TM 9-2320-272-24-2
5 Ton M939 Series Truck
Direct and General Support
Maintenance Manual
Vol 2 of 4
June 1998

This manual contains maintenance information for the 5 ton M939 US Military Trucks. This is volume 2 of 4 in the Direct / General Support Manual Series. M939 series trucks are a 5 ton heavy duty 6x6 truck. Cargo versions were designed to transport 10,000 pounds of cargo in all terrain and all weather conditions. Originally designed in the 1970's to replace the M39 and M809 series of vehicles. 44,590 units were produced. This manual is printed to help private owners in the maintenance of their vehicles.

Should you have suggestions or feedback on ways to improve this book please send email to Books@OcotilloPress.com

---

Edited 2021 Ocotillo Press
ISBN 978-1-954285-64-4

Cover Image Credit: Forest & Kim Starr
Image Source: Public Domain on Wikimedia
Image Caption: Sporobolus virginicus (watering with hydromulcher). Location: Kahoolawe, Honokanaia

Ocotillo Press
Houston, TX 77017
Books@OcotilloPress.com

ARMY TM 9-2320-272-24-2
AIR FORCE TO 36A12-1C-1155-2-2

This publication supersedes TM 9-2320-272-20-1, October 1985, and changes 1 through 4; TM 9-2320-272-20-2, October 1985, and changes 1 through 3; TM 9-2320-272-34- 1, June 1986, and changes 1 through 2; TM 9-2320-272-34-2, June 1986, and changes 1 and 2; and TM 9-2320-358-24&P, October 1992.

TECHNICAL MANUAL
VOLUME 2 OF 4

UNIT, DIRECT SUPPORT, AND GENERAL SUPPORT MAINTENANCE MANUAL FOR

TRUCK, 5-TON, 6X6, M939, M939A1, M939A2 SERIES TRUCKS (DIESEL)

UNIT MAINTENANCE (Condt) 3-615

DIRECT SUPPORT (DS) MAINTENANCE 4-1

TRUCK, CARGO: 5-TON, 6X6, DROPSIDE,
M923 (2320-01-050-2084) (EIC: BRY); M923A1 (2320-01-206-4087) (EIC: BSS); M923A2 (2320-01-230-0307) (EIC: BS7);
M925 (2320-01-047-8769) (EIC: BRT); M925Al (2320-01-206-4088) (EIC: BST); M925A2 (2320-01-230-0308) (EIC: BS8);

TRUCK, CARGO: 5-TON, 6X6 XLWB,
M927 (2320-01-047-8771) (EIC: BRV); M927A1 (2320-01-206-4089) (EIC: BSW); M927A2 (2320-01-230-0309) (EIC: BS9);
M928 (2320-01-047-8770) (EIC: BRU); M928A1 (2320-01-206-4090) (EIC: BSX); M928A2 (2320-01-230-0310) (EIC: BTM);

TRUCK, DUMP: 5-TON, 6X6,
M929 (2320-01447-8756) (EIC: BTH); M929Al (2320-01-206-4079) (EIC: BSY); M929A2 (2320-01-230-0305) (EIC: BTN);
M930 (2320-01-047-8755) (EIC: BTG); M930Al (2320-01-206-4080) (EIC: BSZ); M930A2 (2320-01-230-0306) (EIC: BTO);

TRUCK, TRACTOR: 5-TON, 6X6,
M931 (2320-01-047-8753) (EIC: BTE); M931Al (2320-01-206-4077) (EIC: BS2); M931A2 (2320-01-230-0302) (EIC: BTP);
M932 (2320-01-047-8752) (EIC: BTD); M932Al (2320-01-205-2684) (EIC: BS5); M932A2 (2320-01-230-0303) (EIC: BTQ);

TRUCK, VAN, EXPANSIBLE: 5-TON, 6X6,
M934 (2320-01-047-8750) (EIC: BTB); M934A1 (2320-01-205-2682) (EIC: BS4); M934A2 (2320-01-230-0300) (EIC: BTR);

TRUCK, MEDIUM WRECKER: 5-TON, 6X6,
M936 (2320-01-047-8754) (EIC: BTF); M936A1 (2320-01-206-4078) (EIC: BS6); M936A2 (2320-01-230-0304) (EIC: BTT).

DEPARTMENTS OF THE ARMY AND THE AIR FORCE

JUNE 1998

WARNING

EXHAUST GASES CAN KILL

1. DO NOT operate vehicle engine in enclosed area.
2. DO NOT idle vehicle engine with windows closed.
3. DO NOT drive vehicle with inspection plates or cover plates removed.
4. BE ALERT at all times for odors.
5. BE ALERT for exhaust poisoning symptoms. They are:
   - Headache
   - Dizziness
   - Sleepiness
   - Loss of muscular control
6. IF YOU SEE another person with exhaust poisoning symptoms:
   - Remove person from area
   - Expose to open air
   - Keep person warm
   - Do not permit person to move
   - Administer artificial respiration or CPR, if necessary*

   *For artificial respiration, refer to FM 21-11.
7. BE AWARE: The field protective mask for Nuclear, Biological, or Chemical (NBC) protection will not protect you from carbon monoxide poisoning.

   THE BEST DEFENSE AGAINST EXHAUST POISONING IS ADEQUATE VENTILATION.

WARNING SUMMARY

- Hearing protection is required for the driver and passenger. Hearing protection is also required for all personnel working in and around this vehicle while the engine is running (AR-40-5 and TB MED 501).
- If required to remain inside vehicle during extreme heat, occupants should follow the water intake, work/rest cycle, and other stress preventive measures (FM 21-10, Field Hygiene and Sanitation).
- If NBC exposure is suspected, all air filter media should be handled by personnel wearing protective equipment. Consult with your unit NBC officer or NBC NCO for appropriate handling or disposal instructions.
- This vehicle has been designed to operate safely and efficiently within the limits specified in this TM. Operation beyond these limits is prohibited by IAW AR 70-1 without written approval from the commander, U.S. Army Tank-automotive and Armaments Command, ATTN: AMCPEO-CM-S, Warren, MI 48397-5000.
- Never work under dump body unless safety braces are properly positioned. Failure to do this will result in injury to personnel.
- During winching operation, never stand between vehicles. Assistant must remain in secondary vehicle to engage service brake if cable snaps or automatic brake fails while towing vehicle. Failure to do this may result in injury to personnel.
- Accidental or intentional introduction of liquid contaminants into the environment is in violation of state, federal, and military regulations. Refer to Army POL (para. 1-7) for information concerning storage, use, and disposal of these liquids. Failure to do so may result in injury or death.
- Cleaning solvents are flammable and toxic. Do not use near open flame and always have a fire extinguisher nearby when solvents are used. Use only in well-ventilated places, wear protective clothing, and dispose of cleaning rags in approved container. Failure to do this will result in injury to personnel and/or damage to equipment.
- Eyeshields must be worn when cleaning with compressed air. Compressed air source will not exceed 30 psi (207 kPa). Failure to do so may result in injury to personnel.
- Extreme care should be taken when removing surge tank filler cap if temperature gauge reads above 175°F (79°C). Steam or hot coolant under pressure will cause injury.
- Alcohol used in the alcohol evaporator is flammable, poisonous, and explosive. Do not smoke when removing alcohol evaporator or adding fluid, and do not drink fluid. Failure to do this will result in injury or death.
- Do not perform electrical circuit testing fuel tank with fill cap or sending unit removed. Fuel may ignite, causing injury to personnel.
- When performing battery maintenance, ensure batteries are seated and clamped down, all rubber boots are installed, clamps are well down on battery posts, and all battery cables lie flat against the top of the batteries. Failure to do this may result in injury to personnel and/or damage to equipment.
- Ensure companion seatbelts are not caught inside battery box. This will cause belts to rot which may lead to injury of personnel.
- On M936/A1/A2 model vehicles, remove spare tire prior to changing tire and install tire in spare tire carrier after tire change is complete. Operation of crane and/or vehicle engine while vehicle is on jacks may result in injury to personnel or damage to equipment.
- Never assemble or disassemble tire and rim assembly while inflated, use inflation to seat lockring on split rim or tire on two-piece rim, or inflate a tire without a tire inflation cage. Injury to personnel may result.
- Do not disconnect air lines or hoses, remove safety valves or CTIS components, or perform brake chamber repairs before draining air reservoirs. Small parts under pressure may shoot out with high velocity, causing injury to personnel.

WARNING SUMMARY (Contd)

- Remove all jewelry when working on electrical circuits. Jewelry coming in contact with electrical circuits may produce a short circuit, causing extreme heat, explosions, and fling particles of metal. Failure to do so will result in injury or death and damage to equipment.
- Use eyeshields and follow instructions carefully when performing assembling, disassembling, or maintenance on this device. Components of this device are under spring tension and may shoot out at a high velocity. Failure to do so will result in injury to personnel.
- Do not remove hoses with engine running or start engine with hoses removed. High-pressure fluids may cause hoses to whip violently and spray randomly. Failure to do so may result in injury to personnel.
- Keep hands out from between metal surfaces when removing heavy components. Failure to do so may result in injury to personnel.
- Keep personnel out from under equipment and components of equipment when supported by only a lifting device. Sudden loss of lifting power or shift in load may result in injury or death.
- Do not drain engine, transmission, or radiator fluids, or remove lines containing these fluids, when hot. Doing so may result in injury to personnel.
- Vehicle will become charged with electricity if it contacts or breaks high-voltage wires. Do not leave vehicle while high-voltage lines are in contact with vehicle. Failure to do so may result in injury to personnel.
- Wear hand protection when handling lifting and winching cables, hot exhaust components, and parts with sharp edges. Failure to do so may result in injury to personnel.
- Do not perform fuel system procedures while smoking or within 50 ft (15.2 m) of sparks or open flame. Diesel fuel is highly flammable and can explode easily, causing injury or death to personnel and/or damage to equipment.
- Ensure drainvalve on aftercooler is open when filling cooling system. Failure to do so may result in injury to personnel.
- Turbocharger intake fins are extremely sharp and turn at very high rpm. Keep hands and loose items away from intake openings. Failure to do so may result in injury to personnel.
- Do not place hands between frame and radiator when removing screws from trunnion or lifting radiator. Sudden changes in support may cause the radiator to shift, causing injury to personnel.
- Air pressure may create airborne debris. Use eye protection or injury to personnel may result.
- Air system components are subject to high pressure. Always relieve pressure before loosening or removing air system components.
- Wear safety goggles when using a hammer.
- Ether is extremely flammable. Do not perform ether start system procedures near fire. Injury to personnel may result.

TECHNICAL, MANUAL
NO. 9-2320-272-24-2

TECHNICAL ORDER
NO. 36A12-1C-1155-2-2

HEADQUARTERS
DEPARTMENTS OF THE ARMY AND THE AIR FORCE
Washington, D.C., 30 JUNE 1998

TECHNICAL MANUAL

VOLUME 2 OF 4

UNIT, DIRECT SUPPORT, AND GENERAL SUPPORT MAINTENANCE MANUAL FOR

TRUCK, 5-TON, 6X6, M939, M939A1, M939A2 SERIES TRUCKS (DIESEL)

TRUCK	MODEL	EIC	NSN WITHOUT WINCH	NSN WITH WINCH
Cargo, Dropside	M923	BRY	2320-01-050-2084	
Cargo, Dropside	M923Al	BSS	2320-01-206-4087	
Cargo, Dropside	M923A2	BS7	2320-01-230-0307	
Cargo, Dropside	M925	BRT		2320-01-047-8769
Cargo, Dropside	M925Al	BST		2320-01-206-4088
Cargo, Dropside	M925A2	BS8		2320-01-230-0308
Cargo	M927	BRV	2320-01-047-8771	
Cargo	M927Al	BSW	2320-01-206-4089	
Cargo	M927A2	BS9	2320-01-230-0309	
Cargo	M928	BRU		2320-01-047-8770
Cargo	M928Al	BSX		2320-01-206-4090
Cargo	M928A2	BTM		2320-01-230-0310
Dump	M929	BTH	2320-01-047-8756	
Dump	M929Al	BSY	2320-01-206-4079	
Dump	M929A2	BTN	2320-01-230-0305	
Dump	M930	BTG		2320-01-047-8755
Dump	M930Al	BSZ		2320-01-206-4080
Dump	M930A2	BTO		2320-01-230-0306
Tractor	M931	BTE	2320-01-047-8753	
Tractor	M931Al	BS2	2320-01-206-4077	
Tractor	M931A2	BTP	2320-01-230-0302	
Tractor	M932	BTD		2320-01-047-8752
Tractor	M932Al	BS5		2320-01-205-2684
Tractor	M932A2	BTQ		2320-01-230-0303
Van, Expansible	M934	BTB	2320-01-047-8750	
Van, Expansible	M934Al	BS4	2320-01-205-2682	
Van, Expansible	M934A2	BTR	2320-01-230-0300	
Medium Wrecker	M936	BTF		2320-01-047-8754
Medium Wrecker	M936Al	BS6		2320-01-206-4078
Medium Wrecker	M936A2	BTT		2320-01-230-0304

REPORTING OF ERRORS AND RECOMMENDING IMPROVEMENTS

You can help improve this manual. If you find any mistakes or if you know of a way to improve the procedures, please let us know. Mail your letter or DA Form 2028 (Recommended Changes to Publications and Blank Forms), or DA Form 2028-2 located in back of this manual, directly to: Director, Armament and Chemical Acquisition and Logistics Activity, ATTN: AMSTA-AC-NML, Rock Island, IL 61299-7630. A reply will be furnished to you. You may also provide DA Form 2028-2 information via datafax or e-mail:

- E-mail:amsta-ac-nml.@ria-emh2.army.mil
- Fax: DSN 783-0726 or commercial (309) 782-0726

*This publication supersedes TM 9-2320-272-20-1,24 October 1985, and changes 1 through 4; TM 9-2320-272-20-2,25 October 1985, and changes 1 through 3; TM 9-2320-272-34-1, 10 June 1986, and changes 1 through 2; TM 9-2320-272-34-2 10 June 1986, and changes 1 and 2; and TM 9-2320-358-24&P, 21 October 1992.

This publication is published in four volumes. TM 9-2320-272-24-1 contains chapters 1,2, and 3 (though section IX). TM 9-2320-272-24-2 contains chapters 3 (sections X through XVI) and 4 (sections I through III). TM 9-2320-272-24-3 contains chapter 4 (sections IV through XVI). TM 9-2320-272-24-4 contains chapters 5 and 6 and appendices A through H. Volume 1 contains a table of contents for the entire manual. Volumes 1,2, and 3 contain an alphabetical index covering tasks found in their respective volume. Volume 4 contains an alphabetical index covering all tasks found in the entire manual.

TABLE OF CONTENTS

VOLUME 2 OF 4

		page
CHAPTER 3	UNIT MAINTENANCE	(Contd)
Section X	Wheels, Hubs, Drums, and Steering Maintenance	3-615
Section XI.	FrameMaintenance	3-683
Section XII.	Body, Cab, and Accessories Maintenance	3-730
Section XIII.	Winch and Power Takeoff (FTO) Maintenance	3-823
Section XIV	Special Purpose Bodies Maintenance	3-869
Section XV	Special Purpose Kits	3-972
Section XVI.	Central Tire Inflation System (CTIS) Maintenance	3-1235
CHAPTER 4	DIRECT SUPPORT (DS) MAINTENANCE	4-1
Section I.	Direct Support (DS) and General Support (GS) Mechanical Troubleshooting	4-1
Section II.	Power Plant Maintenance	4-35
Section III.	Engine (M939/Al) Maintenance	4-117

Section X. WHEELS, HUBS, DRUMS, AND STEERING MAINTENANCE

3-217. WHEELS, HUBS, DRUMS, AND STEERING MAINTENANCE INDEX

PARA. NO.	TITLE	PAGE NO.
3-218.	Wheel and Tire (M939) Rotation	3-616
3-219.	Wheel and Tire (M939Al/A2) Rotation	3-622
3-220.	Tire and Tube (M939) Maintenance	3-626
3-221.	Tire and Wheel (M939Al/A2) Maintenance	3-630
3-222.	Wheel Rim Stud Replacement	3-636
3-223.	Front Hub and Drum (M939/Al) Maintenance	3-638
3-224.	Rear Hub and Drum (M939/Al) Maintenance	3-642
3-225.	Wheel Bearing Adjustment	3-648
3-226.	Steering Wheel Replacement	3-651
3-227.	Pitman Arm Replacement (Ross)	3-653
3-228.	Pitman Arm Replacement (Sheppard)	3-654
3-229.	Drag Link Replacement	3-655
3-230.	Steering Pump Drivebelts (M939/Al) Maintenance	3-656
3-231.	Steering Assist Cylinder Stone Shield Replacement	3-658
3-232.	Steering Assist Cylinder Hoses Replacement	3-660
3-233.	Steering Assist Cylinder Maintenance	3-662
3-234.	Power Steering Pump Pressure and Return Hoses Replacement (Ross)	3-668
3-235.	Power Steering Pump Pressure and Return Hoses Replacement (Sheppard)	3-670
3-236.	Power Steering Pump Maintenance	3-672
3-237.	Power Steering Pump Filter and Reservoir (M939A2) Maintenance	3-674
3-238.	Steering Gear Stone Shield Replacement	3-676
3-239.	Steering Gear-to-Assist Cylinder Pressure Lines Replacement	3-678

3-218. WHEEL AND TIRE (M939) ROTATION

THIS TASK COVERS:

a. Wheel and Tire Removal
b. Inspection
c. Wheel and Tire Rotation
d. Inner Rear Wheel Installation
e. Outer Rear Wheel Installation
f. Front Wheel Installation

INITIAL SETUP:

APPLICABLE MODELS
M939

SPECIAL TOOLS
Inner wheel socket (Appendix E, Item 77)

TOOLS
General mechanic's tool kit (Appendix E, Item 1)
Torque wrench (Appendix E, Item 145)
Torque multiplier

REFERENCES (TM)
TM 9-2320-272-10
TM 9-2320-272-24P
TM 9-2610-200-24

EQUIPMENT CONDITION
• Parking brake set (TM 9-2320-272-10).
• Spare tire removed (TM 9-2320-272-10).

GENERAL SAFETY INSTRUCTIONS
Ensure inner and outer lug nuts are properly torqued.

a. Wheel and Tire Removal

1. Loosen sixty stud nuts (3) on four outside rear wheels (5) and two front wheels (2).
2. Using hydraulic jack, raise vehicle and place two jack stands under rear-rear axle (6), two jack stands under forward-rear axle (4), and two jack stands under front axle (1).

NOTE

Tag wheels for rotation.

3. Remove sixty stud nuts (3), two front wheels (2), and four outside rear wheels (5) from vehicle.
4. Using inner wheel socket, remove forty inner wheel spacer nuts (7) and four inside rear wheels (8) from vehicle.

b. Inspection

1. Inspect wheel rims (9) for cracks, stud hole damage, and bends. Replace wheel rims (9), if cracked, damaged, or bent.
2. Inspect tires (10) for tread wear (TM 9-2610-200-24).

3-218. WHEEL AND TIRE (M939) ROTATION (Contd)

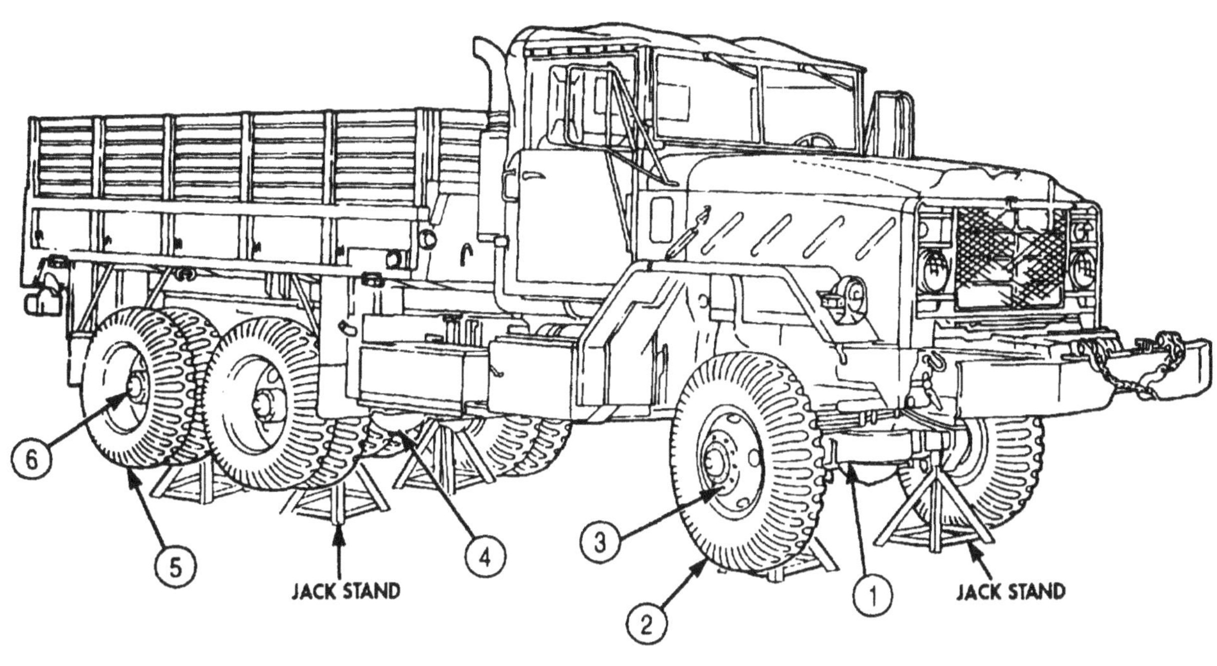

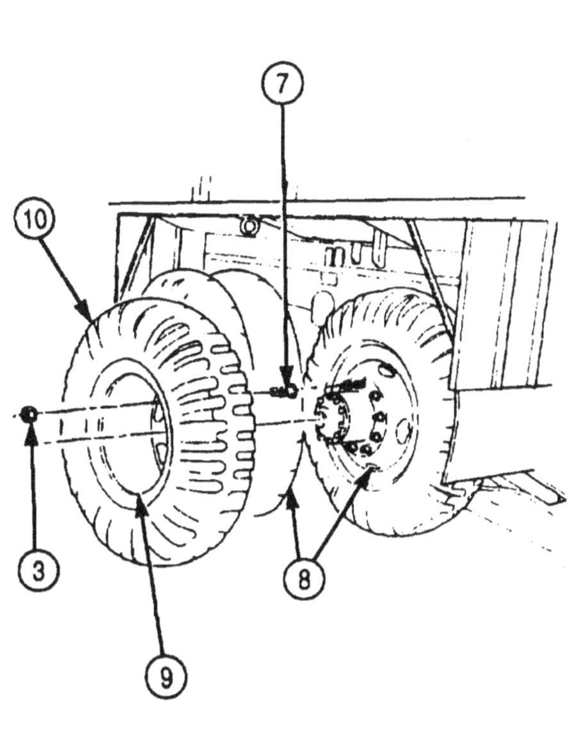

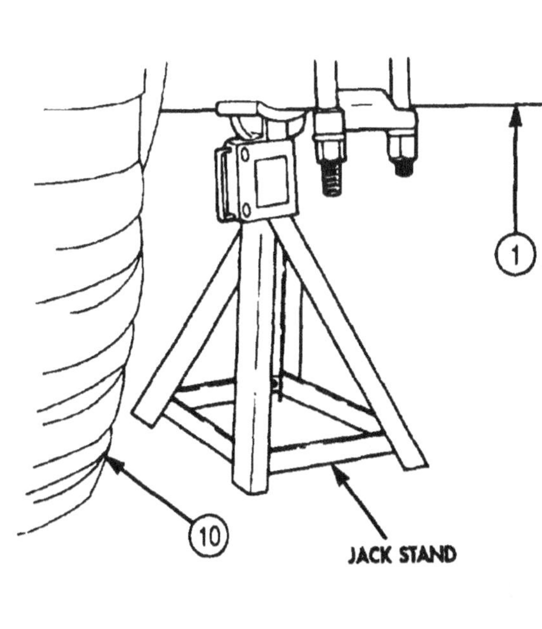

3-218. WHEEL AND TIRE (M939) ROTATION (Contd)

c. Wheel and Tire Rotation

NOTE

- Wheels and tires are rotated the same for all M939 series vehicles.
- To maintain tread depth and pattern of dual tires, tires must be rotated to match tread wear as closely as possible. (TM 9-2610-200-24.)

Rotate wheels and tires as shown.

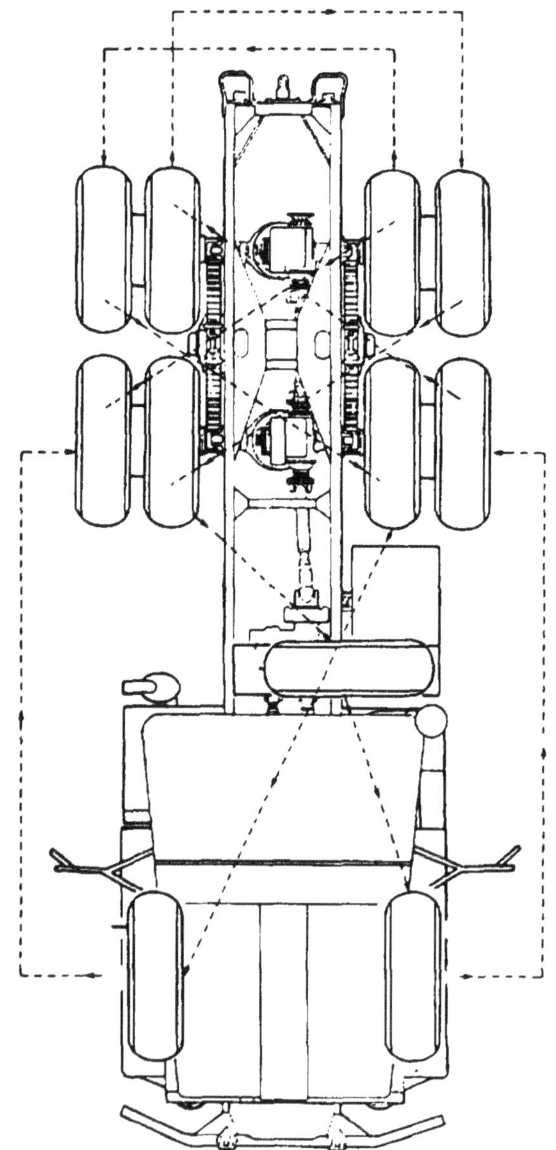

3-218. WHEEL AND TIRE (M939) ROTATION (Contd)

d. Inner Rear Wheel Installation

NOTE

- Ensure inner rear wheels are installed so valve stems to both rear wheels are accessible.
- Start all threaded nuts by hand to prevent cross-threading. Left wheels have left-hand threaded wheel studs and right wheels have right-hand threaded wheel studs.

1. Position four rear inner dual wheels (1) over wheel hub studs (2).

WARNING

Apply proper torque to inner lug nuts. Failure to do so may result in injury to personnel or damage to equipment.

2. Using inner wheel socket, install four rear inner dual wheels (1) on wheel hub (2) with forty inner wheel spacer nuts (3).
3. Tighten spacer nuts (3) 325-355 lb-ft (441-481 N•m) in sequence shown.

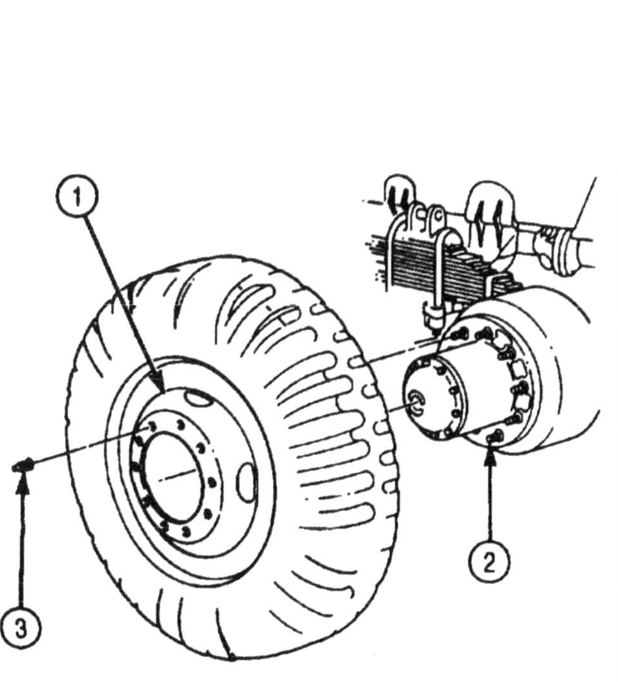

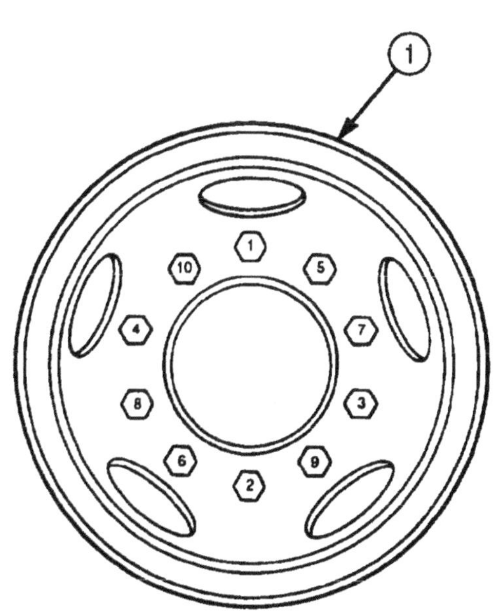

TIGHTENING SEQUENCE

3-218. WHEEL AND TIRE (M939) ROTATION (Contd)

e. Outer Rear Wheel Installation

1. Position four rear outer dual wheels (2) on wheel adapter spacer nuts (1).

WARNING

Apply proper torque to outer lug nuts. Failure to do so may result in injury to personnel or damage to equipment.

2. Install four rear outer dual wheels (2) on vehicle with forty stud nuts (3). Hand-tighten stud nuts (3).
3. Raise rear of vehicle, remove jack stands, and lower vehicle.
4. Tighten forty stud nuts (3) 325-355 lb ft (441-481 N-m) in sequence shown.

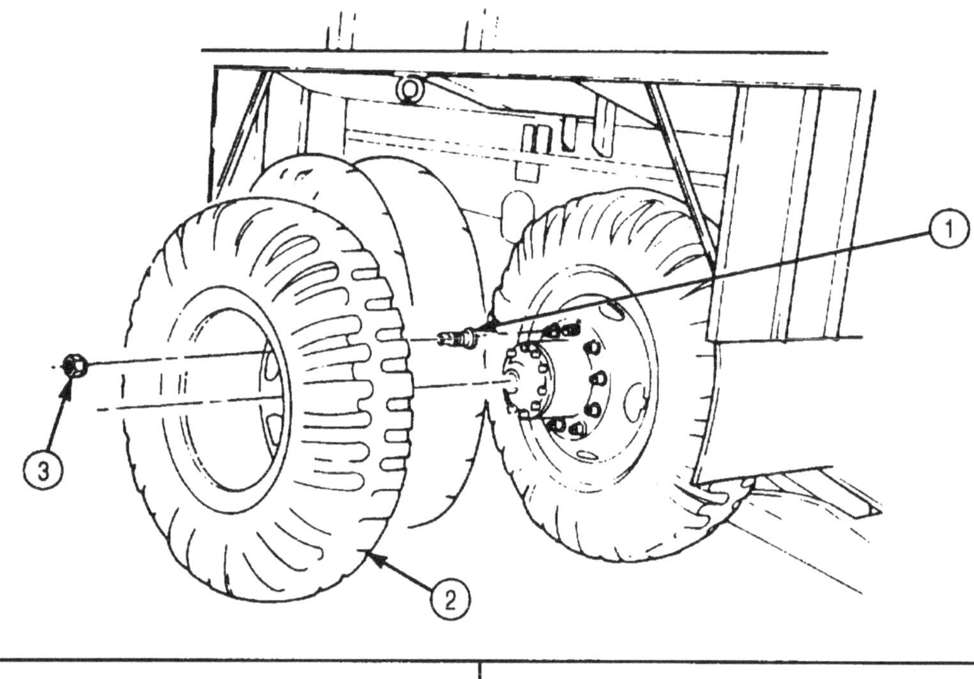

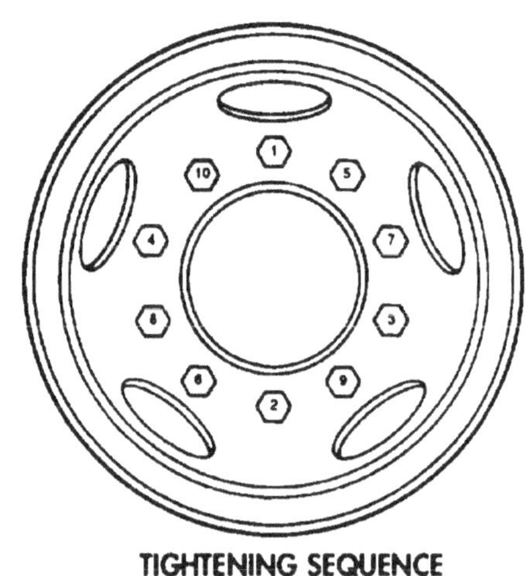

TIGHTENING SEQUENCE

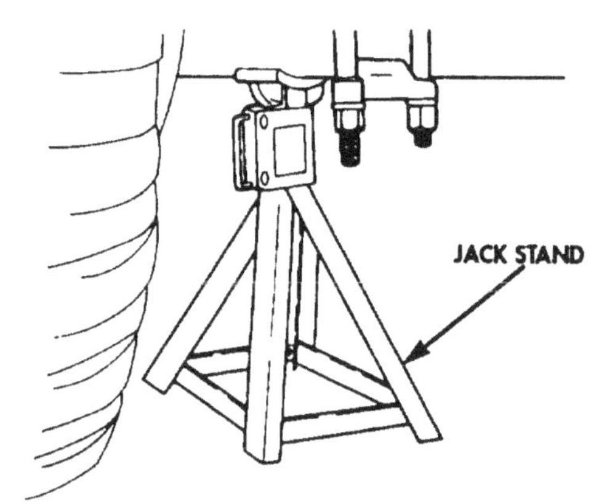

3-218. WHEEL AND TIRE (M939) ROTATION (Contd)

f. Front Wheel Installation

1. Install two front wheels (5) on wheel hub studs (4) with twenty wheel stud nuts (6). Hand-tighten stud nuts (6).
2. Raise front of vehicle clear of jack stands, remove jack stands, and lower vehicle.
3. Tighten twenty wheel stud nuts (6) 325-355 lb-ft (441-481 N•m) in sequence shown.

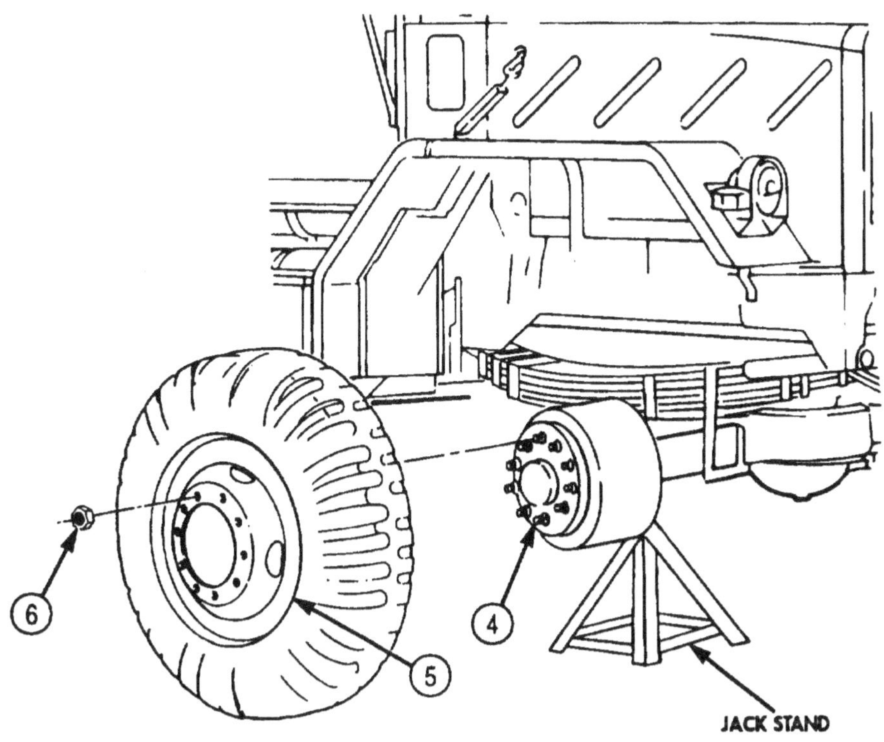

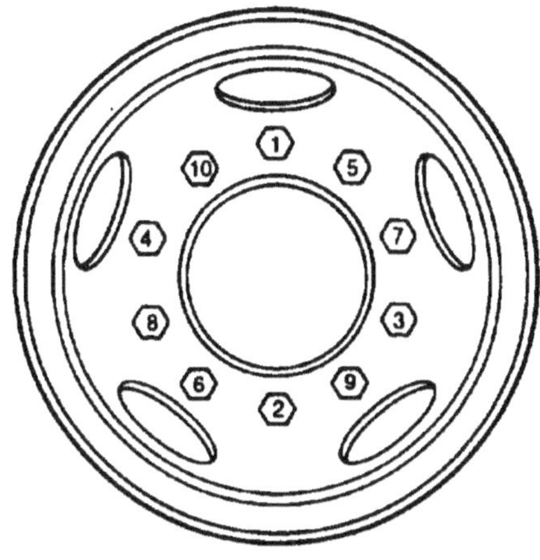

TIGHTENING SEQUENCE

3-219. WHEEL AND TIRE (M939A1/A2) ROTATION

THIS TASK COVERS:

a. **Wheel and Tire Removal**
b. **Inspection**
c. **Wheel and Tire Rotation**
d. **Rear Wheel Installation**
e. **Front Wheel Installation**

INITIAL SETUP:

APPLICABLE MODELS
M939A1/A2

TOOLS
General mechanic's tool kit (Appendix E, Item 1)
Torque wrench (Appendix E, Item 145)

REFERENCES (TM)
TM 9-2320-272-10
TM 9-2320-272-24P
TM 9-2610-200-24

EQUIPMENT CONDITION
- Parking brake set (TM 9-2320-272-10).
- Front wheel valve removed (para. 3-456).
- Rear wheel valve removed (para. 3-457).

GENERAL SAFETY INSTRUCTIONS
Ensure lug nuts are properly torqued.

a. Wheel and Tire Removal

NOTE
Perform step 1 for M939A2 vehicles only.

1. Remove two nuts (6) and wheel weights (5) from two front wheels (2) and studs (4).
2. Loosen sixty stud nuts (7) on four rear wheels (1) and two front wheels (2).
3. Raise vehicle and place two jack stands under rear-rear axle, two jack stands under forward-rear axle (9), and two jack stands under front axle (3).

NOTE
Tag wheels for installation.

4. Remove sixty stud nuts (7), two front wheels (2), and four rear wheels (1) from vehicle.

b. Inspection

1. Inspect wheel rims (8) for cracks, stud hole damage, and bends. Replace wheel rims (8) if cracked, stud holes damaged or bent.
2. Inspect tires for thread wear (TM 9-2610-200-24).

3-219. WHEEL AND TIRE (M939A1/A2) ROTATION (Contd)

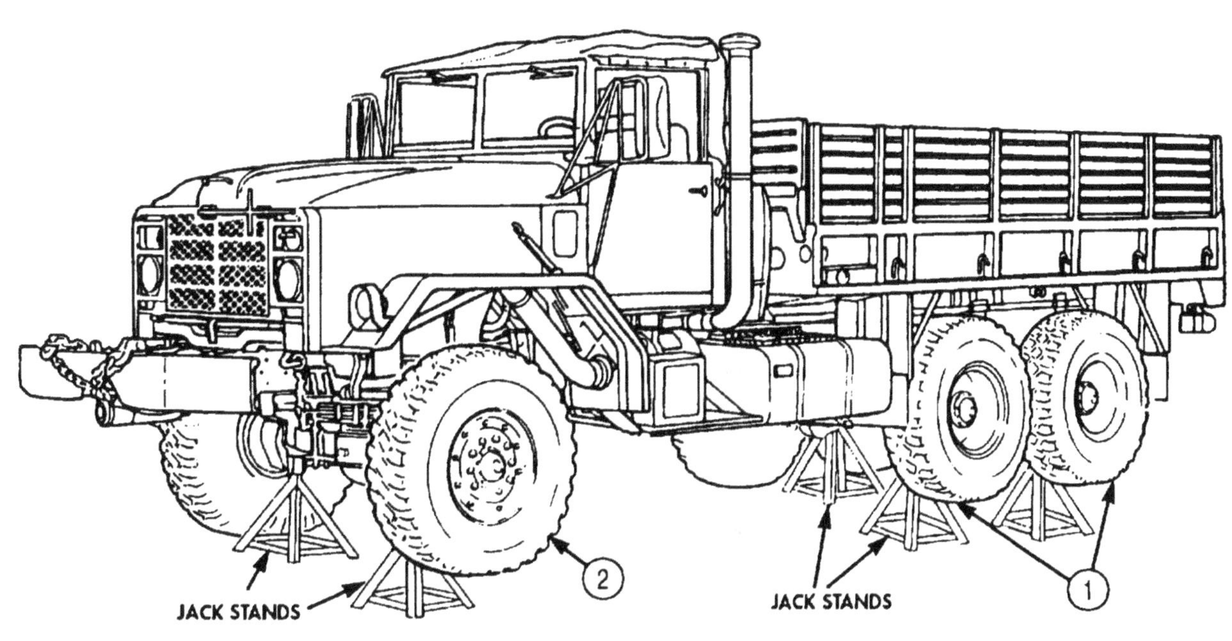

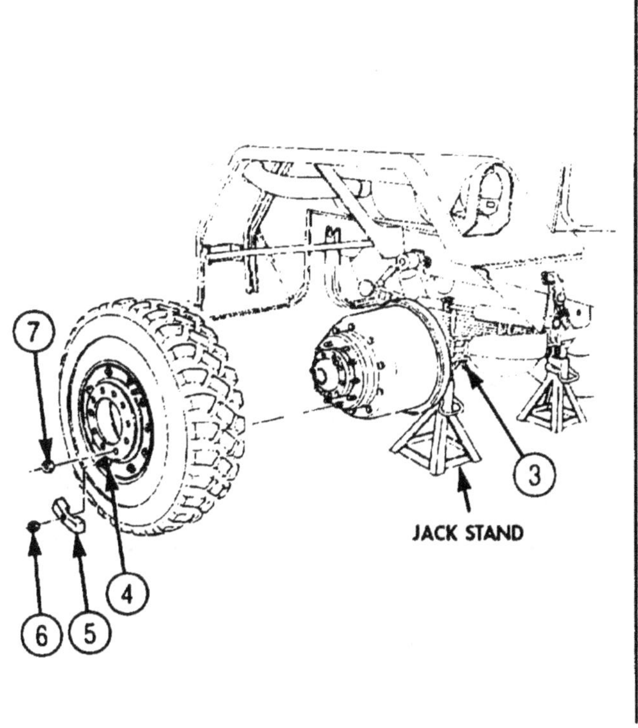

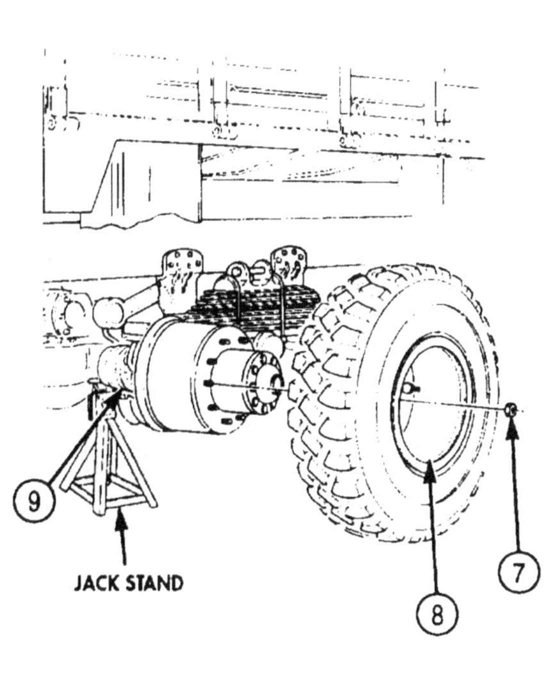

3-219. WHEEL AND TIRE (M939A1/A2) ROTATION (Contd)

c. Wheel and Tire Rotation

NOTE

- Wheels and tires are rotated the same for all M939A1/A2 series vehicles.
- To maintain tread depth, tires must be rotated to match tread wear as closely as possible (TM 9-2610-200-24).
- Tires (14:00xR20) have unidirectional tread design and can be installed in either direction.

Rotate wheels and tires as shown.

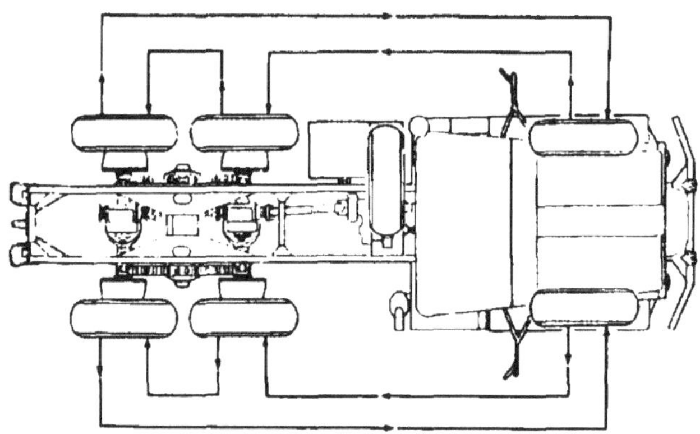

d. Rear Wheel Installation

WARNING

Apply proper torque to outer lug nuts. Failure to do so may result in injury to personnel or damage to equipment.

1. Install four rear wheels (1) on studs (3) with forty stud nuts (2). Hand-tighten stud nuts (2).
2. Raise rear of vehicle, remove jack stands, and lower vehicle.
3. Tighten forty stud nuts (2) 450-500 lb-ft (610-678 N•m) in sequence shown.

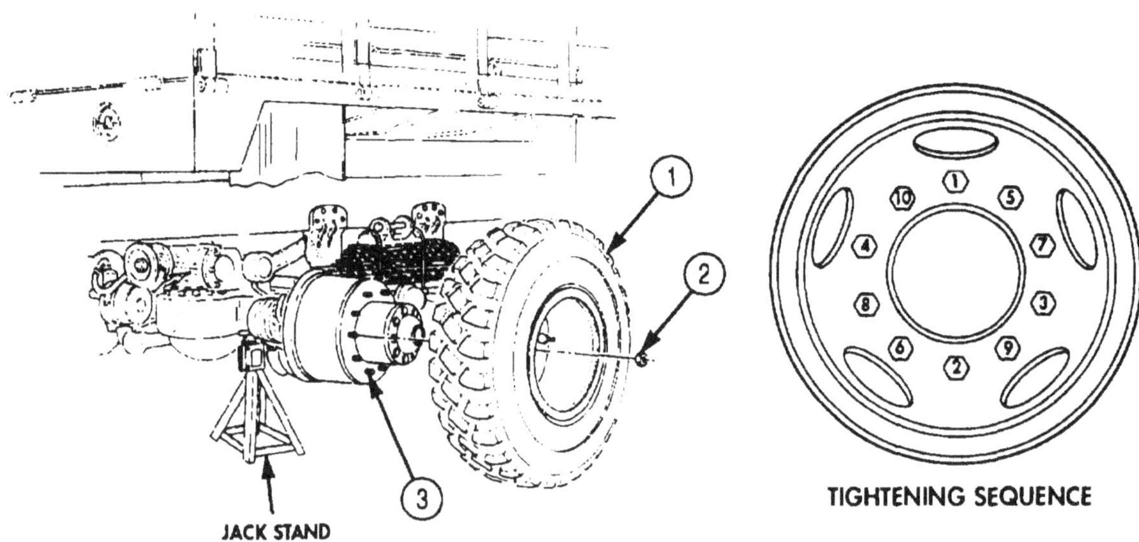

3-219. WHEEL AND TIRE (M939A1/A2) ROTATION (Contd)

e. Front Wheel Installation

WARNING

Apply proper torque to outer lug nuts. Failure to do so may result in injury to personnel or damage to equipment.

1. Install two front wheels (4) on studs (5) with twenty stud nuts (9). Hand tighten stud nuts (9).
2. Raise front of vehicle, remove jack stands, and lower vehicle.
3. Tighten twenty stud nuts (9) 450-500 lb-ft (610-678 N•m) in sequence shown.

NOTE

- Perform step 4 for M939A2 vehicles only.
- Install wheel weight with slot flush against rim nut.

4. Install wheel weights (7) on two front wheels (4) and studs (6) with two nuts (8). Tighten nuts (8) 40 lb-ft (54 N•m).

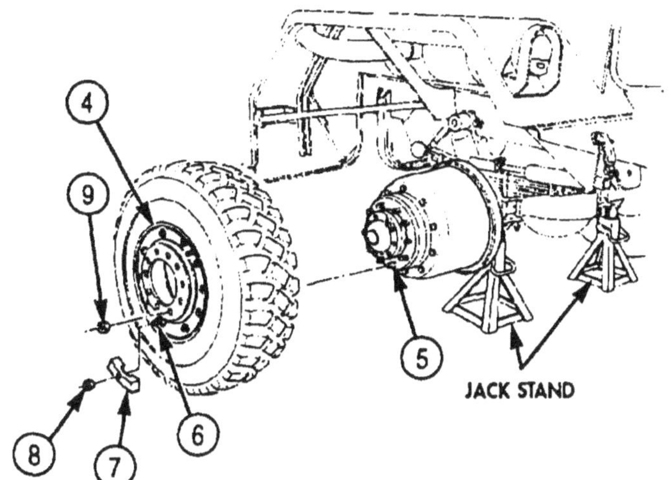

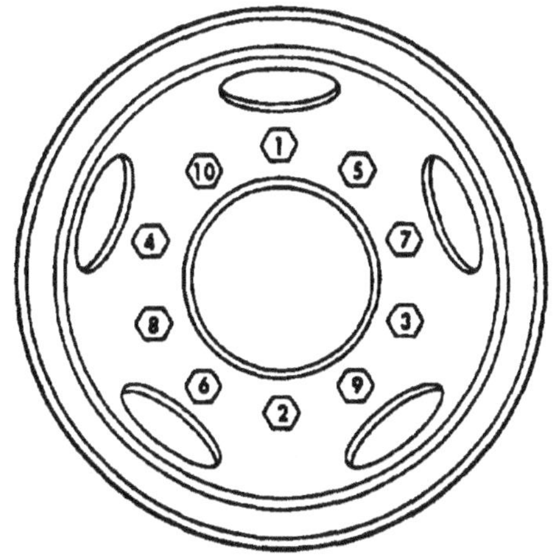

FOLLOW-ON TASKS: • Install front wheel valve (para. 3-456).
• Install rear wheel valve (para. 3-457).

3-220. TIRE AND TUBE (M939) MAINTENANCE

THIS TASK COVERS:

a. Removal
b. Inspection
c. Installation

INITIAL SETUP:

APPLICABLE MODELS
M939

TOOLS
General mechanic's tool kit (Appendix E, Item 1)

REFERENCES (TM)
TM 9-2320-272-10
TM 9-2320-272-24P

EQUIPMENT CONDITION
Wheel(s) removed (para. 3-218).

GENERAL SAFETY INSTRUCTIONS
- Never remove tire lockring without first deflating tire.
- Ensure lockring is properly seated around wheel before inflating tire.
- Never inflate tire with lockring facing personnel.
- Never attempt to seat lockring by striking while tire is inflating.

NOTE

Ensure tire inflation cage does not have cracked welds, cracked or bent components, or pitting from corrosion. If any of these are found, obtain new inflation cage.

a. Removal

1. Remove valve cap (11) and valve core (10) from inner tube valve stem (9).

WARNING

Never remove tire lockring without first deflating tire. Lockring may explode off, causing injury to personnel.

2. Position tire (4) on side, with lockring (1) facing upward.
3. Break tire bead (13) from lockring (1).

NOTE

Assistant will help with step 3.

4. Insert first tire iron (3) in lockring slit (2) and pry upward until slot (5) is exposed.
5. Insert second tire iron (6) into lockring slot (5) and pry upward while running first tire iron (3) completely around and remove lockring (1).
6. Break tire bead (13) on opposite side of wheel (7).
7. Remove tire (4) from wheel (7).
8. Remove tire liner (12) and inner tube (8) from tire (4).

3-220. TIRE AND TUBE (M939) MAINTENANCE (Contd)

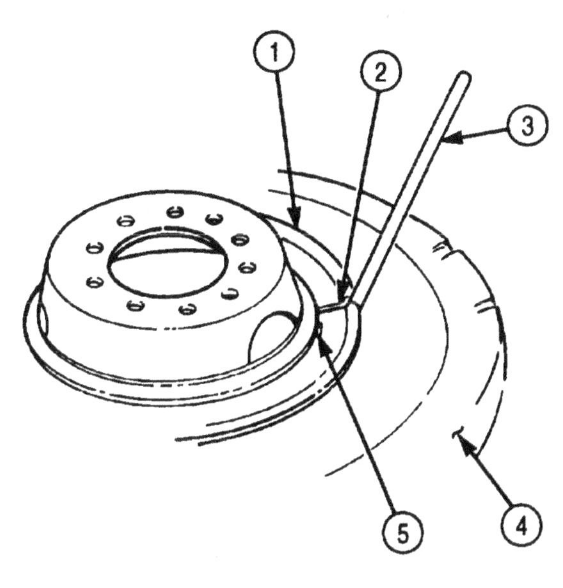

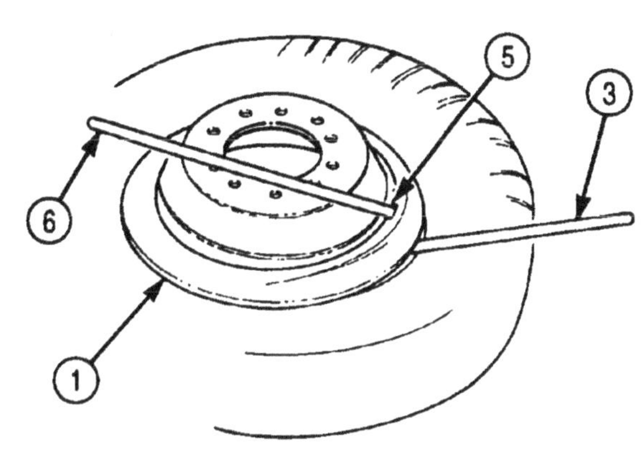

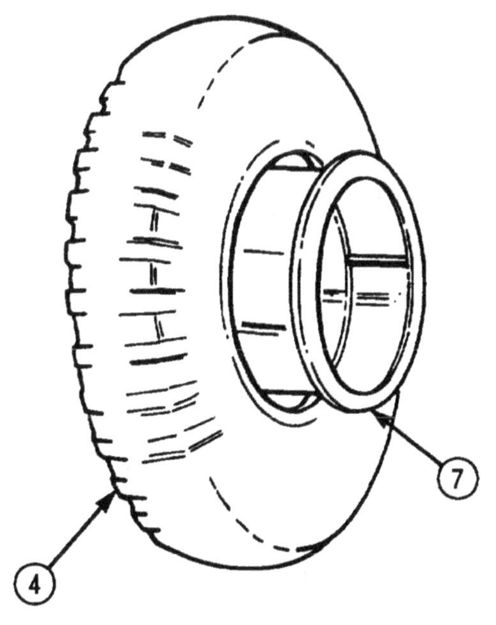

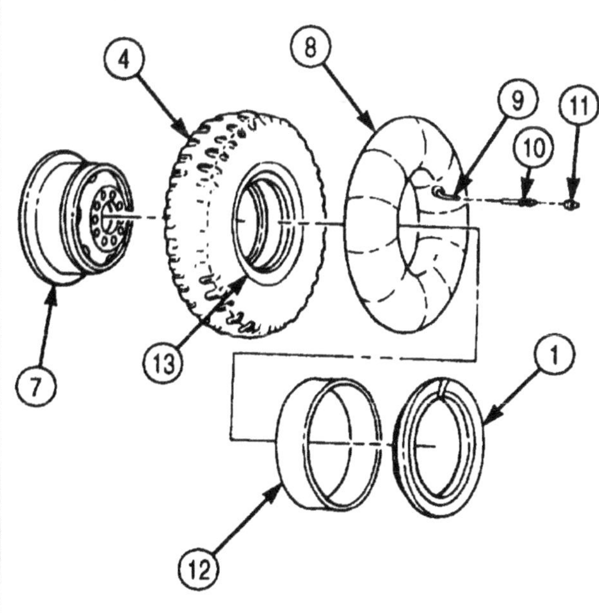

3-220. TIRE AND TUBE (M939) MAINTENANCE (Contd)

b. Inspection

Inspect rim (5) and tire lockring (2) for bends, cracks, breaks, distortion, or pitting from corrosion. Replace rim (5) and tire lockring (2) if bent, cracked, broken, distorted, or pitted.

c. Installation

1. Install inner tube (6) and tire liner (9) inside tire (3).
2. Install valve core (7) on inner tube valve stem (4).
3. Inflate inner tube (6) just enough to prevent puncture when wheel (1) is installed.
4. Place wheel (1) on the ground, rim (5) side down.

WARNING

- Lockring must be properly seated around wheel when installed. If lockring is not correctly installed, it may explode off when tire is inflated, causing injury to personnel.
- Never attempt to correct seating of lockring by hammering, striking, or forcing while tire is inflated. Lockring may explode off, causing injury to personnel.

5. Install tire (3) over wheel (1) and guide stem (4) through hole in wheel (1).
6. Force lockring (2) onto wheel (1), opposite rim (5).

WARNING

Always use a tire inflation cage for inflation purposes. Stand on one side of cage during inflation; never directly in front. Keep hands out of cage during inflation. Inflate assembly to recommended pressure, using a clip-on air chuck. Do not exceed cold inflation pressure. See vehicle operator's manual for recommended tire pressure. Failure to do so may result in injury to personnel.

7. Inflate tire (3) to proper pressure (TM 9-2320-272-10).
8. Observe lockring (2) for proper seating on wheel (1).
9. Install valve cap (8) on inner tube valve stem (4).
10. Inspect rim (5) and lockring (2) for proper seating and locking while tire (3) is still in inflation cage.
11. If lockring (2) adjustment is required, deflate tire (3) completely before adjusting.

3-220. TIRE AND TUBE (M939) MAINTENANCE (Contd)

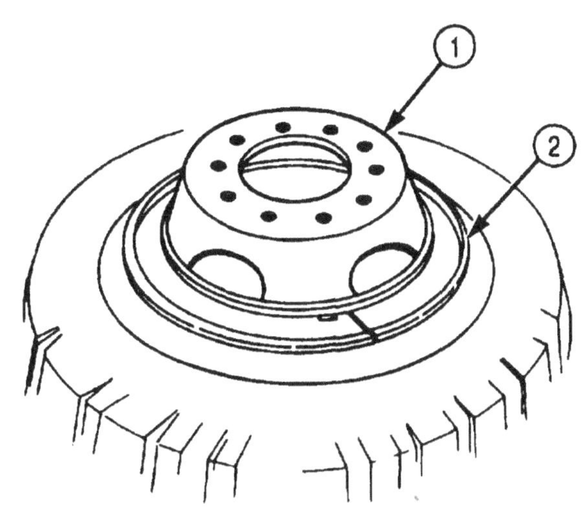

PROPERLY SEATED LOCKRING

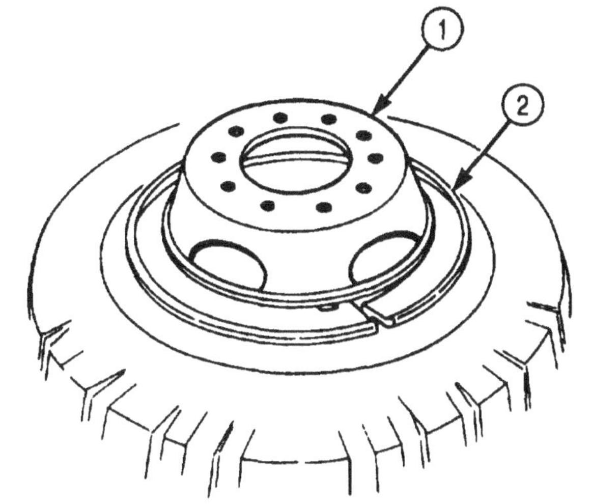

IMPROPERLY SEATED LOCKRING

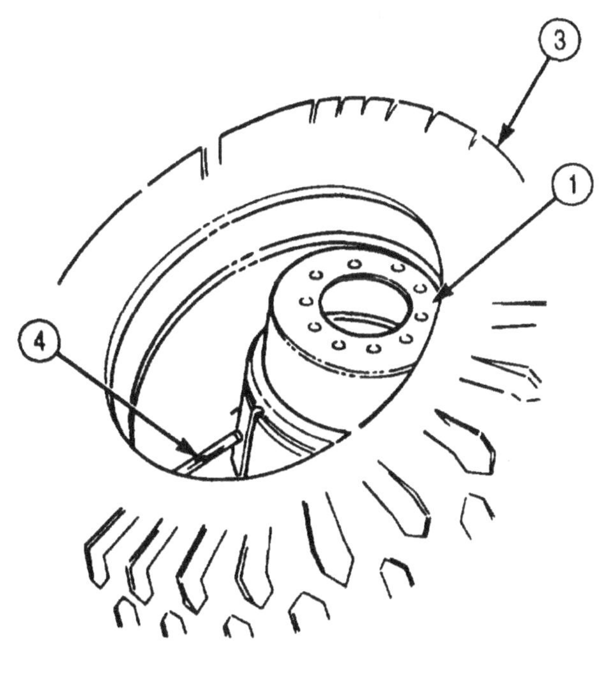

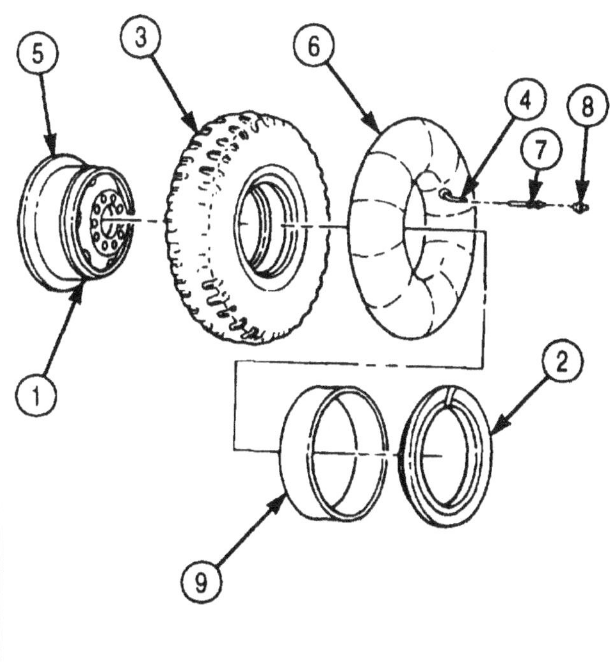

FOLLOW-ON TASK: Install wheel(s) (para. 3-218).

THIS TASK COVERS:

a. Disassembly **b. Assembly**

INITIAL SETUP:

APPLICABLE MODELS
M939A1/A2

SPECIAL TOOLS
Wheel assembly tool (Appendix E, Item 160)

TOOLS
General mechanic's tool kit (Appendix E, Item 1)
Torque wrench (Appendix E, Item 145)

MATERIALS/PARTS
Ten locknuts (Appendix D, Item 290)
O-ring (Appendix D, Item 432)
Lubricant (Appendix C, Item 39)

PERSONNEL REQUIRED
TWO

REFERENCES (TM)
TM 9-2320-272-10
TM 9-2320-272-24P

EQUIPMENT CONDITION
Wheel(s) removed (para. 3-219).

GENERAL SAFETY INSTRUCTIONS
- Never remove tire locknuts before deflating tire.
- Never use tubes in wheel assembly.
- Always use tire cage when inflating tire.
- Mount only tires recommended for vehicle application as specified in TM 9-2320-272-24P.

WARNING

- Never use tubes in wheel assemblies. The wheel is designed for tubeless application. Use of a tube defeats built-in safety features and could allow the wheel to come apart under pressure, resulting in serious injury to personnel.
- Rim surfaces and safety pressure relief grooves in the inner rim must be kept clean and free of dirt. Rust and dirt could clog safety relief grooves, causing the assembly to separate under pressure if improperly disassembled, causing injury to personnel.
- Never use wheel assemblies with studs or flanged locknuts which are cracked, broken, rusted, pitted, bent, or which have loose studs or damaged, mutilated, or deformed threads. Damaged or rusted fasteners can cause improper assembly, which could cause the assembly to separate under pressure or could cause individual fasteners to fail. Any of these conditions could result in serious injury to personnel.
- In all disassembly operations, ensure tire is totally deflated before removing wheel nuts. Remove valve core and exhaust all air from tire. Check the valve stem for obstacles by running a piece of wire through it. Failure to follow proper safety precautions could cause serious injury or death.

a. Disassembly

1. Place tire (1) in tire inflation cage.

NOTE

Run a piece of wire through valve stem to ensure it is not plugged.

3-221. TIRE AND WHEEL (M939A1/A2) MAINTENANCE (Contd)

2. Remove valve cap (4) and valve core (3) from valve stem (5) and allow tire (1) to completely deflate.

NOTE

Assistant will help with step 3.

3. Remove tire (1) from tire inflation cage and place tire (1) flat on floor, with shallow side of rim (2) up.

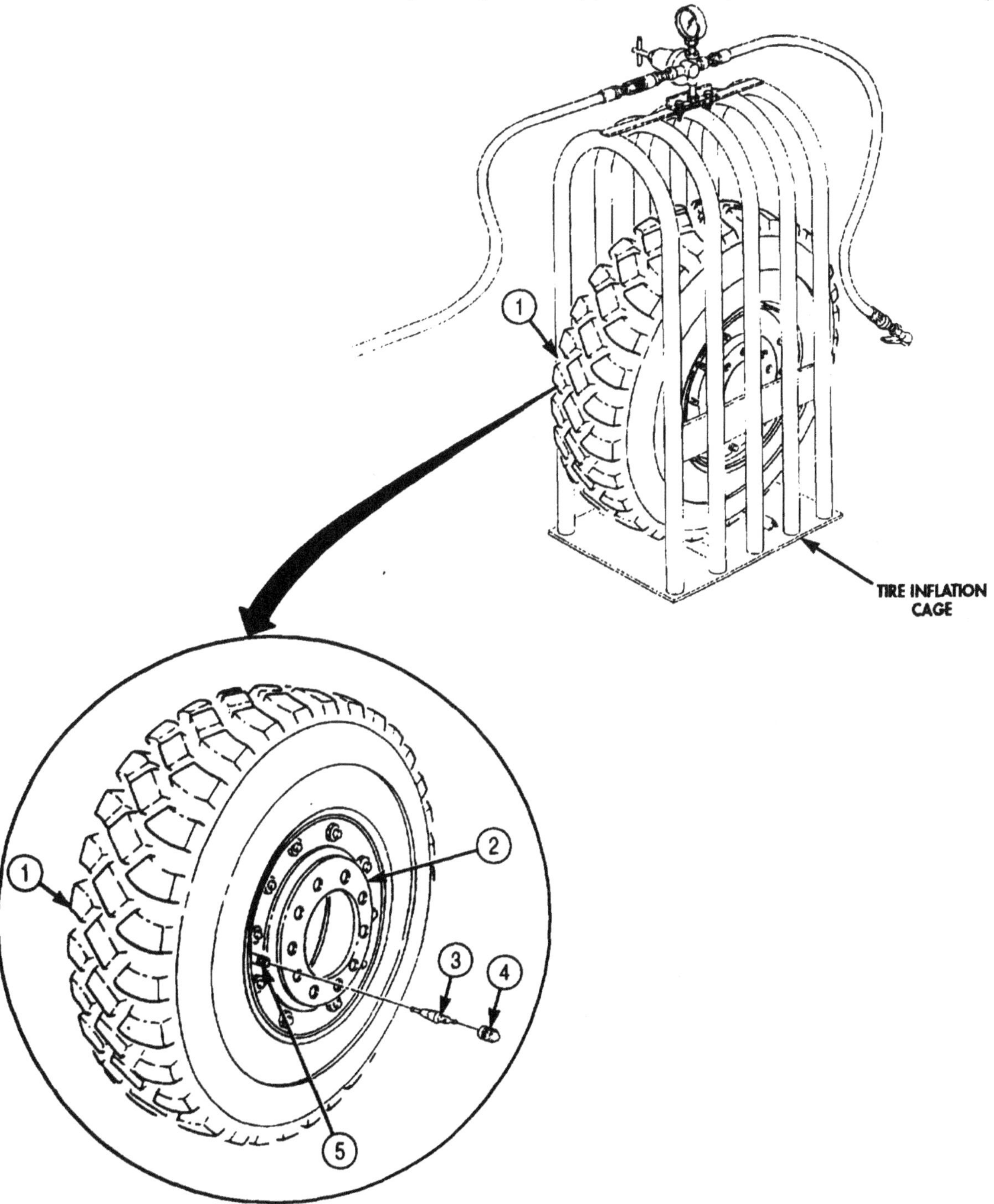

3-221. TIRE AND WHEEL (M939A1/A2) MAINTENANCE (Contd)

WARNING

- Never use a tire mounting/demounting machine on a bolt-together wheel assembly. Failure to comply may result in injury or death.
- Never inflate a wheel assembly with the wheel nuts removed in an attempt to separate inner and outer rim halves. Failure to comply may result in injury or death.
- Never inflate a wheel assembly without first checking wheel nut torques to ensure the wheel nuts are tightened to specifications (use a recently calibrated torque wrench). Failure to comply may result in injury or death.
- Never inflate a wheel assembly assembled with components which are not replacement parts specified in TM 9-2320-272-24P. Wheel assemblies without nuts, with improperly tightened nuts, or with components not to specifications, could cause the assembly to separate under pressure, resulting in serious injury.

CAUTION

Loosen locknuts no more than 1/2 in. (12 mm) at a time. Uneven beadlock pressure on clamp ring may result in damage to wheel rim studs.

4. Remove ten locknuts (1) from wheel rim studs (5). Discard locknuts (1).
5. Remove clamp ring (2) and wheel rim (6) from tire (4).

NOTE

Perform step 6 only if wheel rim is being replaced.

6. Remove nut (10), grommet (9), and turret valve (8) from wheel rim (6).
7. Remove O-ring (3) from wheel rim (6). Discard O-ring (3).
8. Install rope on beadlock (7).
9. Remove beadlock (7) from tire (4).

3-221. TIRE AND WHEEL (M939A1/A2) MAINTENANCE (Contd)

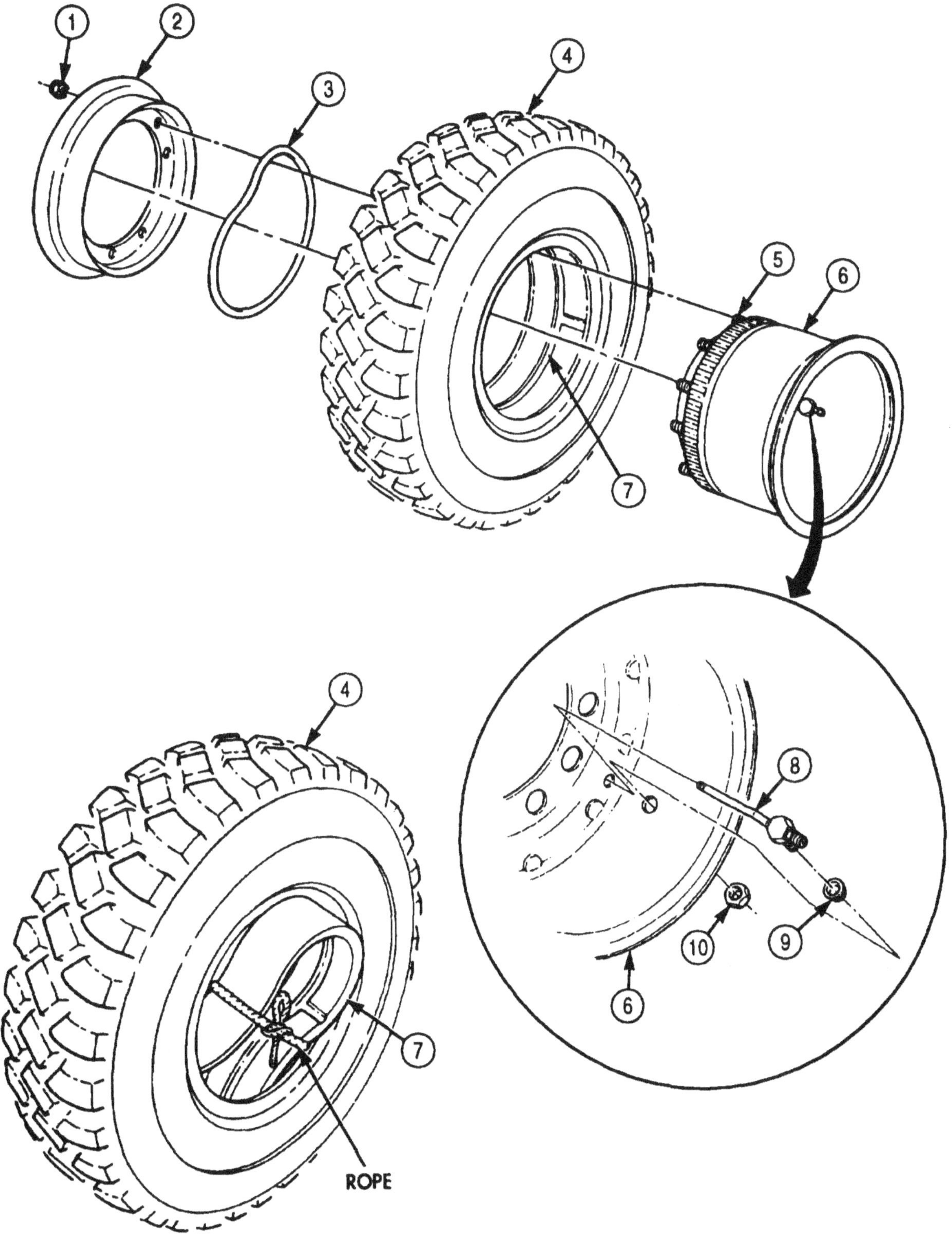

3-221. TIRE AND WHEEL (M939A1 /A2) MAINTENANCE (Contd)

b. Assembly

WARNING

- Mount only replacement tires specified in TM 9-2320-272-24P Do not exceed recommended tire inflation pressure. Do not exceed load stamped on wheel. Wrong size tires, overinflated tires, or overloaded assemblies could case assemblies to separate under pressure, resulting in injury or death.
- Always inflate the assembly in a tire inflation cage, using a clip-on air chuck. Inflation cage can help reduce risk of injury if the assembly should separate during inflation.

NOTE

- Assistant will help when required.
- When new wheel rim is installed, use attaching parts from old wheel rim.
- Ensure O-ring groove, wheel rim, and clamp ring are free of rust, dirt, and other foreign material that may prevent tire from sealing when assembled and inflated.

1. Using rope, install beadlock (5) in tire (1). Ensure beadlock (5) is centered in tire (1). Remove rope.
2. Install grommet (11) on turret valve (3).
3. Install turret valve (3) on wheel rim (2) with nut (12). Tighten nut (12) 40-65 lb-in, (5-7 N-m).
4. Install tire (1) on wheel rim (2).

NOTE

Do not stretch or twist O-ring.

5. Lubricate O-ring (6) and install on O-ring groove (4) of wheel rim (2).
6. Install clamp ring (7) on wheel rim (2), with valve hole aligned with valve (3).
7. Install wheel assembly tool base on wheel rim (2).
8. Install wheel assembly tool plate on wheel assembly tool base with two washers (15) and nuts (16).
9. Install ten new locknuts (14) on wheel rim studs (13). Finger-tighten locknuts (14).
10. Remove two nuts (16), washers (15), tool plate, and tool base from wheel rim (2).
11. Tighten locknuts (14) 210-240 lb-ft (285-325 N•m) in sequence shown.

WARNING

Place tire in tire inflation cage before inflating. Failure to do so may result in injury to personnel.

12. Install valve core (9) on valve stem (8).
13. Inflate tire (1) to proper pressure (TM 9-2320-272-10).
14. Install valve cap (10) on valve stem (8).

3-221. TIRE AND WHEEL (M939A1/A2) MAINTENANCE (Contd)

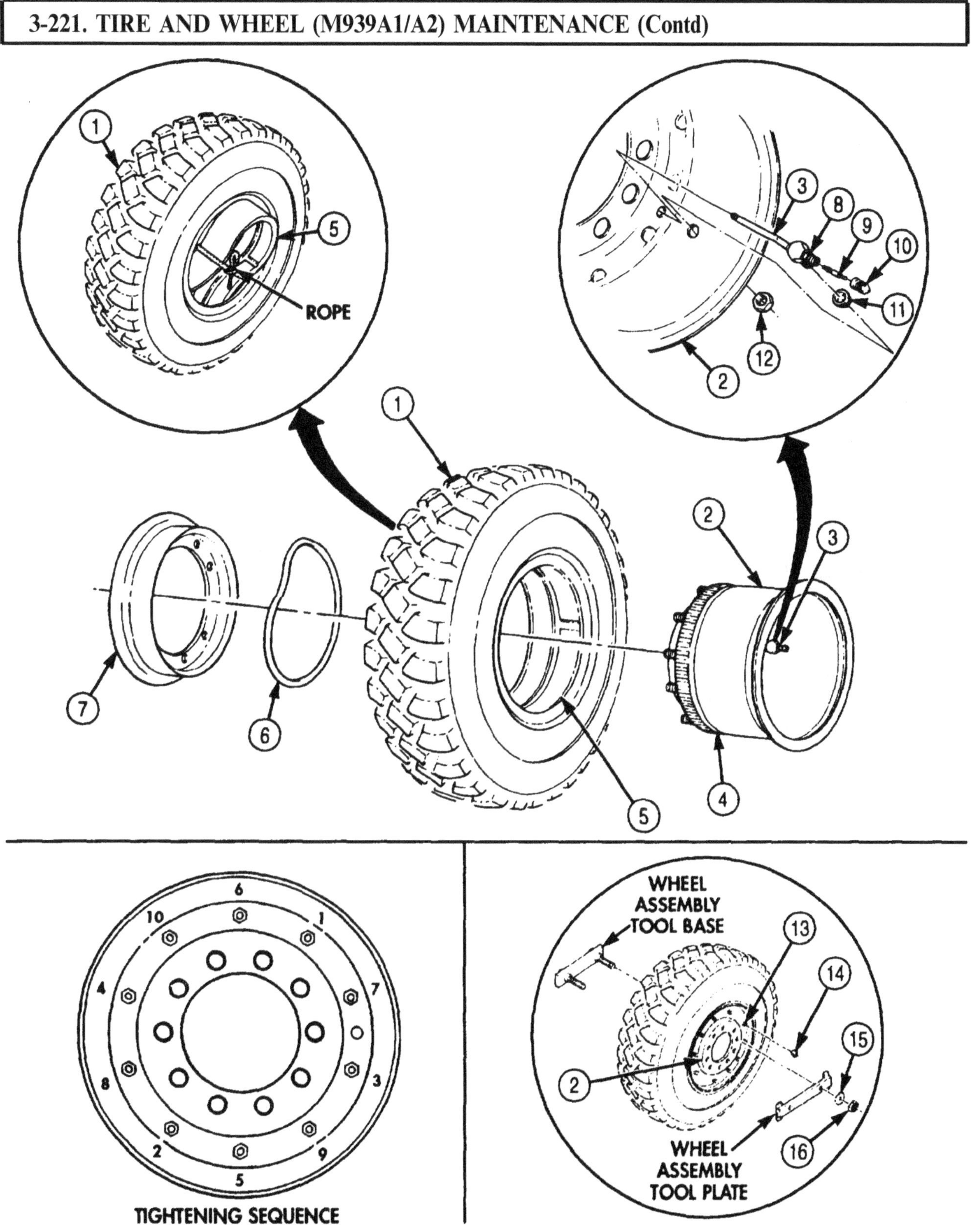

FOLLOW-ON TASK: Install wheel(s) (para. 3-219).

3-222. WHEEL RIM STUD REPLACEMENT

THIS TASK COVERS:

a. Removal **b. Installation**

INITIAL SETUP:

APPLICABLE MODELS
All

SPECIAL TOOLS
Bolt inserting tool (Appendix E, Item 17)

TOOLS
General mechanic's tool kit (Appendix E, Item 1)

REFERENCES (TM)
TM 9-2320 272-10
TM 9-2320-272-24P

EQUIPMENT CONDITION
Wheel(s) disassembled (para. 3-220 or 3-221).

a. Removal

1. Place wheel rim (3) on flat surface, with wheel studs (2) facing up.
2. Install nut (1) on wheel rim stud (2) with six turns.
3. Strike nut (1) with hammer until wheel stud (2) is loose.
4. Remove wheel rim stud (2) from wheel rim (3).

b. Installation

NOTE

Ensure flat side of wheel rim stud head is flat to inside wheel rim.

1. Install wheel rim stud (2) through hole in wheel rim (3).
2. Using bolt inserting tool, tighten nut (1) onto wheel rim stud (2) until wheel rim stud (2) properly seats on wheel rim (3).
3. Remove nut (1) and bolt inserting tool from wheel rim stud (2).

3-222. WHEEL RIM STUD REPLACEMENT (Contd)

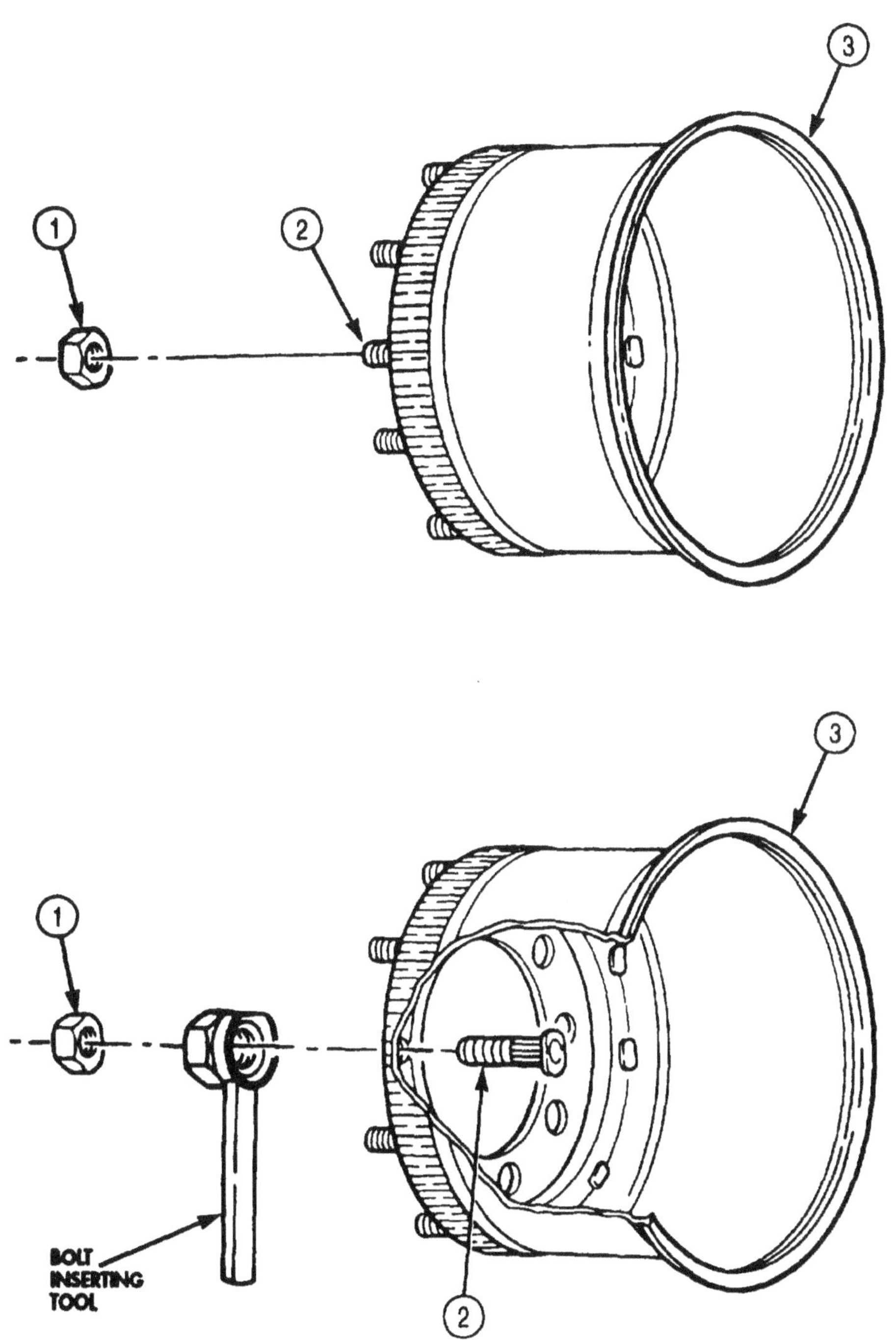

FOLLOW-ON TASK: Assemble wheel(s) (para. 3-220 or 3-221).

3-223. FRONT HUB AND DRUM (M939/A1) MAINTENANCE

THIS TASK COVERS:

a. Removal
b. Cleaning and Inspection
c. Lubrication
d. Installation

INITIAL SETUP:

APPLICABLE MODELS
M939/A1

TOOLS
General mechanic's tool kit (Appendix E, Item 1)
Torque wrench (Appendix E, Item 145)

MATERIAL/PARTS
Inner oil seal (Appendix D, Item 497)
Drive flange gasket (Appendix D, Item 164)
Ten lockwashers (Appendix D, Item 350)
GAA grease (Appendix C, Item 28)
Gasket sealant (Appendix C, Item 30)
Drycleaning solvent (Appendix C, Item 71)

REFERENCES (TM)
TM 9-214
TM 9-2320-272-10
TM 9-2320-272-24P

EQUIPMENT CONDITION
Wheel(s) removed (para. 3-218).

GENERAL SAFETY INSTRUCTIONS
- Drycleaning solvent is flammable and toxic. Do not use near an open flame.
- Keep fire extinguisher nearby when using drycleaning solvent.

a. Removal

1. Remove ten screws (11) and lockwashers (12) from drive flange (10) and hub (2). Discard lockwashers (12).
2. Install two screws (11) in threaded holes (13) to separate drive flange (10) from hub (2).
3. Remove drive flange (10) and gasket (9), if present, from hub (2). Discard gasket (9).
4. Remove two screws (11) from drive flange (10).
5. Remove outer bearing locknut (8), bearing nut washer (7), bearing adjusting nut (6), and outer bearing (5) from spindle (4).
6. Remove hub (2) and drum (1) from spindle (4).
7. Remove inner bearing oil seal (14) and inner bearing (15) from hub (2). Discard inner bearing oil seal (14).
8. Remove ten wheel studs (3) and hub (2) from drum (1).

3-223. FRONT HUB AND DRUM (M939/A1) MAINTENANCE (Contd)

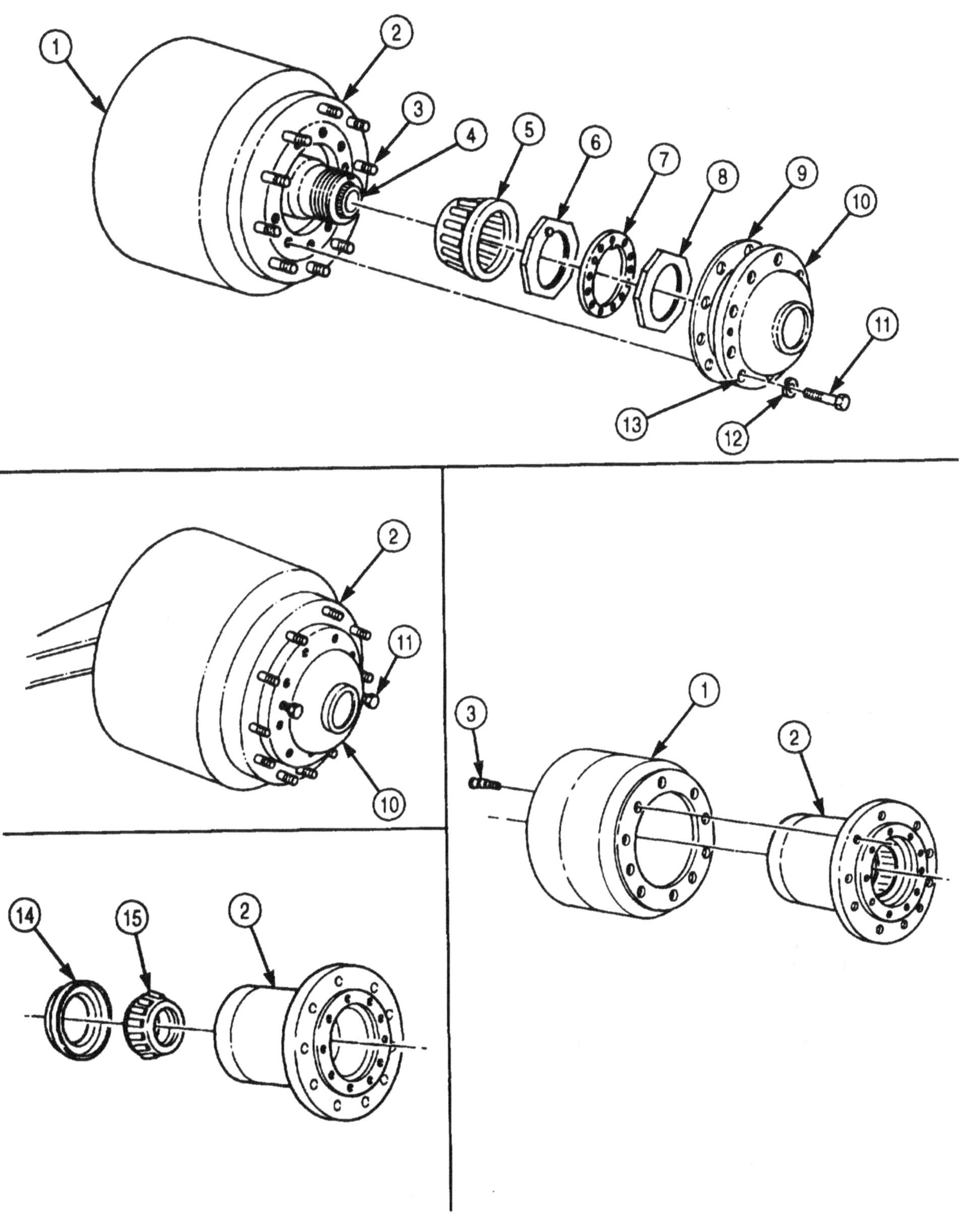

3-223. FRONT HUB AND DRUM (M939/A1) MAINTENANCE (Contd)

b. Cleaning and Inspection

WARNING

Drycleaning solvent is flammable and toxic. Do not use near open flame and always have a fire extinguisher nearby when solvents are used. Use only in well-ventilated places, wear protective clothing, and dispose of cleaning rags in approved container. Failure to do this may result in injury or death to personnel and/or damage to equipment.

1. Clean gasket or sealant remains from mating surfaces of driveshaft flange (14) and hub (2).

NOTE

Do not use compressed air to dry bearings.

2. Clean all hub (2) and drum (1) components in drycleaning solvent.
3. Inspect hub (2) for cracks and breaks. Replace hub (2) if cracked or broken.
4. Inspect inner and outer bearings (4) and (7) (TM 9-214). Replace if damaged.
5. Inspect inner and outer bearing cups (5) and (6) (TM 9-214). Replace if damaged.

NOTE

Perform steps 6 and 7 only if bearing cups are to be replaced.

6. Remove bearing cups (5) and (6) from hub (2) by tapping alternately on outer edge.
7. Press new inner bearing cup (5) and outer bearing cup (6) into hub (2). Ensure bearing cups (5) and (6) are seated.
8. Inspect drum (1) for deep grooves. Replace drum (1) if grooves are deeper than 1/32 in. (0.79 mm).

c. Lubrication

1. Pack inner and outer bearings (4) and (7) with GAA grease (TM 9-214).
2. Pack inner rubber section of new inner oil seal (3) with GAA grease (TM 9-214).

d. Installation

1. Install hub (2) in drum (1).
2. Press wheel studs (8) through drum (1) and into hub (2). Ensure studs (8) are seated.
3. Install inner bearing (4) and inner bearing oil seal (3) into hub (2). Ensure inner oil bearing oil seal (3) is seated.

NOTE

Assistant will help with step 4.

4. Install hub (2) and drum (1) on spindle (9).
5. Install outer bearing (7) on spindle (9) with nut (10). While rotating hub (2), tighten nut (10) to 50 lb-ft (68 N-m).
6. Back out nut (10) 1/6-1/4 turn so washer (12) can be positioned to spindle (9) and adjusting pin (11).
7. Install washer (12) and locknut (13) on spindle (9). Tighten locknut (13) 250-400 lb-ft (339-542 N•m).
8. Apply sealant to mating surfaces of driveshaft flange (14).
9. Install new gasket (17) and driveshaft flange (14) on hub (2) with ten new lockwashers (16) and screws (15). Tighten screws (15) 60-100 lb-ft (81-136 N•m).

3-223. FRONT HUB AND DRUM (M939/A1) MAINTENANCE (Contd)

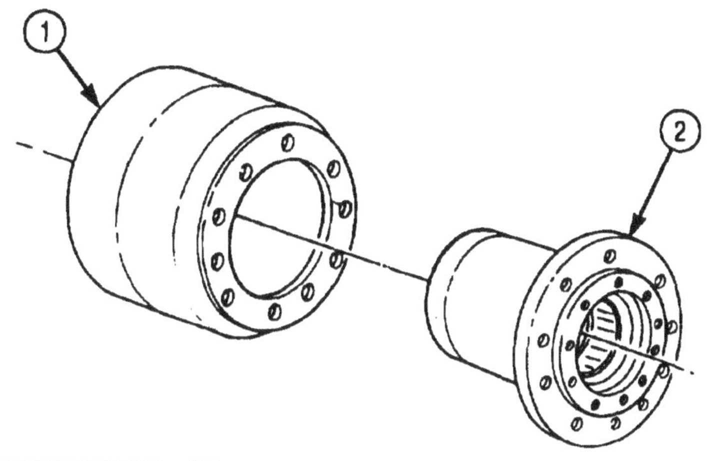

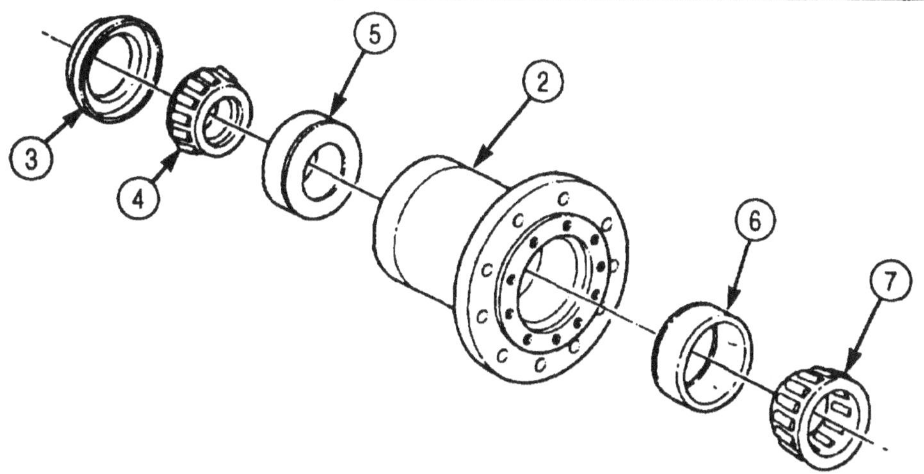

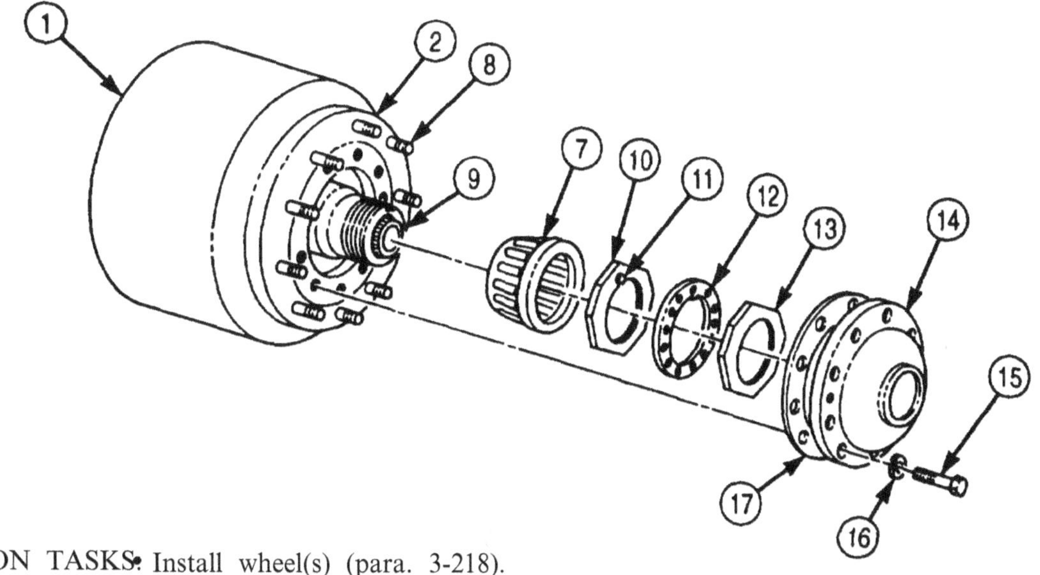

FOLLOW-ON TASKS: • Install wheel(s) (para. 3-218).
• Road test vehicle (TM 9-2320-272-10).

3-224. REAR HUB AND DRUM (M939/A1) MAINTENANCE

THIS TASK COVERS:

a. Removal
b. Cleaning and Inspection
c. Lubrication
d. Installation

INITIAL SETUP:

APPLICABLE MODELS
M939/A1

SPECIAL TOOLS
Wiper replacer (Appendix E, Item 113)

TOOLS
General mechanic's tool kit (Appendix E, Item 1)
Torque wrench (Appendix E, Item 145)

MATERIALS/PARTS
Keyway insert (Appendix D, Item 268)
Wiper (Appendix D, Item 720)
Inner oil seal (Appendix D, Item 634)
Outer oil seal (Appendix D, Item 508)
Axle flange gasket (Appendix D, Item 164)
GAA grease (Appendix C, Item 28)
Gasket sealant (Appendix C, Item 30)
Drycleaning solvent (Appendix C, Item 71)

REFERENCES (TM)
TM 9-214
TM 9-2320-272-10
TM 9-2320-272-24P

EQUIPMENT CONDITION
- Wheels chocked (TM 9-2320-272-10).
- Spring brake(s) caged (TM 9-2320-272-10).
- Wheel(s) removed (para. 3-218 or 3-219).

GENERAL SAFETY INSTRUCTIONS
- Drycleaning solvent is flammable and toxic. Do not use near an open flame.
- Keep fire extinguisher nearby when using drycleaning solvent.

a. Removal

1. Remove ten screws (6) and washers (7) from axle shaft (5).
2. Remove axle shaft (5) and gasket (4), if present, from hub (1). Discard gasket (4).
3. Remove outer bearing locknut (8), bearing nut washer (9), and adjusting nut (10) from axle housing (13).

NOTE

Assistant will help with step 4.

4. Remove hub (1) and drum (2) from axle housing (13). It may be necessary to back off brake adjustment to remove drum (2).
5. Remove outer bearing oil seal (11) and outer bearing (12) from hub (1). Discard oil seal (11).
6. Remove keyway insert (14) from axle housing (13). Discard keyway insert (14).
7. Remove inner bearing oil seal (16) and inner bearing (17) from hub (1). Discard oil seal (16).
8. Remove wiper (15) from axle housing (13). Discard wiper (15).
9. Remove ten wheel studs (3) and hub (1) from drum (2).
10. Remove five screws (20), washers (19), and dustshield (18) from inside drum (2).

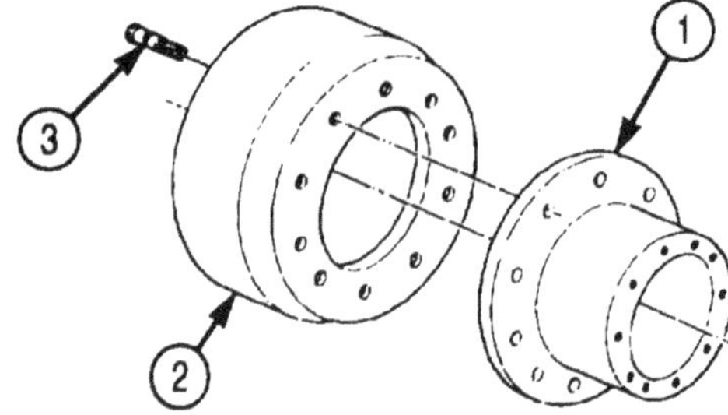

3-224. REAR HUB AND DRUM (M939/A1) MAINTENANCE (Contd)

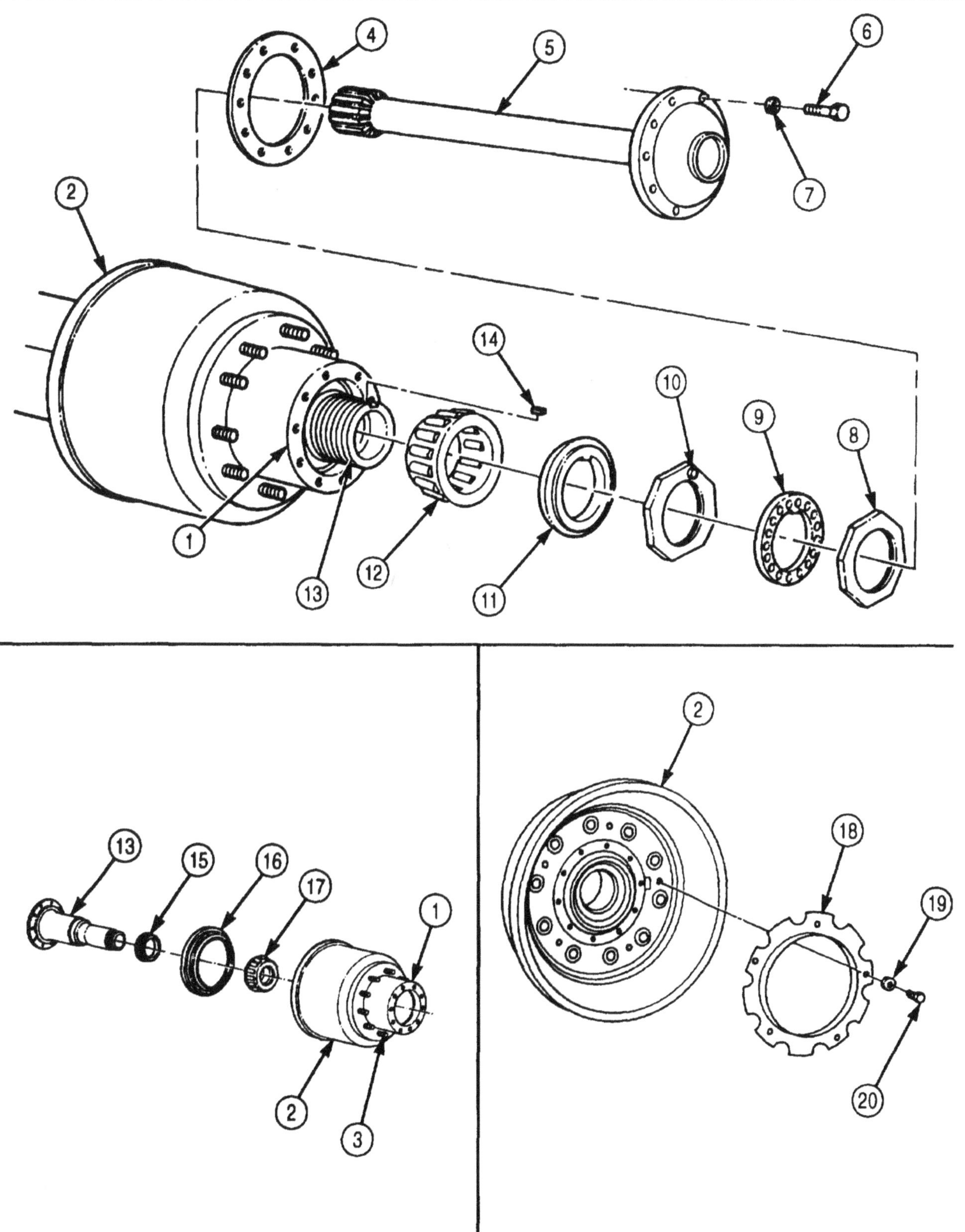

3-224. REAR HUB AND DRUM (M939/A1) MAINTENANCE (Contd)

b. Cleaning and Inspection

WARNING

Drycleaning solvent is flammable and toxic. Do not use near open flame and always have a fire extinguisher nearby when solvents are used. Use only in well-ventilated places, wear protective clothing, and dispose of cleaning rags in approved container. Failure to do this may result in injury or death to personnel and/or damage to equipment.

1. Clean gasket or sealant remains from mating surfaces of axle shaft and hub (2).

NOTE

Do not use compressed air to dry bearings

2. Clean all hub (2) and drum (1) components in drycleaning solvent and allow to air-dry.
3. Inspect hub (2) for cracks and breaks. Replace hub (2) if cracked or broken.
4. Inspect inner bearing (4) and outer bearing (7) (TM 9-214). Replace if damaged.
5. Inspect inner bearing cup (5) and outer bearing cup (6) (TM 9-214). Replace if damaged.

NOTE

Perform steps 6 and 7 only if bearings or bearing cups are to be replaced.

6. Remove bearing cups (5) and (6) from hub (2) by tapping alternately on outer edge.
7. Press new inner bearing cup (5) and new outer bearing cup (6) into hub (2). Ensure bearing cups (5) and (6) are seated.
8. Inspect drum (1) for deep grooves. Replace drum (1) if grooves are deeper than 1/32 in. (0.79 mm).

c. Lubrication

1. Pack inner bearing (4) and outer bearing (7) with GAA grease (TM 9-214).
2. Pack inner rubber section of inner bearing oil seal (3) with GAA grease (TM 9-214).

3-224. REAR HUB AND DRUM (M939/A1) MAINTENANCE (Contd)

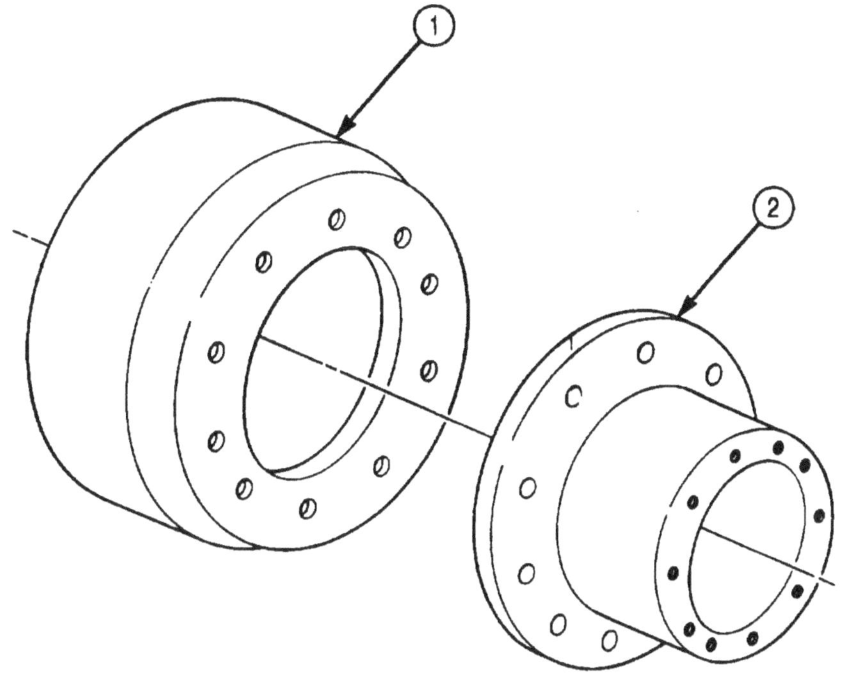

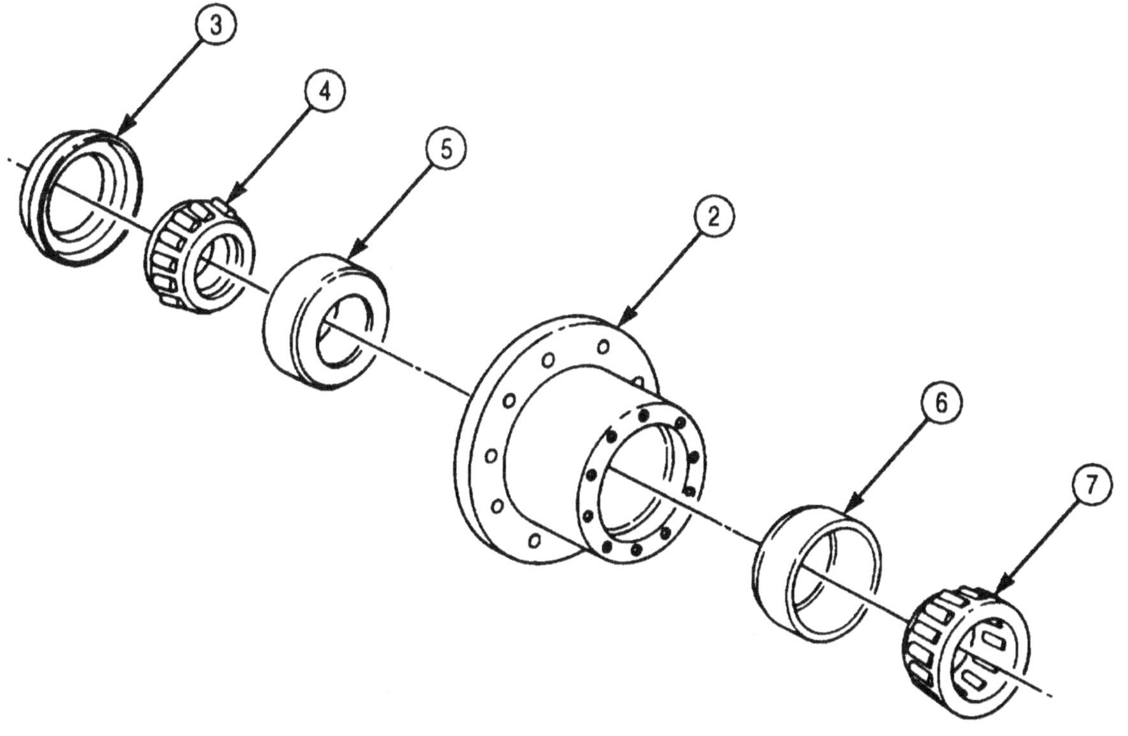

3-224. REAR HUB AND DRUM (M939/A1) MAINTENANCE (Contd)

d. Installation

1. Install dustshield (2) on inside of drum (1) with five washers (3) and screws (4).
2. Position hub (6) on drum (1).
3. Press wheel studs (5) through drum (1) and hub (6). Ensure studs (5) are seated.
4. Install inner bearing (10) and new inner bearing oil seal (9) in hub (6).
5. Using bearing replacer, install new wiper (8) on axle housing (7).

NOTE

Assistant will help with step 6.

6. Install hub (6), drum (1), new keyway insert (21), outer bearing (20), and new outer bearing oil seal (19) on axle housing (7) with bearing adjusting nut (18).
7. Tighten bearing adjusting nut (18) while turning hub (6). Tighten bearing adjusting nut (18) 50 lb-ft (68 N•m).
8. Loosen bearing adjusting nut (18) 1/6-1/4 turn so washer (16) can be positioned to axle housing (7) and adjusting nut pin (17).
9. Install outer bearing locknut (15) on axle housing (7). Tighten outer bearing locknut (15) 250-400 lb-ft (339-542 N•m).
10. Apply sealant to mating surfaces of axle shaft flange (12).
11. Install axle shaft (11) on hub (6) with ten washers (14) and screws (13). Tighten screws (13) 60-100 lb-ft (81-136 N•m).

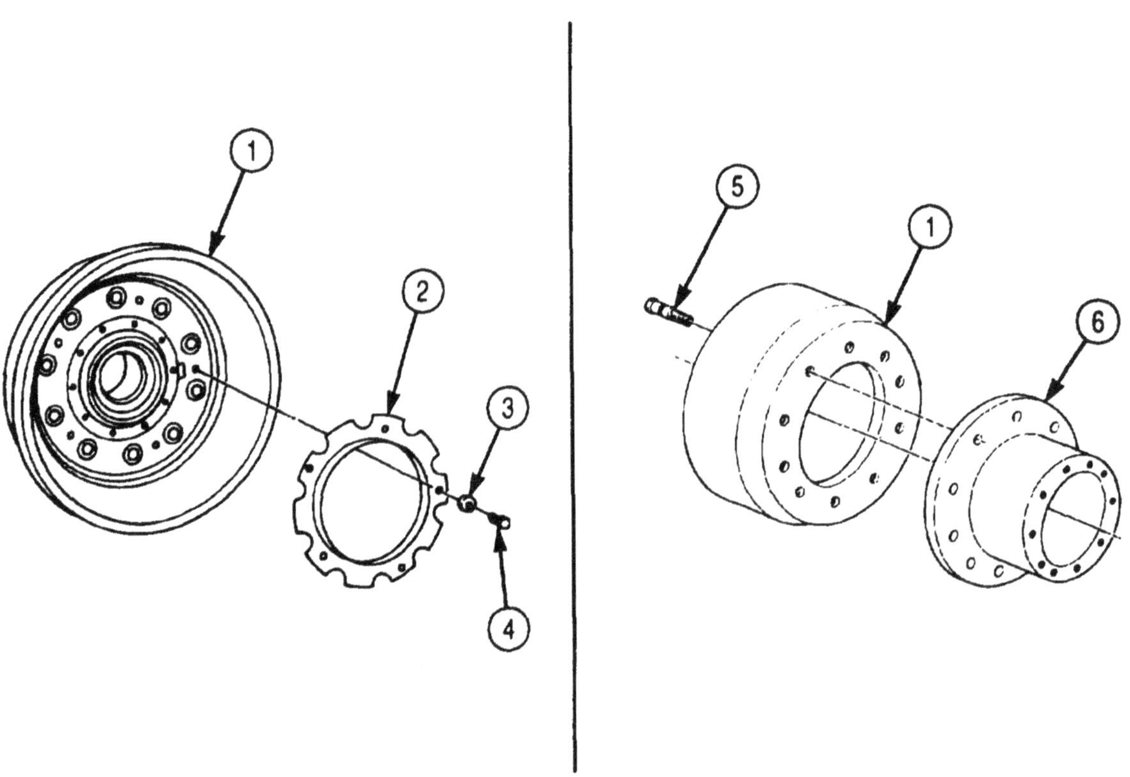

3-224. REAR HUB AND DRUM (M939/A1) MAINTENANCE (Contd)

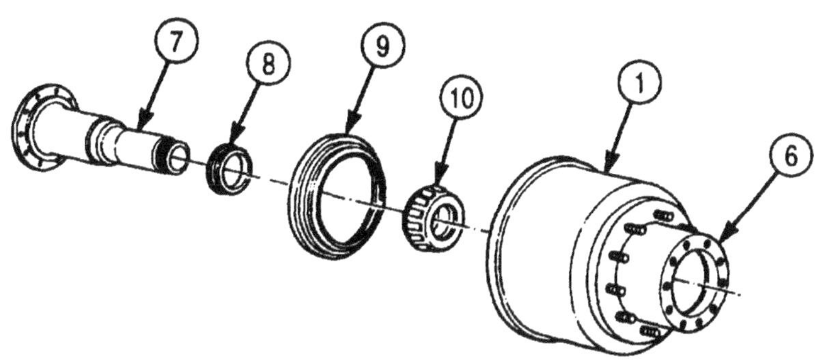

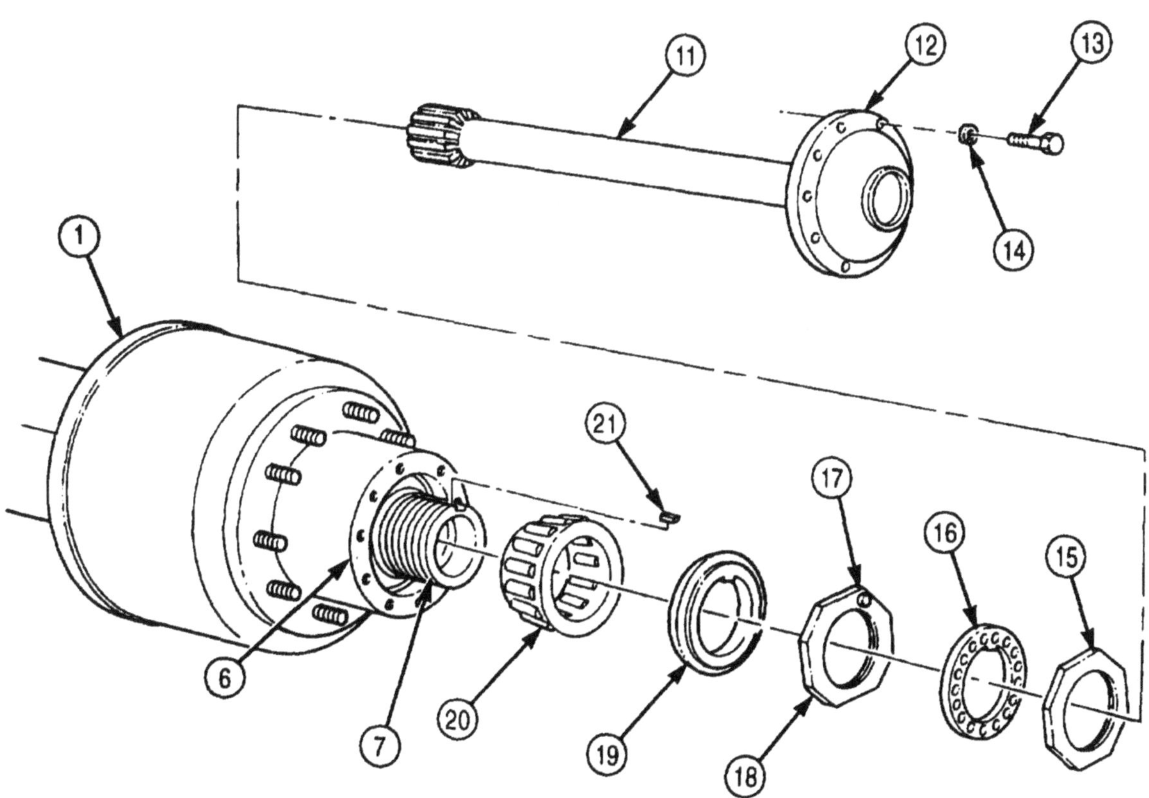

FOLLOW-ON TASKS: • Release caged spring brake(s) (TM 9-2320-272-10).
• Install wheel(s) (para. 3-218 or 3-219).
• Road test vehicle (TM 9-2320-272-10).

3-225. WHEEL BEARING ADJUSTMENT

THIS TASK COVERS:

a. Front Wheel Bearing Adjustment **b. Rear Wheel Bearing Adjustment**

INITIAL SETUP:

APPLICABLE MODELS
All

TOOLS
General mechanic's tool kit (Appendix E, Item 1)
Torque wrench (Appendix E, Item 145)

MATERIAL/PARTS
Two gaskets (Appendix D, Item 164)
Ten lockwashers (Appendix D, Item 350)
Gasket sealant (Appendix C, Item 30)

REFERENCES (TM)
TM 9-2320-272- 10
TM 9-2320-272-24P

EQUIPMENT CONDITION
• Wheels chocked (TM 9-2320-272-10).
• Spring brake(s) caged (TM 9-2320-272-10).

a. Front Wheel Bearing Adjustment

1. Raise vehicle until tire (1) is clear off the ground.
2. Remove ten screws (7) and lockwashers (6) from drive flange (5) and hub (9). Discard lockwashers (6).
3. Install two screws (7) into two threaded holes (8) to separate drive flange (5) from hub (9).
4. Remove drive flange (5) and gasket (4) from hub (9). Discard gasket (4).
5. Remove two screws (7) from drive flange (5).
6. Remove outer bearing locknut (3) and bearing nut washer (2) from spindle (11).
7. While rotating hub (9), tighten outer bearing adjusting nut (12) 50 lb-ft (68 N•m) to set preload.
8. Back outer bearing adjusting nut (12) 1/6-1/4 turn so bearing nut washer (2) can be positioned to spindle (11) and adjusting nut insert (10).
9. Install bearing nut washer (2) and outer bearing locknut (3) on spindle (11). Tighten outer bearing locknut (3) 250-400 lb-ft (339-542 N•m).
10. Install new gasket (4) and drive flange (5) on hub (9) with ten new lockwashers (6) and screws (7). Tighten screws (7) 60-100 lb-ft (81-136 N•m).
11. Lower vehicle to ground.

3-225. WHEEL BEARING ADJUSTMENT (Contd)

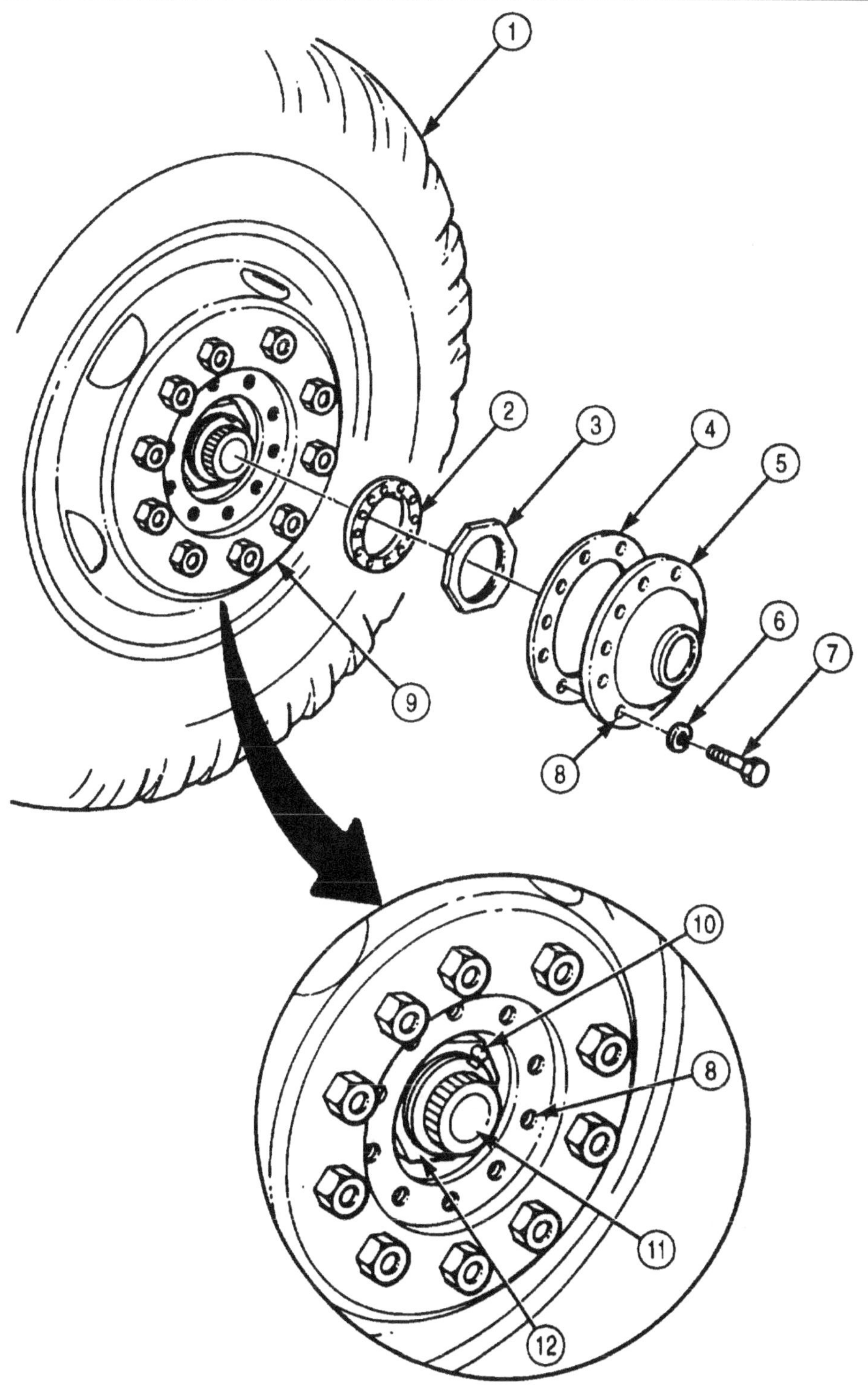

3-225. WHEEL BEARING ADJUSTMENT (Contd)

b. Rear Wheel Bearing Adjustment

1. Raise vehicle until tire (1) is clear off the ground.
2. Remove ten screws (7) and washers (6) from axle shaft (5).
3. Install two screws (7) into threaded holes (8) and remove axle shaft (5) and gasket (4) from hub (9). Discard gasket (4).
4. Remove two screws (7) from axle shaft (5).
5. Remove outer bearing locknut (3) and outer bearing washer (2) from axle housing (9).
6. While turning hub (ll), tighten bearing adjusting nut (10) 50 lb-ft (68 N•m).
7. Back out bearing adjusting nut (10) 1/6 - l/4 turn so outer bearing washer (2) can be positioned to axle housing (9) and adjusting nut insert (11).
8. Install outer bearing washer (2) and outer bearing locknut (3) on hub (11). Tighten outer bearing locknut (3) 250-400 lb-ft (339-542 N•m).
9. Apply sealant to mating surfaces of axle shaft (5).
10. Install new gasket (4) and axle shaft (5) on hub (11) with ten washers (6) and screws (7). Tighten screws (7) 60-100 lb-ft (81-136 N•m).
11. Lower vehicle until tire is on ground.

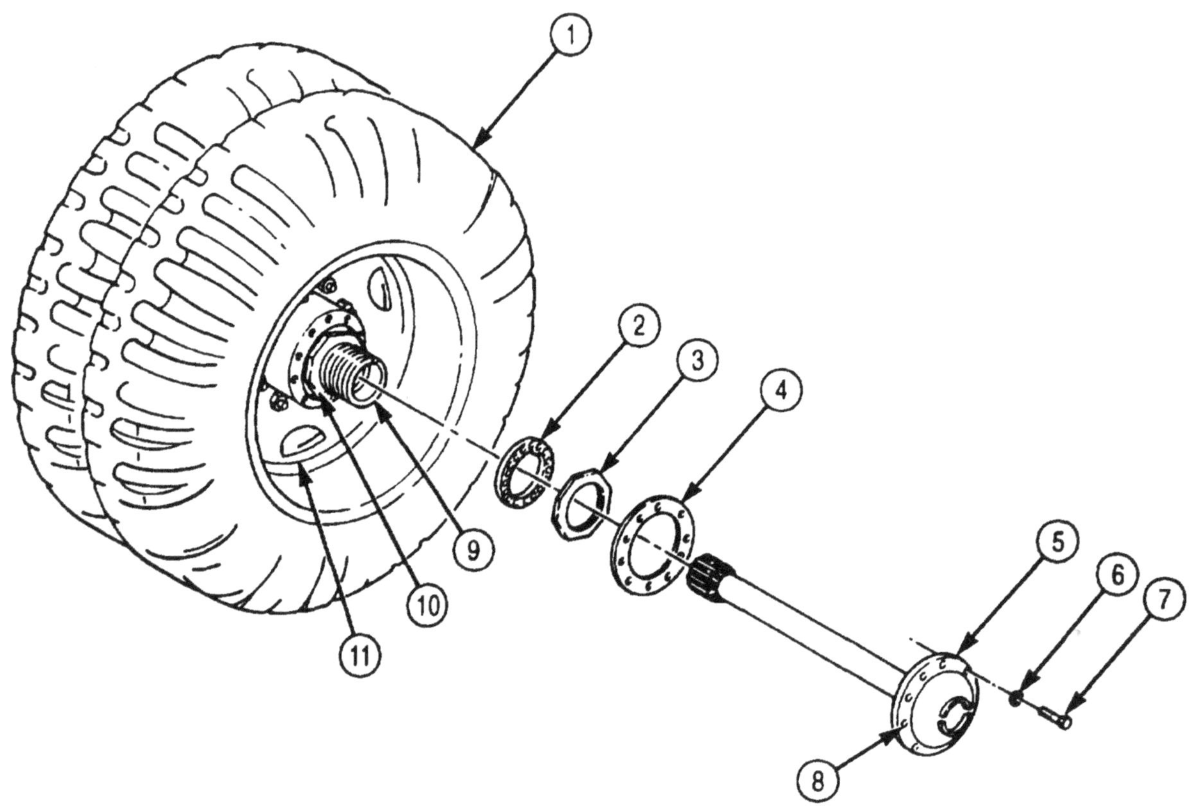

FOLLOW-ON TASKS: • Release caged spring brake(s) (TM 9-2320-272-10).
• Road test vehicle (TM 9-2320-272-10).

3-226. STEERING WHEEL REPLACEMENT

THIS TASK COVERS:

a. Removal b. Installation

APPLICABLE MODELS
All

TOOLS
General mechanic's tool kit (Appendix E, Item 1)
Torque wrench (Appendix E, Item 144)
Puller (Appendix E, Item 110)

REFERENCES (TM)
TM 9-2320-272-10

EQUIPMENT CONDITION
- Parking brake set (TM 9-2320-272-10).
- Wheels straight (TM 9-2320-272-10).
- Horn switch removed (para. 3-104).

a. Removal

1. Loosen clamp (6) and slide turn signal control (8) down steering column (7).

NOTE

Perform step 2 only if vehicle is equipped with hand airbrake control lever.

2. Loosen two screws (5) and slide hand airbrake control lever (4) down steering column (7).
3. Loosen nut (2) until flush with top of steering wheel shaft (3).
4. Install two adapters (14) and (13) on steering column (7) and steering wheel shaft (3).
5. Install puller on adapters (14) and (13) and tighten puller screw until steering wheel (1) is loose.
6. Remove puller, nut (2), two adapters (14) and (13), and steering wheel (1) from steering column (7).

NOTE

Perform steps 7 and 8 on vehicles equipped with turn signal canceling ring.

7. Remove spring (12) from steering wheel shaft (3).
8. Remove three screws (11), turn signal canceling ring 10), and steering wheel ring (9) from steering wheel (1).

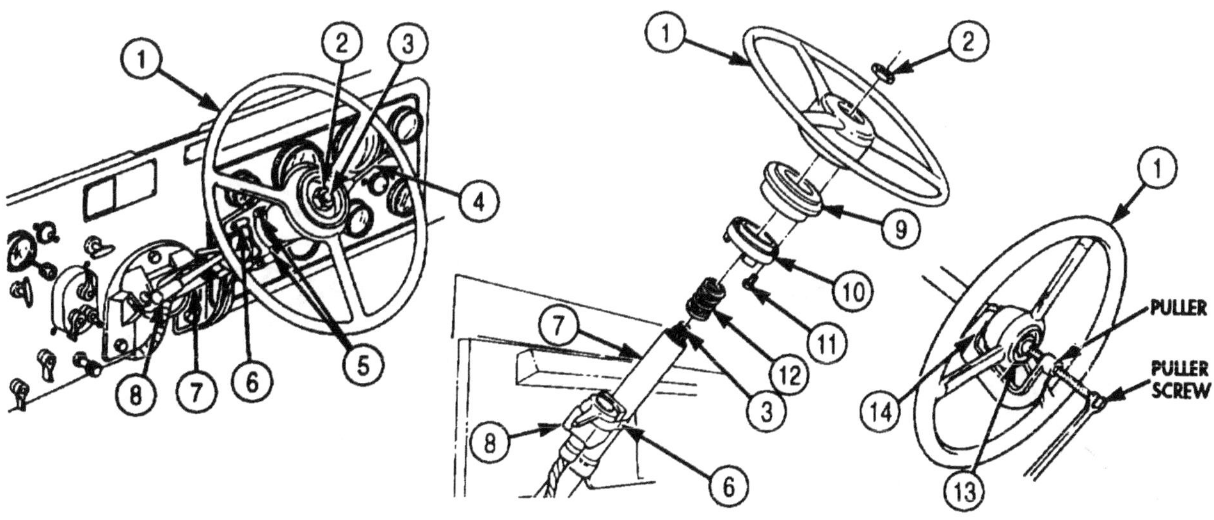

3-226. STEERING WHEEL REPLACEMENT (Contd)

b. Installation

NOTE

Perform steps 1 and 2 on vehicles equipped with turn signal canceling ring only.

1. Install spring (6) on steering column (10).
2. Install turn signal canceling ring (4) and steering wheel ring (3) on steering wheel (1) with three screws (5).
3. Install steering wheel (1) on steering column (10) by evenly tapping until nut (2) can be installed on steering shaft (7). Tighten nut (2) 55-60 lb-ft (75-81 N•m).

NOTE

Perform step 4 on vehicles equipped with hand airbrake control lever.

4. Slide hand airbrake control lever (11) up steering column (10) and tighten two screws (12).
5. Slide turn signal control (9) up steering column (10) and tighten clamp (8).

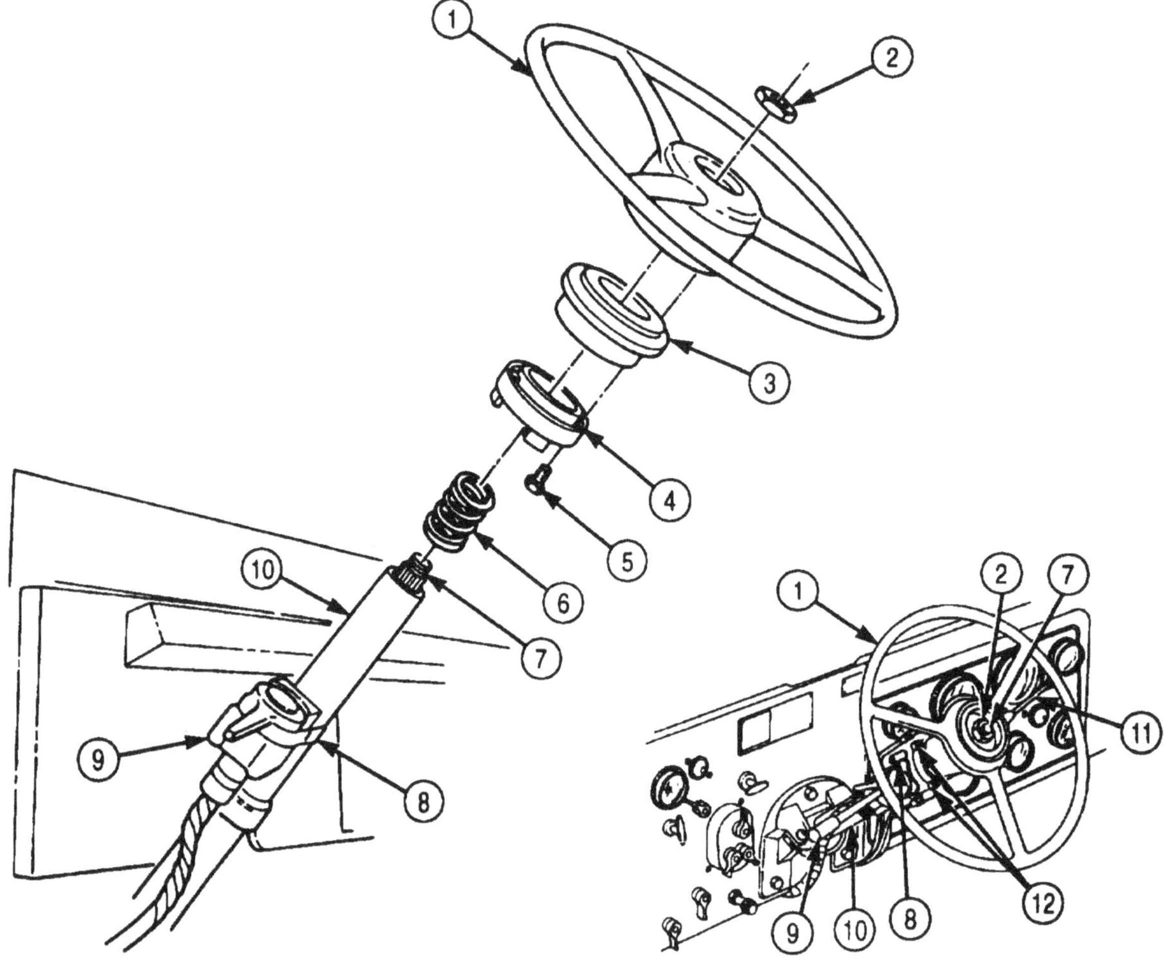

FOLLOW-ON TASK: Install horn switch (para. 3-104).

3-227. PITMAN ARM REPLACEMENT (ROSS)

THIS TASK COVERS:

a. Removal
b. Installation

INITIAL SETUP:

APPLICABLE MODELS
All

TOOLS
General mechanic's tool kit (Appendix E, Item 1)

MATERIALS/PARTS
Cotter pin (Appendix D, Item 85)
Locknut (Appendix D, Item 273)

REFERENCES (TM)
TM 9-2320-272-10
TM 9-2320-272-24P

EQUIPMENT CONDITION
Parking brake set (TM 9-2320-272-10).

a. Removal

1. Remove cotter pin (6) from slotted nut (5). Discard cotter pin (6).
2. Remove slotted nut (5) from drag link (4).
3. Remove locknut (3) and screw (1) from pitman arm (7). Discard locknut (3).
4. Remove pitman arm (7) from steering gear shaft (2) and drag link (4).

b. Installation

NOTE

Ensure pitman arm and steering gear shaft alignment marks match.

1. Install pitman arm (7) on steering gear shaft (2) and drag link (4).
2. Install pitman arm (7) on steering gear shaft (2) with screw (1) and new locknut (3).
3. Install pitman arm (7) on drag link (4) with slotted nut (5). Tighten slotted nut (5) 140 lb-ft (190 N•m).
4. Install cotter pin (6) on slotted nut (5) and drag link (4).

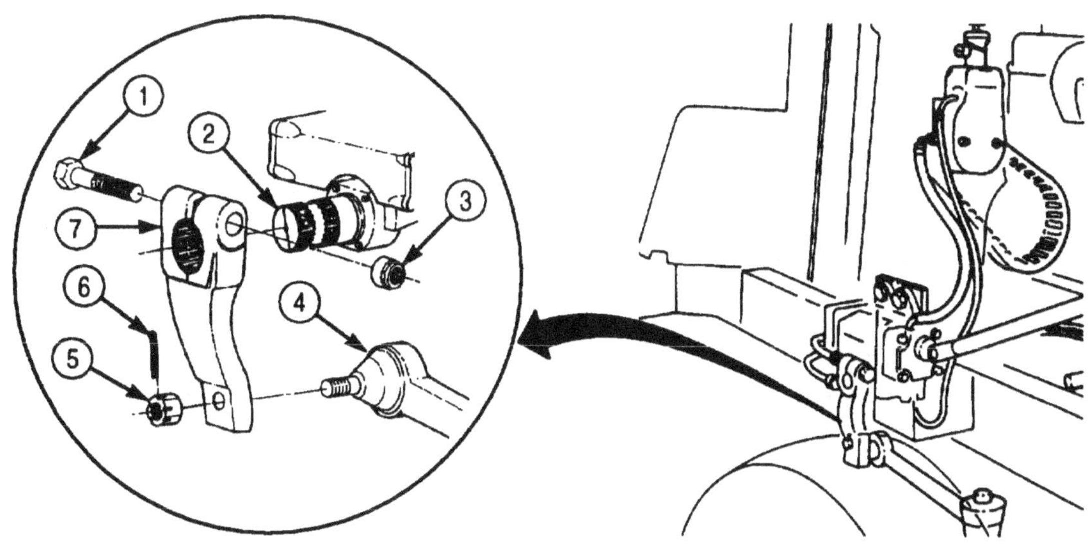

3-228. PITMAN ARM REPLACEMENT (SHEPPARD)

THIS TASK COVERS:

a. Removal **b. Installation**

INITIAL SETUP:

APPLICABLE MODELS
All

TOOLS
General mechanic's tool kit (Appendix E, Item 1)
Torque wrench (Appendix E, Item 145)

MATERIALS/PARTS
Steering parts kit (Appendix D, Item 677)

REFERENCES (TM)
TM 9-2320-272- 10
TM 9-2320-272-24P

EQUIPMENT CONDITION
- Parking brake set (TM 9-2320-272-10).
- Left splash shield removed (TM 9-2320-272-10).
- Drag link disconnected (para. 3-229).

a. Removal

1. Bend two long tabs (5) of retainer (6) out of notches in pitman arm (7).
2. Bend two short tabs (4) out of notches in retainer (6).
3. Remove retainer (6) and pitman arm (7) from shaft (8) of steering gear (1). Discard retainer (6), tab lockwasher (11), and friction washer (9).

b. Installation

1. Press new friction washer (9). and new tab lockwasher (11) into slot on new retainer (6).
2. Press three nylon balls (10) into indentations on retainer (6).
3. Align marks (2) and (3), and position pitman arm (7) on output shaft (8).
4. Install retainer (6) into shaft (8) until drag is felt on friction washer (9).
5. Align long tabs (5) to notches on pitman arm (7) and bend tabs (5) into notches.
6. Tighten retainer (6) 225 lb-ft (305 N•m).
7. Bend two short tabs (4) into notches on retainer (6). Tighten retainer (6), if necessary, to align notches.

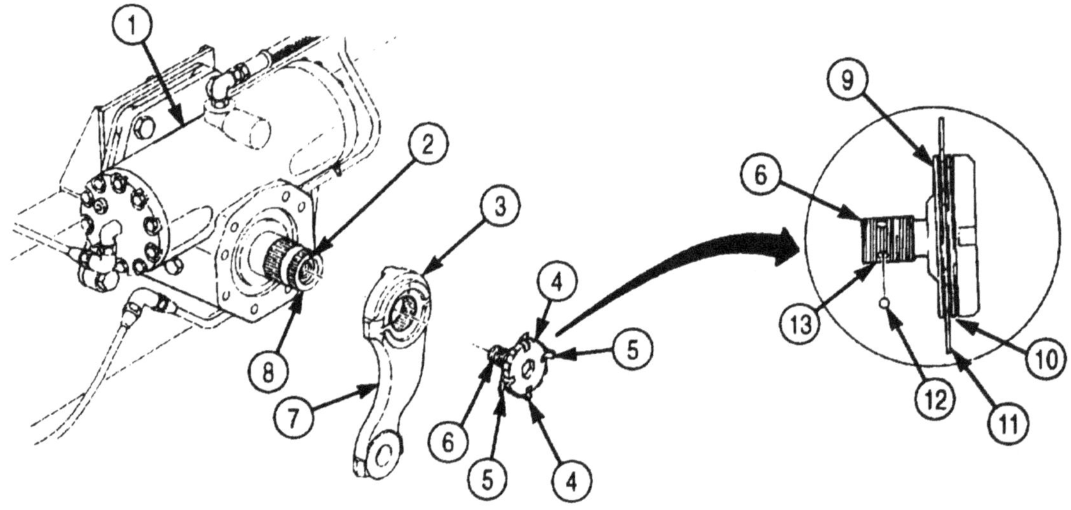

FOLLOW-ON TASKS: • Connect drag link (para. 3-229).
• Install left splash shield (TM 9-2320-272-10).

3-229. DRAG LINK REPLACEMENT

THIS TASKS COVERS:

a. Removal **b. Installation**

INITIAL SETUP:

APPLICABLE MODELS
All

TOOLS
General mechanic's tool kit (Appendix E, Item 1)
Torque wrench (Appendix E, Item 144)

MATERIALS/PARTS
Two cotter pins (Appendix D. Item 85)

REFERENCES (TM)
TM 9-2320-272-10
TM 9-2320-272-24P

EQUIPMENT CONDITION
Parking brake set (TM 9-2320-272-10).

a. Removal

1. Remove two cotter pins (5) from two slotted nuts (6) and drag link (2). Discard cotter pins (5).
2. Remove two slotted nuts (6) from drag link (2).
3. Remove drag link (2) from pitman arm (7) and steering knuckle arm (4).
4. Remove two rubber boots (3) from drag link (2).
5. Remove two grease fittings (1) from drag link (2).

b. Installation

1. Install two grease fittings (1) on drag link (2).
2. Install two rubber boots (3) on drag link (2).
3. Install drag link (2) on steering knuckle arm (4) and pitman arm (7) with two slotted nuts (6). Tighten two slotted nuts (6) 140 lb-ft (190 N•m).
4. Install two new cotter pins (5) in slotted nuts (6) and drag link (2).

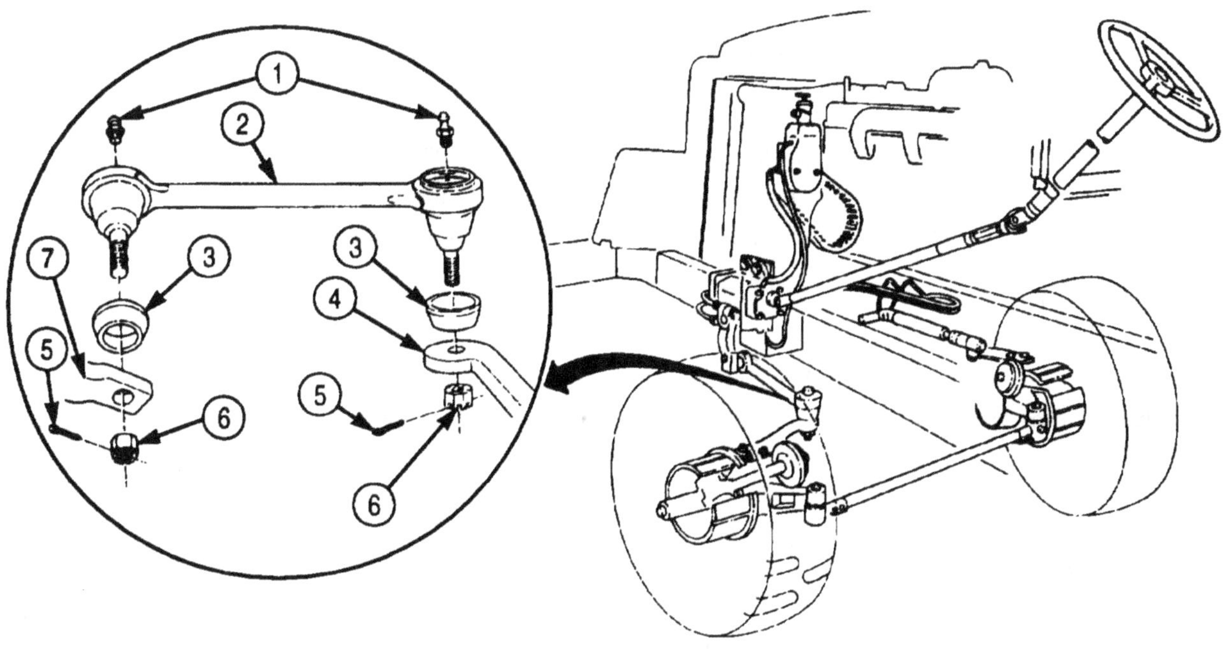

3-230. STEERING PUMP DRIVEBELTS (M939/A1) MAINTENANCE

THIS TASK COVERS:

a. Adjustment
b. Removal
c. Inspection
d. Installation

INITIAL SETUP:

APPLICABLE MODELS
M939/A1

SPECIAL TOOLS
Belt tension gauge (Appendix D, Item 16)

TOOLS
General mechanic's tool kit (Appendix E, Item 1)
Torque wrench (Appendix E, Item 144)
Prybar

REFERENCES (TM)
TM 9-2320-272-10
TM 9-2320-272-24P

EQUIPMENT CONDITION
- Parking brake set (TM 9-2320-272-10).
- Left splash shield removed (TM 9-2320-272-10).
- Fan drivebelts removed (para. 3-70)

a. Adjustment

1. Loosen screw (6) and nut (5) from pump adjusting link (4) and pump bracket (7).
2. Loosen two screws (1) from pump bracket (7) and mounting bracket (10).
3. Place prybar beneath steering pump (2) so end rests on engine (9).
4. Using engine (9) for leverage, push prybar upward until drivebelts (11) appear tight.
5. Tighten screw (6) and nut (5).
6. Tighten two screws (1) 30-40 lb-ft (41-54 N•m).

NOTE

If belt tension cannot be properly adjusted, replace drivebelts.

7. Position belt tension gauge on drivebelt (11) between pump pulley (3) and accessory drive pulley (8).
   a. New belt (12) tension should be 95-105 lb (418-462 N•m).
   b. Used belt (12) tension should be 85-95 lb (378-422 N•m).

b. Removal

1. Loosen nut (5) on pump adjusting link (4) and pump bracket (7).
2. Loosen two screws (1) on pump bracket (7) and mounting bracket (10).
3. While pushing downward on power steering pump (2), remove two drivebelts (11).

c. Inspection

NOTE

Pump drivebelts must be replaced in matched sets.

Inspect two pump drivebelts (11) for cracks, splits, breaks, and wear. Replace both drivebelts (11) if either is cracked, split, broken, or worn.

3-230. STEERING PUMP DRIVEBELTS (M939/A1) MAINTENANCE (Contd)

d. Installation

1. Place two drivebelts (11) over pump pulley (3) and third and fourth slots of accessory drive pulley (8).
2. Perform adjustment, task a.

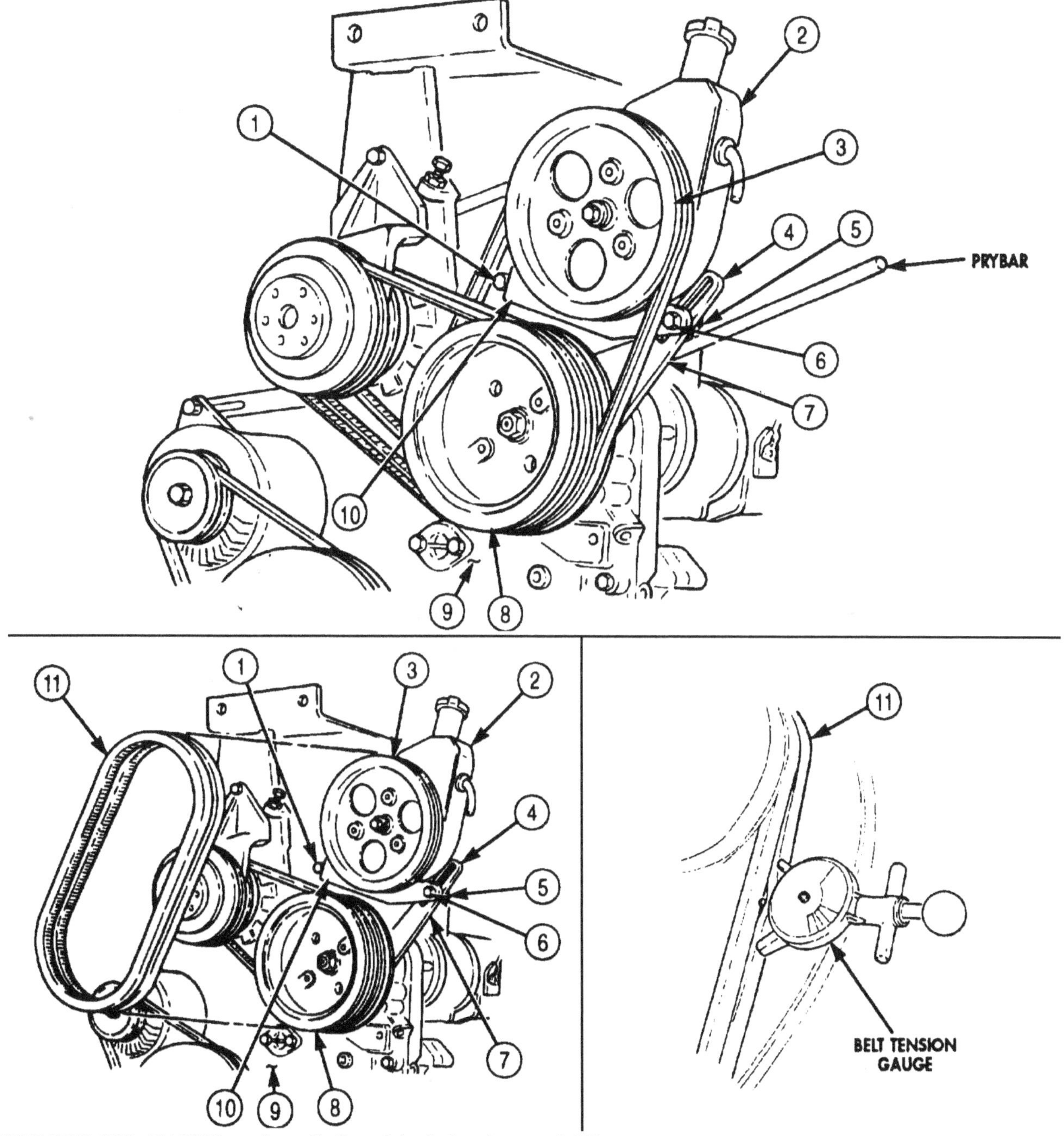

FOLLOW-ON TASKS:
- Install fan drivebelts (para. 3-70).
- Install left splash shield (TM 9-2320-272-10).

3-231. STEERING ASSIST CYLINDER STONE SHIELD REPLACEMENT

THIS TASK COVERS:

a. Removal b. Installation

INITIAL SETUP:

APPLICABLE MODELS:
All

TOOLS
General mechanic's tool kit (Appendix E, Item 1)

MATERIALS/PARTS
Five locknuts (Appendix D, Item 291)

REFERENCES (TM)
TM 9-2320-272-10
TM 9-2320-272-24P

EQUIPMENT CONDITION
- Parking brake set (TM 9-2320-272-10).
- Right splash shield removed (TM 9-2320-272-10).

a. Removal

1. Remove three locknuts (5), screws (l), and washers (2) from stone shield (11), frame rail (13), and transmission oil cooler support (12). Discard locknuts (5).
2. Remove locknut (4), cable clamp (3), screw (8), and washer (9) from stone shield (11) and transmission oil cooler support (12). Discard locknut (4).
3. Remove locknut (4), screw (7), and washer (6) from stone shield (11). Discard locknut (4).
4. Remove two screws (7), washers (6), splash shield guide (10), and stone shield (11) from transmission oil cooler support (12).

b. Installation

1. Install stone shield (11) on frame rail (13) and transmission oil cooler support (12) with three washers (2), screws (1), and new locknuts (5).
2. Install cable clamp (3) and stone shield (11) on transmission oil cooler support (12) with washer (9), screw (8), and locknut (4).
3. Secure stone shield (11) with washer (6), screw (7), and new locknut (4).
4. Install stone shield (11) and splash shield guide (10) on transmission oil cooler support (12) with two washers (6) and screws (7).

3-231. STEERING ASSIST CYLINDER STONE SHIELD REPLACEMENT (Contd)

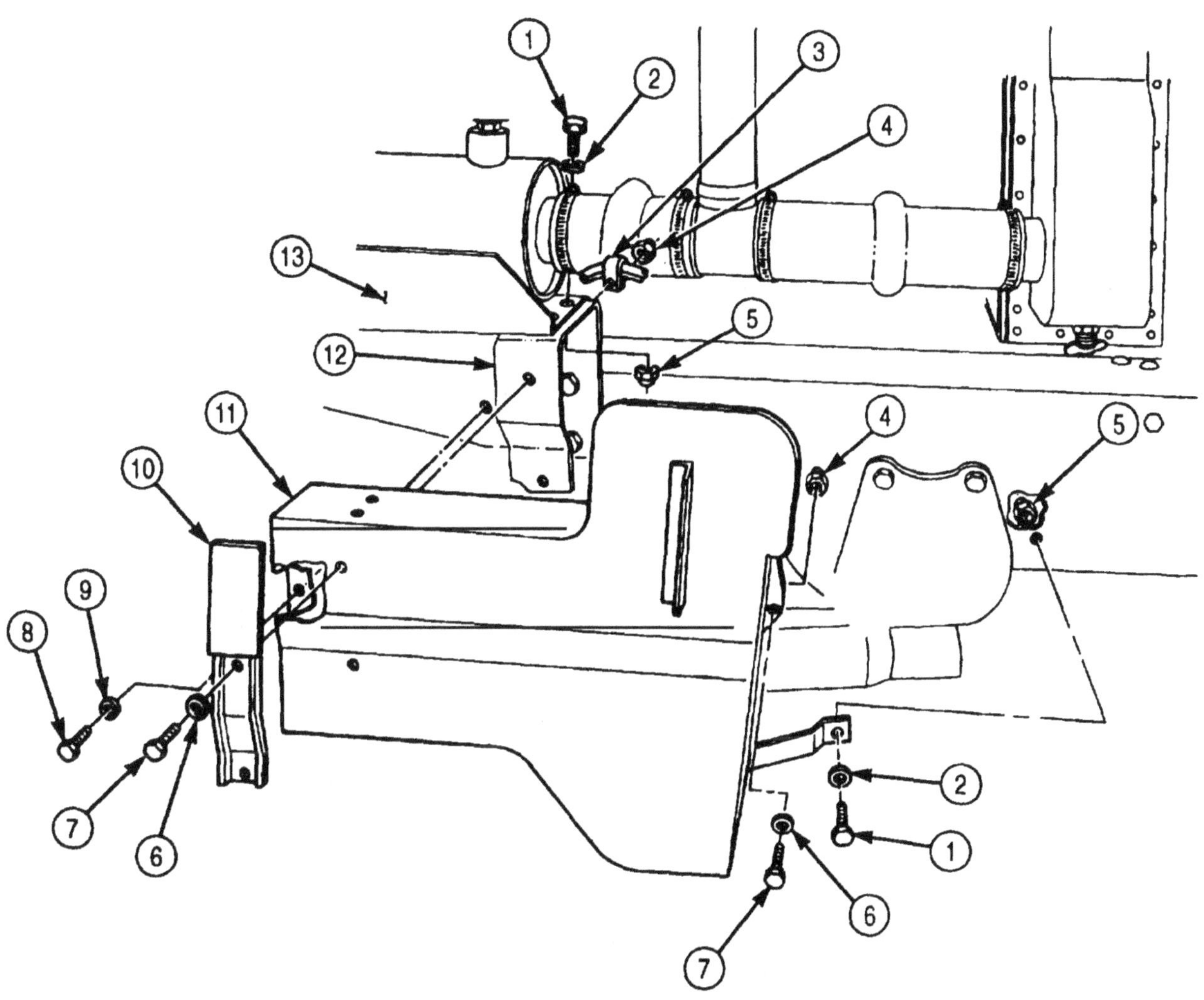

FOLLOW-ON TASK: Install right splash shield (TM 9-2320-272-10).

3-232. STEERING ASSIST CYLINDER HOSES REPLACEMENT

THIS TASK COVERS:

a. Removal b. Installation

INITIAL SETUP:

APPLICABLE MODELS
All

TOOLS
General mechanic's tool kit (Appendix E, Item 1)

MATERIALS/PARTS
Cap and plug set (Appendix C, Item 14)

REFERENCES (TM)
LO 9-2320-272-12
TM 9-2320-272-10
TM 9-2320-272-24P

EQUIPMENT CONDITION
Steering assist cylinder stone shield removed (para. 3-231).

GENERAL SAFETY INSTRUCTIONS
Do not start engine when steering assist cylinder hoses are disconnected.

WARNING

Do not start engine when steering assist cylinder hoses are disconnected. Pressure may whip hoses, causing injury to personnel.

a. Removal

CAUTION

Cap or plug all openings immediately after disconnecting lines and hoses to prevent contamination. Failure to do so may result in steering system damage.

NOTE

- Have container ready to catch hydraulic oil.
- Tag all hydraulic lines and hoses for installation.

Disconnect hoses (1) and (2) from elbows (5) and (6) and adapters (3) and (4).

b. Installation

NOTE

- Do not reuse hydraulic oil.
- Remove all caps and plugs prior to installation.

Install hoses (1) and (2) on adapters (3) and (4) and elbows (5) and (6)1.

3-232. STEERING ASSIST CYLINDER HOSES REPLACEMENT (Contd)

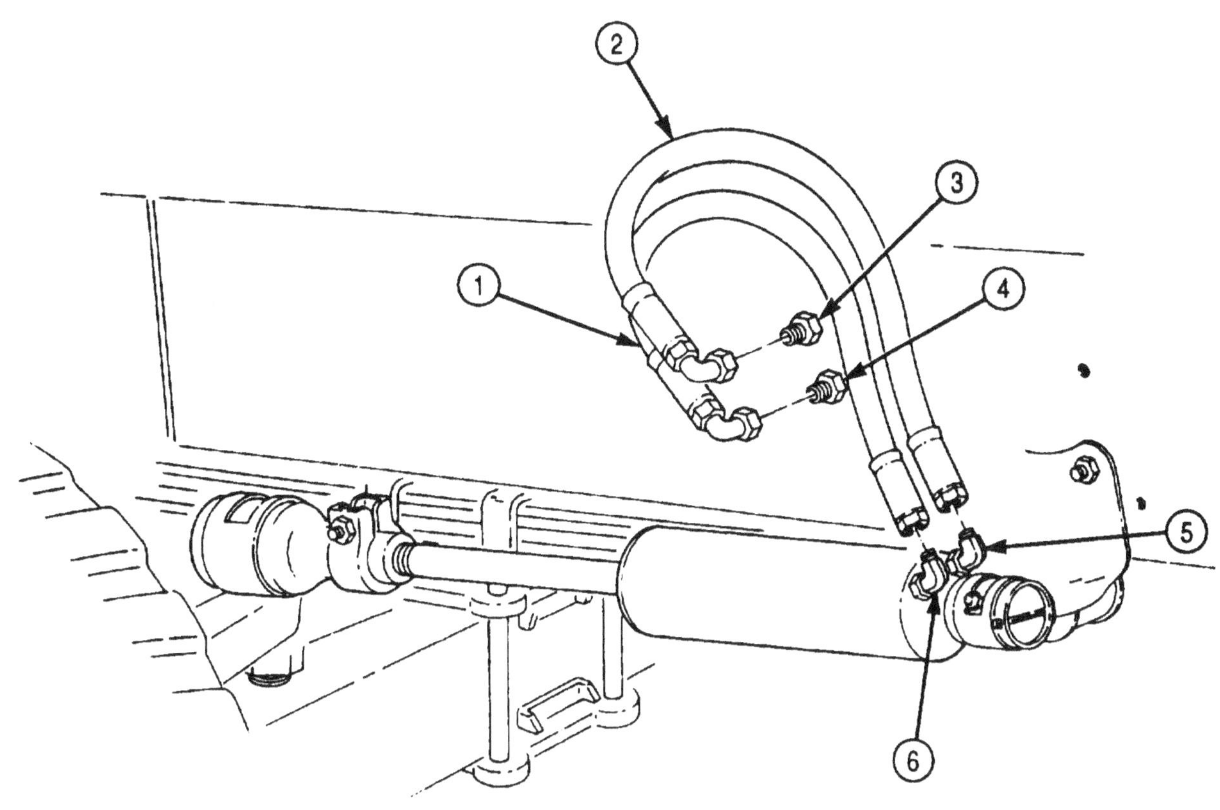

FOLLOW-ON TASKS:
- Fill power steering reservoir to proper level (LO 9-2320-272-12).
- Start engine (TM 9-2320-272-10). Check hoses for leaks and steering for proper operation.
- Install steering assist cylinder stone shield (para. 3-231).

3-233. STEERING ASSIST CYLINDER MAINTENANCE

THIS TASK COVERS:

a. Removal
b. Disassembly
c. Assembly
d. Installation
e. Travel Adjustment

INITIAL SETUP:

APPLICABLE MODELS

All

TOOLS

General mechanic's tool kit (Appendix E. Item 1)
Torque wrench (Appendix E, Item 146)

MATERIALS/PARTS

Two cotter pins (Appendix D, Item 49)
One cotter pin (Appendix D, Item 85)
Cap and plug set (Appendix C, Item 14)

REFERENCES (TM)

TM 9-2320-272-10
TM 9-2320-272-24P

EQUIPMENT CONDITION

- Parking brake set (TM 9-2320-272-10).
- Steering assist cylinder stone shield removed (para. 3-231).

GENERAL SAFETY INSTRUCTIONS

Do not start vehicle when steering hoses are disconnected.

WARNING

Do not start engine when steering hoses are disconnected. Pressure may whip hoses, causing injury to personnel.

a. Removal

CAUTION

Cap or plug all openings immediately after disconnecting hydraulic lines and hoses to prevent contamination. Failure to do so may result in steering system damage.

NOTE

- Have container ready to catch hydraulic oil.
- Tag all lines and hoses for installation.

1. Disconnect return hoses (4) and (5) from elbows (6).
2. Remove two grease fittings (2) from steering assist cylinder (11).
3. Remove two cotter pins (1) from steering assist cylinder (11). Discard cotter pins (1).
4. Loosen adjustable plugs (10) and (3), as far as possible without removing, and loosen two dustcovers (8).
5. Tap on adjustable plugs (10) and (3) to loosen steering assist cylinder (11) from spring shackle ball stud (7) and steering knuckle ball stud (12).
6. Remove steering assist cylinder (11) from ball studs (7) and (12).
7. Remove two dustcovers (8) and felt pads (9) from ball studs (7) and (12).
8. Remove cotter pin (13) and nut (14) from steering knuckle ball stud (12). Discard cotter pin (13).

3-233. STEERING ASSIST CYLINDER MAINTENANCE (Contd)

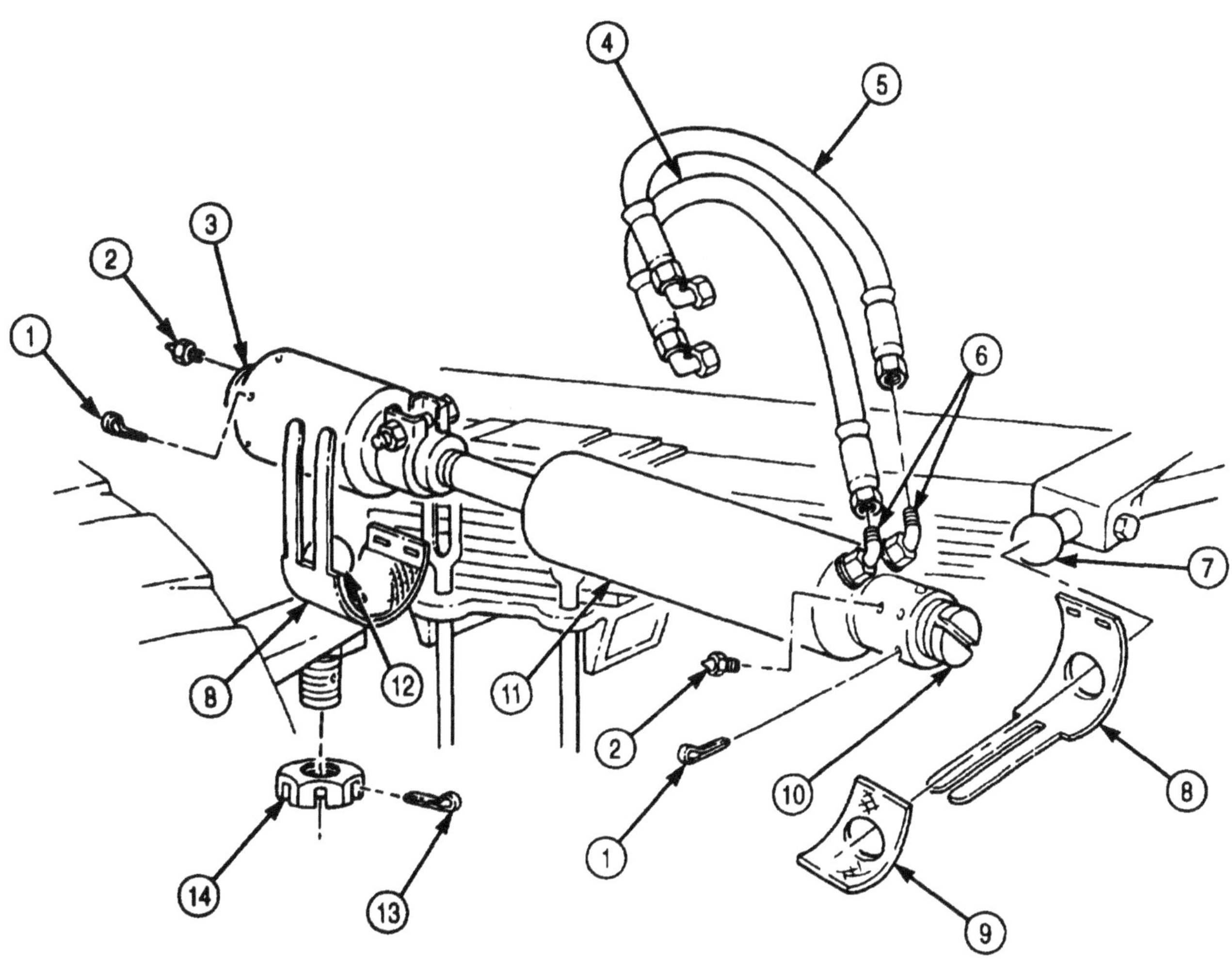

3-233. STEERING ASSIST CYLINDER MAINTENANCE (Contd)

b. Disassembly

NOTE

Mark direction of adapter elbows.

1. Remove two adapter elbows (9) and from steering assist cylinder (8).

NOTE

Mark direction of seat valves.

2. Remove two adjustable plugs (l), seat valves (2), spring (3), and ring (4) from socket assembly (5).

NOTE

Mark socket assembly location on shaft with chalk before removing.

3. Remove nut (11), screw (7), clamp (6), and socket assembly (5) from shaft (10).

c. Assembly

1. Install clamp (6) loosely on socket assembly (5) with screw (7) and nut (11).
2. Install socket assembly (5) on shaft (10).
3. Tighten nut (11) on clamp (6) 30-40 lb-ft (41-54 N•m).

3-233. STEERING ASSIST CYLINDER MAINTENANCE (Contd)

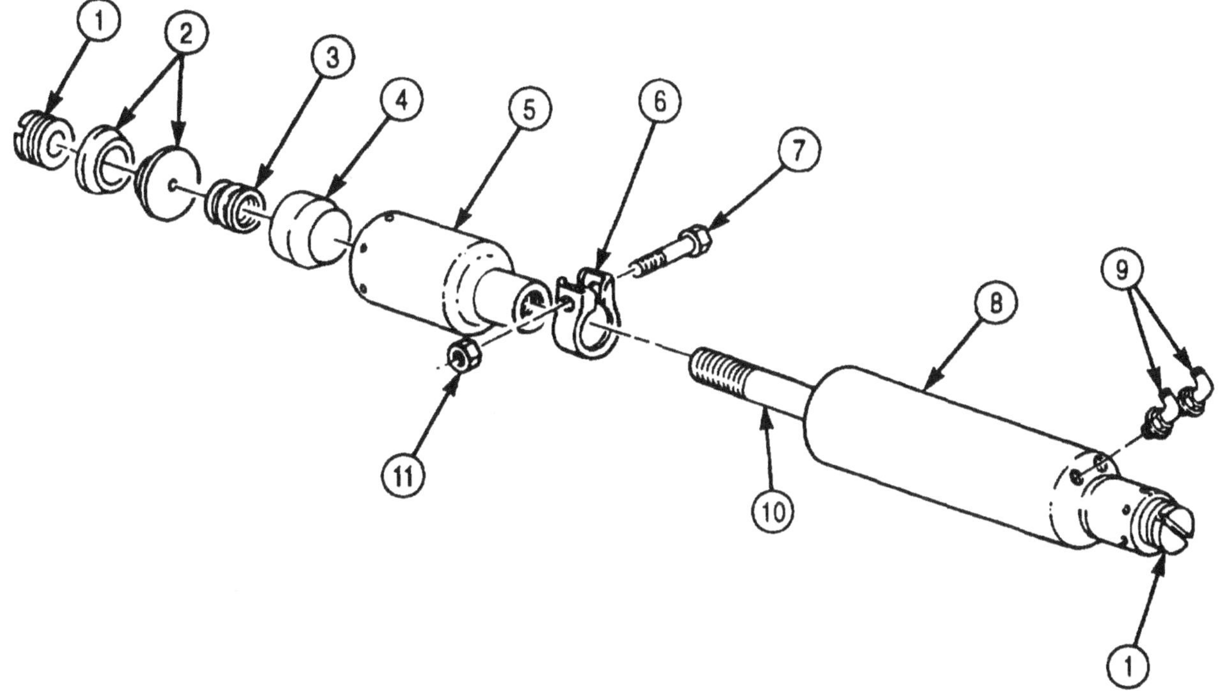

3-233. STEERING ASSIST CYLINDER MAINTENANCE (Contd)

4. Install ring (5), spring (4), and two seat valves (3) in socket assembly (6).
5. Install adjustable plugs (2) and (15) loosely on steering assist cylinder (18) and socket assembly (6).
6. Install grease fittings (1) and (17) on steering assist cylinder (18).
7. Install two adapter elbows (11) on steering assist cylinder (18).

d. Installation

NOTE

- Do not reuse hydraulic oil.
- Remove all caps and plugs prior to installation.

1. Install nut (21) and new cotter pin (19) on ball stud (22).
2. Install dustcovers (13) and (20) and felt pads (14) on ball studs (12) and (22).
3. Position steering assist cylinder (18) on steering knuckle ball stud (22) and spring shackle ball stud (12).
4. Perform travel adjustment task e. on steering assist cylinder (18).
5. Tighten adjustable plugs (2) and (15) until they bottom out, back out one turn, and align and install new cotter pins (24) and (16).
6. Install pressure return hoses (9) and (10) on elbows (11).
7. Fasten dustcovers (13) and (20) around steering assist cylinder (18) and socket assembly (6).

e. Travel Adjustment

NOTE

To check for proper travel adjustment, measure the distance from center of ball stud to center of ball stud for a distance of 25.50 in. (64.8 cm). If adjustment is incorrect, the front wheels must be raised prior to performing the following steps.

1. Loosen nut (23), screw (8), and clamp (7) from socket assembly (6).
2. Turn socket assembly (6) counterclockwise to increase steering assist cylinder (18) travel; clockwise to decrease cylinder (18) travel.
3. Tighten nut (23) 30-40 lb-ft (41-54 N•m).

3-233. STEERING ASSIST CYLINDER MAINTENANCE (Contd)

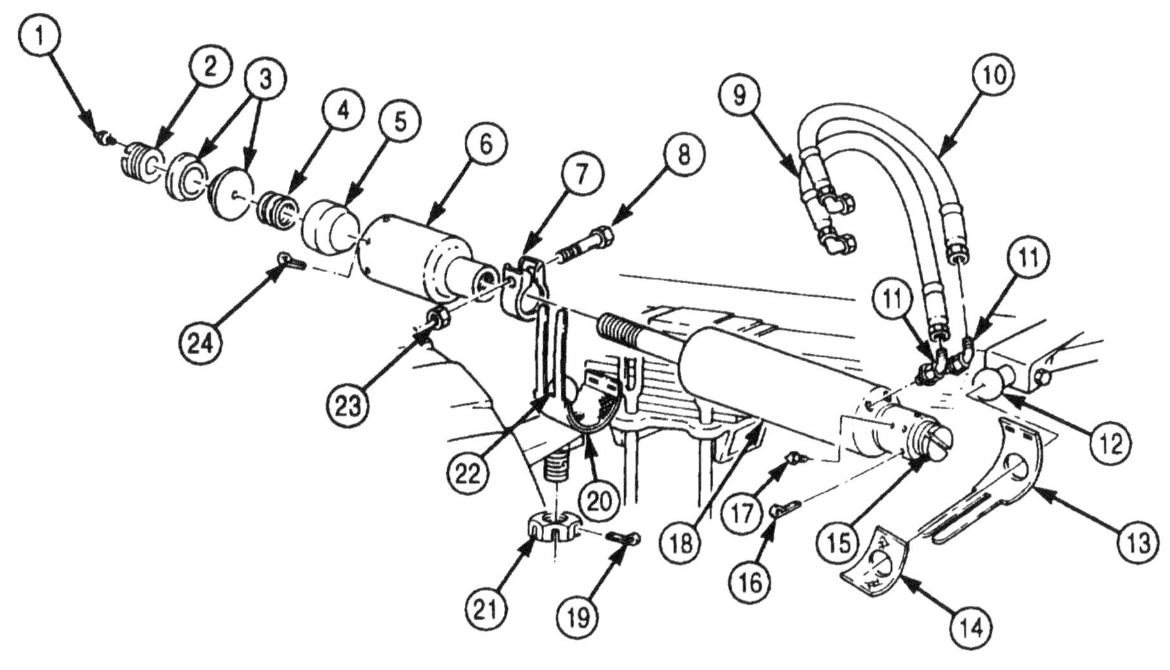

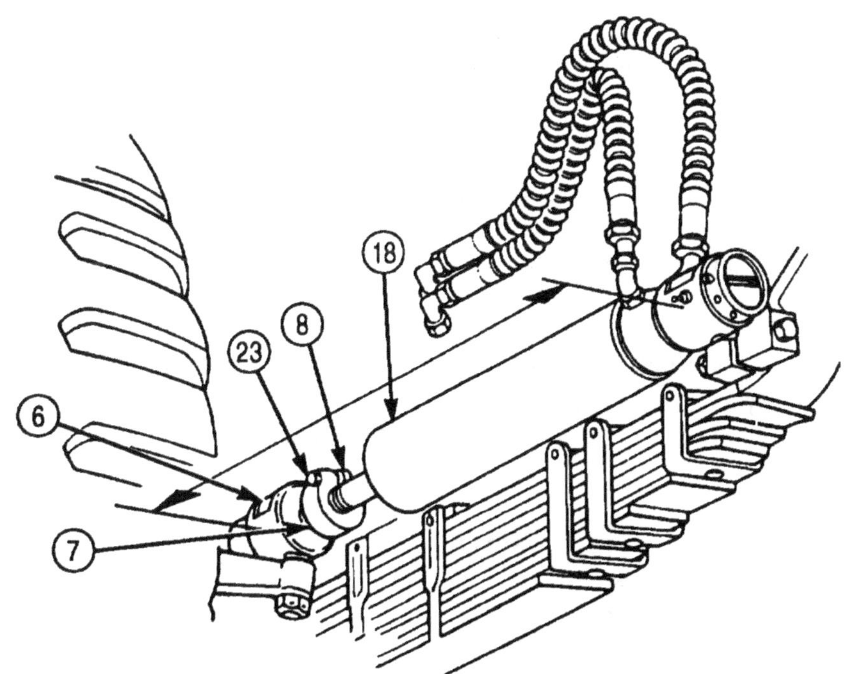

FOLLOW-ON TASK: Install steering assist cylinder stone shield (para. 3-231).

3-234. POWER STEERING PUMP PRESSURE AND RETURN HOSES REPLACEMENT (ROSS)

THIS TASK COVERS:

a. Removal
b. Installation

INITIAL SETUP:

APPLICABLE MODELS
All

TOOLS
General mechanic's tool kit (Appendix E, Item 1)

MATERIAL/PARTS
O-ring (Appendix D, Item 437)
Cap and plug set (Appendix D, Item 14)
Antiseize tape (Appendix C, Item 72)

REFERENCES (TM)
LO 9-2320-272-12
TM 9-2320-272-10
TM 9-2320-272-24P

EQUIPMENT CONDITION
Steering assist cylinder stone shield removed (para. 3-231).

GENERAL SAFETY INSTRUCTIONS
Do not start engine when steering hoses are disconnected.

WARNING

Do not start engine when steering hoses are disconnected. Pressure may whip hoses, causing injury to personnel.

a. Removal

CAUTION

Cap or plug all openings immediately after disconnecting hydraulic lines and hoses to prevent contamination. Failure to do so may result in damage to steering system.

1. Remove nut (13), screw (12), and retaining strap (4) from pump return hose (3) and return tube (5).

NOTE

Have container ready to catch hydraulic oil.

2. Loosen two hose clamps (2) and remove pump return hose (3) from pump nozzle (1) and return tube (5).
3. Disconnect pump return tube (5) from steering gear adapter (6).
4. Disconnect pump pressure hose (11) from pump adapter (14) and steering gear adapter elbow (10).
5. Remove adapter (6) from steering gear (7).
6. Remove elbow (10) and O-ring (8) from steering gear (7). Discard O-ring (8).

b. Installation

NOTE

- Do not reuse hydraulic oil.
- Remove all cap and plugs prior to installation.

1. Install new O-ring (8) and elbow (10) on steering gear (7). Tighten jamnut (9) until new O-ring (8) seats.
2. Wrap male pipe threads of steering gear adapter (6) with antiseize tape and install on steering gear (7).
3. Install pump return tube (5) on adapter (6).

3-234. POWER STEERING PUMP PRESSURE AND RETURN HOSES REPLACEMENT (ROSS) (Contd)

NOTE

Do not slide more than 1 in. (25.4 mm) of return hose onto return tube.

4. Install pump return hose (3) on pump nozzle (1) and return tube (5) with two clamps (2).
5. Install pump pressure hose (11) on pump adapter (14) and steering gear adapter elbow (10).
6. Install retaining strap (4) on hoses (3) and (11) with screw (12) and nut (13).

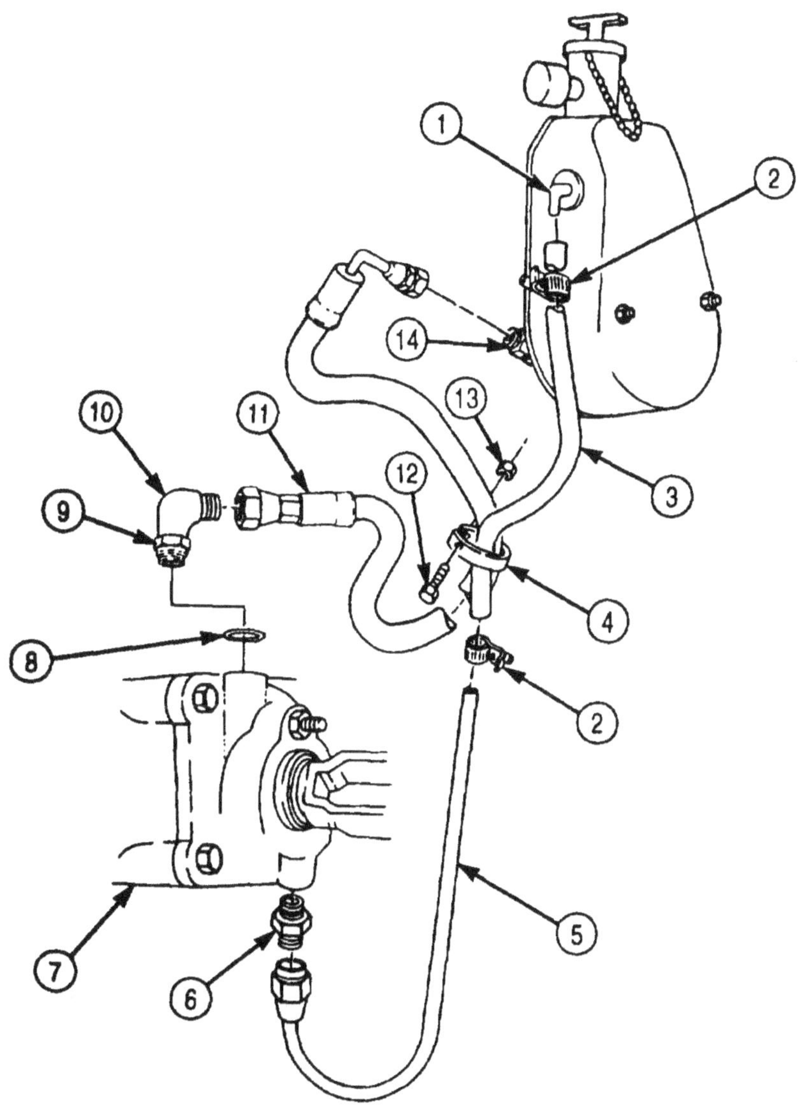

FOLLOW-ON TASKS:
- Fill power steering reservoir to proper level (LO 9-2320-272-12).
- Start engine (TM 9-2320-272-10), check hoses for leaks, and steering for proper operation.
- Install steering assist cylinder stone shield (para. 3-231).

3-235. POWER STEERING PUMP PRESSURE AND RETURN HOSES REPLACEMENT (SHEPPARD)

THIS TASK COVERS:

a. Removal

b. Installation

INITIAL SETUP:

APPLICABLE MODELS

All

TOOLS

General mechanic's tool kit (Appendix E, Item 1)

MATERIALS/PARTS

Two O-rings (Appendix D. Item 437)
Cap and plug set (Appendix C, Item 14)
Antiseize tape (Appendix C, Item 72)

REFERENCES (TM)

TM 9-2320-272-10
TM 9-2320-272-24P

EQUIPMENT CONDITION

- Parking brake set (TM 9-2320-272-10).
- Steering assist cylinder stone shield removed (para. 3-231).

GENERAL SAFETY INSTRUCTIONS

Do not start engine when steering hoses are disconnected.

WARNING

Do not start engine when steering hoses are disconnected. Pressure may whip hoses, causing injury to personnel.

a. Removal

CAUTION

Cap or plug all openings immediately after disconnecting lines and hoses to prevent contamination. Failure to do so may result in steering system damage.

NOTE

- Have container ready to catch hydraulic oil.
- Tag all hydraulic lines and hoses for installation.

1. Remove nut (5), screw (13), and retaining strap (4) from pump return hose (3) and return tube (6).
2. Loosen two hose clamps (2) and remove pump return hose (3) from pump nozzle (1) and return tube (6).
3. Disconnect pump return tube (6) from steering gear adapter (7).
4. Disconnect pump pressure hose (12) from pump adapter (14) and steering gear adapter elbow (11).
5. Remove adapter (7) and O-ring (8) from steering gear (9). Discard O-ring (8).
6. Remove adapter elbow (11) and O-ring (10) from steering gear (9). Discard O-ring (10).

b. Installation

NOTE

- Do not reuse hydraulic oil.
- Remove all cap and plugs prior to removal.

1. Install new O-ring (10) and steering gear adapter elbow (11) on steering gear (9).
2. Install new O-ring (8) and steering gear adapter (7) on steering gear (9).

NOTE

Do not slide more than 1 in. (25.4 mm) of return hose onto return tube.

3. Install pump return hose (3) on pump nozzle (1) and return tube (6) with two hose clamps (2).
4. Install pump pressure hose (12) on pump adapter (14) and elbow (11).
5. Install hose retaining strap (4) on hoses (3) and (6) with screw (13) and nut (5).

3-235. POWER STEERING PUMP PRESSURE AND RETURN HOSES REPLACEMENT (SHEPPARD) (Contd)

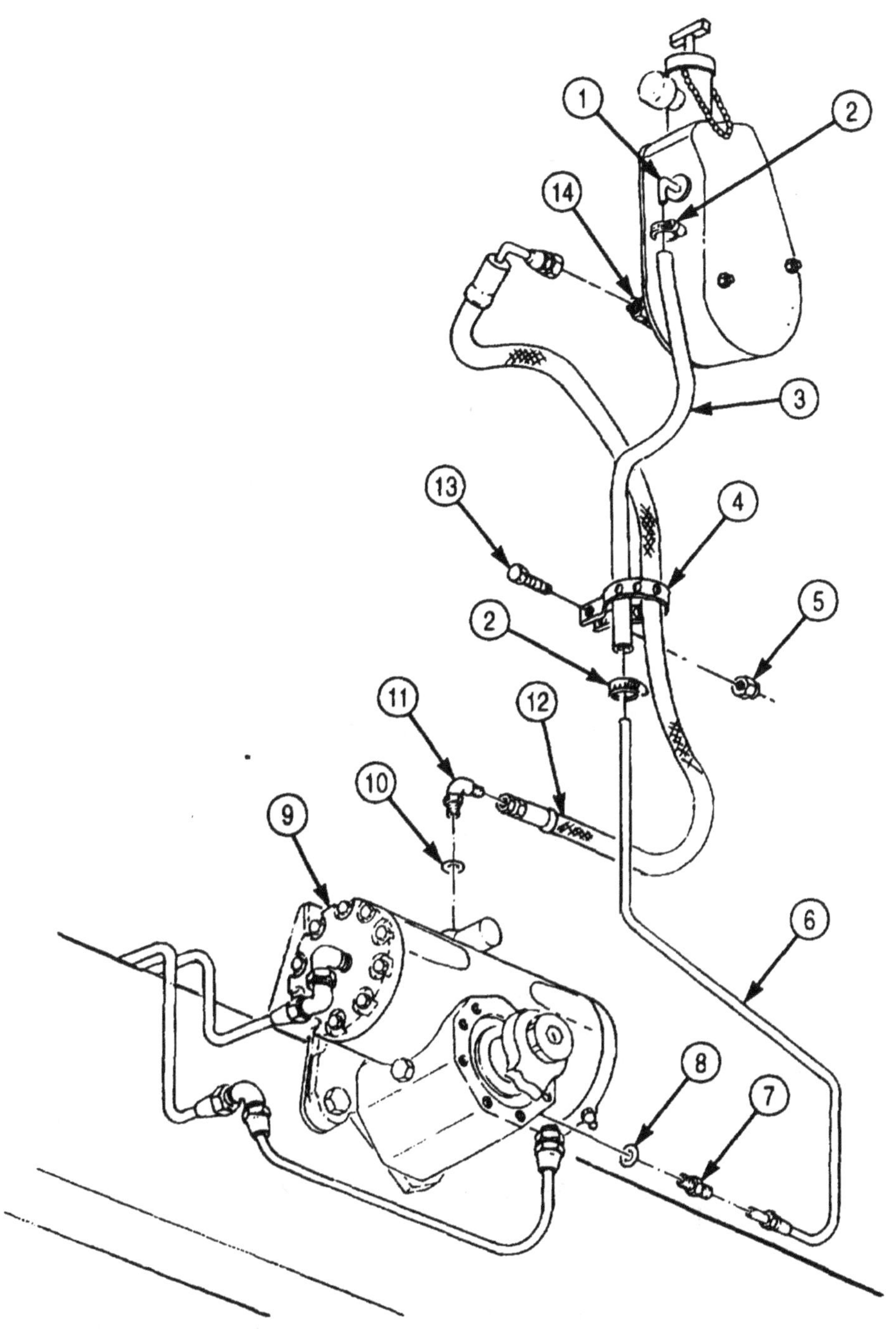

FOLLOW-ON TASKS:
- Install steering assist cylinder stone shield (para. 3-231).
- Check hydraulic oil level (TM 9-2320-272-10).
- Check for oil leaks (TM 9-2320-272-20).

3-236. POWER STEERING PUMP MAINTENANCE

THIS TASK COVERS:

a. Removal
b. Disassembly
c. Assembly
d. Installation

INITIAL SETUP:

APPLICABLE MODELS
All

TOOLS
General mechanic's tool kit (Appendix E, Item 1)

MATERIALS/PARTS
Springtite assembly (Appendix D, Item 675)
Woodruff key (Appendix D, Item 727)
Three lockwashers (Appendix D, Item 382)
Two lockwashers (Appendix D, Item 354)
Locknut (Appendix D, Item 327)
Cap and plug set (Appendix C, Item 14)

MANUAL REFERENCES (TM)
LO 9-2320-272-12
TM 9-2320-272-10
TM 9-2320-272-24P

EQUIPMENT CONDITION
- Parking brake set (TM 9-2320-272-10).
- Hood raised and secured (TM 9-2320-272-10).
- Steering pump drivebelts removed (M939/A1) (para. 3-230).

GENERAL SAFETY INSTRUCTIONS
Do not start engine when power steering hoses are disconnected.

WARNING

Do not start engine when power steering hoses are disconnected. Pressure may whip hoses, causing injury to personnel.

NOTE

M9391A1 and M939A2 power steering pumps are maintained basically the same. This procedure is for M939/A1.

a. Removal

CAUTION

Cap or lug all openings immediately after disconnecting hydraulic lines and hoses to prevent contamination. Failure to do so may result in steering system damage.

NOTE

- Have drainage container ready to catch hydraulic oil.
- Tag all hydraulic lines and hoses for installation.

1. Loosen clamp (4) and remove hose (5) from steering pump return tube (3).
2. Loosen nut (7) and remove high-pressure hose (6) from steering pump high-pressure fitting (8).
3. Remove three screws (18), lockwashers (19), and steering pump (2) from mounting bracket (17). Discard lockwashers (19).
4. Remove locknut (9), washer (10), screw (13), washer (12), screw (14), washer (15), and adjusting link (11) from mounting bracket (17) and engine (20). Discard locknut (9).

b. Disassembly

1. Remove two screws (24), lockwashers (23), washers (22), and mounting bracket (17) from engine mounting bracket (21). Discard lockwashers (23).
2. Remove springtite (28), washer (27), pulley (1), and woodruff key (26) from steering pump (2). Discard woodruff key (26) and springtite (28).
3. Remove breather (25) from steering pump (2).

3-236. POWER STEERING PUMP MAINTENANCE (Contd)

c. Assembly

1. Install breather (25) on steering pump (2).
2. Install new woodruff key (26). Press pulley (1) on pump shaft (29) and install washer (27) and new springtite (28).
3. Install mounting bracket (17) on engine bracket (21) with two washers (22), new lockwashers (23), and screws (24).

d. Installation

1. Install adjusting link (11) on mounting bracket (17) with washer (12), screw (13), washer (10), and new locknut (9).
2. Install adjusting link (11) on engine (20) with washer (15) and screw (14).
3. Install steering pump (2) on mounting bracket (17) with three new lockwashers (19) and screws (18).

NOTE

- Do not reuse hydraulic oil.
- Remove all caps and plugs prior to installation.

4. Install return hose (5) on return tube (3) with clamp (4).
5. Install high-pressure return hose (6) on high-pressure fitting (8) and tighten nut (7).

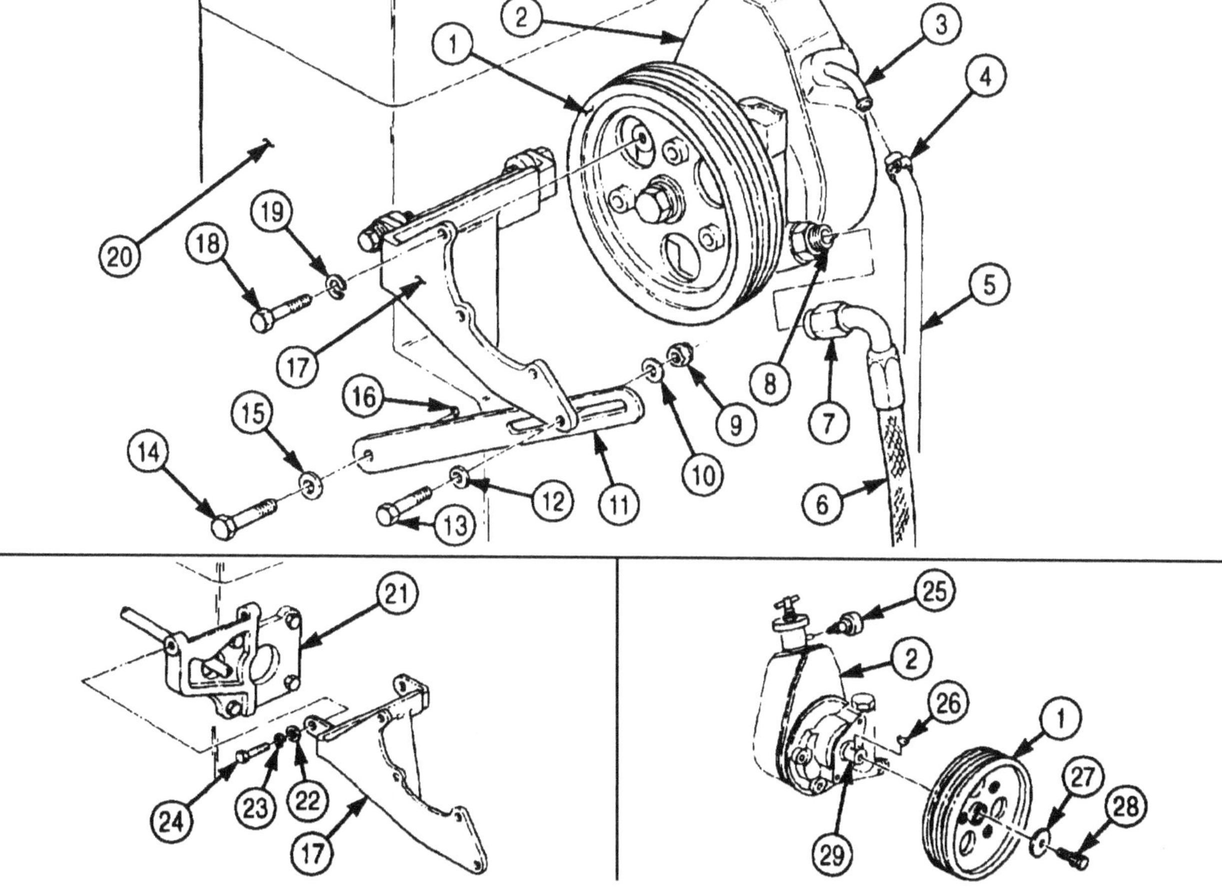

FOLLOW-ON TASKS:
- Install and adjust steering pump drivebelts (para. 3-230).
- Fill power steering reservoir to proper level (LO 9-2320-272-12).
- Start engine (TM 9-2320-272-10), check hoses for leaks and steering for proper operation.

3-237. POWER STEERING PUMP FILTER AND RESERVOIR (M939A2) MAINTENANCE

THIS TASK COVERS:

a. Filter Removal
b. Reservoir Removal
c. Cleaning
d. Reservoir Installation
e. Filter Installation

INITIAL SETUP:

APPLICABLE MODELS
M939A2

TOOLS
General mechanic's tool kit (Appendix E, Item 1)

MATERIALS/PARTS
Filter element kit (Appendix D, Item 123)
Crocus cloth (Appendix C, Item 20)

REFERENCES (TM)
LO 9-2320-272-12
TM 9-2320-272-10
TM 9-2320-272-24P

EQUIPMENT CONDITION
- Parking brake set (TM 9-2320-272-10).
- Hood raised and secured (TM 9-2320-272-10).
- Left splash shield removed (TM 9-2320-272-10).

NOTE

Have container ready to catch hydraulic fluid.

a. Filter Removal

1. Loosen clamp (17) and remove hose (18) from tube (16).
2. Remove wing nut (4), washer (5), gasket (3), lid (2), and gasket (1) from stud (6) and reservoir (12). Discard gaskets (3) and (1).
3. Remove spring (21), filter cap (20), filter (19), and packing (7) from stud (6) and reservoir (12). Discard filter (19) and packing (7).

b. Reservoir Removal

1. Remove stud (6) and valve (8) from plate (11).
2. Remove two screws (9), lockwashers (10), plate (11), and reservoir (12) from power steering pump (15). Discard lockwashers (10).
3. Remove gaskets (13) and (14) from power steering pump (15). Discard gaskets (13) and (14).

c. Cleaning

Clean inside of reservoir (12) with crocus cloth.

d. Reservoir Installation

1. Install new gaskets (13) and (14) on power steering pump (15).
2. Install reservoir (12) and plate (11) on power steering pump (15) with two new lockwashers (10) and screws (9).
3. Install valve (8) and stud (6) on plate (11).

e. Filter Installation

1. Install new packing (7), new filter (19), filter cap (20), and spring (21) on stud (6) and reservoir (12).
2. Install new gasket (1), lid (2), and new gasket (3) on stud (6) and reservoir (12) with washer (5) and wing nut (4).
3. Install hose (18) on tube (16) with clamp (17).

3-237. POWER STEERING PUMP FILTER AND RESERVOIR (M939A2) MAINTENANCE (Contd)

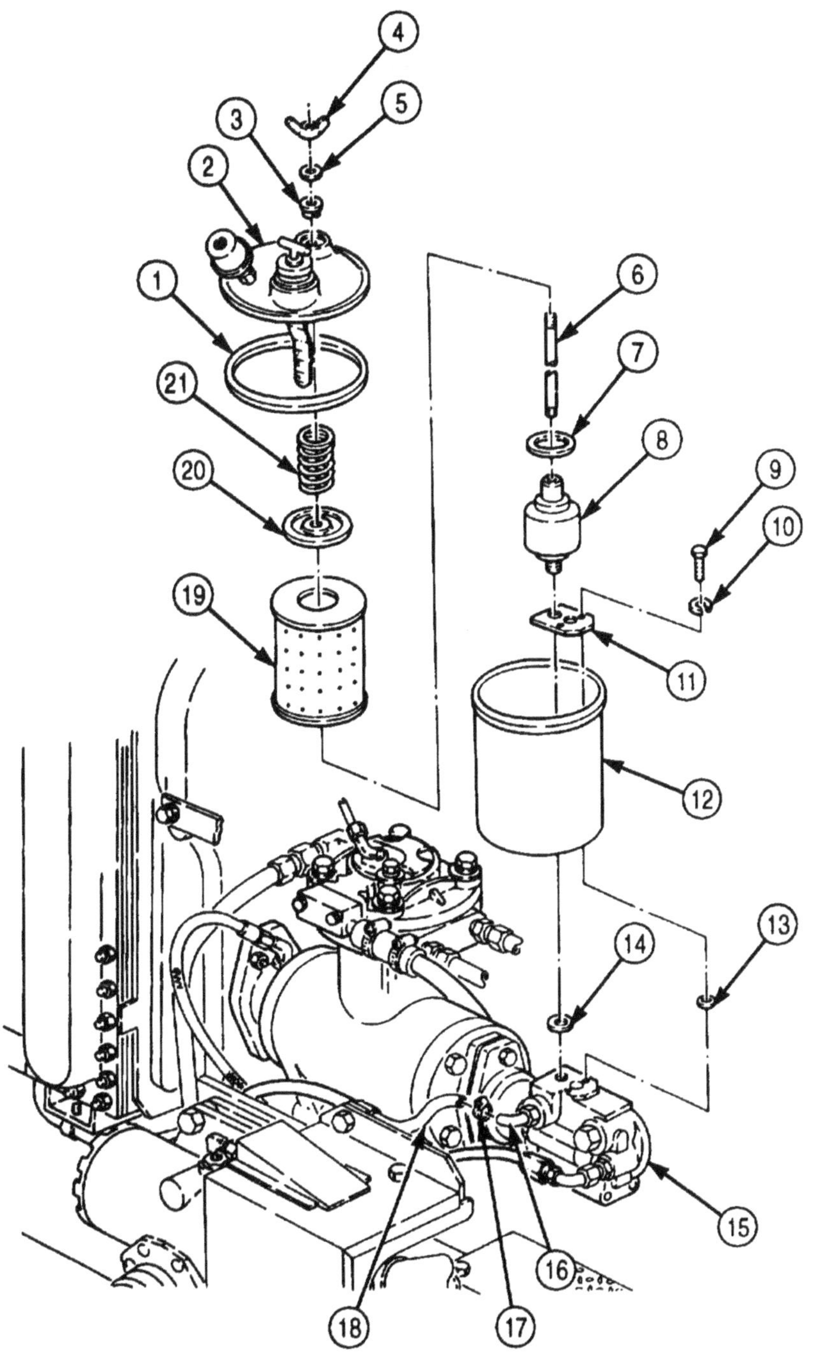

FOLLOW-ON TASKS:
- Fill power steering reservoir (LO 9-2320-272-12).
- Start engine (TM 9-2320-272-10) and check for filter leaks and steering for proper operation.
- Install left splash shield (TM 9-2320-272-10).

3-238. STEERING GEAR STONE SHIELD REPLACEMENT

THIS TASK COVERS:

a. Removal
b. Installation

INITIAL SETUP:

APPLICABLE MODELS
All

TOOLS
General mechanic's tool kit (Appendix E, Item 1)

MATERIALS/PARTS
Two locknuts (Appendix D, Item 291)
Lockwasher (Appendix D, Item 354)

REFERENCES (TM)
TM 9-2320-272-10
TM 9-2320-272-24P

EQUIPMENT CONDITION
Parking brake set (TM 9-2320-272-10).

a. Removal

1. Remove two locknuts (4), washer (5), three screws (8), and lockwasher (1) from stone shield (9) and air check valve (7). Discard locknuts (4) and lockwasher (1).
2. Remove stone shield (9) from splash shield (3) and frame rail (6).

b. Installation

1. Align stone shield (9) with holes in frame rail (6) and splash shield (3).
2. Install stone shield (9) and air check valve (7) on splash shield (3) and frame rail (6) with new lockwasher (1), three screws (8), washer (5), and two new locknuts (4).

3-238. STEERING GEAR STONE SHIELD REPLACEMENT (Contd)

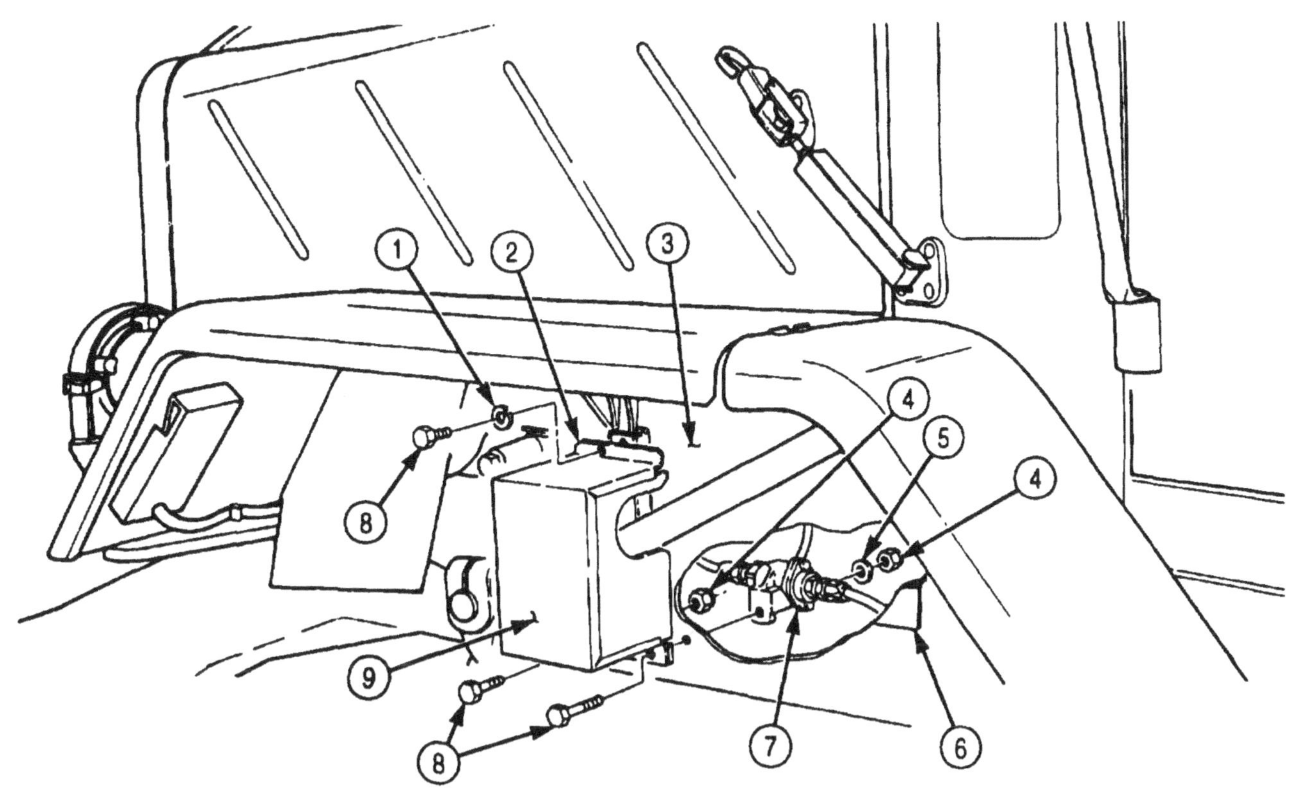

3-239. STEERING GEAR-TO-ASSIST CYLINDER PRESSURE LINES REPLACEMENT

THIS TASK COVERS:

a. Removal

b. Installation

INITIAL SETUP:

APPLICABLE MODELS

All

TOOLS

General mechanic's tool kit (Appendix E, Item 1)

MATERIALS/PARTS

Two O-rings (Appendix D, Item 435)
Two locknuts (Appendix D, Item 313)
Tiedown strap (Appendix D, Item 690)
Cap and plug set (Appendix C, Item 14)
Antiseize tape (Appendix C, Item 72)

REFERENCES (TM)

TM 9-2320-272-10
TM 9-2320-272-24P

EQUIPMENT CONDITION

- Parking brake set (TM 9-2320-272-10).
- Left splash shield removed (TM 9-2320-272-10).

GENERAL SAFETY INSTRUCTIONS

Do not start engine when power steering hoses are disconnected.

WARNING

Do not start engine when power steering hoses are disconnected. Pressure may whip hoses, causing injury to personnel.

a. Removal

CAUTION

Cap or plug all openings immediately after disconnecting lines and hoses to prevent contamination. Failure to do so may result in damage to equipment.

1. Remove two locknuts (17), screws (4), spacers (5), and four clamps (6) from crossmember (16). Discard locknuts (17).

NOTE

- Have drainage container ready to catch hydraulic oil.
- Step 2 applies to Ross steering gears only.
- Steps 3, 4, and 5 apply to Sheppard steering gears only.

2. Loosen nuts (11) and (14) and remove pressure lines (12) and (15) from elbow (13) and adapter (10).
3. Loosen nuts (25) and (27) and remove pressure lines (12) and (15) from elbows (24) and (26).
4. Loosen nut (23) and remove elbows (24) and (26) from extension line (22) and elbow (13).
5. Loosen nut (21) and remove extension line (22) from adapter (10).
6. Remove adapter (10) and O-ring (9) from steering gear (7). Discard O-ring (9).
7. Remove elbow (13) and O-ring (8) from steering gear (7). Discard O-ring (8).
8. Remove tiedown strap (2) from pressure lines (12) and (15). Discard tiedown strap (2).
9. Loosen two nuts (18) and disconnect pressure lines (12) and (15) from two elbows (3).
10. Disconnect two hoses (19) from elbows (3).
11. Remove two nuts (20) and elbows (3) from frame rail (1).

3-239. STEERING GEAR-TO-ASSIST CYLINDER PRESSURE LINES REPLACEMENT (Contd)

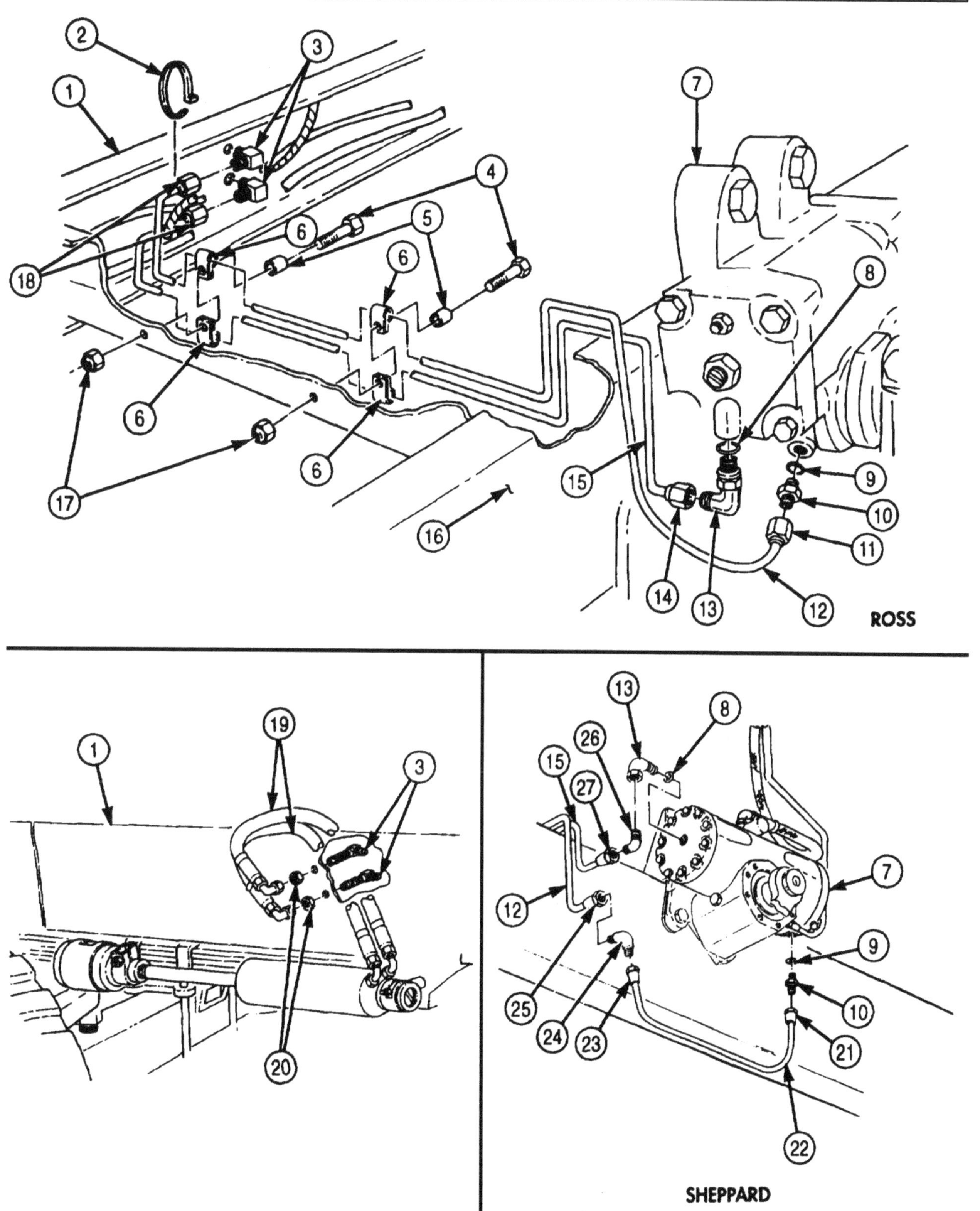

3-239. STEERING GEAR-TO-ASSIST CYLINDER PRESSURE LINES REPLACEMENT (Contd)

b. Installation

NOTE

- Remove all caps and plugs prior to installation.
- Fittings must be cleaned and inspected for cracks or stripped threads.
- Male pipe threads must be wrapped with antiseize tape before installation.
- Do not reuse hydraulic oil.

1. Install new O-ring (8) and elbow (13) on steering gear (7), with opening of elbow (13) facing down.
2. Install new O-ring (9) and adapter (10) on steering gear (7).

NOTE

Steps 3 through 5 apply to Sheppard steering gears only.

3. Install extension line (22) on adapter (10) and tighten nut (211.
4. Install elbows (24) and (26) on extension line (22) and elbow (13) and tighten nut (23).
5. Install pressure lines (12) and (15) on elbows (24) and (26) and tighten nuts (25) and (27).
6. Install two elbows (3) on frame rail (1) with two nuts (20).
7. Connect two hoses (19) to elbows (3).
8. Connect pressure lines (12) and (15) to elbows (3) and tighten nuts (18).

NOTE

Steps 9 through 11 apply to Ross steering gears only.

9. Install new O-ring (8) and elbow (13) on steering gear (7).
10. Install new O-ring (9) and adapter (10) on steering gear (7).
11. Connect pressure lines (12) and (15) to elbow (13) and adapter (10) and tighten nuts (11) and (14).
12. Install new tiedown strap (2) on pressure lines (12) and (15).
13. Install pressure lines (121 and (15) on crossmember (16) with four clamps (6), two spacers (5), screws (4), and new locknuts (17).

3-239. STEERING GEAR-TO-ASSIST CYLINDER PRESSURE LINES REPLACEMENT (Contd)

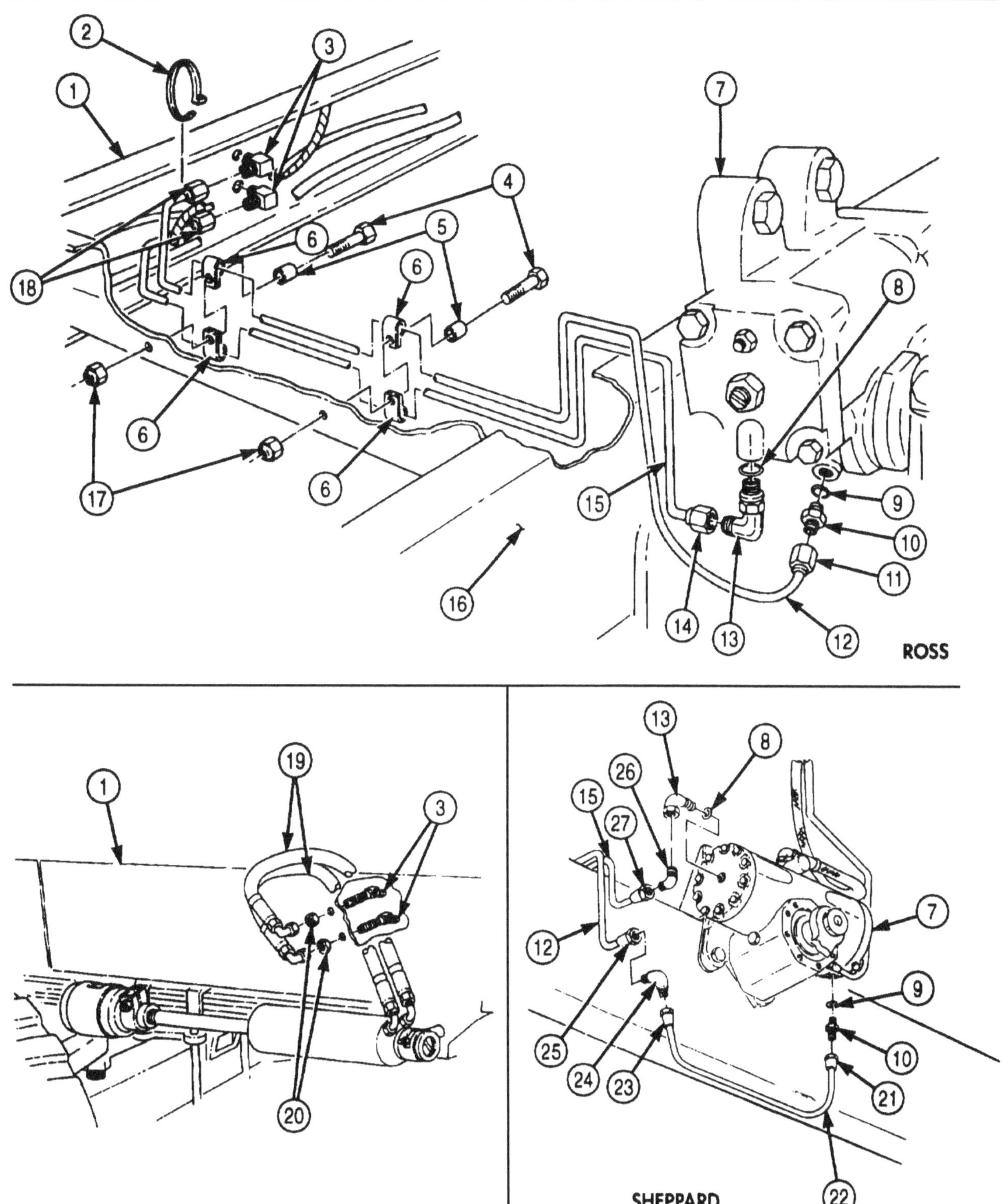

FOLLOW-ON TASK: Install left splash shield (TM 9-2320-272-10).

Section XI. FRAME MAINTENANCE

3-240. FRAME MAINTENANCE INDEX

PARA. NO.	TITLE	PAGE NO.
3-241.	Front and Rear Lifting Shackle and Bracket Replacement	3-684
3-242.	Pintle Hook Maintenance	3-686
3-243.	Front Bumper and Plates Replacement	3-688
3-244.	Front and Rear Field Chock Anchors (M936/A1) Replacement	3-690
3-245.	Winch Frame Extension Replacement	3-692
3-246.	Bumperette Replacement	3-694
3-247	Hood Retaining Bracket Replacement	3-695
3-248.	Tractor Fifth Wheel Replacement	3-696
3-249.	Fifth Wheel Approach Plates Replacement	3-698
3-250.	Tractor Spare Tire Carrier Toolbox Replacement	3-699
3-251.	Fifth Wheel Deck Plate Replacement	3-700
3-252.	Fifth Wheel Spacers Replacement	3-701
3-253.	Tractor Spare Tire Carrier (M931, M932) Replacement	3-702
3-254.	Tractor Spare Tire Carrier (M931A1/A2, M932A1/A2) Replacement	3-704
3-255.	Dump and Tractor Spare Tire Carrier Access Step Replacement	3-710
3-256.	Dump Spare Tire Carrier (M929, M930) Replacement	3-712
3-257.	Dump Spare Tire Carrier (M929A1/A2, M930A1/A2) Replacement	3-713
3-258.	Cargo Spare Tire Carrier (M923, M925, M927, M928) Replacement	3-716
3-259.	Cargo Spare Tire Carrier (M923A1/A2,M925A1/A2, M927A1/A2, M928A1/A2) Replacement	3-718
3-260.	Cargo Spare Tire Carrier Access Step Replacement	3-720
3-261.	Tailgate Bumpers Replacement	3-721
3-262.	Van Spare Tire Carrier (M934) Replacement	3-722
3-263.	Van Spare Tire Carrier (M934A41/A2) Replacement	3-722
3-264.	Van Davit Chain and Wire Rope Replacement	3-724
3-265.	Van Swing Davit and Pulley Replacement	3-726
3-266.	Van Davit Winch (M934A1/A2) Replacement	3-728

3-241. FRONT AND REAR LIFTING SHACKLE AND BRACKET REPLACEMENT

THIS TASK COVERS:

a. Removal
b. Installation

INITIAL SETUP:

APPLICABLE MODELS
All

TOOLS
General mechanic's tool kit (Appendix E, Item 1)

MATERIALS/PARTS
Three locknuts (Appendix D, Item 321)

REFERENCES (TM)
TM 9-2320-272-10
TM 9-2320-272-24P

EQUIPMENT CONDITION
- Parking brake set (TM 9-2320-272-10).
- Winch chain and hook removed from front shackles (TM 9-2320-272-10) (winch models only).

NOTE

- Replacement of front and rear shackles are the same, except rear shackles do not use washers. This procedure covers the front shackle.
- On M936/A1 model vehicles the long screw attaching each front shackle has an additional washer between anchor and locknut.

a. Removal

1. Remove retaining clip (12), S-hooks (1) and (13), and chain (14) from shackle pin (2).
2. Remove shackle pin (2) and shackle (3) from shackle bracket (6).
3. Remove two locknuts (10), screws (4), locknut (9), long screw (5), pipe coupling (8), shackle bracket (6), and washer (7) from bumper (11). Discard locknuts (10) and (9).

b. Installation

1. Install washer (7) and shackle bracket (6) on bumper (11) with long screw (5), pipe coupling (8), new locknut (9), two screws (4), and new locknuts (10).
2. Install shackle (3) on shackle bracket (6) with shackle pin (2).
3. Install retaining clip (12), S-hooks (1) and (13), and chain (14) on shackle pin (2).

3-241. FRONT AND REAR LIFTING SHACKLE AND BRACKET REPLACEMENT (Contd)

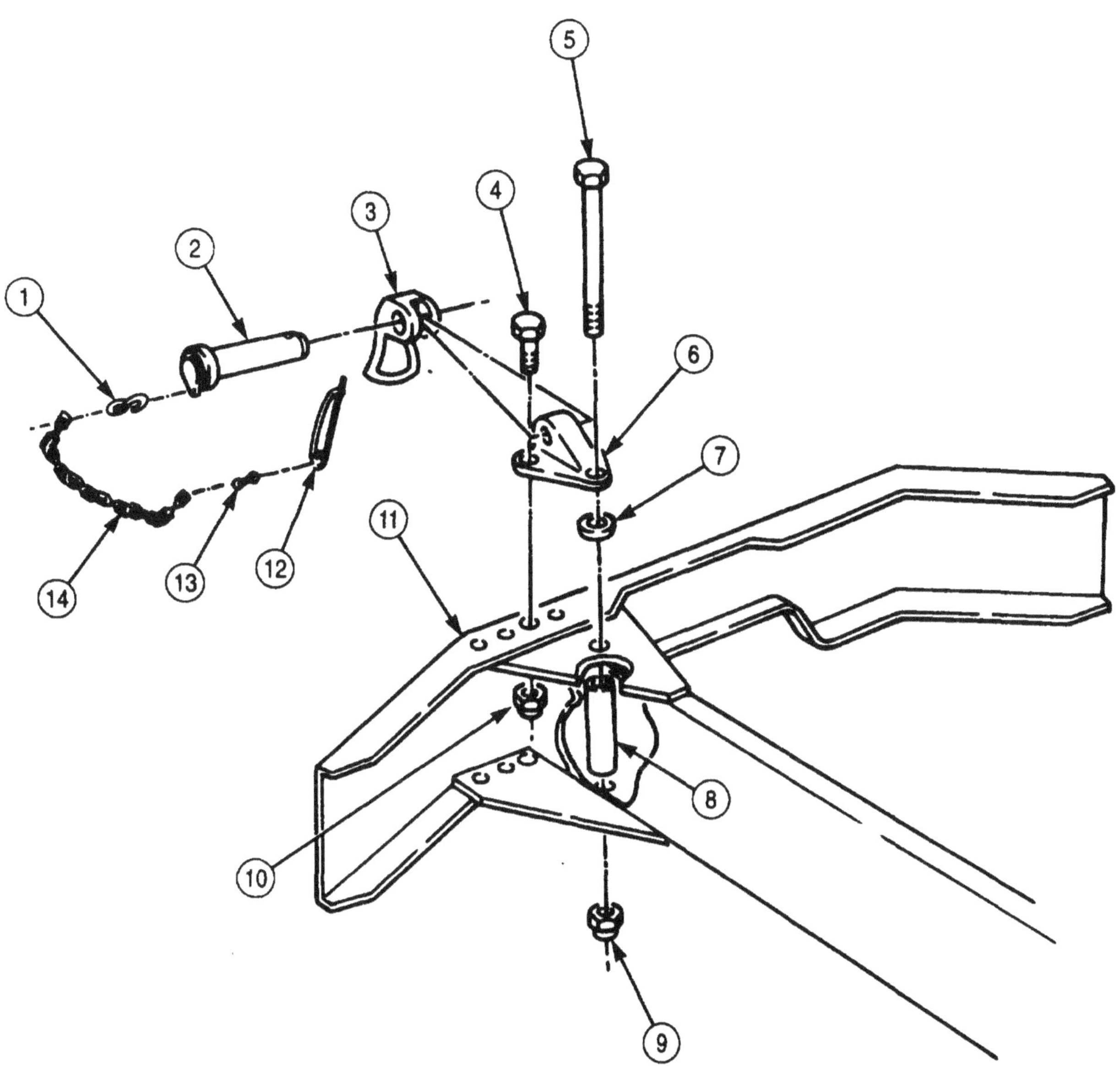

FOLLOW-ON TASK: Stow winch chain and hook on front shackles (TM 9-2320-272-10) (winch models only).

3-242. PINTLE HOOK MAINTENANCE

THIS TASK COVERS:

a. Removal
b. Disassembly
c. Cleaning and Inspection
d. Assembly
e. Installation

INITIAL SETUP:

APPLICABLE MODELS
All

TOOLS
General mechanic's tool kit (Appendix E, Item 1)

MATERIALS/PARTS
Cotter pin (Appendix D, Item 73)
Cotter pin (Appendix D, Item 74)
Cotter pin (Appendix D, Item 76)
Drycleaning solvent (Appendix C, Item 71)
GAA grease (Appendix C, Item 28)

REFERENCES (TM)
LO 9-2320-272-12
TM 9-2320-272-10
TM 9-2320-272-24P

EQUIPMENT CONDITION
Parking brake set (TM 9-2320-272-10).

GENERAL SAFETY INSTRUCTIONS
Keep fire extinguisher nearby when using drycleaning solvent.

---

a. Removal

1. Remove cotter pin (13) from slotted nut (14). Discard cotter pin (13).
2. Remove slotted nut (14), washer (15), and pintle hook (12) from mounting bracket (16).

b. Disassembly

1. Remove two grease fittings (7) from screw (9) and latch shaft (6).
2. Remove drive pin (101, chain (11), and cotter pin (8) from pintle hook (12) and latch (5). Discard cotter pin (8).
3. Remove cotter pin (1), slotted nut (2), screw (9), and latch (5) from pintle hook (12). Discard cotter pin (1).
4. Remove latch shaft (6), latch lock (3), and spring (4) from latch (5).

c. Cleaning and Inspection

WARNING

Drycleaning solvent is flammable and toxic. Do not use near an open flame and always have a fire extinguisher nearby when solvents are used. Use only in well-ventilated places, wear protective clothing, and dispose of cleaning rags in approved container. Failure to do this may result in injury to personnel and/or damage to equipment.

1. Clean pintle hook (12), latch (5), latch lock (3), latch shaft (6), and screw (9) with drycleaning solvent.
2. Inspect pintle hook (12), latch (5), latch lock (3), latch shaft (6), and screw (9) for bends, cracks, and breaks. Replace any damaged parts.

3-242. PINTLE HOOK MAINTENANCE (Contd)

d. Assembly

1. Install latch lock (3) and spring (4) on latch (5) with latch shaft (6).
2. Install latch (5) on pintle hook (12) with screw (9), slotted nut (2), and new cotter pin (1).
3. Install new cotter pin (8) and chain (11) on latch (5) and pintle hook (12) with drive pin (10).
4. Install two grease fittings (7) on screw (9) and latch shaft (6).

e. Installation

1. Install pintle hook (12) on mounting bracket (16) with washer (15) and slotted nut (14).
2. Install new cotter pin (13) through pintle hook (12) and slotted nut (14).

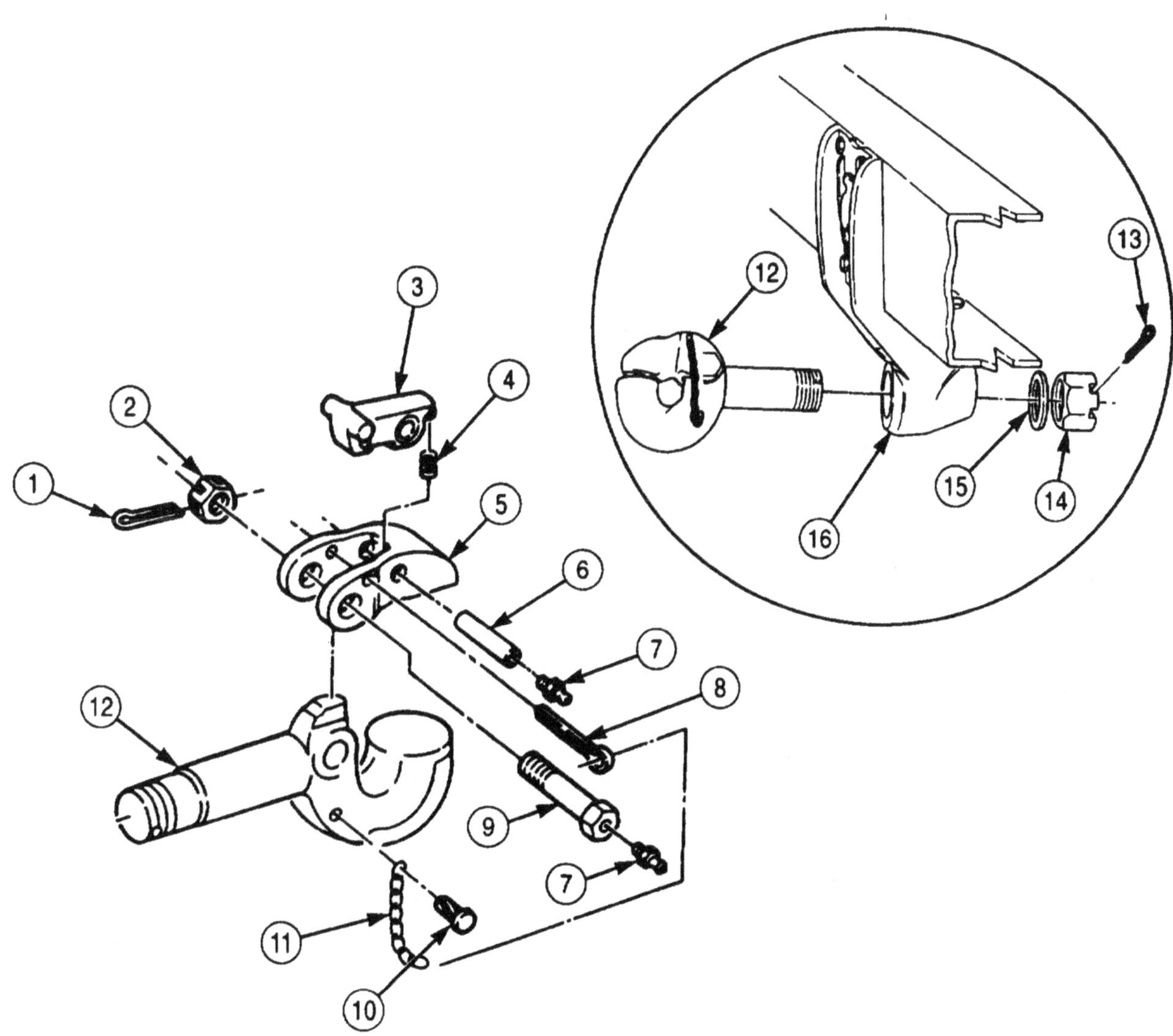

FOLLOW-ON TASK: Lubricate pintle hook (LO 9-2320-272-12).

3-243. FRONT BUMPER AND PLATES REPLACEMENT

THIS TASK COVERS:

a. Removal b. Installation

INITIAL SETUP:

APPLICABLE MODELS

All

TOOLS

General mechanic's tool kit (Appendix E, Item 1)

MATERIALS/PARTS

Ten locknuts (Appendix D, Item 294)
Twelve locknuts- (M936/A1) (Appendix D, Item 294)
Eight locknuts (Appendix D, Item 319)
Four lockwashers (M936/A1) (Appendix D, Item 350)

REFERENCES (TM)

TM 9-2320-272-10
TM 9-2320-272-24P

EQUIPMENT CONDITION

- Hood retaining bracket removed (para. 3-247).
- Front lifting shackle brackets removed (para. 3-241).

a. Removal

1. Remove four locknuts (10) and (12), and screws (5) and (4) from front bumper (2) and two lower plates (13). Discard locknuts (10) and (12).

NOTE

Perform step 2 for M9361A1 model vehicles.

2. Remove six locknuts (9), (11), and (19), screws (14), (6), and (15), four lockwashers (16), two chock anchors (17), and lower plates (13) from front bumper (2) and frame (18). Discard lockwashers (16) and locknuts (9), (11), and (19).
3. Remove two locknuts (9) and (11), screws (14) and (6), and two lower plates (13) from front bumper (2) and frame (18). Discard locknuts (9) and (11).
4. Remove four locknuts (7), screws (3), and bumper (2) from two upper plates (20). Discard locknuts (7).
5. Remove two locknuts (8), screws (1), and upper plates (20) from frame (18). Discard locknuts (8).

b. Installation

1. Install two upper plates (20) on frame (18) with two screws (1) and new locknuts (8).

NOTE

The bumper is inverted on winch model vehicles.

2. Install bumper (2) on two upper plates (20) with four screws (3) and new locknuts (7).

NOTE

Perform step 3 for M936/A1 model vehicles.

3. Install two chock anchors (17) and lower plates (13) on front bumper (2) and frame (18) with four new lockwashers (16), six screws (14), (6), and (15), and new locknuts (9), (11), and (19).
4. Install two lower plates (13) on front bumper (2) and frame (18) with two screws (14) and (6) and new locknuts (9) and (11).
5. Install four screws (4) and (5) in front bumper (2) and two lower plates (13) with four new locknuts (10) and (12).

3-243. FRONT BUMPER AND PLATES REPLACEMENT (Contd)

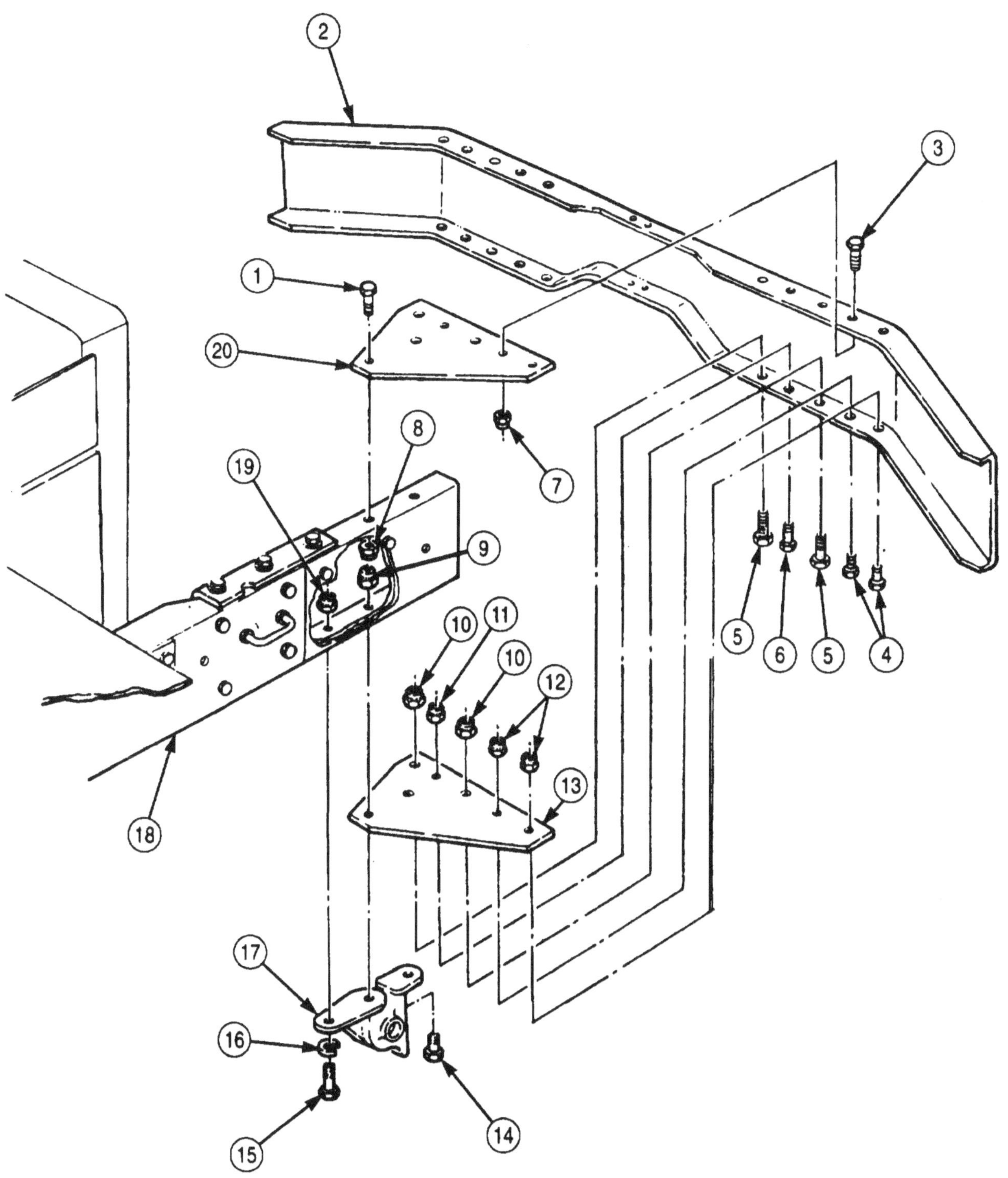

FOLLOW-ON TASKS:
- Install front lifting shackle brackets (para. 3-241).
- Install hood retaining bracket (para. 3-247).

3-244. FRONT AND REAR FIELD CHOCK ANCHORS (M936/A1) REPLACEMENT

THIS TASK COVERS:

a. Removal b. Installation

INITIAL SETUP:

APPLICABLE MODELS

M936/A1

TOOLS

General mechanic's tool kit (Appendix E, Item 1)

MATERIALS/PARTS

Four locknuts (Appendix D, Item 294)
Sixteen locknuts (Appendix D, Item 321)
Two lockwashers (Appendix D, Item 368)

REFERENCES (TM)

LO 9-2320-272-12
TM 9-2320-272-10
TM 9-2320-272-24P

EQUIPMENT CONDITION

Parking brake set (TM 9-2320-272-10).

a. Removal

1. Remove two locknuts (8) and screws (1) from front anchors (4), shackles (9), and frame (3). Discard locknuts (8).
2. Remove four locknuts (2), two screws (6) and (7), lockwashers (5), and front anchors (4) from frame (3). Discard lockwashers (5) and locknuts (2).
3. Remove eight locknuts (11), screws (17), and two side anchors (16) from frame (3). Discard locknuts (11).
4. Remove two locknuts (12), screws (13), four locknuts (14), screws (10), and two rear anchors (15) from frame (3). Discard locknuts (12) and (14).

b. Installation

1. Install two rear anchors (15) on frame (3) with two screws (13), new locknuts (12), four screws (10), and new locknuts (14).
2. Install two side anchors (16) on frame (3) with eight screws (17) and new locknuts (11).
3. Install two front anchors (4) on frame (3) with two new lockwashers (5), screws (6) and (7), and four locknuts (2).
4. Install two screws (1) on front anchors (4), shackles (9), and frame (3) with two new locknuts (8).

3-244. FRONT AND REAR FIELD CHOCK ANCHORS (M936/A1) REPLACEMENT (Contd)

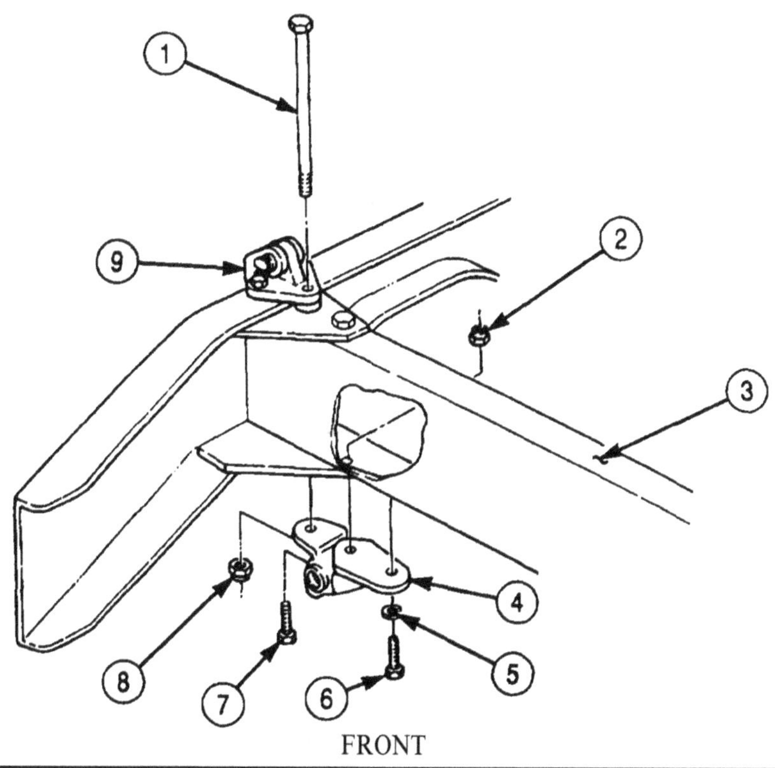

FRONT

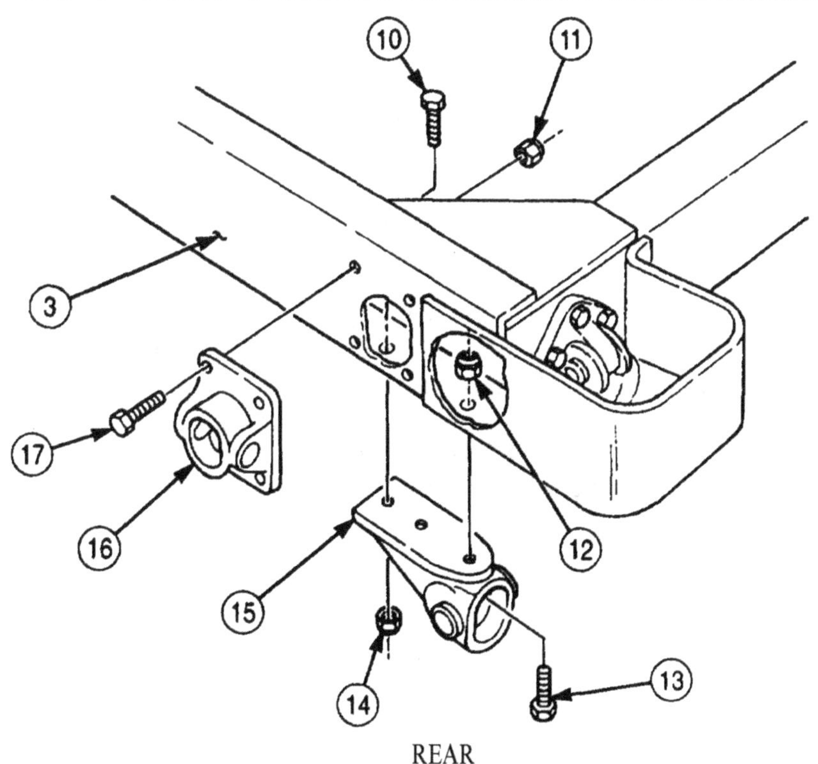

REAR

FOLLOW-ON TASK: Lubricate anchors (LO 9-2320-272-12).

3-245. WINCH FRAME EXTENSION REPLACEMENT

THIS TASK COVERS:

a. Removal b. Installation

INITIAL SETUP:

APPLICABLE MODELS
All w/winch

TOOLS
General mechanic's tool kit (Appendix E, Item 1)
Torque wrench (Appendix E, Item 144)

MATERIALS/PARTS
Four locknuts (Appendix D, Item 309)
Twenty-four locknuts (Appendix D, Item 294)
Six lockwashers (Appendix D. Item 350)

REFERENCES (TM)
TM 9-2320-272-24P

EQUIPMENT CONDITION
- Front bumper and plates removed (para. 3-243).
- Front winch removed para. 3-329).
- Hood removed (para. 3-275).

a. Removal

1. Remove six screws (9), lockwashers (8), washers (7), and two brackets (10) from frame extensions (1). Discard lockwashers (8).
2. Remove four locknuts (12), screws (5), washers (4), and two loop tiedowns (3) from frame extensions (1). Discard locknuts (12).
3. Remove eight locknuts (11), screws (6), and two frame extensions (1) from channel reinforcements (2). Discard locknuts (11).
4. Remove four locknuts (21), screws (13), and washers (14) from two end supports (23) and crossmember (24). Discard locknuts (21).
5. Remove four locknuts (22), screws (17), and two end supports (23) from channel reinforcements (2). Discard locknuts (22).
6. Remove eight locknuts (15), screws (16), two screws (20), washers (19), and two channel reinforcements (2) from frame rails (18). Discard locknuts (15).

b. Installation

1. Install two channel reinforcements (2) on frame rails (18) with eight screws (161, new locknuts (15), two washers (19), and screws (20). Tighten locknuts (15) and screws (20) 85 lb-ft (115 N•m).
2. Install two end supports (231 on channel reinforcements (2) with four screws (17) and new locknuts (22). Tighten locknuts (22) 120 lb-ft (163 N•m).
3. Install two end supports (23) on crossmember (24) with four washers (141, screws (131, and new locknuts (21). Tighten locknut-s (21) 120 lb-ft (163 N•m).
4. Install two frame extensions (1) on channel reinforcements (2) with eight screws (6) and new locknuts (11). Tighten locknuts (12185 lb-ft (115 N•m).
5. Install two loop tiedowns (3) on frame extensions (1) with four washers (4), screws (5), and new locknuts (12).
6. Install two brackets (10) on frame extensions (1) with six washers (7), new lockwashers (8), and screws (9).

3-245. WINCH FRAME EXTENSION REPLACEMENT (Contd)

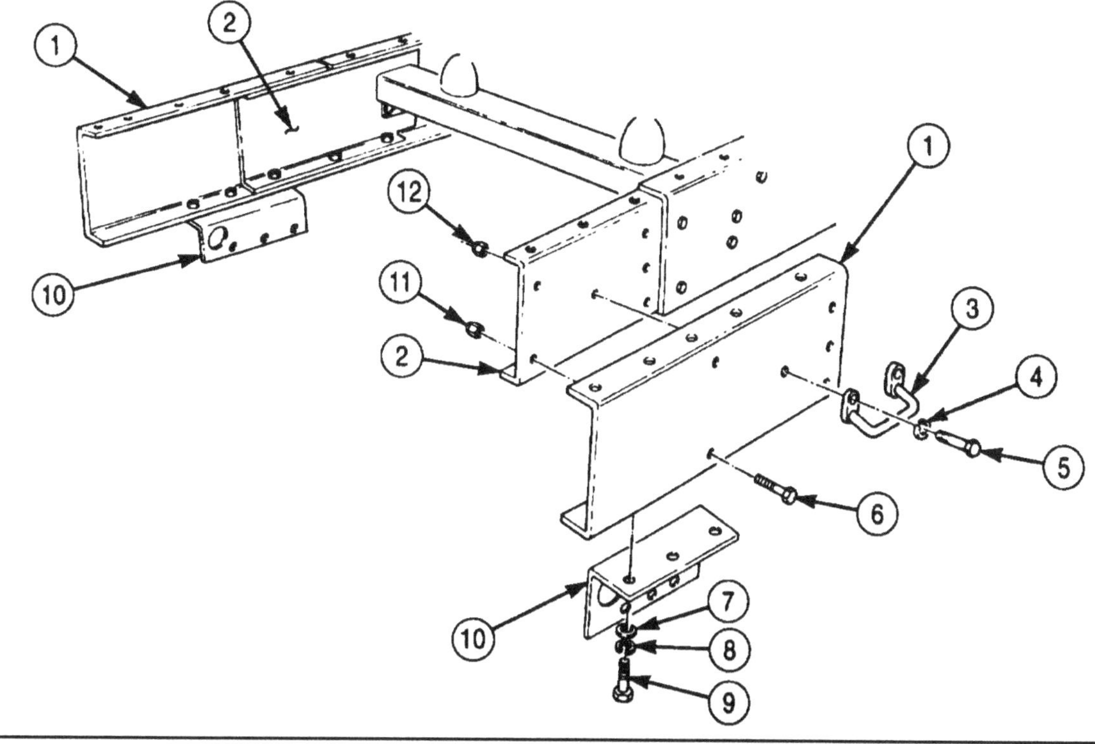

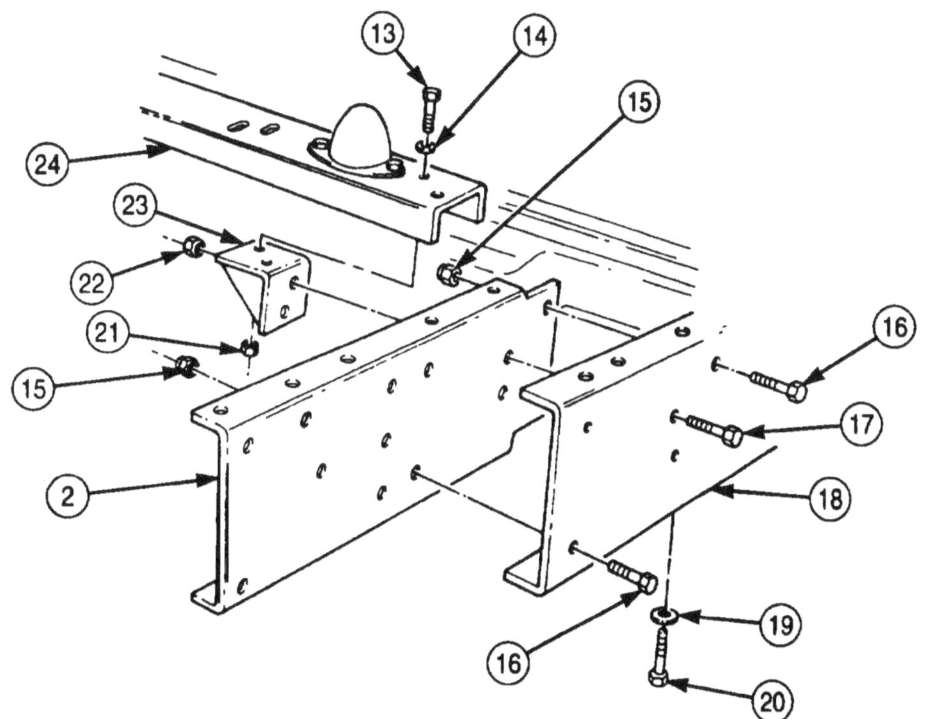

FOLLOW-ON TASKS:
- Install hood (para. 3-275).
- Install front winch (para. 3-329).
- Install bumper and plates (para. 3-243).

3-246. BUMPERETTE REPLACEMENT

THIS TASK COVERS:

a. Removal b. Installation

INITIAL SETUP:

APPLICABLE MODELS
M923/A1/A2, M925/A1/A2, M927/A1/A2, M928/A1/A2

REFERENCES (TM)
TM 9-2320-272-10
TM 9-2320-272-24P

TOOLS
General mechanic's tool kit (Appendix E, Item 1)

EQUIPMENT CONDITION
Parking brake set (TM 9-2320-272-10).

MATERIALS/PARTS
Six locknuts (Appendix D, Item 294)

NOTE

Left and right bumperettes are replaced in the same way. This procedure covers left bumperette.

a. Removal

1. Remove four locknuts (2) and screws (7) from left hand frame rail (1) and bumperette (6). Discard locknuts (2).
2. Remove two locknuts (3), screws (5), and bumperette (6) from rear crossmember (4). Discard locknuts (3).

b. Installation

1. Install bumperette (6) on rear crossmember (4) with two screws (5) and new locknuts (3).
2. Install bumperette (6) on left hand frame rail (1) with four screws (7) and new locknuts (2).

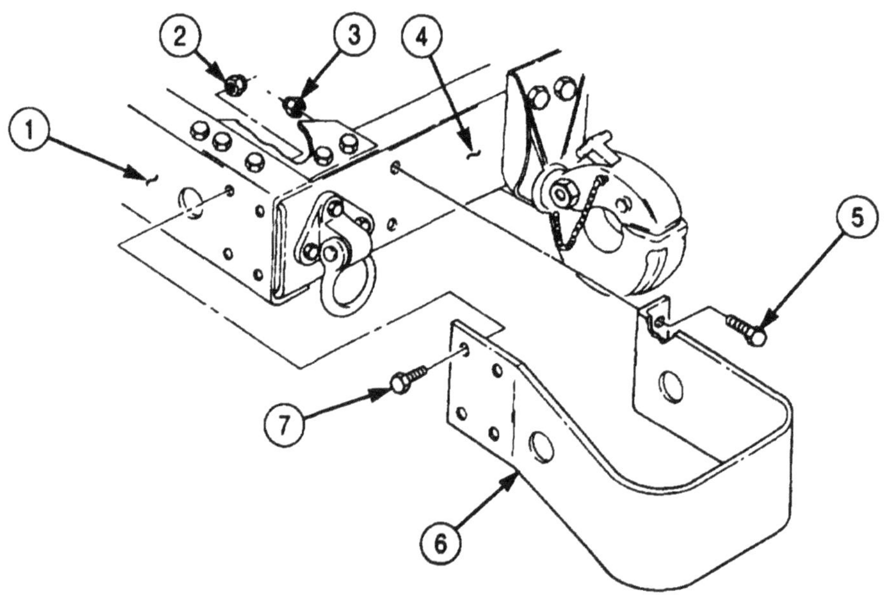

3-247. HOOD RETAINING BRACKET REPLACEMENT

THIS TASK COVERS:

a. Removal b. Installation

INITIAL SETUP:

APPLICABLE MODELS
All

TOOLS
General mechanic's tool kit (Appendix E, Item 1)

MATERIALS/PARTS
Two locknuts (Appendix D, Item 313)

REFERENCES (TM)
TM 9-2320-272-10
TM 9-2320-272-24P

EQUIPMENT CONDITION
Parking brake set (TM 9-2320-272-10).

a. Removal

Remove two locknuts (4), screws (1), and retaining bracket (2) from front bumper (3). Discard locknuts (4).

b. Installation

Install retaining bracket (2) on front bumper (3) with two screws (1) and new locknuts (4).

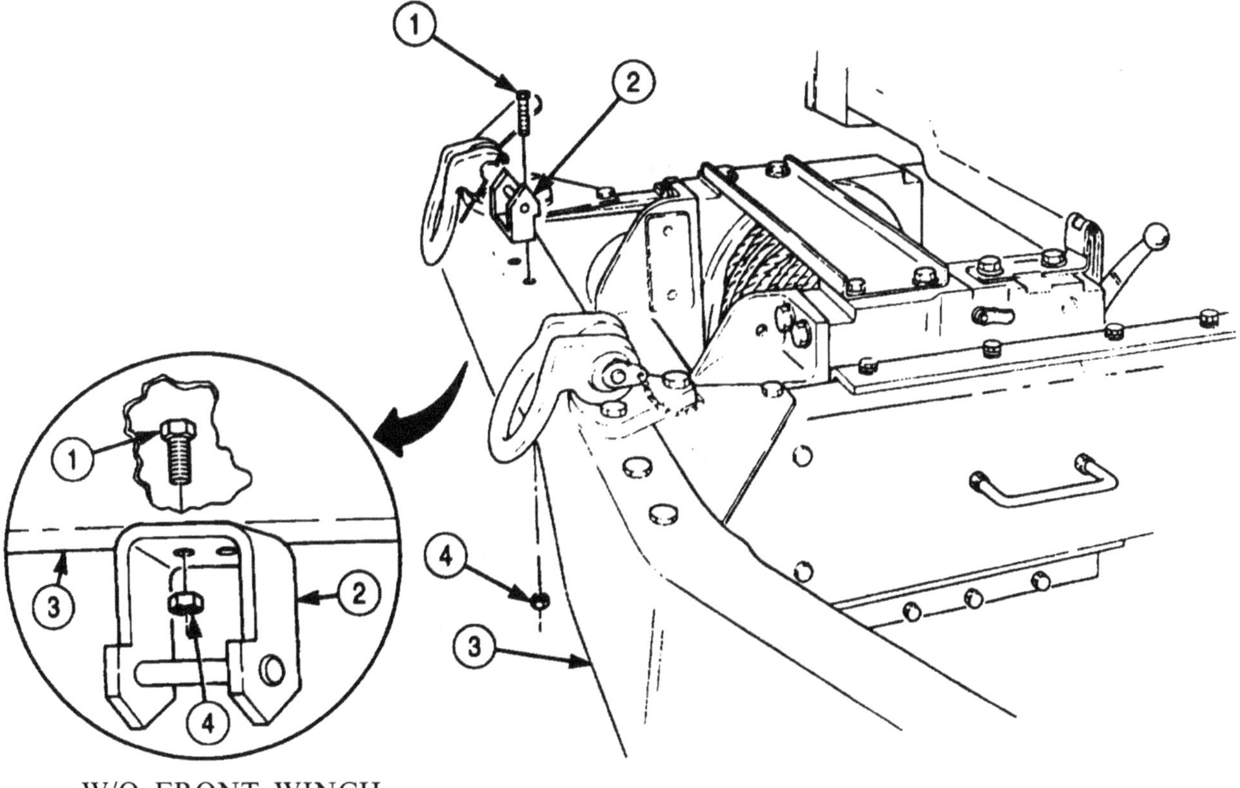

W/O FRONT WINCH

FOLLOW-ON TASK: Check bracket placement with hood retaining latch.

3-248. TRACTOR FIFTH WHEEL REPLACEMENT

THIS TASK COVERS:

a. Removal b. Installation

INITIAL SETUP:

APPLICABLE MODELS
M931/A1/A2, M932/A1/A2

TOOLS
General mechanic's tool kit (Appendix E. Item 1)
Torque wrench (Appendix E, Item 144)
Lifting device
Utility chain

MATERIALS/PARTS
Ten lockwashers (Appendix D, Item 399)

REFERENCES (TM)
LO 9-2320-272-12
TM 9-2320-272-10
TM 9-2320-272-24P

EQUIPMENT CONDITION
Parking brake set (TM 9-2320-272-10).

GENERAL SAFETY CONDITIONS
Personnel must stand clear during lifting operation.

WARNING

All personnel must stand clear during lifting operations. A shifting load may cause injury to personnel.

a. Removal

1. Remove ten screws (7) and lockwashers (6) from fifth wheel (3) and base support (5). Discard lockwashers (6).

NOTE

Ensure washers are between head of screws and fifth wheel, and chain links and nuts.

2. Attach utility chain on fifth wheel (3) with two screws (4), four washers (1), and two nuts (2).

NOTE

Assistant will help with step 3.

3. Attach lifting device to chain, and remove fifth wheel (3) from base support (5).
4. Remove lifting device from chain.
5. Remove two nuts (2), four washers (1), two screws (4). and utility chain from fifth wheel (3).

b. Installation

1. Install utility chain on fifth wheel (3) with two screws (4), four washers (1), and two nuts (2).

NOTE

Assistant will help with step 2.

2. Using lifting device attached to chain, raise fifth wheel (3) and lower onto base support (5). Align fifth wheel (3) with base support (5) and install with ten new lockwashers (6) and screws (7). Tighten screws (7) 160-170 lb-ft (217-231 N•m).
3. Remove lifting device from utility chain, two nuts (2), four washers (1), two screws (4), and utility chain from fifth wheel (3).

3-248. TRACTOR FIFTH WHEEL REPLACEMENT (Contd)

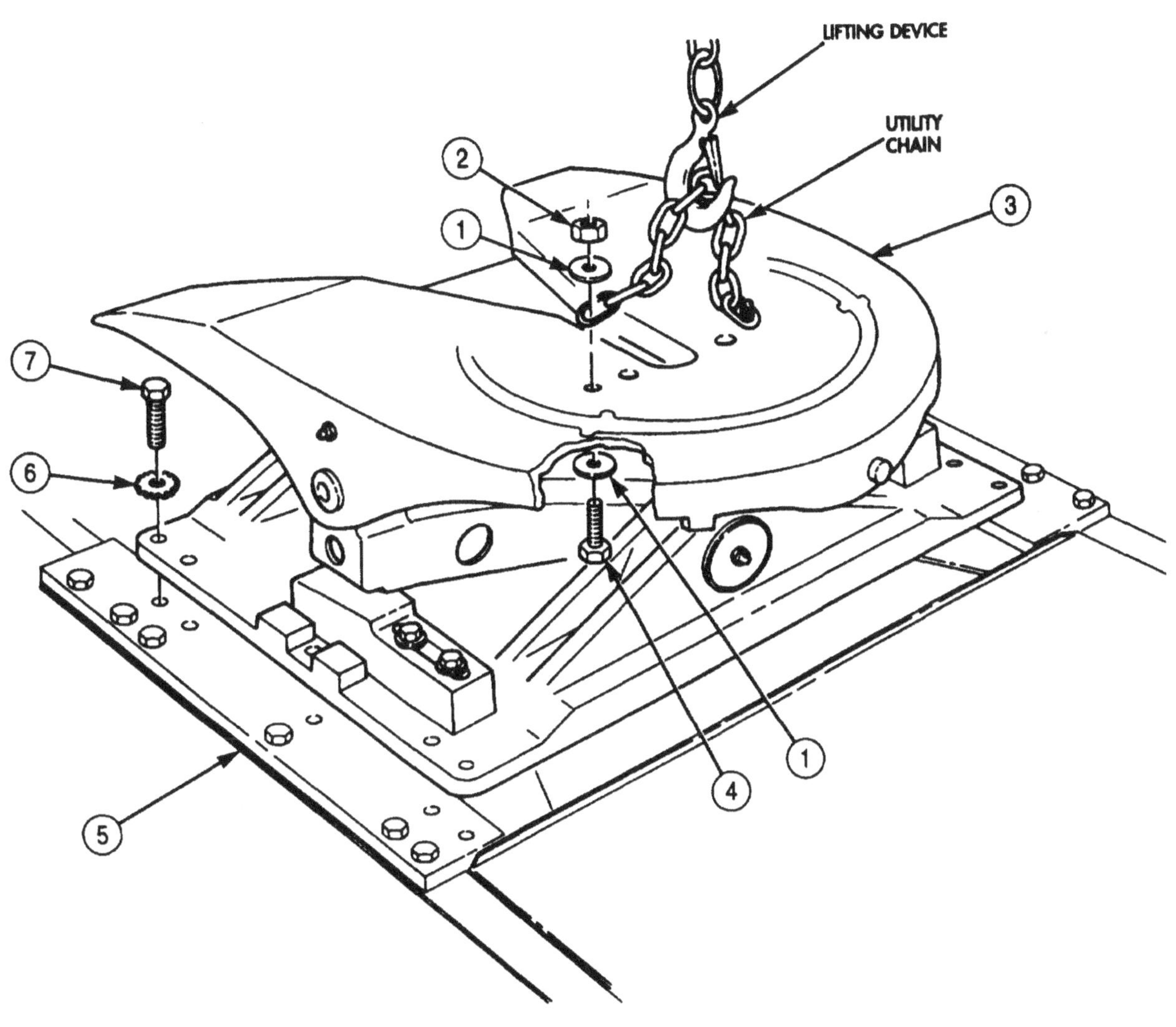

FOLLOW-ON TASK: Lubricate fifth wheel (LO 9-2320-272-12).

3-249. FIFTH WHEEL APPROACH PLATES REPLACEMENT

THIS TASK COVERS:

a. Removal
b. Installation

INITIAL SETUP:

APPLICABLE MODELS
M931/A1/A2, M932/A1/A2

TOOLS
General mechanic's tool kit (Appendix E, Item 1)

MATERIALS/PARTS
Sixteen locknuts (Appendix D, Item 321)
Four lockwashers (Appendix D, Item 400)

REFERENCES (TM)
TM 9-2320-272-10
TM 9-2320-272-24P

EQUIPMENT CONDITION
- Parking brake set (TM 9-2320-272-10).
- Fifth wheel removed (para. 3-248).

a. Removal

1. Remove four locknuts (9), (10), and (5) and screws (7), (2), and (3) from two approach plates (6), channel (4), and frame (8). Discard locknuts (9), (10), and (5).
2. Remove four locknuts (11), screws (1), two approach plates (6), and channel (4) from frame (8). Discard locknuts (11).
3. Remove four screws (12), lockwashers (13), and two reflectors (14) from approach plates (6). Discard lockwashers (13).

b. Installation

1. Install two reflectors (14) on approach plates (6) with four new lockwashers (13) and screws (12).
2. Install two approach plates (6) and channel (4) on frame (8) with four screws (7), (2), and (3) and new locknuts (9), (10), and (5).
3. Install channel (4) on two approach plates (6) and frame (8) with four screws (1) and new locknuts (11).

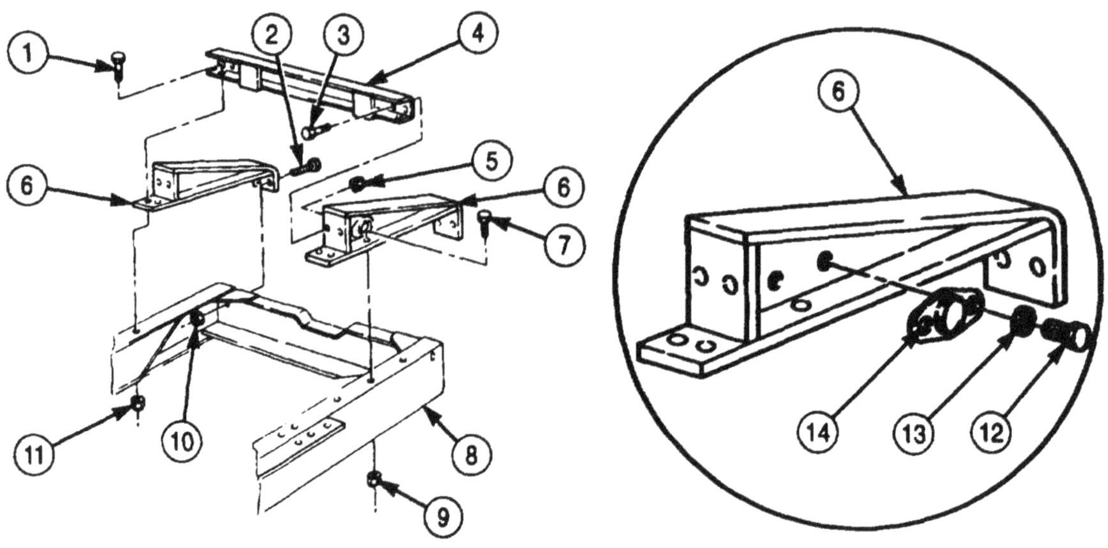

FOLLOW-ON TASK: Install fifth wheel (para. 3-248).

3-250. TRACTOR SPARE TIRE CARRIER TOOLBOX REPLACEMENT

THIS TASK COVERS:

a. Removal b. Installation

INITIAL SETUP:

APPLICABLE MODELS
M931/A1/A2, M932/A1/A2

TOOLS
General mechanic's tool kit (Appendix E, Item 1)

MATERIALS/PARTS
Six locknuts (M931, M932) (Appendix D, Item 291)
One locknut (Appendix D, Item 288)

REFERENCES (TM)
TM 9-2320-272-10
TM 9-2320-272-24P

EQUIPMENT CONDITION
Parking brake set (TM 9-2320-272-10).

a. Removal

NOTE

- Ensure toolbox is empty before starting procedure.
- Assistant will help with step 2.

1. Remove six locknuts (7), screws (4), and washers (5) from carrier base (6) and toolbox (31. Discard locknuts (7).
2. Remove locknut (2), screw (8), and toolbox (3) from support bracket (1) and carrier base (6). Discard locknut (2).

b. Installation

NOTE

Assistant will help with step 2.

1. Install toolbox (3) on carrier base (6) with six screws (4), washers (5), and new locknuts (7).
2. Install toolbox (3) on support bracket (1) with screw (8) and new locknut (2).

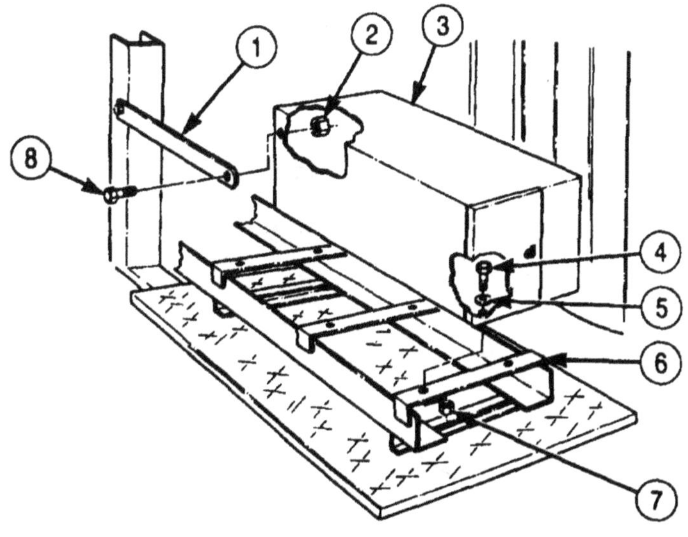

3-251. FIFTH WHEEL DECK PLATE REPLACEMENT

THIS TASK COVERS:

a. Removal

b. Installation

INITIAL SETUP:

APPLICABLE MODELS
M931/A1/A2, M932/A1/A2

TOOLS
General mechanic's tool kit (Appendix E, Item 1)

MATERIALS/PARTS
Six locknuts (Appendix D, Item 291)

REFERENCES (TM)
TM 9-2320-272-10
TM 9-2320-272-24P

EQUIPMENT CONDITION
Parking brake set (TM 9-2320-272-10).

a. Removal

1. Remove four locknuts (9), screws (l), and plate retainers (8) from frame (6) and deck plate (4). Discard locknuts (9).
2. Remove two locknuts (7), screws (3), washers (2), deck plate (4), and two spacers (5) from frame (6). Discard locknuts (7).

b. Installation

1. Install two spacers (5) and deck plate (4) on frame (6) with two washers (2), screws (3), and new locknuts (7).
2. Install four plate retainers (8) on deck plate (4) and frame (6) with four screws (1) and new locknuts (9).

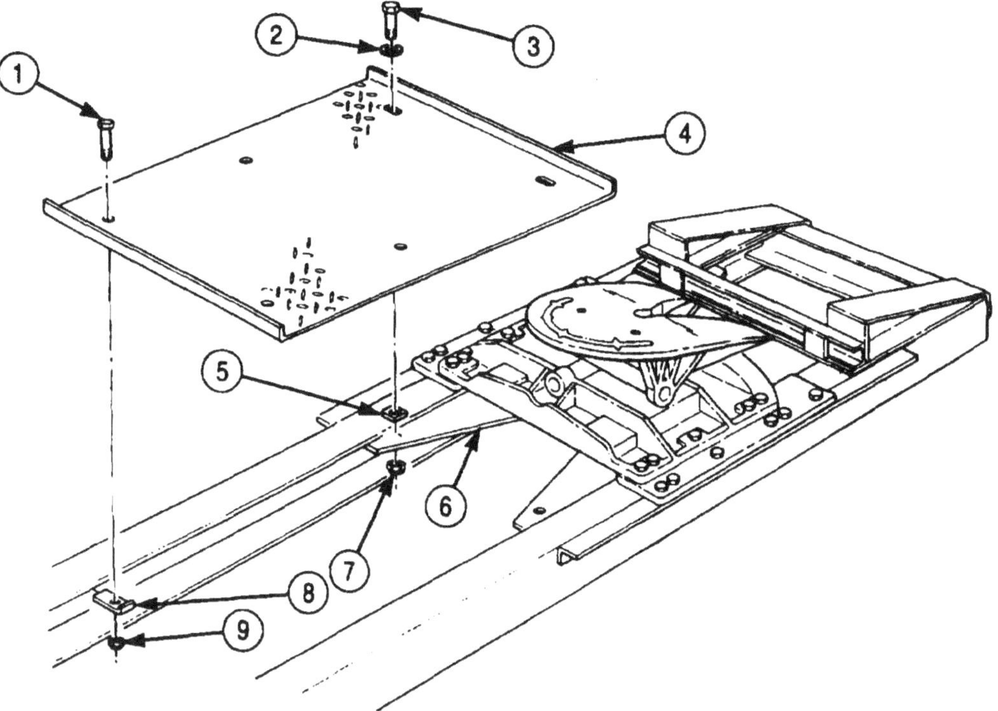

3-252. FIFTH WHEEL SPACERS REPLACEMENT

THIS TASK COVERS:

a. Removal
b. Installation

INITIAL SETUP:

APPLICABLE MODELS
M931/A1/A2, M932/A1/A2

TOOLS
General mechanic's tool kit (Appendix E, Item 1)

MATERIALS/PARTS
Twelve locknuts (Appendix D, Item 321)

REFERENCES (TM)
TM 9-2320-272-10
TM 9-2320-272-24P

EQUIPMENT CONDITION
- Parking brake set (TM 9-2320-272-10).
- Fifth wheel removed (para. 3-248).

a. Removal

Remove twelve locknuts (4), screws (2), and two spacers (1) from frame (3). Discard locknuts (4).

b. Installation

Install two spacers (1) on frame (3) with twelve screws (2) and new locknuts (4).

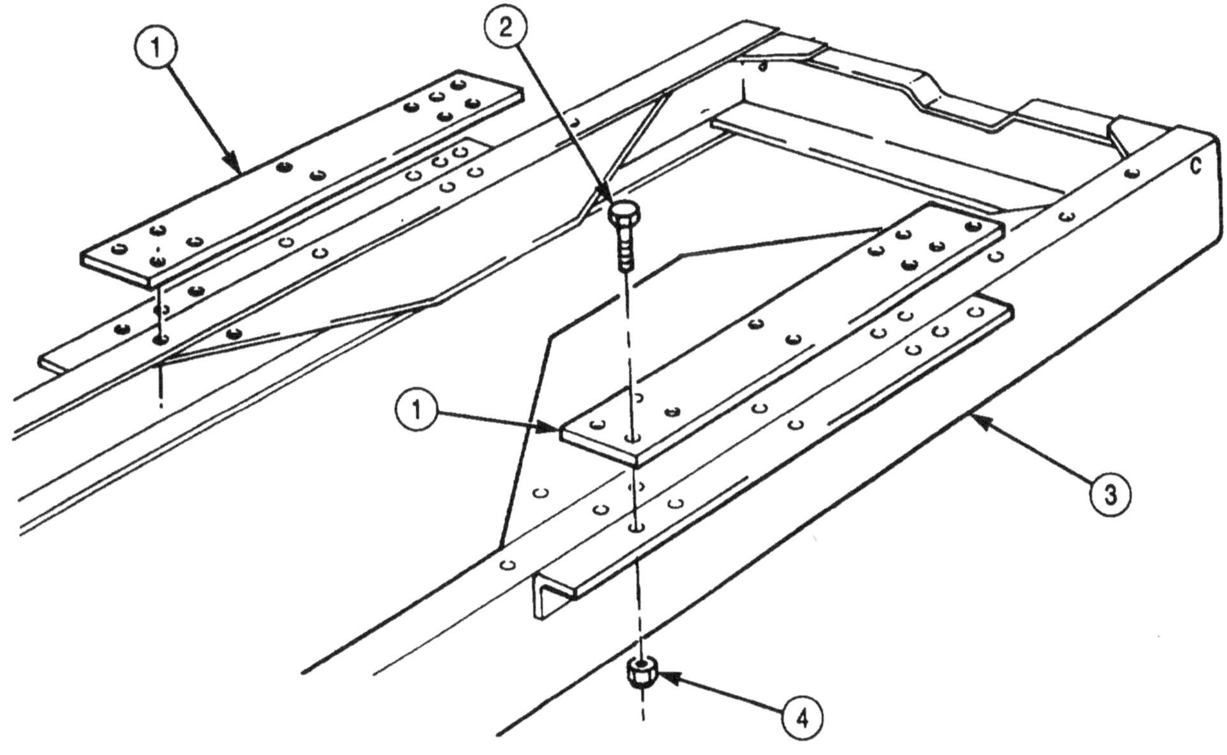

FOLLOW-ON TASK: Install fifth wheel (para. 3-248).

3-253. TRACTOR SPARE TIRE CARRIER (M931, M932) REPLACEMENT

THIS TASK COVERS:

a. Removal
b. Installation

INITIAL SETUP:

APPLICABLE MODELS
M931, M932

TOOLS
General mechanic's tool kit (Appendix E Item 1)

MATERIALS/PARTS
Four locknuts (Appendix D, Item 294)

PERSONNEL REQUIRED
Two

REFERENCES (TM)
TM 9-2320-272-10
TM 9-2320-272-24P

EQUIPMENT CONDITION
- Parking brake set (TM 9-2320-272-10).
- Spare tire removed (TM 9-2320-272-10).
- Toolbox removed (para. 3-250).
- Carrier access steps removed (para. 3-255).

a. Removal

1. Loosen retaining pin (2) and remove boom extension (1) from boom (3).
2. Loosen screw (4) and remove boom (3) from boom support (5).
3. Remove two locknuts (8) and (11) and screws (6) and (12) from frame rail brackets (9) and (10) and carrier base (7). Discard locknuts (8) and (11).
4. Remove carrier base (7) from frame rail brackets (9) and (10).

b. Installation

1. Position carrier base (7) on frame rail brackets (9) and (10).
2. Align carrier base (7) with frame rail brackets (9) and (10) and install two screws (6) and (12) and new locknuts (8) and (11).
3. Install boom (3) on boom support (5).

NOTE

Ensure screw is tightened against preset grove in boom support.

4. Tighten screw (4) on boom (3).
5. Install boom extension (1) on boom (3).
6. Tighten retaining pin (2) in boom extension (1) and boom (3).

3-253. TRACTOR SPARE TIRE CARRIER (M931, M932) REPLACEMENT (Contd)

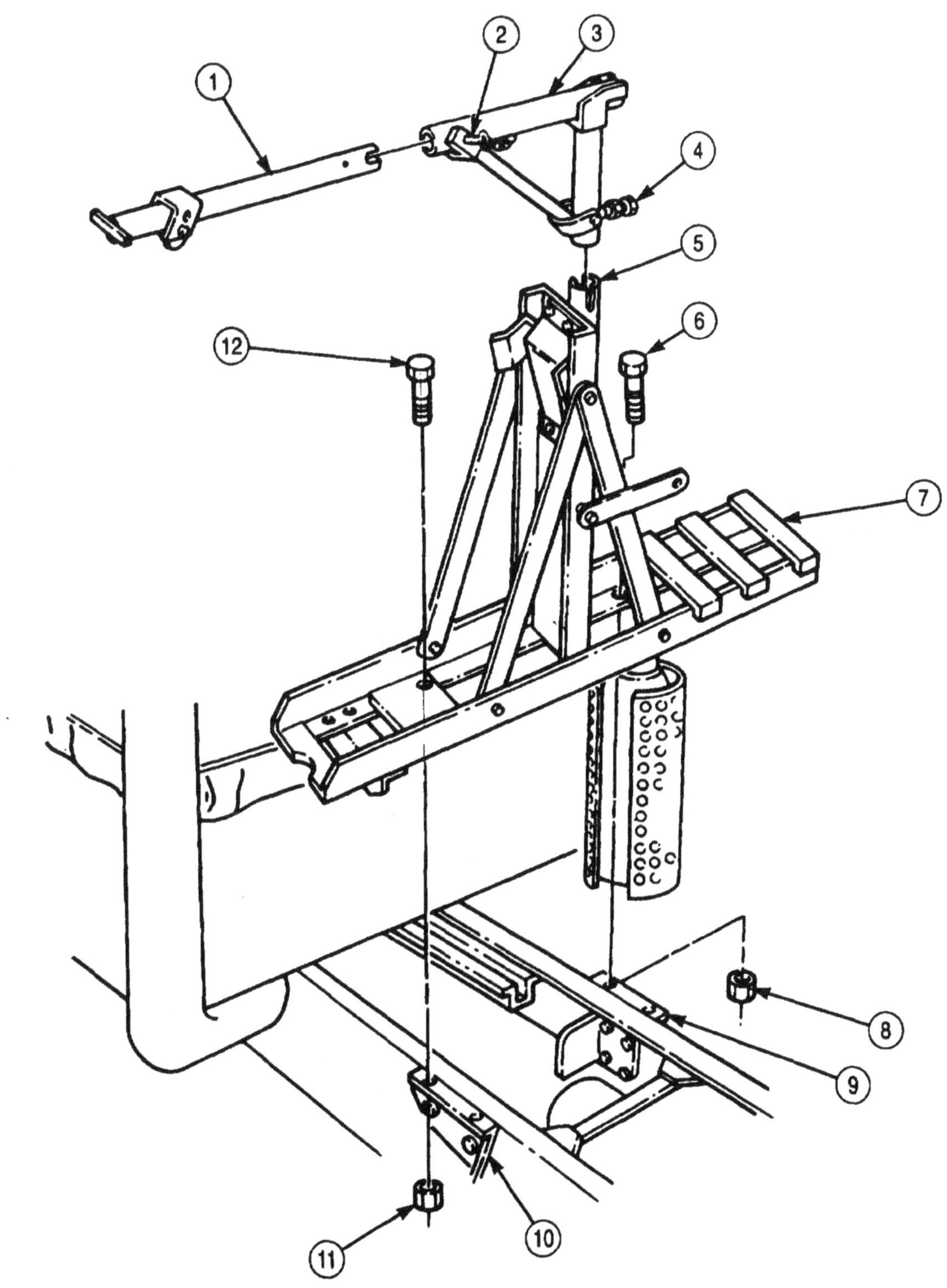

FOLLOW-ON TASKS:
- Install carrier access steps (para. 3-255).
- Install toolbox (para. 3-250).
- Install spare tire (TM 9-2320-272-10).

3-254. TRACTOR SPARE TIRE CARRIER (M931A1/A2, M932A1 /A2) REPLACEMENT

THIS TASK COVERS:

a. Removal b. Installation

INITIAL SETUP:

APPLICABLE MODELS

M931A1/A2, M932A1/A2

TOOLS

General mechanic's tool kit (Appendix E, Item 1)
Four jack stands
Prybar
Lifting device
Utility chain

MATERIALS/PARTS

Twenty locknuts (Appendix D, Item 291)
Eight locknuts (Appendix D, Item 294)
Four locknuts (Appendix D, Item 275)
Six locknuts (Appendix D, Item 274)
Lockwasher (Appendix D, Item 402)
Two lockwashers (Appendix D, Item 345)
One locknut (Appendix D, Item 288)

REFERENCES (TM)

TN 9-2320-272-10
TM 9-2320-272-24P

EQUIPMENT CONDITION

- Parking brake set (TM 9-2320-272-10).
- Spare tire removed (TM 9-2320-272-10).
- Trailer coupling hoses removed (para. 3-209).

a. Removal

1. Remove two locknuts (16), screws (13), and reflector bracket (14) from carrier base (15). Discard locknuts (16).
2. Remove three locknuts (8), washers (7), and screws (5) from trailer harness plug (4) and carrier base (15). Discard locknuts (8).
3. Remove locknut (34), washer (33), lockwashers (32), (31), and (30), two ground straps (6), screw (28), and trailer harness plug (4) from harness base (29). Discard locknut (34) and lockwashers (30), (31), and (32).
4. Remove locknut (37), screw (35), and two clamps (36) from fuel lines (38) and carrier base (15). Discard locknut (37).
5. Remove locking pin (23) and retaining pin (22) from boom extension (1).
6. Remove boom extension (1) from boom (21).
7. Remove four locknuts (3), screws (2), and boom (21) from boom support (20). Discard locknuts (3).
8. Remove four locknuts (26), screws (24), and two reflector brackets (27) from carrier access steps (25). Discard locknuts (26).
9. Remove locknut (10), screw (19), three locknuts (12), screws (18), and washers (17) from frame brackets (9) and (11) and carrier base (15). Discard locknuts (10) and (12).

3-254. TRACTOR SPARE TIRE CARRIER (M931A1/A2, M932A1/A2) REPLACEMENT (Contd)

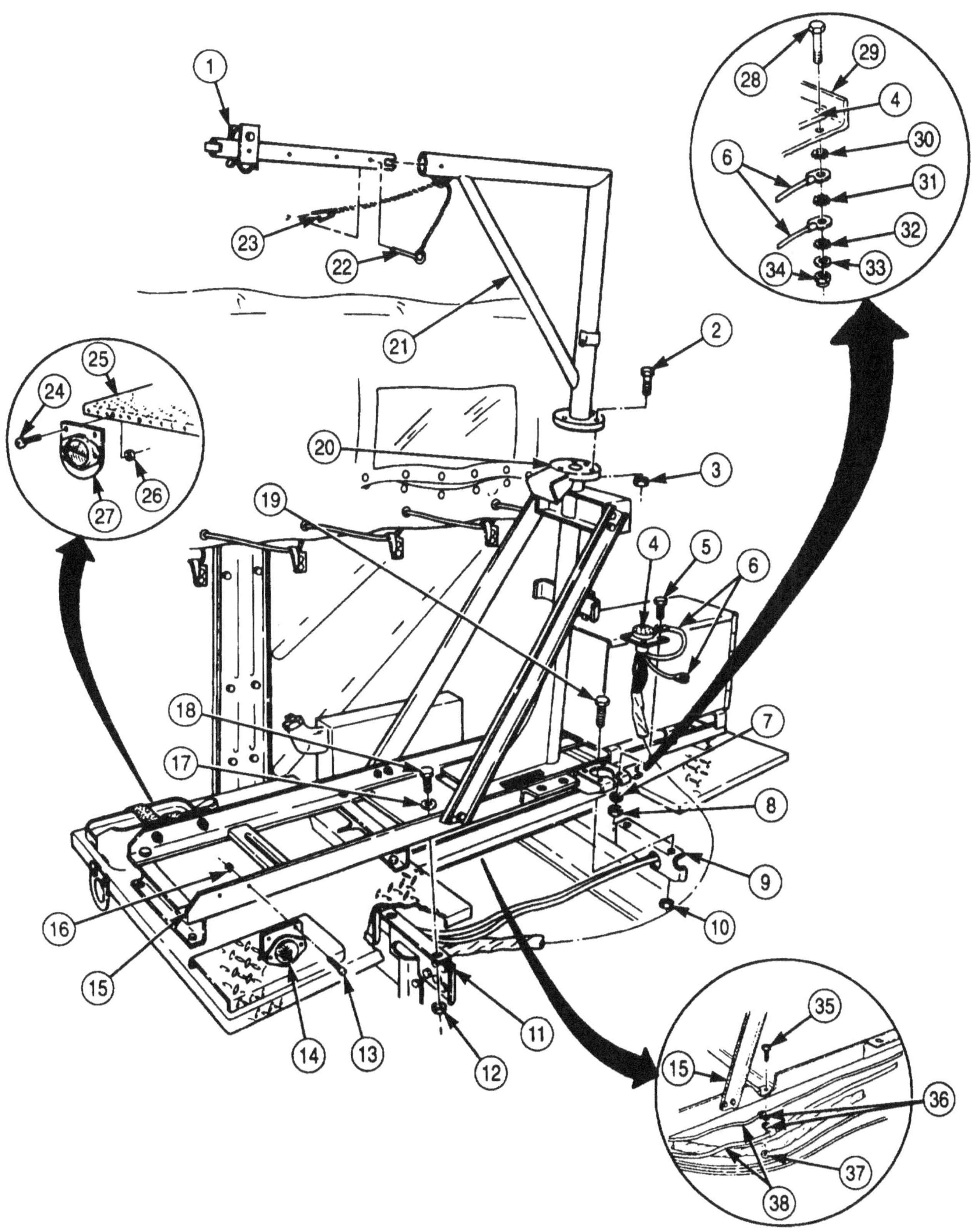

3-254. TRACTOR SPARE TIRE CARRIER (M931A1/A2, M932A1/A2) REPLACEMENT (Contd)

10. Install utility chain around upper carrier (1) and attach to suitable lifting device.
11. Lift carrier base (2) from vehicle and position on four jack stands.
12. Remove utility chain from upper carrier (1).
13. Remove six locknuts (13), screws (11), washers (12), and toolbox (10) from carrier base (2). Discard locknuts (13).
14. Remove six locknuts (5), washers (6), screws (9), and bracket assembly (8) from carrier step (7) and carrier base (2). Discard locknuts (5).

NOTE

Assistant will support access steps during step 15.

15. Remove eight locknuts (4), screws (3), and two carrier steps (7) from carrier base (2). Discard locknuts (4).

b. Installation

NOTE

Assistant will support access steps during step 1.

1. Install two carrier steps (7) on carrier base (2) with eight screws (3) and new locknuts (4).
2. Install bracket assembly (8) on carrier step (7) and carrier base (2) with six screws (9). washers (6), and new locknuts (5).
3. Install toolbox (10) on carrier base (2) with six washers (12), screws (11), and new locknuts (13).
4. Install utility chain around upper carrier (1) and attach to suitable lifting device.
5. Lift carrier base (2) from four jack stands and position on vehicle.

3-254. TRACTOR SPARE TIRE CARRIER (M931A1/A2, M932A1/A2) REPLACEMENT (Contd)

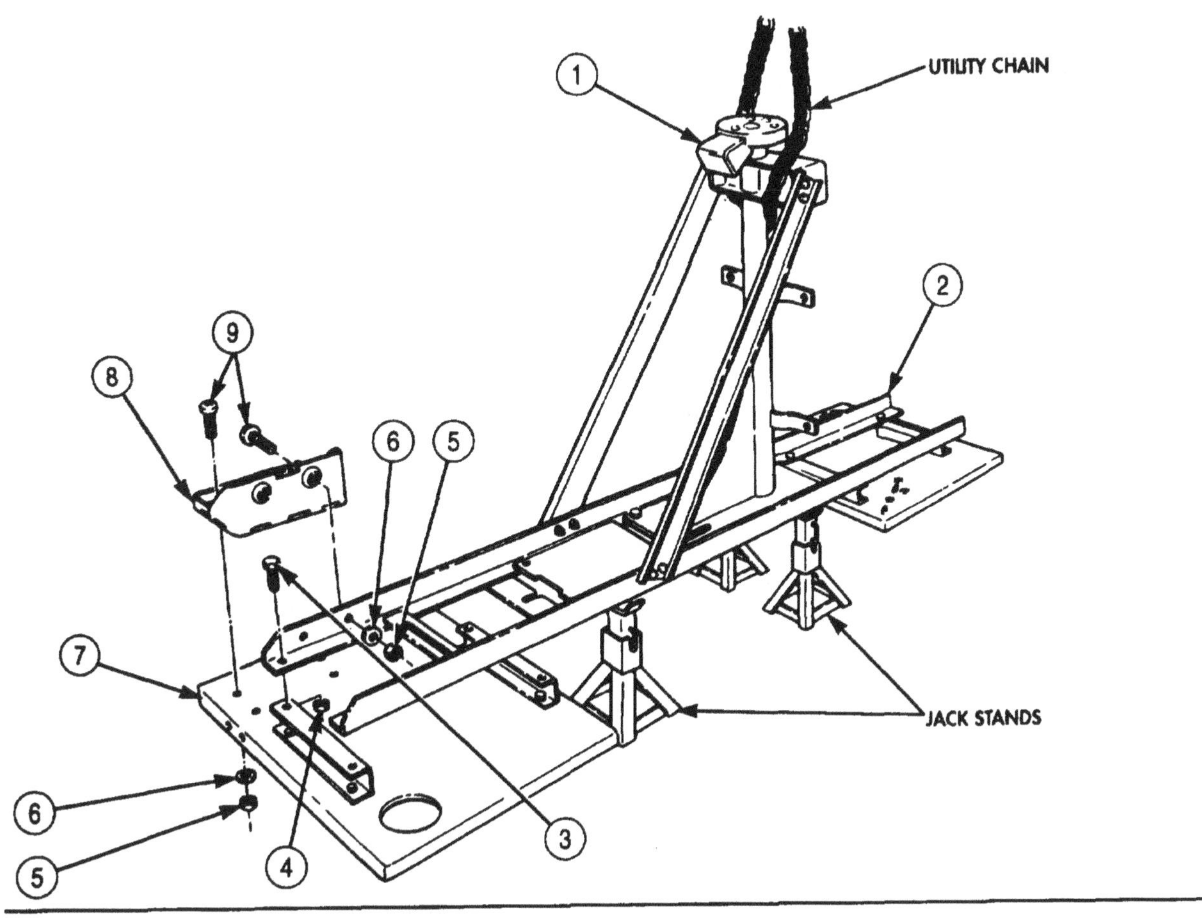

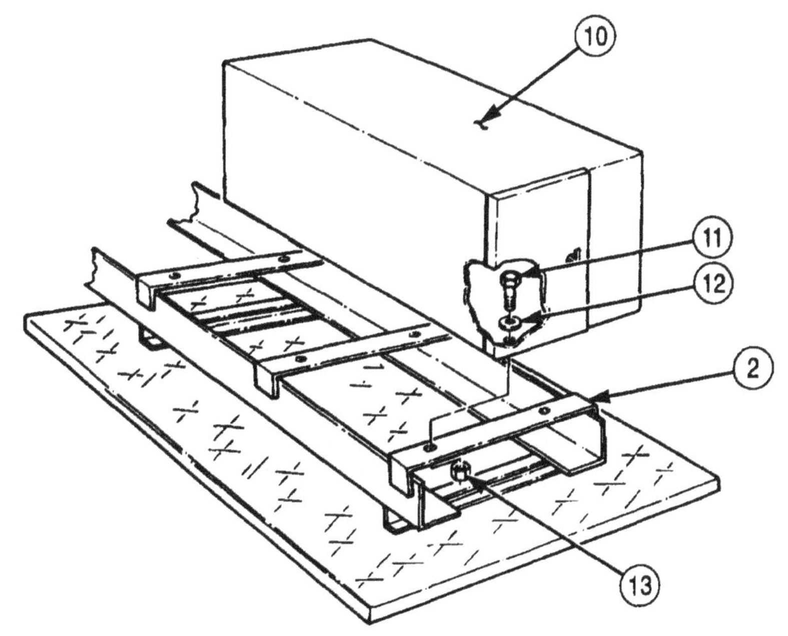

3-254. TRACTOR SPARE TIRE CARRIER (M931A1/A2, M932A1/A2) REPLACEMENT (Contd)

6. Align carrier base (15) on frame brackets (9) and (11) and install screw (19), three washers (17), and screws (18).
7. Remove lifting device and utility chain from carrier base (15).
8. Install new locknut (10) on screw (19) and three new locknuts (12) on screws (18).
9. Install two reflector brackets (27) on carrier access steps (25) with four screws (24) and new locknuts (26).
10. Install boom (21) on boom support (20) with four screws (2) and new locknuts (3).
11. Install boom extension (1) on boom (21).
12. Install retaining pin (22) and locking pin (23) in boom extension (1).
13. Install two fuel lines (38) on carrier base (15) with two clamps (36), screw (35), and new locknut (37).
14. Install trailer harness plug (4) on carrier base (15) with three screws (5), washers (7), and new locknuts (8).
15. Install two ground straps (6) on harness base (29) and trailer harness plug (4) with screw (28), new lockwashers (30), (31), and (32), washer (33), and new locknut (34).
16. Install reflector bracket (14) on carrier base (15) with two screws (13) and new locknuts (16).

3-254. TRACTOR SPARE TIRE CARRIER (M931A1/A2, M932A1/A2) REPLACEMENT (Contd)

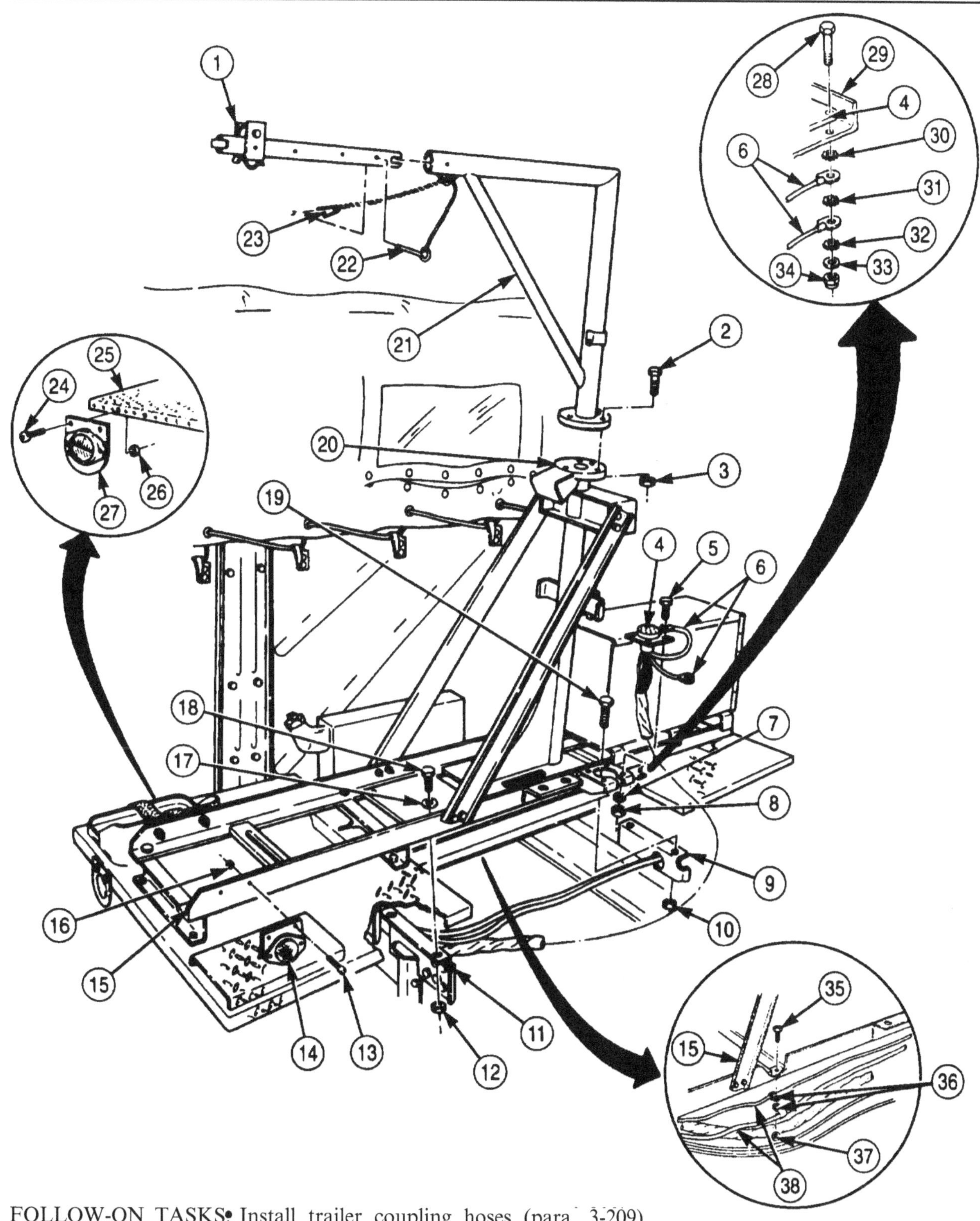

FOLLOW-ON TASKS: • Install trailer coupling hoses (para. 3-209).
• Install spare tire (TM 9-2320-272-10).

3-255. DUMP AND TRACTOR SPARE TIRE CARRIER ACCESS STEP REPLACEMENT

THIS TASK COVERS:

a. Removal b. Installation

INITIAL SETUP:

APPLICABLE MODELS
M929/A1/A2, M930/A1/A2, M931/A1/A2, M932/A1/A2

TOOLS
General mechanic's tool kit (Appendix E, Item 1)

MATERIALS/PARTS
Sixteen locknuts (Appendix D, Item 291)
Two locknuts (Appendix D, Item 313)
Rags (Appendix C, Item 58)

REFERENCES (TM)
TM 9-2320-272-10
TM 9-2320-272-24P

EQUIPMENT CONDITION
- Parking brake set (TM 9-2320-272-10).
- Toolbox removed if required for access (para. 3-250).

GENERAL SAFETY INSTRUCTIONS
Do not perform task near open flame.

WARNING

Diesel fuel is highly flammable. Do not perform this procedure near open flame. Injury to personnel may result.

NOTE

This procedure is the same for right and left spare tire carrier access steps.

a. Removal

1. Remove two locknuts (14), screws (16), and reflector bracket (15) from carrier access step (9). Discard locknuts (14).
2. Remove six locknuts (20), washers (19), screws (18), and fuel can bracket (21) from carrier base (2) and step (9). Discard locknuts (20).

NOTE

- Perform step 3 for vehicles with top fill tank.
- Vehicles may have one or two fuel cap covers. Perform step 3 for both fuel cap covers.

3. Lift fuel cap cover (6) and remove fuel cap (4) and chain and strainer assembly (5) from fuel tank (10). Wrap chain and strainer assembly (5) with rag. Cover fuel tank opening with rags.
4. Remove four locknuts (13), screws (1), two step brackets (17), and carrier access step (9) from carrier base (2). Discard locknuts (13).
5. Remove four locknuts (11), screws (12), and two step brackets (17) from carrier access step (9). Discard locknuts (11).

NOTE

Perform step 6 for M931/A1/A2 and M932/A1/A2 models.

6. Remove two locknuts (3), screws (7), washers (8), and fuel tank cap cover (6) from carrier access step (9). Discard locknuts (3).

3-255. DUMP AND TRACTOR SPARE TIRE CARRIER ACCESS STEP REPLACEMENT (Contd)

NOTE

Perform step 1 for M931/A1/A2 and M932/A1/A2 models.

1. Install fuel tank cap cover (6) on carrier access step (9) with two screws (7), washers (8), and new locknuts (3).
2. Install two step brackets (17) on carrier access step (9) with four screws (12) and new locknuts (11).
3. Install carrier access step (9) and two step brackets (17) on carrier base (2) with four screws (1) and new locknuts (13).

NOTE

- Perform step 4 for vehicles with top fill tank.
- Vehicles may have one or two fuel cap covers. Perform step 4 for both fuel cap covers.

4. Lift fuel cap cover (6), remove rags from fuel tank opening, and install chain and strainer assembly (5) and fuel cap (4) on fuel tank (10).
5. Install fuel can bracket (21) on carrier access step (9) and carrier base (2) with six screws (18), washers (19), and new locknuts (20).
6. Install reflector bracket (15) on carrier access step (9) with two screws (16) and new locknuts (14).

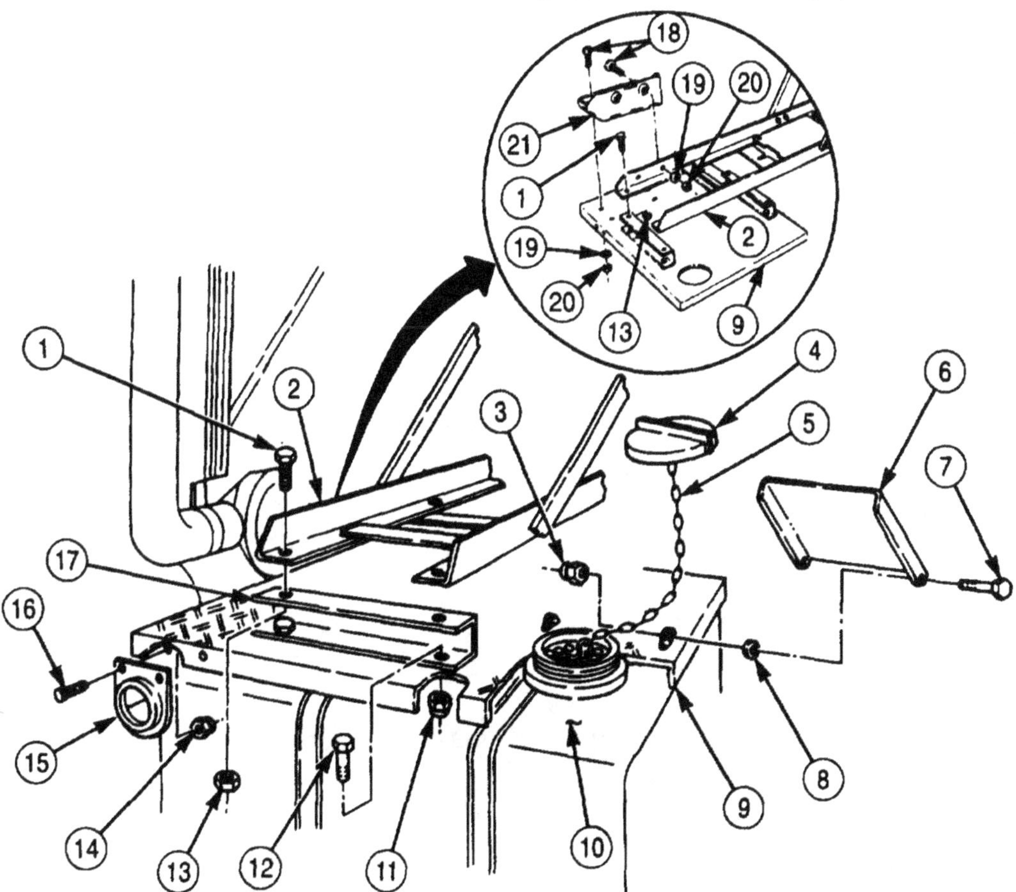

FOLLOW-ON TASK: Install toolbox if removed (para. 3-250).

3-256. DUMP SPARE TIRE CARRIER (M929, M930) REPLACEMENT

THIS TASK COVERS:

a. Removal b. Installation

INITIAL SETUP:

APPLICABLE MODELS
M929, M930

TOOLS
General mechanic's tool kit (Appendix E, Item 1)

MATERIALS/PARTS
Four locknuts (Appendix D, Item 294)

PERSONNEL REQUIRED
Two

REFERENCES (TM)
TM 9-2320-272-10
TM 9-2320-272-24P

EQUIPMENT CONDITION
- Parking brake set (TM 9-2320-272-10).
- Spare tire removed (TM 9-2320-272-10).
- Spare tire carrier access steps removed (para. 3-255).

a. Removal

1. Remove two locknuts (6), screws (3), and washers (4) from carrier base (1) and frame bracket (5). Discard locknuts (6).
2. Remove two locknuts (8), screws (2), and carrier base (1) from frame brackets (5) and (7). Discard locknuts (8).

b. Installation

1. Position carrier base (1) on frame brackets (5) and (7).
2. Install carrier base (1) on frame brackets (5) and (7) with two screws (2), new locknuts (8), washers (4), screws (3), and new locknuts (6).

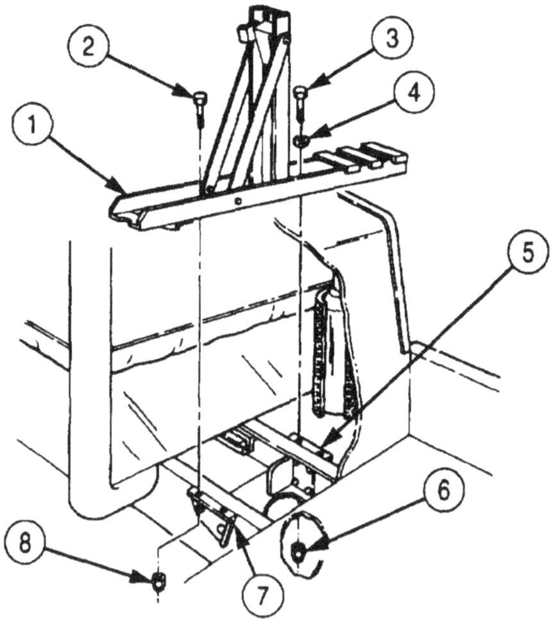

FOLLOW-ON TASKS:
- Install spare tire carrier access steps (para. 3-255).
- Install spare tire (TM 9-2320-272-10).

3-257. DUMP SPARE TIRE CARRIER (M929A1/A2, M930A1/A2) REPLACEMENT

THIS TASK COVERS:

a. Removal

b. Installation

INITIAL SETUP:

APPLICABLE MODELS

M929A1/A2, M930A1/A2

TOOLS

General mechanic's tool kit (Appendix E, Item 1)
Four jack stands
Prybar
Lifting device
Utility chain

MATERIALS/PARTS

Fourteen locknuts (Appendix D, Item 291)
Four locknuts (Appendix D, Item 294)
Two locknuts (Appendix D, Item 297)

REFERENCES (TM)

TM 9-2320-272-10
TM 9-2320-272-24P

EQUIPMENT CONDITION

- Parking brake set (TM 9-2320-272-10).
- Raise and secure dump body (TM 9-2320-272-10).
- Spare tire removed (TM 9-2320-272-10).

GENERAL SAFETY INSTRUCTIONS

Dump body must be raised and secured.

a. Removal

WARNING

Dump body must be raised and secured with safety braces before removal and installation of dump spare tire carrier. Failure to do this may result in injury to personnel.

1. Remove two locknuts (11), screws (10), and reflector bracket (12) from carrier access step (7). Discard locknuts (11).
2. Remove locknut (3), screw (1), three locknuts (6), screws (9), and washers (8) from frame brackets (2) and (5) and carrier base (4). Discard locknuts (3) and (6).

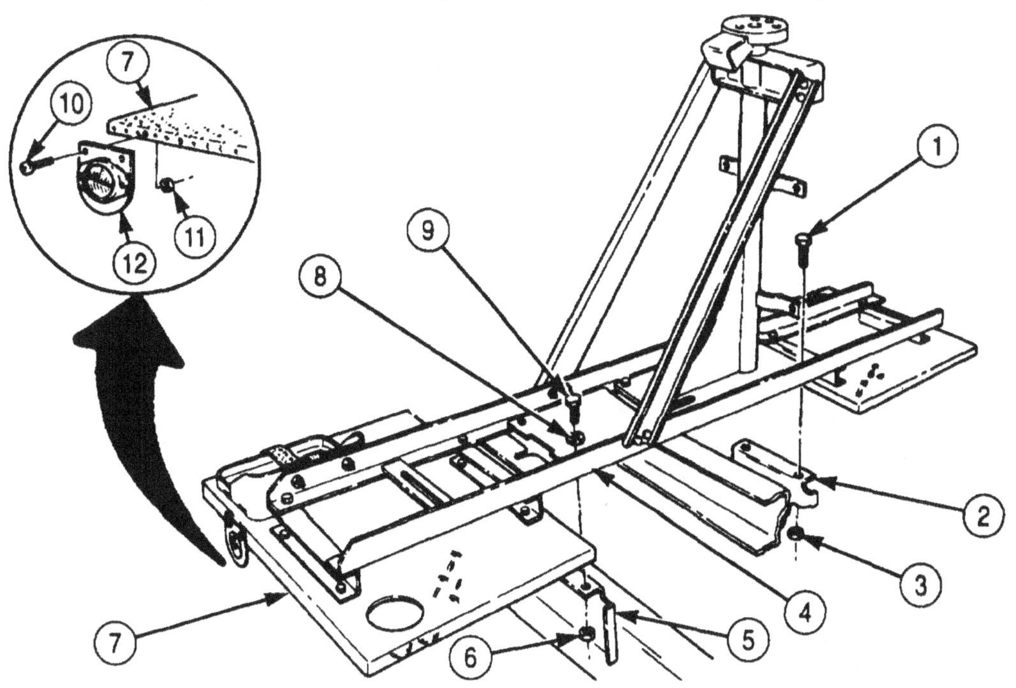

3-257. DUMP SPARE TIRE CARRIER (M929A1/A2, M930A1/A2) REPLACEMENT (Contd)

3. Attach utility chain to upper tire carrier (1) and a suitable lifting device.
4. Lift carrier base (2) from vehicle and position on four jack stands.
5. Remove utility chain from upper tire carrier (1).
6. Remove six locknuts (4), washers (7), screws (8), and bracket assembly (9) from carrier step (3) and carrier base (2). Discard locknuts (4).

NOTE

Assistant will support access steps during step 7.

7. Remove eight locknuts (6), screws (5), and two carrier steps (3) from carrier base (2). Discard locknuts (6).

b. Installation

NOTE

Assistant will support access steps during step 1.

1. Install two carrier steps (3) on carrier base (2) with eight screws (5) and new locknuts (6).
2. Install bracket assembly (9) on carrier step (3) and carrier base (2) with six screws (8), washers (7), and new locknuts (4).
3. Install utility chain to upper carrier (1) and a suitable lifting device.
4. Lift carrier base (2) from four jack stands and position on vehicle.
5. Using lifting device and prybar, align carrier base (2) on frame brackets (11) and (13) and install screw (10), three washers (15), and screws (16).
6. Remove lifting device and utility chain from upper tire carrier (1).
7. Install new locknut (12) and three new locknuts (14) on screw (10) and three screws (16).
8. Install reflector bracket (19) on carrier access step (3) with two screws (17) and new locknuts (18).

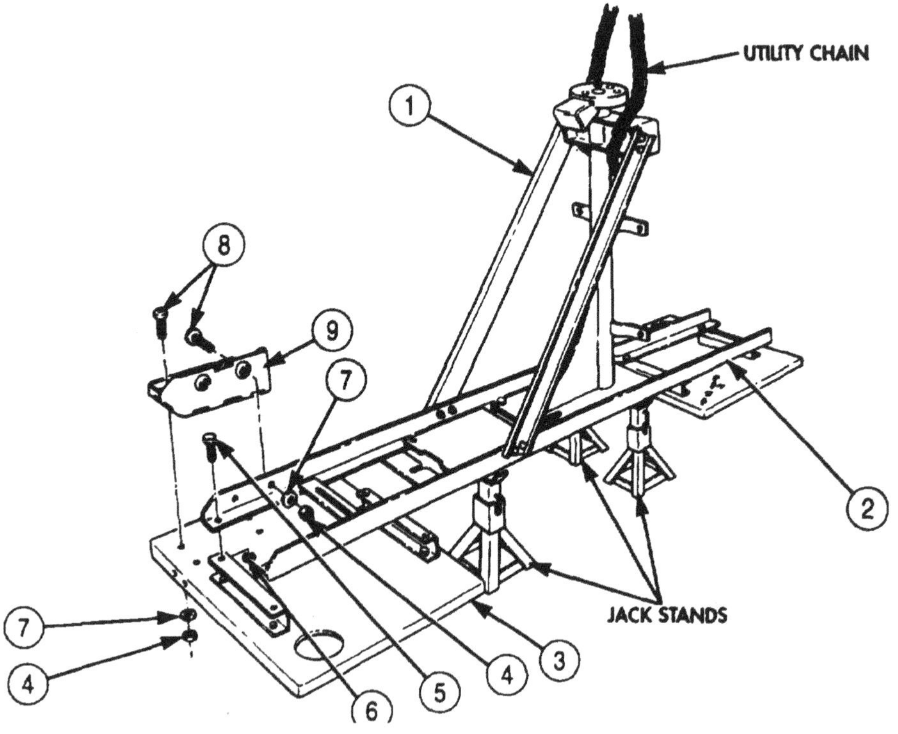

3-257. DUMP SPARE TIRE CARRIER (M929A1/A2, M930A1/A2) REPLACEMENT (Contd)

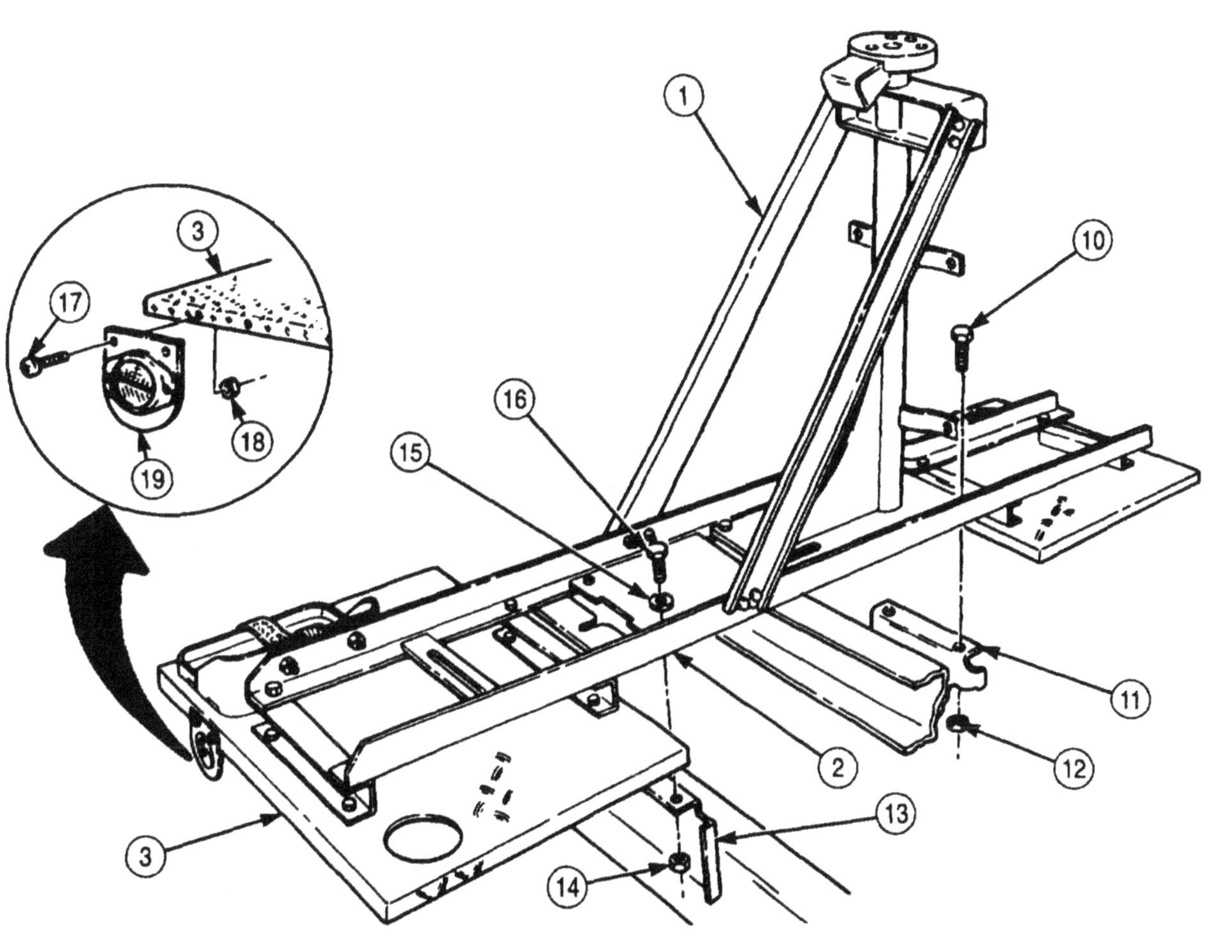

FOLLOW-ON TASK: Install spare tire (TM 9-2320-272-10).

3-258. CARGO SPARE TIRE CARRIER (M923, M925, M927, M928) REPLACEMENT

THIS TASK COVERS:

a. Removal b. Installation

INITIAL SETUP:

APPLICABLE MODELS
M923, M925, M927, M928

TOOLS
General mechanic's tool kit (Appendix E, Item 1)

MATERIALS/PARTS
Four locknuts (Appendix D, Item 291)
Four locknuts (Appendix D, Item 294)

PERSONNEL REQUIRED
Two

REFERENCES (TM)
TM 9-2320-272-10
TM 9-2320-272-24P

EQUIPMENT CONDITION
- Parking brake set (TM 9-2320-272-10).
- Spare tire removed (TM 9-2320-272-10).

a. Removal

1. Remove retaining pin (2) from boom extension (1) and boom (3).
2. Remove boom extension (1) from boom (3).
3. Loosen screw (4) on boom (3).
4. Remove boom (3) from boom support (5).
5. Remove two locknuts (8), washers (7), and screws (16) from muffler support braces (6) and carrier base (13). Discard locknuts (8).
6. Remove four locknuts (14) and screws (10) from muffler support (9) and carrier base (13). Discard locknuts (14).
7. Remove two locknuts (12) and screws (15) from frame rail bracket (11) and carrier base (13). Discard locknuts (12).
8. Remove carrier base (13) from vehicle.

b. Installation

1. Install carrier base (13) on frame rail bracket (11) with two screws (15) and new locknuts (12).
2. Install carrier base (13) on muffler support (9) with four screws (10) and new locknuts (14).
3. Install muffler support braces (6) on carrier base (13) with two screws (16), washers (7), and new locknuts (8).
4. Install boom (3) on boom support (5).

NOTE

Ensure screw is tightened against preset grove in boom support.

5. Tighten screw (4) on boom (3).
6. Install boom extension (1) on boom (3).
7. Install retaining pin (2) in boom extension (1) and boom (3).

3-258. CARGO SPARE TIRE CARRIER (M923, M925, M927, M928) REPLACEMENT (Contd)

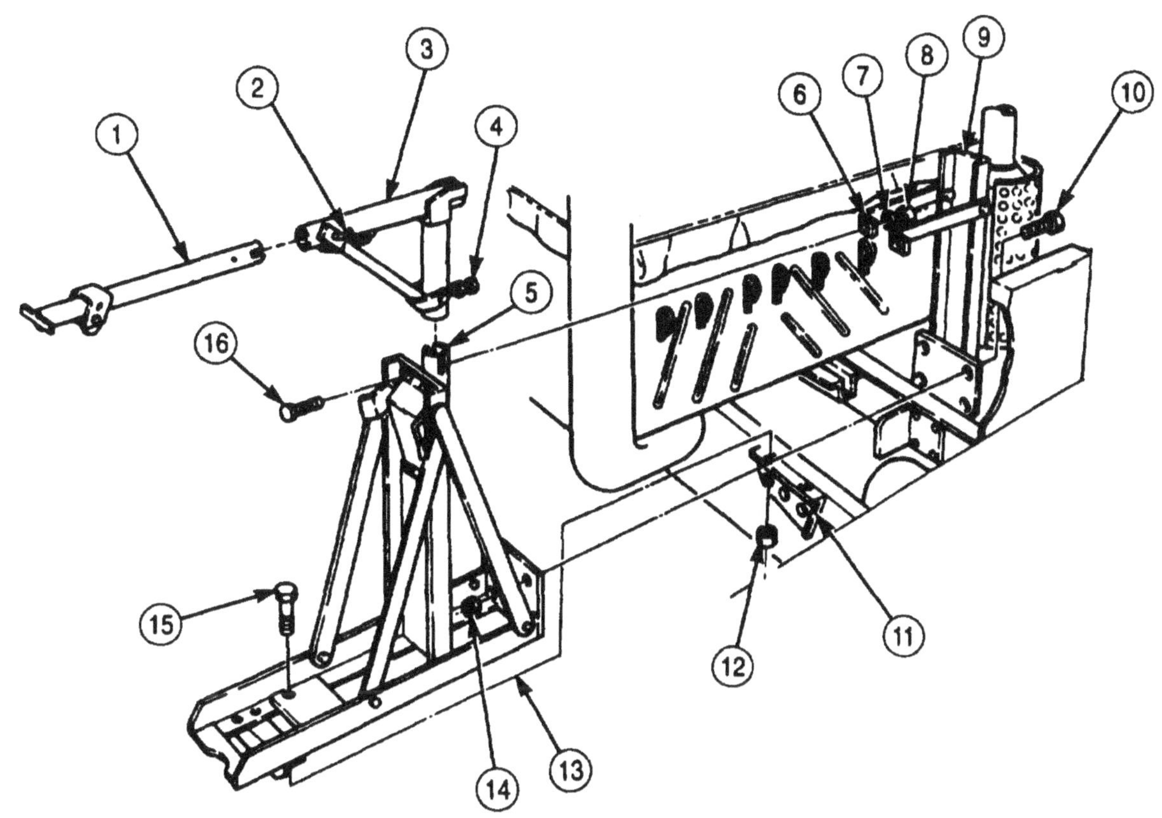

FOLLOW-ON TASK: Install spare tire (TM 9-2320-272-10).

3-259. CARGO SPARE TIRE CARRIER (M923A1/A2, M925A1/A2, M927A1/A2, M928A1/A2) REPLACEMENT

THIS TASK COVERS:

a. Removal b. Installation

INITIAL SETUP:

APPLICABLE MODELS
M923A1/A2, M925A1/A2, M927A1/A2, M928A1/A2

TOOLS
General mechanic's tool kit (Appendix E, Item 1)
Lifting device
Prybar
Utility chain

MATERIALS/PARTS
Ten locknuts (Appendix D, Item 294)
Two locknuts (Appendix D, Item 291)

PERSONNEL REQUIRED
Two

REFERENCES (TM)
TM 9-2320-272-10
TM 9-2320-272-24P

EQUIPMENT CONDITION
- Parking brake set (TM 9-2320-272-10).
- Spare tire removed (TM 9-2320-272-10).
- Carrier access step removed (para. 3-260).
- Forward cargo rack removed (TM 9-2320-272-10).

a. Removal

1. Remove locking pin (18) and retaining pin (19) from boom extension (1).
2. Remove boom extension (1) from boom (2).
3. Remove four locknuts (4), screws (3), and boom (2) from boom support (17). Discard locknuts (4).
4. Remove two locknuts (5), washers (7), screws (6), and washers (7) from muffler support bracket (8) and carrier base (14). Discard locknuts (5).
5. Install utility chain around upper part of carrier base (14) and attach to a suitable lifting device.
6. Remove four locknuts (13) and screws (10) from muffler support (9) and carrier base (14). Discard locknuts (13).
7. Remove two locknuts (12), screws (16), washers (15), and carrier base (14) from left frame rail bracket (11). Discard locknuts (12).
8. Remove utility chain from carrier base (14).

b. Installation

1. Install utility chain around upper part of carrier base (14) and attach to a suitable lifting device.
2. Using lifting device lift carrier base (14), lower onto vehicle, and align carrier base (14) with muffler support (9) while installing four screws (10) and new locknuts (13).
3. Using a prybar to align carrier base (14) with left frame rail bracket (11), install two washers (15), screws (16), and new locknuts (12).
4. Remove lifting device and utility chain from carrier base (14).
5. Install two washers (7), screws (6), washers (7), and new locknuts (5) on muffler support bracket (8) and carrier base (14).
6. Install boom (2) on boom support (17) with four screws (3) and new locknuts (4).
7. Install boom extension (1) on boom (2).
8. Install retaining pin (19) and locking pin (18) in boom extension (1) and boom (2).

3-259. CARGO SPARE TIRE CARRIER (M923A1/A2, M925A1/A2, M927A1/A2, M928A1/A2) REPLACEMENT (Contd)

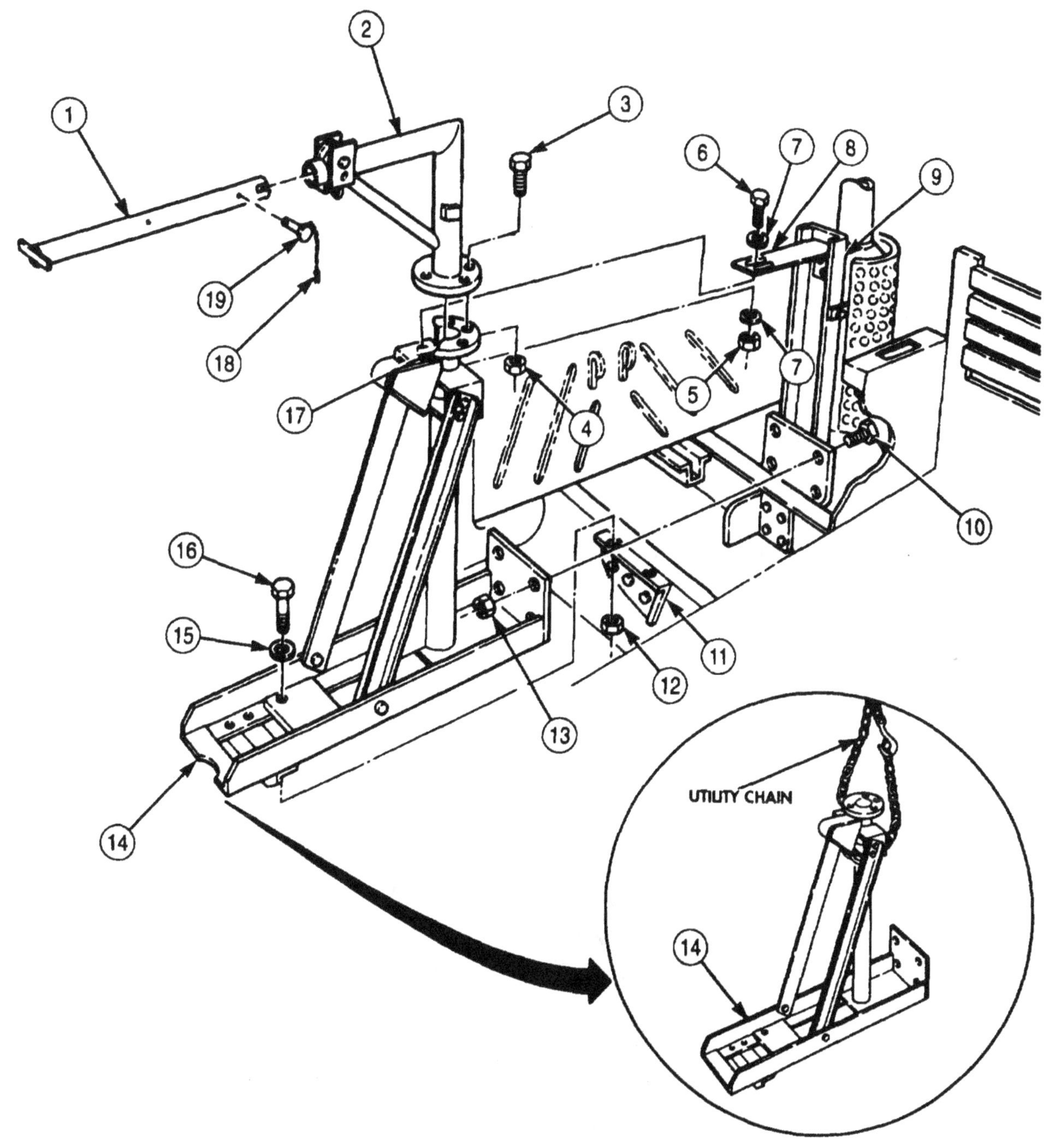

FOLLOW-ON TASKS:
- Install carrier access step (para. 3-260).
- Install spare tire (TM 9-2320-272-10).
- Install forward cargo rack (TM 9-2320-272-10).

3-260. CARGO SPARE TIRE CARRIER ACCESS STEP REPLACEMENT

THIS TASK COVERS:

a. Removal b. Installation

INITIAL SETUP:

APPLICABLE MODELS

M923/A1/A2, M925/A1/A2, M927/A1/A2, M928/A1/A2

TOOLS

General mechanic's tool kit (Appendix E, Item 1)

MATERIALS/PARTS

Four locknuts (Appendix D, Item 296)
Rags (Appendix C, Item 58)

REFERENCES (TM)

TM 9-2320-272-10
TM 9-2320-272-24P

EQUIPMENT CONDITION

Parking brake set (TM 9-2320-272-10).

GENERAL SAFETY INSTRUCTIONS

Do not perform this task near flames.

WARNING

Diesel fuel is highly flammable. Do not perform this procedure near open flame. Injury to personnel may result.

a. Removal

1. Remove fuel cap (2), chain (3), and strainer (5) from fuel tank (6). Wrap strainer (5) with rags,
2. Remove four locknuts (7), screws (l), washers (9), and step bracket (4) from carrier base (8). Discard locknuts (7).

b. Installation

1. Install step bracket (4) on carrier base (8) with four screws (1). washers (9), and new locknuts (7).
2. Install strainer (5), chain (3), and fuel cap (2) in fuel tank (6).

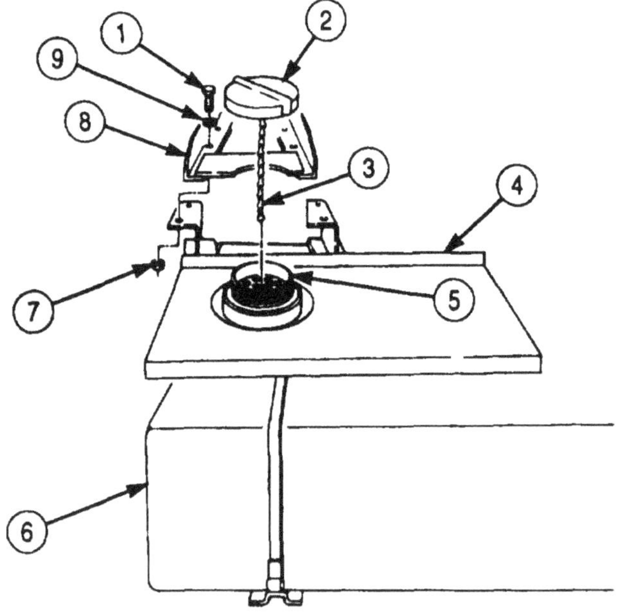

3-261. TAILGATE BUMPERS REPLACEMENT

THIS TASK COVERS:

a. Removal   b. Installation

INITIAL SETUP:

APPLICABLE MODELS
M923/A1/A2, M9251A1/A2, M927/A1/A2, M928/A1/A2

TOOLS
General mechanic's tool kit (Appendix E, Item 1)

MATERIALS/PARTS
Locknut (Appendix D, Item 288)

REFERENCES (TM)
TM 9-2320-272-10
TM 9-2320-272-24P

EQUIPMENT CONDITION
Parking brake set (TM 9-2320-272-10).

NOTE

All tailgate bumpers are replaced the same.

a. Removal

Remove locknut (2) and bumper (3) from tailgate (1). Discard locknut (2).

b. Installation

Install bumper (3) on tailgate (1) with new locknut (2).

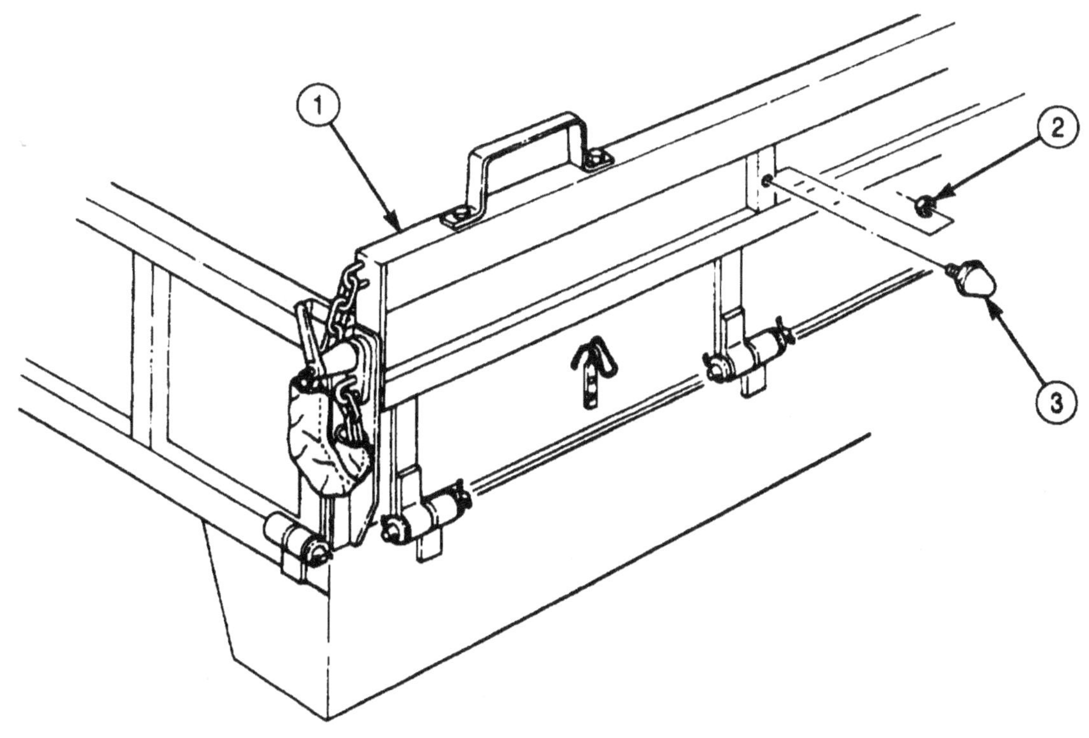

3-262. VAN SPARE TIRE CARRIER (M934) REPLACEMENT

For van spare tire carrier replacement, refer to para. 3-256.

3-263. VAN SPARE TIRE CARRIER (M934A1/A2) REPLACEMENT

THIS TASK COVERS:

a. Removal b. Installation

INITIAL SETUP:

APPLICABLE MODELS
M934A1/A2

TOOLS
General mechanic's tool kit (Appendix E, Item 1)
Jack stands

MATERIALS/PARTS
Ten locknuts (Appendix D, Item 294)

REFERENCES (TM)
TM 9-2320-272-10
TM 9-2320-272-24P

EQUIPMENT CONDITION
- Parking brake set (TM 9-2320-272-10).
- Spare tire removed (TM 9-2320-272-10).

---

a. Removal

1. Remove two locknuts (4), washers (2), screws (l), and washers (2) from muffler support brace (3) and carrier base (11). Discard locknuts (4).
2. Attach davit chain (14) to carrier base (11).
3. Remove four locknuts (12) and screws (5) from muffler support (6) and carrier base (11). Discard locknuts (12).
4. Remove two locknuts (8), screws (10), and washers (9) from carrier base (11) and frame rail bracket (7) Discard locknuts (8).
5. Using davit (13), remove carrier base (11) from frame rail bracket (7), position on jack stands, and remove davit chain (14) from carrier base (11).
6. Remove four locknuts (19), screws (15), washers (16), and carrier access step (18) with step brackets (17) from carrier base (11). Discard locknuts (19).

b. Installation

1. Position step brackets (17) under carrier base (11) with holes aligned and install carrier access step (18) on carrier base (11) with four washers (16), screws (15), and new locknuts (19).
2. Attach davit chain (14) to carrier base (11), and using davit (13), position carrier base (11) on vehicle.
3. Using two washers (9) and screws (10), align carrier base (11) with frame rail bracket (7), and install two carrier base (11) on muffler support (6) with four screws (5) and new locknuts (12).
4. Remove davit chain (14) from carrier base (11).
5. Install two new locknuts (8) on screws (10).
6. Install carrier base (11) on mufller support brace (3) with two washers (2), screws (1), washers (2), and new locknuts (4).

3-263. VAN SPARE TIRE CARRIER (M934A1/A2) REPLACEMENT (Contd)

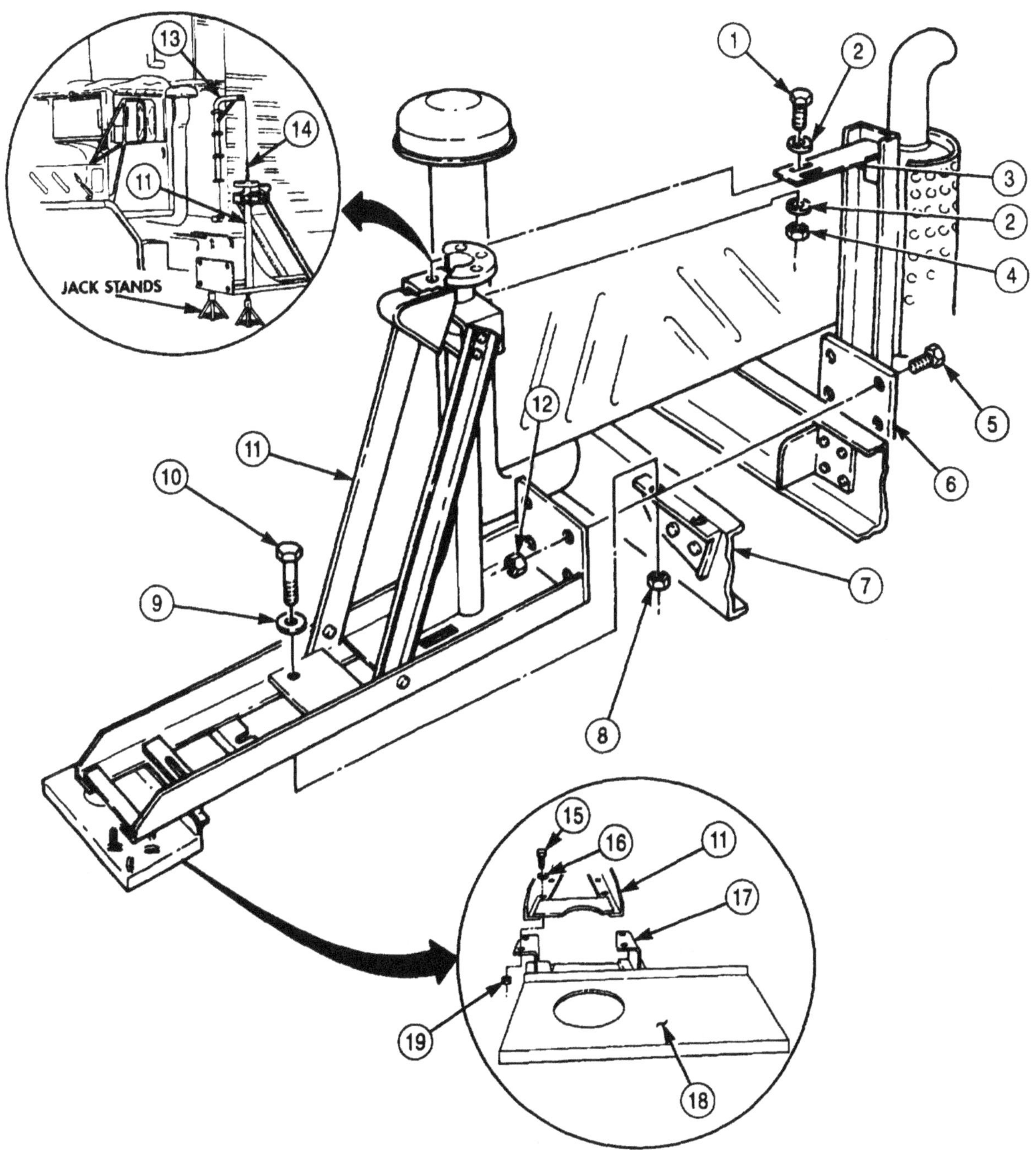

FOLLOW-ON TASK: Install spare tire (TM 9-2320-272-10).

3-264. VAN DAVIT CHAIN AND WIRE ROPE REPLACEMENT

THIS TASK COVERS:

a. Removal
b. Installation

INITIAL SETUP:

APPLICABLE MODELS
M934/A1/A2

TOOLS
General mechanic's tool kit (Appendix E, Item 1)

MATERIALS/PARTS
Two cotter pins (Appendix D, Item 50)
Locknut (Appendix D, Item 299)

REFERENCES (TM)
LO 9-2320-272-12
TM 9-2320-272-10
TM 9-2320-272-24P

EQUIPMENT CONDITION
Parking brake set (TM 9-2320-272-10).

GENERAL SAFETY INSTRUCTIONS
Wear hand protection when handling wire rope.

WARNING

Wear hand protection when handling wire rope. Broken wires may cause injury to personnel.

a. Removal

NOTE

Davit chain must be removed from spare tire before performing step 1.

1. Remove four nuts (l), two U-bolts (3), clamps (2), chain link (4), and thimble (5) from wire rope (6).
2. Remove screw (8) and wire rope (6) from winch barrel (7).
3. Remove wire rope (6) from swing davit (9) and pulleys (12) and (10).

b. Installation

NOTE

Inspect wire rope for cracks, frays, and abrasions, and lubricate as necessary (LO 9-2320-272-12).

1. Install wire rope (6) on winch barrel (7) with screw (8).
2. Remove two cotter pins (16) and davit pin (15) from swing davit (9). Discard cotter pins (16).
3. Remove locknut (14), washer (13), two spacers (11), pulley (12), and screw (17) from swing davit (9). Discard locknut (14).
4. Wind wire rope (6) evenly on winch barrel (7), and thread through pulley (10) and swing davit (9).
5. Loop wire rope (6) over pulley (12) while installing pulley (12) in swing davit (9) with screw (17), two spacers (11), washer (13), and new locknut (14).
6. Install davit pin (15) in swing davit (9) with two new cotter pins (16).
7. Install wire rope (6) on chain link (4) and thimble (5) with two U-bolts (3), clamps (2), and four nuts (1).

3-264. VAN DAVIT CHAIN AND WIRE ROPE REPLACEMENT (Contd)

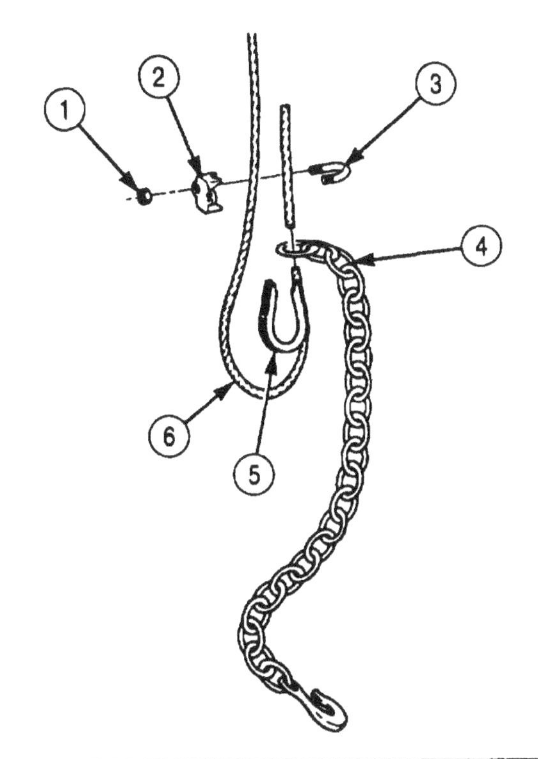

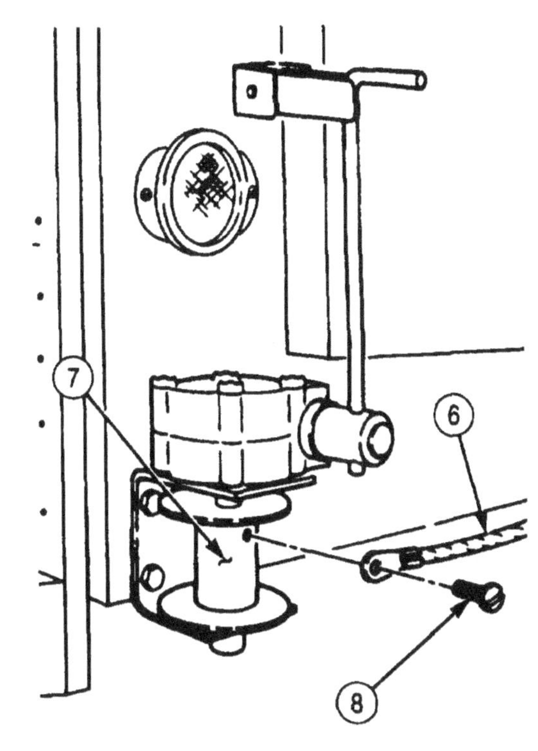

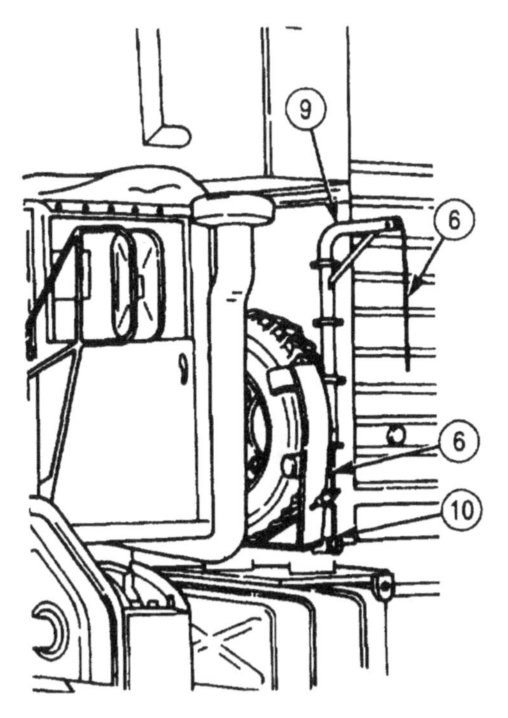

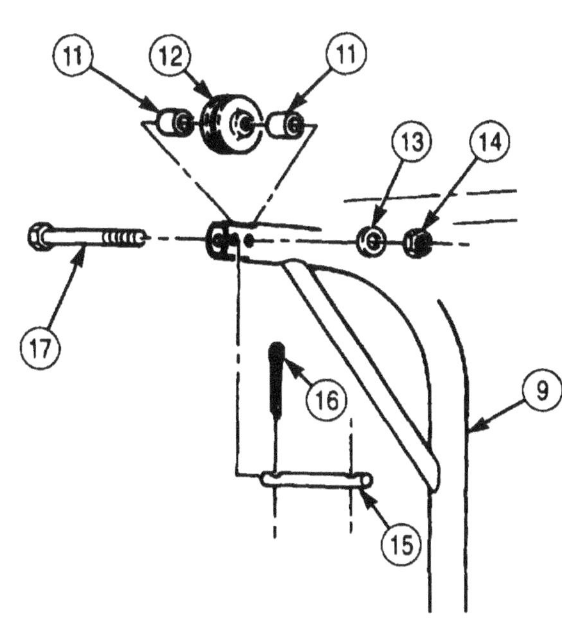

3-265. VAN SWING DAVIT AND PULLEY REPLACEMENT

THIS TASK COVERS:

a. Removal b. Installation

INITIAL SETUP:

APPLICABLE MODELS
M934A1/A2

TOOLS
General mechanic's tool kit (Appendix E, Item 1)

MATERIALS/PARTS
Locknut (Appendix D, Item 299)
Two cotter pins (Appendix D, Item 50)
Cotter pin (Appendix D, Item 51)

REFERENCES (TM)
TM 9-2320-272-10
TM 9-2320-272-24P

EQUIPMENT CONDITION
Parking brake set (TM 9-2320-272-10).
Davit chain and wire rope removed (para. 3-264).

a. Removal

1. Remove six screws (16), three clamps (15), and swing davit (5) from van body (6) and base (7).
2. Remove two cotter pins (18) and davit pin (17) from swing davit (5). Discard cotter pins (18).
3. Remove locknut (4), washer (3), screw (19), two spacers (1), and davit pulley (2) from swing davit (5). Discard locknut (4).
4. Remove three screws (14) and base (7) from van body (6).
5. Remove cotter pin (11), shaft (10), spacer (12), and pulley (13) from pulley bracket (8). Discard cotter pin (11).
6. Remove four screws (9) and pulley bracket (8) from van body (6).

b. Installation

1. Install pulley bracket (8) on van body (6) four screws (9).
2. Install spacer (12) and pulley (13) on pulley bracket (8) with shaft (10) and new cotter pin (11).
3. Install base (7) on van body (6) with three screws (14).
4. Install davit pulley (2) and two spacers (1) on swing davit (5) with screw (19), washer (3), and new locknut (4).
5. Install davit pin (17) in swing davit (5) with two new cotter pins (18).
6. Install swing davit (7) on van body (6) and base (7) with three clamps (15) and six screws (16).

3-265. VAN SWING DAVIT AND PULLEY REPLACEMENT (Contd)

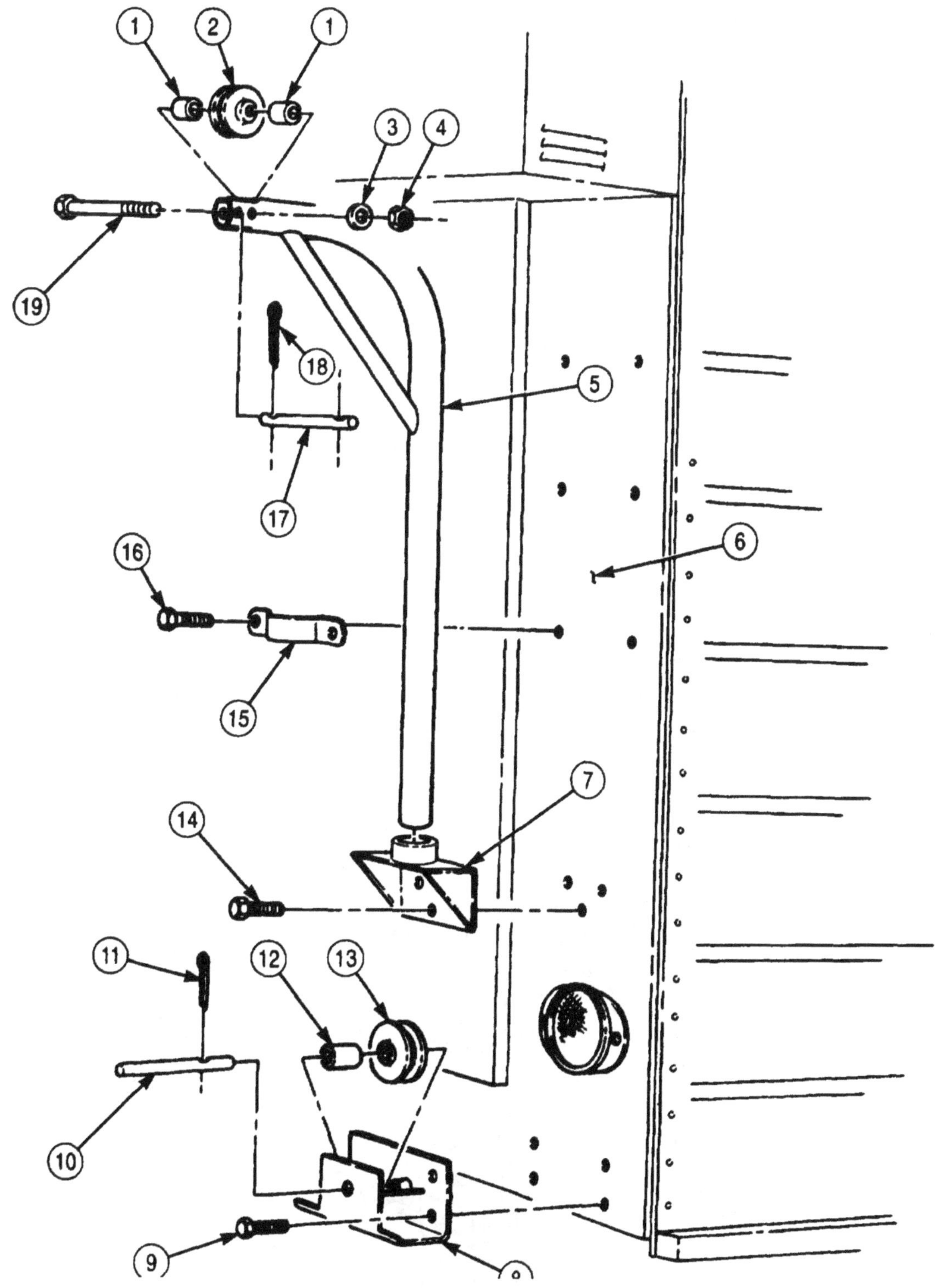

FOLLOW-ON TASK: Install davit chain and wire rope (para. 3-264).

3-266. VAN DAVIT WINCH (M934A1/A2) REPLACEMENT

THIS TASK COVERS:

a. Removal b. Installation

INITIAL SETUP:

APPLICABLE MODELS
M934A1/A2

TOOLS
General mechanic's tool kit (Appendix E, Item 1)

MATERIALS/PARTS
Cotter pin (Appendix D, Item 50)

REFERENCES (TM)
TM 9-2320-272-10
TM 9-2320-272-24P

EQUIPMENT CONDITION
- Parking brake set (TM 9-2320-272-10).
- Davit chain and wire rope removed (para. 3-264).

a. Removal

1. Remove four screws (7) and winch (8) from van body (9).
2. Push button (10) and remove retaining pin (11) from brace (1).
3. Remove cotter pin (2), pin (12), spacer (4), handle lock (5), and spacer (6) from brace (1). Discard cotter pin (2).
4. Remove two screws (3) and brace (1) from van body (9).

b. Installation

1. Install brace (1) on van body (9) with two screws (3).
2. Install spacer (4), handle lock (5), and spacer (6) on brace (1) with pin (12) and new cotter pin (2).
3. Install winch (8) on van body (9) with four screws (7).
4. Push button (10) and install retaining pin (11) on brace (1).

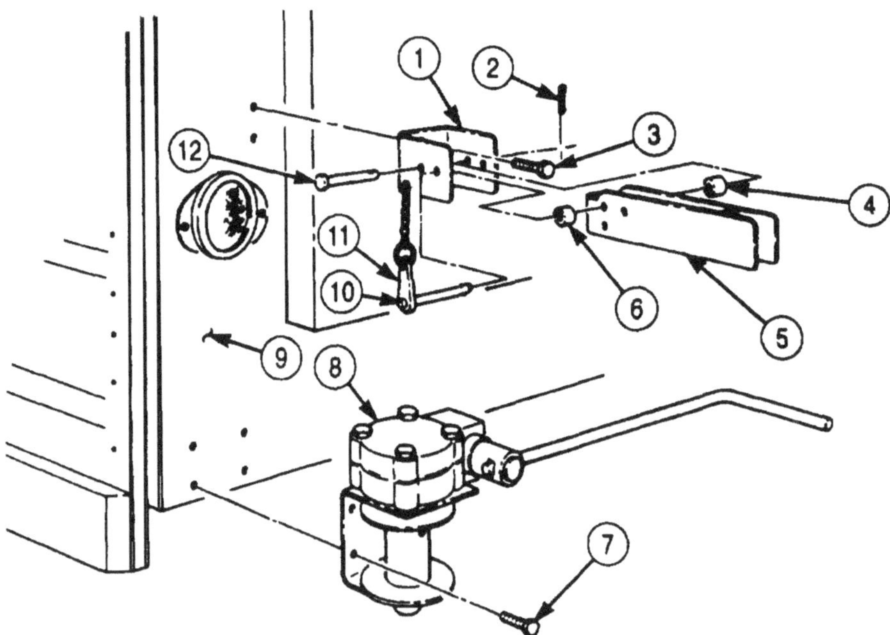

FOLLOW-ON TASK: Install davit chain and wire rope (para. 3-264).

Section XII. BODY, CAB, AND ACCESSORIES MAINTENANCE

3-267. BODY, CAB, AND ACCESSORIES MAINTENANCE INDEX

PARA. NO.	TITLE	PAGE NO.
3-268.	Hood Latch and Bracket Replacement	3-732
3-269.	Hood Support Bar and Bracket Replacement	3-734
3-270.	Hood Stop Cables Replacement	3-736
3-271.	Hood Grab Handle Replacement	3-737
3-272.	Radiator Baffles, Seals, and Plates Replacement	3-738
3-273.	Cab Hood Stop Bracket Replacement	3-740
3-274.	Hood Bumper Replacement	3-741
3-275.	Engine Hood Maintenance	3-742
3-276.	Windshield Stop Bracket and Latch Replacement	3-747
3-277.	Windshield Wiper Blade, Wiper Arm, and Wiper Motor Replacement	3-748
3-278.	Windshield and Outer Frame Assembly Replacement	3-752
3-279.	Windshield Frame Assembly Replacement	3-754
3-280.	Cab Windshield Hinge Assembly Replacement	3-756
3-281.	Windshield Wiper Reservoir, Jet, and Control Replacement	3-758
3-282.	Windshield Washer Hoses Replacement	3-760
3-283.	Driver's Seat Replacement	3-762
3-284.	Driver's Seat Frame and Base Maintenance	3-764
3-285.	Driver's Seat Cushion, Backrest Cushion, and Frame Replacement	3-768
3-286.	Companion Seat Cushion, Backrest Cushion, and Frame Replacement	3-770
3-287.	Map Compartment Replacement	3-772
3-288.	Cab Grab Handle Replacement	3-773
3-289.	Seatbelt Replacement	3-774
3-290.	Rearview Mirror and Brace Replacement	3-778
3-291.	Personnel Heater Inlet and Outlet Hose Replacement	3-780
3-292.	Personnel Hot Water Heater Replacement	3-782
3-293.	Fresh Air Vent Control Assembly Replacement	3-784
3-294.	Defrost and Heat Controls Replacement	3-786
3-295.	Diverter Assembly Replacement	3-788
3-296.	Fresh Air Inlet Ducting Replacement	3-790
3-297.	Cab Heat and Defrost Air Ducting Replacement	3-791
3-298.	Vent Door Weatherseal Replacement	3-792
3-299.	Cab Cowl Vent Screen and Door Replacement	3-793
3-300.	Front Fender Extension Replacement	3-794
3-301.	Fender Splash Shield Replacement	3-795
3-302.	Toolbox and Steps Replacement	3-796
3-303.	Cab Turnbuttons and Lashing Hooks Replacement	3-798
3-304.	Front Cab Mount Replacement	3-800

3-267. BODY, CAB, AND ACCESSORIES MAINTENANCE INDEX (Contd)

PARA. NO.	TITLE	PAGE NO.
3-305.	Rear Cab Mount Replacement	3-802
3-306.	Cab Insulation Replacement	3-804
3-307.	Outside Door Handle Replacement	3-806
3-308.	Window Regulator and Inside Door Handle Replacement	3-807
3-309.	Cab Door Dovetail Wedge Replacement	3-808
3-310.	Cab Door Dovetail Replacement	3-809
3-311.	Cab Door Weatherseal Replacement	3-810
3-312.	Cab Door Inspection Hole Cover Replacement	3-811
3-313.	Cab Door Lock Replacement	3-812
3-314.	Cab Door Glass Maintenance	3-813
3-315.	Window Weatherstripping (Cab Door) Replacement	3-814
3-316.	Cab Top Seal and Retainer Replacement	3-815
3-317.	Cab Door Regulator Assembly Replacement	3-816
3-318.	Cab Door Check Rod Replacement	3-817
3-319.	Cab Door Replacement	3-818
3-320.	Cab Door Hinge Replacement	3-820
3-321.	Cab Door Catch Replacement	3-821

3-268. HOOD LATCH AND BRACKET REPLACEMENT

THIS TASK COVERS:

a. Removal
b. Installation

INITIAL SETUP

APPLICABLE MODELS
All

TOOLS
General mechanic's tool kit (Appendix E, Item 1)

REFERENCES (TM)
TM 9-2320-272-10
TM 9-2320-272-24P

EQUIPMENT CONDITION
Parking brake set (TM 9-2320-272-10).

a. Removal

1. Release hood latch (4).

NOTE

Assistant will help with step 2.

2. Remove three screws (7), nuts (6), and hood latch (4) from hood cowl (5).
3. Remove three screws (1) and upper latch retaining bracket (2) from hood (3).

b. Installation

1. Install upper latch retaining bracket (2) on hood (3) with three screws (1).

NOTE

Assistant will help with step 2.

2. Install hood latch (4) on cab cowl (5) with three screws (7) and nuts (6).
3. Secure hood latch (4).

3-268. HOOD LATCH AND BRACKET REPLACEMENT (Contd)

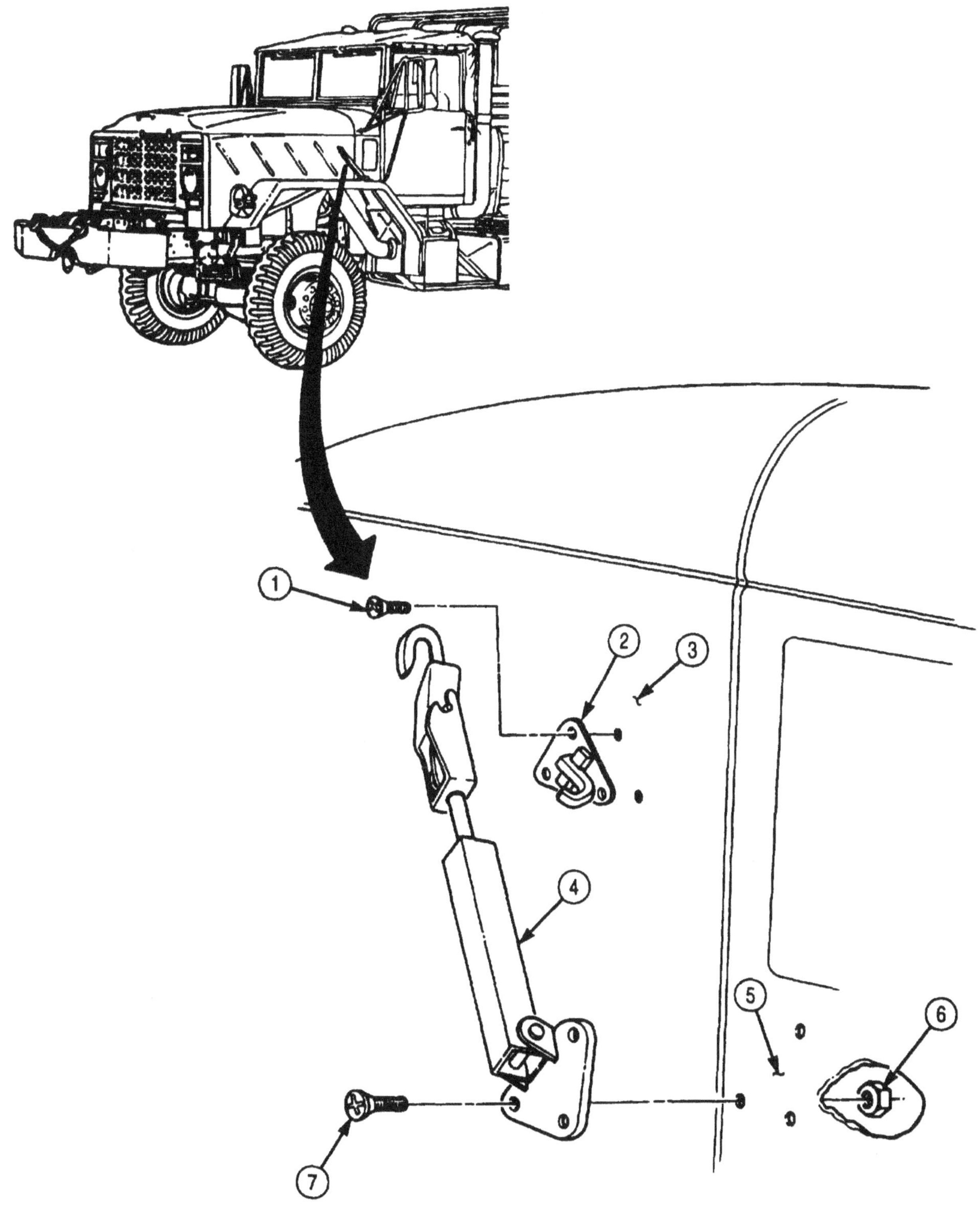

3-269. HOOD SUPPORT BAR AND BRACKET REPLACEMENT

THIS TASK COVERS:

a. Removal b. Installation

INITIAL SETUP:

APPLICABLE MODES
All

TOOLS
General mechanic's tool kit (Appendix E, Item 1)

MATERIALS/PARTS
Cotter pin (Appendix D, Item 52)
Locknut (Appendix D, Item 313)

REFERENCES (TM)
TM 9-2320-272-10
TM 9-2320-272-24P

EQUIPMENT CONDITION
Parking brake set (TM 9-2320-272-10).

GENERAL SAFETY INSTRUCTIONS
Hood must be supported during replacement of hood support bar bracket.

a. Removal

WARNING

Hood must be supported during replacement of hood support bar mounting bracket, or injury to personnel may result.

1. Remove bar support pin (10) from stowage bracket (5).
2. Remove cotter pin (14), washer (13), pin (7), and hood support bar (11) from mounting bracket (12). Discard cotter pin (14).
3. Remove locknut (2), washer (3), and spacer (4) from screw (6). Discard locknut (2).
4. Remove screw (6), washer (15), and mounting bracket (12) from hood (1).
5. Remove loop link (9) and bar support pin (10) from chain (8).

b. Installation

1. Install loop link (9) and bar support pin (10) on chain (8).
2. Install mounting bracket (12) on hood (1) with screw (6), washer (15), spacer (4), washer (3), and new locknut (2).
3. Install hood support bar (11) on mounting bracket (12) with pin (7), washer (13), and new cotter pin (14).
4. Install bar support pin (10) on stowage bracket (5).

3-269. HOOD SUPPORT BAR AND BRACKET REPLACEMENT (Contd)

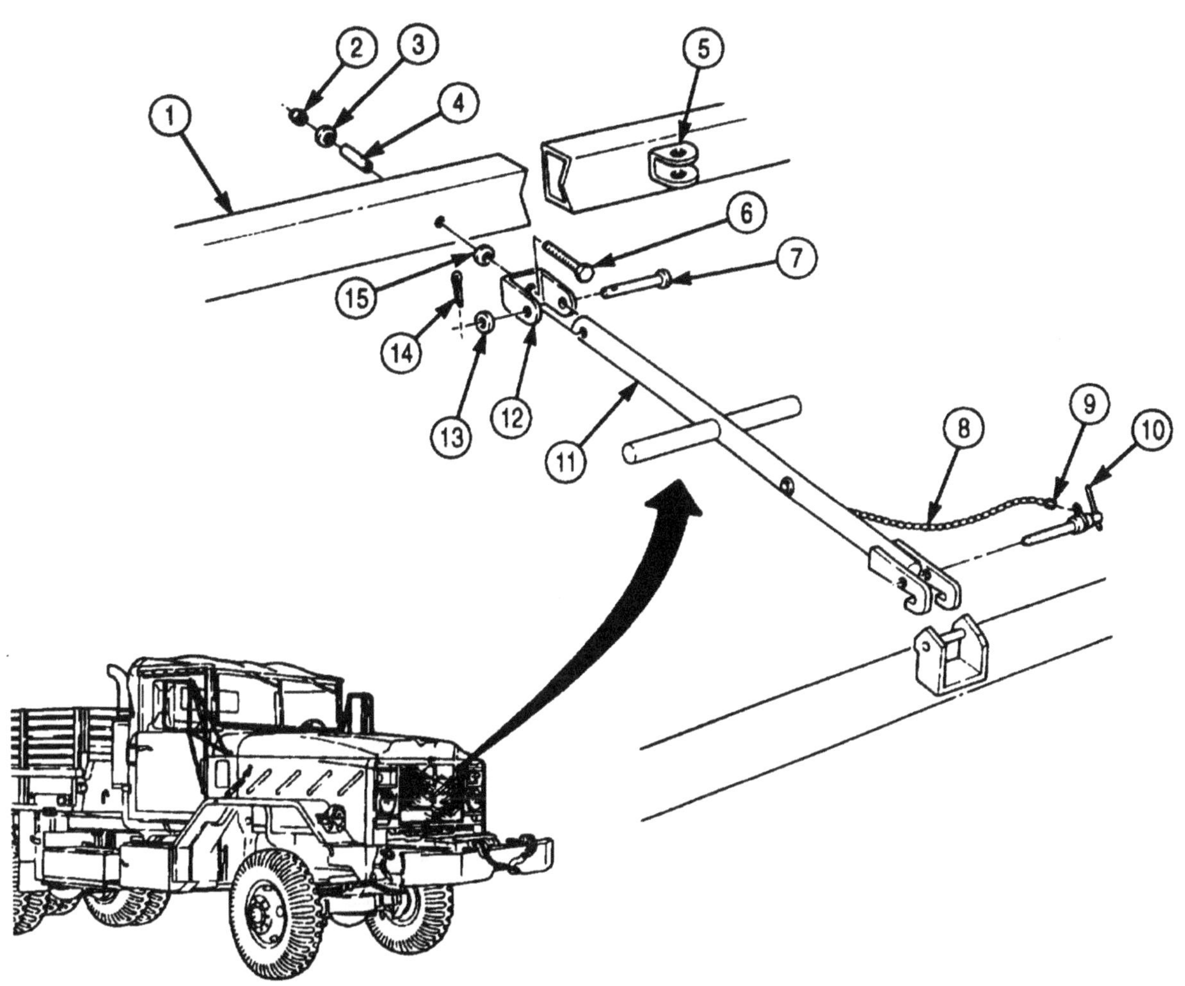

3-270. HOOD STOP CABLES REPLACEMENT

THIS TASK COVERS:

a. Removal

b. Installation

INITIAL SETUP

APPLICABLE MODELS
All

TOOLS
General mechanic's tool kit (Appendix E, Item 1)

MATERIALS/PARTS
Two locknuts (Appendix D, Item 291)

REFERENCES (TM)
TM 9-2320-272-10
TM 9-2320-272-24P

EQUIPMENT CONDITION
- Parking brake set (TM 9-2320-272-10).
- Hood raised and secured (TM 9-2320-272-10).

NOTE

Both hood stop cables are replaced the same way.
This procedure covers one cable.

a. Removal

1. Remove locknut (9), screw (6), washer (7), and hood stop cable (5) from radiator support bracket (8). Discard locknut (9).
2. Remove locknut (4), screw (1), washer (2), and hood stop cable (5) from hood bracket (3). Discard locknut (4).

b. Installation

1. Install hood stop cable (5) on radiator support bracket (8) with washer (7), screw (6), and new locknut (9).
2. Installhood stop cable (5) on hood bracket (3) with washer (2), screw (1), and new locknut (4).

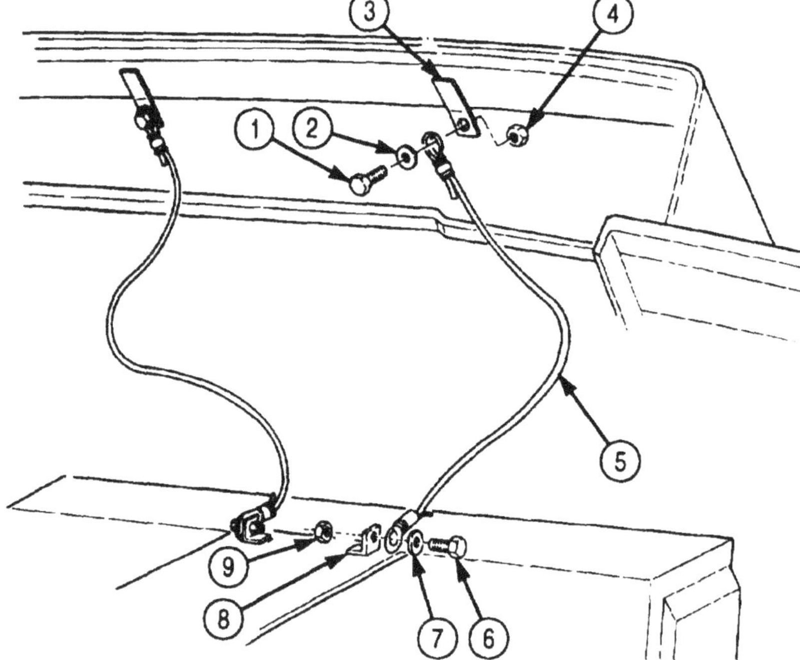

3-271. HOOD GRAB HANDLE REPLACEMENT

THIS TASK COVERS:

a. Removal

b. Installation

INITIAL SETUP:

APPLICABLE MODELS

All

TOOLS

General mechanic's tool kit (Appendix E, Item 1)

MATERIALS/PARTS

Four locknuts (Appendix D, Item 313)

REFERENCES (TM)

TM 9-2320-272-10

TM 9-2320-272-24P

EQUIPMENT CONDITION

- Parking brake set (TM 9-2320-272-10).
- Hood raised and secured (TM 9-2320-272-10).

a. Removal

Remove four screws (5), locknuts (3), two plates (2), and hood grab handle (1) from hood (4). Discard locknuts (3).

b. Installation

Install hood grab handle (1) on hood (4) with four screws (5), two plates (2), and four new locknuts (3).

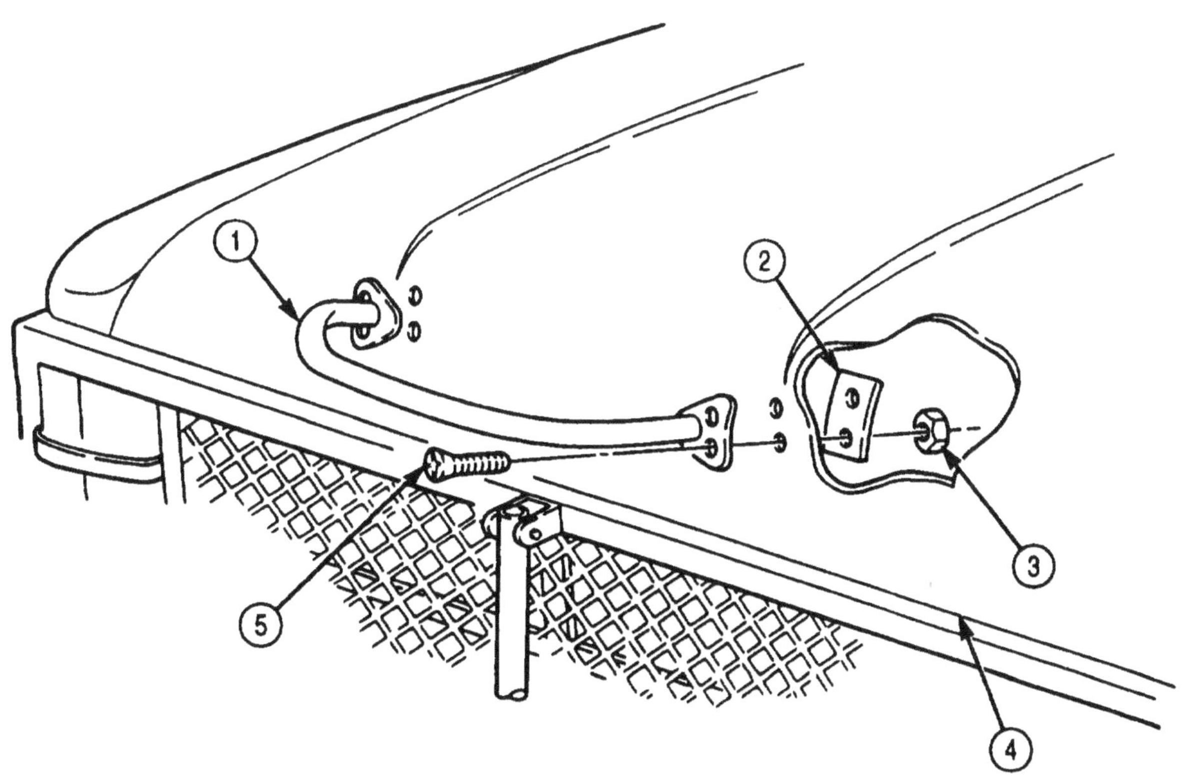

3-272. RADIATOR BAFFLES, SEALS, AND PLATES REPLACEMENT

THIS TASK COVERS:

a. Removal b. Installation

ININITAL SETUP:

APPLICABLE MODELS
All

Tools
General mechanic's tool kit (Appendix E, Item 1)

MATERIALS/PARTS
Twelve locknuts (Appendix D, Item 276)
Twelve lockwashers (Appendix D, Item 400)

REFERENCES (TM)
TM 9-2320-272-10
TM 9-2320-272-24P

EQUIPMENT CONDITION
• Parking brake set (TM 9-2320-272-10).
• Hood raised and secured (TM 9-2320-272-10).

NOTE

Upper baffle consists of a seal and plate as one unit. Side baffles consist of seals and plates as separate units.

a. Removal

1. Remove four screws (12) and lockwashers (11) from upper radiator baffle seal and plate (3) and brush guard (13). Discard lockwashers (11).
2. Remove two screws (2), lockwashers (l), and upper radiator baffle seal and plate (3) from left and right baffle plates (7). Discard lockwashers (1).

NOTE

Steps 3 and 4 apply to both left and right baffle seals and plates.

3. Remove six locknuts (5), washers (6), screws (8), and baffle seal (4) from radiator baffle plate (7). Discard locknuts (5).
4. Remove three screws (9), lockwashers (10), and baflle plate (7) from brush guard (13). Discard lockwashers (10).

b. Installation

NOTE

Steps 1 and 2 apply to both left and right baffle seals and plates.

1. Install radiator baffle plate (7) on brush guard (13) with three new lockwashers (10) and screws (9).
2. Install baffle seal (4) on baffle plate (7) with six screws (8), washers (6), and new locknuts (5).
3. Install upper radiator baffle seal and plate (3) on left and right baffle plates (7) with two new lockwashers (1) and screws (2).
4. Install upper radiator baffle seal and plate (3) on brush guard (13) with four new lockwashers (11) and screws (12).

3-272. RADIATOR BAFFLES, SEALS, AND PLATES REPLACEMENT (Contd)

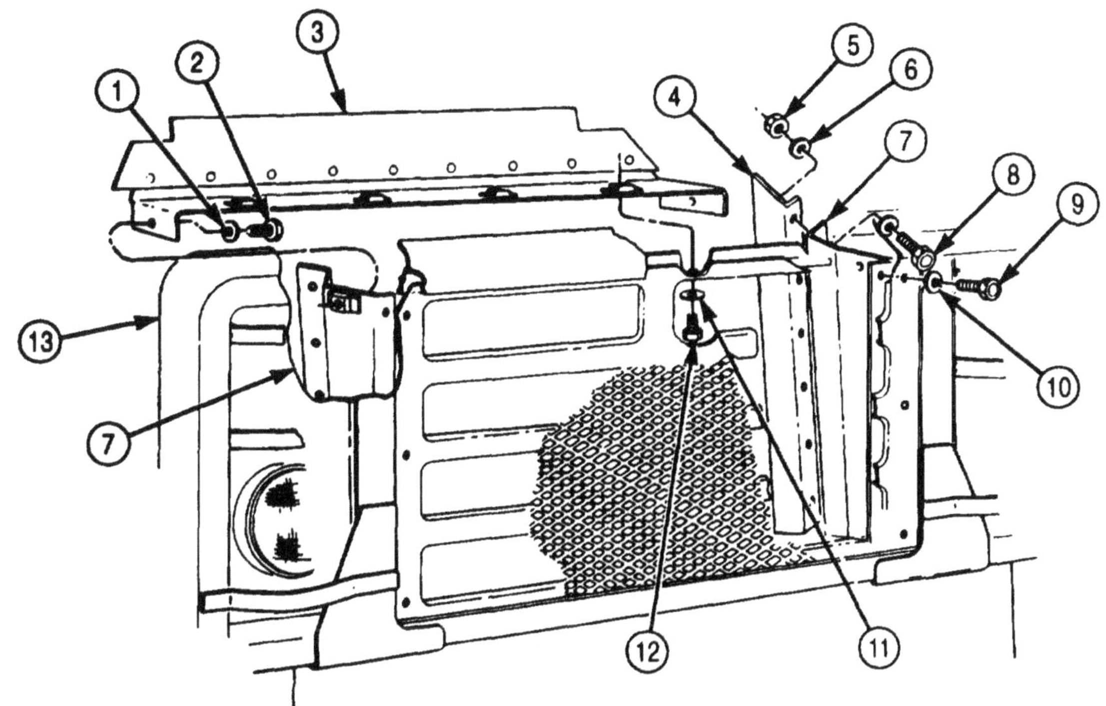

3-273. CAB HOOD STOP BRACKET REPLACEMENT

THIS TASK COVERS:

a. Removal b. Installation

INITIAL SETUP

APPLICABLE MODELS
All

TOOLS
General mechanic's tool kit (Appendix E, Item 1)

MATERIALS/PARTS
Two locknuts (Appendix D. Item 291)
Shims (Appendix D, Item 652)

REFERENCES (TM)
TM 9-2320-272-10
TM 9-2320-272-24P

EQUIPMENT CONDITION
- Parking brake set (TM 9-2320-272-10).
- Hood raised and secured (TM 9-2320-272-10).

a. Removal

NOTE

This procedure applies to both left and right stop brackets.

Remove two locknuts (5), stop bracket (4), and shim (3) from firewall (2). Discard locknuts (5).

b. Installation

NOTE

Install original shim or new shims as required to establish proper alignment between vehicle and stop bracket.

Install shim (3) and stop bracket (4) on firewall (2) with two new locknuts (5).

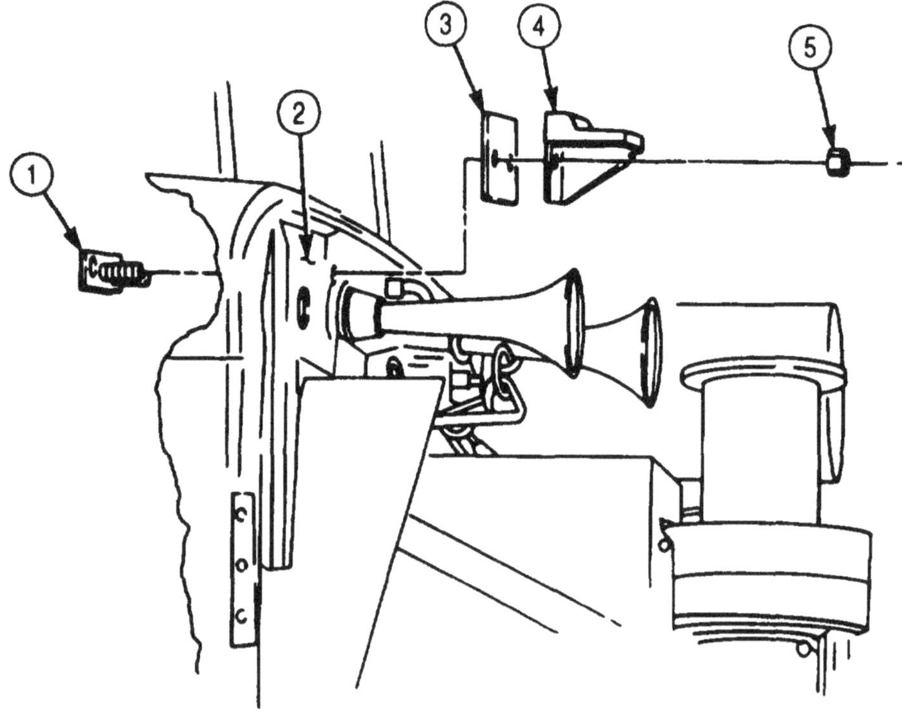

3-274. HOOD BUMPER REPLACEMENT

THIS TASK COVERS:

a. Removal b. Installation

INITIAL SETUP:

APPLICABLE MODELS

All

TOOLS

General mechanic's tool kit (Appendix E, Item 1)

MATERIALS/PARTS

Three locknuts (Appendix D, Item 274)

REFERENCES (TM)

TM 9-2320-272-10

TM 9-2320-272-24P

EQUIPMENT CONDITION

- Parking brake set (TM 9-2320-272-10).
- Hood raised and secured (TM 9-2320-272-10).

a. Removal

NOTE

This procedure applies to both left and right hood bumpers.

Remove three locknuts (5), screws (1), washers (2), and hood bumper (3) from cab body (4). Discard locknuts (5).

b. Installation

Install hood bumper (3) on cab body (4) with three washers (2), screws (1), and new locknuts (5).

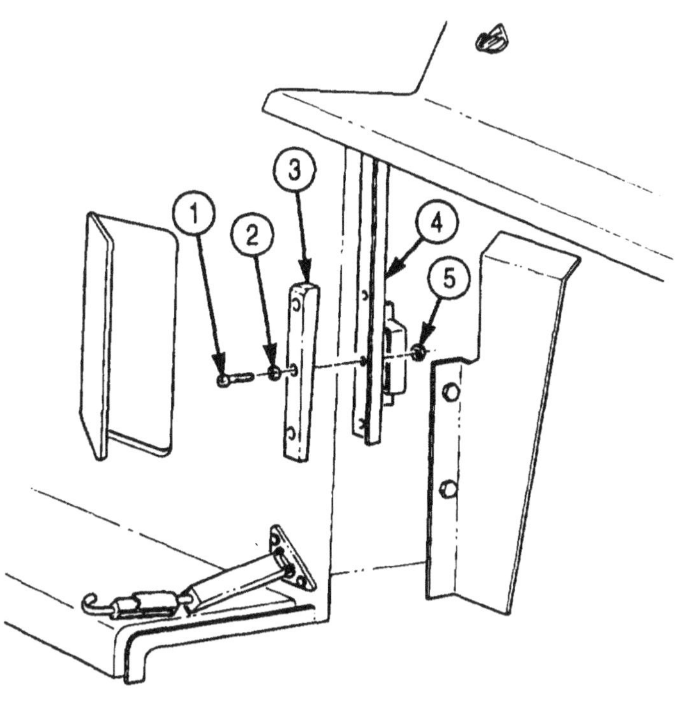

3-275. ENGINE HOOD MAINTENANCE

THIS TASK COVERS:

a. Removal
b. Installation
c. Adjustment

INITIAL SETUP:

APPLICABLE MODELS
All

TOOLS
General mechanic's tool kit (Appendix E, Item 1)
Torque wrench (Appendix E, Item 144)

MATERIALS/PARTS
Two locknuts (Appendix D. Item 291)
Four lockwashers (Appendix D, Item 345)
Locknut (Appendix D, Item 294)
Lockwasher (Appendix D, Item 372)

REFERENCES (TM)
TM 9-2320-272-10
TM 9-2320-272-24P

EQUIPMENT CONDITION
- Parking brake set (TM 9-2320-272-10).
- Hood raised and secured (TM 9-2320-272-10).
- Left and right splash shields removed (TM 9-2320-272-10).

GENERAL SAFETY INSTRUCTIONS
All personnel must stand clear during lifting operations.

---

a. Removal

1. Remove two locknuts (3), washers (6), screws (5), washers (6), and hood stop cables (4) from radiator support brackets (7). Discard locknuts (3).
2. Disconnect wiring harness quick-disconnect (2) from harness plug (1).
3. Remove locknut (8), washer (9), screw (12), washer (9), ground strap (11), and lockwasher (10) from inside left fender (13). Discard locknut (8) and lockwasher (10).

3-275. ENGINE HOOD MAINTENANCE (Contd)

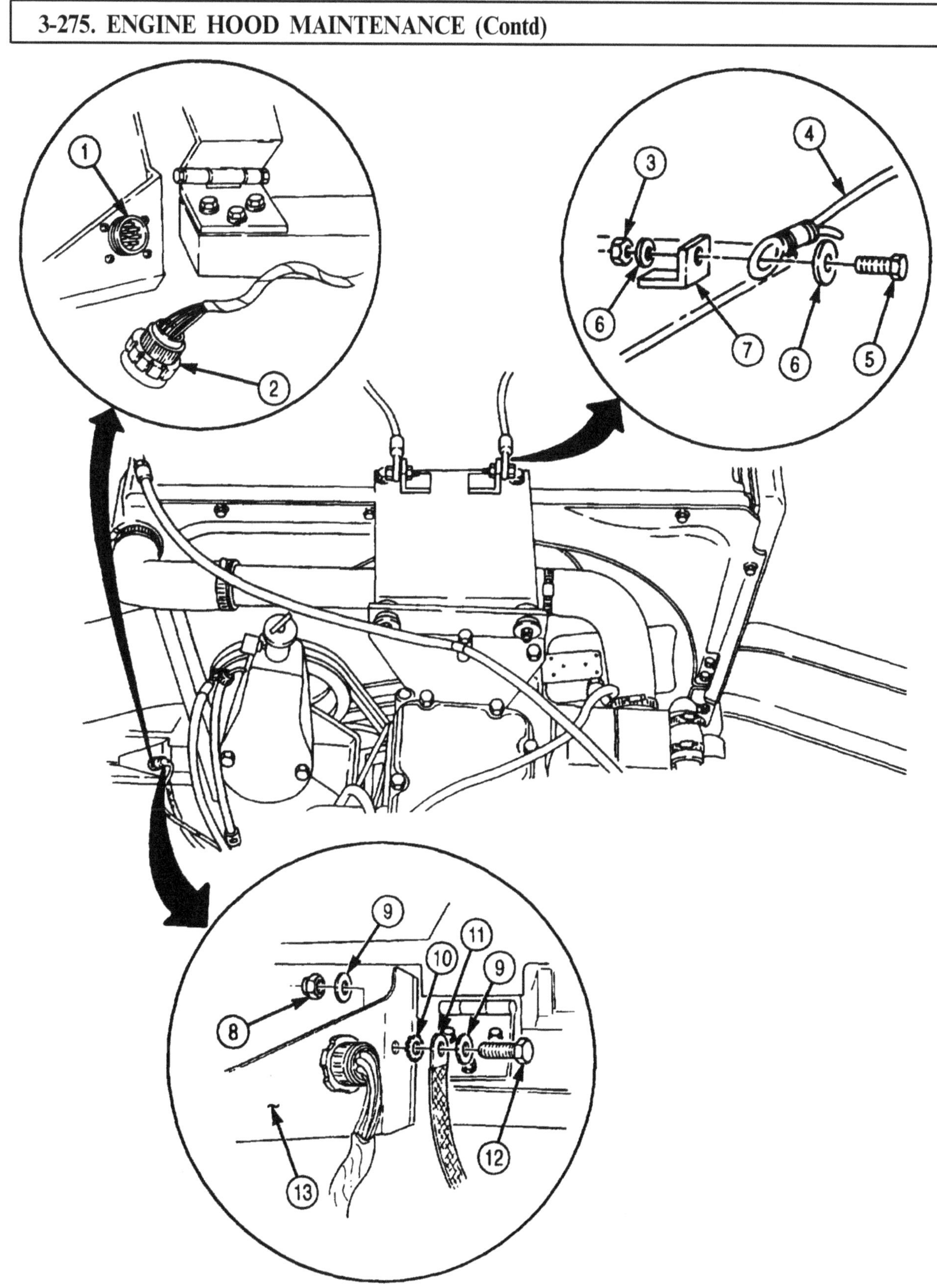

3-275. ENGINE HOOD MAINTENANCE (Contd)

4. Attach chain to hood support brackets (2) and lifting device and raise hood (1) until slack is removed.
5. Remove hood retaining bar (3) from bumper bracket (7) and stow on hood stowage bracket (4) (TM 9-2320-272-10).
6. Remove four screws (8), lockwashers (9), and mounting plate (10) from front crossmember (5). Discard lockwashers (9).

WARNING

All personnel must stand clear during hoisting operation. A shifting or swinging load may cause injury or death to personnel.

7. Lift hood (1) straight up and away from bumper (6).
8. Lower and remove hood (1) from lifting device.

b. Installation

1. Attach chain to hood support brackets (2) and lifting device and hoist onto front crossmember (5).

WARNING

All personnel must stand clear during hoisting operation. A shifting or swinging load may cause injury or death to personnel.

2. Install hood (1) and mounting plate (10) on front crossmember (5) with four new lockwashers (9) and screws (8).
3. Secure hood (1) to bumper bracket (7) with hood retaining bar (3) (TM 9-2320-272-10).
4. Remove lifting device and chain from hood (1).

3-275. ENGINE HOOD MAINTENANCE (Contd)

5. Connect wiring harness quick-disconnect (12) to harness plug (11)

NOTE

Ensure ground strap is between washer and lockwasher.

6. Install new lockwasher (20) and ground strap (21) on inside left fender (24) with washer (22), screw (23), washer (19), and new locknut (25).
7. Install two hood stop cables (14) on radiator support brackets (17) with two washers (16), screws (15), washers (18), and new locknut (13).

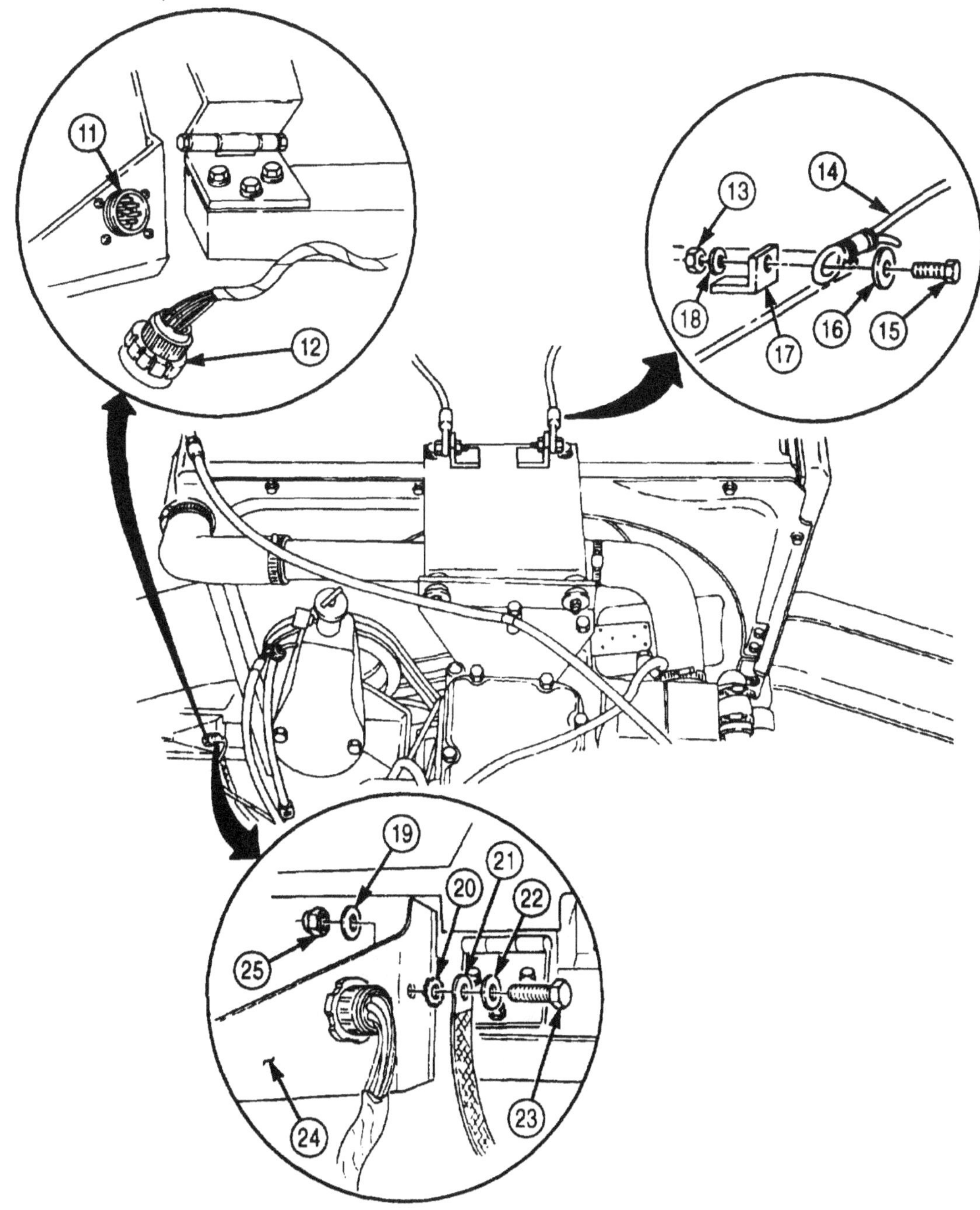

3-275. ENGINE HOOD MAINTENANCE (Contd)

c. Adjustment

1. Close hood (1) and check clearance at top and sides of hood (1) and cowl (2). Clearance should be 1/4-3/4 in. (0.64-1.9 cm) and equal at both top and bottom of hood (1).
2. Loosen four screws (4) at frame crossmember (3).
3. Position hood (1) until proper clearance is obtained.
4. Tighten four screws (4) 100 lb-ft (136 N•m).

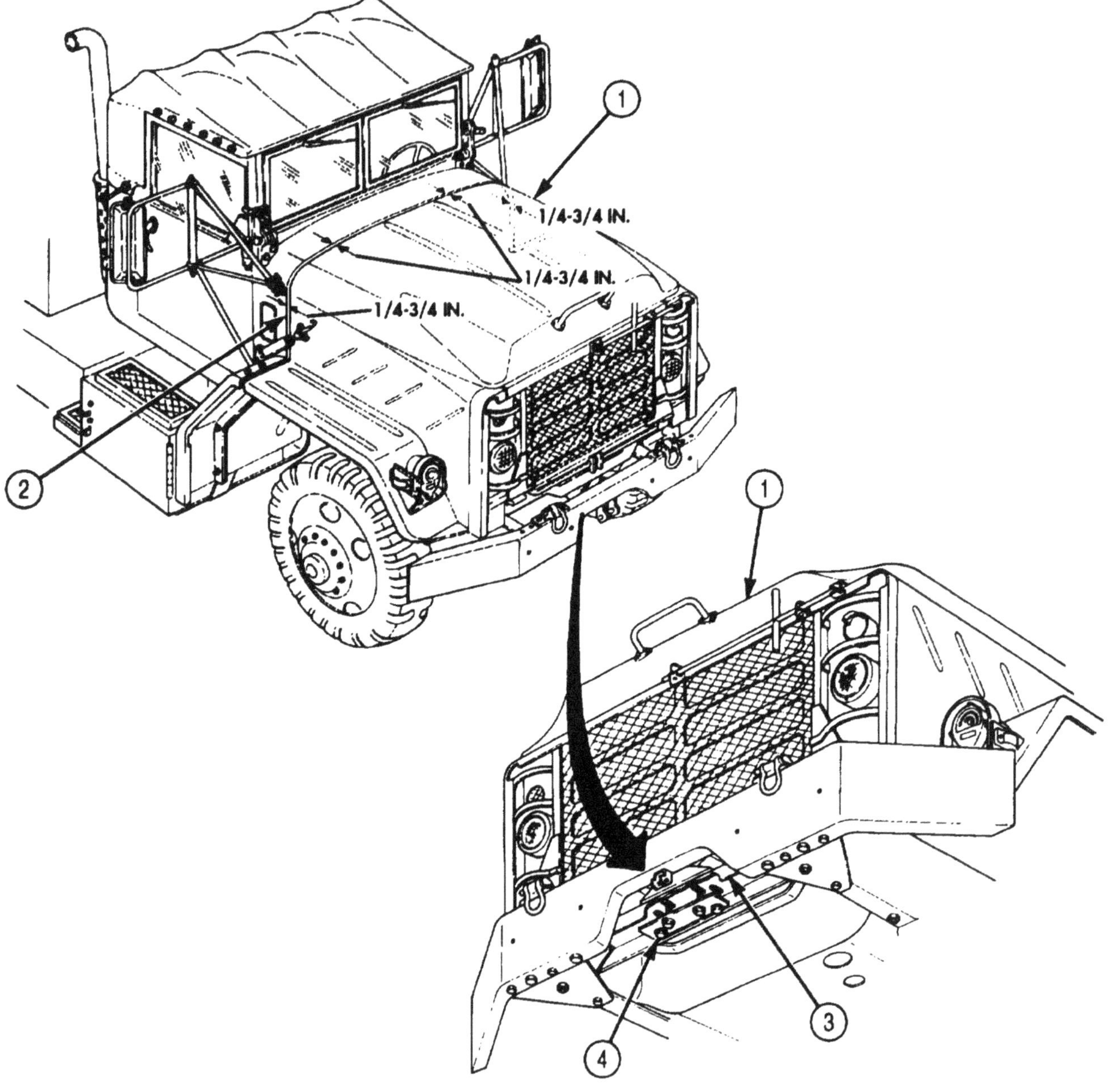

FOLLOW-ON TASK: Install left and right splash shields (TM 9-2320-272-10).

3-276. WINDSHIELD STOP BRACKET AND LATCH REPLACEMENT

THIS TASK COVERS:

a. Removal   b. Installation

INITIAL SETUP:

APPLICABLE MODELS
All

TOOLS
General mechanic's tool kit (Appendix E, Item 1)

MATERIALS/PARTS
Four locknuts (Appendix D, Item 313)

PERSONNEL REQUIRED
Two

REFERENCES (TM)
TM 9-2320-272-10
TM 9-2320-272-24P

EQUIPMENT CONDITION
- Parking brake set (TM 9-2320-272-10).
- Hood raised and secured (TM 9-2320-272-10)

a. Removal

1. Remove two locknuts (8), screws (3), and windshield latch (1) from hood (2). Discard locknuts (8).

NOTE

Hood insulation must be pulled back to gain access to nuts and reinforcement plates.

2. Remove two locknuts (7), screws (4), reinforcement plate (6), and windshield stop bracket (5) from hood (2). Discard locknuts (7).

b. Installation

1. Install windshield stop bracket (5) and reinforcement plate (6) on hood (2) with two screws (4) and new locknuts (7).
2. Install windshield latch (1) on hood (2) with two screws (3) and new locknuts (8).

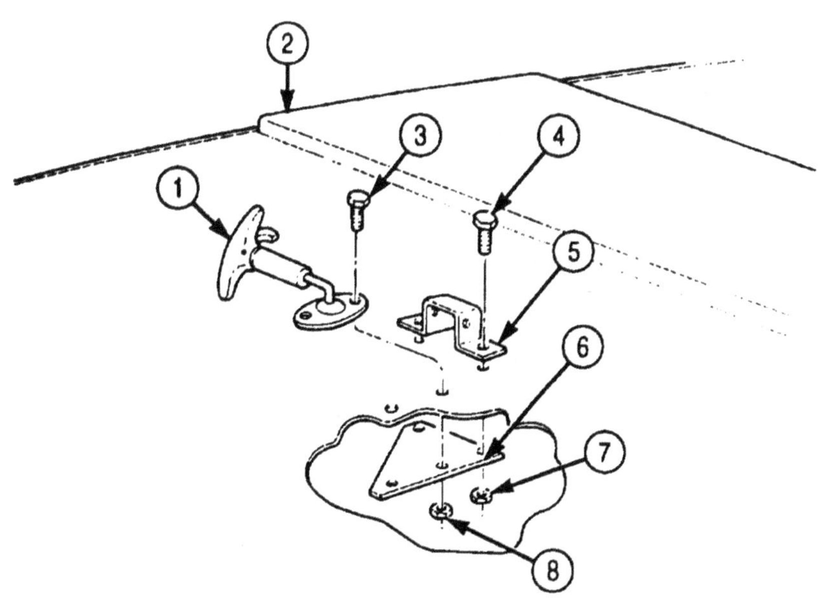

3-277. WINDSHIELD WIPER BLADE, WIPER ARM, AND WIPER MOTOR REPLACEMENT

THIS TASK COVERS:

a. Removal b. Installation

INITIAL SETUP:

APPLICABLE MODELS
All

TOOLS
General mechanic's tool kit (Appendix E, Item 1)
Torque wrench (Appendix E, Item 146)

MATERIALS/PARTS
Two locknuts (Appendix D, Item 292)
Locknut (Appendix D, Item 293)
Leather washer (Appendix D, Item 269)
Lockwasher (Appendix D, Item 367)
Antiseize tape (Appendix C, Item 72)
Cap and plug set (Appendix C, Item 14)

REFERENCES (TM)
TM 9-2320-272-10
TM 9-2320-272-24P

EQUIPMENT CONDITION
- Parking brake set (TM 9-2320-272-10).
- Air reservoir drained (TM 9-2320-272-101.

GENERAL SAFETY INSTRUCTIONS
Do not disconnect air lines before draining air reservoirs.

a. Removal

WARNING

Do not disconnect air lines before draining air reservoirs. Small parts under pressure may shoot out with high velocity, causing injury to personnel.

CAUTION

Cap or plug all openings immediately after disconnecting lines or hoses to prevent contamination. Failure to do so may result in system damage or injury to personnel.

1. Remove locknut (10), screw (11), and wiper blade (12) from wiper arm (7). Discard locknut (10).

NOTE

Mark position of wiper arm for installation.

2. Remove nut (9), lockwasher (8), and wiper arm (7) from wiper motor shaft (1). Discard lockwasher (8).
3. Remove knurled drive (6) from wiper motor shaft (1).
4. Remove nut (5), washer (4), and leather washer (3) from wiper motor shaft (1). Discard leather washer (3).
5. Remove hose clamps (14) and (17) and hoses (15) and (16) from adapter fittings (22) and (21).
6. Remove two locknuts (20), screws (18), wiper motor (13), and wiper motor bracket (19) from windshield frame (2). Discard locknuts (20).
7. Remove two screws (25), bracket (19), and two spacers (24) from wiper motor (13).
8. Remove adapter fittings (21) and (22) from wiper motor (13).
9. Remove muffler (23) from wiper motor (13).

3-277. WINDSHIELD WIPER BLADE, WIPER ARM, AND WIPER MOTOR REPLACEMENT (Contd)

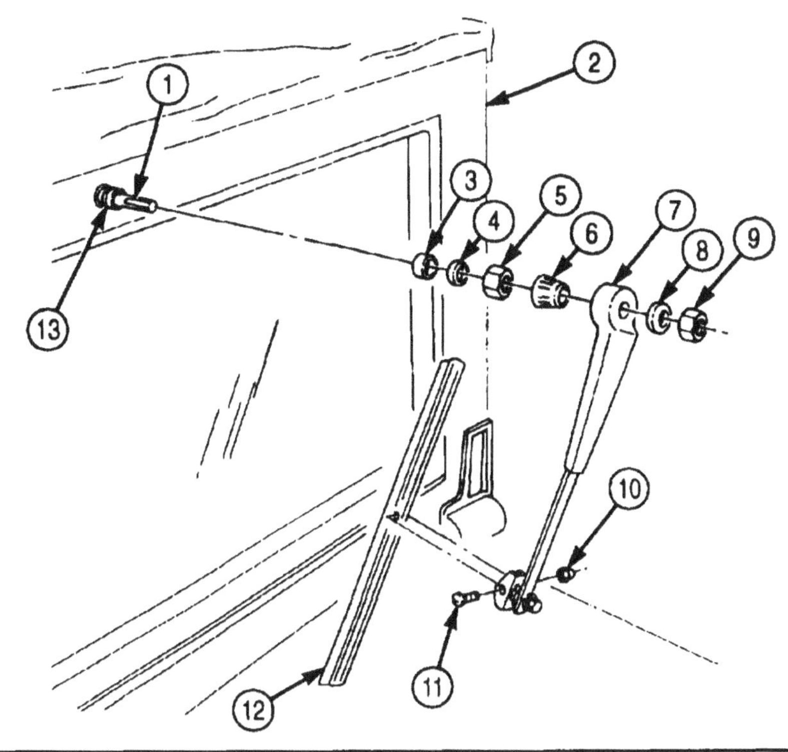

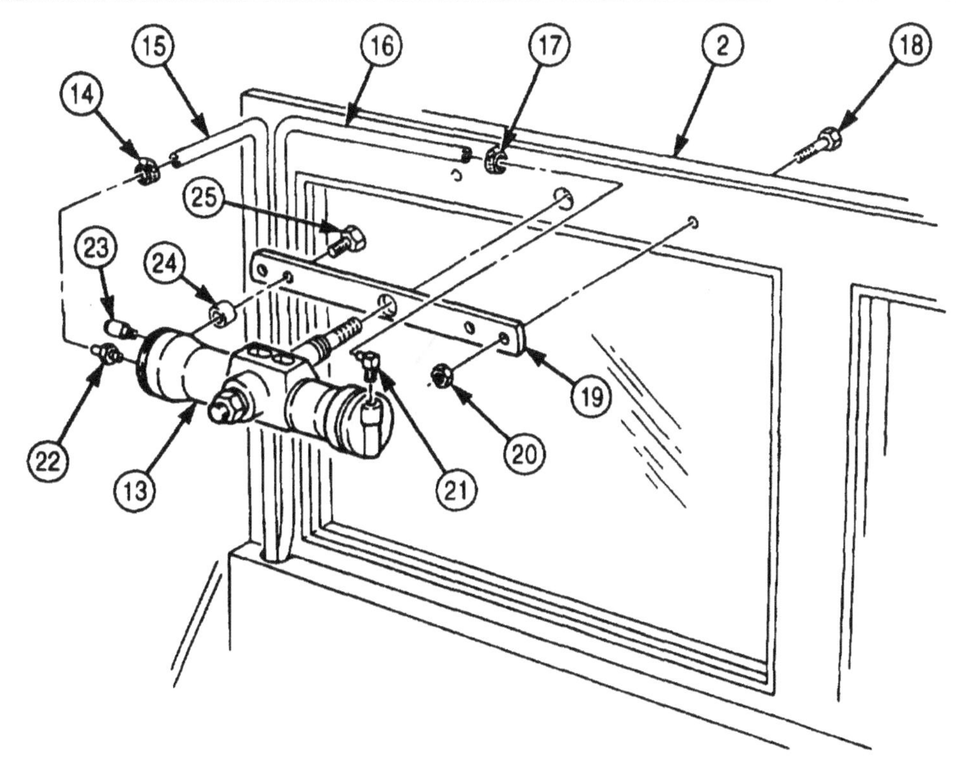

3-277. WINDSHIELD WIPER BLADE, WIPER ARM, AND WIPER MOTOR REPLACEMENT (Contd)

b. Installation

1. Install muffler (12) on wiper motor (10).

NOTE

Wrap male pipe threads with antiseize tape before installation.

2. Install adapter fittings (9) and (11) in wiper motor (10).
3. Install bracket (7) on wiper motor (10) with two spacers (13) and screws (14).
4. Install wiper motor (10) and bracket (7) on windshield frame (5) with two screws (6) and new locknuts (8).
5. Install hoses (2) and (3) on adapter fittings (11) and 9) with hose clamps (1) and 4).
6. Install new leather washer (16), washer (17), and nut (18) on wiper motor shaft (15).
7. Install knurled drive (19) on wiper motor shaft (15).
8. Install wiper arm (20) on wiper motor shaft (15) with new lockwasher (21) and nut (22). Tighten nut (22) 15-20 lb-ft (20-27 N•m).
9. Install wiper blade (25) on wiper arm (20) with screw (24) and new locknut (23). Tighten locknut (23) 15-20 lb-ft (20-27 N•m).

3-277. WINDSHIELD WIPER BLADE, WIPER ARM AND WIPER MOTOR REPLACEMENT (Contd)

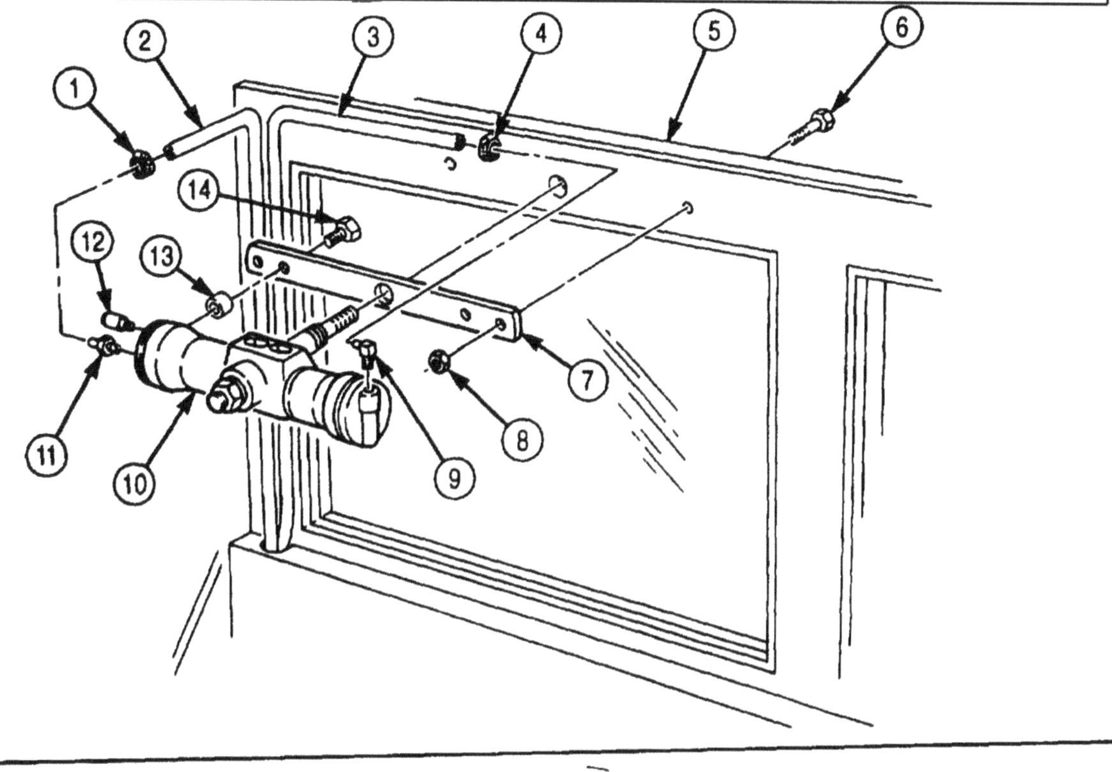

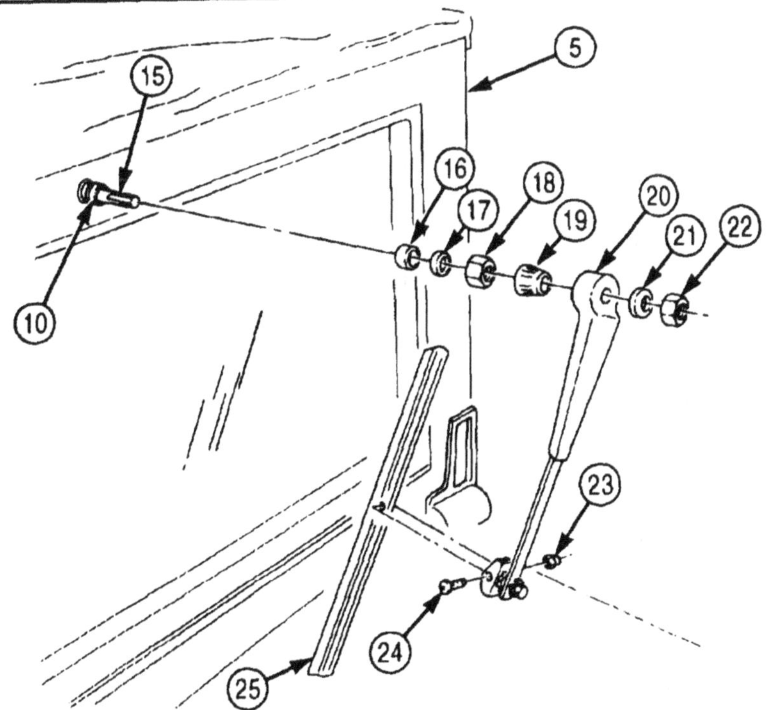

FOLLOW-ON TASK: Start engine (TM 9-2320-272-10) and check for air leaks and proper operation.

3-278. WINDSHIELD AND OUTER FRAME ASSEMBLY REPLACEMENT

THIS TASK COVERS:

a. Removal b. Installation

INITIAL SETUP:

APPLICABLE MODELS

All

TOOLS

General mechanic's tool kit (Appendix E, Item 1)

MATERIALS/PARTS

Six lockwashers (Appendix D, Item 389)
Two lockwashers (Appendix D, Item 369)

REFERENCES (TM)

TM 9-2320-272-10
TM 9-2320-272-24P

EQUIPMENT CONDITION

- Parking brake set (TM 9-2320-272-10).
- Cab top removed from windshield (TM 9-2320-272-10)
- Wiper blade, wiper motor, and wiper arm removed (para 3-277).

a. Removal

NOTE

Perform steps 1 and 2 only on vehicles with hard top kit.

1. Remove two screws (3) and lockwashers (2) from roof (7) and corner post (5). Discard lockwashers (2).
2. Remove four nuts (1) and hook bolts (4) from roof (7) and windshield frame (6).
3. Remove hose clamp (9) and wiper hose (8) from copper air line (10).
4. Remove six screws (13) and lockwashers (14) from two windshield hinges (11) and windshield frame (6). Discard lockwashers (14).
5. Remove windshield frame (6) from two windshield hinges (11).

b. Installation

1. With doors (12) opened, align windshield frame (6) on two windshield hinges (11) and install with six new lockwashers (14) and screws (13).
2. Install wiper hose (8) on copper air line (10) with hose clamp (9).

NOTE

Perform steps 3 and 4 on vehicles with hardtop kit.

3. Install roof (7) on windshield frame (6) with four hook bolts (4) and nuts (1).
4. Install roof (7) on comer post (5) with two new lockwashers (2) and screws (3).

3-278. WINDSHIELD AND OUTER FRAME ASSEMBLY REPLACEMENT (Contd)

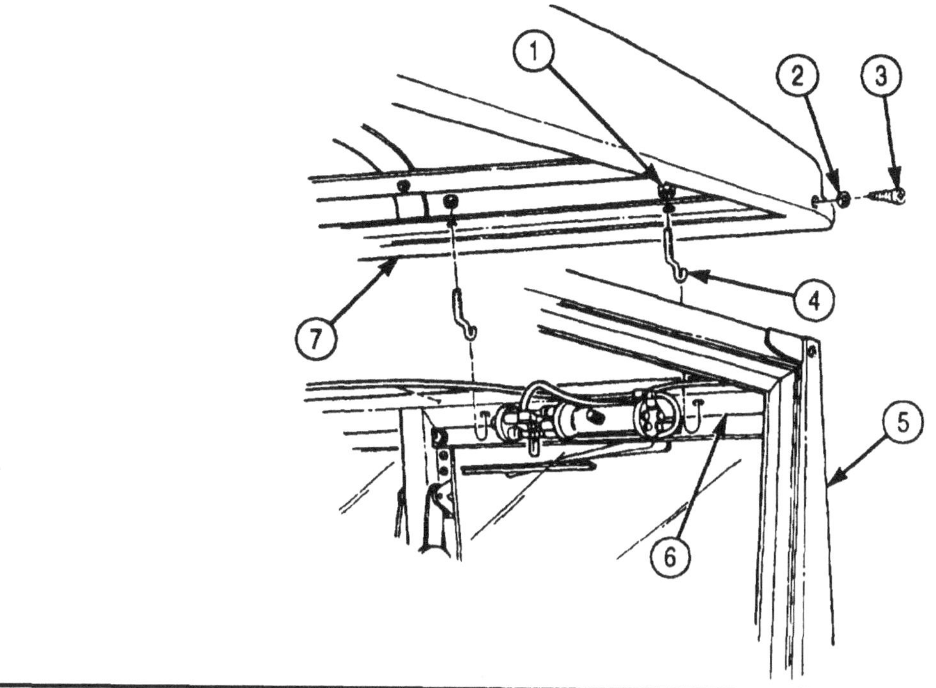

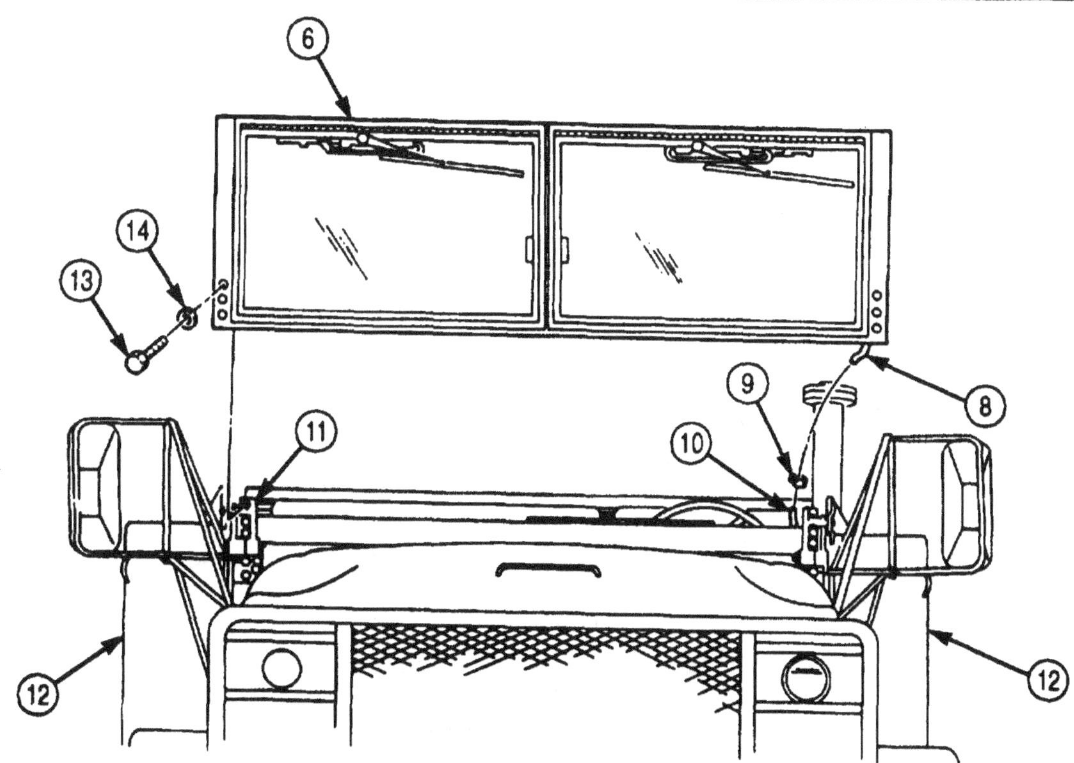

FOLLOW-ON TASKS: • Install cab top on windshield (TM 9-2320-272-10).
• Install wiper blade, wiper motor, and wiper arm (para. 3-277).

3-279. WINDSHIELD FRAME ASSEMBLY REPLACEMENT

THIS TASK COVERS:

a. Removal
b. Installation

INITIAL SETUP:

APPLICABLE MODELS
All

TOOLS
General mechanic's tool kit (Appendix E, Item 1)

MATERIALS/PARTS
Two lockwashers (Appendix D. Item 389)
Eight screw-assembled washers (Appendix D, Item 593)
Sealing compound (Appendix C. Item 62)

REFERENCES (TM)
TM 9-2320-272-10
TM 9-2320-272-24P

EQUIPMENT CONDITION
- Parking brake set (TM 9-2320-272-10).
- Windshield wiper blade, wiper arm, and wiper motor removed (para. 3-277).

NOTE

Left and right windshield frame assemblies are replaced the same. This procedure covers right windshield frame assembly.

a. Removal

1. Secure windshield frame assembly (6) in open position.
2. Remove eight screw-assembled washers (1) from windshield hinge (5) and windshield Frame (2). Discard screw-assembled washers (1).
3. Lower windshield frame assembly (6).
4. Remove two screws (12) and spring washers (14) from two friction lockarms (11).
5. Remove windshield frame assembly (6) from windshield frame (2).
6. Remove two capnuts (8), lockwashers (7), screws (10), and lock handle (9) from windshield frame assembly (6). Discard lockwashers (7).
7. Remove seven screws (4). windshield hinge (5), and hinge seal (3) from windshield frame assembly (6).
8. Remove two screws (13), spring washers (15), and friction lockarms (11) from windshield frame assembly (6).

b. Installation

1. Install two friction lockarms (11) on windshield frame assembly (6) with two spring washers (15) and screws (13).
2. Apply sealing compound to hinge seal (3) and install on hinge (5).
3. Install windshield hinge (5) on windshield frame assembly (6) with seven screws (4).
4. Install lock handle (9) on windshield frame assembly (6) with two screws (10), new lockwashers (7), and capnuts (8).
5. Install two friction lockarms (11) on windshield outer frame (6) with two spring washers (14) and screws (12).
6. Install windshield hinge (5) on windshield frame (2) with eight new screw-assembled washers (1).
7. Close and lock windshield frame assembly (6).

3-279. WINDSHIELD FRAME ASSEMBLY REPLACEMENT (Contd)

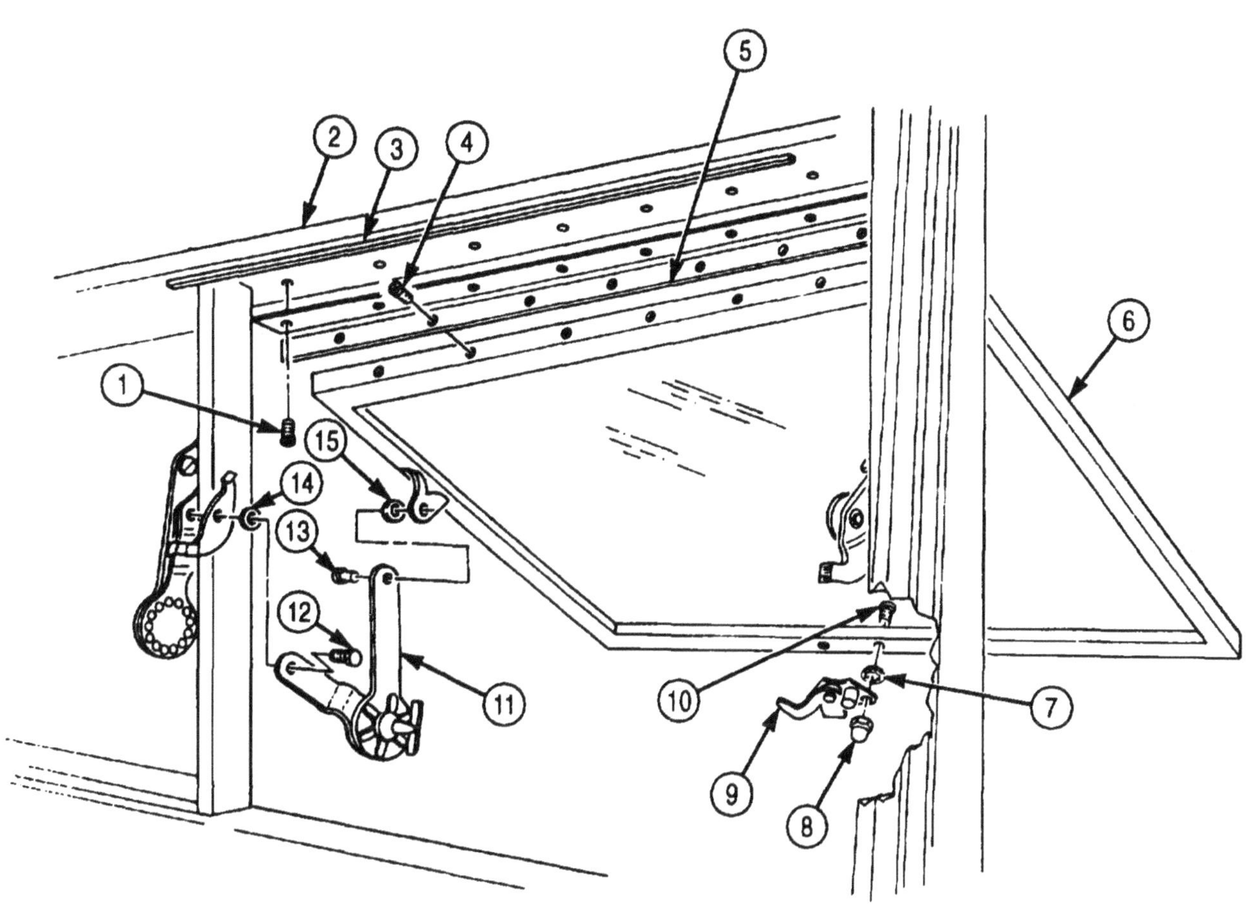

FOLLOW-ON TASK: Install windshield wiper blade, wiper arm, and wiper motor (para. 3-277).

3-280. CAB WINDSHIELD HINGE ASSEMBLY REPLACEMENT

THIS TASK COVERS:

a. Removal b. Installation

INITIAL SETUP

APPLICABLE MODELS
All

TOOLS
General mechanic's tool kit (Appendix E, Item 1)

REFERENCES (TM)
TM 9-2320-272-10
TM 9-2320-272-24P

EQUIPMENT CONDITION
Parking brake set (TM 9-2320-272-10).

NOTE

Left and right hinge assemblies are replaced the same. This procedure covers right side.

a. Removal

1. Remove three screws (1) from windshield hinge (3) and windshield frame (2).
2. Remove four screws (5) and windshield hinge (3) from cab (4).

b. Installation

1. Install windshield hinge (3) on cab (4) with four screws (5).
2. Install windshield hinge (3) on windshield frame (2) with three screws (1).

3-280. CAB WINDSHIELD HINGE ASSEMBLY REPLACEMENT (Contd)

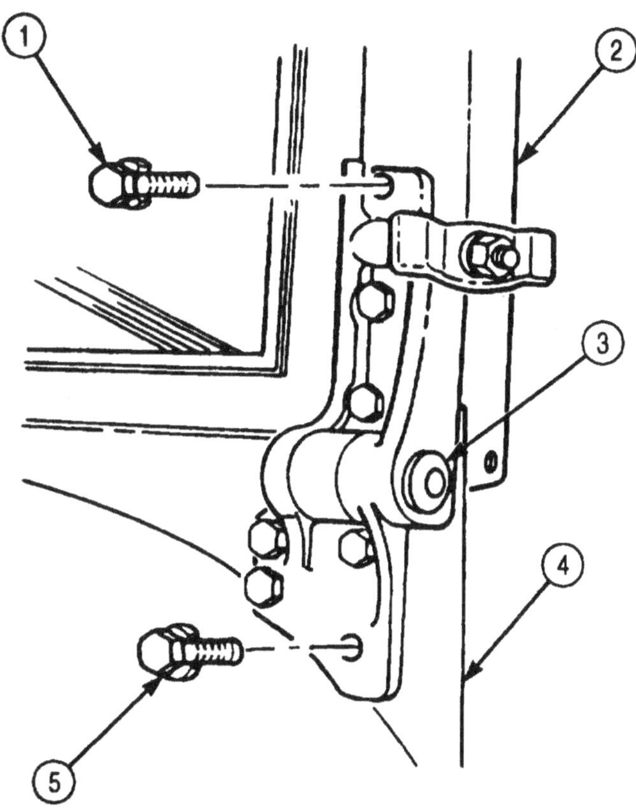

3-281. WINDSHIELD WIPER RESERVOIR, JET, AND CONTROL REPLACEMENT

THIS TASK COVERS:

a. Removal **b. Installation**

INITIAL SETUP:

APPLICABLE MODELS
All

TOOLS
General mechanic's tool kit (Appendix E, Item 1)

MATERIALS/PARTS
Three locknuts (Appendix D, Item 283)
Lockwasher (Appendix D, Item 370)
Antiseize tape (Appendix C, Item 72)

REFERENCES (TM)
TM 9-2320-272-10
TM 9-2320-272-24P

EQUIPMENT CONDITION
- Parking brake set (TM 9-2320-272-10).
- Air reservoir drained (TM 9-2320-272-10).
- Wheels chocked (TM 9-2320-272-10).

GENERAL SAFETY INSTRUCTIONS
Do not disconnect air lines before draining air reservoirs.

a. Removal

WARNING

Do not disconnect air lines before draining air reservoirs. Small parts under pressure may shoot out with high velocity, causing injury to personnel.

NOTE

Tag all air lines for installation.

1. Remove setscrew (13) and knob (14) from windshield wiper control (11).
2. Remove nut (1), lockwasher (2), and windshield wiper control (11) from instrument panel (4). Discard lockwasher (2).
3. Remove clamp (10) and hose (12) from windshield wiper control (11).
4. Remove hose (9) and elbow (8) from windshield wiper control (11).
5. Remove clamp (6) and duct hose (5) from defrost duct (3). Pull duct hose (5) from firewall (7).
6. Disconnect hoses (17) and (31) from jets (18) and (15).
7. Remove wingnuts (19) and (33) and jets (18) and (15) from cab (32).
8. Remove hoses (17), (30), and (31) from tee (16).
9. Remove clamps (22) and (23) and hoses (12) and (30) from windshield wiper reservoir cap (24).
10. Remove windshield wiper reservoir cap (24) from windshield wiper reservoir (25).
11. Remove windshield wiper reservoir (25) from bracket (26).
12. Remove three locknuts (29), screws (27), and bracket (26) from mounting bracket (28). Discard locknuts (29).
13. Remove grommets (20) and (21) and hoses (12) and (30) from firewall (7).

b. Installations

NOTE

Wrap all male threads with antiseize tape prior to installation.

1. Install bracket (26) on mounting bracket (28) with three screws (27) and new locknuts (29).
2. Install windshield wiper reservoir (25) on bracket (26).
3. Install windshield wiper reservoir cap (24) on reservoir (25).
4. Install hoses (12) and (30) on windshield wiper reservoir cap (24) with clamps (22) and (23).
5. Insert hoses (12) and (30) through holes in firewall (7).

3-281. WINDSHIELD WIPER RESERVOIR, JET, AND CONTROL REPLACEMENT (Contd)

6. Install hoses (17), (30), and (31) on tee (16).
7. Install jets (15) and (18) on cab (32) with wing nuts (33) and (19).
8. Connect hoses (31) and (17) on jets (15) and (18).
9. Place grommets (20) and (21) on hoses (30) and (12) and install on firewall (7).
10. Install elbow (8) and hose (9) on windshield wiper control (11).
11. Install hose (12) on windshield wiper control (11) with clamp (10).
12. Install windshield wiper control (11) on instrument panel (4) with new lockwasher (2) and nut (1).
13. Install knob (14) on windshield wiper control (11) with setscrew (13).
14. Install duct hose (5) on defrost duct (3) with clamp (6).

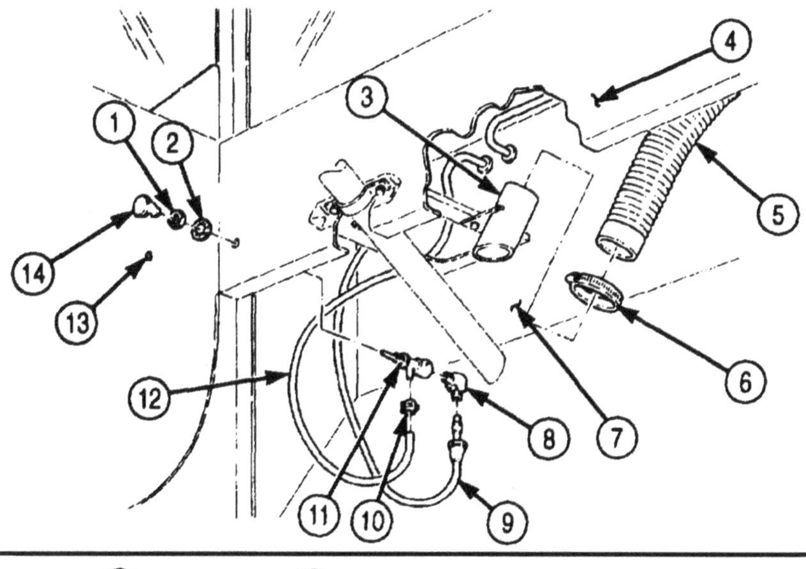

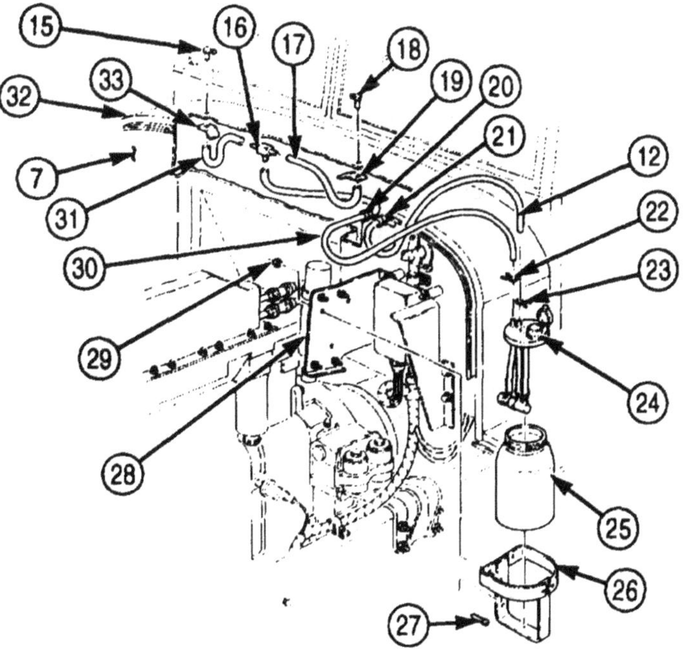

FOLLOW-ON TASK: Start engine (TM 9-2320-272-10) and check for leaks and proper operation.

3-282. WINDSHIELD WASHER HOSES REPLACEMENT

THIS TASK COVERS:

a. Removal b. Installation

INITIAL SETUP:

APPLICABLE MODELS
All

TOOLS
General mechanic's tool kit (Appendix E, Item 1)

REFERENCES (TM)
TM 9-2320-272-10
TM 9-2320-272-24P

EQUIPMENT CONDITION
- Parking brake set (TM 9-2320-272-10).
- Hood raised and secured (TM 9-2320-272-10).
- Instrument cluster removed (para. 3-83).

a. Removal

1. Remove wire clamp (6) and air supply line (7) from washer control valve outlet adapter (5).
2. Remove wire clamp (11) and air supply line (7) from washer bottle air inlet adapter (9).
3. Remove air supply line (7) from grommet (8) and firewall (14).
4. Disconnect washer jet supply tube (12) from outlet adapter (10).
5. Disconnect washer jet supply tube (12) from tee (13), grommet (3), and firewall (14).
6. Disconnect two windshield washer jet supply tubes (2) from washer jet adapters (1).
7. Remove two windshield washer jet supply tubes (2) from tee (13).

b. Installation

1. Connect two windshield washer jet supply tubes (2) to tee (13) under cowl (4).
2. Connect two windshield washer jet supply tubes (2) to washer jet adapters (1).
3. Push washer jet supply tube (12) through grommet (3) in firewall (14) and connect to tee (13).
4. Connect washer jet supply tube (12) to outlet adapter (10).
5. Install air supply line (7) on washer bottle air inlet adapter (9) with wire clamp (11).
6. Push air supply line (7) through grommet (8) in firewall (14).
7. Install air supply line (7) on washer control valve outlet adapter (5) with wire clamp (6).

3-282. WINDSHIELD WASHER HOSES REPLACEMENT (Contd)

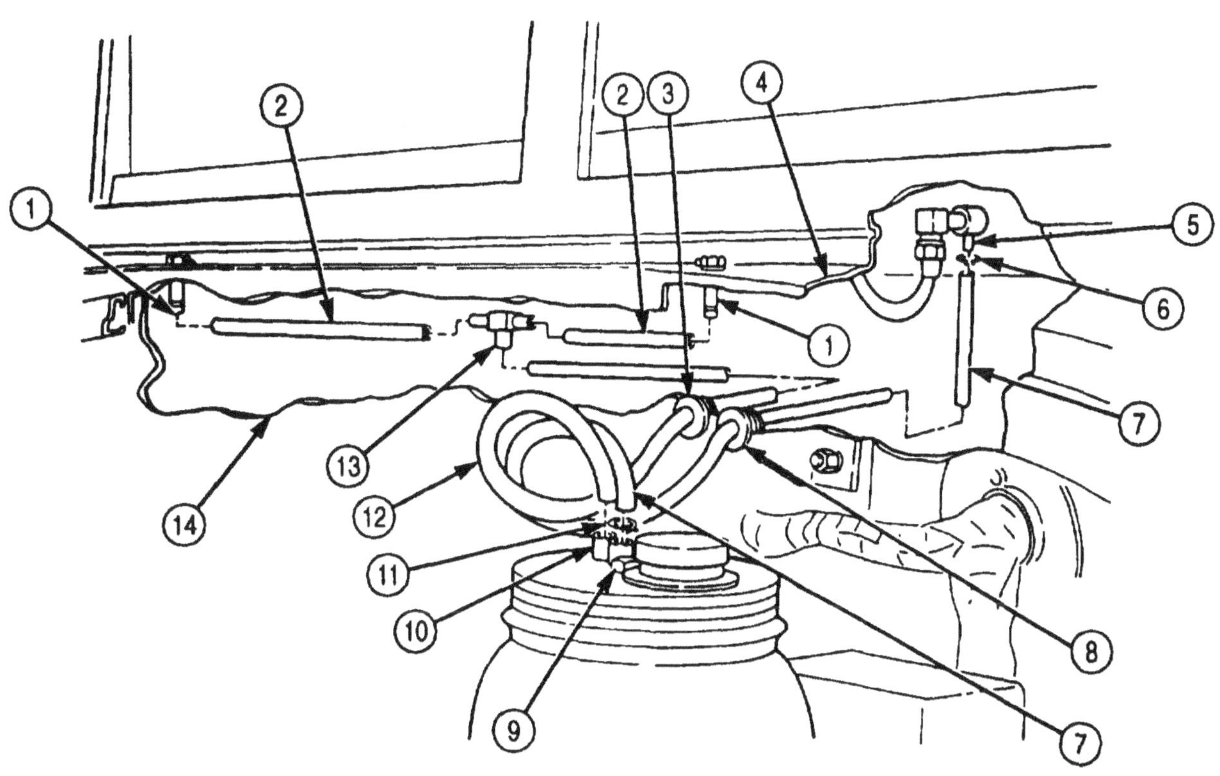

FOLLOW-ON TASK: Install instrument cluster (para. 3-83).

3-283. DRIVER'S SEAT REPLACEMENT

THIS TASK COVERS:

a. Removal **b. Installation**

INITIAL SETUP:

APPLICABLE MODELS
All

TOOLS
General mechanic's tool kit (Appendix E, Item 1)

MATERIALS/PARTS
Six screw-assembled lockwashers (Appendix D, Item 584)

REFERENCES (TM)
TM 9-2320-272-10
TM 9-2320-272-24P

EQUIPMENT CONDITION
- Parking brake set (TM 9-2320-272-10).
- Driver's seatbelts removed (para. 3-289).

a. Removal

1. Remove six screw-assembled lockwashers (1) from seat base (3) and cab floor (2). Discard screw-assembled lockwashers (1).
2. Slide driver's seat (4) close to door opening.

NOTE

Assistance will help with step 3.

3. Tilt driver's seat (4) carefully out of door opening and remove.

b. Installation

NOTE

Assistance will help with step 1.

1. Position driver's seat (4) so back faces rear of cab.
2. Install seat base (3) on cab floor (2) with six new screw-assembled lockwashers (1).

3-283. DRIVER'S SEAT REPLACEMENT (Contd)

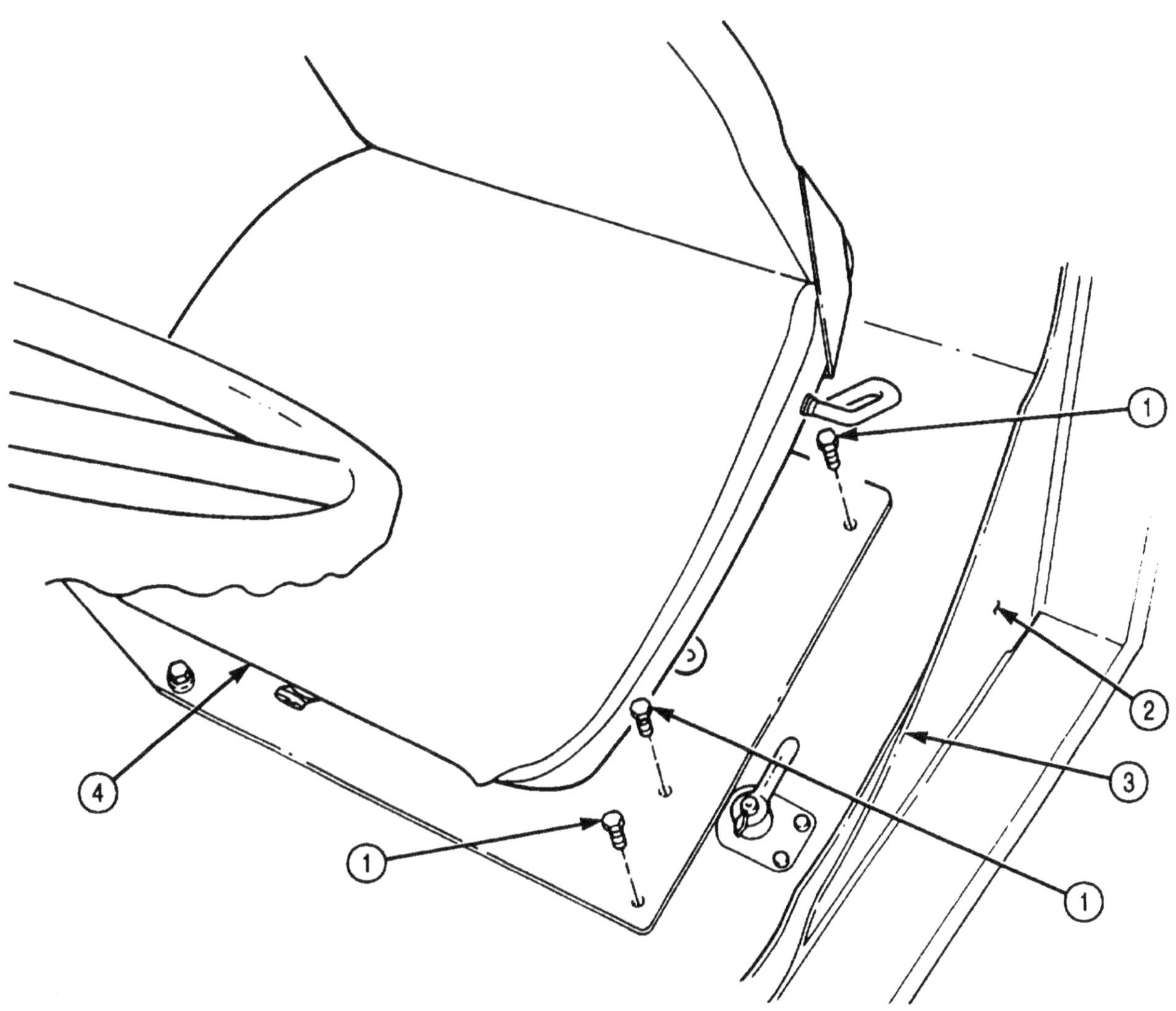

FOLLOW-ON TASK: Install driver's seatbelts (para. 3-289).

3-284. DRIVER'S SEAT FRAME AND BASE MAINTENANCE

THIS TASK COVERS:

a. Removal
b. Inspection and Repair
c. Installation

INITIAL SETUP:

APPLICABLE MODELS
All

TOOLS
General mechanic's tool kit (Appendix E, Item 1)

MATERIALS/PARTS
Eight locknuts (Appendix D, Item 277)
Eight lockwashers (Appendix D, Item 350)
Four lockwashers (Appendix D, Item 371)
GAA grease (Appendix C, Item 28)

REFERENCES (TM)
TM 9-2320-272-10
TM 9-2320-272-24P
TM 9-237

EQUIPMENT CONDITION

- Parking brake set (TM 9-2320-272-10).
- Driver's seat removed (para. 3-283).
- Driver's seat cushion and backrest cushion removed (para. 3-285).

a. Removal

1. Remove four locknuts (2) and seat frame (1) from adjusters (3). Discard locknuts (2).
2. Remove two screws (17), four locknuts (7), washers (5), adjusters (3), and washers (15) from top frame (14). Discard locknuts (7).
3. Remove two nuts (16), screws (4), and shock absorber (6) from seat base (13).
4. Remove four nuts (19), lockwashers (20), screws (11), and two brackets (21) from seat base (13). Discard lockwashers (20).
5. Remove spring (18) from seat base (13).
6. Turn crank (12) fully clockwise to remove tension from torque springs.
7. Remove four nuts (8), lockwashers (9), and screws (10) from seat base (13). Discard lockwashers (9).
8. Remove top frame (14) from seat base (13).

3-284. DRIVER'S SEAT FRAME AND BASE MAINTENANCE (Contd)

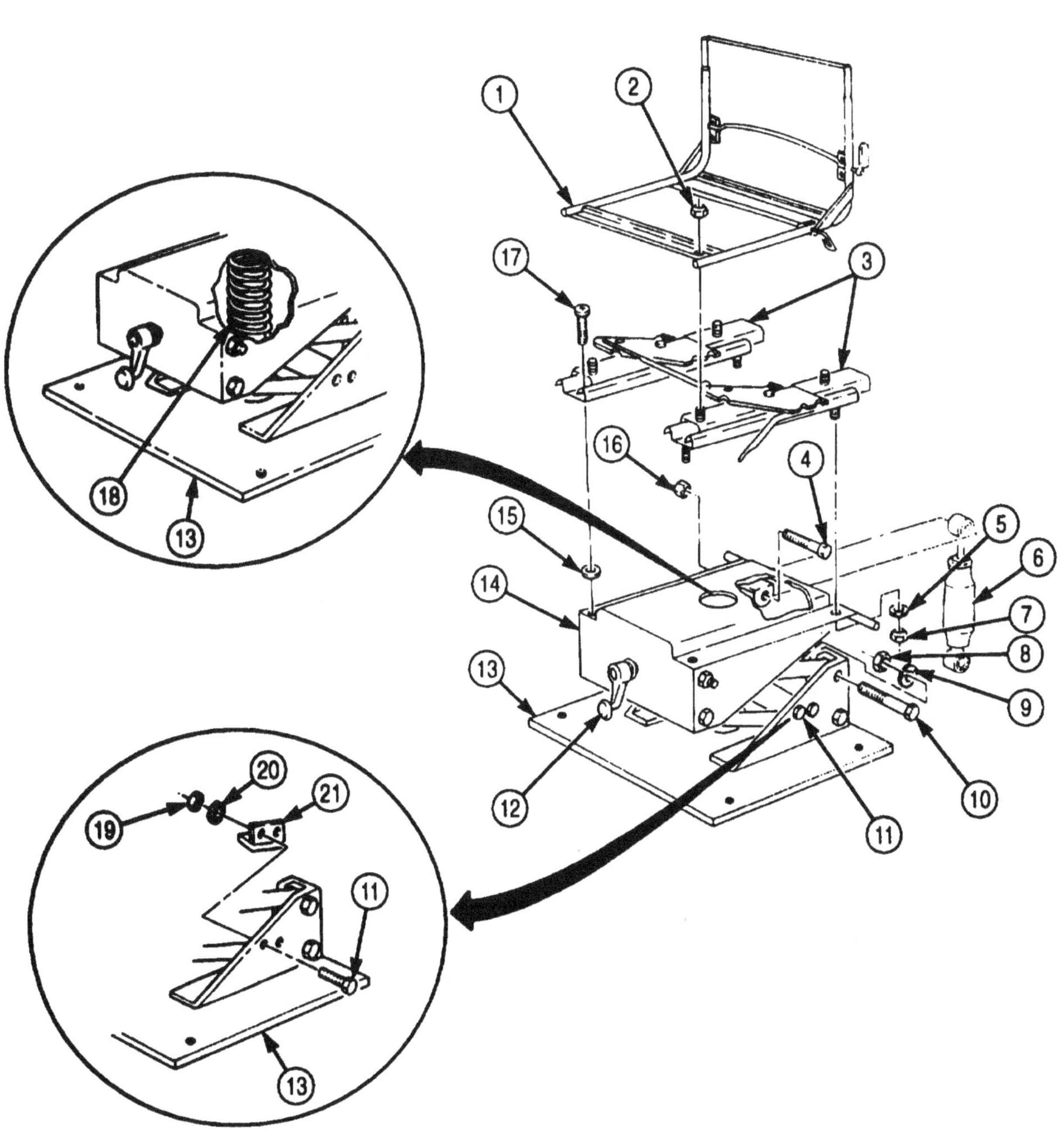

3-284. DRIVER'S SEAT FRAME AND BASE MAINTENANCE (Contd)

NOTE

Tag strut for installation.

9. Remove two nuts (10), lockwashers (9), screws (5), and lower strut (4) from top frame (11). Discard lockwashers (9).
10. Remove two nuts (6), lockwashers (7), torque rod (8), sleeve (12), two springs (2), and upper strut (1) from top frame (11). Discard lockwashers (7).
11. Remove pin (16) from crank (14).
12. Remove crank (14) and washer (13) from swivel nut (17) on bracket (15).

b. Inspection and repair

1. Inspect sheet metal parts, springs, brackets, struts, and pins for breaks, bends, and cracks. Replace if broken, bent, or cracked (TM 9-237).
2. Inspect crank (14), crank adjuster swivel nut (17), torque rod (8), and screws (5) for damaged threads. Replace if damaged.
3. Inspect shock absorber (18) for damage. Replace if damaged.
4. Inspect adjusters (19) and crank (14) for breaks, bends, and cracks. Replace if broken, bent, or cracked (TM 9-237).

c. Installation

1. Install washer (13) and crank (14) on top frame (11) and swivel nut (17) with pin (16).
2. Apply light coat of GAA grease to eight strut bushings (3) and torque rod (8).
3. Install lower strut (4) on top frame (11) with two screws (5), new lockwashers (9), and nuts (10).

NOTE

Torque rod can be installed through one side only.

4. Install sleeve (12), two springs (2), and upper strut (1) on top frame (11) with torque rod (8), two new lockwashers (7), and nuts (6).

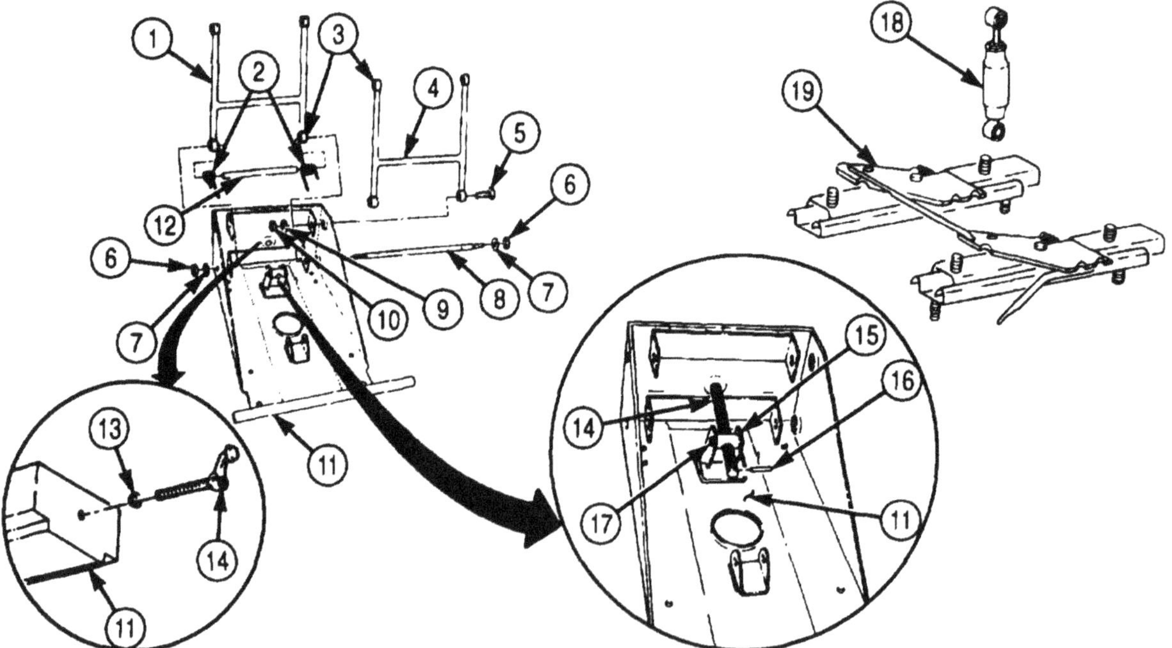

3-284. DRIVER'S SEAT FRAME AND BASE MAINTENANCE (Contd)

5. Install lower strut (4) and upper strut (1) on seat base (31) with four screws (30), new lockwashers (29), and nuts (27).
6. Position spring (33) on seat base (31).
7. Install two brackets (36) on seat base (31) with four screws (35), new lockwashers (37), and nuts (34).
8. Install shock absorber (18) on top frame (11) with screw (24) and nut (23).
9. Install fixed end (28) of shock absorber (18) on seat base (31) with screw (24) and nut (23).
10. Install adjusters (19) on top frame (11) with two washers (32), screws (20), four washers (25), and new locknuts (26).
11. Install seat frame (21) on adjusters (19) with four new locknuts (22).

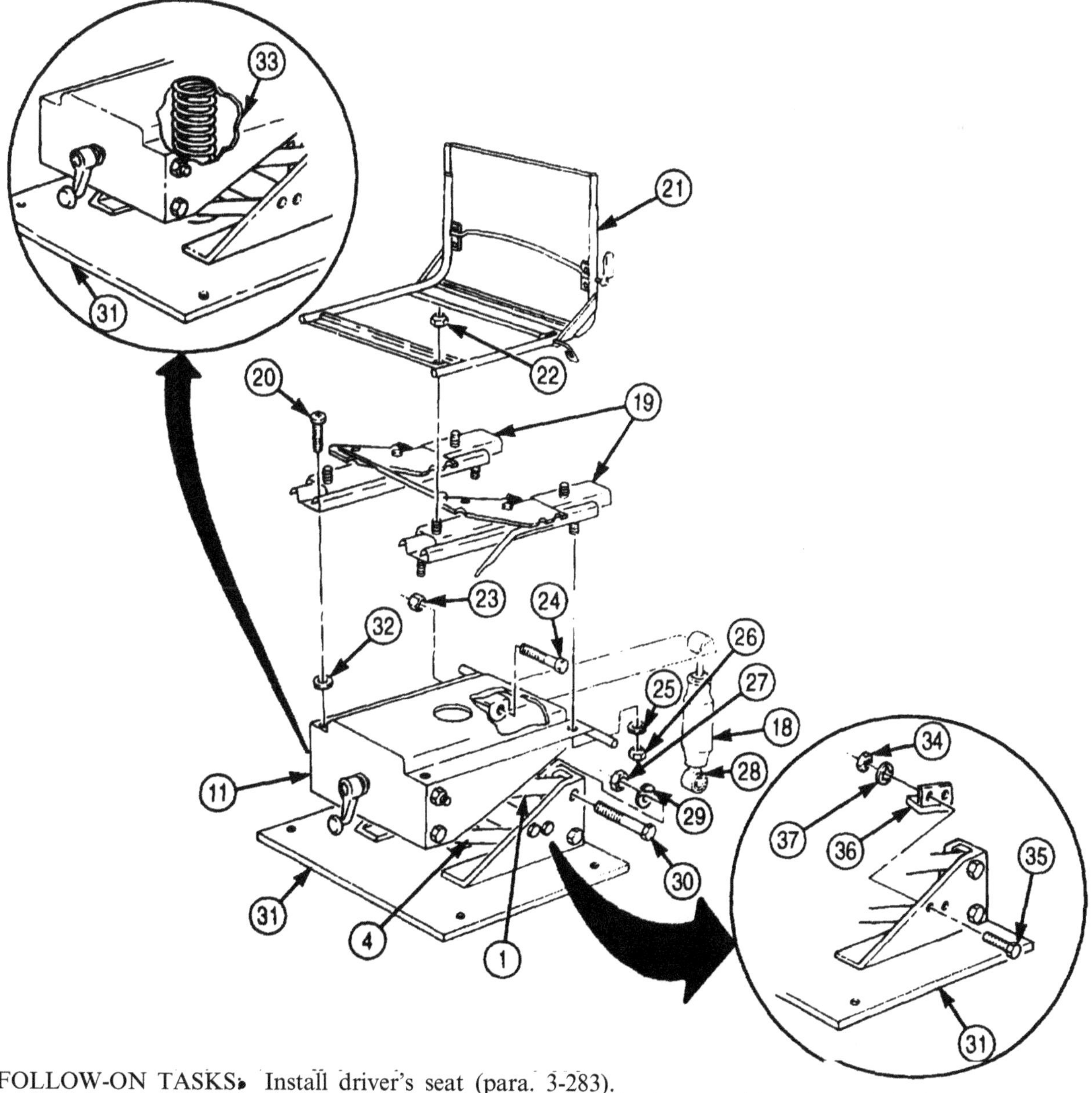

FOLLOW-ON TASKS:
- Install driver's seat (para. 3-283).
- Install driver's seat cushion and backrest cushion (para. 3-285).

3-285. DRIVER'S SEAT CUSHION, BACKREST CUSHION, AND FRAME REPLACEMENT

THIS TASK COVERS:

a. Removal   b. Installation

INITIAL SETUP:

APPLICABLE MODELS
All

TOOLS
General mechanic's tool kit (Appendix E, Item 1)

MATERIALS/PARTS
Four lockwashers (Appendix D. Item 371)

REFERENCES (TM)
TM 9-2320-272-10
TM 9-2320-272-24P

EQUIPMENT CONDITION
Parking brake set (TM 9-2320-272-10).

a. Removal

1. Remove four screws (5) and lockwashers (6) from two brackets (7) and seat cushion (1). Discard lockwashers (6).
2. Remove two seat brackets (7) and washers (2) from seat frame pins (3).
3. Remove seat cushion (1) from seat frame (4).
4. Remove two screws (12) from adjuster rod brackets (13) and remove adjuster rod brackets (13).
5. Remove two screws (11) and upper mounting brackets (10) from seat frame (4).
6. Remove backrest cushion (8) and wear plate (9) from seat frame (4).

b. Installation

1. Install wear plate. (9), backrest cushion (8), and two adjuster rod brackets (13) on seat frame (4) with two screws (12).
2. Install two upper mounting brackets (10) on seat frame (4) and backrest cushion (8) with two screws (11).
3. Install two seat brackets (7) on seat cushion (1) with four new lockwashers (6) and screws (5).
4. Position seat cushion (1) on seat frame (4).
5. Install two seat brackets (7) and washers (2) on seat frame pins (3).

3-285. DRIVER'S SEAT CUSHION, BACKREST CUSHION, AND FRAME REPLACEMENT (Contd)

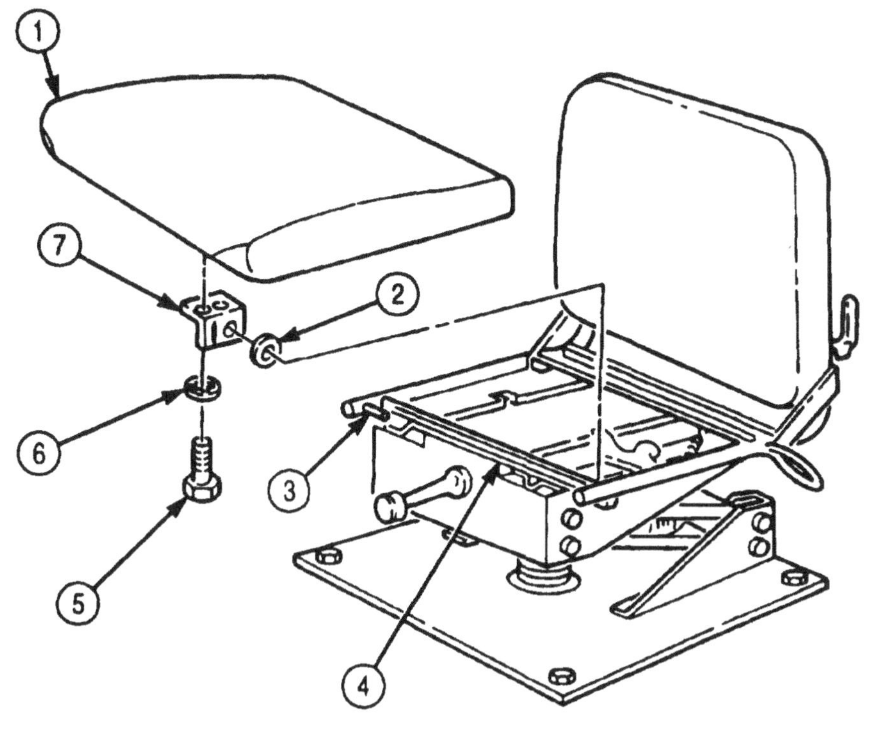

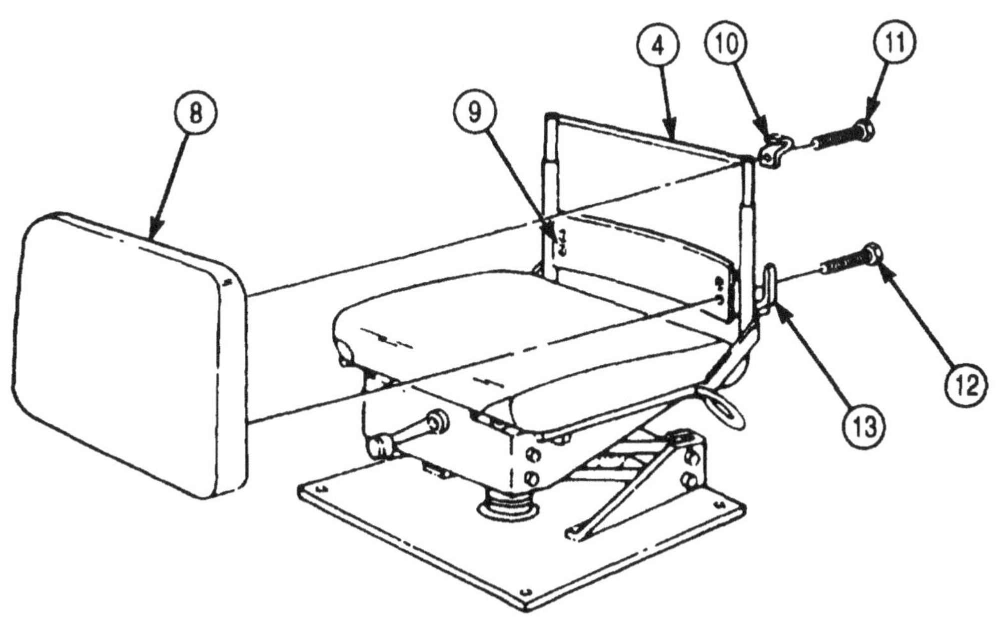

3-286. COMPANION SEAT CUSHION, BACKREST CUSHION, AND FRAME REPLACEMENT

THIS TASK COVERS:

a. Removal **b. Installation**

INITIAL SETUP:

APPLICABLE MODELS
All

TOOLS
General mechanic's tool kit (Appendix E, Item 1)

MATERIALS/PARTS
Two cotter pins (Appendix D, Item 46)
Fourteen lockwashers (Appendix D, Item 400)

REFERENCES (TM)
TM 9-2320-272-10
TM 9-2320-272-24P

EQUIPMENT CONDITION
Parking brake set (TM 9-2320-272-10).

a. Removal

1. Fold down backrest cushion frame (2).
2. Remove eight screws (1), lockwashers (6), washers (5), and backrest cushion (3) from backrest cushion frame (2). Discard lockwashers (6).
3. Remove six screws (8), lockwashers (9), washers (10), and seat cushion (7) from battery box cushion (4). Discard lockwashers (9).
4. Remove two springs (13) from two battery box cover extensions (15) and backrest cushion frame (2).
5. Remove two cotter pins (11), four washers (12), two pins (14), and backrest cushion frame (2) from two battery box cover extensions (15). Discard cotter pins (11).

b. Installation

1. Install backrest cushion frame (2) on battery box cover extensions (15) with two pins (14), four washers (12), and two new cotter pins (11).
2. Install two springs (13) on battery box cover extensions (15) and backrest cushion frame (2).
3. Install seat cushion (7) on battery box cushion (4) with six washers (10), lockwashers (9), and screws (8).
4. Install backrest cushion (3) on backrest cushion frame (2) with eight washers (5), new lockwashers (6), and screws (1).

3-286. COMPANION SEAT CUSHION, BACKREST CUSHION, AND FRAME REPLACEMENT (Contd)

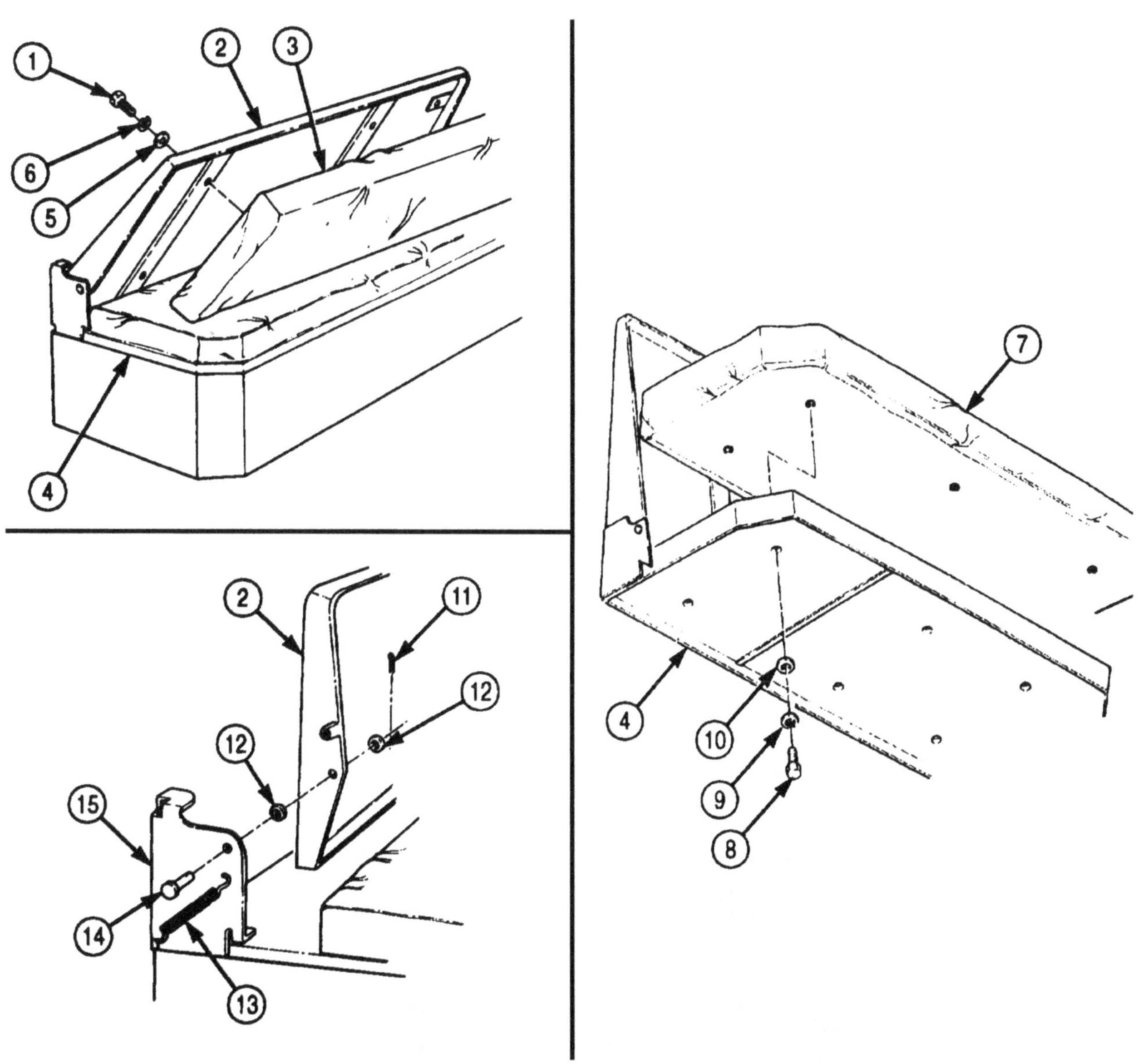

3-287. MAP COMPARTMENT REPLACEMENT

THIS TASK COVERS:

a. Removal **b. Installation**

INITIAL SETUP:

APPLICABLE MODELS
All

TOOLS
General mechanic's tool kit (Appendix E, Item 1)

MATERIALS/PARTS
Four locknuts (Appendix D. Item 291)

REFERENCES (TM)
TM 9-2320-272-10
TM 9-2320-272-24P

EQUIPMENT CONDITION
Parking brake set (TM 9-2320-272-10).

a. Removal

1. Fold companion seat backrest (3) forward.
2. Release two latches (5) on battery box (4).
3. Raise battery box cover (2) and secure with support rod (1).
4. Remove four locknuts (10), screws (7), and washers (6) from map compartment risers (9) and battery box cover (2). Discard locknuts (10).
5. Remove map compartment (8) from battery box (4).

b. Installation

1. Install map compartment (8) on battery box cover (2) with four washers (6), screws (7), and four new locknuts (10).
2. Secure support rod (1) in battery box (4), lower battery box cover (2), and secure with two latches (5).
3. Lower companion seat backrest (3).

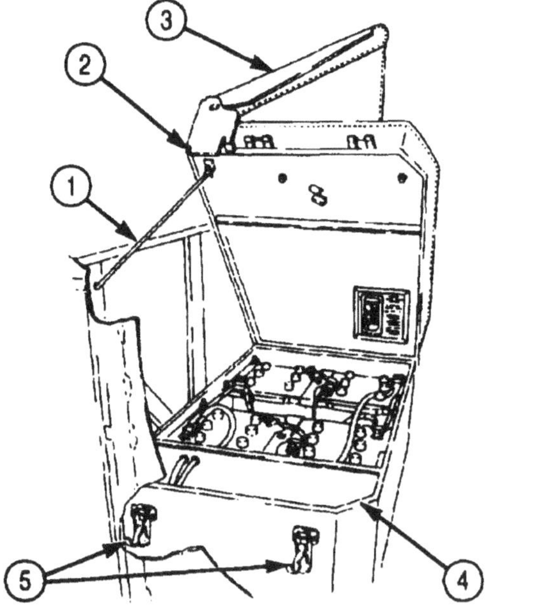

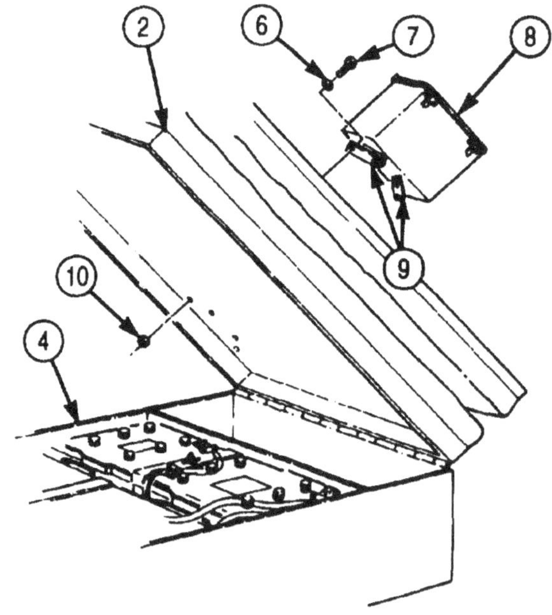

3-288. CAB GRAB HANDLE REPLACEMENT

THIS TASK COVERS:

a. Removal **b. Installation**

INITIAL SETUP:

APPLICABLE MODELS
All

TOOLS
General mechanic's tool kit (Appendix E, Item 1)

MATERIALS/PARTS
Four locknuts (Appendix D, Item 313)
Adhesive (Appendix C, Item 7)

REFERENCES (TM)
TM 9-2320-272-10
TM 9-2320-272-24P

EQUIPMENT CONDITION
Parking brake set (TM 9-2320-272-10).

a. Removal

1. Lift insulation and cut two square patches to gain access to locknuts (5) and reinforcing plates (4). Save patches for installation.
2. Remove four locknuts (5), screws (1), two reinforcing plates (4), and grab handle (2) from cab (3). Discard locknuts (5).

b. Installation

1. Install grab handle (2) on cab (3) with four screws (1), two reinforcing plates (4), and four new locknuts (5).
2. Install insulation on cab (3) with adhesive.

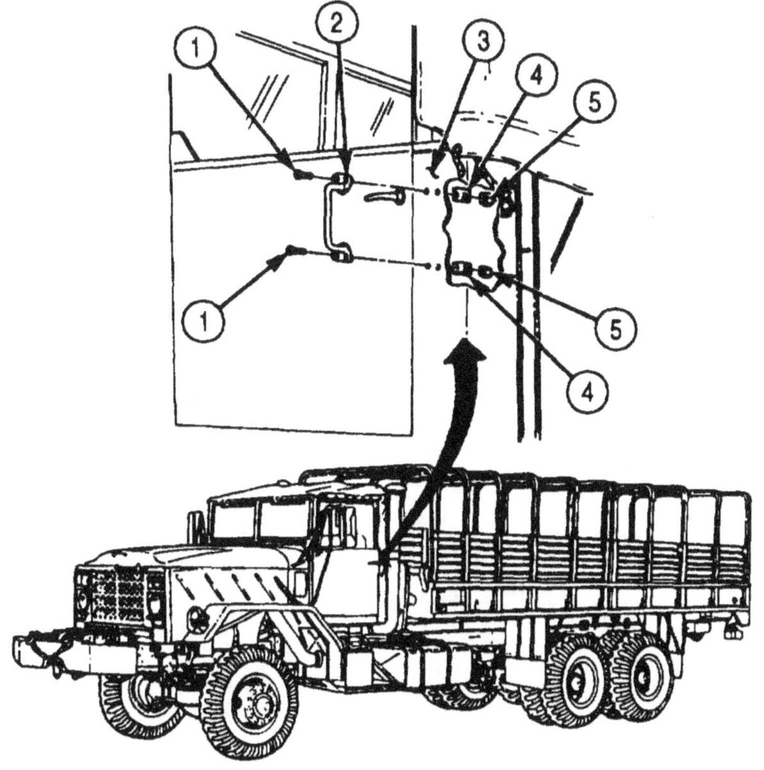

3-289. SEATBELT REPLACEMENT

THIS TASK COVERS:

a. Driver's Seatbelt Removal
b. Driver's Seatbelt Installation
c. Companion Seatbelts Removal
d. Companion Seatbelts Installation

INITIAL SETUP:

APPLICABLE MODELS
All

TOOLS
General mechanic's tool kit (Appendix E, Item 1)
Torque wrench (Appendix E, Item 144)

REFERENCES (TM)
TM 9-2320-272-10
TM 9-2320-272-24

EQUIPMENT CONDITION
- Parking brake set (TM 9-2320-272-10).
- Companion seat removed (para. 3-286).
- Battery box removed (para. 3-128).

a. Driver's Seatbelt Removal

1. Remove screw (7), washer (6), retractor (5), washer (4), tether assembly (3), and spacer (2) from seat rod (1).
2. Remove nut (8), washer (9), screw (13), washer (12), tether assembly (3), and washer (11) from vertical channel (10).
3. Remove nut (14), washer (15), bracket (16), screw (19), washer (20), tether assembly (21), washer (18), and sleeve (17) from left diagonal channel (22).
4. Remove screw (29), washer (28), seatbelt buckle (26), washer (23), and tether assembly (21) from seat rod (24).
5. Remove seatbelt buckle (26) from between driver's seat (27) and seat backrest support (25).

b. Driver's Seatbelt Installation

1. Position seatbelt buckle (26) between seat backrest support (25) and driver's seat (27).
2. Install tether assembly (21), washer (23), seatbelt buckle (26), washer (28), and screw (29) on seat rod (24).
3. Install sleeve (17), washer (18), tether assembly (21), washer (20), screw (19), bracket (16), washer (15), and nut (14) on left diagonal channel (22). Tighten nut (14) 20-25 lb-ft (27-34 N.m).
4. Install spacer (2), tether assembly (3), washer (4), retractor (5), washer (6), and screw (7) on seat rod (1).
5. Install washer (11), tether assembly (3), washer (12), screw (13), washer (9), and nut (8) on vertical channel (10). Tighten nut (8) 40-45 lb-ft (54-61 N.m).

3-289. SEATBELT REPLACEMENT (Contd)

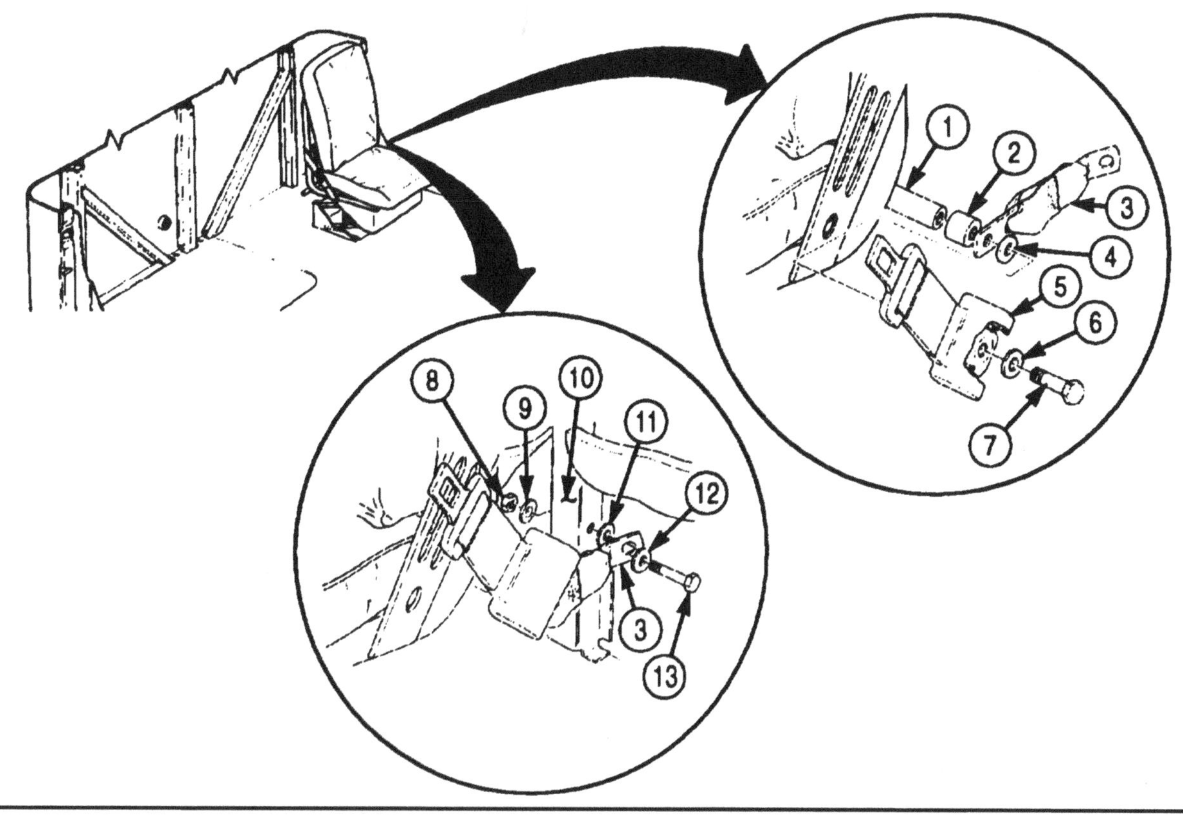

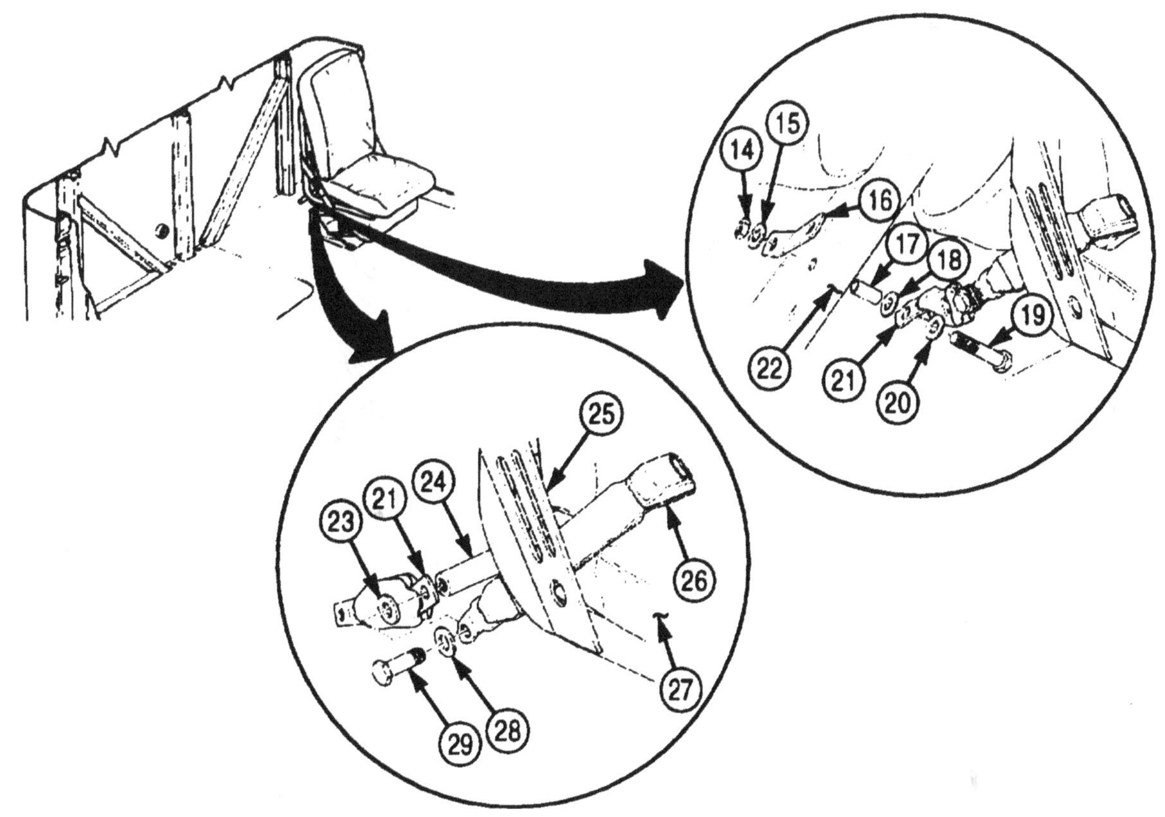

3-289. SEATBELT REPLACEMENT (Contd)

c. Companion Seatbelts Removal

1. Remove nut (1), washer (2), screw (5), washer (6), left seatbelt buckle (4), and washer (7) from bracket (3).
2. Remove nut (9), washer (10), middle companion seatbelt (8), washer (11), screw (12), washer (13), and sleeve (14) from right diagonal channel (15).
3. Remove nut (17), washer (18), screw (21), washer (22), right seatbelt buckle (16), and washer (20) from B-pillar (19).

d. Companion Seatbelts Installation

1. Install washer (20), right seatbelt buckle (16), washer (22), screw (21), washer (18), and nut (17) on B-pillar channel (19). Tighten nut (17) 20-25 lb-ft (27-34 N•m).
2. Install sleeve (14), washer (13), screw (12), washer (11), middle companion seatbelt (8), washer (10), and nut (9) on right diagonal channel (15). Tighten nut (9) 20-25 lb-ft (27-34 N•m).
3. Install washer (7), left seatbelt buckle (4), washer (6), screw (5), washer (2), and nut (1) on bracket (3). Tighten nut (1) 20-25 lb-ft (27-34 N.m).

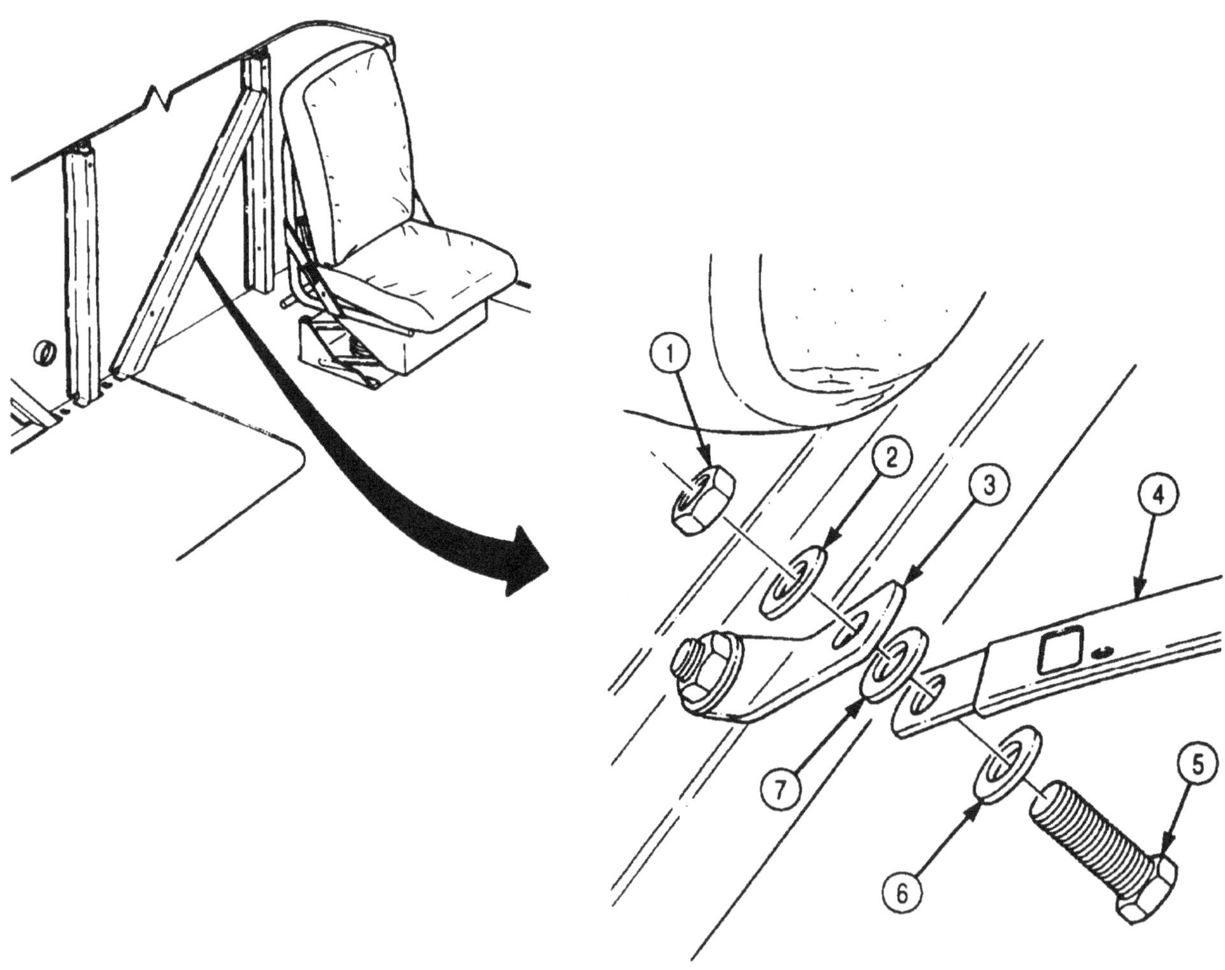

3-289. SEATBELT REPLACEMENT (Contd)

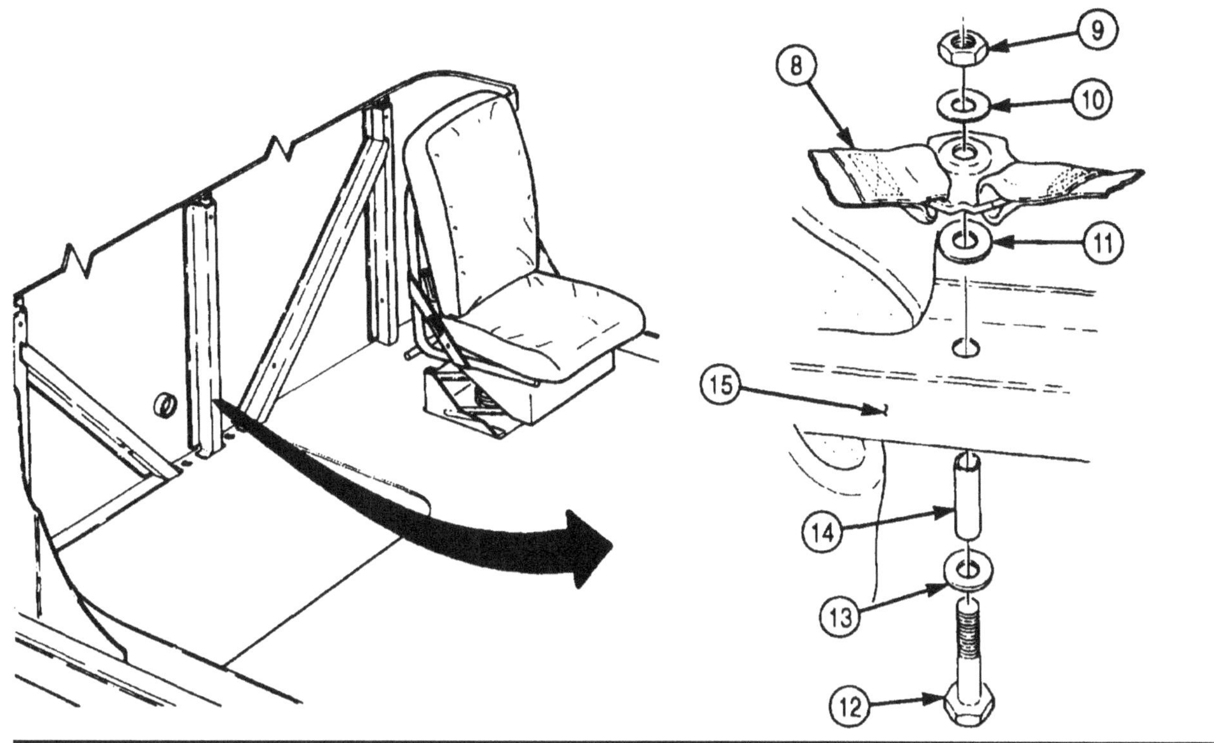

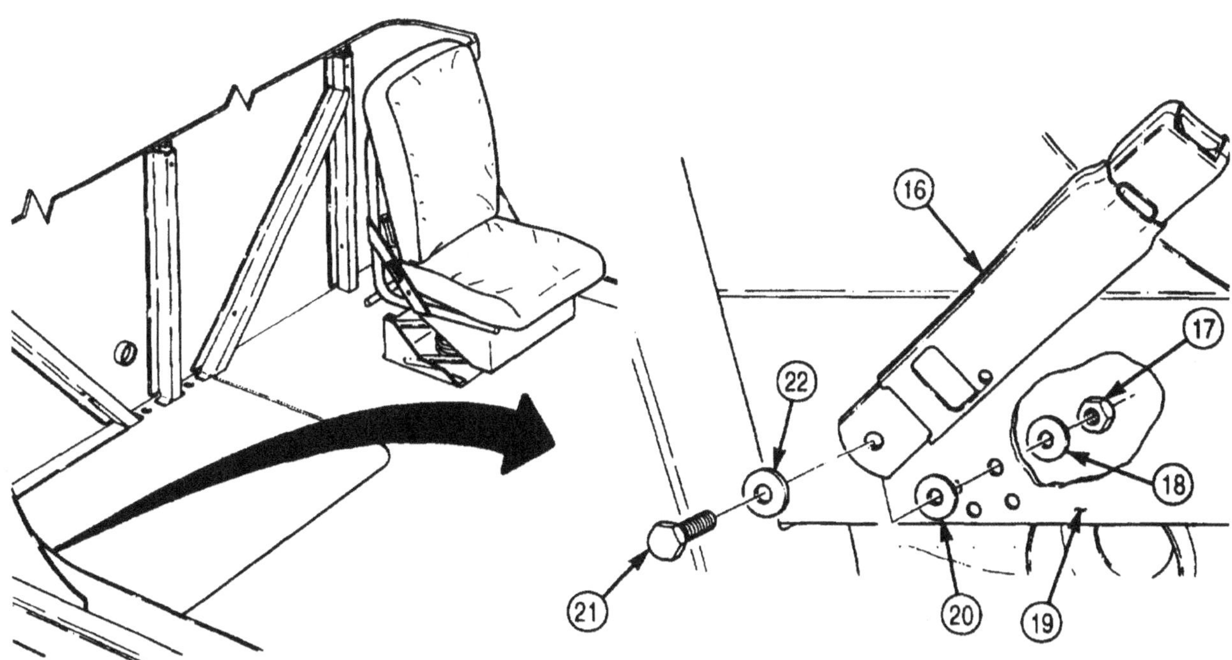

FOLLOW-ON TASKS:
- Install companion seat (para. 3-286).
- Install battery box (para. 3-128).

3-290. REAR VIEW MIRROR AND BRACE REPLACEMENT

THIS TASK COVERS:

a. Removal **b. Installation**

INITIAL SETUP:

APPLICABLE MODELS
All

TOOLS
General mechanic's tool kit (Appendix E, Item 1)

MATERIALS/PARTS
Five locknuts (Appendix D, Item 294)
Two locknuts (Appendix D, Item 272)
Lockwasher (Appendix D, Item 400)
Screw-assembled lockwasher (Appendix D, Item 572)
Two locknuts (Appendix D, Item 275)

REFERENCES (TM)
TM 9-2320-272-10
TM 9-2320-272-24P

EQUIPMENT CONDITION
Parking brake set (TM 9-2320-272-10).

a. Removal

1. Remove two locknuts (6) and washers (7) from mirror (9). Discard locknuts (6).
2. Remove rearview mirror (9) and two spacers (8) from rearview mirror brace (4).
3. Remove nut (10), lockwasher (11), and screw (15) from two clamp halves (12). Discard lockwasher (11).
4. Remove screw-assembled lockwasher (14), two clamp halves (12), and convex mirror (5) from upper brace (13). Discard screw-assembled lockwasher (14).
5. Remove locknut (24), washer (23), and hinge screw (36) from upper mirror braces (16) and (38) and upper cab door hinge (22). Discard locknut (24).
6. Push upper mirror braces (16) and (38) away from cab door hinge (22).
7. Insert hinge screw (36) into upper cab door hinge (22). Do not tighten hinge screw (36).
8. Remove locknut (30), washer (29), and hinge screw (19) from lower mirror brace (18) and lower cab door hinge (28). Discard locknut (30).
9. Push lower mirror brace (18) away from cab door hinge (28).
10. Insert hinge screw (19) into lower cab door hinge (28). Do not tighten hinge screw (19).
11. Remove locknut (27), washer (29), and screw (21) from front mirror brace (37) and (20) and cab cowl bracket (25). Discard locknut (27).
12. Remove two locknuts (2), washers (1), gaskets (3), braces (18), (20), (16), (4), (37), and (38) from rod (17) and upper brace (13). Discard locknuts (2).
13. Open air vent door (31).
14. Remove two locknuts (34), screws (32), cowl plate (33), and bracket (25) from cowl side panel (35). Discard locknuts (34).

3-290. REAR VIEW MIRROR AND BRACE REPLACEMENT (Contd)

b. Installation

1. Install bracket (25) and cowl plate (33) on cowl side panel (35) with two screws (32) and two new locknuts (34).
2. Install braces (13), (4), (16), (37), (38), (20), (18), and two gaskets (3) on rod (17) with two washers (1) and new locknuts (2).
3. Install mirror braces (20) and (37) on cab cowl bracket (25) with screw (21), washer (26), and new locknut (27).
4. Remove hinge screw (36) from top door hinge (22).
5. Position upper mirror brace (38) and mirror brace (16) over upper door hinge (22) and install hinge screw (36) through brace (38) and hinge (22) with washer (23) and new locknut (24).
6. Remove hinge screw (19) from lower door hinge (28).
7. Position lower mirror brace (18) over lower door hinge (28) and insert hinge screw (19) through brace (18) and hinge (28) and secure with washer (29) and new locknut (30).
8. Install two clamp halves (12) on upper brace (13) with screw (15), new lockwasher (11), and nut (10).
9. Install convex mirror (5) on clamp halves (12) with new screw-assembled lockwasher (14).
10. Position rearview mirror (9) and two spacers (8) on mirror brace (4) and install with two washers (7) and new locknuts (6).

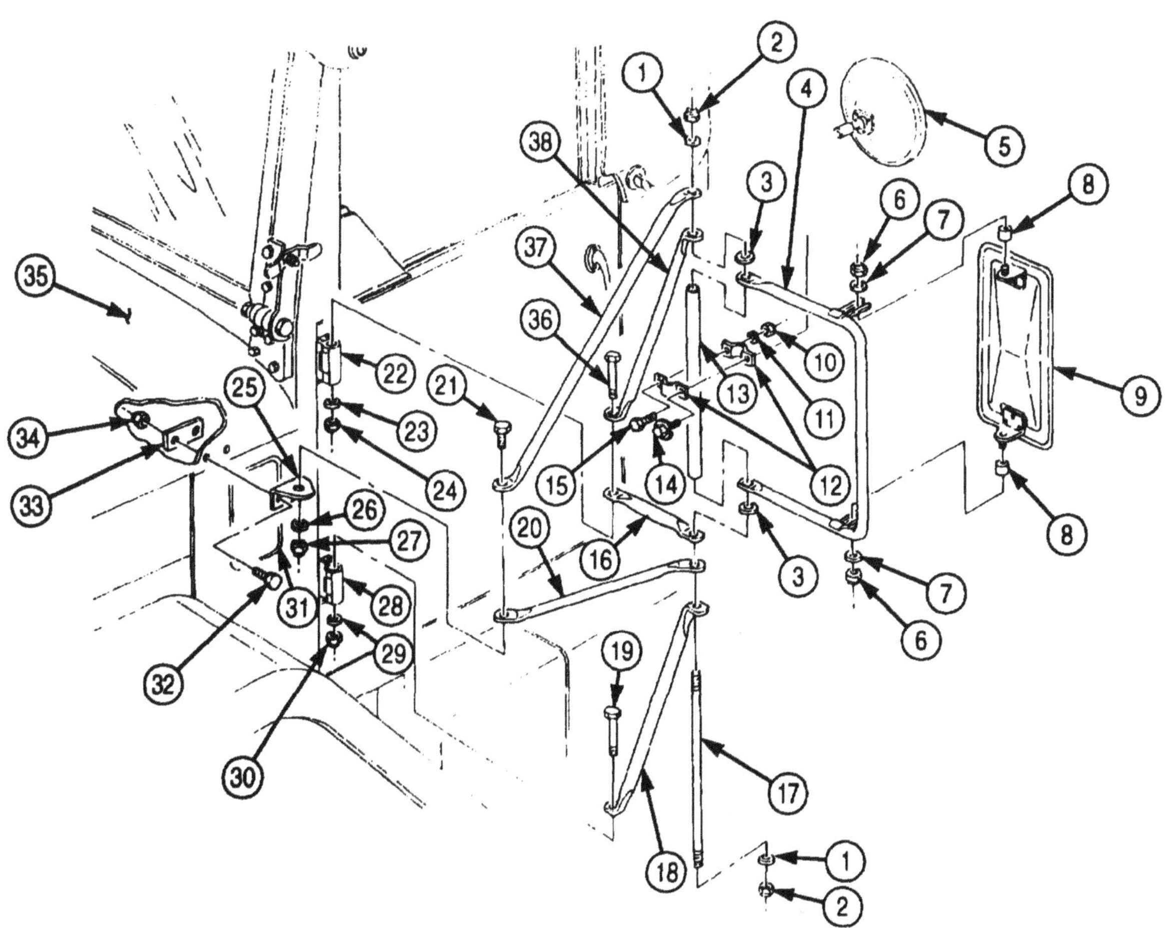

3-291. PERSONNEL HEATER INLET AND OUTLET HOSE REPLACEMENT

THIS TASK COVERS:

a. Removal **b. Installation**

INITIAL SETUP:

APPLICABLE MODELS
All

TOOLS
General mechanic's tool kit (Appendix E, Item 1)

REFERENCES (TM)
TM 9-2320-272-10
TM 9-2320-272-24P

EQUIPMENT CONDITION
- Parking brake set (TM 9-2320-272-10).
- Right fender splash shield removed (para. 3-301).

NOTE

- Have drainage container ready to catch coolant.
- The personnel heater inlet hose and outlet hose are removed the same. This procedure covers the inlet hose.

a. Removal

1. Loosen hose clamp (4) and remove heater inlet hose (2) from elbow (1).
2. Remove nut (7), screw (9), clamp (8), and heater inlet hose (2) from heater (3).
3. Loosen hose clamp (5) and remove heater inlet hose (2) from oil cooler shutoff valve (6).

b. Installation

1. Install heater inlet hose (2) on elbow (1) with clamp (4).
2. Install heater inlet hose (2) on oil cooler shutoff valve (6) with hose clamp (5).
3. Install heater inlet hose (2) on heater (3) with clamp (8), screw (9), and nut (7).

3-291. PERSONNEL HEATER INLET AND OUTLET HOSE REPLACEMENT (Contd)

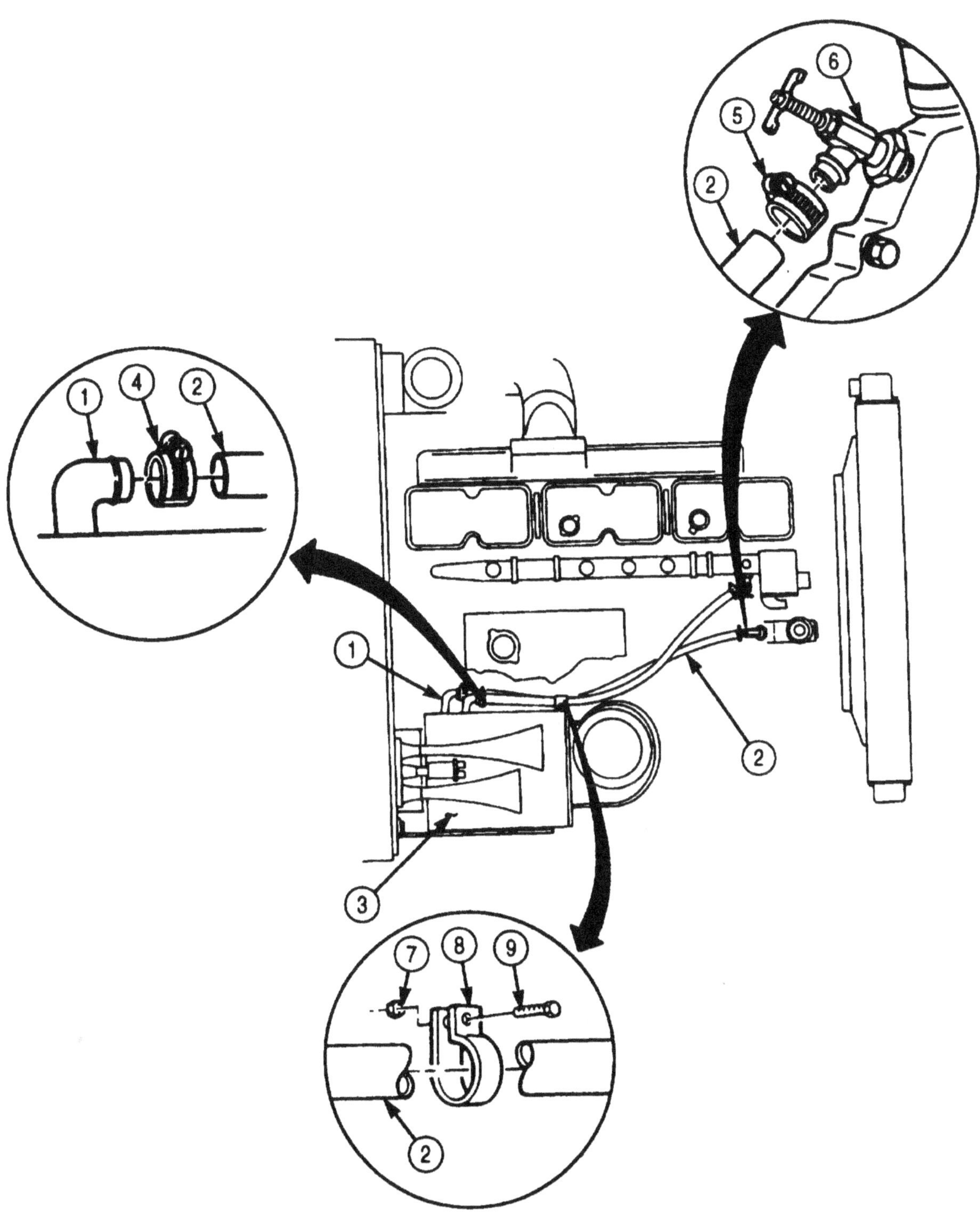

FOLLOW-ON TASKS:
- Fill cooling system to proper level and test antifreeze (para. 3-53).
- Install right fender splash shield (para. 3-301).
- Check heater for proper operation (TM 9-2320-272-10).

3-292. PERSONNEL HOT WATER HEATER REPLACEMENT

THIS TASK COVERS:

a. Removal **b. Installation**

INITIAL SETUP:

APPLICABLE MODELS
All

TOOLS
General mechanic's tool kits (Appendix E Item 1)

MATERIALS/PARTS
Cotter pin (Appendix D, Item 53)

REFERENCES (TM)
TM 9-2320-272-10
TM 9-2320-272-24P

EQUIPMENT CONDITION
- Parking brake set (TM 9-2320-272-10).
- Personnel heater inlet and outlet hoses removed (para 3-291).

a. Removal

1. Remove screw (3), sheet spring nut (7), and clamp bracket (4) from bracket (6).
2. Remove cotter pin (8) from fresh air shutoff rod (2). Discard cotter pin (8).
3. Remove fresh air cable (5) and clip (9) from fresh air shutoff rod (2).
4. Remove four screws (17), washers (16), and canister (1) from personnel hot water heater (13).
5. Disconnect lead (34) from heater blower motor lead (35).
6. Remove setscrew (15) and fan/impeller (14) from heater blower motor (12).
7. Remove four screws (11), washers (10), and heater blower motor (12) from personnel hot water heater (13).
8. Remove nut (29), washer (24), screw (23), and washer (24) from rear heater support (25).
9. Remove nut (33), washer (18), screw (19), and washer (18) from forward heater support (26).
10. Remove four screws (32) and washers (31) from forward and rear heater supports (26) and (25).
11. Remove two screws (28), washers (27), and left mounting bracket (30) from firewall (21).

CAUTION

Do not twist or bend elbows. They are welded to personnel hot water heater.

12. Remove personnel hot water heater (13) from diverter duct (22) and right mounting bracket (20).

b. Installation

CAUTION

During installation of personnel water heater, do not twist or bend elbows. They are welded and may be damaged.

1. Position personnel hot water heater (13) on diverter duct (22) and right mounting bracket (20).
2. Install left mounting bracket (30) on firewall (21) with two washers (27) and screws (28).
3. Install forward and rear supports (26) and (25) on personnel hot water heater (13) with four washers (31) and screws (32).
4. Install forward and rear supports (26) and (25) on left mounting bracket (30) and right mounting bracket (20) with washer (18), screw (19), washer (18), nut (33), washer (24), screw (23), washer (24), and nut (29).
5. Install heater blower motor (12) on personnel hot water heater (13) with four washers (10) and screws (11).

3-292. PERSONNEL HOT WATER HEATER REPLACEMENT (Contd)

6. Install fan/impeller (14) on heater blower motor (12) with setscrew (15).
7. Connect lead (34) to heater blower motor lead (35).
8. Install canister (1) on personnel hot water heater (13) with four washers (16) and screws (17).
9. Install clip (9) and fresh air control cable (5) on fresh air shutoff rod (2) with new cotter pin (8).
10. Install clamp bracket (4) and fresh air control cable (5) on bracket (6) with screw (3) and spring nut (7).

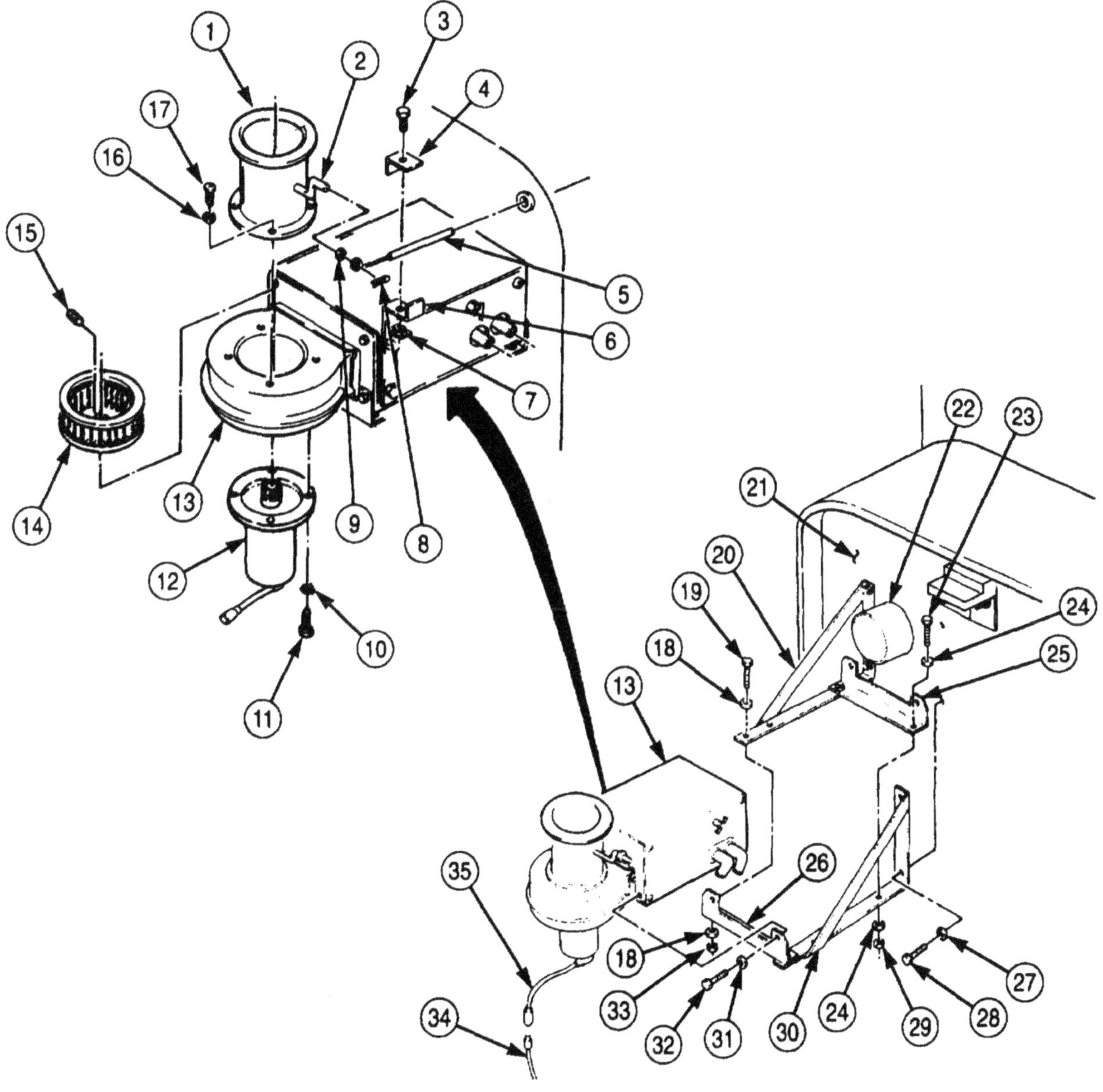

FOLLOW-ON TASKS:
- Install personnel heater inlet and outlet hoses (para. 3-291).
- Check heater for proper operation (TM 9-2320-272-10).

3-293. FRESH AIR VENT CONTROL ASSEMBLY REPLACEMENT

THIS TASK COVERS:

a. Removal   b. Installation

INITIAL SETUP:

APPLICABLE MODELS
All

TOOLS
General mechanics tool kit (Appendix E, Item 1)

MATERIALS/PARTS
Cotter pin (Appendix D, Item 53)
Lockwasher (Appendix D, Item 354)

REFERENCES (TM)
TM 9-2320-272-10
TM 9-2320-272-24P

EQUIPMENT CONDITION
- Parking brake set (TM 9-2320-272-10).
- Hood raised and secured (TM 9-2320-272-10).

a. Removal

1. Remove screw (4), sheet spring nut (7), and clamp bracket (5) from personnel hot water heater (9).
2. Remove cotter pin (8), fresh air control cable (3), and clip (2) from fresh air shutoff rod (10). Discard cotter pin (8).
3. Remove eight screws (11) and pull instrument cluster (16) from instrument panel (12).
4. Remove nut (13) and lockwasher (14) from fresh air control handle (15) and slide off fresh air control cable (3) at coiled end. Discard lockwasher (14).
5. Remove fresh air control handle (15) and fresh air control cable (3) from instrument cluster (16).

b. Installation

1. Install fresh air control handle (15) and fresh air control cable (3) through hole in instrument cluster (16).
2. Position fresh air control handle (15) against instrument cluster (16) and install with new lockwasher (14) and nut (13).
3. Push fresh control cable (3) through firewall (6).
4. Install instrument cluster (16) on instrument panel (12) with eight screws (11).
5. Push fresh air control handle (15) in all the way.
6. Close fresh air shutoff rod (10) before connecting fresh air control cable (3).
7. Install clip (2) and fresh air control cable (3) on fresh air shutoff rod (10) with new cotter pin (8).
8. Install fresh air control cable (3) on personnel hot water heater (9) with clamp bracket (5), spring nut (7), and screw (4).
9. Adjust air intake flap (1) to the closed position.

3-293. FRESH AIR VENT CONTROL ASSEMBLY REPLACEMENT (Contd)

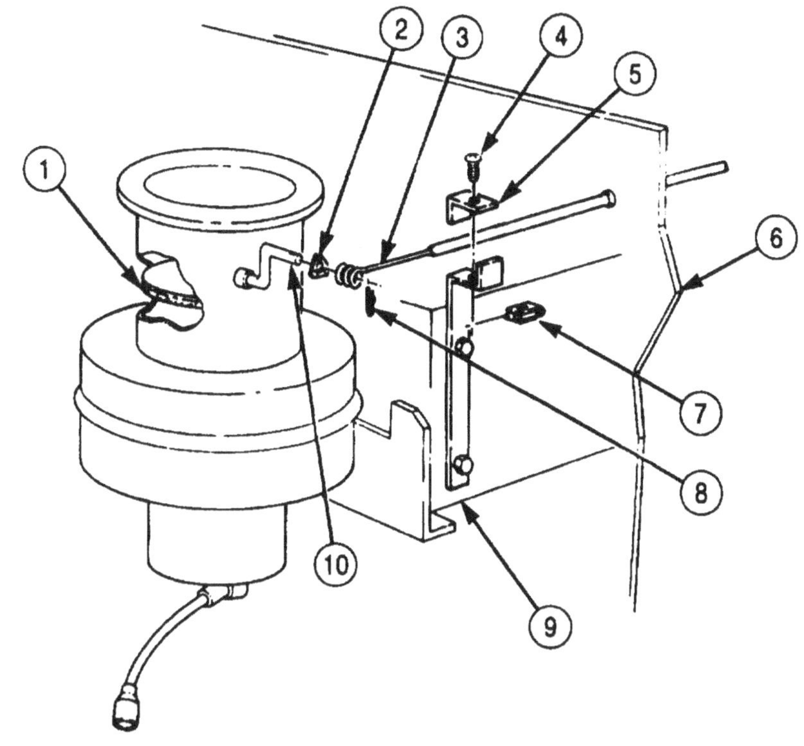

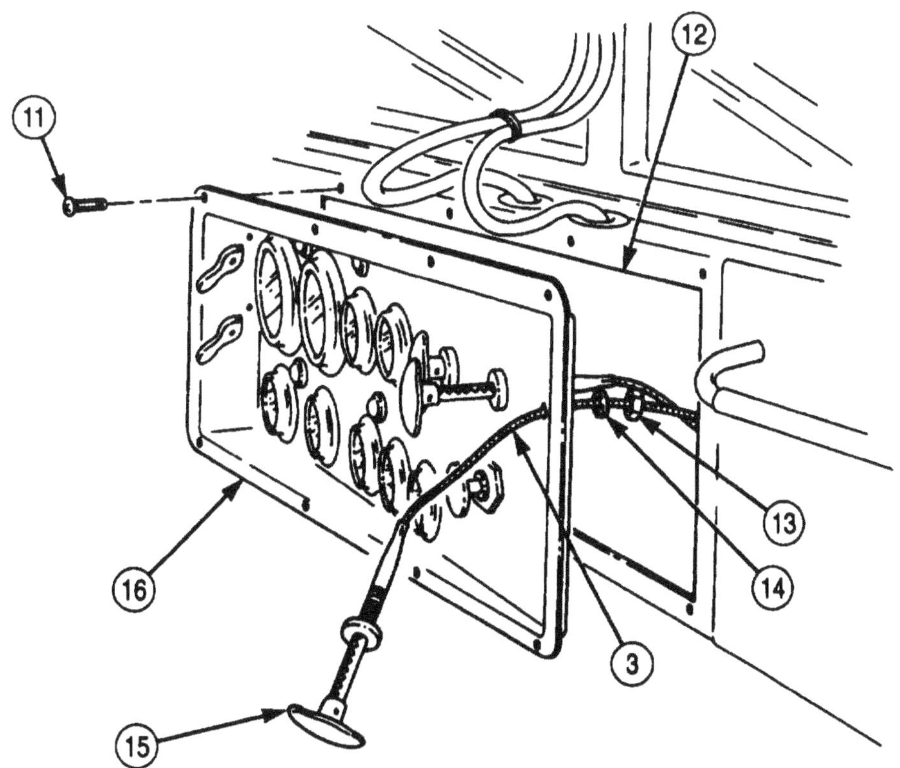

FOLLOW-ON TASK: Check fresh air vent for proper operation (TM 9-2320-272-10).

3-294. DEFROST AND HEAT CONTROLS REPLACEMENT

THIS TASK COVERS:

a. Removal **b. Installation**

INITIAL SETUP:

APPLICABLE MODELS
All

TOOLS
General mechanic's tool kit (Appendix E, Item 1)

MATERIALS/PARTS
Two cotter pins (Appendix D, Item 53)
Two spring nuts (Appendix D, Item 672)

REFERENCES (TM)
TM 9-2320-272-10
TM 9-2320-272-24P

EQUIPMENT CONDITION
Parking brake set (TM 9-2320-272-10).

NOTE
Tag all cables for installation.

a. Removal

1. Remove screw (15), clamp (3), defrost cable (1), and retaining clip (4) from diverter bracket (6).
2. Remove cotter pin (16), defrost cable (1), and spring nut (2) from control rod (5). Discard cotter pin (16) and spring nut (2).
3. Remove screw (14), clamp (10), heat control cable (13), and retaining clip (8) from diverter bracket (7).
4. Remove cotter pin (12), heat control cable (13), and spring nut (11) from control rod (9). Discard cotter pin (12) and spring nut (11).
5. Remove eight screws (23) and pull instrument cluster (22) away from instrument panel (17).
6. Remove two nuts (19), washers (18), and defrost cable (1) from instrument cluster (22).
7. Remove heat control cable handle (20), heat control cable (13), defrost cable (1), and defrost control handle (21) from instrument cluster (22).

b. Installation

1. Install heat control cable handle (20), heat control cable (13), defrost control handle (21), and defrost cable (1) on instrument cluster (22) with two washers (18) and nuts (19).
2. Route cables (1) and (13) to diverter brackets (6) and (7).
3. Install instrument cluster (22) on instrument panel (17) with eight screws (23).
4. Push heat control handle (20) and defrost handle (21) in all the way.
5. Install heat control cable (13) on control rod (9) with new spring nut (11) and new cotter pin (12).
6. Install heat control cable (13) on diverter bracket (7) with retaining clip (8), clamp (10), and screw (14).
7. Install defrost control cable (1) on control rod (5) with new spring nut (2) and new cotter pin (16).
8. Install defrost control cable (1) on diverter bracket (6) with retaining clip (4), clamp (3), and screw (15).

3-294. DEFROST AND HEAT CONTROLS REPLACEMENT (Contd)

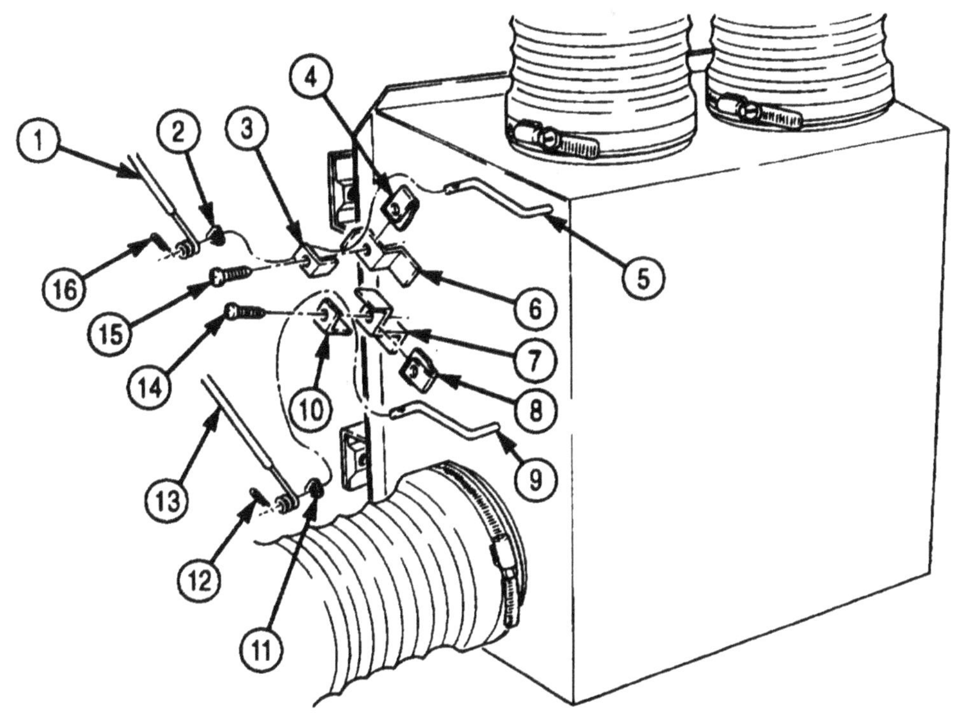

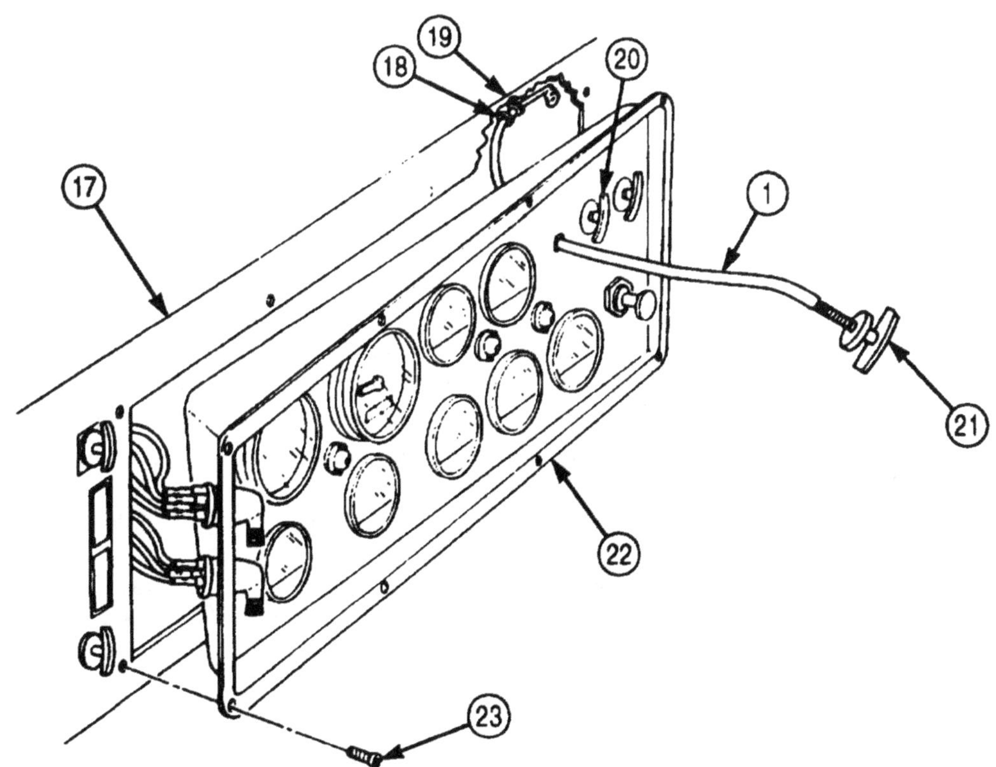

FOLLOW-ON TASK: Check defrost and heat controls for proper operation (TM 9-2320-272-10).

3-295. DIVERTER ASSEMBLY REPLACEMENT

THIS TASK COVERS:

a. Removal b. Installation

INITIAL SETUP:

APPLICABLE MODELS
All

TOOLS
General mechanic's tool kit (Appendix E, Item 1)

MATERIALS/PARTS
Two cotter pins (Appendix D, Item 53)
Two spring nuts (Appendix D, Item 672)

REFERENCES (TM)
TM 9-2320-272-10
TM 9-2320-272-24P

EQUIPMENT CONDITION
- Parking brake set (TM 9-2320-272-10).
- Defrost and heat control handles closed (TM 9-2320-272-10).
- Right fender splash shield removed (para. 3-301).

a. Removal

1. Remove two screw-assembled washers (3) and bracket (2) from right side of engine compartment cowl (1).
2. Remove three clamps (10) and ducting hoses (7) from diverter (9) and three adapter flanges (8).
3. Remove screw (23), clamp (22), and retaining clip (12) from defroster control cable bracket (13).
4. Remove cotter pin (24), defroster control cable (25), and spring nut (26) from control rod (11). Discard cotter pin (24) and spring nut (26).
5. Remove screw (21), clamp (20), and retaining clip (15) from heater control cable bracket (14).
6. Remove cotter pin (18), heater control cable (19), and spring nut (17) from control rod (16). Discard cotter pin (18) and spring nut (17).

CAUTION

Hold personnel hot water heater and brackets in place during removal of diverter. Both mount to firewall with same screws.

7. Remove four screw-assembled washers (27), cage nuts (6), personnel hot water heater bracket (4), and diverter (9) from firewall (5).

b. Installation

1. Install diverter (9) on hot water heater bracket (4) and firewall (5) with four screw-assembled washers (27) and cage nuts (6).
2. Install bracket (2) to right side of engine compartment cowl (1) with two screw-assembled washers (3).
3. Install three ducting hoses (7) on diverter (9) and adapter flanges (8) with three clamps (10).
4. Install heater control cable (19) on control rod (16) with new spring nut (17) and new cotter pin (18).
6. Install heater control cable (19) on heater control cable bracket (14) with clamp (20), screw (21), and retaining clip (15).
7. Install defroster control cable (25) on control rod (11) with new spring nut (26) and new cotter pin (24).
8. Install defroster control cable (25) on defroster control cable bracket (13) with clamp (22), screw (23), and retaining clip (12).

3-295. DIVERTER ASSEMBLY REPLACEMENT (Contd)

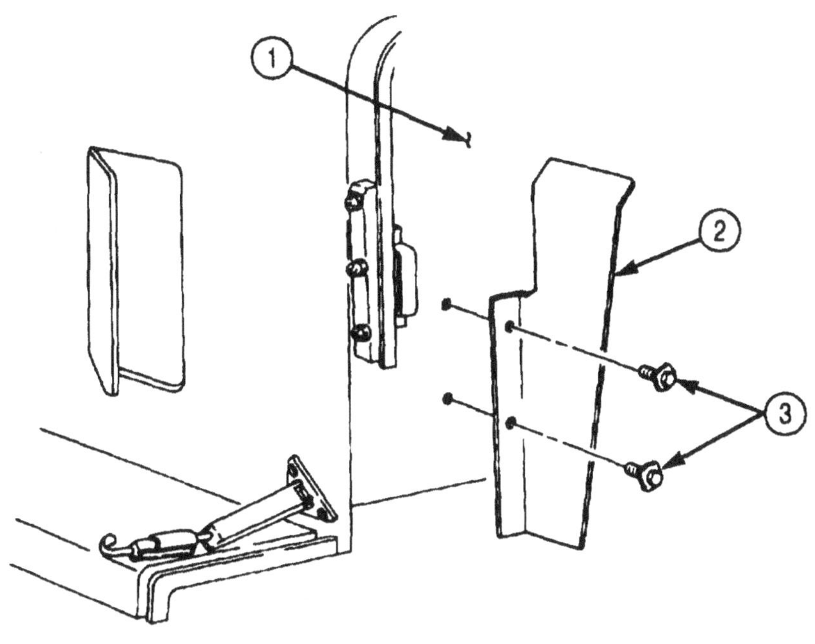

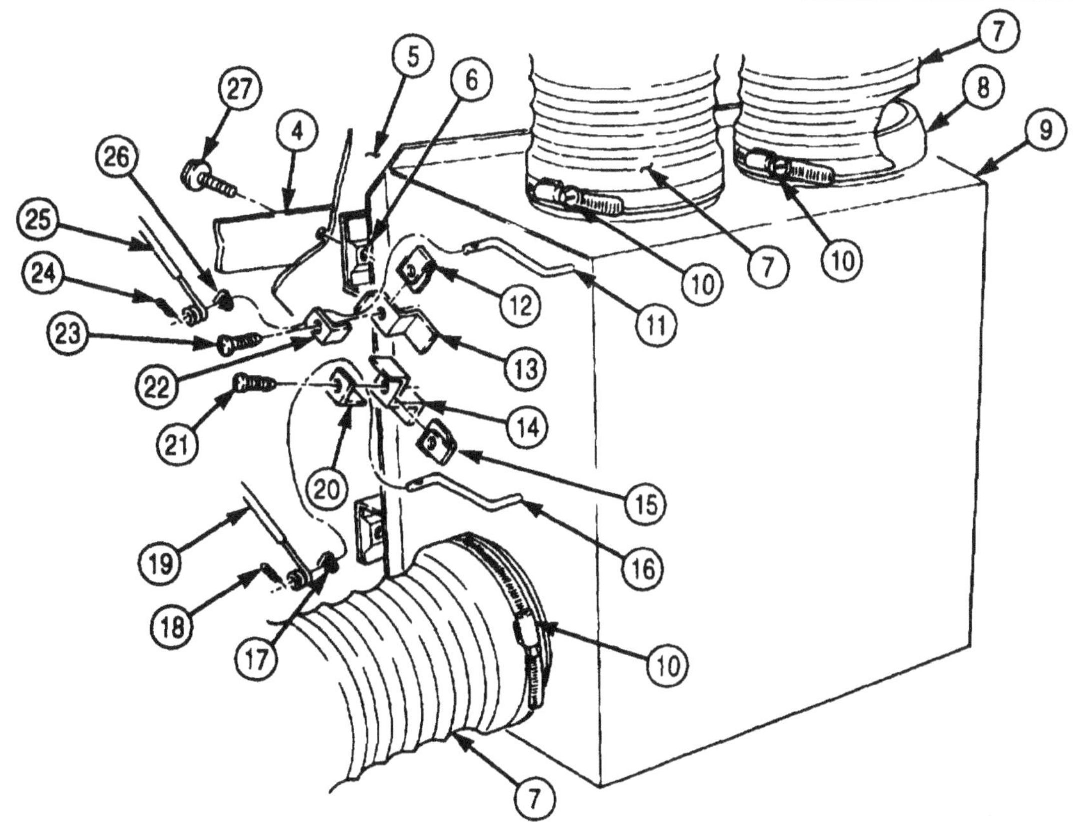

FOLLOW-ON TASKS:

- Install right fender splash shield (para. 3-301).
- Check heater for proper operation (TM 9-2320-272-10).

3-296. FRESH AIR INLET DUCTING REPLACEMENT

THIS TASK COVERS:

a. Removal **b. Installation**

INITIAL SETUP:

APPLICABLE MODELS
All

TOOLS
General mechanic's tool kit (Appendix E, Item 1)

REFERENCES (TM)
TM 9-2320-272-10
TM 9-2320-272-24P

EQUIPMENT CONDITION
- Parking brake set (TM 9-2320-272-10).
- Defrost and heat control levers closed (TM 9-2320-272-10).
- Hood raised and secured (TM 9-2320-272-10).

a. Removal

1. Loosen clamp (1) and remove fresh air ducting (5) from adapter flange (2).
2. Remove two nuts (6), screws (9), and clamps (3) from fresh air ducting (5) and hood (4).
3. Loosen clamp (8) and remove fresh air ducting (5) from flange (7).

b. Installation

1. Install fresh air ducting (5) on flange (7) with clamp (8).
2. Install fresh air ducting (5) on hood (4) with two clamps (3), screws (9), and nuts (6).
3. Install fresh air ducting (5) on adapter flange (2) with clamp (1).

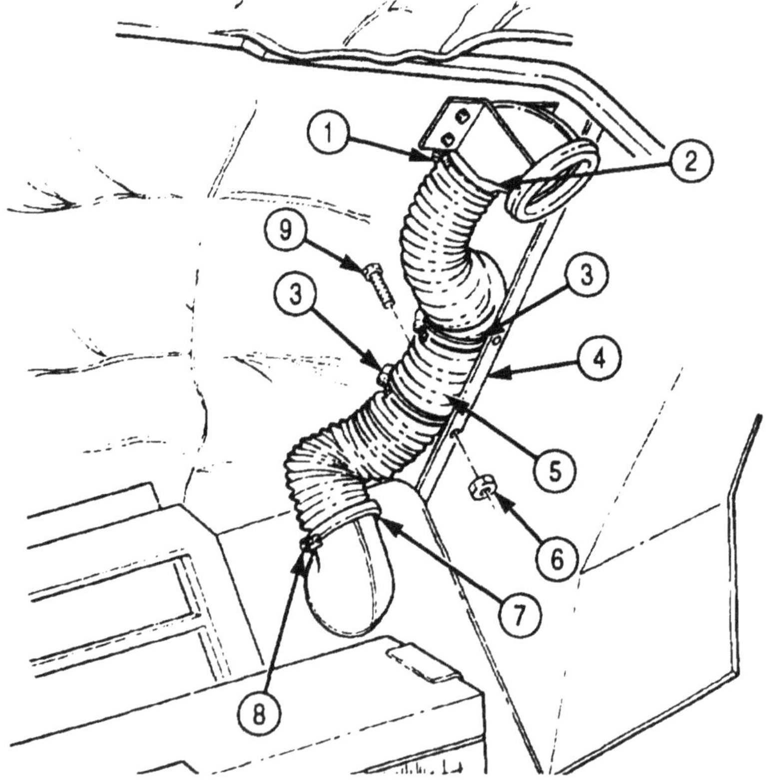

3-297. CAB HEAT AND DEFROST AIR DUCTING REPLACEMENT

THIS TASK COVERS:

a. Removal **b. Installation**

INITIAL SETUP:

APPLICABLE MODELS
All

TOOLS
General mechanic's tool kit (Appendix E, Item 1)

REFERENCES (TM)
TM 9-2320-272-10
TM 9-2320-272-24P

EQUIPMENT CONDITION
Parking brake set (TM 9-2320-272-10).

a. Removal

1. Remove eight screws (3) and instrument cluster (2) from instrument panel (1).
2. Loosen six clamps (10) and remove cab heat ducting hose (7) and defrost ducting hoses (4) and (6) from diverter (5), exhaust flange (8), and two defrost air flanges (9).

b. Installation

1. Install cab heat ducting hose (7) on diverter (5) and exhaust flange (8) with clamps (10).
2. Install defrost ducting hoses (4) and (6) on defrost air flanges (9) with clamps (10).
3. Install instrument cluster (2) on instrument panel (1) with eight screws (3).

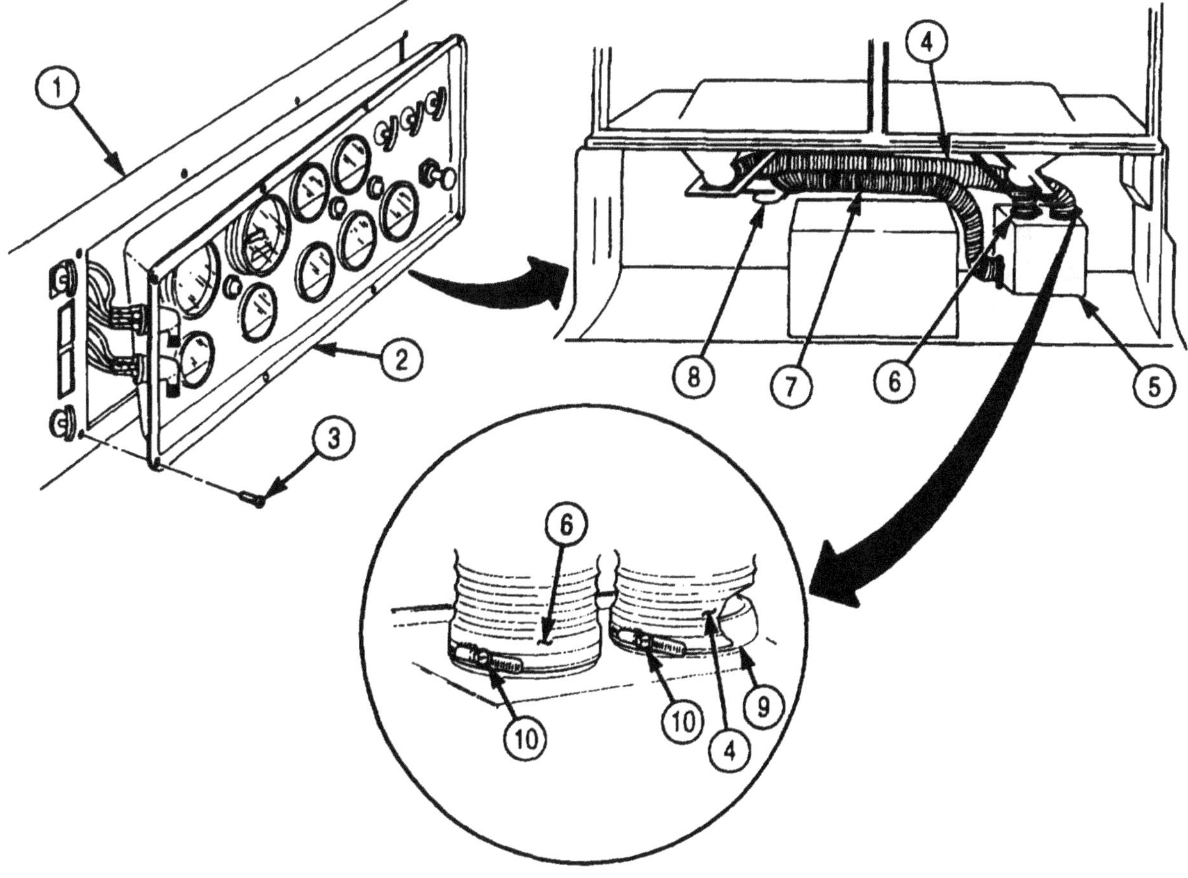

3-298. VENT DOOR WEATHERSEAL REPLACEMENT

THIS TASK COVERS:

a. Removal **b. Installation**

INITIAL SETUP:

APPLICABLE MODELS
All

TOOLS
General mechanic's tool kit (Appendix E, Item 1)

MATERIALS/PARTS
Vent door weatherseal (Appendix D, Item 710)
Adhesive (Appendix C, Item 7)

REFERENCES (TM)
TM 9-2320-272-10
TM 9-2320-272-24P

EQUIPMENT CONDITION
Parking brake set (TM 9-2320-272-10).

a. Removal

1. Open vent door (2).
2. Remove vent door weatherseal (3) from vent door (2). Discard weatherseal (3).

b. Installation

NOTE

Surface of cowl vent door opening must be clean, dry, and free of oil and grease before seal is installed.

1. Apply thin coat of adhesive to mating surfaces of vent door (2) and new weatherseal (3) and allow to dry until tacky.
2. Position weatherseal (3) into channel (1) of vent door (2) and press firmly into place.

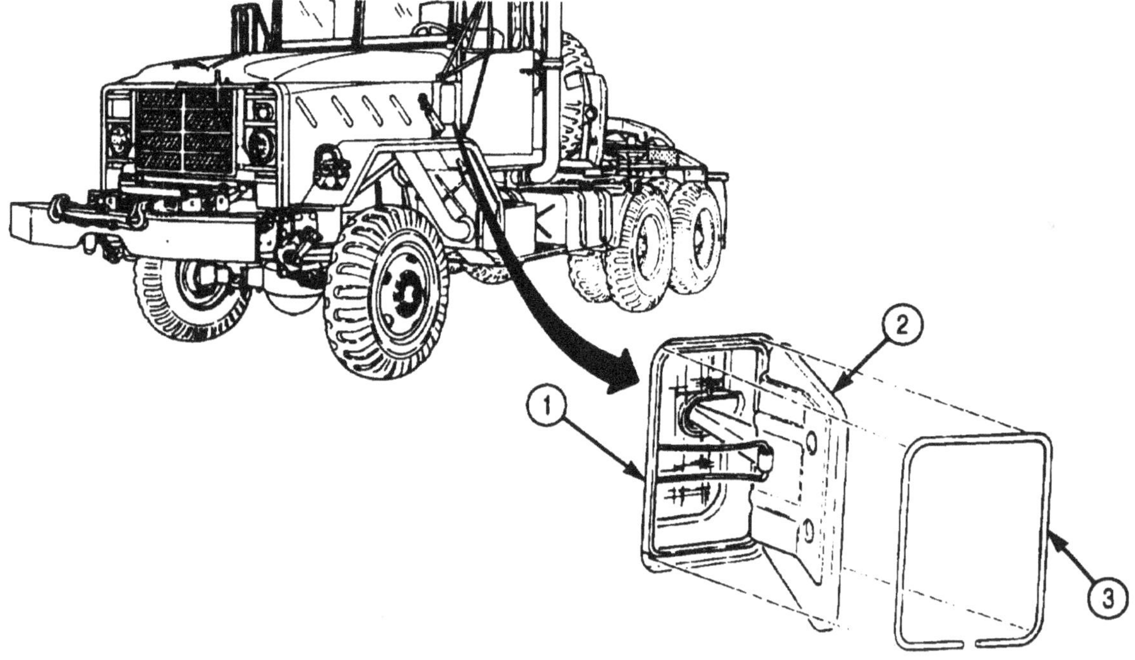

3-299. CAB COWL VENT SCREEN AND DOOR REPLACEMENT

THIS TASK COVERS:

a. Removal
b. Installation

INITIAL SETUP:

APPLICABLE MODELS
All

TOOLS
General mechanic's tool kit (Appendix E, Item 1)

MATERIALS/PARTS
Cotter pin (Appendix D, Item 50)

REFERENCES (TM)
TM 9-2320-272-10
TM 9-2320-272-24P

EQUIPMENT CONDITION
Parking brake set (TM 9-2320-272-10).

a. Removal

1. Remove nine screws (1) and vent screen (2) from cab cowl (3).
2. Spread vent door spring (9) apart and disconnect vent door bracket (7).
3. Remove cotter pin (8) from door pivot pin (4). Discard cotter pin (8).
4. Remove door pivot pin (4) and vent door (6) from hinge (5) and cab cowl (3).

b. Installation

1. Install vent door (6) on cab cowl (3) and hinge (5) with door pivot pin (4) and new cotter pin (8).
2. Spread vent door spring (9) apart and clamp to door hinge bracket (7).
3. Install vent screen (2) on cab cowl (3) with nine screws (1).

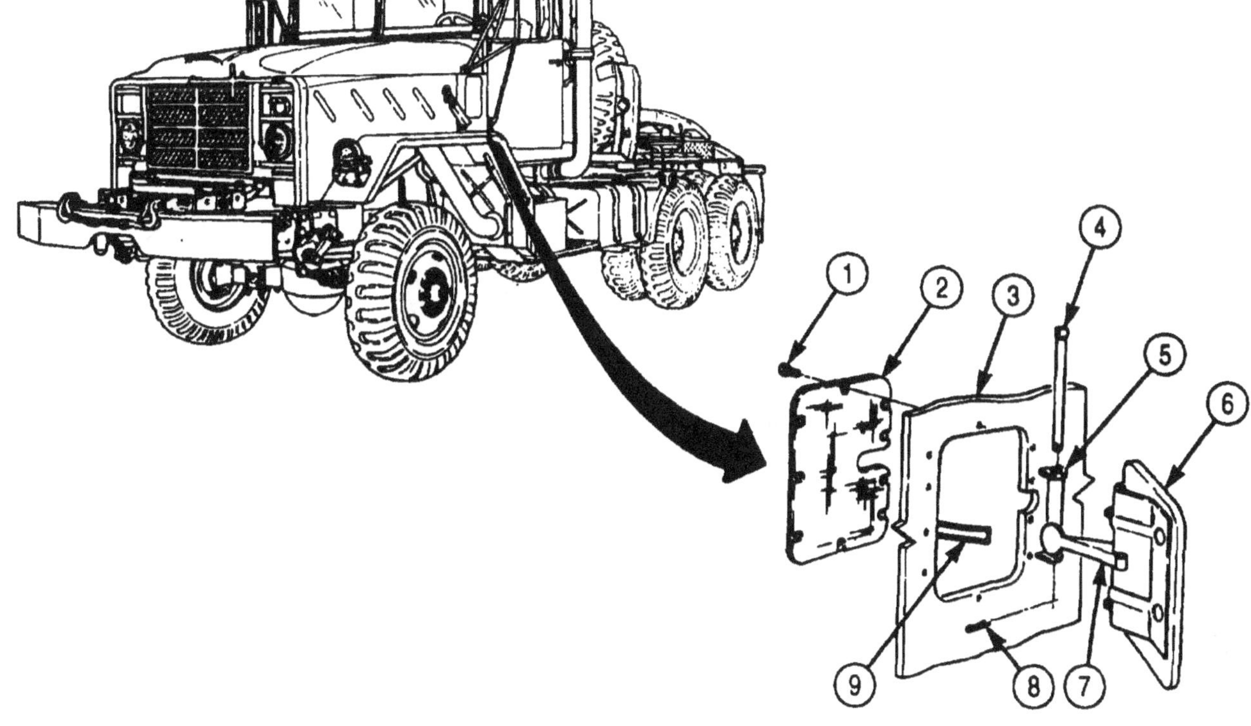

3-300. FRONT FENDER EXTENSION REPLACEMENT

THIS TASK COVERS:

a. Removal **b. Installation**

INITIAL SETUP:

APPLICABLE MODELS
All

TOOLS
General mechanic's tool kit (Appendix E, Item 1)

MATERIALS/PARTS
Two locknuts (Appendix D, Item 291)

REFERENCES (TM)
TM 9-2320-272-10
TM 9-2320-272-24P

EQUIPMENT CONDITION
- Parking brake set (TM 9-2320-272-10).
- Hood raised and secured (TM 9-2320-272-10).

NOTE

Front extension replacement is the same for both left and front fender extensions. This procedure covers the right fender extension.

a. Removal

1. Remove five screw-assembled washers (5) from fender extension (6) and splash shield (4).
2. Remove two locknuts (3), screws (1), and fender extension (6) from support bracket (2). Discard locknuts (3).

b. Installation

1. Install fender extension (6) on support bracket (2) with two screws (1) and new locknuts (3).
2. Install fender extension (6) on splash shield (4) with five screw-assembled washers (5).

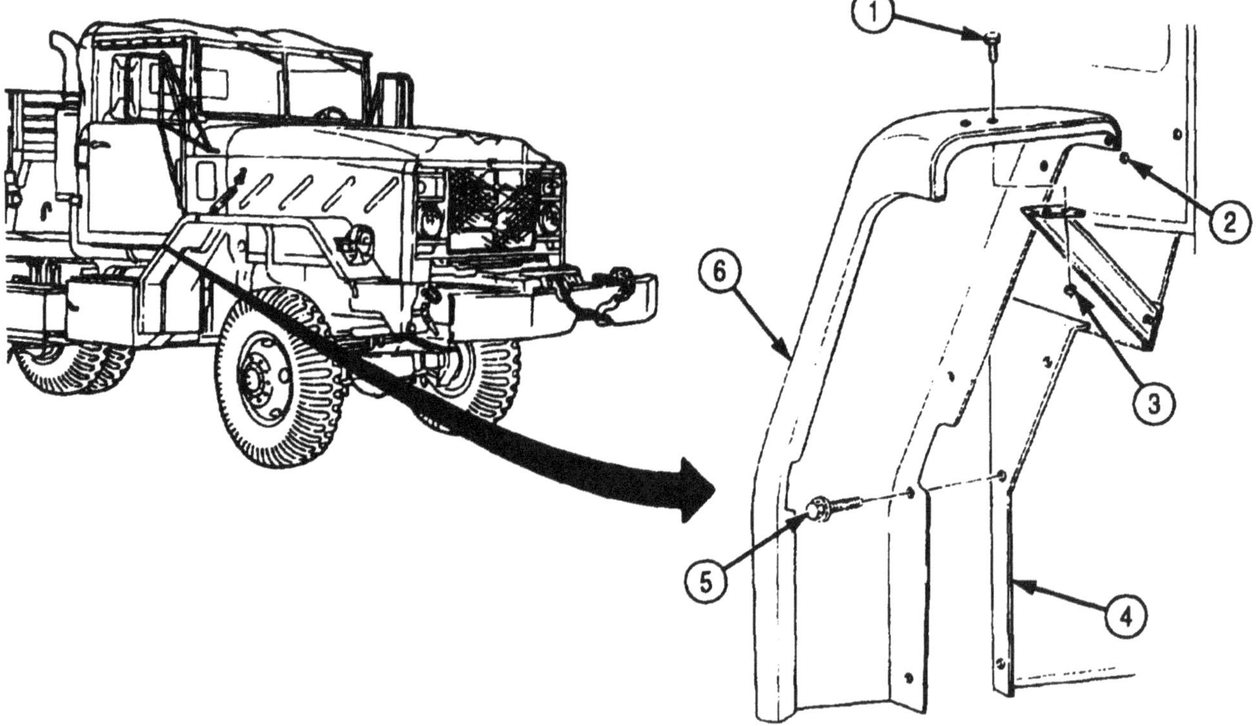

3-301. FENDER SPLASH SHIELD REPLACEMENT

THIS TASK COVERS:

a. Removal **b. Installation**

INITIAL SETUP:

APPLICABLE MODELS
All

TOOLS
General mechanic's tool kit (Appendix E, Item 1)

MATERIALS/PARTS
Four locknuts (Appendix D, Item 291)
Three lockwashers (Appendix D. Item 358)

REFERENCES (TM)
TM 9-2320-272-10
TM 9-2320-272-24P

EQUIPMENT CONDITION
- Parking brake set (TM 9-2320-272-10).
- Fender extension removed (para. 3-300).
- Air cleaner intake pipe hump hose removed (para. 3-13).

NOTE

Fender splash shield replacement is the same for both left and right side. This procedure covers the right fender splash shield.

a. Removal

1. Remove locknut (8), screw (6), and washer (7) from splash shield (5) and support brace (11). Discard locknut (8).
2. Remove three screws (9) and lockwashers (10) from splash shield (5) and cab extension (12). Discard lockwashers (11).
3. Remove three locknuts (4), washers (3), and splash shield (5) from screws (2) and cab floor (1). Discard locknuts (4).

b. Installation

1. Install splash shield (5) on cab floor (1) and three screws (2) with three washers (3) and new locknuts (4).
2. Install splash shield (5) on cab extension (12) with three new lockwashers (10) and screws (9).
3. Install splash shield (5) on support brace (11) with washer (7), screw (6), and new locknut (8).

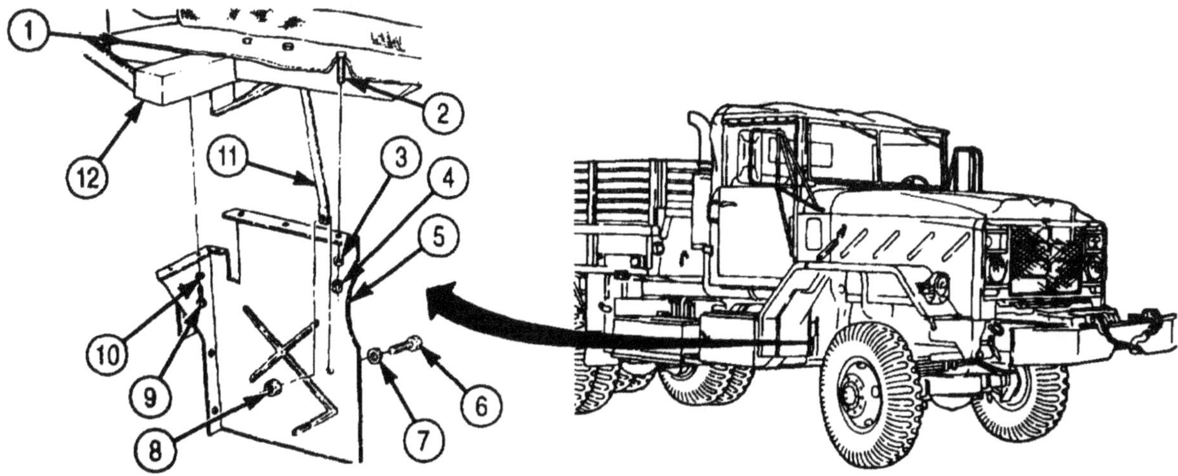

FOLLOW-ON TASKS:
- Install air cleaner intake pipe hump hose (para. 3-13).
- Install fender extension (para. 3-300).

3-302. TOOLBOX AND STEPS REPLACEMENT

THIS TASK COVERS:

a. Removal **b. Installation**

INITIAL SETUP:

APPLICABLE MODELS
All

TOOLS
General mechanic's tool kit (Appendix E, Item 1)

MATERIALS/PARTS
Fourteen locknuts (Appendix D, Item 313)
Two locknuts (Appendix D, Item 291)

REFERENCES (TM)
TM 9-2320-272-10
TM 9-2320-272-24P

EQUIPMENT CONDITION
Parking brake set (TM 9-2320-272-10).

a. Removal

1. Remove two locknuts (2), screws (13), and drainvalve bracket (1) from toolbox with step (3). Discard locknuts (2).
2. Remove two locknuts (10), screws (12), and toolbox support bracket (11) from toolbox with step (3). Discard locknuts (10).
3. Release latch (4) and open toolbox door (9).
4. Remove four locknuts (7), screws (5), washers (6), toolbox (18), and step (3) from hangers (8). Discard locknuts (7).
5. Remove two screws (17) and washers (16) from step (3) and toolbox (18).
6. Remove eight locknuts (14), screws (15), and step (3) from toolbox (18). Discard locknuts (14).

b. Installation

1. Install step (3) on toolbox (18) with eight screws (15) and new locknuts (14).
2. Install step (3) on toolbox (18) with two screws (17) and washers (16).
3. Install toolbox (18) with step (3) on top of hangers (8) with four washers (6), screws (5), and new locknuts (7).
4. Install toolbox support bracket (11) on toolbox (18) with step (3) with two screws (12) and new locknuts (10).
5. Install drainvalve bracket (1) on toolbox (18) with step (3) with two screws (13) and new locknuts (2).

3-302. TOOLBOX AND STEPS REPLACEMENT (Contd)

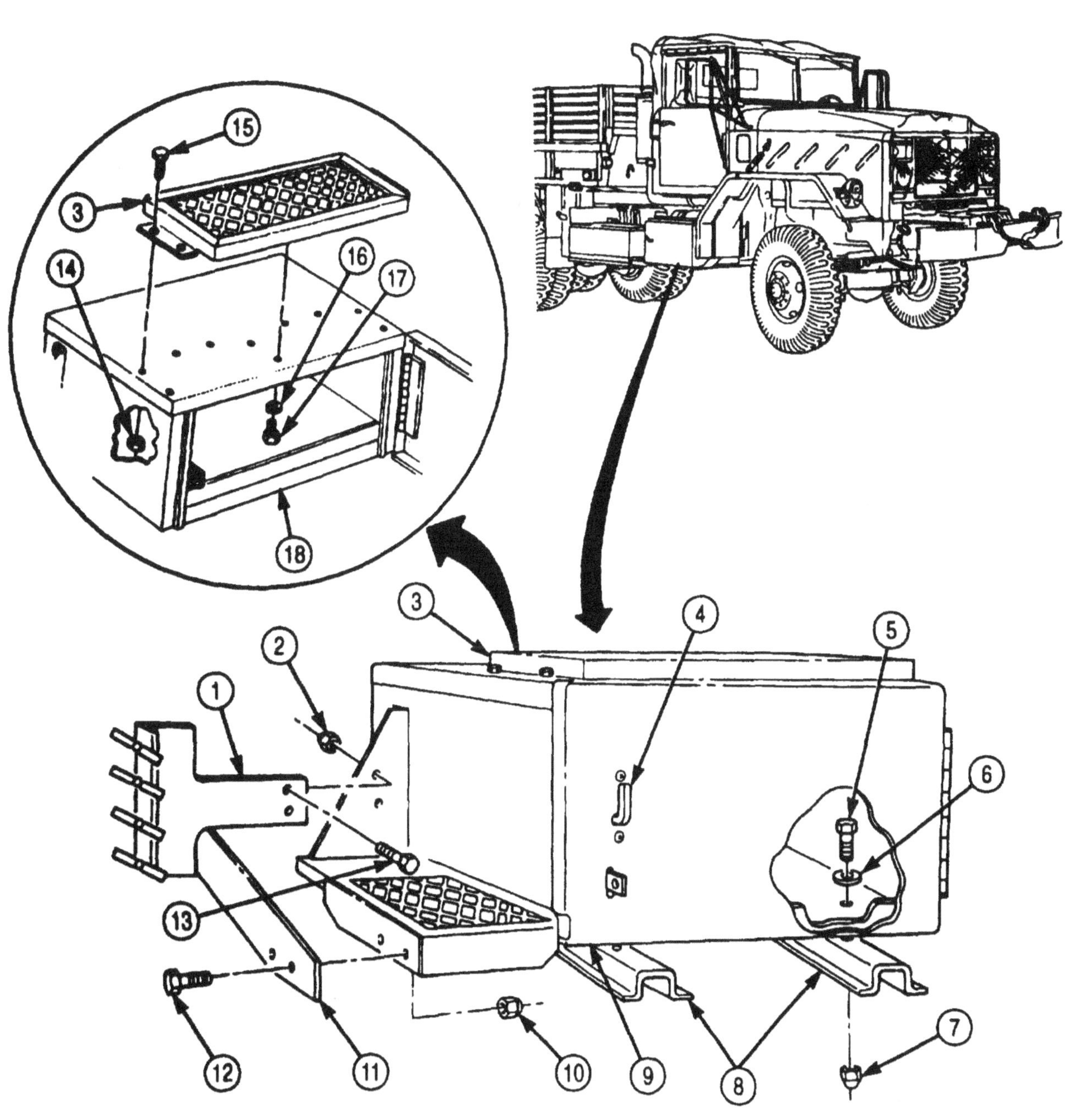

3-303. CAB TURNBUTTONS AND LASHING HOOKS REPLACEMENT

THIS TASK COVERS:

a. Removal **b. Installation**

INITIAL SETUP:

APPLICABLE MODELS
All

TOOLS
General mechanic's tool kit (Appendix E, Item 1)

MATERIALS/PARTS
Two locknuts (Appendix D, Item 277)

REFERENCES (TM)
TM 9-2320-272-10
TM 9-2320-272-24P

EQUIPMENT CONDITION
Parking brake set (TM 9-2320-272-10).

a. Removal

1. Unsnap snapbutton (1) on cab top (3).
2. Remove turnbutton (2) from cab side rail (5).

NOTE

Center lashing hooks have only two screws.

3. Remove two locknuts (8), screws (6), and forward lashing hook (4) from rear of cab (7). Discard locknuts (8).

b. Installation

NOTE

Center lashing hooks have only two screws.

1. Install forward lashing hook (4) on rear of cab (7) with two screws (6) and new locknuts (8).
2. Install turnbutton (2) on cab side rail (5).
3. Install snapbutton (1) on cab top (3).

3-303. CAB TURNBUTTONS AND LASHING HOOKS REPLACEMENT (Contd)

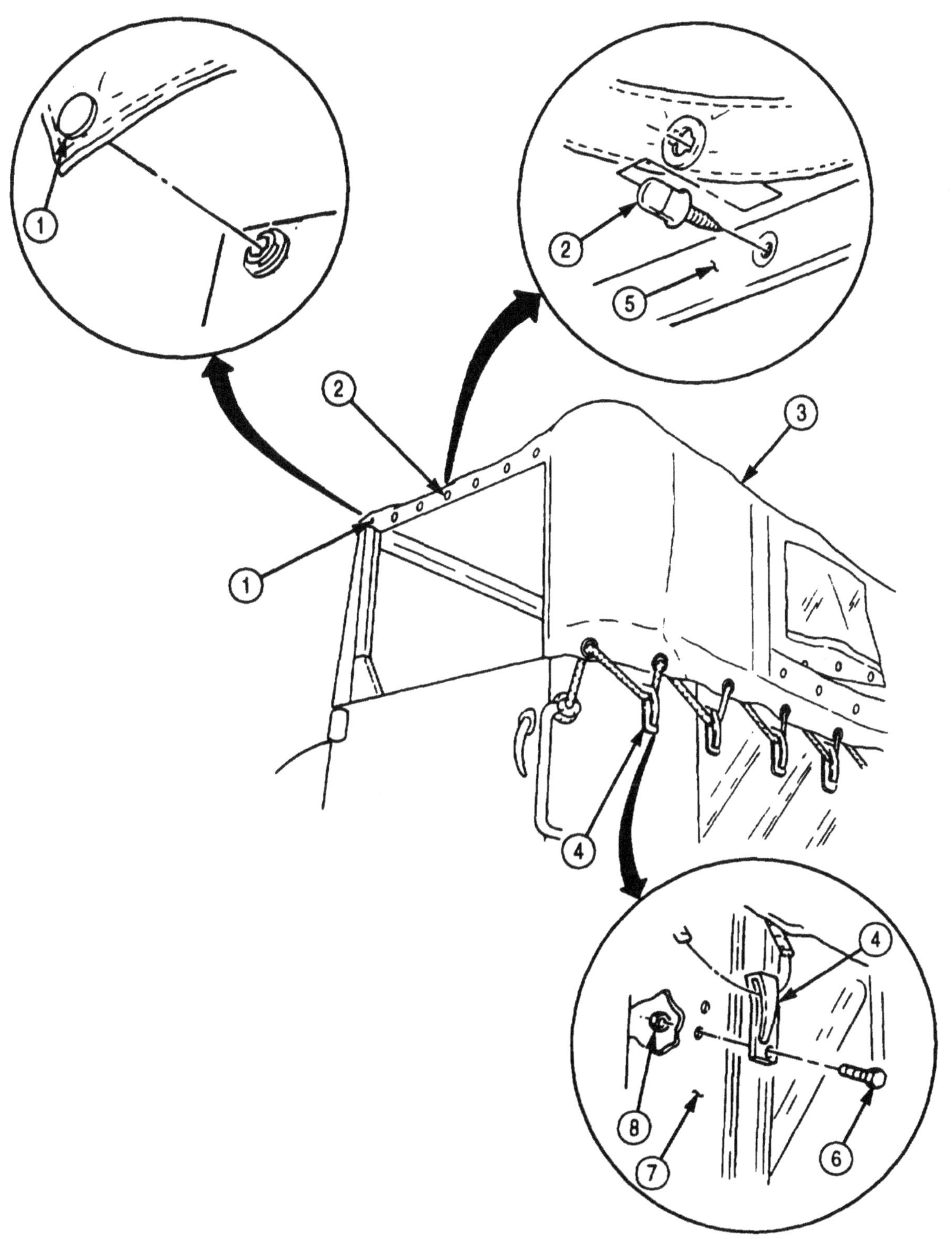

3-304. FRONT CAB MOUNT REPLACEMENT

THIS TASK COVERS:

a. Removal

b. Installation

MATERIAL SETUP:

APPLICABLE MODELS

All

TOOLS

General mechanic's tool kit (Appendix E, Item 1)

MATERIALS/PARTS

Locknut (Appendix D, Item 294)
Two insulators (Appendix D, Item 261)

REFERENCES (TM)

TM 9-2320-272-10
TM 9-2320-272-24P

EQUIPMENT CONDITION

Parking brake set (TM 9-2320-272-10).

NOTE

This procedure applies to both right and left front cab mounts.

a. Removal

1. Turn jacking screw (9) until cab weight is supported.
2. Remove locknut (7), washer (6), and insulator lower half (5) from cab bracket (1). Discard locknut (7) and insulator (5).
3. Remove screw (2), washer (3), and insulator upper half (4). Discard insulator (4).

b. Installation

1. Place new insulator upper half (4) between cab bracket (1) and frame bracket (8).
2. Install washer (3) and screw (2) on insulator upper half (4), cab bracket (1), and frame bracket (8).
3. Install new insulator lower half (5) on screw (2) with washer (6) and new locknut (7).
4. Turn out jacking screw (9) until it is secured all the way down.

3-304. FRONT CAB MOUNT REPLACEMENT (Contd)

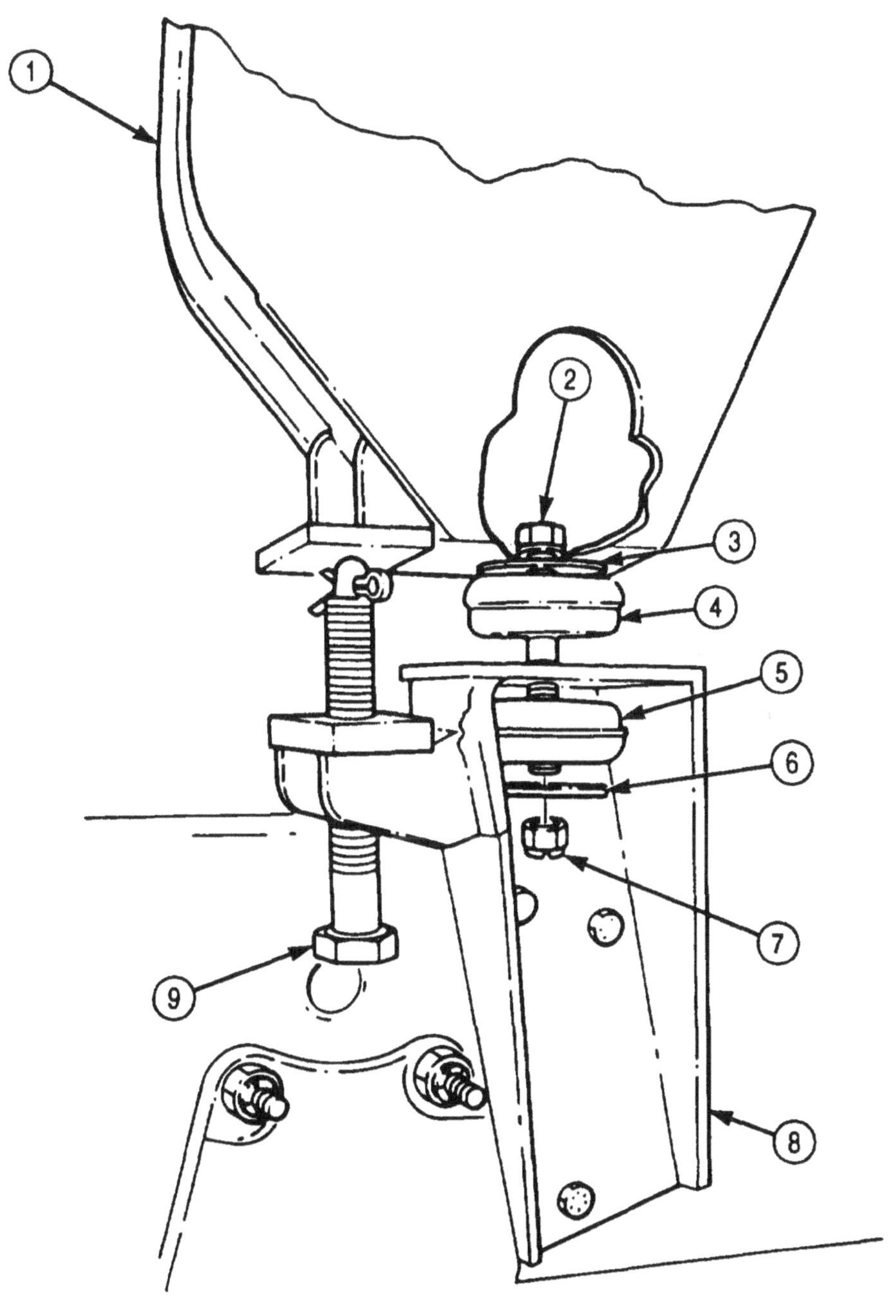

3-305. REAR CAB MOUNT REPLACEMENT

THIS TASK COVERS

a. Removal **b. Installation**

INITIAL SETUP:

APPLICABLE MODELS
All

TOOLS
General mechanic's tool kit (Appendix E, Item 1)

MATERIALS/PARTS
Two locknuts (Appendix D, Item 294)
Two insulators (Appendix D, Item 261)

REFERENCES (TM)
TM 9-2320-272-10
TM 9-2320-272-24P

EQUIPMENT CONDITION
- Parking brake set (TM 9-2320-272-10).
- Dump body raised (M929/A1/A2, M930/A1/A2) (TM 9-2320-272-10).
- Cab top removed (TM 9-2320-272-10).
- Spare tire removed (TM 9-2320-272-10).

a. Removal

1. Pull back insulation and remove two locknuts (7) and washers (6) from cab (9) and frame (8). Discard locknuts (7).

CAUTION

Raise rear of cab enough to remove insulators. Damage to cab will result if raised too high.

2. Using overhead lifting device, raise cab (9) until clear of insulators (5).
3. Remove two screws (4), washers (3), and springs (2) from inside cab (9).
4. Remove two insulators (5). Discard insulators (5).

b. Installation

1. Position two new insulators (5) between cab (9) and frame (8).
2. Install two washers (3), springs (2), and screws (4) through insulators (5).
3. Lower cab (9) to frame (8) and remove lifting device
4. Install two washers (6) and new locknuts (7) on screws (4) and tighten until washers (3) are even with cab (9) floor.

3-305. REAR CAB MOUNT REPLACEMENT (Contd)

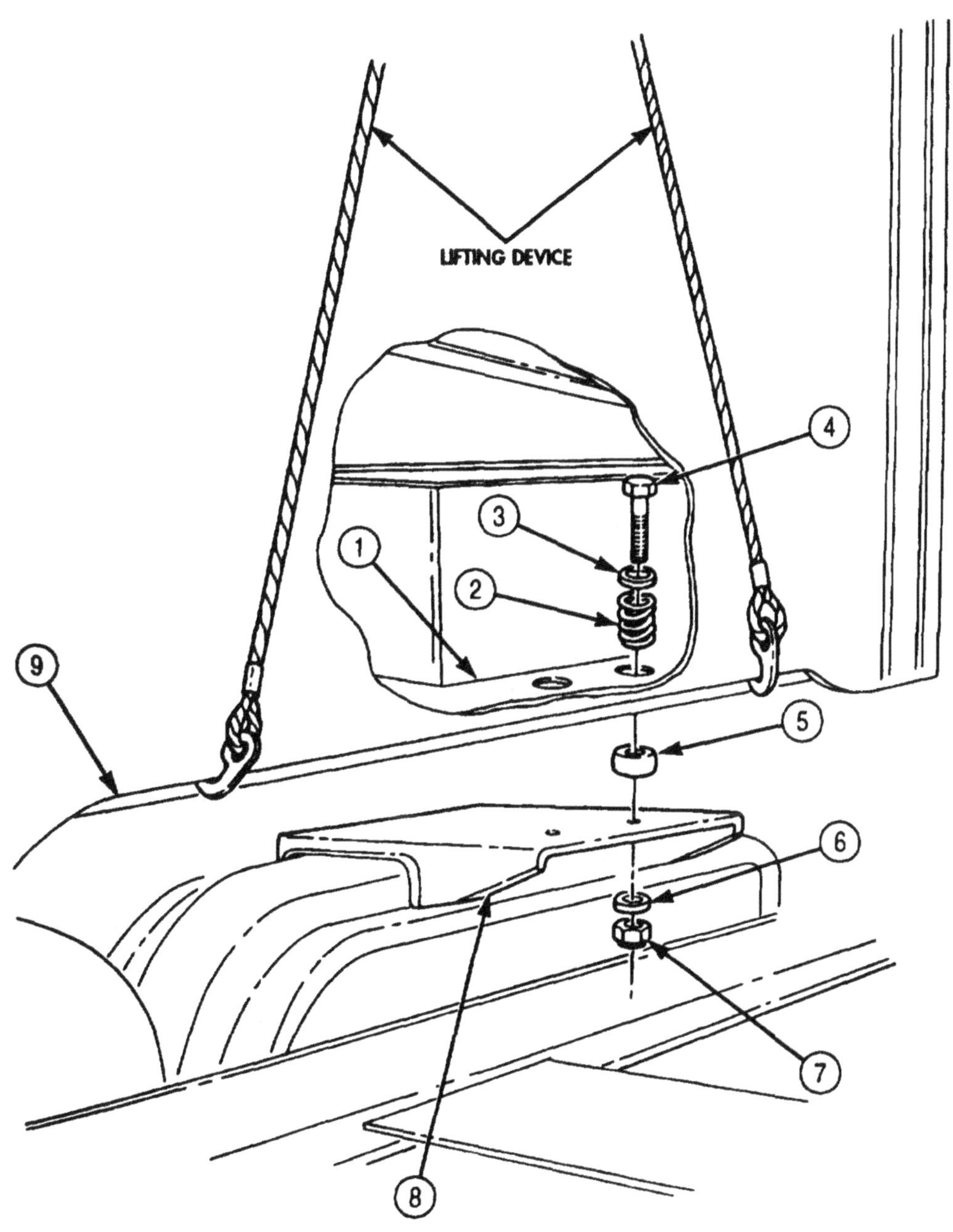

FOLLOW-ON TASKS:
- Install spare tire (TM 9-2320-272-10).
- Lower dump body (M929/A1/A2, M930/A1/A2) (TM 9-2320-272-10).
- Install cab top (TM 9-2320-272-10).

3-306. CAB INSULATION REPLACEMENT

THIS TASK COVERS:

a. Removal **b. Installation**

INITIAL SETUP:

APPLICABLE MODELS
All

TOOLS
General mechanic's tool kit (Appendix E, Item 1)

MATERIALS/PARTS
Adhesive (Appendix C, Item 6)

REFERENCES (TM)
TM 9-2320-272-10
TM 9-2320-272-24P

EQUIPMENT CONDITION
- Parking brake set (TM 9-2320-272-10).
- Driver's seat removed (left rear upper insulation only) (para. 3-283).

NOTE

- All insulation is removed the same way except where noted. This procedure covers replacement of left upper and engine access cover panels only.
- Clean all insulating material and adhesive from contact area.

a. Removal

1. Pull left rear upper panel insulation (2) away from cab (1) interior.
2. Remove four screws (6) and washers (5) from engine access cover (4).
3. Pull engine access cover insulation (3) away from engine access cover (4).

b. Installation

1. Apply adhesive to foam side of insulation (2) and (3) and install on cab (1) interior and engine access cover (4).
2. Install four washers (5) and screws (6) on engine access cover (4).

3-306. CAB INSULATION REPLACEMENT (Contd)

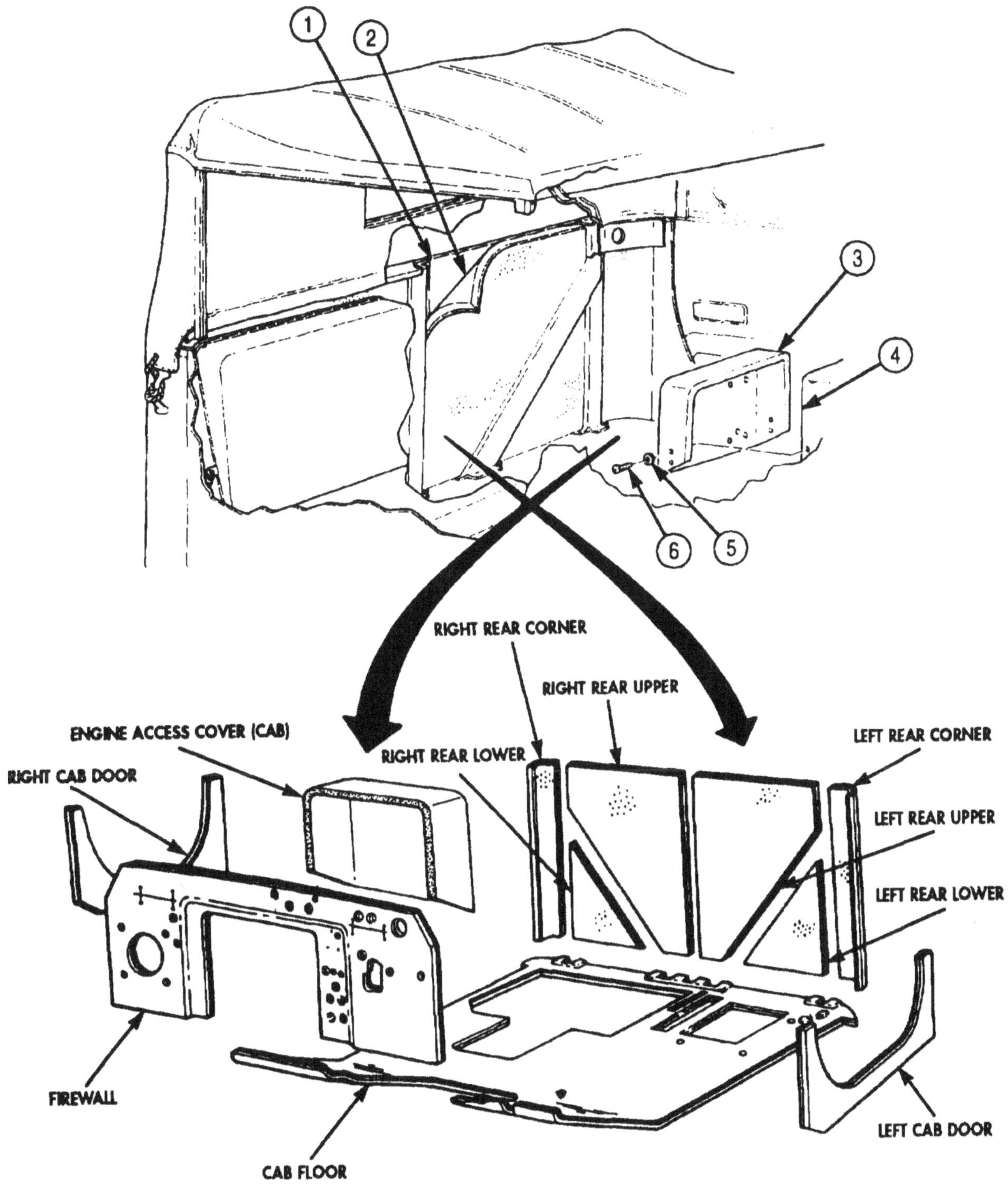

FOLLOW-ON TASK: Install driver's seat, if removed (para. 3-283).

3-307. OUTSIDE DOOR HANDLE REPLACEMENT

THIS TASK COVERS:

a. Removal **b. Installation**

INITIAL SETUP:

APPLICABLE MODELS
All

TOOLS
General mechanic's tool kit (Appendix E, Item 1)

REFERENCES (TM)
TM 9-2320-272-10
TM 9-2320-272-24P

EQUIPMENT CONDITION
Parking brake set (TM 9-2320-272-10).

a. Removal

1. Remove two screws (1) from door handle bracket (2) and cab door (3).
2. Remove door handle (5) and bracket (2) from cab door (3) by rotating door handle (5) 1/4-turn counterclockwise and pulling out at the same time.

b. Installation

1. Install door handle (5) by rotating 1/4-turn clockwise to insert shaft (4) in cab door (3).
2. Install door handle bracket (2) on cab door (3) with two screws (1).

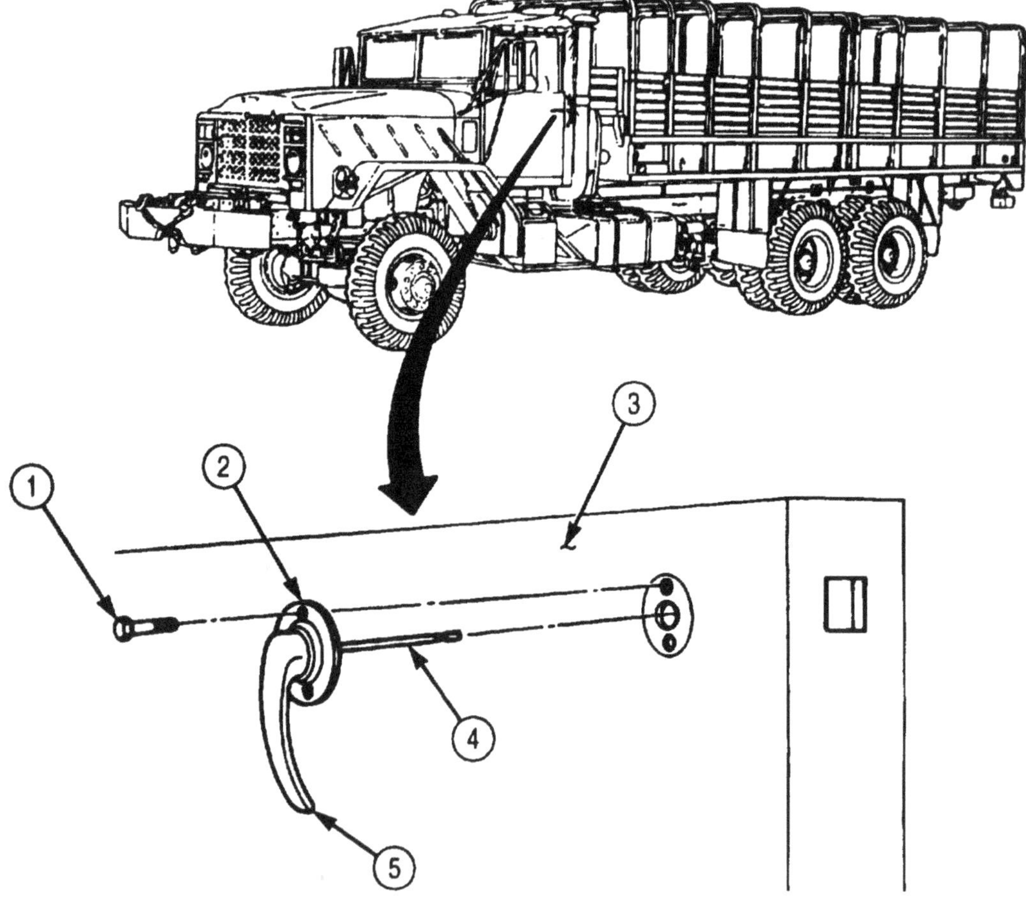

3-308. WINDOW REGULATOR AND INSIDE DOOR HANDLE REPLACEMENT

THIS TASK COVERS:

a. Removal **b. Installation**

INITIAL SETUP:

APPLICABLE MODELS
All

TOOLS
General mechanic's tool kit (Appendix E, Item 1)

REFERENCES (TM)
TM 9-2320-272-10
TM 9-2320-272-24P

EQUIPMENT CONDITION
Parking brake set (TM 9-2320-272-10).

NOTE

The steps for replacing window handles and door handles are the same; this procedure covers the window handles.

a. Removal

Remove screw (1), window regulator handle (2), and washer (3) from shaft (4).

b. Installation

Install washer (3) and window regulator handle (2) on shaft (4) with screw (1).

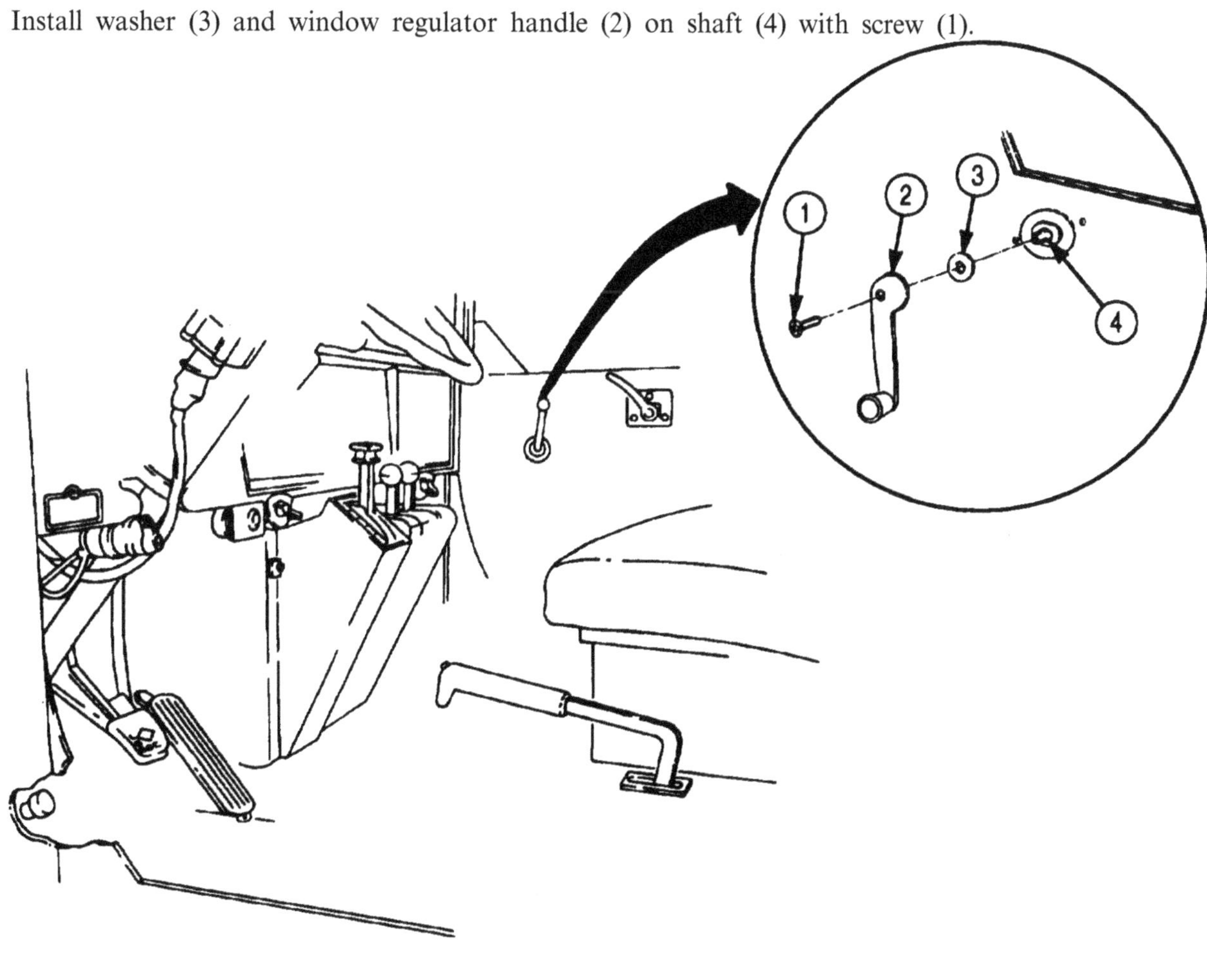

3-309. CAB DOOR DOVETAIL WEDGE REPLACEMENT

THIS TASK COVERS:

a. Removal **b. Installation**

INITIAL SETUP:

APPLICABLE MODELS
All

TOOLS
General mechanic's tool kit (Appendix E, Item 1)

MATERIALS/PARTS
Gasket (Appendix D, Item 235)
Two screw-assembled lockwashers (Appendix D, Item 573)

REFERENCES (TM)
TM 9-2320-272-10
TM 9-2320-272-24P

EQUIPMENT CONDITION
Parking brake set (TM 9-2320-272-10).

a. Removal

Remove two screw-assembled lockwashers (4), dovetail wedge (3), and gasket (2) from door (1). Discard gasket (2) and screw-assembled lockwashers (4).

b. Installation

Install new gasket (2) and dovetail wedge (3) on door (1) with two new screw-assembled lockwashers (4).

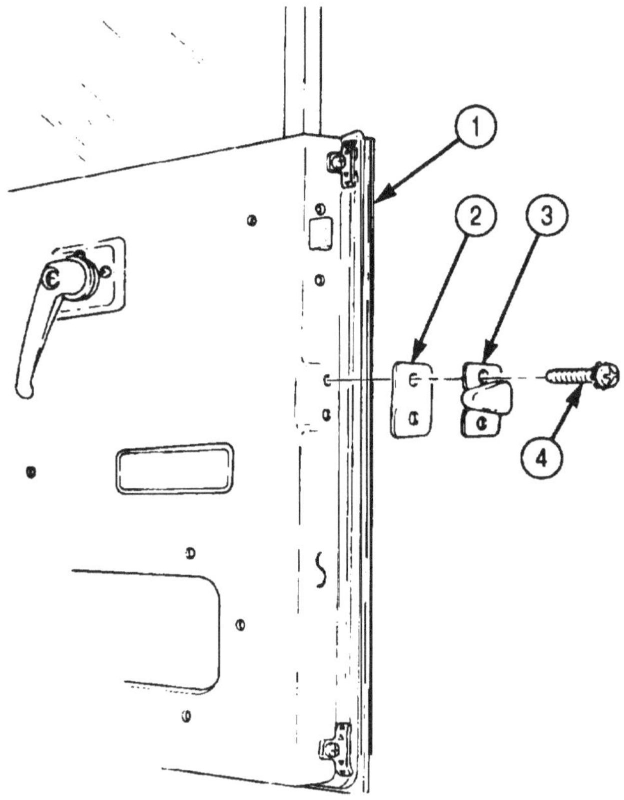

FOLLOW-ON TASK: Check cab door for proper operation (TM 9-2320-272-10).

3-310. CAB DOOR DOVETAIL REPLACEMENT

THIS TASK COVERS:

a. Removal **b. Installation**

INITIAL SETUP:

APPLICABLE MODELS
All

TOOLS
General mechanic's tool kit (Appendix E, Item 1)

REFERENCES (TM)
TM 9-2320-272-10
TM 9-2320-272-24P

EQUIPMENT CONDITION
Parking brake set (TM 9-2320-272-10).

a. Removal

Remove dovetail (3) and spring (2) from door post (1) by inserting screwdriver in notch (4) at end of dovetail (3) and push in and up.

b. Installation

Install spring (2) and dovetail (3) in door post (1) by inserting screwdriver in notch (4) of dovetail (3) and pushing down. Release pressure to snap dovetail (3) in place.

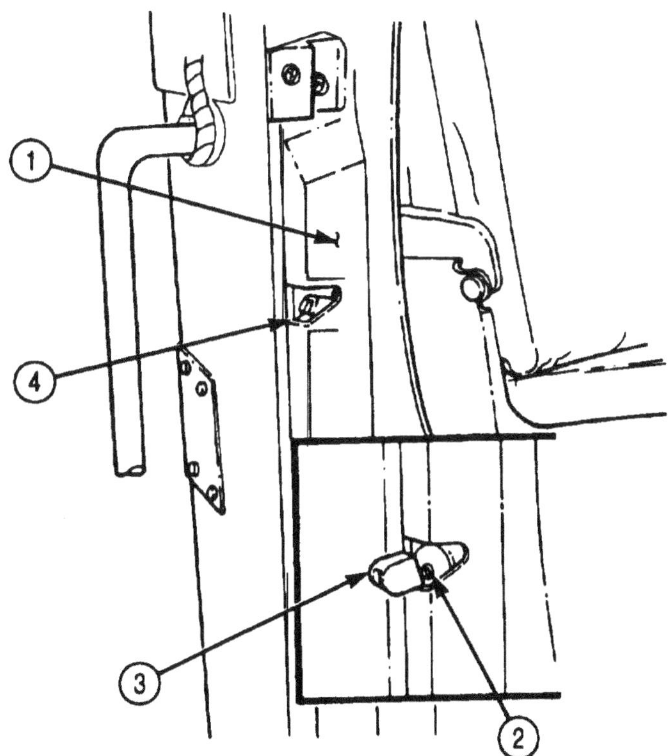

FOLLOW-ON TASK: Check cab door for proper operation (TM 9-2320-272-10).

3-311. CAB DOOR WEATHERSEAL REPLACEMENT

THIS TASK COVERS:

a. Removal **b. Installation**

INITIAL SETUP:

APPLICABLE MODELS
All

TOOLS
General mechanic's tool kit (Appendix E, Item 1)

MATERIALS/PARTS
Adhesive (Appendix C, Item 7)

REFERENCES (TM)
TM 9-2320-272-10
TM 9-2320-272-24P

EQUIPMENT CONDITION
Parking brake set (TM 9-2320-272-10).

a. Removal

1. Remove five screws (3) and retainer (2) from weatherseal (4) and door (1).
2. Remove two weatherseals (4) and (5) from door (1).

b. Installation

1. Apply light coat of adhesive to weatherseals (4) and (5) and door (1).
2. Position weatherseals (4) and (5) in place and install with five retainers (2) and screws (3).
3. Keep door (1) open until adhesive dries.

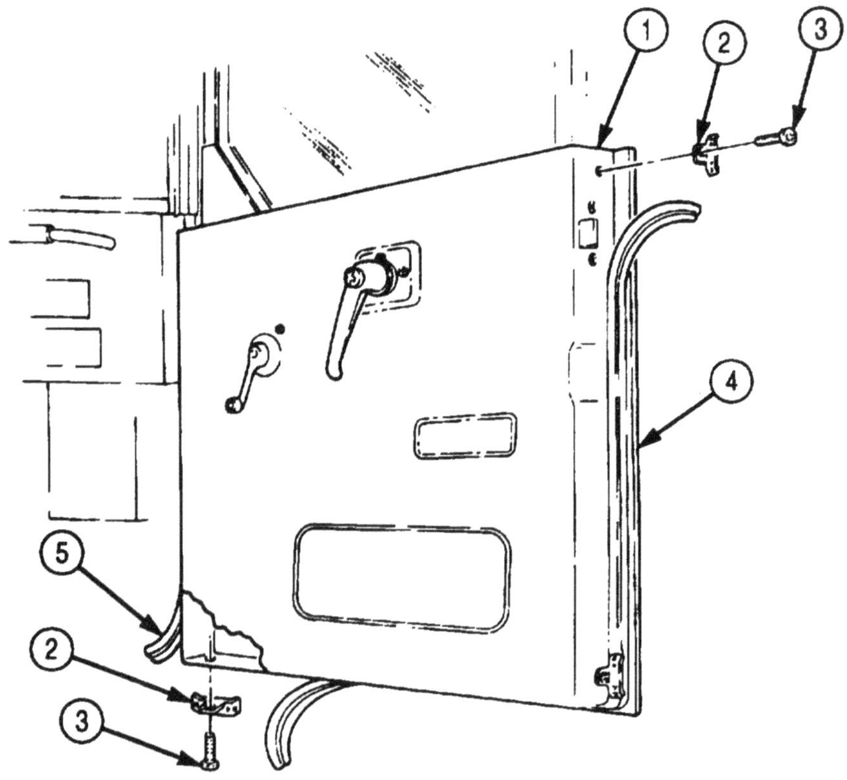

3-312. CAB DOOR INSPECTION HOLE COVER REPLACEMENT

THIS TASK COVERS:

a. Removal **b. Installation**

INITIAL SETUP

APPLICABLE MODELS
All

TOOLS
General mechanic's tool kit (Appendix E, Item 1)

REFERENCES (TM)
TM 9-2320-272-10
TM 9-2320-272-24P

EQUIPMENT CONDITION
Parking brake set (TM 9-2320-272-10).

a. Removal

Remove six screws (2) and inspection hole cover (3) from door (1).

b. Installation

Install inspection hole cover (3) on door (1) with six screws (2).

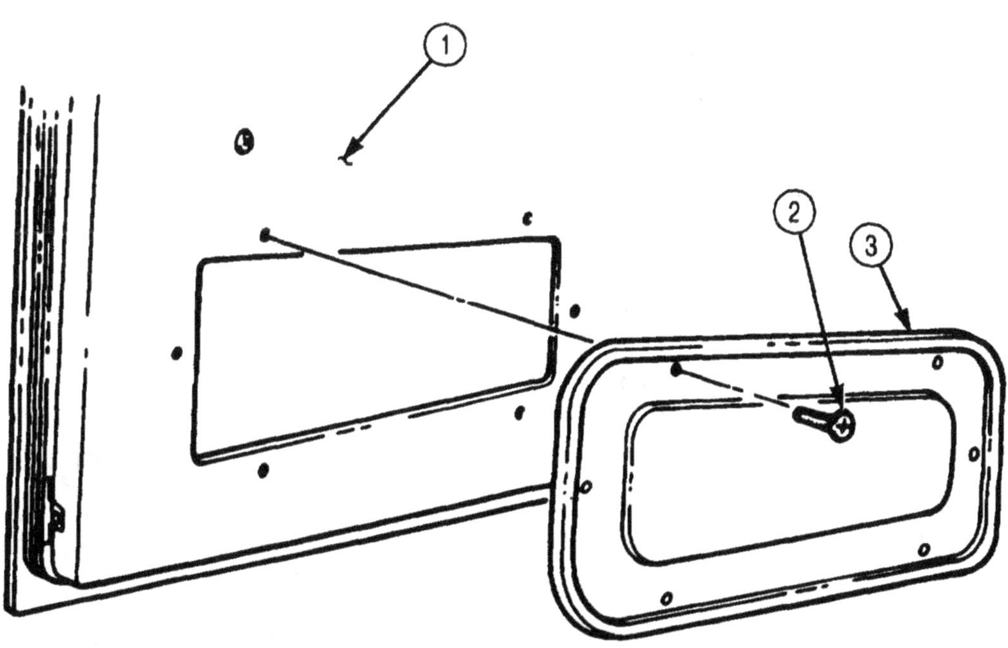

3-313. CAB DOOR LOCK REPLACEMENT

THIS TASK COVERS:

a. Removal b. Installation

INITIAL SETUP:

APPLICABLE MODELS

All

TOOLS

General mechanic's tool kit (Appendix E, Item 1)

MATERIALS/PARTS

Six screw-assembled lockwashers (Appendix D, Item 574)

REFERENCES (TM)

TM 9-2320-272-10
TM 9-2320-272-24P

EQUIPMENT CONDITION

- Parking brake set (TM 9-2320-272-10).
- Outside door handle removed (para. 3-307).
- Cab door inspection hole cover removed (para. 3-312).
- Inside door handle removed (para. 3-308).

a. Removal

1. Remove three screw-assembled lockwashers (2) from cab door (3). Discard screw-assembled lockwashers (2).
2. Remove three screw-assembled lockwashers (1) from cab door (3). Discard screw-assembled lockwashers (1).
3. Remove door lock assembly (6) through door inspection hole (8).

b. Installation

1. Position door lock assembly (6) with latch (7) inserted through hole (8) in cab door (3) and lock shaft (5) through inner door panel (4).
2. Install door lock assembly (6) on cab door (3) with three new screw-assembled lockwashers (2) and (1).

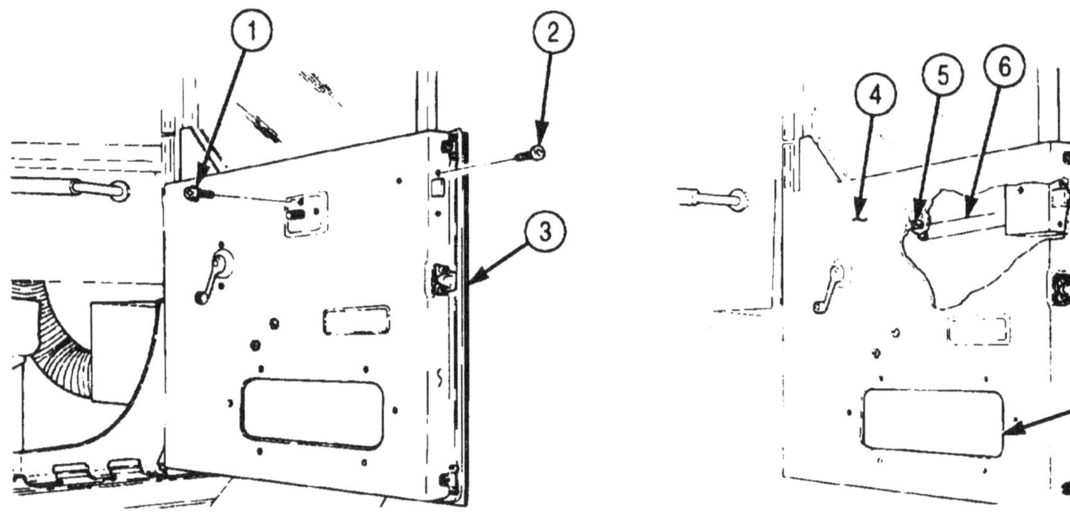

FOLLOW-ON TASKS:
- Install door inspection hole cover (para. 3-312).
- Install outside door handle (para. 3-307).
- Install inside door handle (para. 3-308).
- Check cab door lock for proper operation (TM 9-2320-272-10).

3-314. CAB DOOR GLASS MAINTENANCE

THIS TASK COVERS:

a. Removal
b. Installation
c. Adjustment

INITIAL SETUP:

APPLICABLE MODELS
All

TOOLS
General mechanic's tool kit (Appendix E, Item 1)

MATERIALS/PARTS
Two fasteners (Appendix D, Item 109)
Four lockwashers (Appendix D, Item 373)

REFERENCES (TM)
TM 9-2320-272-10
TM 9-2320-272-24P

EQUIPMENT CONDITION
- Parking brake set (TM 9-2320-272-10).
- Cab door inspection hole cover removed (para. 3-312).

a. Removal

1. Position door glass (7) in cab door (4) so regulator channel (6) is accessible through door inspection hole (5).
2. Remove four screws (1), lockwashers (2), and two window regulator stop brackets (3) from regulator channel (6). Discard lockwashers (2).
3. Remove two fasteners (8) and arm studs (9) from regulator channel (6). Discard fasteners (8).
4. Pull door glass (7) up and out of cab door (4).

b. Installation

1. Position door glass (7) in cab door (4) so regulator channel (6) are accessible through door inspection hole (5).
2. Place two window regulator arm studs (9) in regulator channel (6) and install with two new fasteners (8)
3. Install two window regulator stop brackets (3) on regulator channel (6) with four new lockwashers (2) and screws (1).

c. Adjustment

1. Raise door glass (7) to full up position.
2. Lower door glass (7) and tighten four screws (1).

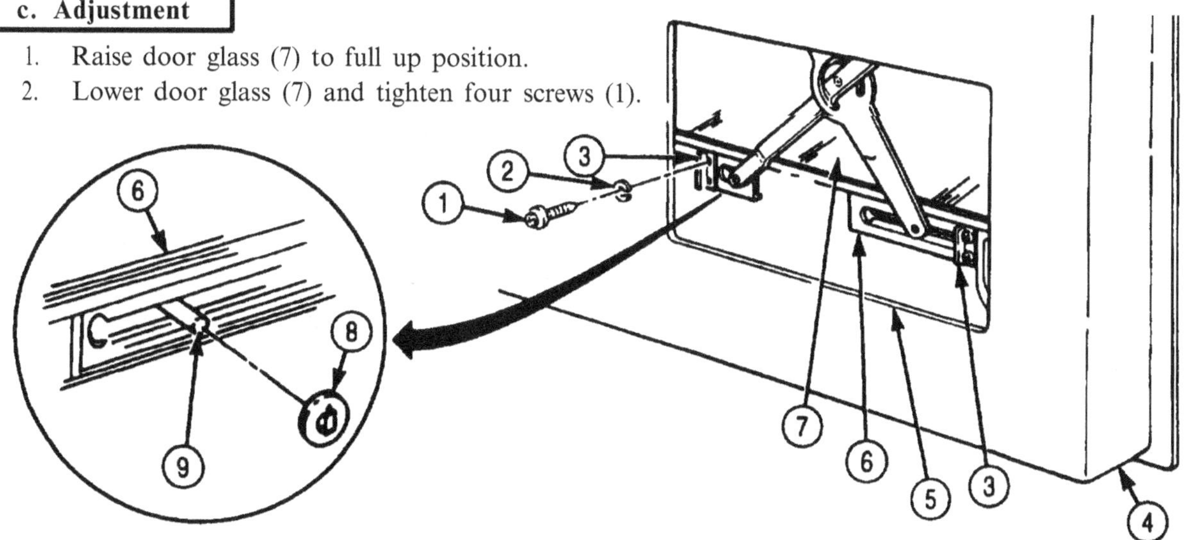

FOLLOW-ON TASK: Install door inspection hole cover (para. 3-312).

3-315. WINDOW WEATHERSTRIPPING (CAB DOOR) REPLACEMENT

THIS TASK COVERS:

a. Removal **b. Installation**

INITIAL SETUP:

APPLICABLE MODELS
All

TOOLS
General mechanic's tool kit (Appendix E, Item 1)

MATERIALS/PARTS
Rivet (Appendix D, Item 543)

REFERENCES (TM)
TM 9-2320-272-10
TM 9-2320-272-24P

EQUIPMENT CONDITION
- Parking brake set (TM 9-2320-272-10).
- Cab door glass removed (para. 3-314).

a. Removal

1. Remove screw (6) and channel (4) from cab door (5).
2. Remove rivet (1) and channel (2) from cab door (5). Discard rivet (1).
3. Remove weatherstripping (3) from cab door (5).

b. Installation

Position weatherstripping (3) on cab door (5) and install channels (2) and (4) with screw (6) and new rivet (1).

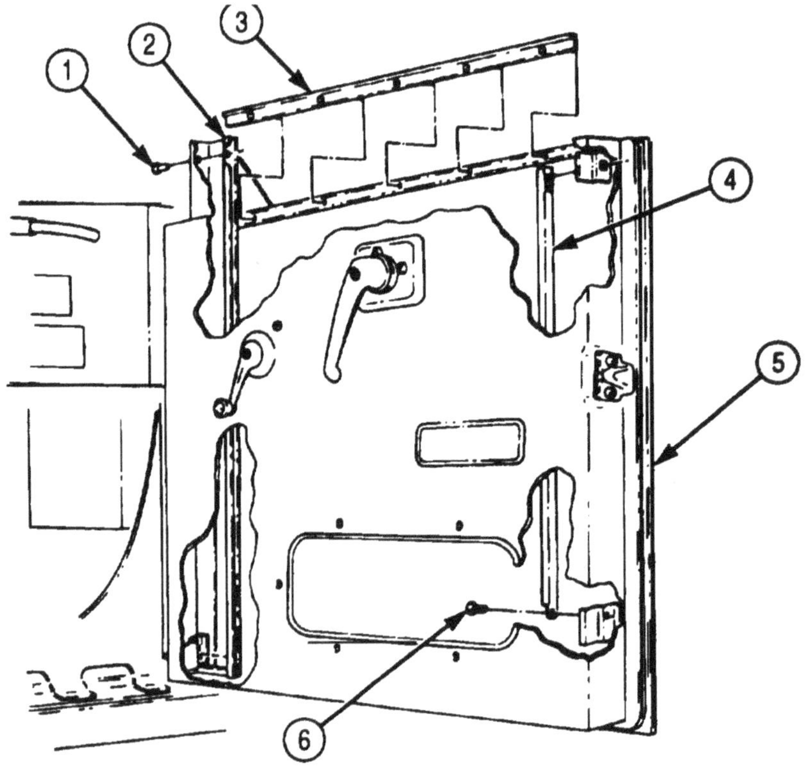

FOLLOW-ON TASK: Install cab door glass (para. 3-314).

3-316. CAB TOP SEAL AND RETAINER REPLACEMENT

THIS TASK COVERS:

a. Removal **b. Installation**

INITIAL SETUP:

APPLICABLE MODELS
All

TOOLS
General mechanic's tool kit (Appendix E, Item 1)

MATERIALS/PARTS
Adhesive (Appendix C, Item 7)

REFERENCES (TM)
TM 9-2320-272-10
TM 9-2320-272-24P

EQUIPMENT CONDITION
Parking brake set (TM 9-2320-272-10).

a. Removal

1. Remove rubber seals (1) and (6) from cab pillar (3).
2. Remove nine screws (7) and retainer (2) from cab pillar (3).
3. Remove six screws (5) and retainer (4) from cab pillar (3).

b. Installation

1. Install retainers (2) and (4) on cab pillar (3) with nine screws (7) and six screws (5).
2. Apply light coat of adhesive to rubber seals (6) and (1) and install on cab pillar (3).

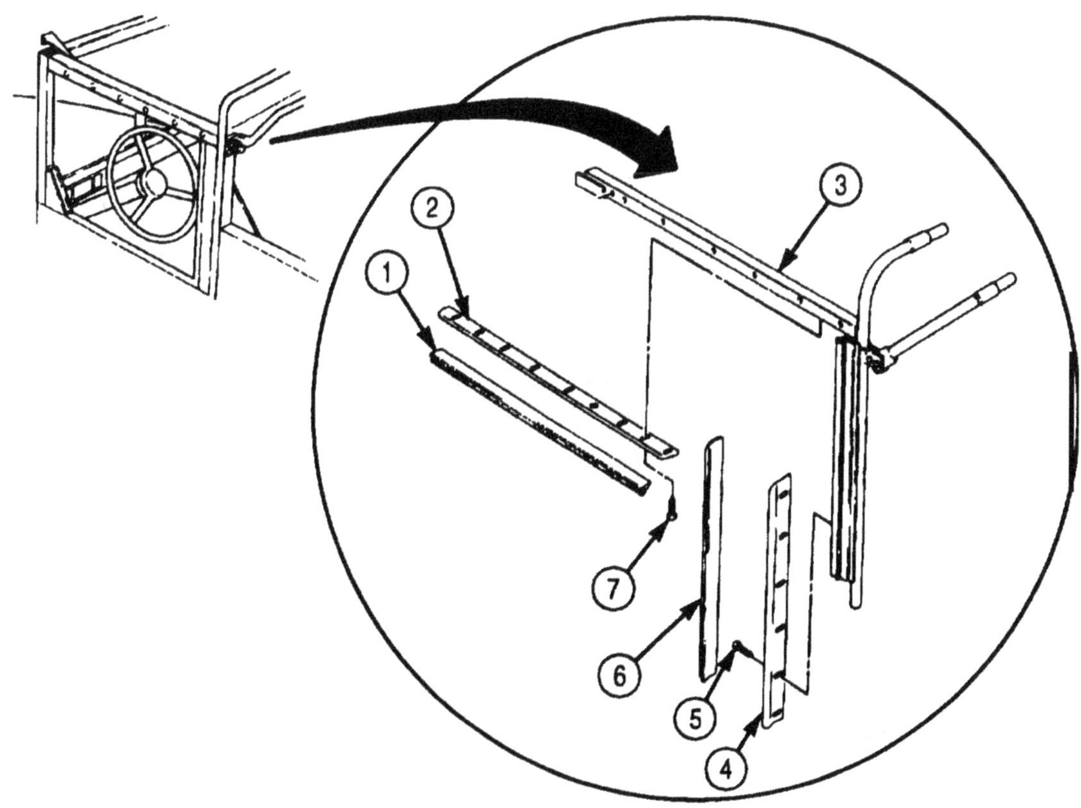

3-317. CAB DOOR REGULATOR ASSEMBLY REPLACEMENT

THIS TASK COVERS:

a. Removal b. Installation

INITIAL SETUP:

APPLICABLE MODELS
All

TOOLS
General mechanic's tool kit (Appendix E, Item 1)

REFERENCES (TM)
TM 9-2320-272-10
TM 9-2320-272-24P

EQUIPMENT CONDITION
- Parking brake set (TM 9-2320-272-10).
- Cab door glass removed (para. 3-314).
- Window regulator handle removed (para. 3-308).

a. Removal

1. Remove four screws (1) and regulator (6) from inner door panel (2).
2. Allow regulator arm stud (3) to slide out of stationary track (4) and remove through door inspection hole (5).

b. Installation

1. Slide regulator arm stud (3) into stationary track (4) on regulator (6).
2. Install regulator (6) on inner door (2) with four screws (1).

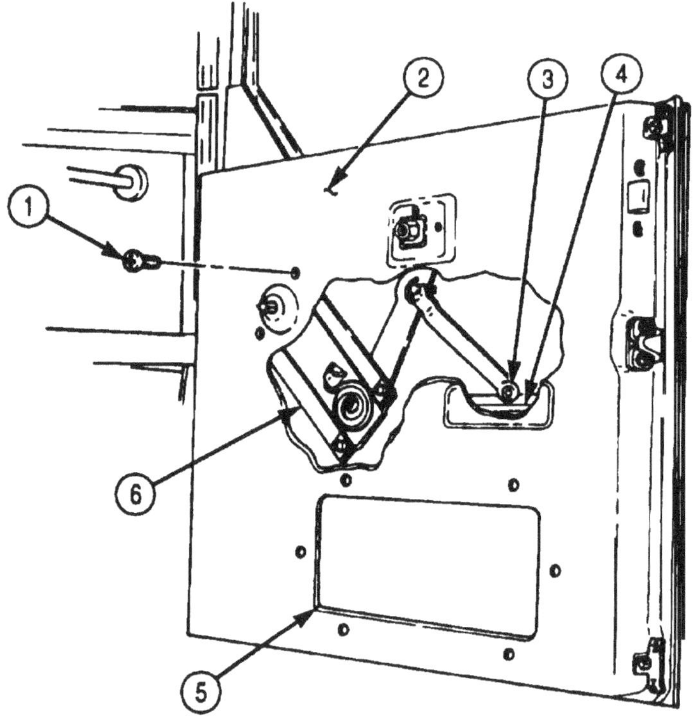

FOLLOW-ON TASKS:
- Install window regulator handle (para. 3-308).
- Install cab door glass (para. 3-314).

3-318. CAB DOOR CHECK ROD REPLACEMENT

THIS TASK COVERS

a. Removal **b. Installation**

INITIAL SETUP:

APPLICABLE MODELS
All

TOOLS
General mechanic's tool kit (Appendix E, Item 1)

MATERIAL/PARTS
Cotter pin (Appendix D, Item 54)

REFERENCES (TM)
TM 9-2320-272-10
TM 9-2320-272.24P

EQUIPMENT CONDITION
Parking brake set (TM 9-2320-272-10).

a. Removal

1. Remove cotter pin (2) from check rod (1) and cab door (3). Discard cotter pin (2).
2. Open vent door (4) and remove check rod (1) and pad (5).

b. Installation

1. Install check rod (1) and pad (5) on vent door (4).
2. Install check rod (1) on cab door (3) with new cotter pin (2).

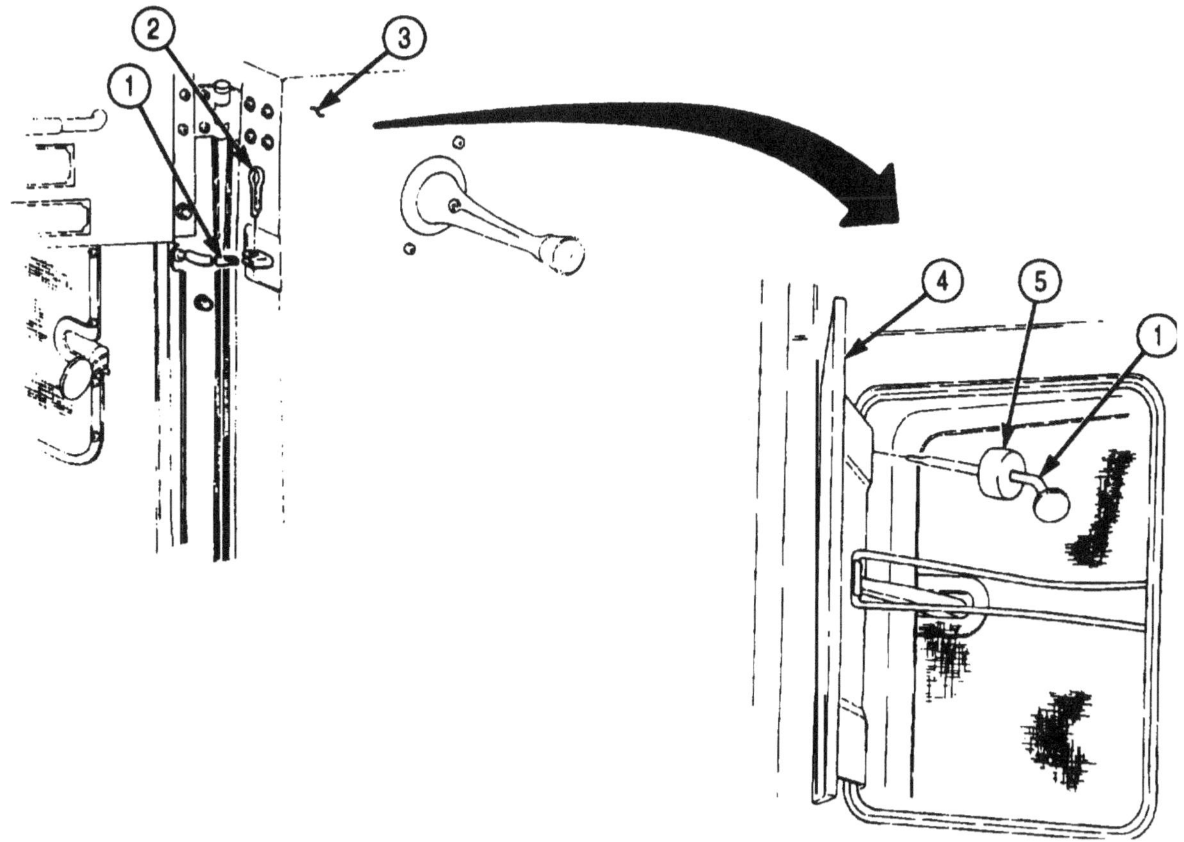

3-319. CAB DOOR REPLACEMENT

THIS TASK COVERS:

a. Removal

b. Installation

INITIAL SETUP

APPLICABLE MODELS

A11

TOOLS

General mechanic's tool kit (Appendix E, Item 1)

MATERIALS/PARTS

Three locknuts (Appendix D, Item 272)

REFERENCES (TM)

TM 9-2320-272-10
TM 9-2320-272-24P

EQUIPMENT CONDITION

- Parking brake set (TM 9-2320-272-10).
- Cab door lock removed (para. 3-313).
- Cab door dovetail wedge removed (para. 3-309).
- Window weatherstripping (cab door) removed (para. 3-315).
- Cab door window regulator removed tpara. 3-308).
- Cab door check rod removed (para. 3-318).
- Cab door weatherseal removed (para. 3-311).

a. Removal

1. Remove three locknuts (6), washers (5), and screws (1) from three mirror braces (2) and two door hinges (7). Discard three locknuts (6).
2. Remove cab door (4) from two door hinges (3).

b. Installation

1. Position cab door (4) and two door hinges (3) on three mirror braces (2) and two door hinges (7).
2. Install cab door on three. mirror braces (2) and two door hinges (7) with three screws (l), washers (5), and new locknuts (6).

3-319. CAB DOOR REPLACEMENT (Contd)

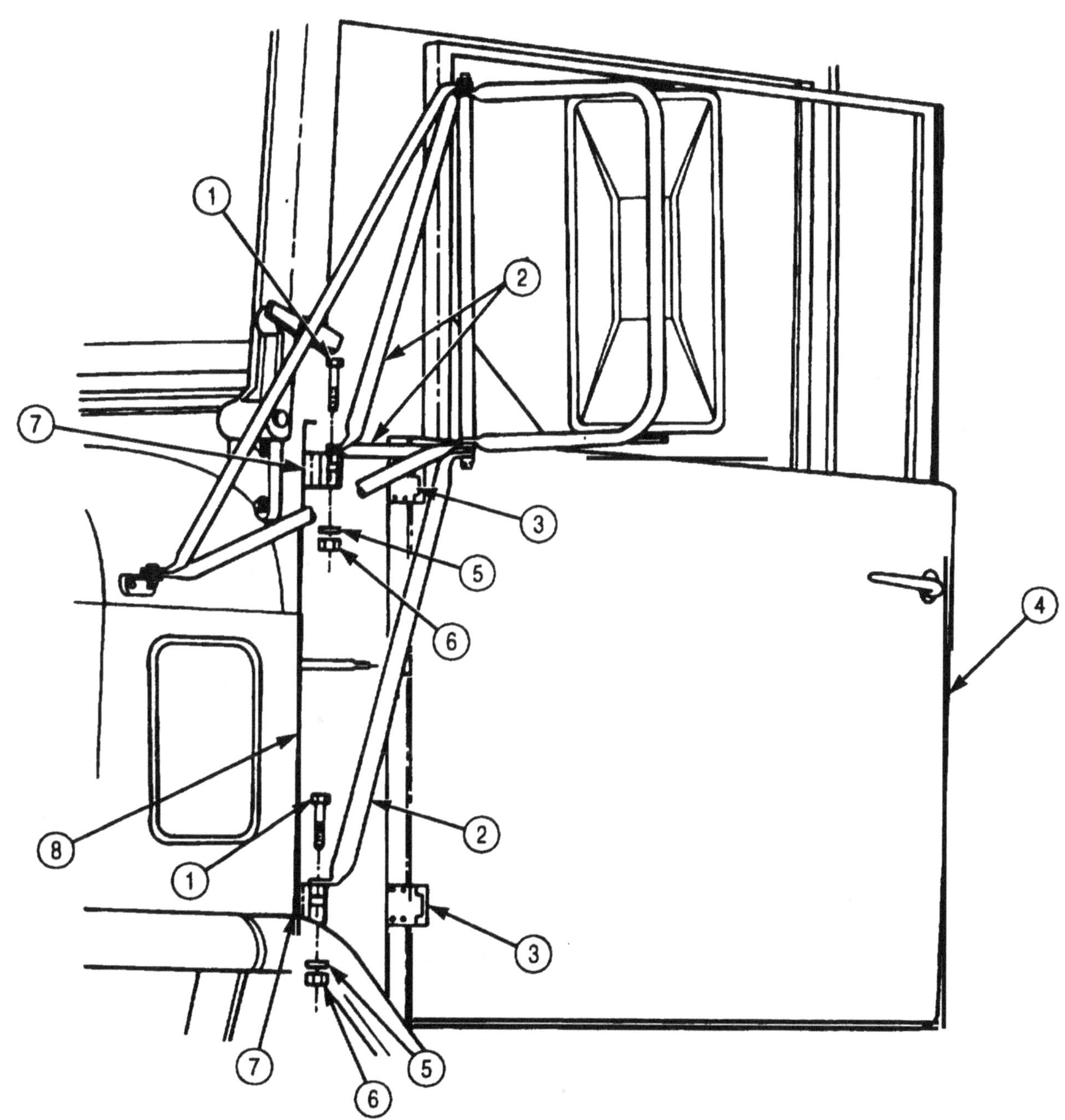

FOLLOW-ON TASKS:
- Install cab door weatherseal (para. 3-311).
- Install cab door check rod (para. 3-318).
- Install cab door window regulator (para. 3-308).
- Install window weatherstripping (cab door) (para. 3-315).
- Install cab door dovetail wedge (para. 3-309).
- Install cab door lock (para. 3-313).

3-320. CAB DOOR HINGE REPLACEMENT

THIS TASK COVERS:

a. Removal

b. Installation

INITIAL SETUP:

APPLICABLE/MODELS
All

TOOLS
General mechanic's tool kit (Appendix E, Item 1)

REFERENCES (TM)
TM 9-2320-272-10
TM 9-2320-272-24P

EQUIPMENT CONDITION
- Parking brake set (TM 9-2320-272-10).
- Cab door removed (para. 3-319).

a. Removal

Remove eight screws (4) and hinges (1) and (2) from cab door (3) and cab body (5).

b. Installation

Install hinges (1) and (2) on cab body (5) and cab door (3) with eight screws (4) .

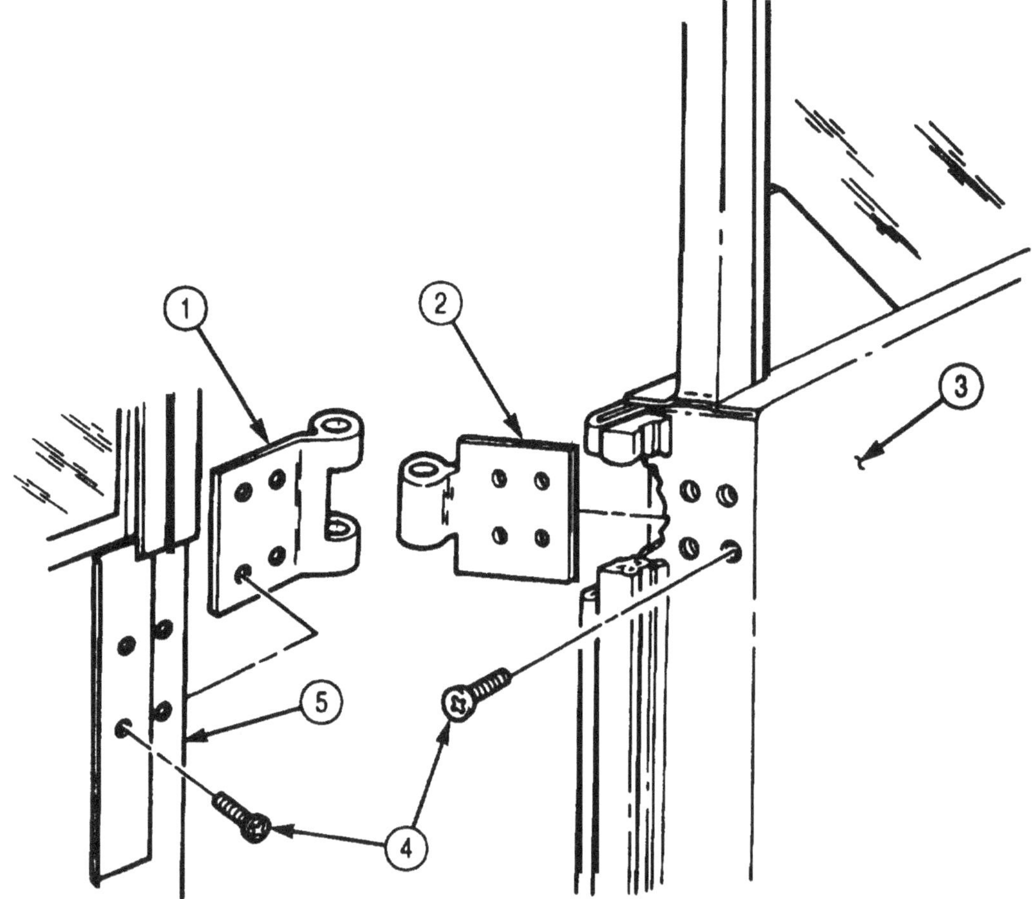

FOLLOW-ON TASK: Install cab door (para. 3-319).

3-321. CAB DOOR CATCH REPLACEMENT

THIS TASK COVERS:

a. Removal **b. Installation**

INITIAL SETUP

APPLICABLE MODELS
All

TOOLS
General mechanic's tool kit (Appendix E, Item 1)

MATERIALS/PARTS
Two screw-assembled lockwashers (Appendix D, Item 573)

REFERENCES (TM)
TM 9-2320-272-10
TM 9-2320-272-24P

EQUIPMENT CONDITION
Parking brake set (TM 9-2320-272-10).

a. Removal

Remove two screw-assembled lockwashers (3) and cab door catch (2) from cab body (1). Discard screw-

b. Installation

Install cab door catch (2) on cab body (1) with two new screw-assembled lockwashers (3).

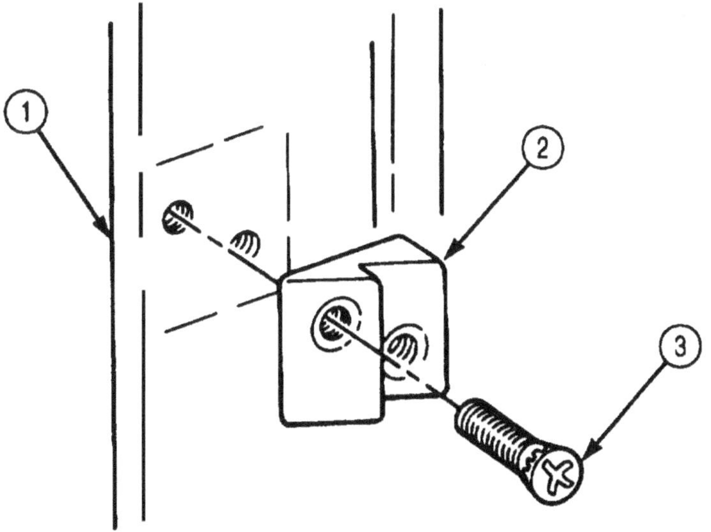

FOLLOW-ON TASK: Check cab door for proper operation (TM 9-2320-272-10).

Section XIII. WINCH AND POWER TAKEOFF MAINTENANCE

3-322. WINCH AND POWER TAKEOFF MAINTENANCE INDEX

PARA. NO.	TITLE	PAGE NO.
3-323.	Front Winch Automatic Brake Adjustment	3-824
3-324.	Front Winch Drag Brake Adjustment	3-828
3-325.	Front Winch Cable Chain and Hook Replacement	3-830
3-326.	Winch Cable Clevis Replacement	3-831
3-327.	Front Winch Cable Replacement	3-834
3-328.	Front Winch Motor Replacement	3-836
3-329.	Front Winch Replacement	3-840
3-330.	Rear Winch Adjustment	3-844
3-331.	Rear Winch Cable Replacement	3-846
3-332.	Rear Winch Replacement	3-847
3-333.	Housing Assembly Cover Replacement	3-852
3-334.	Transmission PTO-to-Hydraulic Pump Propeller Shaft Replacement	3-853
3-335.	Transmission PTO-to-Hydraulic Pump Propeller Shaft Universal Joint Maintenance	3-854
3-336.	Winch Hydraulic Oil Reservoir Filter Replacement	3-858
3-337.	Winch Hydraulic Oil Reservoir Replacement	3-860
3-338.	Tractor Winch Hydraulic Oil Reservoir (M932/A1/A2) Replacement	3-864

3-323. FRONT WINCH AUTOMATIC BRAKE ADJUSTMENT

THIS TASK COVERS:

a. Testing **b. Adjustment**

INITIAL SETUP:

APPLICABLE MODELS
All with winch

TOOLS
General mechanic's tool kit (Appendix E, Item 1)

PERSONNEL REQUIRED
Two

REFERENCES (TM)
TM 9-2320-272-10

EQUIPMENT CONDITION
Parking brake set (TM 9-2320-272-10).

GENERAL SAFETY INSTRUCTIONS
- Wear hand protection when handling winch cable.
- Never stand between vehicles during test.
- Assistant must remain in second vehicle to engage service brakes if cable snaps or winch automatic brake fails.
- A minimum of four turns of cable must remain on winch drive at all times.

NOTE

The procedures for testing and adjustment of front and rear winch automatic brakes are the same. This procedure covers the front winch automatic brake. Refer to TM 9-2320-272-10 for winch operation.

a. Testing

CAUTION

Selection of grade used in this procedure should be within the tolerance capabilities of the second vehicle. Failure to comply may result in damage to equipment.

1. Position test vehicle (1) at top of steep grade, facing downhill, with engine running (TM 9-2320-272-10).
2. Position second vehicle (4) at bottom of steep grade, facing uphill, with engine running. Refer to operator's manual for second vehicle (4).

WARNING

- Never stand between vehicles during test. Assistant must remain in second vehicle to engage service brakes if cable snaps or winch automatic brake fails while towing vehicle. Failure to comply may result in injury to personnel.
- Wear hand protection when handling winch cable. Broken wires may cause injury to personnel.
- A minimum of four turns of cable must remain on winch drum at all times. Failure to do this may result in injury to personnel or damage to equipment.

3. Release winch cable (3) (TM 9-2320-272-10) from test vehicle (1) and rig to second vehicle (4).
4. Refer to operator's manual for second vehicle (4) and prepare vehicle (1) as follows:
   a. Place transmission lever (5) in NEUTRAL position.
   b. Disengage parking brake (TM 9-2320-272-10).
   c. Disengage front wheel drive lever if engaged (TM 9-2320-272-10).
5. Prepare front winch (2) of vehicle (1) for winding (TM 9-2320-272-10).

3-323. FRONT WINCH AUTOMATIC BRAKE ADJUSTMENT (Contd)

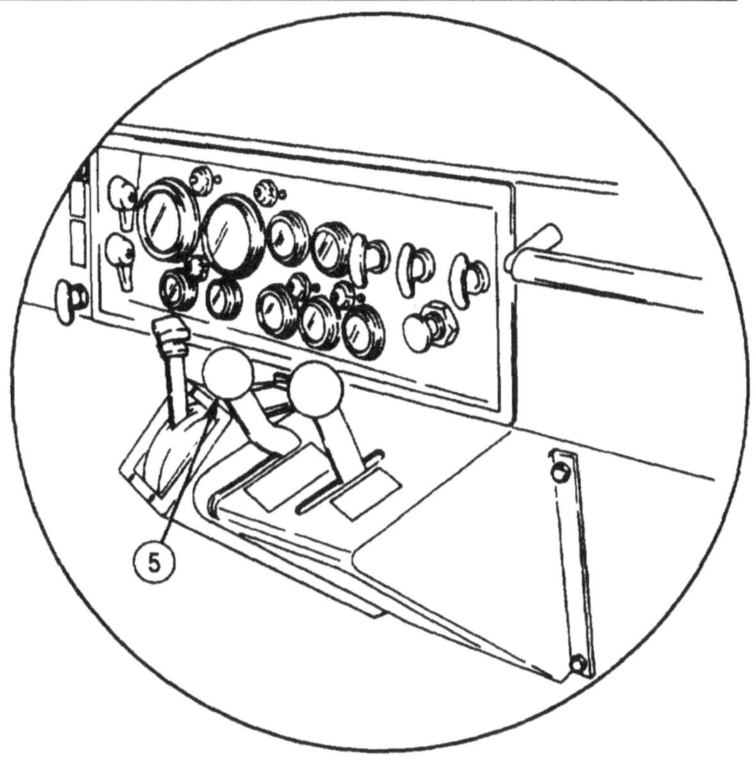

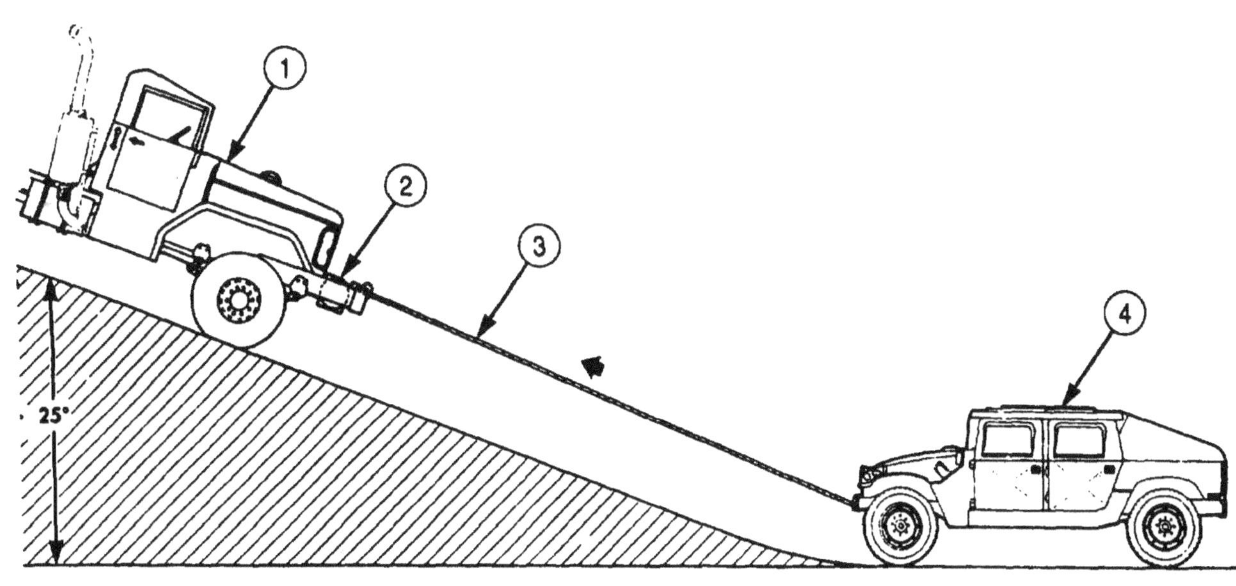

3-323. FRONT WINCH AUTOMATIC BRAKE ADJUSTMENT (Contd)

6. Pull back winch control lever (4) to WIND position, pull second vehicle (3) part way up grade, and observe movement of second vehicle (3).
   a. If second vehicle (3) rolls backward, go to task b. and adjust automatic winch brake of vehicle (1).
   b. If second vehicle (3) holds steady on the incline, no adjustment is required.
7. Push winch control lever (4) forward to WINCH position and unwind winch cable (2) until second vehicle (3) is on level ground.
8. Remove winch cable (2) from second vehicle (3) if no adjustment is required.

b. Adjustment

1. Adjust brake band by turning adjustment screw (5) clockwise in one-half turn increments to increase brake action enough to hold second vehicle (3) steady on incline.
2. Repeat task a., steps 4 through 8 until adjustment is correct.

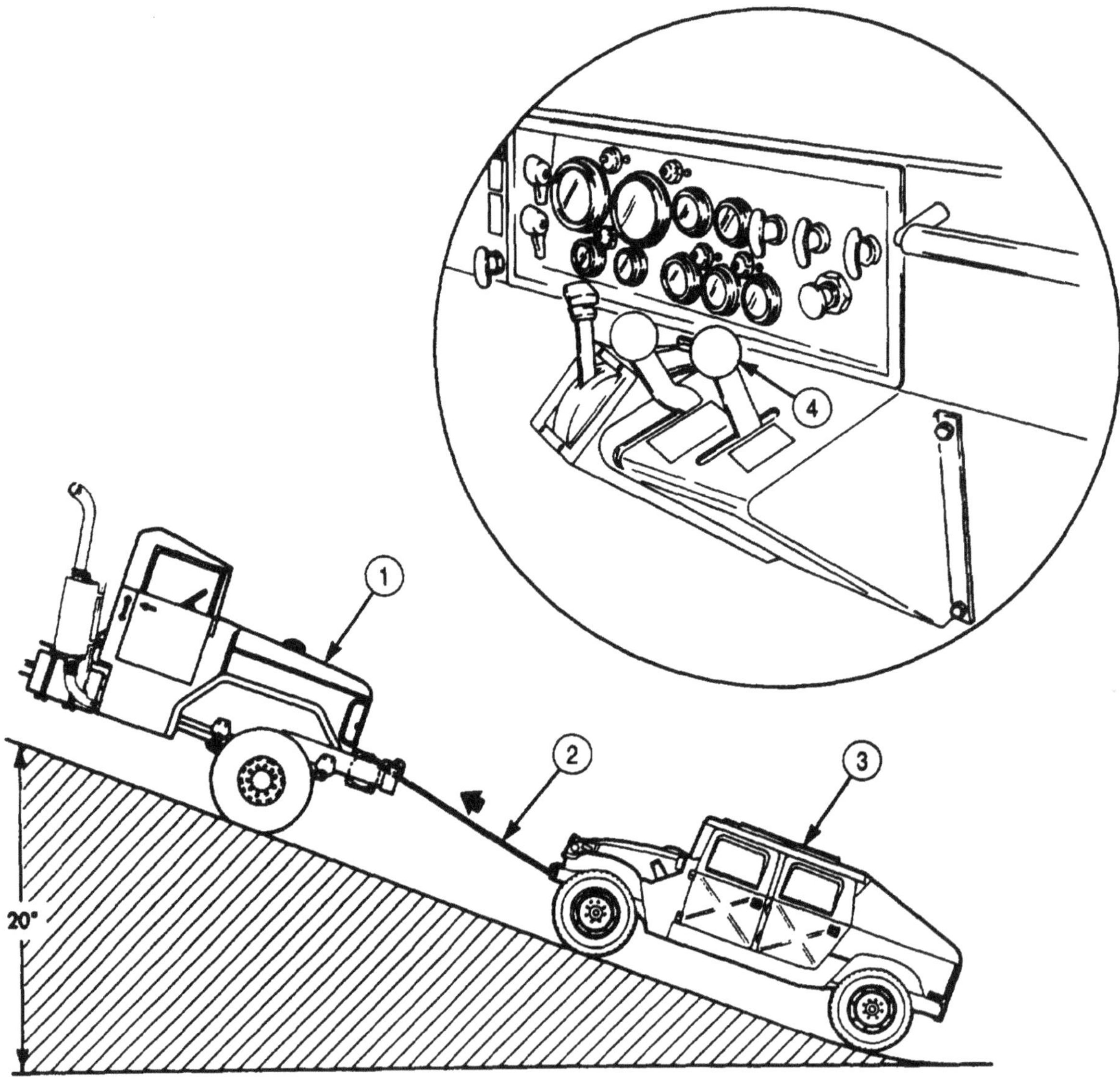

3-323. FRONT WINCH AUTOMATIC BRAKE ADJUSTMENT (Contd)

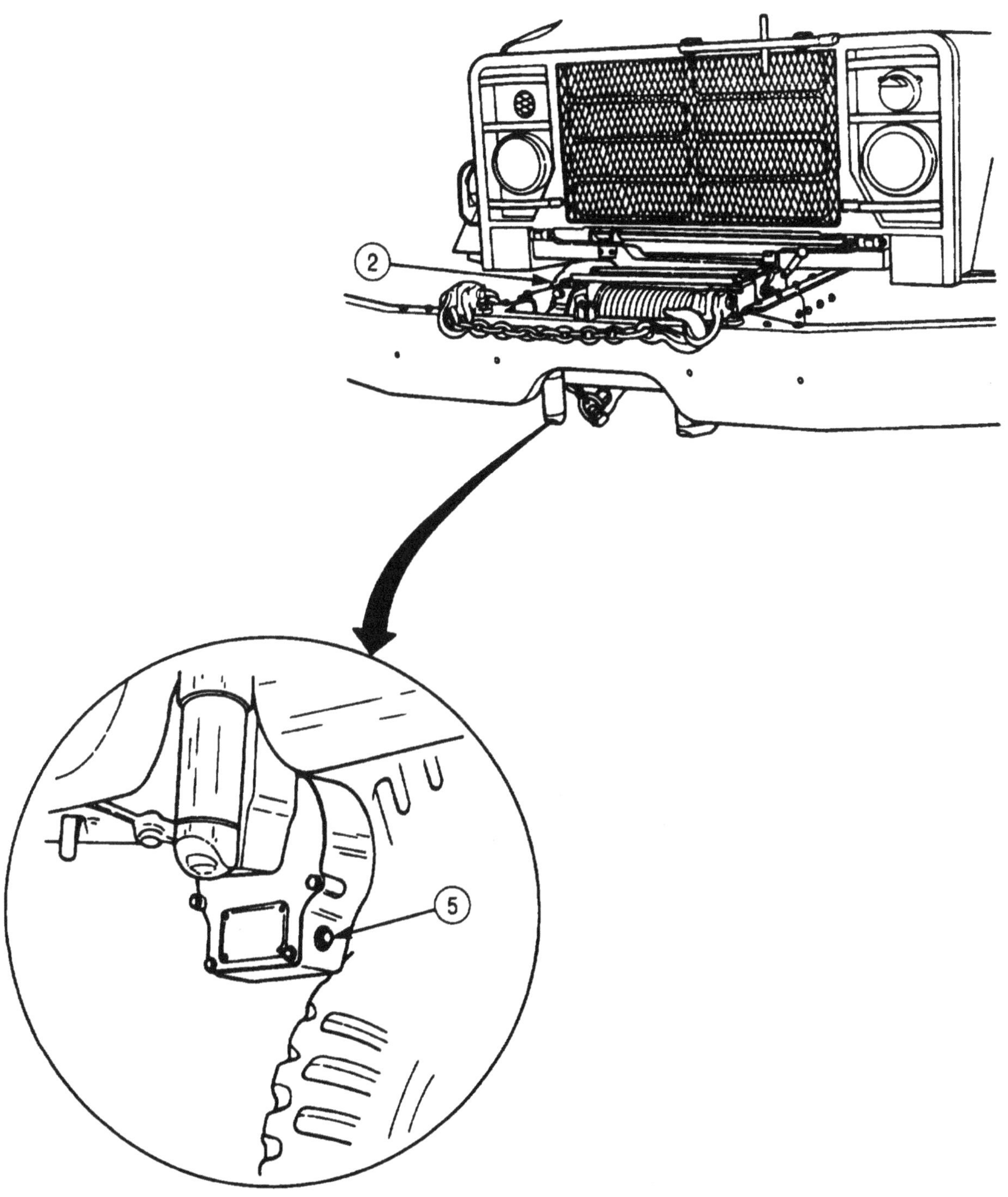

FOLLOW-ON TASK: Rewind winch cable (TM 9-2320-272-10).

3-324. FRONT WINCH DRAG BRAKE ADJUSTMENT

THIS TASK COVERS:

a. Testing **b. Adjustment**

INITIAL SETUP:

APPLICABLE MODELS
All with winch

TOOLS
General mechanic's tool kit (Appendix E, Item 1)

REFERENCES (TM)
TM 9-2320-272-10

EQUIPMENT CONDITION
Parking brake set (TM 9-2320-272-10).

GENERAL SAFETY INSTRUCTIONS
Wear hand protection when handling winch cable.

a. Testing

WARNING

Wear hand protection when handling winch cable. Broken wires may cause injury to personnel.

1. Pull out drum lock knob (2) on left side of front winch (1), rotate 90 degrees, and release.

NOTE

Perform steps 2 and 3 for M936/A1 model vehicles equipped with level wind. M936A2 model vehicles do not have a level wind.

2. Pull out level wind lock knob (6) on level winch frame (5), rotate one-quarter turn, and release.
3. Pull out cable tensioner lock knob (8) and release tensioner lever (7).
4. Push clutch lever (3) toward front of vehicle to disengage clutch.
5. Pull winch cable (4) 3-4ft. (.9-1.2 m) off drum. Drum will stop turning as soon as pulling has stopped if drag brake is adjusted properly If drum continues to turn after pulling has stopped, perform task b. to adjust drag brake.

b. Adjustment

1. Turn drag brake adjusting screw (9), located under front winch (1), one-quarter turn clockwise to increase drag; counterclockwise to decrease drag.
2. Repeat task a, step 5, until drag brake is properly adjusted.

3-324. FRONT WINCH DRAG BRAKE ADJUSTMENT (Contd)

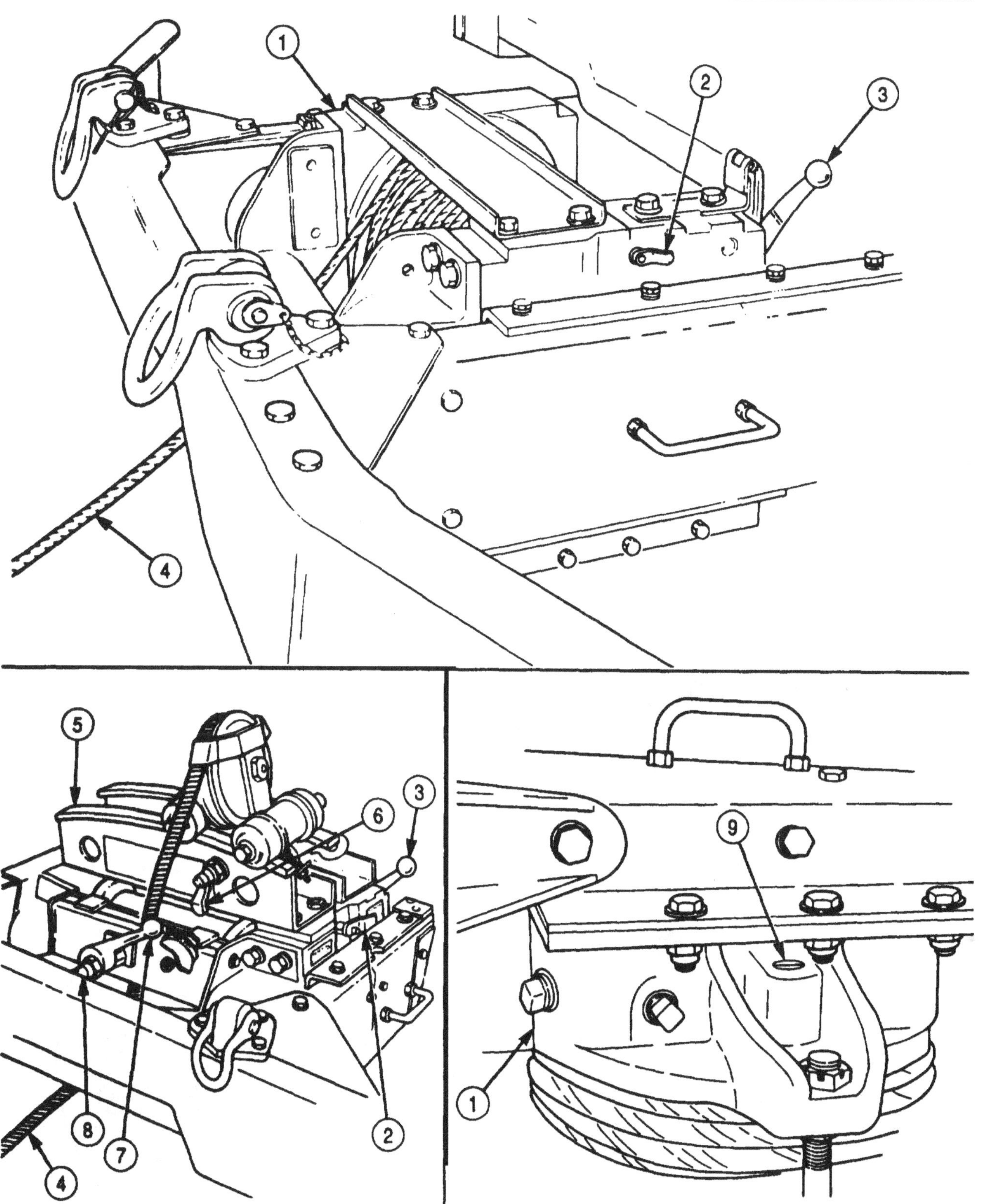

FOLLOW-ON TASK: Rewind winch cable (TM 9-2320-272-10)

3-325. FRONT WINCH CABLE CHAIN AND HOOK REPLACEMENT

THIS TASK COVERS:

a. Removal **b. Installation**

INITIAL SETUP:

APPLICABLE MODELS
All with winch

TOOLS
General mechanic's tool kit (Appendix E, Item 1)

REFERENCE (TM)
TM 9-2320-272-10
TM 9-2320-272-24P

EQUIPMENT CONDITION
Parking brake set (TM 9-2320-272-10)

a. Removal

1. Remove cable chain and hook (1) from front bumper lifting shackle (2).
2. Remove nut (5), clevis pin (3), and clevis (4) from cable chain and hook (1).

b. Installation

1. Install cable chain and hook (1) on clevis (4) with clevis pin (3) and nut (5).
2. Store cable chain and hook (1) on front bumper lifting shackle (2) for travel.

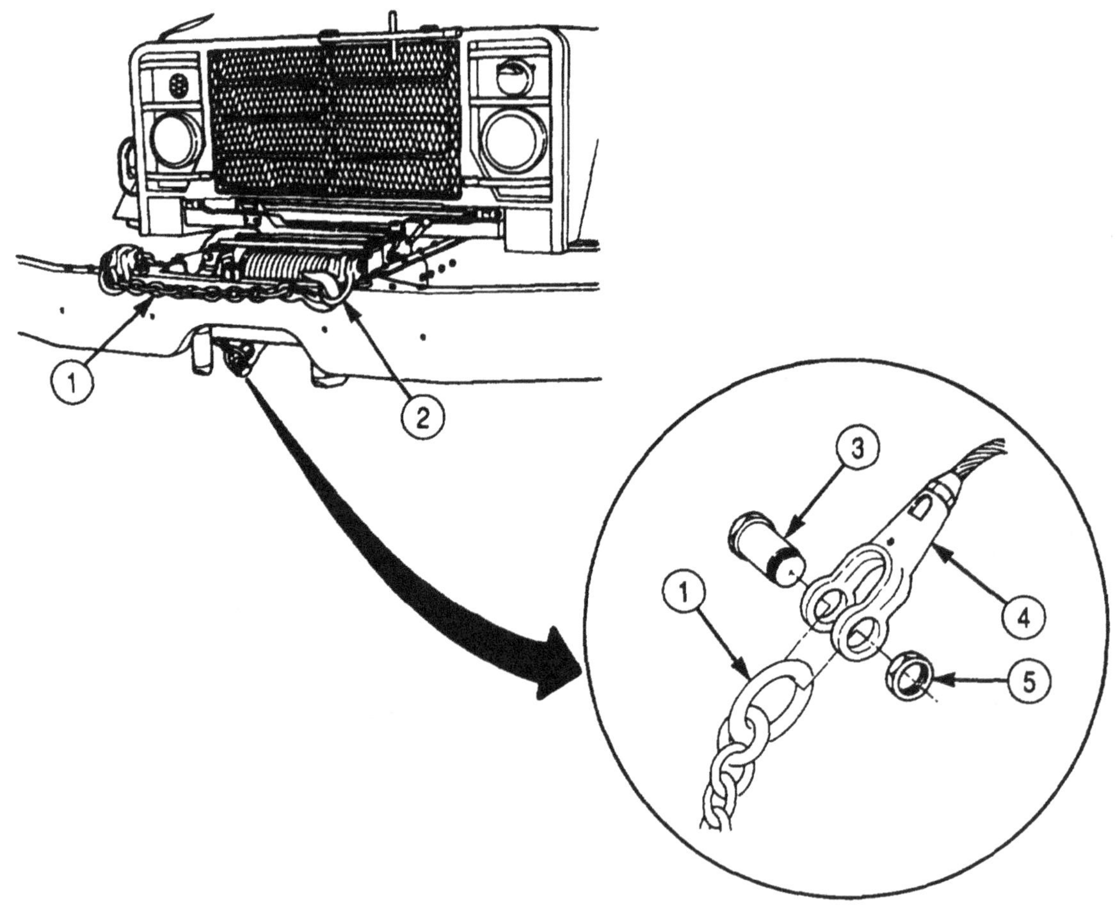

3-326. WINCH CABLE CLEVIS REPLACEMENT

THIS TASK COVERS:

a. Removal **b. Installation**

INITIAL SETUP:

APPLICABLE MODELS:
All with winch

TOOLS
General mechanic's tool kit (Appendix E, Item 1)

REFERENCES (TM)
TB 43-0142
TM 9-2320-272-10
TM 9-2320-272-24P

EQUIPMENT CONDITION
- Parking brake set (TM 9-2320-272-10).
- Front winch cable unwound (TM 9-2320-272-10).
- Cable chain and hook removed (para. 3-325).

GENERAL SAFETY INSTRUCTIONS
- Wear hand protection when handling winch cable.
- Top seizing must not be less than 5 in. (12.7 cm) from end of cable.

WARNING

Wear hand protection when handling winch cable. Broken wires may cause injury to personnel.

a. Removal

1. Place threaded sleeve (1) in vise.
2. Remove clevis socket (2) from threaded sleeve (1).
3. Remove threaded sleeve (1) from vise, and clamp cable (5) in vise below seizing wire (4), if present.
4. Remove plug (3), threaded sleeve (1), and seizing wire (4), if present, from cable (5).
5. Remove cable (5) from vise and trim to provide new end.

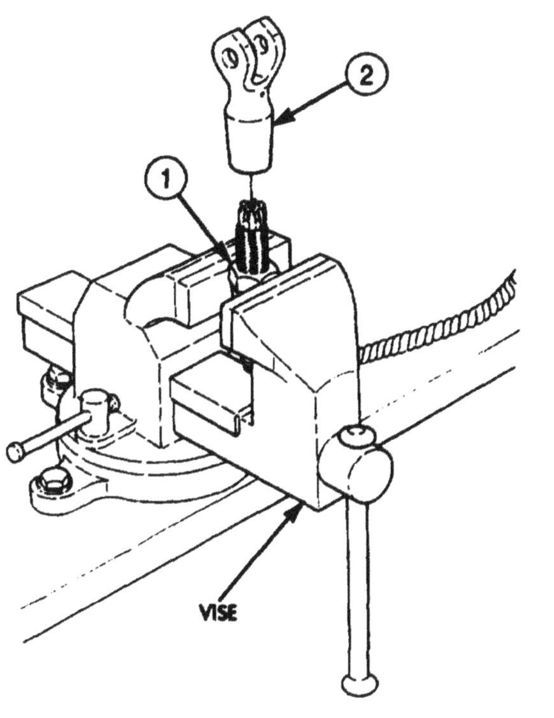

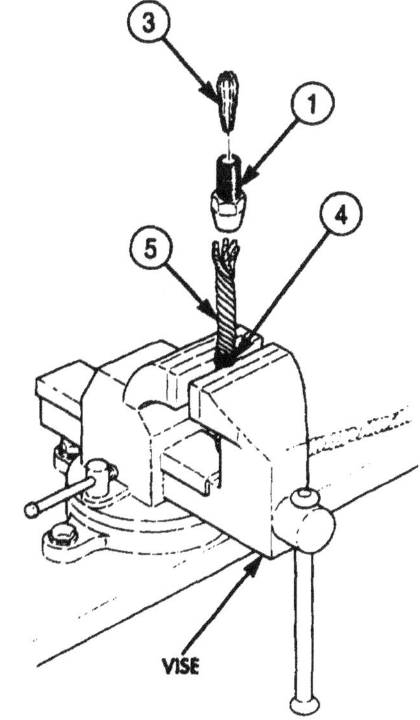

3-326. WINCH CABLE CLEVIS REPLACEMENT (Cont)

b. Installation

WARNING

Top seizing must not be less than 5.0 in. (12.7 cm) from end of cable. Faulty installation will cause cable failure and may result in injury or death to personnel.

1. Place cable (1) in vise with 5-1/2 in. (14.0 cm) of cable (1) above jaws of vise.
2. Measure 5.0 in. (12.7 cm) below end of cable (1). Beginning at this point, wrap 1/2 in. (12.7 mm) of mechanic's wire (2) around cable (1) toward vise. Twist ends of mechanic's wire (2) together and bend flat in a groove of cable (1).

NOTE

If cable is wire-core type, proceed to step 5.

3. Unravel six strands (3) of cable (1) above vise.
4. Cut off hemp core (4) as close to vise as possible.
5. Wrap upper ends of strands (3) with mechanic's wire (2).

NOTE

Approximately 1-3/8 in. (3.5 cm) of cable should extend above threaded sleeve.

6. Place threaded sleeve (5) over end of cable (1) and seat against mechanic's wire (2).
7. Remove cable (1) from vise and reclamp on the hex-flats of threaded sleeve (5).
8. Remove upper wire (6) from cable (1).

NOTE

- Use P/N 7071906 plug for hemp core cable. Use P/N 7071871 plug for wire core cable.
- Perform step 10 for wire-core type cable.

9. Drive plug (7) into center of cable (1) and into threaded sleeve (5). Align strands (3) with grooves in plug (7).
10. Slide plug (8) over core wire (9) and down onto cable (1), align strands (3) with grooves of plug (8), and drive plug (8) into cable (1) and threaded sleeve (5).
11. Place clevis socket (10) over end of cable (1) and install on threaded sleeve (5). Tighten clevis socket (10) on threaded sleeve (5) until only 3 to 5 threads are exposed. Ensure cable (1) is visible through inspection hole (11) of clevis socket (10).

NOTE

Refer to TB 43-0142 for cable and clevis assemblies that must be proof-load tested.

3-326. WINCH CABLE CLEVIS REPLACEMENT (Contd)

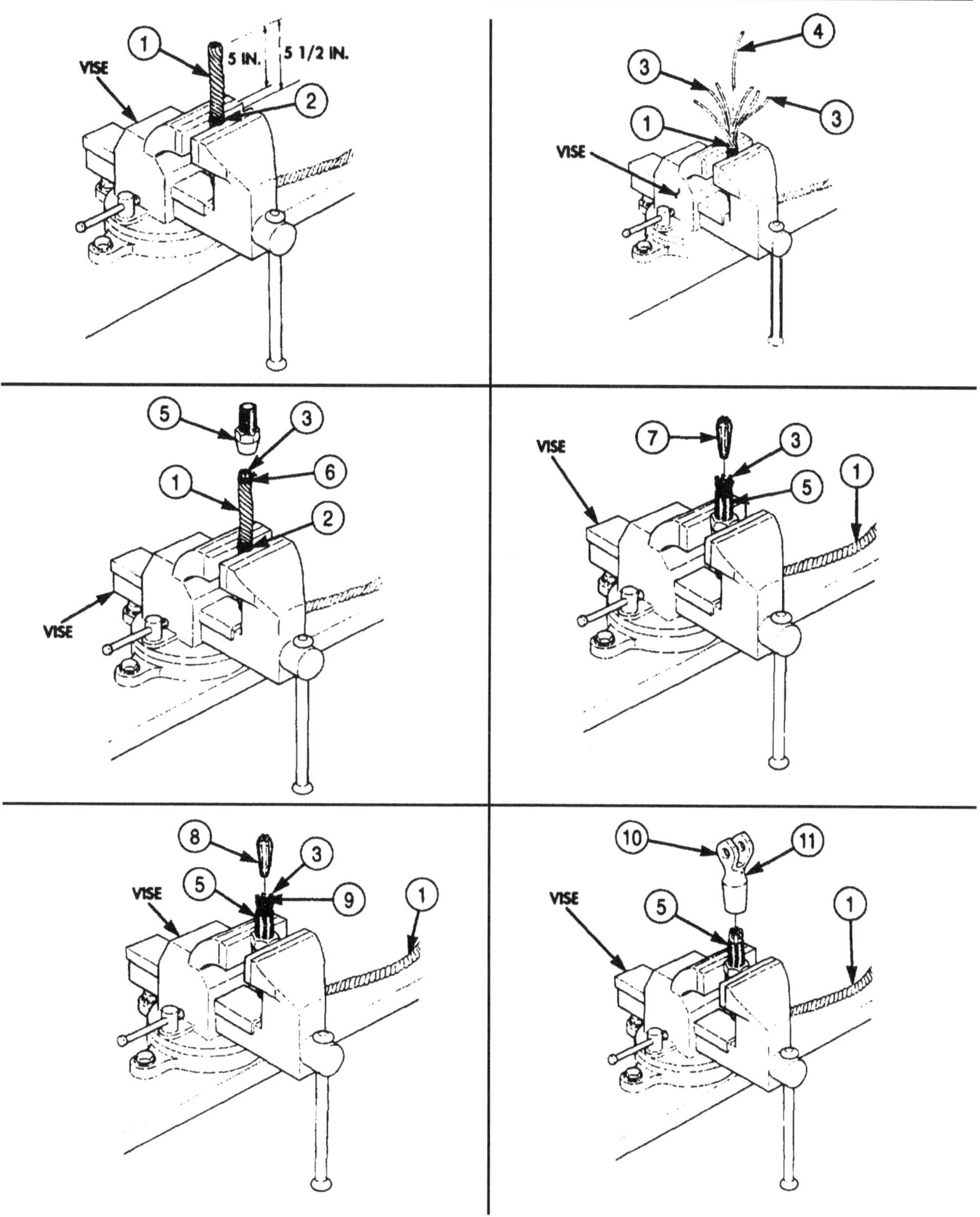

FOLLOW-ON TASKS:
- Install cable chain and hook (para. 3-325).
- Wind front winch cable (TM 9-2320-272-10).

3-327. FRONT WINCH CABLE REPLACEMENT

THIS TASK COVERS:

a Removal **b. Installation**

INITIAL SETUP:

APPLICABLE MODELS
All with winch

TOOLS
General mechanic's tool kit (Appendix E, Item 1)

REFERENCES (TM)
TM 9-2320-272-10
TM 9-2320-272-24P

EQUIPMENT CONDITION
- Parking brake set (TM 9-2320-272-10).
- Front winch cable unwound (TM 9-2320-272-10).
- Cable chain and hook removed (para. 3-325).

GENERAL SAFETY INSTRUCTIONS
Wear hand protection when handling winch cable.

WARNING

Wear hand protection when handling winch cable. Broken wires may cause injury to personnel.

a. Removal

1. Remove setscrew (1) and winch cable (4) from hole (3) in winch drum (2).

NOTE

Perform step 2 if front winch is equipped with level wind and tensioner.

2. Pull winch cable (4) out at side roller (9), over level wind pulley (6), down through tensioner sheaves (7), and out of rollers (8) and (9).

b. Installation

NOTE

Perform step 1 if front winch is equipped with level wind and tensioner.

1. Thread winch cable (4) into center of rollers (8) and (9), tensioner sheaves (7), under level wind pulley plate (5), and over level wind pulley (6).
2. Insert winch cable (4) in hole (3) of winch drum (2) and install setscrew (1).

3-327. FRONT WINCH CABLE REPLACEMENT (Contd)

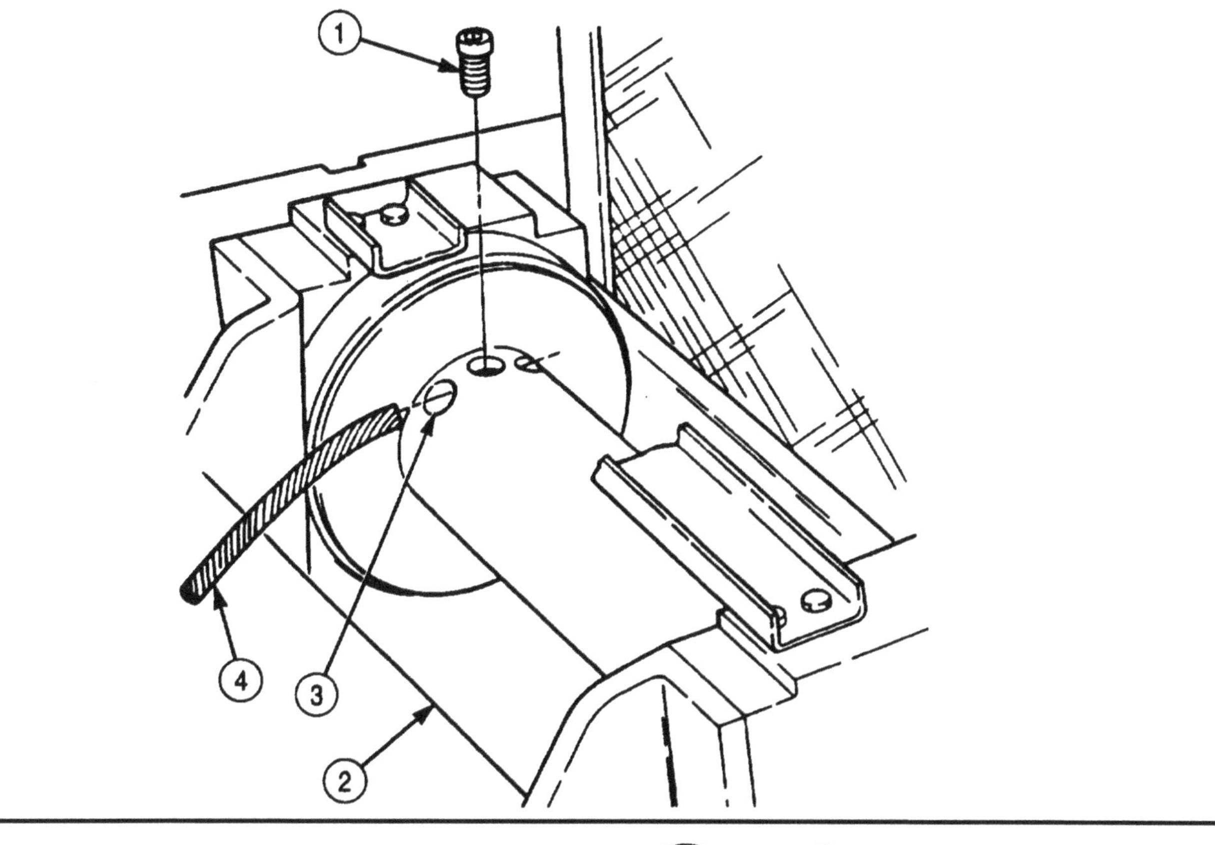

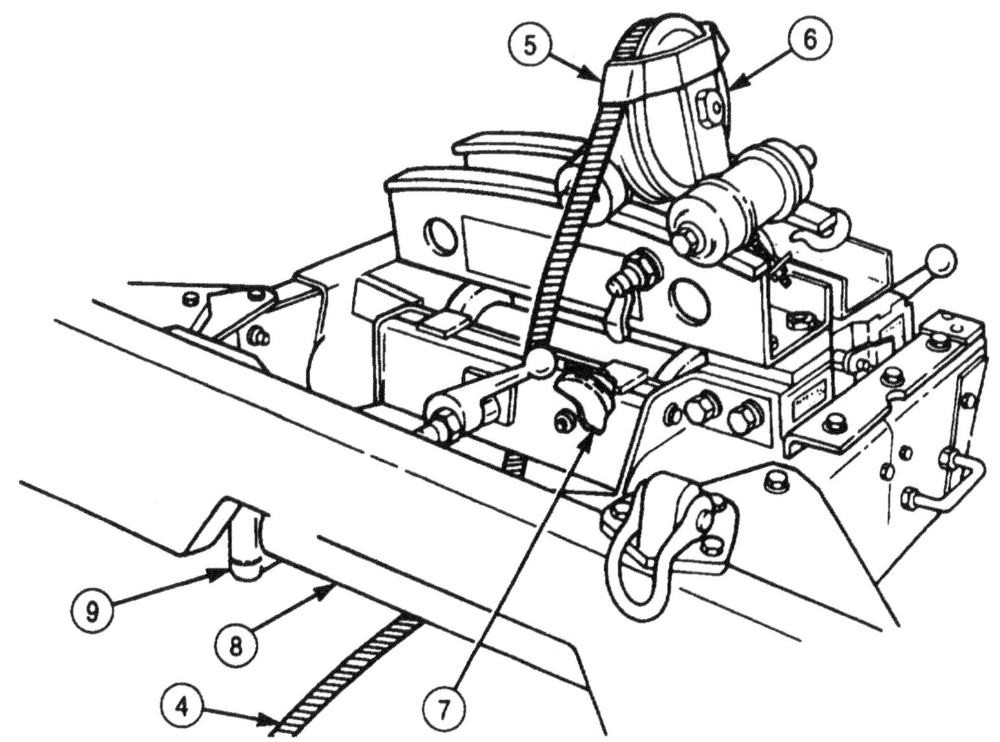

FOLLOW-ON TASKS: • Wind cable on drum (TM 9-2320-272-10).
• Install cable chain and hook (para. 3-325).

3-328. FRONT WINCH MOTOR REPLACEMENT

THIS TASK COVERS:

a. Removal **b. Installation**

INITIAL SETUP:

APPLICABLE MODELS
All with winch

TOOLS
General mechanic's tool kit (Appendix E, Item 1)
Lifting device
Chain

MATERIALS/PARTS
Gasket (Appendix D, Item 165)
Ten lockwashers (Appendix D, Item 350)
Six locknuts (Appendix D, Item 294)
Cap and plug set (Appendix C, Item 14)
Antiseize tape (Appendix C, Item 72)

REFERENCES (TM)
LO 9-2320-272-12
TM 9-2320-272-10
TM 9-2320-272-24P

EQUIPMENT CONDITION
- Parking brake set (TM 9-2320-272-10).
- Hydraulic oil reservoir drained (LO 9-2320-272-12).

a. Removal

1. Wrap chain around front of winch (11) and install on lifting device.
2. Remove six screws (1), lockwashers (2), and washers (3) from left and right support plates (5) and frame rails (10). Discard lockwashers (2).
3. Loosen two screws (4) on left and right support plates (5).
4. Remove six locknuts (8). screws (6), and washers (7) from left and right support plates (5) and support plate brackets (9). Discard locknuts (8).
5. Lift and tilt front winch (11) until winch motor (16) clears crossmember (12).

CAUTION

When disconnecting hydraulic lines and hoses, plug all openings to prevent dirt from entering and causing damage to internal parts.

NOTE

- Have drainage container ready to catch oil.
- Tag all hydraulic lines for installation.

6. Disconnect hydraulic return hose (14) and supply hose (15) from elbows (13) on winch motor (16).
7. Remove four screws (18), lockwashers (19), front winch motor (16), and gasket (17) from winch motor adapter (20). Discard lockwashers (19) and gasket (17).

3-328. FRONT WINCH MOTOR REPLACEMENT (Contd)

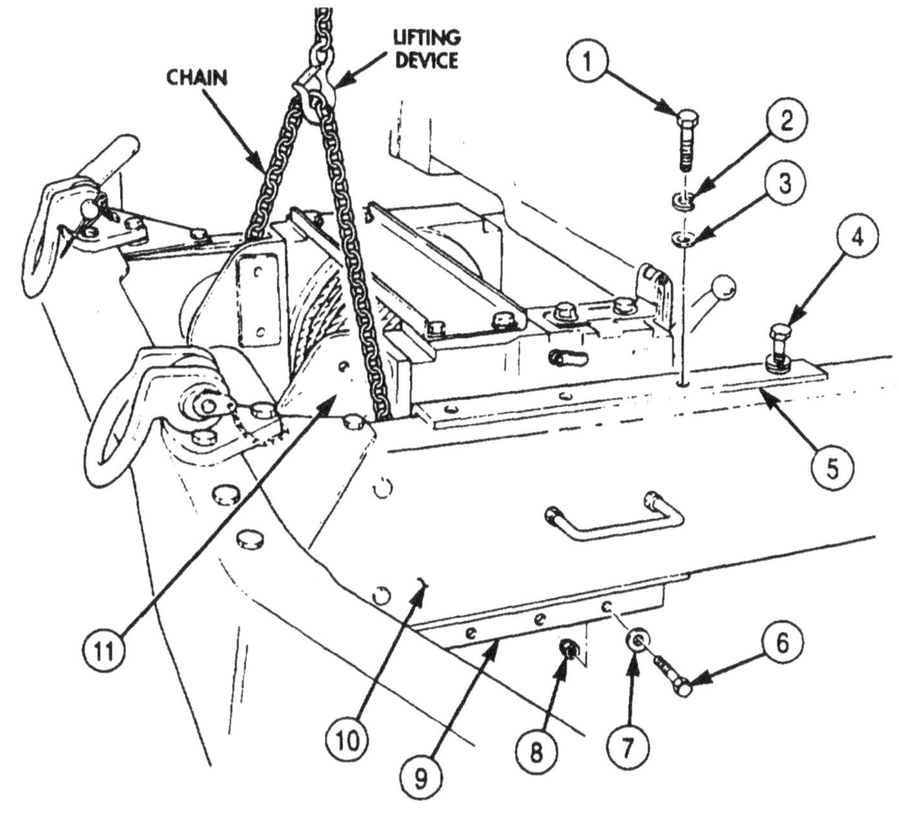

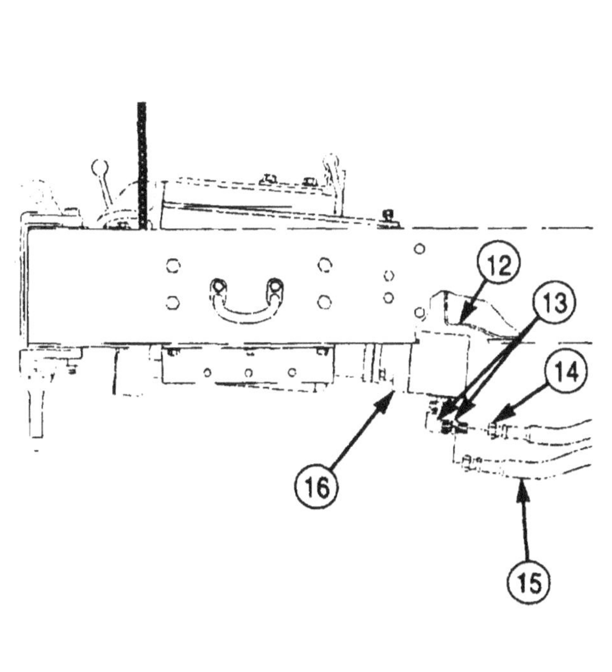

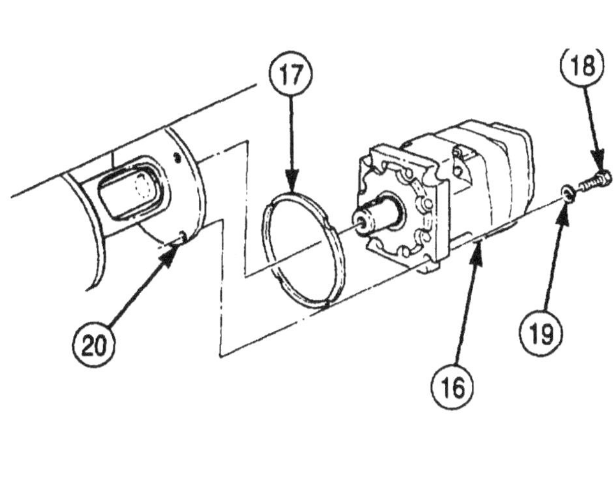

3-328. FRONT WINCH MOTOR REPLACEMENT (Contd)

b. Installation

NOTE

- If new winch motor is being installed, transfer fittings from old winch motor. Fittings must be clean and free of defects.
- Wrap all male threads with antiseize tape before installation.

1. Align winch motor shaft key (3) with keyway of winch gear shaft (1) and install new gasket (2) and winch motor (6) on winch motor adapter (7) with four new lockwashers (5) and screws (4).
2. Connect hydraulic return hose (9) and supply hose (10) to elbows (8) on winch motor (6).
3. Lower front of winch (11) and align holes of left and right support plates (16) with holes of frame rails (21) and support plate brackets (20).
4. Install six washers (14), new lockwashers (13), and screws (12) on left and right support plates (16) and frame rails (21). Finger tighten screws (12).
5. Install six washers (18), screws (17), and new locknuts (19) on left and right support plates (16) and support plate brackets (20).
6. Tighten screws (12), (15), and (17).
7. Remove lifting device and chain from front winch (11).

3-328. FRONT WINCH MOTOR REPLACEMENT (Contd)

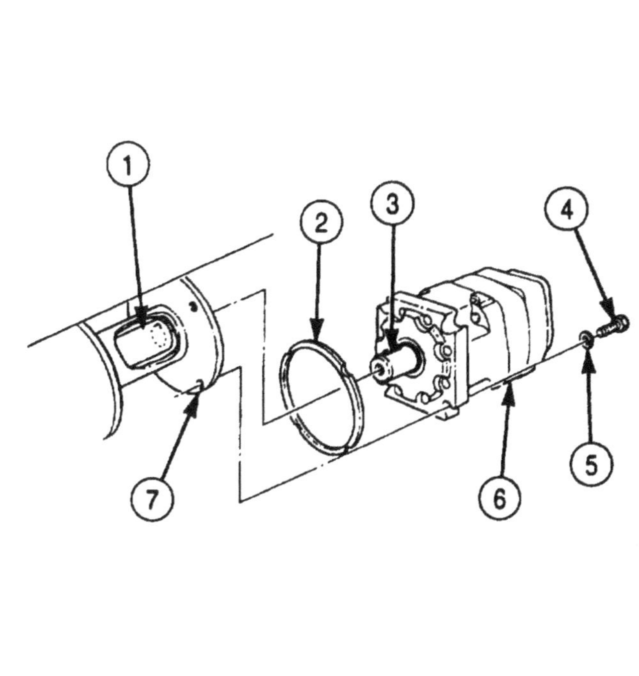

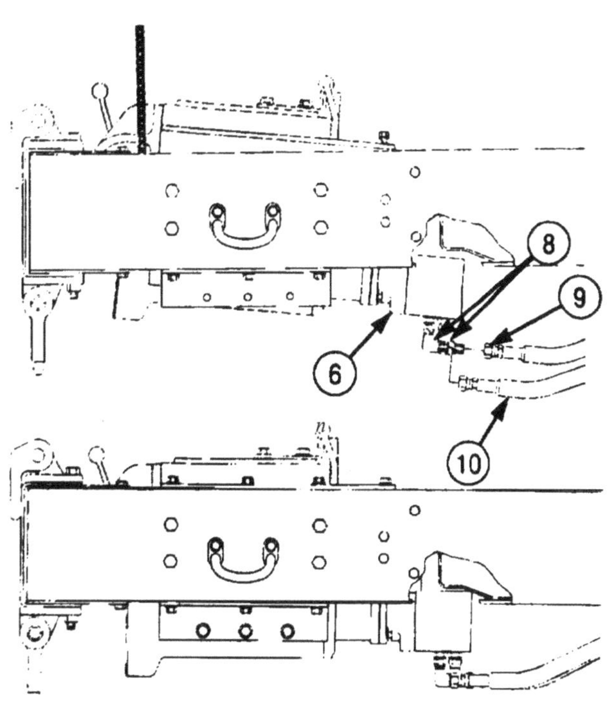

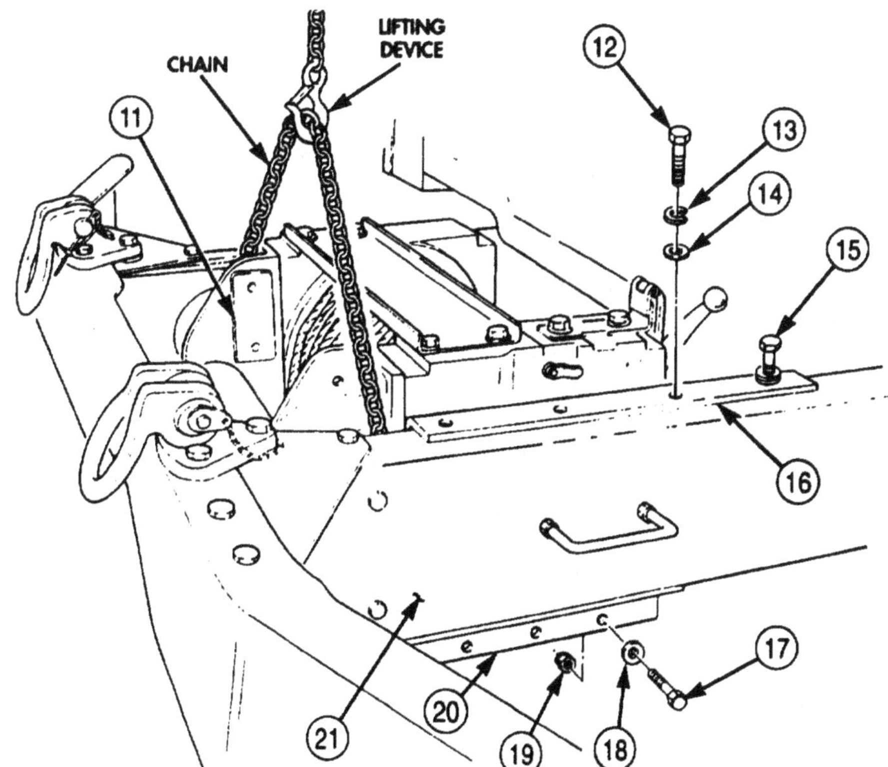

FOLLOW-ON TASKS:
- Fill hydraulic oil reservoir to proper fluid level (LO 9-2320-272-12).
- Start engine (TM 9-2320-272-10) and check winch for oil leaks and proper operation.

3-329. FRONT WINCH REPLACEMENT

THIS TASK COVERS:

a. Removal **b. Installation**

INITIAL SETUP:

APPLICABLE MODELS
All with winch

TOOLS
General mechanic's tool kit (Appendix E, Item 1)
Lifting device
Chains

MATERIALS/PARTS
Six locknuts (Appendix D, Item 294)
Eight lockwashers (Appendix D, Item 350)
Cap and plug set (Appendix C, Item 14)
Antiseize type (Appendix C, Item 72)

REFERENCES (TM)
LO 9-2320-272-12
TM 9-2320-272-10
TM 9-2320-272-24P

EQUIPMENT CONDITION
- Parking brake set (TM 9-2320-272-10)
- Front winch oil drained (LO 9-2320-272-12)
- Front bumper removed (para. 3-243).
- Front winch cable removed (para. 3-327).

GENERAL SAFETY INSTRUCTIONS
All personnel must stand clear during lifting operations.

a. Removal

CAUTION

When disconnecting hydraulic lines and hoses, plug all openings to prevent dirt from entering and causing damage to internal parts.

NOTE

- Have drainage container ready to catch oil.
- Tag all hydraulic lines for installation.

1. Disconnect oil supply hose (26) from rear elbow (27) on winch motor (28).
2. Disconnect oil return hose (25) from front elbow (24) on winch motor (28).
3. Remove four screws (2), lockwashers (3), and washers (4) from left mounting support plate (5) and left frame rail extension (11). Discard lockwashers (3).
4. Remove three locknuts (12), screws (8), and washers (9) from left inner support plate bracket (10) and left mounting support plate (5). Discard locknuts (12).
5. Remove four screws (1), lockwashers (23), and washers (22) from right mounting support plate (21) and right frame rail extension (15). Discard lockwashers (23).
6. Remove three locknuts (13), screws (17), and washers (16) from right inner support plate bracket (14) and right mounting support plate (21). Discard locknuts (13).
7. Position chains on front winch (18) and attach to lifting device.

WARNING

All personnel must stand clear during lifting operations. A snapped cable, or swinging or shifting load, may result in injury to personnel.

NOTE

Assistant will help with step 8.

8. Remove front winch (18) from frame rail extensions (11) and (15).

3-329. FRONT WINCH REPLACEMENT (Contd)

9. Remove four screws (7), washers (6), and left mounting support plate (5) from front winch (18).
10. Remove four screws (20), washers (19), and right mounting support plate (21) from front winch (18).

NOTE

Perform step 11 if replacing with new front winch.

11. Install four washers (6) and (19) and screws (7) and (20) on front winch (18).
12. Lower front winch (18) and remove chains and lifting device.

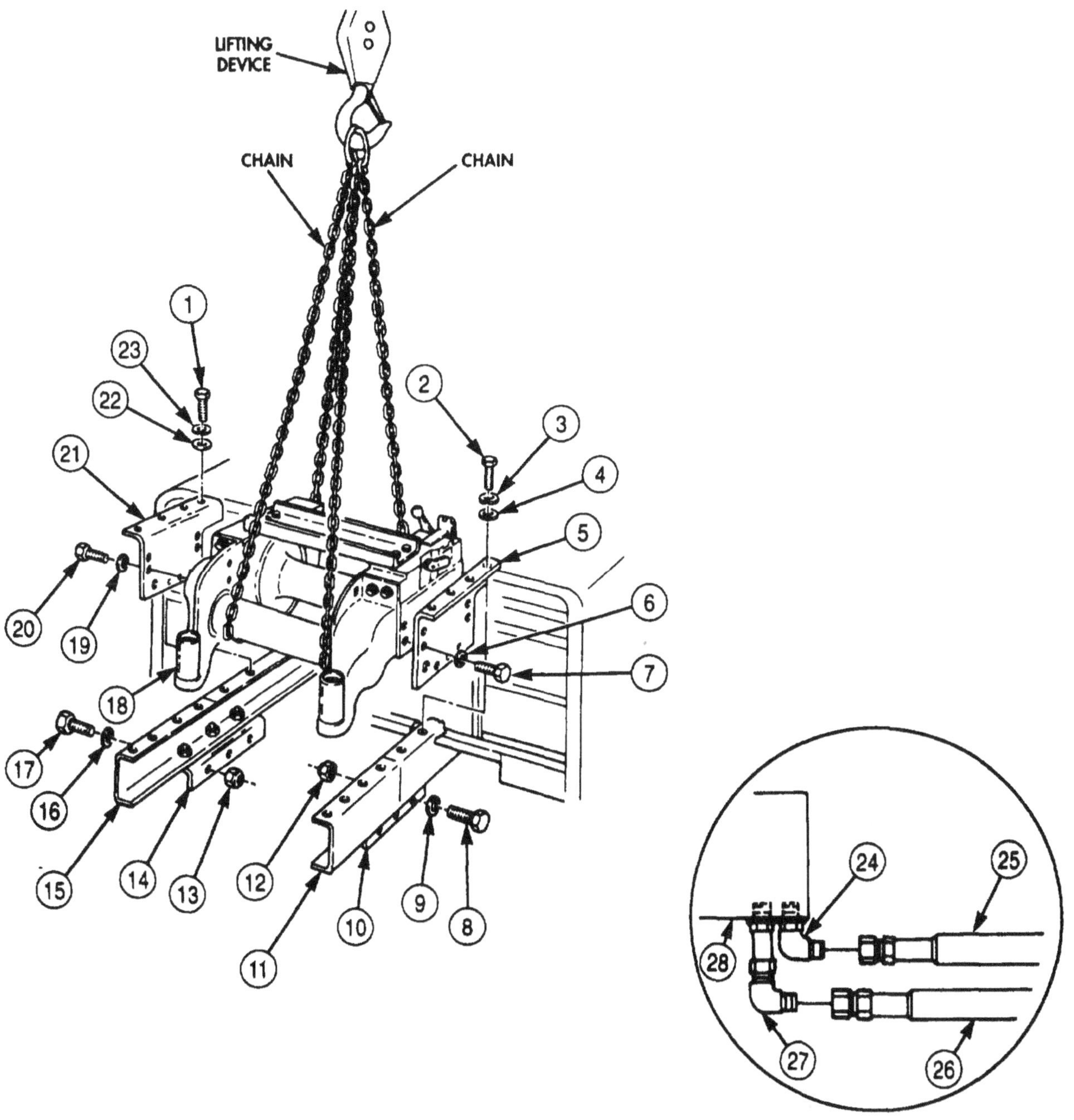

3-329. FRONT WINCH REPLACEMENT (Contd)

b. Installation

NOTE

Perform step 1 if installing new front winch.

1. Remove four screws (7) and (20) and washers (6) and (19) from front winch (18).
2. Install right mounting support plate (21) on front winch (18) with four washers (19) and screws (20).
3. Install left mounting support plate (5) on front winch (18) with four washers (6) and screws (7).
4. Position chains on front winch (18) and attach to lifting device.

WARNING`

All personnel must stand clear during lifting operations. A snapped cable, or swinging or shifting load, may result in injury to personnel.

NOTE

Assistant will help with step 5.

5. Position front winch (18) on frame rails extensions (11) and (15). Do not remove chains or lifting device until front winch (18) is secured.
6. Install right mounting support plate (21) on right inner support plate (14) with three washers (16), screws (17), and new locknuts (13).
7. Install right mounting support plate (21) on right frame rail extension (15) with four washers (22), new lockwashers (23), and screws (1).
8. Install left mounting support plate (5) on left inner support plate (10) with three washers (9), screws (8), and new locknuts (12).
9. Install install left mounting support plate (5) on left frame rail extension (11) with four washers (4), new lockwashers (3), and screws (2).
10. Remove lifting device and chains from front winch (18).

NOTE

Wrap all male threads with antiseize type before installation.

11. Connect oil supply hose (26) to rear elbow (27) on winch motor (28).
12. Connect oil return hose (25) to front elbow (24) on winch motor (28).

3-329. FRONT WINCH REPLACEMENT (Contd)

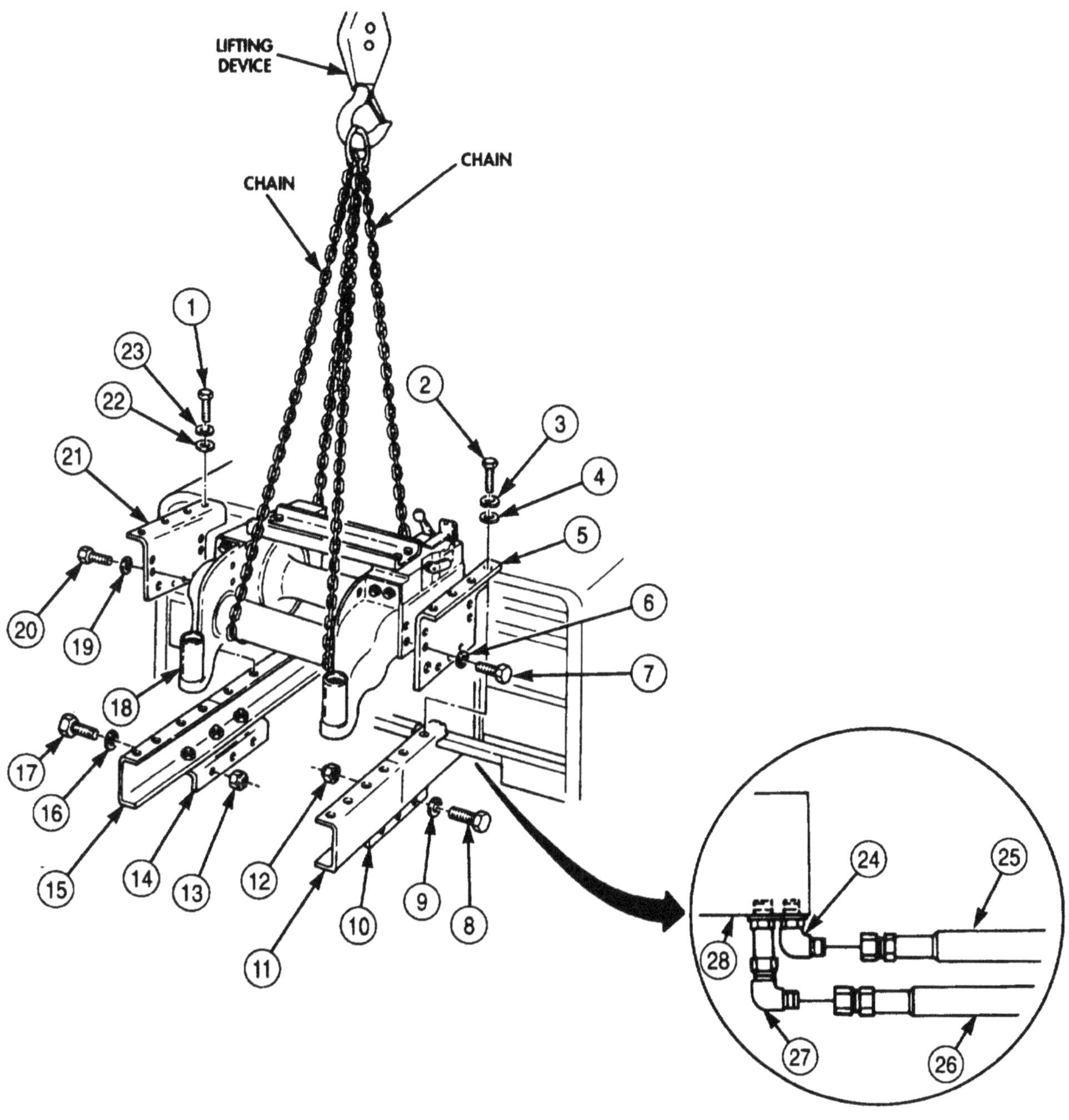

FOLLOW-ON TASKS:
- Fill hydraulic reservoir to proper fluid level (LO 9-2320-272-12).
- Fill front winch to proper fluid level (LO 9-2320-272-12).
- Install front winch cable (para. 3-327).
- Operate front winch (TM 9-2320-272-10) and check for leaks.
- Install front bumper (para. 3-243).

3-330. REAR WINCH ADJUSTMENT

THIS TASK COVERS:

a. Cable Tensioner Check **b. Cable Tensioner Adjustment**

INITIAL SETUP:

APPLICABLE MODELS
M936/A1/A2

TOOLS
General mechanic's tool kit (Appendix E, Item 1)
Test rod. 5/8 in. diameter

REFERENCES (TM)
TM 9-2320-272-10
TM 9-2320-272-24P

EQUIPMENT CONDITION
Rear winch cable removed (para. 3-331).

a. Cable Tensioner Check

1. Start engine (TM 9-2320-272-10) and allow air system to build to normal operating pressure.
2. Place cable tensioner control valve lever (1) in ON position.
3. Place test rod (3) between sheaves (2). If test rod (3) cannot be inserted or fits loosely, adjust cable tension (task b).
4. Stop engine (TM 9-2320-272-10) and remove test rod (3) if adjustment is not required.

b. Cable Tensioner Adjustment

1. Remove cotter pin (6), pin (8), and pushrod yoke (7) from pivot arm (5).
2. Place test rod (3) between sheaves (2) on cable tensioner (4).
3. Loosen jamnut (9) on pushrod (10).
4. Position pivot arm (5) so sheaves (2) are against test rod (3).
5. Turn pushrod yoke (7) until holes in pushrod (10) and pivot arm (5) align.
6. Install pushrod yoke (7) on pivot arm (5) with pin (8) and pin (6). Tighten jamnut (9).
7. Place tensioner control valve lever (1) in OFF position and remove test rod (3) from sheaves (2).
8. Stop engine (TM 9-2320-272-10).

3-330. REAR WINCH ADJUSTMENT (Contd)

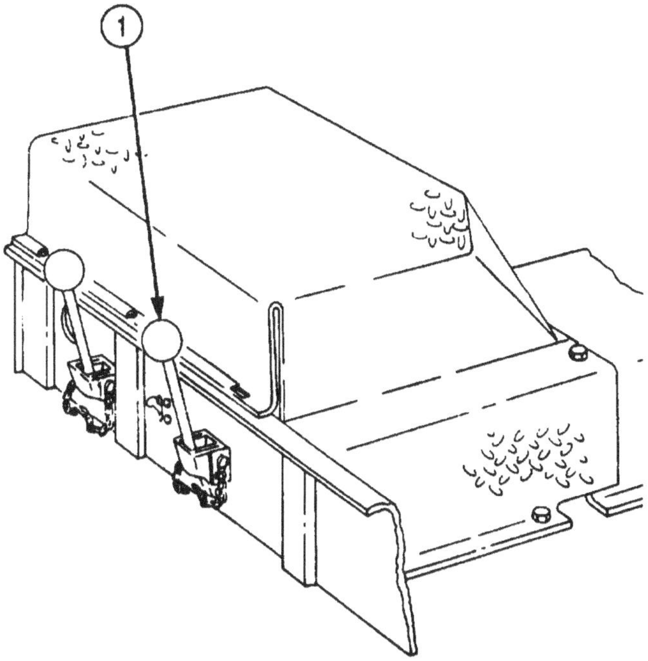

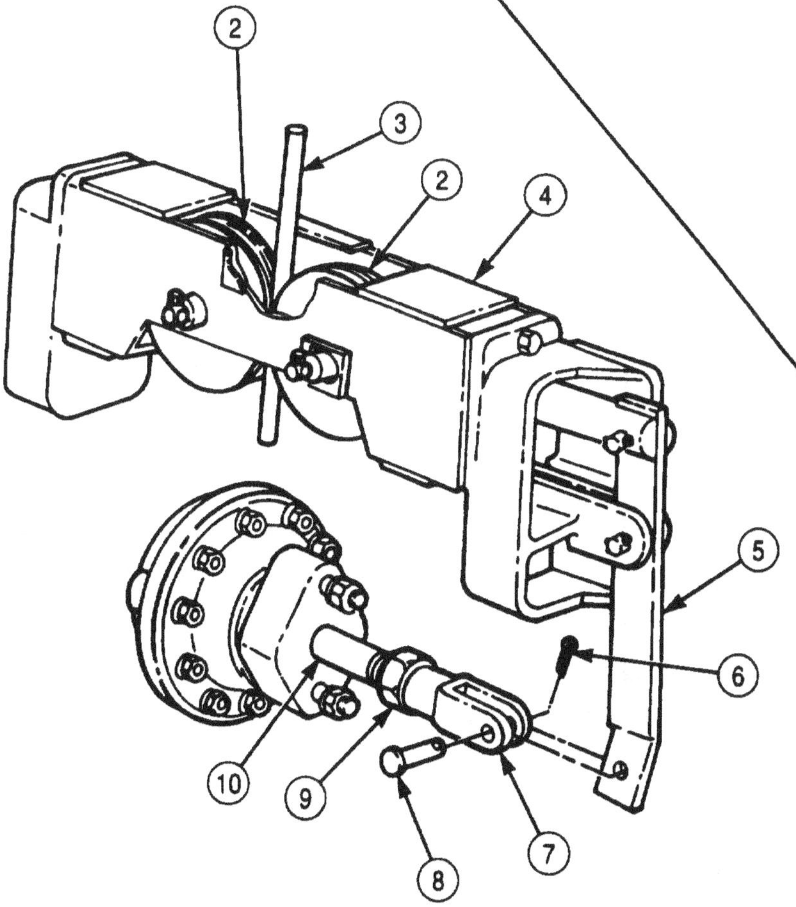

FOLLOW-ON TASK: Install rear winch cable (para. 3-331).

3-331. REAR WINCH CABLE REPLACEMENT

THIS TASK COVERS:

a. Removal **b. Installation**

INITIAL SETUP:

APPLICABLE MODELS
M936/A1/A2

TOOLS
General mechanic's tool kit (Appendix E, Item 1)

REFERENCES (TM)
TM 9-2320-272-10
TM 9-2320-272-24P

EQUIPMENT CONDITION
- Parking brake set (TM 9-2320-272-10).
- Rear winch cable unwound (TM 9-2320-272-10).

GENERAL SAFETY INSTRUCTIONS
Wear hand protection when handling winch cable.

a. Removal

WARNING

Wear hand protection when handling winch cable. Broken wires may cause injury to personnel.

1. Remove setscrew (4) and cable (2) from hole (3) in winch drum (1).
2. Pull cable (2) out over level wind pulley (9), down through tensioner sheaves (6), and out of rollers (8), and side of rollers (7).

b. Installation

1. Thread cable (2) into center of side rollers (7), rollers (8), tensioner sheaves (6), under level wind pulley plate (5), and over level wind pulley (9).
2. Insert cable (2) in hole (3) of winch drum (1) and install setscrew (4).

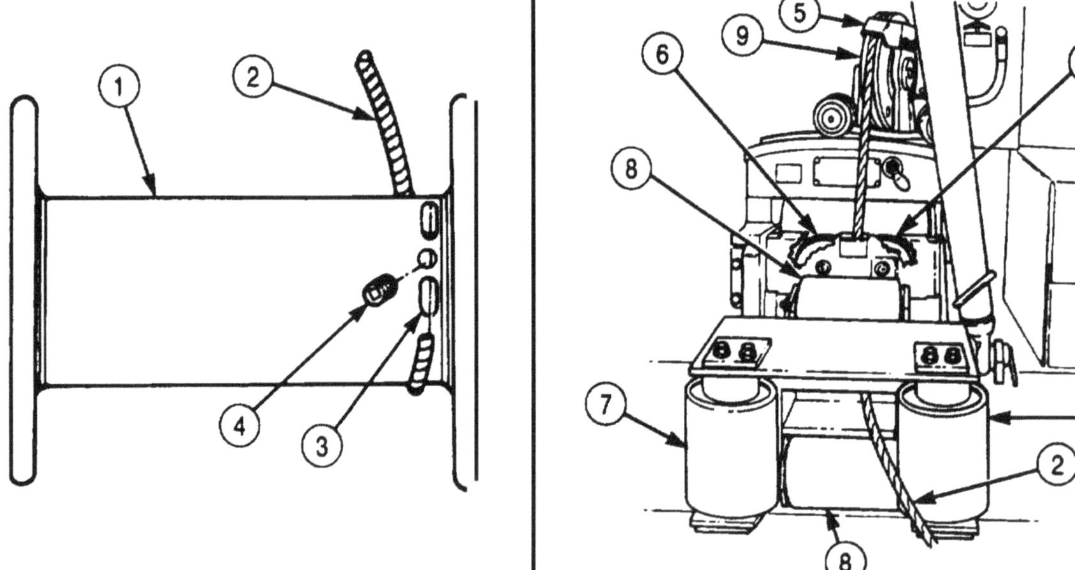

FOLLOW-ON TASK: Wind winch cable on drum (TM 9-2320-272-10).

3-332. REAR WINCH REPLACEMENT

THIS TASK COVERS:

a. Removal **b. Installation**

INITIAL SETUP:

APPLICABLE MODELS
M936/A1/A2

TOOLS
General mechanic's tool kit (Appendix E, Item 1)
Chain

MATERIALS/PARTS
Seven lockwashers (Appendix D, Item 354)

PERSONNEL REQUIRED
Two

REFERENCES (TM)
LO 9-2320-272-12
TM 9-2320-272-10
TM 9-2320-272-24P

EQUIPMENT CONDITION
- Parking brake set (TM 9-2320-272-10).
- Boom jack base plates removed (TM 9-2320-272-10).
- Rear winch oil drained (LO 9-2320-272-12).
- Rear winch cable removed (para. 3-331).
- Housing assembly cover removed (para. 3-333).

GENERAL SAFETY INSTRUCTIONS
Direct all personnel to stand clear during lifting operation.

a. Removal

1. Remove pin (1), pin (4), and pushrod yoke (3) from pivot arm (2).
2. Remove seven screws (7), lockwashers (8), washers (9), cover (6), and plate (5) from crane body (10). Discard lockwashers (8).

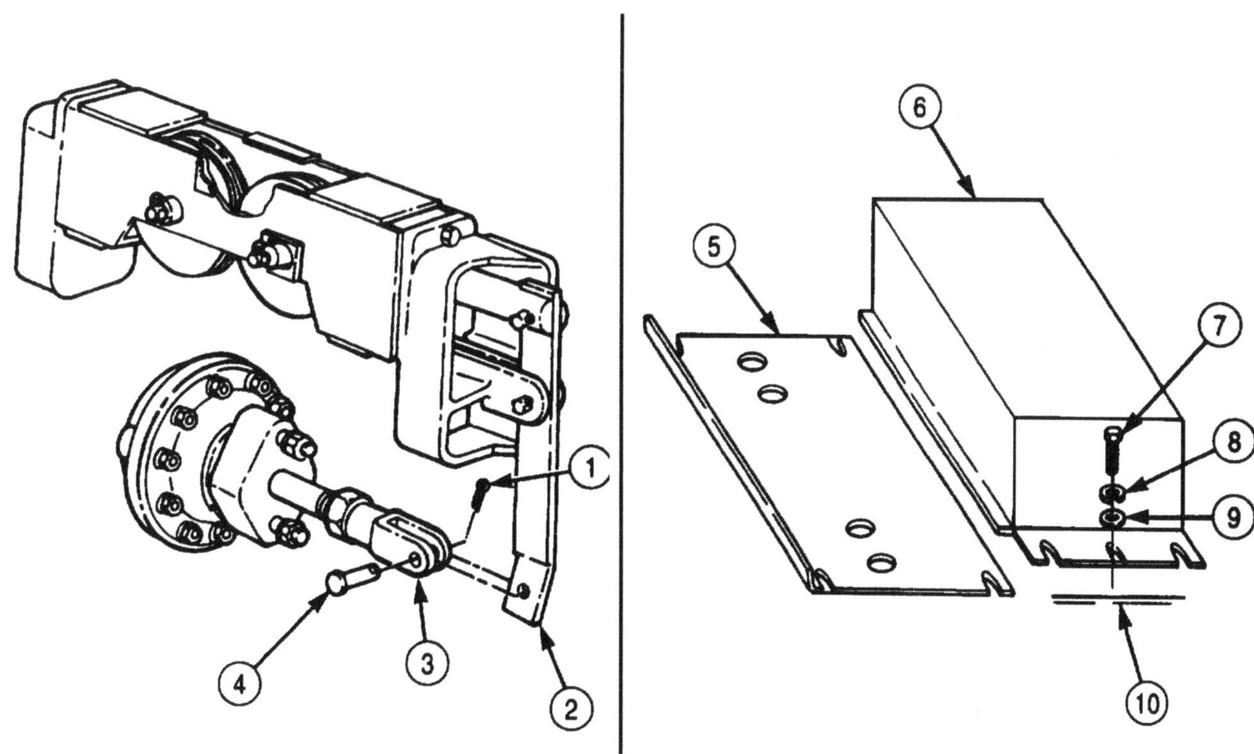

3-332. REAR WINCH REPLACEMENT (Contd)

3. Rotate drive coupling (9) to access pin (4). Remove pin (4) and shear pin (2) from drive coupling (9).
4. Remove two nuts (10), screws (14), and washers (13) from front motor bracket (12) and crane body (11).
5. Remove two nuts (5), washers (6), screws (8), and washers (6) from rear motor bracket (7) and crane body (11).
6. Slide winch motor (1) forward until clear of winch drive shaft (3).
7. Secure chain around rear winch (15) and attach to hoist hook. Raise hoist (TM 9-2320-272-10) until slack is removed from chain.

WARNING

All personnel must stand clear during lifting operations. A snapped cable, or swinging or shifting load, may result in injury to personnel.

NOTE

Mechanic will direct hoisting operation. Assistant will operate hoist. Refer to TM 9-2320-272-10 for crane operating instructions.

8. Remove four nuts (16), screws (19). and washers (18) from winch frame mounts (17) and crane body (11).
9. Remove rear winch (15) from vehicle.
10. Remove chains from hoist hook and rear winch (15).

b. Installation

1. Secure chain around rear winch (15) and attach to hoist hook. Raise hoist until slack is removed from chain.

WARNING

All personnel must stand clear during lifting operations. A snapped cable, or swinging or shifting load, may result in injury to personnel.

NOTE

Mechanic will direct hoisting operation. Assistant will operate hoist. Refer to TM 9-2320-272-10 for crane operating instructions.

2. Lift rear winch (15) into position on crane body (11) and align winch drive shaft (3) and winch frame mounts (17).
3. Remove chain from hoist hook and rear winch (11).
4. Install four washers (18), screws (19), and nuts (16) on winch frame mounts (17) and crane body (11).

3-332. REAR WINCH REPLACEMENT (Contd)

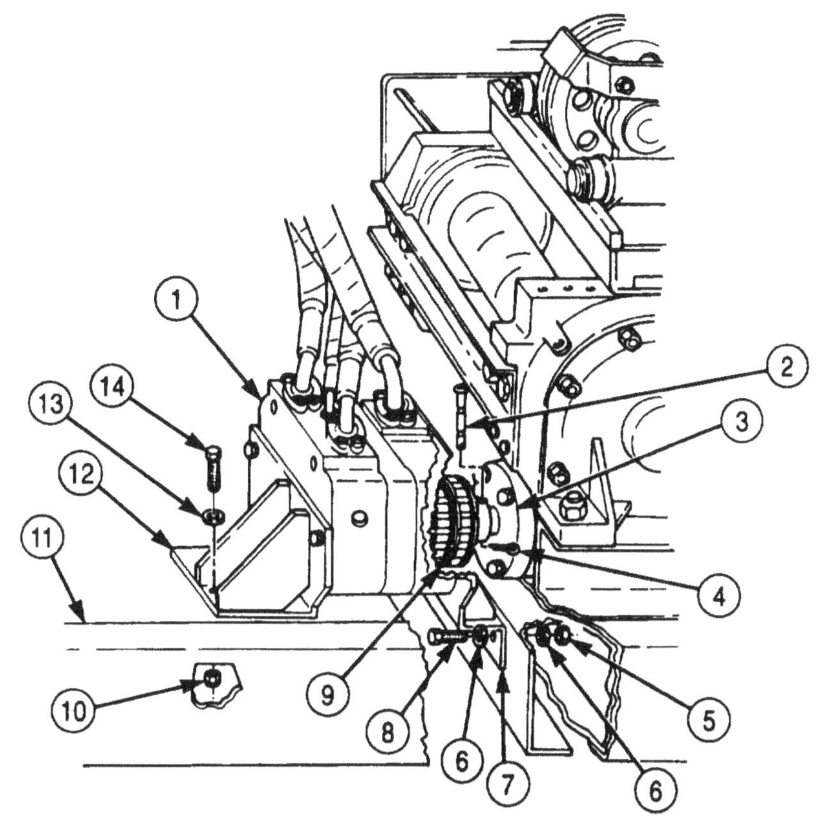

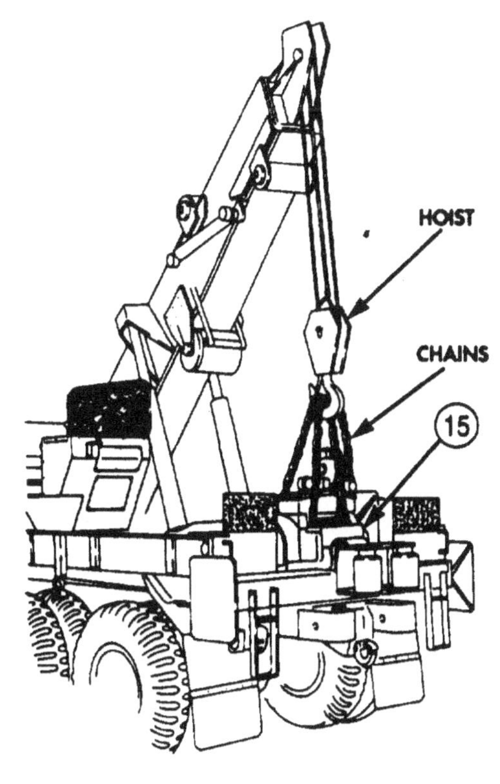

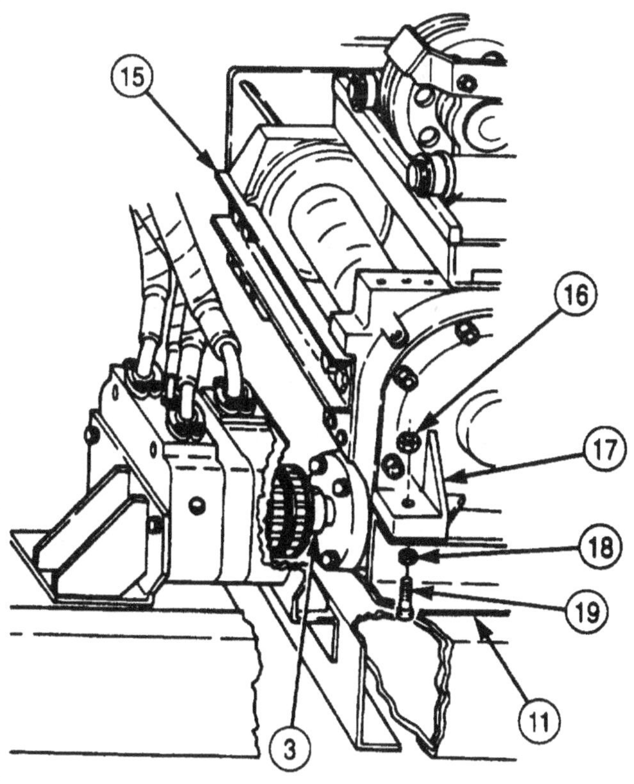

3-332. REAR WINCH REPLACEMENT (Contd)

5. Slide winch motor (16) backward position at rear winch (1), align holes in drive coupling (10) and winch drive shaft (3) and install shear pin (2) and pin (4).
6. Install front mounting bracket (13) on crane body (12) with two washers (14), screws (15), and nuts (11).
7. Install rear mounting bracket (7) on crane body (12) with two washers (8), screws (9), washers (6), and nuts (5).
8. Position plate (17) and cover (18) over winch motor (16) and install on crane body (12) with seven washers (21), new lockwashers (20), and screws (19).
9. Install pushrod yoke (24) on pivot arm (23) with pin (25) and pin (22).

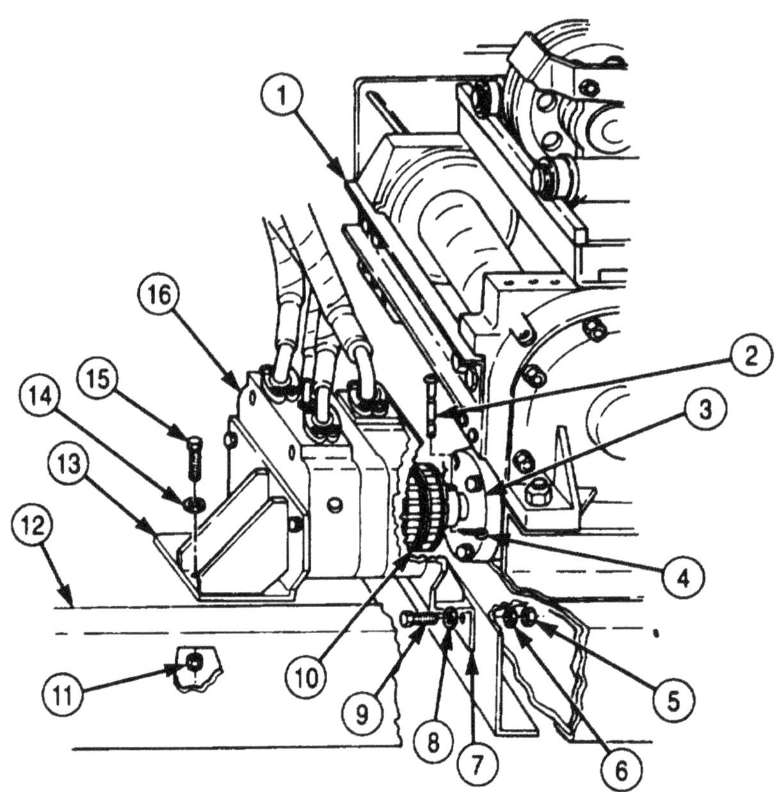

3-332. REAR WINCH REPLACEMENT (Contd)

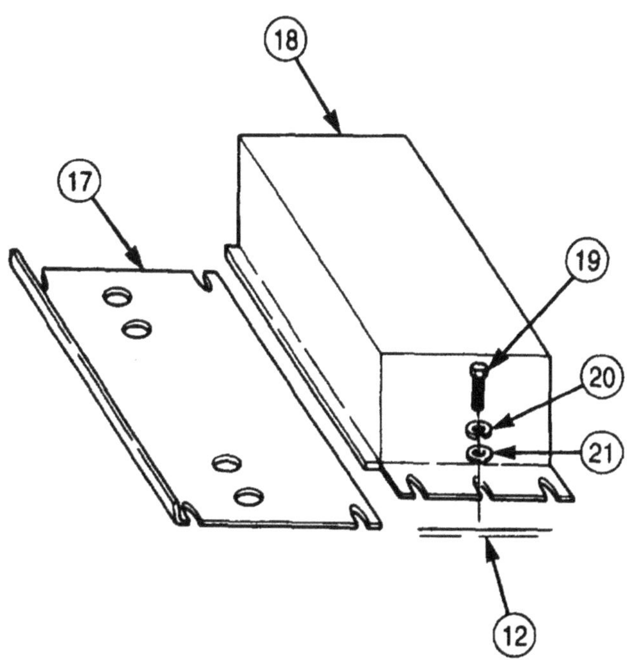

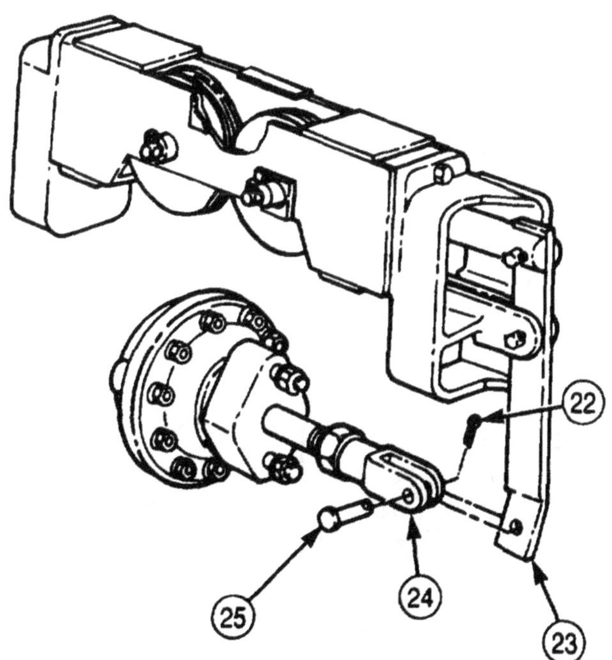

FOLLOW-ON TASKS:
- Fill rear winch to proper fluid level (LO 9-2320-272-12).
- Adjust rear winch cable tensioner (para. 3-330).
- Install rear winch cable (para. 3-331).
- Operate rear winch (TM 9-2320-272-10) and check for leaks.
- Install boom jack base plates (TM 9-2320-272-10).
- Install housing assembly cover (para. 3-333).
- Lubricate winch (LO 9-2320-272-12).

3-333. HOUSING ASSEMBLY COVER REPLACEMENT

THIS TASK COVERS:

a. Removal

b. Installation

INITIAL SETUP:

APPLICABLE MODELS

M936/A1/A2

TOOLS

General mechanic's tool kit (Appendix E, Item 1)

MATERIALS/PARTS

Nine lockwashers (Appendix D, Item 354)

REFERENCES (TM)

TM 9-2320-272-10

TM 9-2320-272-24P

EQUIPMENT CONDITION

Parking brake set (TM 9-2320-272-10).

a. Removal

Remove nine screws (1), lockwashers (3), washers (4), bracket (5), and housing assembly cover (2) from crane body (6). Discard lockwashers (3).

b. Installation

Install housing assembly cover (2) on crane body (6) with bracket (5), nine washers (4), new lockwasher (3), and screws (1).

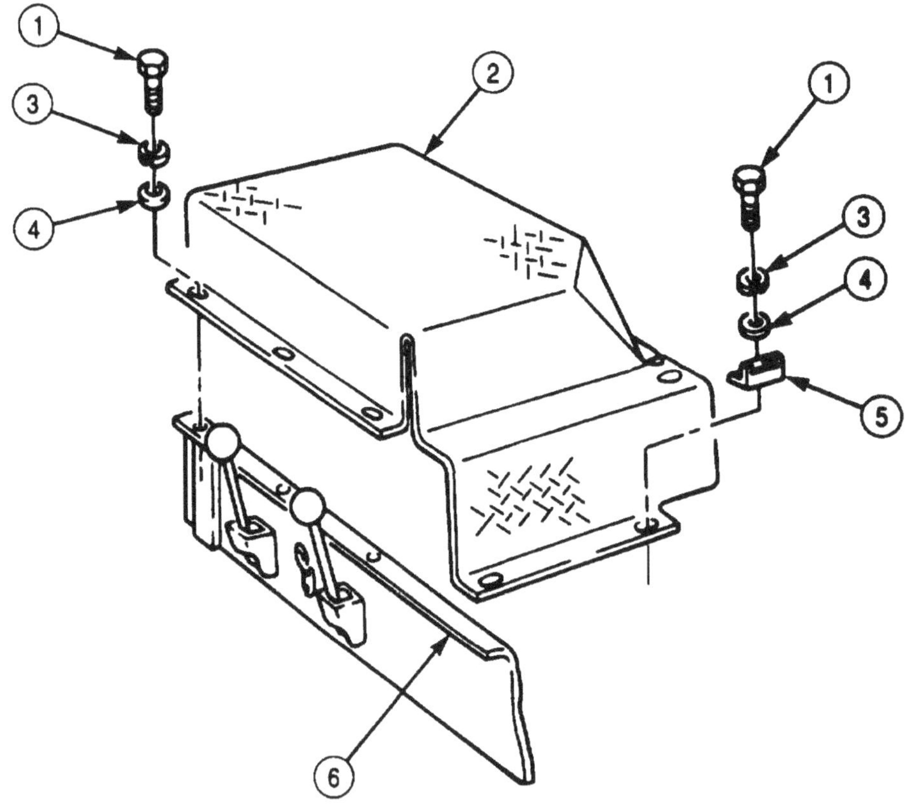

3-334. TRANSMISSION PTO-TO-HYDRAULIC PUMP PROPELLER SHAFT REPLACEMENT

THIS TASK COVERS:

a. Removal b. Installation

INITIAL SETUP:

APPLICABLE MODELS
M925/A1/A2, M928/A1/A2, M929/A1/A2,
M930/A1/A2, M932/A1/A2, M936/A1/A2

TOOLS
General mechanic's tool kit (Appendix E, Item 1)

REFERENCES (TM)
TM 9-2320-272-10
TM 9-2320-272-24P

EQUIPMENT CONDITION
Parking brake set (TM 9-2320-272-10).

a. Removal

Remove setscrews (5) and (9) and propeller shaft (4) from PTO shaft (2) and hydraulic pump shaft (8).

b. Installation

1. Position universal joint ends (3) of propeller shaft (4) over PTO shaft (2) and hydraulic pump shaft (8). Ensure universal joint ends (3) are seated properly over woodruff keys (6) of PTO shaft (2) and hydraulic pump shaft (8).
2. Install propeller shaft (4) on PTO shaft (2) and hydraulic pump shaft (8) with setscrews (5) and (9). Ensure setscrews (5) and (9) seat properly in shaft channels (1) and (7).

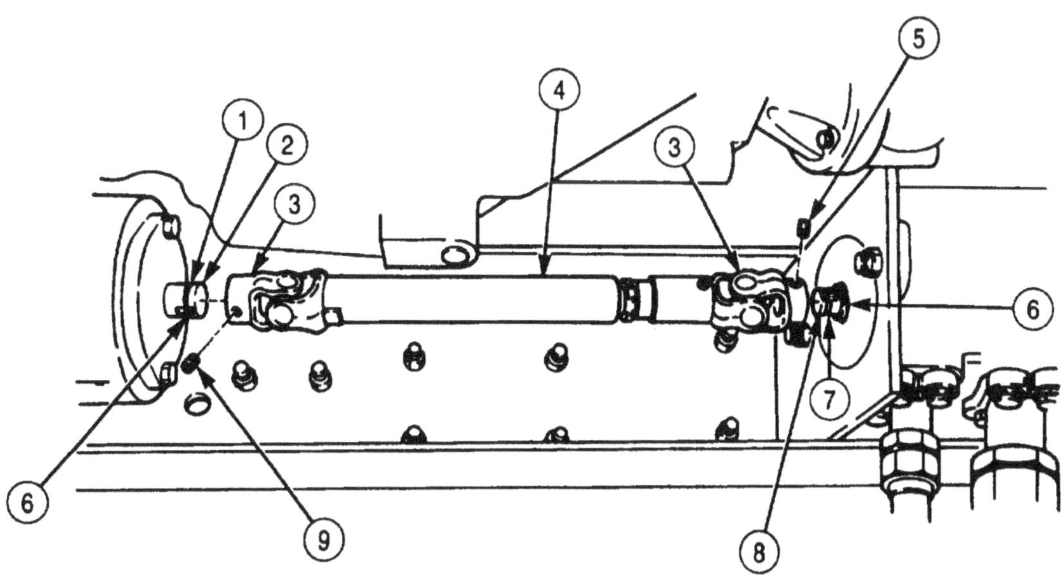

FOLLOW-ON TASK: Start engine (TM 9-2320-272-10) and operate any accessory driven by the transmission PTO-to-hydraulic pump propeller shaft. Stop engine (TM 9-2320-272-10) and check shaft for looseness.

3-335. TRANSMISSION PTO-TO-HYDRAULIC PUMP PROPELLER SHAFT UNIVERSAL JOINT MAINTENANCE

THIS TASK COVERS:

a. Disassembly
b. Inspection
c. Assembly

INITIAL SETUP:

APPLICABLE MODELS
M925/A1/A2, M928/A1/A2, M929/A1/A2, M930/A1/A2, M932/A1/A2, M936/A1/A2

TOOLS
General mechanic's tool kit (Appendix E, Item 1)

MATERIALS/PARTS
Dust cap assembly (M939/A1) (Appendix D, Item 103)
Dust cap assembly (M939A2) (Appendix D, Item 104)
Lubricating oil (Appendix C, Item 50)

REFERENCES (TM)
TM 9-2320-272-24P

EQUIPMENT CONDITION
Transmission PTO-to-hydraulic pump propeller shaft removed (para. 3-334).

---

a. Disassembly

NOTE

This procedure covers maintenance for both universal joints,

1. Position propeller shaft (1) in soft-jawed vise so end yoke (2) moves freely.
2. Remove four lockrings (6) from universal joint (5).

NOTE

Do not drop bearing cups. Needle bearings inside are very small and can easily be lost.

3. Remove two bearing cups (4) and end yoke (2) from propeller shaft yoke (3).
4. Remove two bearing cups (7) and universal joint (5) from propeller shaft yoke (3).
5. Remove grease fitting (8) from universal joint (5).
6. Remove dust cap (10) and slip yoke (13) from propeller shaft splines (9).
7. Remove dust cap (10), nylon washer (11), and felt washer (12) from slip yoke (13). Discard felt washer (12), nylon washer (11), and dust cap (10).
8. Remove grease fitting (14) from slip yoke (13).

b. Inspection

1. Inspect bearing cups (4) and (7) for worn or missing needle bearings. Replace needle bearings if worn or missing.
2. Apply a few drops of lubricating oil in bearing cups (4) and (7). Using finger, roll needle bearings around to check for free movement. Replace bearing cup(s) (4) or (7) if needle bearing movement is rough or uneven.
3. Inspect end yoke (2), propeller shaft yoke (3), slip yoke (13), and universal joint (5) for scoring, burrs, cracks, and bends. Replace part(s) if defective.

3-335. TRANSMISSION PTO-TO-HYDRAULIC PUMP PROPELLER SHAFT UNIVERSAL JOINT MAINTENANCE (Contd)

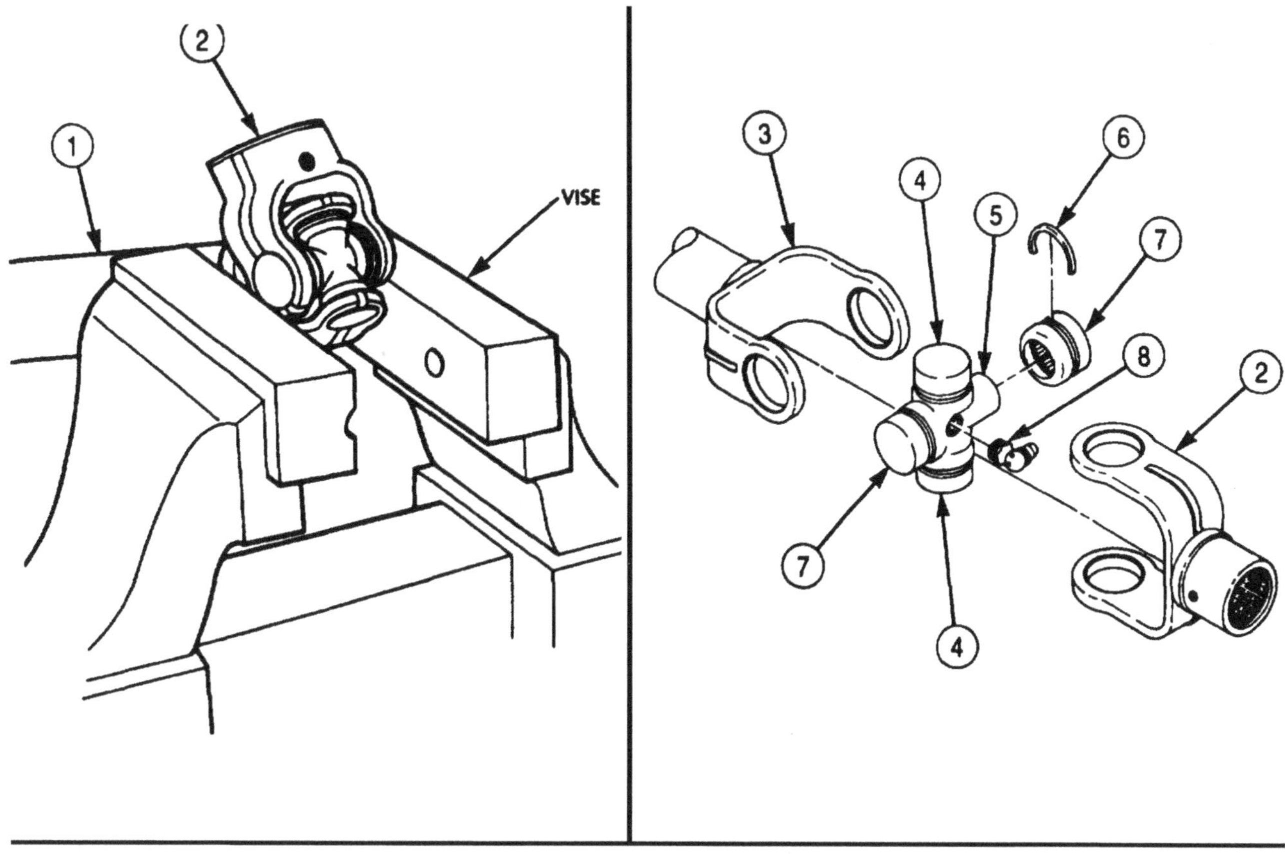

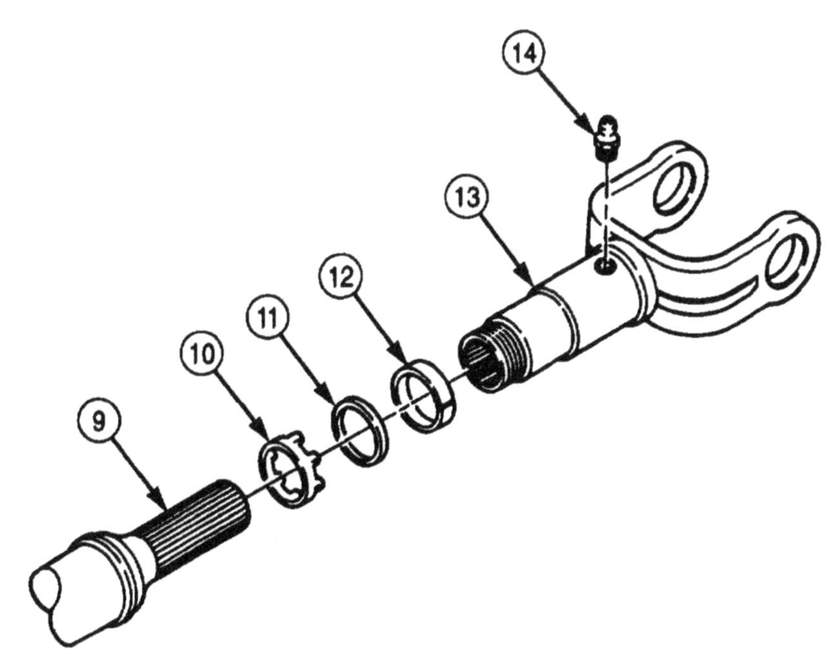

3-335. TRANSMISSION PTO-TO-HYDRAULIC PUMP PROPELLER SHAFT UNIVERSAL JOINT MAINTENANCE (Contd)

c. Assembly

1. Install grease fitting (6) on slip yoke (5).

NOTE

Soak new felt washer in lubricating oil prior to installation.

2. Slide new dust cap (2), new nylon washer (3), new felt washer (4), and slip yoke (5) over propeller shaft splines (1).
3. Install nylon washer (3) and felt washer (4) in dust cap (2).
4. Install dust cap (2) on slip yoke (5).
5. Place propeller shaft (14) in soft-jawed vise.
6. Install grease fitting (12) on universal joint (9).

NOTE

- Lubricate new or used universal joint before assembly.
- Press bearing cups into yoke far enough to install lockrings.

7. Place universal joint (9) in propeller shaft yoke (7).
8. Install two bearing cups (8) on universal joint (9) and end yoke (13) with two lockrings (10).
9. Install two bearing cups (11) on universal joint (9) and propeller shaft yoke (7) with two lockrings (10).
10. Remove propeller shaft (14) from soft-jawed vise.

3-335. TRANSMISSION PTO-TO-HYDRAULIC PUMP PROPELLER SHAFT UNIVERSAL JOINT MAINTENANCE (Contd)

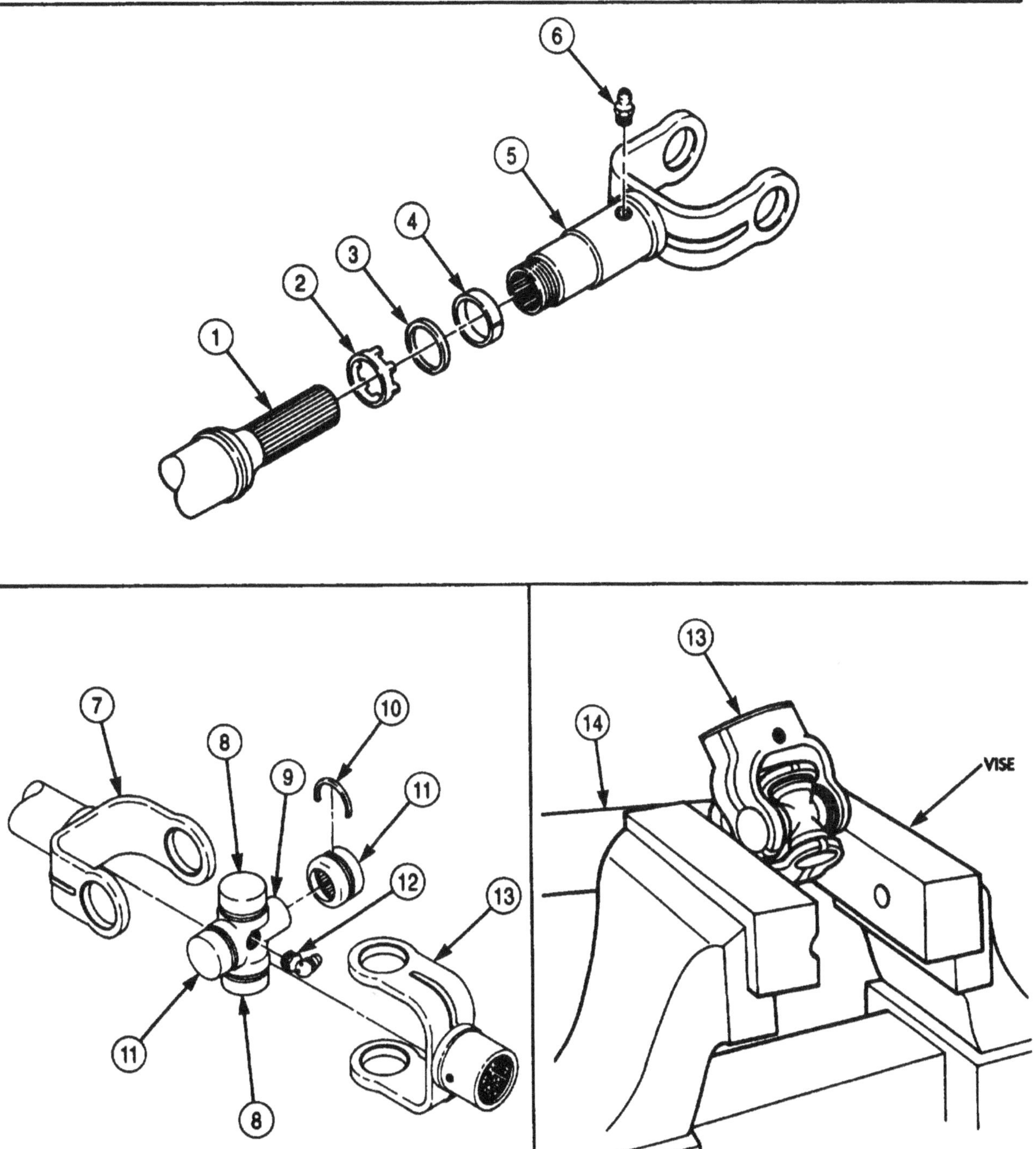

FOLLOW-ON TASK: Install transmission PTO-to-hydraulic pump propeller shaft (para. 3-334).

3-336. WINCH HYDRAULIC OIL RESERVOIR FILTER REPLACEMENT

THIS TASK COVERS:

a. Removal b. Installation

INITIAL SETUP:

APPLICABLE MODELS
M925/A1/A2, M928/A1/A2, M929/A1/A2,
M930/Al/A2, M932/A1/A2, M936/A1/A2

TOOLS
General mechanic's tool kit (Appendix E, Item 1)
Torque wrench (Appendix E, Item 146)

MATERIALS/PARTS
Oil filter and gasket (Appendix D, Item 492)

REFERENCES (TM)
LO 9-2320-272-12
TM 9-2320-272-10
TM 9-2320-272-24P

EQUIPMENT CONDITION
Parking brake set (TM 9-2320-272-10).

a. Removal

NOTE

- The oil filter is located on the right frame rail above the wet tank air reservoir.
- Have drainage container ready to catch oil.

1. Loosen center bolt (6) and remove filter housing (7), spring (3), and gasket (2) from filter base (1). Discard gasket (2).
2. Remove center bolt (6), washer (5), and oil filter (4) from filter housing (7). Discard oil filter (4).

b. Installation

1. Install washer (5), center bolt (6), new oil filter (4), and spring (3) in filter housing (7).
2. Position new gasket (2) on filter base (1).
3. Install filter housing (7) on gasket (2) and filter base (1) with center bolt (6). Tighten center bolt (6) 30-35 lb-ft (41-47 N•m).

3-336. WINCH HYDRAULIC OIL RESERVOIR FILTER REPLACEMENT (Contd)

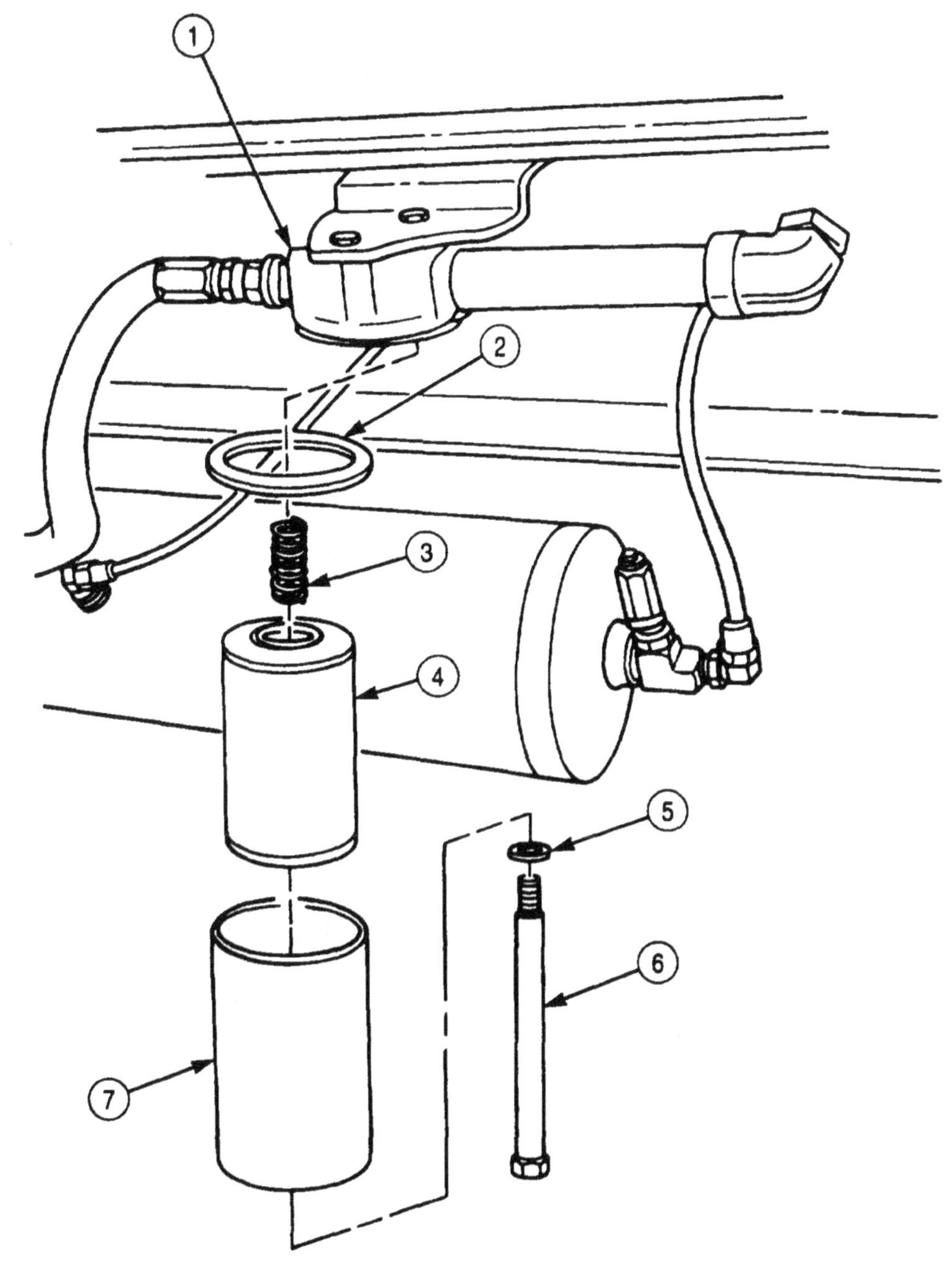

FOLLOW-ON TASKS: • Fill hydraulic oil reservoir to proper fluid level (LO 9-2320-272-12).
• Operate hydraulic system (TM 9-2320-272-10) and check for leaks.

3-337. WINCH HYDRAULIC OIL RESERVOIR REPLACEMENT

THIS TASK COVERS:

a. Removal b. Installation

INITIAL SETUP:

APPLICABLE MODELS

M925/A1/A2, M928/A1/A2, M929/A1/A2, M930/A1/A2, M932/A1/A2, M936/A1/A2

TOOLS

General mechanic's tool kit (Appendix E, Item 1)

MATERIALS/PARTS

Eight locknuts (Appendix D. Item 291)
Tiedown straps (Appendix D, Item 695)
Cap and plug set (Appendix C, Item 14)
Antiseize tape (Appendix C, Item 72)

REFERENCES (TM)

LO 9-2320-272-12
TM 9-2320-272-10
TM 9-2320-272-24P

EQUIPMENT CONDITION

- Parking brake set (TM 9-2320-272-10).
- Hydraulic oil reservoir drained (LO 9-2320-272-12).

GENERAL SAFETY INSTRUCTIONS

Store or dispose of used oil properly.

WARNING

Accidental or intentional introduction of liquid contaminants into the environment is in violation of state, federal, and military regulations. Refer to Army POL (para. 1-8) for information concerning storage, use, and disposal of these liquids. Failure to do so may result in injury or death.

a. Removal

CAUTION

When disconnecting hydraulic lines and hoses, plug all openings to prevent dirt from entering and causing damage to internal parts.

NOTE

- Have drainage container ready to catch oil.
- Tag all hydraulic lines for installation.

1. Remove tiedown straps from oil supply hose (6) as required. Discard tiedown straps.
2. While holding nut (5), loosen nut (4) and remove oil supply hose (6) from elbow (1).
3. Loosen nut (10) and remove oil return hose (9) from elbow (11).
4. Remove elbow (1) and nipple (2) from reservoir (3).
5. Remove nut (18), screw (8), clamp (7), and oil return hose (9) from reservoir (3).
6. Remove eight locknuts (17) and screws (16) from frame bracket (19) and reservoir (3). Discard locknuts (17).
7. Remove and pull reservoir (3) out about 2 in. (5 cm) from frame bracket (19).
8. Remove reservoir (3) from frame bracket (19).
9. Remove elbow (11) from reservoir (19).
10. Remove filler cap (15), dipstick (14), spacer (13), and strainer (12) from reservoir (3).

3-337. WINCH HYDRAULIC OIL RESERVOIR REPLACEMENT (Contd)

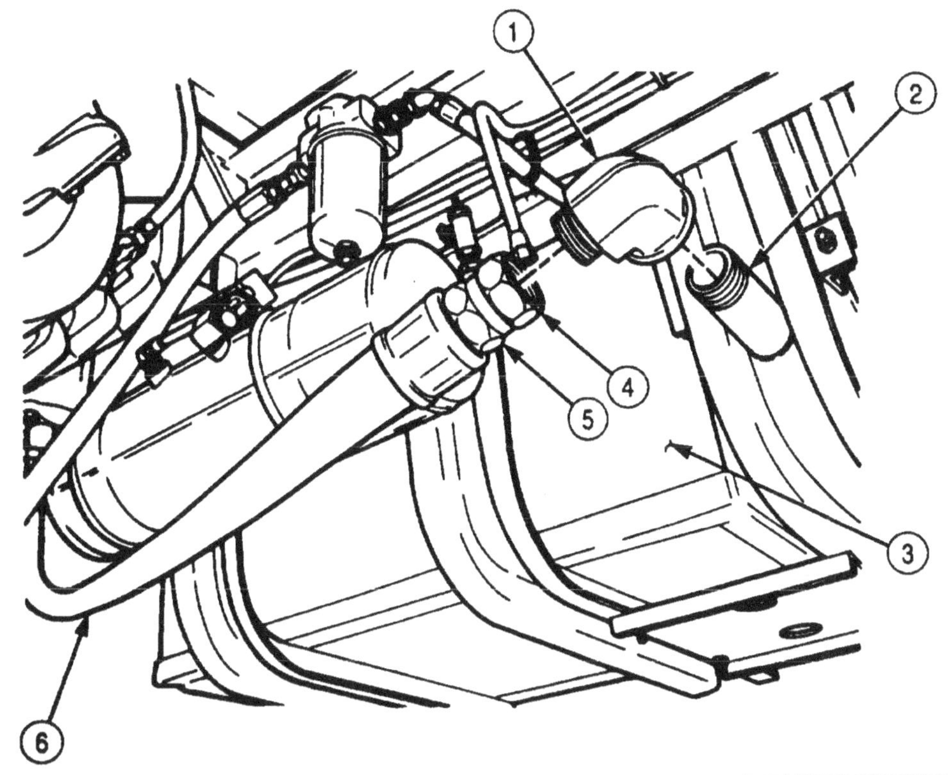

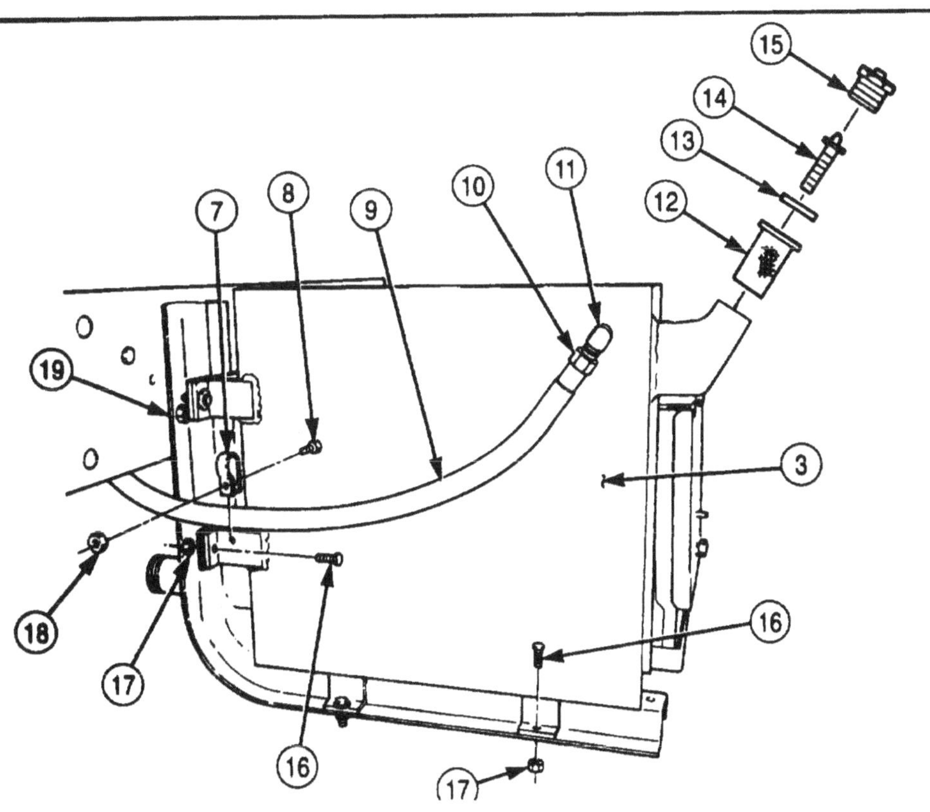

3-337. WINCH HYDRAULIC OIL RESERVOIR REPLACEMENT (Contd)

b. Installation

NOTE

- When new hydraulic oil reservoir is installed, use attaching parts and fittings from old hydraulic oil reservoir.
- Wrap all male threads with antiseize tape before installation.

1. Install strainer (5), spacer (6), dipstick (7), and filler cap (8) on reservoir (9).
2. Install elbow (4) on reservoir (9).
3. Position reservoir (9) on frame bracket (14).
4. Install nipple (16) and elbow (15) on reservoir (9).
5. Install reservoir (9) on frame bracket (14) with eight screws (10) and new locknuts (11).
6. Install oil return hose (12) on elbow (4) and tighten nut (3).
7. Install oil supply hose (19) on elbow (15), and while holding nut (18), tighten nut (17).
8. Install oil return hose (12) on reservoir (9) with clamp (1), screw (2), and nut (13).
9. Install new tiedown straps on oil supply hose (19) as required.

3-337. WINCH HYDRAULIC OIL RESERVOIR REPLACEMENT (Contd)

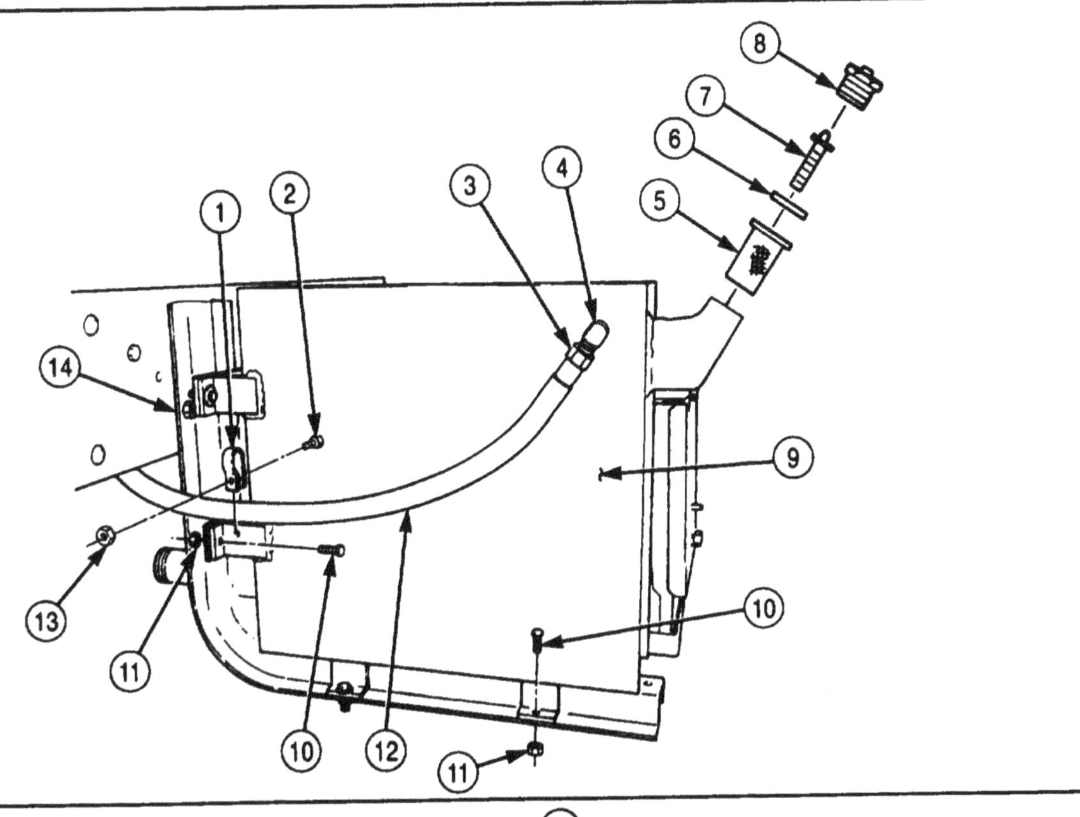

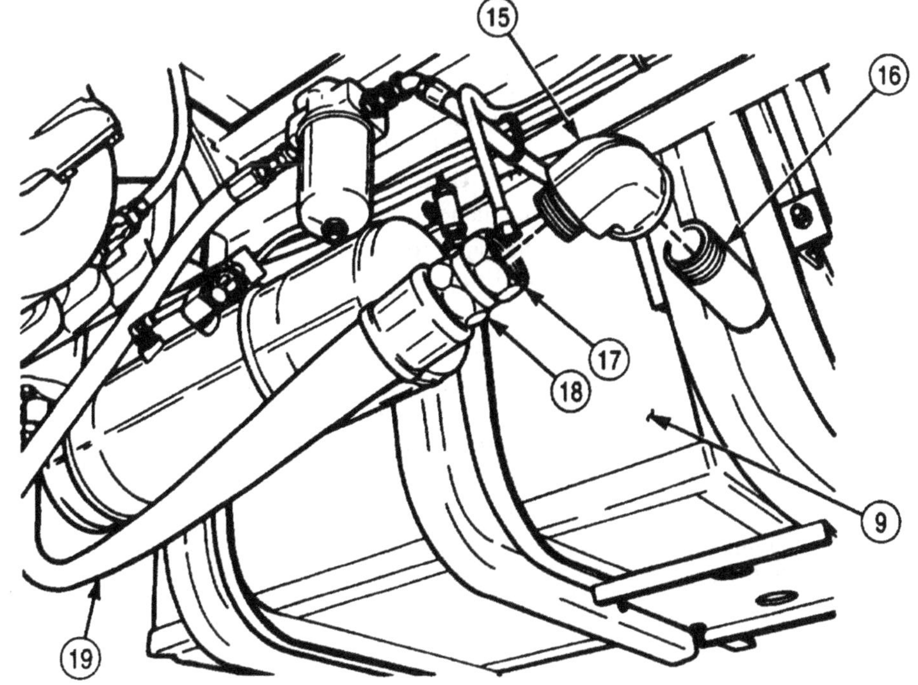

FOLLOW-ON TASKS:
- Fill hydraulic oil reservoir to proper fluid level (LO 9-2320-272-12).
- Operate hydraulic system (TM 9-2320-272-10) and check for leaks.

3-338. TRACTOR WINCH HYDRAULIC OIL RESERVOIR (M932/A1/A2) REPLACEMENT

THIS TASK COVERS:

a. Removal | b. Installation

INITIAL SETUP:

APPLICABLE MODELS
M932/A1/A2

TOOLS
General mechanic's tool kit (Appendix E, Item 1)

MATERIALS/PARTS
Six locknuts (Appendix D, Item 294)
Four locknuts (Appendix D, Item 295)
Locknut (Appendix D, Item 274)
Two locknuts (Appendix D, Item 300)
Four lockwashers (Appendix D, Item 354)
O-ring (Appendix D, Item 455)
Cap and plug set (Appendix C, Item 14)
Antiseize tape (Appendix C, Item 72)

PERSONNEL REQUIRED
TWO

REFERENCES (TM)
LO 9-2320-272-12
TM 9-2320-272-10
TM 9-2320-272-24P

EQUIPMENT CONDITION
- Parking brake set (TM 9-2320-272-10).
- Spare tire removed (TM 9-2320-272-10).
- Hydraulic oil reservoir drained (LO 9-2320-272-12).

GENERAL SAFETY INSTRUCTIONS
Store or dispose of used oil properly.

WARNING

Accidental or intentional introduction of liquid contaminants into the environment is in violation of state, federal, and military regulations. Refer to Army POL (para. 1-8) for information concerning storage, use, and disposal of these liquids. Failure to do so may result in injury or death.

a. Removal

CAUTION

When disconnecting hydraulic lines and hoses, plug all openings to prevent dirt from entering and causing damage to internal parts.

NOTE

- Have drainage container ready to catch oil.
- Tag all hydraulic lines for installation.

1. Remove two locknuts (4), screws (2), and brace (5) from muffler support (1) and reservoir base (3). Discard locknuts (4).
2. While holding nut (20), disconnect oil supply line (19) from elbow (13).
3. Remove four screws (18), lockwashers (17), two split flanges (16), oil supply hose (19), and O-ring (15) from hydraulic oil pump (14). Discard lockwashers (17) and O-ring (15).
4. Remove two locknuts (9), washers (8), and U-bolt (22) from nipple support (7). Discard locknuts (9).
5. Remove elbow (13) and nipple (12) from reservoir (6).
6. Remove two locknuts (21), screws (10), washers (11), and nipple support (7) from reservoir base (3). Discard locknuts (21).

3-338. TRACTOR WINCH HYDRAULIC OIL RESERVOIR (M932/A1/A2) REPLACEMENT (Contd)

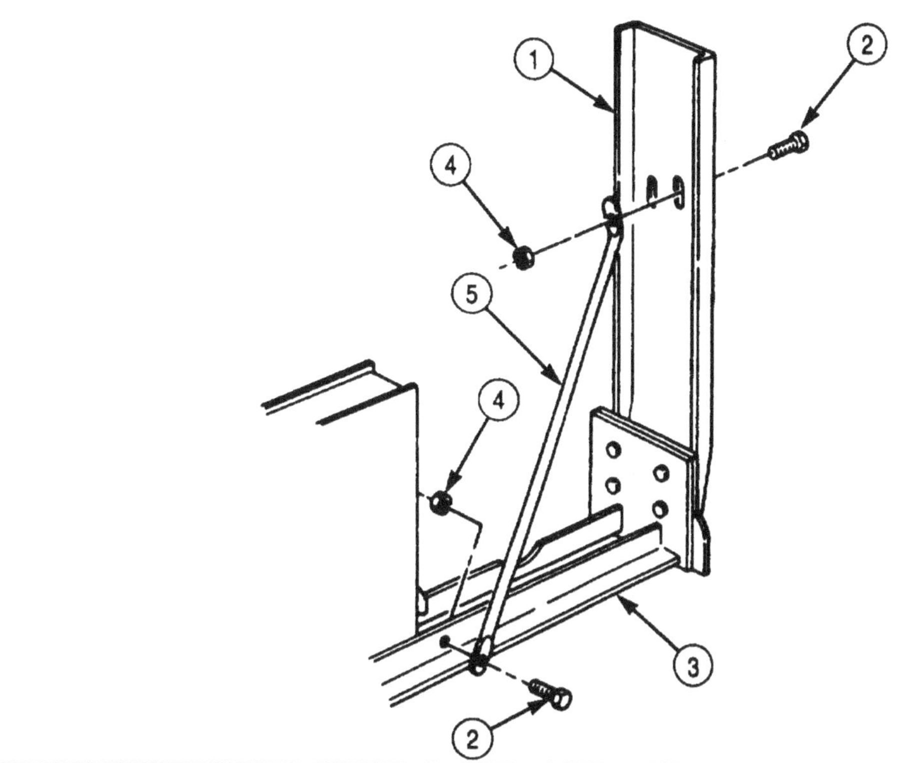

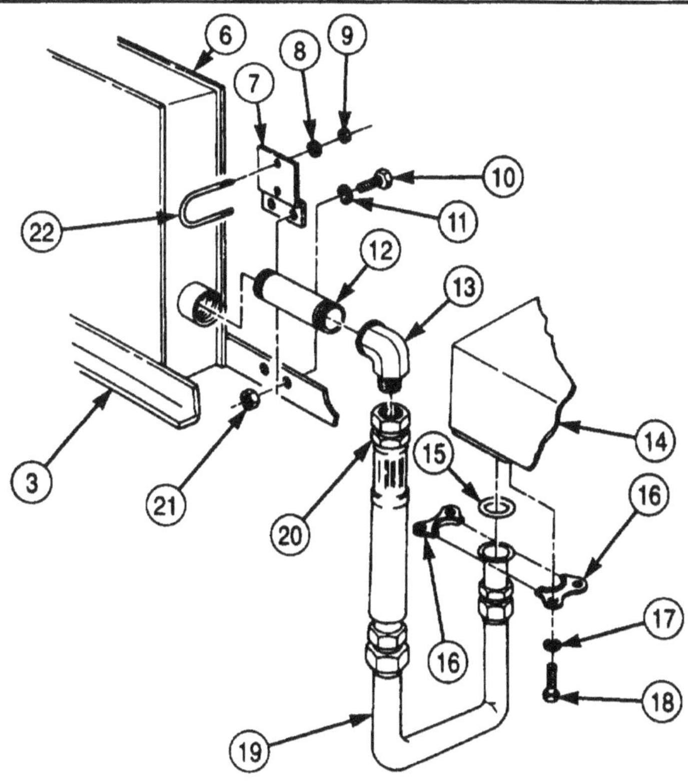

3-338. TRACTOR WINCH HYDRAULIC OIL RESERVOIR (M932/A1/A2) REPLACEMENT (Contd)

7. While holding nut (16), disconnect oil return line (6) from elbow (15).
8. Remove locknut (19), screw (17), clamp (18), and oil return hose (6) from reservoir base (5). Discard locknut (191.
9. Remove four locknuts (3), screws (4), and muffler support (2) from reservoir base (5). Discard locknuts (3).
10. Remove two locknuts (8), washers (9), screws (10), washers (9), and reservoir (1) from reservoir base (5) and frame bracket (7). Discard locknuts (8).
11. Remove elbow (15) from reservoir (1).
12. Remove filler cap (11), dipstick (12), spacer (13), and strainer (14) from reservoir (1).

b. Installation

NOTE

- When new hydraulic oil reservoir is installed, use attaching parts and fittings from old hydraulic oil reservoir.
- Wrap all male threads with antiseize tape before installation.

1. Install strainer (14), spacer (13), dipstick (12), and filler cap (11) on reservoir (1).
2. Install elbow (15) on reservoir (1).
3. Position reservoir (1) on reservoir base (5) and frame bracket (7).
4. Install reservoir (1) on frame bracket (7) with two washers (9), screws (10), washers (9), and new locknuts (8).
5. Install muffler support (2) on reservoir base (5) with four screws (4) and new locknuts (3).
6. Install oil return hose (6) on elbow (15), holding nut (16) to prevent turning.
7. Install oil return hose (6) on reservoir base (5) with clamp (18), screw (17), and new locknut (19).
8. Install nipple support (21) on reservoir base (5) with two washers (25), screws (24), and new locknuts (35).
9. Install nipple (26) and elbow (27) on reservoir (1).
10. Install U-bolt (36) on nipple support (21) with two washers (22) and new locknuts (23).
11. Install new O-ring (29) and oil supply hose (33) on hydraulic oil pump (28) with two split flanges (30), four new lockwashers (31), and screws (32).
12. Install oil supply hose (33) on elbow (27), holding nut (34) to prevent turning.
13. Install brace (39) on muffler support (2) and reservoir base (5) with two screws (37) and new locknuts (38).

3-338. TRACTOR WINCH HYDRAULIC OIL RESERVOIR (M932/A1/A2) REPLACEMENT (Contd)

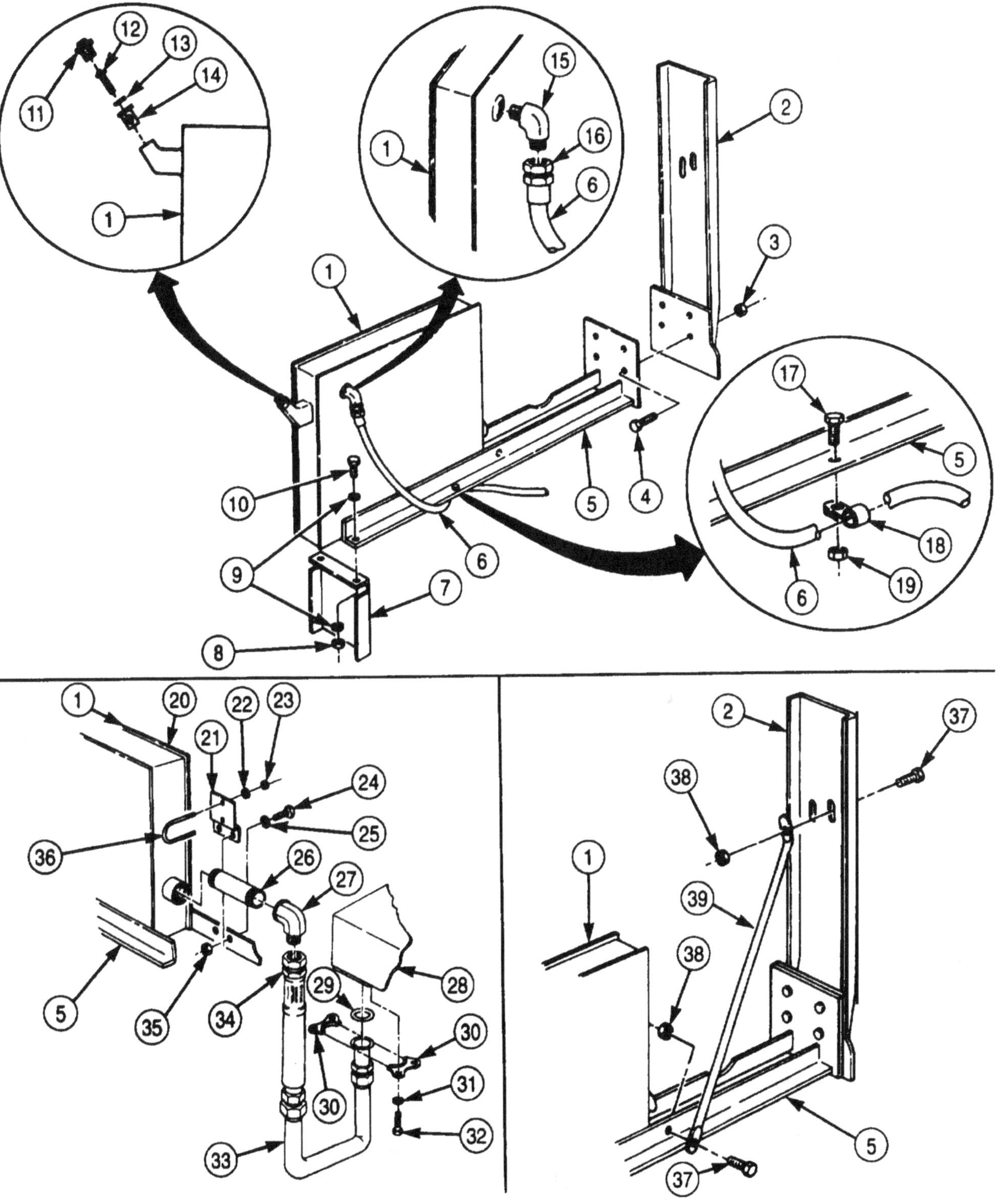

FOLLOW-ON TASKS:
- Install spare tire (TM 9-2320-272-10).
- Fill hydraulic oil reservoir to proper fluid level (LO 9-2320-272-12).
- Operate hydraulic system (TM 9-2320-272-10) and check for leaks.

Section XIV. SPECIAL PURPOSE BODIES MAINTENANCE

3-339. SPECIAL PURPOSE BODIES MAINTENANCE INDEX

PARA. NO.	TITLE	PAGE NO.
3-340.	Cargo Troop Seat Replacement	3-870
3-341.	Cargo Body Cover Bows Replacement	3-872
3-342.	Cargo Upper and Lower Wheel Splash Guard Replacement	3-874
3-343.	Cargo Tailgate Replacement	3-876
3-344.	Side Locking Pin Retaining Clip Replacement	3-878
3-345.	Reflectors Replacement	3-879
3-346.	Cargo Storage Box Replacement	3-880
3-347.	Tailgate Personnel Step Replacement	3-881
3-348.	Dump Tailgate Assembly Replacement	3-882
3-349.	Dump Tailgate Control Linkage Replacement	3-884
3-350.	Van Rear Door and Side Door Window Replacement	3-888
3-351.	Retractable Window Replacement	3-889
3-352.	Window Blackout Panel Replacement	3-890
3-353.	Window Screen Replacement	3-891
3-354.	Retractable Window Regulator Replacement	3-892
3-355.	Window Brush Guard Replacement	3-893
3-356.	Hinged Roof and Floor Counterbalance Cable Maintenance	3-894
3-357.	Side Panel-to-Roof Toggle Clamp Replacement	3-896
3-358.	Toggle Clamp Anchor Post Replacement	3-897
3-359.	Side Panel Roof Swivel Hook Replacement	3-898
3-360.	Ladder Locking Clamp Replacement	3-899
3-361.	Bonnet Handle and Control Rod Replacement	3-900
3-362.	Door Hinge and Seal Replacement	3-902
3-363.	Panel Seals Replacement	3-904
3-364.	Door Handle and Lock Replacement	3-905
3-365.	Door Checks Replacement	3-906
3-366.	Ladder Rack Bumpers Replacement	3-907
3-367.	Side Panel Rubber Bumpers Replacement	3-908
3-368.	Side Panel Rear Lock Replacement	3-909
3-369.	Side Panel Front Lock and Hinged-Type Roof Lock Replacement	3-911
3-370.	Side Panel Exterior Lock Replacement	3-912
3-371.	Fluorescent Light Tube Replacement	3-914
3-372.	Emergency/Blackout Light Lamp and Lens Replacement	3-915
3-373.	Blackout Light Switch and 110-Volt Receptacle Replacement	3-916
3-374.	Inside Telephone Jack Post Replacement	3-918
3-375.	Outside Telephone Jack Post Replacement	3-920
3-376.	Side and Rear Door Blackout Light Switch Maintenance	3-922

3-339. SPECIAL PURPOSE BODIES MAINTENANCE INDEX (Contd)

PARA. NO.	TITLE	PAGE NO.
3-377.	Hinged Roof-Operated Blackout Circuit Plungers Replacement	3-924
3-378.	Expanding and Retracting Mechanism Locks Replacement	3-926
3-379.	Van Heater and Exhaust Replacement	3-928
3-380.	Van Power Cable Reel (M934A1/A2) Replacement	3-932
3-381.	Van Heater Fuel Pump (M934A1/A2) Replacement	3-934
3-382.	Air Conditioner Drain Tube Replacement	3-938
3-383.	Automatic Brake (Hoist Winch) Adjustment	3-940
3-384.	Hoist Winch Cable Replacement	3-942
3-385.	Boom Floodlight Wire Replacement	3-946
3-386.	Crane Wiring Harness Replacement	3-948
3-387.	Wrecker Crane Hydraulic Hose and Tube Replacement	3-952
3-388.	Wrecker Crane Hydraulic Pump Replacement	3-954
3-389.	Pressure Relief Valve Maintenance	3-956
3-390.	Snubber Valve Assembly Replacement	3-958
3-391.	Crane Hydraulic Filter Maintenance	3-960
3-392.	Forward Deck Plate Replacement	3-964
3-393.	Transfer PTO-to-Hydraulic Pump Propeller Shaft Replacement	3-966

3-340. CARGO TROOP SEAT REPLACEMENT

THIS TASK COVERS:

a. Removal   b. Installation

INITIAL SETUP:

APPLICABLE MODELS
M923/A1/2, M925/A1/A2, M927/A1/A2, M928/A1/A2

TOOLS
General mechanic's tool kit (Appendix E, Item 1)

MATERIALS/PARTS
Five cotter pins (Appendix D, Item 46)
One cotter pin (Appendix D, Item 55)

PERSONNEL REQUIRED
TWO

REFERENCES (TM)
TM 9-2320-272-10
TM 9-2320-272-24P

EQUIPMENT CONDITION
Parking brake set (TM 9-2320-272-10).

NOTE

M927/A1/A2 and M928/A1/A2 model trucks have six hinges per troop seat. The M923/A1/A2 and M925/A1/A2 have five hinges per troop seat as shown in this procedure.

a. Removal

1. Remove five cotter pins (4) and hinge pins (2) from hinges (5) and (6). Discard cotter pins (4).
2. Remove cotter pin (11), washer (10), and stabilizer bar (9) from side rack (1) and cargo bed (8). Discard cotter pin (11).
3. Remove two seat latches (3), release, and lower troop seat (7) from side rack (1).

b. Installation

1. Align hinges (6) of troop seat (7) with hinges (5) on side rack (1) and install with five hinge pins (2) and new cotter pins (4).
2. Install stabilizer bar (9) on side rack (1) and cargo bed (8) with washer (10) and new cotter pin (11).
3. Raise troop seat (7) to stow position on side rack (1) and install with two seat latches (3).

3-340. CARGO TROOP SEAT REPLACEMENT (Contd)

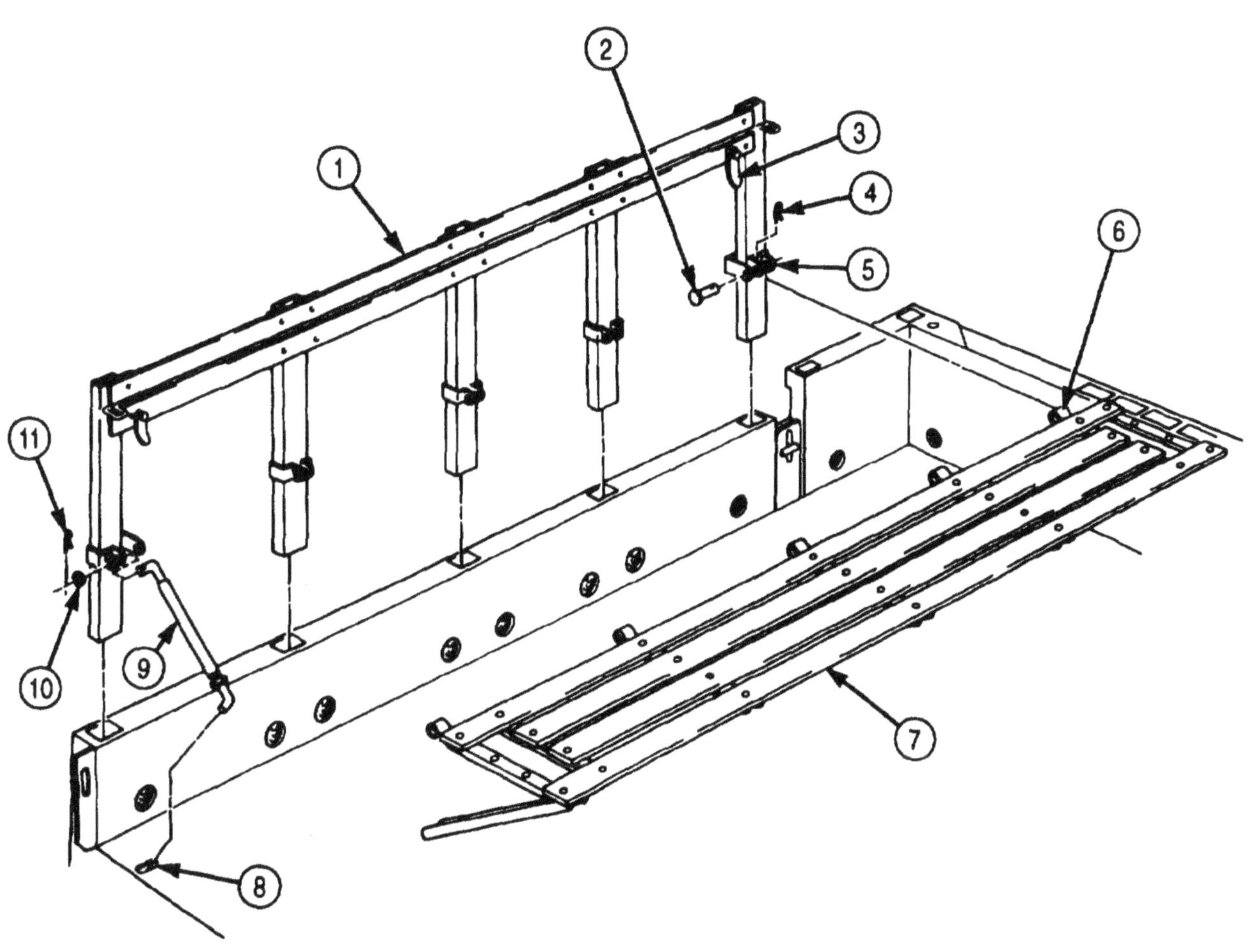

3-341. CARGO BODY COVER BOWS REPLACEMENT

THIS TASK COVERS:

a. Removal **b. Installation**

INITIAL SETUP:

APPLICABLE MODELS
M923/A1/A2 M925/A1/A2, M927/A1/A2,
M928/A1/A2

TOOLS
General mechanic's tool kit (Appendix E, Item 1)

REFERENCES (TM)
TM 9-2320-272-10
TM 9-2320-272-24P

EQUIPMENT CONDITION
- Parking brake set (TM 9-2320-272-10).
- Tarpaulin, curtains, and extensions removed (TM 9-2320-272-10).

a. Removal

1. Remove two screws (6) and stake (7) from corner sections (1).
2. Rotate two latches (2) from rivets (8) and remove comer sections (1) from bow (3).
3. Remove screw (4) and strap (5) from comer sections (1).

1. Install strap (5) on comer sections (1) with screw (4).
2. Insert long end of corner sections (1) over bow (3) and rotate two latches (2) over rivets (8).
3. Install stake (7) on comer sections (1) with two screws (6).

3-341. CARGO BODY COVER BOWS REPLACEMENT (Contd)

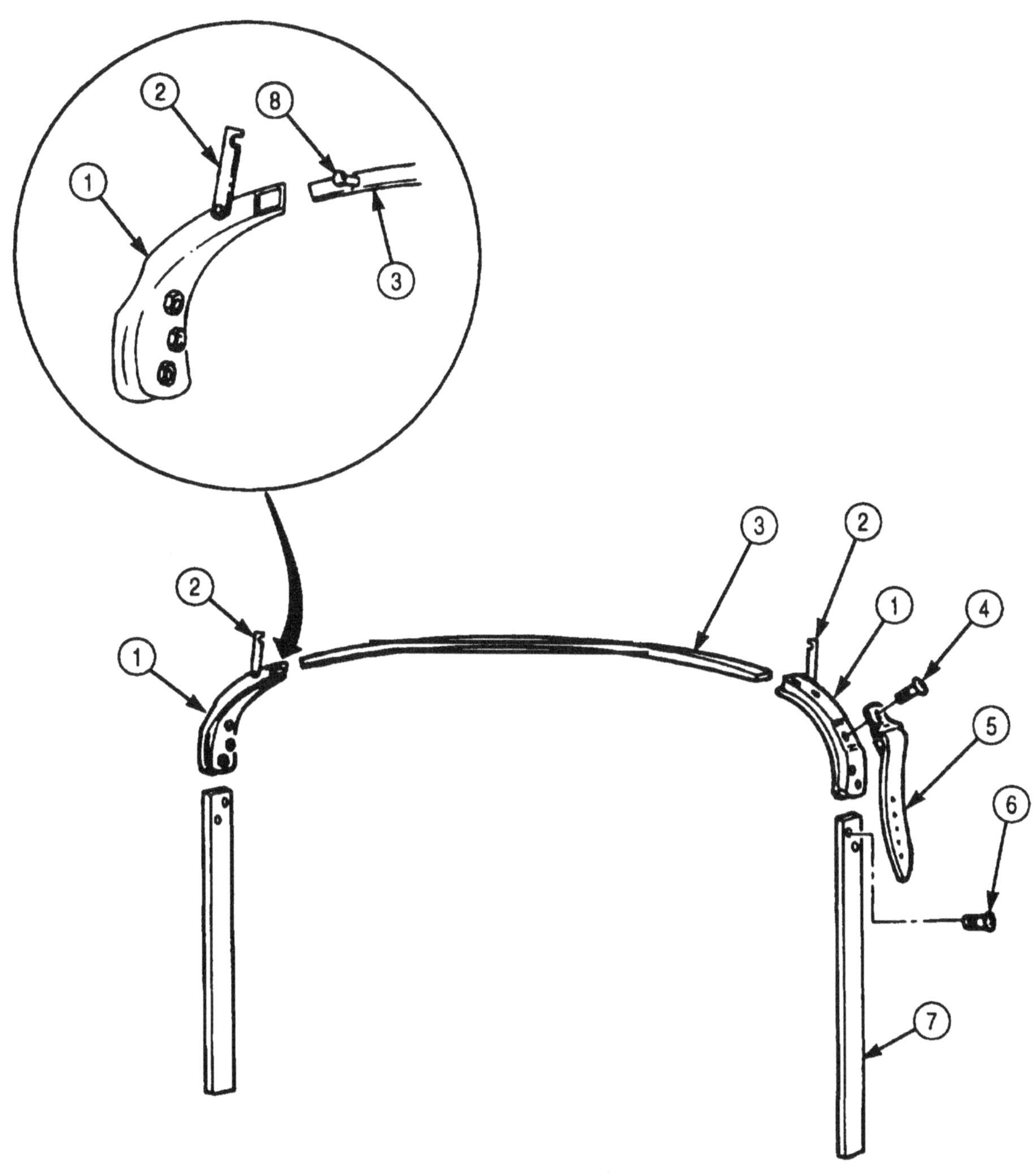

FOLLOW-ON TASK: Install tarpaulin, curtains, and extensions (TM 9-2320-272-10).

3-342. CARGO UPPER AND LOWER WHEEL SPLASH GUARD REPLACEMENT

THIS TASK COVERS:

a. Removal b. Installation

INITIAL SETUP:

APPLICABLE MODELS
M923/A1/42, M925/A1/A2, M927/A1/A2, M928/A1/A2

REFERENCES (TM)
TM 9-2320-272-10
TM 9-2320-272-24P

TOOLS
General mechanic's tool kit (Appendix E, Item 1)

EQUIPMENT CONDITION
Parking brake set (TM 9-2320-272-10).

MATERIALS/PARTS
Ten locknuts (Appendix D, Item 291)

a. Removal

1. Remove five locknuts (5), screws (8), lower splash guard (6), and retainer (7) from upper splash guard (3). Discard locknuts (5).
2. Remove three locknuts (1), screws (4), and upper splash guard (3) from weld bracket (2). Discard locknuts (1).
3. Remove two locknuts (11), screws (10), and braces (9) from weld bracket (2). Discard locknuts (11).

b. Installation

1. Install two braces (9) on weld bracket (2) with two screws (10) and new locknuts (11).
2. Install upper splash guard (3) on weld bracket (2) with three screws (4) and new locknuts (1).
3. Install retainer (7) and lower splash guard (6) on upper splash guard (3) with five screws (8) and new locknuts (5).

3-342. CARGO UPPER AND LOWER WHEEL SPLASH GUARD REPLACEMENT (Contd)

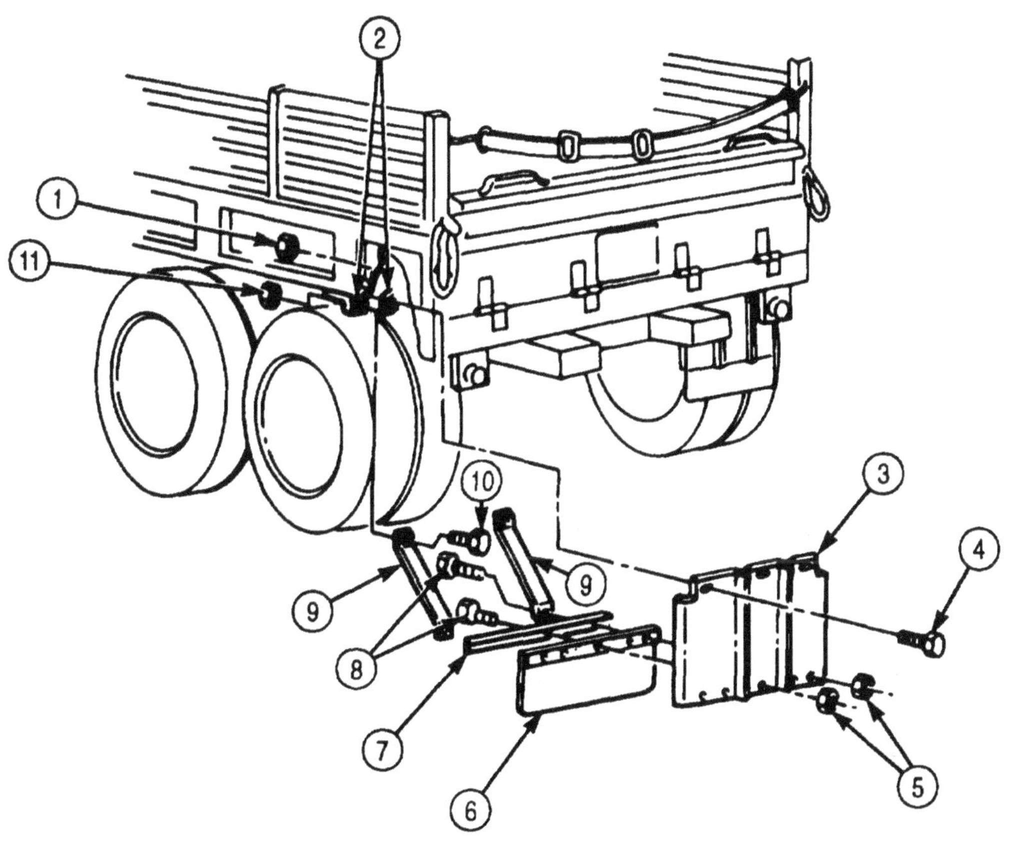

3-343. CARGO TAILGATE REPLACEMENT

THIS TASK COVERS:

a. Removal b. Installation

INITIAL SETUP:

APPLICABLE MODELS

M923/A1/A2, M925/A1/A2, M927/A1/A2, M928/A1/A2

TOOLS

General mechanic's tool kit (Appendix E, Item 1)
Lifting device
Chain

MATERIALS/PARTS

Four cotter pins (Appendix D, Item 56)
GM grease (Appendix C, Item 28)

REFERENCES (TM)

TM 9-2320-272-10
TM 9-2320-272-24P

EQUIPMENT CONDITION

- Parking brake set (TM 9-2320-272-10).
- Tailgate closed and secured (TM 9-2320-272-10).

GENERAL SAFETY INSTRUCTIONS

All personnel must stand clear during lifting operations.

a. Removal

1. Install chain on two steps (2) and lifting device.
2. Remove four cotter pins (6), pins (3), and eight washers (4) from tailgate hinges (5). Discard cotter pins (6).

WARNING

All personnel must stand clear during lifting operations. A snapped cable, or swinging or shifting load, may result in injury to personnel.

3. Remove two hooks (8) from tailgate latches (9) and raise tailgate (1) away from cargo body (7).
4. Lower tailgate (1) onto supports and remove chain and lifting device from two steps (2).

b. Installation

1. Install chain on two steps (2) and lifting device.

WARNING

All personnel must stand clear during lifting operations. A snapped cable, or swinging or shifting load, may result in injury to personnel.

2. Raise tailgate (1) and position on cargo body (7).
3. Install two hooks (8) in tailgate latches (9).
4. Install four pins (3) on tailgate hinges (5) with eight washers (4) and four new cotter pins (6).
5. Remove chain and lifting device from two steps (2).

3-343. CARGO TAILGATE REPLACEMENT (Contd)

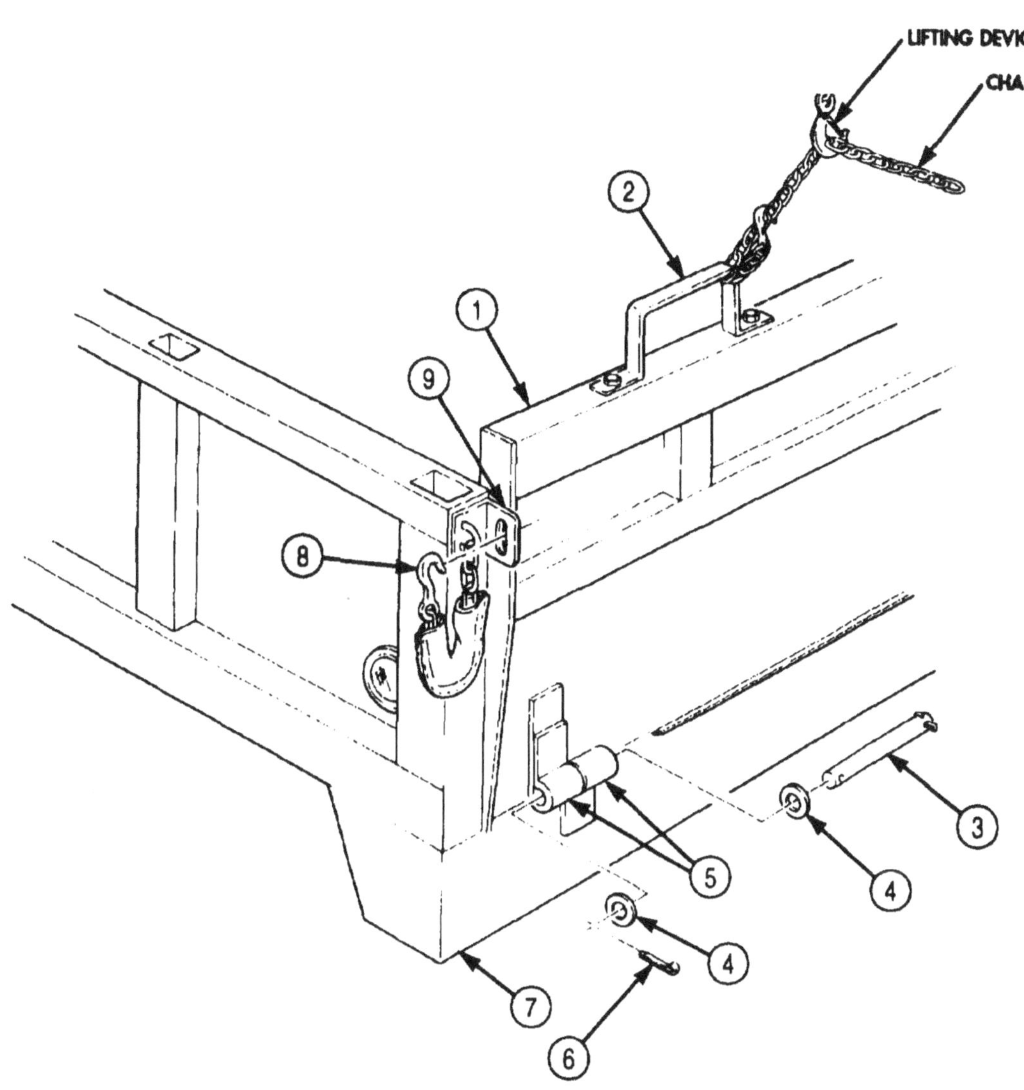

3-344. SIDE LOCKING PIN RETAINING CLIP REPLACEMENT

THIS TASK COVERS:

a. Removal | b. Installation

INITIAL SETUP:

APPLICABLE MODELS
M923/A1/A2, M925/A1/A2

TOOLS
General mechanic's tool kit (Appendix E, Item 1)

MATERIALS/PARTS
Locknut (Appendix D, Item 329)

REFERENCES (TM)
TM 9-2320-272-10
TM 9-2320-272-24P

EQUIPMENT CONDITION
- Parking brake set (TM 9-2320-272-10).
- Side gates up and locked (TM 9-2320-272-10).

a. Removal

Remove locknut (2), screw (4), and retaining clip (3) from welded bracket (5) on dropside (1). Discard locknut (2).

b. Installation

Install retaining clip (3) on welded bracket (5) on dropside (1) with screw (4) and new locknut (2).

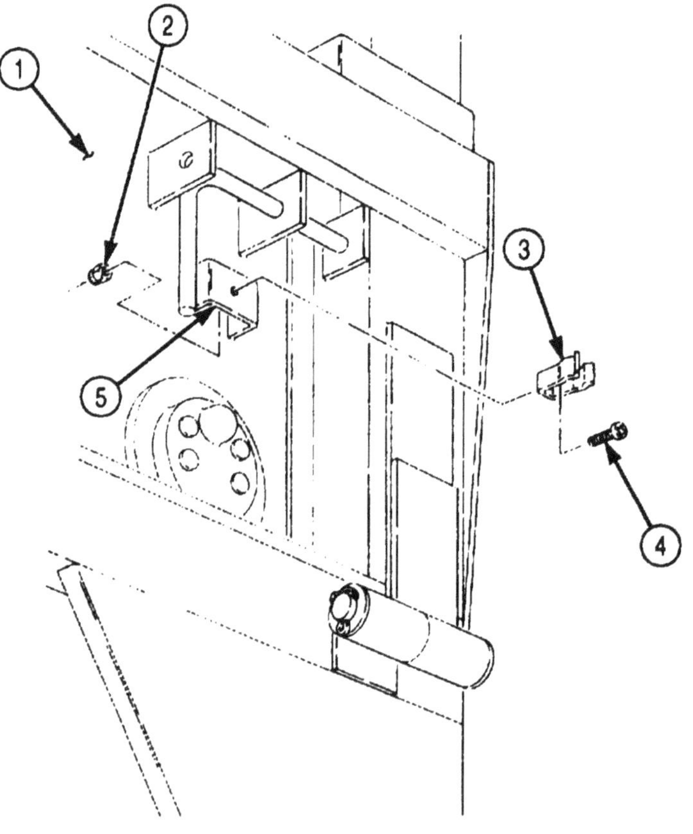

3-345. REFLECTORS REPLACEMENT

THIS TASK COVERS:

a. Removal | b. Installation

INITIAL SETUP:

APPLICABLE MODELS
All

TOOLS
General mechanic's tool kit (Appendix E, Item 1)

REFERENCES (TM)
TM 9-2320-272-10
TM 9-2320-272-24P

EQUIPMENT CONDITION
Parking brake set (TM 9-2320-272-10).

NOTE

Reflectors on all vehicles are replaced basically the same. Mounting hardware may differ. This procedure covers replacement of reflectors on cargo dropside vehicle.

a. Removal

Remove two nuts (4), screws (1), and reflector (2) from cargo body (3).

b. Installation

Install reflector (2) on cargo body (3) with two screws (1) and nuts (4).

3-346. CARGO STOWAGE BOX REPLACEMENT

THIS TASK COVERS:

a. Removal | b. Installation

INITIAL SETUP:

APPLICABLE MODELS
M923/A1/A2, M925/A1/A2, M927/A1/A2, M92/A1/A2

TOOLS
General mechanic's tool kit (Appendix E, Item 1)

MATERIALS/PARTS
Four locknuts (Appendix D, Item 294)

REFERENCES (TM)
TM 9-2320-272-10
TM 9-2320-272-24P

EQUIPMENT CONDITION
Parking brake set (TM 9-2320-272-10).

a. Removal

1. Release two latches (7) and open stowage box door (6).
2. Remove four locknuts (4), screws (2), washers (3), and stowage box (1) from box hangers (5). Discard locknuts (4).

b. Installation

1. Install stowage box (1) on box hangers (5) with four washers (3), screws (2), and new locknuts (4).
2. Close stowage box door (6) and secure with two latches (7).

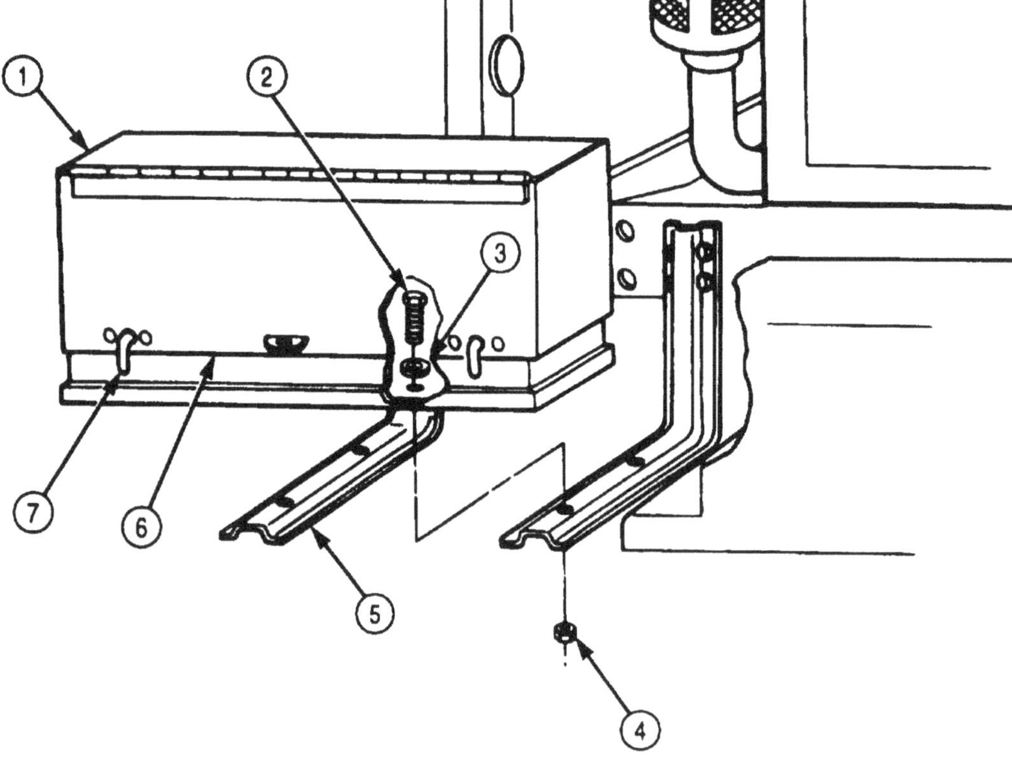

3-347. TAILGATE PERSONNEL STEP REPLACEMENT

THIS TASK COVERS:

a. Removal   b. Installation

INITIAL SETUP:

APPLICABLE MODELS
M923/A1/A2, M925/A1/A2, M927/A1/A2, M923/A1/A2, M929/A1/A2, M930/A1/A2

TOOLS
General mechanic's tool kit (Appendix E, Item 1)

MATERIALS/PARTS
Two locknuts (Appendix D, Item 294)

REFERENCES (TM)
TM 9-2320-272-10
TM 9-2320-272-24P

EQUIPMENT CONDITION
- Parking brake set (TM 9-2320-272-10).
- Tailgate closed and secured (TM 9-2320-272-10).

NOTE

Left and right tailgate personnel steps are replaced the same. This procedure covers the left tailgate personnel step.

a. Removal

Remove two locknuts (5), screws (2), plate (4), and personnel step (3) from tailgate (1). Discard locknuts (5).

b. Installation

Install personnel step (3) and plate (4) on tailgate (1) with two screws (2) and new locknuts (5).

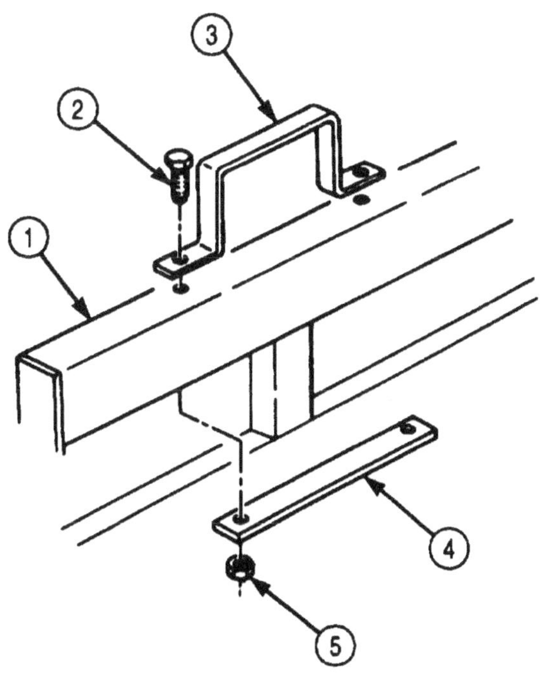

3-348. DUMP TAILGATE ASSEMBLY REPLACEMENT

THIS TASK COVERS:

a. Removal b. Installation

INITIAL SETUP:

APPLICABLE MODELS
M929/A1/A2, M930/A1/A2

TOOLS
General mechanic's tool kit (Appendix E, Item 1)
Lifting device
Chain

PERSONNEL REQUIRED
TWO

REFERENCES (TM)
TM 9-2320-272-10
TM 9-2320-272-24P

EQUIPMENT CONDITION
Parking brake set (TM 9-2320-272-10).

GENERAL SAFETY INSTRUCTIONS
All personnel must stand clear during lifting operations.

a. Removal

1. Install chain on two steps (3) and attach chain to lifting device.
2. Place tailgate control lever in OPEN position (TM 9-2320-272-10).
3. Remove two retaining pins (2) from dump body (1).

WARNING

All personnel must stand clear during lifting operations. A snapped cable, or swinging or shifting load, may result in injury to personnel.

4. Remove tailgate (4) from dump body (1) and lower tailgate (4) onto supports.
5. Remove lifting device and chain from steps (3).

b. Installation

1. Install chain on two steps (3) and attach chain to lifting device.

WARNING

All personnel must stand clear during lifting operations. A snapped cable, or swinging or shifting load, may result in injury to personnel.

2. Raise tailgate (4) and position on dump body (1).
3. Install two retaining pins (2) on dump body (1).
4. Place tailgate control lever to LOCKED position (TM 9-2320-272-10).
5. Remove lifting device and chain from steps (3).

3-348. DUMP TAILGATE ASSEMBLY REPLACEMENT (Contd)

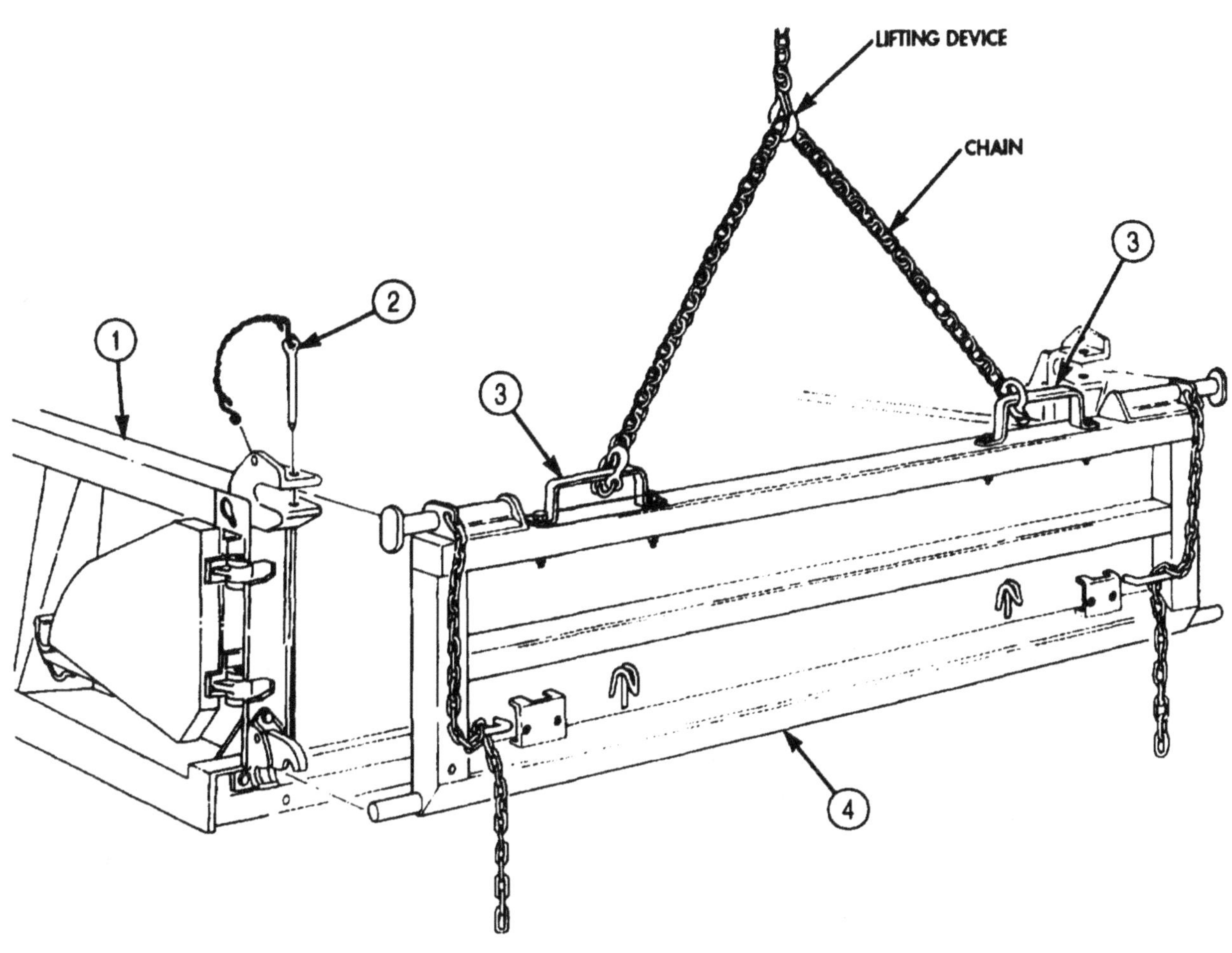

3-349. DUMP TAILGATE CONTROL LINKAGE REPLACEMENT

THIS TASK COVERS:

a. Removal b. Installation

INITIAL SETUP:

APPLICABLE MODELS
M929/A1/A2, M930/A1/A2

TOOLS
General mechanic's tool kit (Appendix E, Item 1)

MATERIALS/PARTS
Fourteen locknuts (Appendix D. Item 309)
Two locknuts (Appendix D, Item 296)
Four cotter pins (Appendix D, Item 85)
Two woodruff keys (Appendix D, Item 728)

REFERENCES (TM)
TM 9-2320-272-10
TM 9-2320-272-24P

EQUIPMENT CONDITION
• Parking brake set (TM 9-2320-272-10).
• Dump body in lowered position (TM 9-2320-272-10).

a. Removal

1. Place tailgate control lever (2) in OPEN position.
2. Remove adjusting nut (10) from threaded end of control rod (8).
3. Remove locknut (12), screw (7), clevis (9), and washer (11) from tailgate control lever (2). Discard locknut (12).
4. Remove locknut (13) and screw (14) and slide tailgate control lever (2) off control rod (8). Discard locknut (13).
5. Remove woodruff key (15) from control lever (2). Discard woodruff key (15).

NOTE

Repeat steps 2 through 5 for removal of control linkage at opposite side. Tailgate control linkage is identical on each side except for tailgate control lever. Tailgate control lever on left side has a hand control. Tailgate control lever on vehicle right side does not.

6. Remove six locknuts (4), screws (6), and three crossshaft bearings (3) from bearing supports (5). Discard locknuts (4).
7. Remove two cotter pins (22), nuts (23), link plate (24), and link (16) from control rod (8), tailgate latch (20), and dump body (1). Discard cotter pins (22).
8. Slide control rod (8) from front of dump body (1).
9. Remove three locknuts (17), washers (18), screws (21), and tailgate latch (20) from hole (19) in rear of dump body (1). Discard locknuts (17).

NOTE

Repeat steps 7 through 9 for removal of control linkage at opposite side of dump body.

3-349. DUMP TAILGATE CONTROL LINKAGE REPLACEMENT (Contd)

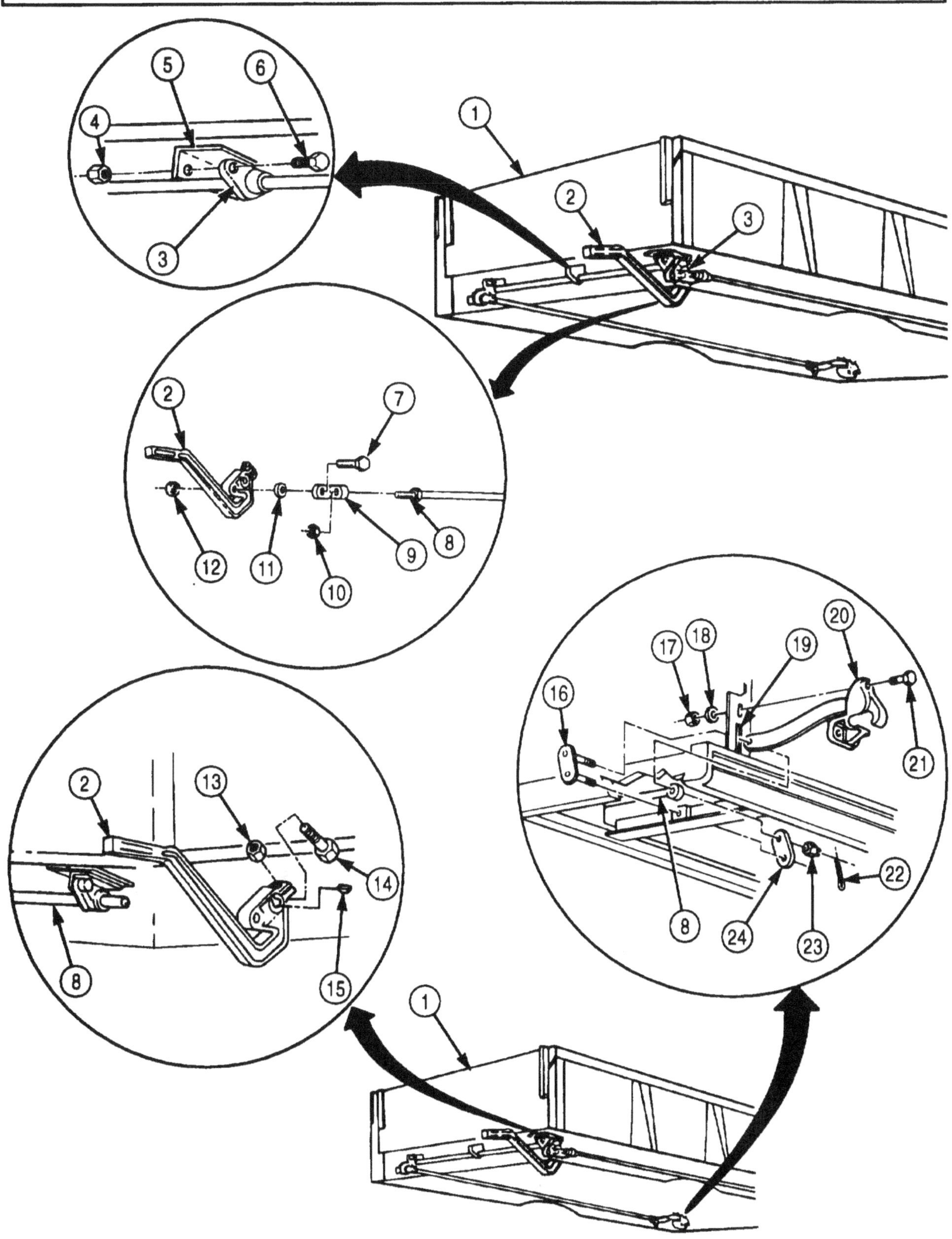

3-349. DUMP TAILGATE CONTROL LINKAGE REPLACEMENT (Contd)

b. Installation

1. Insert bar end of tailgate latch (8) through hole (7) in dump body (1) and install with three screws (9), washers (6), and new locknuts (5).
2. Insert yoke end of control rod (2) through slots (14) in crossmembers (15) of dump body (1) to tailgate latch (8).
3. Insert upper stud (4) of link (3) through hole in tailgate latch (8) and yoke side of control rod (2).
4. Insert lower stud (13) of link (3) into hole in dump body (1).
5. Install link plate (12) on link (3) with two nuts (11) and new cotter pins (10).
6. Install three crossshaft bearings (20) on control rod (2).
7. Position three crossshaft bearings (20) against outer bearing brackets (26) and install with six screws (27) and new locknuts (28).
8. Place tailgate control lever (16) in OPEN position.
9. Place new woodruff key (19) on tailgate control lever (16).
10. Install tailgate control lever (16) on control rod (2) with screw (18) and new locknut (17).
11. Place clevis (22) on threaded end of control rod (2) and install washer (24) and clevis (22) on tailgate control lever (16) with screw (21) and new locknut (25).
12. Install adjusting nut (23) on threaded end of control rod (2).

NOTE

Repeat steps 8 through 12 for tailgate control linkage on opposite side.

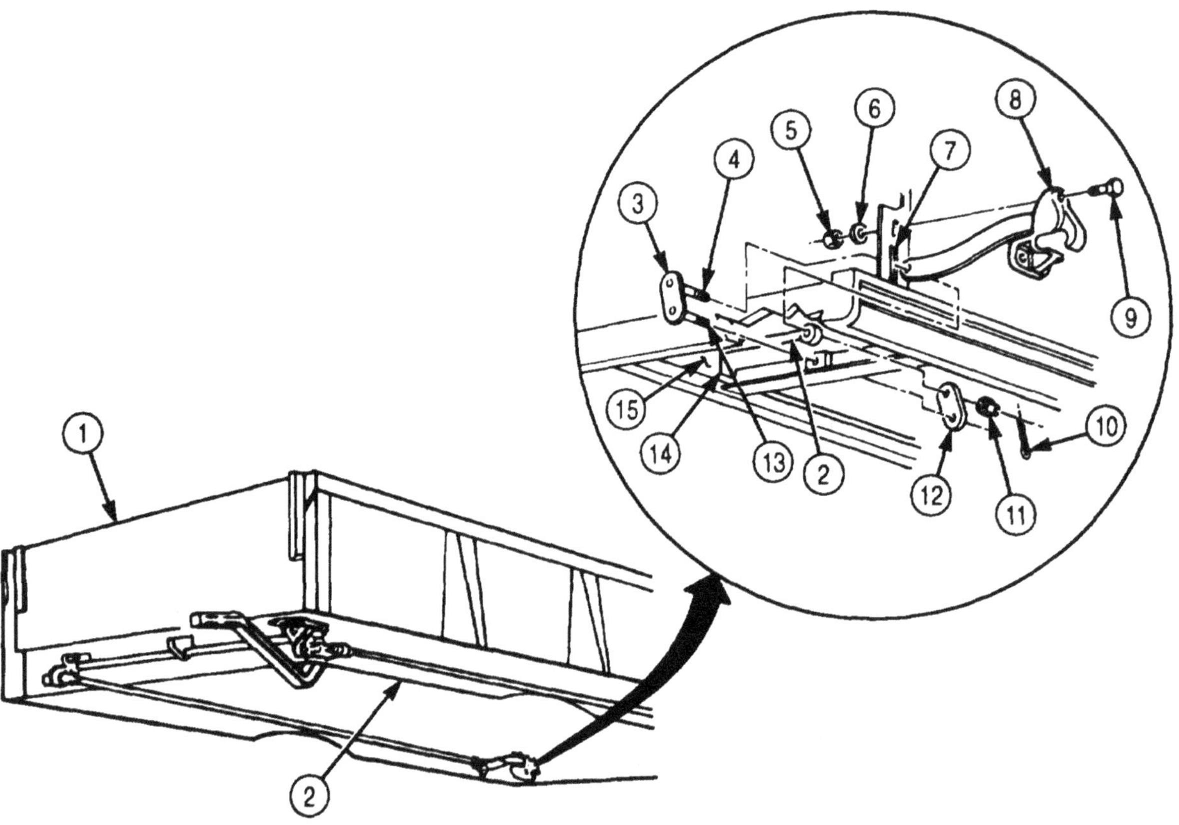

3-349. DUMP TAILGATE CONTROL LINKAGE REPLACEMENT (Contd)

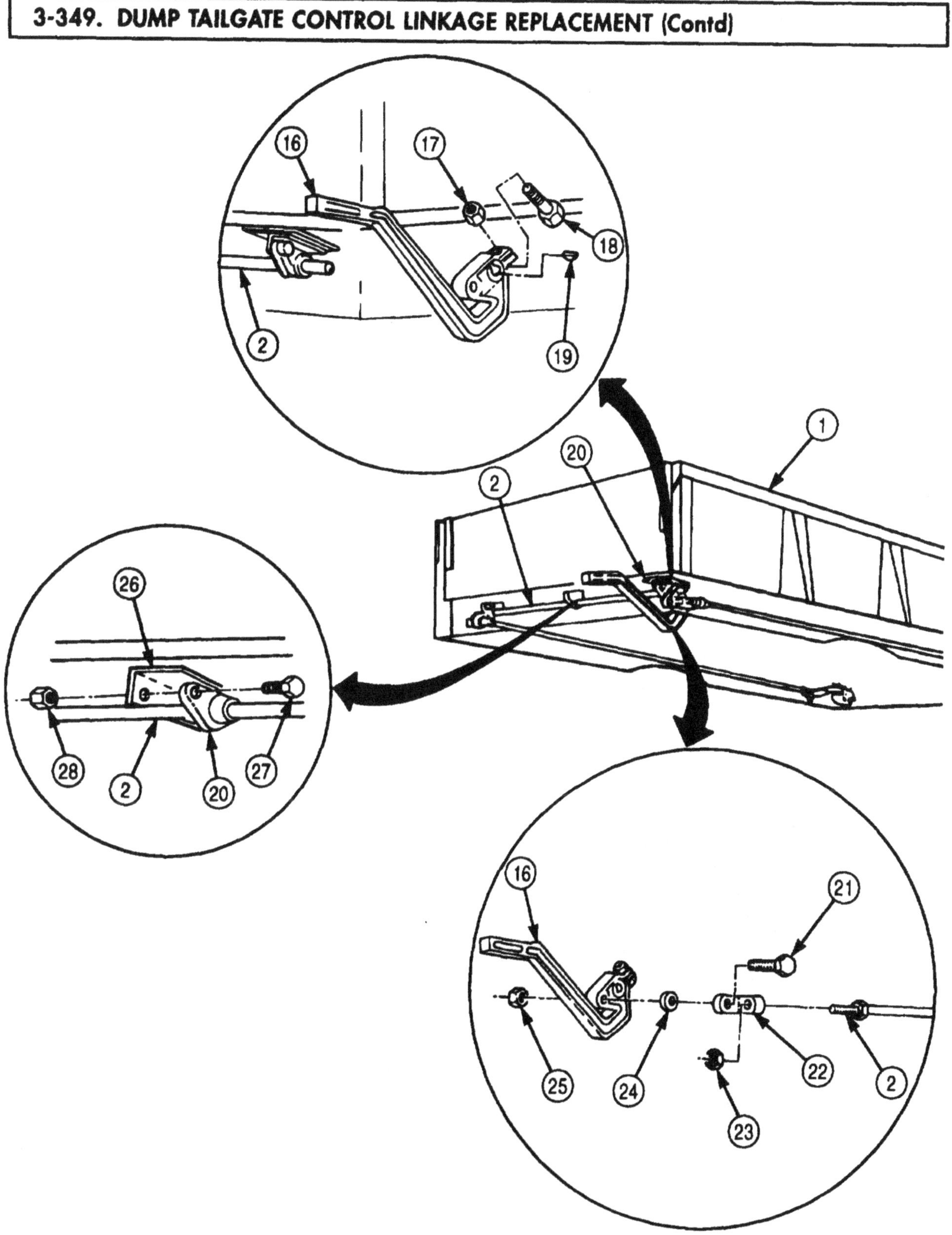

3-350. VAN REAR DOOR AND SIDE DOOR WINDOW REPLACEMENT

THIS TASK COVERS:

a. Removal

b. Installation

INITIAL SETUP:

APPLICABLE MODELS
M934/A1/A2

TOOLS
General mechanic's tool kit (Appendix E, Item 1)

REFERENCES (TM)
TM 9-2320-272-10
TM 9-2320-272-24P

EQUIPMENT CONDITION
- Parking brake set (TM 9-2320-272-10).
- Van body fully expanded and secured (side door window only) (TM 9-2320-272-10).

a. Removal

Remove twenty-one screws (3) and door window frame (1) from door (2).

b. Installation

Install door window frame (1) on door (2) with twenty-one screws (3).

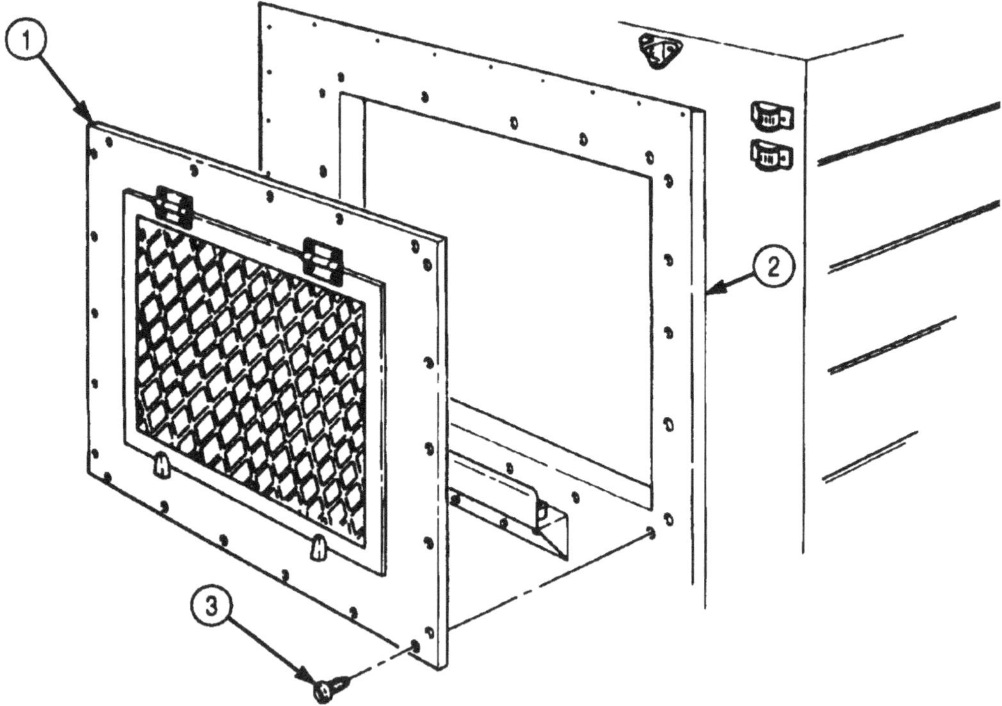

FOLLOW-ON TASK: Retract and secure van body (side door window only) (TM 9-2320-272-10).

3-351. RETRACTABLE WINDOW REPLACEMENT

THIS TASK COVERS:

a. Removal b. Installation

INITIAL SETUP:

APPLICABLE MODELS
M934/A1/A2

TOOLS
General mechanic's tool kit (Appendix E, Item 1)

MATERIALS/PARTS
Cotter pin (Appendix D, Item 57)

REFERENCES (TM)
TM 9-2320-272-10
TM 9-2320-272-24P

EQUIPMENT CONDITION
- Parking brake set (TM 9-2320-272-10).
- Van body fully expanded and secured (side door window only) (TM 9-2320-272-10).

a. Removal

1. Open retractable window (1).
2. Remove cotter pin (5) and regulator arm (4) from frame bracket (2). Discard cotter pin (5).
3. Remove five screws (7) and retractable window (1) from outer frame (3) and van body side panel (6).

b. Installation

1. Install retractable window (1) on outer frame (3) and van body side panel (6) with five screws (7).
2. Install regulator arm (4) on frame bracket (2) with new cotter pin (5).
3. Close retractable window (1).

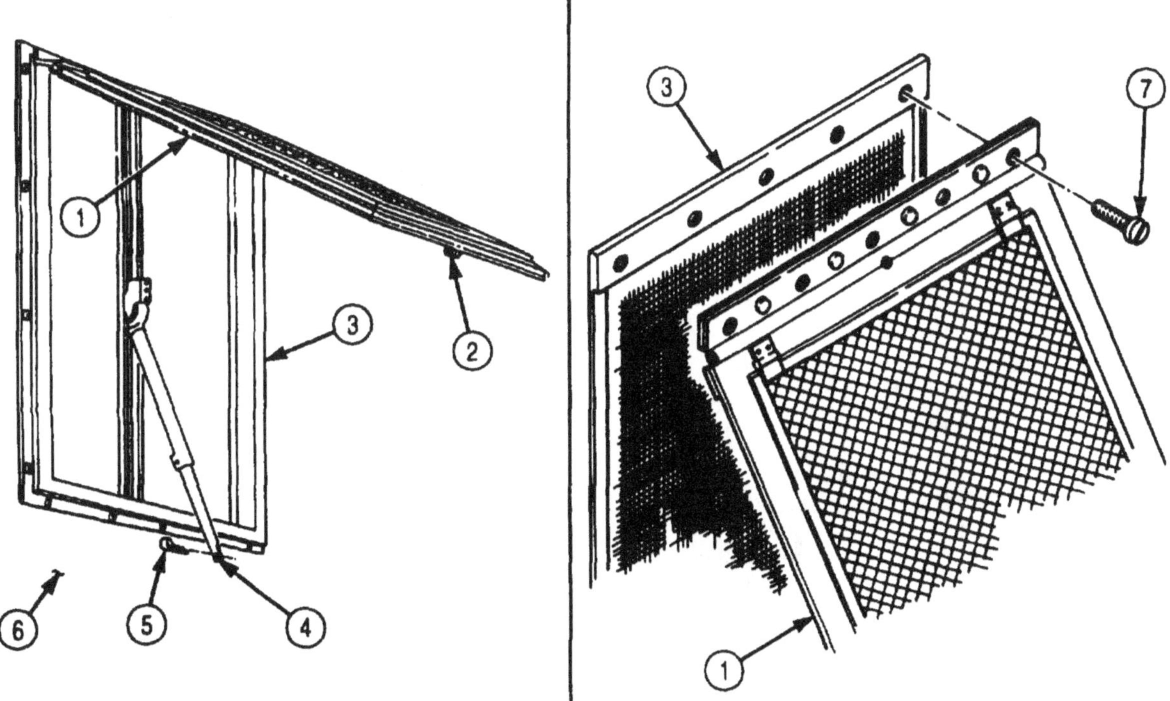

FOLLOW-ON TASK: Retract and secure van body (side door window only) (TM 9-2320-272-10).

3-352. WINDOW BLACKOUT PANEL REPLACEMENT

THIS TASK COVERS:

a. Removal b. Installation

INITIAL SETUP:

APPLICABLE MODELS
M934/A1/A2

TOOLS
General mechanic's tool kit (Appendix E, Item 1)

REFERENCES (TM)
TM 9-2320-272-10
TM 9-2320-272-24P

EQUIPMENT CONDITION
- Parking brake set (TM 9-2320-272-10)
- Van body fully expanded and secured (side door window only) (TM 9-2320-272-10).

---

a. Removal

Remove fifteen screws (4), guide frame (3), and blackout panel (2) from van body side panel (1).

b. Installation

Position blackout panel (2) in guide frame (3) and install on van body side panel (1) with fifteen screws (4).

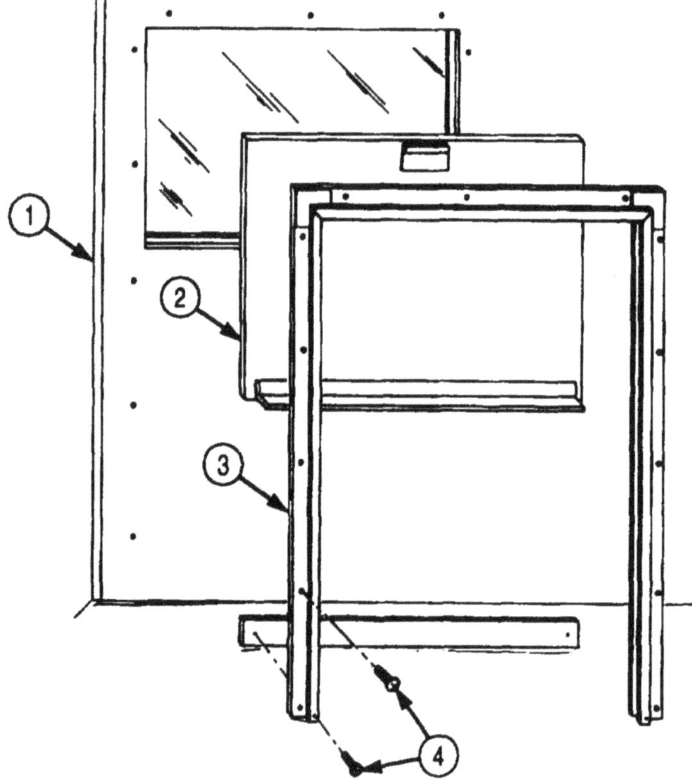

FOLLOW-ON TASK: Retract and secure van body (side door window only) (TM 9-2320-272-10).

3-353. WINDOW SCREEN REPLACEMENT

THIS TASK COVERS:

a. Removal b. Installation

INITIAL SETUP:

APPLICABLE MODELS
M934/A1/A2

TOOLS
General mechanic's tool kit (Appendix E, Item 1)

REFERENCES (TM)
TM 9-2320-272-10
TM 9-2320-272-24P

EQUIPMENT CONDITION
- Parking brake set (TM 9-2320-272-10).
- Van body fully expanded and secured (side door window only) (TM 9-2320-272-10).

a. Removal

Carefully remove screen retainer cord (3) and screen (2) from window frame (1).

b. Installation

Position screen (2) to window frame (1) and install with screen retainer cord (3).

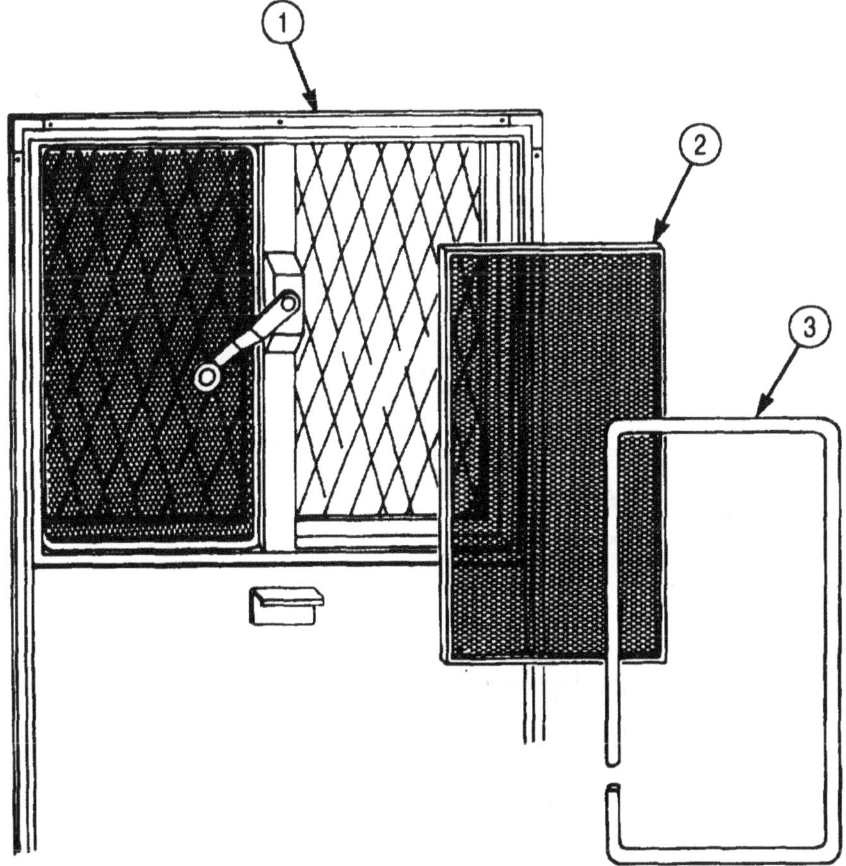

FOLLOW-ON TASK: Retract and secure van body (side door window only) (TM 9-2320-272-10).

3-354. RETRACTABLE WINDOW REGULATOR REPLACEMENT

THIS TASK COVERS:

a. Removal | b. Installation

INITIAL SETUP:

APPLICABLE MODELS
M934/A1/A2

TOOLS
General mechanic's tool kit (Appendix E, Item 1)

MATERIALS/PARTS
Cotter pin (Appendix D, Item 57)

REFERENCES (TM)
TM 9-2320-272-10
TM 9-2320-272-24P

EQUIPMENT CONDITION
- Parking brake set (TM 9-2320-272-10).
- Van body fully expanded and secured (side door window only) (TM 9-2320-272-10).

a. Removal

1. Open retractable window (2) and prop open.
2. Remove cotter pin (6) and regulator arm (5) from frame bracket (3). Discard cotter pin (6).
3. Remove screw (9) and crank handle (8) from window regulator (4).
4. Remove three screws (7) and window regulator (4) from window frame (1).

b. Installation

1. Install window regulator (4) on window frame (1) with three screws (7).
2. Install crank handle (8) on window regulator (4) with screw (9).
3. Install regulator arm (5) on frame bracket (3) with new cotter pin (6).
4. Close retractable window (2).

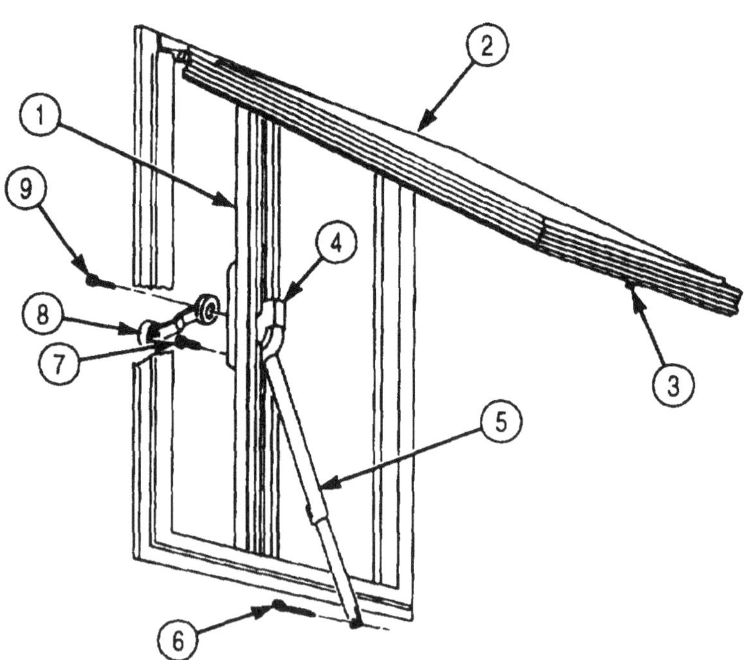

FOLLOW-ON TASK: Retract and secure van body (side door window only) (TM 9-2320-272-10).

3-355. WINDOW BRUSH GUARD REPLACEMENT

THIS TASK COVERS:

a. Removal — b. Installation

INITIAL SETUP:

APPLICABLE MODELS
M934/A1/A2

TOOLS
General mechanic's tool kit (Appendix E, Item 1)

REFERENCES (TM)
TM 9-2320-272-10
TM 9-2320-272-24P

EQUIPMENT CONDITION
- Parking brake set (TM 9-2320-272-10).
- Van body fully expanded and secured (side door window only) (TM 9-2320-272-10).

a. Removal

1. Release two latches (4) and brush guard (3) from window frame (1).
2. Remove two setscrews (6) and hinge pins (5) from hinges (2) and brush guard (3) from window frame (1).

b. Installation

1. Position brush guard (3) on window frame (1) and two hinges (2).
2. Insert two hinge pins (5) through hinges (2) and install with two setscrews (6).
3. Secure brush guard (3) on window frame (1) with two latches (4).

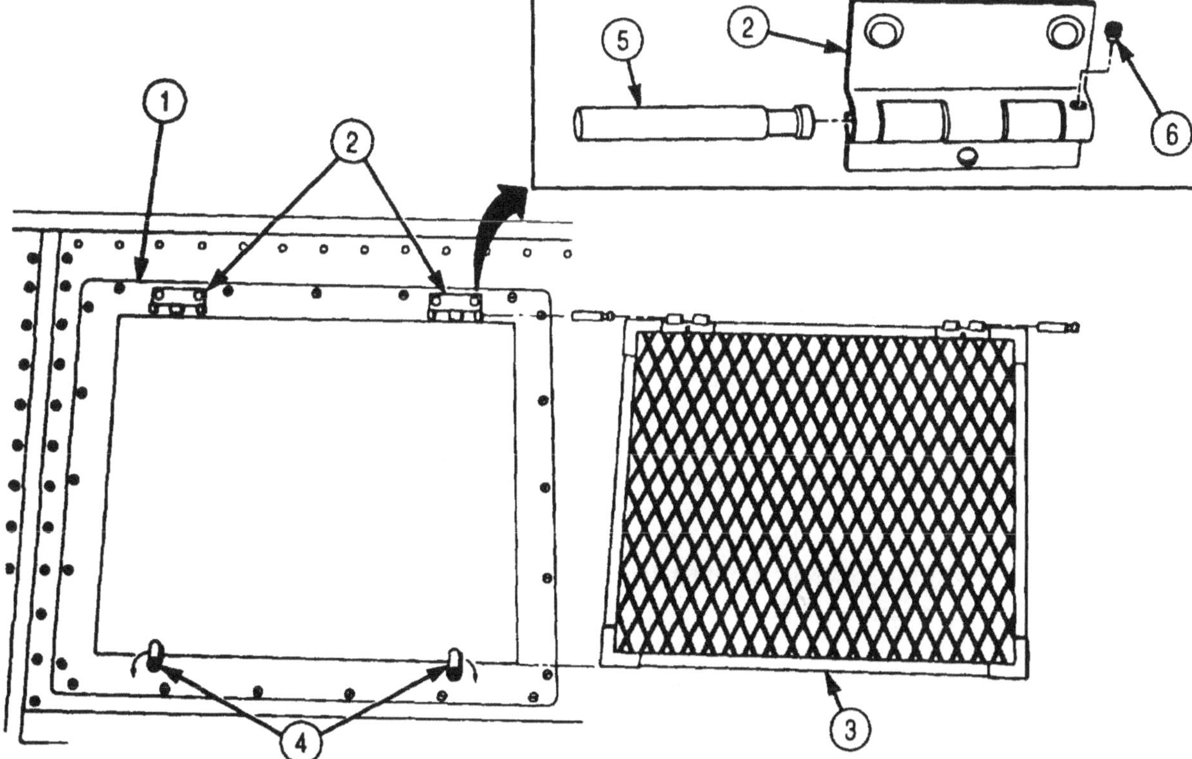

FOLLOW-ON TASK: Retract and secure van body (side door window only) (TM 9-2320-272-10).

3-356. HINGED ROOF AND FLOOR COUNTERBALANCE CABLE MAINTENANCE

THIS TASK COVERS:

a. Removal
b. Installation
c. Cable Adjustment

INITIAL SETUP:

APPLICABLE MODELS
M934/A1/A2

TOOLS
General mechanic's tool kit (Appendix E, Item 1)

MATERIALS/PARTS
'Iwo cotter pins (Appendix D, Item 57)
Cotter pin (Appendix D, Item 66)
Wood block

REFERENCES (TM)
TM 9-2320-272-10
TM 9-2320-272-24P

EQUIPMENT CONDITION
- Parking brake set (TM 9-2320-272-10).
- Van body fully expanded and secured (side door window only) (TM 9-2320-272-10).

GENERAL SAFETY INSTRUCTIONS
Always wear hand protection when handling cable.

WARNING

Wear hand protection when handling cable. Broken wires may cause injury to personnel.

a. Removal

1. Place a l-in. (2.54-cm) block of wood between end of swivel hook (1) and hinged roof (2).
2. Turn turnbuckle (5) on hinged floor (6) counterclockwise to decrease cable tension.
3. Remove cotter pin (8), pin (10), and cable clevis (9) from drop arm (11). Discard cotter pin (8).
4. Remove cotter pin (12), pin (13), and upper roller (15) from mounting plate (14). Discard cotter pin (12).
5. Remove cotter pin (18), pin (20), and lower roller (17) from folding arm (19). Discard cotter pin (18).
6. Disconnect lower cable end (16) from turnbuckle (5).
7. Pull cable (4) upward through two cable guides (3) and remove from vehicle.

b. Installation

NOTE

Perform step 1 if installing new counterbalance cable.

1. Remove eye of turnbuckle (5) from replacement cable end (16).
2. Thread cable (4) through two cable guides (3) and connect to turnbuckle (5).
3. Align lower folding arm (19) with holes in comer post (7) and install lower roller (17) on folding arm (19) with pin (20) and new cotter pin (18). Ensure cable (4) is behind lower roller (17).
4. Install upper roller (15) on mounting plate (14) with pin (13) and new cotter pin (12). Ensure cable (4) is behind upper roller (15).
5. Install cable clevis (9) on drop arm (11) with pin (10) and new cotter pin (8).
6. Remove block of wood from swivel hook (1) and hinged roof (2).
7. Adjust cable tension of turnbuckle (5) if necessary (task c.).

3-356. HINGED ROOF AND FLOOR COUNTERBALANCE CABLE MAINTENANCE (Contd)

c. Cable Adjustment

Turn turnbuckle (5) clockwise to increase cable tension; counterclockwise to decrease cable extension.

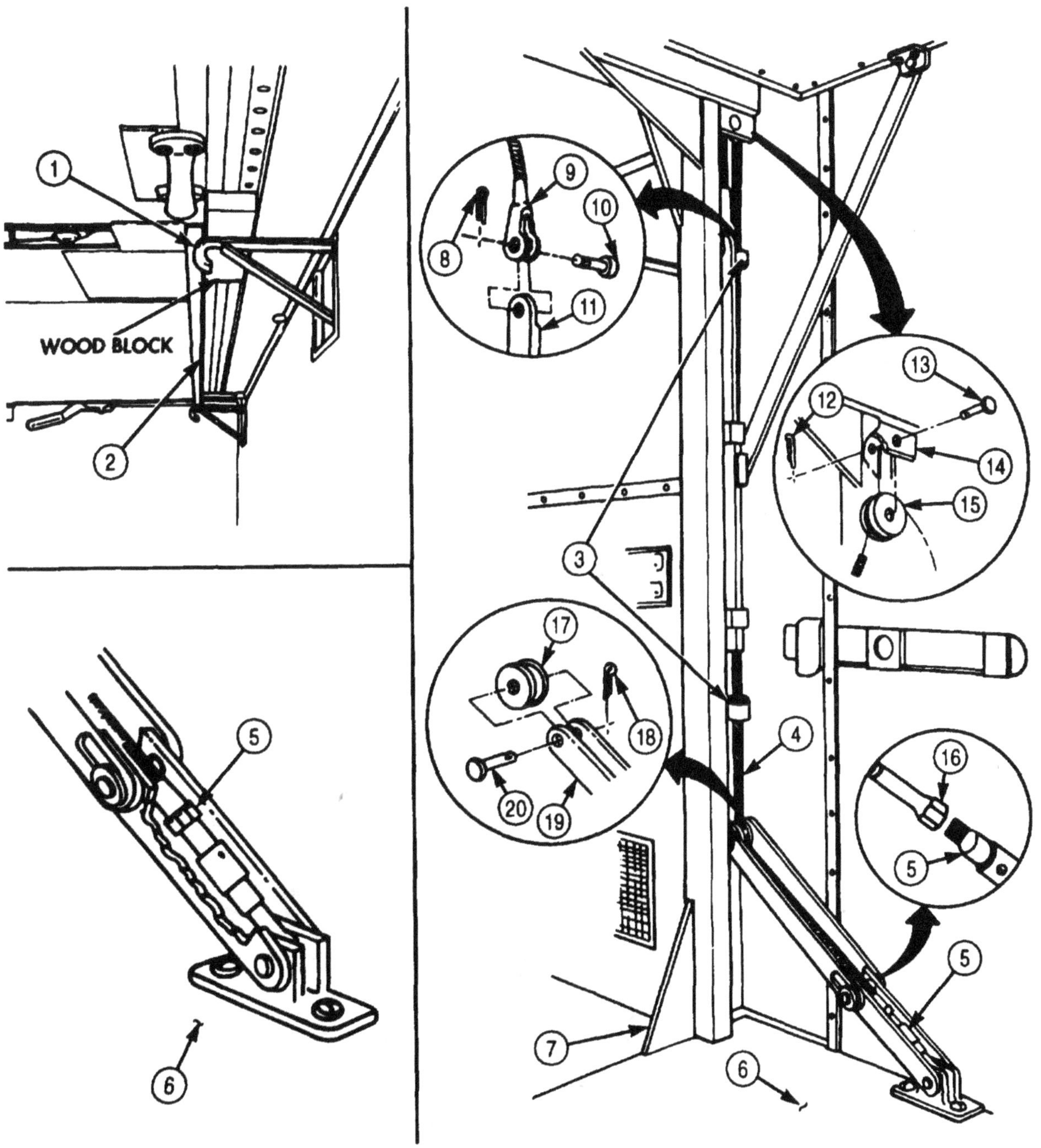

FOLLOW-ON TASK: Retract and secure van body (side door window only) (TM 9-2320-272-10).

3-357. SIDE PANEL-TO-ROOF TOGGLE CLAMP REPLACEMENT

THIS TASK COVERS:

a. Removal

b. Installation

INITIAL SETUP

APPLICABLE MODELS
M934/A1/A2

TOOLS
General mechanic's tool kit (Appendix E, Item 1)

REFERENCES (TM)
TM 9-2320-272-10
TM 9-2320-272-24P

EQUIPMENT CONDITION
Parking brake set (TM 9-2320-272-10).

a. Removal

1. Release toggle clamp (5) and remove eyebolt (3) from swivel hook (2).
2. Remove four screws (4) and toggle clamp (5) from hinged roof (1).

b. Installation

1. Install toggle clamp (5) on hinged roof (1) with four screws (4). Ensure eyebolt (3) faces swivel hook (2).
2. Attach eyebolt (3) of toggle clamp (5) on swivel hook (2) and close toggle clamp (5).

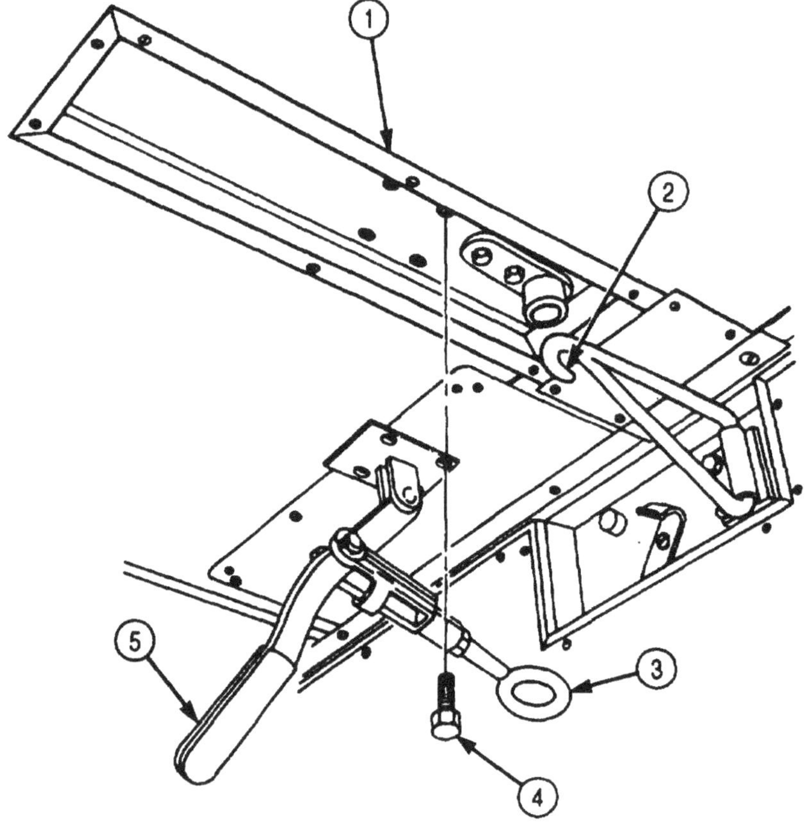

3-358. TOGGLE CLAMP ANCHOR POST REPLACEMENT

THIS TASK COVERS:

a. Removal b. Installation

INITIAL SETUP:

APPLICABLE MODELS
M934/A1/A2

TOOLS
General mechanic's tool kit (Appendix E, Item 1)

REFERENCES (TM)
TM 9-2320-272-10
TM 9-2320-272-24P

EQUIPMENT CONDITION
Parking brake set (TM 9-2320-272-10).

a. Removal

Remove two screws (3) and anchor post (2) from hinged roof (1).

b. Installation

Install anchor post (2) on hinged roof (1) with two screws (3).

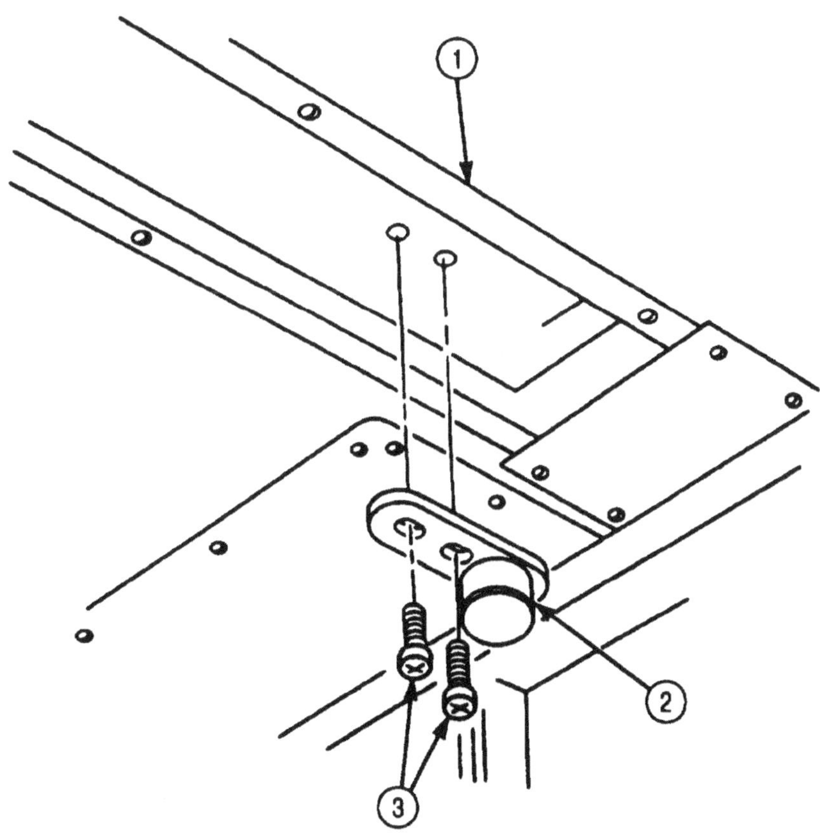

3-359. SIDE PANEL ROOF SWIVEL HOOK REPLACEMENT

THIS TASK COVERS:

a. Removal

b. Installation

INITIAL SETUP:

APPLICABLE MODELS

M934/A1/A2

TOOLS

General mechanic's tool kit (Appendix E, Item 1)

REFERENCES (TM)

TM 9-2320-272-10

TM 9-2320-272-24P

EQUIPMENT CONDITION

- Parking brake set (TM 9-2320-272-10).
- Van body fully expanded and secured (side door window only) (TM 9-2320-272-10).

a. Removal

1. Release toggle clamp (5) and remove eyebolt (4) from swivel hook (2).
2. Remove two screws (1) and swivel hook (2) from side panel (3).

b. Installation

1. Install swivel hook (2) on side panel (3) with two screws (1).
2. Attach eyebolt (4) of toggle clamp (5) on swivel hook (2) and close toggle clamp (5).

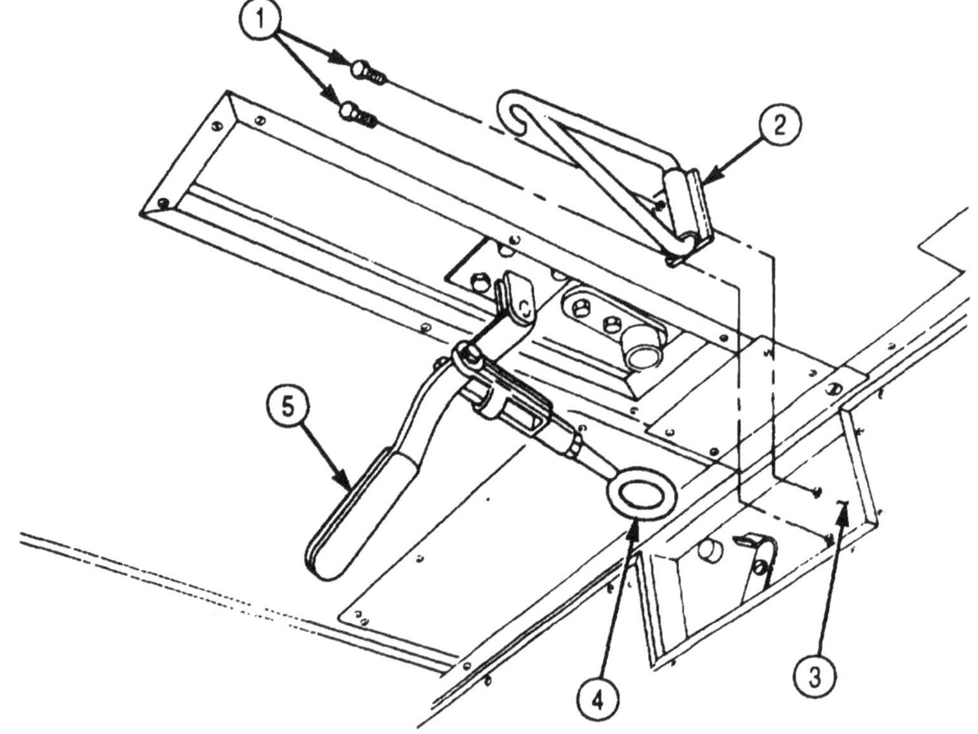

FOLLOW-ON TASK: Retract and secure van body (side door window only) (TM 9-2320-272-10).

3-360. LADDER LOCKING CLAMP REPLACEMENT

THIS TASK COVERS:

a. Removal
b. Installation

INITIAL SETUP:

APPLICABLE MODELS
M934/A1/A2

TOOLS
General mechanic's tool kit (Appendix E, Item 1)

REFERENCES (TM)
TM 9-2320-272-10
TM 9-2320-272-24P

EQUIPMENT CONDITION
- Parking brake set (TM 9-2320-272-10).
- Ladders removed (TM 9-2320-272-10).

a. Removal

Remove four screws (3) and locking clamp (2) from door (1).

b. Installation

Install locking clamp (2) on door (1) with four screws (3).

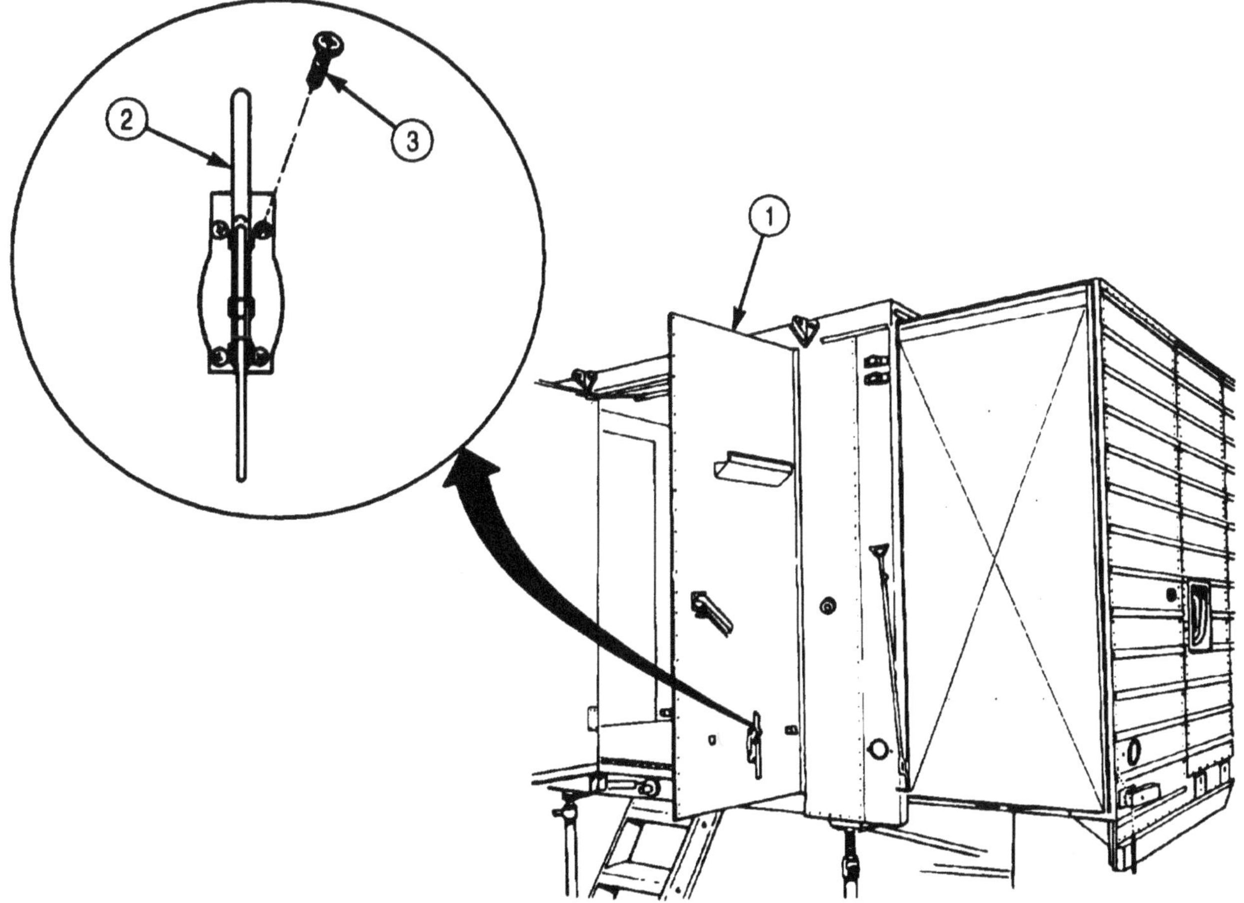

FOLLOW-ON TASK: Install ladders (TM 9-2320-272-10).

3-361. BONNET HANDLE AND CONTROL ROD REPLACEMENT

THIS TASK COVERS:

a. Removal
b. Installation

INITIAL SETUP:

APPLICABLE MODELS
M934/A1/A2

TOOLS
General mechanic's tool kit (Appendix E, Item 1)

MATERIALS/PARTS
Two cotter pins (Appendix D, Item 57)

REFERENCES (TM)
TM 9-2320-272-10
TM 9-2320-272-24P

EQUIPMENT CONDITION
- Parking brake set (TM 9-2320-272-10).
- Bonnet front door opened and braced (TM 9-2320-272-10).

a. Removal

1. Remove two cotter pins (3) and control rod (2) from handle (5) and door bracket (1). Discard cotter pins (3).
2. Remove nut (4), handle (5), and spacer (6) from mounting bracket (7).

b. Installation

1. Install spacer (6) and handle (5) on mounting bracket (7) with nut (4).
2. Install control rod (2) on door bracket (1) and handle (5) with two new cotter pins (3).

3-361. BONNET HANDLE AND CONTROL ROD REPLACEMENT (Contd)

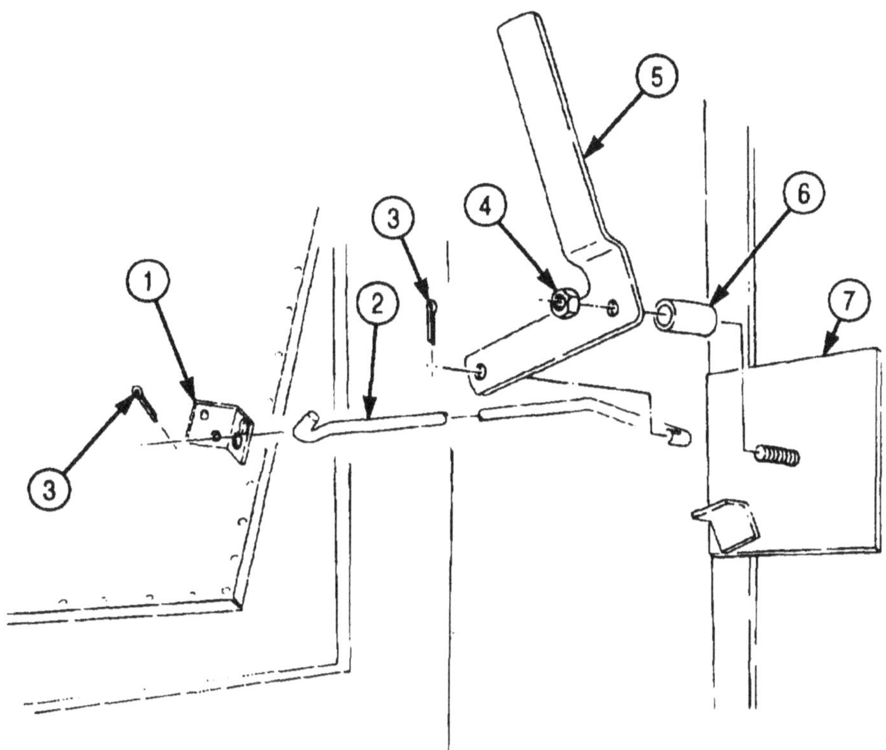

FOLLOW-ON TASK: Close bonnet front door (TM 9-2320-272-10).

3-362. DOOR HINGE AND SEAL REPLACEMENT

THIS TASK COVERS:

a. Removal b. Installation

INITIAL SETUP:

APPLICABLE MODELS
M934/A1/A2

TOOLS
General mechanic's tool kit (Appendix E, Item 1)

MATERIALS/PARTS
Cotter pin (Appendix D, Item 51)

REFERENCES (TM)
TM 9-2320-272-10
TM 9-2320-272-24P

EQUIPMENT CONDITION
Parking brake set (TM 9-2320-272-10).

a. Removal

1. Open door (6).
2. Remove cotter pin (4), pin (2), and door check arm (3) from bracket (1). Discard cotter pin (4).

NOTE

Assistant will help with step 3.

3. Remove fifteen screws (5) and door (6) from van body (8).
4. Remove sixteen screws (12), hinge seal retainer (11), weather stripping (9), hinge (7), and outer hinge seal (10) from door (6).

b. Installation

1. Install outer hinge seal (10), hinge (7), weather stripping (9), and hinge seal retainer (11) on door (6) with sixteen screws (12).

NOTE

Assistant will help with step 2.

2. Install door (6) on van body (8) with fifteen screws (5).
3. Install door check arm (3) on bracket (1) with pin (2) and new cotter pin (4).
4. Close door (6).

3-362. DOOR HINGE AND SEAL REPLACEMENT (Contd)

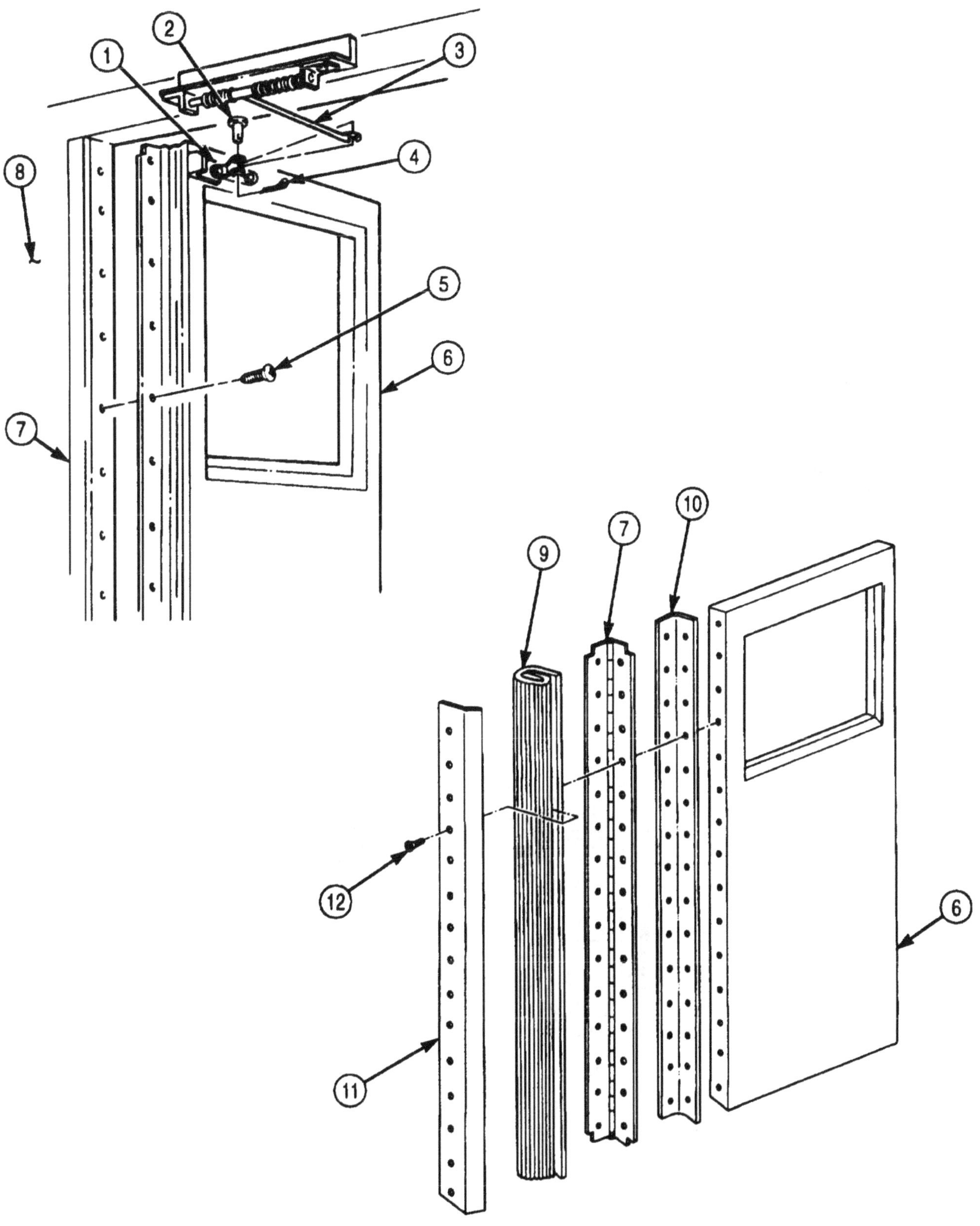

3-363. PANEL SEALS REPLACEMENT

THIS TASK COVERS:

a. Removal b. Installation

INITIAL SETUP:

APPLICABLE MODELS
M934/A1/A2

TOOLS
General mechanic's tool kit (Appendix E, Item 1)

REFERENCES (TM)
TM 9-2320-272-10
TM 9-2320-272-24P

EQUIPMENT CONDITION
Parking brake set, (TM 9-2320-272-10).

NOTE

All panel seals are installed by retainers with either screws or screws and nuts. This procedure shows screws only. The quantity of screws also may differ.

a. Removal

1. Raise panel seal (4) to expose screws (2).
2. Remove screws (2), as required, retainer (3), and panel seal (4) from van body (1).

b. Installation

1. Position retainer (3) in flap of panel seal (4) and align holes.
2. Install panel seal (4) and retainer (3) on van body (1) with screws (2), as required.

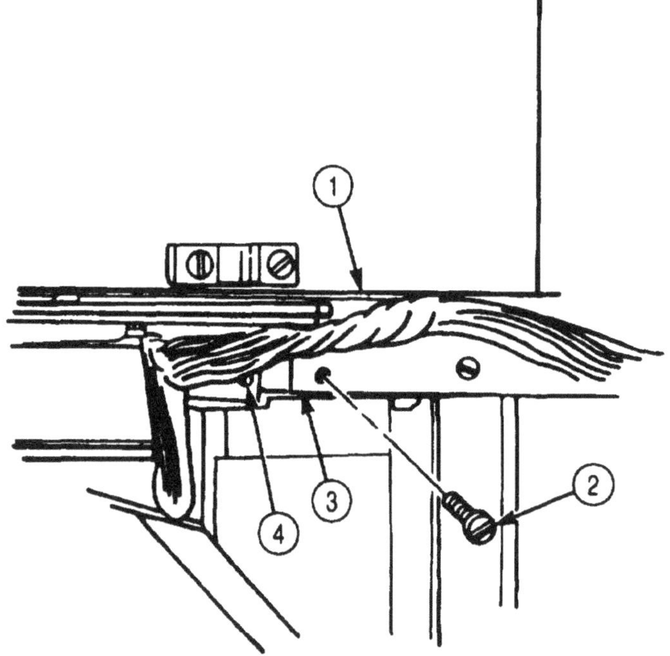

3-364. DOOR HANDLE AND LOCK REPLACEMENT

THIS TASK COVERS:

a. Removal

b. Installation

INITIAL SETUP:

APPLICABLE MODELS
M934/A1/A2

TOOLS
General mechanic's tool kit (Appendix E, Item 1)

MATERIALS/PARTS
Two cotter pins (Appendix D, Item 51)

REFERENCES (TM)
TM 9-2320-272-10
TM 9-2320-272-24P

EQUIPMENT CONDITION
- Parking brake set (TM 9-2320-272-10).
- Rear doors open (rear door handle and lock replacement only) (TM 9-2320-272-10).

NOTE
This procedure applies to van side and rear doors.

1. Remove pin (10) and inner handle (8) from handle shank (9).
2. Remove two cotter pins (1), vertical bars (2), and washers (3) from lockpins (4). Discard cotter pins (1).
3. Remove four screws (7) and door lock (6) from door (5).

1. Install door lock (6) on door (5) with four screws (7).
2. Install two vertical bars (2) on lockpins (4) with two washers (3) and new cotter pins (1).
3. Install inner handle (8) on handle shank (9) with pin (10).

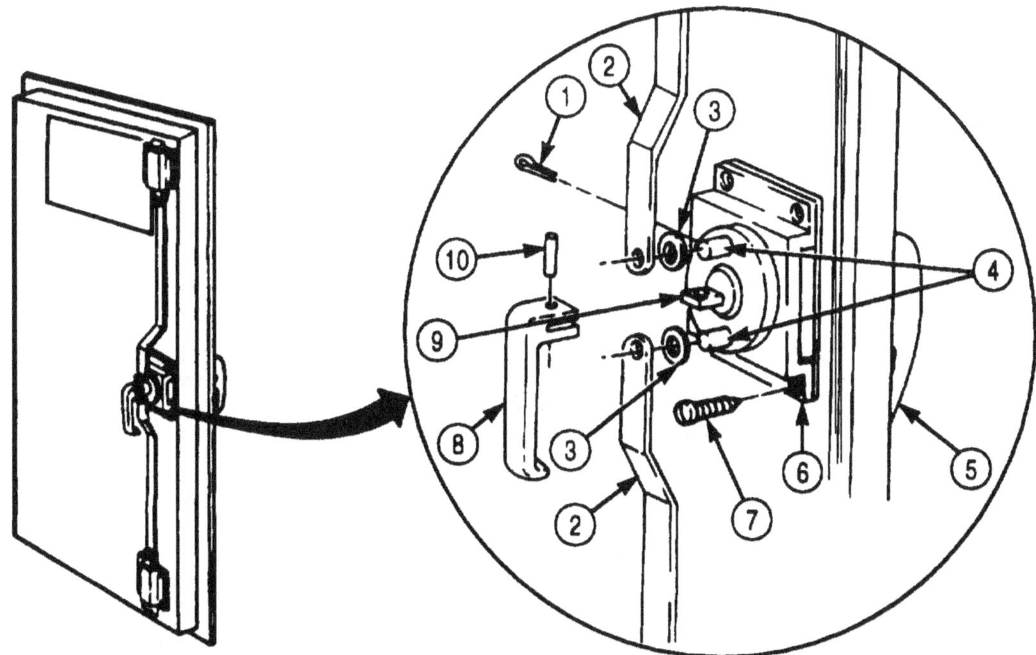

FOLLOW-ON TASK: Close rear doors (rear door handle and lock replacement only) (TM-9-2320-272-10).

3-365. DOOR CHECKS REPLACEMENT

THIS TASK COVERS:

a. Removal
b. Installation

INITIAL SETUP:

APPLICABLE MODELS
M934/A1/A2

TOOLS
General mechanic's tool kit (Appendix E, Item 1)

MATERIALS/PARTS
Four lockwashers (Appendix D, Item 400)

REFERENCES (TM)
TM 9-2320-272-10
TM 9-2320-272-24P

EQUIPMENT CONDITION
- Parking brake set (TM 9-2320-272-10).
- Rear doors open (TM 9-2320-272-10).

NOTE

Blackout switch striker is removed with door check arm bracket on all doors except left rear.

1. Remove two screws (9), lockwashers (8), door check arm bracket (12), and two washers (2) from edge of door (1) and blackout switch striker (13). Discard lockwashers (8).
2. Remove screw (10), washer (11), and blackout switch striker (13) from edge of door (1).
3. Remove two nuts (5), lockwashers (6), screws (3), and door check (7) from bracket (4). Discard lockwashers (6).

b. Installation

1. Install door check (7) on bracket (4) with two screws (3), new lockwashers (6), and nuts (5).
2. Install blackout switch striker (13) on edge of door (1) with washer (11) and screw (10).
3. Install two washers (2) and door check arm bracket (12) on edge of door (1) with two new lockwashers (8) and screws (9).

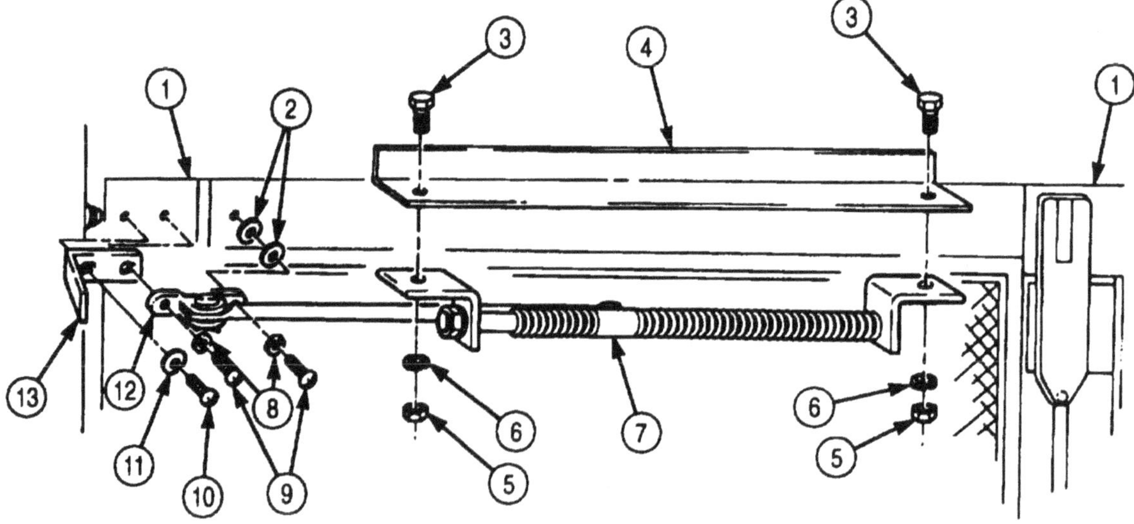

FOLLOW-ON TASK: Close rear doors (TM 9-2320-272-10).

3-366. LADDER RACK BUMPERS REPLACEMENT

THIS TASK COVERS:

a. Removal **b. Installation**

APPLICABLE MODELS
M934/A1/A2

TOOLS
General mechanic's tool kit (Appendix E, Item 1)

REFERENCES (TM)
TM 9-2320-272-10
TM 9-2320-272-24P

EQUIPMENT CONDITION
- Parking brake set (TM 9-2320-272-10).
- Ladders removed (TM 9-2320-272-10)

a. Removal

Remove two nuts (3), screws (1), washers (4), and bumpers (2) from ladder rack (5).

b. Installation

Install two bumpers (2) on ladder rack (5) with two washers (4), screws (1). and nuts (3).

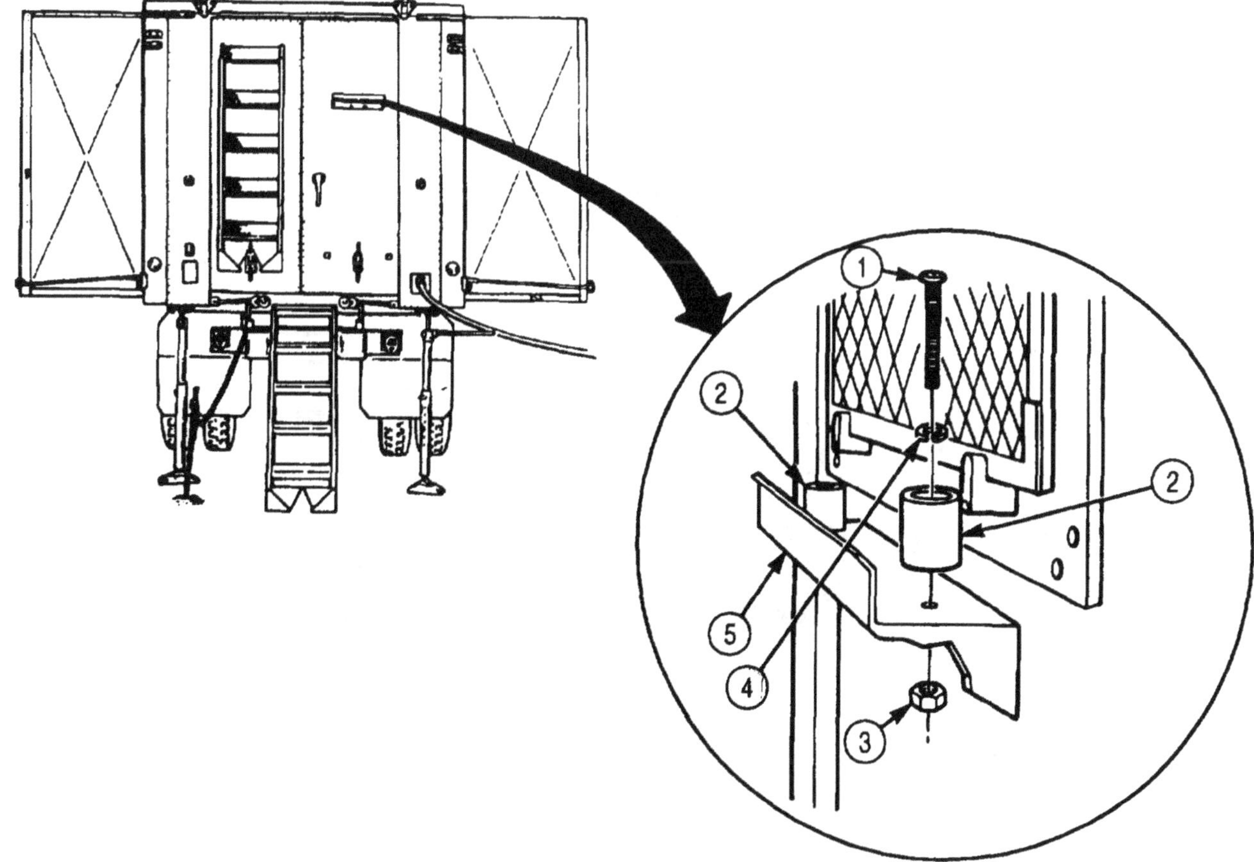

FOLLOW-ON TASK: Install ladders (TM 9-2320-272-10).

3-367. SIDE PANEL RUBBER BUMPERS REPLACEMENT

THIS TASK COVERS:

a. Removal b. Installation

INITIAL SETUP:

APPLICABLE MODELS
M934/A1/A2

TOOLS
General mechanic's tool kit (Appendix E, Item 1)

REFERENCES (TM)
TM 9-2320-272-10
TM 9-2320-272-24P

EQUIPMENT CONDITION
- Parking brake set (TM 9-2320-272-10).
- Van side panel fully expanded and secured (TM 9-2320-272-10).

NOTE

Removal procedures for side panel bumpers, rubber bumpers (rear interior wall), and swivel hood rubber bumpers are the same.

a. Removal

Remove screw (2), washer (3), and bumper (4) from side panel (1).

b. Installation

Install bumper (4) on side panel (1) with washer (3) and screw (2).

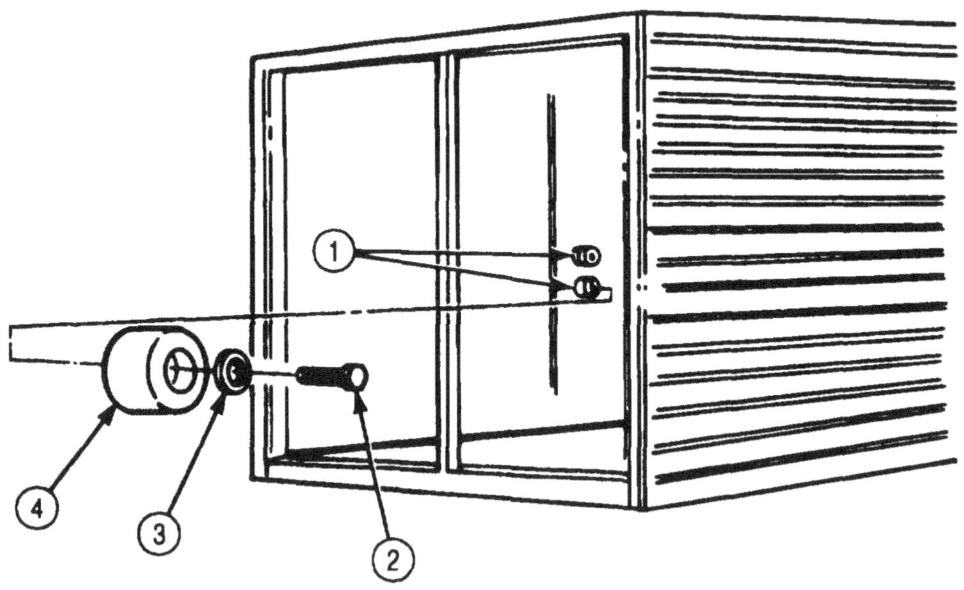

FOLLOW-ON TASK: Retract and secure van side panel (TM 9-2320-272-10).

3-368. SIDE PANEL REAR LOCK REPLACEMENT

THIS TASK COVERS:

a. Removal   b. Installation

INITIAL SETUP:

APPLICABLE MODELS
M934/A1/A2

TOOLS
General mechanic's tool kit (Appendix E, Item 1)

MATERIALS/PARTS
Two cotter pins (Appendix D, Item 51)
Four lockwashers (Appendix D, Item 364)

REFERENCES (TM)
TM 9-2320-272-10
TM 9-2320-272-24P

EQUIPMENT CONDITION
- Parking brake set (TM 9-2320-272-10).
- Van side panel fully expanded and secured (TM 9-2320-272-10).

a. Removal

1. Remove eight screws (5) and lock cover plate with insulation (6) from side panel (13).
2. Remove two screws (17) from lock bolt retainer (16) and door frame (18).
3. Remove two cotter pins (1), pins (4), and clevises (2) from locking arms (3). Discard cotter pins (1).
4. Remove nut (7) and washer (8) from lock (11) and lock handle shank (14).
5. Remove four nuts (9), lockwashers (10), lock (11). lockbolt (12), and lockbolt retainer (16) from side panel (13). studs (15), and door frame (18). Discard lockwashers (10).

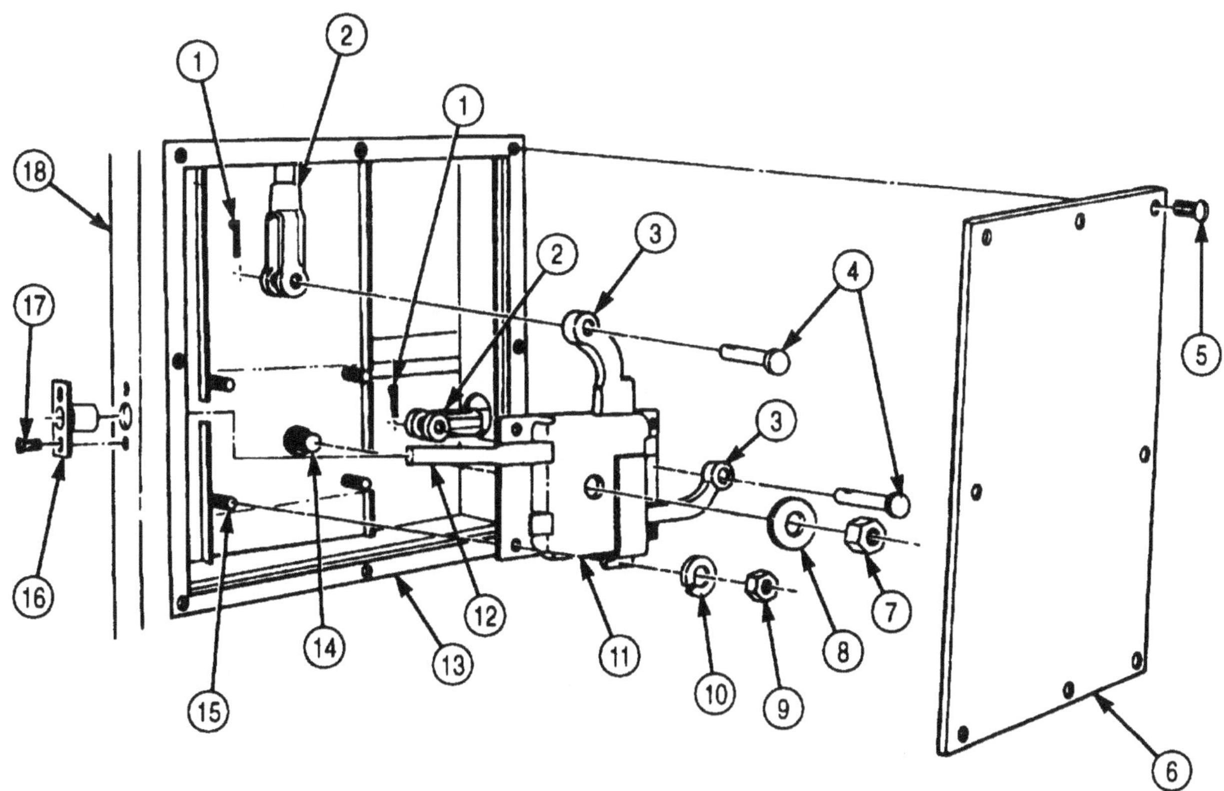

3-368. SIDE PANEL REAR LOCK REPLACEMENT (Contd)

b. Installation

1. Position lock (11) over lock handle shank (14) and studs (15).
2. Slide lockbolt retainer (16) over lockbolt (12) and install on door frame (18) with two screws (17).
3. Install lock (11) on side panel (13) and studs (15) with four new lockwashers (10) and nuts (9).
4. Install washer (8) and nut (7) on lock (11) and lock handle shank (14).
5. Install two clevises (2) on locking arms (3) with two pins (4) and new cotter pins (1).
6. Install lock cover plate with insulation (6) on side panel (13) with eight screws (5).

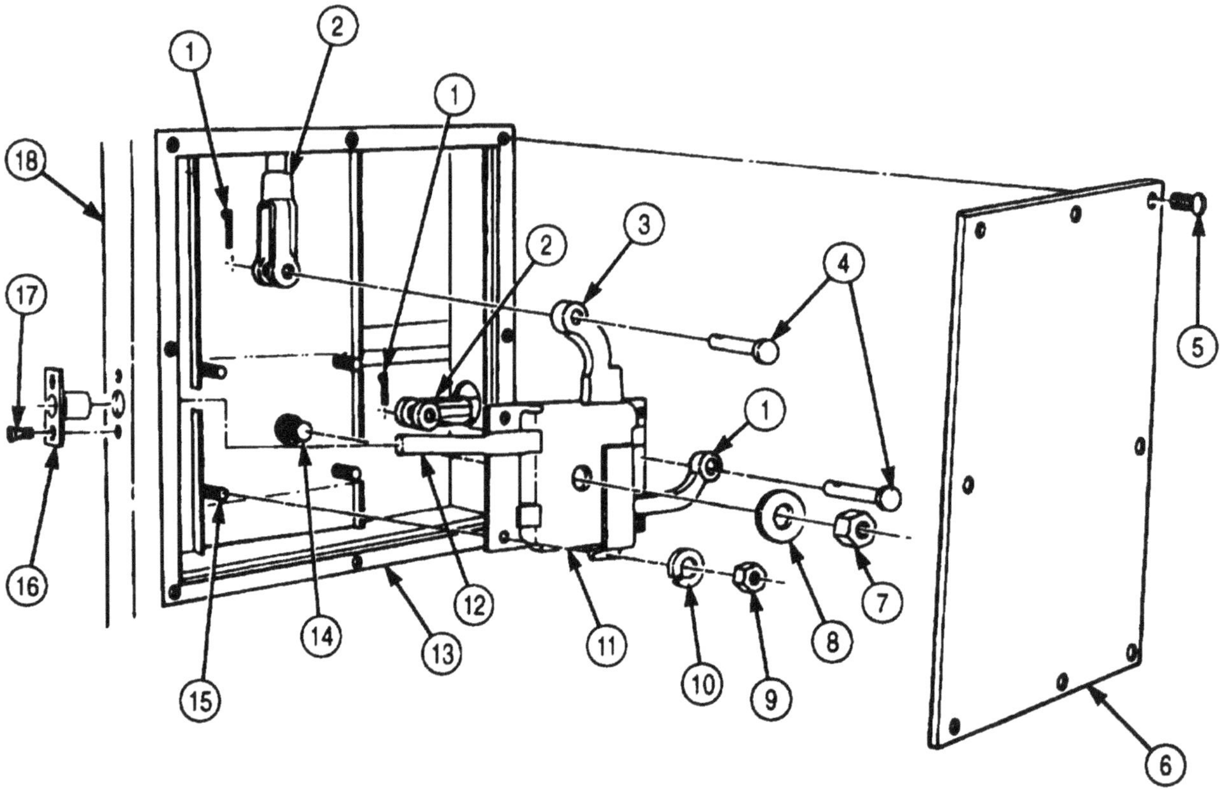

FOLLOW-ON TASK: Retract and secure van side panel (TM 9-2320-272-10).

3-369. SIDE PANEL FRONT LOCK AND HINGED-TYPE ROOF LOCK REPLACEMENT

THIS TASK COVERS:

a. Removal

b. Installation

INITIAL SETUP

APPLICABLE MODELS
M934/A1/A2

TOOLS
General mechanic's tool kit (Appendix E, Item 1)

MATERIALS/PARTS
Four lockwashers (Appendix D, Item 364)
Two cotter pins (Appendix D, Item 51)
GAA grease (Appendix C, Item 28)

REFERENCES (TM)
TM 9-2320-272-10
TM 9-2320-272-24P

EQUIPMEMT CONDITION
Parking brake set (TM 9-2320-272-10).

a. Removal

1. Remove ten screws (10) and cover plate (11) from side panel (14).
2. Remove two cotter pins (1), pins (7), and clevises (2) from lock arms (6). Discard cotter pins (1).
3. Remove nut (12) and washer (13) from lock (3) and lock handle shank (4).
4. Remove four nuts (9), lockwashers (8), and lock (3) from four studs (5) and side panel (14). Discard lockwashers (8).

b. Installation

1. Position lock (3) on side panel (14), four studs (5), and lock handle shank (4) and install washer (13) and nut (12).
2. Install four new lockwashers (8) and nuts (9) on lock (3) and four studs (5).
3. Install two clevises (2) on lock arms (6) with two pins (7) and new cotter pins (1).
4. Install cover plate (11) on side panel (14) with ten screws (10).

3-370. SIDE PANEL EXTERIOR LOCK REPLACEMENT

THIS TASK COVERS:

a. Removal b. Installation

INITIAL SETUP:

APPLICABLE MODELS
M934/A1/A2

TOOLS
General mechanic's tool kit (Appendix E, Item 1)

REFERENCES (TM)
TM 9-2320-272-10
TM 9-2320-272-24P

EQUIPMENT CONDITION
Parking brake set (TM 9-2320-272-10).

a. Removal

1. Remove locking pin (4) and release lock handle (1).
2. Remove four screws (2) and lock (3) from side panel (5).

b. Installation

1. Install lock (3) on side panel (5) with four screws (2).
2. Close lock handle (1) and insert locking pin (4).

3-370. SIDE PANEL EXTERIOR LOCK REPLACEMENT (Contd)

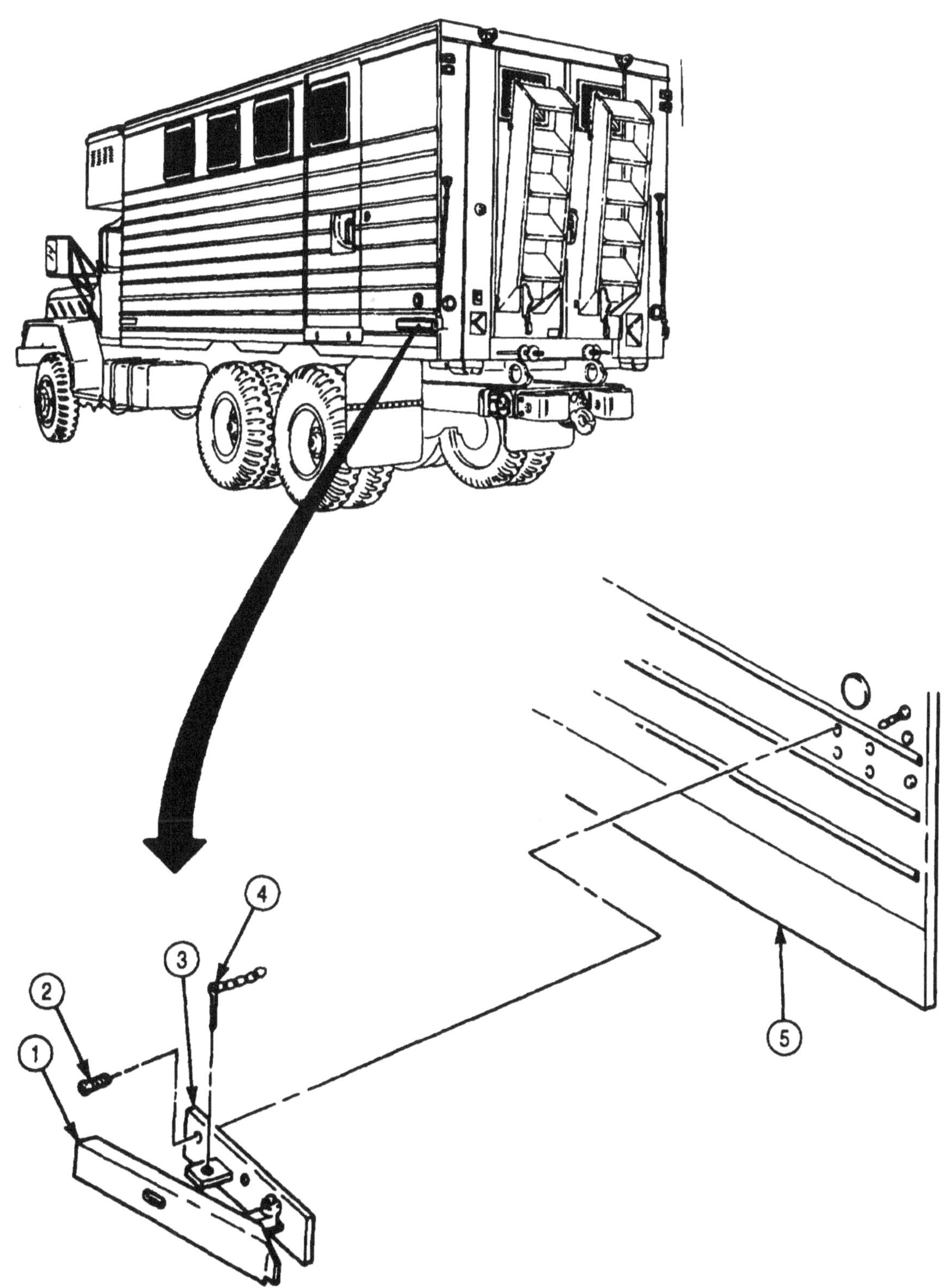

3-371. FLUORESCENT LIGHT TUBE REPLACEMENT

THIS TASK COVERS:

a. Removal
b. Installation

INITIAL SETUP:

APPLICABLE MODELS
M934/A1/A2

TOOLS
General mechanic's tool kit (Appendix E, Item 1)

REFERENCES (TM)
TM 9-2320-272-10
TM 9-2320-272-24P

EQUIPMENT CONDITIONS
- Parking brake set (TM 9-2320-272-10).
- Fluorescent ceiling light switch off (TM 9-2320-272-10).

a. Removal

1. Remove two screws (3) and lower mesh guard (2) from light fixture (1).
2. Remove three fluorescent tubes (4) from light fixture (1).

b. Installation

1. Install three fluorescent tubes (4) on light fixture (1).
2. Position mesh guard (2) on light fixture (1) and install with two screws (3).

FOLLOW-ON TASK: Check operation of fluorescent lights TM 9-2320-272-10).

3-372. EMERGENCY/BLACKOUT LIGHT LAMP AND LENS REPLACEMENT

THIS TASK COVERS:

a. Removal

b. Installation

INITIAL SETUP:

APPLICABLE MODELS
M934/A1/A2

TOOLS
General mechanic's tool kit (Appendix E, Item 1)

REFERENCES (TM)
TM 9-2320-272-10
TM 9-2320-272-24P

EQUIPMENT CONDITION
- Parking brake set (TM 9-2320-272-10).
- Emergency light switch off (TM 9-2320-272-10)
- Blackout light switch off (TM 9-2320-272-10).

NOTE

Emergency light and blackout light lamps and lenses are replaced the same. This procedure is for the emergency light lamps and lenses.

a. Removal

1. Loosen lockscrew (2) and open light door (1).
2. Remove lamp (4) from lamp socket (3).
3. Turn lens retaining clip (5) and remove lens (6) from light door (1).

b. Installation

NOTE

White lens is installed in emergency light door and blue lens is installed in blackout light door.

1. Position lens (6) in light door (1) and install with lens retaining clip (5).
2. Install lamp (4) in lamp socket (3).
3. Close light door (1) and tighten lockscrew (2).

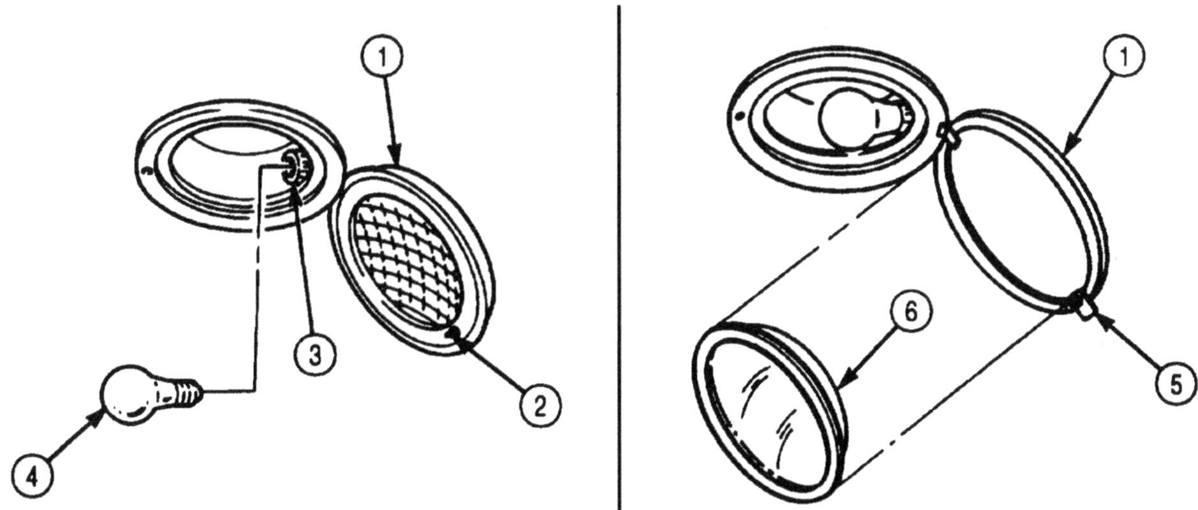

FOLLOW-ON TASK: Check operation of emergency/blackout light (TM-9-2320-272-10).

3-373. BLACKOUT LIGHT SWITCH AND 110-VOLT RECEPTACLE REPLACEMENT

THIS TASK COVERS:

a Removal | b. Installation

INITIAL SETUP:

APPLICABLE MODELS
M934/A1/A2

TOOLS
General mechanic's tool kit (Appendix E, Item 1)

REFERENCES (TM)
TM 9-2320-272- 10
TM 9-2320-272-24P

EQUIPMENT CONDITION
- Parking brake set TM 9-2820-272-10).
- Auxiliary A/C power source disconneted (TM 9-2320-272-10).

a. Removal

1. Remove three screws (7) and cover plate (6) from box (1).
2. Remove two screws (8) and pull switch (9) away from box (1).
3. Remove two screws (5) and pull receptacle (4) away from box (1).

NOTE

Tag all leads for installation.

4. Remove five screws (3) and leads (2) from pull switch (9).
5. Remove five screws (3) and leads (2) from receptacle (4).

b. Installation

1. Install five leads (2) on receptacle (4) with five screws (3).
2. Install five leads (2) on pull switch (9) with five screws (3).
3. Install receptacle (4) on box (1) with two screws (5).
4. Install pull switch (9) on box (1) with two screws (8).
5. Install cover plate (6) on box (1) with three screws (7).

3-373. BLACKOUT LIGHT SWITCH AND 110-VOLT RECEPTACLE REPLACEMENT (Contd)

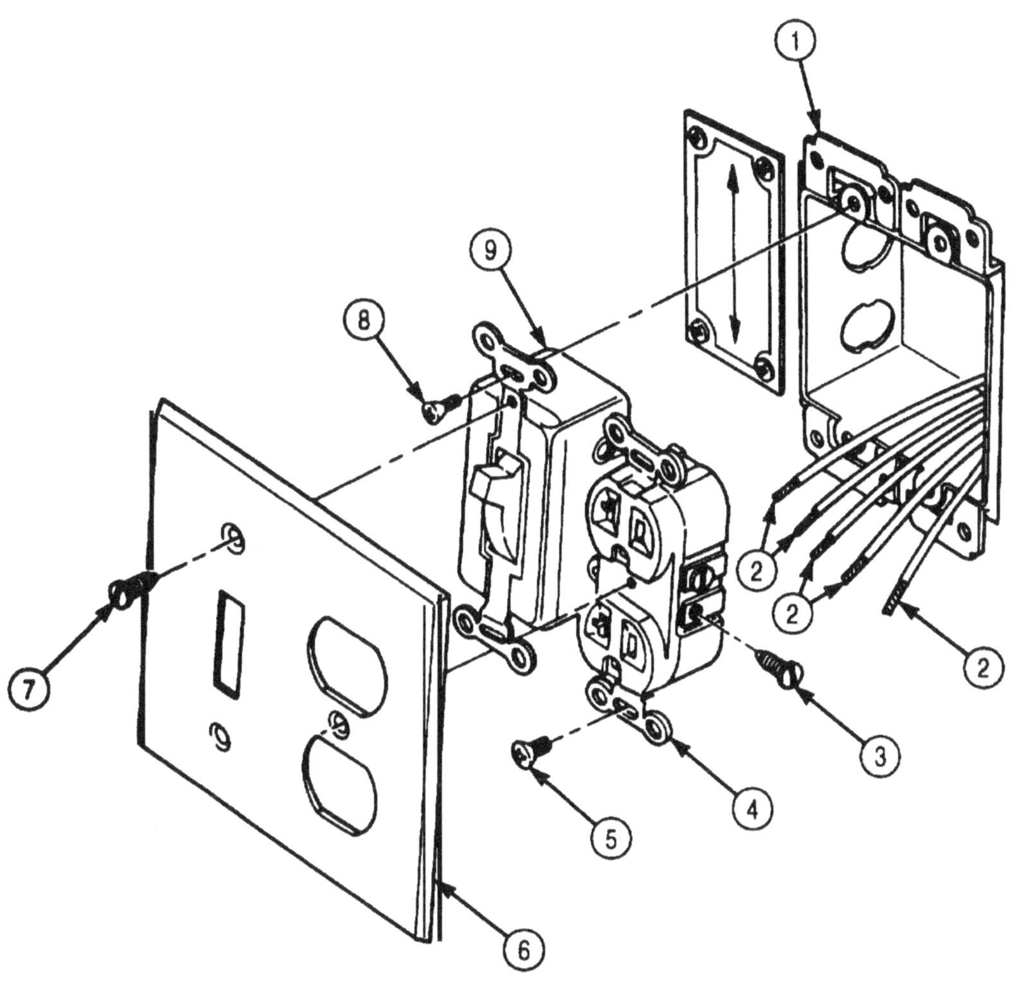

FOLLOW-ON TASK: Connect auxiliary A/C power source and check operation of light switch and 110-volt receptacle (TM 9-2320-272-10).

3-374. INSIDE TELEPHONE JACK POST REPLACEMENT

THIS TASK COVERS:

a. Removal **b. Installation**

APPLICABLE MODELS:
M934/A1/A2

TOOLS
General mechanic's tool kit (Appendix E, Item 11

REFERENCES (TM)
TM 9-2320-272-10
TM 9-2320-272-24P

EQUIPMENT CONDITION
Parking brake set (TM 9-2320-272-10).

NOTE

Inside telephone jack posts are located on van ceiling next to emergency light and on left rear panel below fire extinguisher. Both telephone jack posts are replaced the same. This procedure is for rear panel telephone jack post.

a. Removal

1. Remove four screws (6) and pull junction box (4) away from panel (7).

NOTE

- Ceiling telephone jack posts are connected with one wire.
- Tag all leads for installation.

2. Remove screw (1), two leads (2), nut (8), and telephone jack post (5) from grommet (3) and junction box (4).
3. Remove grommet (3) from junction box (4).

b. Installation

1. Install grommet (3) on junction box (4).
2. Install telephone jack post (5) on grommet (3) and junction box (4) with nut (8).
3. Install two leads (2) on telephone jack post (5) with screw (1).

3-374. INSIDE TELEPHONE JACK POST REPLACEMENT (Contd)

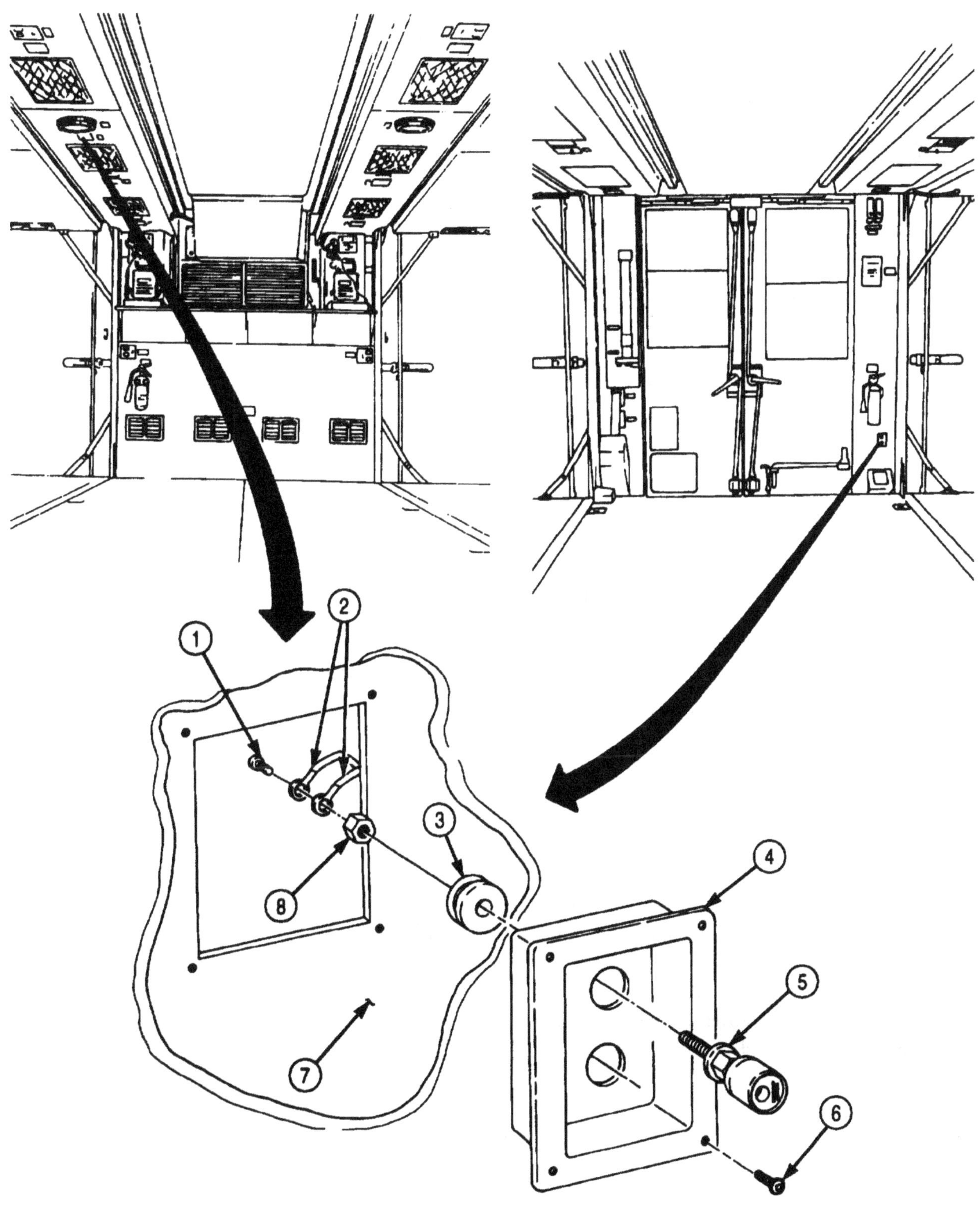

3-375. OUTSIDE TELEPHONE JACK POST REPLACEMENT

THIS TASK COVERS

a. Removal b. Installation

INITIAL SETUP:

APPLICABLE MODEL
M934/A1/A2

TOOLS
General mechanic's tool kit (Appendix E, Item 1)

MATERIAL/PARTS
Seal (Appendix D, Item 601)

REFERENCES (TM)
TM 9-2320-272-10
TM 9-2320-272-24P

EQUIPMENT CONDITION
- Parking brake set (TM 9-2320-272-10).
- Fire extinguisher removed (TM 9-2320-272-10).
- Inside telephone jack posts removed (para. 3-374).

a. Removal

1. Remove four screws (11), cover (12). and seal (19) from van body (13). Discard seal (19).
2. Remove two screws (5) and thermostat cover (4) from two thermostats (2) and inside panel (10).
3. Remove eight screws (3) and two thermostats (2) from inside panel (10). Thermostats (2) remain connected to wires.
4. Remove twenty-three screws (8) and panel retainer (7) from inside panel (10).
5. Pry inside panel (10) away from frame (1) enough to clear studs (6) and gain access to rear of junction box (9).

NOTE
- Tag all leads for installation.
- Assistant will help with step 6.

6. Remove screws (17), lead (16), nut (15), and telephone jack post (18) from grommet (14) and junction
7. Remove grommet (14) from junction box (9).

b. Installation

1. Install grommet (14) on junction box (9).

NOTE
Assistant will help with stop 2.

2. Install telephone jack post (18) on grommet (14) and junction box (9) with screw (15).
3. Install lead (16) on telephone jack post (18) with nut (17).
4. Pry inside panel (10) over studs (6) and onto frame (1).
6. Install panel retainer (7) on inside panel (10) with twenty-three screws (8).
6. Install two thermostats (2) on inside panel (10) with eight screws (3).
7. Install thermostat cover (4) on two thermostats (2) and inside panel (10) with two screws (6).
8. Install seal (19) and cover (12) on van body (13) with four screws (11).

3-375. OUTSIDE TELEPHONE JACK POST REPLACEMENT (Contd)

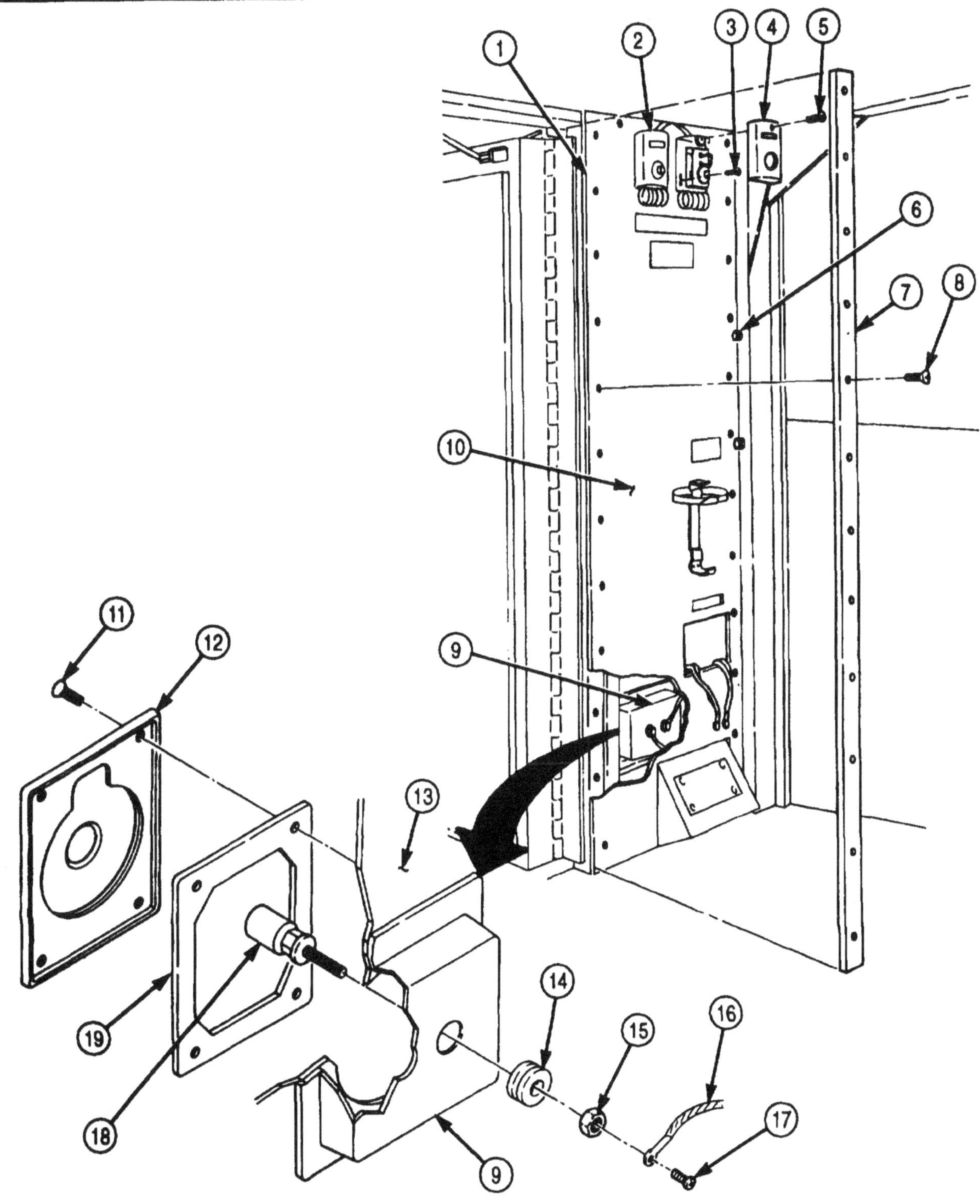

FOLLOW-ON TASKS: • Install inside panel telephone jack posts (para. 3-374).
• Install fire extinguisher (TM 9-2320-272-10).

3-376. SIDE AND REAR DOOR BLACKOUT LIGHT SWITCH MAINTENANCE

THIS TASK COVERS:

a. Removal
b. Installation
c. Adjustment

INITIAL SETUP:

APPLICABLE MODEL
M934/A1/A2

TOOLS
General mechanic's tool kit (Appendix E, Item 1)

MATERIALS/PARTS
Solder (Appendix C, Item 70)

REFERENCES (TM)
TM 9-2320-272-10
TM 9-2320-272-24P

EQUIPMENT CONDITION
Parking brake set (TM 9-2320-272-10)
Main power switch off (TM 9-2320-272-10).
Van body side panel fully expanded and secured (TM 9-2320-272-10).

a. Removal

1. Remove two screws (4) and blackout light switch (2) from electrical panel (8) or switch box (1).
2. Remove solder from terminals of blackout light switch (2) and disconnect two wires (7).

b. Installation

1. Connect two wires (7) to terminals of blackout light switch (2) and install with solder.
2. Install blackout light switch (2) on electrical panel (8) or switch box (1) with two screws (4).

c. Adjustment

NOTE

With main power switch, interior lights, and blackout light circuit on, interior lights must turn off when door starts to open, and turn on when door is closed.

1. Remove blackout light switch (2) (task a.).
2. Remove nut (5) and plate (6) from blackout light switch (2).
3. Rotate ring (3) down farther on threaded shaft of blackout light switch (2).
4. Install plate (6) on blackout light switch (2) with nut (5).
5. Install blackout light switch (2) (task b.).

3-376. SIDE AND REAR DOOR BLACKOUT LIGHT SWITCH MAINTENANCE (Contd)

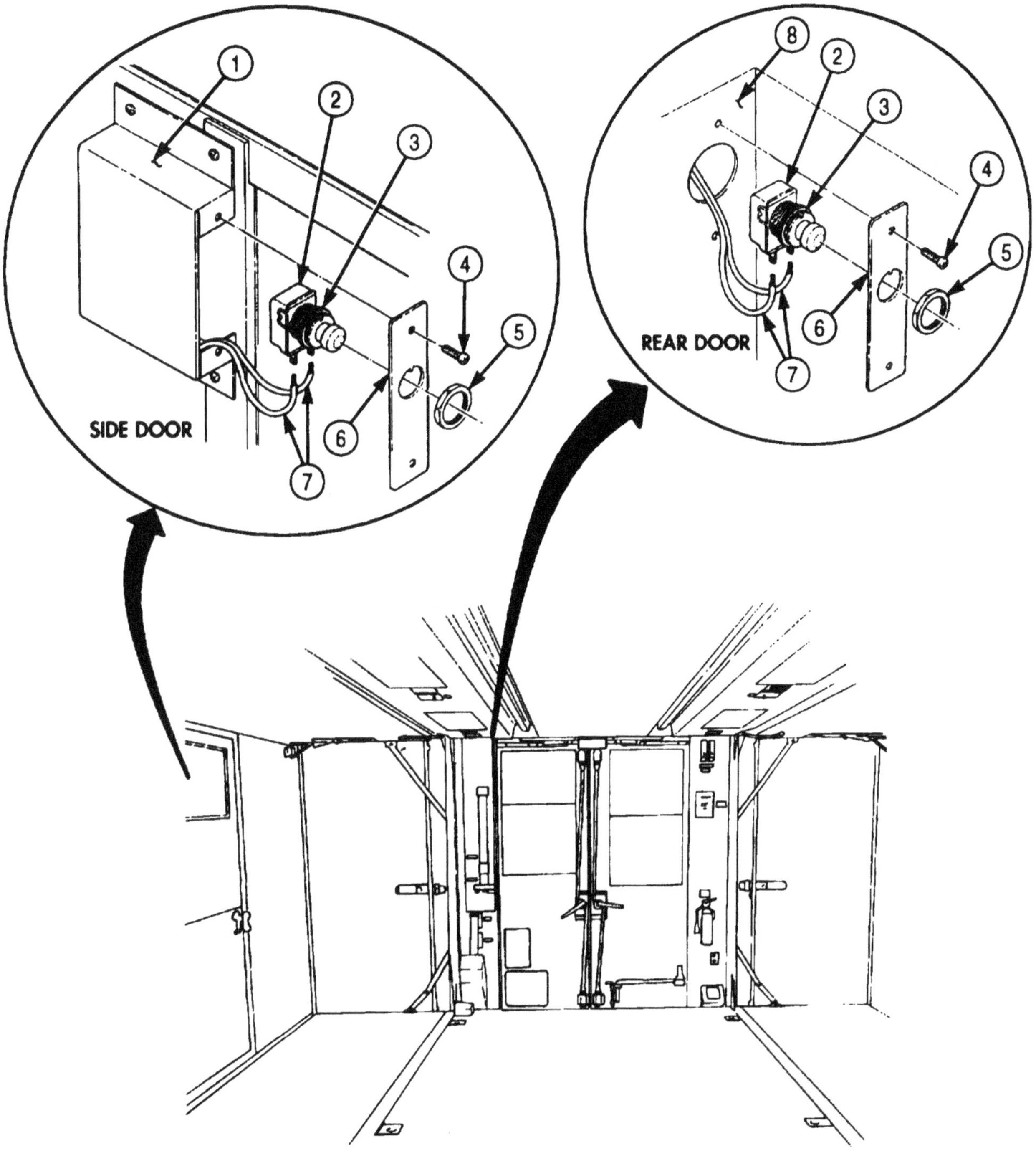

FOLLOW-ON TASKS:
- Turn on main power switch and check operation of side and rear door blackout switches (TM 9-2320-272-10).
- Retract and secure van body side panel (TM 9-2320-272-10).

3-377. HINGED ROOF-OPERATED BLACKOUT CIRCUIT PLUNGERS REPLACEMENT

THIS TASK COVERS:

a. Removal b. Installation

INITIAL SETUP:

APPLICABLE MODELS

M934/A1/A2

TOOLS

General mechanic's tool kit (Appendix E, Item 1)

REFERENCES (TM)

TM 9-2320-272-10
TM 9-2320-272-24P

EQUIPMENT CONDITION

- Parking brake set (TM 9-2320-272-10).
- Main power switch off TM 9-2320-272-10).
- Van body side panel fully expanded and secured (TM 9-2320-272-10).

a. Removal

1. Remove four screws (4) and pull plunger plate (5) from hinged roof (1).

NOTE

Tag all leads for installation.

2. Remove two screws (2) and leads (3) from plunger plate (5).

b. Installation

1. Install two leads (3) on plunger plate (5) with two screws (2).
2. Install plunger plate (5) on hinged roof (1) with four screws (4).

3-377. HINGED ROOF-OPERATED BLACKOUT CIRCUIT PLUNGERS REPLACEMENT (Contd)

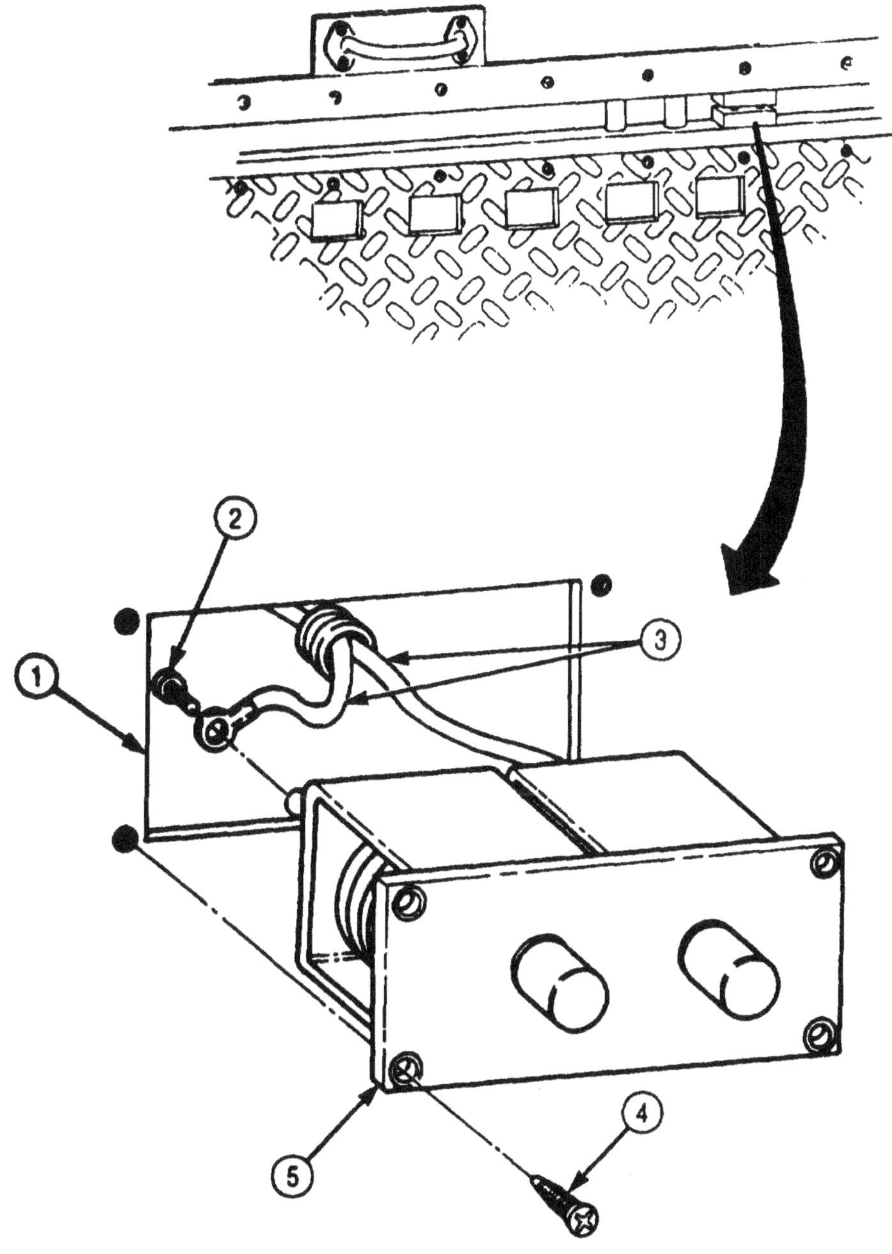

FOLLOW-ON TASKS: • Turn on main power switch and check operation of side and rear door blackout switches (TM 9-2320-272-10)
• Retract and secure van body side panel (TM 9-2320-272-10)

3-378. EXPANDING AND RETRACTING MECHANISM LOCKS REPLACEMENT

THIS TASK COVERS:

a. Removal

b. Installation

INITIAL SETUP:

APPLICABLE MODELS

M934/A1/A2

TOOLS

General mechanic's tool kit (Appendix E, Item 1)

MATERIALS/PARTS

Locknut (Appendix D, Item 297)

REFERENCES (TM)

TM 9-2320-272-10

TM 9-2320-272-24P

EQUIPMENT CONDITION

Parking brake set (TM 9-2320-272-10).

a. Removal

Remove locknut (4), screw (6), lock plunger (2), and lock pawl (5) from crossmember (1) and stud (3). Discard locknut (4).

b. Installation

1. Position lock plunger (2) in slot of lock pawl (5).
2. Install lock plunger (2) and lock pawl (5) on crossmember (1) and stud (3) with screw (6) and new locknut (4).

3-378. EXPANDING AND RETRACTING MECHANISM LOCKS REPLACEMENT (Contd)

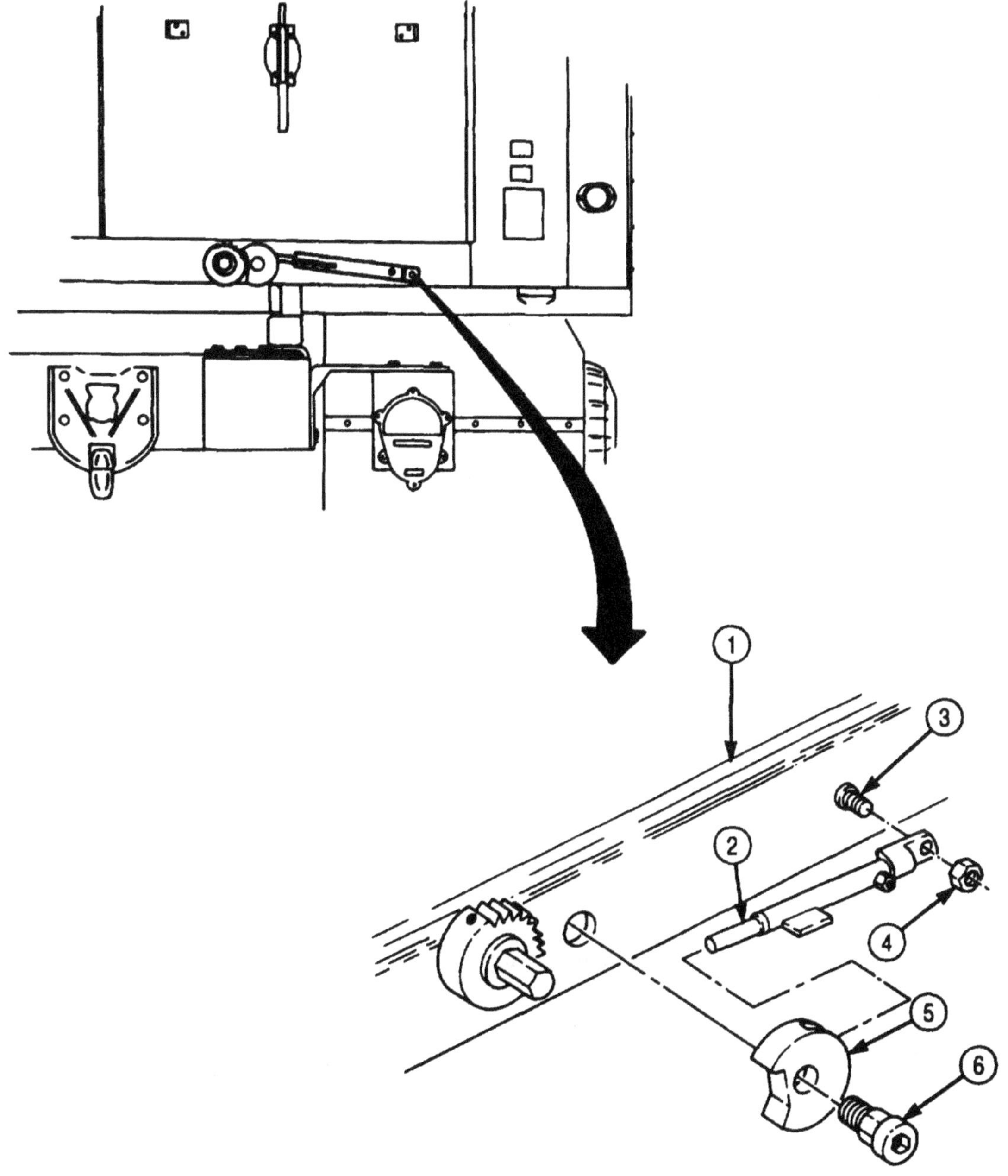

FOLLOW-ON TASK: Check expanding and retracting mechanism for proper operation (TM 9-2320-272-10).

3-379. VAN HEATER AND EXHAUST REPLACEMENT

THIS TASK COVERS:

a. Removal   b. Installation

APPLICABLE MODELS
M934/A1/A2

TOOLS
General mechanic's tool kit (Appendix E, Item 1)

MATERIALS/PARTS
Seven lockwashers (Appendix D, Item 400)
Lockwasher (Appendix D, Item 418)
Four lockwashers (Appendix D, Item 348)
Antiseize tape (Appendix C, Item 72)

REFERENCES (TM)
TM 9-2320-272-10
TM 9-2320.272-24P

EQUIPMENT CONDITION
- Parking brake set (TM 9-2320-272-10).
- Battery ground cables disconnected (para. 3-126).

GENERAL SAFETY INSTRUCTIONS
Do not perform this procedure near open flames.

a. Removal

1. Remove eight screws (3), exhaust pipe (4), and two clamps (15) from van body (6) and heater exhaust pipe (12).
2. Remove two screws (14). lockwashers (2), nuts (1), and clamps (15) from exhaust pipe (4). Discard lockwashers (2).

NOTE

Tag all leads and lines for installation.

3. Remove nut (19) and fuel overflow line (23) from fitting (18).
4. Remove nut (24) and fuel receptacle line (25) from heater fuel valve (17).
5. Disconnect power receptacle connector (16) from heater (7).
6. Disconnect external fuel pump receptacle connector (26) from heater (7).
7. Disconnect room thermostat connector (27) from heater (7).
8. Remove four screws (8) and lockwashers (9) from heater bracket (10). Discard lockwashers (9).
9. Remove screw (22), ground strap (21), and lockwasher (20) from bracket (10) and heater (7). Discard lockwasher (20).

NOTE

Assistant will help with step 10.

10. Remove heater (7) from van body (6).
11. I&move five screws (22) and lockwashers (20) from two brackets (10) and heater (7). Discard lockwashers (20).
12. Remove heater exhaust pipe (12) from heater (7).
13. Remove four screws (5) and (13) and two seals (11) from van body (6).

3-379. VAN HEATER AND EXHAUST REPLACEMENT (Contd)

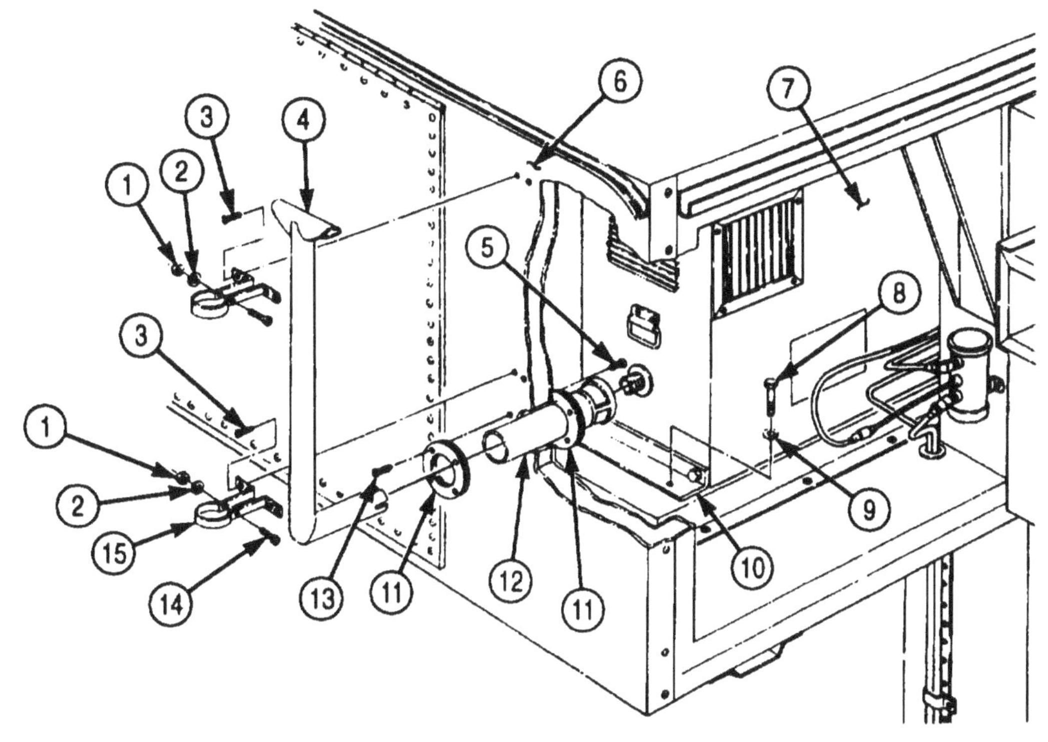

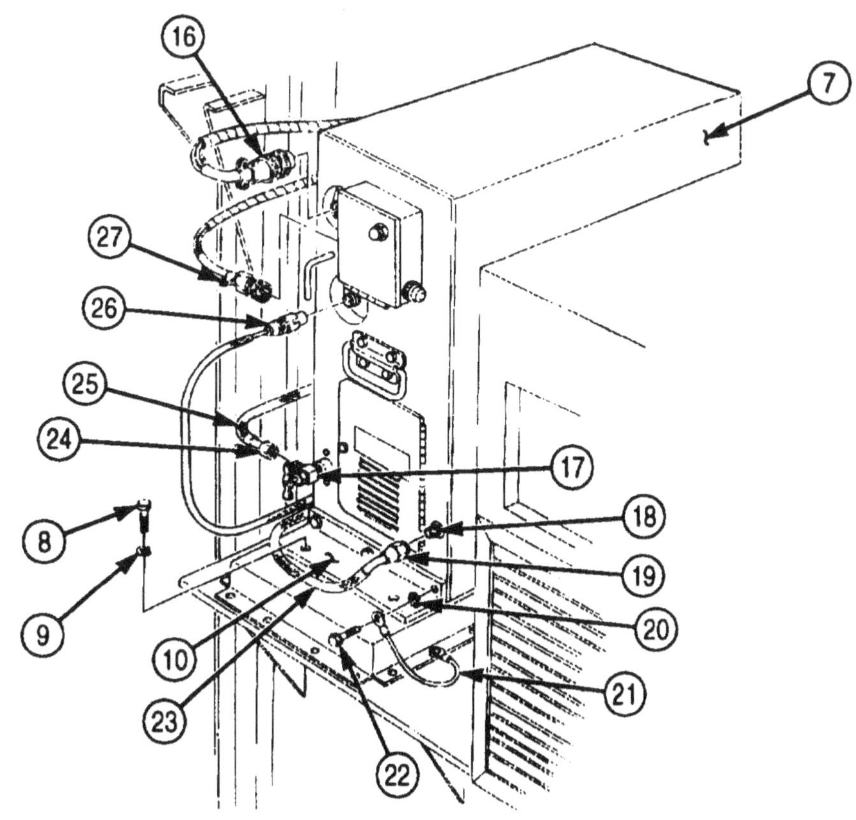

3-379. VAN HEATER AND EXHAUST REPLACEMENT (Contd)

b. Installation

1. Apply antiseize tape to male threads of heater fuel valve (16) and heater nipple (17).
2. Install two seals (9) on van body (3) with eight screws (4).
3. Install two brackets (8) on heater (5) with five screws (21) and new lockwashers (19).
4. Install heater exhaust pipe (10) on heater (5).

NOTE

Assistant will help with step 5.

5. Place heater (5) in van body (3).
6. Install bracket (8) on van body (3) with four screws (6) and new lockwashers (7).
7. Install ground strap (20) on bracket (8) and heater (5) with screws (21) and new lockwasher (19).
8. Install fuel overflow line (22) on fitting (17) with nut (18).
9. Install fuel receptacle line (24) on heater fuel valve (16) with nut (23).
10. Connect external fuel pump receptacle connector (25) to heater (5).
11. Connect power receptacle connector (15) to heater (5).
12. Connect room thermostat connector (26) to heater (5).
13. Install exhaust pipe (2) on heater exhaust pipe (10) with two clamps (12), eight screws (1), two screws (11), new lockwashers (13), and nuts (14).

3-379. VAN HEATER AND EXHAUST REPLACEMENT (Contd)

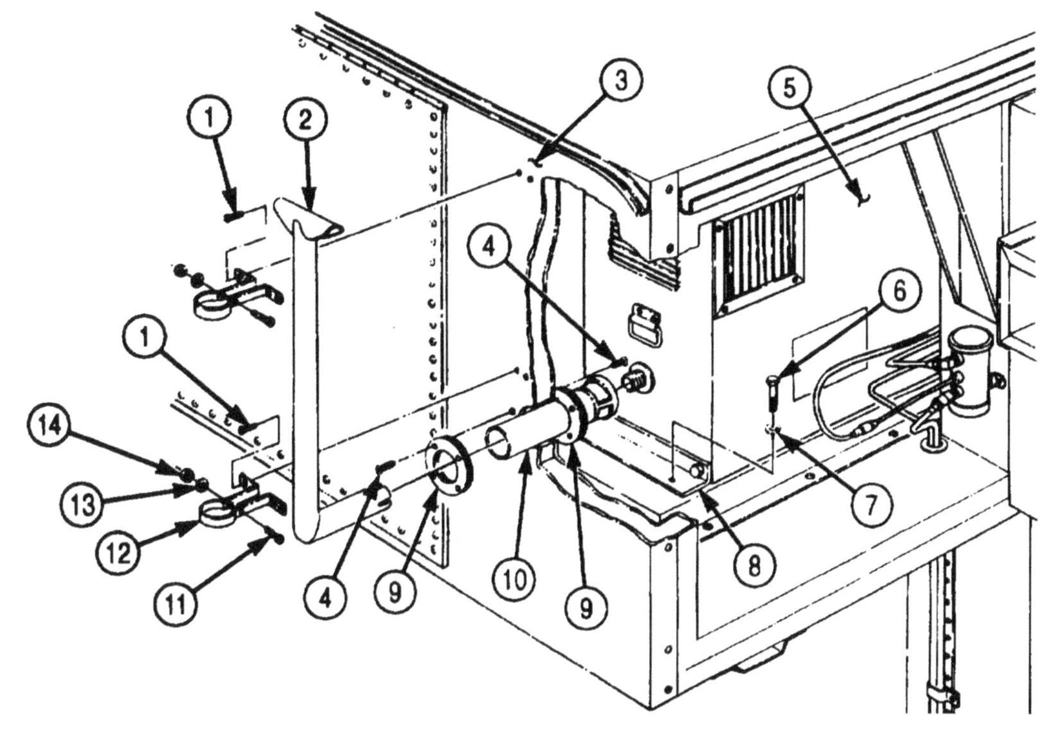

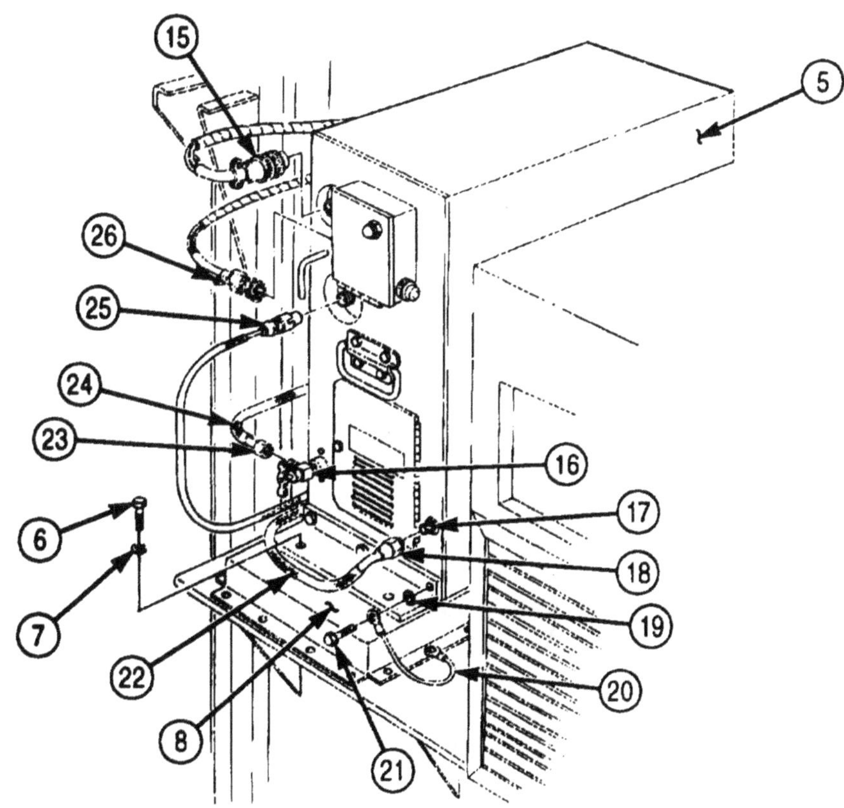

FOLLOW-ON TASK: Connect battery ground cables (para 3-126).

3-380. VAN POWER CABLE REEL (M934A1/A2) REPLACEMENT

THIS TASK COVERS:FX3:

a. Removal b. Installation

APPLICABLE MODELS
M934/A1/A2

TOOLS
General mechanic's tool kit (Appendix E, Item 1)

MATERIAL/PARTS
Four locknuts (Appendix D, Item 291)

REFERENCES (TM)
TM 9-2320-272-10
TM 9-2320-272-24P

EQUIPMENT CONDITION
- Parking brake set (TM 9-2320-272-10).
- Power cable removed (TM 9-2320-272-10).

a. Removal

1. Remove retaining pin (2) from power cable reel (5).
2. Remove four locknuts (1), screws (4), and power cable reel (5) from mud flap support brackets (3). Discard locknuts (1).

b. Installation

1. Install power cable reel (5) on mud flap support brackets (3) with four screws (4) and new locknuts (1).
2. Install retaining pin (2) on power cable reel (5).

3-380. VAN POWER CABLE REEL (M934A1/A2) REPLACEMENT (Contd)

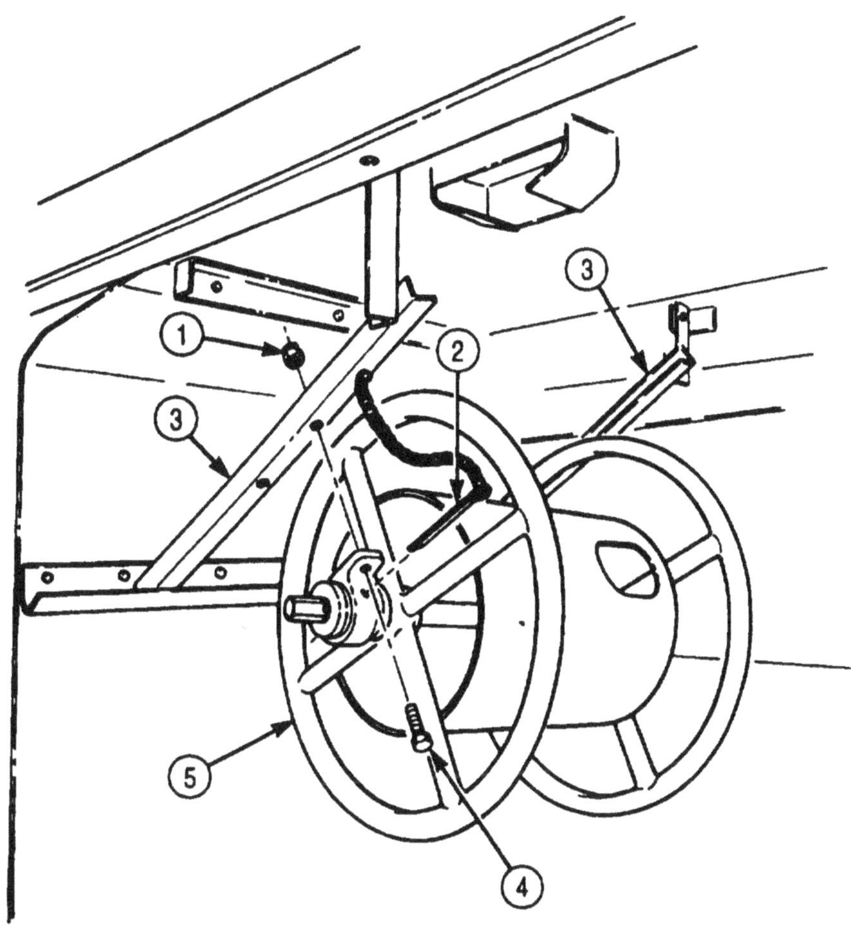

FOLLOW-ON TASK: Install power cable (TM 9-2320-272-10).

3-381. VAN HEATER FUEL PUMP (M934A1/A2) REPLACEMENT

THIS TASK COVERS:

a. Removal (M934A1/A2)
b. Installation (M934A1/A2
c. Removal (M934)
d. Installation (M934)

INITIAL SETUP:

APPLICABLE MODELS
M934/A1/A2

TOOLS
General mechanic's tool kit (Appendix E, Item 1)

MATERIAL/PARTS
Antiseize tape (Appendix C, Item 72)

REFERENCES (TM)
TM 9-2320-272-10
TM 9-2320-272-24P

EQUIPMENT CONDITION
- Parking brake set (TM 9-2320-272-10).
- Battery ground cables disconnected (para. 3-126).

GENERAL SAFETY INSTRUCTIONS
Diesel fuel is highly flammable. Do not perform fuel system procedures near open flame.

a. Removal (M934A1/A2)

1. Loosen two locking studs (8) from turnbuckles (3) and open access door (9).
2. Disconnect fuel pump lead (10) from harness lead (11).
3. Disconnect output line (4) from elbow (5).
4. Disconnect input line (1) from elbow (2).
5. Remove two locknuts (12)1, screws (6), clamp (7) with fuel pump lead (10), and fuel pump (14) from hinge plate (13). Discard locknuts (12).
6. Remove elbows (2) and (5) from fuel pump (14).

b. Installation (M934A1/A2)

NOTE

Wrap male threads with antiseize tape prior to installation.

1. Install elbows (2) and (5) on fuel pump (14).
2. Install fuel pump (14) and clamp (7) with fuel pump lead (10) on hinge plate (13) with two screws (6) and new locknuts (12).
3. Connect input line (1) to elbow (2).
4. Connect output line (4) to elbow (5).
5. Connect fuel pump lead (10) to harness lead (11).
6. Close access door (9) and tighten two locking studs (8) in turnbuckles (3).

3-381. VAN HEATER FUEL PUMP (M934A1/A2) REPLACEMENT (Contd)

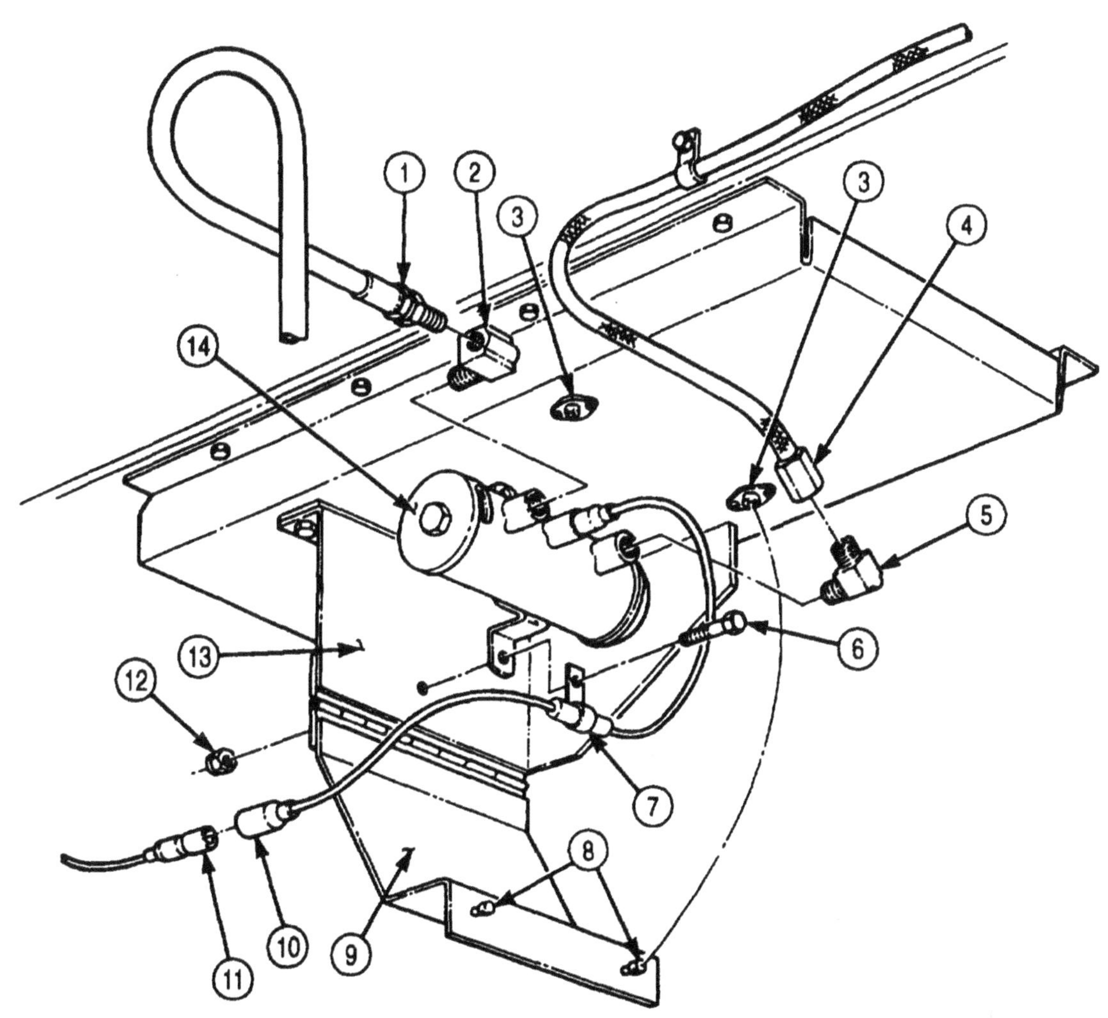

3-381. VAN HEATER FUEL PUMP (M934A1/A2) REPLACEMENT (Contd)

WARNING

Diesel fuel is highly flammable. Do not perform fuel system procedures near open flame. Injury to personnel may result.

c. Removal (M934)

1. Remove seven screws (13) and cover (12) from van body (5).
2. Disconnect fuel pump lead (11) from harness lead (10).
3. Remove outlet line (3) and bushing (2) from elbow (1).
4. Disconnect inlet line (9) from adapter (8).
5. Remove two screws (7), clamp (14), and fuel pump (4) from van body (5).
6. Remove elbow (1), adapter (8), and elbow (6) from fuel pump (4).

d. Installation(M934)

NOTE

Wrap male threads with antiseize tape prior to installation.

1. Install elbow (6), adapter (8), and elbow (1) on fuel pump (4).
2. Install fuel pump (4) and clamp (14) on van body (5) with two screws (7).
3. Connect inlet line (9) on adapter (8).
4. Install bushing (2) and outlet line (3) on elbow (1).
5. Connect fuel pump lead (11) to harness lead (10).
6. Install cover (12) on van body (5) with seven screws (13).

3-381. VAN HEATER FUEL PUMP (M934A1/A2) REPLACEMENT (Contd)

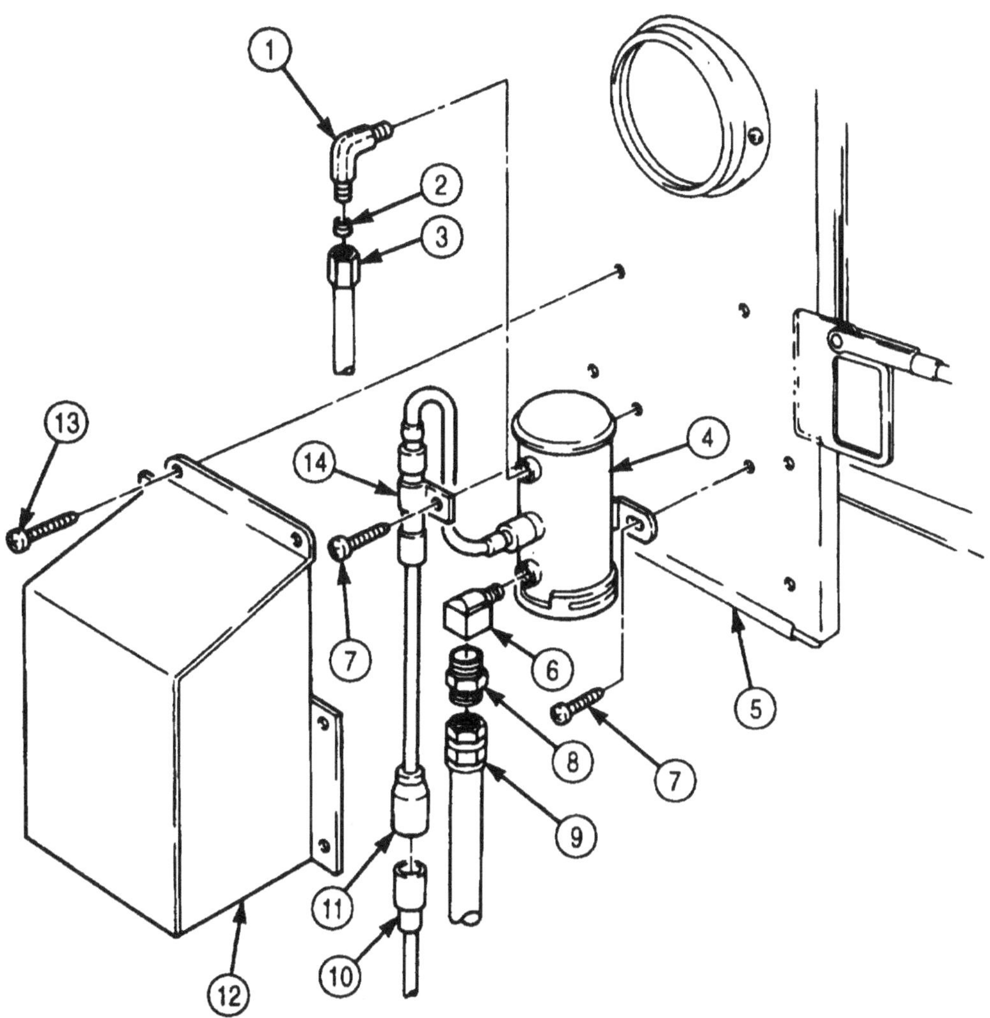

FOLLOW-ON TASK: Connect battery ground cables (para. 3-126).

3-382. AIR CONDITIONER DRAIN TUBE REPLACEMENT

THIS TASK COVERS:

a. Removal **b. Installation**

APPLICABLE MODELS
M934/A1

TOOLS
General mechanic's tool kit (Appendix E, Item 1)

MATERIALS/PARTS
Cotter pin (Appendix D, Item 57)
Sleeve (Appendix D, Item 654)
Antiseize tape (Appendix C, Item 72)
Pipe sealant (Appendix C, Item 5)

REFERENCES (TM)
TM 9-243
TM 9-2320-272-10
TM 9-2320-272-24P

EQUIPMENT CONDITION
Parking brake set (TM 9-2320-272-10).

a. Removal

1. Open bonnet door (5) and support with support rod (6).
2. Remove cotter pin (13) and door rod (2) from bonnet (1). Discard cotter pin (13).
3. Remove drain tube (7) and sleeve (4) from elbow (3). Discard sleeve (4).
4. Remove four screws (11), clamps (12), and drain tube (7) from van body (10).
5. Remove drain tube (7) from two grommets (8) and bonnet (1).
6. Remove two grommets (8) from bonnet holes (9).
7. Remove elbow (3) from air conditioner (14).

b. Installation

NOTE

Wrap male threads with antiseize tape prior to installation.

1. Install elbow (3) on air conditioner (14).
2. Install two grommets (8) in bonnet holes (9).

CAUTION

Do not crimp drain tube when routing through bonnet holes; drainage will stop.

NOTE

If tube was damaged, refer to TM 9-243 for tube fabrication.

3. Insert drain tube (7) through grommets (8).
4. Install new sleeve (4) and drain tube (7) on elbow (3).
5. Install drain tube (7) on van body (10) with four clamps (12) and screws (11).
6. Release bonnet door (5) support rod (6).
7. Install door rod (2) on bonnet door (5) with new cotter pin (13).

3-382. AIR CONDITIONER DRAIN TUBE REPLACEMENT (Contd)

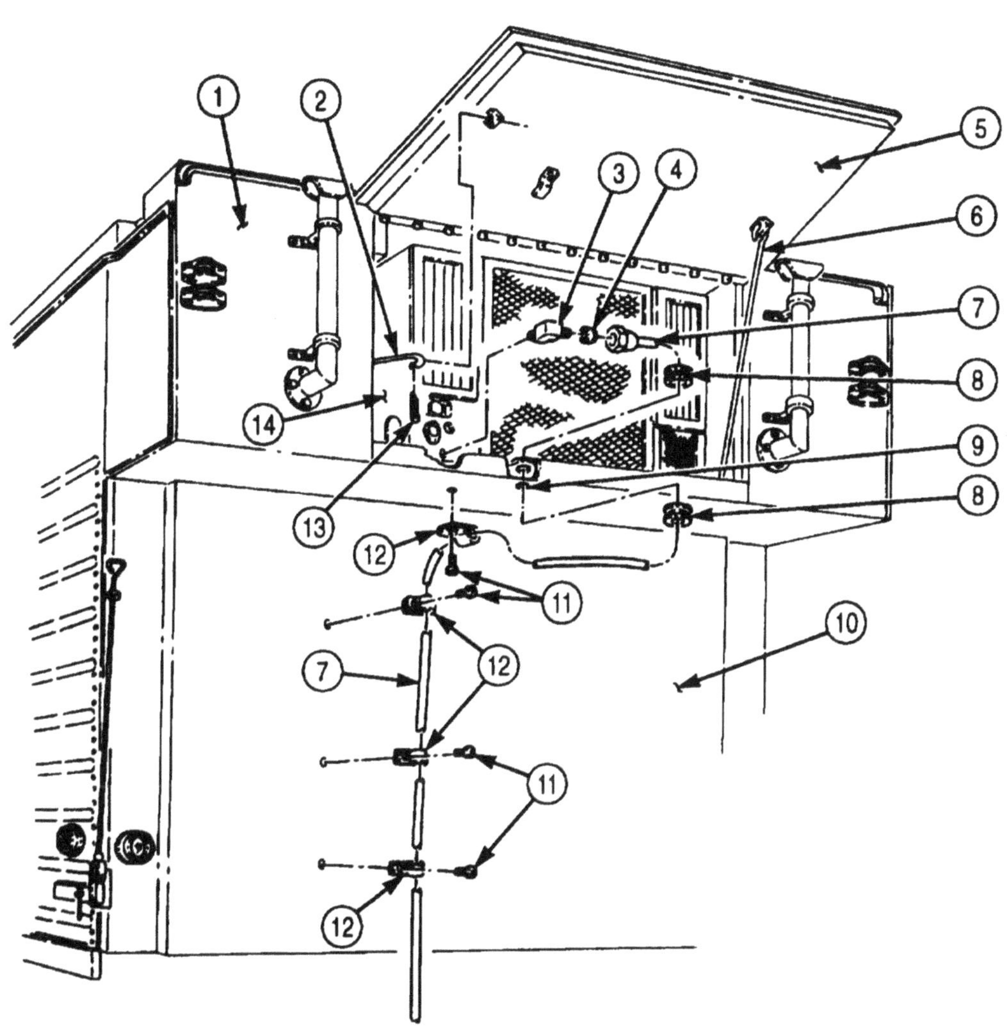

3-383. AUTOMATIC BRAKE (HOIST WINCH) ADJUSTMENT

THIS TASK COVERS:

Adjustment

INITIAL SETUP:

APPLICABLE MODELS
M936/A1/A2

TOOLS
General mechanic's tool kit (Appendix E, Item 1)

MATERIALS/PARTS
Six lockwashers (Appendix D, Item 377)
Gaskets (Appendix D, Item 166)
Grease

REFERENCES (TM)
TM 9-2320-272-10
TM 9-2320-272-24P

EQUIPMENT CONDITION
Parking brake set (TM 9-2320-272-10).

Adjustment

1. Remove six screws (2), lockwashers (1), automatic brake cover (3), and gasket (5) from brake case (6). Discard lockwashers (1) and gasket (5).
2. Measure distance between two ears (8) of brake band (7). Distance should be 1-7/32 in. ± 1/32 in. (31 mm ± 1 mm).
3. Turn adjusting screw (4) clockwise to tighten and counterclockwise to loosen until proper distance between band ears (8) is reached.
4. Position new gasket (5) on brake case (6). Apply a light coat of grease on brake case (6) to hold gasket (5) in place.
5. Position cover (3) over gasket (5) and brake case (6) and install with six new lockwashers (1) and screws (2).

3-383. AUTOMATIC BRAKE (HOIST WINCH) ADJUSTMENT (Contd)

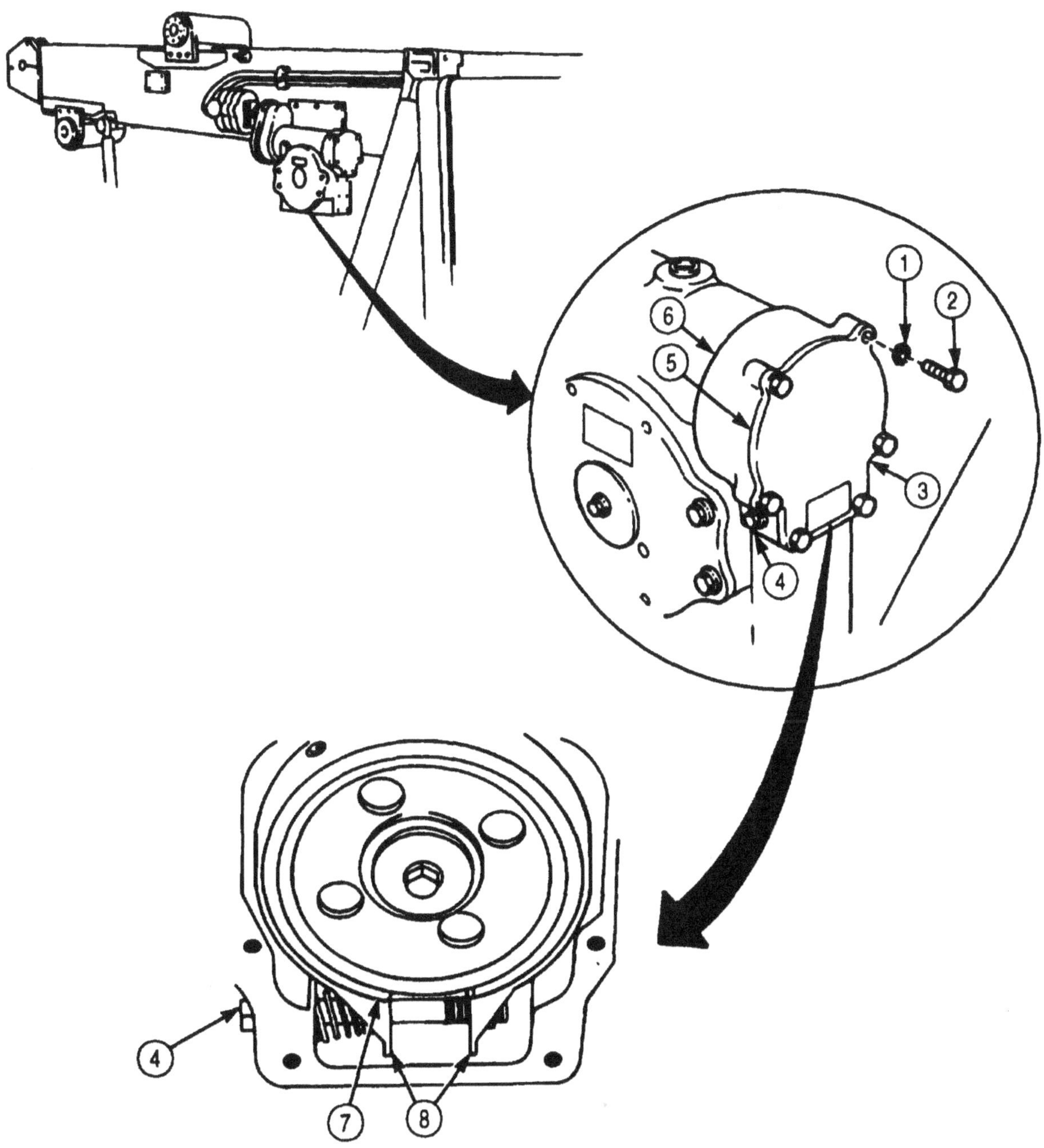

FOLLOW-ON TASK: Lift a heavy load with crane (TM 9-2320-272-10) and check adjustment by observing slippage when trying to sustain load. If crane does not hold load, notify DS maintenance.

3-384. HOIST WINCH CABLE REPLACEMENT

THIS TASK COVERS:

a. Removal **b. Installation**

INITIAL SETUP:

APPLICABLE MODELS:
M936/A1/A2

TOOLS
General mechanic's tool kit (Appendix E, Item 1)

MATERIALS/PARTS
Four lockwashers (Appendix D, Item 311)
Two locknuts (Appendix D, Item 354)

PERSONNEL REQUIRED
Two

REFERENCES (TM)
TM 9-2320-272-10
TM 9-2320-272-24P

EQUIPMENT CONDITION
Parking brake set (TM 9-2320-272-10).

GENERAL SAFETY INSTRUCTIONS
Wear hand protection when handling winch cable.

a. Removal

1. Remove four screws (5), lockwashers (4), washers (3), and rear cable guard (2) from boom (1). Discard lockwashers (4).
2. Remove two locknuts (7), screws (9), and spacers (6) from inner boom (8). Discard locknuts (7).

WARNING

Wear hand protection when handling winch cable. Broken wires may cause injury to personnel.

NOTE

Maintain manual tension on hoist cable when removing block from cable clevis.

3. Unwind hoist cable (12) (TM 9-2320-272-10) until snatch block (13) touches ground. Place hoist control in NEUTRAL position.
4. Remove nut (11), anchor bolt (14), and cable clevis (10) from snatch block (13).
5. Install anchor bolt (14) and nut (11) on snatch block (13) for storage.

3-384. HOIST WINCH CABLE REPLACEMENT (Contd)

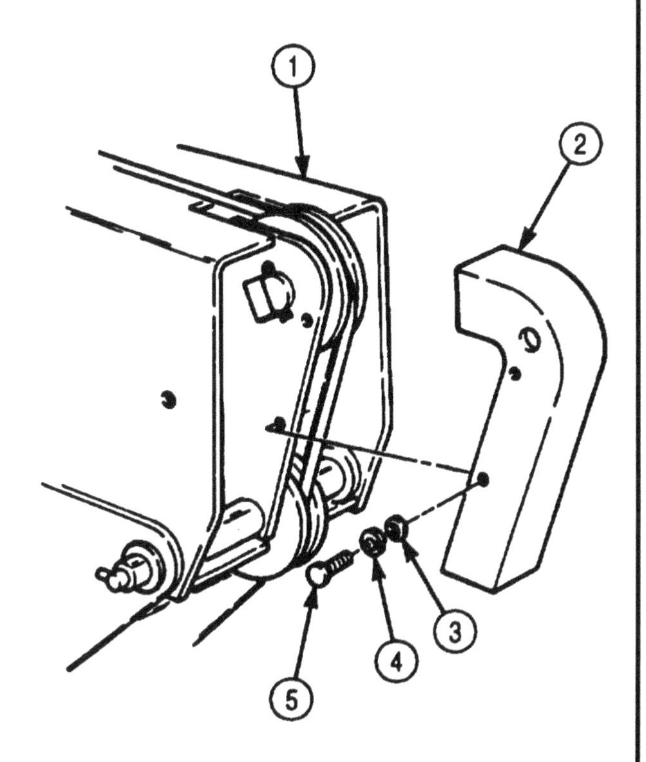

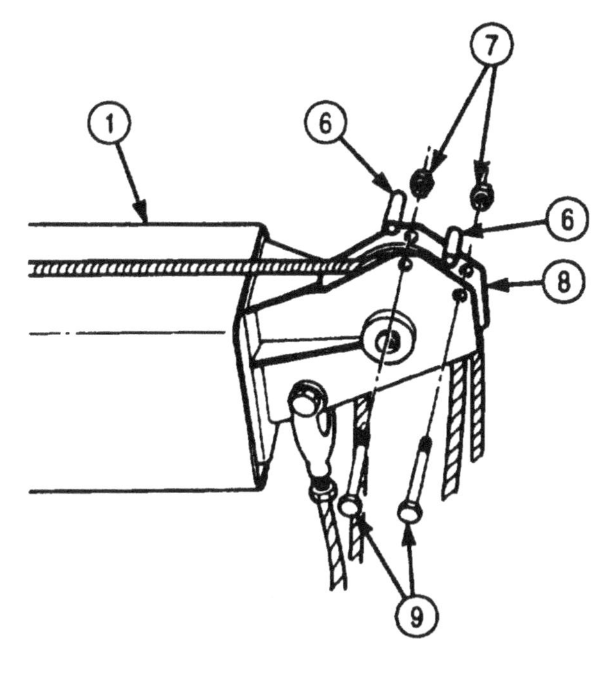

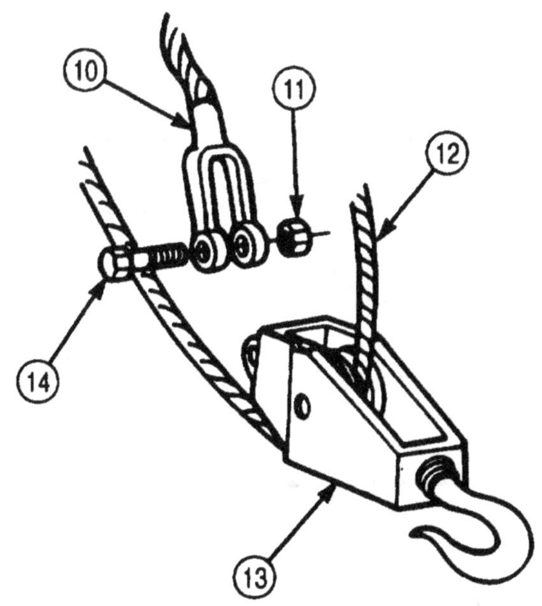

3-384. HOIST WINCH CABLE REPLACEMENT (Contd)

6. Thread cable clevis (3) through forward boom sheaves (2) until only one part of cable (4) extends from forward boom sheaves (2).

NOTE

Direct assistant to maintain tension on cable while crane hoist is in operation.

7. Continue to unwind cable (4) until screw (5) on hoist winch drum (6) is visible. Place hoist control in NEUTRAL position and shut down hoist winch operation (TM 9-2320-272-10).
8. Remove screw (5) and cable (4) from hole (7) in hoist winch drum (6).
9. Install screw (6) in hoist winch drum (6) for storage.
10. Pull cable (4) from boom (1).

b. Installation

1. Thread cable end of cable (4) over forward boom sheaves (2) and through upper boom roller (9).
2. Thread cable (4) around upper rear sheaves (8) and lower rear sheaves (10).
3. Feed cable (4) under boom (1) to hoist winch drum (6).
4. Remove screw (5), insert cable (4) through hole (7) in hoist winch drum (6), and install screw (5).

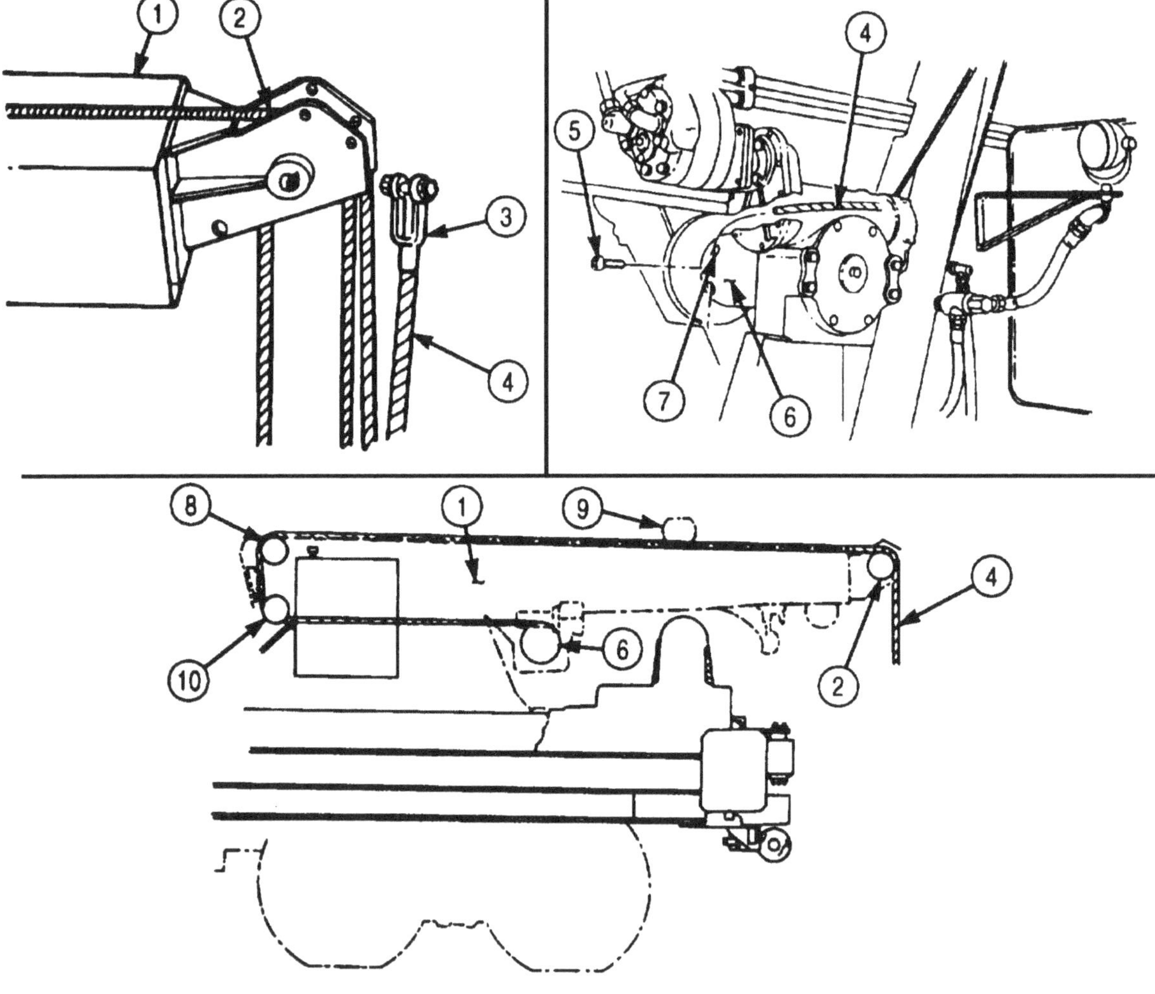

3-384. HOIST WINCH CABLE REPLACEMENT (Contd)

NOTE

Direct assistant to maintain tension on hoist cable and observe that cable is winding properly.

5. Wind cable (4) on hoist winch drum (6) until clevis (3) end of cable (4) leaves ground.
6. Thread cable clevis (3) through snatch block (14), over forward boom sheave (11), and back to snatch block (14).
7. Remove nut (13) and anchor bolt (12) from snatch block (14) and install cable clevis (3) with anchor bolt (12) and nut (13).
8. Install two spacers (15) on forward boom (18) with two screws (17) and new locknuts (16).
9. Install rear cable guard (20) over rear sheaves (19) on boom (1) with four washers (21), new lockwashers (22), and screws (23).
10. Raise snatch block (14) and place in stowage position (TM 9-2320-272-10).
11. Place hoist control in NEUTRAL position and stop vehicle (TM 9-2320-272-10).

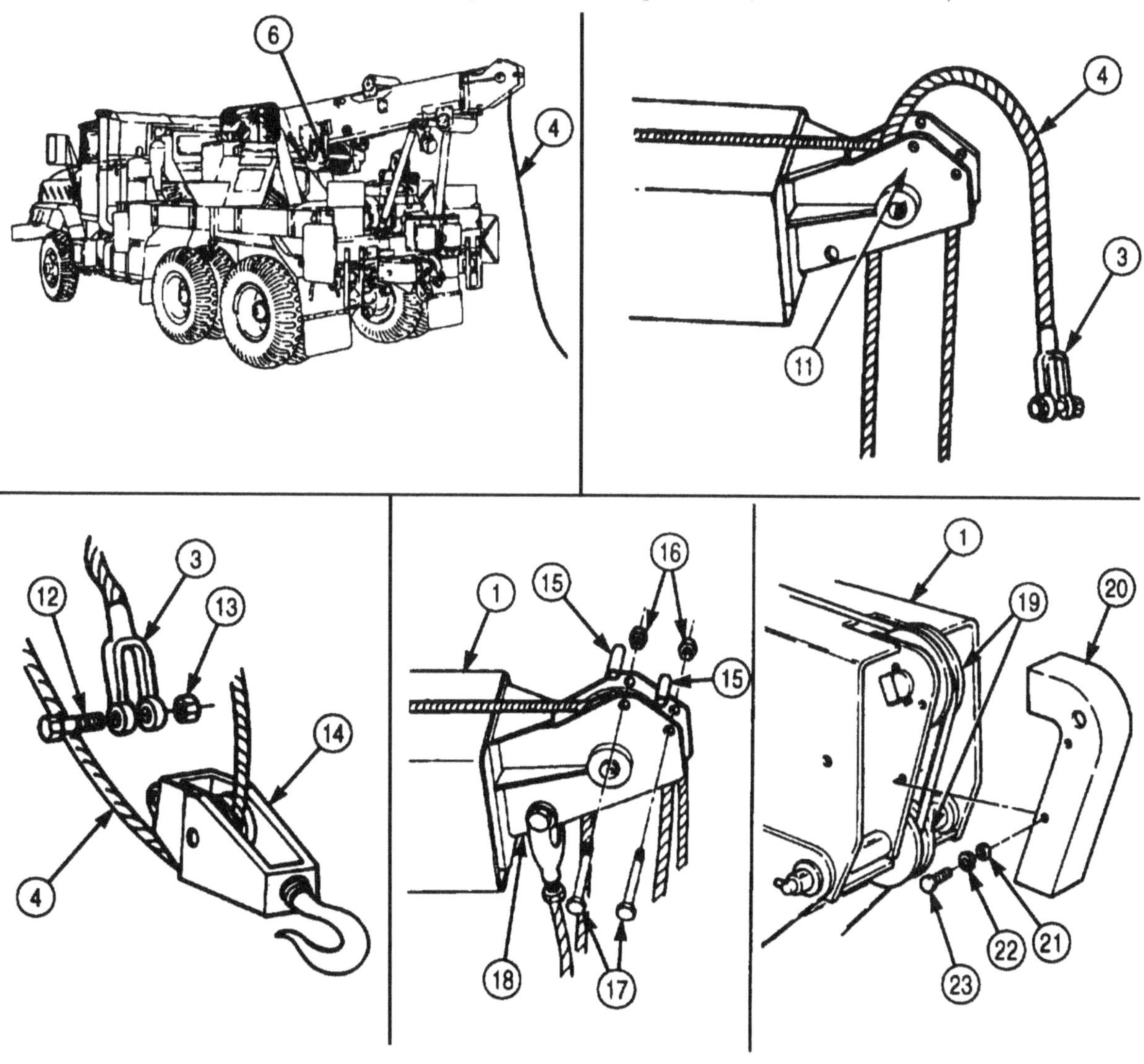

3-385. BOOM FLOODLIGHT WIRE REPLACEMENT

THIS TASK COVERS:

a. Removal **b. Installation**

INITIAL SETUP:

APPLICABLE MODELS
M936/A1/A2

TOOLS
General mechanic's tool kit (Appendix E, Item 1)

PERSONNEL REQUIRED
TWO

REFERENCES (TM)
TM 9-2320-272-10
TM 9-2320-272-24P

EQUIPMENT CONDITION
- Parking brake set (TM 9-2320-272-10).
- Battery ground cables disconnected (para. 3-126).

GENERAL SAFETY INSTRUCTIONS
- All personnel must stand clear during raising and lowering of boom.
- Assistant must remain at crane controls while work is being performed beneath boom.

NOTE

Tag all leads for installation.

1. Disconnect floodlight wire leads (7) from floodlight (8).
2. Remove seven screws (6), clamps (5), and floodlight wire (1) from boom (2).

WARNING

- Assistant must remain at crane controls until removal operation is completed. Injury to personnel may result if boom control lever is accidentally engaged while work is being done between raised boom and swivel base.
- All personnel must stand clear during hoisting operations. A snapped cable, or swinging or shifting load, may result in injury to personnel.

NOTE

Assistant will operate crane. Mechanic will continue with removal operation after boom has been raised.

3. Raise boom (2) to a 45-degree position (TM 9-2320-272-10).
4. Remove remaining screws (6), clamps (5), and floodlight wire (1) from boom (2).
5. Disconnect floodlight wire leads (3) from crane harness leads (4) and remove floodlight cable (1) from vehicle.

b. Installation

WARNING

Assistant must remain at crane controls until installation operation is completed. Injury to personnel may result if boom control lever is accidentally engaged while work is being done between raised boom and swivel base.

All personnel must stand clear during hoisting operations. A snapped cable, or swinging or shifting load, may result in injury to personnel.

3-385. BOOM FLOODLIGHT WIRE REPLACEMENT (Contd)

1. Spread out floodlight wire (1) on vehicle along general lines of installation.
2. Connect floodlight wire leads (7) to floodlight (8).
3. Connect floodlight wire leads (3) to crane harness leads (4).
4. Install floodlight wire (1) on boom (2) with ten clamps (5) and screws (6).
5. Lower boom (2) (TM 9-2320-272-10).

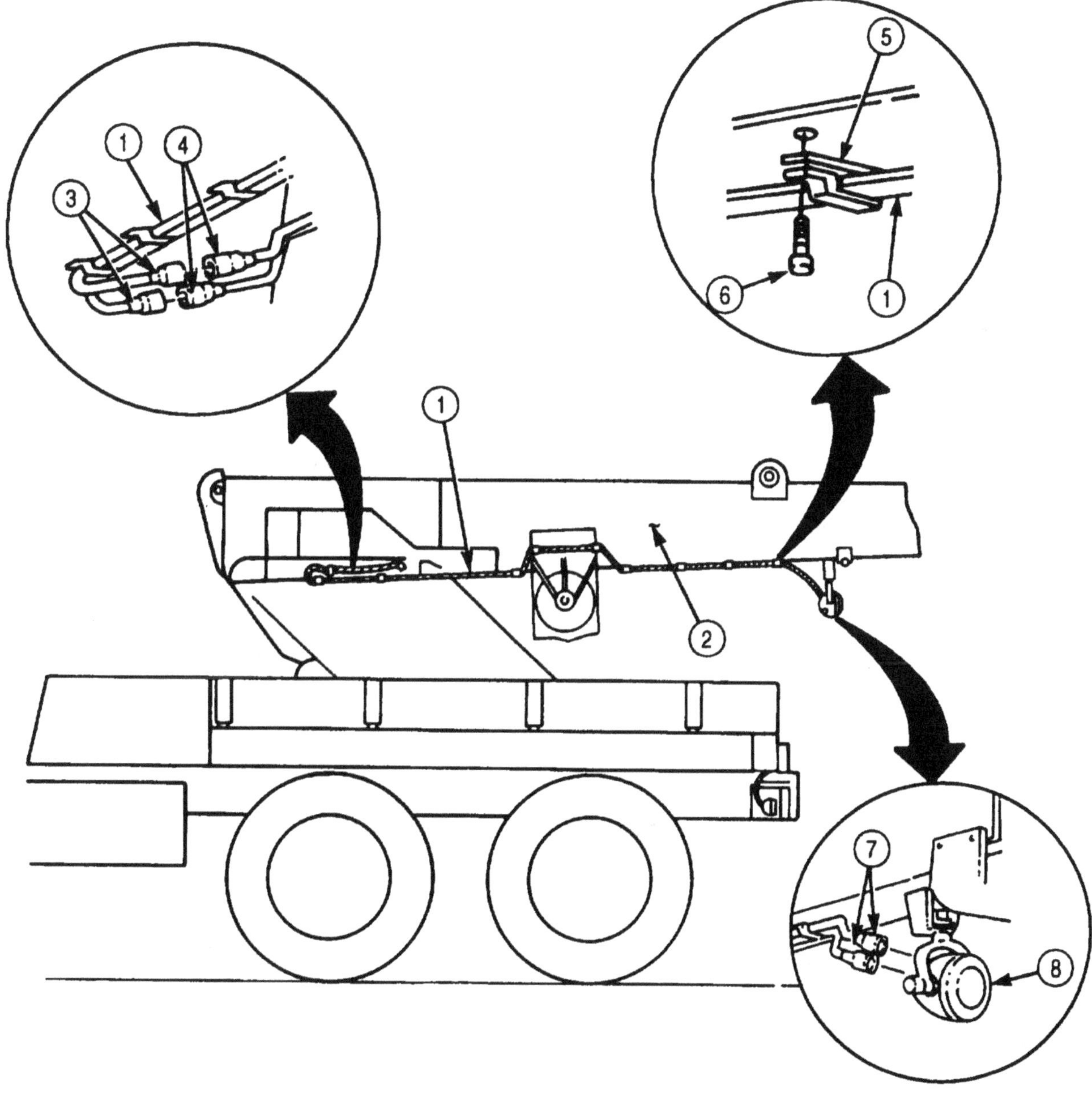

FOLLOW-ON TASK: Connect battery g-round cables (para. 3-126).

3-386. CRANE WIRING HARNESS REPLACEMENT

THIS TASK COVERS:

a. Removal **b. Installation**

APPLICABLE MODELS
M936/A1/A2

TOOLS
General mechanic's tool kit (Appendix E, Item 1)

PERSONNEL REQUIRED
Two

REFERENCES (TM)
TM 9-2320-272-10
TM 9-2320-272-24P

EQUIPMENT CONDITION
- Parking brake set (TM 9-2320-272-10).
- Battery ground cables disconnected (para. 3-126).

GENERAL SAFETY INSTRUCTIONS
- Stand clear of boom during raising and lowering.
- Assistant must remain at crane controls while work is being done under boom.

a. Removal

1. Disconnect gondola floodlight wire leads (2) from floodlight (1).
2. Remove three screws (19), clamps (3), and gondola floodlight wire (18) from gondola (20).
3. Pull gondola floodlight wire (11) through side plate (4).
4. Disconnect leads (9) of oil reservoir floodlight wire (11) from oil reservoir floodlight (10).
5. Remove three screws (8), clamps (12), and oil reservoir floodlight wire (11) from side plate (13).
6. Pull gondola floodlight wire (11) through side plate (13).

WARNING

- Assistant must remain at crane controls until removal operation is completed. Injury to personnel may result if boom control lever is accidentally engaged while work is being done between raised boom and swivel base.
- All personnel must stand clear during hoisting operations. A snapped cable, or swinging or shifting load, may result in injury to personnel.

NOTE

Assistant will operate crane. Mechanic will continue with removal operation after boom has been raised.

7. Raise boom (5) to a 45-degree position (TM 9-2320-272-10).
8. Disconnect crane harness leads (6) from floodlight wire leads (7).
9. Remove six screws (21), clamps (15), and crane harness (14) from turntable (22) and side plates (4) and (13).
10. Disconnect crane harness leads (17) from swivel leads (16) and remove crane harness (14) from vehicle.

3-386. CRANE WIRING HARNESS REPLACEMENT (Contd)

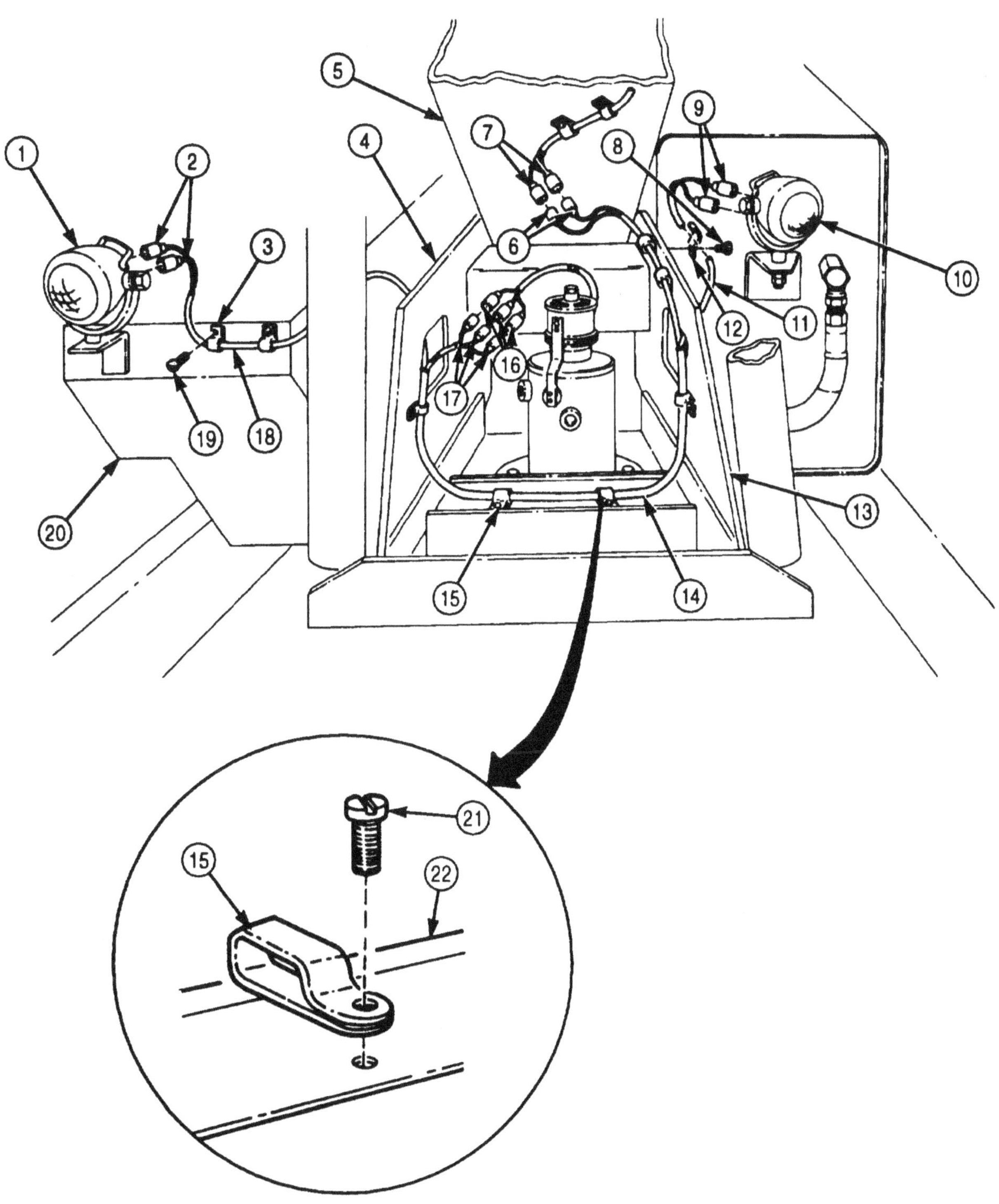

3-386. CRANE WIRING HARNESS REPLACEMENT (Contd)

b. Installation

WARNING

- Assistant must remain at crane controls until removal operation is completed. Injury to personnel may result if boom control lever is accidentally engaged while work is being done between raised boom and swivel base.
- All personnel must stand clear during hoisting operations. A snapped cable, or swinging or shifting load, may result in injury to personnel.

1. Raise boom (5) to a 45-degree position (TM 9-2320-272-10).
2. Spread out crane harness (14) on vehicle along general lines of installation.
3. Thread gondola floodlight wire (18) through side plate (4) and connect leads (2) to floodlight (1).
4. Connect crane harness leads (17) to swivel leads (16).
5. Connect crane harness leads (6) to floodlight wire leads (7).
6. Thread oil reservoir floodlight wire (11) through side plate (13) and connect leads (9) to oil reservoir floodlight (10).
7. Install oil reservoir floodlight wire (11) on side plate (13) with three clamps (12) and screws (8).
8. Install crane harness (14) on turntable (22) and side plates (4) and (13) with six clamps (15) and screws (21).
9. Install gondola floodlight wire (18) on gondola (20) with three clamps (3) and screws (19).
10. Lower boom (5) (TM 9-2320-272-102

3-386. CRANE WIRING HARNESS REPLACEMENT (Contd)

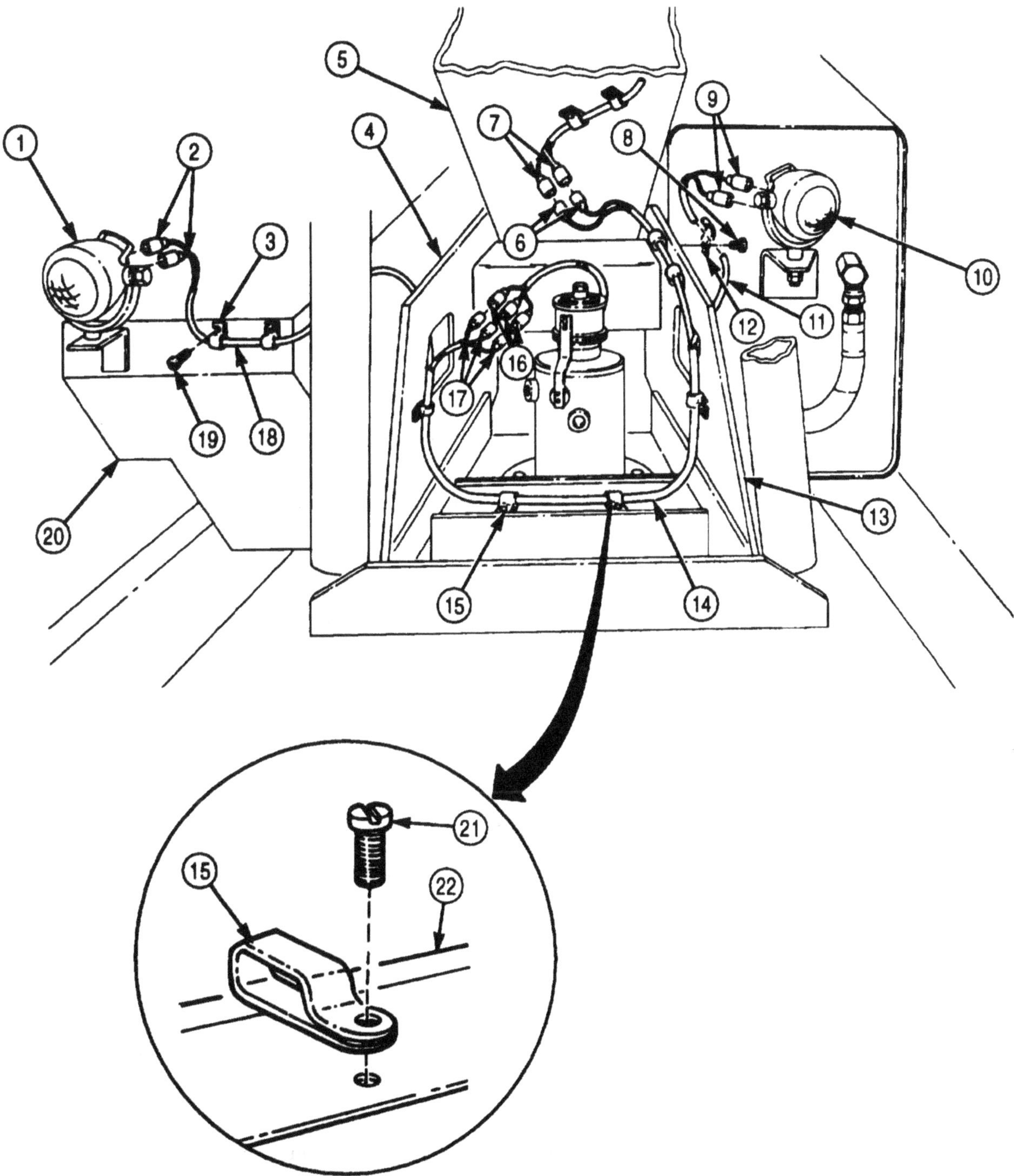

FOLLOW-ON TASKS:
- Connect battery ground cables (para. 3-128).
- Check boom floodlight for proper operation (TM 9-2320-272-10).
- Lower boom and secure for travel (TM 9-2320-272-10).

3-387. WRECKER CRANE HYDRAULIC HOSE AND TUBE REPLACEMENT

THIS TASK COVERS:

a. Removal b. Installation

INITIAL SETUP:

APPLICABLE MODELS
M936/A1/A2

TOOLS
General mechanic's tool kit (Appendix E, Item 1)

MATERIALS/PARTS
Six locknuts (Appendix D, Item 291)
Cap and plug set (Appendix C. Item 14)

REFERENCES (TM)
LO 9-2320-272-12
TM 9-2320-272-10
TM 9-2320-272-24P

EQUIPMENT CONDITION
- Parking brake set (TM 9-2320-272-10).
- Shipper brace in travel position (TM 9-2320-272-10).
- Hydraulic oil reservoir drained (LO 9-2320-272-12).

a. Removal

CAUTION

When disconnecting hydraulic hoses and tubes, plug all openings to prevent dirt from entering and causing internal parts damage.

NOTE

- Do not twist hose during removal or attempt to remove hose with one wrench.
- Hose fitting ends connected by a single hexagonal nut cannot be disconnected until the flare nut connected at the opposite end is removed. The entire hose must be free to turn whenever removing hose connected by a single hexagonal nut.
- Tag all hoses and lines for installation.
- Have drainage container ready to catch oil.

1. Remove locknut (8), clamp (3), hose (4), and/or tube (1) from bracket (6) and wrecker body (7). Discard locknut (8).
2. Position wrenches on hexagonal fitting (2) and flare nut (5).
3. Holding hexagonal fitting (2) in place, loosen flare nut (5), and disconnect hose (4) and/or tube (1) from hexagonal fitting (2).
4. Repeat steps 1 through 3 for remaining hoses (4) and/or tubes (1).

b. Installation

CAUTION

Ensure no particles of plugging become trapped in crane hydraulic system during installation of hoses or tubes. Failure to do so may result in damage to equipment.

NOTE

- Do not twist hose or attempt to install hose with one wrench.
- Hose fitting ends connected by a single hexagonal nut cannot be connected until the hexagonal nut end is connected at the opposite end. The entire hose must be free to turn whenever installing hose connected by a single hexagonal nut.

3-387. WRECKER CRANE HYDRAULIC HOSE AND TUBE REPLACEMENT (Contd)

1. Install hose (4) and/or tube (1) on wrecker body (7) and bracket (6) with clamp (3) and new locknut (8).
2. Position wrenches on hexagonal fitting (2) and flare nut (5).
3. Holding hexagonal fitting (2) in place, tighten flare nut (5).
4. Repeat steps 1 through 3 for remaining hoses (4) and/or tubes (1).

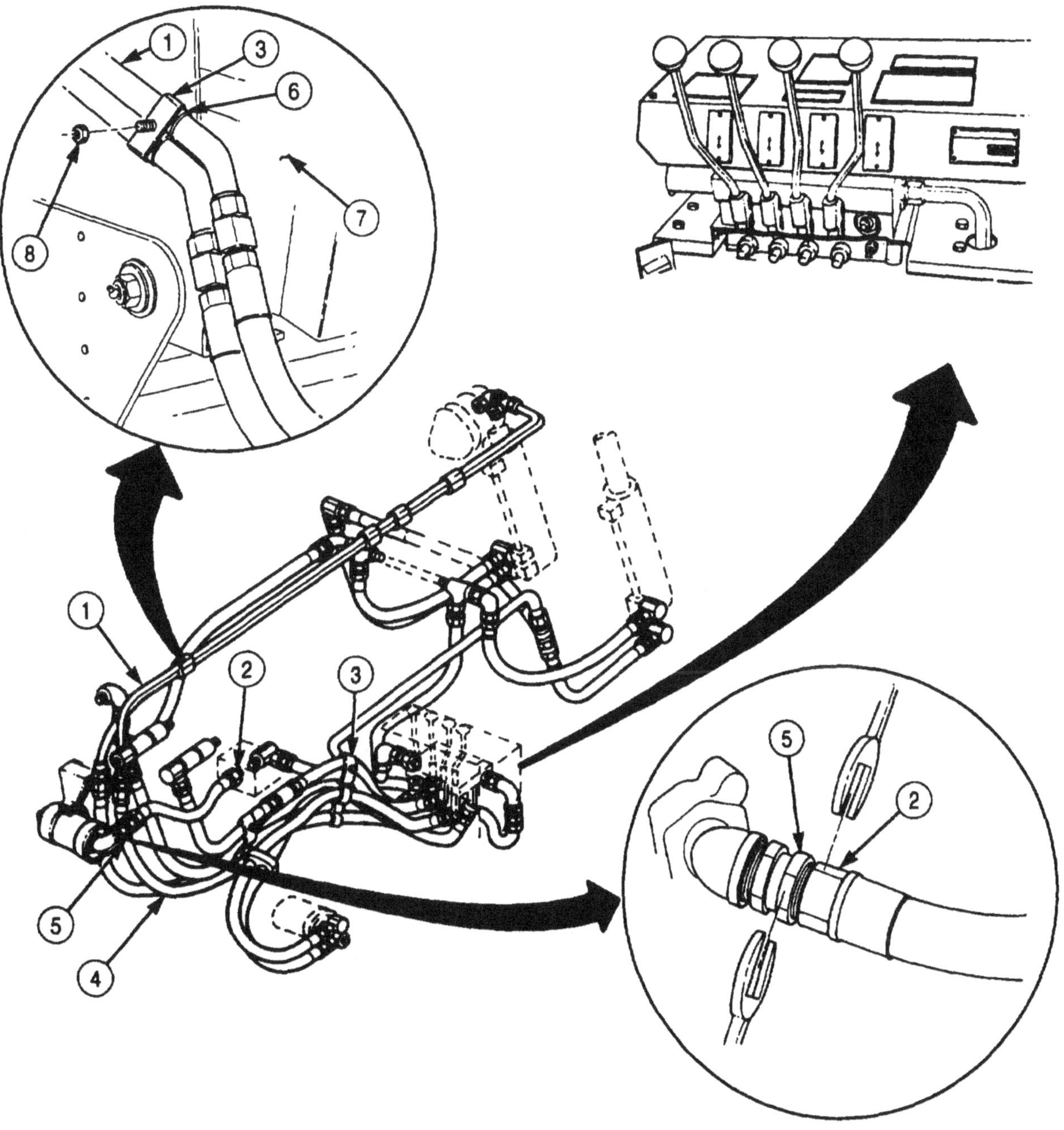

FOLLOW-ON TASK: Fill hydraulic oil reservoir to proper fluid level (LO 9-2320-272-12).

3-388. WRECKER CRANE HYDRAULIC PUMP REPLACEMENT

THIS TASK COVERS:

a. Removal **b. Installation**

INITIAL SETUP:

APPLICABLE MODELS
M936/A1/A2

TOOLS
General mechanic's tool kit (Appendix E, Item 1)
Torque wrench (Appendix E, Item 144)

MATERIALS /PARTS
Sixteen lockwashers (Appendix D, Item 350)
O-ring (Appendix D, Item 456)
O-ring (Appendix D, Item 471)
Cap and Plug set. (Appendix C, Item 14)

PERSONNEL REQUIRED
Two

REFERENCES (TM)
LO 9-2320-272-12
TM 9-2320-272-10
TM 9-2320-272-24P

EQUIPMENT CONDITION
- Parking brake set (TM 9-2320-272-10).
- Hydraulic oil reservoir drained (LO 9-2320-272-12).

a. Removal

CAUTION

When disconnecting hydraulic hoses and lines, plug all openings to prevent dirt from entering and causing internal parts damage.

NOTE

- Tag all hoses and lines for installation.
- Have drainage container ready to catch oil.

1. Disconnect hose (16) from hose (15).
2. Disconnect hoses (13) and (15) from inlet tee (14).
3. Disconnect hose (7) from outlet elbow (6).
4. Remove four screws (21), lockwashers (22), and mount (19) with crane pump (10) from bracket (1), lower crane pump (10), and slide from driveshaft hub (20). Discard lockwashers (22).
5. Remove four screws (17), lockwashers (18), two split flanges (12), inlet tee (14), and O-ring (11) from crane pump (10). Discard lockwashers (18) and O-ring (11).
6. Remove four screws (4), lockwashers (3), two split flanges (5), outlet elbow (6), and O-ring (8) from crane pump (10). Discard lockwashers (3) and O-ring (8).
7. Remove four screws (9), lockwashers (2), and mount (19) from crane pump (10). Discard lockwashers (2).

b. Installation

1. Install mount (19) on crane pump (10) with four new lockwashers (2) and screws (9).
2. Install new O-ring (8) and outlet elbow (6) on crane pump (10) with two split flanges (5), four new lockwashers (3), and screws (4). Tighten screws (4) 50-60 lb-R (68-81 N•m).
3. Install new O-ring (11) and inlet tee (14) on crane pump (10) with two split flanges (12), four new lockwashers (18), and screws (17). Tighten screws (17) 50-60 lb-ft (68-81 N•m).
4. Position shaft of crane pump (10) on driveshaft hub (20).
5. Position mount (19) with crane pump (10) on bracket (1) and install with four new lockwashers (22) and screws (21).
6. Connect hose (7) to outlet elbow (6).
7. Connect hoses (13) and (15) to inlet tee (14).
8. Connect hose (16) to hose (15).

3-388. WRECKER CRANE HYDRAULIC PUMP REPLACEMENT (Contd)

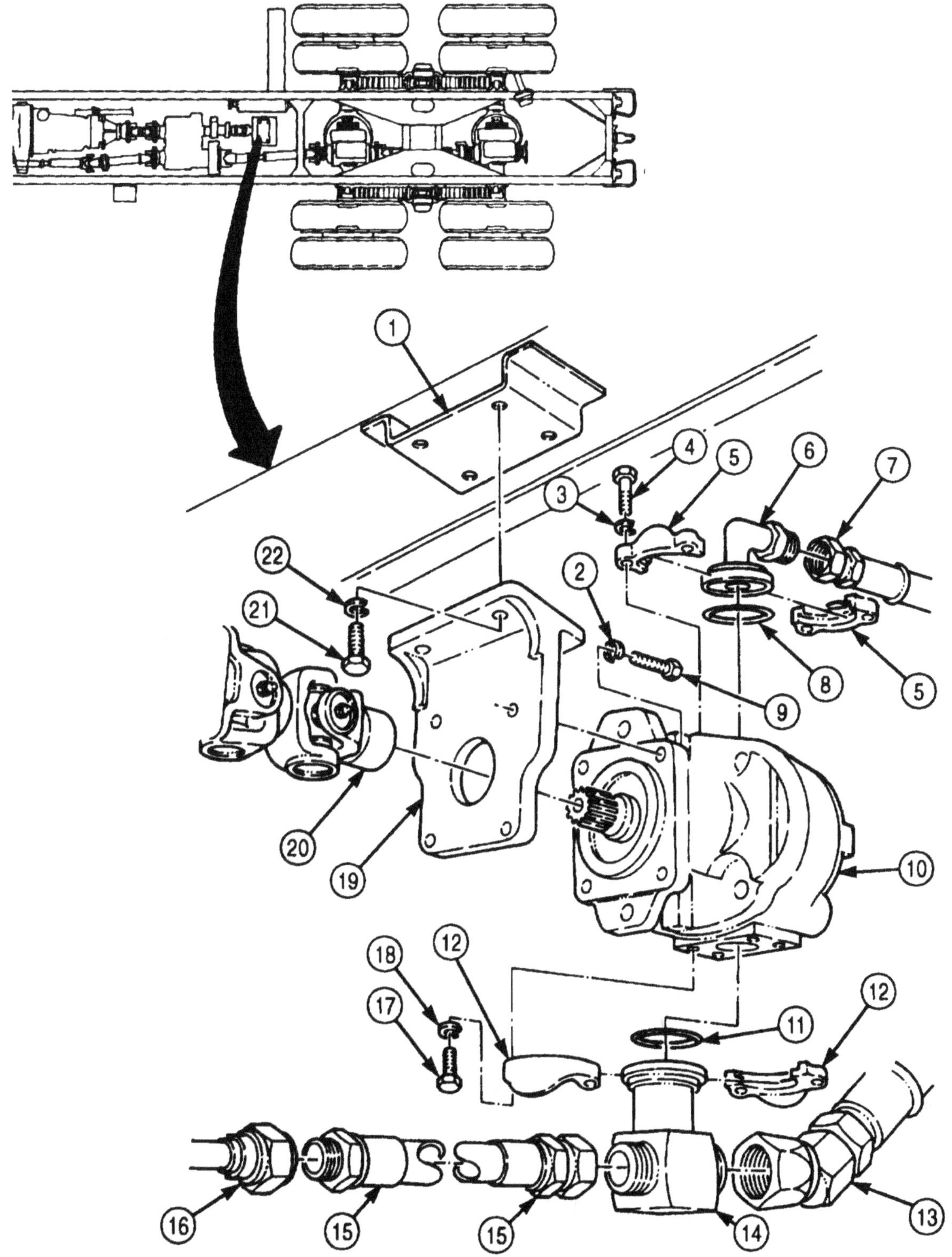

FOLLOW-ON TASKS:
- Fill hydraulic oil reservoir to proper fluid level (LO 9-2320-272-12).
- Check crane hydraulic pump for proper operation (TM 9-2320-272-10).

3-389. PRESSURE RELIEF VALVE MAINTENANCE

THIS TASK COVERS:

a. Testing b. Adjustment

INITIAL SETUP:

APPLICABLE MODELS

M936/A1/A2

TOOLS

General mechanic's tool kit (Appendix E, Item 1)

Hydraulic pressure gauge (Appendix E, Item 51)

REFERENCES (TM)

TM 9-2320-272-10

TM 9-2320-272-24P

EQUIPMENT CONDITION

- Parking brake set (TM 9-2320-272-10).
- Boom secured in travel position (TM 9-2320-272-101.

a. Testing

1. Remove five screws (1) and cover (2) from gondola (8).
2. Remove valve plug (3) from valve bank (4) and install hydraulic pressure gauge.
3. Start engine and engage crane hydraulic system (TM 9-2320-272-10).
4. Raise engine idle to 1,250 rpm and observe hydraulic pressure gauge reading. Hydraulic pressure gauge should read 1,350 psi ± 25 psi (9,308 kPa ± 172 kPa).
5. If reading is within limits, go to step 7.
6. If reading is not within limits, perform task b.
7. Remove hydraulic pressure gauge from valve bank (4) and install plug (3).
8. Install cover (2) on gondola (8) with five screws (1).

b. Adjustment

1. Remove acorn nut (7) from valve bank (4).
2. While holding adjusting screw (6), loosen jamnut (5)
3. Turn adjusting screw (6) until hydraulic pressure gauge reads 1,350 psi ± 25 psi (9,308 kPa ± 172 kPa).
4. While holding adjusting screw (6), tighten jamnut (5).
5. If correct pressure cannot be achieved, notify DS maintenance.

3-389. PRESSURE RELIEF VALVE MAINTENANCE (Contd)

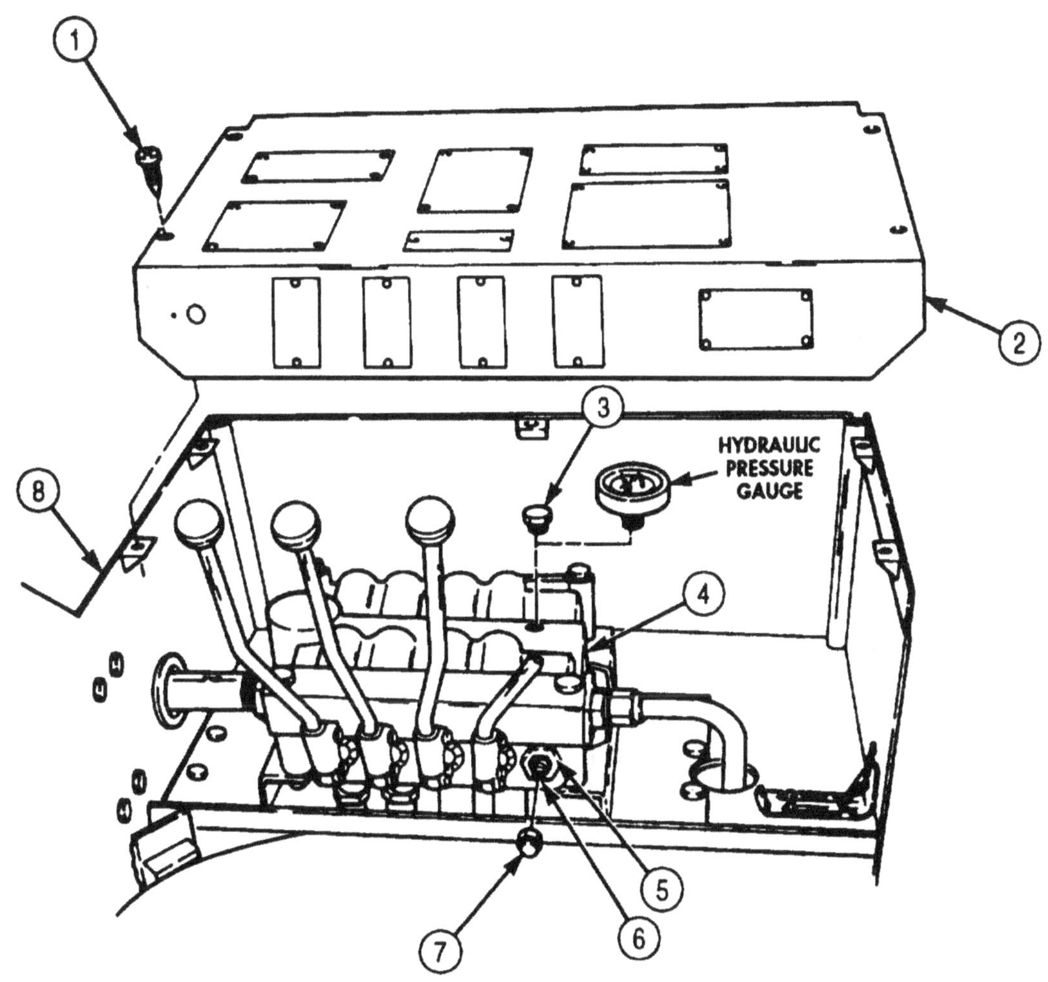

3-390. SNUBBER VALVE ASSEMBLY REPLACEMENT

THIS TASK COVERS:

a. Removal **b. Installation**

INITIAL SETUP:

APPLICABLE MODELS
M936/A1/A2

TOOLS
Ganeral mechanic's tool kit (Appendix E, Item 1)

REFERENCES (TM)
LO 9-2320-272-12
TM 9-2320-272-10
TM 9-2320-272-24P

EQUIPMENT CONDITION
- Boom in down position (TM 9-2320-272-10).
- Shipper braces in travel position (TM 9-2320-272-10).

GENERAL SAFETY INSTRUCTIONS
Snubber valve must be removed as an assembly. Do not disassemble.

WARNING

Replace snubber valve and lift cylinder adapter cap as an assembly. Do not disconnect adapter cap from snubber valve. Valve and cap are locked against spring tension. Improper removal of valve from adapter cap may cause injury to personnel.

CAUTION

When disconnecting hydraulic hoses and lines, plug all openings to prevent dirt from entering and causing internal parts damage.

NOTE

- Tag all hoses and lines for installation.
- Have drainage container ready to catch oil.

a. Removal

1. Disconnect hose (1) from snubber valve (2).
2. Remove snubber valve (2) from lift cylinder fitting (3) and install plug (4) to prevent excessive oil loss.

b. Installation

CAUTION

- Ensure no particles of plugging become trapped in crane hydraulic system during installation of hoses or tubes. Failure to do so may result in damage to equipment.
- Ensure plug is removed from lift cylinder fitting before installation.

NOTE

- Do not twist hose or attempt to install hose with one wrench.
- Hose fitting ends connected by a single hexagonal nut cannot be installed until the hexagonal nut end is connected at the opposite end. The entire hose must be free to turn whenever installing hose connected by a single hexagonal nut.

1. Remove plug (4) from lift cylinder fitting (3) and install snubber valve (2).
2. Connect hose (1) to snubber valve (2).

3-390. SNUBBER VALVE ASSEMBLY REPLACEMENT (Contd)

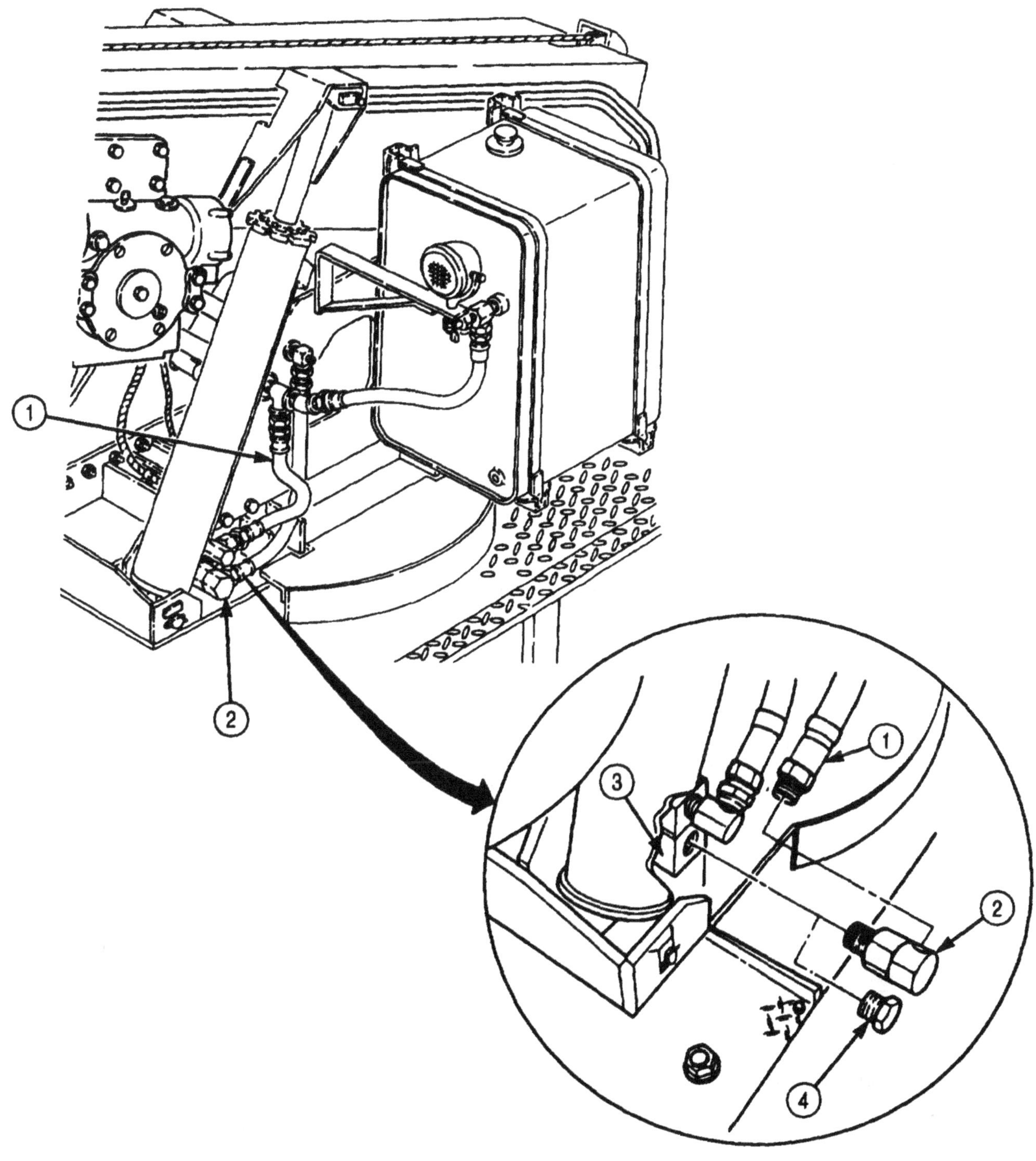

FOLLOW-ON TASK: Fill hydraulic oil reservoir to proper fluid level (LO 9-2320-272-12).

3-391. CRANE HYDRAULIC FILTER MAINTENANCE

THIS TASK COVERS:

a. Removal
b. Disassembly
c. Cleaning and Inspection
d. Assembly
e. Installation

INITIAL SETUP:

APPLICABLE MODELS
M936/A1/A2

TOOLS
General mechanic's tool kit (Appendix E, Item 1)

MATERIALS/PARTS
O-ring (Appendix D, Item 435)
O-ring (Appendix D, Item 434)
Ring seal (Appendix D, Item 541)
Drycleaning solvent (Appendix C, Item 71)

PERSONNEL REQUIRED
TWO

REFERENCES (TM)
LO 9-2320-272-12
TM 9-2320-272-10
TM 9-2320-272-24P

EQUIPMENT CONDITION
Parking brake set (TM 9-2320-272-10).

GENERAL SAFETY INSTRUCTIONS
- Wear eyeshields during replacement of crane hydraulic filter.
- Keep fire extinguisher nearby when using drycleaning solvent.
- When cleaning with compressed air, wear eyeshields and ensure source pressure does not exceed 30 psi (207 kPa).

a. Removal

WARNING

Hydraulic filter assembly is under great pressure and oil will spurt out from housing during removal. Wear eyeshields during removal of assembly, Failure to do this may cause injury to personnel.

NOTE

- Mechanic must hold filter cover firmly in place while assistant removes screws. Cover and attached filter assembly must be pulled quickly from housing. A shutoff valve inside housing stops oil flow immediately after filter assembly is removed.
- Have drainage container ready to catch oil.

1. While holding filter cover (2) in place, direct assistant to remove four screws (1) from filter cover (2).
2. Pull filter cover (2) with filter assembly (4) quickly from filter housing (3).
3. Remove O-ring (5) from rear of filter housing (3). Discard O-ring (5).

b. Disassembly

1. Place filter cover (2) with filter assembly (4) on flat surface.
2. Pull filter assembly (4) straight up to separate from filter cover (2).
3. Remove O-ring (6) from filter cover (2). Discard O-ring (6).
4. Remove three screws (7), rear cap (12), front cap (11), and ring seal (10) from filter element (8). Discard ring seal (10).
5. Remove filter element (8) from shroud (9).

3-391. CRANE HYDRAULIC FILTER MAINTENANCE (Contd)

3-391. CRANE HYDRAULIC FILTER MAINTENANCE (Contd)

c. Cleaning and Inspection

WARNING

- Drycleaning solvent is flammable and toxic. Do not use near open flame and always have a fire extinguisher nearby when solvents are used. Use only in well-ventilated places, wear protective clothing, and dispose of cleaning rags in approved container. Failure to do this may result in injury to personnel and damage to equipment.
- Eyeshields must be worn when cleaning with compressed air. Compressed air source will not exceed 30 psi (207 kPa). Failure to do so may result in injury to personnel.

1. Using drycleaning solvent, clean filter cover (12), rear cap (1), front cap (8), filter element (4), and shroud (5) and dry with compressed air.
2. Inspect filter cover (12), rear cap (1), front cap (8), filter element (4), and shroud (5) for cracks, holes, and excessive wear. Replace damaged or worn parts.

d. Assembly

1. Install filter element (4) over shroud (5).
2. Position new ring seal (7) on front cap (8) and install front cap (8) on filter element (4).
3. Position rear cap (1) on opposite end of filter element (4), align holes (3) and (6), and install with three screws (2).
4. Install new O-ring (13) on filter cover (12).
5. Place filter cover (12) on flat surface.
6. Align key (11) of filter cover (12) with slot (10) of filter assembly (9) and install filter assembly (9) on filter cover (12).

e. Installation

1. Install new O-ring (14) sideways in filter housing (15). Rotate and seat into position when O-ring (14) contacts rear of filter housing (15).

WARNING

Hydraulic filter assembly is under great pressure and oil will spurt out from housing during installation. Wear eyeshields during installation of assembly. Failure to do this may cause injury to personnel.

CAUTION

- Indicator must be level and on the right side of filter housing during installation. Filter assembly will not seat if indicator is improperly seated.
- Do not reuse hydraulic oil from drainage container. Damage to equipment may result if drained oil is used.

NOTE

- Mechanic must seat filter cover quickly and hold in place while assistant installs screws.
- Have drainage container ready to catch oil.

2. Position filter cover (12) with filter assembly (9) just inside filter housing (15).
3. Push filter cover (12) quickly into filter housing (15) and install with four screws (16).

3-391. CRANE HYDRAULIC FILTER MAINTENANCE (Contd)

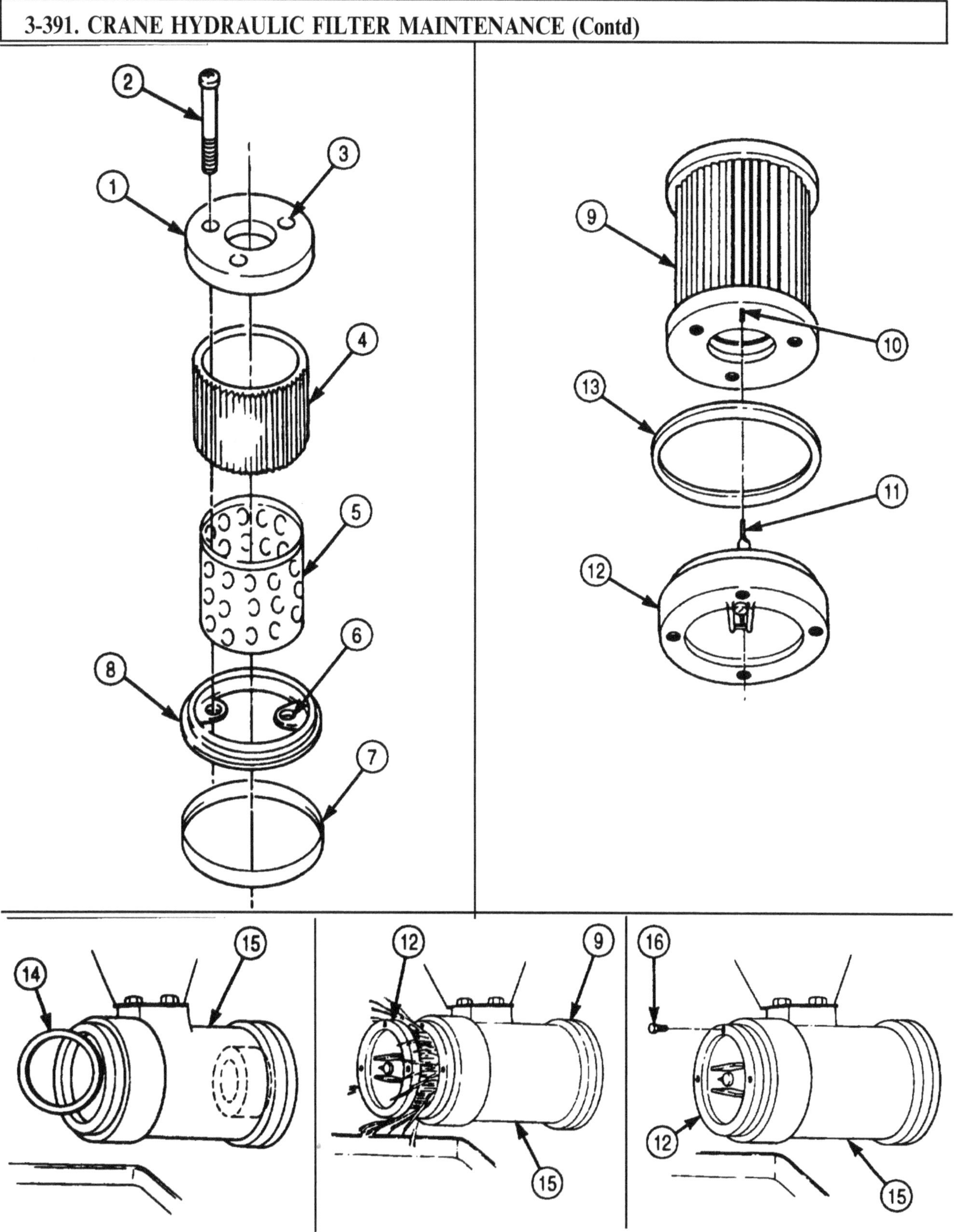

FOLLOW-ON TASK: Fill hydraulic oil reservoir to proper fluid level (LO 9-2320-272-12).

3-392. FORWARD DECK PLATE REPLACEMENT

THIS TASK COVERS:

a Removal b. Installation

INITIAL SETUP

APPLICABLE MODELS
M936/A1/A2

TOOLS
General mechanic's tool kit (Appendix E, Item 1)

MATERIALS/PARTS
Six lockwashers (Appendix D, Item 354)

REFERENCES (TM)
TM 9-2320-272-10
TM 9-2320-272-24P

EQUIPMENT CONDITION
Parking brake set (TM 9-2320-272-10).

a. Removal

Remove six screws (2), lockwashers (3), washers (4), and deck plate (1) from wrecker body (5). Discard lockwashers (3).

b. Installation

Install deck plate (1) on wrecker body (5) with six washers (4), new lockwashers (3). and screws (2).

3-392. FORWARD DECK PLATE REPLACEMENT (Contd)

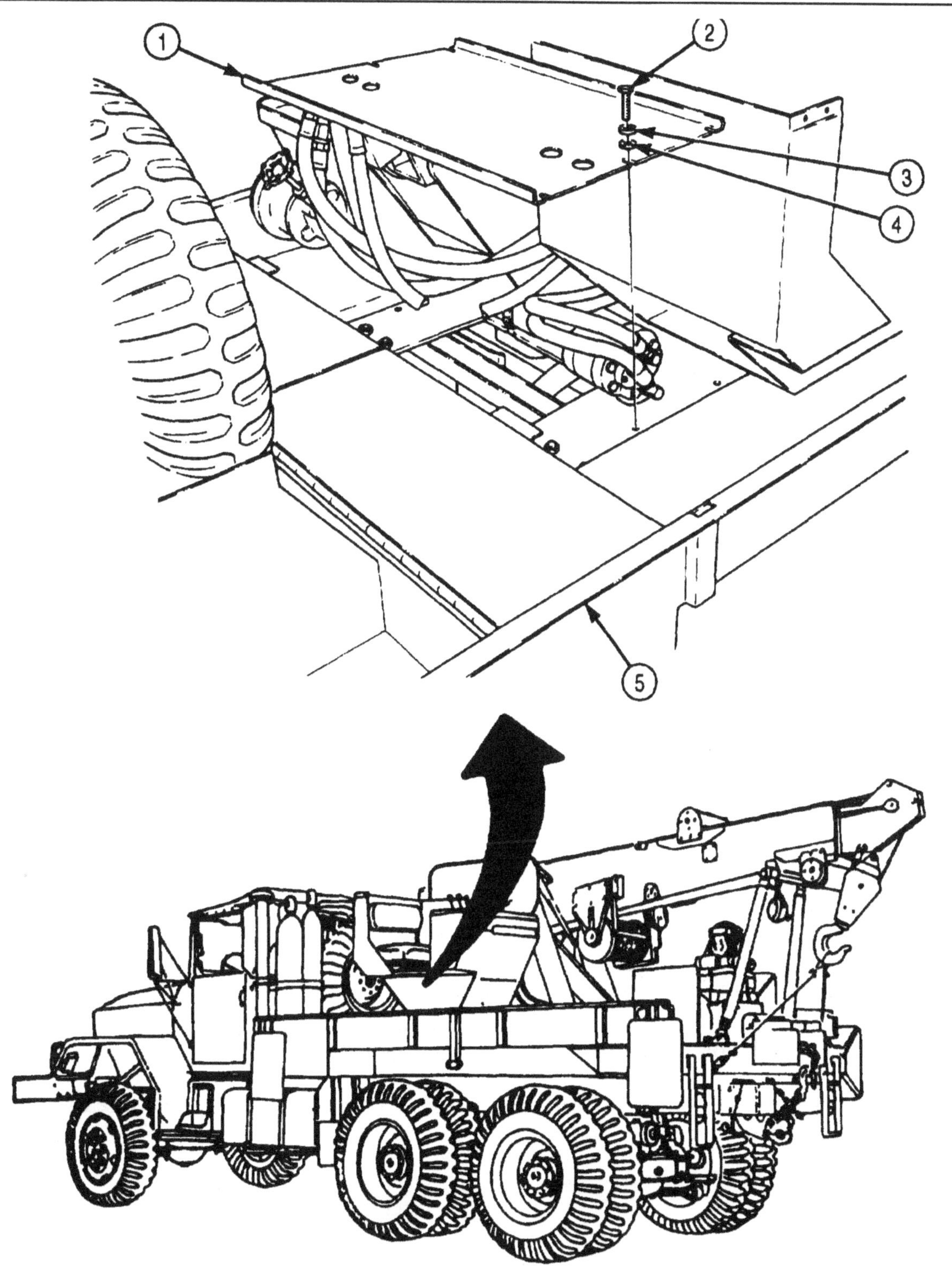

3-393. TRANSFER PTO-TO-HYDRAULIC PUMP PROPELLER SHAFT REPLACEMENT

THIS TASK COVERS:

a. Removal   b. Installation

INITIAL SETUP:

APPLICABLE MODELS
M936/A1/A2

TOOLS
General mechanic's tool kit (Appendix E, Item 1)
Hydraulic jack

MATERIALS/PARTS
Four lockwashers (Appendix D, Item 409)
Four lockwashers (Appendix D, Item 350)

REFERENCES (TM)
TM 9-2320-272-10
TM 9-2320-272-24P

EQUIPMENT CONDITION
Parking brake set (TM 9-2320-272-10).

a. Removal

NOTE

Transfer PTO-to-crane hydraulic pump propeller shaft and adapter come assembled as one unit but adapter must be replaced separately.

1. Remove four nuts (3), lockwashers (4), and screws (5) from propeller shaft (14) and driveshaft flange (17). Discard lockwashers (4).
2. Rotate propeller shaft (14) to position lubrication fitting (6) at top.
3. Support weight of pump (9) with hydraulic jack.
4. Remove four screws (8) and lockwashers (7) from pump (9) and mount (12). Discard lockwashers (7).
5. Separate pump (9) from mount (12) and slide toward rear of vehicle until pump boss (10) clears mount (12).
6. Lower hydraulic jack enough to allow separation and removal of propeller shaft (14) from pump shaft (11).
7. Remove nut (15) and washer (16) from transfer PTO shaft (2) and slide driveshaft flange (17) off transfer PTO shaft (1).

b. Installation

1. Install driveshaft flange (17) on transfer PTO shaft (2) with washer (16) and nut (15).
2. Raise hydraulic jack enough to position pump shaft (11) and boss (10) of pump (9) through mount (12). Ensure rear hub of propeller shaft (14) slides on pump shaft (11).
3. Position propeller shaft (14) on PTO shaft (1) with lubrication fitting (6) positioned at top.
4. Install propeller shaft (14) on driveshaft flange (17) with four screws (5), new lockwashers (4), and nuts (3).
5. Install pump (9) on mount (12) with four new lockwashers (7) and screws (8). Tighten screws (8) 32-40 lb ft (43-54 N•m).

3-393. TRANSFER PTO-TO-HYDRAULIC PUMP PROPELLER SHAFT REPLACEMENT (Contd)

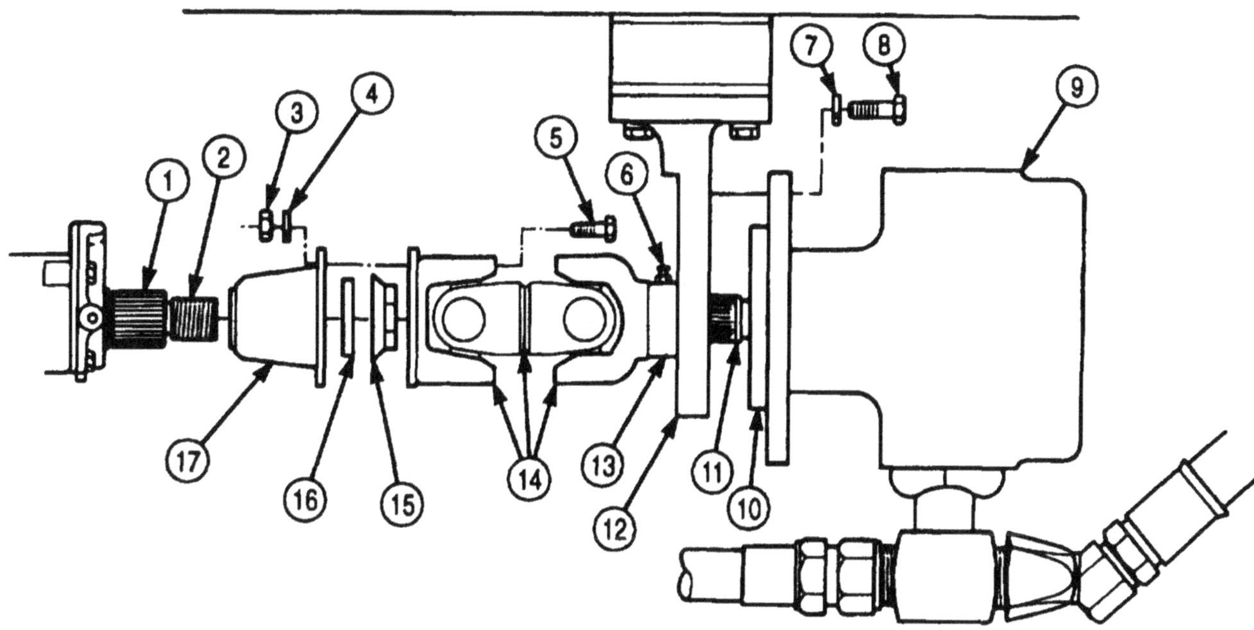

FOLLOW-ON TASK: Check operation of crane hydraulic system (TM 9-2320-272-10).

Section XV. SPECIAL PURPOSE KITS

3-394. SPECIAL PURPOSE KITS INDEX

PARA. NO.	TITLE	PAGE NO.
3-395.	Radiator Cover Kit Replacement	3-970
3-396.	Hardtop Kit Replacement	3-972
3-397.	Engine Coolant Heater Replacement	3-974
3-398.	Engine Coolant Heater Pump Replacement	3-976
3-399.	Engine Coolant Heater Control Box Replacement	3-978
3-400.	Engine Coolant Heater Harness (M939/A1) Replacement	3-980
3-401.	Engine Coolant Heater Harness (M939A2) Replacement	3-982
3-402.	Engine Coolant Oil Pan Shroud and Exhaust Tube Replacement	3-984
3-403.	Engine Coolant Heater Hose Replacement	3-986
3-404.	Engine Coolant Battery Box Heater Pad Replacement	3-994
3-405.	Swingfire Heater Pump Replacement	3-996
3-406.	Swingfire Heater and Mounting Bracket Replacement	3-1000
3-407.	Swingfire Heater Electrical Components Replacement	3-1004
3-408.	Swingfire Heater Harness Replacement	3-1010
3-409.	Swingfire Heater Oil Pan Shroud and Exhaust Tube Replacement	3-1016
3-410.	Swingfire Heater Battery Box Heater Pad Replacement	3-1018
3-411.	Swingfire Heater Water Jacket Replacement	3-1020
3-412.	Air Dryer Kit (M923/A1/A2, M925/A1/A2, M927/A1/A2, M928/A1/A2, M934/A1/A2) Replacement	3-1022
3-413.	Air Dryer Kit (M929/A1/A2, M930/A1/A2, M931/A1/A2, M932/A1/A2, M936/A1/A2) Replacement	3-1032
3-414.	A-frame Kit Maintenance	3-1046
3-415.	Pioneer Tool Kit Mounting Bracket Replacement	3-1052
3-416.	Fire Extinguisher Mounting Bracket Kit Replacement	3-1056
3-417.	Chemical Agent Alarm Mounting Bracket Kit Replacement	3-1058
3-418.	Machine Gun Mounting Kit Maintenance	3-1066
3-419.	Decontamination (M13) Apparatus Mounting Bracket Kit Replacement	3-1074
3-420.	Mud Guard Kit (M931/A1/A2, M932/A1/A2) Replacement	3-1078
3-421.	Rifle Mounting Kit Replacement	3-1080
3-422.	Hand Airbrake Air Supply Valve Replacement	3-1084
3-423.	Hand Airbrake Controller Valve Replacement	3-1086
3-424.	Hand Airbrake Doublecheck Valves Replacement	3-1088
3-425.	Hand Airbrake Tractor Protection Valve Replacement	3-1092
3-426.	100-Amp Alternator Replacement	3-1094
3-427.	100-Amp Alternator Harness Replacement	3-1096
3-428.	100-Amp Voltage Regulator Replacement	3-1100
3-429.	Troop Seat and Siderack Kit (M929/A1/A2, M930/A1/A2) Maintenance	3-1102
3-430.	Convoy Warning Light Mount Replacement	3-1104

3-394. SPECIAL PURPOSE KITS INDEX (Contd)

PARA. NO.	TITLE	PAGE NO.
3-431.	Convoy Warning Light Mount (M934/A1/A2) Replacement	3-1108
3-432.	Convoy Warning Light Mount (M929/A1/A2, M930/A1/A2) Replacement	3-1112
3-433.	Convoy Warning Light Harness Replacement	3-1114
3-434.	Convoy Warning Light Harness (M929/A1/A2, M930/A1/A2) Replacement	3-1116
3-435.	Convoy Warning Light Harness (M934/A1/A2) Replacement	3-1120
3-436.	Convoy Warning Light Resistor and Leads Replacement	3-1124
3-437.	Convoy Warning Light Replacement	3-1128
3-438.	Convoy Warning Light Switch Replacement	3-1130
3-439.	European Mini-lighting Kit Replacement	3-1132
3-440.	Automatic Throttle Kit (M936/A1) Replacement	3-1134
3-441.	Atmospheric Fuel Tank Vent System Kit Replacement	3-1140
3-442.	Vehicle Tiedown Kit Replacement	3-1146
3-443.	Hydraulic Hose Chafe Guard Kit Replacement	3-1148
3-444.	Hydraulic Reservoir Drain Kit Replacement	3-1152
3-445.	Hydraulic Reservoir Shutoff Modification Kit Replacement	3-1154
3-446.	Lightweight Weapon Station Modification Kit Maintenance	3-1160
3-447.	Engine Exhaust Brake Modification Kit (M939/A1) Replacement	3-1186
3-448.	Engine Exhaust Brake Modification Kit (M939A2) Replacement	3-1204
3-449.	Exhaust Heat Shield Accessory Kit Replacement	3-1224
3-450.	Van Handrail Modification Kit (M934/A1/A2) Replacement	3-1225

3-395. RADIATOR COVER KIT REPLACEMENT

THIS TASK COVERS:

a Removal | b. Installation

INITIAL SETUP:

APPLICABLE MODELS
All

TOOLS
General mechanic's tool kit (Appendix E, Item 1)

REFERENCES (TM)
TM 9-2320-272-10
TM 9-2320-272-24P

EQUIPMENT CONDITION
Parking brake set (TM 9-2320-272-10).

a. Removal

1. Untie ends of tiedown strap (2) and remove from twenty-eight tiedown loops (3).
2. Remove radiator cover (1), fifty-six screws (4), and twenty-eight tiedown loops (3) from hood (5).

b. Installation

1. Install twenty-eight tiedown loops (3) on hood (5) with fifty-six screws (4).
2. Position radiator cover (1) on twenty-eight tiedown loops (3).
3. Thread tiedown strap (2) through twenty-eight tiedown loops (3) and tie ends together.

3-395. RADIATOR COVER KIT REPLACEMENT (Contd)

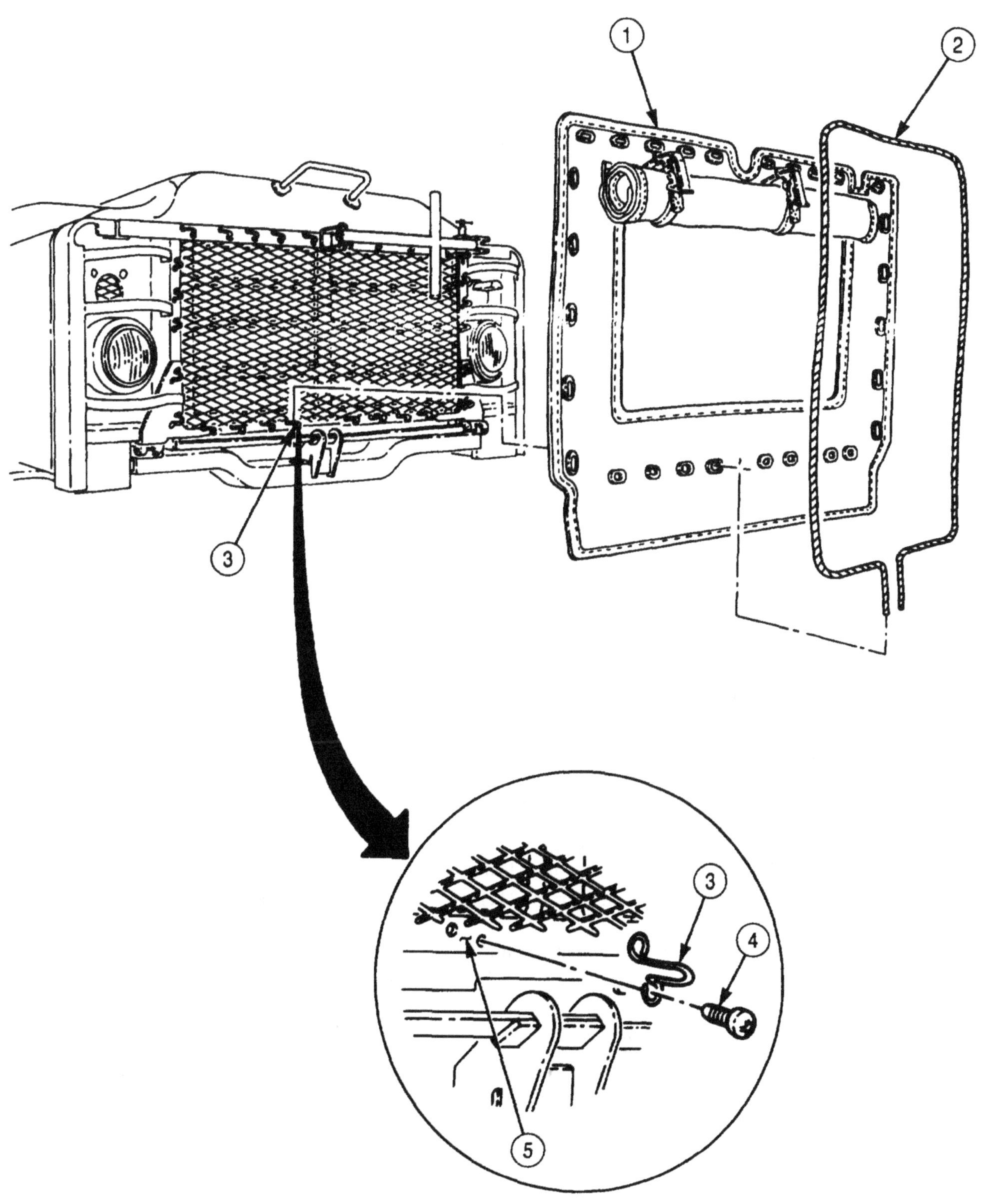

3-396. HARDTOP KIT REPLACEMENT

THIS TASK COVERS:

a. Removal b. Installation

INITIAL SETUP:

APPLICABLE MODELS
All

TOOLS
General mechanic's tool kit (Appendix E, Item 1)

MATERIALS/PARTS
Thirty-two locknuts (Appendix D, Item 276)
Two lockwashers (Appendix D, Item 369)
Two rubber seals (Appendix D, Item 566)

REFERENCES (TM)
TM 9-2320-272-10
TM 9-2320-272-24P

EQUIPMENT CONDITION
Parking brake set (TM 9-2320-272-10).

NOTE
Assistant will help when required.

a. Removal

1. Remove two screws (2) and lockwashers (1) from roof assembly (3). Discard lockwashers (1).

NOTE
Note position of hook bolts for installation.

2. Remove four nuts (4) and hook bolts (10) from top of roof assembly (3) and windshield frame (11).
3. Remove sixteen locknuts (5), washers (6), screws (9), and washers (6) from roof assembly (3). Discard locknuts (5).
4. Remove roof assembly (3) and rubber seal (7) from back panel assembly (8) and windshield frame (11). Discard rubber seal (7).
5. Remove sixteen locknuts (13), screws (12), washers (16), back panel assembly (8), and rubber seal (15) from cab body (14). Discard locknuts (13) and rubber seal (15).

b. Installation

1. Install new rubber seal (15) and back panel assembly (8) on cab body (14) with sixteen washers (16), screws (12), and new locknuts (13). Do not tighten locknuts (13).
2. Install new rubber seal (7) and roof assembly (3) on back panel assembly (8) with sixteen washers (6), screws (9), washers (6), and new locknuts (5). Do not tighten locknuts (5).
3. Install roof assembly (3) on windshield frame (11) with four hook bolts (10) and nuts (4). Do not tighten nuts (4).
4. Inspect all hardtop kit panels for alignment and seating. Adjust as required.
5. Close cab windows and inspect for alignment with roof assembly (3). Adjust as required.
6. Tighten sixteen new locknuts (13) and (5) and four nuts (4).
7. Install two new lockwashers (1) and screws (2) on roof assembly (3).

3-396. HARDTOP KIT REPLACEMENT (Contd)

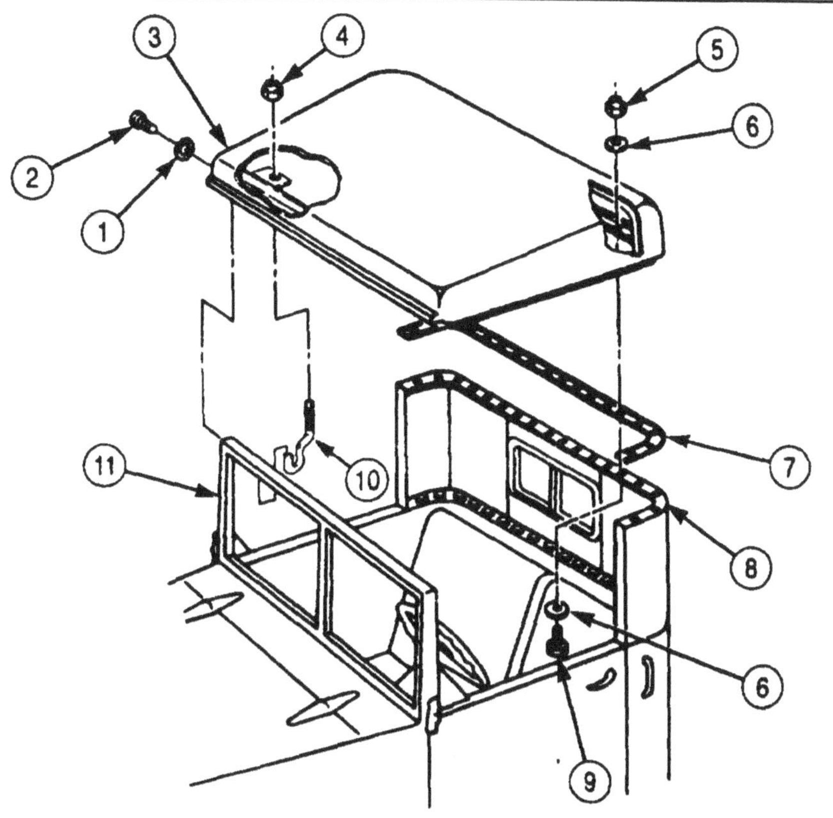

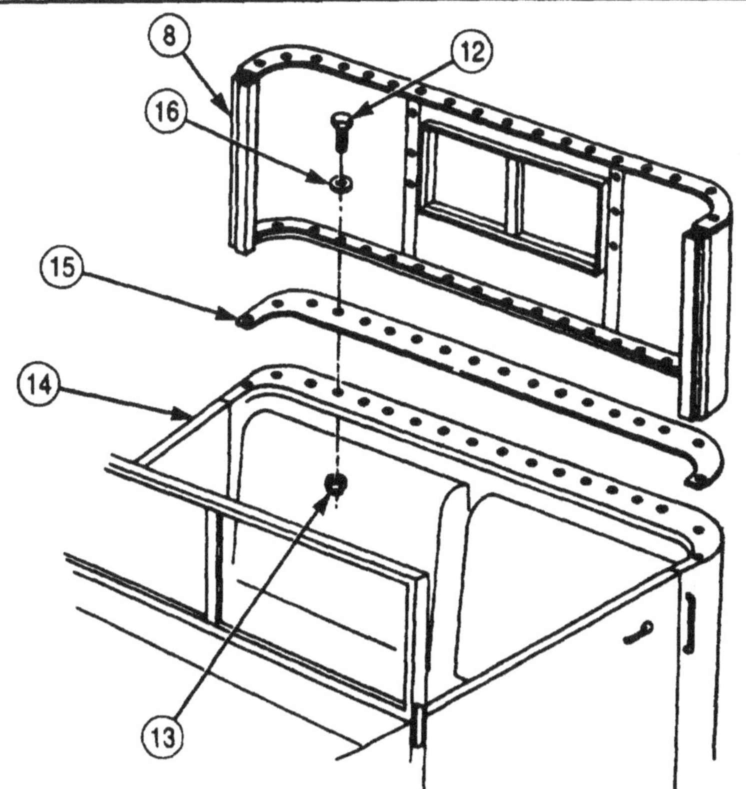

3-397. ENGINE COOLANT HEATER REPLACEMENT

THIS TASK COVERS:

a. Removal b. Installation

INITIAL SETUP:

APPLICABLE MODELS
All

TOOLS
General mechanic's tool kit (Appendix E, Item 1)

MATERIALS/PARTS
Cotter pin (Appendix D, Item 58)
Four locknuts (Appendix D. Item 291)

REFERENCES (TM)
TM 9-2320-272-10
TM 9-2320-272-24P

EQUIPMENT CONDITION
Engine coolant heater pump removed (para. 3-398).

a. Removal

1. Disconnect heater harness (3) from engine coolant heater (2).
2. Disconnect fuel line (8) and elbow (9) from engine coolant heater (2).
3. Disconnect hose (7), adapter (6), elbow (5), union (4), and nipple (10) from engine coolant heater (2).
4. Remove cotter pin (19) and exhaust tube (18) from engine coolant heater (2). Discard cotter pin (19).
5. Remove hose (13), adapter (15), and elbow (16) from engine coolant heater (2).
6. Remove two clamps (11) from engine coolant heater (2).
7. Remove engine coolant heater (2) from bracket (12).
8. Remove four locknuts (17) and screws (14) from bracket (12). Discard locknuts (17).
9. Remove bracket (12) from toolbox (1).

b. Installation

1. Install bracket (12) on toolbox (1).
2. Install four screws (14) and new locknuts (17) on bracket (12).
3. Install engine coolant heater (2) on bracket (12).
4. Install two clamps (11) on engine coolant heater (2).
5. Install elbow (16), adapter (15), and hose (13) on engine coolant heater (2).
6. Install exhaust tube (18) on engine coolant heater (2) with new cotter pin (19).
7. Install nipple (10), union (4), elbow (5), adapter (6), and hose (7) on engine coolant heater (2).
8. Install elbow (9) and fuel line (8) on engine coolant heater (2).
9. Connect heater harness (3) to engine coolant heater (2).

3-397. ENGINE COOLANT HEATER REPLACEMENT (Contd)

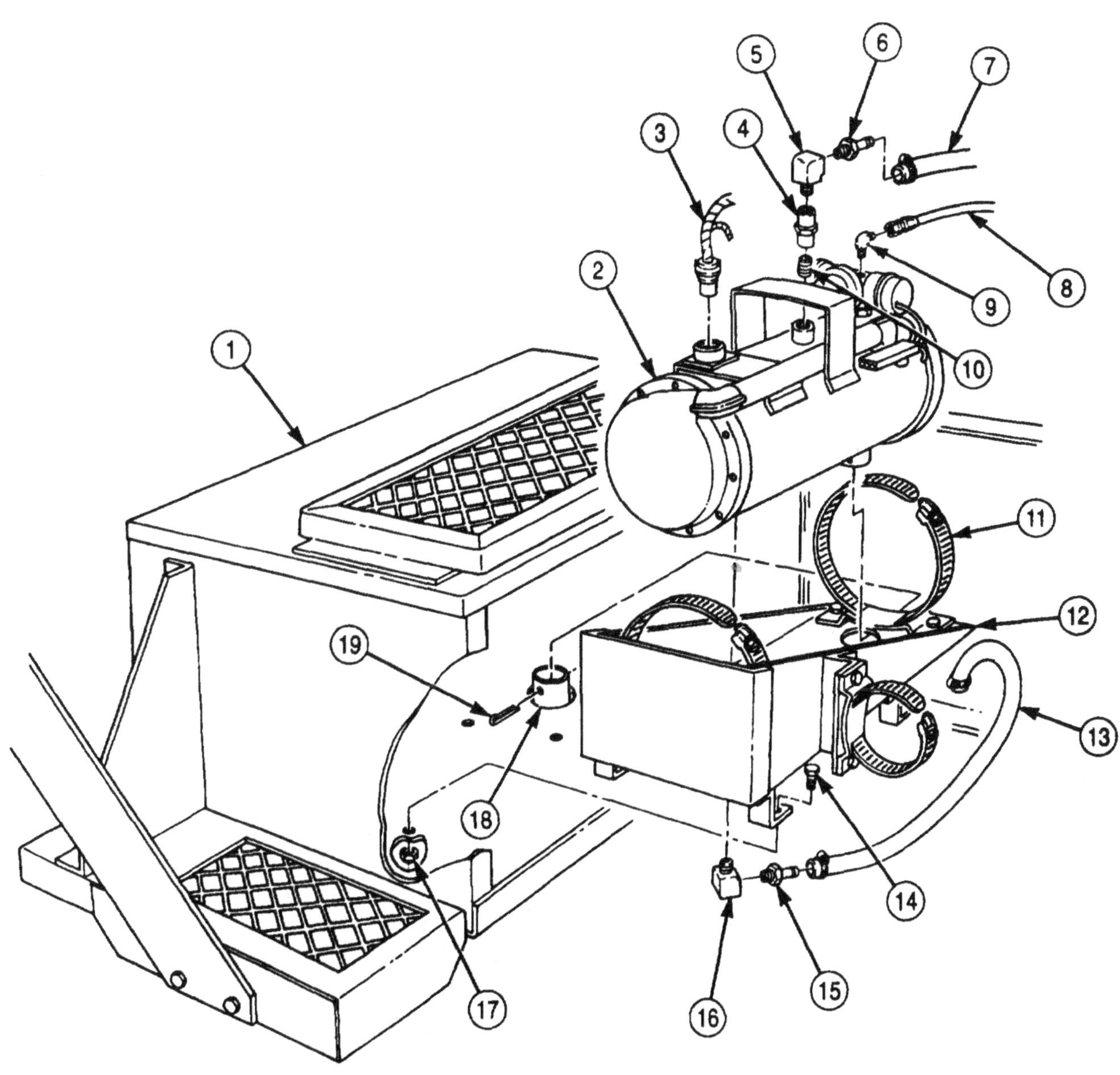

FOLLOW-ON TASK: Install engine coolant heater pump (para. 3-398).

3-398. ENGINE COOLANT HEATER PUMP REPLACEMENT

THIS TASK COVERS:

a. Removal b. Installation

INITIAL SETUP:

APPLICABLE MODELS
All

TOOLS
General mechanic's tool kit (Appendix E, Item 1)

MATERIALS/PARTS
Two lockwashers (Appendix D, Item 299)

REFERENCES (TM)
TM 9-2320-272-10
TM 9-2320-272-24P

EQUIPMENT CONDITION
- Parking brake set (TM 9-2320-272-10).
- Hood raised and secured (TM 9-2320-272-10).

a. Removal

1. Close water manifold drainvalve (1) and engine oil cooler drainvalve (2).
2. Open toolbox door (6).
3. Close two engine coolant heater drainvalves (3).
4. Loosen clamp (16) and remove manifold inlet hose (19) from heater pump elbow (15).
5. Loosen clamp (18) and remove pump outlet hose (4) from heater pump adapter (17).
6. Remove nut (5) and lockwasher (7) from terminal stud (9) and disconnect ground wire (8) from heater pump (14). Discard lockwasher (7).
7. Remove nut (13) and lockwasher (12) from terminal stud (10) and disconnect wire (11) from heater pump (14). Discard lockwasher (12).
8. Remove clamp (21) and heater pump (14) from pump bracket (20).
9. Remove heater pump elbow (15) and adapter (22) from heater pump (14).
10. Remove heater pump adapter (17) from port (23).

b. Installation

1. Install heater pump adapter (17) in port (23).
2. Install adapter (22) and heater pump elbow (15) in heater pump (14).
3. Install heater pump (14) on pump bracket (20) with clamp (21).
4. Install wire (11) on terminal stud (10) with new lockwasher (12) and nut (13).
5. Install ground wire (8) on terminal stud (9) with new lockwasher (7) and nut (5).
6. Install pump outlet hose (4) on heater pump adapter (17) and tighten clamp (18).
7. Install manifold inlet hose (19) on heater pump elbow (15) and tighten clamp (16).
8. Open two engine coolant heater drainvalves (3).
9. Open water manifold drainvalve (1) and engine oil cooler drainvalve (2).
10. Close toolbox door (6).

3-398. ENGINE COOLANT HEATER PUMP REPLACEMENT (Contd)

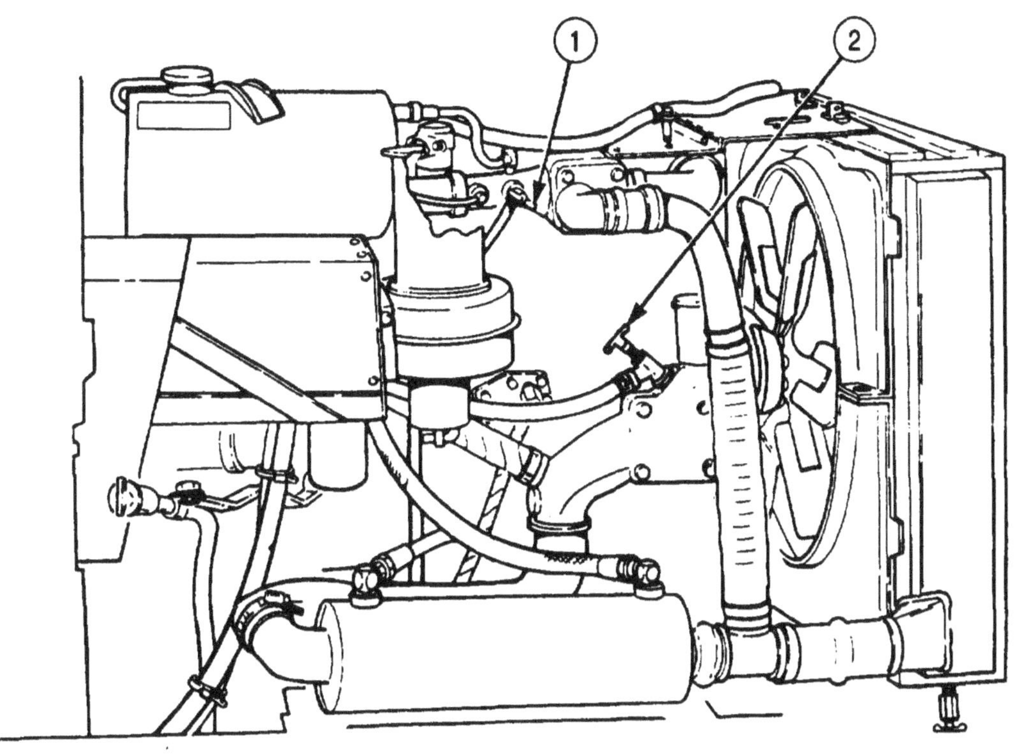

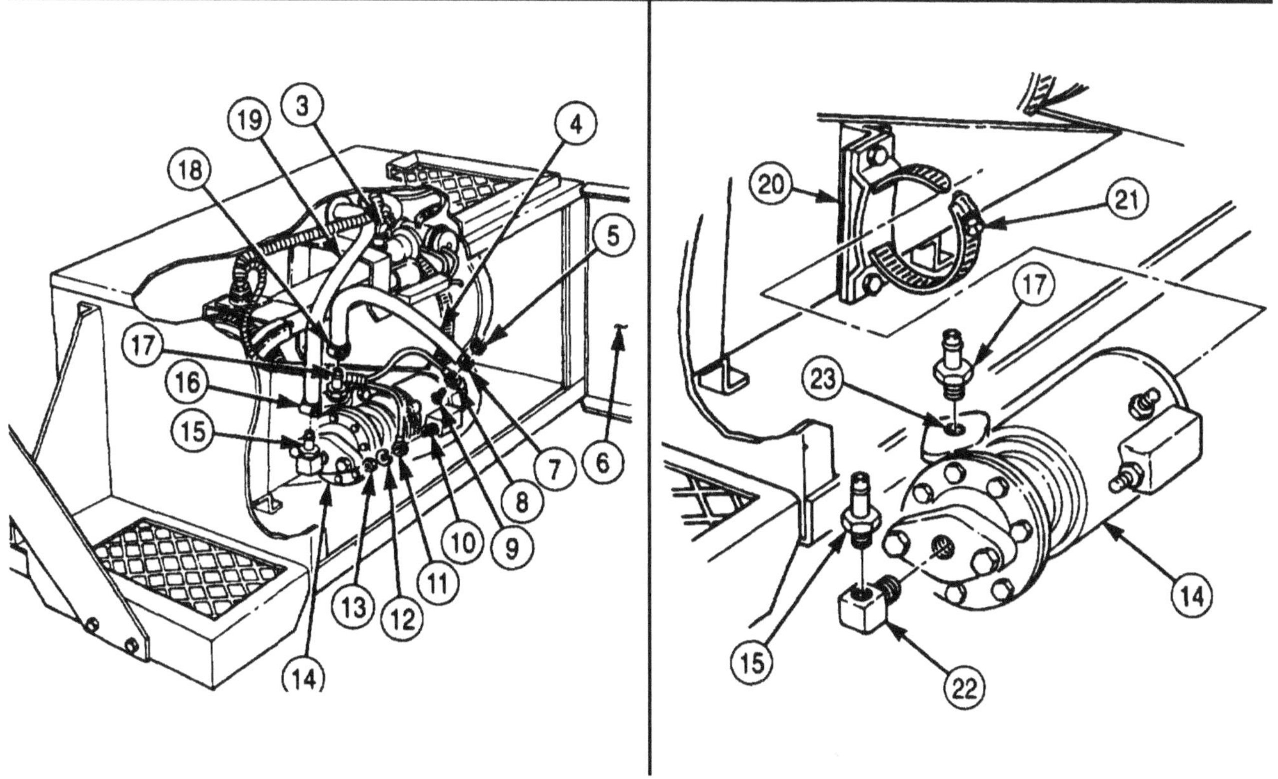

FOLLOW-ON TASK: Check engine coolant heater for proper operation (TM 9-2320-272-10).

3-399. ENGINE COOLANT HEATER CONTROL BOX REPLACEMENT

THIS TASK COVERS:

a. Removal b. Installation

INITIAL SETUP:

APPLICABLE MODELS
All

TOOLS
General mechanic's tool kit (Appendix E, Item 1)

MATERIALS/PARTS
Two lockwashers (Appendix D, Item 400)
Two locknuts (Appendix D. Item 313)

REFERENCES (TM)
TM 9-2320-272-10
TM 9-2320-272-24P

EQUIPMENT CONDITION
- Parking brake set (TM 9-2320-272-10).
- Battery ground cables disconnected (para. 3-126).

a. Removal

1. Disconnect coolant heater harness connector (5) from coolant heater control box (8).
2. Disconnect connector (6) from control box wire (7).
3. Remove two nuts (2), lockwashers (3), and coolant heater control box (8) from control mounting bracket (4). Discard lockwashers (3).
4. Remove two locknuts (1), screws (9), and control mounting bracket (4) from instrument panel (10). Discard locknuts (1).

b. Installation

1. Install control mounting bracket (4) on instrument panel (10) with two screws (9) and new locknuts (1).
2. Install coolant heater control box (8) on control mounting bracket (4) with two new lockwashers (3) and nuts (2).
3. Connect connector (6) to control box wire (7).
4. Connect coolant heater harness connector (5) to coolant heater control box (8).

3-399. ENGINE COOLANT HEATER CONTROL BOX REPLACEMENT (Contd)

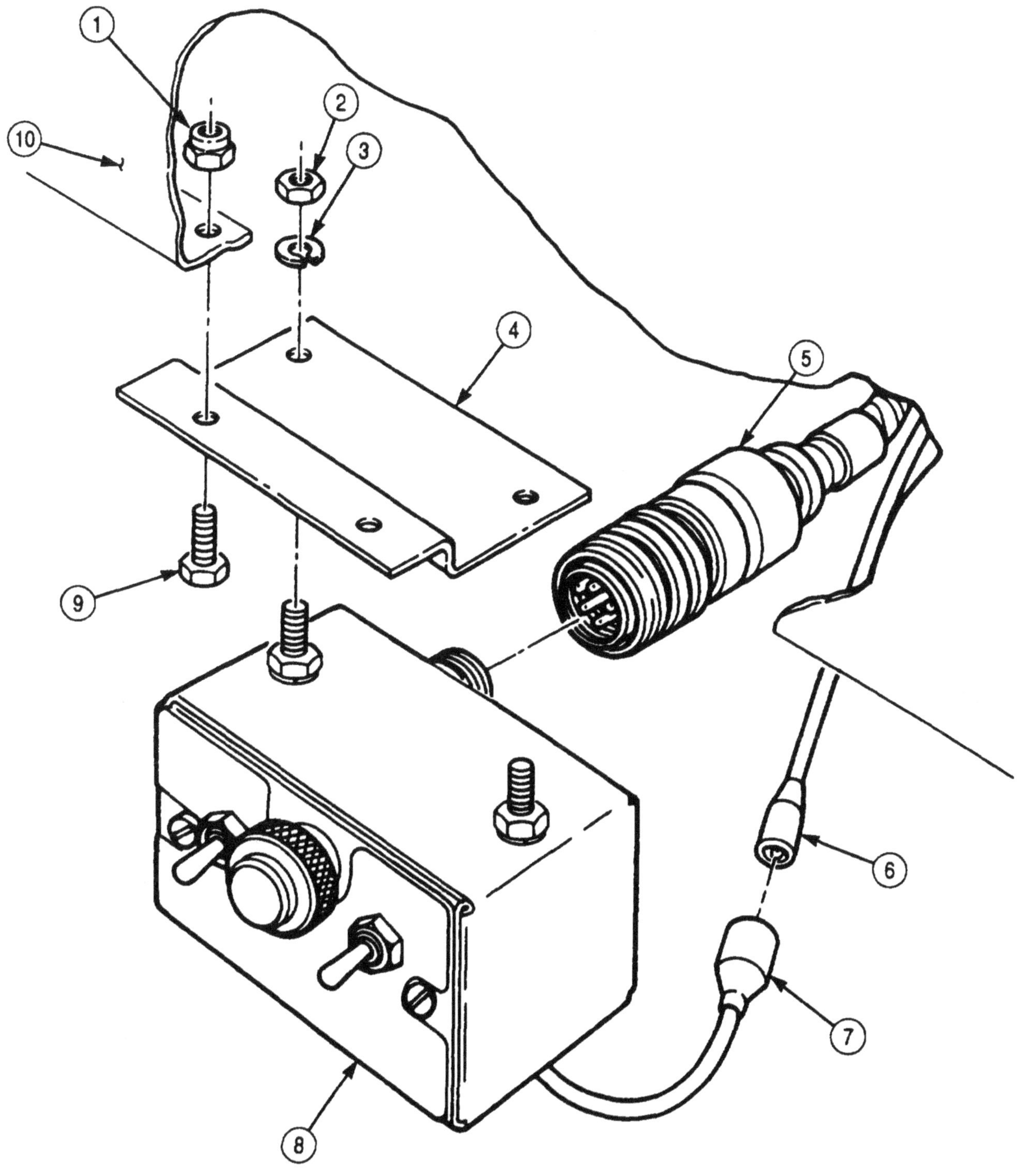

FOLLOW-ON TASKS: •Connect battery ground cables (para. 3-126).
• Check engine coolant heater for proper operation (TM 9-2320-272-10).

3-400. ENGINE COOLANT HEATER HARNESS (M939/A1) REPLACEMENT

THIS TASK COVERS:

a. Removal   b. Installation

INITIAL SETUP:

APPLICABLE MODELS
M939/A1

TOOLS
General mechanic's tool kit (Appendix E, Item 1)

MATERIALS/PARTS
Two lockwashers (Appendix D, Item 400)
Ten tiedown straps (Appendix D, Item 685)

REFERENCES (TM)
TM 9-2320-272-10
TM 9-2320-272-24P

EQUIPMENT CONDITION
- Parking brake set (TM 9-2320-272-10).
- Hood raised and secured (TM 9-2320-272-10).
- Battery ground cables disconnected (para. 3-126).

a. Removal

1. Open toolbox door (9).
2. Disconnect engine coolant heater harness (2) from engine coolant heater (20).
3. Remove nut (10), lockwasher (11), and ground wire (12) from terminal stud (13) on heater pump (18). Discard lockwasher (11).
4. Remove nut (17), lockwasher (16), and ground wire (15) from terminal stud (14) on heater pump (18). Discard lockwasher (16).
5. Remove ten tiedown straps (5) from engine coolant heater harness (2), fuel pump cable (6), hose (7), and lead (22). Discard tiedown straps (5).
6. Disconnect lead (22) from engine coolant heater harness lead (4).
7. Disconnect engine coolant heater harness (2) from control box (1), slide through grommet (3) in firewall (21) and grommet (8) in toolbox (19), and remove from vehicle.

b. Installation

1. Slide engine coolant heater harness (2) through grommet (8) in toolbox (19) and grommet (3) in firewall (21) and connect to control box (1).
2. Connect lead (22) to engine coolant heater harness lead (4).
3. Install ten new tiedown straps (5) on engine coolant heater harness (2), fuel pump cable (6), hose (7), and lead (22).
4. Connect ground wire (15) to terminal stud (14) on heater pump (18) with new lockwasher (16) and nut (17).
5. Connect ground wire (12) to terminal stud (13) on heater pump (18) with new lockwasher (11) and nut (10).
6. Connect engine coolant heater harness (2) to engine coolant heater (20).
7. Close toolbox door (9).

3-400. ENGINE COOLANT HEATER HARNESS (M939/A1) REPLACEMENT (Contd)

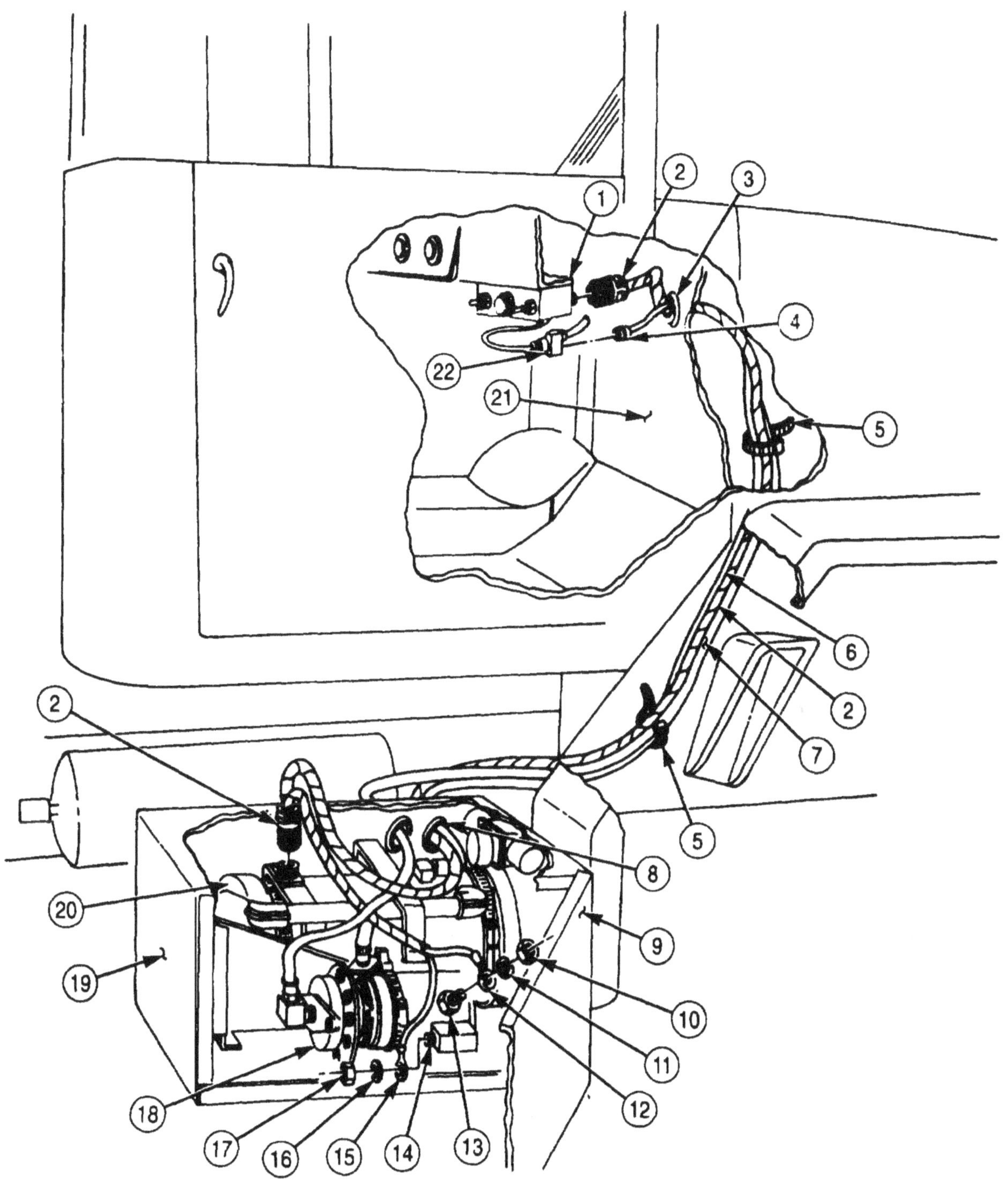

FOLLOW-ON TASKS:

- Connect battery ground cables (para. 3-126).
- Check engine coolant heater for proper operation (TM 9-2320-272-10).

3-401. ENGINE COOLANT HEATER HARNESS (M939A2) REPLACEMENT

THIS TASK COVERS:

a. Removal

b. Installation

INITIAL SETUP:

APPLICABLE MODELS
M939A2

TOOLS
General mechanic's tool kit (Appendix E, Item 1)

MATERIALS/PARTS
Two lockwashers (Appendix D, Item 400)
Tiedown strap (Appendix D, Item 685)

REFERENCES (TM)
TM 9-2320-272-10
TM 9-2320-272-24P

EQUIPMEMT CONDITION
- Parking brake set (TM 9-2320-272-102
- Battery ground cables disconnected (para. 3-126).

a. Removal

1. Open toolbox door (12).
2. Disconnect engine coolant heater harness (3) from engine coolant heater (22).
3. Remove nut (13), lockwasher (14), and ground wire (21) from terminal stud (15) on heater pump (19). Discard lockwasher (14).
4. Remove nut (18), lockwasher (201, and wire (17) from terminal stud (16) on heater pump (19). Discard lockwasher (20).
5. Remove screw (7), nut (6), and clamp (8) from bracket (9).
6. Remove tiedown strap (5) from engine coolant heater harness (3). Discard tiedown strap (5).
7. Disconnect engine coolant heater harness (3) from control box (1). and slide harness (3) through grommet (4) in firewall (2) and grommet (10) in toolbox (11).

b. Installation

1. Slide engine coolant heater harness (3) through grommet (4) in firew(2l) and grommet (10) in toolbox (11) and connect to control box (1).
2. Install new tiedown strap (5) on engine coolant heater harness (3).
3. Install engine coolant heater harness (3) on bracket (9) with clamp (8), screw (7), and nut (6).
4. Connect engine coolant heater harness (3) to engine coolant heater (22).
5. Install wire (17) on terminal stud (16) with new lockwasher (20) and nut (18).
6. Install ground wire (21) on terminal stud (15) with new lockwasher (14) and nut (13).
7. Close toolbox door (12).

3-401. ENGINE COOLANT HEATER HARNESS (M939A2) REPLACEMENT (Contd)

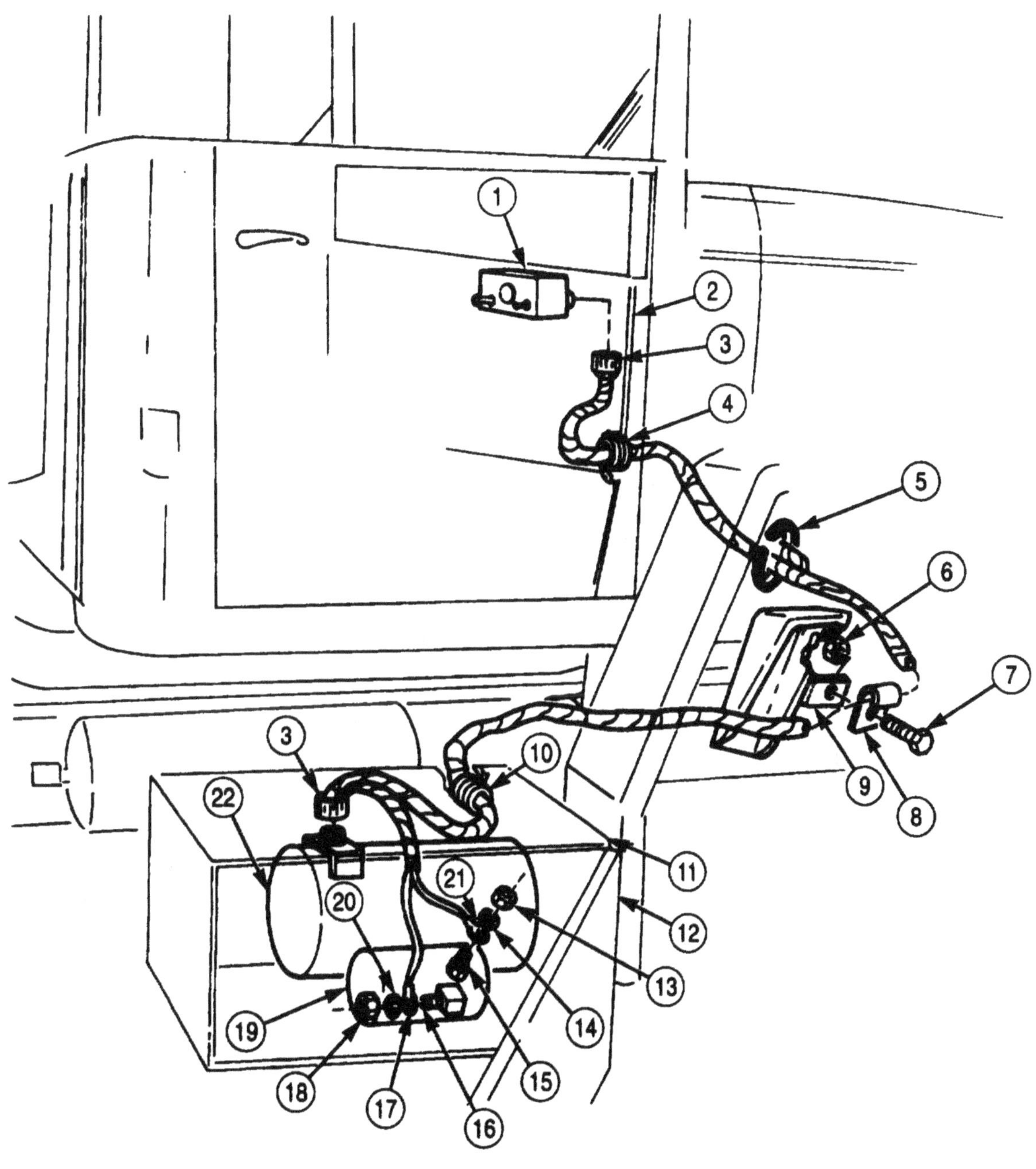

FOLLOW-ON TASKS: • Connect battery ground cables (para. 3-126).
• Check engine coolant heater for proper operation (TM 9-2320-272-10).

3-402. ENGINE COOLANT OIL PAN SHROUD AND EXHAUST TUBE REPLACEMENT

THIS TASK COVERS:

a. Removal　　b. Installation

INITIAL SETUP:

APPLICABLE MODELS
All

TOOLS
General mechanic's tool kit (Appendix E, Item 1)

MATERIALS/PARTS
Three cotter pins (Appendix D, Item 59)
Two locknuts (Appendix D, Item 291)

REFERENCES (TM)
TM 9-2320-272-10
TM 9-2320-272-24P

EQUIPMEMT CONDITION
Parking brake set (TM 9-2320-272-10).

a. Removal

1. Remove two locknuts (4), screws (10), and clamps (9) from exhaust tube (13), crossmember (11), and bracket (12). Discard locknuts (4).
2. Open toolbox door (1).
3. Remove two cotter pins (31, exhaust tube (13). and elbow (14) from engine coolant heater (2) and oil pan shroud (6). Discard cotter pins (3).
4. Remove cotter pin (3) and elbow (14) from exhaust tube (13). Discard cotter pin (3).
5. Remove four nuts (8), washers (7), and oil pan shroud (6) from oil pan studs (5).

b. Installation

1. Install oil pan shroud (6) on oil pan studs (5) with four washers (7) and nuts (8).
2. Install elbow (14) on exhaust tube (13) with new cotter pin (3).
3. Install exhaust tube (13) and elbow (14) on engine coolant heater (2) and oil pan shroud (6) with two new cotter pins (3).
4. Close toolbox door (1).
5. Install exhaust tube (13) on crossmember (11) and bracket (12) with two clamps (9), screws (10), and new locknuts (4).

3-402. ENGINE COOLANT OIL PAN SHROUD AND EXHAUST TUBE REPLACEMENT (Contd)

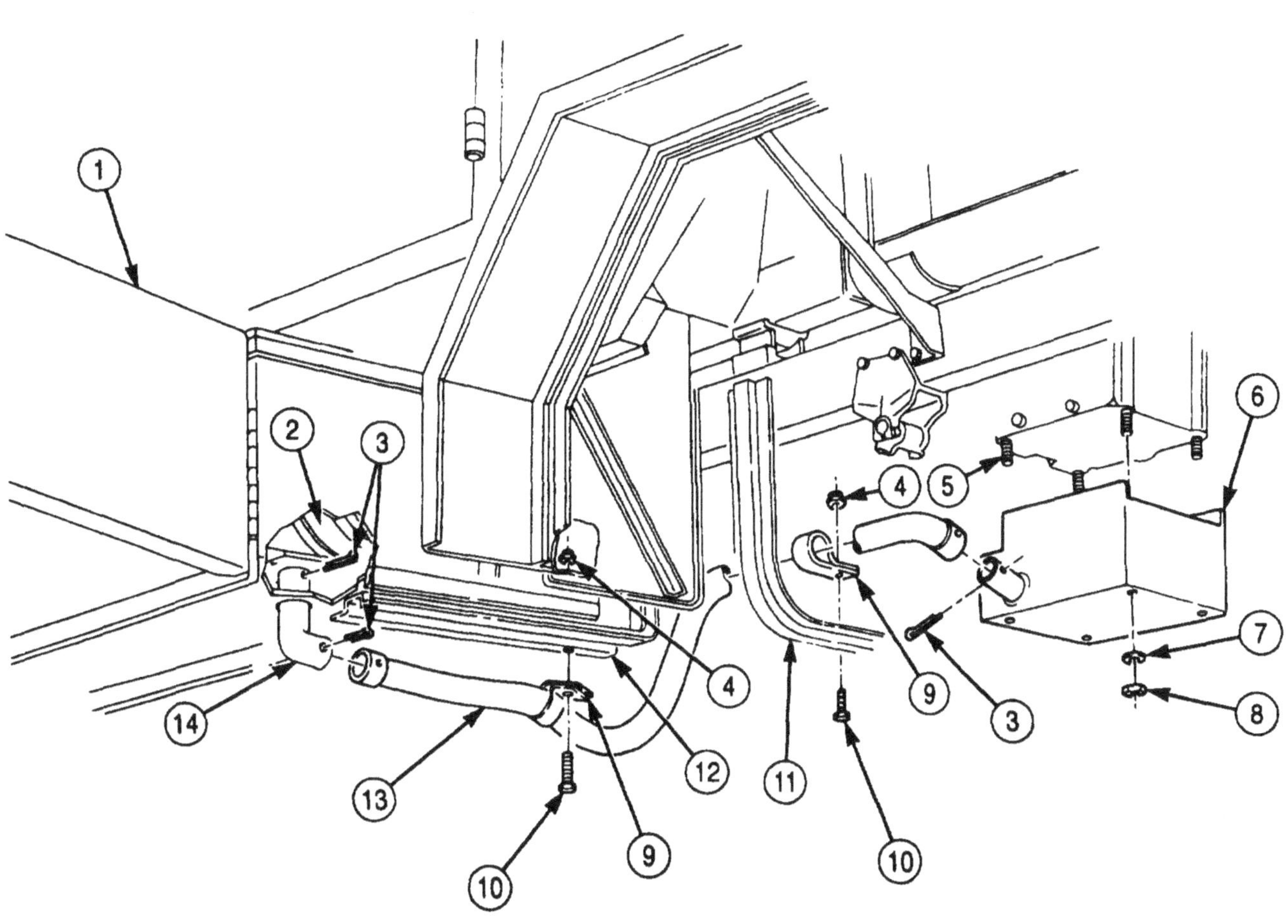

3-403. ENGINE COOLANT HEATER HOSE REPLACEMENT

THIS TASK COVERS:

a. Removal
b. Installation

INITIAL SETUP:

APPLICABLE MODELS
All

TOOLS
General mechanic's tool kit (Appendix E, Item 1)

MATERIALS/PARTS
Lockwasher (Appendix D, Item 354)
Nine tiedown straps (Appendix D. Item 685)

REFERENCES (TM)
TM 9-2320-272-10
TM 9-2320-272-24P

EQUIPMENT CONDITION
- Parking brake set (TM 9-2320-272-10).
- Hood raised and secured (TM 9-2320-272-10).
- Right splash shield removed (TM 9-2320-272-10).

a. Removal

1. Close water manifold drainvalve (1) on right side of engine (3).
2. Close coolant outlet drainvalve (2) on right side of engine (3).

NOTE

Have drainage container ready to catch excess coolant.

3. Loosen clamp (27) and remove hose (20) from water manifold drainvalve (1).
4. Remove nut (10), lockwasher (9), clamps (12) and (24), washer (22), and screw (23) from oil dipstick tuba bracket (8) and oil dipstick tube (11). Discard lockwasher (9).
5. Remove screw (18) and clamp (19) from cab support (13).
6. Remove three tiedown straps (21) from hose (20), fuel line (26), and electrical harness (25). Discard tiedown straps (21).
7. Loosen clamp (7) and remove hose (6) from coolant outlet drainvalve (2).
8. Remove screw (16), washer (15), and clamp (14) from engine access cover (5).
9. Remove two tiedown straps (17) from hose (6) and two transmission oil cooler lines (4). Discard tiedown straps (17).

3-403. ENGINE COOLANT HEATER HOSE REPLACEMENT (Contd)

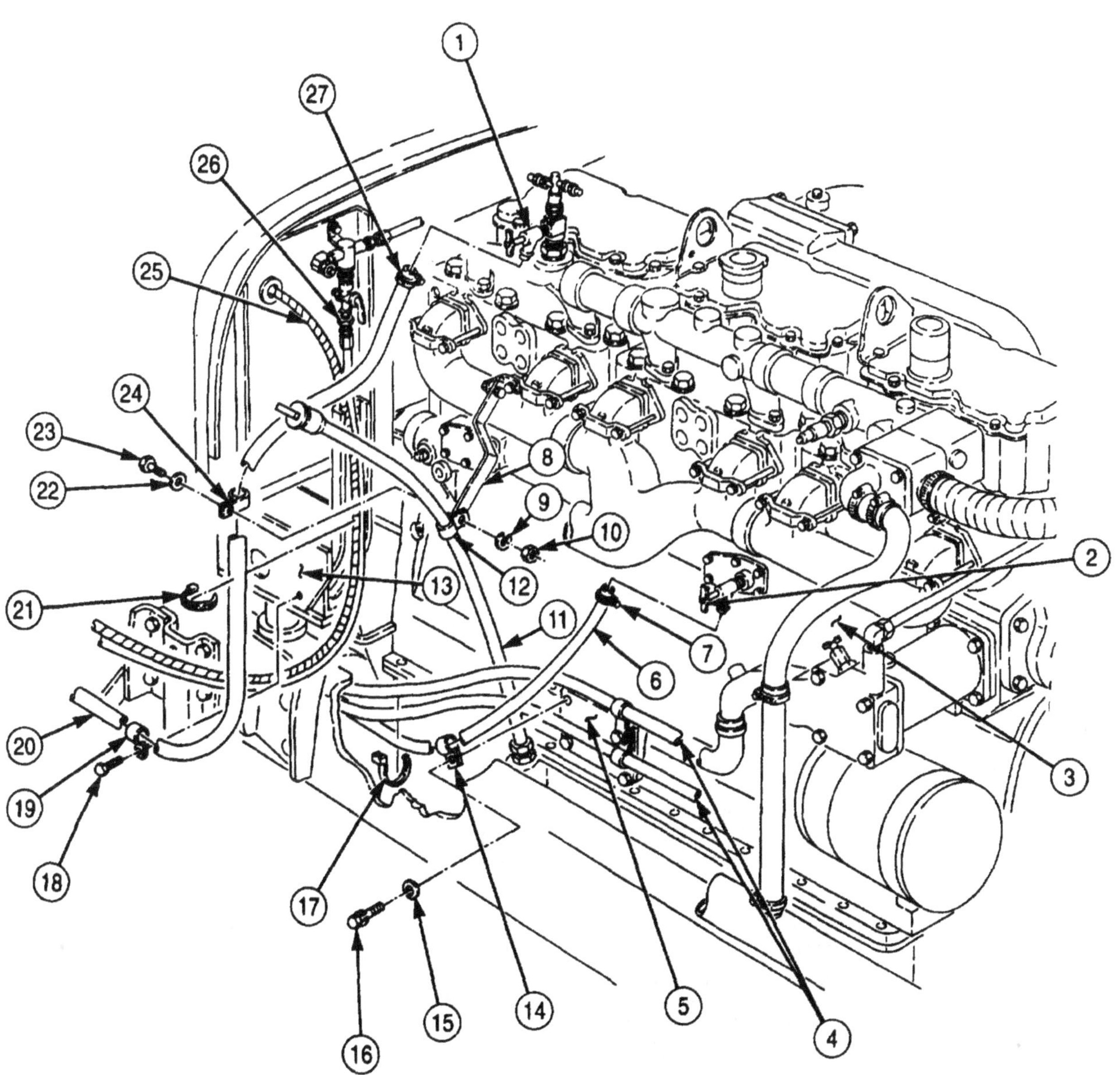

3-403. ENGINE COOLANT HEATER HOSE REPLACEMENT (Contd)

10. Remove two tiedown straps (3) from hose (7), fuel line (14), and electrical harness (13). Discard tiedown straps (3).
11. Remove screw (8), clamp (5), and hose (7) from crossmember (6).

NOTE

Note routing of hose for installation.

12. Loosen clamp (12) and remove hose (7) from coolant heater (15), adapter (11), and vehicle.
13. Remove nut (4), screw (10), clamp (2), and hose (1) from air tank bracket (9).
14. Remove two tiedown straps (24) from hose (1) and electrical harness (13). Discard tiedown straps (24).
15. Loosen clamp (23) and remove hose (1) from adapter (22) on coolant heater pump (21).
16. Loosen clamp (19) and remove hose (17) from adapter (20) on coolant heater pump (21).
17. Loosen clamp (18) and remove hose (17) from adapter (16) on coolant heater (15).
18. Loosen clamp (29) and remove hose (1) from adapter (25) and vehicle.
19. Loosen clamp (28) and remove hose (27) from nipple (26) and vehicle.

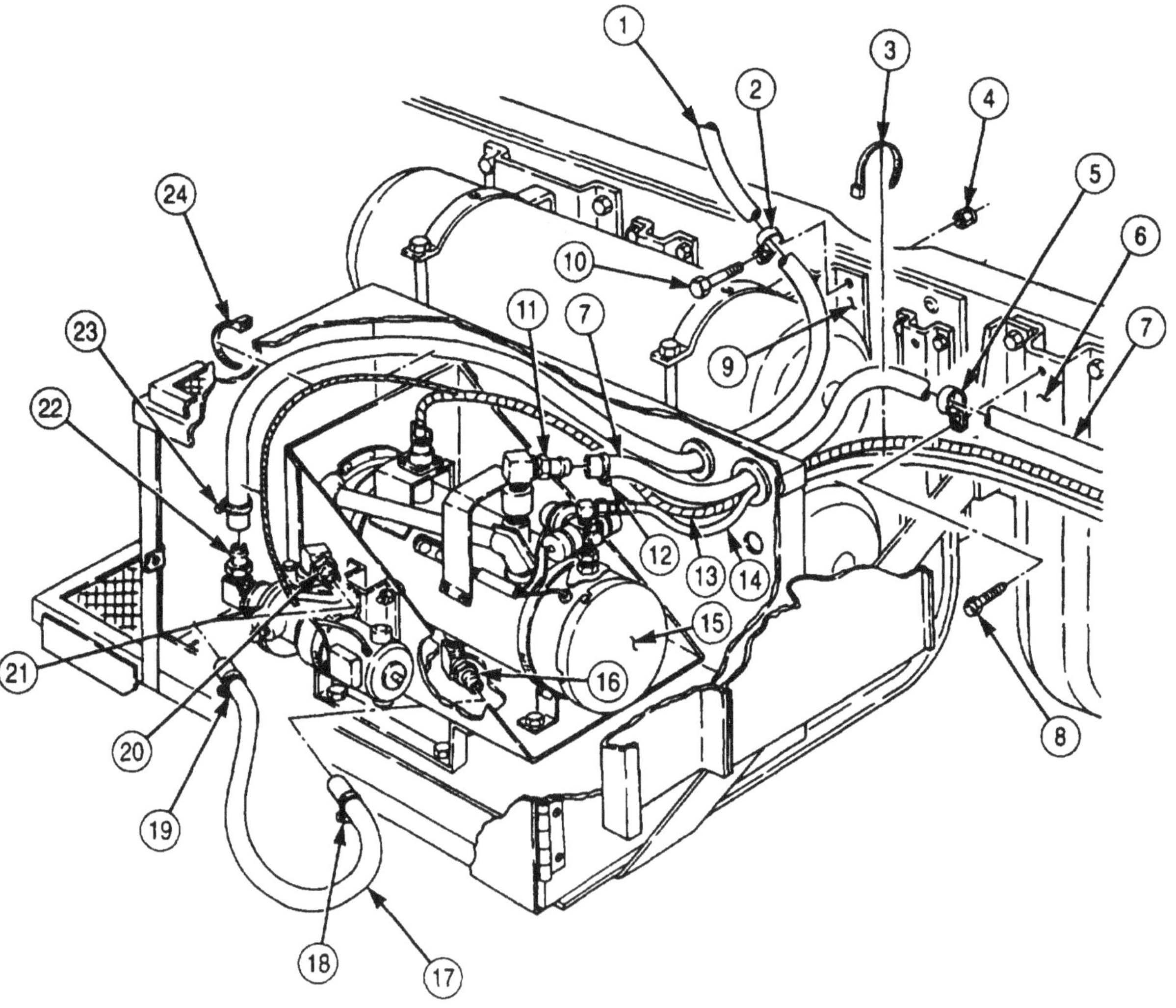

3-403. ENGINE COOLANT HEATER HOSE REPLACEMENT (Contd)

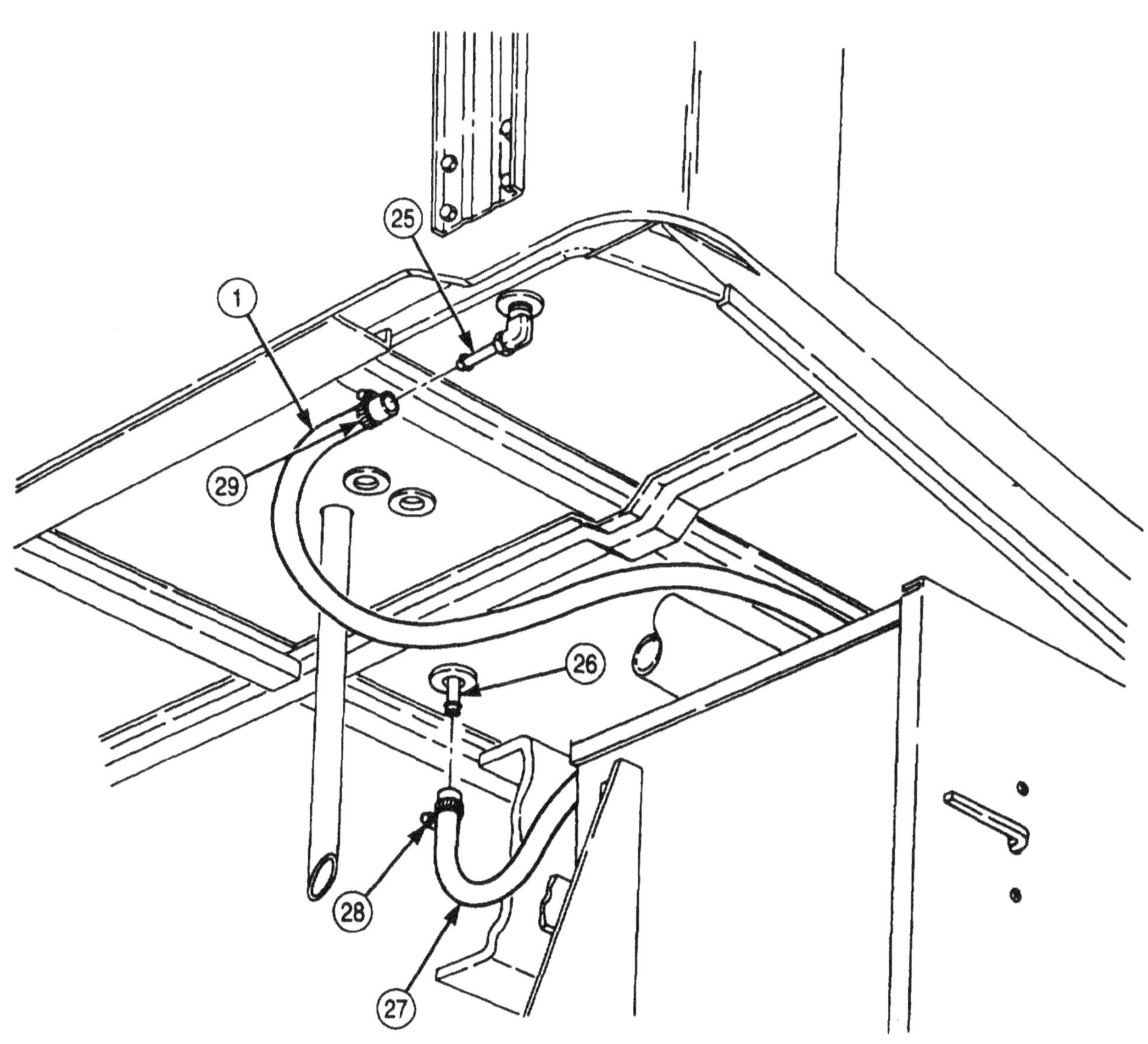

3-403. ENGINE COOLANT HEATER HOSE REPLACEMENT (Contd)

b. Installation

1. Route hose (5) into position on vehicle and install hose (5) on nipple (4) and tighten clamp (6).
2. Route hose (1) into position on vehicle and install hose (1) on adapter (3) and tighten clamp (2).
3. Install hose (19) on adapter (18) and tighten clamp (20).
4. Install hose (19) on adapter (22) and tighten clamp (21).
5. Install hose (1) on adapter (24) and tighten clamp (25).
6. Install two new tiedown straps (26) on hose (1) and electrical harness (27).
7. Install hose (1) on air tank bracket (30) with clamp (7), screw (29), and nut (9).
8. Route hose (12) into position on vehicle and install hose (12) and adapter (28) on coolant heater (17) and tighten clamp (16).
9. Install hose (12) on crossmember (11) with clamp (10) and screw (13).
10. Install two new tiedown straps (8) on hose (12), fuel line (14), and electrical harness (15).

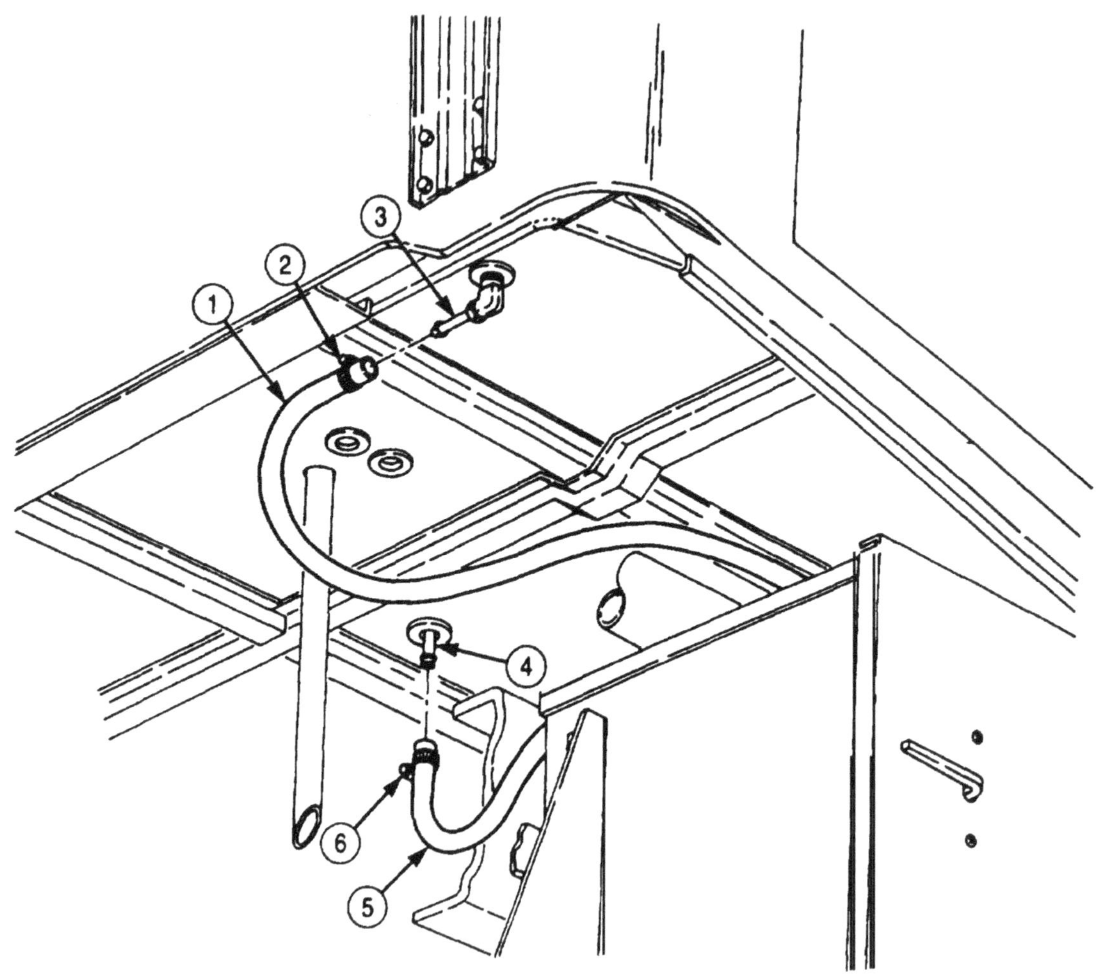

3-403. ENGINE COOLANT HEATER HOSE REPLACEMENT (Contd)

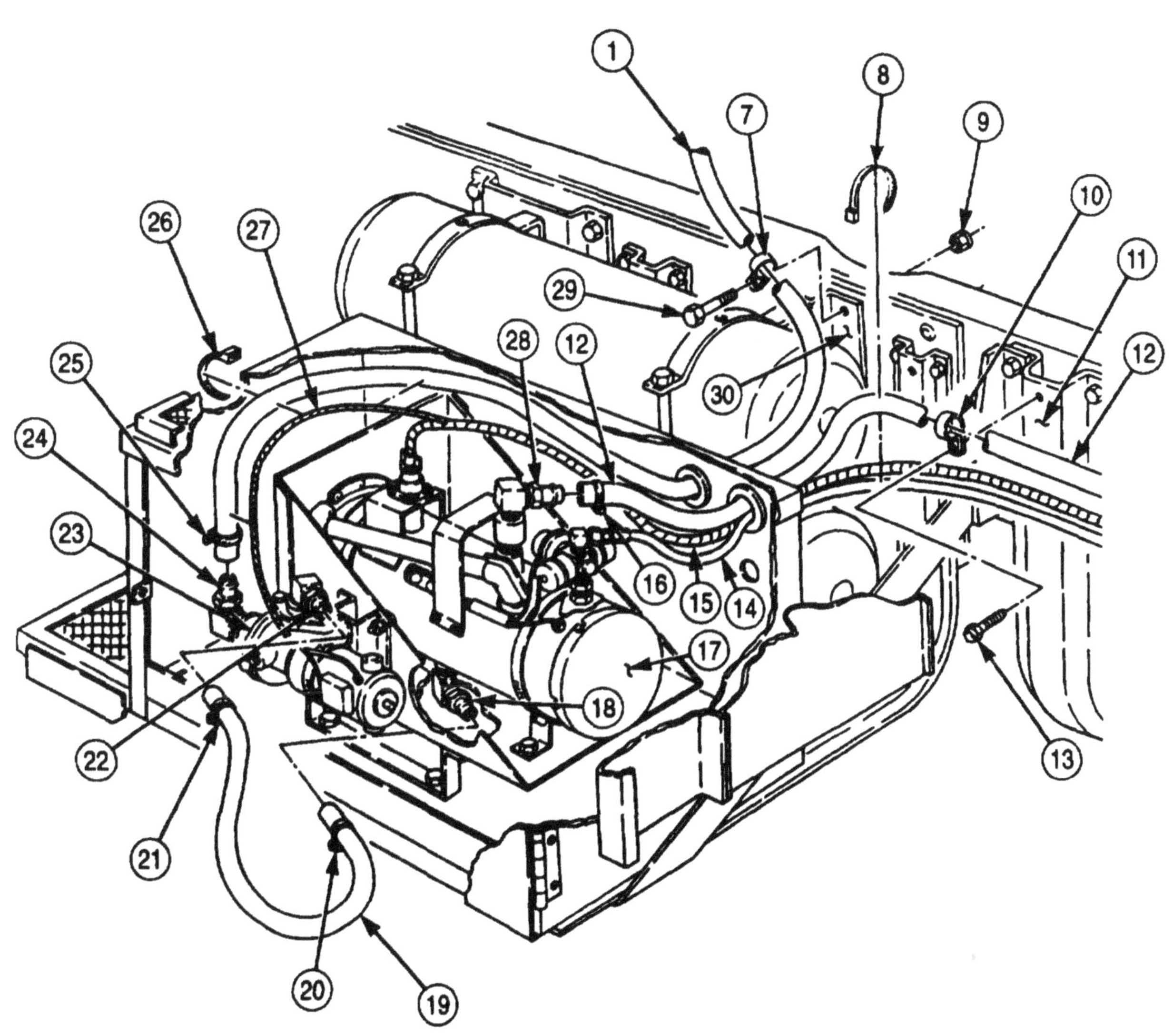

3-403. ENGINE COOLANT HEATER HOSE REPLACEMENT (Contd)

11. Install two new tiedown straps (17) on hose (7) and two transmission oil cooler lines (5).
12. Install hose (7) on engine access cover (6) with clamp (14), washer (15), and screw (16).
13. Install hose (7) on coolant outlet drainvalve (3) and tighten clamp (8).
14. Install three new tiedown straps (22) on hose (21), fuel line (27), and electrical harness (26).
15. Install hose (21) on cab support (18) with clamp (20) and screw (19).
16. Install hose (21) and oil dipstick tube (13) on oil dipstick tube bracket (9) with clamps (23) and (10), washer (24), screw (25), new lockwasher (12), and nut (11).
17. Install hose (21) on water manifold drainvalve (2) and tighten clamp (1).
18. Open coolant outlet drainvalve (2) on right side of engine (4).
19. Open water manifold drainvalve (3) on right side of engine (4).

3-403. ENGINE COOLANT HEATER HOSE REPLACEMENT (Contd)

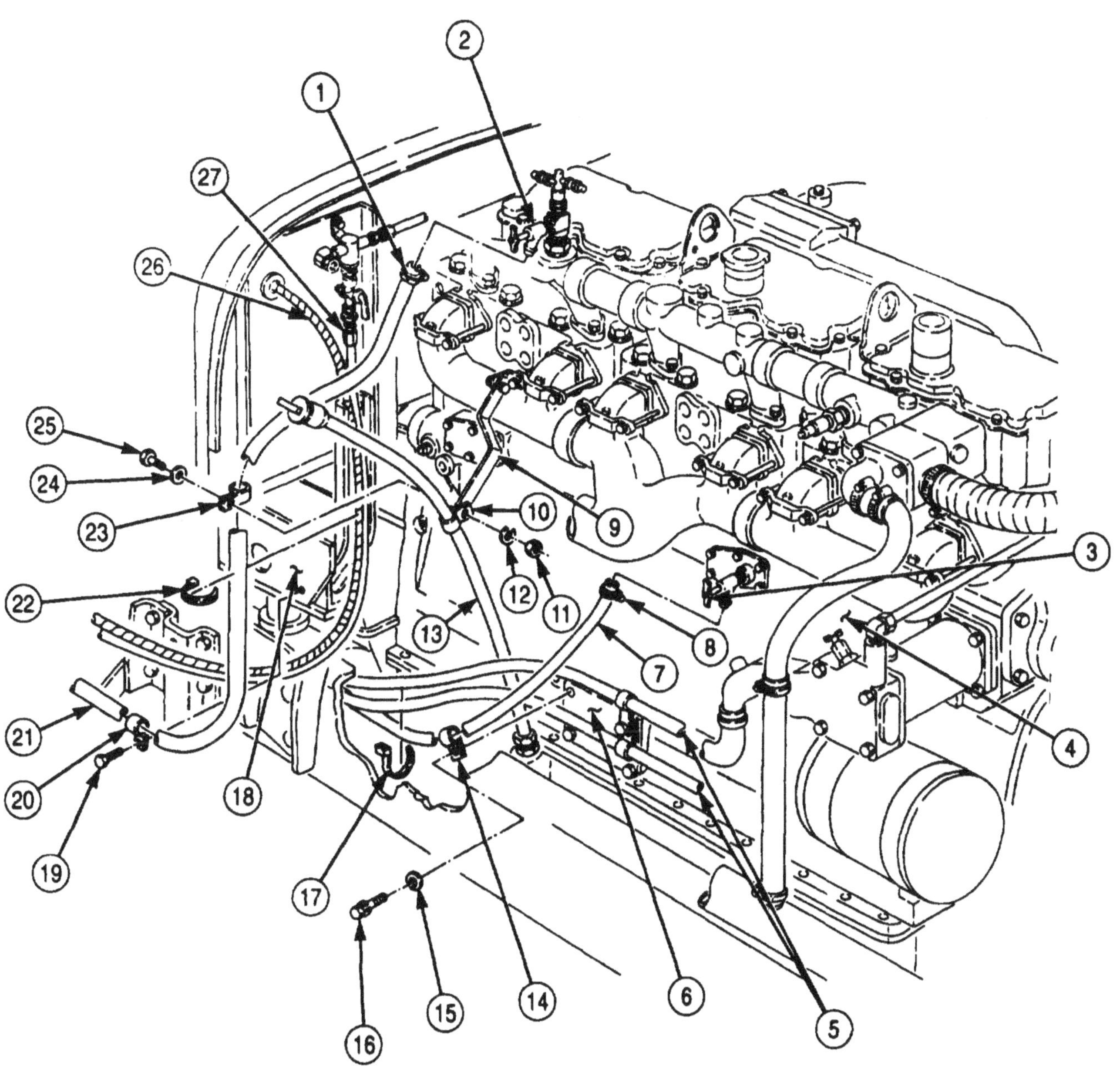

FOLLOW-ON TASKS: • Fill coolant to proper level (para. 3-53).

- Start engine (TM 9-2320-272-10) and check hose connections for leaks.
- Install right splash shield (TM 9-2320-272-10).

3-404. ENGINE COOLANT BATTERY BOX HEATER PAD REPLACEMENT

THIS TASK COVERS:

a Removal   b. Installation

INITIAL SETUP:

APPLICABLE MODELS
All

TOOLS
General mechanic's tool kit (Appendix E, Item 1)

MATERIALS/PARTS
Antiseize tape (Appendix C, Item 72)

REFERENCES (TM)
TM 9-2320-272-10
TM 9-2320-272-24P

EQUIPMENT CONDITION
- Parking brake set (TM 9-2320-272-10).
- Batteries removed (para. 3-125).

a. Removal

1. Loosen two hose clamps (3) and remove inlet hose (5) and outlet hose (4) from engine coolant heater pad (1) and adapter (6).
2. Remove adapter (6) from elbow (7).

NOTE

Mark position of elbow for installation.

3. Remove elbow (7) and nipple (12) from heater pad (1) at floor panel (8).
4. Remove heater pad (1), two blocks (10), four blocks (9), and two blocks (11) from battery box (2).

b. Installation

NOTE

Wrap all male pipe threads with antiseize tape before installation.

1. Position two blocks (11), four blocks (9), two blocks (10), and heater pad (1) in battery box (2).
2. Install nipple (12) and elbow (7) on heater pad (1) at floor panel (8).
3. Install adapter (6) on elbow (7).
4. Install inlet hose (5) and outlet hose (4) on engine coolant heater pad (1) and adapter (6) and tighten two hose clamps (3).

3-404. ENGINE COOLANT BATTERY BOX HEATER PAD REPLACEMENT (Contd)

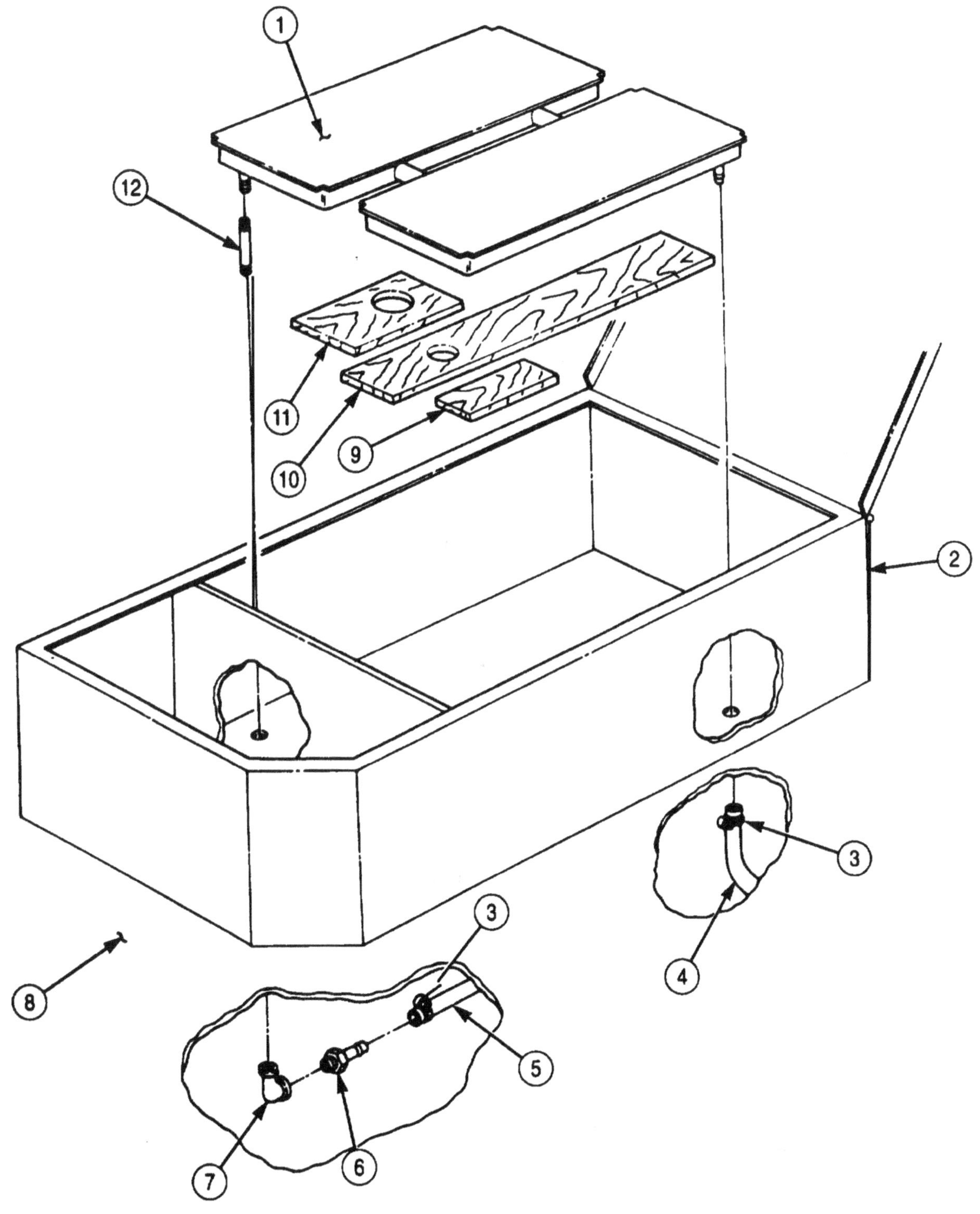

FOLLOW-ON TASK: Install batteries (para. 3-125).

3-405. SWINGFIRE HEATER PUMP REPLACEMENT

THIS TASK COVERS:

a Removal b. Installation

INITIAL SETUP:

APPLICABLE MODELS
All

TOOLS
General mechanic's tool kit (Appendix E, Item 1)

MATERIALS/PARTS
Two key washers (Appendix D, Item 267)
Antiseize tape (Appendix C, Item 72)

REFERENCES (TM)
TM 9-2320-272-10
TM 9-2320-272-24P

EQUIPMENT CONDITION
- Parking brake set (TM 9-2320-272-10).
- Hood raised and secured (TM 9-2320-272-10).
- Battery ground cables disconnected (para. 3-126).

a. Removal

1. Remove nut (28) and wire (29) from heater pump (6) and ground terminal (27).
2. Remove nut (26) and wire (25) from heater pump relay (24).

NOTE

Have drainage container ready to catch excess coolant.

3. Remove clamp (17), clamp (21), and hose (20) from connector (9) and pump outlet tube (18).
4. Remove connector (9) and adapter (22) from elbow (23).

NOTE

Mark position of elbow for installation.

5. Remove elbow (23) from heater pump (6).
6. Remove heater pump (6) and nipple (4) from tee (1).
7. Remove clamp (5) and heater pump (6) from bracket (8) and support (10).
8. Remove two nuts (12), washers (11), screws (7), and bracket (8) from support (10).
9. Remove two screws (13), key washers (14), bracket (15), and support (10) from exhaust port (16). Discard key washers (14).
10. Remove reducer bushing (2), thermal close valve (3), and tee (1) from water manifold (19).

3-405. SWINGFIRE HEATER PUMP REPLACEMENT (Contd)

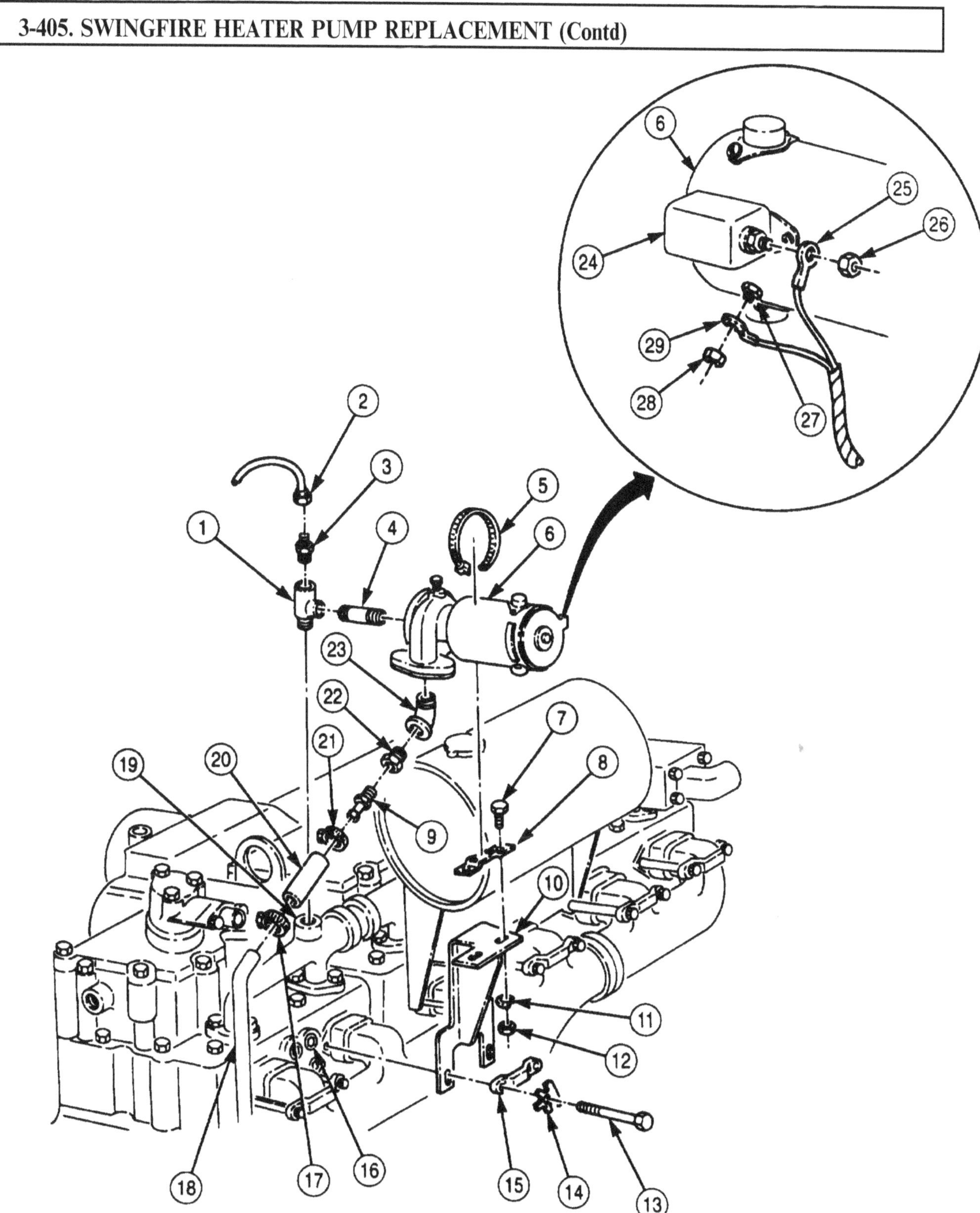

3-405. SWINGFIRE HEATER PUMP REPLACEMENT (Contd)

b. Installation

NOTE

Wrap all male pipe threads with antiseize tape before installation.

1. Install tee (1), thermal close valve (3), and reducer bushing (2) on water manifold (19).
2. Install support (10) on exhaust port (16) with bracket (15), two new key washers (14), and screws (13).
3. Install bracket (8) on support (10) with two screws (7), washers (11), and nuts (12).
4. Install heater pump (6) on bracket (8) and support (10) with clamp (5).
5. Install nipple (4) and heater pump (6) on tee (1).
6. Install elbow (23) on heater pump (6).
7. Install adapter (22) and connector (9) on elbow (23).
8. Install hose (20) on connector (9) and pump outlet tube (18) with clamp (17) and clamp (21).
9. Install wire (25) on heater pump relay (24) with nut (26).
10. Install wire (29) on heater pump (6) and ground terminal (27) with nut (28).

3-405. SWINGFIRE HEATER PUMP REPLACEMENT (Contd)

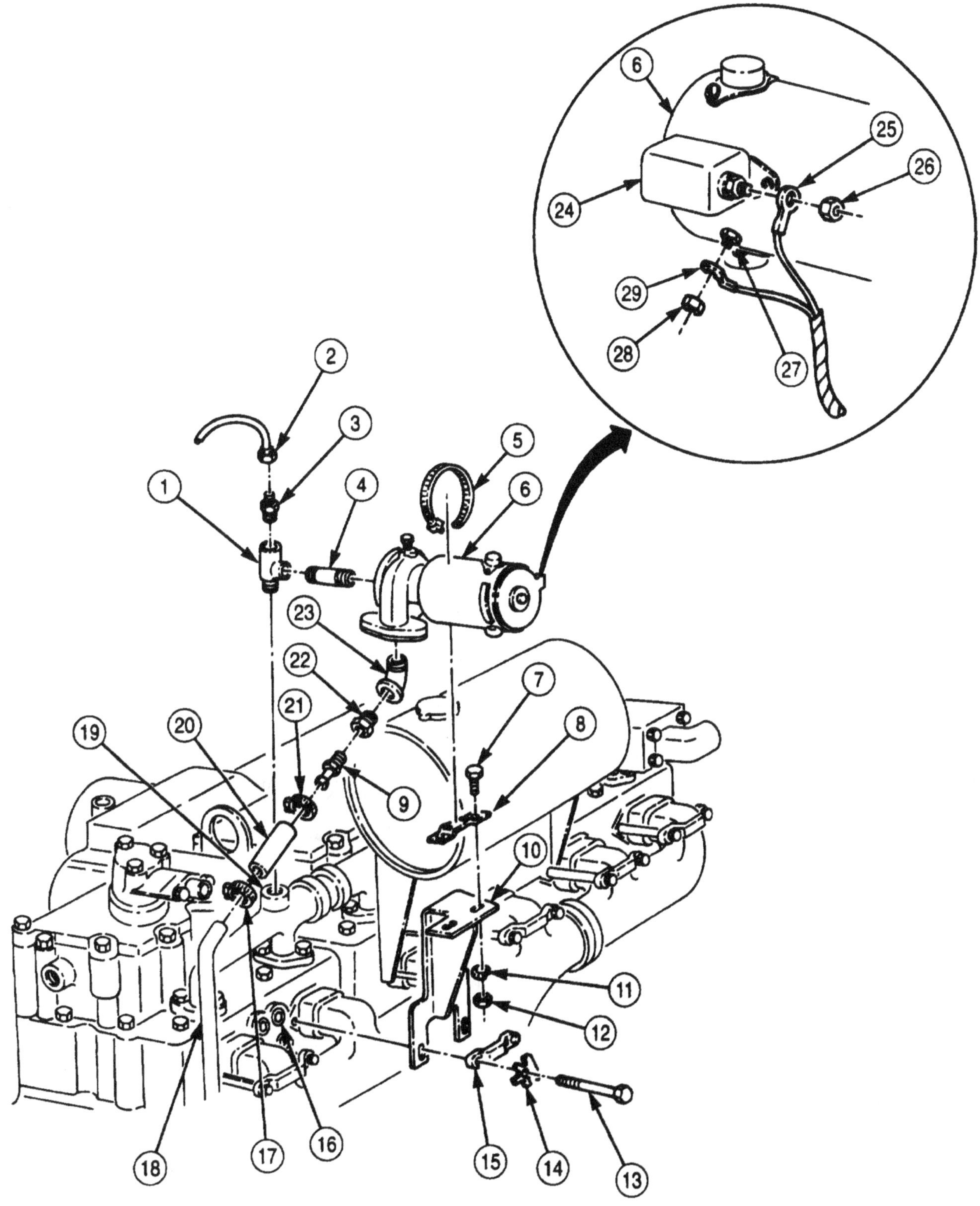

FOLLOW-ON TASKS: • Fill cooling system to proper level (para. 3-53).
• Connect battery ground cables (para. 3-126).

3-406. SWINGFIRE HEATER AND MOUNTING BRACKET REPLACEMENT

THIS TASK COVERS:

a. Removal | b. Installation

INITIAL SETUP:

APPLICABLE MODELS
ALL

TOOLS
General mechanic's tool kit (Appendix E, Item 1)

MATERIALS/PARTS
Seventeen lockwashers (Appendix D, Item 374)
Two locknuts (Appendix D, Item 291)
Six locknuts (Appendix D, Item 313)

REFERENCES (TM)
TM 9-2320-272-10
TM 9-2320-272-24P

EQUIPMENT CONDITION
Parking brake set (TM 9-2320-272-10).

a. Removal

1. Turn three turnbuttons (22) and eight turnbuttons (4) to release cover (23) from right front fender (18) and bracket (7). Position cover (23) to gain access to hardware.
2. Open clamp (16) on bracket (20).
3. Remove clamp (17) from bracket (7).
4. Remove swingfire heater (21) from bracket (7).
5. Remove six nuts (13), lockwashers (12), and screws (1) from right front fender (18) and bracket (7). Discard lockwashers (12).
6. Remove three nuts (3), lockwashers (2), screws (5), and cover (23) from right front fender (18) and bracket (7). Discard lockwashers (2).
7. Remove four locknuts (15), washers (14), screws (19), and bracket (20) from right front fender (18). Discard locknuts (15).
8. Remove two locknuts (11), washers (10), screws (6), and bracket (7) from right front fender (18). Discard locknuts (11).
9. Remove three turnbuttons (22) from right front fender (18).
10. Remove six nuts (9), lockwashers (8), and turnbuttons (4) from right front fender (18). Discard lockwashers (8).
11. Remove two nuts (9), lockwashers (8), and turnbuttons (4) from bracket (7). Discard lockwashers (8).

3-406. SWINGFIRE HEATER AND MOUNTING BRACKET REPLACEMENT (Contd)

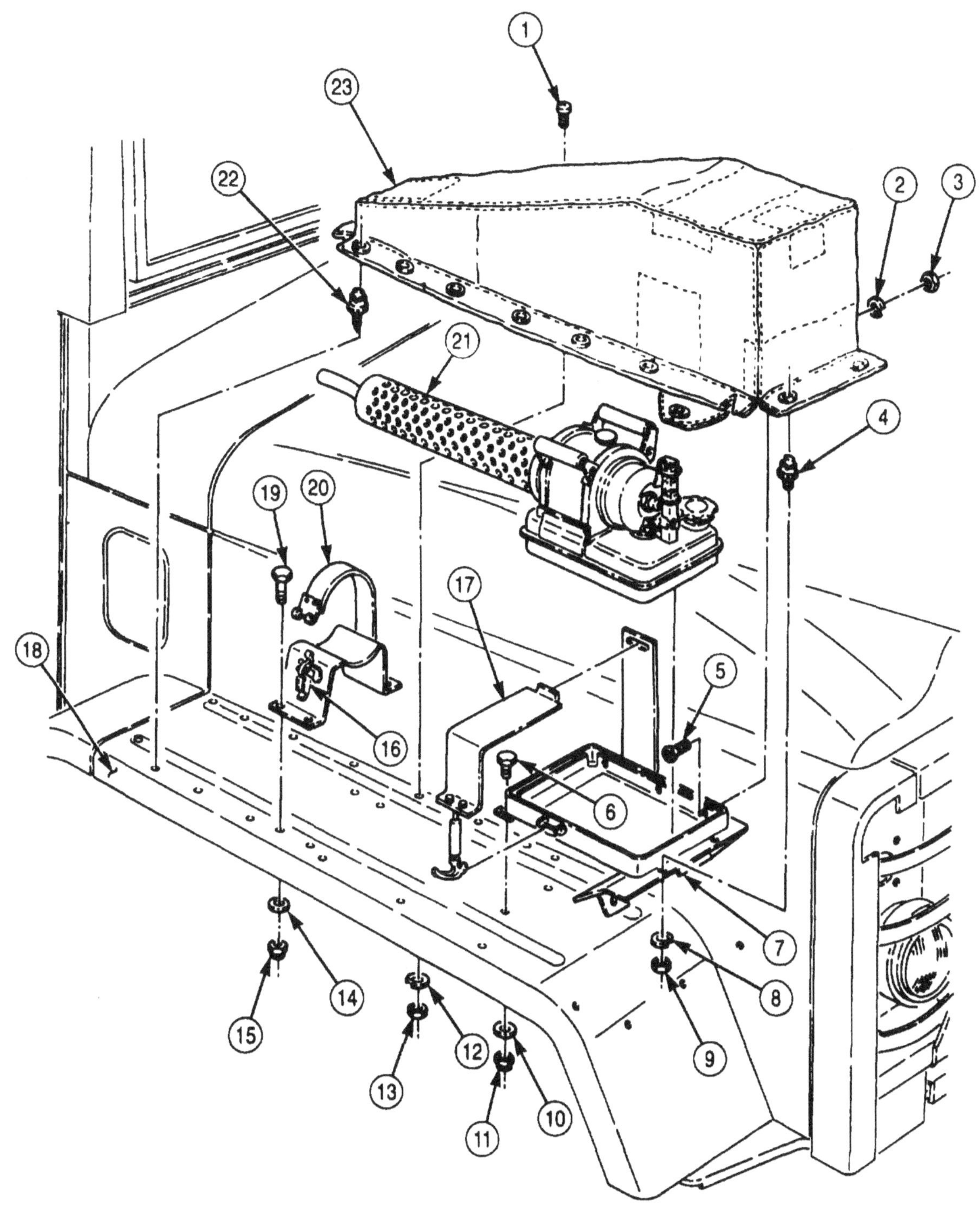

3-406. SWINGFIRE HEATER AND MOUNTING BRACKET REPLACEMENT (Contd)

b. Installation

1. Install two turnbuttons (4) on bracket (7) with two new lockwashers (8) and nuts (9).
2. Install six turnbuttons (4) on right front fender (18) with six new lockwashers (8) and nuts (9).
3. Install three turnbuttons (22) on right front fender (18).
4. Install bracket (7) on right front fender (18) with two screws (6). washers (10), and new locknuts (11).
5. Install bracket (20) on right front fender (18) with four screws (19), washers (14). and new locknuts (15).
6. Install cover (23) on right front fender (18) and bracket (7) with three screws (5), new lockwashers (2), and nuts (3).
7. Secure cover (23) to right front fender (18) and bracket (7) with six screws (1), new lockwashers (12), and nuts (13).
8. Install swingfire heater (21) on bracket (7).
9. Install clamp (17) on bracket (7).
10. Close clamp (16) on bracket (20).
11. Position cover (23) over swingfire heater (21) and install on right front fender (18) with three turnbuttons (22) and eight turnbuttons (4).

3-406. SWINGFIRE HEATER AND MOUNTING BRACKET REPLACEMENT (Contd)

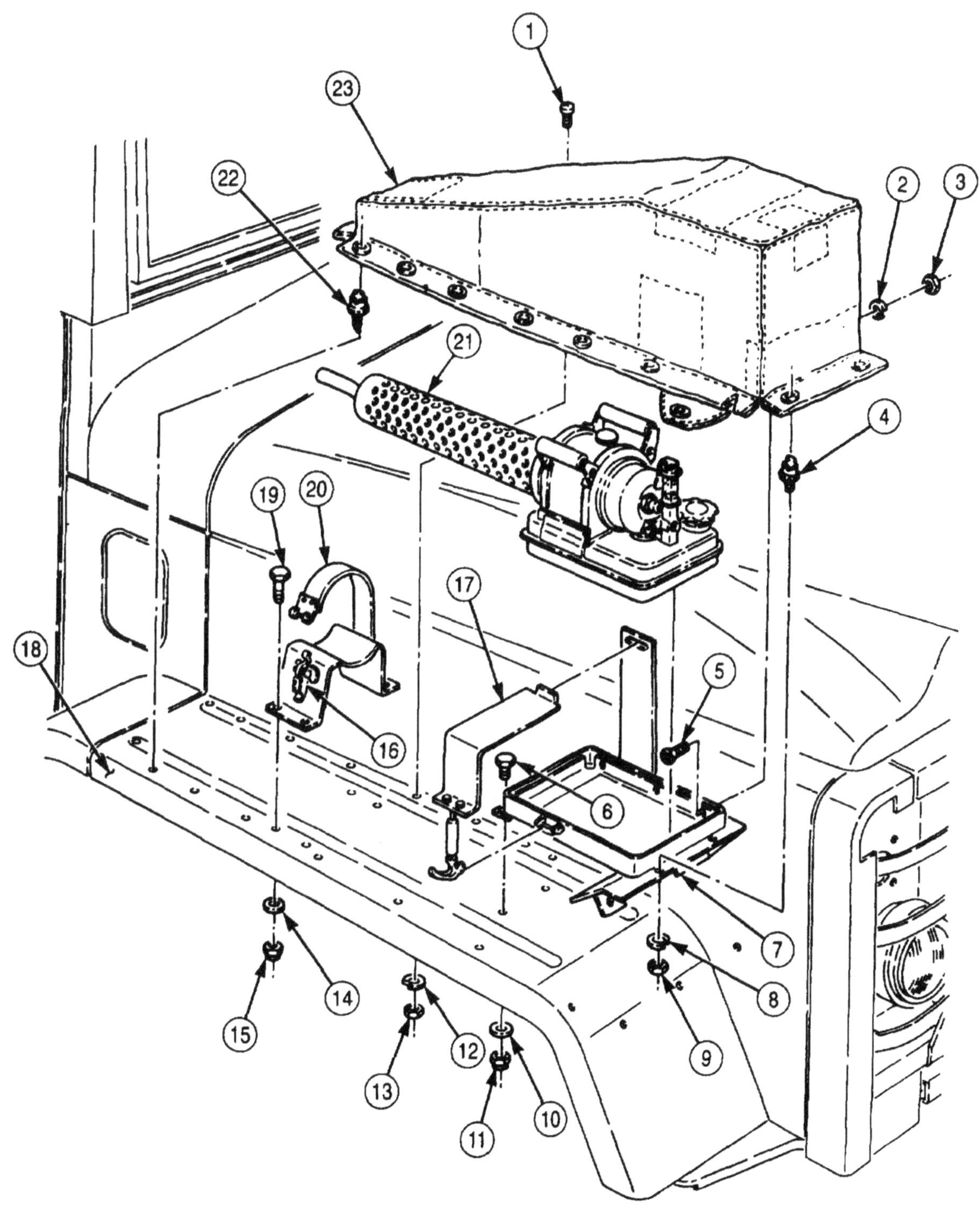

3-407. SWINGFIRE HEATER ELECTRICAL COMPONENTS REPLACEMENT

THIS TASK COVERS:

a. Electrical Connector Removal
b. Relay Removal
c. Circuit Breaker Removal
d. Thermal Switch Removal
e. Thermal Switch Installation
f. Circuit Breaker Installation
g. Relay Installation
h. Electrical Connector Installation

INITIAL SETUP:

APPLICABLE MODELS
All

TOOLS
General mechanic's tool kit (Appendix E, Item 1)

MATERIALS/PARTS
Lockwasher (Appendix D, Item 375)
Locknut (Appendix D, Item 291)
Three locknuts (Appendix D, Item 313)
Two locknuts (Appendix D, Item 276)
Antiseize compound (Appendix C, Item 62)

REFERENCES (TM)
TM 9-2320-272-10
TM 9-2320-272-24P

EQUIPMEMT CONDITION
- Parking brake set (TM 9-2320-272-10).
- Battery ground cables disconnected (para. 3-126).

a. Electrical Connector Removal

1. Loosen screw (10) and remove wire (9) from electrical connector (11).
2. Remove nut (13), lockwasher (12), and electrical connector (11) from cab panel (14). Discard lockwasher (12).

b. Relay Removal

NOTE

Tag all wires for installation.

1. Remove four screws (8), clips (15), and wires (2) from relay (3) between dash panel (1) and firewall (6).
2. Remove two locknuts (4), screws (7), and relay (3) from bracket (5). Discard locknuts (4).

3-407. SWINGFIRE HEATER ELECTRICAL COMPONENTS REPLACEMENT (Contd)

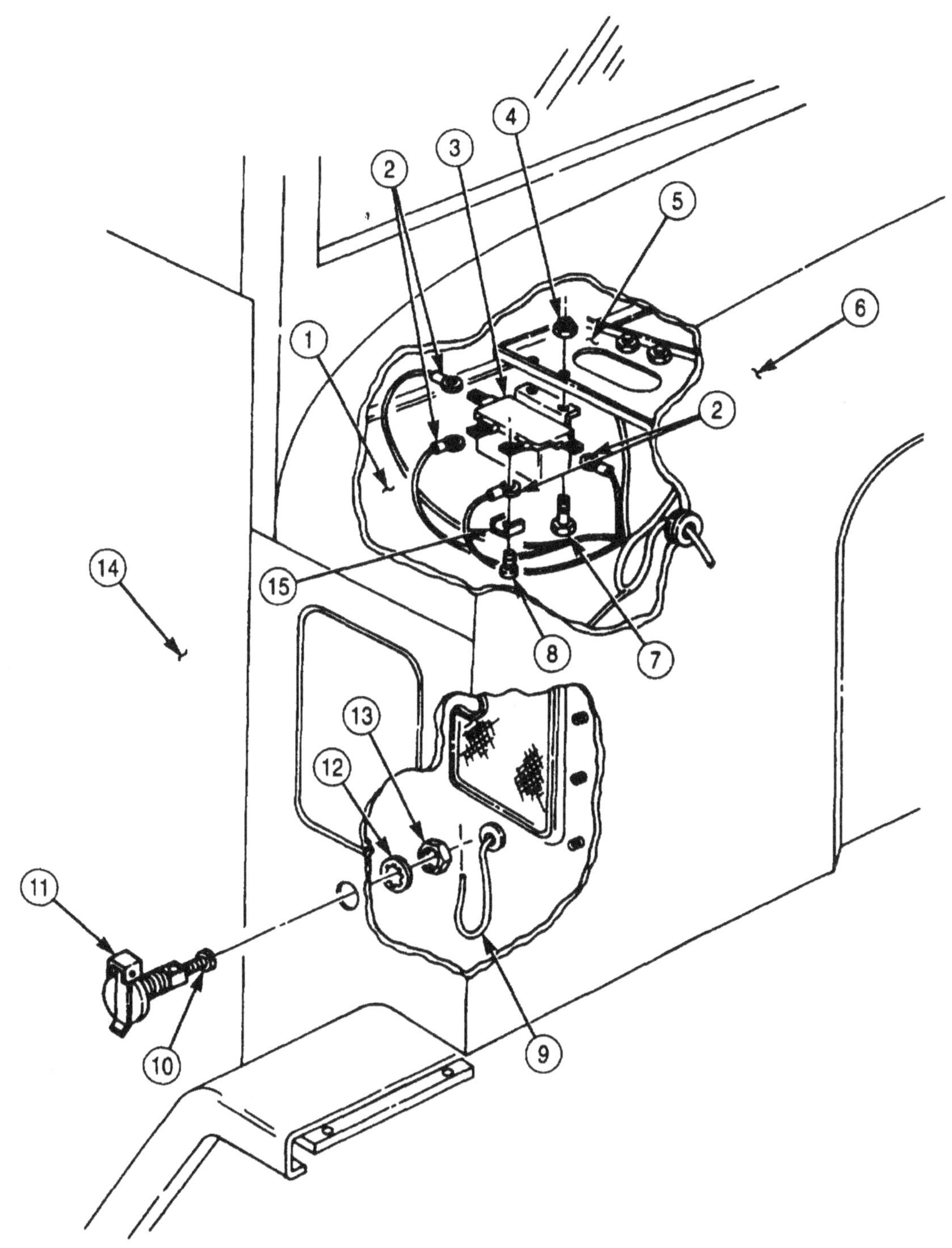

3-407. SWINGFIRE HEATER ELECTRICAL COMPONENTS REPLACEMENT (Contd)

c. Circuit Breaker Removal

1. Disconnect two wires (6) from circuit breaker (4) between dash panel (2) and firewall (1).
2. Remove two screws (5) and circuit breaker (4) from bracket (3).

d. Thermal Switch Removal

1. Remove locknut (9), clamp (8), wire (7), and screw (12) from thermal switch shield (15). Discard locknut (9).
2. Remove two locknuts (18) and screws (13) from thermal switch shield (15) and bracket (17). Discard locknuts (18).
3. Remove locknut (16), screw (14), and thermal switch shield (15) from bracket (11). Discard locknut (16).
4. Remove two screws (21) and wires (20) from thermal switch (19).
5. Remove thermal switch (19) from tube (10).

e. Thermal Switch Installation

1. Install thermal switch (19) on tube (10).

NOTE

Apply liberal coating of antiseize compound to both terminals.

2. Install two wires (20) on thermal switch (19) with two screws (21).
3. Install shield (15) on bracket (11) with screw (14) and new locknut (16).
4. Install two screws (13) and new locknuts (18) on shield (15) and bracket (17).
5. Install clamp (8) and wire (7) on thermal switch shield (15) with screw (12) and new locknut (9).

f. Circuit Breaker Installation

1. Install circuit breaker (4) on bracket (3) with two screws (5).
2. Connect two wires (6) to circuit breaker (4) between dash panel (2) and firewall (1).

3-407. SWINGFIRE HEATER ELECTRICAL COMPONENTS REPLACEMENT (Contd)

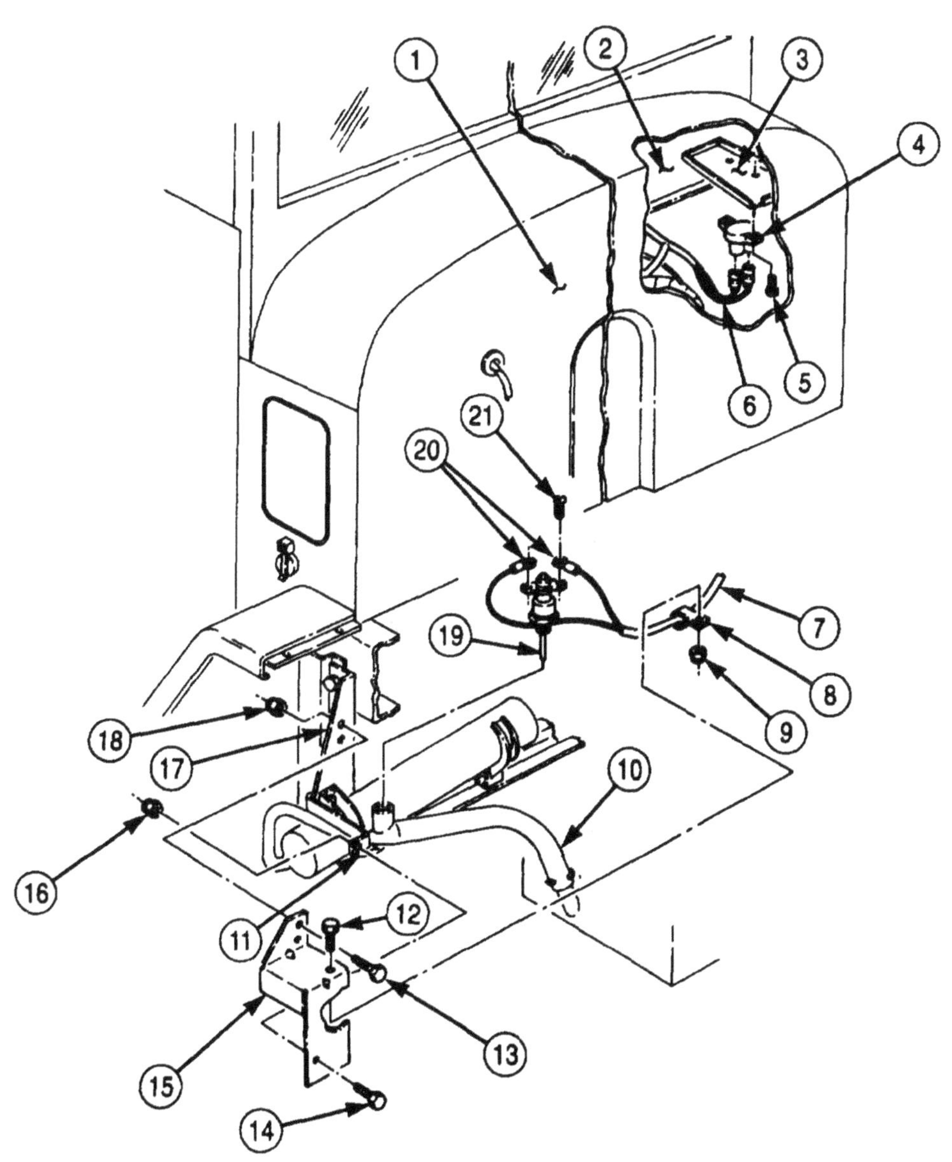

3-407. SWINGFIRE HEATER ELECTRICAL COMPONENTS REPLACEMENT (Contd)

g. Relay Installation

1. Install relay (3) on bracket (5) with two screws (7) and new locknuts (4).
2. Install four wires (2) on relay (3) between dash panel (1) and firewall (6) with four clips (15) and screws (8).

h. Electrical Connector Installation

1. Install electrical connector (11) on cab panel (14) with new lockwasher (12) and nut (13).
2. Install wire (9) on electrical connector (11) and tighten screw (10).

3-407. SWINGFIRE HEATER ELECTRICAL COMPONENTS REPLACEMENT (Contd)

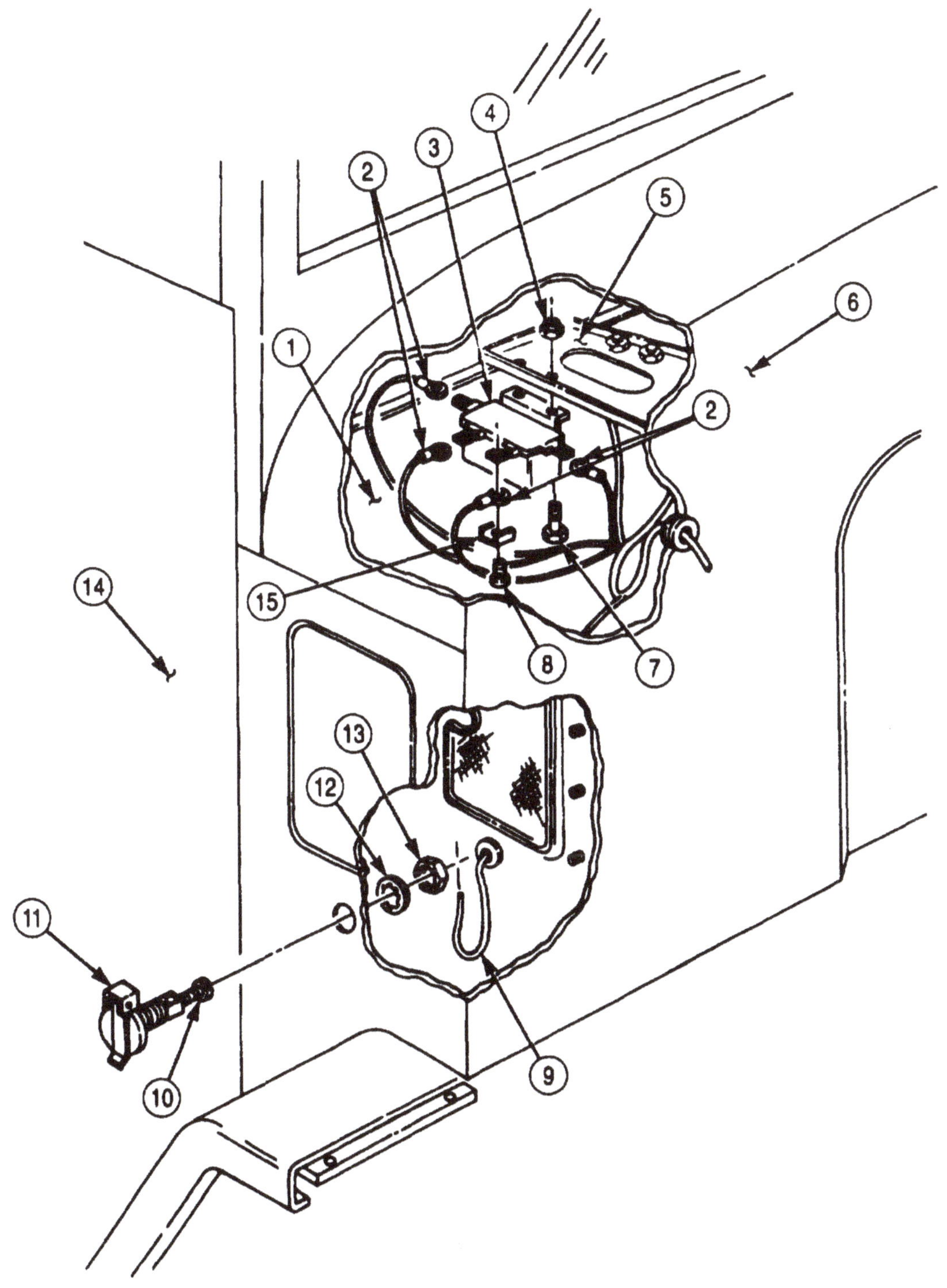

FOLLOW-ON TASK: Connect battery ground cables (para. 3-126).

3-408. SWINGFIRE HEATER HARNESS REPLACEMENT

THIS TASK COVERS:

a. Removal b. Installation

INITIAL SETUP:

APPLICABLE MODELS
All

TOOLS
General mechanic's tool kit (Appendix E, Item 1)

MATERIALS/PARTS
Three lockwashers (Appendix D, Item 391)
Three locknuts (Appendix D. Item 313)
Locknut (Appendix D, Item 291)
Five tiedown straps (Appendix D, Item 686)
Six tiedown straps (Appendix D, Item 684)
Eleven tiedown straps (Appendix D, Item 690)
Sealing compound (Appendix C, Item 62)

REFERENCES (TM)
TM 9-2320-272-10
TM 9-2320-272-24P

EQUIPMENT CONDITION
- Parking brake set (TM 9-2320-272-10).
- Battery ground cables disconnected (para. 3-126).

a. Removal

NOTE

If vehicle is equipped with a personnel heater, perform step 1.

1. Disconnect personnel heater wiring harness wires (27) and (28) behind cab dash panel (26).
2. Disconnect two wires (11) and (13) from circuit breaker (12).
3. Disconnect wire (7) and front wiring harness lead (10) behind cab dash panel (26).
4. Disconnect wire (9) from battery switch (8).
5. Remove six tiedown straps (15) from front wiring harness (14). Discard tiedown straps (15).
6. Remove nut (5), lockwasher (4), screw (16), and wire (3) from diagnostic connector (6). Discard lockwasher (4).
7. Remove four screws (24), clips (25), and wires (1) from relay (2).
8. Remove screw (17), clamp (18), and wire (19) from cab (23).
9. Loosen screw (20) and remove wire (19) from heater electrical connector (21) and grommet (22).

3-408. SWINGFIRE HEATER HARNESS REPLACEMENT (Contd)

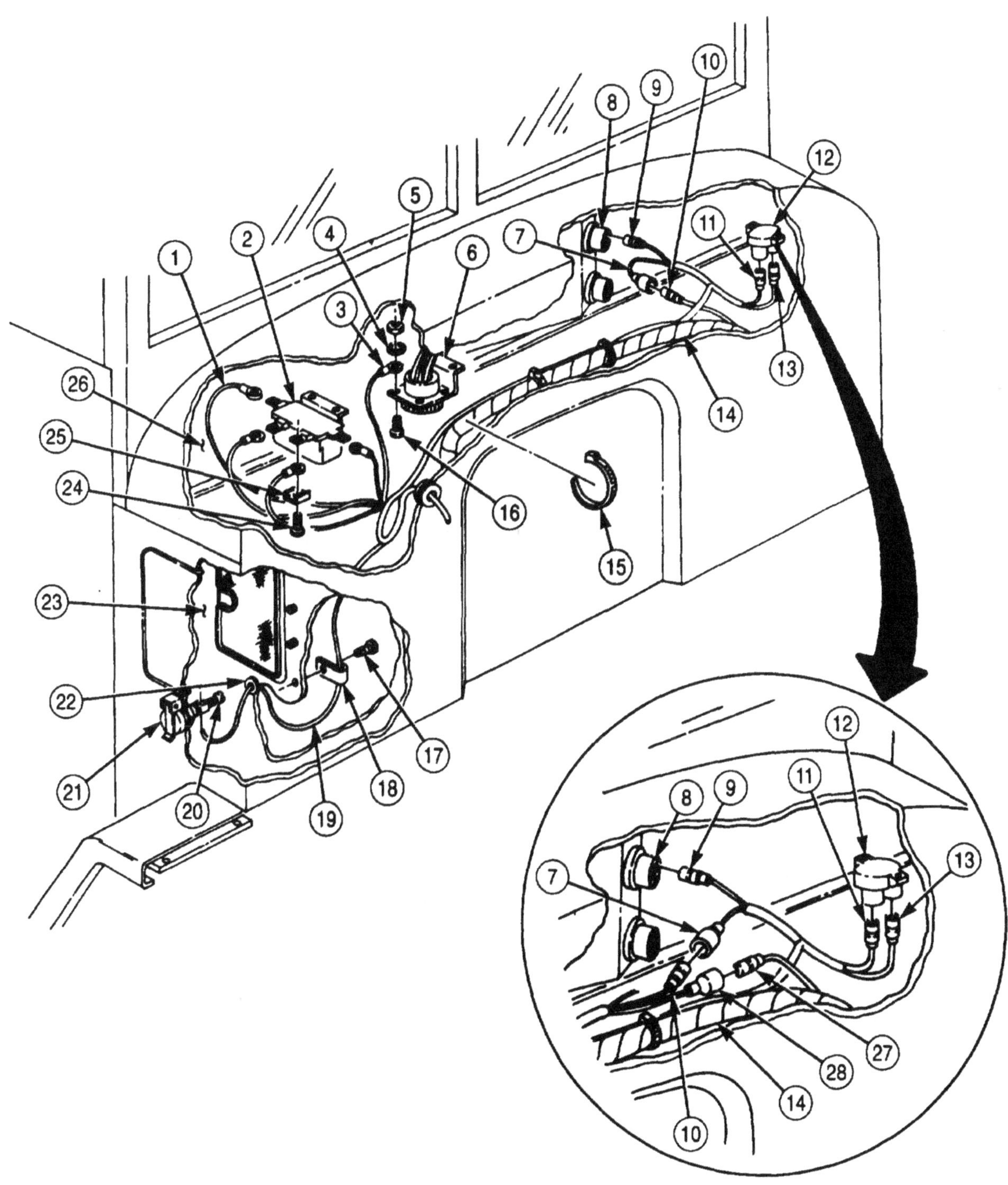

3-408. SWINGFIRE HEATER HARNESS REPLACEMENT (Contd)

10. Remove five tiedown straps (1) from heater harness (10). Discard tiedown straps (1).
11. Remove nut (5) and wire (4) from heater fuel pump solenoid (7).
12. Remove nut (8) and wire (9) from heater fuel pump (6).
13. Remove two screws (24), lockwashers (23), and wires (22) and (13) from thermal switch (20). Discard lockwashers (23).
14. Remove two locknuts (21) and screws (15) from bracket (25) and shield (16). Discard locknuts (21).
15. Remove locknut (19), screw (17), and shield (16) from bracket (18). Discard locknut (19).
16. Remove locknut (12), clamp (11), heater harness (10), and screw (14) from shield (16). Discard locknut (12).
17. Remove grommet (3) from firewall (2) and heater harness (10).
18. Remove clamp (11) from heater harness (10) and push heater harness (10) through firewall (2).

b. Installation

1. Push heater harness (10) through firewall (2) and install clamp (11) on heater harness (10).
2. Install grommet (3) on heater harness (10) and firewall (2).
3. Install heater harness (10) and clamp (11) on shield (16) with screw (14) and new locknut (12).
4. Install shield (16) on bracket (18) with screw (17) and new locknut (19).
5. Install two screws (15) and new locknuts (21) on bracket (25) and shield (16).

NOTE

Apply sealing compound to both thermal switch terminals.

6. Install wires (22) and (13) on thermal switch (20) with two new lockwashers (23) and screws (24).
7. Install wire (9) on heater fuel pump (6) with nut (8).
8. Install wire (4) on heater fuel pump solenoid (7) with nut (5).
9. Install five new tiedown straps (1) on heater harness (10).

3-408. SWINGFIRE HEATER HARNESS REPLACEMENT (Contd)

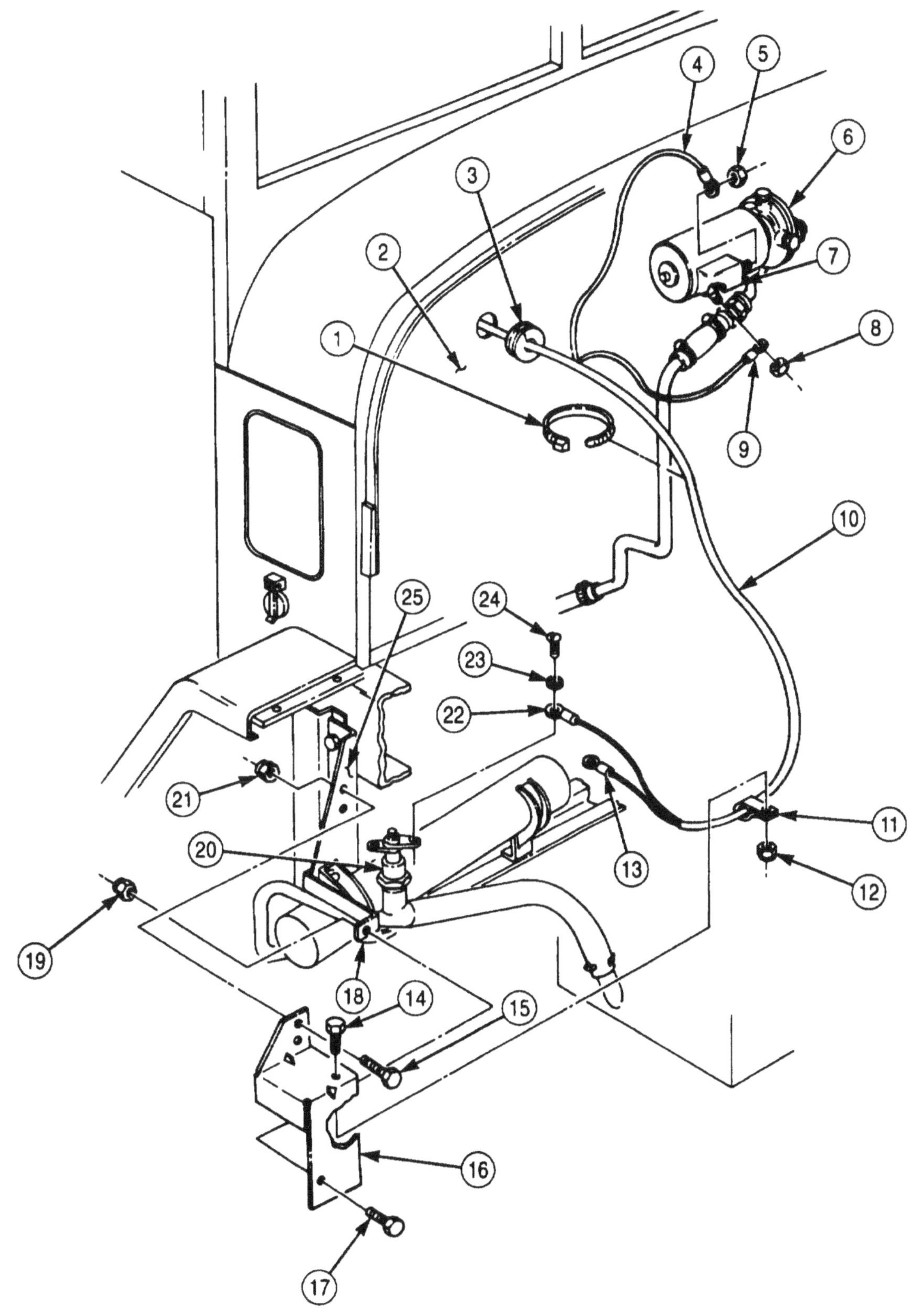

3-408. SWINGFIRE HEATER HARNESS REPLACEMENT (Contd)

10. Push wire (19) through grommet (22) and install on heater electrical connector (21) and tighten screw (20).
11. Install wire (19) on cab (23) with clamp (18) and screw (17).
12. Install four wires (1) on relay (2) with four clips (25) and screws (24).
13. Install wire (3) on diagnostic connector (6) with screw (16), new lockwasher (4), and nut (5).
14. Install six new tiedown straps (15) on front wiring harness (14).
15. Connect wire (9) to battery switch (8).
16. Connect wire (7) to front wiring harness lead (10) behind cab dash panel (26).
17. Connect wires (11) and (13) to circuit breaker (12).

NOTE

If vehicle is equipped with a personnel heater, perform step 18.

18. Connect personnel heater wiring harness wire (27) to wire (28) behind cab dash panel (26).

3-408. SWINGFIRE HEATER HARNESS REPLACEMENT (Contd)

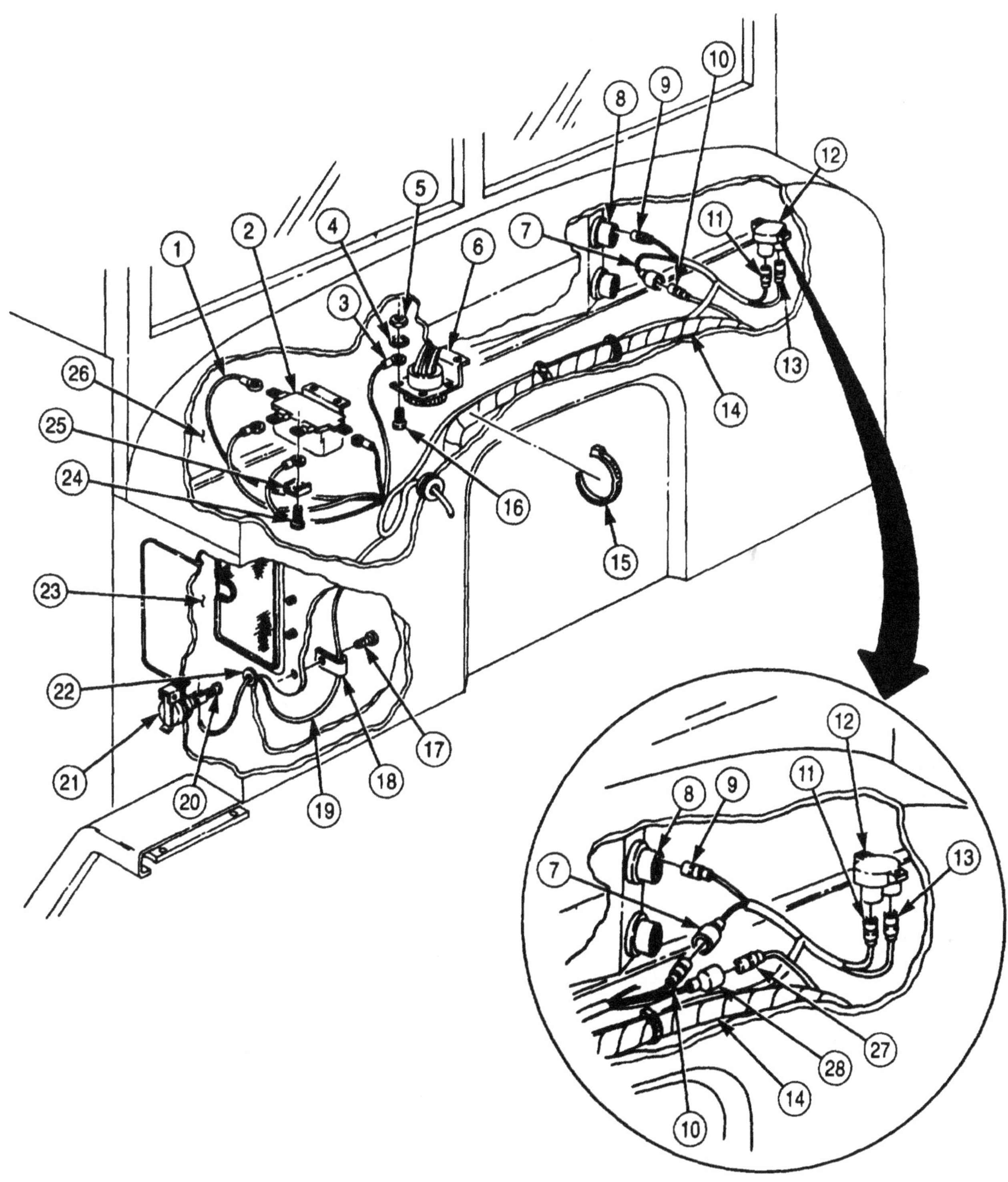

FOLLOW-ON TASK: Connect battery ground cables (para. 3-126).

3-409. SWINGFIRE HEATER OIL PAN SHROUD AND EXHAUST TUBE REPLACEMENT

THIS TASK COVERS:

a. Removal
b. Installation

INITIAL SETUP:

APPLICABLE MODELS
All

TOOLS
General mechanic's tool kit (Appendix E, Item 1)

MATERIALS/PARTS
Cotter pin (Appendix D, Item 49)
Four lockwashers (Appendix D, Item 400)

REFERENCES (TM)
TM 9-2320-272-10
TM 9-2320-272-24P

EQUIPMENT CONDITION
- Parking brake set (TM 9-2320-272-10).
- Thermal switch removed (para. 3-407).

a. Removal

1. Remove cotter pin (10) from oil pan shroud (6) and exhaust tube (1). Discard cotter pin (10).
2. Remove two nuts (12), screws (11), and clamp halves (2) from exhaust tube (1) and water jacket outlet (3).
3. Remove exhaust tube (1) from water jacket outlet (3) and oil pan shroud (6).
4. Remove four nuts (7), lockwashers (9), oil pan shroud (6), and four washers (4) from oil pan (5). Discard lockwashers (9).
5. Remove four studs (8) from oil pan (5).

b. Installation

1. Install four studs (8) in oil pan (5).
2. Install four washers (4) and oil pan shroud (6) on oil pan (5) with four new lockwashers (9) and nuts (7).
3. Install exhaust tube (1) on water jacket outlet (3) and oil pan shroud (6).
4. Install two clamp halves (2) on exhaust tube (1) and water jacket outlet (3) with two screws (11) and nuts (12).
5. Install new cotter pin (10) on oil pan shroud (6) and exhaust tube (1).

3-409. SWINGFIRE HEATER OIL PAN SHROUD AND EXHAUST TUBE REPLACEMENT (Contd)

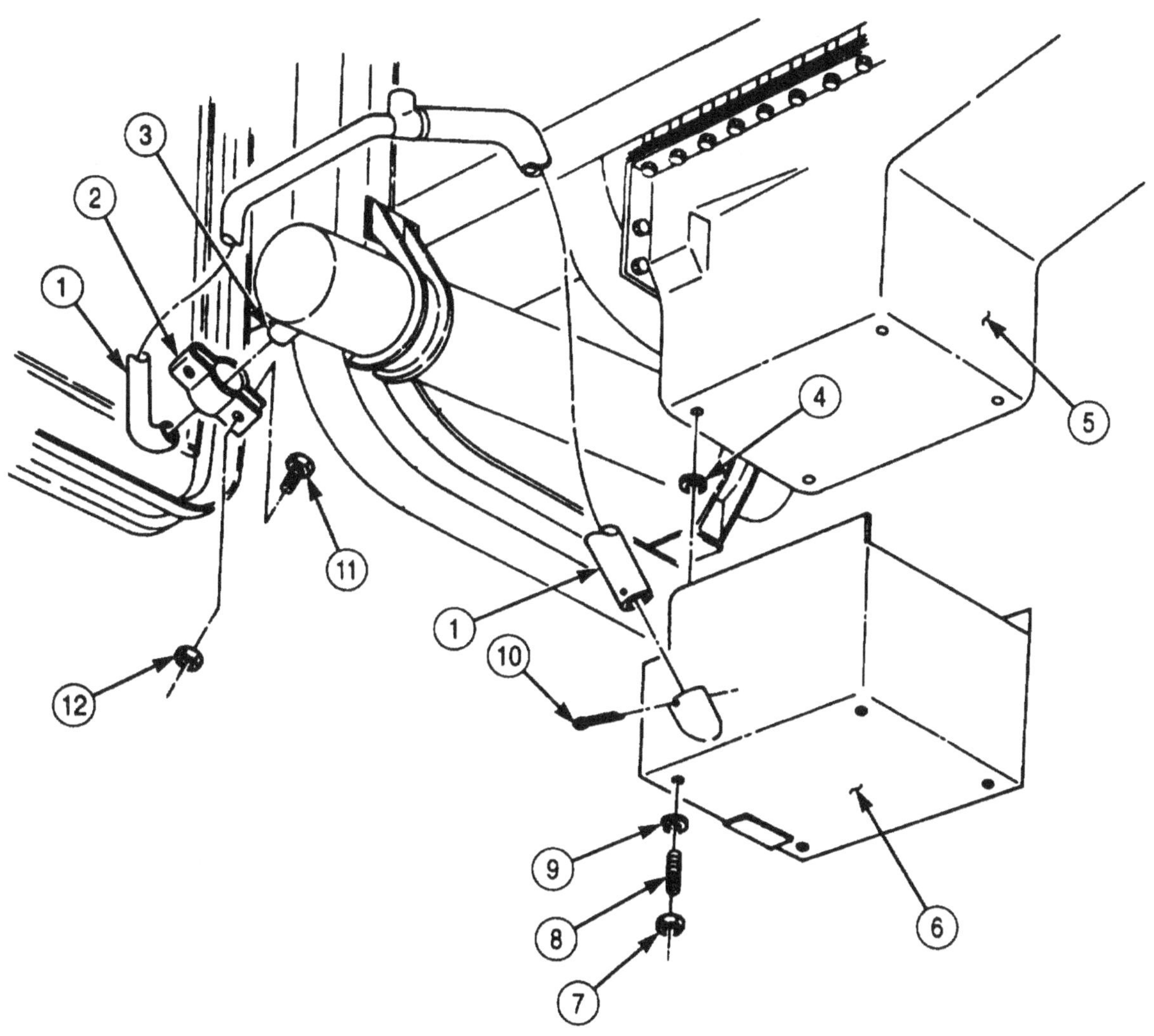

FOLLOW-ON TASK: Install thermal switch (para. 3-407).

3-410. SWINGFIRE HEATER BATTERY BOX HEATER PAD REPLACEMENT

THIS TASK COVERS:

a. Removal
b. Installation

INITIAL SETUP:

APPLICABLE MODELS
All

TOOLS
General mechanic's tool kit (Appendix E, Item 1)

MATERIALS/PARTS
Antiseize tape (Appendix C, Item 72)

REFERENCES (TM)
TM 9-2320-272-10
TM 9-2320-272-24P

EQUIPMENT CONDITION
• Parking brake set (TM 9-2320-272-10).
• Batteries removed (para. 3-125).

---

a. Removal

1. Loosen two hose clamps (2) and remove inlet hose (4) and outlet hose (3) from swingtire heater pad (1) and adapter (5).
2. Remove adapter (5) from elbow (6).

NOTE

Mark position of elbow for installation.

3. Remove elbow (6) and nipple (11) from swingfire heater pad (1).
4. Remove swingfire heater pad (1), two blocks (10), four blocks (8), and two blocks (9) from battery box (7).

b. Installation

NOTE

Wrap all male pipe threads with antiseize tape before installation.

1. Position two blocks (9), four blocks (8), two blocks (10), and swingfire heater pad (1) in battery box (7).
2. Install nipple (11) and elbow (6) on swingfire heater pad (1).
3. Install adapter (5) on elbow (6).
4. Install inlet hose (4) and outlet hose (3) on swingfire heater pad (1) and adapter (5) and tighten two hose clamps (2).

3-410. SWINGFIRE HEATER BATTERY BOX HEATER PAD REPLACEMENT (Contd)

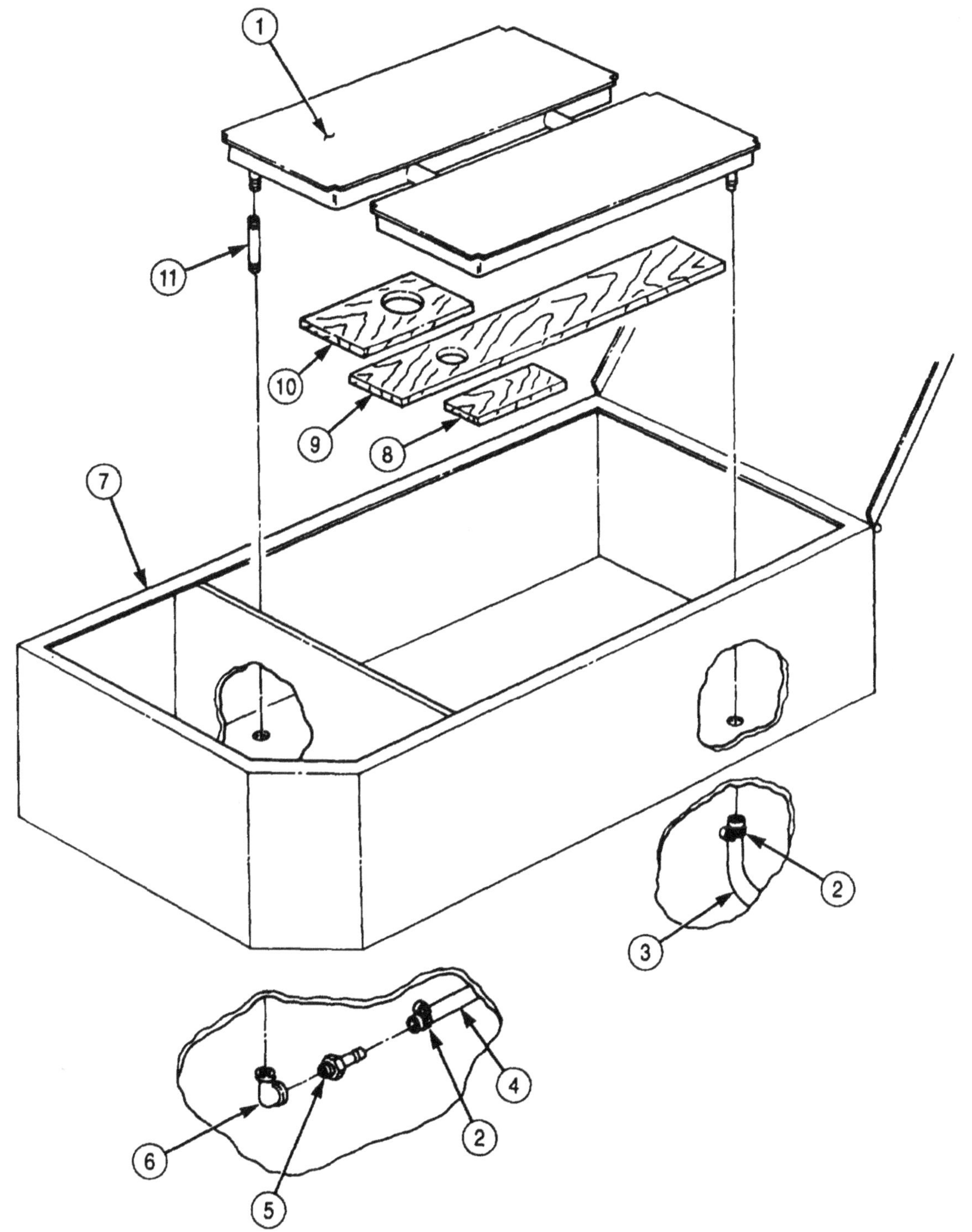

FOLLOW-ON TASK: Install batteries (para. 3-125).

3-411. SWINGFIRE HEATER WATER JACKET REPLACEMENT

THIS TASK COVERS:

a. Removal b. Installation

INITIAL SETUP:

APPLICABLE MODELS

All

TOOLS

General mechanic's tool kit (Appendix E, Item 1)

MATERIALS/PARTS

Four locknuts (Appendix C, Item 291)

REFERENCES (TM)

TM 9-2320-272-10
TM 9-2320-272-24P

EQUIPMENT CONDITION

- Parking brake set (TM 9-2320-272-10).
- Swingfire heater oil pan shroud and exhaust tube removed (para. 3-409).

a. Removal

NOTE

Have drainage container ready to catch excess coolant.

1. Loosen clamps (5) and (13) and remove hoses (4) and (14) from swingfire heater water jacket (9).
2. Remove locknut (1), shield support (15), and screw (11) from bracket (2) and water jacket (9). Discard locknut (1).
3. Remove locknut (6), screw (10), and water jacket (9) from bracket (8). Discard locknut (6).
4. Remove two locknuts (12), screws (7), and bracket (8) from crossmember (3). Discard locknuts (12).

b. Installation

1. Install bracket (8) on crossmember (3) with two screws (7) and new locknuts (12).
2. Install water jacket (9) on bracket (8) with screw (10) and new locknut (6).
3. Install shield support (15) on bracket (2) and water jacket (9) with screw (11) and new locknut (1).
4. Install hoses (4) and (14) on water jacket (9) with clamps (5) and (13).

3-411. SWINGFIRE HEATER WATER JACKET REPLACEMENT (Contd)

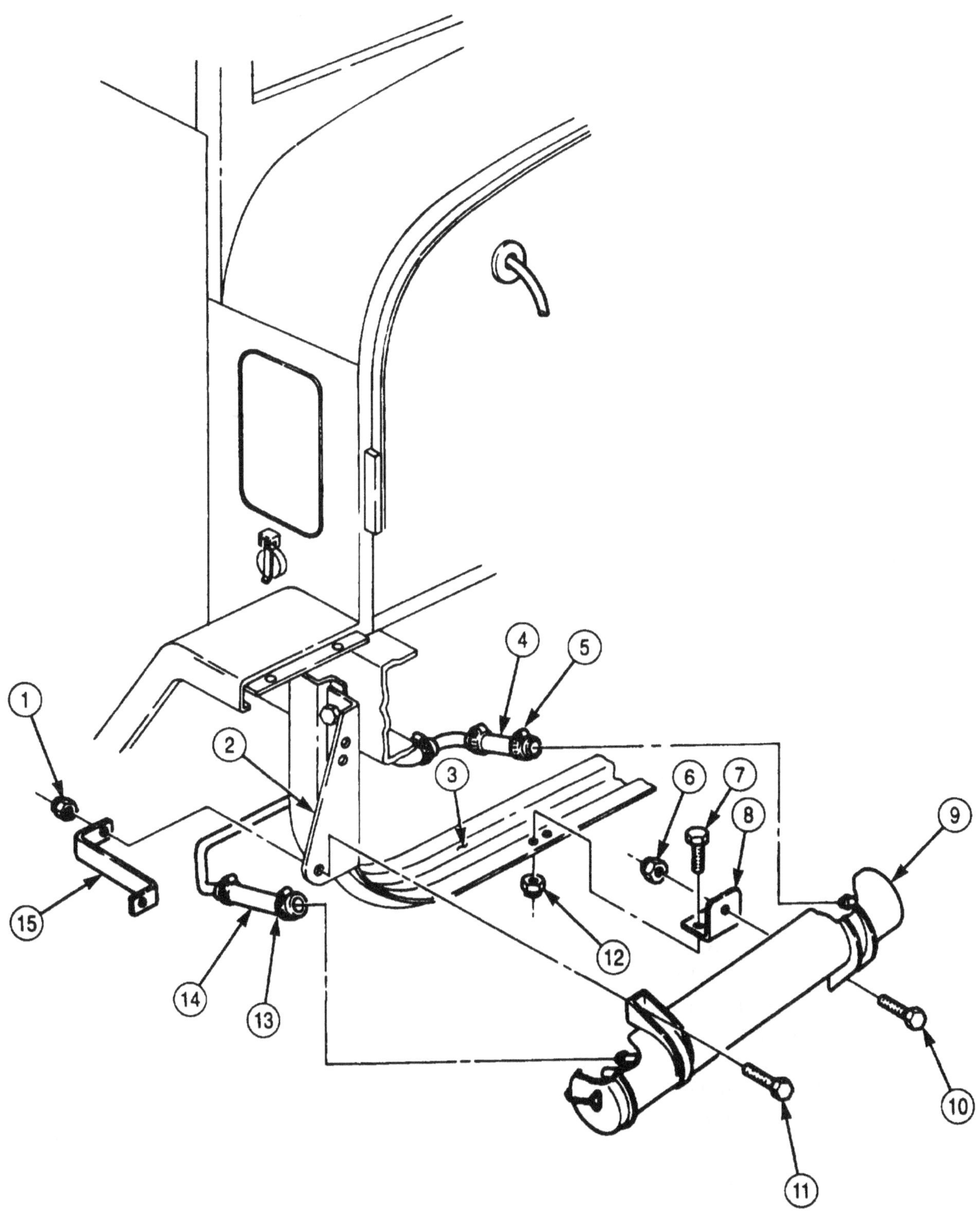

FOLLOW-ON TASK: Install swingfire heater oil pan shroud and exhaust tube (para. 3-409).

3-412. AIR DRYER KIT (M923/A1/A2, M925/A1/A2, M927/A1/A2, M928/A1/A2, M934/A1/A2) REPLACEMENT

THIS TASK COVERS:

a. Removal   b. Installation

INITIAL SETUP:

APPLICABLE MODELS
M923/A1/A2 M925/A1/A2, M927/A1/A2, M928/A1/A2, M934/A1/A2

TOOLS
General mechanic's tool kit (Appendix E, Item 1)

MATERIALS/PARTS
Nine locknuts (M923/A1/A2, M925/A1/A2) (Appendix D, Item 327)
Lockwasher (Appendix D, Item 413)
Nine locknuts (M927/A1/A2, M928/A1/A2, M934/A1/A2) (Appendix D, Item 327)
Locknut (M927/A1/A2, M928/A1/A2, M934/A1/A2) (Appendix D, Item 328)
Five tiedown straps (M923/A1/A2, M925/A1/A2) (Appendix D, Item 694)
Seven tiedown straps (M927/A1/A2, M928/A1/A2, M934/A1/A2) (Appendix D, Item 694)
Antiseize tape (Appendix C, Item 72)

REFERENCES (TM)
TM 9-2320-272-10
TM 9-2320-272-24P

EQUIPMENT CONDITION
- Parking brake set (TM 9-2320-272-10).
- Air reservoirs drained (TM 9-2320-272-10).
- Battery ground cables disconnected (para. 3-126).

a. Removal

1. Disconnect wire (16) from thermostat connector (17) on air dryer (6).

NOTE

Perform steps 2, 3, and 4 for M923/A1/A2 and M925/A1/A2 vehicles.

2. Remove tube (20) from elbow (21).
3. Remove elbow (21) from bushing (22).
4. Remove bushing (22) from wet reservoir adapter (23).

NOTE

Perform steps 5 and 6 for M927/A1/A2, M928/A1/A2, and M934/A1/A2 vehicles.

5. Remove tube (20) from adapter (25).
6. Remove adapter (25) from wet reservoir adapter (24).
7. Remove tube (20) from elbow (18).
8. Remove tiedown strap (19) from tube (20) and ground wire (4). Discard tiedown strap (19).
9. Remove locknut (5), ground wire (4), lockwasher (3), and screw (2) from right-side frame rail (1). Discard locknut (5) and lockwasher (3).
10. Remove tube (10) from elbow (13).
11. Remove line (9) from elbow (15).
12. Remove four locknuts (12), washers (11), screws (7), washers (8), and air dryer (6) from mounting bracket (14). Discard locknuts (12).

3-412. AIR DRYER KIT (M923/A1/A2, M925/A1/A2, M927/A1/A2, M928/A1/A2, M934/A1/A2) REPLACEMENT (Contd)

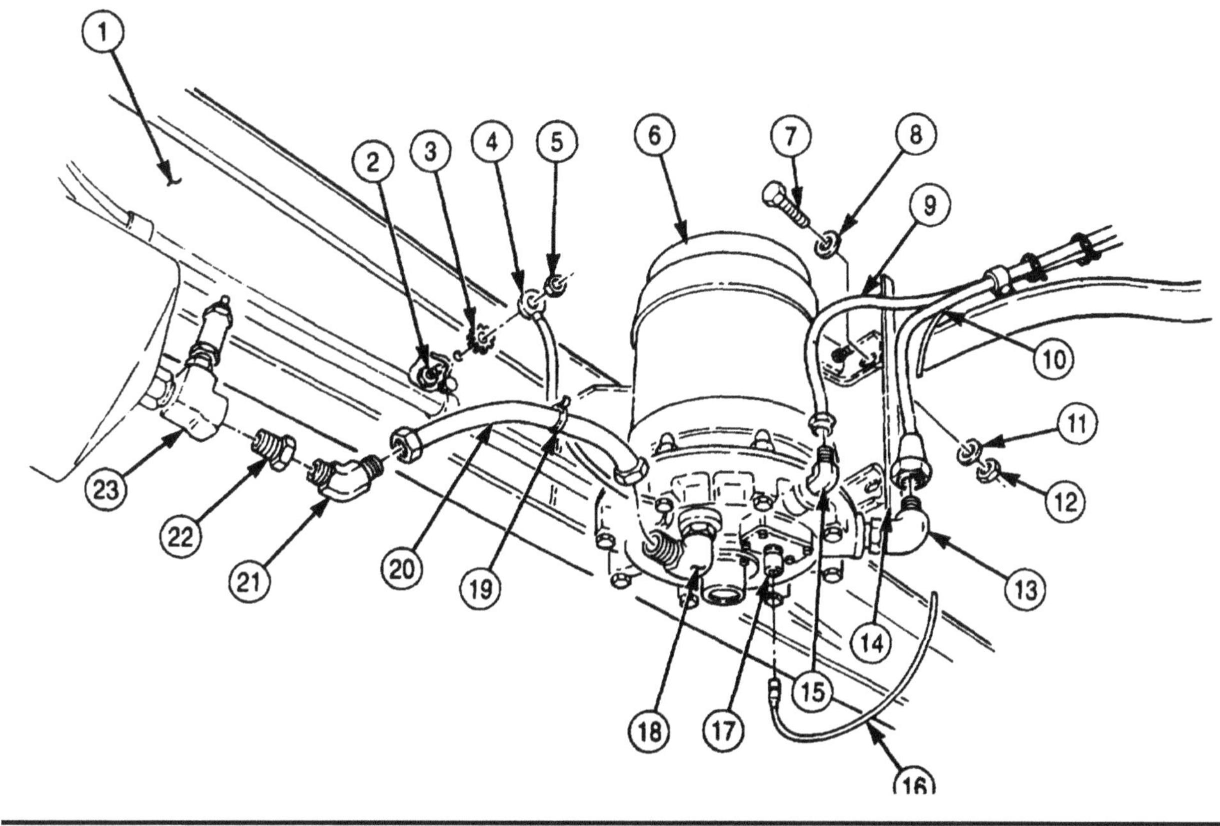

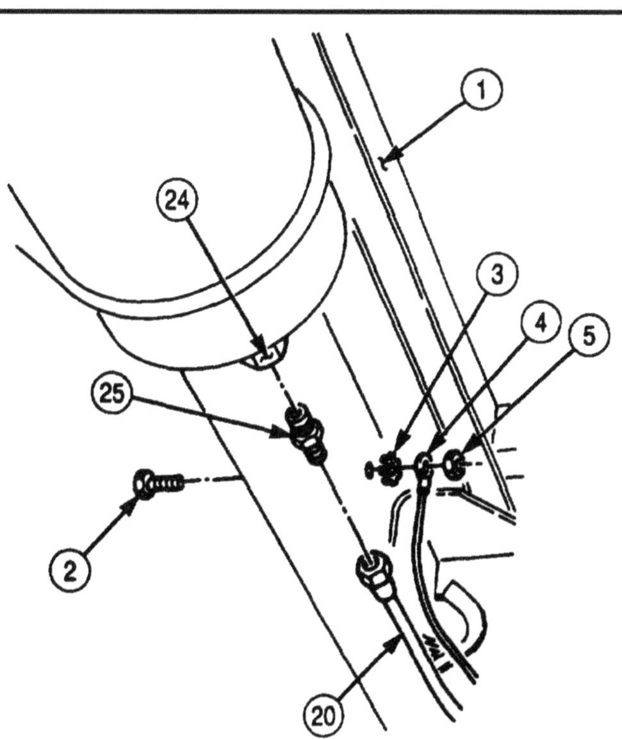

3-412. AIR DRYER KIT (M923/A1/A2, M925/A1/A2, M927/A1/A2, M928/A1/A2, M934/A1/A2) REPLACEMENT (Contd)

13. Remove three locknuts (7), washers (6), clamp (4), three screws (3), mounting bracket (2), and two air lines (5) from frame crossmember (1). Discard locknuts (7).
14. Remove elbow (16) from bushing (15) on air dryer (9).
15. Remove bushing (15) from outlet port (8).
16. Remove elbow (12) from bushing (13).
17. Remove bushing (13) from inlet port (14).
18. Remove elbow (10) from control port (11).

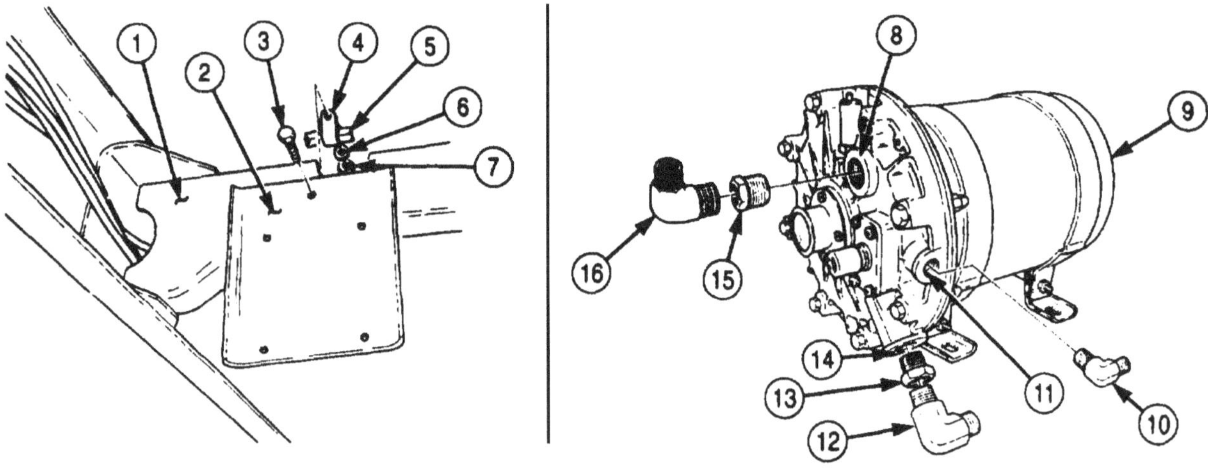

3-412. AIR DRYER KIT (M923/A1/A2, M925/A1/A2, M927/A1/A2, M928/A1/A2, M934/A1/A2) REPLACEMENT (Contd)

NOTE

Perform steps 19 through 21 for M923/A1/A2 and M925/A2/A2 vehicles.

19. Remove locknut (23), screw (19), clamp (24), wire (25), and tube (18) from frame crossmember (22). Discard locknut (23).
20. Remove two tiedown straps (20) from tube (18), line (17), and wire (25). Discard tiedown straps (20).
21. Remove two tubes (18) from nipple (21).

NOTE

Perform steps 22 through 25 for M927/A1/A2, M928/A1/A2, and M934/A1/A2 vehicles.

22. Remove locknut (31), screw (34), clamp (32), tube (18), and washer (33) from mounting bracket (30). Discard locknut (31).
23. Remove locknut (27), screw (35), washer (29), clamp (28), and tube (16) from left-side frame rail (26). Discard locknut (27).
24. Remove four tiedown straps (20) from tube (18), line (17), and wire (25). Discard tiedown straps (20).
25. Remove two tubes (18) from nipple (21).

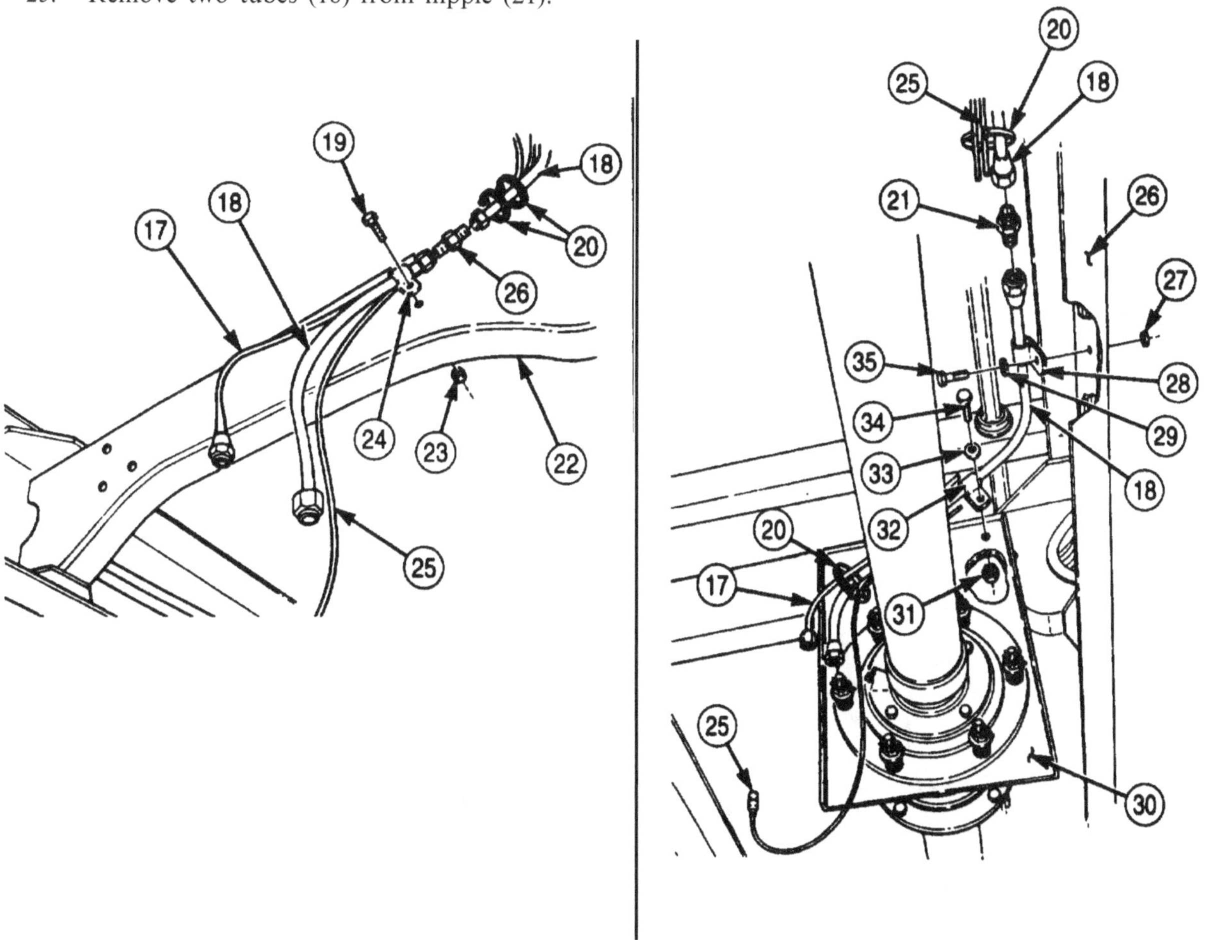

3-412. AIR DRYER KIT (M923/A1/A2, M925/A1/A2, M927/A1/A2, M928/A1/A2, M934/A1/A2) REPLACEMENT (Contd)

26. Disconnect wire (1) from wire (4) on frame crossmember (3).
27. Remove two tiedown straps (8) from lines (9) and (7). Discard tiedown straps (8).
28. Remove line (7) from elbow (5).
29. Remove tube (2) from elbow (6).
30. Remove elbows (5) and (6) from tee (11).
31. Remove tee (11) from air compressor (10).
32. Remove tube (2) from frame crossmember (3).

b. Installation

NOTE

Clean all male pipe threads and wrap with antiseize tape before installation.

1. Install tube (2) on frame crossmember (3).
2. Install tee (11) on air compressor (10).
3. Install elbows (5) and (6) on tee (11).
4. Install tube (2) on elbow (6).
5. Install line (7) on elbow (5).
6. Install two new tiedown straps (8) on lines (9) and (7).
7. Connect wire (1) to wire (4) on frame crossmember (3).

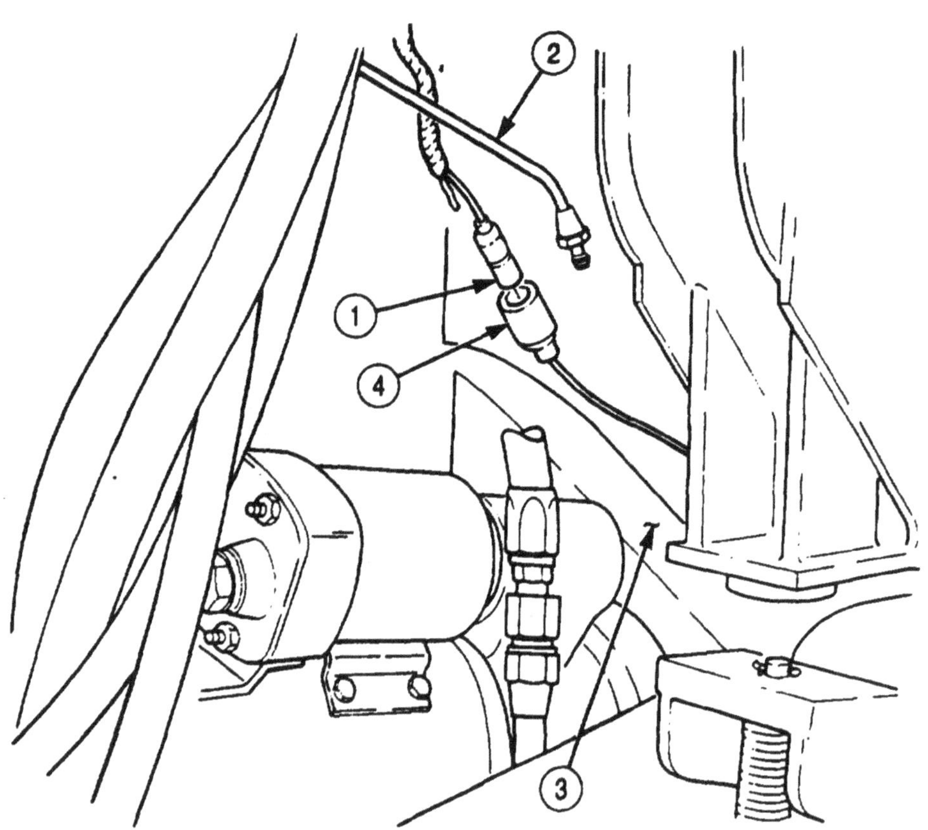

3-412. AIR DRYER KIT (M923/A1/A2, M925/A1/A2, M927/A1/A2, M928/A1/A2, M934/A1/A2) REPLACEMENT (Contd)

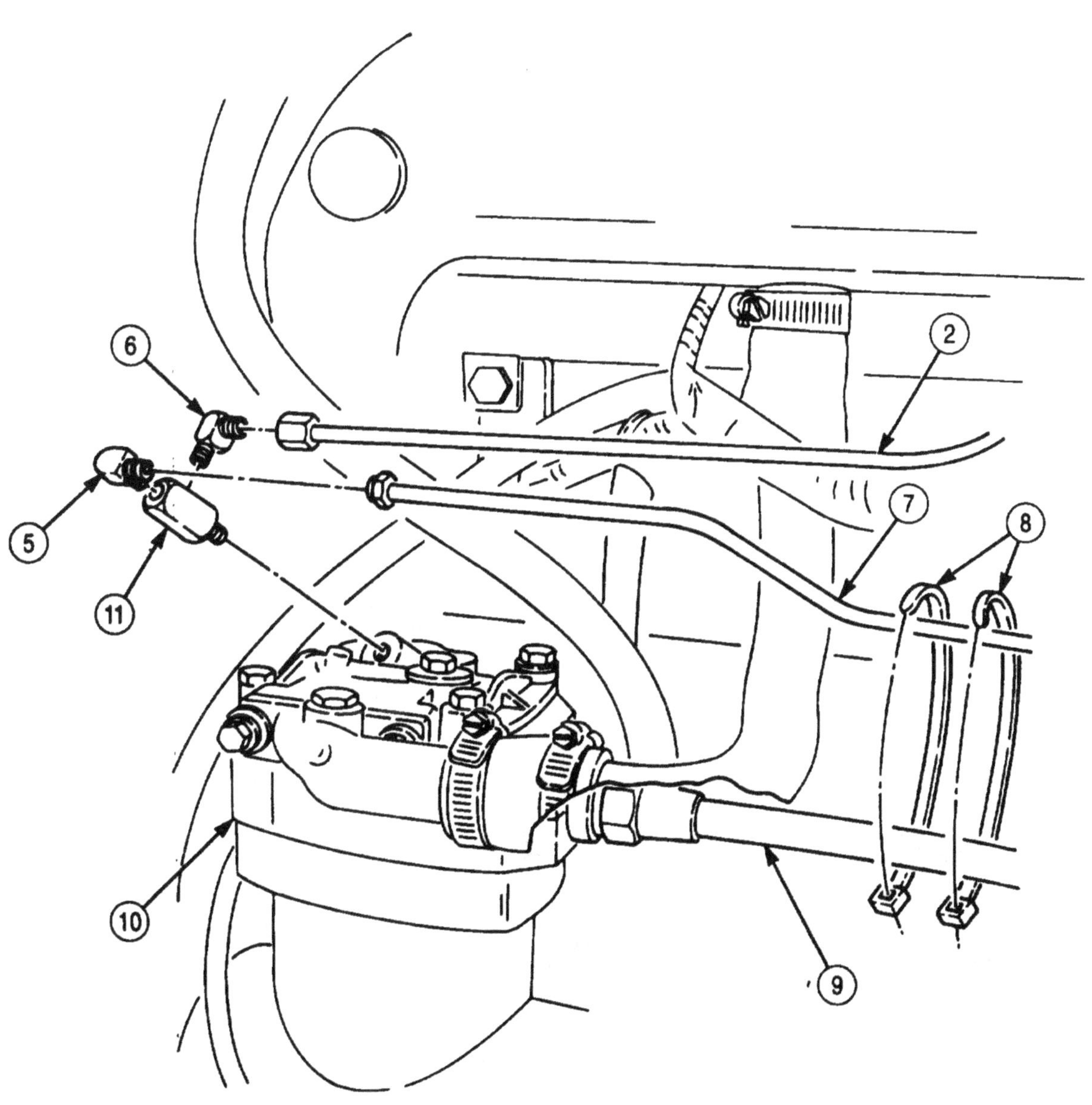

3-412. AIR DRYER KIT (M923/A1/A2, M925/A1/A2, M927/A1/A2, M928/A1/A2, M934/A1/A2) REPLACEMENT (Contd)

NOTE

Perform steps 8 through 11 for M927/A1/A2, M928/A1/A2, and M934/A1/A2 vehicles.

8. Install two tubes (2) on nipple (5).
9. Install four new ticdown straps (4) on tube (2). line (1), and wire (9).
10. Install clamp (12) and tube (2) on left-side frame rail (10) with washer (13), screw (19), and new locknut (11).
11. Install clamp (16) and tube (2) on mounting bracket (14) with washer (17), screw (18), and new locknut (15).

NOTE

Perform steps 12 through 14 for M923/A1/A2 and M925/A1/A2 vehicles.

12. Install two tubes (2) on nipple (5).
13. Install two new tiedown straps (4) on tube (2), line (1), and wire (9).
14. Install clamp (8), wire (9), and tube (2) on frame crossmember (6) with screw (3) and new locknut (7).

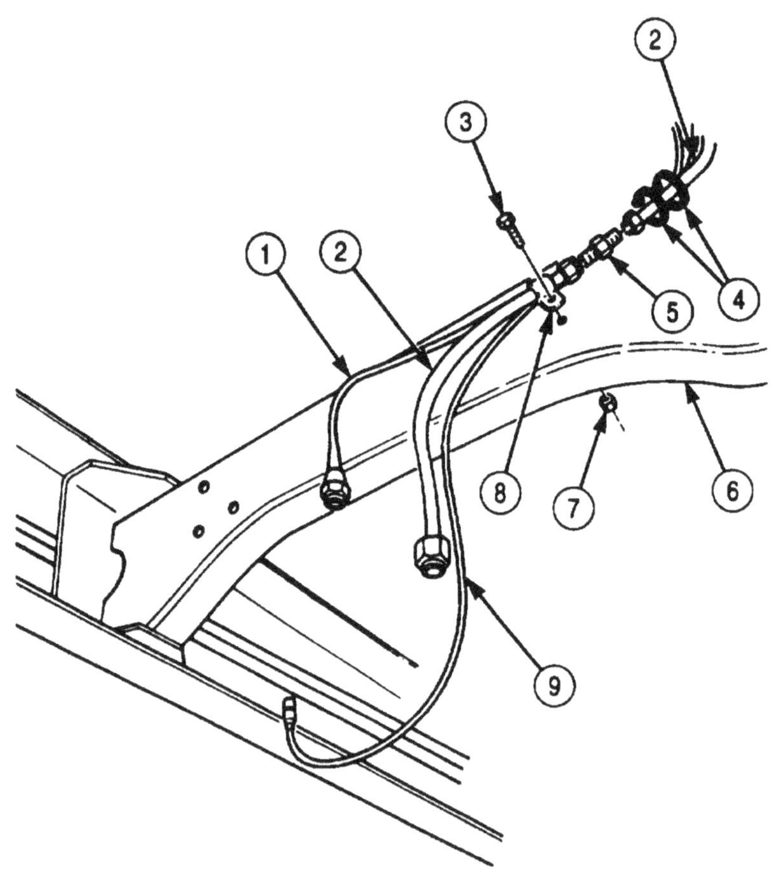

3-412. AIR DRYER KIT (M923/A1/A2, M925/A1/A2, M927/A1/A2, M928/A1/A2, M934/A1/A2) REPLACEMENT (Contd)

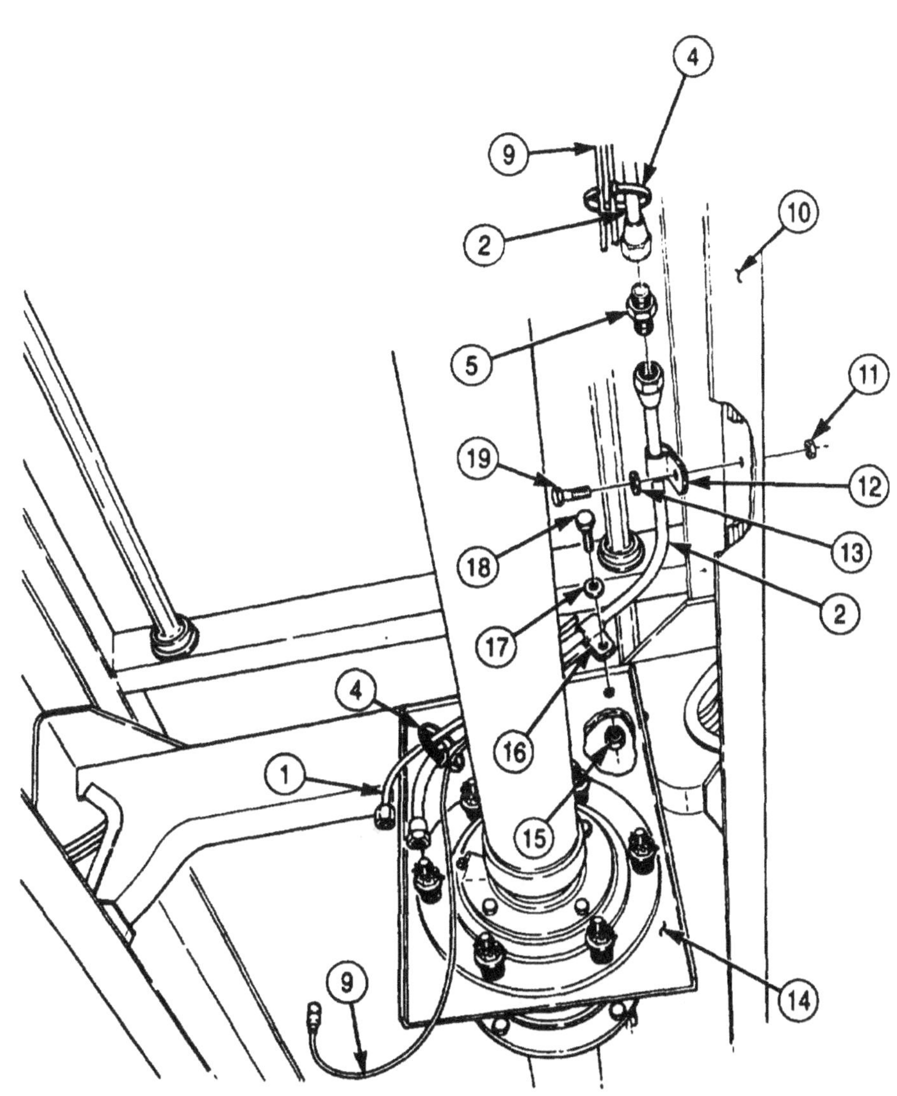

3-412. AIR DRYER KIT (M923/A1/A2, M925/A1/A2, M927/A1/A2, M928/A1/A2, M934/A1/A2) REPLACEMENT (Contd)

15. Install elbow (3) on control port (4).
16. Install bushing (6) on inlet port (7).
17. Install elbow (5) on bushing (6).
18. Install bushing (8) on outlet port (1).
19. Install elbow (9) on bushing (8).
20. Install mounting bracket (11), clamp (13), and two air lines (14) on frame crossmember (10) with three screws (12), washers (15), and new locknuts (16).

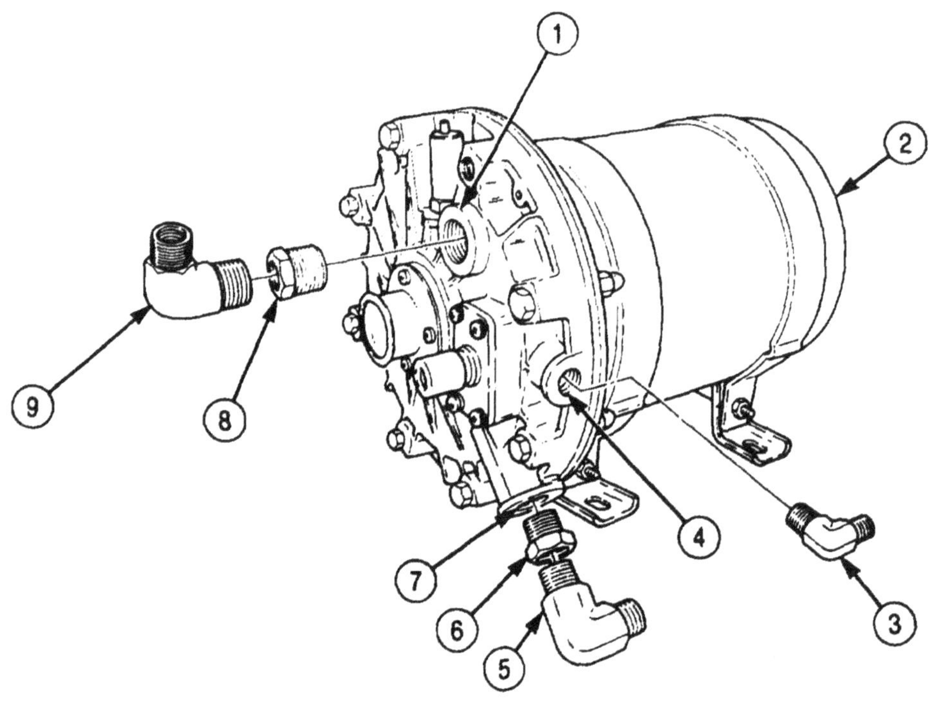

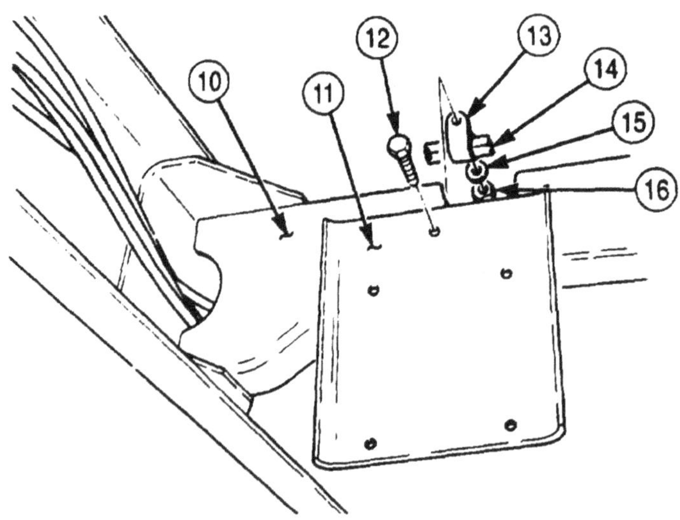

3-412. AIR DRYER KIT (M923/A1/A2, M925/A1/A2, M927/A1/A2, M928/A1/A2, M934/A1/A2) REPLACEMENT (Contd)

21. Install air dryer (2) on mounting bracket (11) with four washers (23), screws (22), washers (261, and new locknuts (27).
22. Install line (24) on elbow (3).
23. Install tube (23) on elbow (5).
24. Install screw (18), new lockwasher (19), ground wire (20), and new locknut (21) on right-side frame rail (17).
25. Install new tiedown strap (30) on tube (31) and ground wire (20).
26. Install tube (31) on elbow (9).

NOTE

Perform steps 27 and 28 for M927/A1/2, M928/A1/A2, and M934/A1/A2 vehicles.

27. Install adapter (35) on wet reservoir adapter (36).
28. Install tube (31) on adapter (35).

NOTE

Perform steps 29 through 31 for M923/A1/A2 and M925/A1/A2 vehicles.

29. Install bushing (33) on wet reservoir adapter (34).
30. Install elbow (32) on bushing (33).
31. Install tube (31) on elbow (32).
32. Connect wire (28) to thermostat connector (29) on air dryer (2).

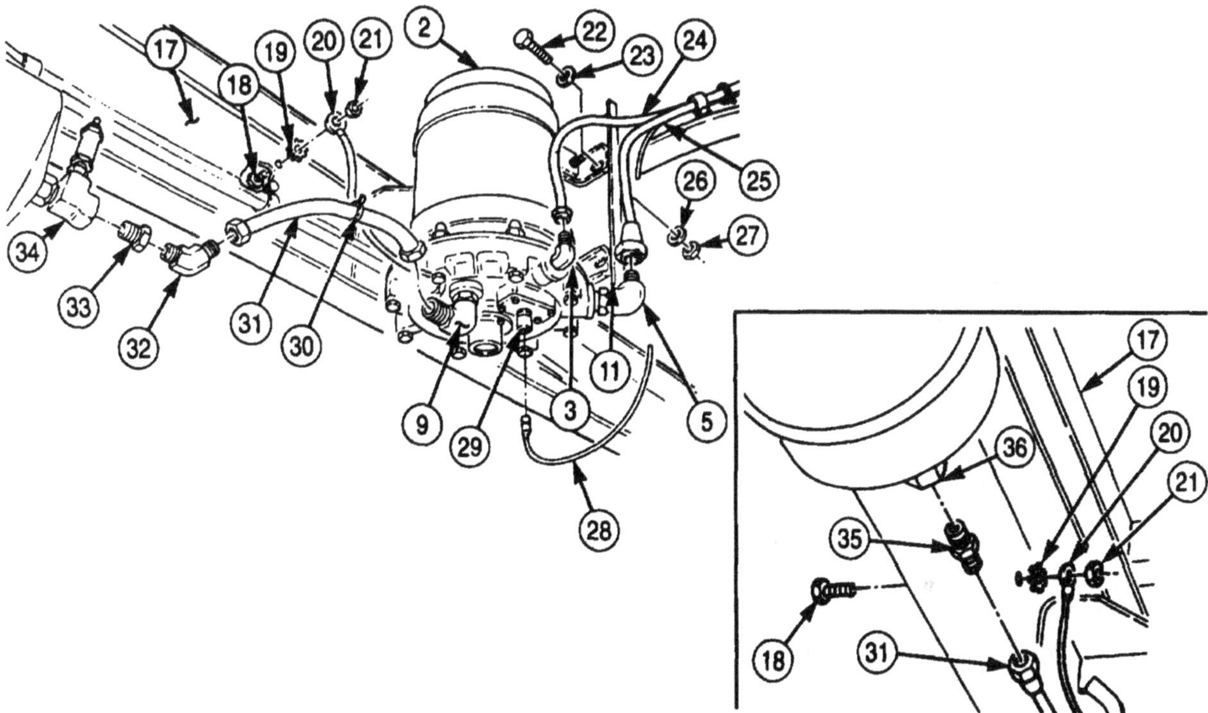

FOLLOW-ON TASKS: • Connect battery ground cables (para. 3-126).
• Start engine, allow air system to reach normal operating pressure, and check air system for leaks (TM 9-2320-272-10).

3-413. AIR DRYER KIT (M929/A1/A2, M930/A1/A2, M931/A1/A2, M932/A1/A2, M936/A1/A2) REPLACEMENT

THIS TASK COVERS:

a. Removal b. Installation

INITIAL SETUP:

APPLICABLE MODELS
M929/A1/A2, M930/A1/A2, M931/A1/A2, M932/A1/A2, M936/A1/A2

TOOLS
General mechanic's tool kit (Appendix E, Item 1)

MATERIALS/PARTS
Nine locknuts (M929/A1/A2, M930/A1/A2, M931/A1/A2, M932/A1/A2) (Appendix D, Item 327)
Five locknuts (M936/A1/A2) (Appendix D, Item 327)
Locknut (M936/A1/A2) (Appendix D, Item 328)
Lockwasher (Appendix D, Item 413)
Ten tiedown straps (M936/A1/A2 (Appendix D, Item 694)
Five tiedown straps (M929/A1/A2, M930/A1/A2, M931/A1/A2, M932/A1/A2) (Appendix D, Item 694)
Antiseize tape (Appendix C, Item 72)

REFERENCES (TM)
TM 9-2320-272-10
TM 9-2320-272-24P

EQUIPMENT CONDITION
- Parking brake set (TM 9-2320-272-10).
- Air reservoirs drained (TM 9-2320-272-10).
- Fifth wheel deck plate removed (M931/A1/A2, M932/A1/A2 only) (para. 3-251).
- Battery ground cables disconnected (para. 3-126).

---

a. Removal

NOTE

Perform steps 1, 2, and 3 for M929/A1/A2, M930/A1/A2, M931/A1/A2, and M932/A1/A2 vehicles.

1. Remove tiedown strap (7) from tube (8) and ground wire (6). Discard tiedown strap (7).
2. Remove tube (8) from elbow (9).
3. Remove elbow (9) from tee (10) on wet reservoir (1).
4. Disconnect wire (4) from thermostat connector (5) on air dryer (3).

NOTE

Perform steps 5, 6, and 7 for M936/A1/A2 vehicles.

5. Remove tube (16) from wet reservoir adapter (18).
6. Remove wet reservoir adapter (18) from wet reservoir (1).
7. Remove three tiedown straps (15) from tube (16) and winch hydraulic line (17). Discard tiedown straps (15).
8. Remove locknut (14), clamp (13), ground wire (6), lockwasher (12), and screw (11) from right-side frame rail (2). Discard locknut (14) and lockwasher (12).

3-413. AIR DRYER KIT (M929/A1/A2, M930/A1/A2, M931/A1/A2, M932/A1/A2, M936/A1/A2) REPLACEMENT (Contd)

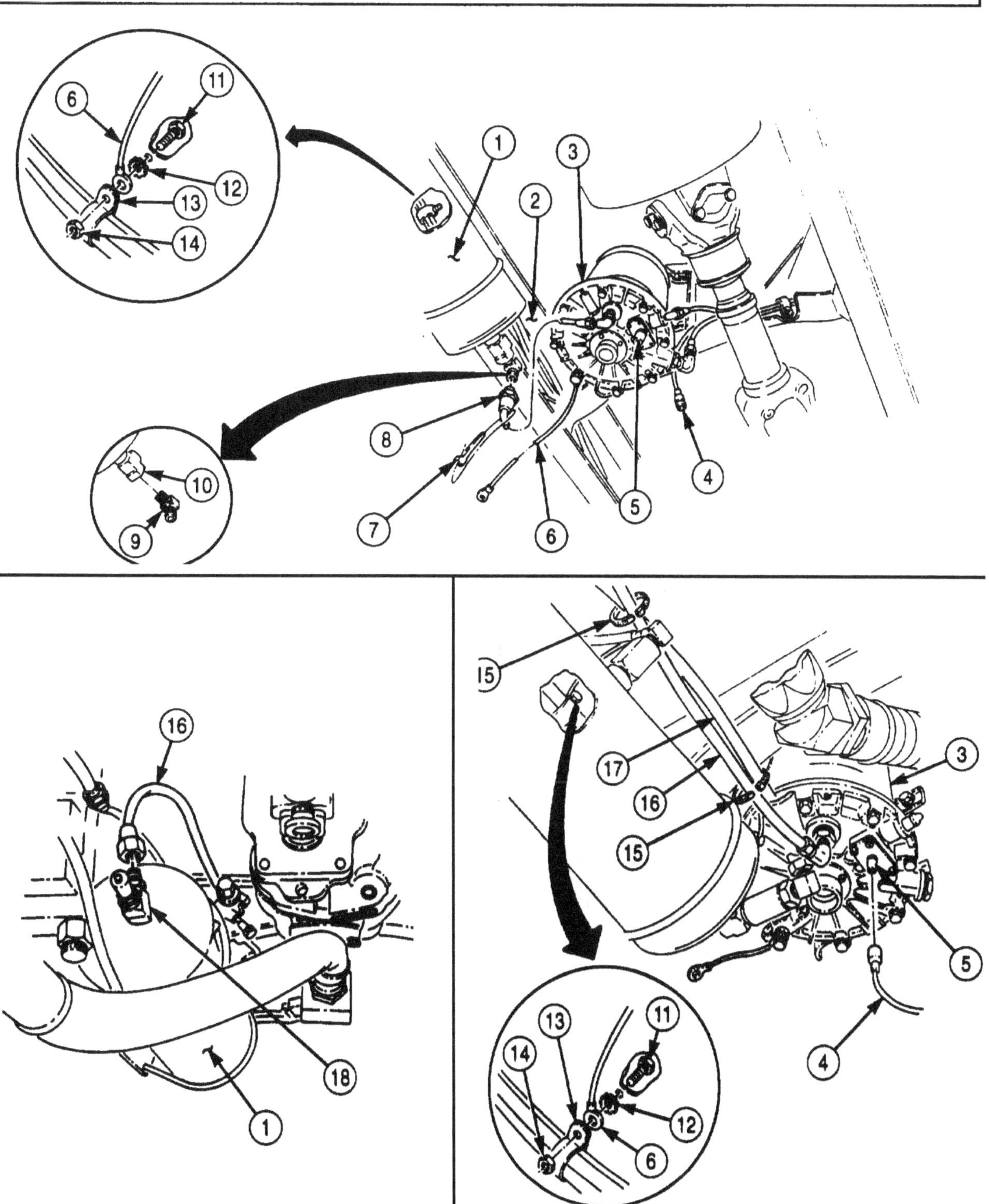

3-413. AIR DRYER KIT (M929/A1/A2, M930/A1/A2, M931/A1/A2, M932/A1/A2, M936/A1/A2) REPLACEMENT (Contd)

9. Remove tube (1) from elbow (2).
10. Remove tube (9) from elbow (11).
11. Remove tube (8) from elbow (4).
12. Remove four locknuts (10), washers (6), screws (5), washers (6), and air dryer (3) from mounting bracket (7). Discard locknuts (10).
13. Remove elbow (2) from bushing (13).
14. Remove bushing (13) from air dryer (3).
15. Remove elbow (11) from bushing (12).
16. Remove bushing (12) from air dryer (3).
17. Remove elbow (4) from air dryer (3).

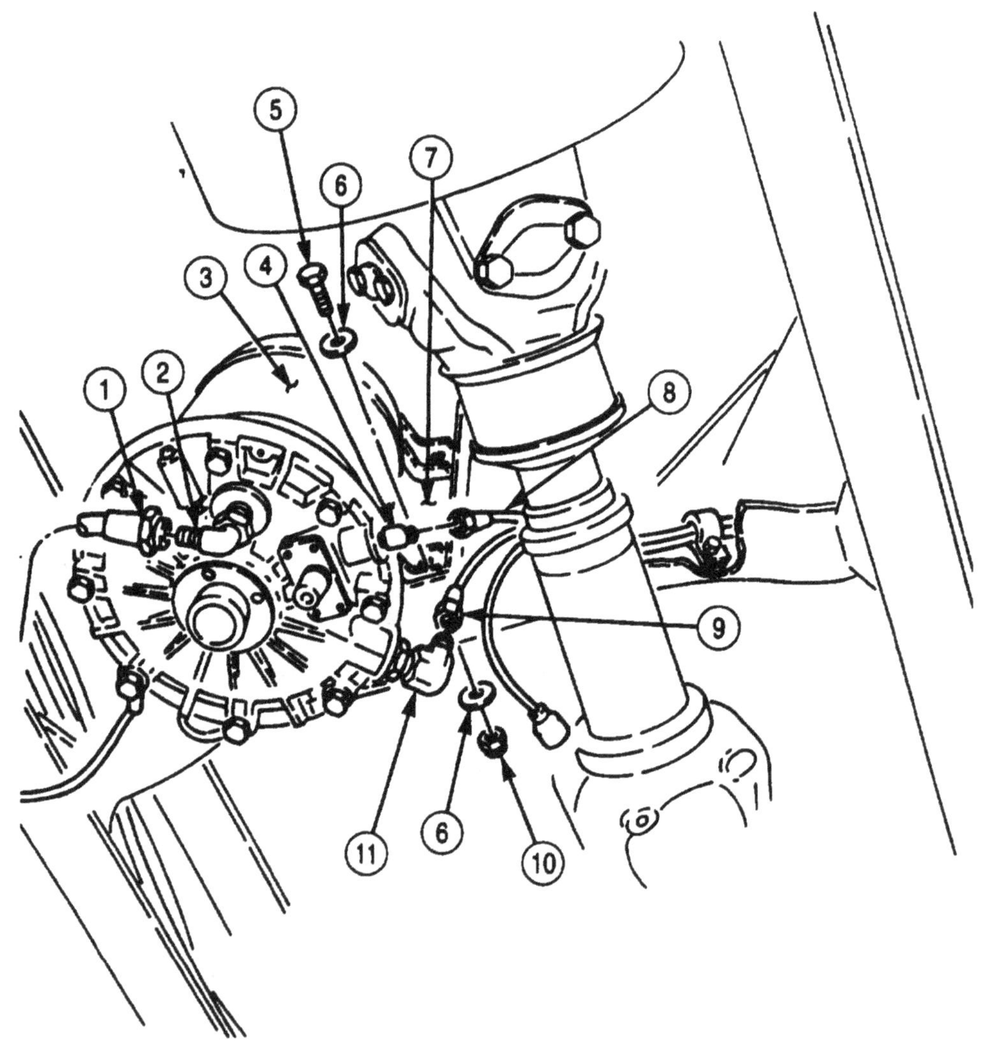

3-413. AIR DRYER KIT (M929/A1/A2, M930/A1/A2, M931/A1/A2, M932/A1/A2, M936/A1/A2) REPLACEMENT (Contd)

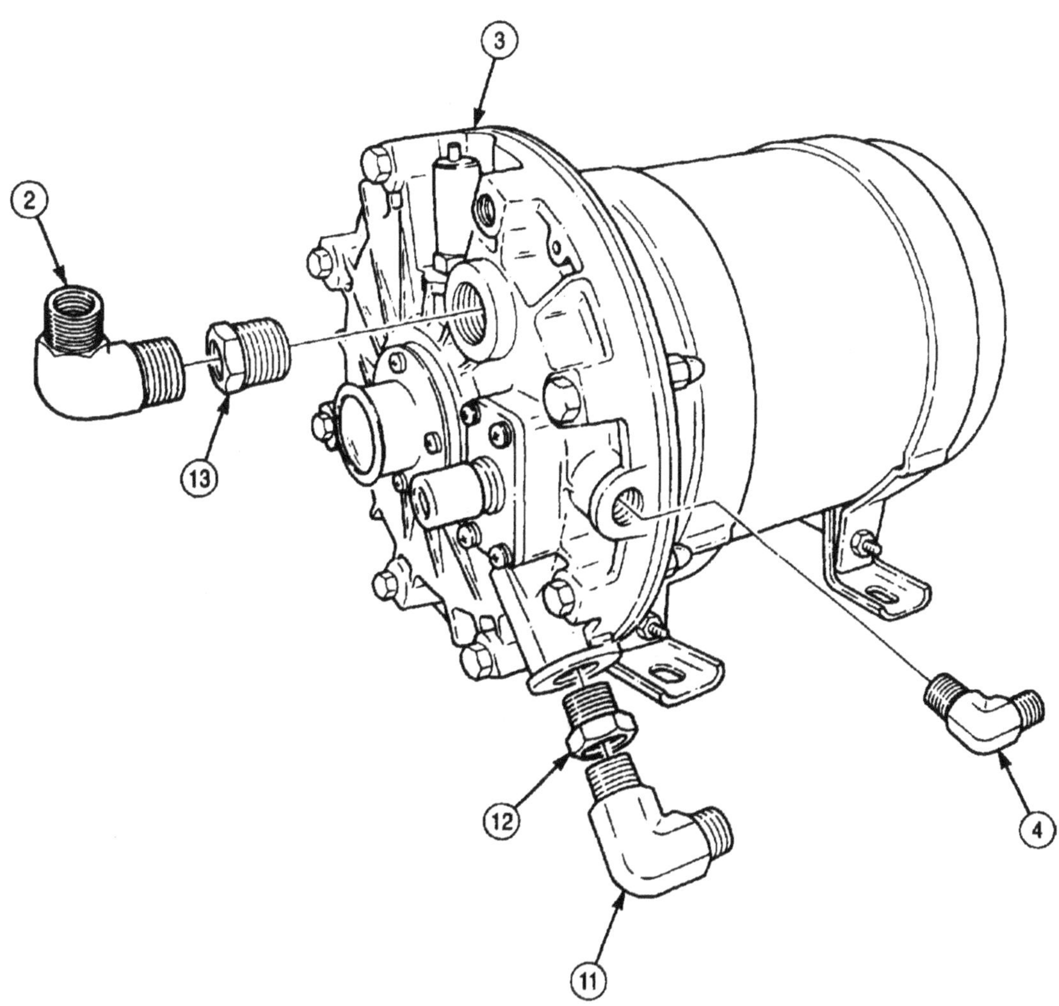

3-413. AIR DRYER KIT (M929/A1/A2, M930/A1/A2, M931/A1/A2, M932/A1/A2, M936/A1/A2) REPLACEMENT (Contd)

18. Disconnect two wires (13) on frame crossmember (1).

NOTE

Perform steps 19 through 22 for M929/A1/A2, M930/A1/A2, M931/A1/A2, and M932/A1/A2 vehicles.

19. Remove three locknuts (7). washers (6), clamp (4), three screws (3), mounting bracket (2), and two air lines (5) from frame crossmember (1). Discard locknuts (7).
20. Remove two tiedown straps (9) from tube (14), line (8), and wire (13). Discard tiedown straps (9).
21. Remove locknut (11), screw (10), clamp (12), wire (13), tube (14), and line (8) from frame crossmember (1). Discard locknut (11).
22. Remove two tubes (14) from nipple (15).

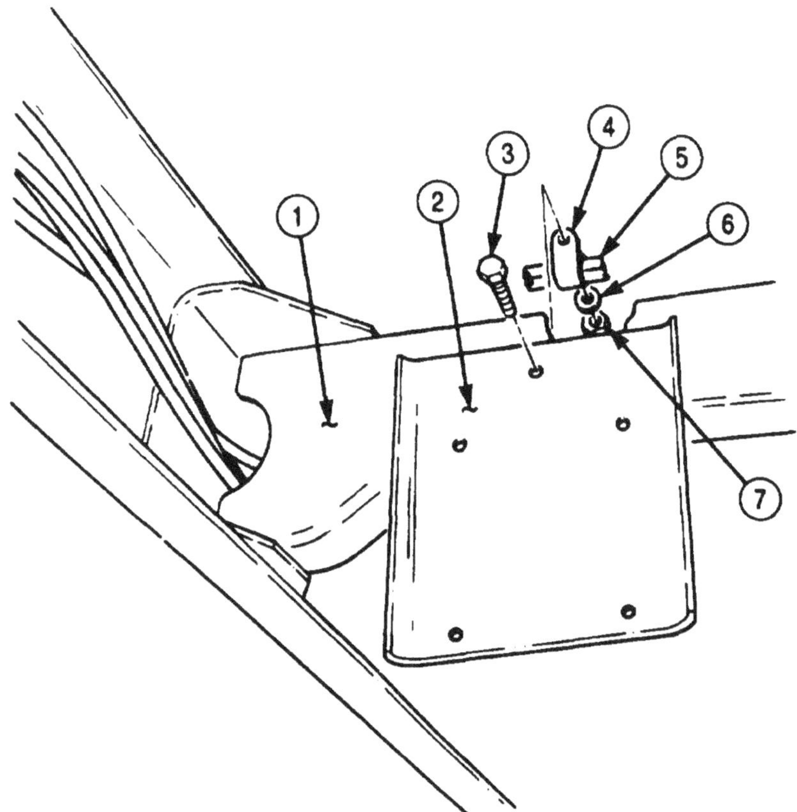

3-413. AIR DRYER KIT (M929/A1/A2, M930/A1/A2, M931/A1/A2, M932/A1/A2, M936/A1/A2) REPLACEMENT (Contd)

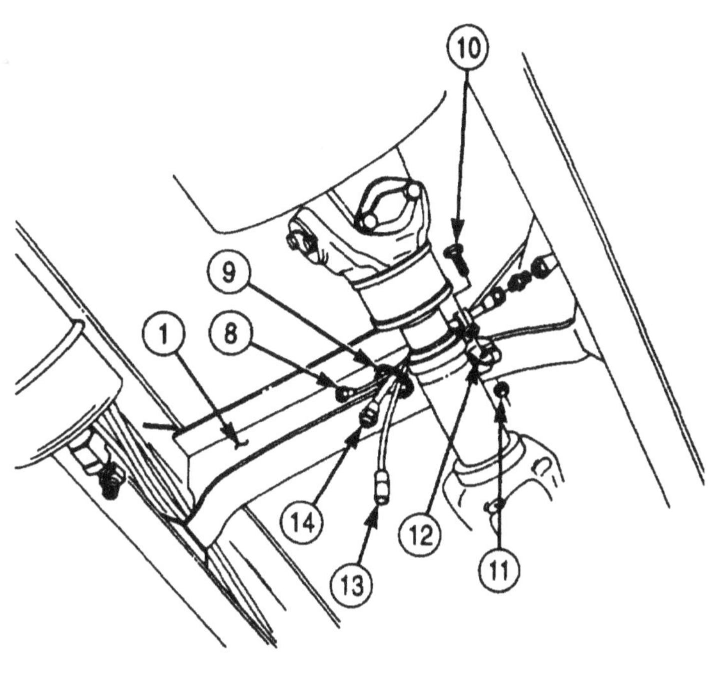

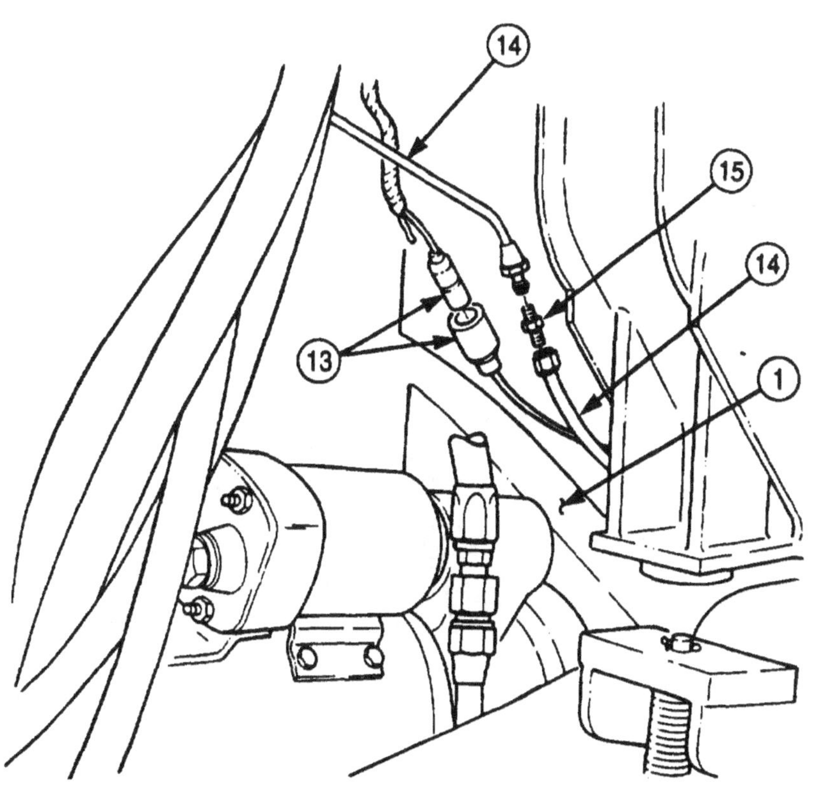

3-413. AIR DRYER KIT (M929/A1/A2, M930/A1/A2, M931/A1/A2, M932/A1/A2, M936/A1/A2) REPLACEMENT (Contd)

NOTE

Perform steps 23 through 26 for M936/A1/A2 vehicles.

23. Remove four tiedown straps (4) from frame crossmember (1). Discard tiedown straps (4).
24. Remove tiedown strap (7) from hose (5), line (6), tube (3), and wire (8). Discard tiedown strap (7).
25. Remove tube (3) from elbow (2).
26. Remove locknut (11), screw (9), clamp (10), and tube (3) from frame crossmember (1). Discard locknut (11).

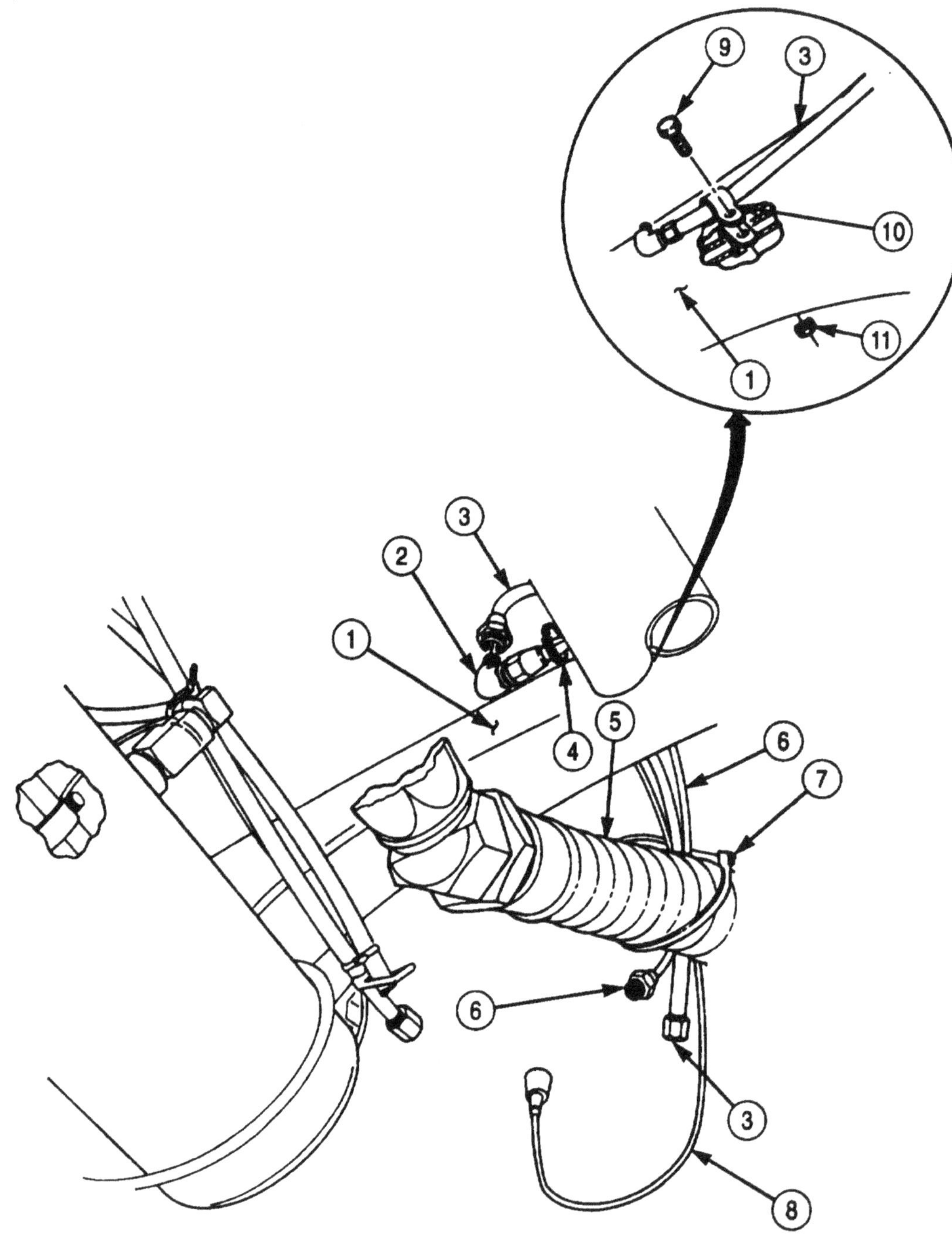

3-413. AIR DRYER KIT (M929/A1/A2, M930/A1/A2, M931/A1/A2, M932/A1/A2, M936/A1/A2) REPLACEMENT (Contd)

27. Remove two tiedown straps (16) from lines (15) and (17). Discard tiedown straps (16).
28. Remove line (15) from elbow (12).
29. Remove tube (14) from elbow (13).
30. Remove elbows (12) and (13) from tee (19).
31. Remove tee (19) from air compressor (18).

b. Installation

NOTE

Clean all male pipe threads and wrap with antiseize tape before installation.

1. Install tee (19) on air compressor (18).
2. Install elbows (12) and (13) on tee (19).
3. Install tube (14) on elbow (13).
4. Install line (15) on elbow (12).
5. Install two new tiedown straps (16) on lines (15) and (17).

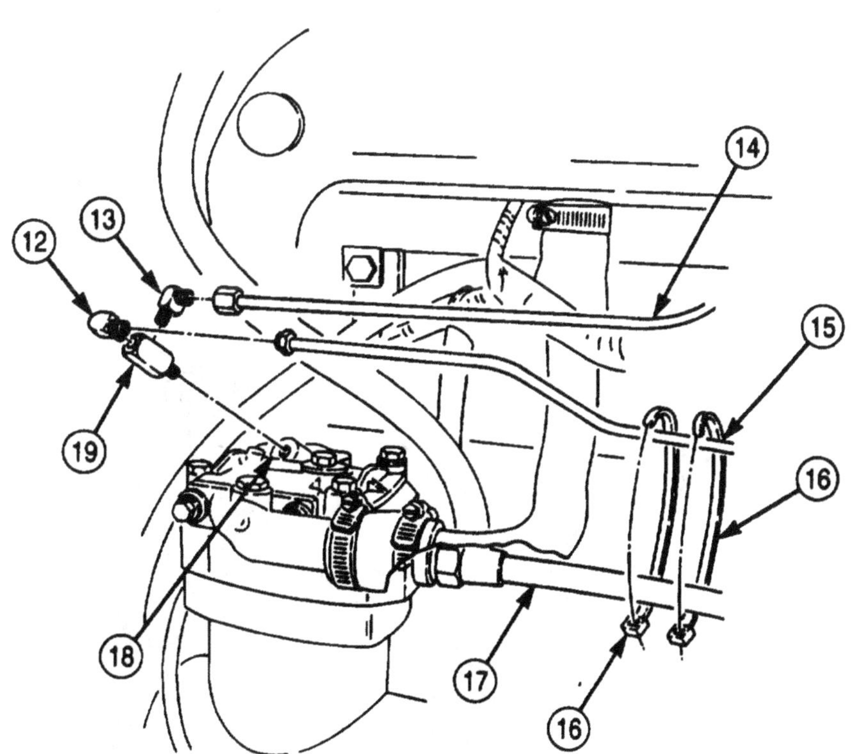

3-413. AIR DRYER KIT (M929/A1/A2, M930/A1/A2, M931/A1/A2, M932/A1/A2, M936/A1/A2) REPLACEMENT (Contd)

NOTE

Perform steps 6 through 9 for M936/A1/A2 vehicles.

6. Install clamp (10) and tube (3) on frame crossmember (1) with screw (9) and new locknut (11).
7. Install tube (3) on elbow (2).
8. Install new tiedown strap (7) on hose (5), line (6), tube (3), and wire (8).
9. Install four new tiedown straps (4) on frame crossmember (1).

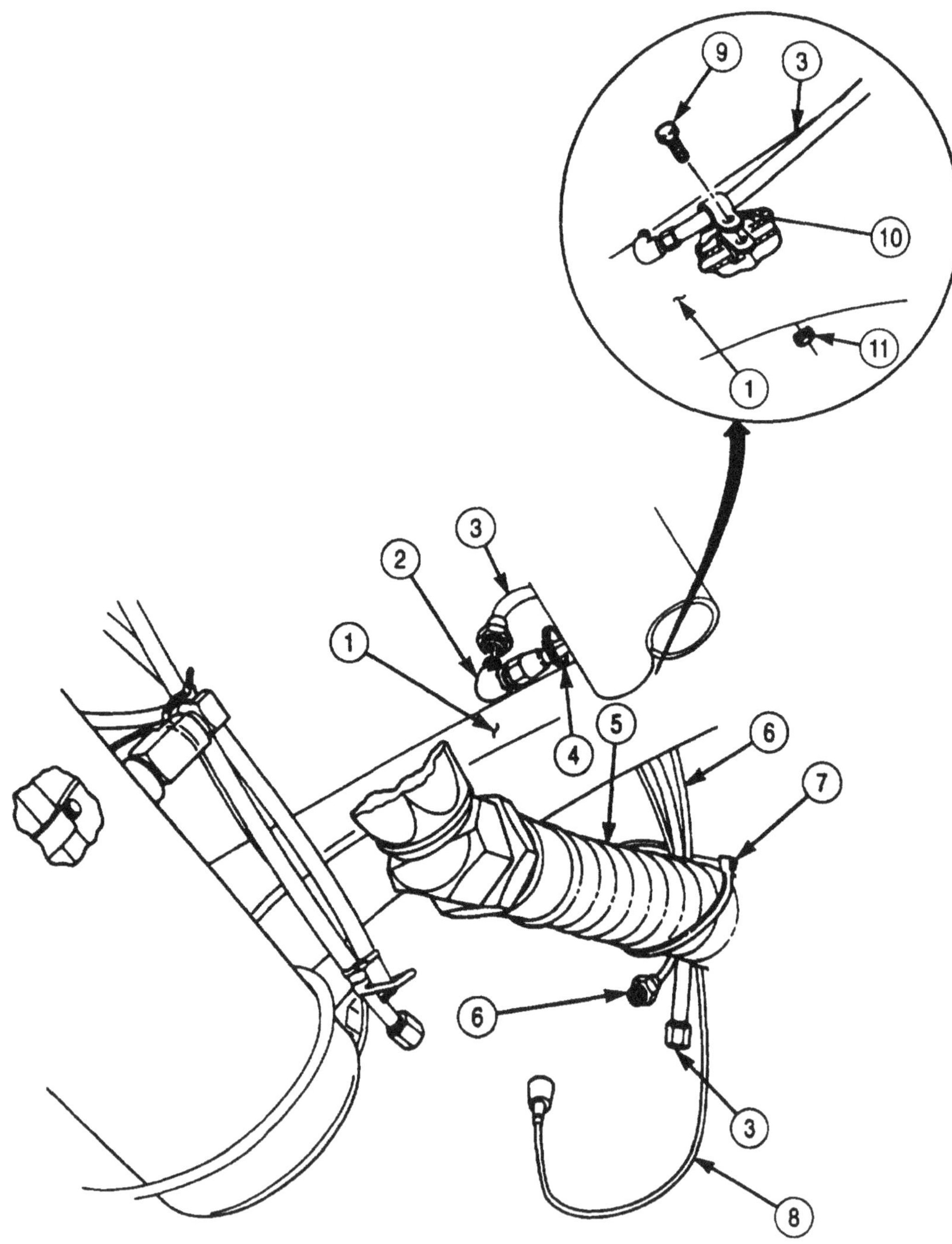

3-413. AIR DRYER KIT (M929/A1/A2, M930/A1/A2, M931/A1/A2, M932/A1/A2, M936/A1/A2) REPLACEMENT (Contd)

NOTE

Perform steps 10 through 12 for M929/A1/A2, M930/A1/A2, M931/A1/A2, and M932/A1/A2 vehicles.

10. Install clamp (20), wire (21), tube (22), and line (23) on frame crossmember (1) with screw (18) and new locknut (19).
11. Install two new tiedown straps (24) on tube (22), line (23), and wire (21).
12. Install mounting bracket (12), two air lines (15), and clamp (14) on frame crossmember (1) with three screws (13), washers (16), and new locknuts (17).

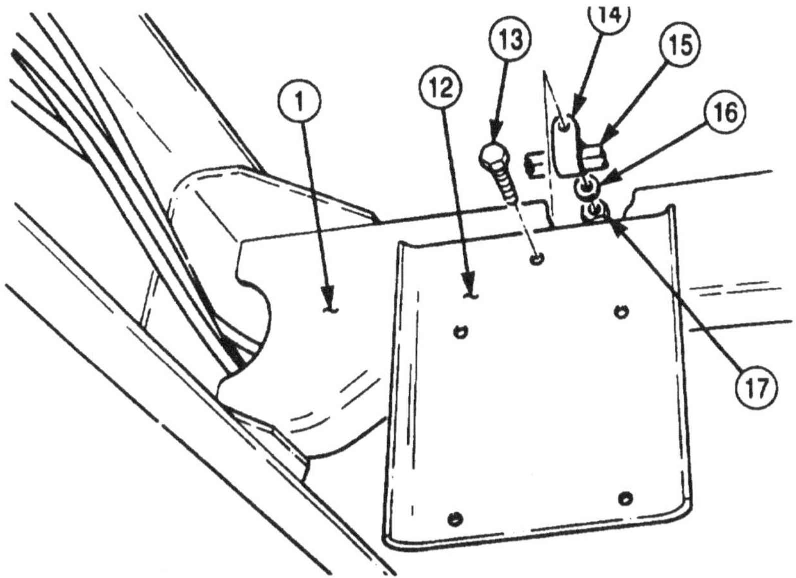

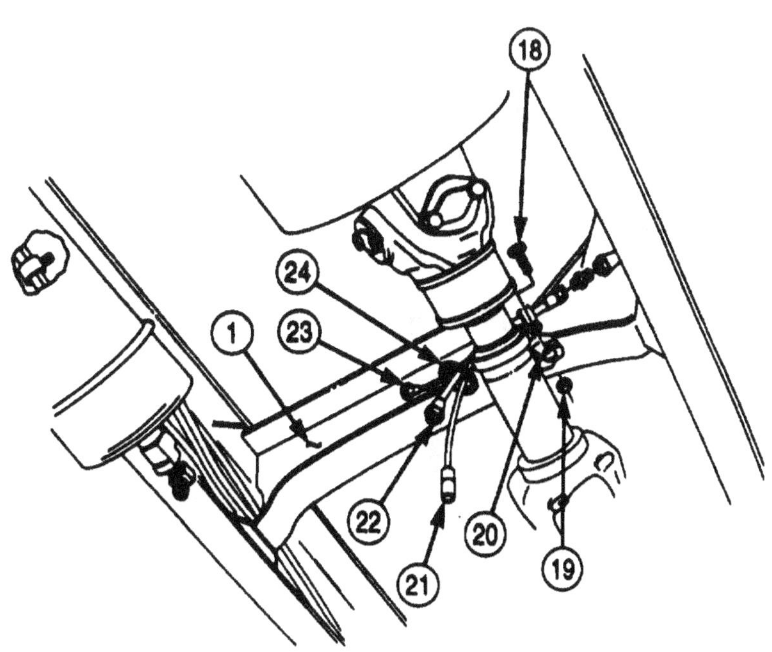

3-413. AIR DRYER KIT (M929/A1/A2, M930/A1/A2, M931/A1/A2, M932/A1/A2, M936/A1/A2) REPLACEMENT (Contd)

13. Install elbow (4) on air dryer (3).
14. Install bushing (12) on air dryer (3).
15. Install elbow (11) on bushing (12).
16. Install bushing (13) on air dryer (3).
17. Install elbow (2) on bushing (13).
18. Install air dryer (3) on mounting bracket (7) with four washers (6), screws (5), washers (6), and new locknuts (10).
19. Install tube (8) on elbow (4).
20. Install tube (9) on elbow (11).
21. Install tube (1) on elbow (2).

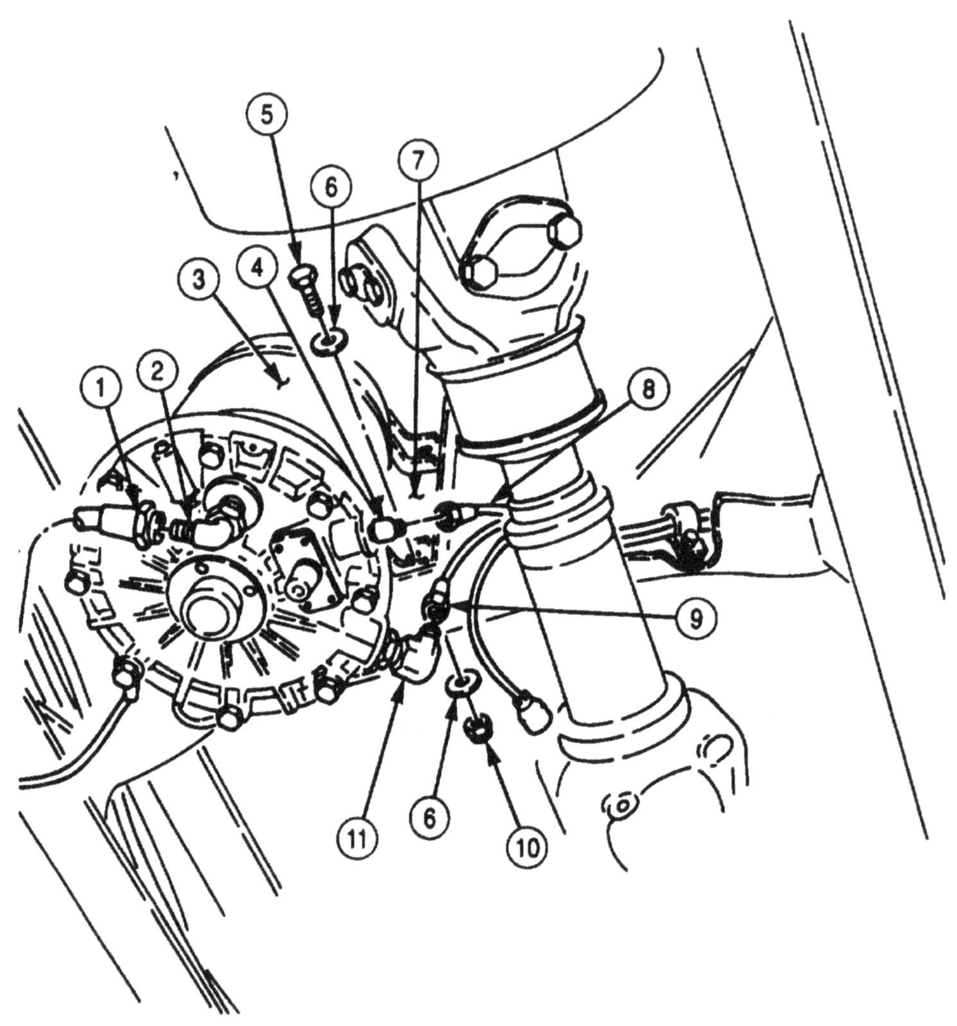

3-413. AIR DRYER KIT (M929/A1/A2, M930/A1/A2, M931/A1/A2, M932/A1/A2, M936/A1/A2) REPLACEMENT (Contd)

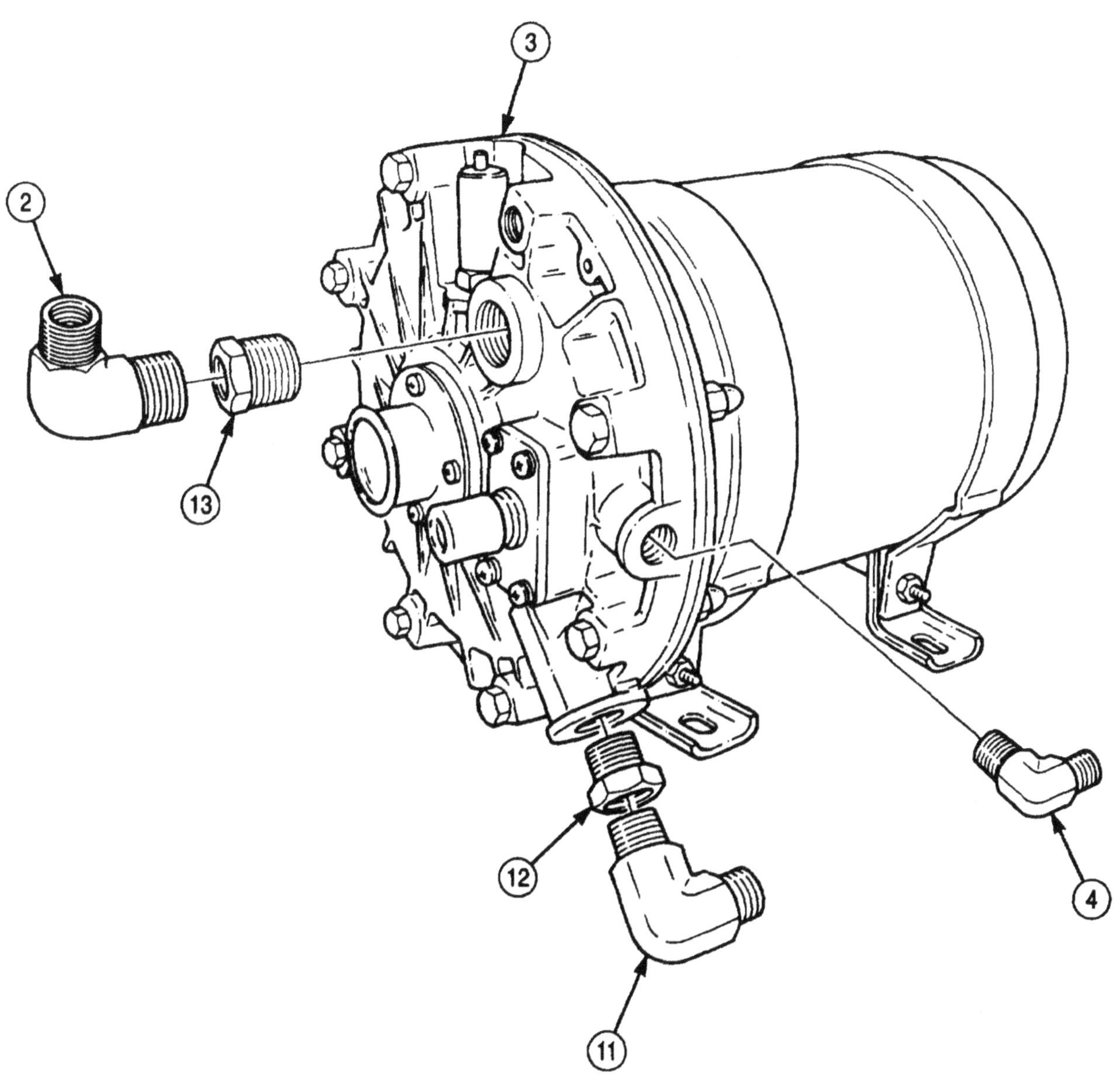

3-413. AIR DRYER KIT (M929/A1/A2, M930/A1/A2, M931/A1/A2, M932/A1/A2, M936/A1/A2) REPLACEMENT (Contd)

22. Install screw (11), new lockwasher (12), ground wire (6), clamp (13), and new locknut (14) on right-side frame rail (2).

NOTE

Perform steps 23 through 25 for M936/A1/A2 vehicles.

23. Install three new tiedown straps (15) on tube (16) and winch hydraulic line (17).
24. Install wet reservoir adapter (18) on wet reservoir (1).
25. Install tube (16) on wet reservoir adapter (17).

NOTE

Perform steps 26 through 28 for M929/A1/A2, M930/A1/A2, M931/A1/A2, and M932/A1/A2 vehicles.

26. Install new tiedown strap (7) on tube (8) and ground wire (6).
27. Install elbow (9) on tee (10).
28. Install tube (8) on elbow (9).
29. Connect wire (4) to thermostat connector (5) on air dryer (3).

3-413. AIR DRYER KIT (M929/A1/A2, M930/A1/A2, M931/A1/A2, M932/A1/A2, M936/A1/A2) REPLACEMENT (Contd)

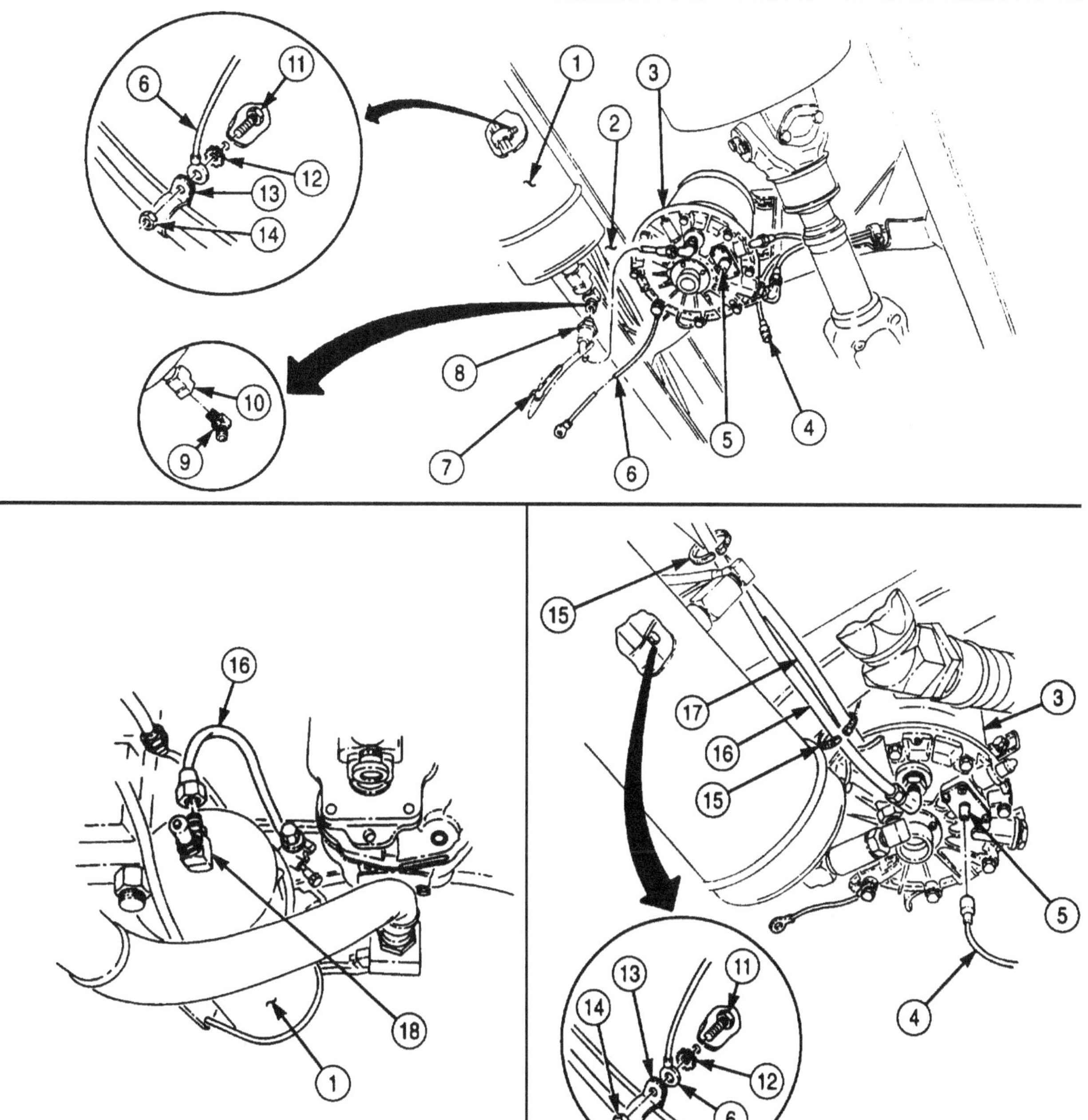

FOLLOW-ON TASKS:
- Connect battery ground cables (para. 3-126).
- Install fifth wheel deck plate (M931/A1/A2, M932/A1/A2 only) (para. 3-251).
- Start engine, allow air system to reach normal operating pressure, and check air system for leaks (TM 9-2320-272-10).

3-414. A-FRAME KIT MAINTENANCE

THIS TASK COVERS:

a. Removal
b. Inspection
c. Installation

INITIAL SETUP:

APPLICABLE MODELS
M925/A1/A2, M928/A1/A2, M932/A1/A2

TOOLS
General mechanic's tool kit (Appendix E, Item 1)

MATERIALS/PARTS
Lockwasher (Appendix D, Item 416)

PERSONNEL REQUIRED
TWO

REFERENCES (TM)
TM 9-2320-272-10
TM 9-2320-272-24P

EQUIPMENT CONDITION
- Parking brake set (TM 9-2320-272-10).
- Cab tarpaulin removed (task c. only) (TM 9-2320-272-10).
- Windshield lowered (task c. only) (TM 9-2320-272-10).
- Tailgate removed (except M932/A1/A2) (task c. only) (para. 3-343).
- Front lifting shackles removed (task c. only) (para. 3-241).

GENERAL SAFETY INSTRUCTION
Do not perform this procedure near high-voltage wires.

a. Removal

WARNING

Do not perform this procedure near high-voltage wires. Vehicle will become charged with electricity if A-frame contacts or breaks high-voltage wire. Do not leave vehicle while high-voltage line is in contact with A-frame or vehicle. Failure to comply may result in injury to personnel.

1. Remove snatch block (2) from A-frame spreader tube (1).
2. Remove setscrew (10) from each A-frame leg (9) and bracket (11).

NOTE

Assistant will push A-frame toward cab during steps 3 and 4.

3. Remove shackle pin (7) from shackle (6) and separate cable (3) and harness (8) from shackle (6).
4. Remove cable (3) from pintle hook (5) and eyebolt (4).

3-414. A-FRAME KIT MAINTENANCE (Contd)

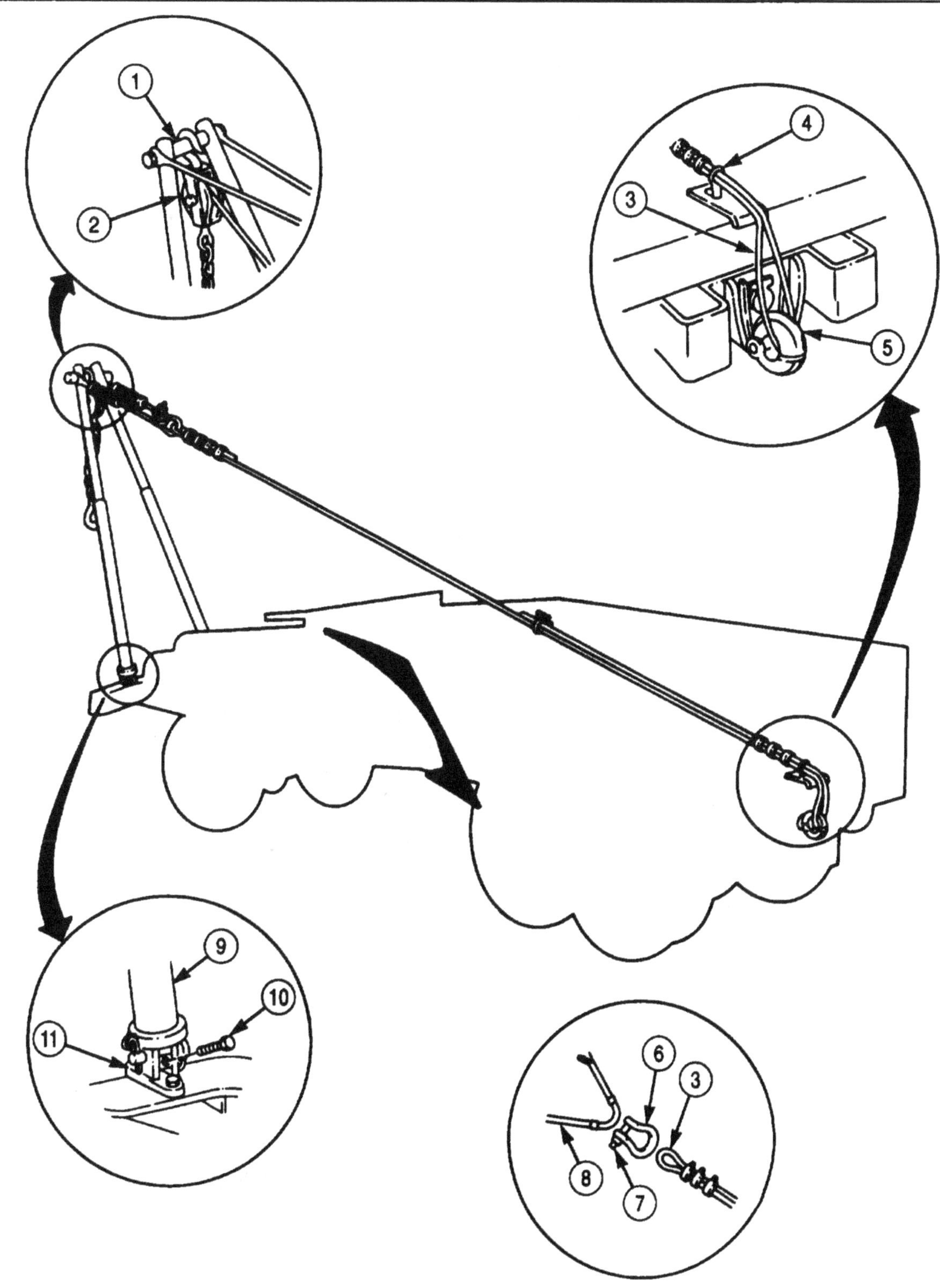

3-414. A-FRAME KIT MAINTENANCE (Contd)

NOTE

Step 5 is not required for M932/A1/A2 vehicles.

5. Remove nut (4), lockwasher (3), washer (2), eyebolt (1), washer (2), and eyebolt plate (6) from rear cargo bed (5). Discard lockwasher (3).
6. Lower A-frame legs (13) and (18) on front bumper (14) to ground.
7. Remove nut (22), nut (7), washer (8), harness (21), and leg spacer (9) from each end of spreader tube stud (10).
8. Remove spreader tube stud (10) and spreader tube (12) from two leg extension tubes (11).
9. Remove two safety pins (19) from pins (20).
10. Remove two pins (20) and detach leg extension tubes (11) from A-frame legs (13) and (18).
11. Remove two safety pins (16) from pins (17).
12. Remove two pins (17) from two front brackets (15).
13. Remove A-frame legs (13) and (18) from two front brackets (15).

b. Inspection

1. Inspect all metal components of A-frame kit. Replace if bent, cracked, or broken.
2. Inspect harness (21). Replace frayed, broken, or loose or missing clamps.

NOTE

If A-frame kit is not to be installed, perform follow-on tasks. Do not perform installation, task c.

WARNING

Do not perform this procedure near high voltage wires. Vehicle will become charged with electricity if A-frame contacts or breaks high voltage wire. Do not leave vehicle while high voltage line is in contact with A-frame or vehicle. Failure to do so may result in injury to personnel.

1. Install A-frame legs (13) and (18) on two front brackets (15).
2. Install two pins (17) on front brackets (15).
3. Install two safety pins (16) on pins (17).
4. Install leg extension tubes (11) on A-frame legs (13) and (18) with two pins (20).
5. Install two safety pins (19) on pins (20).
6. Install spreader tube (12) and spreader tube stud (10) on two leg extension tubes (11).
7. Install leg spacer (9), harness (21), washer (8), nut (7), and nut (22) on each end of spreader tube stud (10).
8. Raise A-frame legs (13) and (18) on front bumper (14) to original position.

NOTE

Step 9 is not required for M932/A1/A2 vehicles.

9. Install eyebolt plate (6) on rear cargo bed (5) with washer (2), eyebolt (1), washer (2), new lockwasher 31, and nut (4).

3-414. A-FRAME KIT MAINTENANCE (Contd)

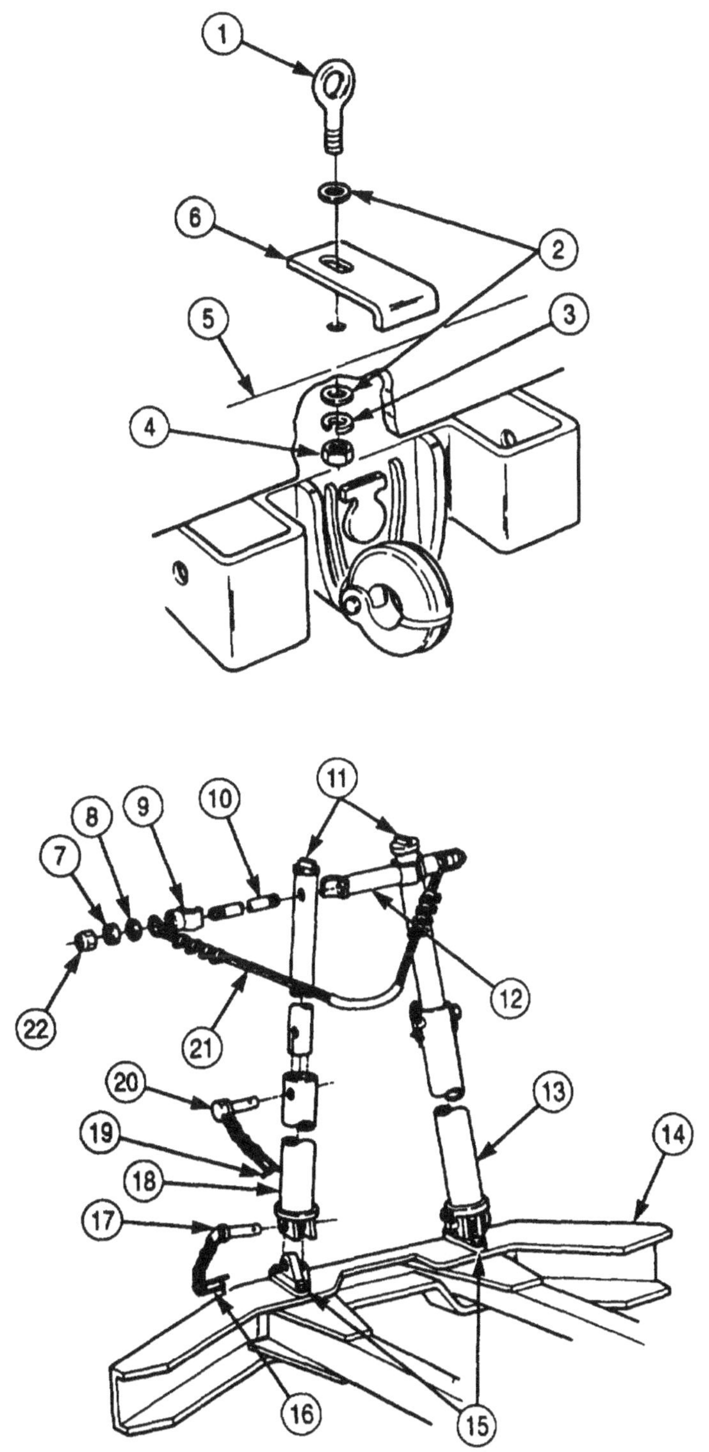

3-414. A-FRAME KIT MAINTENANCE (Contd)

NOTE

- Assistant will push A-frame toward cab during steps 10 and 11.
- On M932/A1/A2 vehicles, cable is passed directly over rear crossmember and attached to pintle hook because vehicle is not equipped with eyebolt assembly.

10. Thread cable (3) through eyebolt (4) and install cable (3) on pintle hook (5).
11. Install shackle (7) on cable (3) and harness (9) with shackle pin (8).

NOTE

A-frame must be angled approximately 60 degrees from horizontal. Do not insert setscrews until adjustment is made.

12. Loosen clamp (6) at cut end of cable (3) and position A-frame legs (10) at an angle of approximately 60 degrees from horizontal. Tighten clamp (6).
13. Install setscrew (11) on each A-frame leg (10) and bracket (12). Tighten setscrew (11) until cable (3) slack is taken up.
14. Install snatch block (2) on A-frame spreader tube (1).

NOTE

Do not perform follow-on tasks if A-frame kit has been installed.

3-414. A-FRAME KIT MAINTENANCE (Contd)

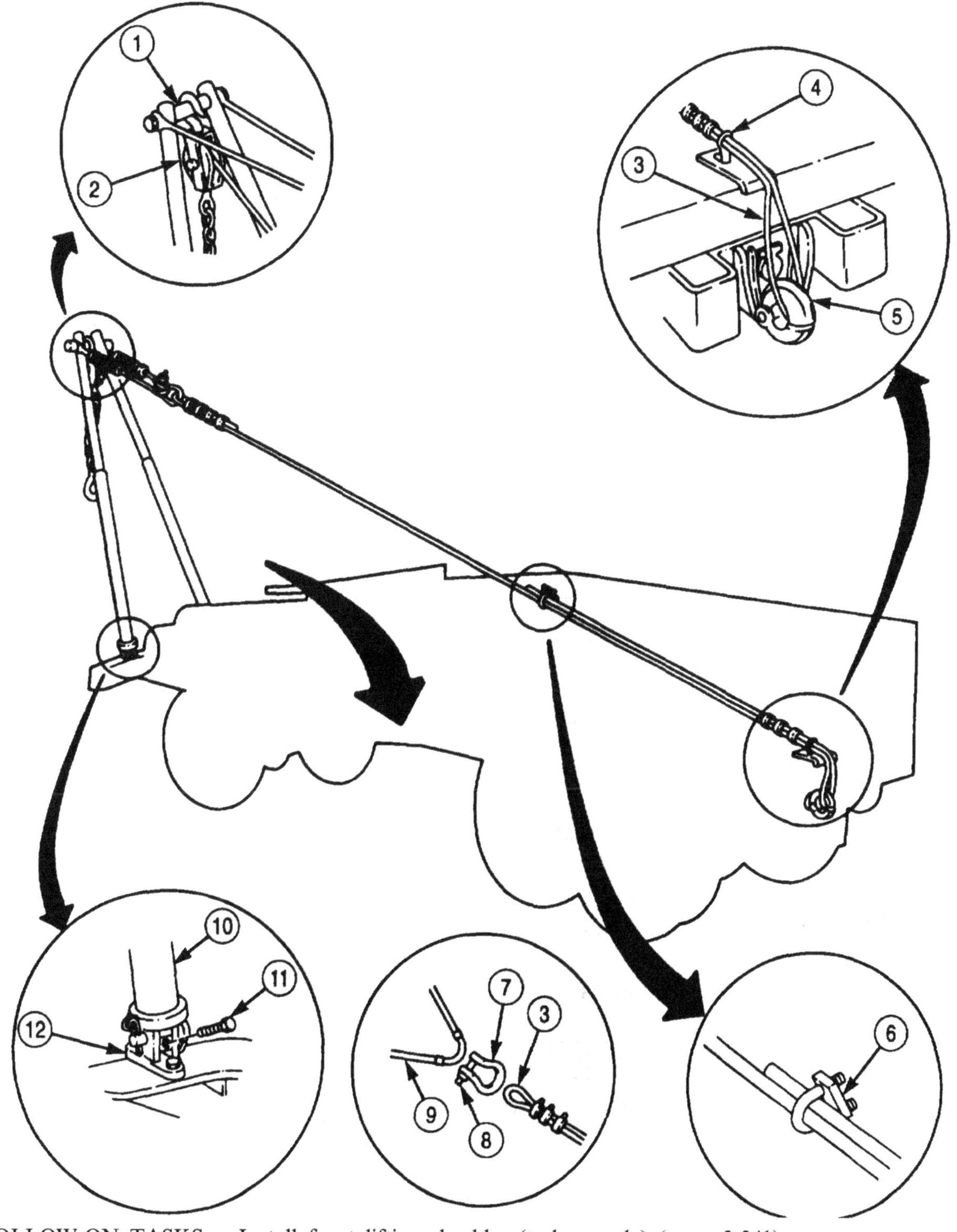

FOLLOW-ON TASKS:
- Install front lifting shackles (task c. only) (para. 3-241).
- Install tailgate (except M932/A1/A2) (task c. only) (para. 3-343).
- Raise windshield (task c. only) (TM 9-2320-272-10).
- Install cab tarpaulin (task c. only) (TM 9-2320-272-10).

3-415. PIONEER TOOL KIT MOUNTING BRACKET REPLACEMENT

THIS TASK COVERS:

a. Bracket Removal (M923, M925, M927, M928)
b. Bracket Installation (M923, M925, M927, M928)
c. Bracket Removal (M931, M932)
d. Bracket Installation (M931, M932)
e. Bracket Removal (M929, M930)
f. Bracket Installation (M929, M930)
g. Bracket Removal (M929A1, M930A1)
h. Bracket Installation (M929A1, M930A1)

INITIAL SETUP:

APPLICABLE MODELS
M923, M925, M927, M928, M929/A1, M930/A2, M931, M932

REFERENCES (TM)
TM 9-2320-272-10
TM 9-2320-272-24P

TOOLS
General mechanic's tool kit (Appendix E, Item 1)

EQUIPMENT CONDITION
Parking brake set (TM 9-2320-272-10).

MATERIALS/PARTS
Four locknuts (M923, M925, M927, M928, M929A1, M930A1, M931, M9321 (Appendix D, Item 291)
Five locknuts (M931, M932) (Appendix D, Item 291)
Eight locknuts (M929, M930) (Appendix D, Item 376)

---

a. Bracket Removal (M923, M925, M927, M928)

1. Open toolbox door (3).
2. Remove four locknuts (5), washers (6), screws (1), and bracket (2) from toolbox (4). Discard locknuts (5).

b. Bracket Installation (M923, M925, M927, M928)

1. Install bracket (2) on toolbox (4) with four screws (1), washers (6), and new locknuts (5).
2. Close toolbox door (3).

c. Bracket Removal (M931, M932)

1. Open toolbox door (11).
2. Remove five locknuts (10), washers (9), screws (8), and bracket (7) from toolbox (4). Discard locknuts (10).

d. Bracket Installation (M931, M932)

1. Install bracket (7) on toolbox (4) with five screws (8), washers (9), and new locknuts (10).
2. Close toolbox door (11).

3-415. PIONEER TOOL KIT MOUNTING BRACKET REPLACEMENT (Contd)

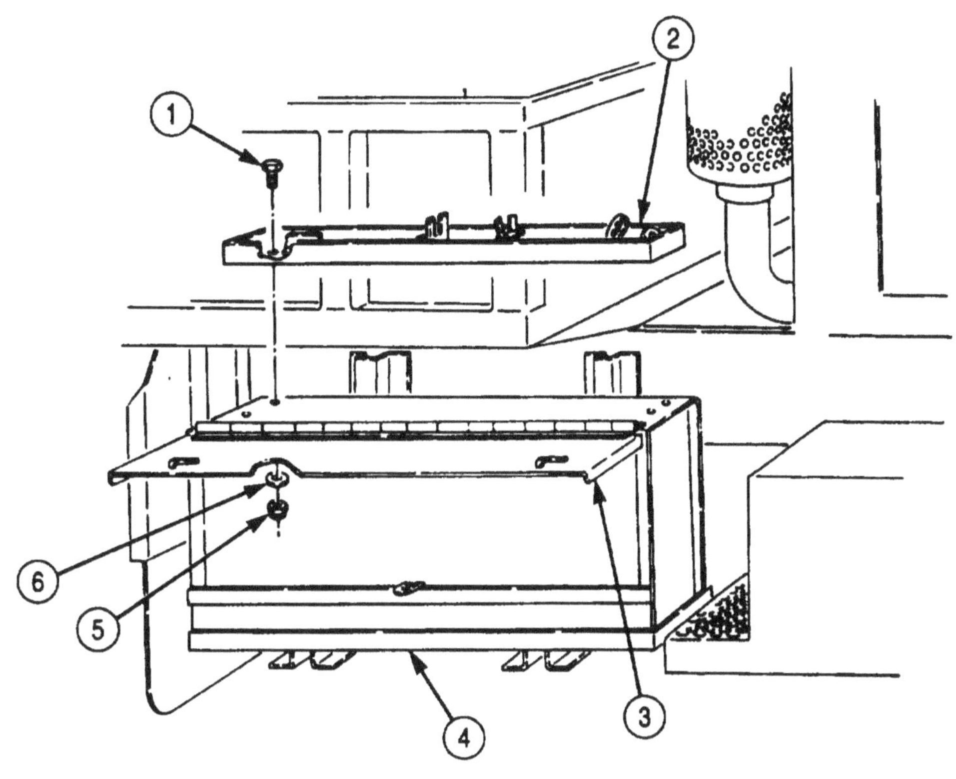

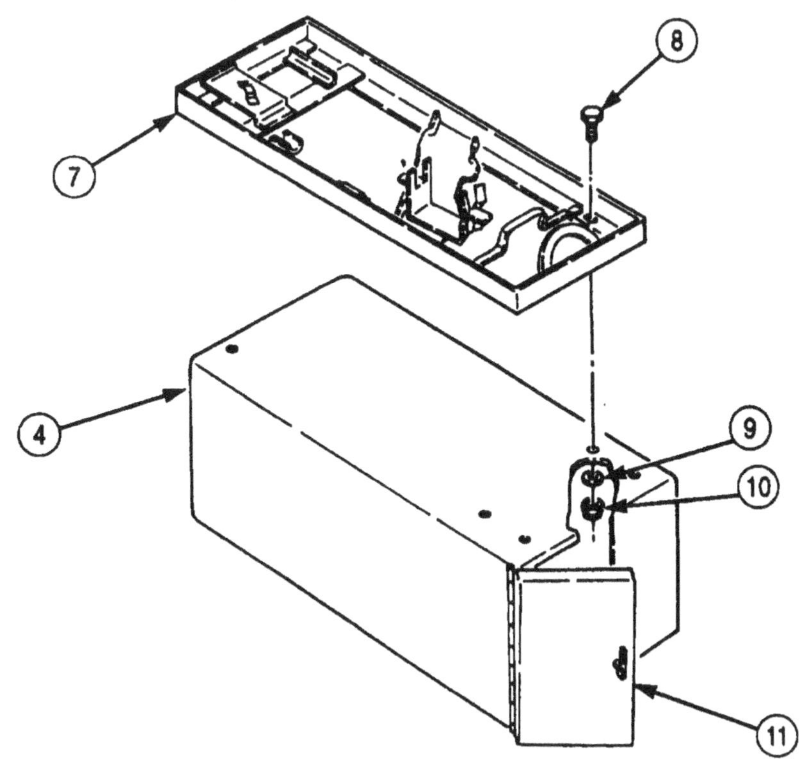

3-415. PIONEER TOOL KIT MOUNTING BRACKET REPLACEMENT (Contd)

e. Bracket Removal (M929, M930)

1. Remove four locknuts (1) and screws (4) from spare tire support (2) and bracket (3). Discard locknuts (1).
2. Remove two locknuts (6), washers (5), screws (10), and bracket (3) from spare tire support (2) and brackets (7). Discard locknuts (6).
3. Remove two locknuts (8), washers (9), screws (12), and two angle brackets (11) from spare tire support (2) and brackets (7). Discard locknuts (8).

f. Bracket Installation (M929, M930)

1. Install two angle brackets (11) on spare tire support (2) and brackets (7) with two screws (12), washers (9), and new locknuts (8).
2. Install bracket (3) on spare tire support (2) and brackets (7) with two screws (10), washers (5), and new locknuts (6).
3. Install four screws (4) and new locknuts (1) on spare tire support (2) and bracket (3).

g. Bracket Removal (M929A1, M930A1)

Remove four locknuts (14), screws (15), and bracket (16) from lower davit (13). Discard locknuts (14).

h. Bracket Installation (M929A1, M930A1)

Install bracket (16) on lower davit (13) with four screws (15) and new locknuts (14).

3-415. PIONEER TOOL KIT MOUNTING BRACKET REPLACEMENT (Contd)

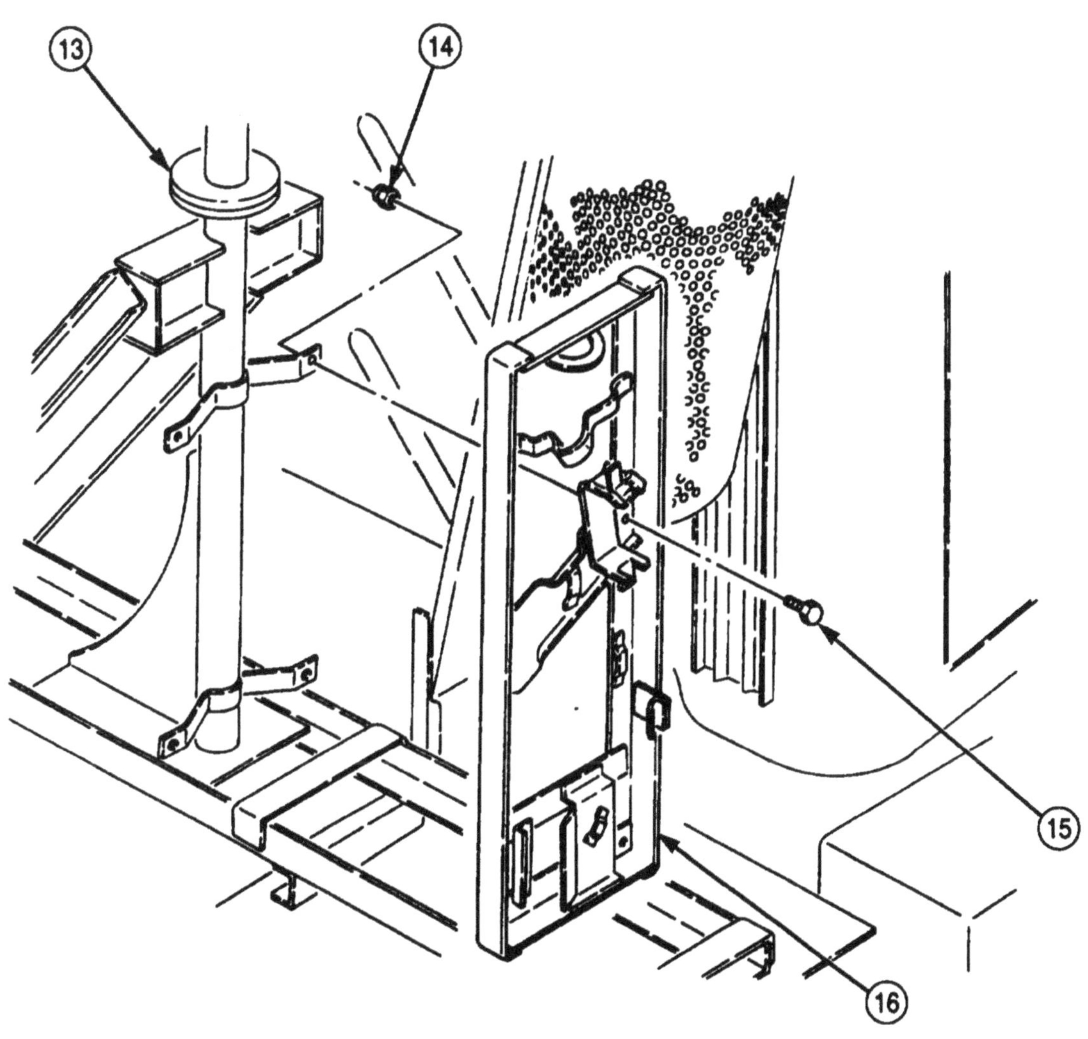

3-416. FIRE EXTINGUISHER MOUNTING BRACKET KIT REPLACEMENT

THIS TASK COVERS:

a. Removal b. Installation

INITIAL SETUP:

APPLICABLE MODELS
All

TOOLS
General mechanic's tool kit (Appendix E, Item 1)

MATERIALS/PARTS
Four locknuts (Appendix D, Item 313)

REFERENCES (TM)
TM 9-2320-272-10
TM 9-2320-272-24P

EQUIPMENT CONDITION
Parking brake set (TM 9-2320-272-10).

a. Removal

1. Loosen two clamps (2) on bracket (3).
2. Remove fire extinguisher (1) from bracket (3).
3. Remove four locknuts (5), washers (4), screws (8), bracket (3), and two spacers (6) from engine cover (7). Discard locknuts (5).

b. Installation

1. Install two spacers (6) and bracket (3) on engine cover (7) with four screws (8), washers (4), and new locknuts (5).
2. Install fire extinguisher (1) on bracket (3).
3. Tighten two clamps (2) on bracket (3).

3-416. FIRE EXTINGUISHER MOUNTING BRACKET KIT REPLACEMENT (Contd)

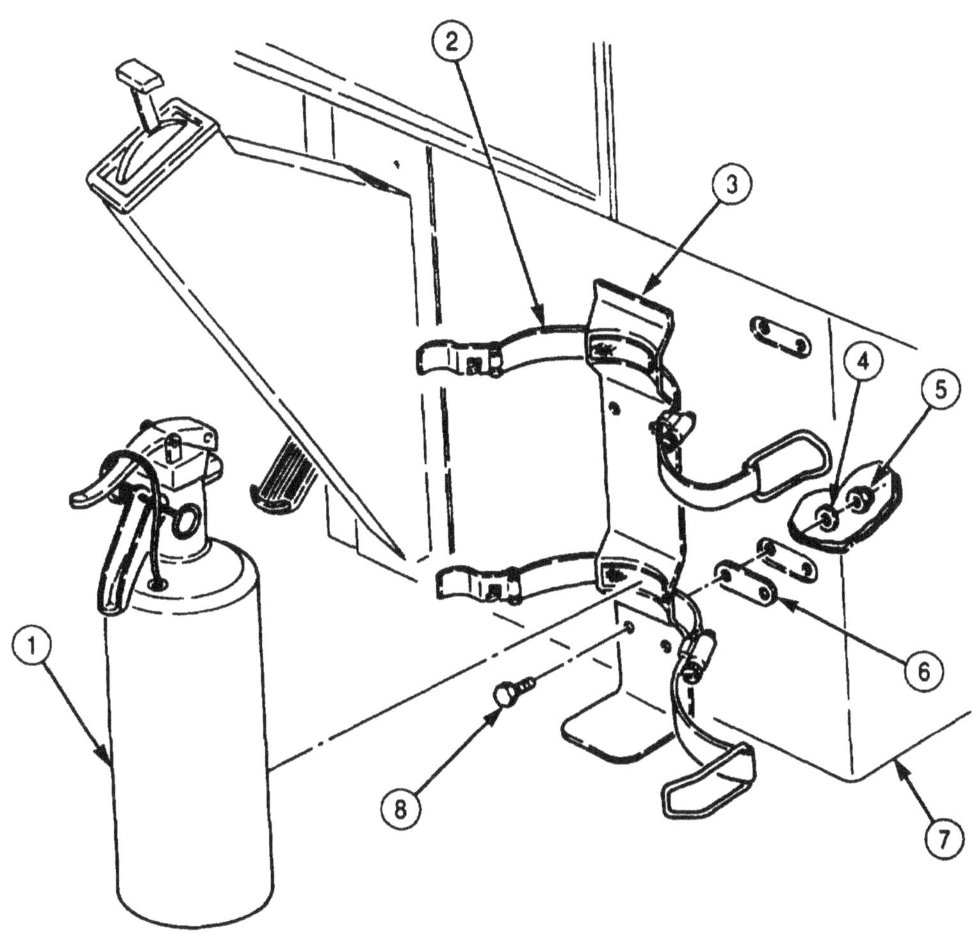

3-417. CHEMICAL AGENT ALARM MOUNTING BRACKET KIT REPLACEMENT

THIS TASK COVERS:

a Chemical Alarm Wiring Harness Removal
b. Detector and Alarm Bracket Removal
cDetector and Alarm Bracket Installation
d. Chemical Alarm Wiring Harness Installation

INITIAL SETUP:

APPLICABLE MODELS
All

TOOLS
General mechanic's tool kit (Appendix E, Item 1)

MATERIALS/PARTS
Six locknuts (Appendix D, Item 299)
Lockwasher (Appendix D, Item 353)
Electrical tape (Appendix C, Item 73)

REFERENCES (TM)
TM 9-2320-272-10
TM 9-2320-272-24P

EQUIPMENT CONDITION
- Parking brake set (TM 9-2320-272-10).
- Hood raised and secured (TM 9-2320-272-10).
- Battery ground cables disconnected (para. 3-126).
- Driver's seat removed (para. 3-283).

a. Chemical Alarm Wiring Harness Removal

NOTE

The cable is provided in two sections. A short wire harness with a screw-on receptacle completes the circuit for the detector and alarm units to the main power source. This harness is only supplied when both chemical detector and alarm units are issued to the field.

1. Remove nut (4) and receptacle (1) from detector bracket (3).
2. Remove screw (6), clamp (7), and cable (2) from locations A, B, C, D, E, and F along left side fender (5) and underside of vehicle.
3. Remove six clamps (7) from cable (2).

3-417. CHEMICAL AGENT ALARM MOUNTING BRACKET KIT REPLACEMENT (Contd)

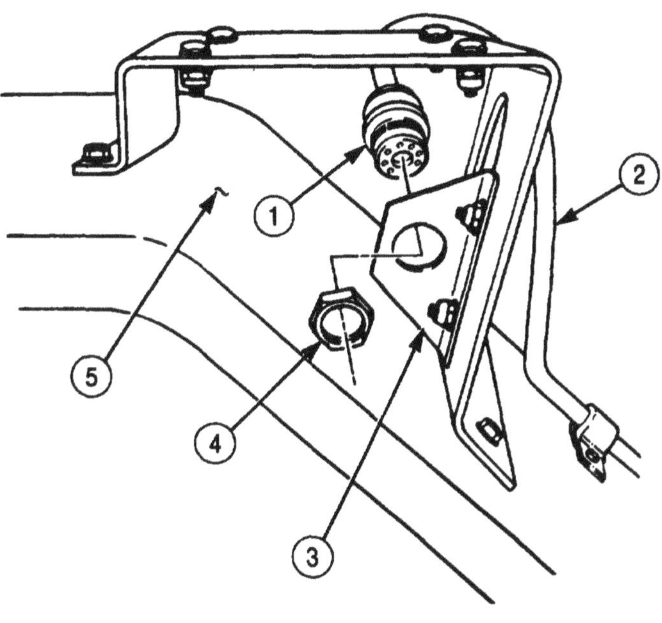

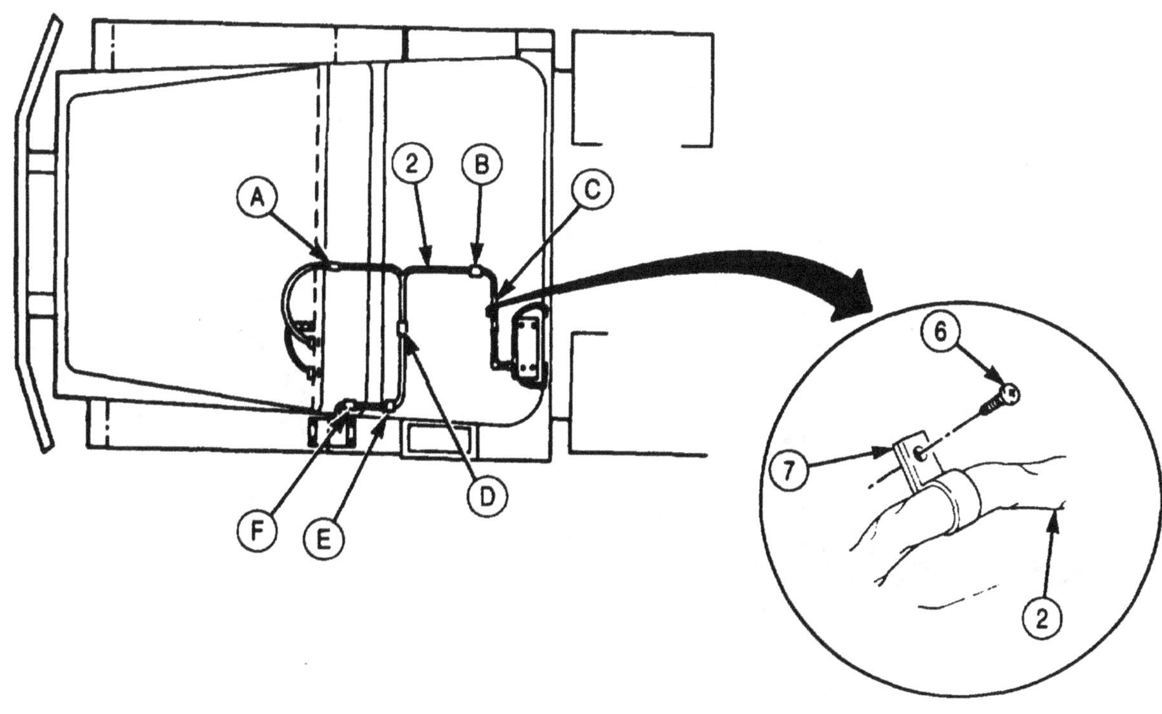

3-417. CHEMICAL AGENT ALARM MOUNTING BRACKET KIT REPLACEMENT (Contd)

4. Disconnect harness cable positive wire (4) from power circuit wire (3) in engine compartment at wiring harness (1).
5. Remove screw (9), clamp (5), lockwasher (6). and ground wires (7) and (8) from protective control box (2). Discard lockwasher (6).
6. Remove wire (11) from grommet (12) and floor (10) behind driver's seat.

b. Detector and Alarm Bracket Removal

1. Remove four locknuts (16), support plate (17), four screws (14), and bracket (15) from left front fender (13). Discard locknuts (16).

NOTE

Assistant will help with step 2.

2. Remove two locknuts (21), mounting bracket (18), two screws (20), and washers (19) from cab body (22). Discard locknuts (21).

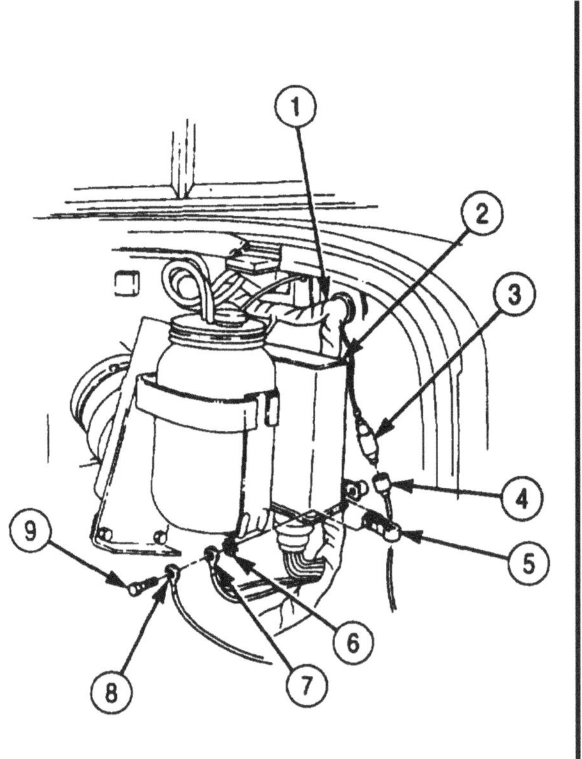

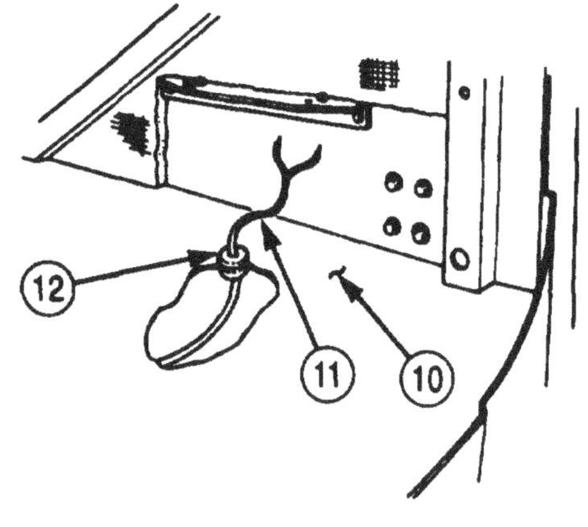

3-417. CHEMICAL AGENT ALARM MOUNTING BRACKET KIT REPLACEMENT (Contd)

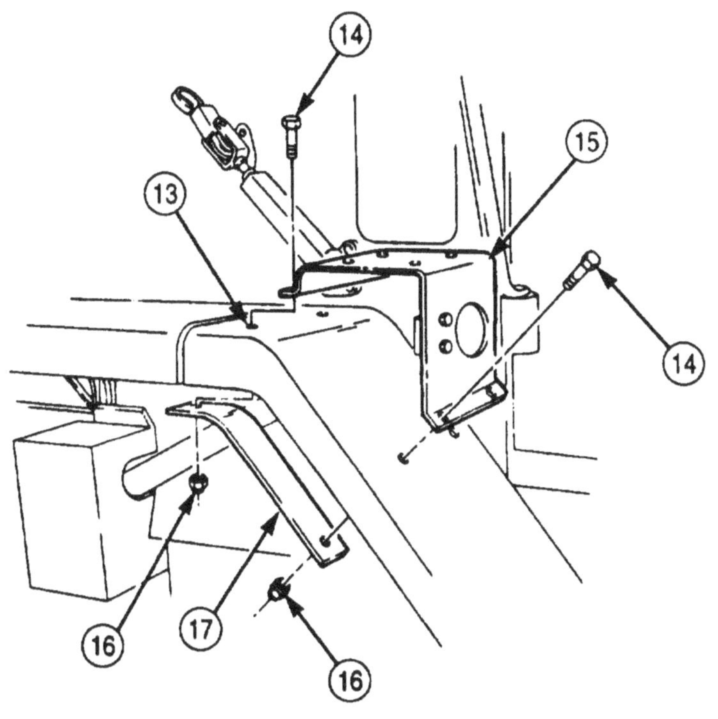

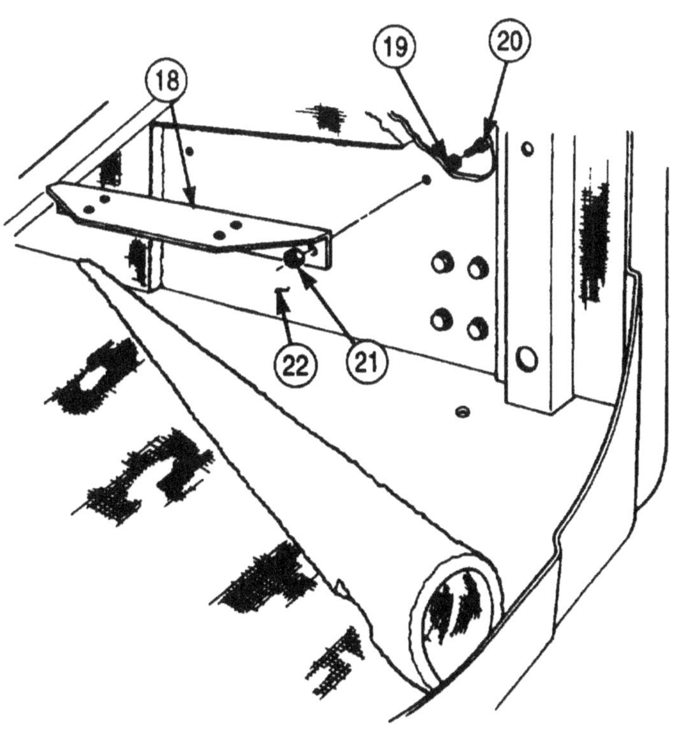

3-417. CHEMICAL AGENT ALARM MOUNTING BRACKET KIT REPLACEMENT (Contd)

c. Detector and Alarm Bracket Installation

1. Install bracket (2) and support plate (5) on left front fender (3) with four screws (1) and new locknuts (4).

NOTE

- Alarm unit bracket is mounted to the cab behind driver's seat.
- Assistant will help with step 2.

2. Install mounting bracket (6) on cab body (9) with two washers (7), screws (8), and new locknuts (10).

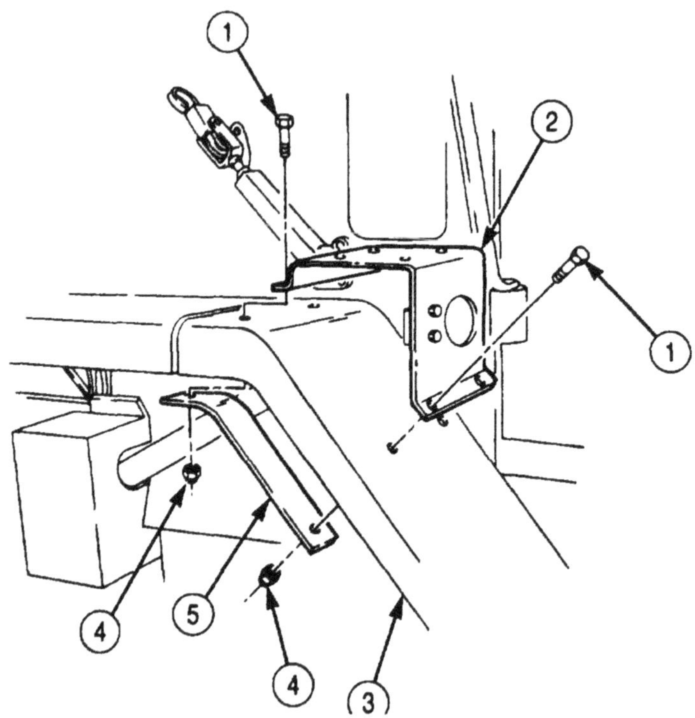

3-417. CHEMICAL AGENT ALARM MOUNTING BRACKET KIT REPLACEMENT (Contd)

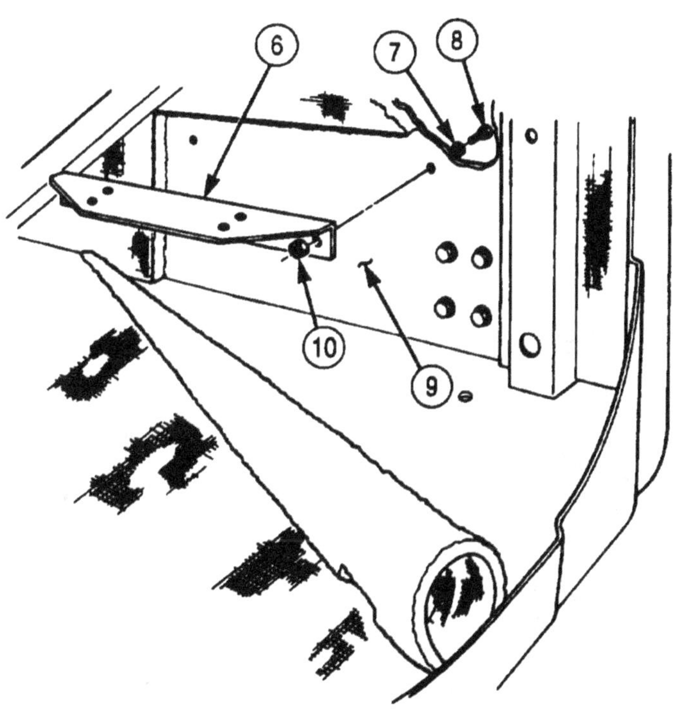

3-417. CHEMICAL AGENT ALARM MOUNTING BRACKET KIT REPLACEMENT (Contd)

d. Chemical Alarm Wiring Harness Installation

1. Insert wiring harness receptacle (1) into detector bracket (3) and install with nut (4).
2. Insert split, tinned ends of alarm unit connector wire (8) up through grommet (6) and floor (7) and tape tinned ends together for protection. Tinned ends will connect to alarm unit.
3. Connect harness cable positive wire (11) to wire (9) from wiring harness (2).
4. Position clamp (12) around harness cable positive wire (11).
5. Install chemical detector ground wire (15), main harness ground wire (14), and clamp (12) on protective control box (10) with new lockwasher (13) and screw (16).

NOTE

Six clamp positions are provided to support a split cable harness routing to the underside of the vehicle.

6. Position six clamps (18) over harness cable (2) and install at cable clamp locations A, B, C, D, E, and F along left side fender (5) with six screws (17). Remove slack in cable (2) for tightening.

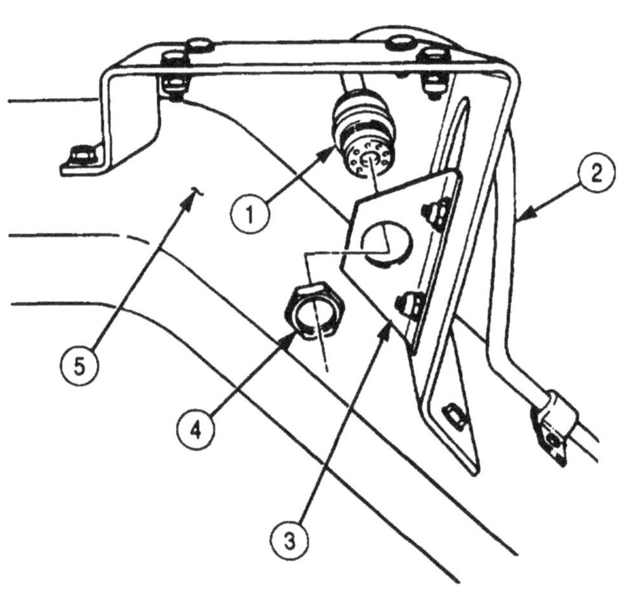

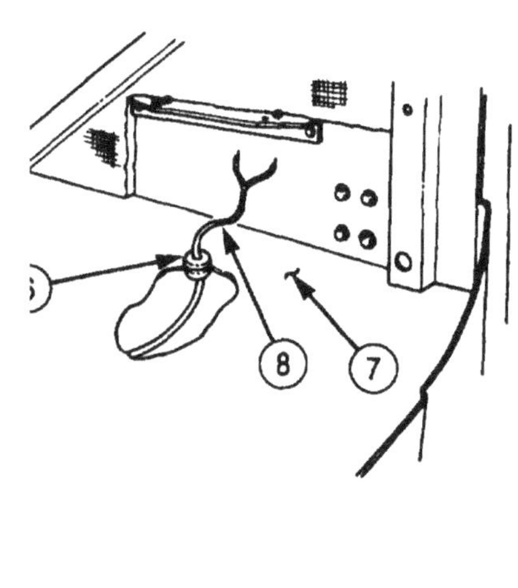

3-417. CHEMICAL AGENT ALARM MOUNTING BRACKET KIT REPLACEMENT (Contd)

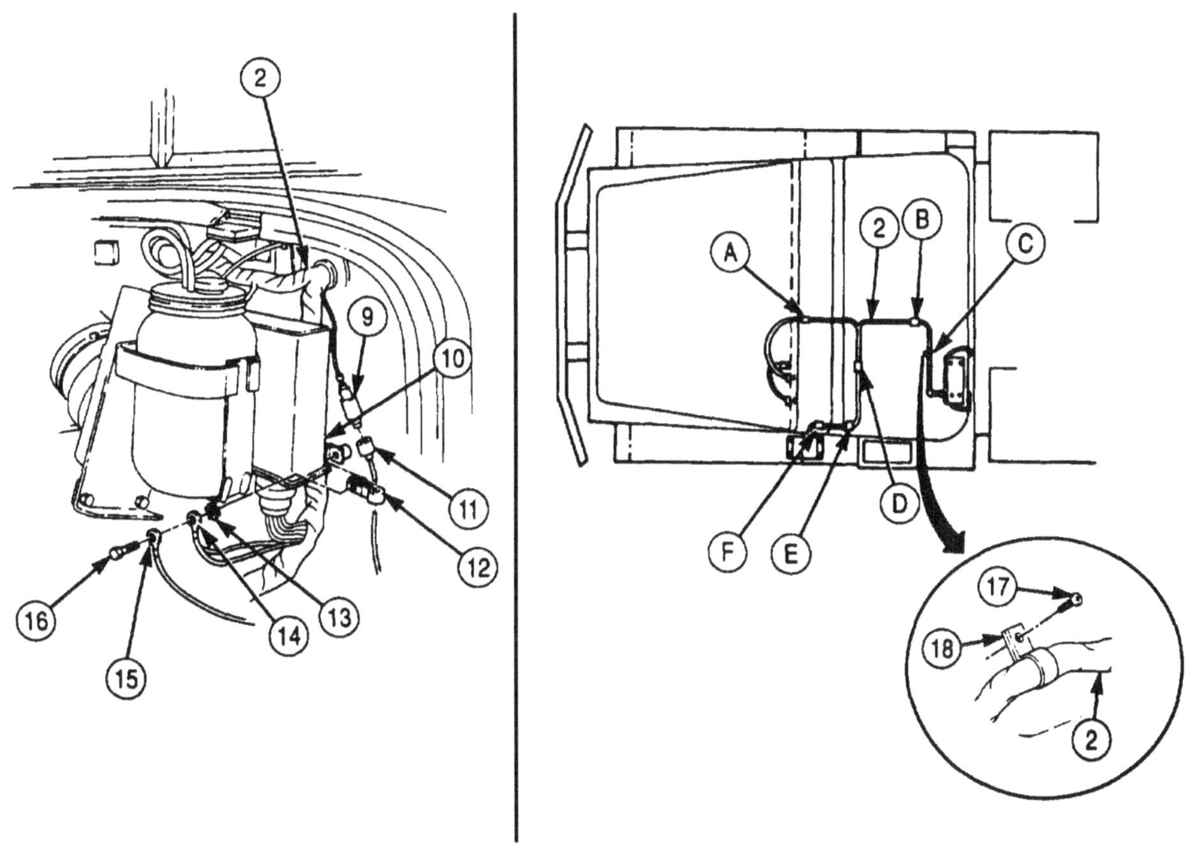

FOLLOW-ON TASKS: • Install driver's seat (para. 3-283).
• Connect battery ground cables (para. 3-126).

3-418. MACHINE GUN MOUNTING KIT MAINTENANCE

THIS TASK COVERS:

a. Removal
b. Disassembly
c. Assembly
d. Installation

INITIAL SETUP:

APPLICABLE MODELS
M923/A1/A2, M925/A1/A2, M927/A1/A2, M928/A1/A2, M931/A1/A2, M932/A1/A2, M936/A1/A2

TOOLS
General mechanic's tool kit (Appendix E, Item 1)

MATERIALS/PARTS
Twenty-four locknuts (Appendix D, Item 294)
Two cotter pins (Appendix D, Item 46)
Eight locknuts (Appendix D, Item 298)
Eight locknuts (Appendix D, Item 309)

REFERENCES (TM)
TM 9-2320-272-10
TM 9-2320-272-24P

EQUIPMENT CONDITION
- Parking brake set (TM 9-2320-272-10).
- Cab top removed (task d. only) (TM 9-2320-272-10).
- Windshield lowered (task d. only) (TM 9-2320-272-10).

a. Removal

1. Remove twelve locknuts (7), washers (3), screws (2), washers (3), and ring mount (1) from adapters (4), (5), and (9). Discard locknuts (7).
2. Loosen four screws (16) and remove right front bracket and post (6) from front gun mount bracket (14). Tighten four screws (16).
3. Remove eight locknuts (13), washers (12), four U-bolts (11), left rear bracket and post (10), and right rear bracket and post (8) from rear gun mount brackets (15). Discard locknuts (13).

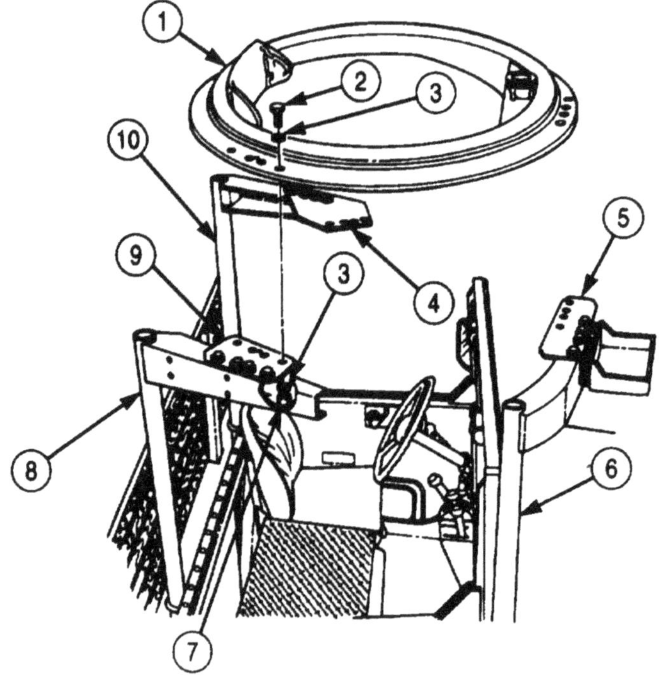

3-418. MACHINE GUN MOUNTING KIT MAINTENANCE (Contd)

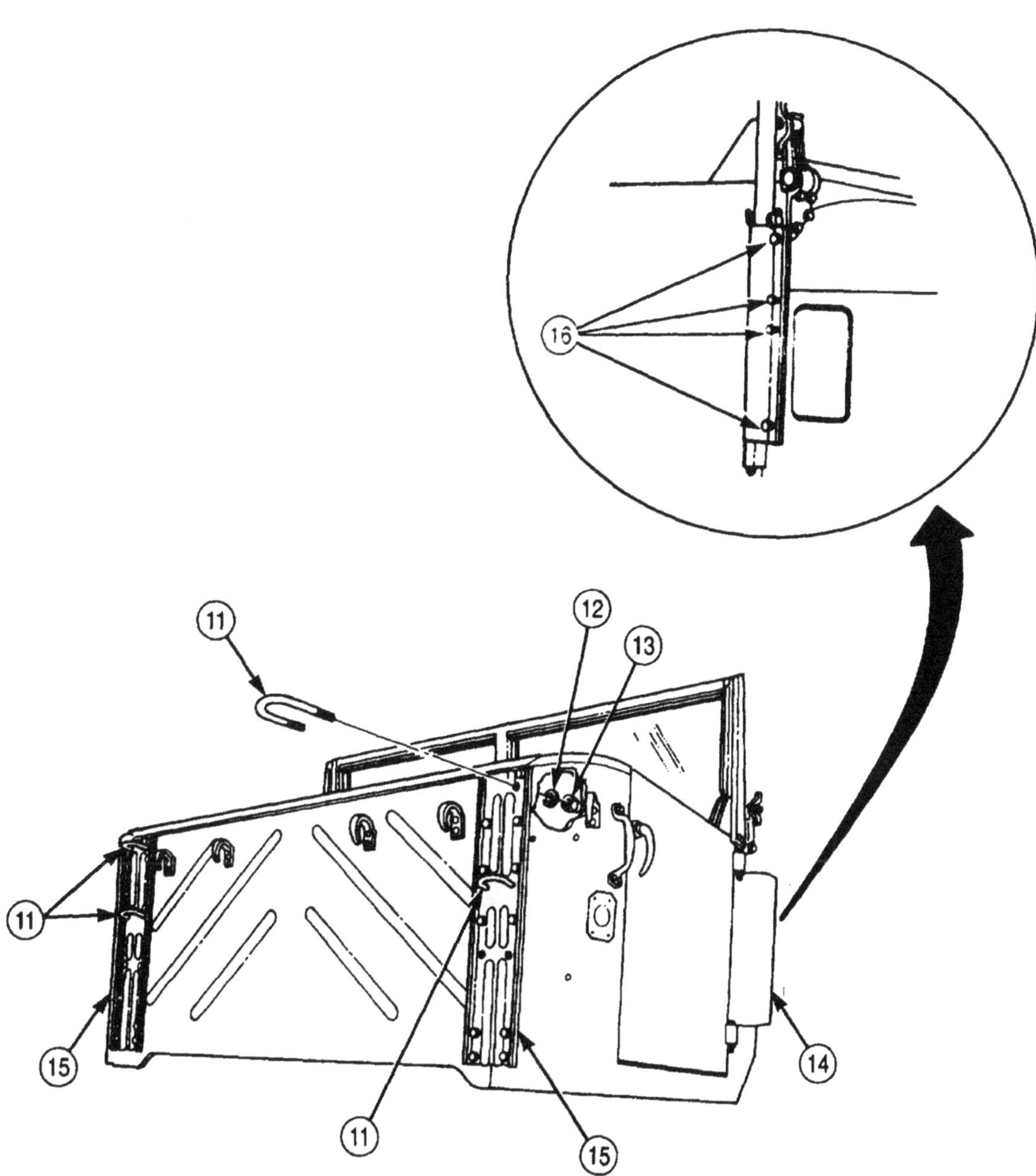

3-418. MACHINE GUN MOUNTING KIT MAINTENANCE (Contd)

b. Disassembly

NOTE

Note position of adapters for installation.

1. Remove four locknuts (7), (11), and (16), washers (11, (9), and (14), screws (2), (8), and (13), washers (l), (9), and (14), and adapters (3), (10), and (15) from bracket and post (6), (12), and (17). Discard locknuts (7), (11), and (16).
2. Remove cotter pin (5) and pin (4) from right front bracket and post (6). Discard cotter pin (5).
3. Remove cotter pin (19) and pin (18) from left rear bracket and post (17). Discard cotter pin (19).

NOTE

Ammunition trays on right front bracket and post and left rear bracket and post are replaced the same. Steps 4, 5, and 6 cover the right front bracket and post.

4. Remove two straps (25) from ammunition tray (26).
5. Remove four locknuts (20), washers (21), screws (23), and ammunition tray (26) from right front bracket and post (6). Discard locknuts (20).

NOTE

Note position of screws and tray brackets for installation.

6. Remove four locknuts (28), washers (27), screws (29), and two tray brackets (22) from ammunition tray extensions (24). Discard locknuts (28).

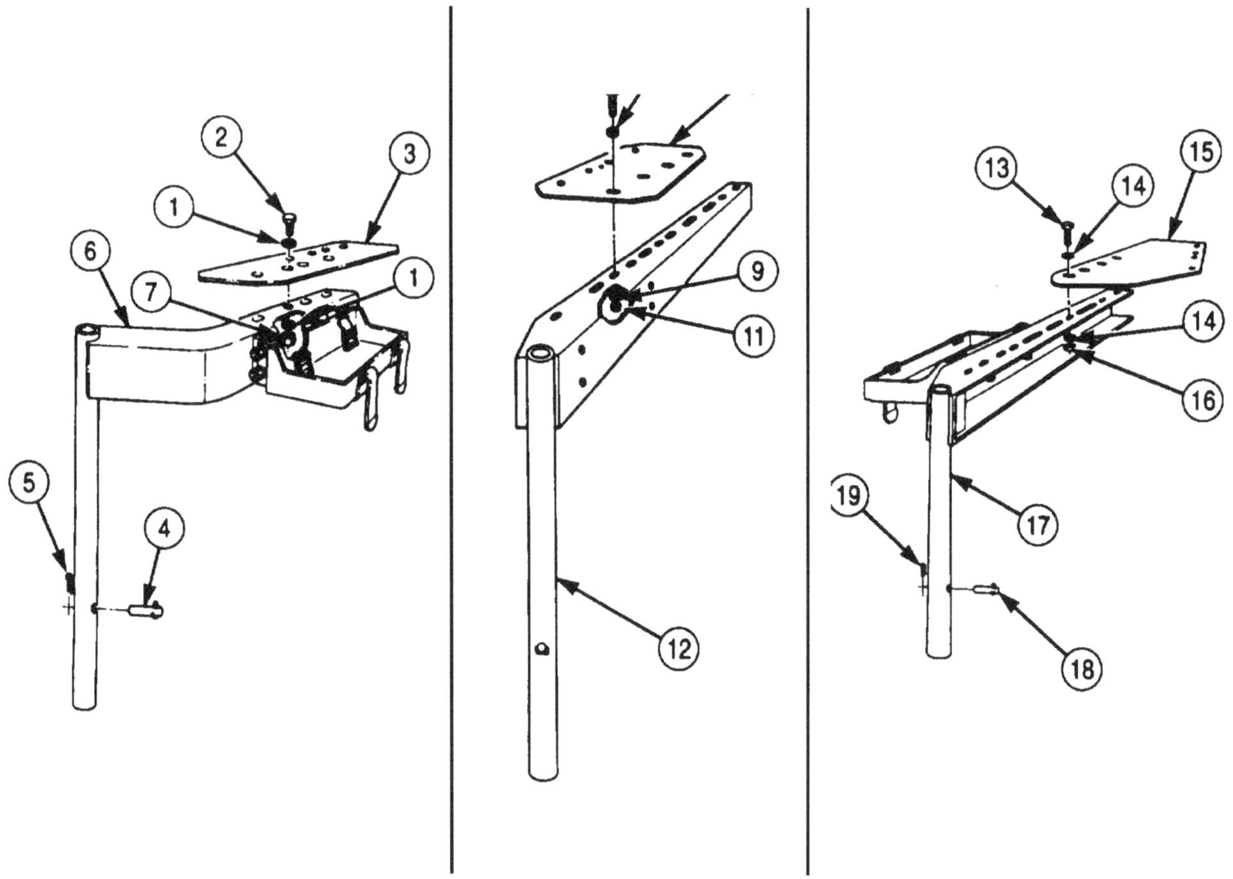

3-418. MACHINE GUN MOUNTING KIT MAINTENANCE (Contd)

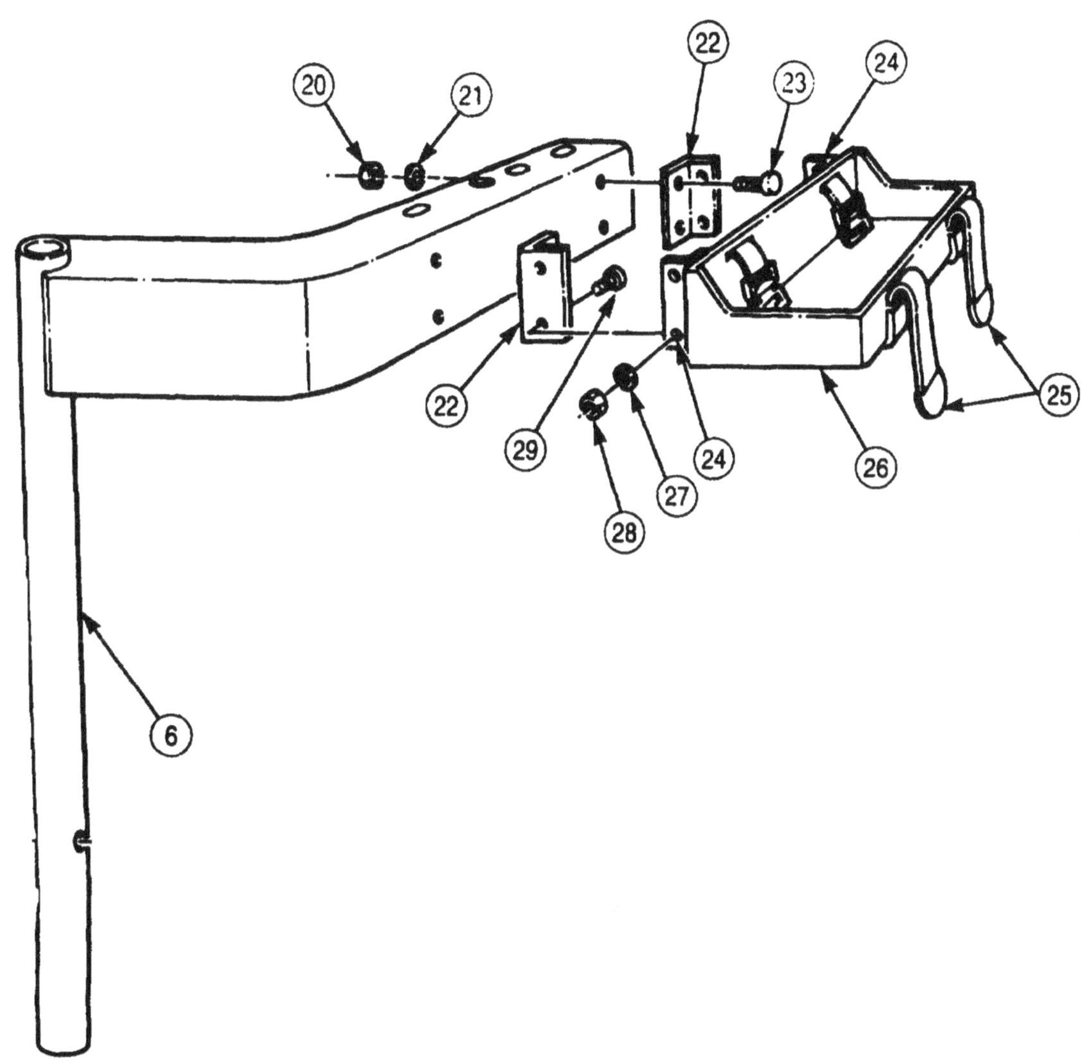

3-418. MACHINE GUN MOUNTING KIT MAINTENANCE (Contd)

c. Assembly

NOTE

Ammunition trays on both right front bracket and post and left rear bracket and post are replaced the same. Steps 1, 2, and 3 cover the right front bracket and post.

1. Install two tray brackets (22) on ammunition tray extensions (24) with four screws (29), washers (27), and new locknuts (28).
2. Install ammunition tray (26) on right front bracket and post (6) with four screws (23), washers (21), and new locknuts (20).
3. Install two straps (25) on ammunition tray (26).
4. Install pin (18) and new cotter pin (19) on left rear bracket and post (17).
5. Install pin (4) and new cotter pin (5) on right front bracket and post (6).
6. Install adapters (3), (10), and (15) on bracket and post (6), (12), and (17) with four washers (l), (9), and (14), screws (2), (8), and (13), washers (l), (9), and (14), and new locknuts (7), (11), and (16).

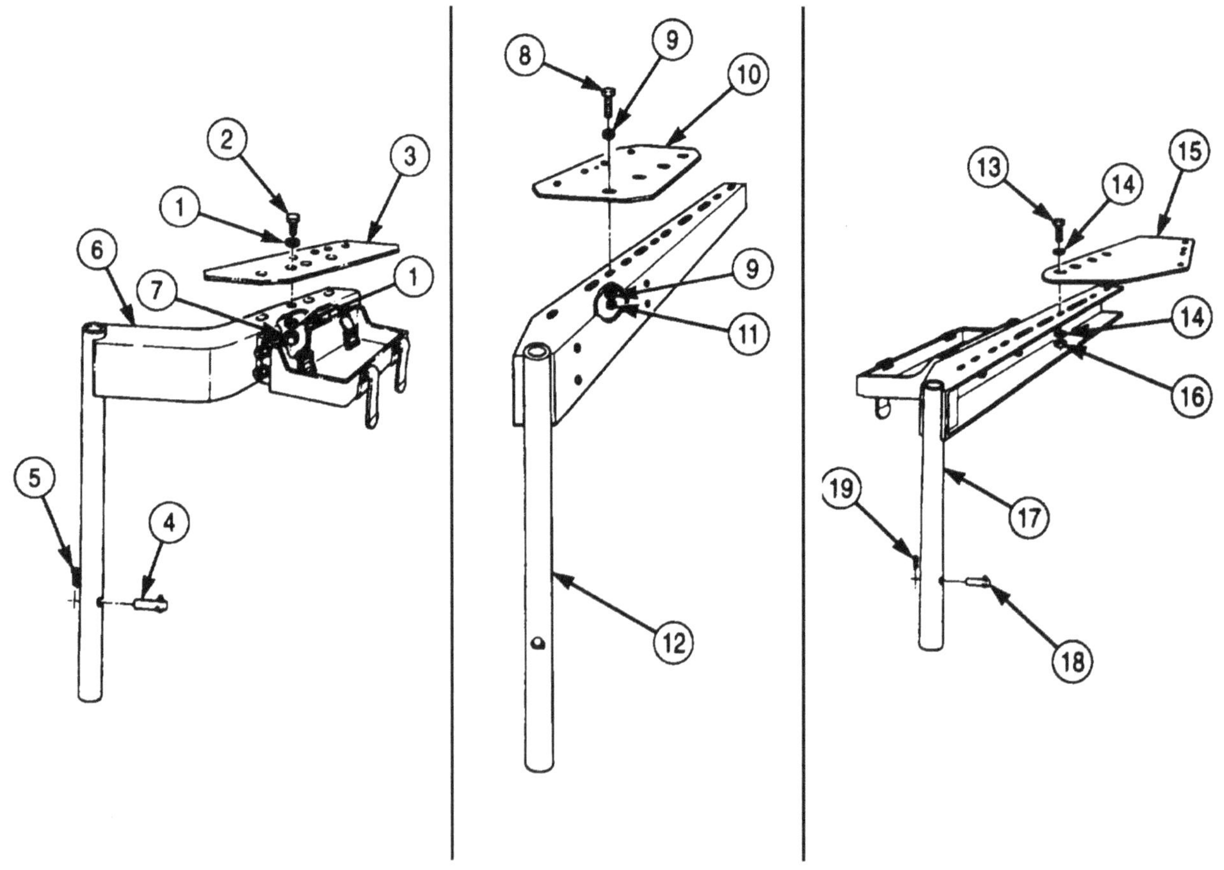

3-418. MACHINE GUN MOUNTING KIT MAINTENANCE (Contd)

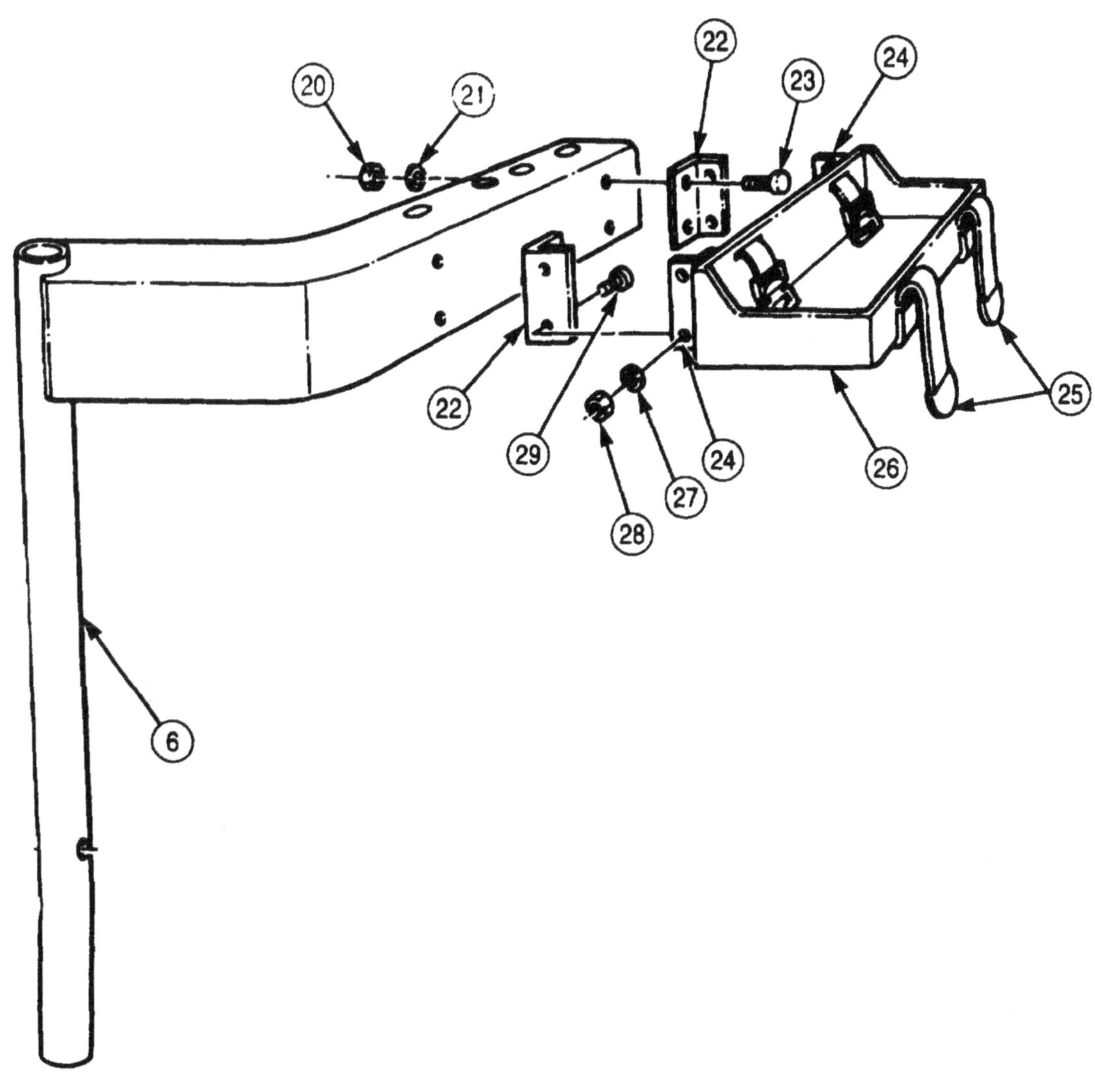

3-418. MACHINE GUN MOUNTING KIT MAINTENANCE (Contd)

d. Installation

NOTE

Step 1 may require the removal of some insulation material from inside of cab.

1. Install two U-bolts (9) in each rear gun mount bracket (7) with four washers (5) and new locknuts (8). Finger-tighten locknuts (8).
2. Install right rear bracket and post (1) through two U-bolts (9) and rear gun mount bracket (7). Ensure welded post nib (2) rests on top of U-bolt (9) and bracket and post (1) turns freely.
3. Install left rear bracket and post (10) through two U-bolts (9) and rear gun mount bracket (7). Ensure welded post nib (2) rests on top of U-bolt (9) and bracket and post (10) turns freely.
4. Loosen four screws (11) on front gun mount bracket (6) and install right front bracket and post (3) in front gun mount bracket (6). Ensure pin (4) rests on top of front gun mount bracket (6).
5. Position ring mount (12) to adapters (15), (18), and (16) and align locating hole (19) in rear adapter (18) and locating hole (20) in ring mount (12).
6. Install ring mount (12) on adapters (15), (18), and (16) with twelve washers (14), screws (13), washers (14), and new locknuts (17). Tighten screws (11).

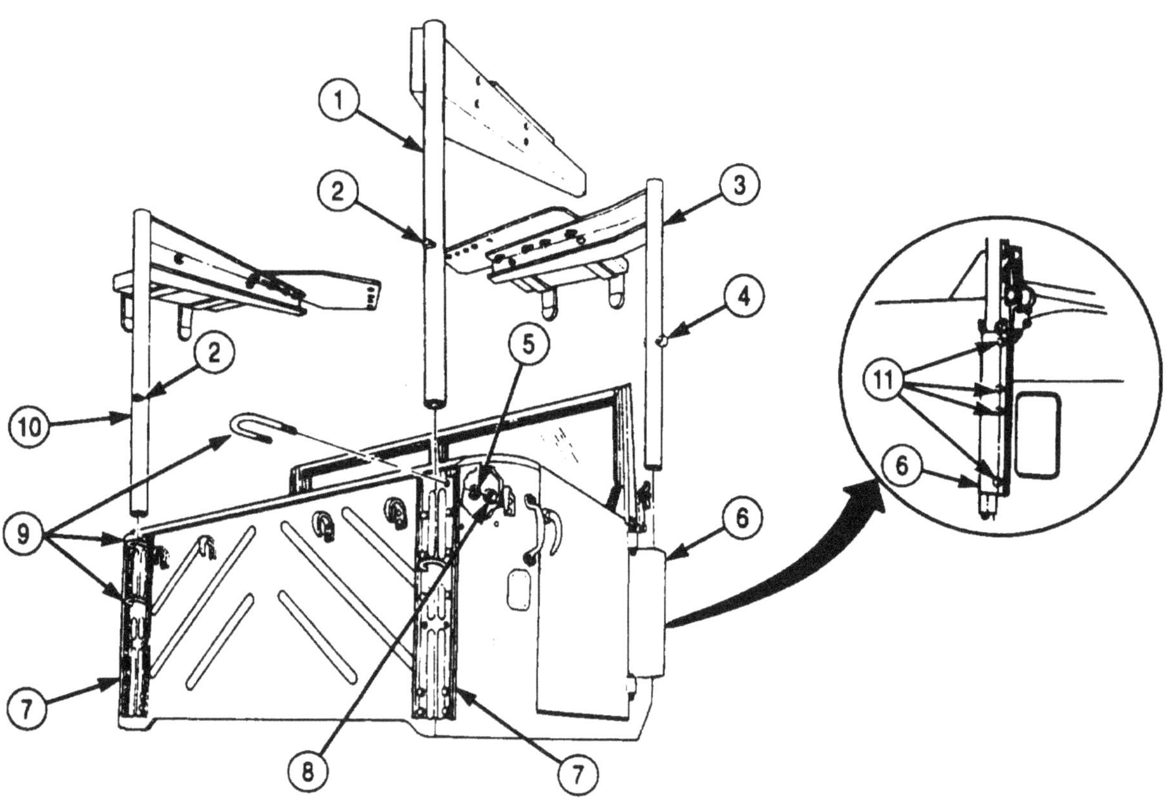

3-418. MACHINE GUN MOUNTING KIT MAINTENANCE (Contd)

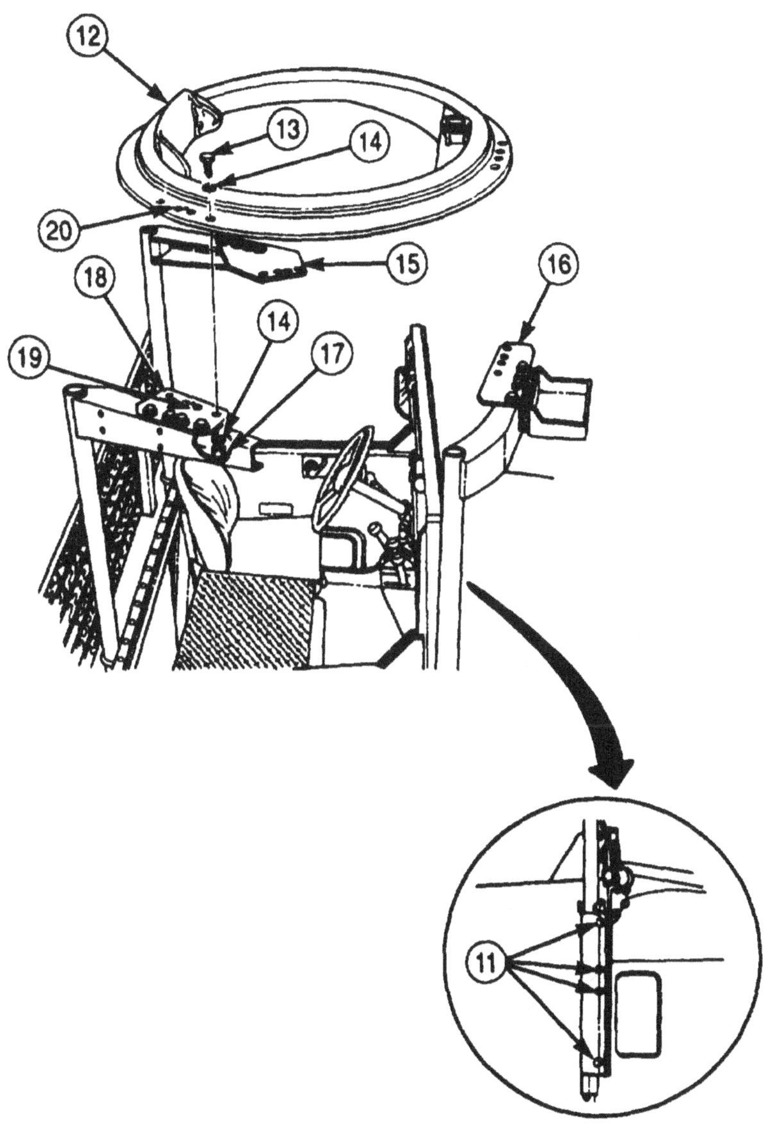

3-419. DECONTAMINATION (M13) APPARATUS MOUNTING BRACKET KIT REPLACEMENT

THIS TASK COVERS:

a. Removal (M929/A1/A2, M930/A1/A2, M931/A1/A2, M932/A1/A2)
b. Installation (M929/A1/A2, M930/A1/A2, M931/A1/A2, M932/A1/A2)
c. Removal (M934/A1/A2)
d. Installation (M934/A1/A2)
e. Removal (M936/A1/A2)
f. Installation (M936/A1/A2)
g. Removal (M923/A1/A2, M925/A1/A2, M927/A1/A2, M928/A1/A2)
h. Installation (M923/A1/A2, M925/A1/A2, M927/A1/A2, M928/A1/A2)

INITIAL SETUP:

APPLICABLE MODELS
AI1

TOOLS
General mechanic's tool kit (Appendix E, Item 1)

MATERIALS/PARTS
Four locknuts (M929/A1/A2, M930lA1/A2, M93l/A1/A2, M932/A1/A2) (Appendix D, Item 291)
Eight locknuts (M934/A1/A2, M936/A1/A2) (Appendix D, Item 291)
Four locknuts (M934/A1/A2) (Appendix D, Item 294)
Four locknuts (M939/A1/A2) (Appendix D, Item 291)
Seventeen locknuts (M923/A1/A2, M925/A1/A2, M927/A1/A2, M928/A1/A2) (Appendix D, Item 291)

REFERENCES (TM)
TM 9-2320-272-10
TM 9-2320-272-24P

EQUIPMENT CONDITION
- Parking brake set (TM 9-2320-272-10)
- Spare tire removed (TM 9-2320-272-10).

---

a. Removal (M929/A1/A2, M930/A1/A2, M93l/A1/A2, M932/Al/A2)

Remove four locknuts (5), screws (2), washers (3), mounting bracket (l), and two supports (6) from deck plate (4). Discard locknuts (5).

b. Installation (M929/A1/A2, M930/A1/A2, M931/A1/A2, M932/A1/A2)

Install two supports (6) and mounting bracket (1) on deck plate (4) with four washers (3), screws (2), and new locknuts (5).

c. Removal (M934/A1/A2)

1. Remove four locknuts (10), screws (9), washers (8), mounting bracket (11, and four washers (8) from support bracket (7). Discard locknuts (10).
2. Remove four locknuts (12), screws (11), and support bracket (7) from right frame rail (13). Discard locknuts (12).

d. Installation (M934/A1/A2)

1. Install support bracket (7) on right frame rail (13) with four screws (11) and new locknuts (12).
2. Install four washers (8) and mounting bracket (1) on support bracket (7) with four washers (8), screws (9), and new locknuts (10).

3-419. DECONTAMINATION (M13) APPARATUS MOUNTING BRACKET KIT REPLACEMENT (Contd)

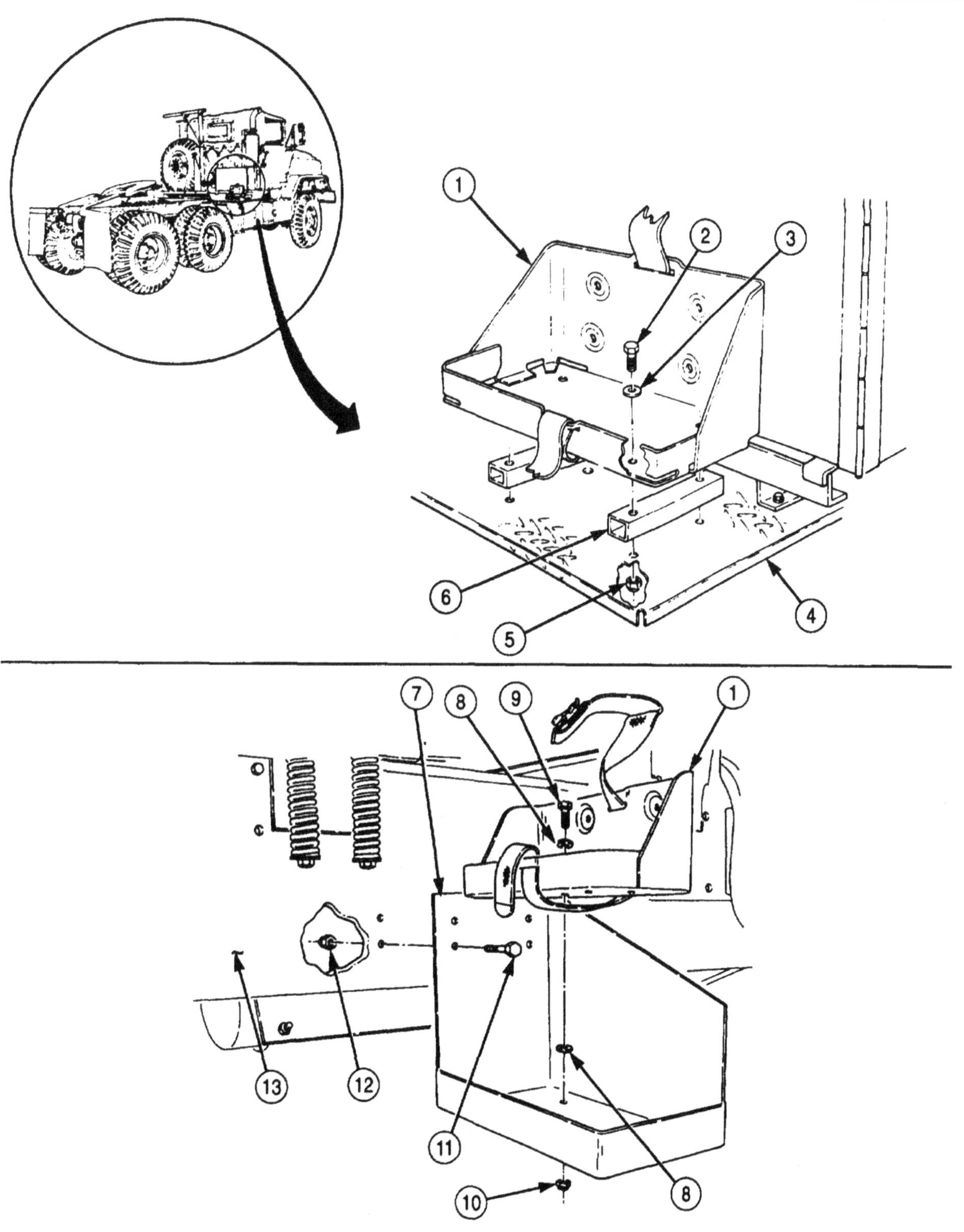

3-419. DECONTAMINATION (M13) APPARATUS MOUNTING BRACKET KIT REPLACEMENT (Contd)

e. Removal (M936/A1/A2)

1. Remove four locknuts (4), washers (3), screws (2), washers (3), and mounting bracket (1) from support bracket (9). Discard locknuts (4).
2. Remove four locknuts (8), two washers (7), four screws (6), washers (7), and support bracket (9) from body (5). Discard locknuts (8).

f. Installation (M936/A1/A2)

1. Install support bracket (9) on body (5) with four washers (7), screws (6), two washers (7), and four new locknuts (8).
2. Install mounting bracket (1) on support bracket (9) with four washers (3), screws (2), washers (3), and new locknuts (4).

g. Removal (M923/A1/A2, M925/A1/A2, M927/A1/A2, M928/A1/A2)

1. Remove six locknuts (25), screws (11), and support bracket (21) with mounting bracket (1) from upper splash guard (28). Discard locknuts (25).
2. Remove four locknuts (22), screws (23), washers (24), and mounting bracket (1) from support bracket (21). Discard locknuts (22).
3. Remove three locknuts (20), screws (16), brace (17), retainer (18), and lower splash guard (19) from upper splash guard (28). Discard locknuts (20).
4. Remove locknut (15), screw (12), and brace (17) from cargo body (13). Discard locknut (15).
5. Remove three locknuts (26), washers (27), upper splash guard (28), reinforcement plate (10), and three screws (14) from cargo body (13). Discard locknuts (26).

h. Installation (M923/A1/A2, M925/A1/A2, M927/A1/A2, M928/A1/A2)

1. Install reinforcement plate (10) and upper splash guard (28) on cargo body (13) with three screws (14), washers (27), and new locknuts (26).
2. Install brace (17) on cargo body (13) with screw (12) and new locknut (15).
3. Install lower splash guard (19), retainer (18), and brace (17) on upper splash guard (28) with three screws (16) and new locknuts (20).
4. Install mounting bracket (1) on support bracket (21) with four washers (24), screws (23), and new locknuts (22).
5. Install support bracket (21) with mounting bracket (1) on upper splash guard (28) with six screws (11) and new locknuts (25).

3-419. DECONTAMINATION (M 13) APPARATUS MOUNTING BRACKET KIT REPLACEMENT (Contd)

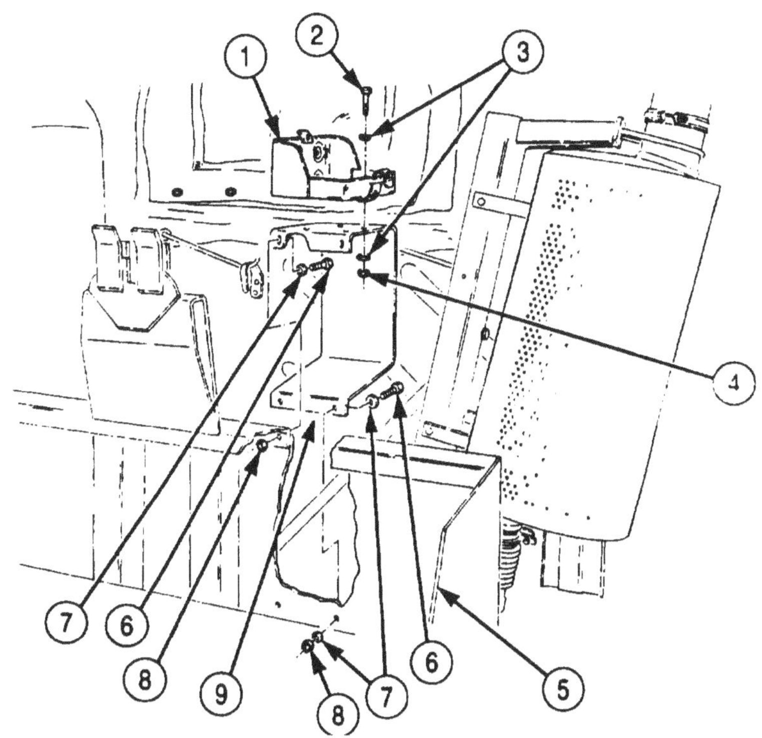

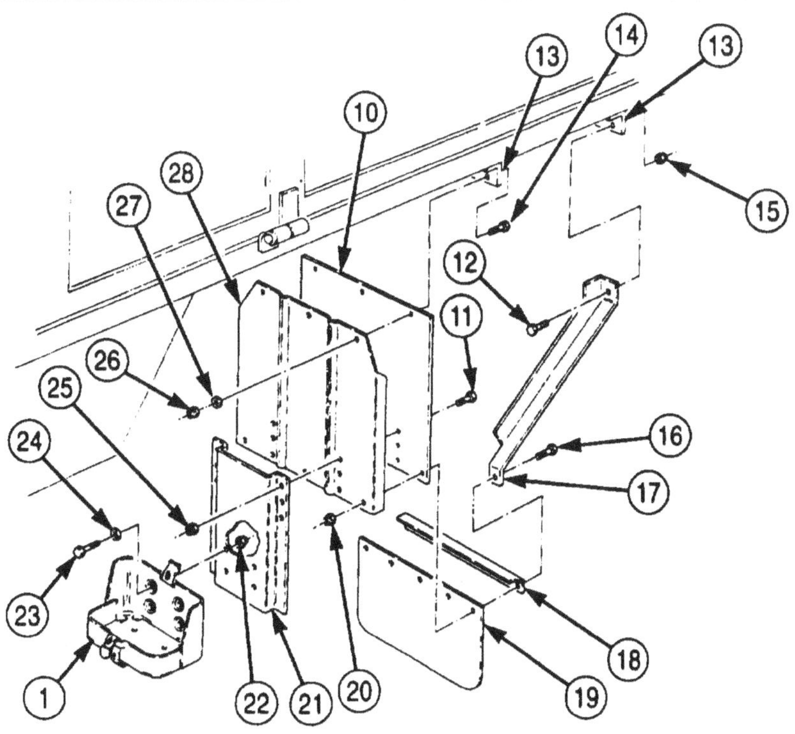

FOLLOW-ON TASK: Install spare tire (TM 9-2320-272-10).

3-420. MUD GUARD KIT (M931/A1/A2, M932/A1/A2) REPLACEMENT

THIS TASK COVERS:

a. Removal b. Installation

INITIAL SETUP:

APPLICABLE MODELS
M931/A1/A2, M932/A1/A2

TOOLS
General mechanic's tool kit (Appendix E, Item 1)

MATERIALS/PARTS
Five locknuts (Appendix D, Item 294)

REFERENCES (TM)
TM 9-2320-272-10
TM 9-2320-272-24P

EQUIPMENT CONDITION
- Parking brake set (TM 9-2320-272-10).
- Pioneer tool kit mounting bracket removed (para. 3-415).

NOTE

Left and right mud guards are replaced the same way. This procedure covers the left mud guard.

a. Removal

1. Remove pin (6) and mud guard (1) from bracket (8).
2. Remove two locknuts (3), screws (7), and bracket (8) from plate (4). Discard locknuts (3).
3. Remove three locknuts (2), screws (5), and plate (4) from frame rail (9). Discard locknuts (2).

b. Installation

1. Install plate (4) on frame rail (9) with three screws (5) and new locknuts (2).
2. Install bracket (8) on plate (4) with two screws (7) and new locknuts (3).
3. Install mud guard (1) on bracket (8) with pin (6).

3-420. MUD GUARD (M931/A1/A2, M932/A1/A2) REPLACEMENT (Contd)

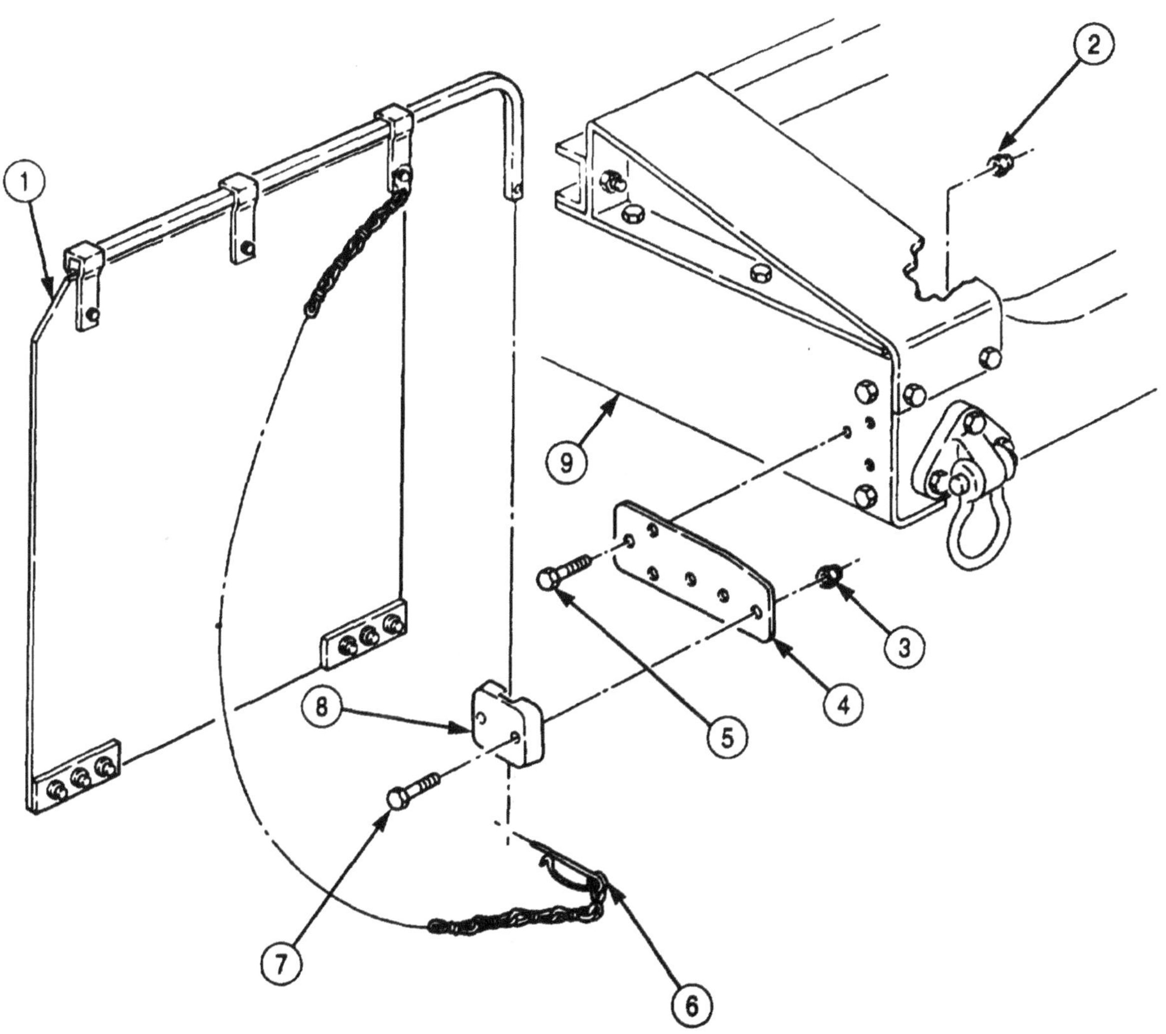

FOLLOW-ON TASK: Install pioneer tool kit mounting bracket (para. 3-415).

3-421. RIFLE MOUNTING KIT REPLACEMENT

THIS TASK COVERS:

a. Removal from Left Door
b. Removal from Dash and Floor
c. Installation on Left Door
d. Installation on Dash and Floor

INITIAL SETUP:

APPLICABLE MODELS
All

TOOLS
General mechanic's tool kit (Appendix E, Item 1)

MATERIALS/PARTS
Three locknuts (left door)
(Appendix D, Item 299)
Four locknuts (dash and floor)
(Appendix D, Item 299)

REFERENCES (TM)
TM 9-2320-272-10
TM 9-2320-272-24P

EQUIPMENT CONDITION
Parking brake set (TM 9-2320-272-10).

a. Removal from Left Door

1. Remove four screws (1) and bracket (2) from door (3).
2. Remove six screws (7) and inspection cover (4) from door (3).
3. Remove three locknuts (8), screws (6), and rifle support (5) from inspection cover (4). Discard locknuts (8).

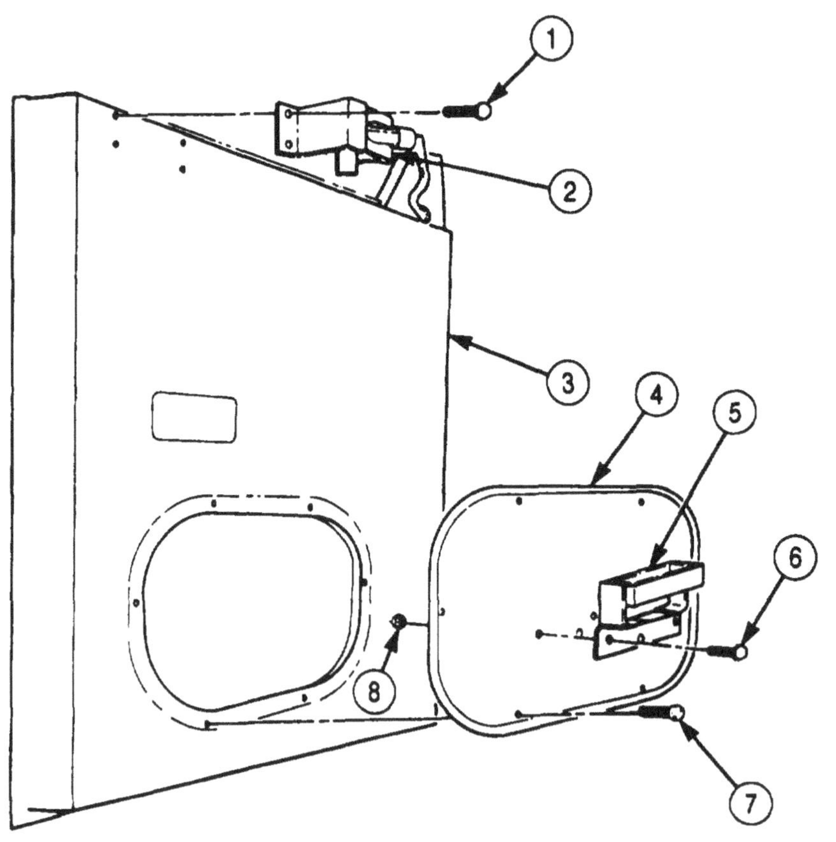

3-421. RIFLE MOUNTING KIT REPLACEMENT (Contd)

b. Removal from Dash and Floor

1. Remove four screws (2), two reinforcements (6), and rifle catch brackets (3) from dash (1).

NOTE

Assistant will help with step 2.

2. Remove two locknuts (8), screws (7) or (9), and rifle supports (5) from floor (4). Discard locknuts (8).

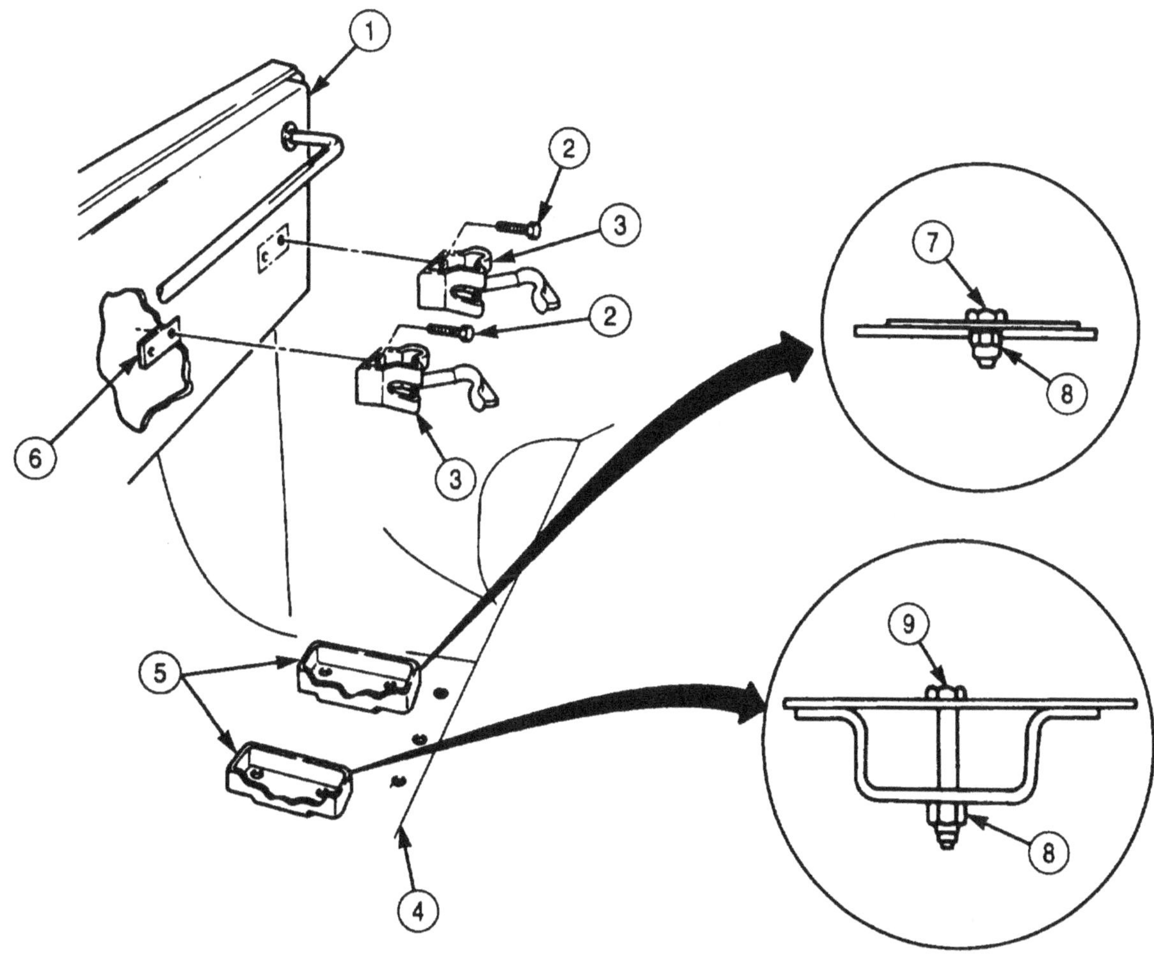

3-421. RIFLE MOUNTING KIT REPLACEMENT (Contd)

c. Installation on Left Door

1. Position bracket (3) to holes (1) in door (4) and install with four screws (2).
2. Install rifle support (6) on inspection cover (5) with three screws (7) and new locknuts (9).
3. Install inspection cover (5) on door (4) with six screws (8).

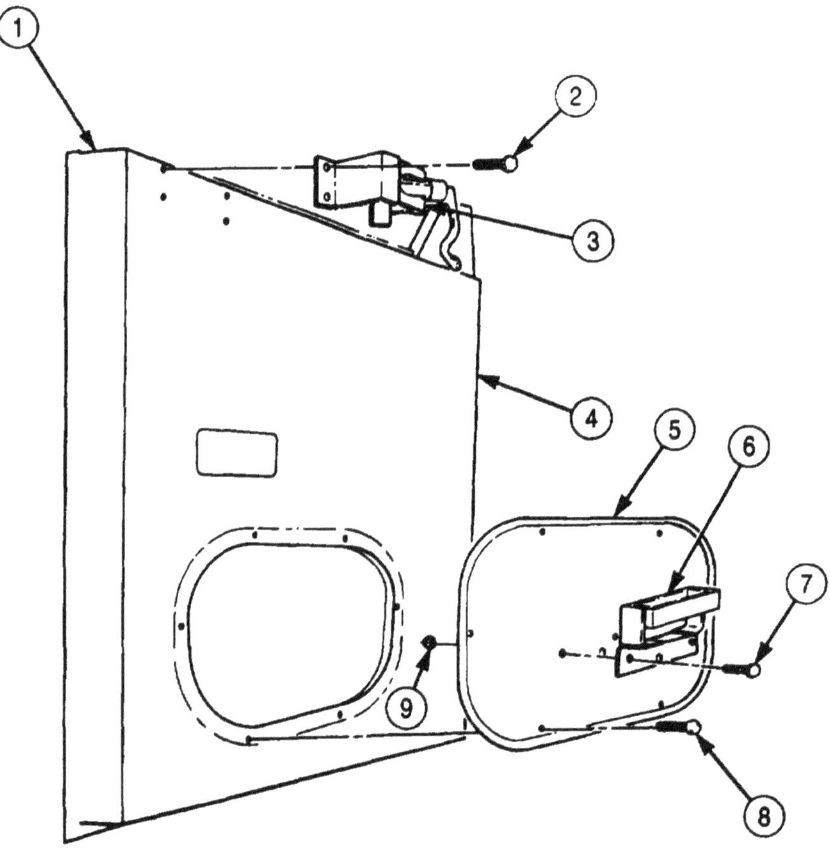

3-421. RIFLE MOUNTING KIT REPLACEMENT (Contd)

d. Installation on Dash and Floor

1. Position two reinforcements (6) behind dash (1) and align holes.
2. Install two rifle catch brackets (3) on dash (1) with four screws (2).

NOTE

Assistant will help with step 3.

3. Install two rifle supports (5) on floor (4) with two screws (7) or (9) and new locknuts (8).

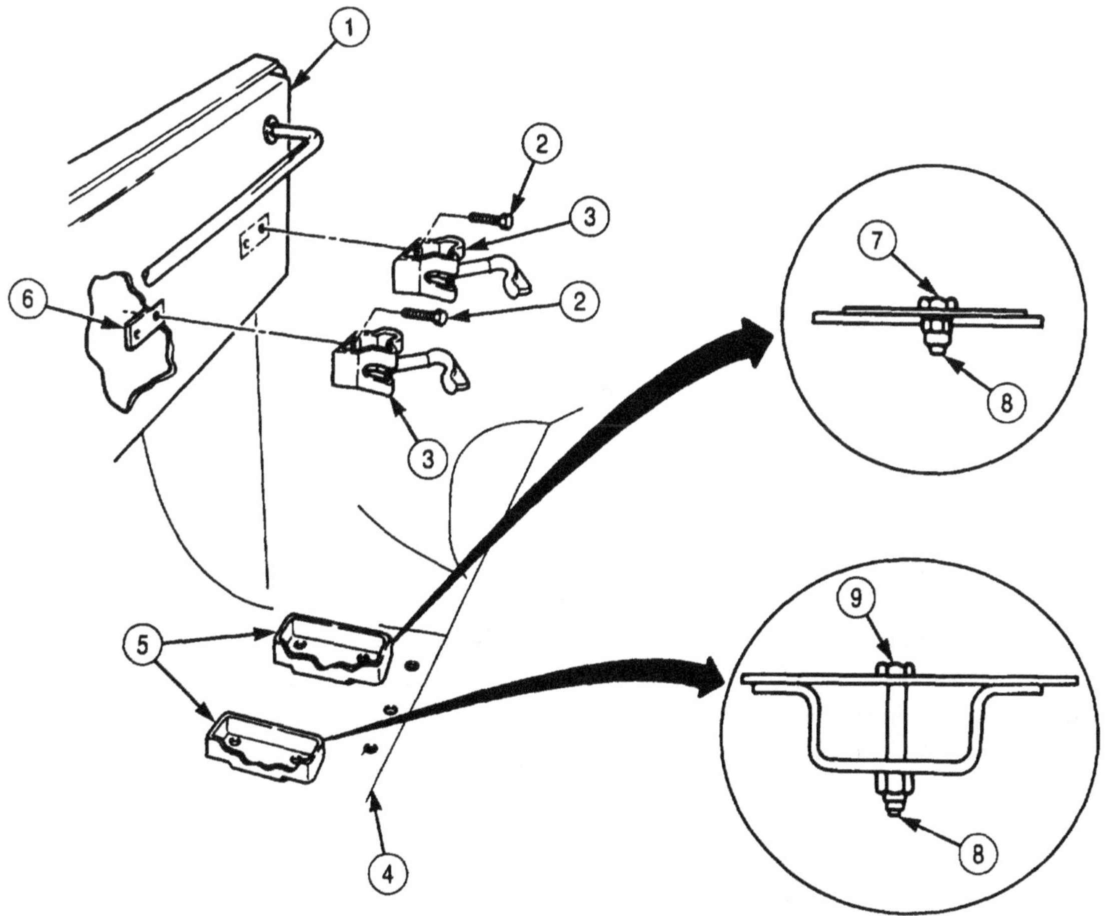

3-422. HAND AIRBRAKE AIR SUPPLY VALVE REPLACEMENT

THIS TASK COVERS:

a. Removal **b. Installation**

INITIAL SETUP:

APPLICABLE MODELS
All (except M931/A1/A2, M932/A1/A2, M933/A1/Al, M936/A1/A2)

TOOLS
General mechanic's tool kit (Appendix E, Item 1)

MATERIALS/PARTS
Two locknuts (Appendix D. Item 291)
Antiseize tape (Appendix C, Item 72)

REFERENCES (TM)
TM 9-2320-272-10
TM 9-2320-272-24P

EQUIPMENT CONDITION
- Parking brake set (TM 9-2320-272-10).
- Air reservoirs drained (TM 9-2320-272-10).

GENERAL SAFETY INSTRUCTIONS
Do not disconnect air lines before draining air reservoirs.

a. Removal

WARNING

Do not disconnect air lines before draining air reservoirs. Small parts under pressure may shoot out with high velocity, causing injury to personnel.

NOTE

Tag air lines for installation.

1. Disconnect two air lines (9) from elbows (8).
2. Remove two elbows (8) from air supply valve (10).
3. Remove pin (3) from button (4).
4. Remove button (4), nut (5), and air supply valve (10) from mounting bracket (6).
5. Remove two locknuts (2), screws (7), and mounting bracket (6) from instrument panel (1). Discard locknuts (2).

b. Installation

1. Install mounting bracket (6) on instrument panel (1) with two screws (7) and new locknuts (2).

NOTE

Wrap all male pipe threads with antiseize tape before installation.

2. Install air supply valve (10) on mounting bracket (6) with nut (5) and button (4).
3. Install pin (3) on button (4).
4. Install two elbows (8) on air supply valve (10).
5. Connect two air lines (9) to elbows (8).

3-422. HAND AIRBRAKE AIR SUPPLY VALVE REPLACEMENT (Contd)

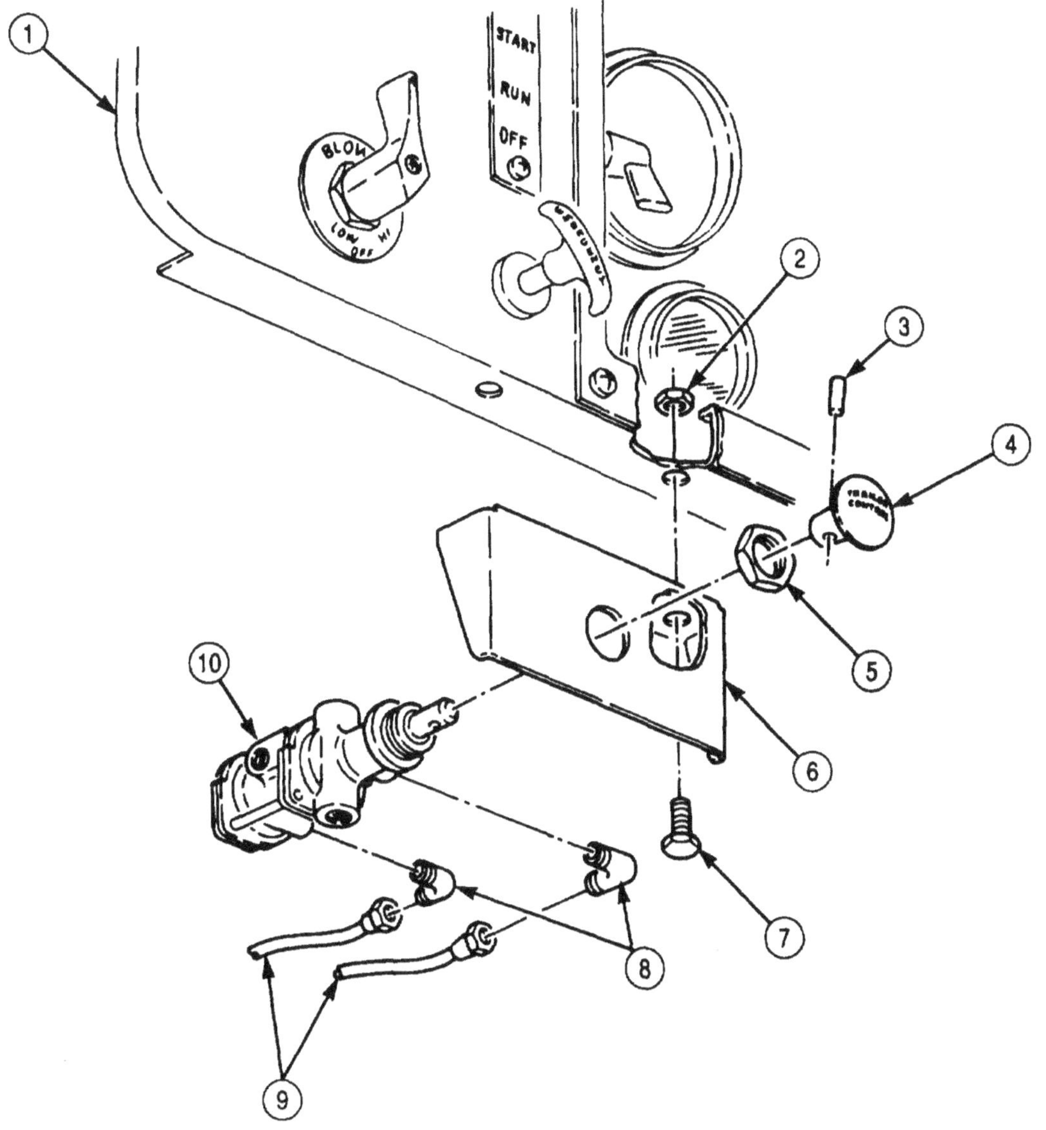

FOLLOW-ON TASK: Start engine (TM 9-2320-272-10) and allow air pressure to build to normal operating range. Check air supply valve for leaks.

3-423. HAND AIRBRAKE CONTROLLER VALVE REPLACEMENT

THIS TASK COVERS:

a. Removal **b. Installation**

INITIAL SETUP:

APPLICABLE MODELS
All (except M931/A1/A2, M932/A1\A2, M933/A1/A2, M936/A1/A2)

TOOLS
General mechanic's tool kit (Appendix E, Item 1)

MATERIALS/PARTS
Antiseize tape (Appendix C, Item 72)

REFERENCES (TM)
TM 9-2320-272-10
TM 9-2320-272-24P

EQUIPMENT CONDITION
- Parking brake set (TM 9-2320-272-10).
- Air reservoirs drained (TM 9-2320- 272-10).

GENERAL SAFETY INSTRUCTIONS
Do not disconnect air lines before draining air reservoirs.

a. Removal

WARNING

Do not disconnect air lines before draining air reservoirs. Small parts under pressure may shoot out with high velocity, causing injury to personnel.

NOTE

Tag air lines for installation.

1. Disconnect two air lines (7) from adapters (6).
2. Remove two adapters (6) from controller valve (5).
3. Remove two nuts (1) from retainer bracket (2) and controller valve (5).
4. Remove retainer bracket (2), strap (3), and controller valve (5) from steering column (4).

b. Installation

NOTE

Wrap all male pipe threads with antiseize tape before installation.

1. Install controller valve (5), strap (3), and retainer bracket (2) on steering column (4).
2. Install retainer bracket (2) on controller valve (5) with two nuts (1).
3. Install two adapters (6) on controller valve (5).
4. Connect two air lines (7) to adapters (6).

3-423. HAND AIRBRAKE CONTROLLER VALVE REPLACEMENT (Contd)

FOLLOW-ON TASK: Start engine (TM 9-2320-272-10) and allow air pressure to build to normal operating range. Check controller valve for leaks.

3-424. HAND AIRBRAKE DOUBLECHECK VALVES REPLACEMENT

THIS TASK COVERS:

a. Removal (Forward-Rear Axle Doublecheck Valve)
b. Installation (Forward-Rear Axle Doublecheck Valve)
c. Removal (Rear-Rear Axle Doublecheck Valve)
d. Installation (Rear-Rear Axle Doublecheck Valve)

INITIAL SETUP:

APPLICABLE MODELS
All (except M931/A1/A2, M932/A1/A2, M933/A1/A2, M936/A1/A2)

TOOLS
General mechanic's tool kit (Appendix E, Item 1)

MATERIALS/PARTS
Locknut (Appendix D, Item 291)
Locknut (Appendix D, Item 299)
Antiseize tape (Appendix C, Item 72)

REFERENCES (TM)
TM 9-2320-272-10
TM 9-2320-272-24P

EQUIPMEMT CONDITION
- Parking brake set (TM 9-2320-272-101.
- Air reservoirs drained (TM 9-2320-272-10).

GENERAL SAFETY INSTRUCTIONS
Do not disconnect air lines before draining air reservoirs.

a. Removal(Forward-Rear Axle Doublecheck Valve)

WARNING

Do not disconnect air lines before draining air reservoirs. Small parts under pressure may shoot out with high velocity, causing injury to personnel.

NOTE

Tag air lines for installation.

1. Disconnect air line (10) from adapter (11).
2. Disconnect two air lines (1) from elbows (2).
3. Remove two elbows (2) from doublecheck valve (8).
4. Remove switch (3) and adapter (11) from tee (9).
5. Remove tee (9) from doublecheck valve (8).
6. Remove locknut (4), washer (5), doublecheck valve (8), and screw (7) from frame rail (6). Discard locknut (4).

b. Installation (Forward-Rear Axle Doublecheck Valve

1. Install doublecheck valve (8) on frame rail (6) with screw (7), washer (5), and new locknut (4).

NOTE

Wrap all male pipe threads with antiseize tape before installation.

2. Install tee (9) on doublecheck valve (8).
3. Install adapter (11) and switch (3) on tee (9).
4. Install two elbows (2) on doublecheck valve (8).
5. Connect two air lines (1) to elbows (2).
6. Connect air line (10) to adapter (11).

3-424. HAND AIRBRAKE DOUBLECHECK VALVES REPLACEMENT (Contd)

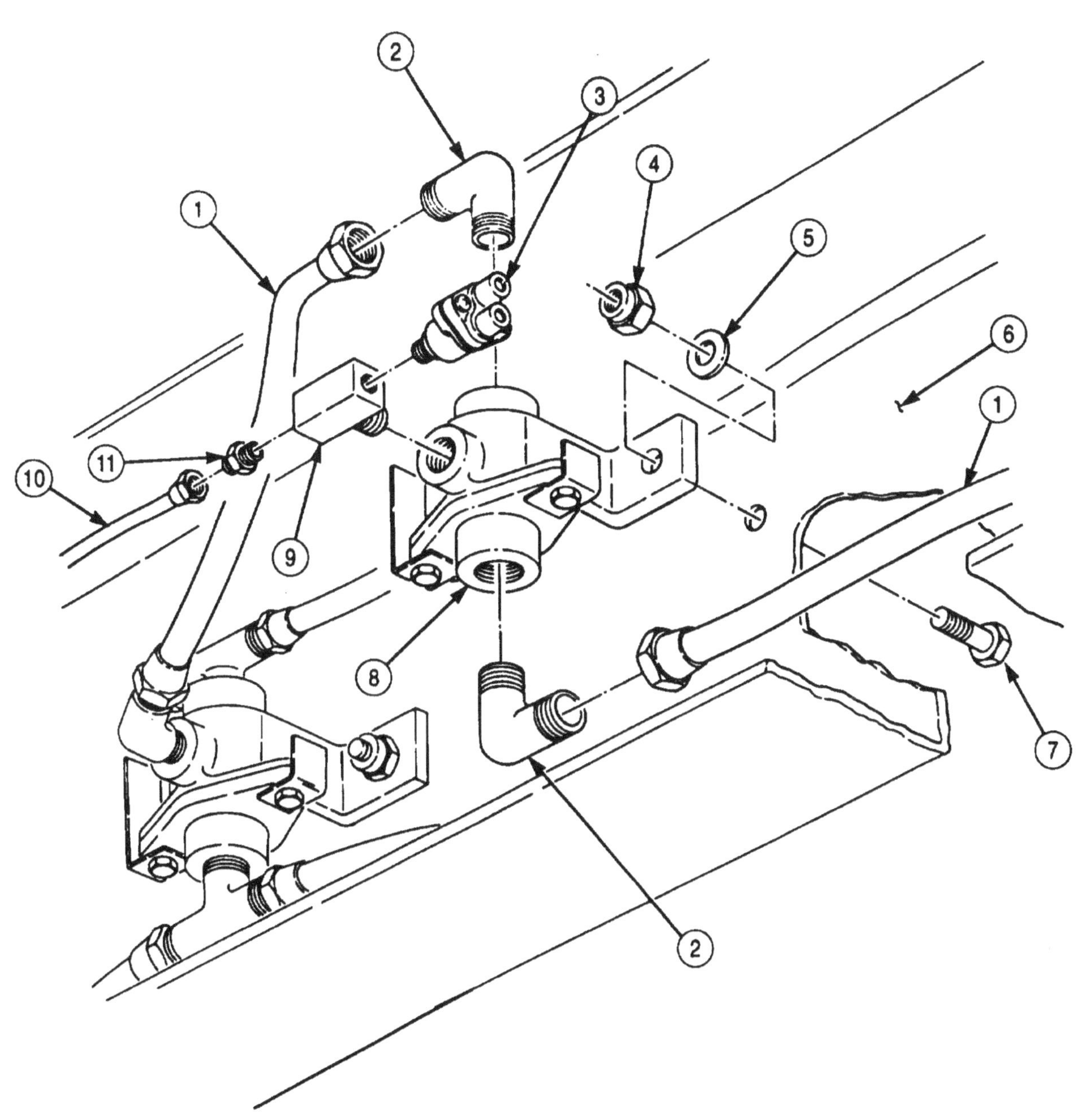

3-424. HAND AIRBRAKE DOUBLECHECK VALVES REPLACEMENT (Contd)

c. Removal (Rear-Rear Axle Doublecheck Valve)

WARNING

Do not disconnect air lines before draining air reservoirs. Small parts under pressure may shoot out with high velocity, causing injury to personnel.

1. Disconnect air line (2) from adapter (3).
2. Disconnect two air lines (1) from elbows (4).
3. Remove two elbows (4) and adapter (3) from doublecheck valve (5).
4. Remove locknut (8), doublecheck valve (5), and screw (6) from frame rail (7). Discard locknut (8).

d. Installation (Rear-Rear Axle Doublecheck Valve)

1. Install doublecheck valve (5) on frame rail (7) with screw (6) and new locknut (8).
2. Install two elbows (4) and adapter (3) on doublecheck valve (5).
3. Connect two air lines (1) to elbows (4).
4. Connect air line (2) to adapter (3).

3-424. HAND AIRBRAKE DOUBLECHECK VALVES REPLACEMENT (Contd)

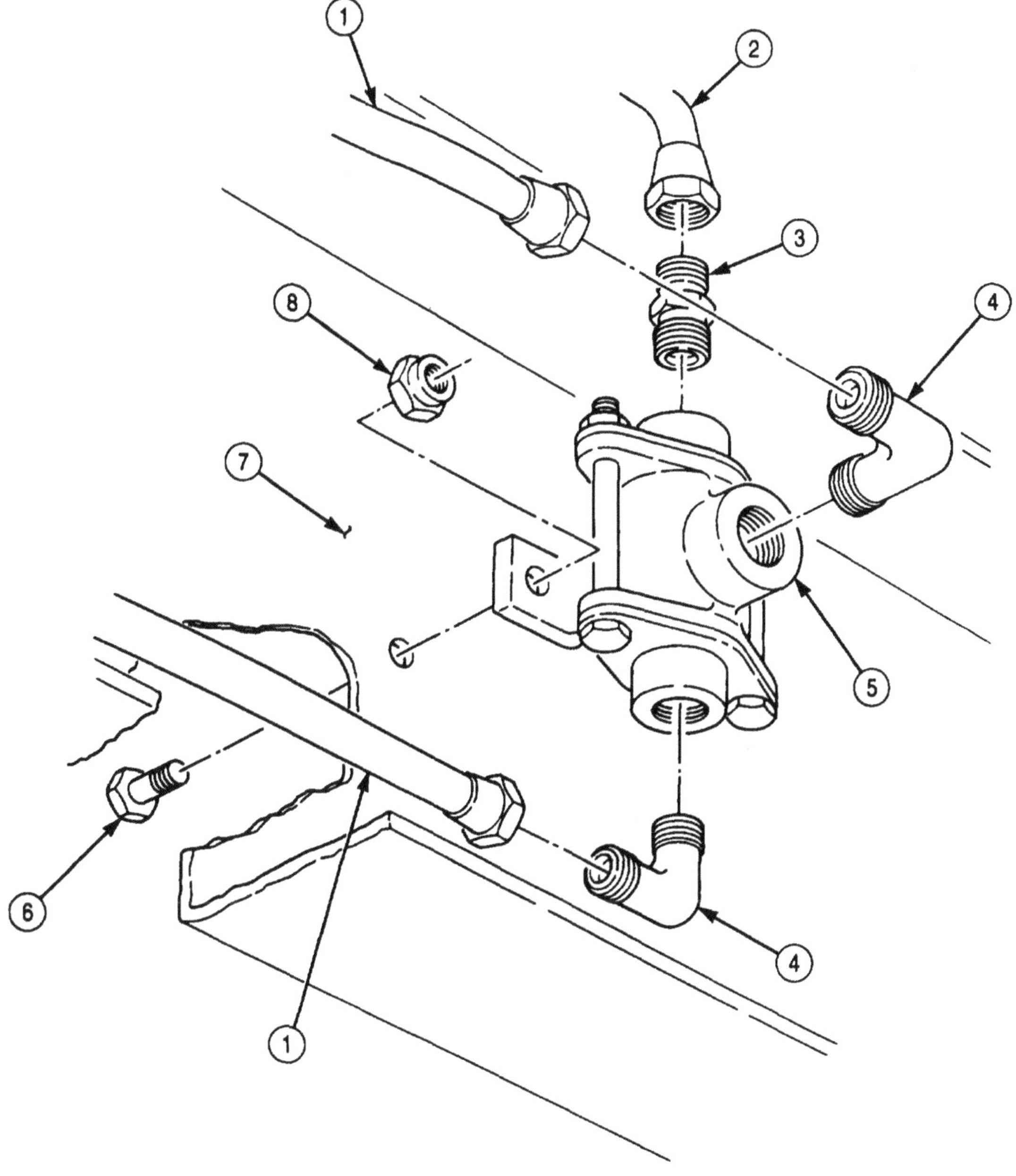

FOLLOW-ON TASK: Start engine (TM 9-2320-272-10) and allow air pressure to build to normal operating range. Check doublecheck valve for leaks.

3-425. HAND AIRBRAKE TRACTOR PROTECTION VALVE REPLACEMENT

THIS TASK COVERS:

a. Removal **b. Installation**

INITIAL SETUP:

APPLICABLE MODELS
All (except M931/A1/A2, M932/A1/A2, M933/A1/A2, M936/A1/A2)

TOOLS
General mechanic's tool kit (Appendix E, Item 1)

MATERIALS/PARTS
Two locknuts (Appendix D, Item 299)
Antiseize tape (Appendix C, Item 72)

REFERENCES (TM)
TM 9-2320-272- 10
TM 9-2320-272-24P

EQUIPMENT CONDITION
• Parking brake set (TM 9-2320-272-10).
• Air reservoirs drained (TM 9-2320-272-10).

GENERAL SAFETY INSTRUCTIONS
Do not disconnect air lines before draining air reservoirs.

a. Removal

WARNING

Do not disconnect air lines before draining air reservoirs. Small parts under pressure may shoot out with high velocity, causing injury to personnel.

NOTE

Tag air lines for installation.

1. Disconnect four air lines (2) from elbows (1).
2. Remove four elbows (1) from valve (3).
3. Remove two locknuts (6), valve (3), and two screws (5) from frame rail (4). Discard locknuts (6).

b. Installation

1. Install valve (3) on frame rail (4) with two screws (5) and new locknuts (6).

NOTE

Wrap all male pipe threads with antiseize tape before installation.

2. Install four elbows (1) on valve (3).
3. Connect four air lines (2) to elbows (1).

3-425. HAND AIRBRAKE TRACTOR PROTECTION VALVE REPLACEMENT (Contd)

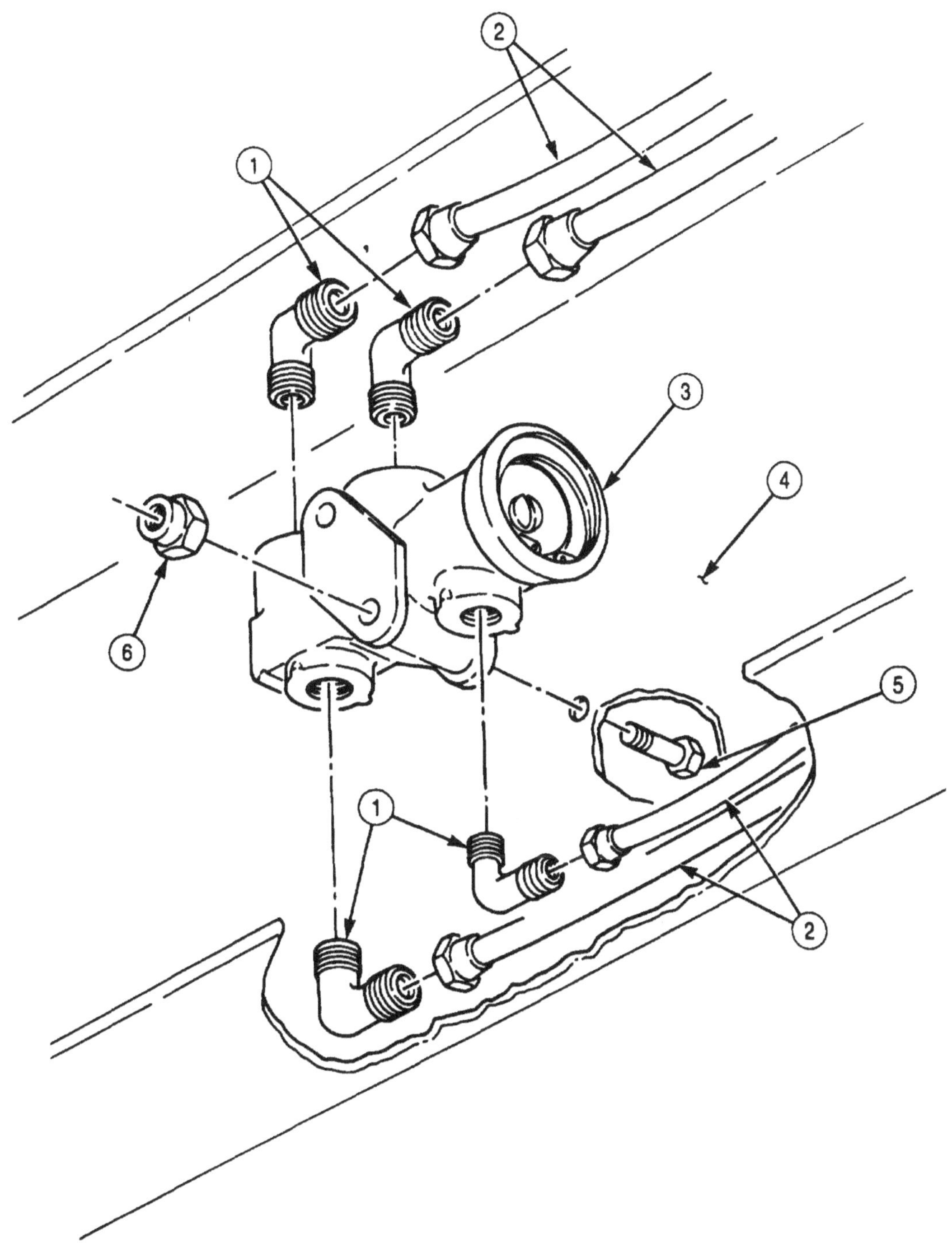

FOLLOW-ONTASK Start engine (TM 9-2320-272-10) and allow air pressure to build to normal operating range. Check valve for leaks.

3-426. 100-AMP ALTERNATOR REPLACEMENT

THIS TASK COVERS:

a. Removal **b. Installation**

INITIAL SETUP:

APPLICABLE MODELS
All

TOOLS
General mechanic's tool kit (Appendix E, Item 1)

MATERIALS/PARTS
Lockwasher (Appendix D, Item 354)

REFERENCES (TM)
TM 9-2320-272-10
TM 9-2320-272-24P

EQUIPMENT CONDITION
- Parking brake set (TM 9-2320-272-10).
- Hood raised and secured (TM 9-2320-272-10).
- Left and right splash shields removed (TM 9-2320-272-10).
- Battery ground cables disconnected (para. 3-126).
- Alternator drivebelts removed (para. 3-78).

a. Removal

1. Disconnect harness connector (12) from alternator (11).
2. Remove two nuts (9), washers (8), and screws (7) from alternator (11) and mounting bracket (10).

NOTE

Assistant will help support alternator.

3. Remove screw (l), washers (4) and (3), lockwasher (2), and alternator (11) from adjusting arm (5). Discard lockwasher (2).
4. Remove alternator pulley (6) from alternator (11) (para. 3-73).

b. Installation

1. Install alternator pulley (6) on alternator (11) (para. 3-73).

NOTE

Assistant will help support alternator.

2. Install alternator (11) on adjusting arm (5) with new lockwasher (2), washers (3) and (4), and screw (1).
3. Install alternator (11) on mounting bracket (10) with two screws (7), washers (8), and nuts (9).
4. Connect harness connector (12) to alternator (11).

3-426. 100-AMP ALTERNATOR REPLACEMENT (Contd)

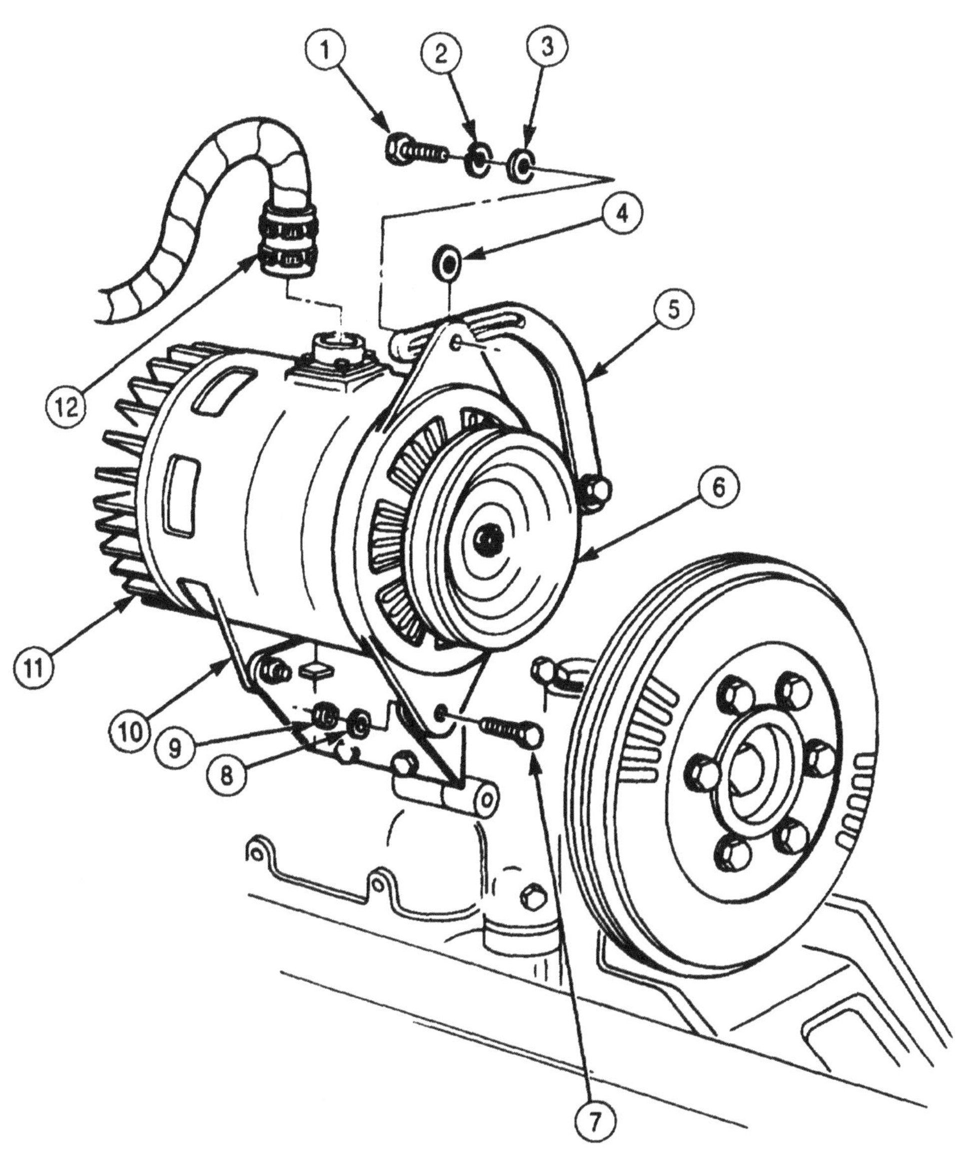

FOLLOW-ON TASKS: 1 Install alternator drivebelts (para. 3-78).
1 Connect battery ground cables (para. 3-126).
1 Install left and right splash shields (TM 9-2320-272-10).

3-427. 100-AMP ALTERNATOR HARNESS REPLACEMENT

THIS TASK COVERS:

a. Removal **b. Installation**

INITIAL SETUP:

APPLICABLE MODELS
All

TOOLS
General mechanic's tool kit (Appendix E, Item 1)

MATERIALS/PARTS
Five tiedown straps (Appendix D, Item 696)
Two lockwashers (Appendix D, Item 379)
Lockwasher (Appendix D, item 416)

REFERENCES (TM)
TM 9-2320-272-10
TM 9-2320-272-24P

EQUIPMENT CONDITION
- Parking brake set (TM 9-2320-272-10).
- Hood raised and secured (TM 9-2320-272-10).
- Left and right splash shields removed (TM 9-2320-272-10).
- Battery ground cables disconnected (para. 3-126).

a. Removal

NOTE

Tag wires, connectors, and cables for installation.

1. Remove three screws (6), washers (5), and clamps (4) from firewall (2), front wiring harness (22), and voltage regulator wiring harness (3).
2. Remove four tiedown straps (7) from front wiring harness (22) and voltage regulator wiring harness (3). Discard tiedown straps (7).
3. Remove two nuts (18), washers (17), and clamps (16) from personnel heater (1).
4. Remove tiedown strap (15) from wiring harness (10). Discard tiedown strap (15).
5. Disconnect harness connector (8) from alternator (9).
6. Disconnect connectors (12) and (13) from wires (11) and (14).
7. Disconnect harness connectors (19) and (21) from regulator (20).

3-427. 100-AMP ALTERNATOR HARNESS REPLACEMENT (Contd)

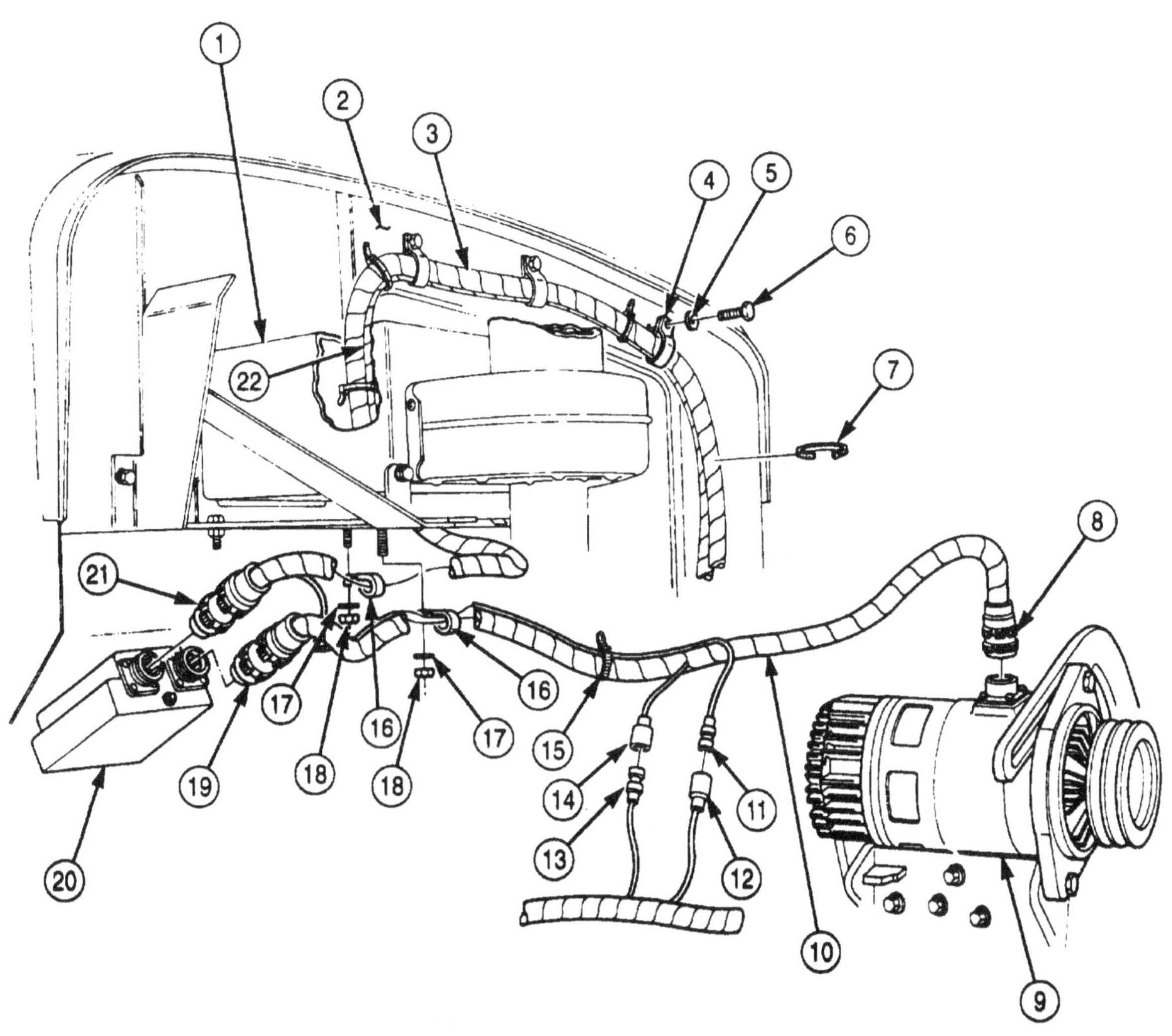

3-427. 100-AMP ALTERNATOR HARNESS REPLACEMENT (Contd)

8. Remove screw (5), lockwasher (8), ground strap (10), lockwasher (8), ground wire (7), and washer (6) from engine (9). Discard lockwasher (8).
9. Remove nut (4), lockwasher (3), and wire (2) from starter solenoid (1). Discard lockwasher (3).

b. Installation

1. Install wire (2) on starter solenoid (1) with new lockwasher (3) and nut (4).
2. Install new washer (6), ground wire (7), new lockwasher (8), ground strap (10), new lockwasher (8), and screw (5) on engine (9).
3. Connect harness connectors (29) and (31) to regulator (30).
4. Connect connectors (22) and (23) to wires (21) and (24).
5. Connect harness connector (18) to alternator (19).
6. Install new tiedown strap (25) on wiring harness (20).
7. Install two clamps (26) on personnel heater (11) with two washers (27) and nut; (28).
8. Install four new tiedown straps (17) on front wiring harness (32) and voltage regulator wiring harness (13).
9. Install three clamps (14) on firewall (12), front wiring harness (32), and voltage regulator wiring harness (13) with three washers (15) and screws (16).

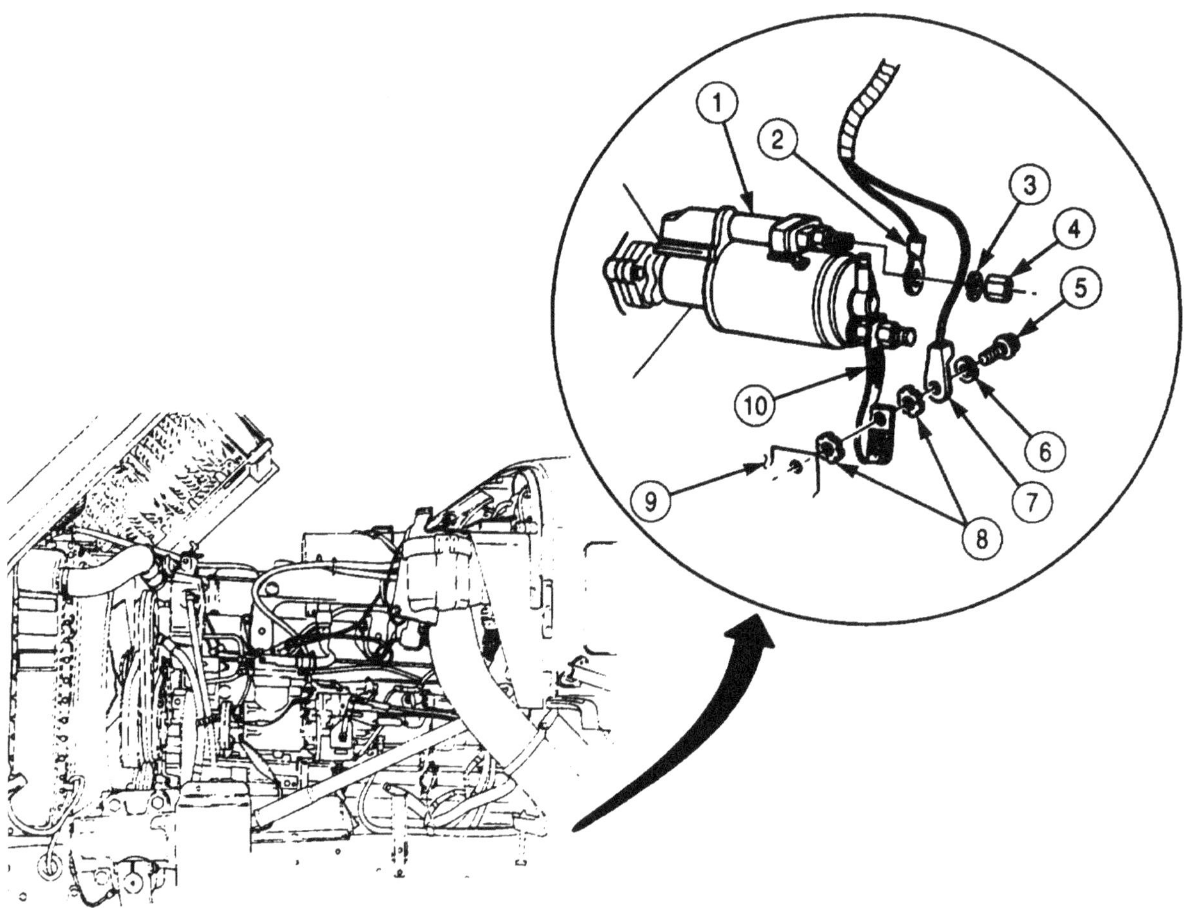

3-427. 100-AMP ALTERNATOR HARNESS REPLACEMENT (Contd)

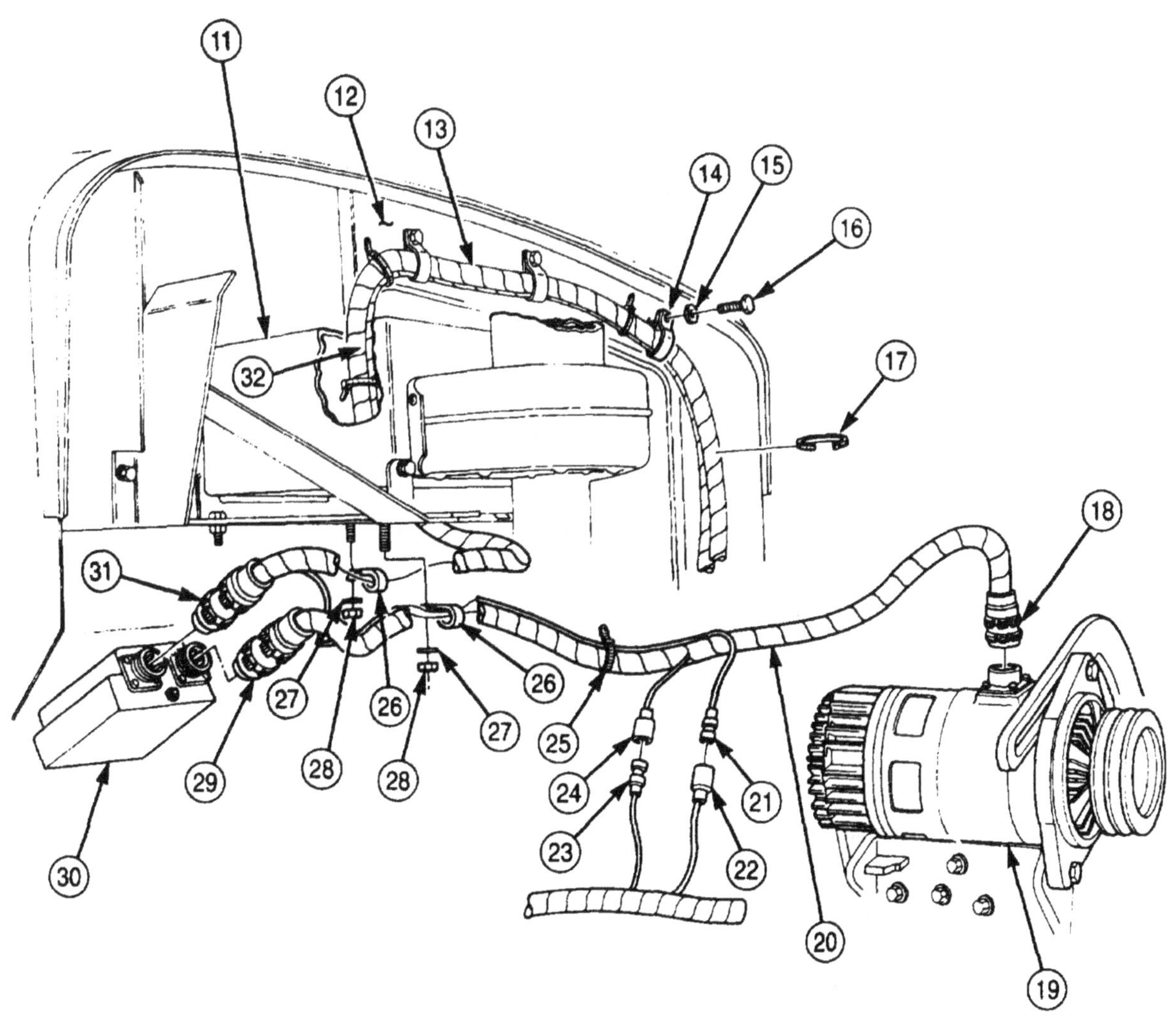

FOLLOW-ON TASKS: • Connect battery ground cables (para. 3-126).
• Install left and right splash shields (TM 9-2320-272-10).

3-428. 100-AMP VOLTAGE REGULATOR REPLACEMENT

THIS TASK COVERS:

a. Removal   **b. Installation**

INITIAL SETUP:

APPLICABLE MODELS
All

TOOLS
General mechanic's tool kit (Appendix E, Item 1)

REFERENCES (TM)
TM 9-2320-272-10
TM 9-2320-272-24P

EQUIPMENT CONDITION
- Parking brake set (TM 9-2320-272-10).
- Hood raised and secured (TM 9-2320-272-10).
- Right splash shield removed (para. 3-301).
- Battery ground cables disconnected (para. 3-126).

a. Removal

NOTE

Tag wires, connectors, and cables for installation.

1. Disconnect harness connectors (6) and (7) from voltage regulator (8).

NOTE

Assistant will help with step 2.

2. Remove four nuts (5), washers (2), screws (1), washers (2), and voltage regulator (8) from floorboard (10).
3. Remove four screws (3), washers (4), and two brackets (9) from voltage regulator (8).

b. Installation

1. Install two brackets (9) on voltage regulator (8) with four washers (4) and screws (3).

NOTE

Assistant will help with step 2.

2. Install voltage regulator (8) on floorboard (10) with four washers (2), screws (1), washers (2), and nuts (5).
3. Connect harness connectors (6) and (7) to voltage regulator (8).
4. Connect battery ground cables (para. 3-126).
5. Start engine (TM 9-2320-272-10).
6. Check battery generator indicator on instrument panel to ensure it is in green area. If adjustment is necessary, notify your supervisor.

3-428. 100-AMP VOLTAGE REGULATOR REPLACEMENT (Contd)

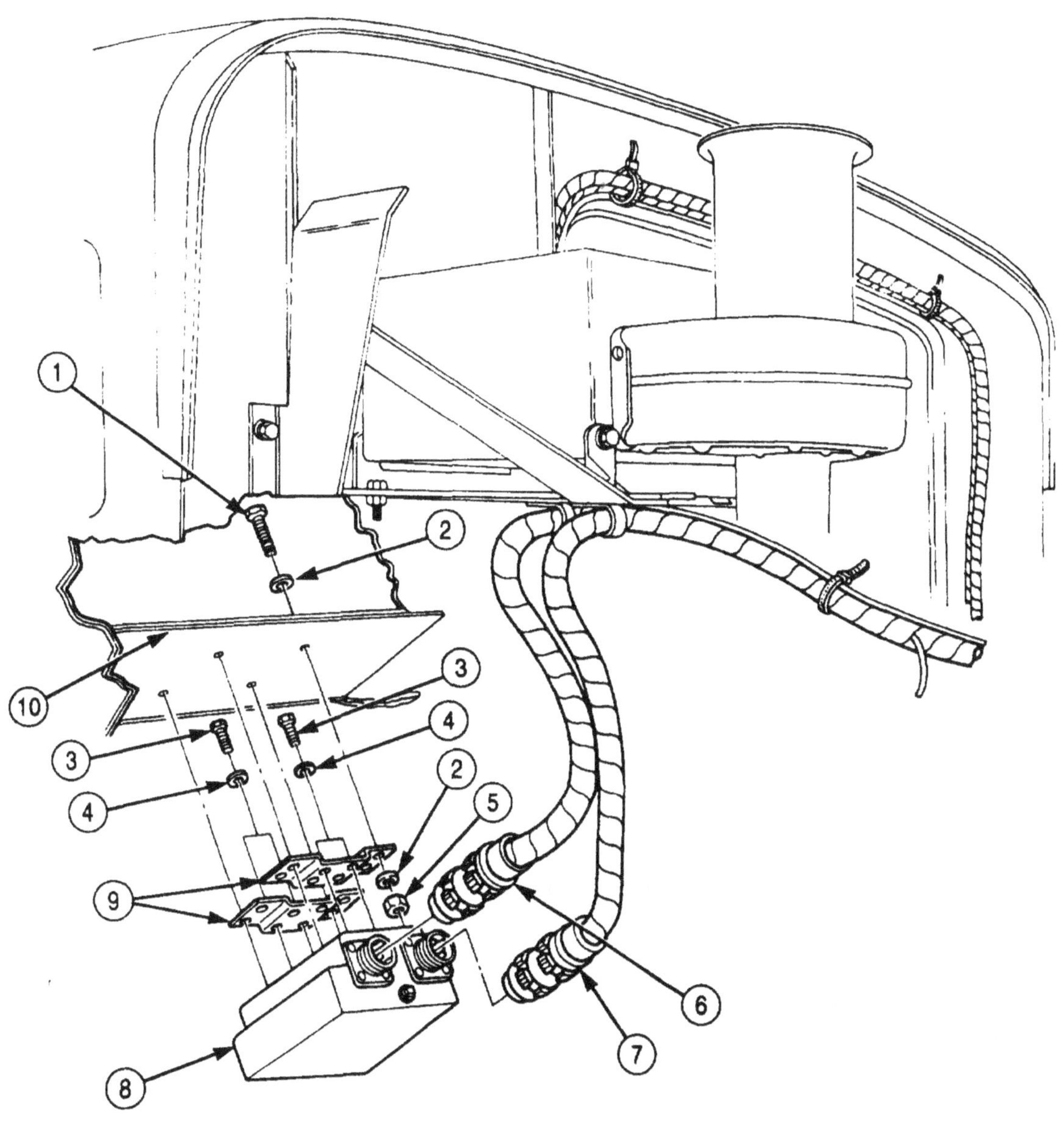

FOLLOW-ON TASK: Install right splash shield (para. 3-301).

3-429. TROOP SEAT AND SIDERACK KIT (M929/A1/A2, M930/A1/A2) MAINTENANCE

THIS TASK COVERS:

a. Troop Seat Disassembly
b. Siderack Disassembly
c. Siderack Assembly
d. Troop Seat Assembly

INITIAL SETUP:

APPLICABLE MODELS
M929/A1/A2, M930/A1/A2

TOOLS
General mechanic's tool kit (Appendix E, Item 1)

MATERIALS/PARTS
Five locknuts (Appendix D, Item 288)

REFERENCES (TM)
TM 9-2320-272-10
TM 9-2320-272-24P

EQUIPMENT CONDITION
- Parking brake set (TM 9-2320-272-10).
- Troop seat and side rack removed (TM 9-2320-272-10).

NOTE

- All troop seats and sideracks are replaced the same. This procedure covers the right side only.
- Assistant will help when necessary.

a. Troop Seat Disassembly

1. Remove five locknuts (7), screws (8), and legs (6) from channels (3). Discard locknuts (7).
2. Remove eight nuts (5), screws (2), and four hinges (4) from channels (3).
3. Remove twelve nuts (9), screws (1), and five channels (3) from four boards (10).

b. Siderack Disassembly

1. Remove eight nuts (14), screws (16), and five pockets (15) from board (18).
2. Remove two nuts (13), washers (12), screws (17), and retainers (11) from board (18).

c. Siderack Assembly

1. Install two retainers (11) on board (18) with two screws (17), washers (12), and nuts (13).
2. Install five pockets (15) on board (18) with eight screws (16) and new nuts (14).

d. Troop Seat Assembly

1. Install five channels (3) on four boards (10) with twelve screws (1) and nuts (9).
2. Install four hinges (4) on channels (3) with eight screws (2) and nuts (5).
3. Install five legs (6) on channels (3) with five screws (8) and new locknuts (7).

3-429. TROOP SEAT AND SIDERACK KIT (M929/A1/A2, M930/A1/A2) MAINTENANCE (Contd)

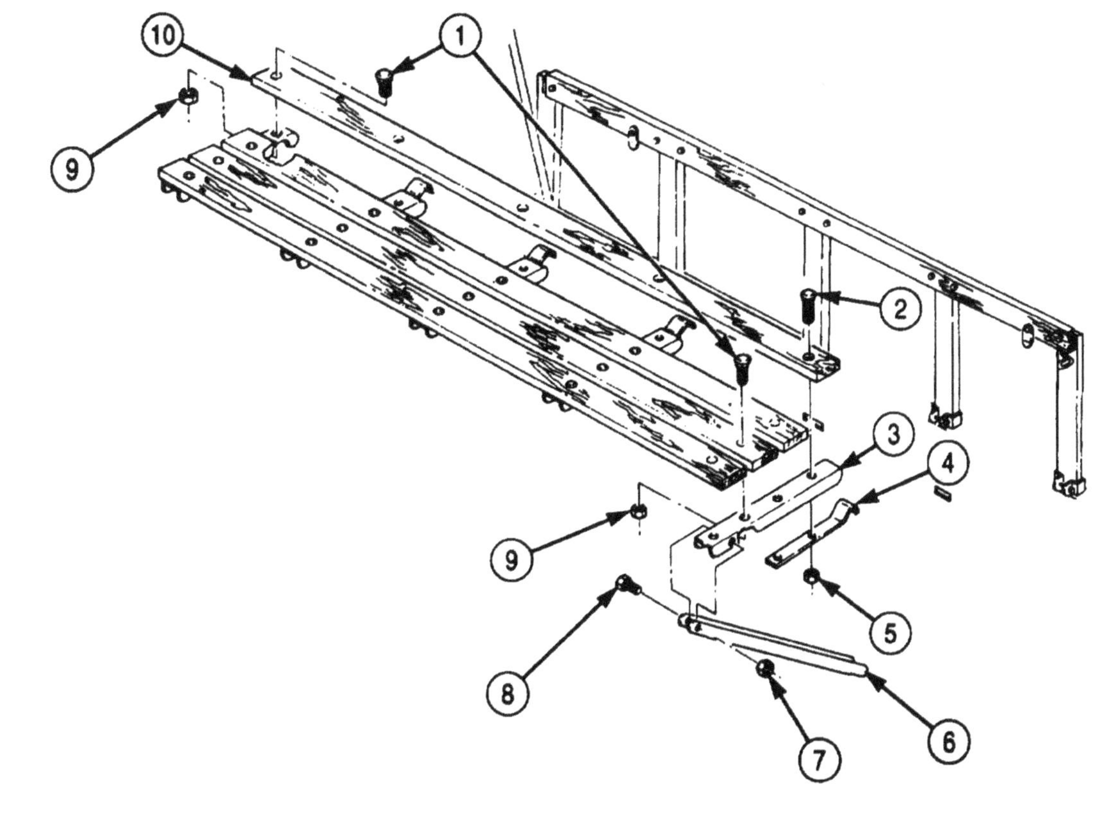

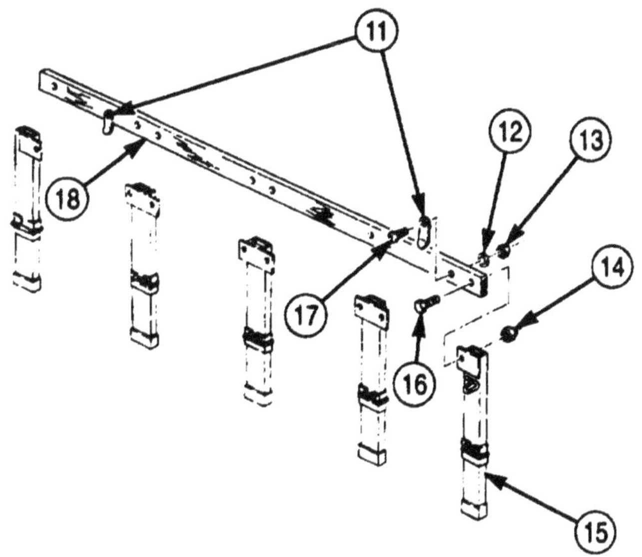

FOLLOW-ON TASK: Install troop seat and siderack (TM 9-2320-272-10).

3-430. CONVOY WARNING LIGHT MOUNT REPLACEMENT

THIS TASK COVERS:

a. Removal **b. Installation**

INITIAL SETUP:

APPLICABLE MODELS
All except M929/A1/A2, M930/A1/A2, M934/A1/A2

TOOLS
General mechanic's tool kit (Appendix E, Item 1)

MATERIALS/PARTS
Five locknuts (Appendix D, Item 309)
Two locknuts (Appendix D, Item 277)
Screw-assembled lockwasher (Appendix D, Item 575)

REFERENCES (TM)
TM 9-2320-272-10
TM 9-2320-272-24P

EQUIPMENT CONDITION
- Parking brake set (TM 9-2320-272-10).
- Convoy warning light removed (para. 3-437).

NOTE

Left and right convoy warning light mounts are replaced the same. This procedure covers left convoy waning light mount only.

a. Removal

1. Disconnect two leads (13) from connectors (14).
2. Remove screw-assembled lockwasher (12), washer (11), and clamp (10) with cable (22) from bracket (9). Discard screw-assembled lockwasher (12).
3. Remove mounting plate (1) and cable (22) from support tube (2).
4. Remove wing screw (20), pin (21), and support tube (2) from bracket tube (16) and bracket (19).
5. Remove two locknuts (18), washers (17), U-bolt (6), locknut (5), washer (4), screw (3), and bracket (19) from bracket tube (16). Discard locknuts (5) and (18).

NOTE

Perform step 6 for vehicles equipped with machine gun mount kit.

6. Remove two locknuts (18), washers (17), U-bolt (6), locknut (5), washer (4), screw (3), and bracket (19) from bracket post (23). Discard locknuts (5) and (18).
7. Remove four locknuts (7), washers (8), two U-bolts (15), and bracket tube (16) from bracket ,(9). Discard locknuts (7).

3-430. CONVOY WARNING LIGHT MOUNT REPLACEMENT (Contd)

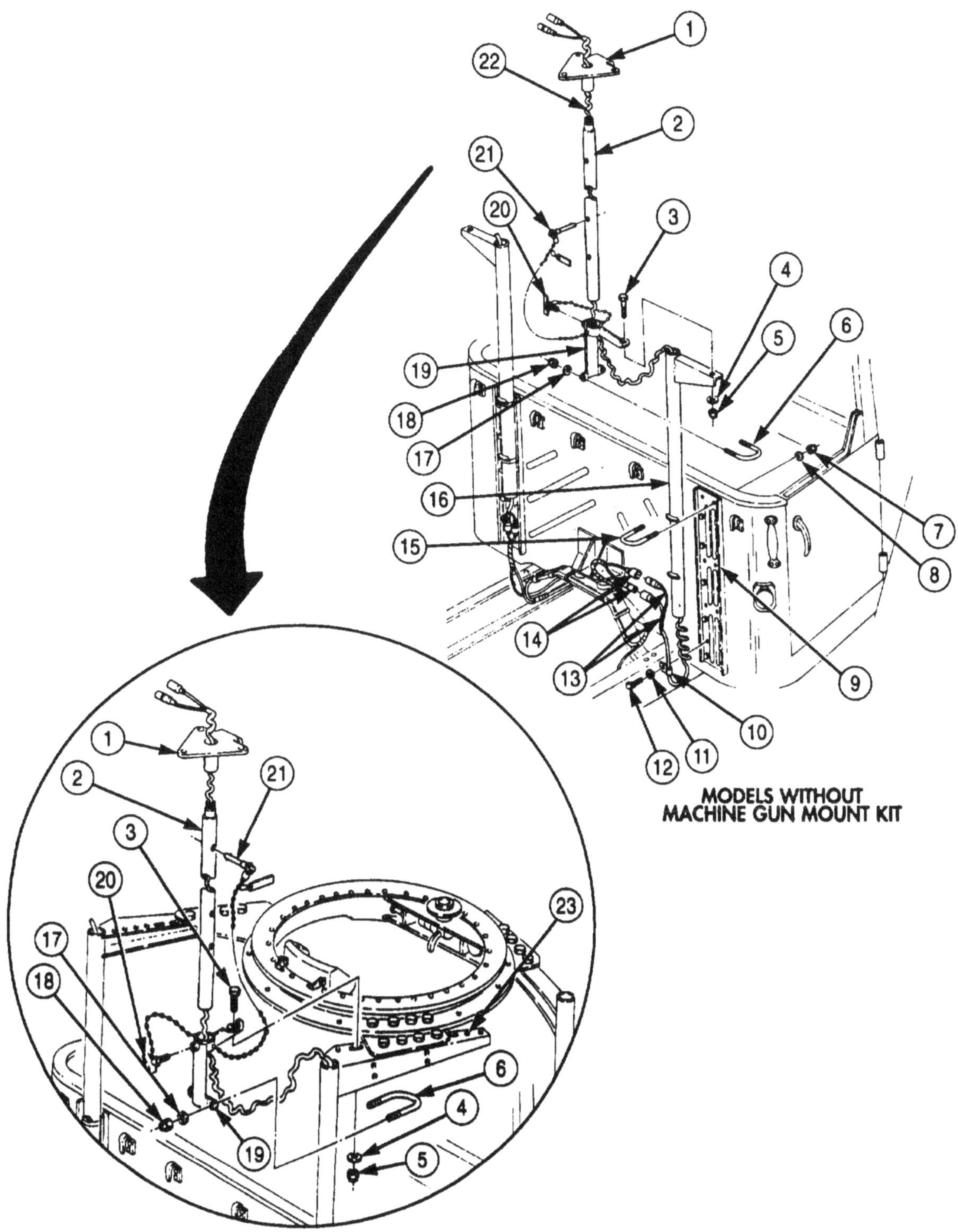

MODELS WITH MACHINE GUN MOUNT KIT

3-430. CONVOY WARNING LIGHT MOUNT REPLACEMENT (Contd)

b. Installation

1. Install bracket tube (16) on bracket (9) with two U-bolts (15), four washers (8), and new locknuts (7).

NOTE

Perform step 2 for vehicles equipped with machine gun mount kit.

2. Install bracket (19) on bracket post (23) with screw (3), washer (4), new locknut (5), U-bolt (6), two washers (17), and new locknuts (18).
3. Install bracket (19) on bracket tube (16) with screw (3), washer (4), new locknut (5), U-bolt (6), two washers (17), and new locknuts (18).
4. Install support tube (2) on bracket tube (16) and bracket (19) with pin (21) and wing screw (20).
5. Install cable (22) and mounting plate (1) on support tube (2).
6. Install washer (11), clamp (10) with cable (22), and new screw-assembled lockwasher (12) on bracket (9).
7. Connect two leads (13) to connectors (14).

3-430. CONVOY WARNING LIGHT MOUNT REPLACEMENT (Contd)

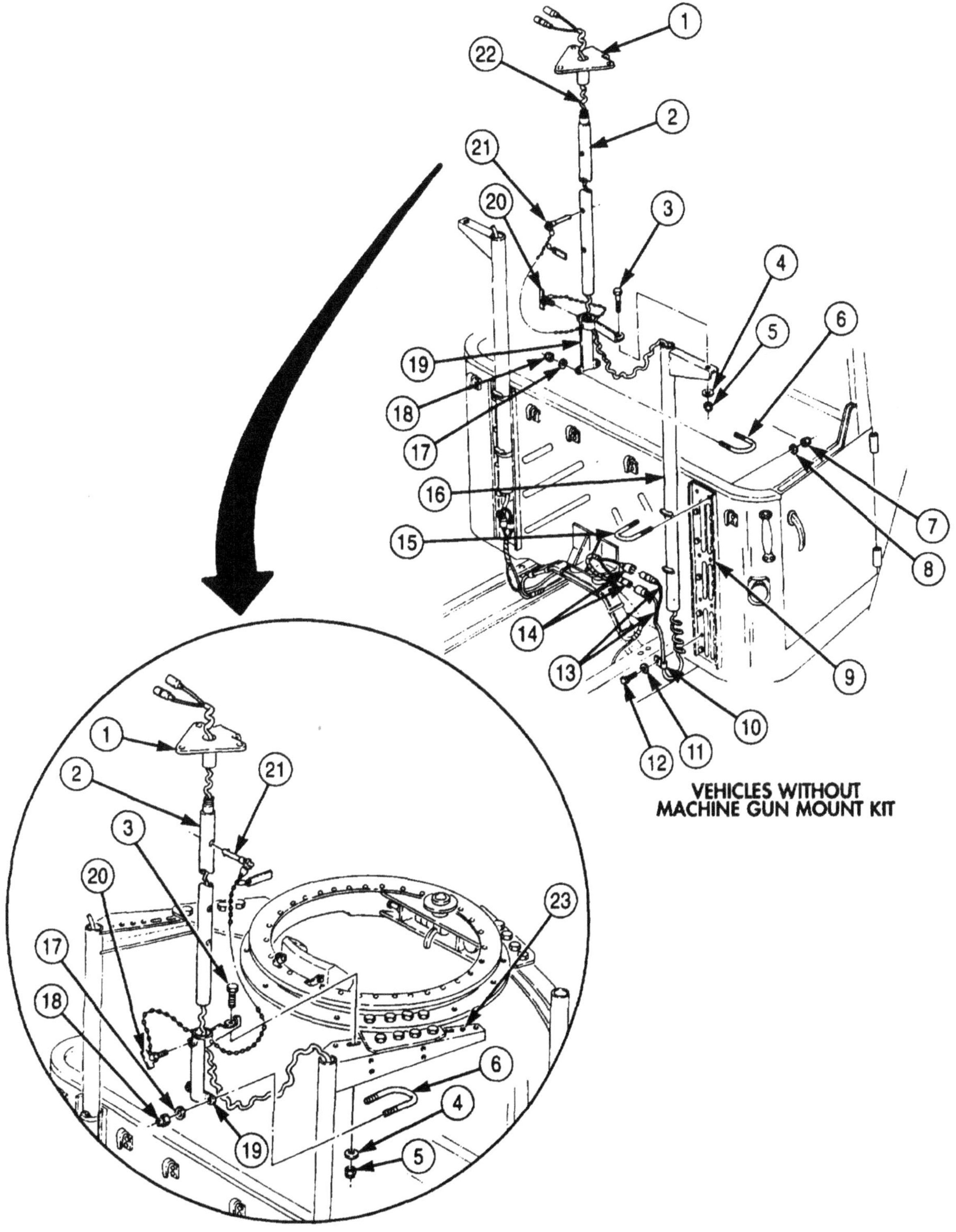

FOLLOW-ON TASK: Install convoy warning light (para. 3-437).

3-431. CONVOY WARNING LIGHT MOUNT M934/A1/A2 REPLACEMENT

THIS TASK COVERS:

a. Removal **b. Installation**

APPLICABLE MODELS
M934/A1/A2

TOOLS
General mechanic's tool kit (Appendix E, Item 1)

MATERIALS/PARTS
Six lockwashers (Appendix D, Item 377)
Two tiedown straps (Appendix D. Item 684)

REFERENCES (TM)
TM 9-2320-272-10
TM 9-2320-272-24P

EQUIPMENT CONDITION
- Parking brake set (TM 9-2320-272-10).
- Convoy warning light removed (para. 3-437).

a. Removal

1. Remove tiedown strap (3) and harness (4) from support tube (10). Discard tiedown strap (3).
2. Remove wing screw (5) and support tube (10) from mounting bracket (9).
3. Remove mounting plate (2) from support tube (10).
4. Remove three screws (6), lockwashers (7), lifting bracket (8), and mounting bracket (9) from van body (1). Discard lockwashers (7).
5. Remove tiedown strap (14) and harness (4) from support tube (12). Discard tiedown strap (14).
6. Remove wing screw (15) and support tube (12) from mounting bracket (20).
7. Remove mounting plate (13) from support tube (12).
8. Remove two screws (17) from heater access door (18) and open heater access door (18).
9. Remove three nuts (11), lockwashers (22), screws (16), lifting bracket (21), mounting bracket (20), and six washers (19) from van body (1). Discard lockwashers (22).

3-431. CONVOY WARNING LIGHT MOUNT (M934/A1/A2) REPLACEMENT (Contd)

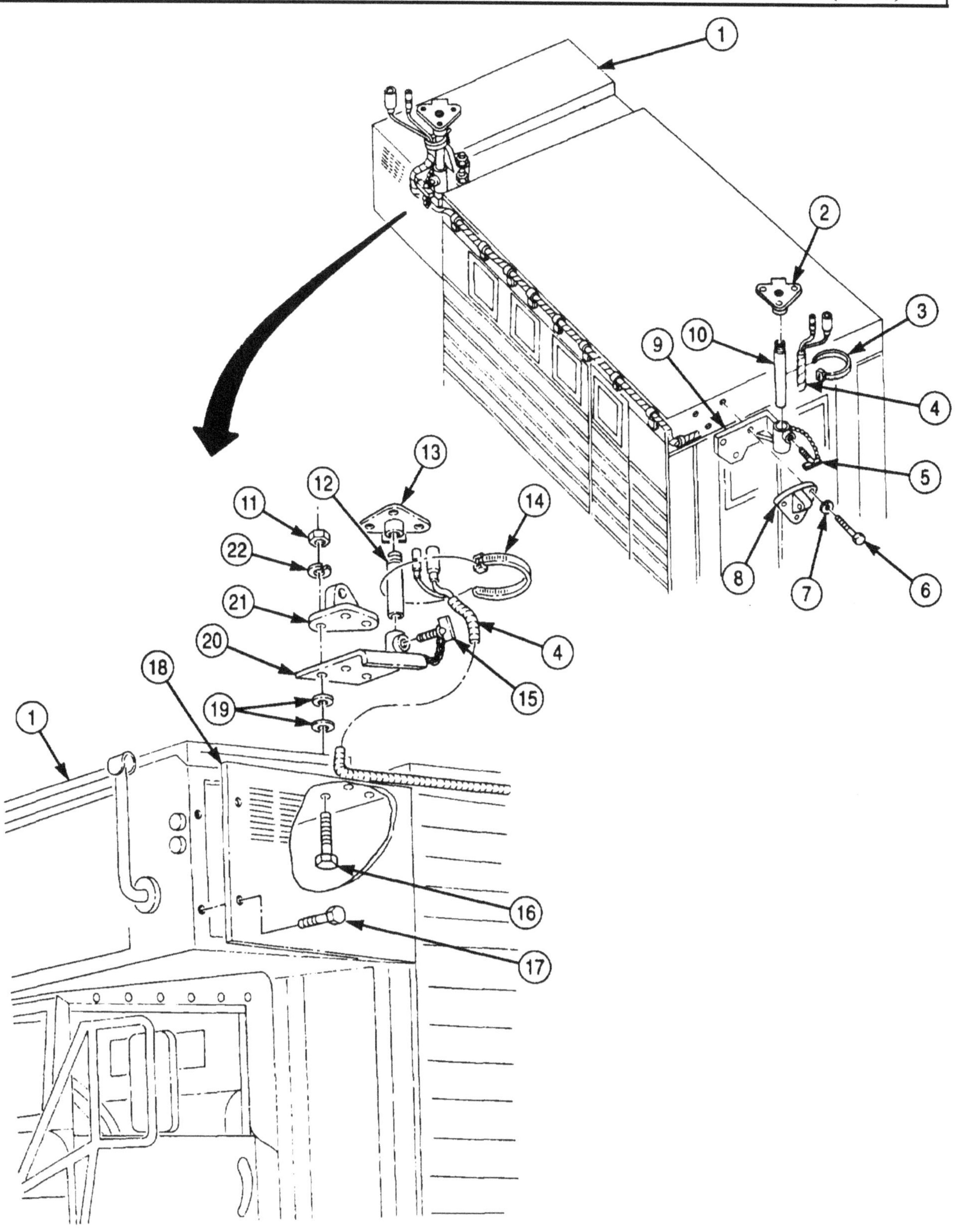

3-431. CONVOY WARNING LIGHT MOUNT (M934/A1/A2) REPLACEMENT (Contd)

b. Installation

1. Install six washers (19), mounting bracket (20), and lifting bracket (21) on van body (1) with three screws (16), new lockwashers (22), and nuts (11).
2. Close heater access door (18) and install two screws (17) on heater access door (18).
3. Install mounting plate (13) on support tube (12).
4. Install support tube (12) on mounting bracket (20) with wing screw (15).
5. Install harness (4) on support tube (12) with new tiedown strap (14).
6. Install mounting bracket (9) and lifting bracket (8) on van body (1) with three new lockwashers (7) and screws (6).
7. Install mounting plate (2) on support tube (10).
8. Install support tube (10) on mounting bracket (9) with wing screw (5).
9. Install harness (4) on support tube (10) with new tiedown strap (3).

3-431. CONVOY WARNING LIGHT MOUNT (M934/A1/A2) REPLACEMENT (Contd)

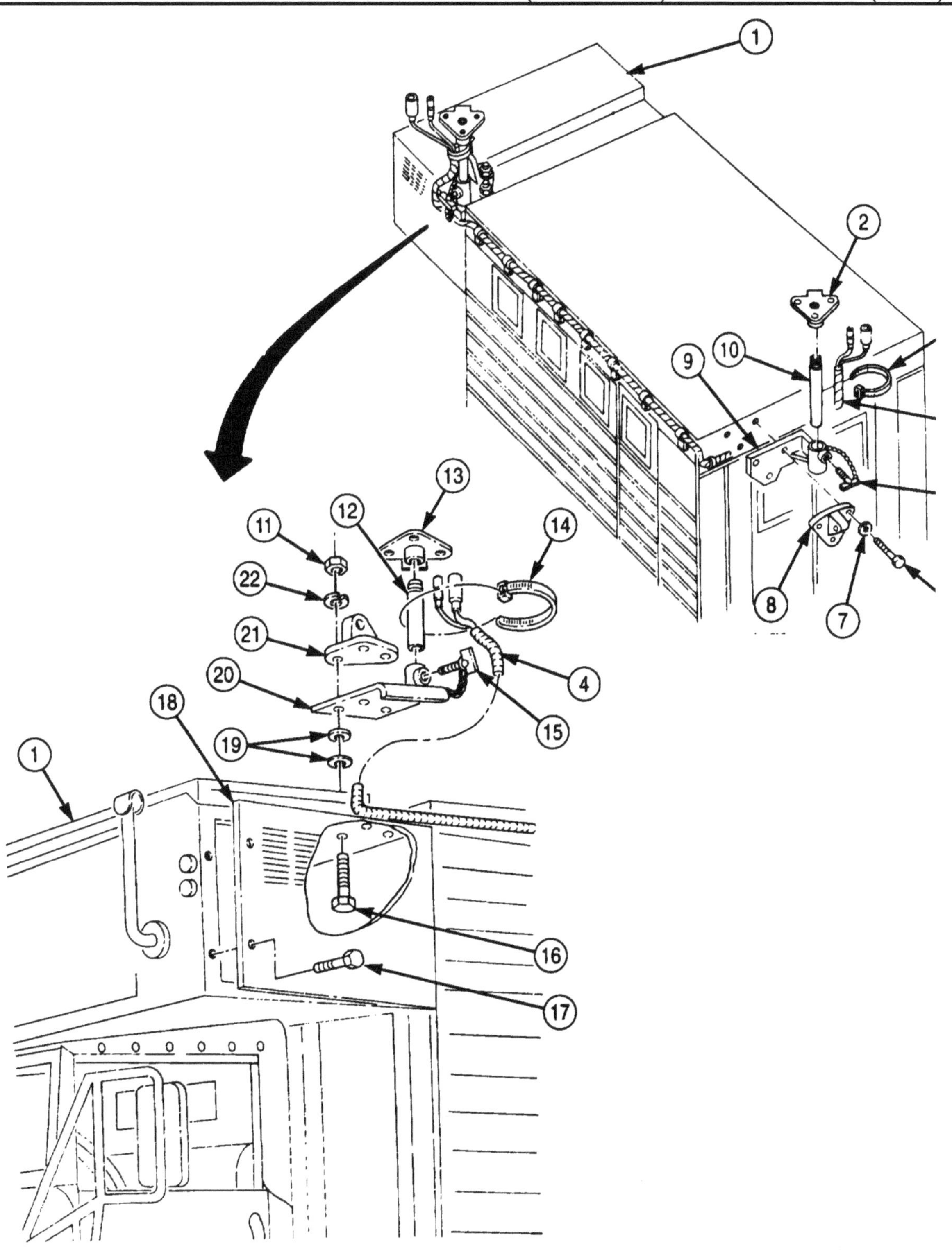

FOLLOW-ON TASK: Install convoy warning light (para. 3-437).

3-432. CONVOY WARNING LIGHT MOUNT (M929/A1/A2, M930/A1/A2) REPLACEMENT

THIS TASK COVERS:

a. Removal **b. Installation**

INITIAL SETUP:

APPLICABLE MODELS
M929/A1/A2, M930/A1/A2

TOOLS
General mechanic's tool kit (Appendix E, Item 1)

MATERIALS/PARTS
Four locknuts (Appendix D, Item 288)

REFERENCES (TM)
TM 9-2320-272-10
TM 9-2320-272-24P

EQUIPMENT CONDITION
- Parking brake set (TM 9-2320-272-10).
- Convoy warning light removed (para. 3-437).

NOTE

Right and left convoy warning light mounts are replaced the same. This procedure covers the right mount only

a. Removal

1. Disconnect two cable leads (8) from harness leads (7).
2. Remove two cable leads (8) from support tube (2) and mounting bracket (11).
3. Remove wing screw (3) and support tube (2) from mounting bracket (11).
4. Remove mounting plate (1) from support tube (2).
5. Remove four locknuts (5), washers (4), screws (10), washers (9), and mounting bracket (11) from cab protector (6). Discard locknuts (5).

b. Installation

1. Install mounting bracket (11) on cab protector (6) with four washers (9), screws (10), washers (4), and new locknuts (5).
2. Install mounting plate (1) on support tube (2).
3. Install support tube (2) on mounting bracket (11) with wing screw (3).
4. Install two cable leads (8) on support tube (2) and mounting bracket (11).
5. Connect two cable leads (8) to harness leads (7).

3-432. CONVOY WARNING LIGHT MOUNT (M929/A1/A2, M930/A1/AZ) REPLACEMENT (Contd)

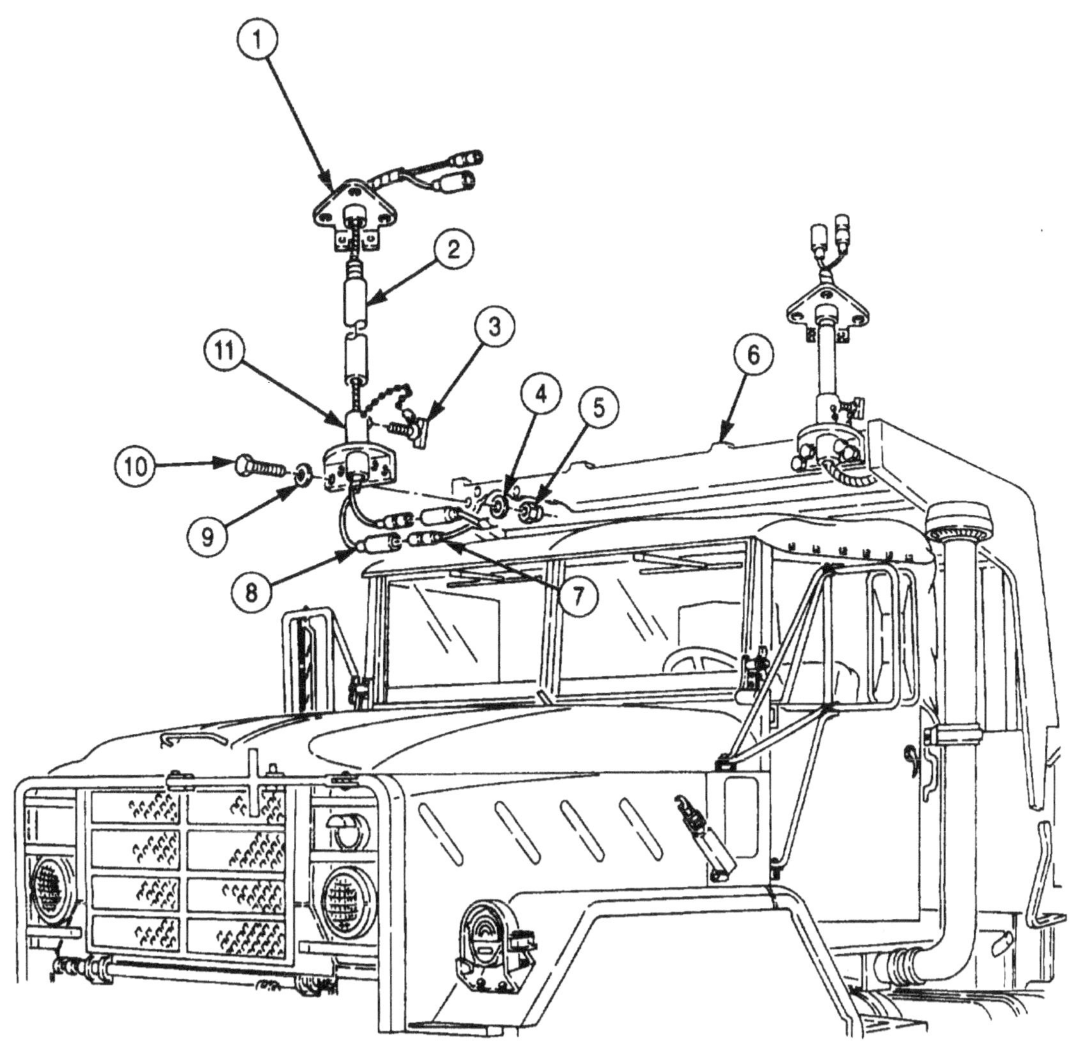

FOLLOW-ON TASK: Install convoy warning light (para. 3-437).

3-433. CONVOY WARNING LIGHT HARNESS REPLACEMENT

THIS TASK COVERS:

a. Removal **b. Installation**

INITIAL SETUP:

APPLICABLE MODELS
All (except M929/A1/A2, M930/A1/A2, M934/A1/A2)

TOOLS
General mechanic's tool kit (Appendix E, Item 1)

MATERIALS/PARTS
Locknut (Appendix D. Item 288)
Lockwasher (Appendix D, Item 405)
Fifteen tiedown straps (Appendix D, Item 687)
Six tiedown straps (Appendix D, Item 685)

REFERENCES (TM)
TM 9-2320-272-10
TM 9-2320-272-24P

EQUIPMENT CONDITION
- Parking brake set (TM 9-2320-272-10).
- Hood raised and secured (TM 9-2320-272-10).
- Toolbox removed (para. 3-302).
- Battery ground cables disconnected (para. 3-126).

a. Removal

NOTE

Perform steps 1 and 2 for left and right rear of cab.

1. Disconnect four cable leads (6) from harness (7).
2. Remove locknut (5), ground lead (4), washer (3), lockwasher (2), and screw (10) from frame (11) and rear step hanger (9). Discard locknut (5) and lockwasher (2).
3. Remove six tiedown straps (1) and harness (7) from cab crossmember (8). Discard tiedown straps (1).

NOTE

Note routing of harness lead for installation.

4. Remove fifteen tiedown straps (17) and harness lead (15) from front main wiring harness (16). Discard tiedown straps (17).
5. Remove screw (14), clip (13), and harness lead (15) from relay (12).

b. Installation

1. Install harness lead (15) on relay (12) with clip (13) and screw (14) .

NOTE

Perform steps 2 and 3 for left and right rear of cab.

2. Install screw (10), new lockwasher (2), washer (3), ground lead (4), and new locknut (5) on rear step hanger (9) and frame (11).
3. Connect four cable leads (6) to harness (7).

NOTE

Tiedown straps must be installed 10 in. (25 cm) apart.

4. Install harness (7) on cab crossmember (8) with six new tiedown straps (1).
5. Install harness lead (15) on front main wiring harness (16) with fifteen new tiedown straps (17).

3-433. CONVOY WARNING LIGHT HARNESS REPLACEMENT (Contd)

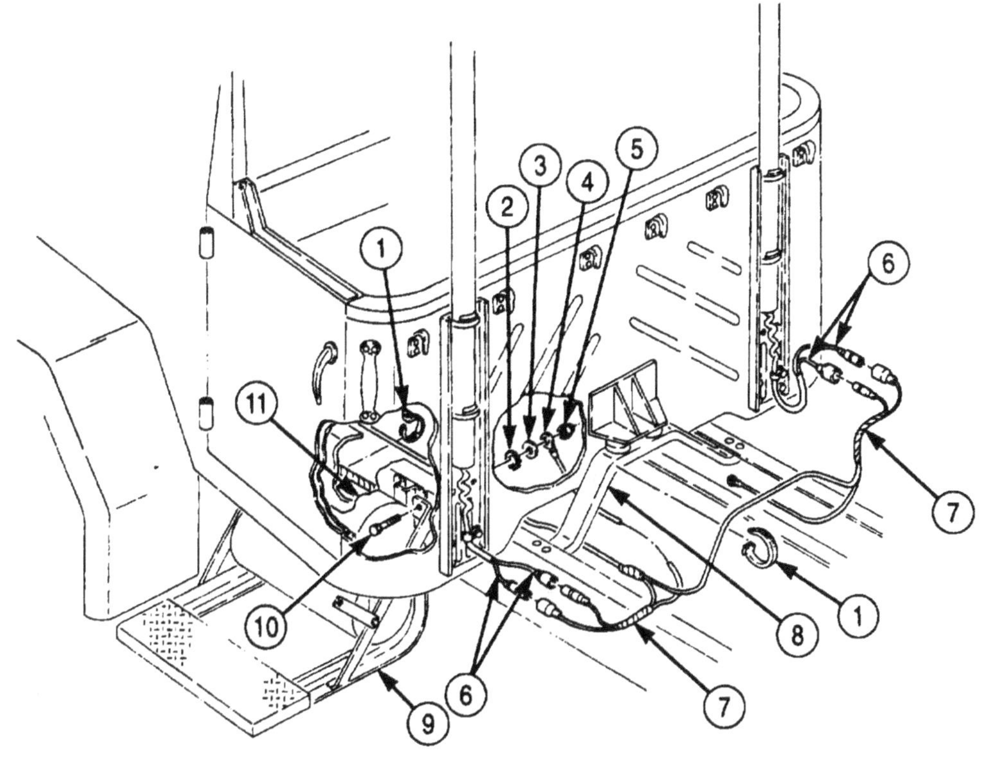

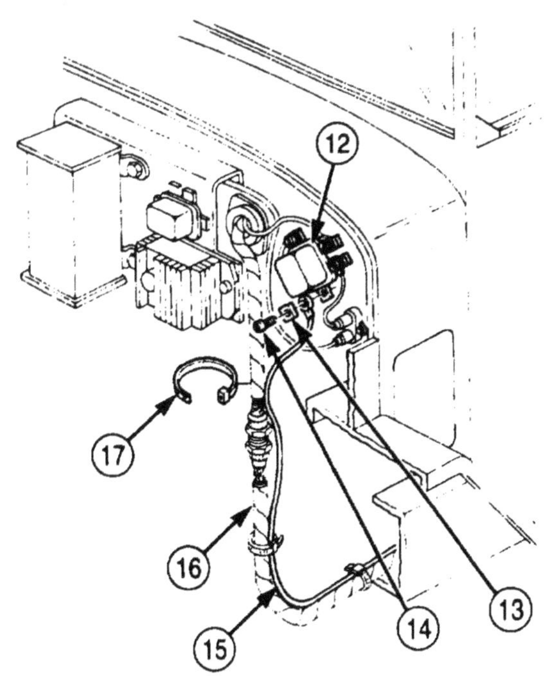

FOLLOW-ON TASKS: • Connect battery ground cables (para. 3-126).
• Install toolbox (para. 3-302).

3-434. CONVOY WARNING LIGHT HARNESS (M929/A1/A2, M930/A1/A2) REPLACEMENT

THIS TASK COVERS:

a. Removal **b. Installation**

INITIAL SETUP:

APPLICABLE MODELS
M929/A1A2, M930/A1/A2

TOOLS
General mechanic's tool kit (Appendix E, Item 1)

MATERIALS/PARTS
Locknut (Appendix D, Item 288)
Twenty-nine locknuts (Appendix D, Item 276)
Lockwasher (Appendix D, Item 379)
Two tiedown straps (Appendix D, Item 684)

REFERENCES (TM)
TM 9-2320-272-10
TM 9-2320-272-24P

EQUIPMENT CONDITION
Parking brake set (TM 9-2320-272-10).

a. Removal

1. Disconnect four cable leads (8) from two convoy warning lights (7).
2. Remove eighteen locknuts (5), screws (2), clamps (3), and conduit (6) from cab protector (4). Discard locknuts (5).
3. Remove ten locknuts (11), screws (9), clamps (10), and conduit (6) from dump body frame (12) and cab protector (4). Discard locknuts (11).
4. Disconnect lead (22) from cable (20).
5. Remove locknut (23), washer (24), ground lead (28), lockwasher (25), screw (27), and washer (26) from crossmember (21). Discard locknut (23) and lockwasher (25).
6. Remove locknut (16), screw (29), clamp (14), and conduit (6) from crossmember (17). Discard locknut (16).
7. Remove screw (13), clamp (15), and conduit (6) from frame (18).
8. Remove two tiedown straps (19) from conduit (6). Discard tiedown straps (19).

NOTE

Note routing of conduit for installation.

9. Remove conduit (6) from front of cab protector (4) and dump body (1).
10. Remove conduit (6) from harness (30).

3-434. CONVOY WARNING LIGHT HARNESS (M929/A1/A2, M930/A1/A2) REPLACEMENT (Contd)

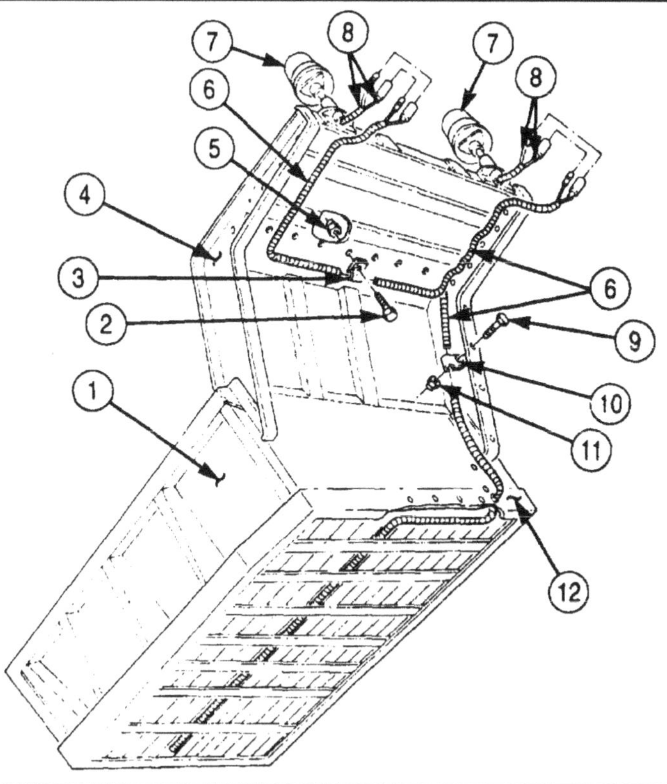

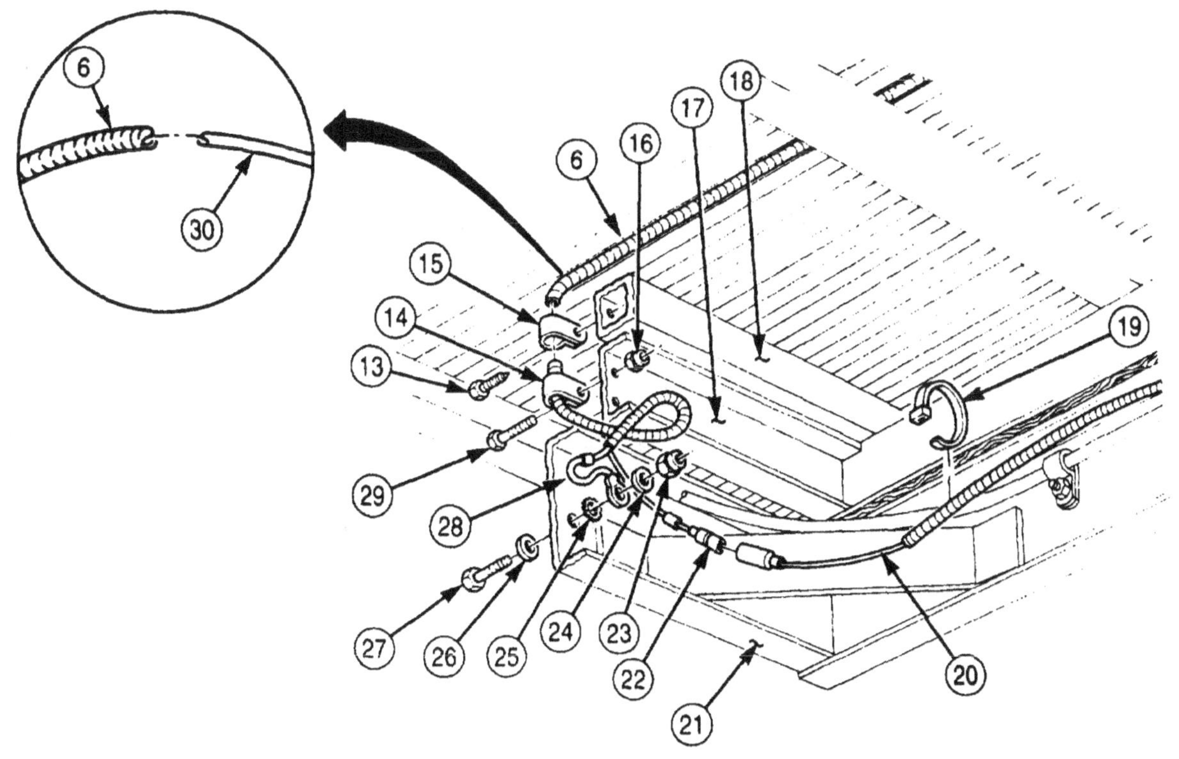

3-434. CONVOY WARNING LIGHT HARNESS (M929/A1/A2, M930/A1/A2) REPLACEMENT (Contd)

b. Installation

1. Install conduit (6) on harness (30).
2. Install conduit (6) on front of cab protector (4) and dump body (1).
3. Install two new tiedown straps (19) on conduit (6).
4. Install conduit (6) on frame (18) with clamp (15) and screw (13).
5. Install conduit (6) on crossmember (17) with clamp (14), screw (29), and new locknut (16).
6. Install washer (261, screw (27), new lockwasher (25), ground lead (28), washer (24), and new locknut (23) on crossmember (21).
7. Connect lead (22) to cable (20).
8. Install conduit (6) on dump body frame (12) and cab protector (4) with ten clamps (10), screws (9), and new locknuts (11).
9. Install conduit (6) on cab protector (4) with eighteen clamps (3), screws (2), and new locknuts (5).
10. Connect four cable leads (8) to two convoy warning lights (7).

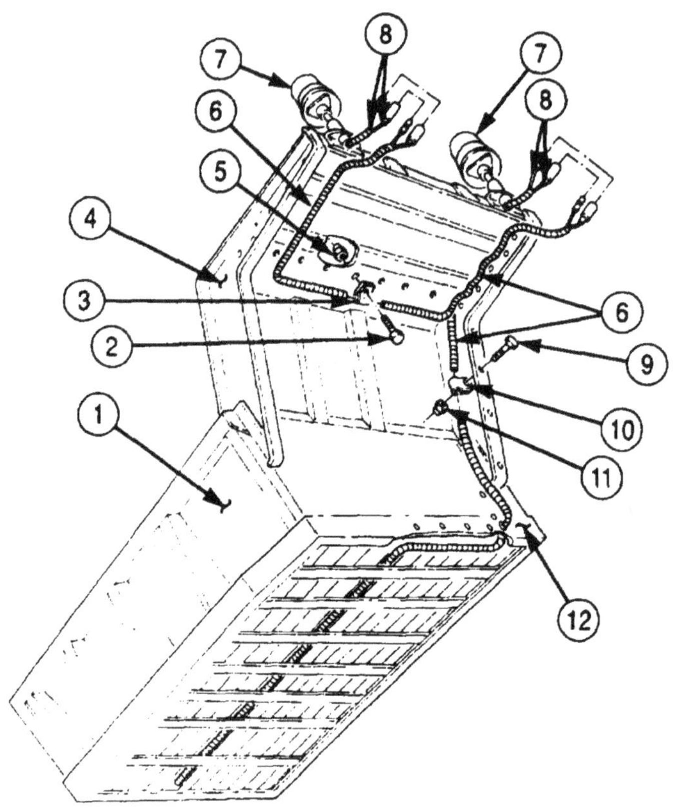

3-434. CONVOY WARNING LIGHT HARNESS (M929/A1/A2, M930/A1/A2) REPLACEMENT (Contd)

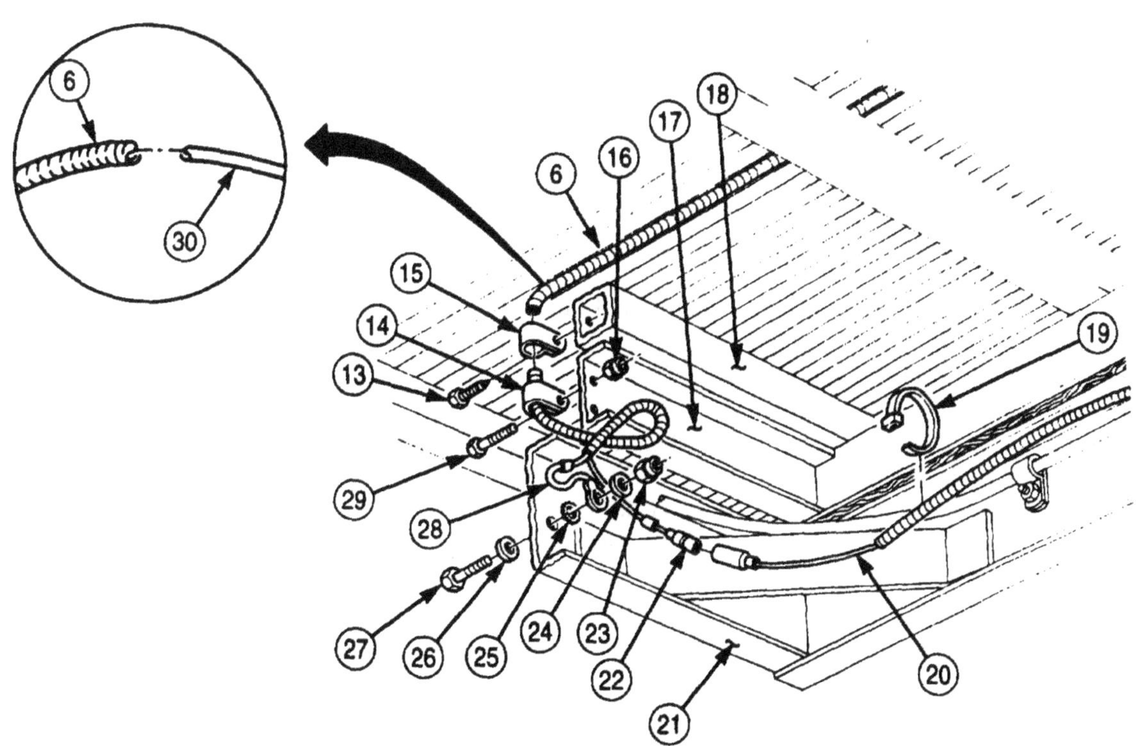

3-435. CONVOY WARNING LIGHT HARNESS (M934A1/A2) REPLACEMENT

THIS TASK COVERS:

a. Removal **b. Installation**

INITIAL SETUP:

APPLICABLE MODELS
M934A1/A2

TOOLS
General mechanic's tool kit (Appendix E, Item 1)

MATERIALS/PARTS
Seven screw-assembled lockwashers (Appendix D, Item 576)
Locknut (Appendix D, Item 288)
Two tiedown straps (Appendix D, Item 684)

REFERENCES (TM)
TM 9-2320-272-10
TM 9-2320-272.24P

EQUIPMENT CONDITION
- Parking brake set (TM 9-2320-272-10).
- Convoy warning lights removed (para. 3-437).

a. Removal

NOTE

Note routing of harness for installation.

1. Remove two tiedown straps (3) and harness (4) from two support tubes (2) and van body (1). Discard tiedown straps (3).
2. Remove twenty-five screws (5), clamps (6), and harness (4) from van body (1).
3. Remove four screw-assembled lockwashers (19), washers (20), clamps (21), and harness (4) from van body (1). Discard screw-assembled lockwashers (19).
4. Remove five screws (18), clamps (17), and harness (4) from van body (1).
5. Remove three screw-assembled lockwashers (14), washers (15), clamps (16), and harness (4) from van body (1). Discard screw-assembled lockwashers (14).
6. Disconnect harness lead (13) from lead (12).
7. Remove locknut (11), washer (9), screw (7), washer (9), and ground lead (8) from fuel tank support (10). Discard locknut (11).

3-435. CONVOY WARNING LIGHT HARNESS (M934A1/A2) REPLACEMENT (Contd)

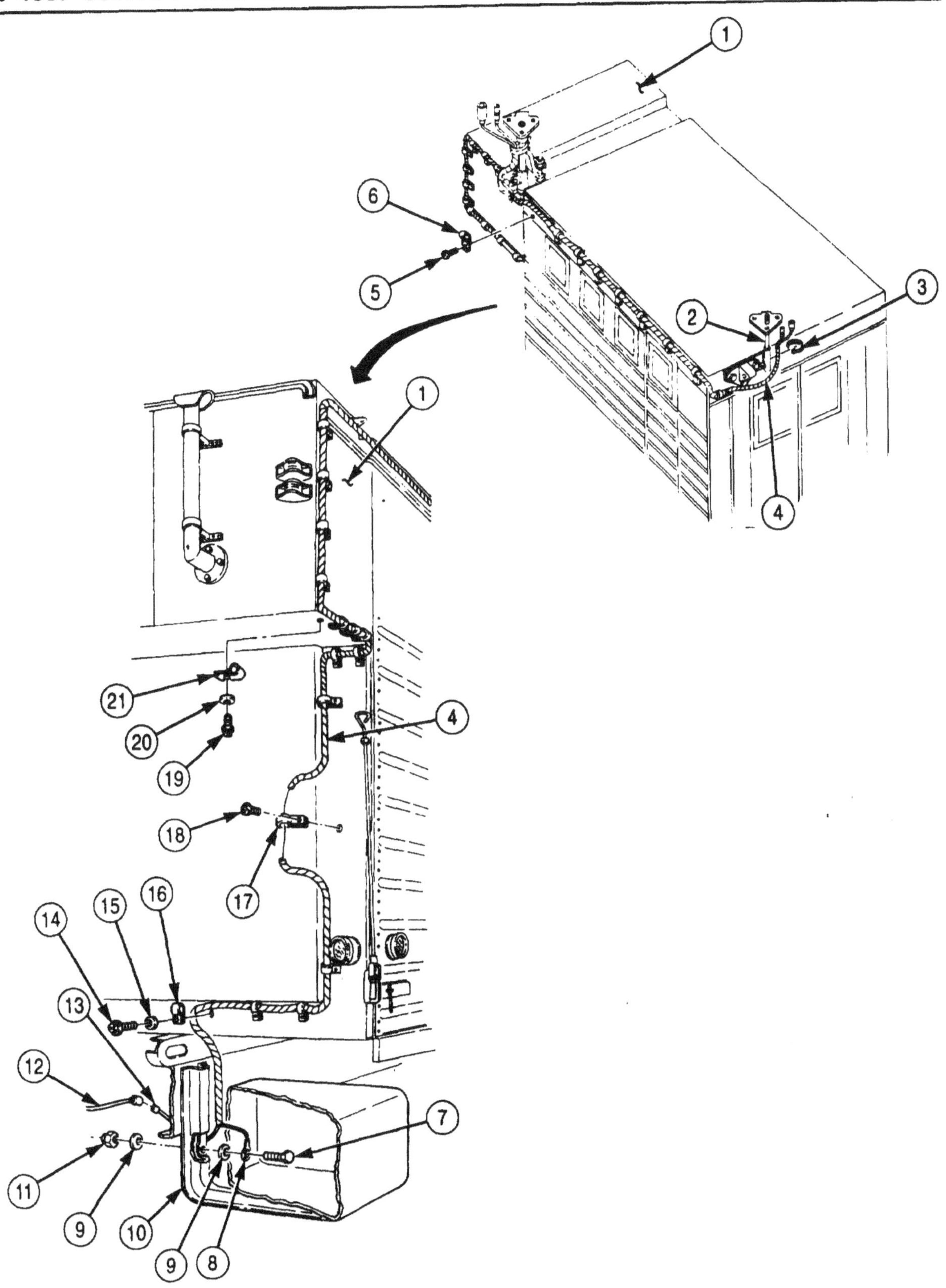

3-435. CONVOY WARNING LIGHT HARNESS (M934A1/A2) REPLACEMENT (Contd)

b. Installation

1. Install ground lead (8) on fuel tank support (10) with washer (9), screw (7), washer (9), and new locknut (11).
2. Connect harness lead (13) to lead (12).
3. Install harness (4) on van body (1) with three clamps (16), washers (15), and new screw-assembled lockwashers (14).
4. Install harness (4) on van body (1) with five clamps (17) and screws (18).
5. Install harness (4) on van body (1) with four clamps (21), washers (20), and new screw-assembled lockwashers (19).
6. Install harness (4) on van body (1) with twenty-five clamps (6) and screws (5).
7. Install harness (4) on two support tubes (2) and van body (1) with two new tiedown straps (3).

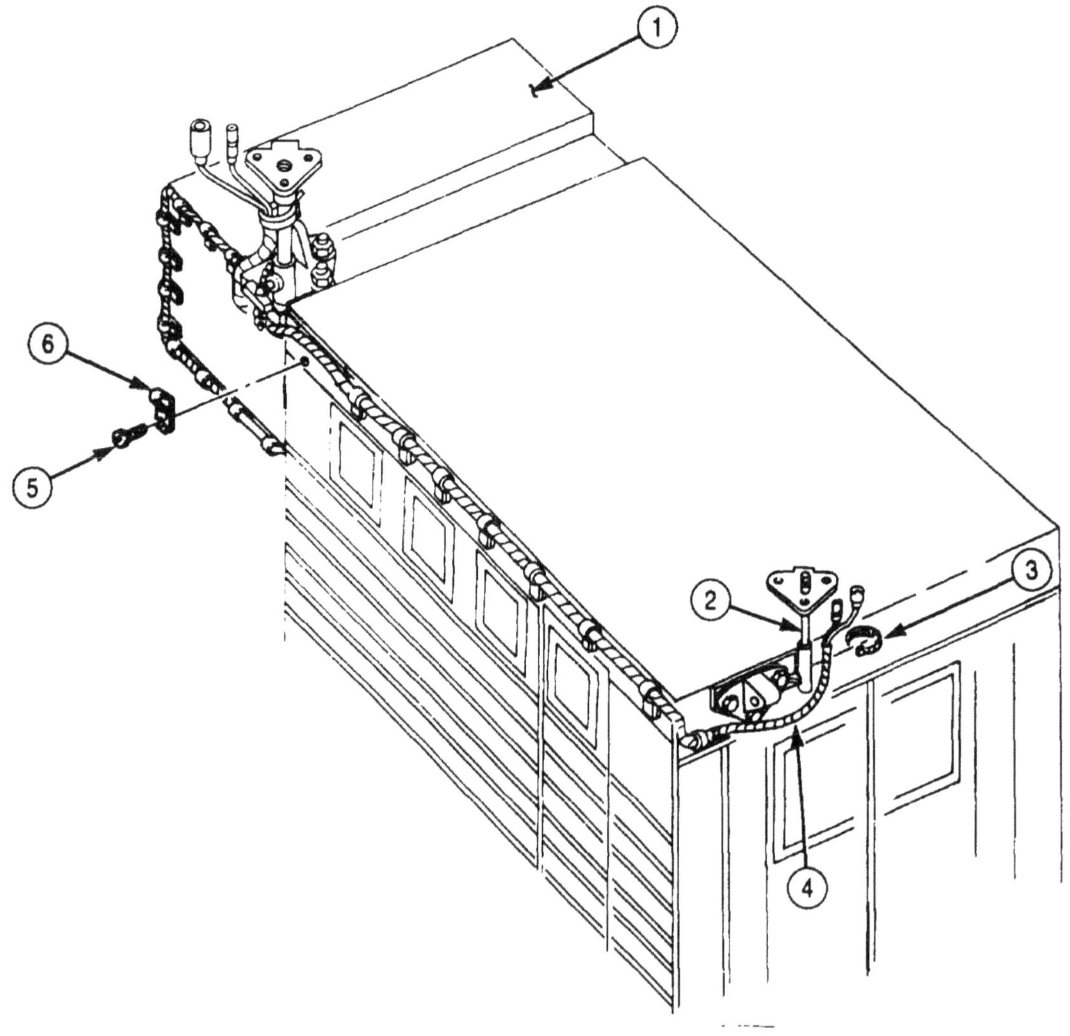

3-435. CONVOY WARNING LIGHT HARNESS (M934A1/A2) REPLACEMENT (Contd)

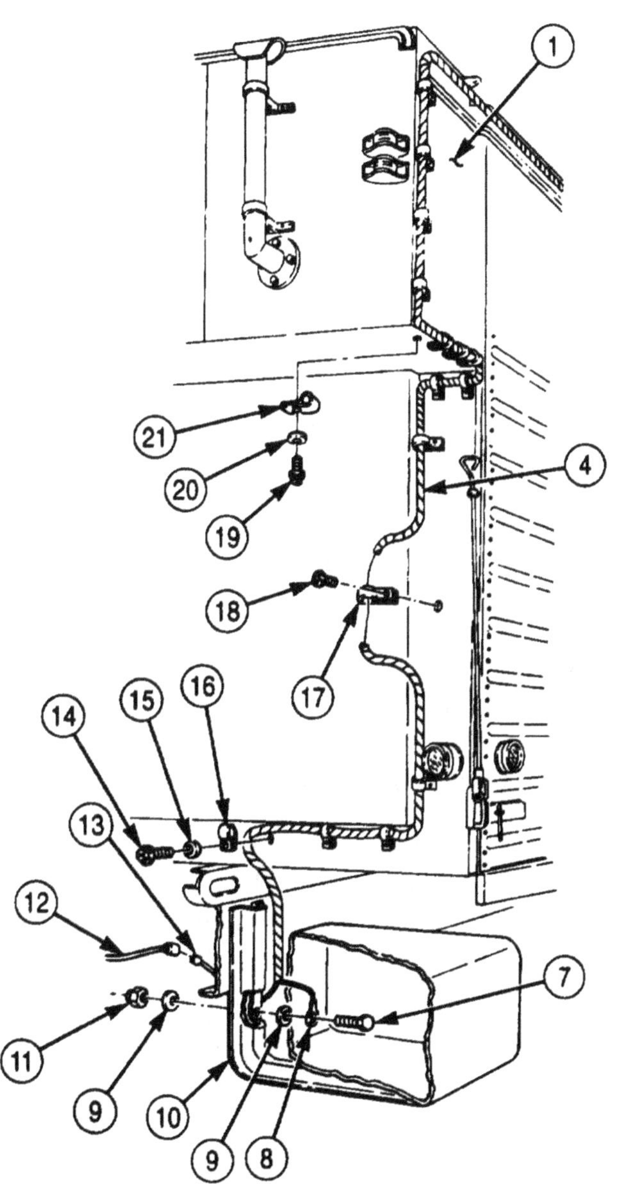

FOLLOW-ON TASK: Install convoy warning lights (para. 3-437).

3-436. CONVOY WARNING LIGHT RESISTOR AND LEADS REPLACEMENT

THIS TASK COVERS:

a. Removal **b. Installation**

INITIAL SETUP:

APPLICABLE MODELS
All

TOOLS
General mechanic's tool kit (Appendix E, Item 1)

MATERIALS/PARTS
Two lockwashers (Appendix D, Item 406)
Tiedown strap (Appendix D. Item 691)
Two lockwashers (Appendix D, Item 418)

REFERENCES (TM)
TM 9-2320-272-10
TM 9-2320-272-24P

EQUIPMENT CONDITION
- Parking brake set (TM 9-2320-272-10).
- Hood raised and secured (TM 9-2320-272-10).
- Battery ground cables disconnected (para. 3-126).

a. Removal

NOTE

- Tag all leads for installation.
- Note routing of all leads for installation.

1. Disconnect leads (1) and (2) from warning light switch (3).
2. Remove tiedown strap (27) from front main wiring harness (21) and leads (17) and (20). Discard tiedown strap (27).

NOTE

Assistant will help with step 3.

3. Remove two screws (30), washers (31), lockwashers (4), retainer (29), and grommet (7) from firewall (10). Discard lockwashers (4).
4. Pull leads (1) and (2) through hole in firewall (10).
5. Remove four screws (5), clips (6), and leads (2), (9), (16), and (28) from resistor (8).
6. Remove two screws (15), lockwashers (14), lead (28), and resistor (8) from tirewall (10). Discard lockwashers (14).
7. Disconnect lead (17) from high-beam selector switch (18).
8. Disconnect leads (11) and (13) from circuit breaker (12).
9. Remove nut (24), lockwasher (23), and wires (25) and (26) from starter solenoid (22). Discard lockwasher (23).
10. Disconnect lead (19) from connector (20).

3-436. CONVOY WARNING LIGHT RESISTOR AND LEADS REPLACEMENT (Contd)

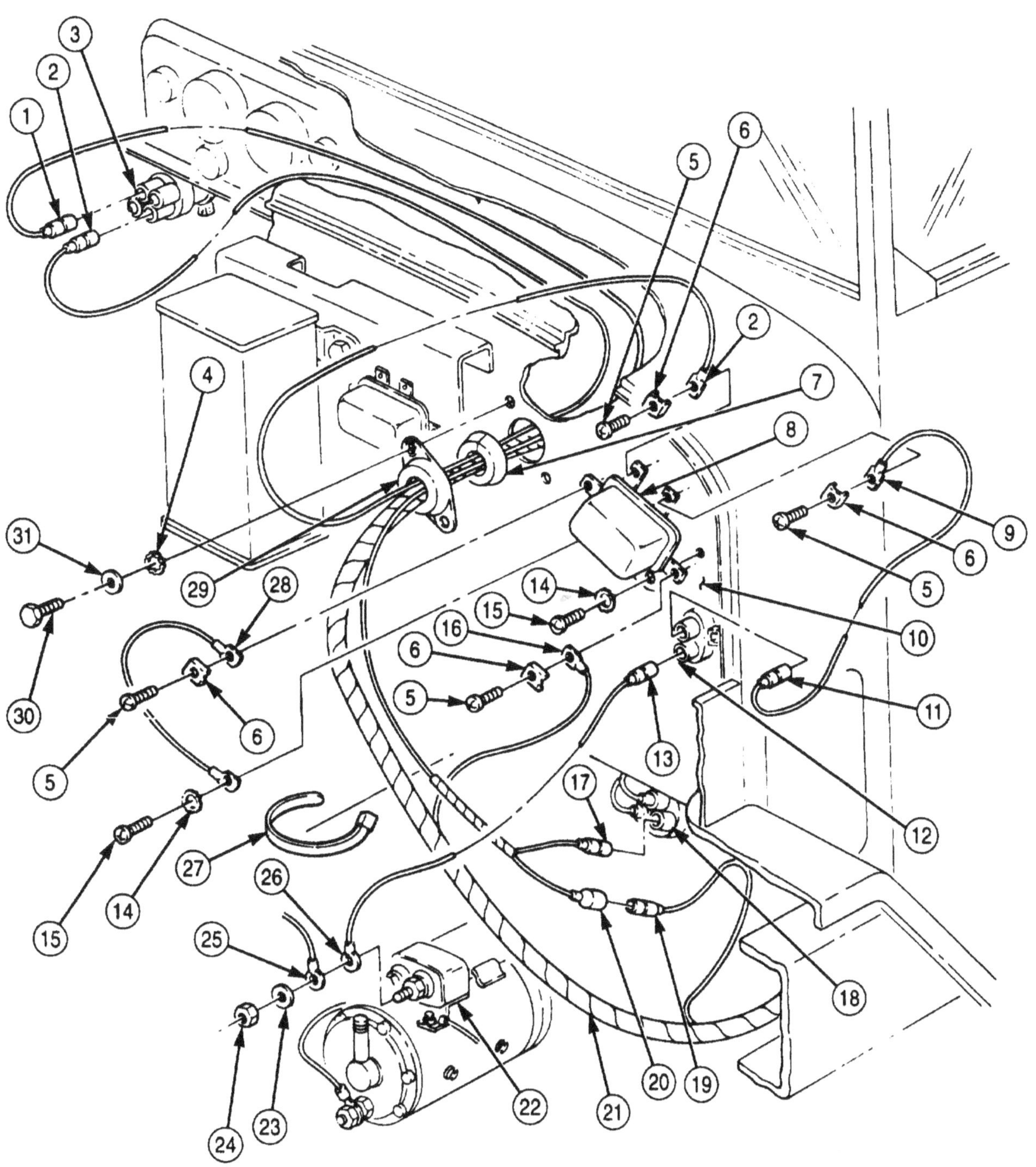

3-436. CONVOY WARNING LIGHT RESISTOR AND LEADS REPLACEMENT (Contd)

b. Installation

1. Connect lead (19) to connector (20).
2. Install wires (25) and (26) on starter solenoid (22) with new lockwasher (23) and nut (24).
3. Connect leads (11) and (13) to circuit breaker (12).
4. Connect lead (17) to high-beam selector switch (18).
5. Install lead (28) and resistor (8) on firewall (10) with two new lockwashers (14) and screws (15).
6. Install four leads (2), (9), (16), and (28) on resistor (8) with clips (6) and screws (5).

NOTE

Assistant will help with step 7.

7. Install grommet (7) and retainer (29) on firewall (10) with two new lockwashers (4), washers (31), and screws (30).
8. Pull leads (1) and (2) through hole in firewall (10).
9. Install new tiedown strap (27) on front main wiring harness (21) and leads (17) and (20).
10. Connect leads (1) and (2) to warning light switch (3).

3-436. CONVOY WARNING LIGHT RESISTOR AND LEADS REPLACEMENT (Contd)

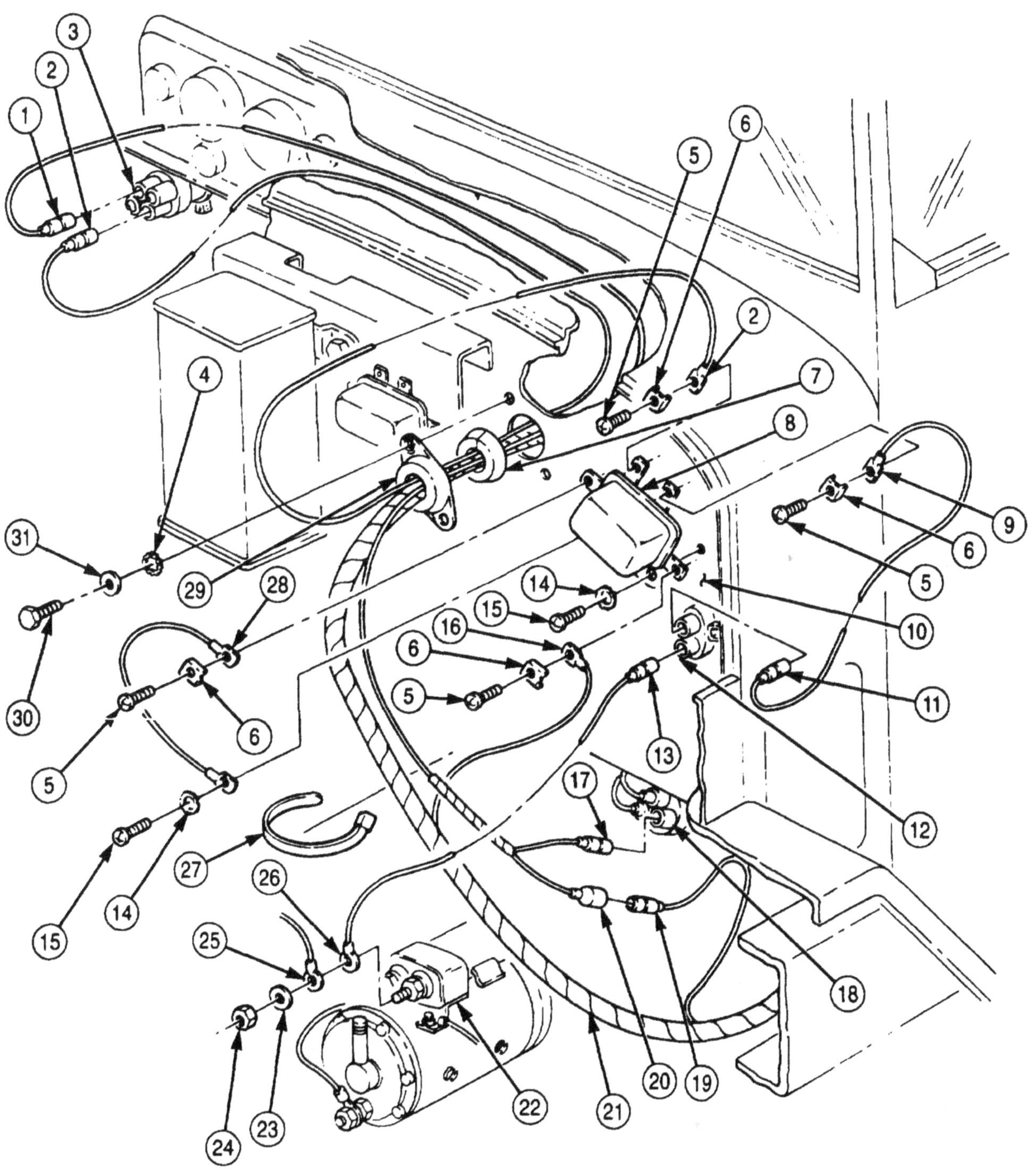

FOLLOW-ON TASK: Connect battery ground cables (para. 3-126).

3-437. CONVOY WARNING LIGHT REPLACEMENT

THIS TASK COVERS:

a. Removal b. Installation

APPLICABLE MODELS
All

TOOLS
General mechanic's tool kit (Appendix E, Item 1)

MATERIALS/PARTS
Three locknuts (Appendix D, Item 297)

REFERENCES (TM)
TM 9-2320-272-10
TM 9-2320-272-24P

EQUIPMENT CONDITION
- Parking brake set (TM 9-2320-272-10).
- Battery ground cables disconnected (para. 3-126).

a. Removal

1. Disconnect two leads (13) from cable leads (12).
2. Loosen three screws (4), rotate clamps (5) 1/2-turn counterclockwise, and remove dome (1) from base (6).
3. Remove three locknuts (2), washers (3), and base (6) from support plate (9).
4. Remove nut (16), washer (15), clamp (14), and two leads (13) from toggle bolt (11).
5. Remove two leads (13) and seal (7) from support plate (9).
6. Remove three oval nuts (8), support plate (9), and three toggle bolts (11) from mounting plate (10).

b. Installation

1. Install support plate (9) on mounting plate (10) with three toggle bolts (11) and oval nuts (8).
2. Install two leads (13) and seal (7) on support plate (9).
3. Install two leads (13) on toggle bolt (11) with clamp (14), washer (15), and nut (16).
4. Install base (6) on support plate (9) with three washers (3) and new locknuts (2).
5. Install dome (1) on base (6), rotate clamps (5) 1/2-turn clockwise, and tighten three screws (4).
6. Connect two leads (13) to cable leads (12).

3-437. CONVOY WARNING LIGHT REPLACEMENT (Contd)

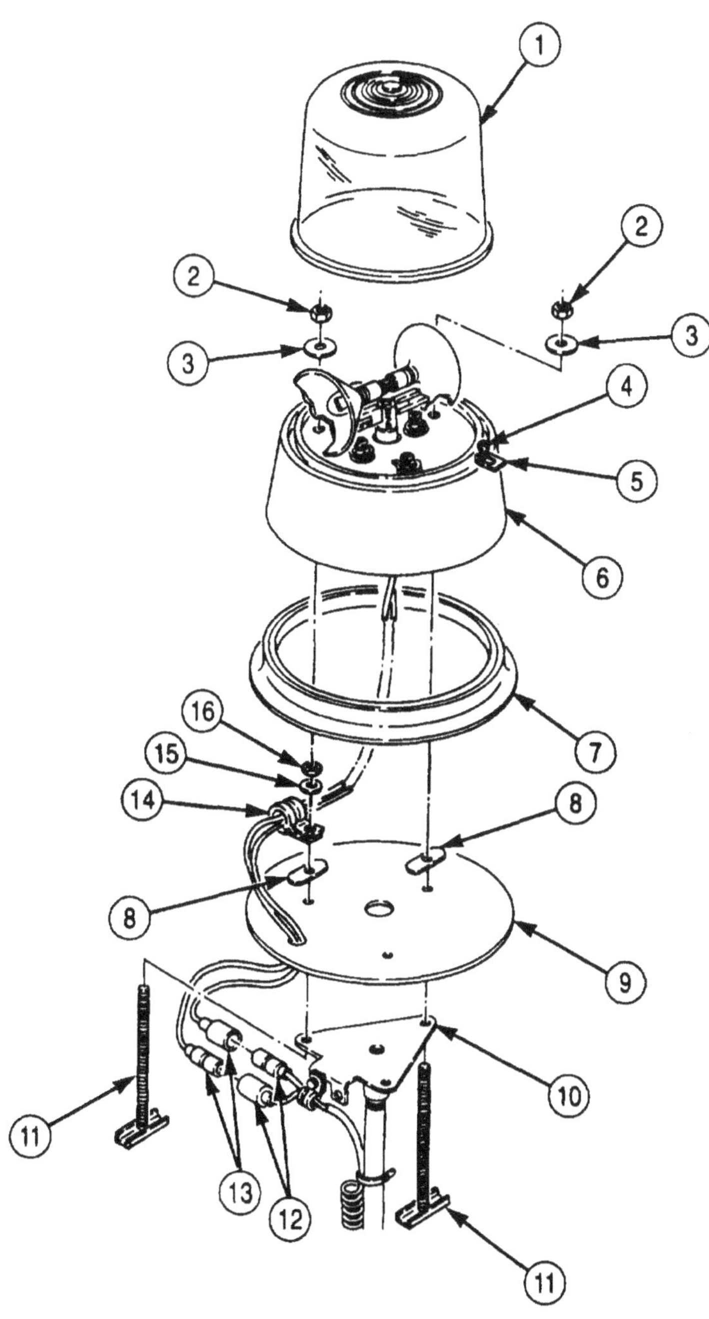

FOLLOW-ON TASK: Connect battery ground cables (para. 3-126).

3-438. CONVOY WARNING LIGHT SWITCH REPLACEMENT

THIS TASK COVERS:

a. Removal **b. Installation**

INITIAL SETUP:

APPLICABLE MODELS
All

TOOLS
General mechanic's tool kit (Appendix E, Item 1)

MATERIALS/PARTS
Two locknuts (Appendix D, Item 283)

REFERENCES (TM)
TM 9-2320-272-10
TM 9-2320-272-24P

EQUIPMENT CONDITION
Parking brake set (TM 9-2320-272-10).
Battery ground cables disconnected (para. 3-126).

a. Removal

1. Remove screw (8), lockwasher (7), and lever (6) from switch (11). Discard lockwasher (7).
2. Remove nut (5), lockwasher (4), identification plate (3), and switch (11) from bracket (9). Discard lockwasher (4).
3. Disconnect two leads (12) from switch (11).
4. Remove two locknuts (21, screws (10), and bracket (9) from instrument panel (1). Discard locknuts (2).

b. Installation

1. Install bracket (9) on instrument panel (1) with two screws (10) and new locknuts (2).
2. Connect two leads (12) to switch (11).
3. Install switch (11) and identification plate (3) on bracket (9) with new lockwasher (4) and nut (5).
4. Install lever (6) on switch (11) with new lockwasher (7) and screw (8).

3-438. CONVOY WARNING LIGHT SWITCH REPLACEMENT (Contd)

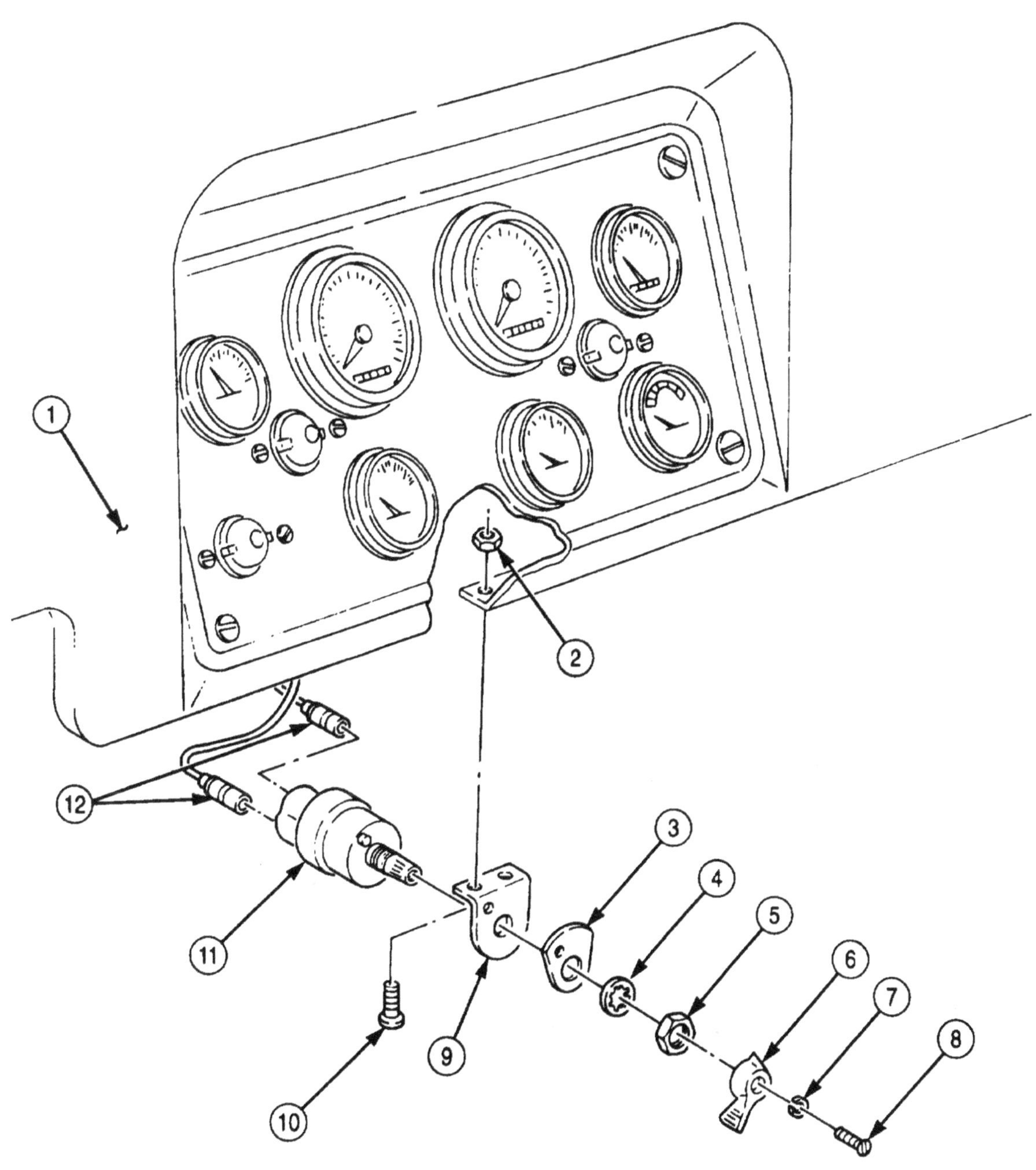

FOLLOW-ON TASK: Connect battery ground cables (para. 3-126).

3-439. EUROPEAN MINI-LIGHTING KIT REPLACEMENT

THIS TASK COVERS:

a. Removal **b. Installation**

INITIAL SETUP:

APPLICABLE MODELS
All

TOOLS
General mechanic's tool kit (Appendix E, Item 1)

MATERIALS/PARTS
Locknut (Appendix D, Item 294)
Four locknuts (Appendix D, Item 299)
Lockwasher (Appendix D, Item 378)
Two screw-assembled lockwashers (Appendix D, Item 577)
Two O-rings (Appendix D, Item 436)

REFERENCES (TM)
TM 9-2320-272-10
TM 9-2320-272-24P

EQUIPMENT CONDITION
- Parking brake set (TM 9-2320-272-10).
- Battery ground cable disconnected (para. 3-126).

NOTE

Right and left European mini-lights are replaced the same. This procedure covers the right side.

a. Removal

NOTE

Note routing of leads through grommet and fender for installation.

1. Remove two lenses (2), O-rings (3), and lamp (1) from lamp housing (4). Discard O-rings (3).
2. Remove four locknuts (13) and protector box (14) from four screws (10) on fender (11). Discard locknuts (13).
3. Disconnect lead (16) and cable (12) from connector (15).
4. Disconnect cable (12) from lead (9).
5. Remove grommet (7) from leads (16) and (9) and fender (11).
6. Remove locknut (18), lockwasher (19), lamp housing (4), and washer (5) from brush guard (6). Discard locknut (18) and lockwasher (19).
7. Remove two screw-assembled lockwashers (17) and brush guard (6) from brush guard (8). Discard screw-assembled lockwashers (17).

b. Installation

1. Install brush guard (6) on brush guard (8) with two new screw-assembled lockwashers (17).
2. Install washer (5) and lamp housing (4) on brush guard (6) with new lockwasher (19) and new locknut (18).
3. Install grommet (7) on leads (16) and (9) and fender (11).
4. Connect cable (12) to lead (9).
5. Connect lead (16) and cable (12) to connector (15).
6. Install protector box (14) on fender (11) and four screws (10) with four new locknuts (13).

NOTE

Ensure red lens of light assembly is installed facing rear of vehicle.

7. Install lamp (11, two new O-rings (3), and lenses (2) on lamp housing (4).

3-439. EUROPEAN MINI-LIGHTING KIT REPLACEMENT (Contd)

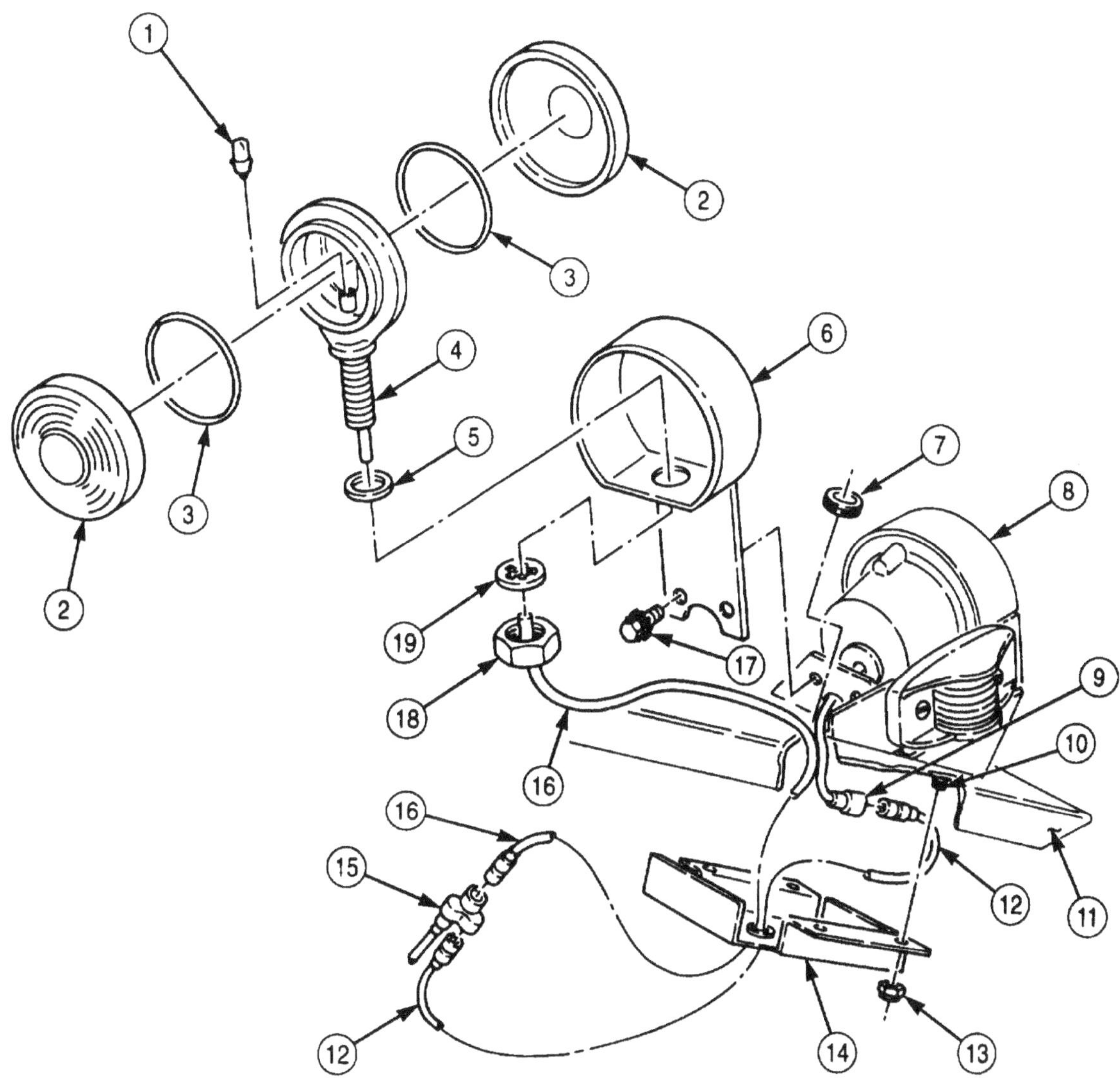

FOLLOW-ON TASK: Connect battery ground cables (para. 3-126).

3-440. AUTOMATIC THROTTLE KIT (M936/A1) REPLACEMENT

THIS TASK COVERS:

a. Removal **b. Installation**

INITIAL SETUP:

APPLICABLE MODELS
M936/A1

TOOLS
General mechanic's tool kit (Appendix E, Item 1)

MATERIALS/PARTS
Cotter pin (Appendix D, Item 78)
Three cotter pins (Appendix D, Item 66)
Antiseize tape (Appendix D, Item 72)

REFERENCES (TM)
TM 9-2320-272-10
TM 9-2320-272-24P

EQUIPMENT CONDITION
- Parking brake set (TM 9-2320-272-10).
- Air reservoirs drained (TM 9-2320-272-10).
- Air cleaner, air cleaner hose, air intake pipe, and hump hose removed (para. 3-13).

GENERAL SAFETY INSTRUCTIONS
Do not disconnect air lines before draining air reservoirs.

a. Removal

WARNING

Do not disconnect air lines before draining air reservoirs. Small parts under pressure may shoot out with high velocity, causing injury to personnel.

1. Remove cotter pin (12), washer (13), and pin (14) from throttle lever (1). Discard cotter pin (12).
2. Remove tube (10) from elbow (11).
3. Remove cotter pin (8), washer (9), pin (5), and air cylinder (2) from bracket (6). Discard cotter pin (8).

NOTE

Assistant will help with step 4.

4. Remove four screws (3), washers (4), and bracket (6) from cab floor (7).
5. Remove clevis (16), jamnut (15), and elbow (11) from air cylinder (2).

3-440. AUTOMATIC THROTTLE KIT (M936/A1) REPLACEMENT (Contd)

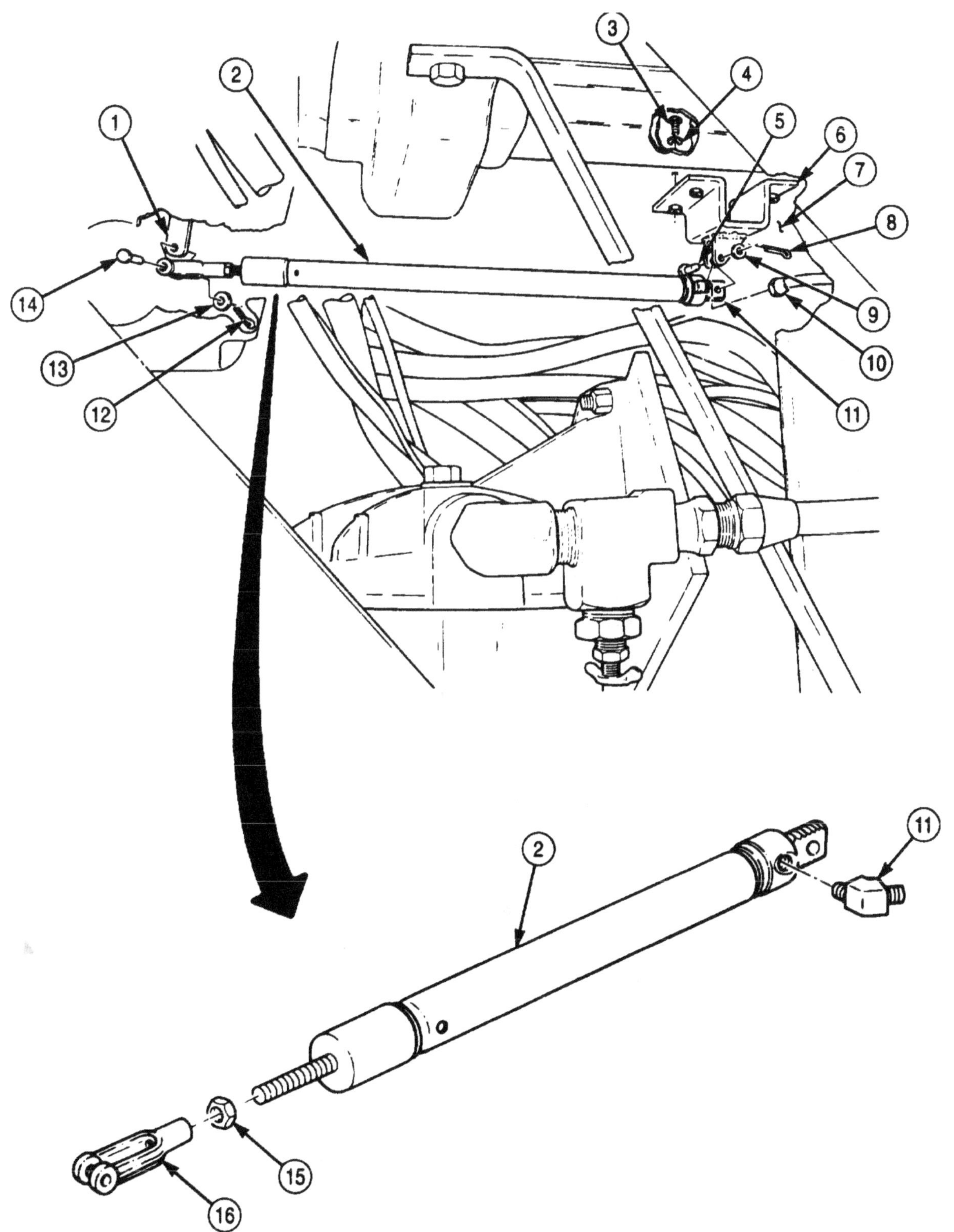

3-440. AUTOMATIC THROTTLE KIT (M936/A1) REPLACEMENT (Contd)

6. Remove cotter pin (6), washer (5), and accelerator pedal pushrod (8) from bellcrank link (4). Discard cotter pin (6).
7. Remove bushing (9), throttle lever (10), and bushing (9) from bellcrank link (4).
8. Remove cotter pin (3), washer (2), throttle shaft (7), and bellcrank link (4) from throttle bracket (1). Discard cotter pin (3).
9. Remove tubes (13) and (14) from tee (12).
10. Remove tee (12) from air pressure switch tee (11).

b. Installation

NOTE

Clean all male pipe threads and wrap with antiseize tape before installation.

1. Install tee (12) on air pressure switch tee (11).
2. Install tubes (13) and (14) on tee (12).
3. Install bellcrank link (4) and throttle shaft (7) on throttle bracket (1) with washer (2) and new cotter pin (3).
4. Install bushing (9), throttle lever (10), and bushing (9) on bellcrank link (4).
5. Install accelerator pedal pushrod (8) on bellcrank link (4) with washer (5) and new cotter pin (6).

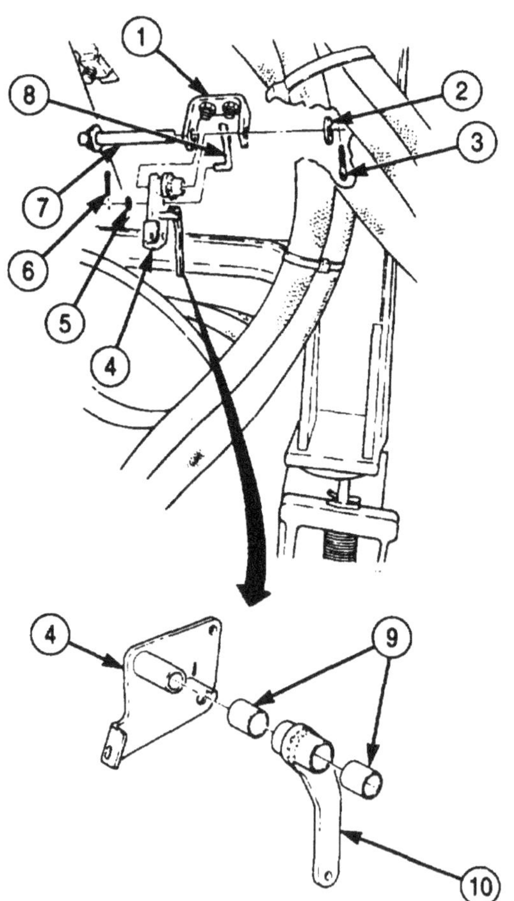

3-440. AUTOMATIC THROTTLE KIT (M936/A1) REPLACEMENT (Contd)

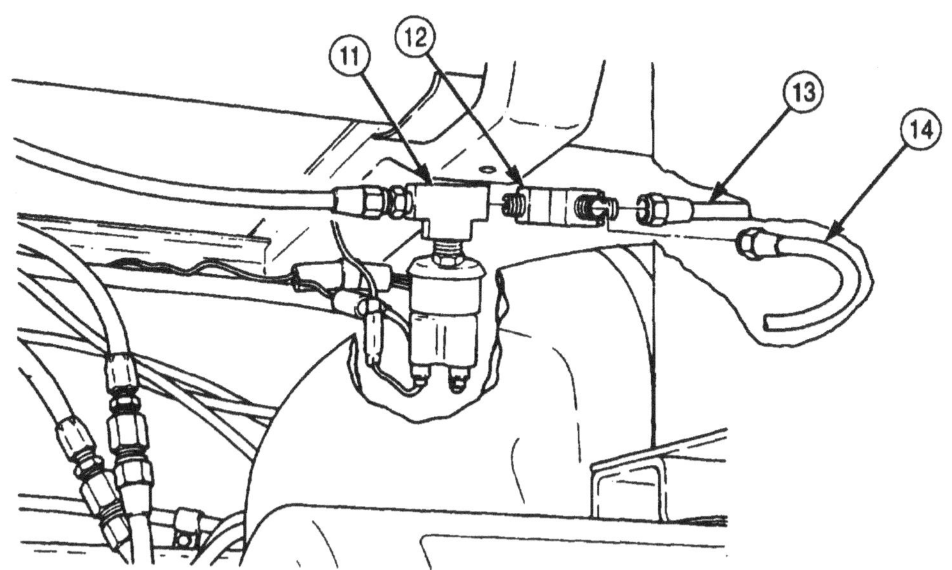

3-440. AUTOMATIC THROTTLE KIT (M936/A1) REPLACEMENT (Contd)

6. Install elbow (11), jamnut (15), and clevis (16) on air cylinder (2). Tighten jamnut (15) a distance of 0.25 in. (6.35 mm) from air cylinder (2) on air cylinder stud (17).

NOTE

Assistant will help with step 7.

7. Install bracket (6) on cab floor (7) with four washers (4) and screws (3).
8. Install air cylinder (2) on bracket (6) with pin (5), washer (9), and new cotter pin (8).
9. Install tube (10) on elbow (11).
10. Install pin (14) on throttle lever (1) with washer (13) and new cotter pin (12).

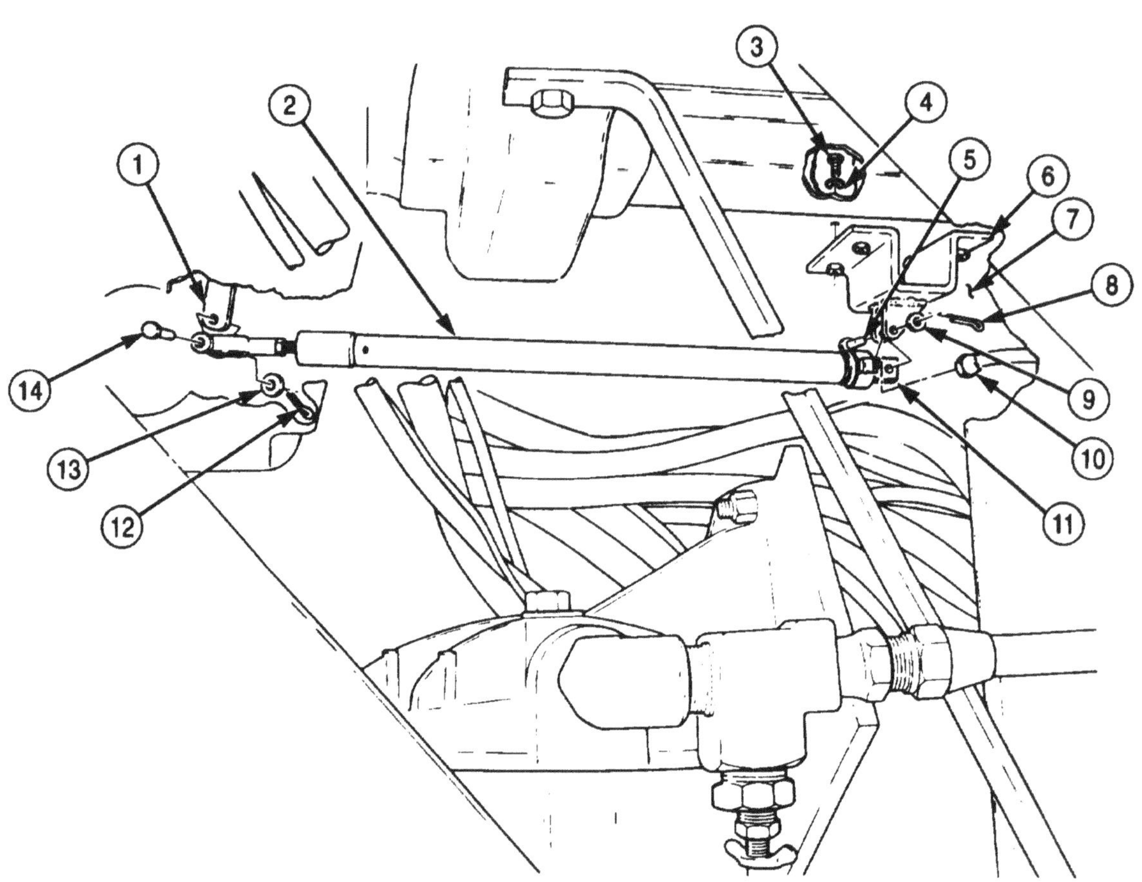

3-440. AUTOMATIC THROTTLE KIT (M936/A1) REPLACEMENT (Contd)

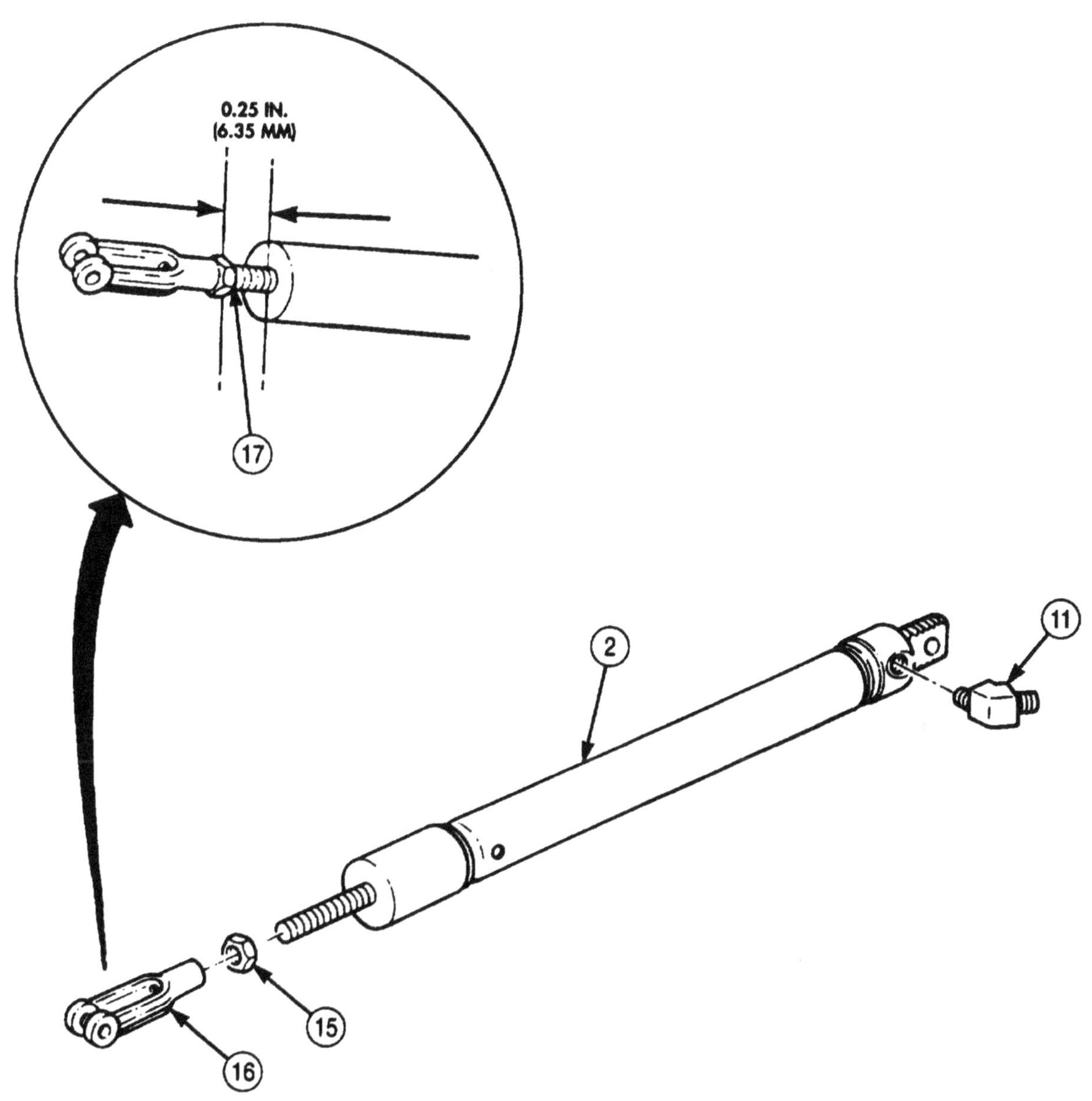

FOLLOW-ON TASKS:
- Install air cleaner, air cleaner hose, air intake pipe, and hump hose (para. 3-13).
- Start engine (TM 9-2320-272-10) and allow air pressure to build to normal operating range. Check for air leaks at service brake chamber.

3-441. ATMOSPHERIC FUEL TANK VENT SYSTEM KIT REPLACEMENT

THIS TASK COVERS:

a. Removal **b. Installation**

APPLICABLE MODELS
All

TOOLS
General mechanic's tool kit (Appendix E, Item 1)

MATERIALS/PARTS
Two locknuts (Appendix D, Item 294)
Antiseize tape (Appendix D, Item 72)

REFERENCES (TM)
TM 9-2320-272-10
TM 9-2320-272-24P

EQUIPMENT CONDITION
- Parking brake set (TM 9-2320-272-10).
- Hood raised and secured (TM 9-2320-272-10).
- Battery ground cables disconnected (para. 3-126).

a. Removal

NOTE

Perform steps 1 through 3 for single fuel tank system only.

1. Remove three fuel lines (1) from adapters (2).
2. Remove locknut (5), screw (4), and tee (3) from rail (6). Discard locknut (5).
3. Remove three adapters (2) from tee (3).
4. Remove three fuel lines (1) from adapters (2).
5. Remove locknut (5), screw (4), and tee (3) from rail (6). Discard locknut (5).
6. Remove three adapters (2) from two elbows (16) and tee (3).
7. Remove two elbows (16) from tee (3).
8. Remove vent line (10) from adapter (11).
9. Remove adapter (11) from adapter (12).
10. Remove adapter (12) from tube (13).
11. Remove four clamps (7), tubes (15) and (13), and hose (9) from air intake pipe (8).

3-441. ATMOSPHERIC FUEL TANK VENT SYSTEM KIT REPLACEMENT (Contd)

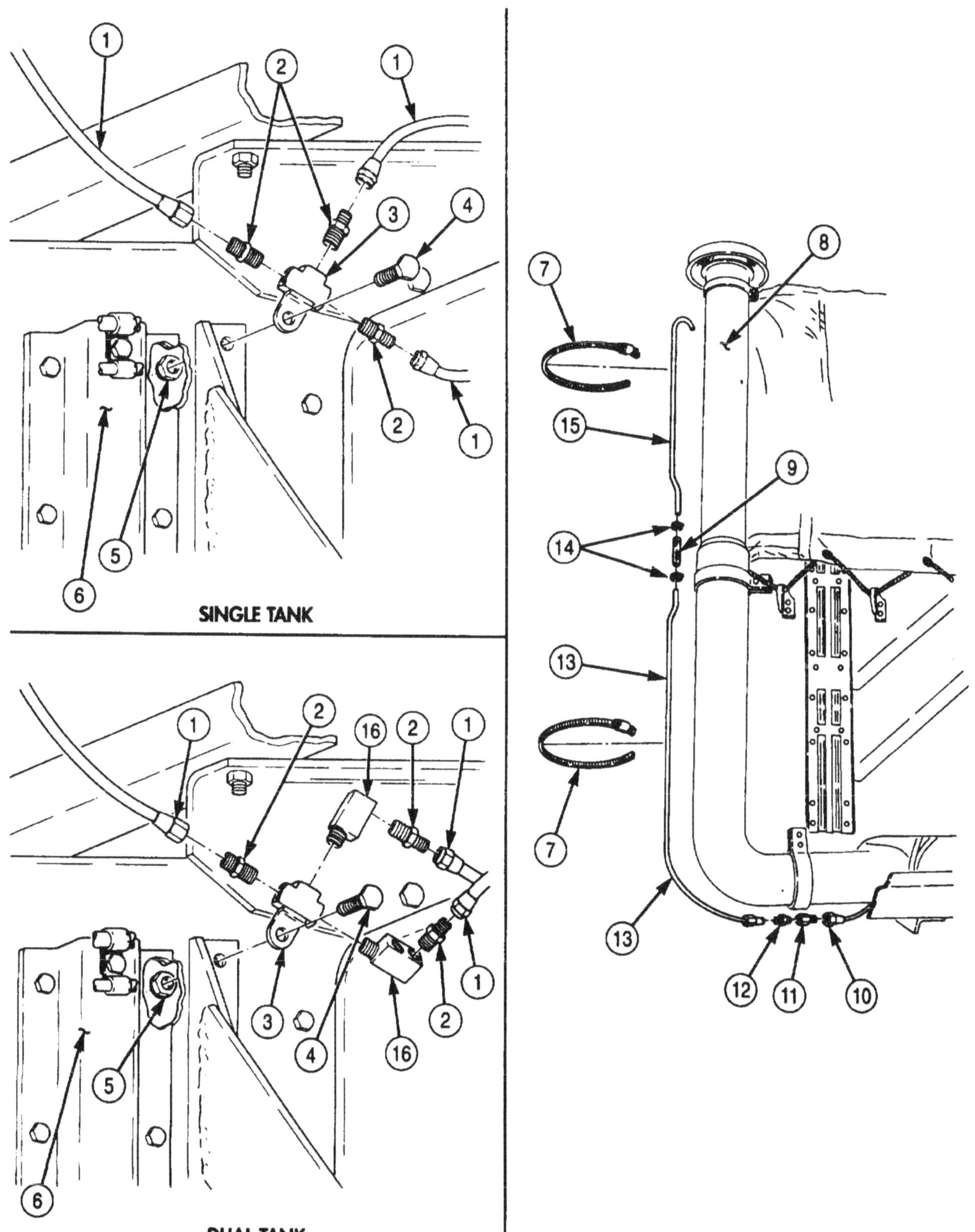

3-441. ATMOSPHERIC FUEL TANK VENT SYSTEM KIT REPLACEMENT (Contd)

12. Remove two clamps (14) and tubes (15) and (13) from hose (9).
13. Remove two nuts (13), clamps (12), screws (11), and clamps (12) from fuel return line (1) and fuel supply line (14).
14. Disconnect fuel return line (1) from elbows (16) and (9).
15. Disconnect fuel supply line (14) from elbow (15) and fuel pump (6).
16. Disconnect tube (4) from elbow (3) and tee (5).
17. Remove screw (7) and bracket (8) from engine (10).
18. Remove elbows (3) and (9) and two bushings (18) from check valve (2).
19. Remove locknut (21), washer (20), clamp (171, bracket (8), washer (20), and screw (19) from check valve (2). Discard locknut (21).

b. Installation

NOTE

Wrap all male pipe threads with antiseize tape before installation.

1. Install clamp (17) and bracket (8) on check valve (2) with washer (20), screw (19), washer (201, and new locknut (21).
2. Install two bushings (18) and elbows (3) and (9) on check valve (2).
3. Install bracket (8) on engine (10) with screw (7).
4. Connect tube (4) to elbow (3) and tee (5).
5. Connect fuel supply line (14) to elbow (15) and fuel pump (6).
6. Connect fuel return line (1) to elbows (16) and (9).
7. Install two clamps (12), screws (11), clamps (12), and nuts (13) on fuel return line (1) and fuel supply line (14).

3-441. ATMOSPHERIC FUEL TANK VENT SYSTEM KIT REPLACEMENT (Contd)

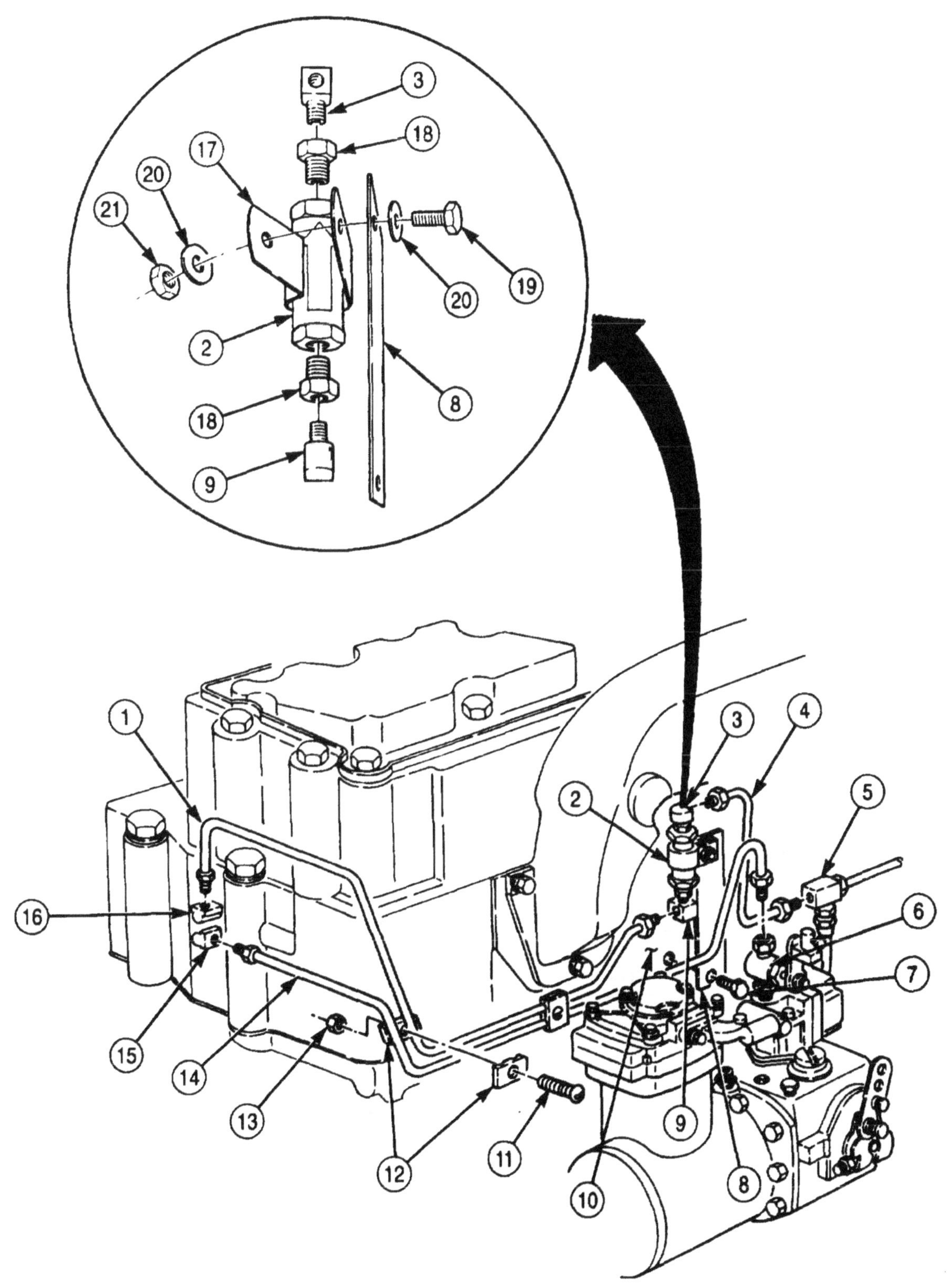

3-441. ATMOSPHERIC FUEL TANK VENT SYSTEM KIT REPLACEMENT (Contd)

8. Install tubes (15) and (13) on hose (9) with two clamps (14).
9. Install tubes (15) and (13) and hose (9) on air intake pipe (8) with four clamps (7).
10. Install adapter (12) on tube (13).
11. Install adapter (11) on adapter (12).
12. Install vent line (10) on adapter (11).
13. Install two elbows (16) on tee (3).
14. Install three adapters (2) on two elbows (16) and tee (3).
15. Install tee (3) on rail (6) with screw (4) and new locknut (5).
16. Install three fuel lines (1) on three adapters (2).

NOTE

Perform steps 17 through 19 for single fuel tank system only.

17. Install three adapters (2) on tee (3).
18. Install tee (3) on rail (6) with screw (4) and new locknut (5).
19. Install three fuel lines (1) on three adapters (2).

3-441. ATMOSPHERIC FUEL TANK VENT SYSTEM KIT REPLACEMENT (Contd)

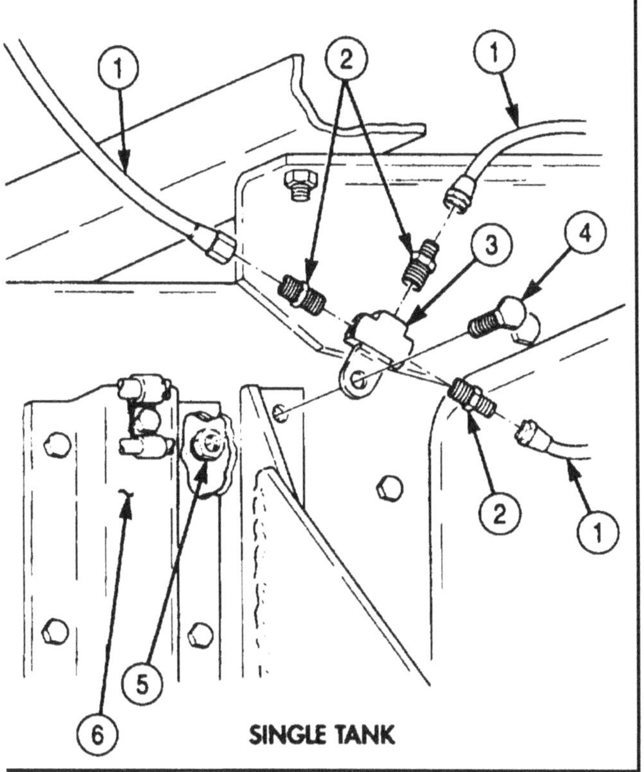

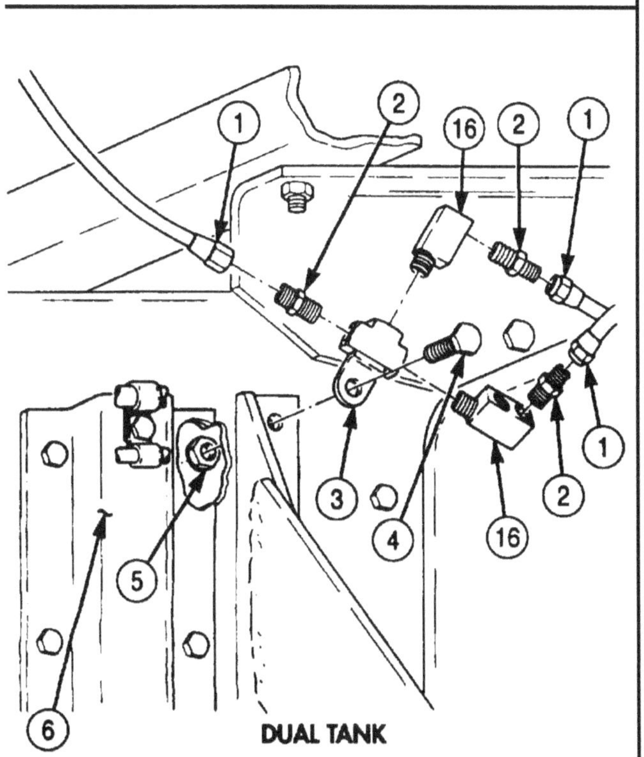

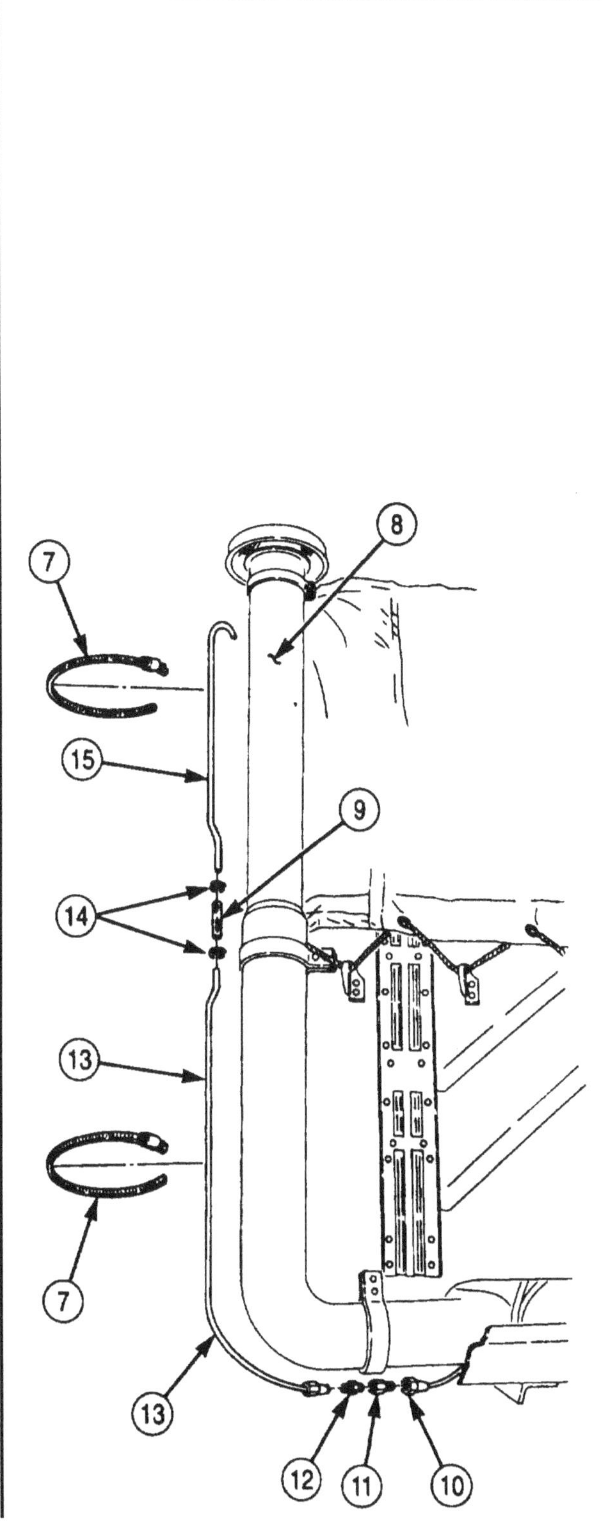

SINGLE TANK

DUAL TANK

FOLLOW-ON TASK: Connect battery ground cables (para. 3-126).

3-442. VEHICLE TIEDOWN KIT REPLACEMENT

THIS TASK COVERS:

a. Removal **b. Installation**

INITIAL SETUP:

APPLICABLE MODELS
All

TOOLS
General mechanic's tool kit (Appendix E, Item 1)

MATERIAL/PARTS
Twenty-four locknuts (M923, M924, M929, M931, M936) (Appendix D. Item 309)
Twenty-eight locknuts (M925, M926, M930, M932) (Appendix D, Item 309)
Thirty-two locknuts (M927, M934) (Appendix D, Item 309)
Thirty-six locknuts (M928) (Appendix D, Item 309)

REFERENCES (TM)
TM 9-2320-272-10
TM 9-2320-272-24P

EQUIPMENT CONDITION
Parking brake set (TM 9-2320-272-10).

NOTE

Left and right side tiedowns are replaced the same. This procedure covers left-side tiedowns.

a. Removal

NOTE

- Depending on vehicle model, as few as 12, or as many as 18, tiedowns may exist.
- Tiedowns are located alongside vehicle frame rail.
- All vehicles have two tiedowns installed with spacers. Mark position for installation.

Remove nine locknuts (2), washers (1), screws (5), spacers (3), and tiedown (6) from left side frame rail (4). Discard locknuts (2) and remove spacer (3) as required.

b. Installation

Install spacers (3) as required and tiedowns (6) on left-side frame rail (4) with two screws (5), washers (1), and new locknuts (2).

3-451. VEHICLE TIEDOWN KIT REPLACEMENT (Contd)

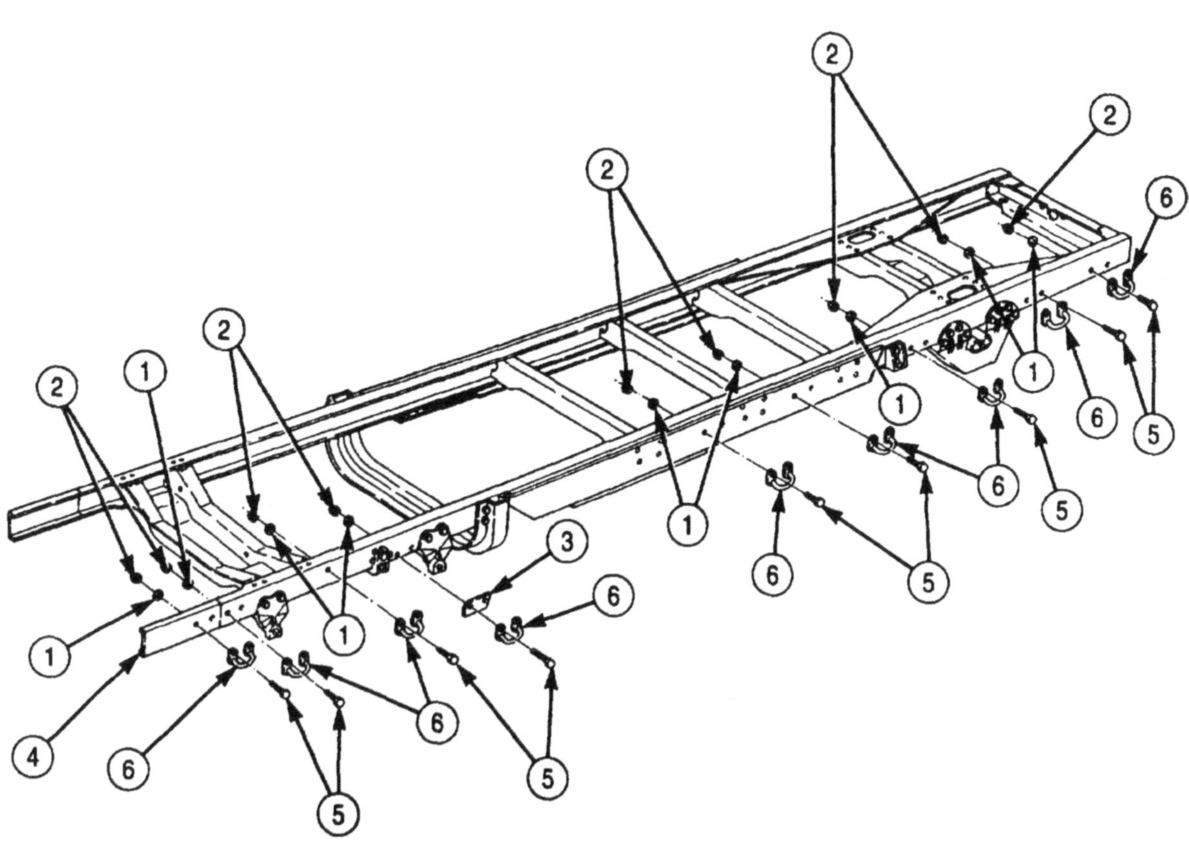

3-443. HYDRAULIC HOSE CHAFE GUARD KIT REPLACEMENT

THIS TASK COVERS:

a. Removal b. Installation

INITIAL SETUP:

APPLICABLE MODELS
M929/A1/A2, M930/A1/A2

TOOLS
General mechanic's tool kit (Appendix E, Item 1)

MATERIALS/PARTS
Two packings (Appendix D, Item 437)
Two locknuts (Appendix D, Item 294)
Tiedown strap (Appendix D, Item 694)
Sealant (Appendix C, Item 62)
Cap and plug set (Appendix C, Item 14)

REFERENCES (TM)
LO 9-2320-272-12
TM 9-2320-272-10
TM 9-2320-272-24P

EQUIPMENT CONDITION
Parking brake set (TM 9-2320-272-10).
Dump body raised and support braces in position (TM 9-2320-272-10).
Drain hydraulic oil reservoir (LO 9-2320-272-12).

GENERAL SAFETY INSTRUCTIONS
Dump body must be raised and secured with safety braces.

WARNING

Dump body must be raised and secured with safety braces before removal and installation of chafe guard kit. Failure to do this may result in injury to personnel.

CAUTION

Plug all hydraulic openings and hoses to prevent contamination.

a. Removal

NOTE

Tag all hoses and tubes for installation.

1. Disconnect hoses (5) and (6) from elbows (9) and (10).
2. Remove hoses (5) and (6) and packings (7) from safety lock cylinder (8). Discard packings (7).
3. Remove hoses (2) and (3) from crosses (1) and (4).
4. Remove hoses (11), (12), (16), and (15) from four connectors (20).
5. Remove four connectors (20) from four cylinder ports (18), (21), (19), and (22).
6. Remove hoses (11) and (16) from cross (4).
7. Remove hoses (12) and (15) from cross (1).
8. Remove two locknuts (17), plate (14), cross (4), plate (14), cross (1), and plate (14) from two screws (13). Discard locknuts (17).
9. Remove elbows (9) and (10) from crosses (1) and (4).

3-443. HYDRAULIC HOSE CHAFE GUARD KIT REPLACEMENT (Contd)

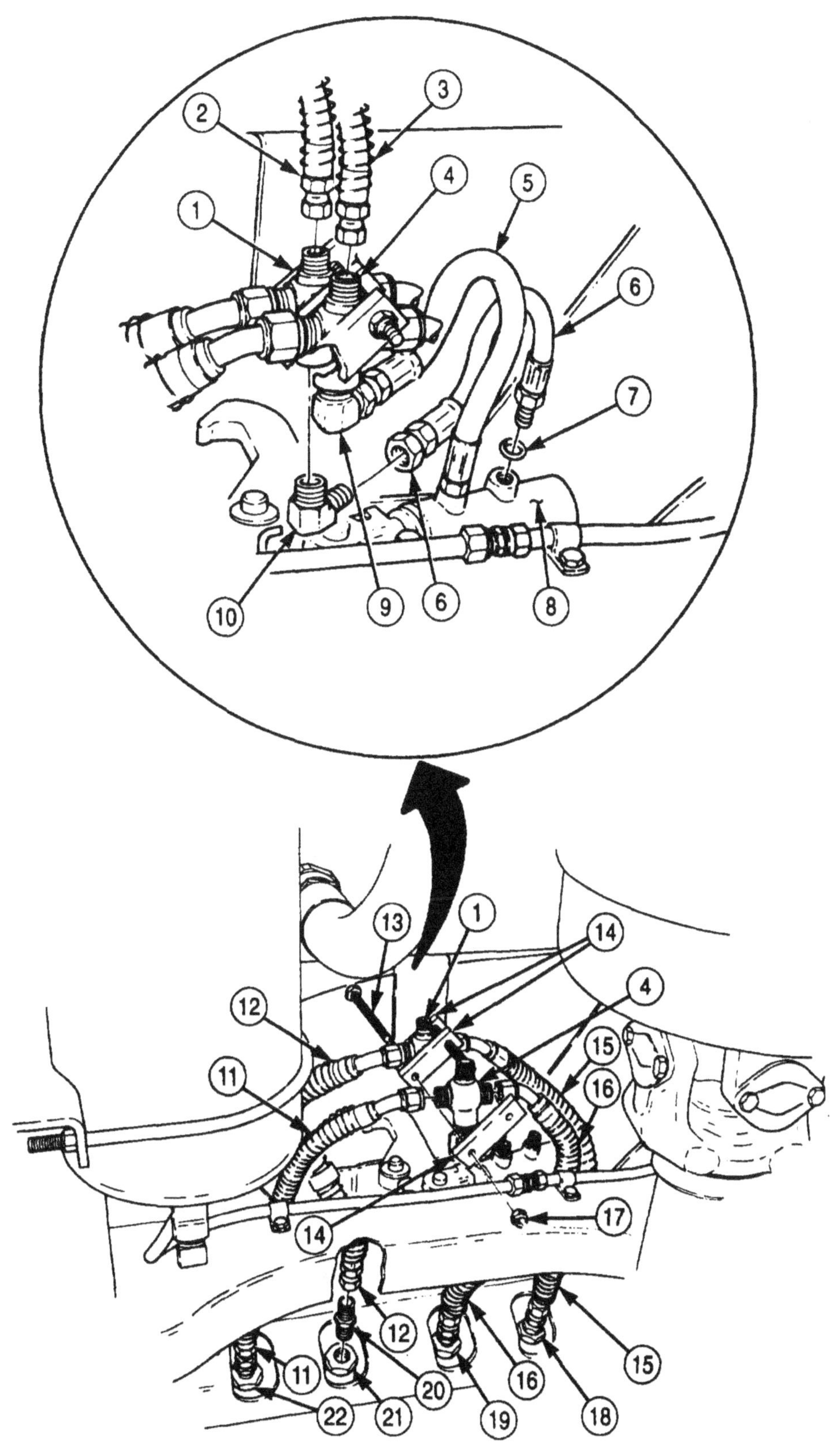

3-443. HYDRAULIC HOSE CHAFE GUARD KIT REPLACEMENT (Contd)

10. Remove hoses (1) and (3) from control valve ports (5) and (4).
11. Remove tiedown strap (2) from hoses (1) and (3). Discard tiedown strap (2).

b. Installation

NOTE

Clean all male pipe threads and apply pipe sealant before installation.

1. Install new tiedown strap (2) on hoses (1) and (3).
2. Install hoses (1) and (3) on control valve ports (5) and (4).
3. Install elbows (24) and (25) on crosses (9) and (11).
4. Install plate (10), cross (9), plate (10), crosses (11), plate (10), and two new locknuts (14) on screws (8).
5. Install hoses (7) and (12) on cross (9).
6. Install hoses (6) and (13) on cross (11).
7. Install four connecters (17) on cylinder ports (15), (16), (18), and (19).
8. Install hoses (6), (7), (12), and (13) on connectors (17).
9. Install hoses (1) and (3) on crosses (9) and (11).
10. Install two new packings (22) and hoses (20) and (21) on safety lock cylinder (23).
11. Connect hoses (20) and (21) to elbows (24) and (25).

3-443. HYDRAULIC HOSE CHAFE GUARD KIT REPLACEMENT (Contd)

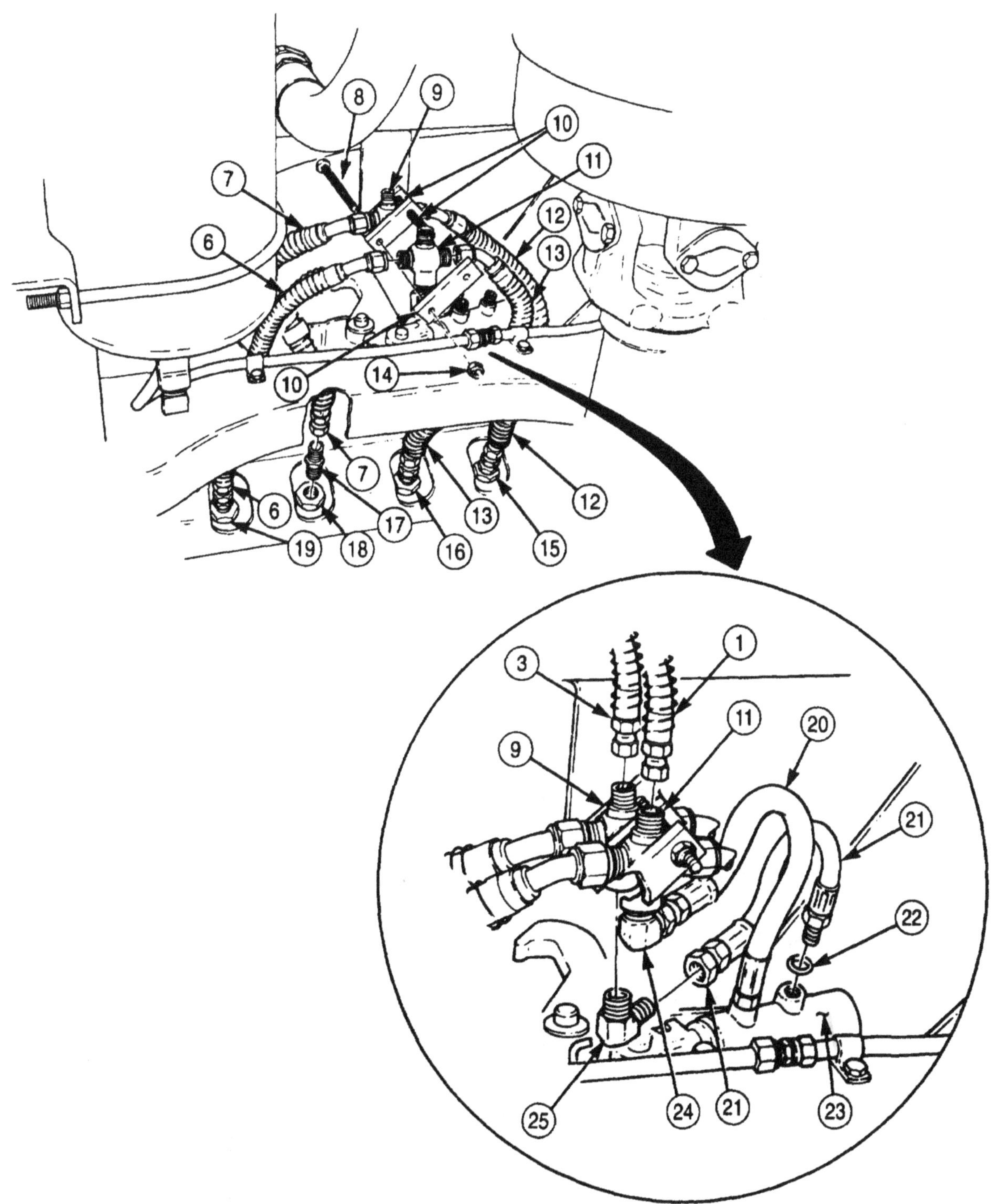

FOLLOW-ON TASKS:
- Fill hydraulic oil reservoir (LO 9-2320-272-12).
- Remove support braces and lower dump body (TM 9-2320-272-10).
- Check for proper operation of hydraulic system (TM 9-2320-272-10).

3-444. HYDRAULIC RESERVOIR DRAIN KIT REPLACEMENT

THIS TASK COVERS:

a. Removal
b. Installation

INITIAL SETUP:

APPLICABLE MODELS
M929/A1, M930/A1

TOOLS
General mechanic's tool kit (Appendix E, Item 1)
Torque wrench (Appendix E, Item 144)

MATERIALS/PARTS
Teflon pipe sealant (Appendix C, Item 67)
Antiseize tape (Appendix C, Item 72)
Cap and plug set (Appendix C, Item 14)

REFERENCES (TM)
TM 9-2320-272-10
TM 9-2320-272-24P

EQUIPMENT CONDITION
- Parking brake set (TM 9-2320-272-10).
- Dump body raised and support braces in position (TM 9-2320-272-10).

GENERAL SAFETY INSTRUCTIONS
Dump body must be raised and secured with safety braces.

WARNING

Dump body must be raised and secured with safety braces before working under dump body. Failure to do this may result in injury to personnel.

CAUTION

- Wrap or plug hydraulic line openings to prevent dirt entering and causing damage.
- Position of transfer case oil pump should be marked to align with transfer case for installation. Improper alignment of transfer case oil pump housing to transfer case may damage components.

a. Removal

NOTE

- Steps 1 and 2 provide clearance for hydraulic reservoir drain removal.
- Have drainage container ready to catch oil.

1. Remove six screws (1) and washers (2) from transfer case oil pump (4).
2. Remove transfer case oil pump (4) from transfer case (3).
3. Disconnect reservoir hydraulic tube (7) from adapter (9).
4. Remove adapter (9) from Y-branch (8).
5. Remove hex plug (13), ball valve (12), pipe nipple (11), and reducer bushing (10) from Y-branch (8).
6. Remove Y-branch (8) and pipe nipple (6) from hydraulic reservoir (5).

b. Installation

NOTE

Apply antiseize tape to all male threads prior to installation.

1. Position pipe nipple (6) in hydraulic reservoir (5) and install Y-branch (8) to pipe nipple (6).
2. Install reducer bushing (10), pipe nipple (11), ball valve (12), and hex plug (13) on Y-branch (8).
3. Install adapter (9) in Y-branch (8).

3-444. HYDRAULIC RESERVOIR DRAIN KIT REPLACEMENT (Contd)

4. Connect reservoir hydraulic tube (7) to adapter (9).
5. Install transfer case oil pump (4) on transfer case (3) with six washers (2) and screws (1). Tighten screws (1) 40-65 lb-ft (54-88 N•m).

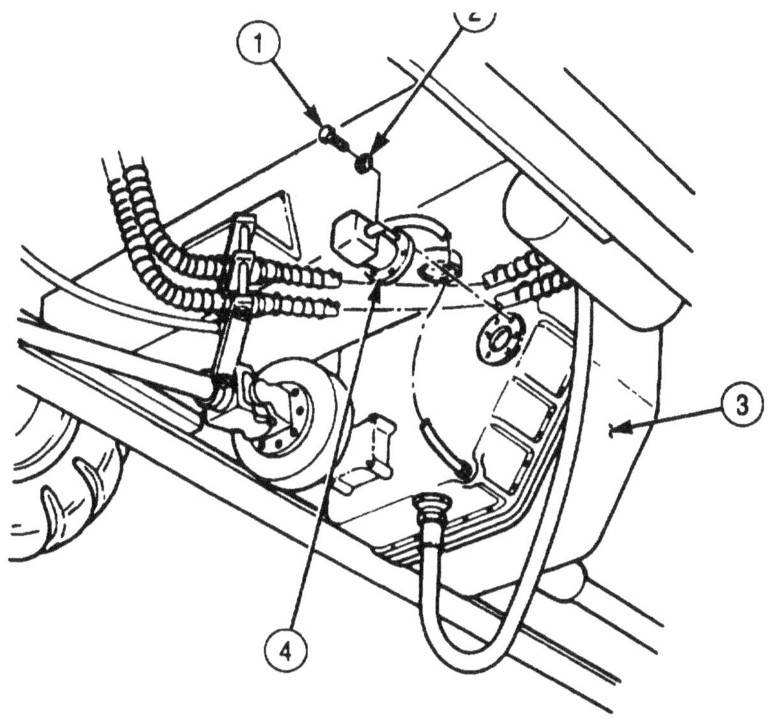

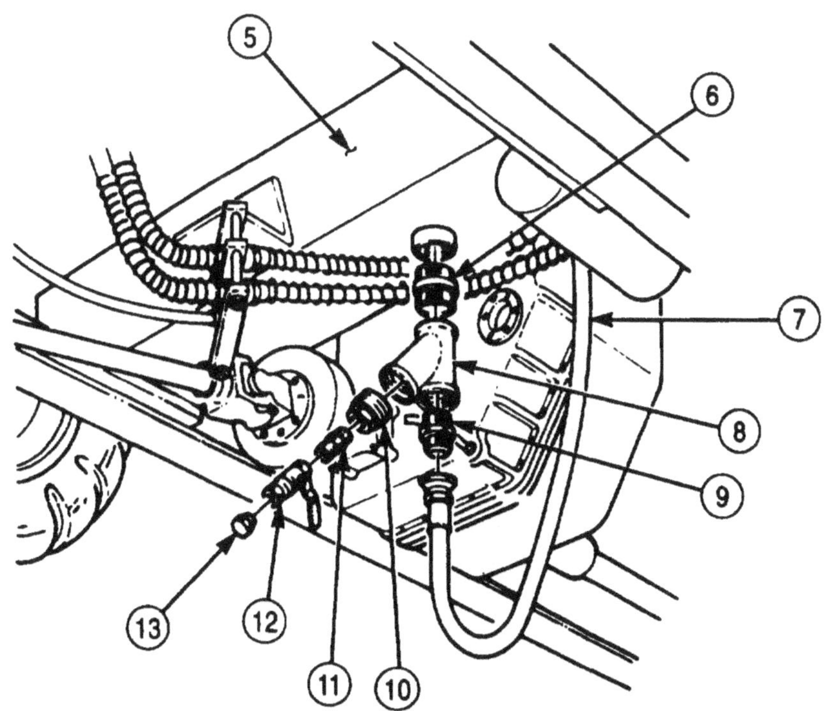

FOLLOW-ON TASK: Fill hydraulic reservoir to proper level (LO 9-2320-272-12).

3-445. HYDRAULIC RESERVOIR SHUTOFF MODIFICATION KIT REPLACEMENT

THIS TASK COVERS:

a. Removal   b. Installation

INITIAL SETUP:

APPLICABLE MODELS

M936/A1/A2

TOOLS

General mechanic's tool kit (Appendix E, Item 1)
Torque wrench (Appendix E, Item 146)

MATERIALS/PARTS

Two O-rings (Appendix D, Item 489)
O-ring (Appendix D, Item 490)
Sealing compound (Appendix C, Item 63)

REFERENCES (TM)

LO 9-2320-272-12
TM 9-2320-272-20
TM 9-2320-272-24P

EQUIPMENT CONDITION

- Battery ground cables disconnected (para. 3-1261.
- Forward deck plate removed (para. 3-392).
- Hydraulic oil reservoir drained (LO 9-2320-272-12).

GENERAL SAFETY INSTRUCTIONS

Directional arrow on check valve must point toward hydraulic reservoir. Failure may result in personnel injury and damage to equipment.

NOTE

Have drainage container ready to catch oil.

a. Removal

1. Remove hose assembly (1) from check valve (10) and adapter (11).
2. Remove check valve (10), O-ring (9), and adapter assembly (8) from adapter (7). Discard O-ring (9).
3. Remove four screws (2), washers (3), two swivel flanges (4), adapter (7), and O-ring (6) from front winch pump outlet port (5). Discard O-ring (6).

NOTE

From under wrecker pull hose assembly far enough to provide clearance for check valve removal.

4. Disconnect hose assembly (13) from check valve (14).
5. Remove adapter assembly (15) and check valve (14) from elbow (12).

3-445. HYDRAULIC RESERVOIR SHUTOFF MODIFICATION KIT REPLACEMENT (Contd

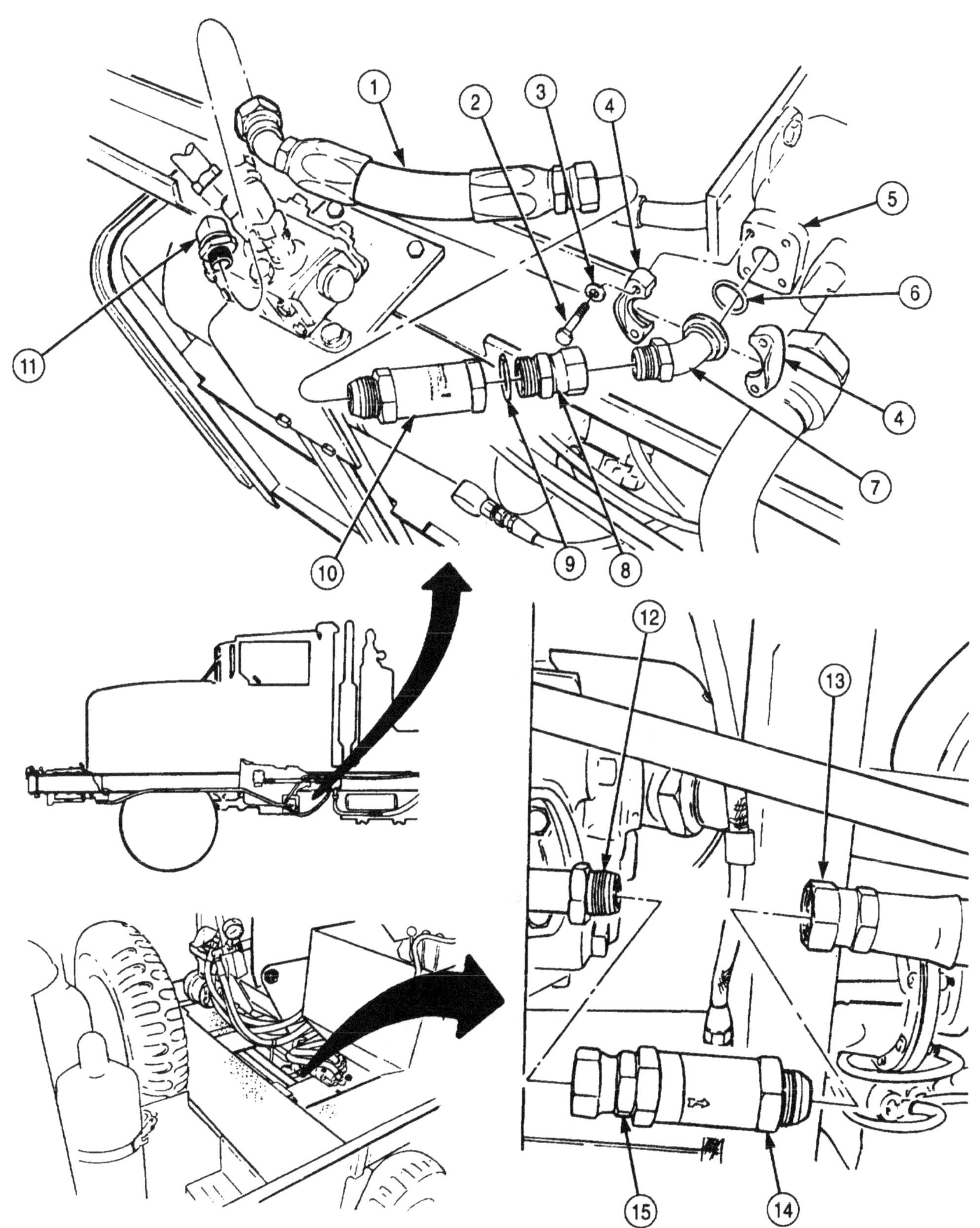

3-445. HYDRAULIC RESERVOIR SHUTOFF MODIFICATION KIT REPLACEMENT (Contd)

6. Remove adapter assembly (8) and O-ring (10) from check valve (7). Discard O-ring (10).
7. Remove hose assembly (1), union assembly (5), check valve (2), and nipple (3) from hydraulic oil reservoir (4).

b. Installation

NOTE

- When applying pipe sealant to male threads, leave first two starter threads clear of sealing compound before installing.
- Sealing compound is not necessary on adapter assemblies with O-rings.

1. Install nipple (3) on hydraulic oil reservoir (4).

WARNING

Ensure directional arrow on check valve points toward the hydraulic oil reservoir. Failure to do this may result in injury to personnel and damage to equipment.

2. Install check valve (2) on nipple (3). Ensure directional arrow on check valve (2) points toward hydraulic reservoir (4).
3. Install union assembly (5) and hose assembly (1) on check valve (2).
4. Position new O-ring (10) on threaded end of adapter assembly (8).
5. Install check valve (7) and adapter assembly (8) on elbow (9).
6. Connect hose assembly (6) to check valve (7).

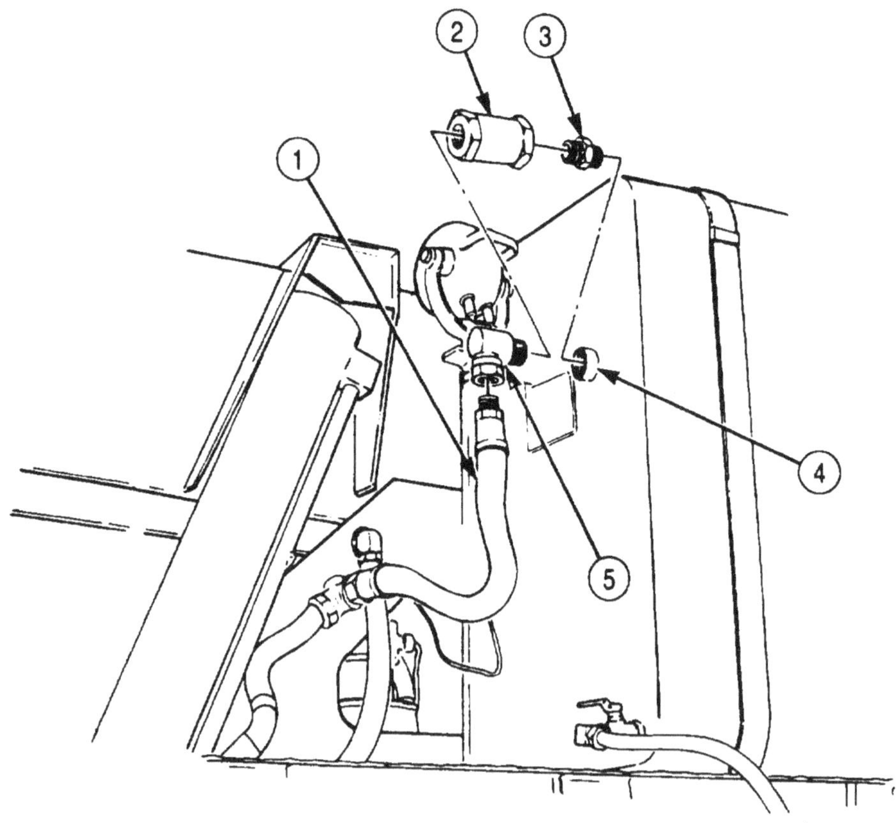

3-445. HYDRAULIC RESERVOIR SHUTOFF MODIFICATION KIT REPLACEMENT (Contd)

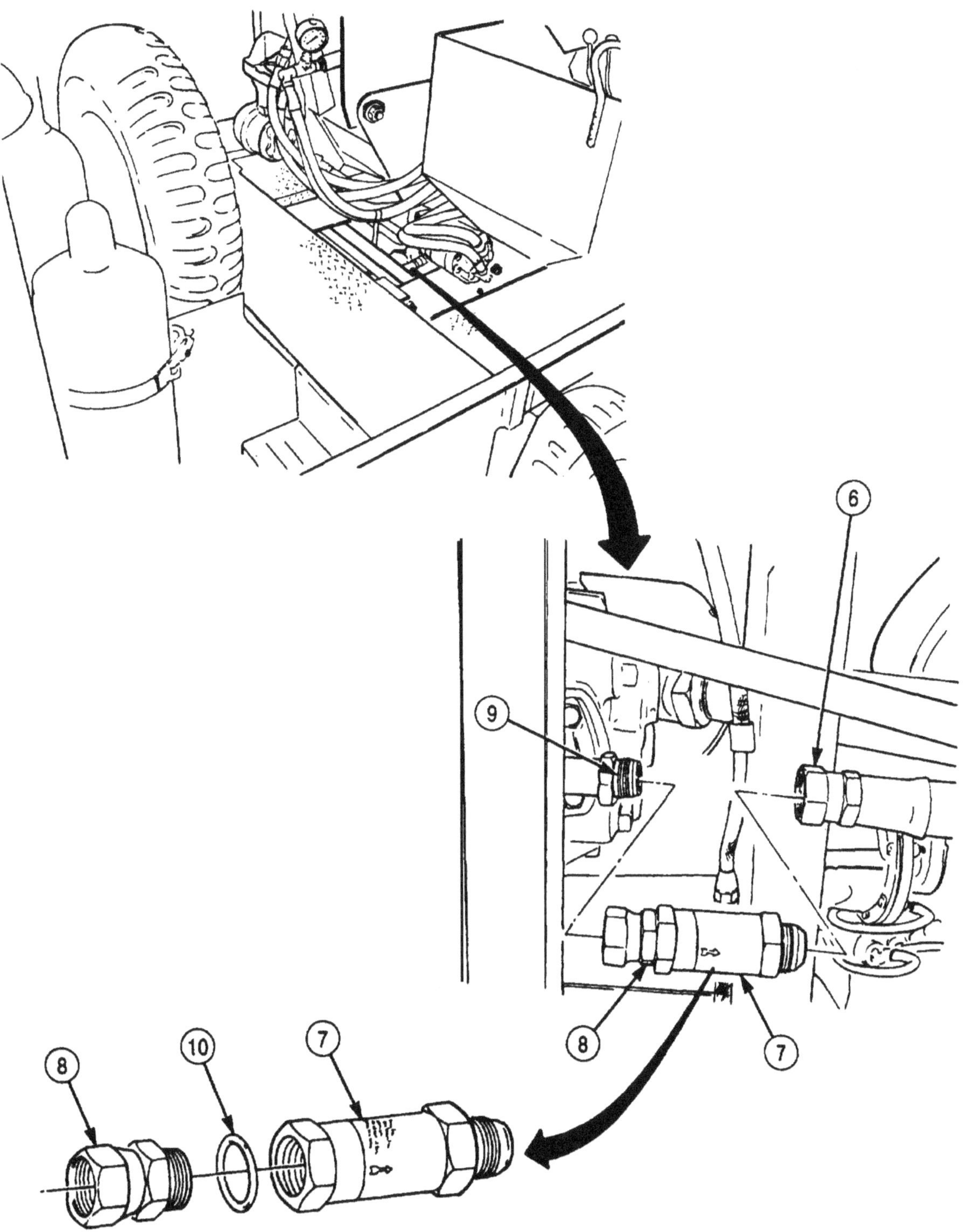

3-445. HYDRAULIC RESERVOIR SHUTOFF MODIFICATION KIT REPLACEMENT (Contd)

7. Position new O-ring (6) and adapter (7) on front winch pump outlet port (5) and install two swivel flanges (4), four washers (3), and screws (2). Tighten screws (2) 31-35 lb-ft (42-47 N•m).
8. Install adapter assembly (8) on adapter (7).
9. Position new O-ring (9) over threaded end of adapter assembly (8) and install check valve (10) on adapter assembly (8).
10. Install hose assembly (1) on adapter (11).
11. Install hose assembly (1) on check valve (10).

3-445. HYDRAULIC RESERVOIR SHUTOFF MODIFICATION KIT REPLACEMENT (Contd)

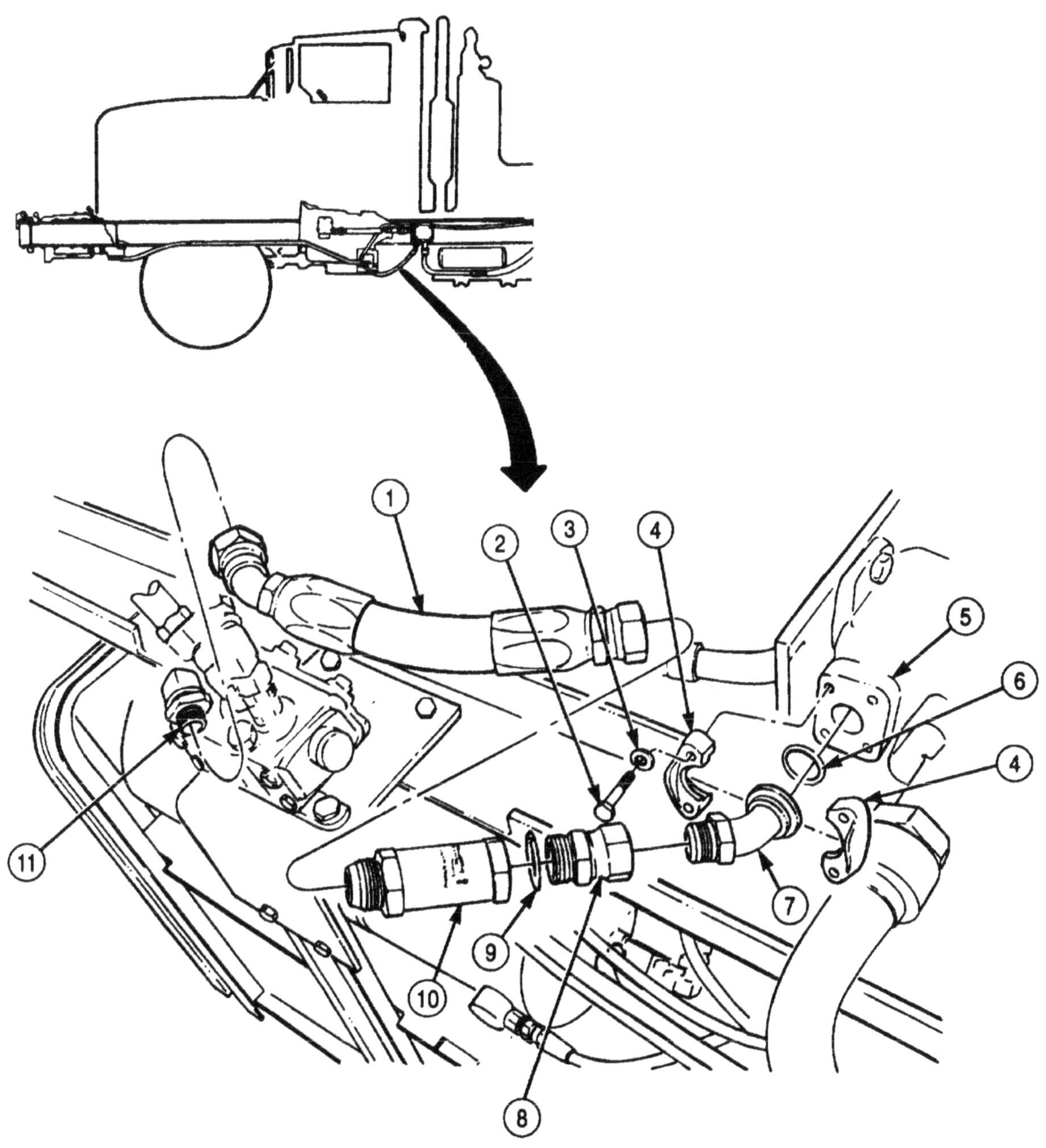

FOLLOW-ON TASKS:
- Install forward deck plate (para. 3-392).
- Connect battery ground cables (para. 3-126).
- Fill hydraulic oil reservoir (LO 9-2320-272-12).
- Start engine and operate front winch and rear boom in all directions until air is completely purged from all lines. Check all hoses and fittings for leaks and tighten if necessary (TM 9-2320-272-10).
- Add additional hydraulic fluid as necessary (LO 9-2320-272-12).

3-446. LIGHTWEIGHT WEAPON STATION MODIFICATION KIT MAINTENANCE

THIS TASK COVERS:

a. Removal
b. Disassembly
c. Assembly
d. Installation

INITIAL SETUP:

APPLICABLE MODELS

M939/A1/A2

TOOLS

General mechanic's tool kit (Appendix E, Item 1)
Torque wrench (Appendix E, Item 146)
Lifting device
Four utility chains

MATERIALS/PARTS

One hundred eight locknuts (Appendix D, Item 337)
Eight locknuts (Appendix D, Item 338)
Locknut (Appendix D, Item 335)
Four locknuts (Appendix D, Item 336)
Eight lockwashers (Appendix D, Item 414)
Six cotter pins (Appendix D, Item 79)

REFERENCES (TM)

TM 9-2320-272-10
TM 9-2320-272-24P

EQUIPMENT CONDITION

- Parking brake set (TM 9-2320-272-10).
- Cab top removed (TM 9-2320-272-10).
- Spare tire removed (TM 9-2320-272-10).
- Backrest cushion on companion seat assembly folded down (TM 9-2320-272-10).
- Companion seat and battery box moved forward to gain access to cab (TM 9-2320-272-10).
- Batteries removed (para. 3-125).
- Muffler shield assembly removed (para. 3-51).
- Left and right cab door check rods removed (para. 3-318).
- Left and right door assemblies removed (para. 3-319).
- Left and right mirror assemblies removed (para. 3-319).
- Driver's seat removed (para. 3-283).
- Right and left cowl ventilation door screens removed (para. 3-299).

GENERAL SAFETY INSTRUCTIONS

All personnel must stand clear during lifting onerations.

a. Removal

WARNING

All personnel must stand clear during lifting operations. A snapped chain, or shifting or swinging load, may result in injury to personnel.

NOTE

It may be necessary to remove left and right reinforcement panels from rear of cab to gain access to U-bolt nuts.

1. Remove four locknuts (7), eight washers (4), four screws (3), and braces (6) from support assembly (10). Discard locknuts (7).
2. Remove eight locknuts (8), sixteen washers (1), eight screws (2), four ammo tray supports (5) and ammo box tray (9) from under left and right front sides of support assembly (10). Discard locknuts (8).

3-446. LIGHTWEIGHT WEAPON STATION MODIFICATION KIT MAINTENANCE (Contd)

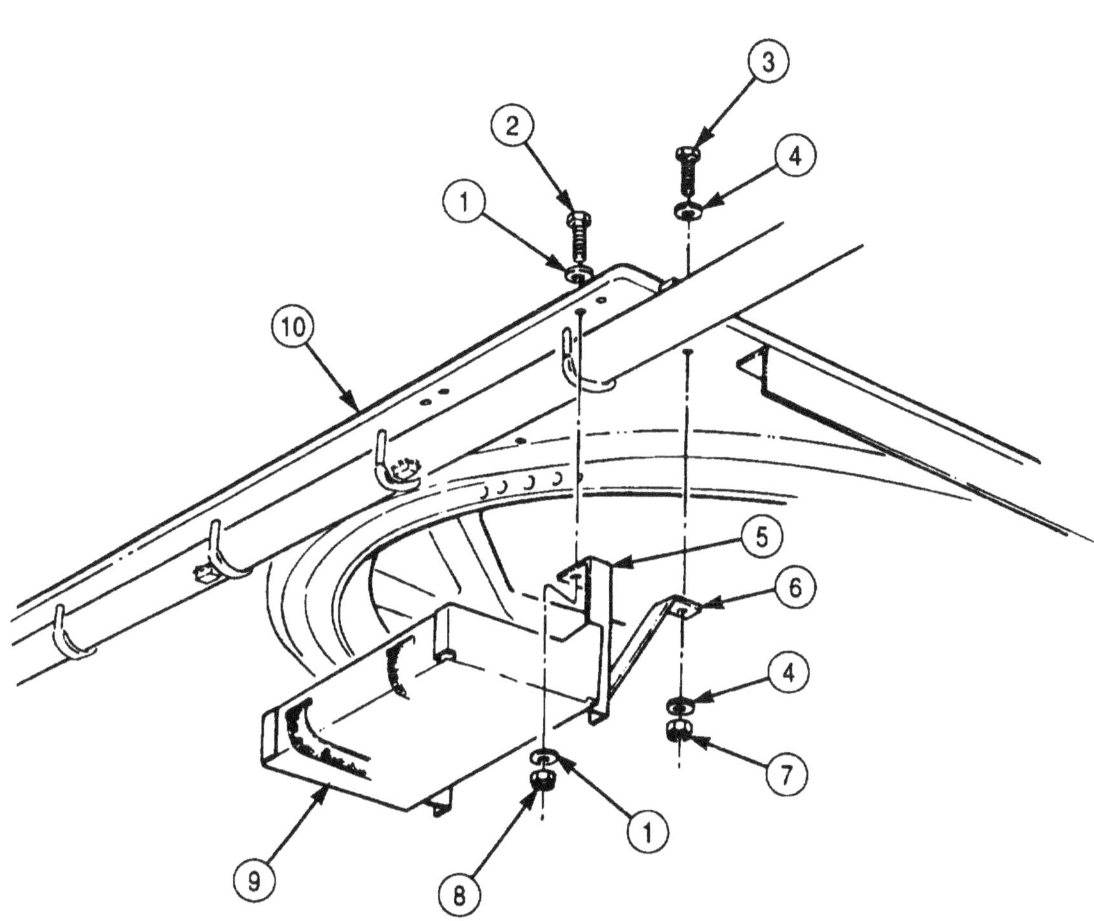

3-446. LIGHTWEIGHT WEAPON STATION MODIFICATION KIT MAINTENANCE (Contd)

3. Attach four utility chains and lifting device to lightweight weapon station platform (11).
4. Remove four locknuts (9). washers (8), plates (7), roof mounts (6), and spacers (5) from screws (1), lightweight weapon station platform (11), and four post assemblies (10). Discard locknuts (9).
5. Remove four screws (1) and washers (2) from lightweight weapon station platform (11).
6. Raise lightweight weapon station platform (11) and remove four plates (3) and roof mounts (4) from four post assemblies (10).
7. Remove fourteen locknuts (14), washers (13), seven U-bolts (17), short crossmember assembly (18), and long crossmember assembly (12) from lightweight weapon station platform (11). Discard locknuts (14).
8. Place lightweight weapon station platform (11) on blocking so inner ring (16) of support assembly (15) is accessible and remove lifting device from lightweight weapon station platform (11).

NOTE

Left and right rear post assemblies are removed the same way. This procedure is for the right rear post assembly.

9. Remove two locknuts (20) and U-bolt (24) from right rear cab support (21). Discard locknuts (20).
10. Remove right rear post assembly (10) from U-bolts (23) and (22) on right rear cab support (21).
11. Remove two cotter pins (19) and pin (25) from bottom hole in right rear post assembly (10). Discard cotter pins (19).
12. Remove four locknuts (20) and U-bolts (23) and (22) from right rear cab support (21). Discard locknuts (20).

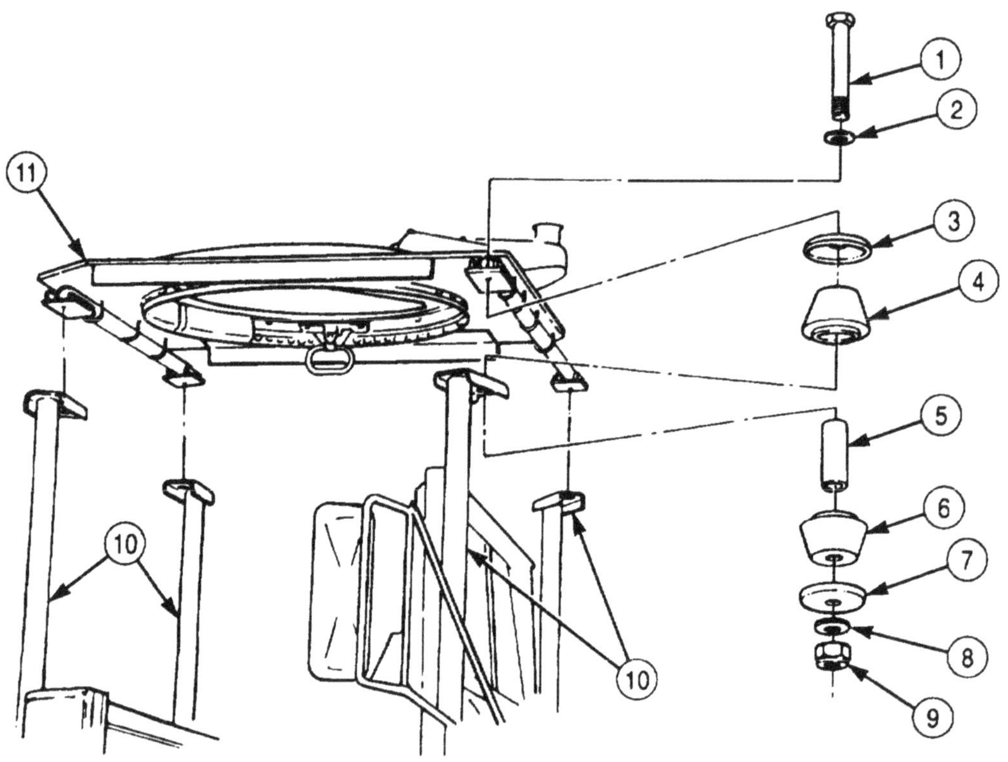

3-446. LIGHTWEIGHT WEAPON STATION MODIFICATION KIT MAINTENANCE (Contd)

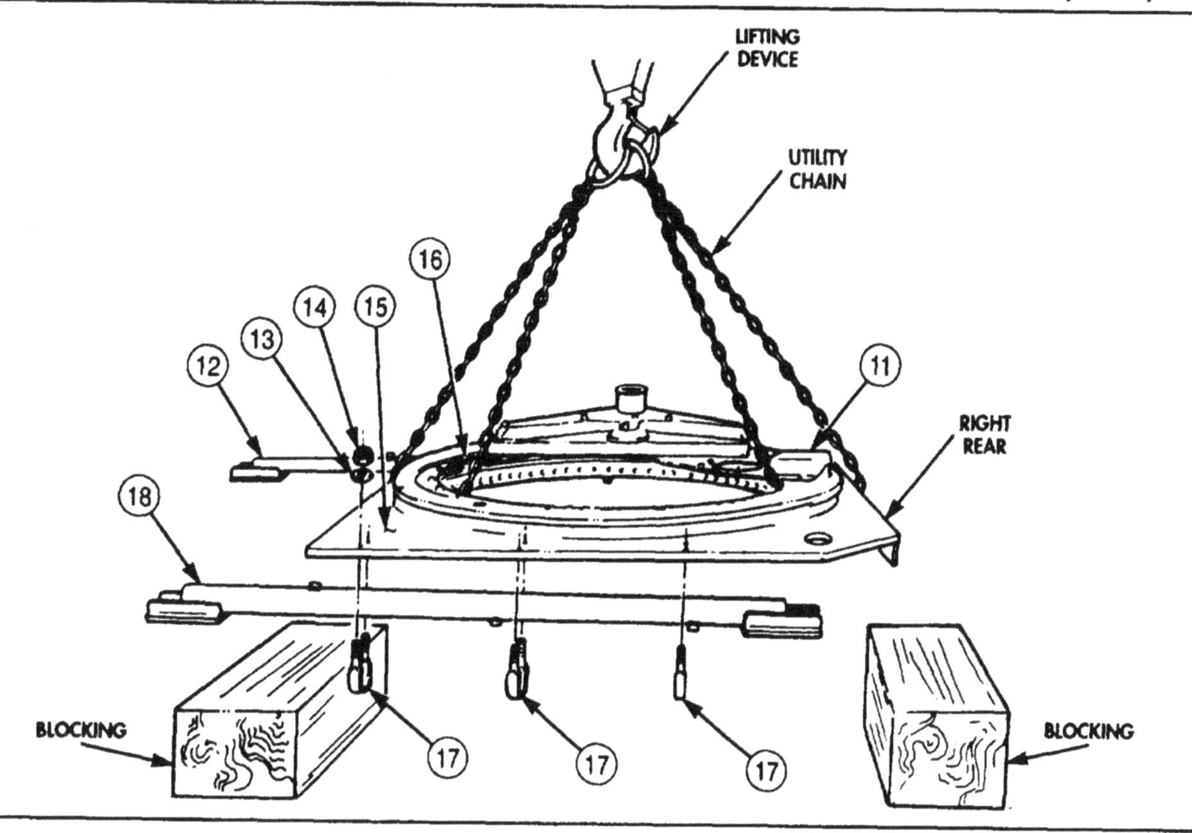

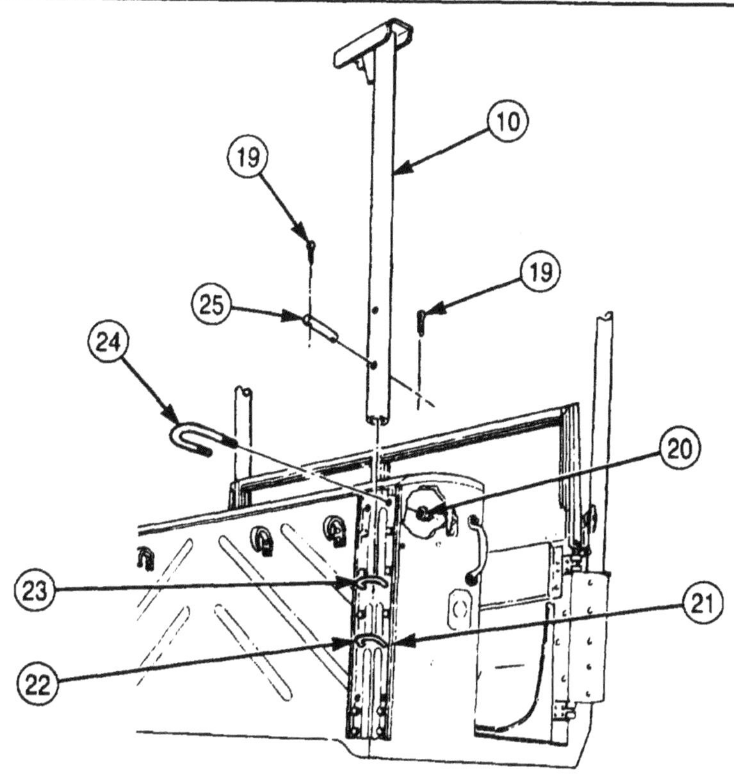

3-446. LIGHTWEIGHT WEAPON STATION MODIFICATION KIT MAINTENANCE (Contd)

13. Remove two cotter pins (2) and pin (6) from top hole in right front post assembly (1). Discard cotter pins (2).
14. Remove three locknuts (3) and screws (5) from right front post assembly (1) and right post support (4). Lift right front post assembly (1) out of right front post support (4). Discard locknuts (3).
15. Remove five locknuts (9) and screws (10) from right post support (8) and right front brace assembly (7). Discard locknuts (9).
16. Remove two screws (15) and screw (14) from instrument panel (11) and right door pillar (12).
17. Remove four screws (13) and right post support (8) from right door pillar (12).
18. Remove four screws (19), washers (18), right front brace assembly (20), and tapping plate (16) from outer cowl wall (17) and inside of door pillar wall (21).
19. Remove two cotter pins (24) and pin (23) from top hole of left front post assembly (22). Discard cotter pins (24).
20. Remove three locknuts (27) and screws (25) from left front post assembly (22) and left post support (26) and lift left front post assembly (22) out of left post support (26). Discard locknuts (27).

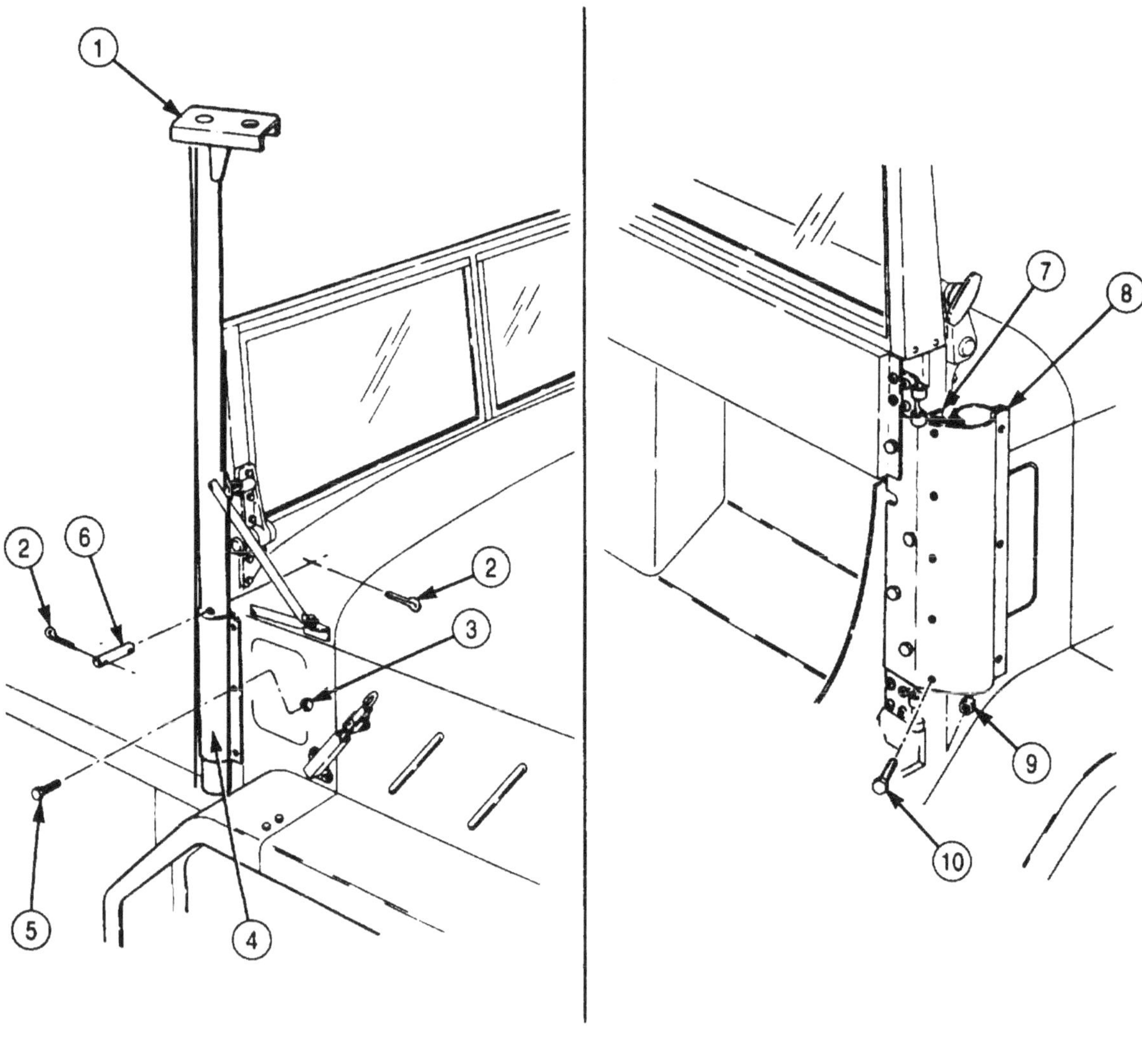

3-446. LIGHTWEIGHT WEAPON STATION MODIFICATION KIT MAINTENANCE (Contd)

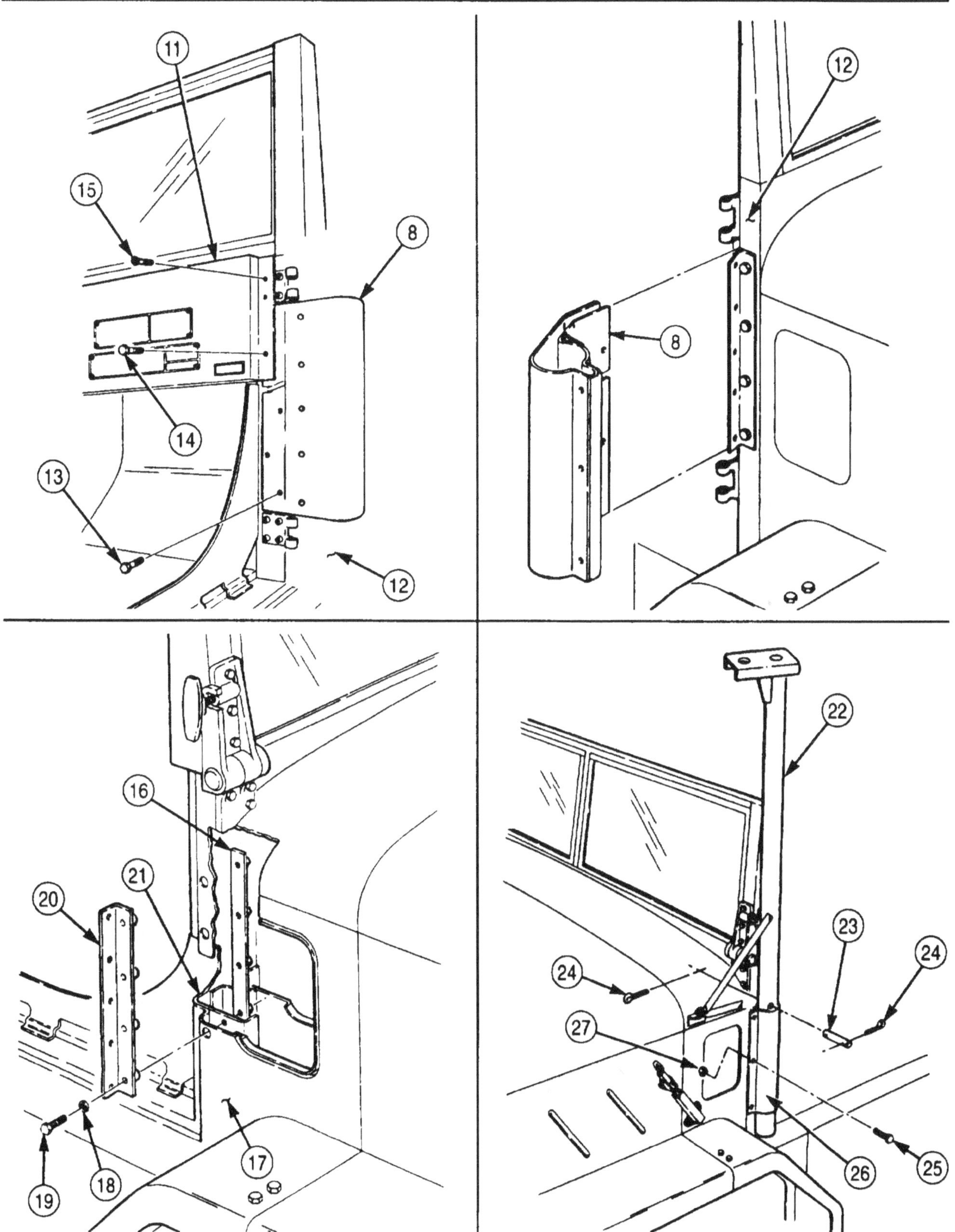

3-446. LIGHTWEIGHT WEAPON STATION MODIFICATION KIT MAINTENANCE (Contd)

21. Remove three nut assemblies (3) and screws (1) from backside of left front pillar wall (2).
22. Remove five locknuts (9), screws (7), and left post support bracket (8) from left front brace assembly (4) and left post support (6). Discard locknuts (9).
23. Remove eight door hinge screws (21), left post support (6), and two door hinges (20) from left door pillar (5).
24. Remove four screws (12), washers (13), left front brace assembly (4), and tapping plate (10) from left door pillar wall (2) and outer cowl wall (11).
25. Remove eight screws (14), lockwashers (15), washers (16), and seat back panel (17) from backrest cushion frame (18) and companion seat assembly (19). Discard lockwashers (15).

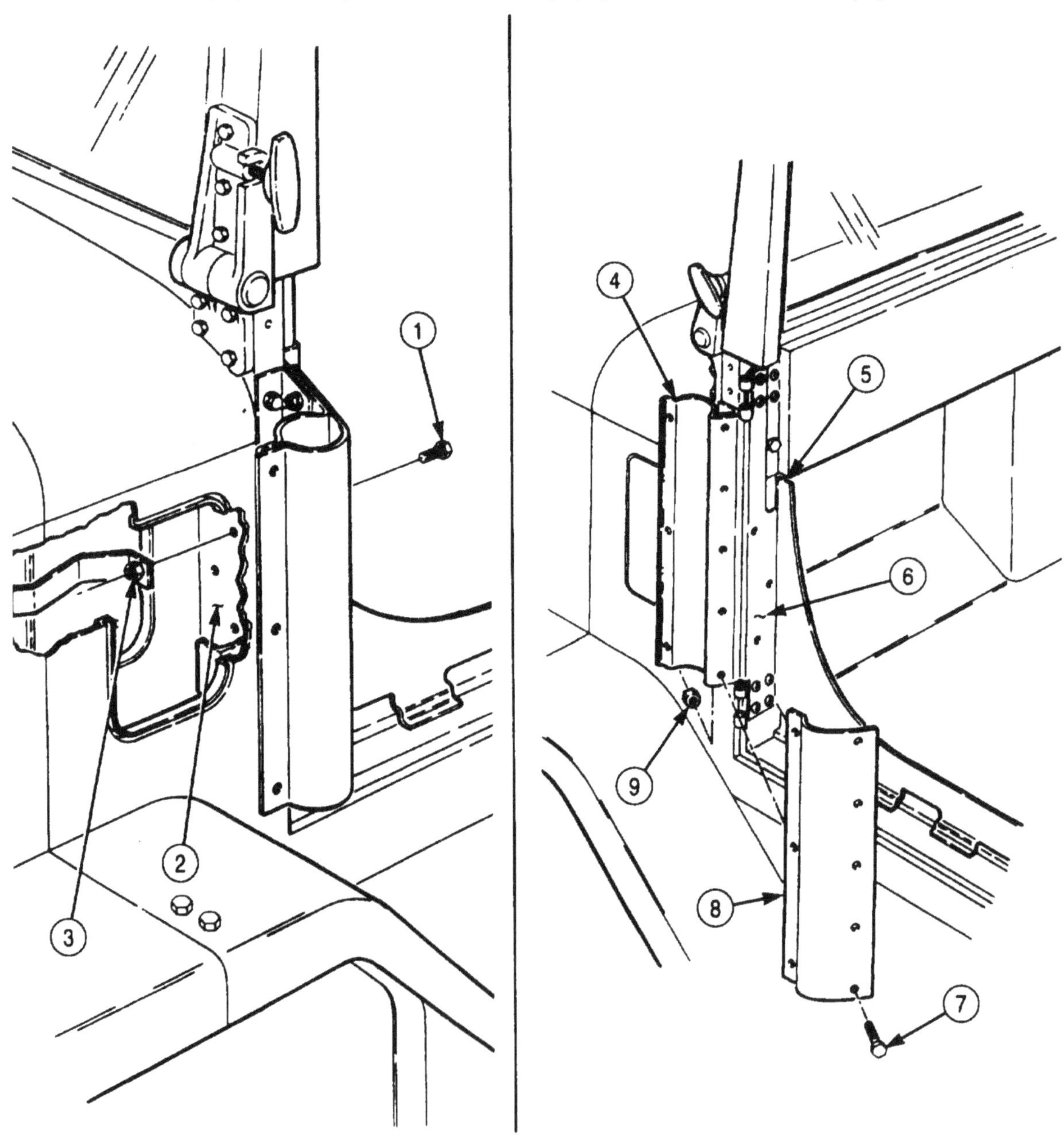

3-446. LIGHTWEIGHT WEAPON STATION MODIFICATION KIT MAINTENANCE (Contd)

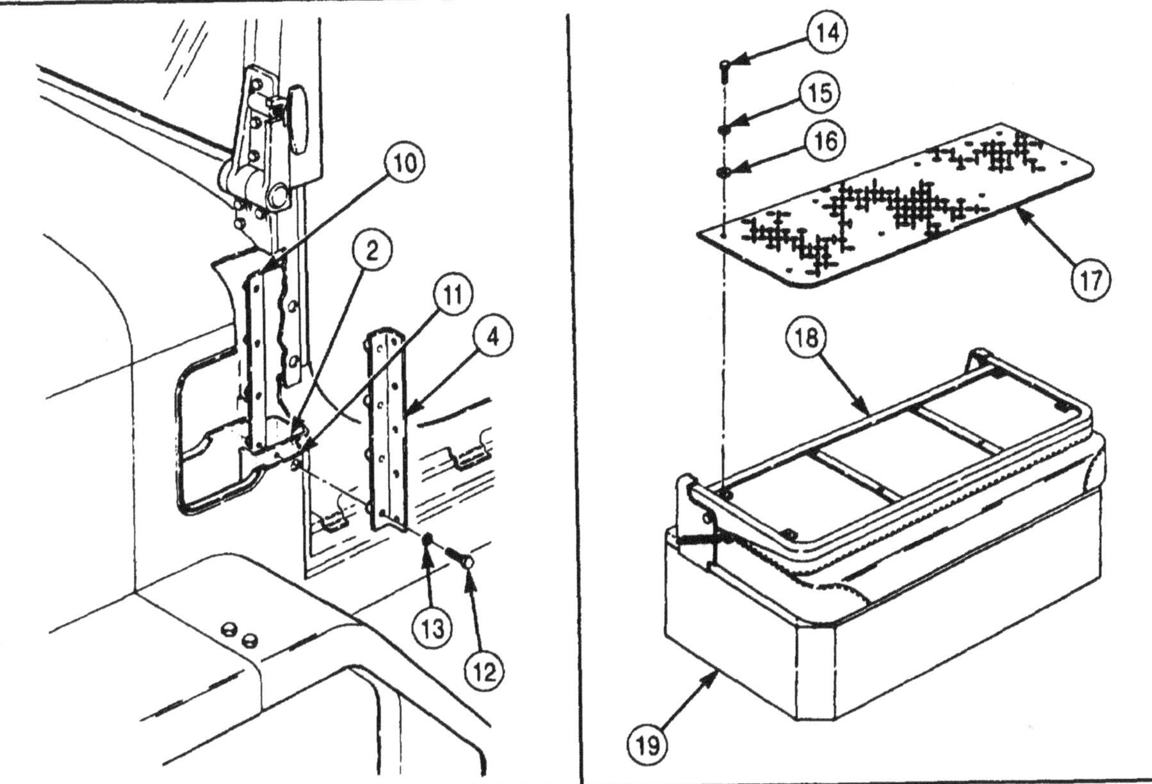

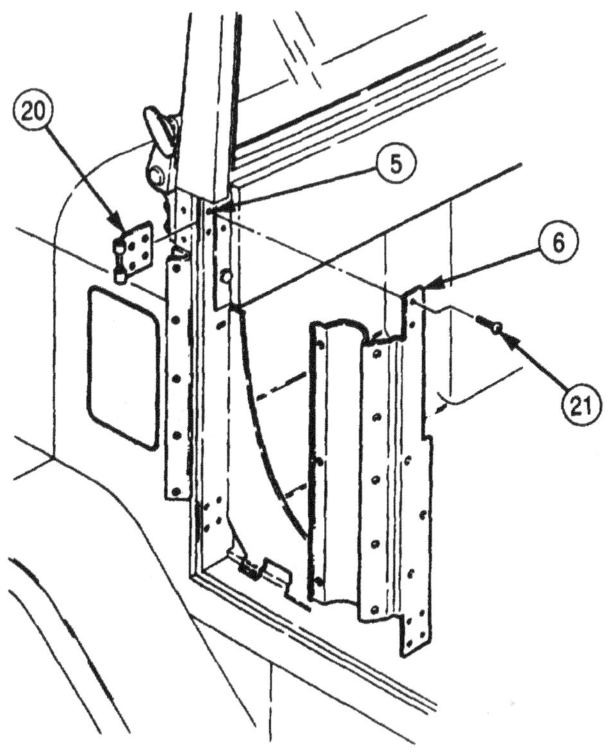

3-446. LIGHTWEIGHT WEAPON STATION MODIFICATION KIT MAINTENANCE (Contd)

b. Disassembly

1. Remove twelve locknuts (3), washers (4), screws (7), washers (8), four braces (2), and ammo tray supports (1) from two ammo box trays (5). Discard locknuts (3).
2. Remove four strap assemblies (6) from two ammo box trays (5).

NOTE

Spacers were used as needed to center lockpin in lockring. Number of spacers in bearing assembly and lockring may vary.

3. Remove two locknuts (41), washers (42), screws (39), lock assembly (43), and spacers (44) from bearing assembly (20) and lockring (40). Discard locknuts (41).
4. Remove two locknuts (26), washers (27), screws (11), four locknuts (24), washers (25), screws (12), and armament mount panel (10) from armament support assembly (28). Discard locknuts (26) and (24).
5. Remove locknut (29), washer (30), screw (9), and pin assembly (31) from armament mount panel (10). Discard locknut (29).

NOTE

Tag all hardware for assembly.

6. Remove six locknuts (21), washers (22), and screws (13) from tube (23), bearing assembly (20), and armament support assembly (28). Discard locknuts (21).
7. Remove two locknuts (37), washers (38), screws (35), and backrests (36) from bearing assembly (20) and armament support assembly (28). Discard locknuts (37).
8. Remove seven locknuts (34), washers (33), and screws (32) from armament support assembly (28) and bearing assembly (20). Discard locknuts (34).
9. Remove locknut (19), washer (18), screw (17), and armament support assembly (28) from bearing assembly (20). Discard locknut (19).
10. Remove four locknuts (16), screws (14), and handle (15) from armament support assembly (28). Discard locknuts (16).

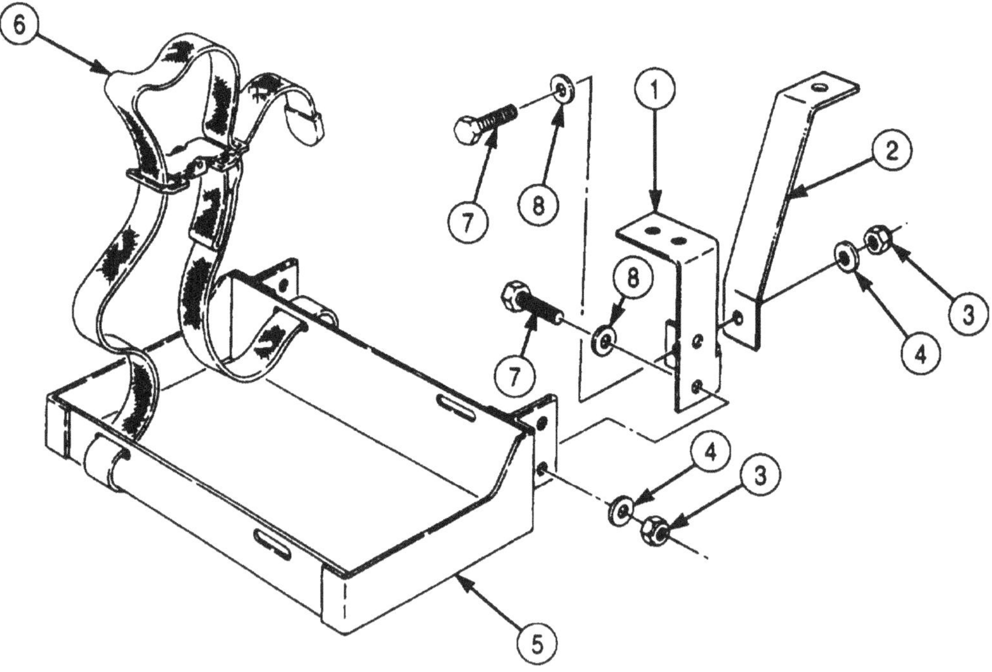

3-446. LIGHTWEIGHT WEAPON STATION MODIFICATION KIT MAINTENANCE (Contd)

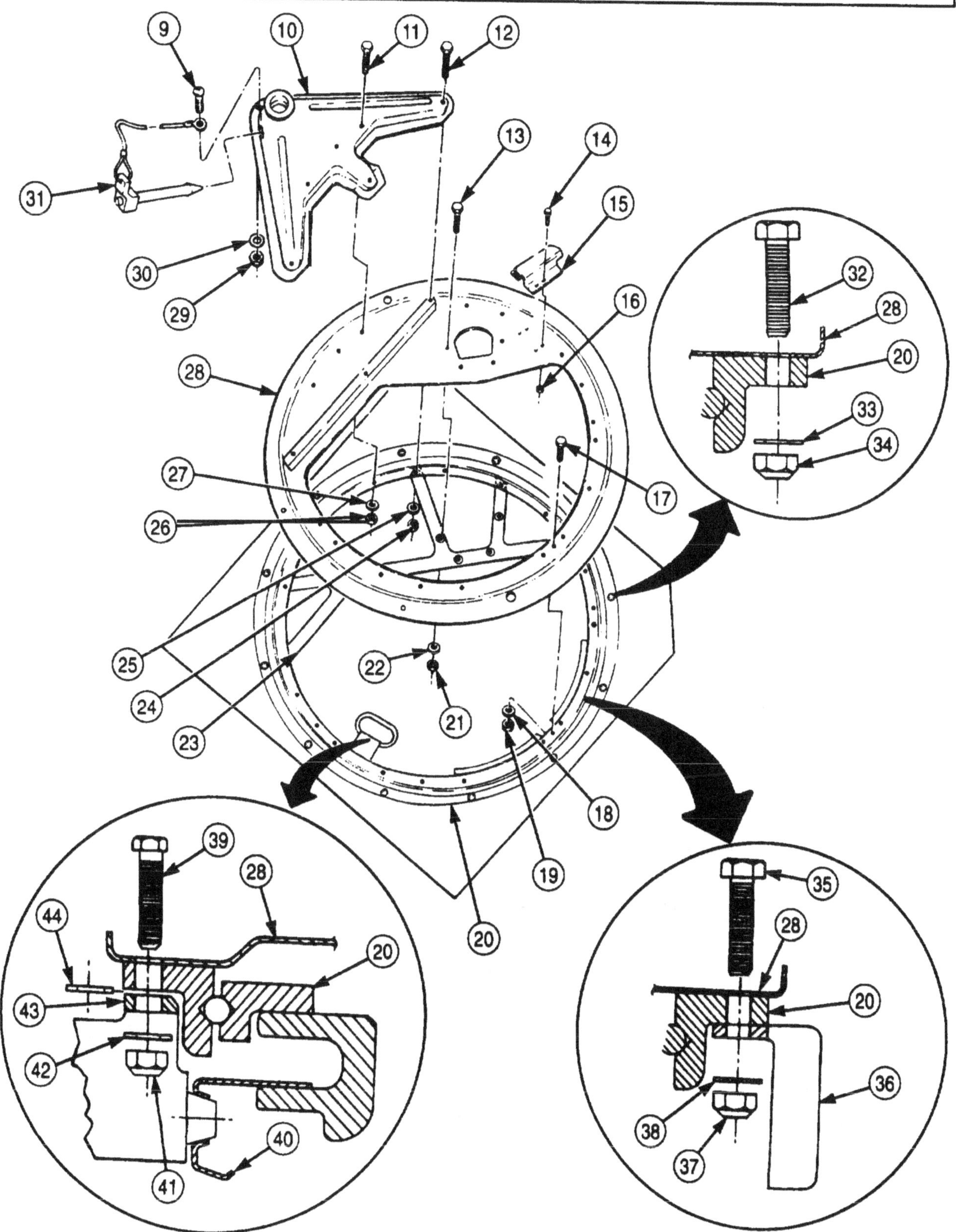

3-446. LIGHTWEIGHT WEAPON STATION MODIFICATION KIT MAINTENANCE (Contd)

11. Remove eighteen screws (14), washers (15), and three lockrings (12) from inner ring of support assembly (13).
12. Remove twelve locknuts (11), washers (10), screws (8), washers (9), and bearing assembly (7) from support assembly (1). Discard locknuts (11).
13. Remove twelve locknuts (2), washers (3), screws (5), washers (6), and two reinforcements (4) from support assembly (1). Discard locknuts (2).

c. Assembly

1. Install two reinforcements (4) on support assembly (1) with twelve washers (6), screws (5), washers (3), and new locknuts (2).
2. Install bearing assembly (7) on support assembly (1) with twelve washers (9), screws (8), washers (10), and new locknuts (11).
3. Install three lockrings (12) to inner ring of support assembly (13) with eighteen washers (15) and screws (14).

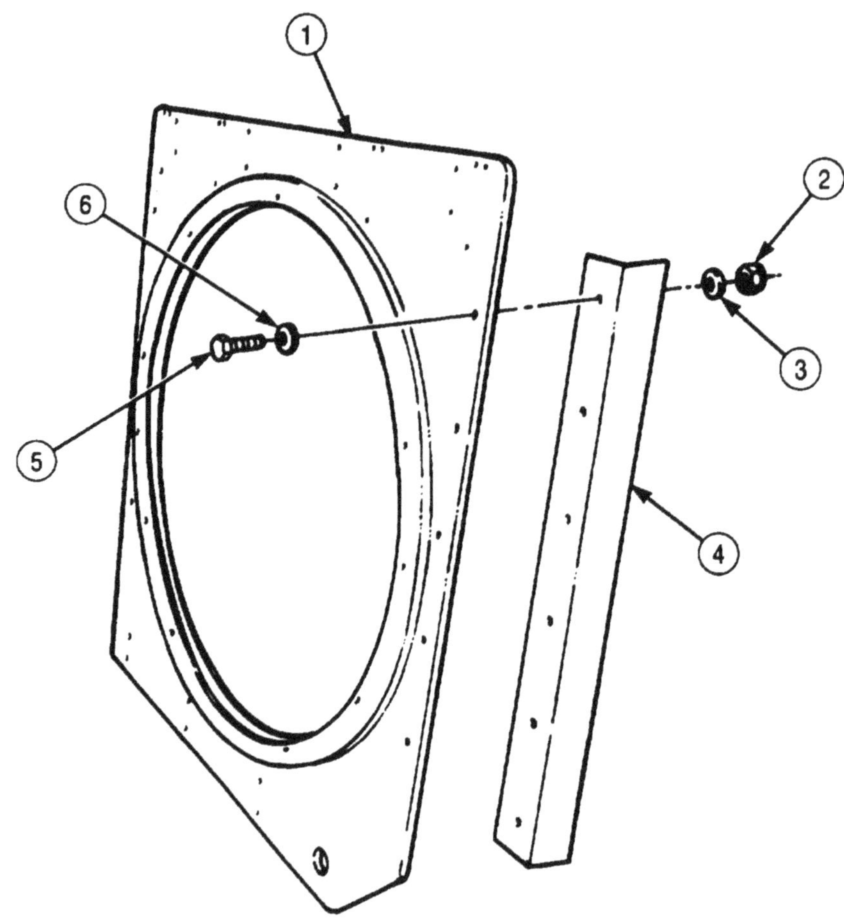

3-446. LIGHTWEIGHT WEAPON STATION MODIFICATION KIT MAINTENANCE (Contd)

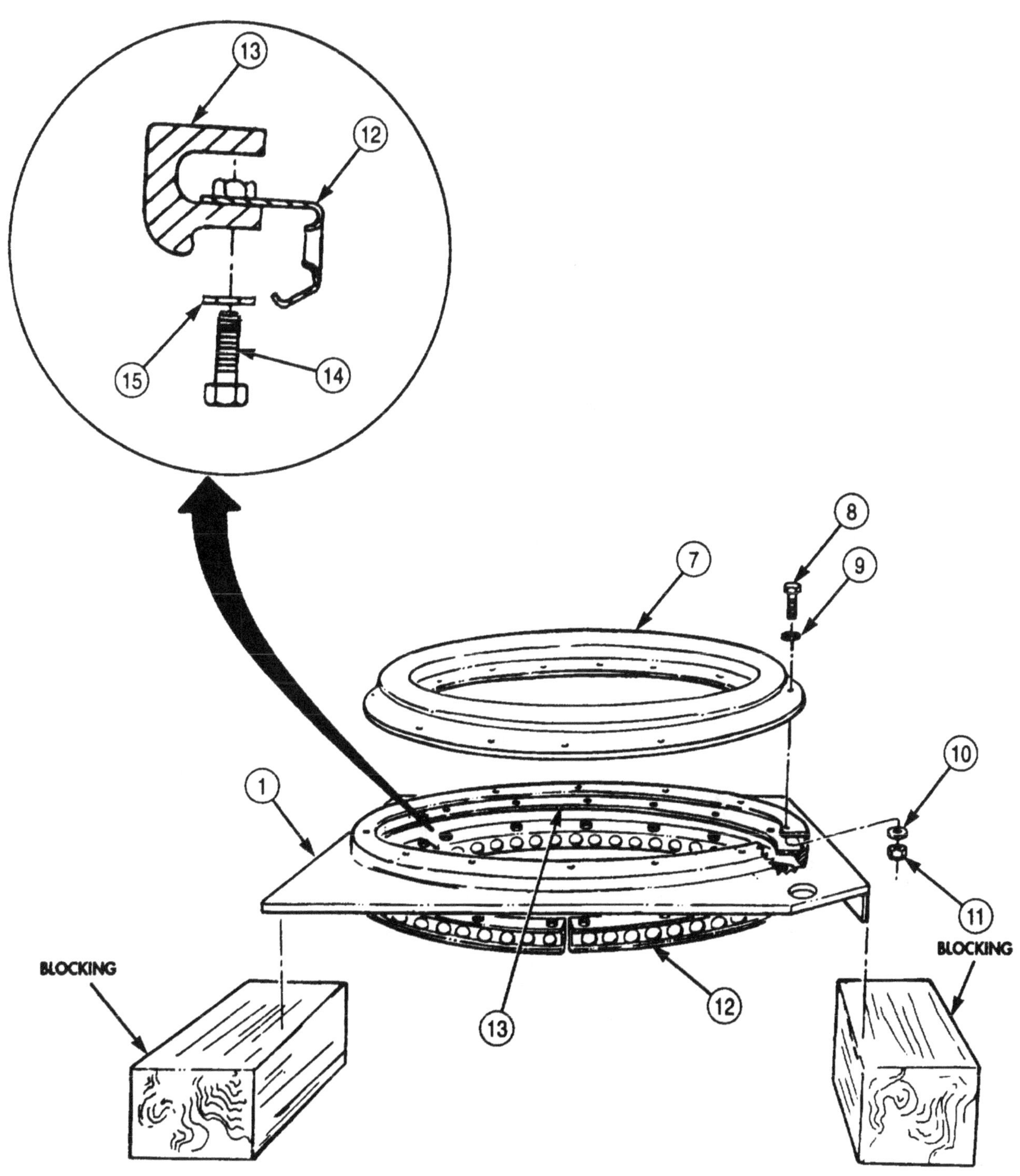

3-446. LIGHTWEIGHT WEAPON STATION MODIFICATION KIT MAINTENANCE (Contd)

4. Install handle (15) on armament support assembly (29) with four screws (14) and new locknuts (16).

NOTE

Finger-tighten locknuts in steps 5 through 10.

5. Align three-hole pattern in armament support assembly (29) with three-hole pattern of bearing assembly (21) and install screw (17), washer (19), and new locknut (20) in center hole (18).
6. Install two backrests (37) on bearing assembly (21) and armament support assembly (29) with four screws (36), washers (39), and new locknuts (38).
7. Install tube (24) on bearing assembly (21) and armament support assembly (29) with six screws (13), washers (23), and new locknuts (22).
8. Install pin assembly (32) on armament mount panel (10) with screw (9), washer (31) and new locknut (30).
9. Install armament mount panel (10) on armament support assembly (29) with four screws (12), washers (26), new locknuts (25), two screws (11), washers (28), and new locknuts (27).
10. Install seven screws (33), washers (34), and new locknuts (35) on armament support assembly (29) and bearing assembly (21).
11. Tighten locknuts (20), (38), (22), (30), (25), (27), and (35).

NOTE

Use spacers as needed to center lockpin in lockring.

12. Position lock assembly (45) between bearing assembly (21) and lockring (41), centering lockpin (42) in lockring (41) using spacers (46) as needed.
13. Install lock assembly (45) on bearing assembly (21) with two screws (40), washers (44), and new locknuts (43).
14. Install four strap assemblies (6) on two ammo box trays (5).
15. Install four ammo tray supports (1) and braces (2) on two ammo box trays (5) with twelve washers (8), screws (7), washers (4), and new locknuts (3). Do not tighten locknuts (3).

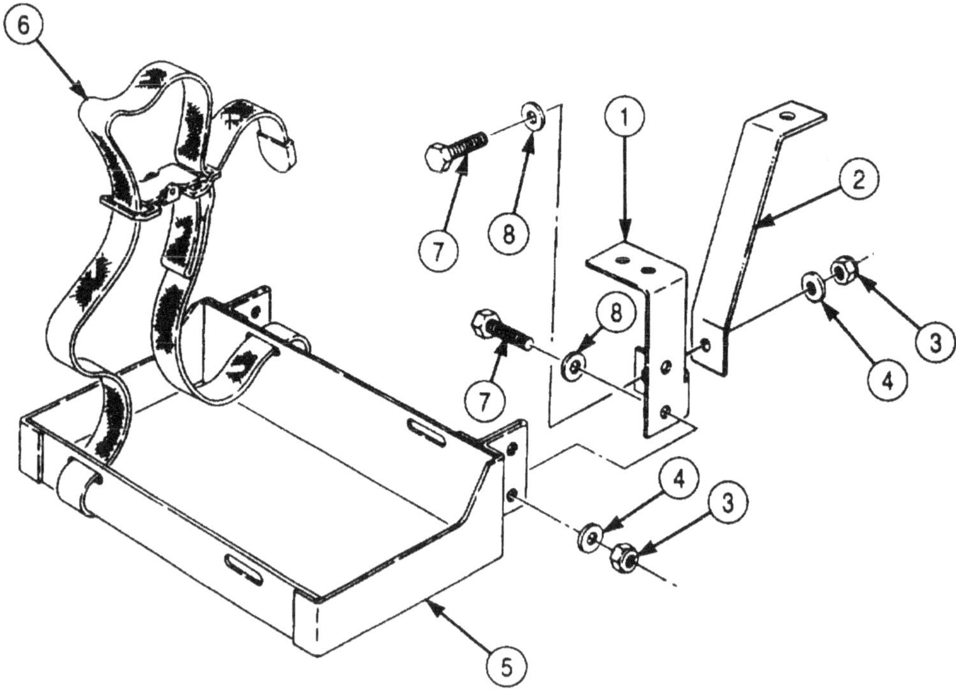

3-446. LIGHTWEIGHT WEAPON STATION MODIFICATION KIT MAINTENANCE (Contd)

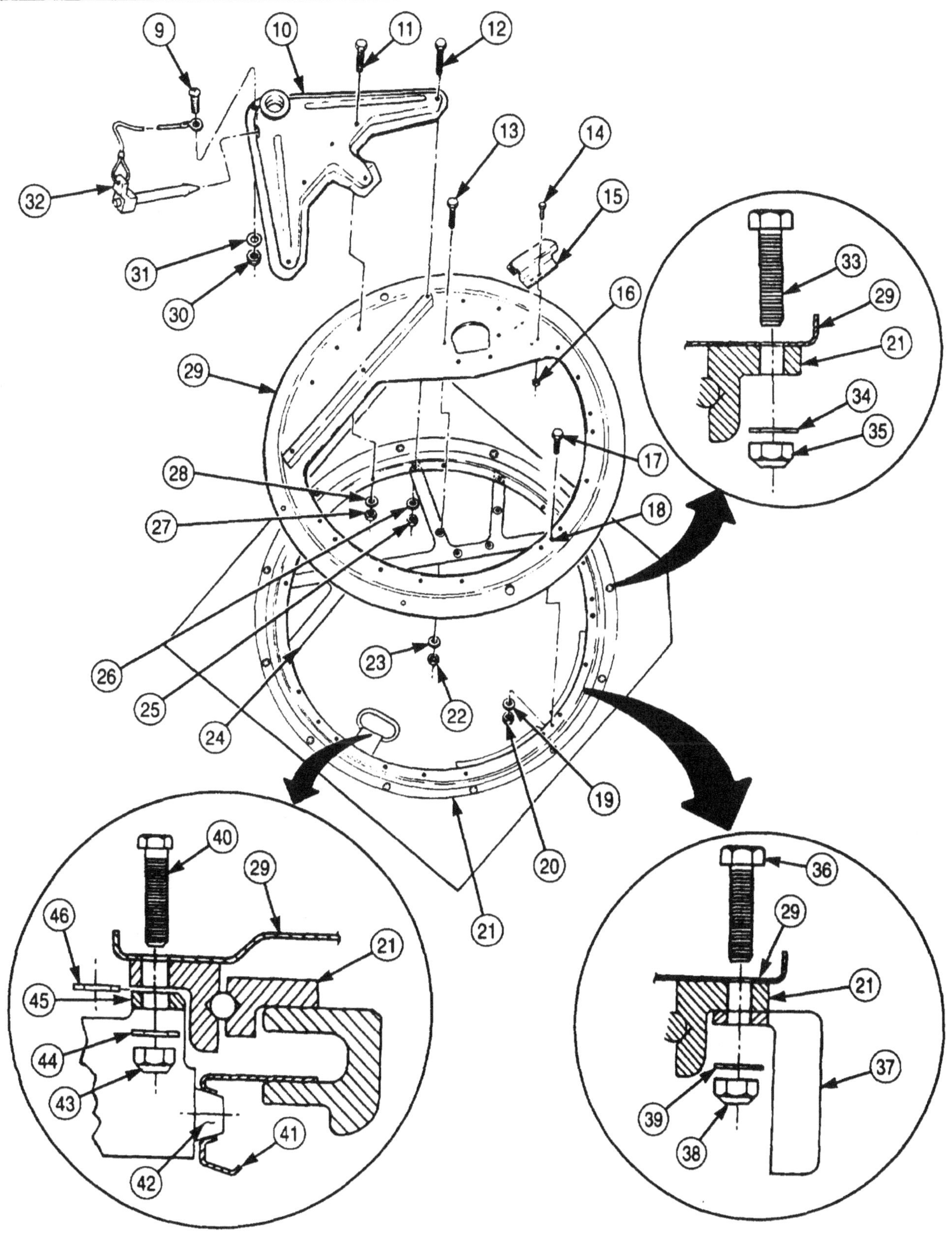

3-446. LIGHTWEIGHT WEAPON STATION MODIFICATION KIT MAINTENANCE (Contd)

d. Installation

1. Install seat back panel (4) on backrest cushion frame (6) and companion seat assembly (5) with eight washers (3), new lockwashers (2), and screws (1).
2. Position tapping plate (16) inside left door pillar wall (11) and install left front brace assembly (13) on outer cowl wall (12) with four washers (15) and screws (14).
3. Install left post support (9) and two door hinges (7) on left door pillar (8) with eight door hinge screws (10). Do not tighten door hinge screws (10).
4. Position left post support bracket (18) on left post support (9) and install to left front brace assembly (13) with five screws (17) and new locknuts (19). Do not tighten locknuts (19).
5. Tighten screws (10) and locknuts (19).

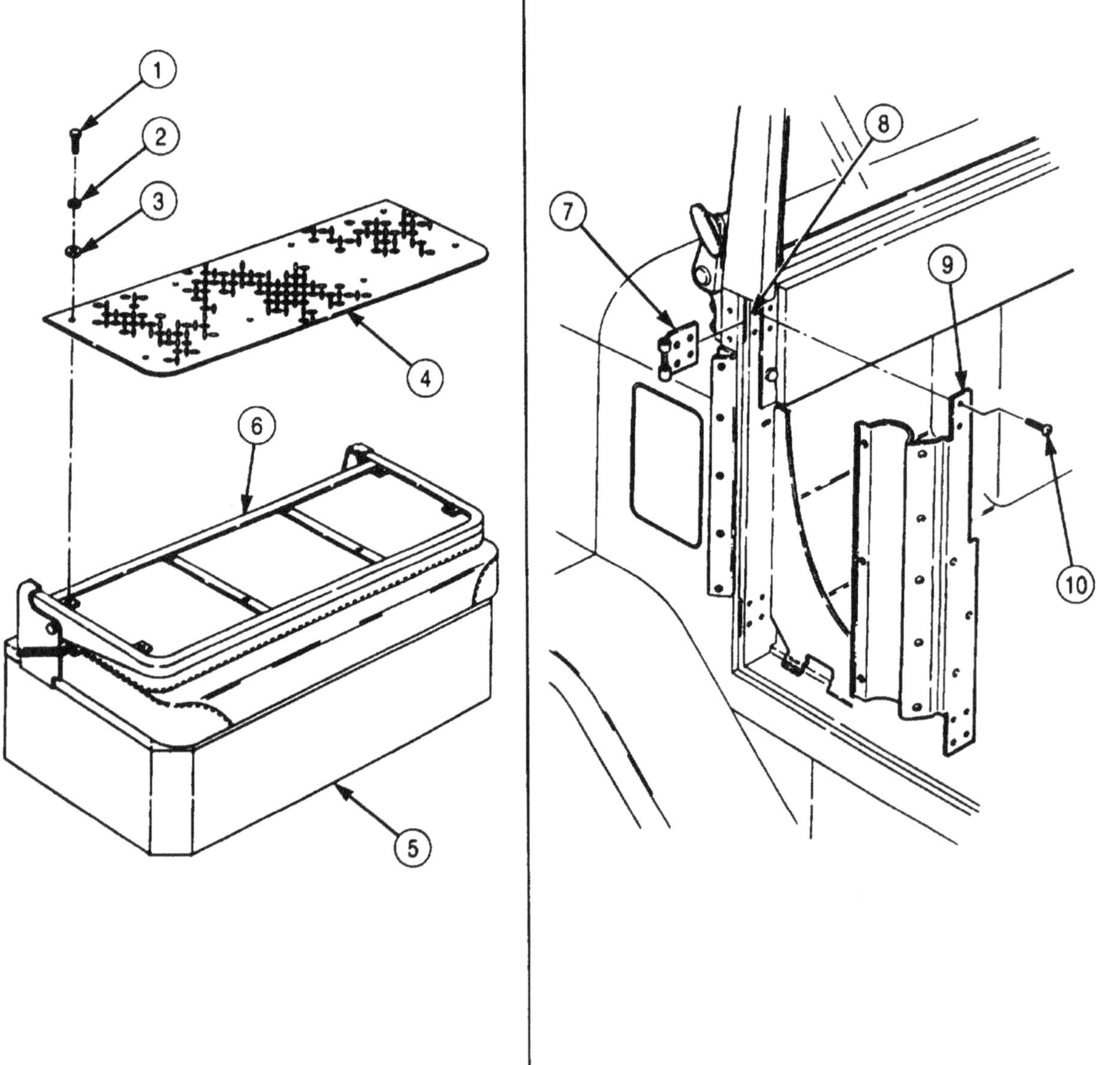

3-446. LIGHTWEIGHT WEAPON STATION MODIFICATION KIT MAINTENANCE (Contd)

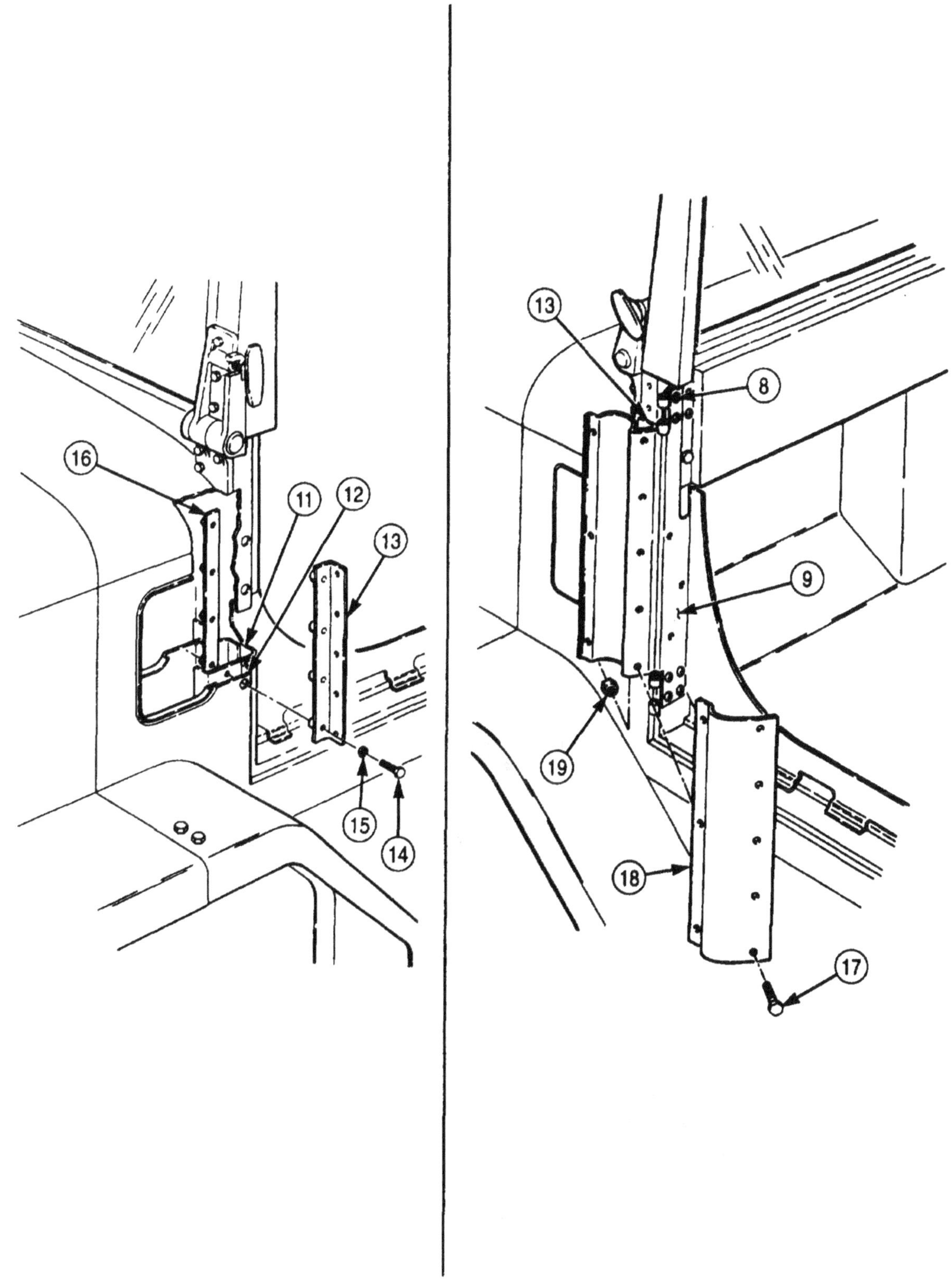

3-446. LIGHTWEIGHT WEAPON STATION MODIFICATION KIT MAINTENANCE (Contd)

NOTE

Tabs on assemblies may require minor bending to fit during installation.

6. Install three screws (8) and nut assemblies (7) to backside of left pillar wall (9).
7. Install pin (11) through top in left front post assembly (10) with two new cotter pins (12).
8. Install left front post assembly (10) on left post support (14) with pin (11) resting on on left post support (14) and install with three screws (13) and new locknuts (15).
9. Position tapping plate (1) inside right door pillar wall (2) and install right front brace assembly (5) on outer cowl wall (6) with four washers (4) and screws (3).

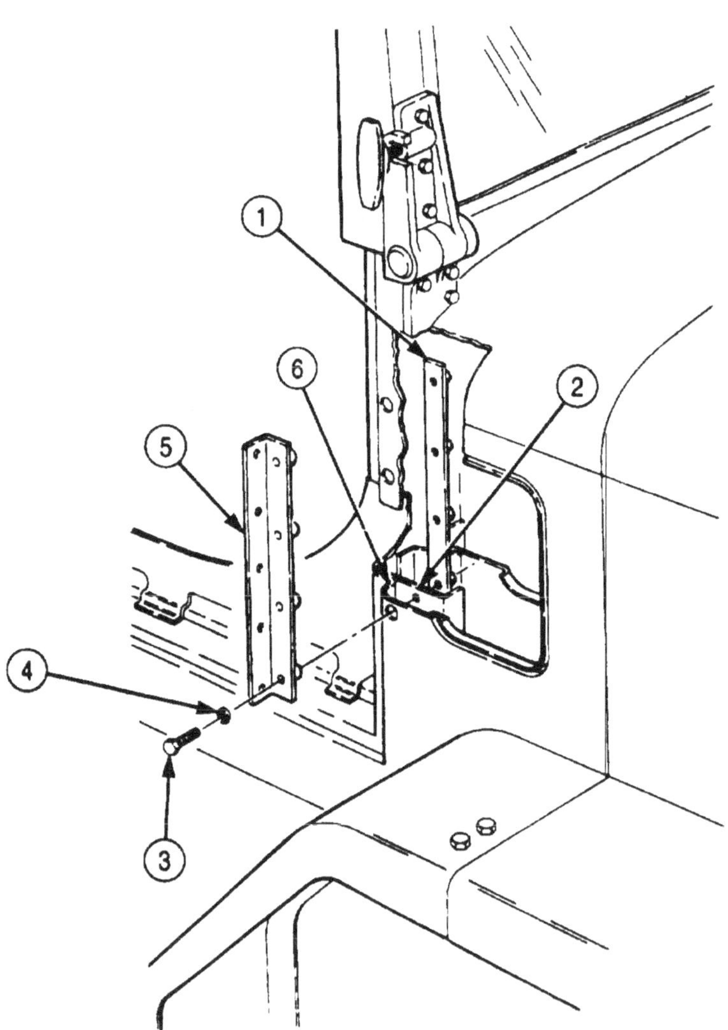

3-446. LIGHTWEIGHT WEAPON STATION MODIFICATION KIT MAINTENANCE (Contd)

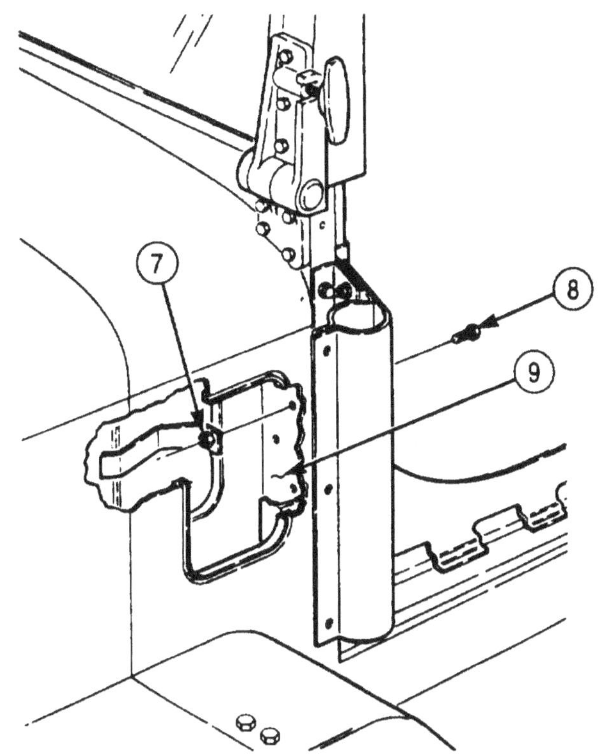

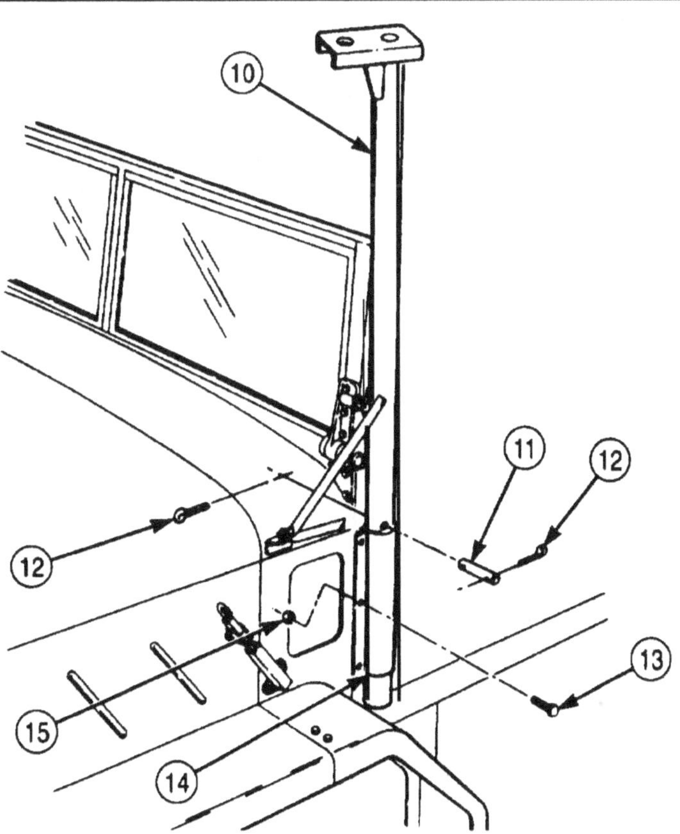

3-446. LIGHTWEIGHT WEAPON STATION MODIFICATION KIT MAINTENANCE (Contd)

10. Align right post support brackets (1) and (2) and install right post support assembly (7) on right door pillar (3) with four screws (11). Do not tighten screws (11).
11. Install instrument panel (10) on right door pillar (2) with two screws (10). Do not tighten screws (10).
12. Loosely install right post support assembly (7) to right front brace assembly (12) with five screws (14) and new locknuts (13). Do not tighten locknuts (13).
13. Install pin (9) in right post assembly (4) with two new cotter pins (5).
14. Install right post assembly (4) on post support assembly (7) with pin (9) resting on top of support assembly (7) and install three screws (8) and new locknuts (6).
15. Tighten screws (11) and (10) and locknut (13).

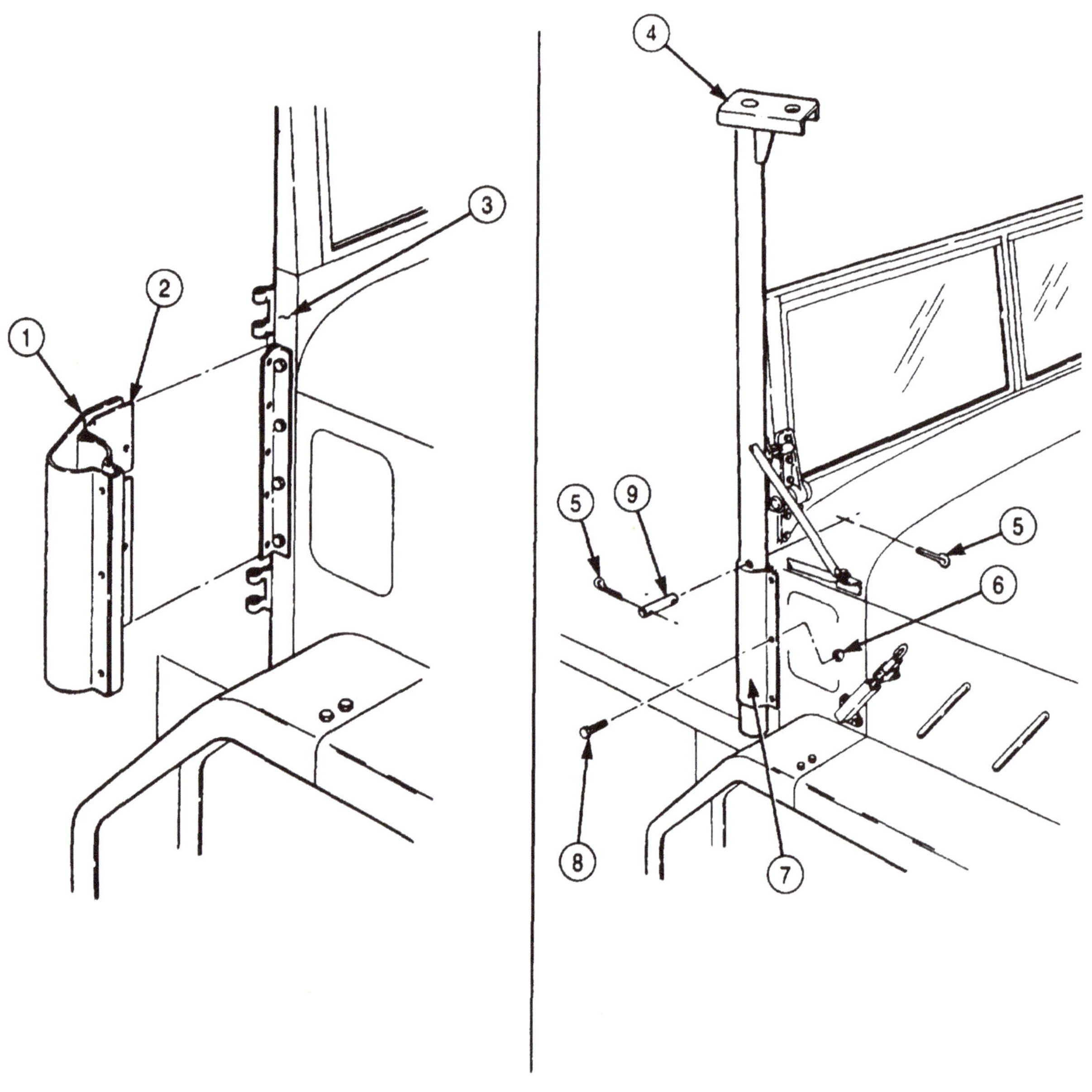

3-446. LIGHTWEIGHT WEAPON STATION MODIFICATION KIT MAINTENANCE (Contd)

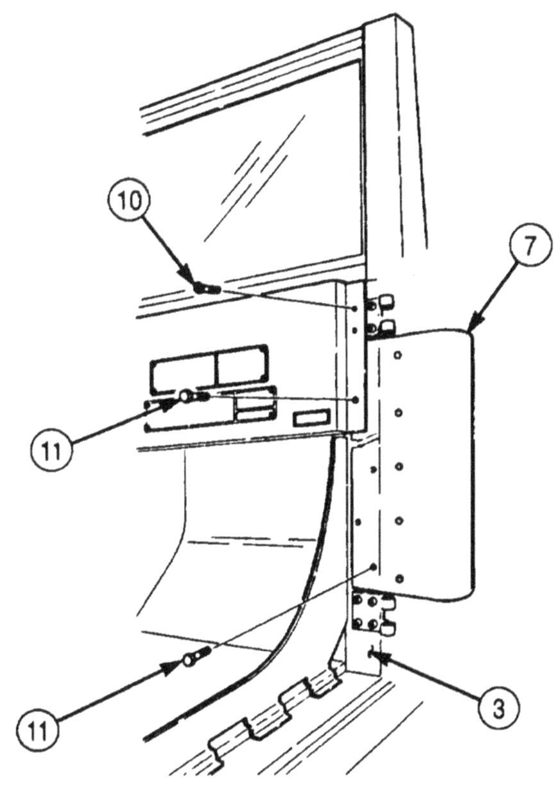

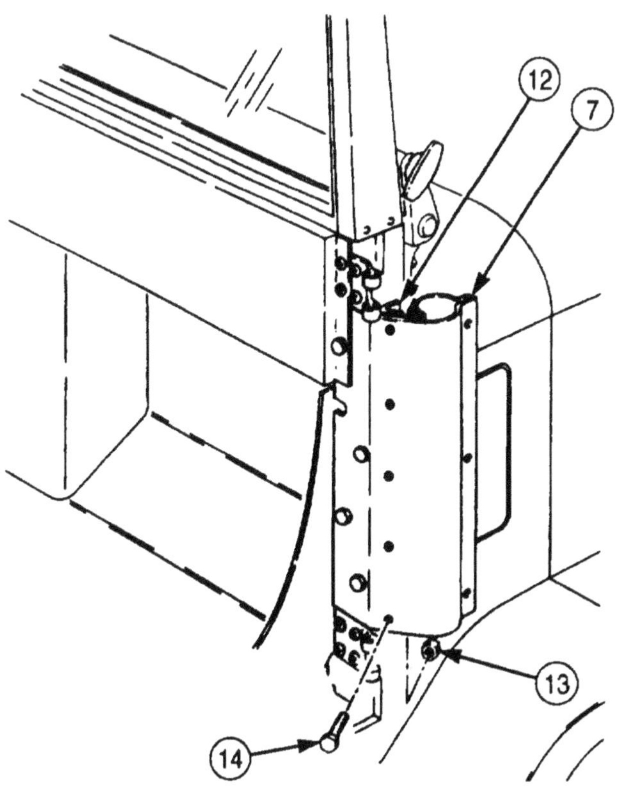

3-446. LIGHTWEIGHT WEAPON STATION MODIFICATION KIT MAINTENANCE (Contd)

NOTE

- Left and right rear mounting posts are installed the same way. This procedure is for the right rear mounting post.
- It may be necessary to remove left and right reinforcement panels from rear of cab to gain access to U-bolt nuts.

16. Install U-bolts (5) and (6) in right rear cab support (4) and cab (7) with two new locknuts (11) and (10). Do not tighten locknuts (11) and (10).
17. Install pin (9) in right rear mounting post assembly (1) with two new cotter pins (2).

NOTE

Ensure pins installed on post assemblies do not slip through U-bolts.

18. Install right rear post assembly (1) through U-bolts (6) and (5) with pin (9) resting on U-bolt (6).
19. Install U-bolt (8) through right rear cab support (4) and cab (7) with two new locknuts (3). Do not tighten locknuts (3).
20. Tighten locknuts (11), (10), and (3).

3-446. LIGHTWEIGHT WEAPON STATION MODIFICATION KIT MAINTENANCE (Contd)

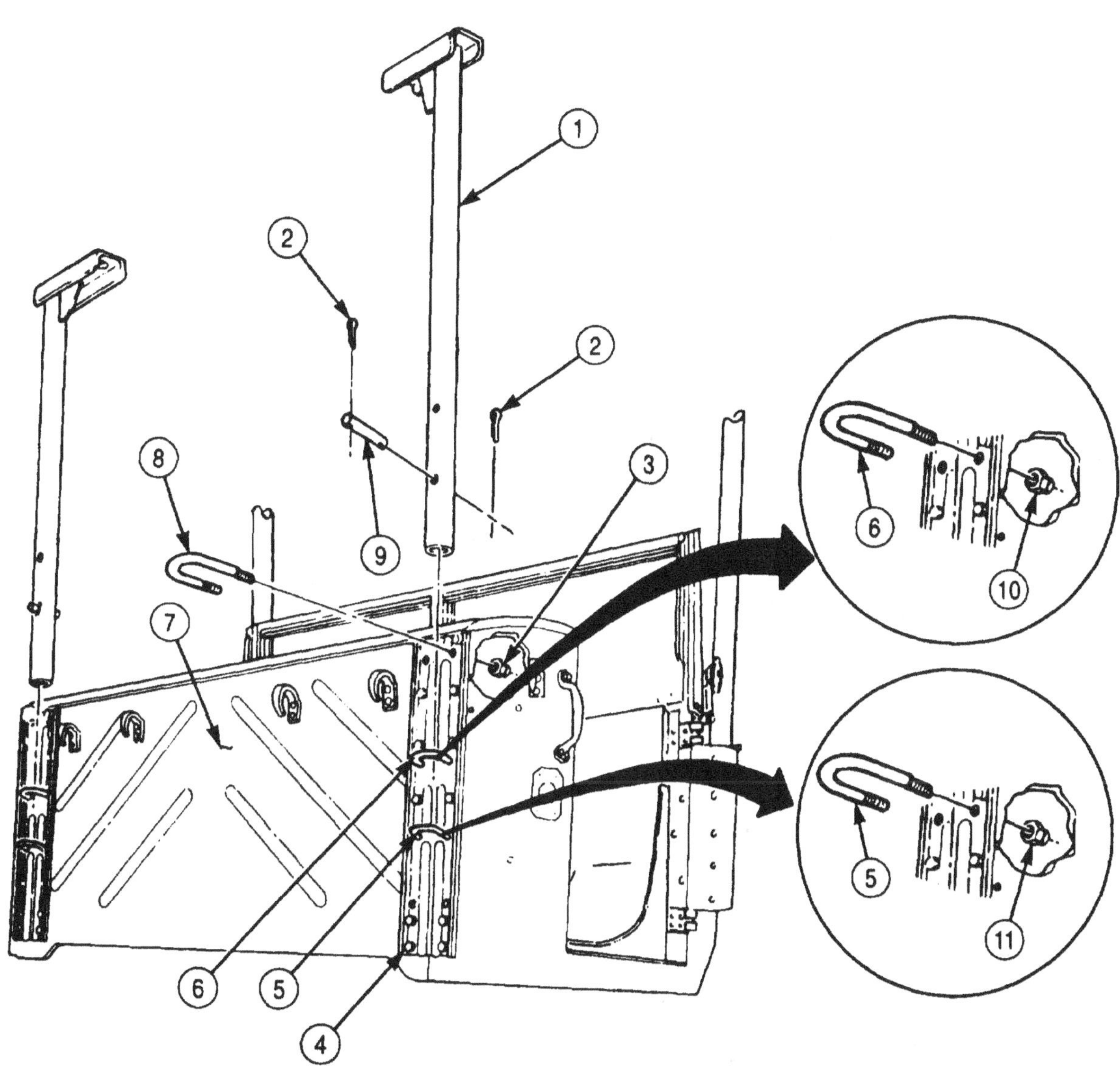

3-446. LIGHTWEIGHT WEAPON STATION MODIFICATION KIT MAINTENANCE (Contd)

WARNING

All personnel must stand clear during lifting operations. A snapped chain, shifting or swinging load, may result in injury to personnel.

21. Attach lifting device to lightweight weapon station platform (1) with four utility chains and raise approximately 12 in.
22. Install long crossmember assembly (6), with locator tab (8) to left, on front of lightweight weapon station platform (1) with four U-bolts (10), eight washers (11), and new locknuts (9). Do not tighten locknuts (9).
23. Install short crossmember assembly (4), with locator tab (3) to left, on rear of lightweight weapon station platform (1) with three U-bolts (2), six washers (5), and new locknuts (7). Do not tighten locknuts (7).
24. Position four roof mounts (15) and plates (14) on top of post assemblies (22).

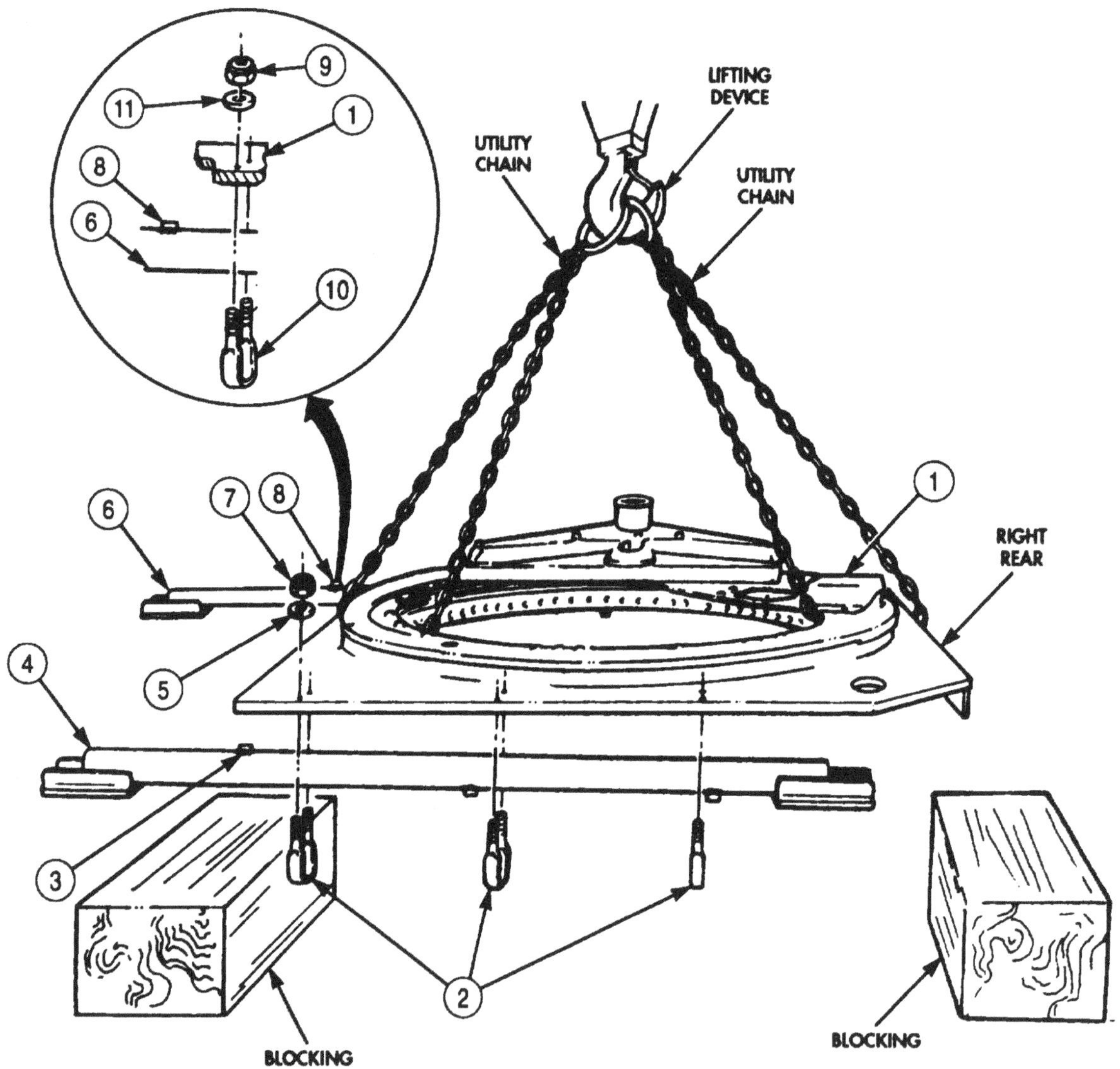

3-446. LIGHTWEIGHT WEAPON STATION MODIFICATION KIT MAINTENANCE (Contd)

25. Align holes in post assemblies (22) with holes in crossmember pads (17) and install lightweight weapon station platform (1) on four post assemblies (22) with four washers (13) and screws (12).
26. Install four spacers (16), roof mounts (18), plates (19), washers (20), and new locknuts (21) on four screws (12). Do not tighten locknuts (21).

NOTE

Ensure locator tabs and lightweight weapon station platform are aligned before tightening U-bolts.

27. Tighten locknuts (9) and (7) 32-40 lb-ft (43-54 N•m).
28. Tighten four locknuts (21) 15-20 lb-ft (20-27 N•m).

CAUTION

U-bolts and pins installed in steps 18 through 22 must be aligned before tightening. Do not over-tighten or damage to cab will result.

29. Tighten all screws and nuts that remain loosely installed from steps 2 through 19.

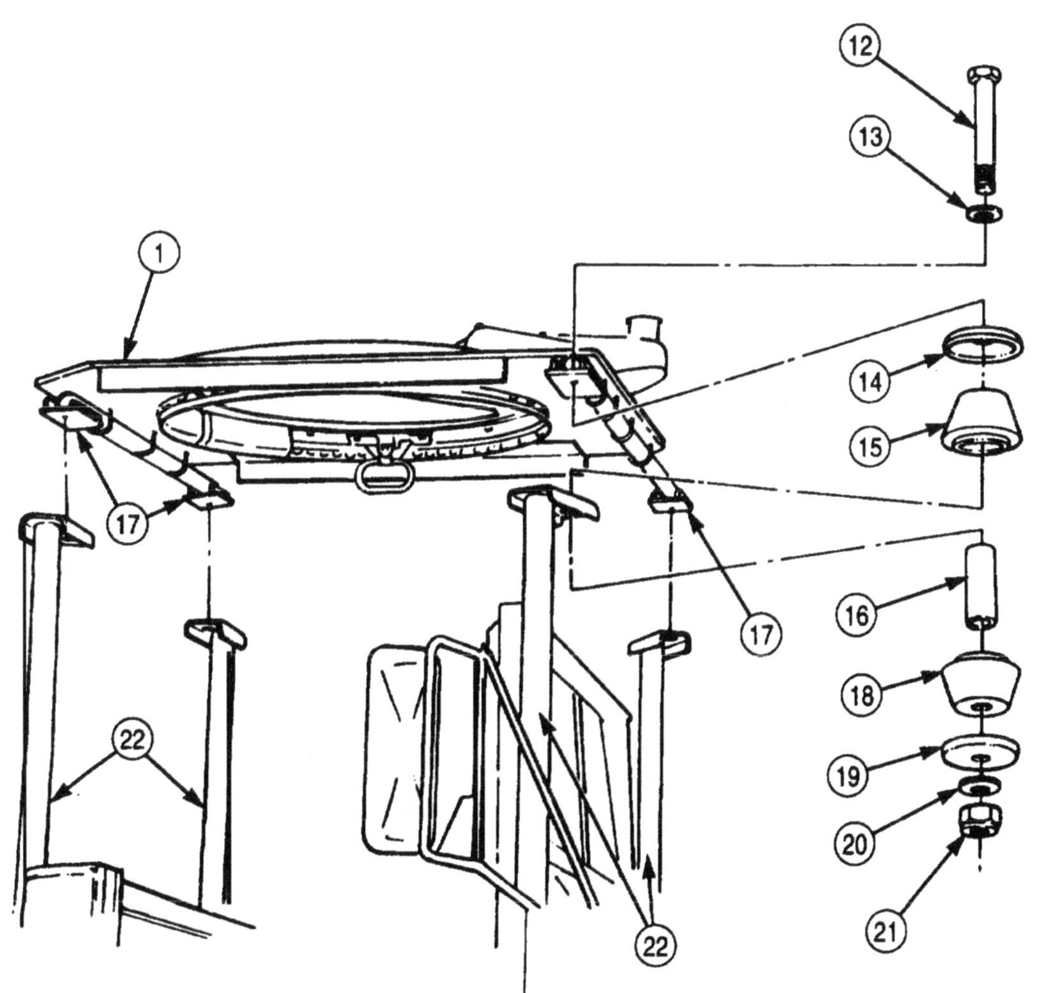

3-446. LIGHTWEIGHT WEAPON STATION MODIFICATION KIT MAINTENANCE (Contd)

30. Install four ammo tray supports (11) and ammo box tray (9) on left and right front sides of support assembly (10) with eight washers (12), screws (1), washers (8), and new locknuts (7). Do not tighten locknuts (7).
31. Install four braces (4) to support assembly (10) with four washers (3), screws (2), washers (5), and new locknuts (6). Do not tighten locknuts (6).
32. Tighten locknuts (7) and (6).
33. Tighten locknuts, installed in subtask c., connecting four ammo tray supports (11) and four braces (4) to two ammo box trays (9).

3-446. LIGHTWEIGHT WEAPON STATION MODIFICATION KIT MAINTENANCE (Contd)

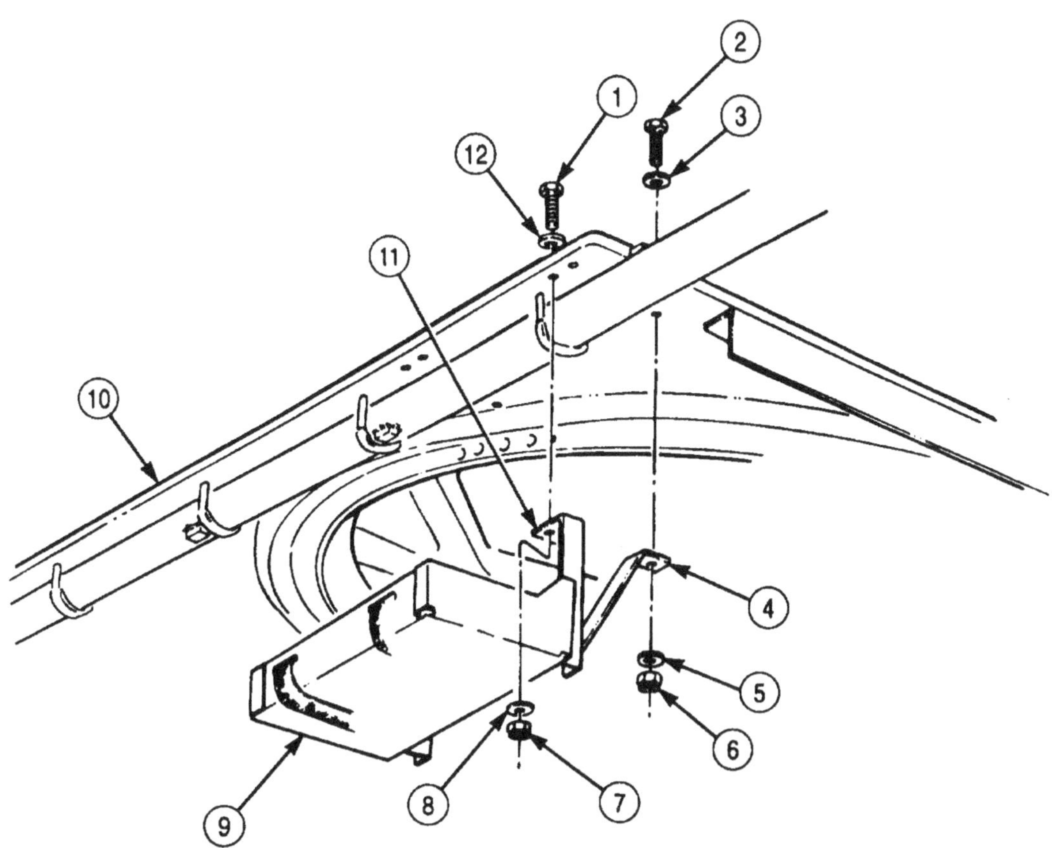

FOLLOW-ON TASKS:
- Reposition companion seat and battery box (TM 923200-272-10).
- Reposition backrest cushion on companion seat assembly (TM 92320-272-10).
- Install right and left cowl ventilation door screens (para. 3-299).
- Install driver's seat (para. 2-283).
- Install left and right mirror assemblies (para. 3-319).
- Install left and right cab door check rods (para. 3-318).
- Install muffler shield assembly (para. 3-51).
- Install batteries (para. 3-125).
- Install spare tire (TM 9-2320-272-10).
- Install cab top (TM 9-2320-272-10).
- Adjust left side door hinges and install left and right door assemblies (para. 3-319).

3-447. ENGINE EXHAUST BRAKE MODIFICATION KIT (M939/A1) REPLACEMENT

THIS TASK COVERS:

a. Removal b. Installation

INITIAL SETUP:

APPLICABLE MODELS
M939/A1

TOOLS
General mechanic's tool kit (Appendix E, Item 1)

MATERIALS/PARTS
Two locknuts (Appendix D, Item 274)
tie lockwashers (Appendix D, Item 346)
Locknut (Appendix D, Item 275)
Gasket (Appendix D, Item 170)
Two gaskets (Appendix D, Item 244)
Gasket (Appendix D, Item 175)
Tiedown straps (Appendix D, Item 697)
Antiseize tape (Appendix C, Item 72)
Sealant (Appendix C, Item 63)

REFERENCES (TM)
TM 9-2320-272-10
TM 9-2320-272-24P

EQUIPMENT CONDITION
- Parking brake set (TM 9-2320-272-10).
- Air reservoir drained (TM 9-2320-272-10).
- Battery ground cables disconnected (para. 3-126).
- Left and right engine splash shields removed (TM 9-2320-272-10).
- Exhaust pipe heat shield if needed removed (para. 3-49).
- Right hood bumper removed (para. 3-274).

a. Removal

NOTE

Remove all tiedown straps as required and note location and position for installation.

1. Remove two toggle switch leads (1) from cable (4) and lead assembly (3) behind instrument panel (8).
2. Remove lead (2) from lead assembly (3).
3. Remove lead assembly (3) from circuit breaker (5).
4. Remove tiedown strap (6) and cable (4) from wiring harness (7). Discard tiedown strap (6).
5. Connect lead (2) to circuit breaker (5).

NOTE

- Instrument cluster may have to be removed for switch removal (para. 3-83).
- Ensure that toggle switch is in OFF position.

6. Remove nut (10), washer (9), plate (12), and toggle switch (13) from instrument cluster (11).
7. Install instrument cluster (11) if required (para. 3-83).

3-447. ENGINE EXHAUST BRAKE MODIFICATION KIT (M939/A1) REPLACEMENT (Contd)

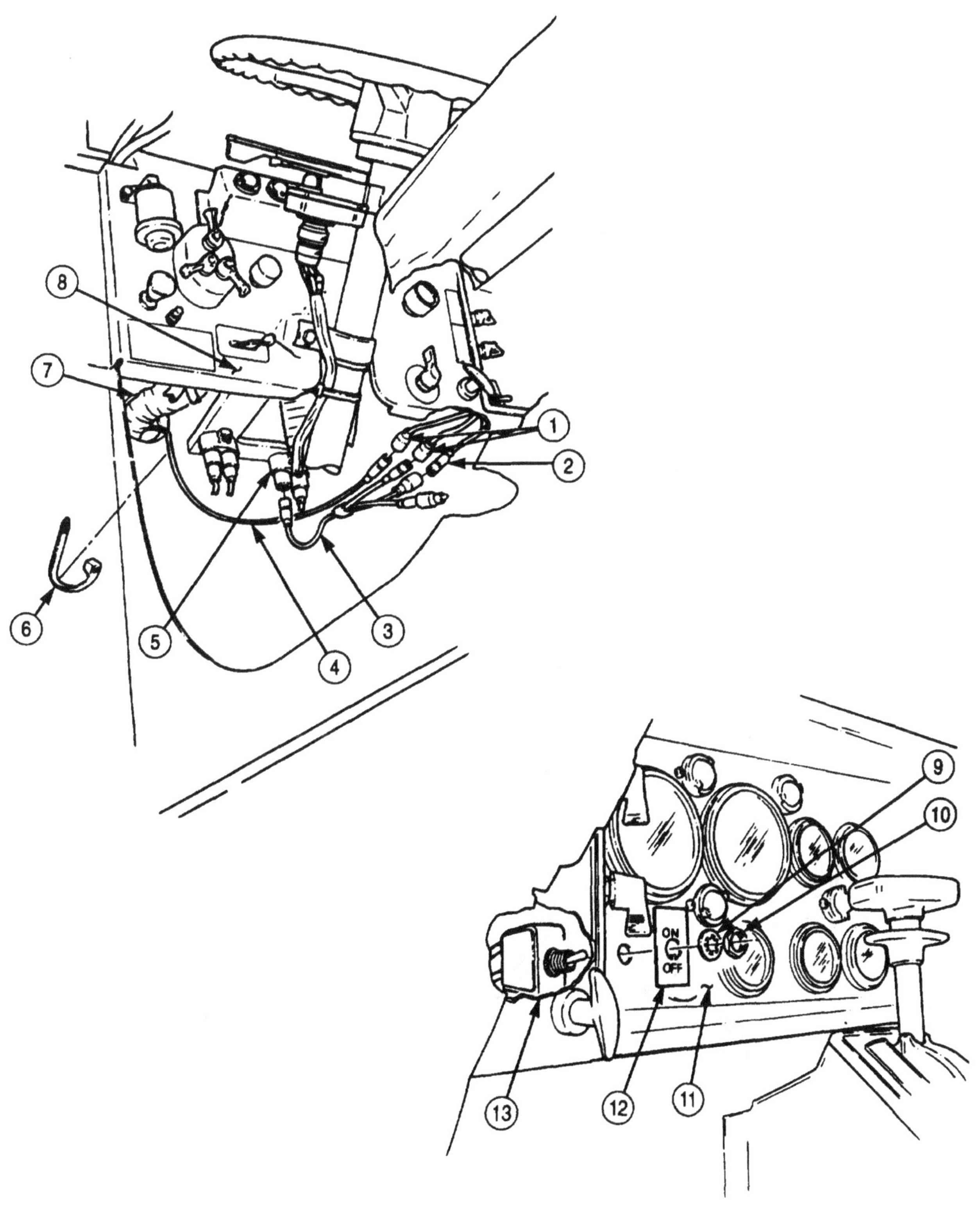

3-447. ENGINE EXHAUST BRAKE MODIFICATION KIT (M939/A1) REPLACEMENT (Contd)

8. Disconnect cable (4) from lead (19) on lever switch bracket (16) and remove cable (4) from firewall.
9. Disconnect cable (7) from lead (19) on lever switch bracket (16).
10. Remove locknut (11), washer (12), lever (14), and T-bolt (13) from fuel pump throttle control lever (10) and bracket assembly (15). Discard locknut (11).
11. Remove two screws (8), washers (9), and lever switch bracket (16) from bracket assembly (15).
12. Loosen two fuel pump screws (17) and remove bracket assembly (15) from fuel pump (18). Tighten fuel pump screws (17).
13. Remove hose assembly (21) and elbow (22) from intake manifold (20).
14. Remove tiedown strap (5) and disconnect cable (4) from valve assembly lead (6) and remove cable (4) from firewall. Discard tiedown strap (5).
15. Remove nut (3) and valve assembly ground lead (2) from valve assembly screw (1).

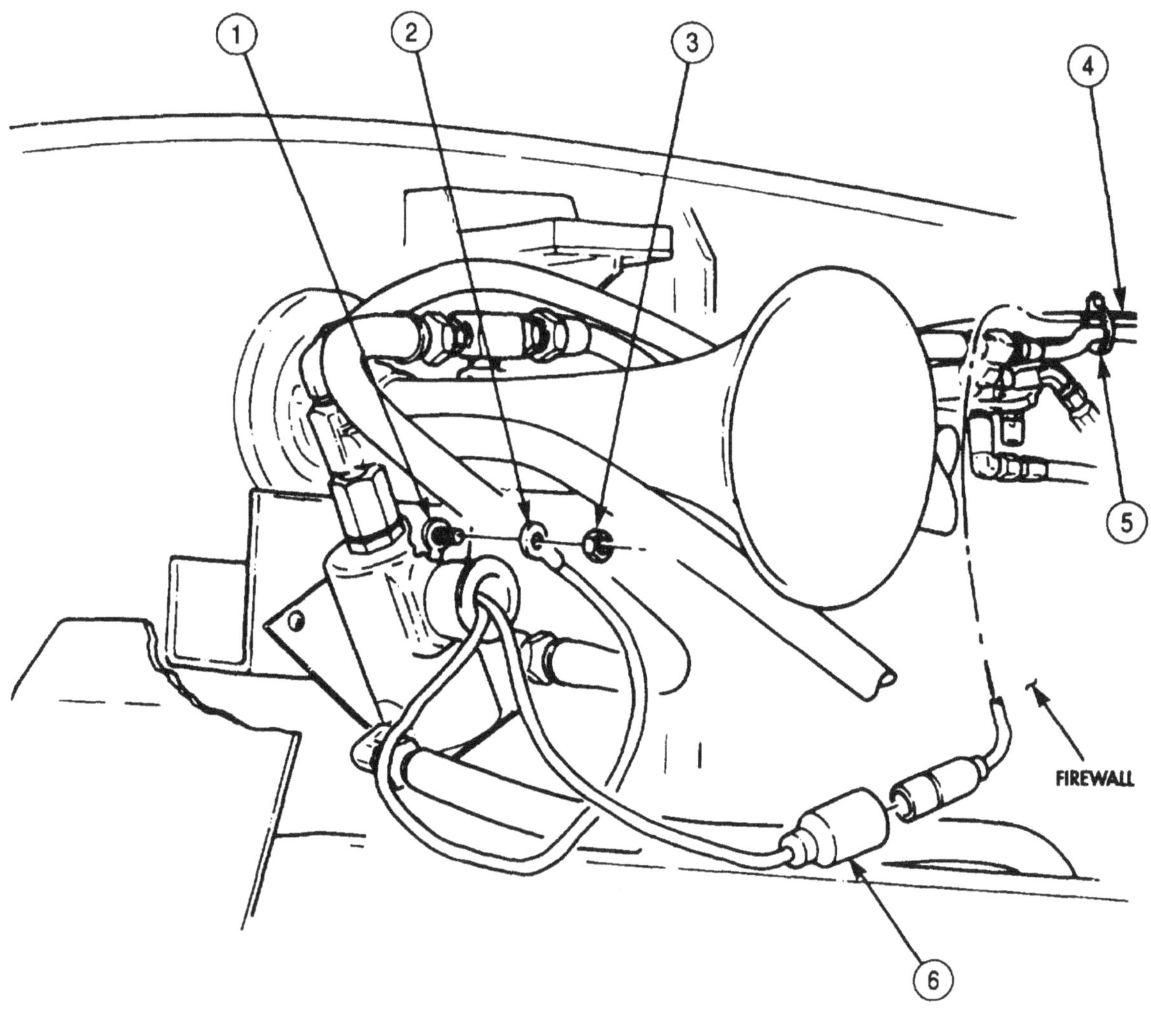

3-447. ENGINE EXHAUST BRAKE MODIFICATION KIT (M939/A1) REPLACEMENT (Contd)

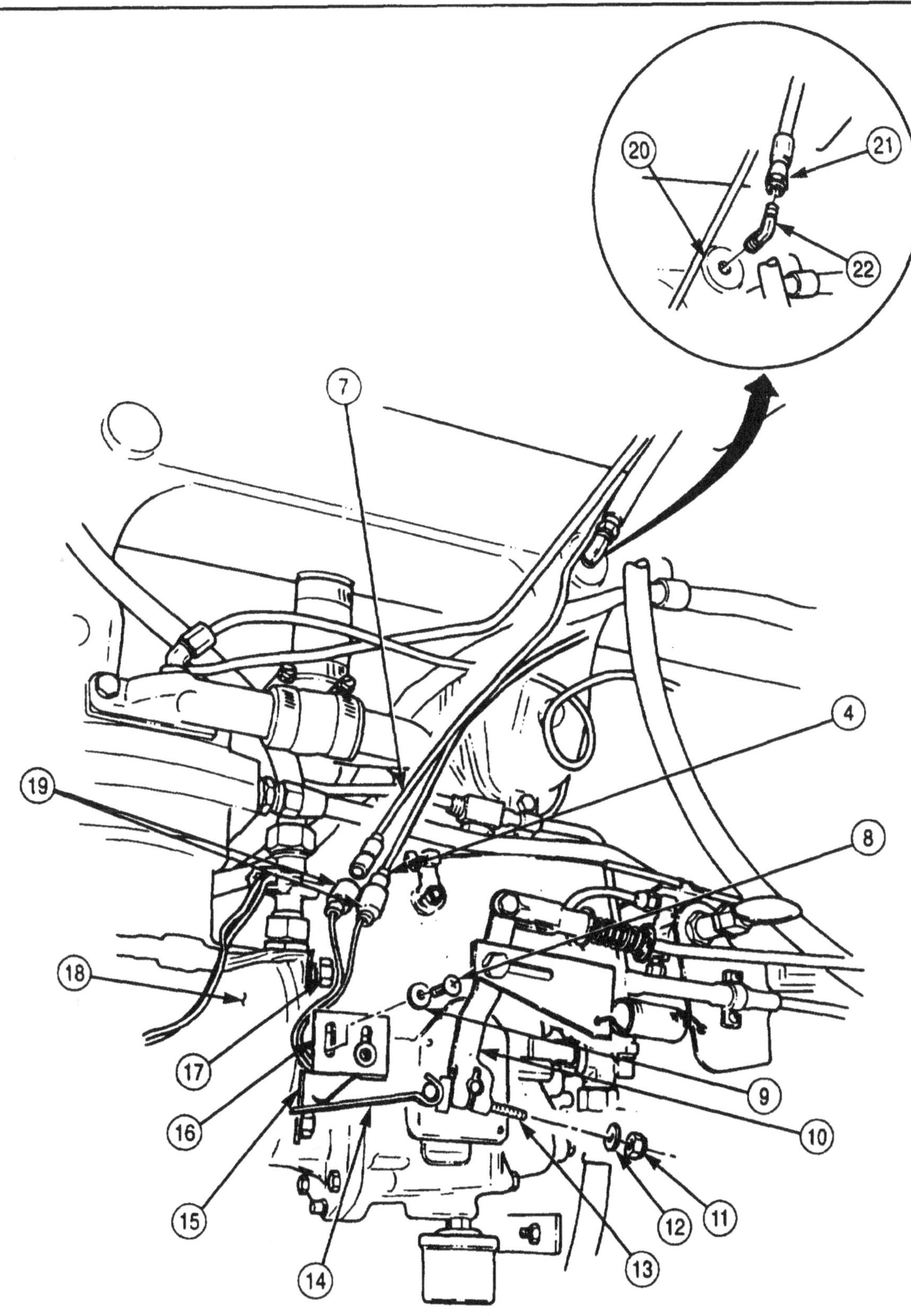

3-447. ENGINE EXHAUST BRAKE MODIFICATION KIT (M939/A1) REPLACEMENT (Contd)

16. Remove air supply line (6) from connectors (5) and (8).
17. Disconnect hose assembly (4) from elbow (1).
18. Disconnect hose assembly (10) from connector (2).
19. Disconnect hose assembly (11) from elbow (13).
20. Remove hose assembly (9) from elbow (12) and connector (3).
21. Remove nut (16), screw (14), washer (15), and valve assembly (18) from horn bracket (17).
22. Remove horn bell (7) from horn assembly (23).
23. Remove horn assembly (para. 3-103).
24. Remove two screws (19), lockwashers (20), and bracket (21) from valve assembly (18) Discard lockwashers (20).
25. Remove elbows (12) and (13) from valve assembly (18).
26. Remove connector (2), elbow (1), and tee (22) from valve assembly (18).
27. Remove connectors (3) and (5) and tee (24) from horn assembly (23).

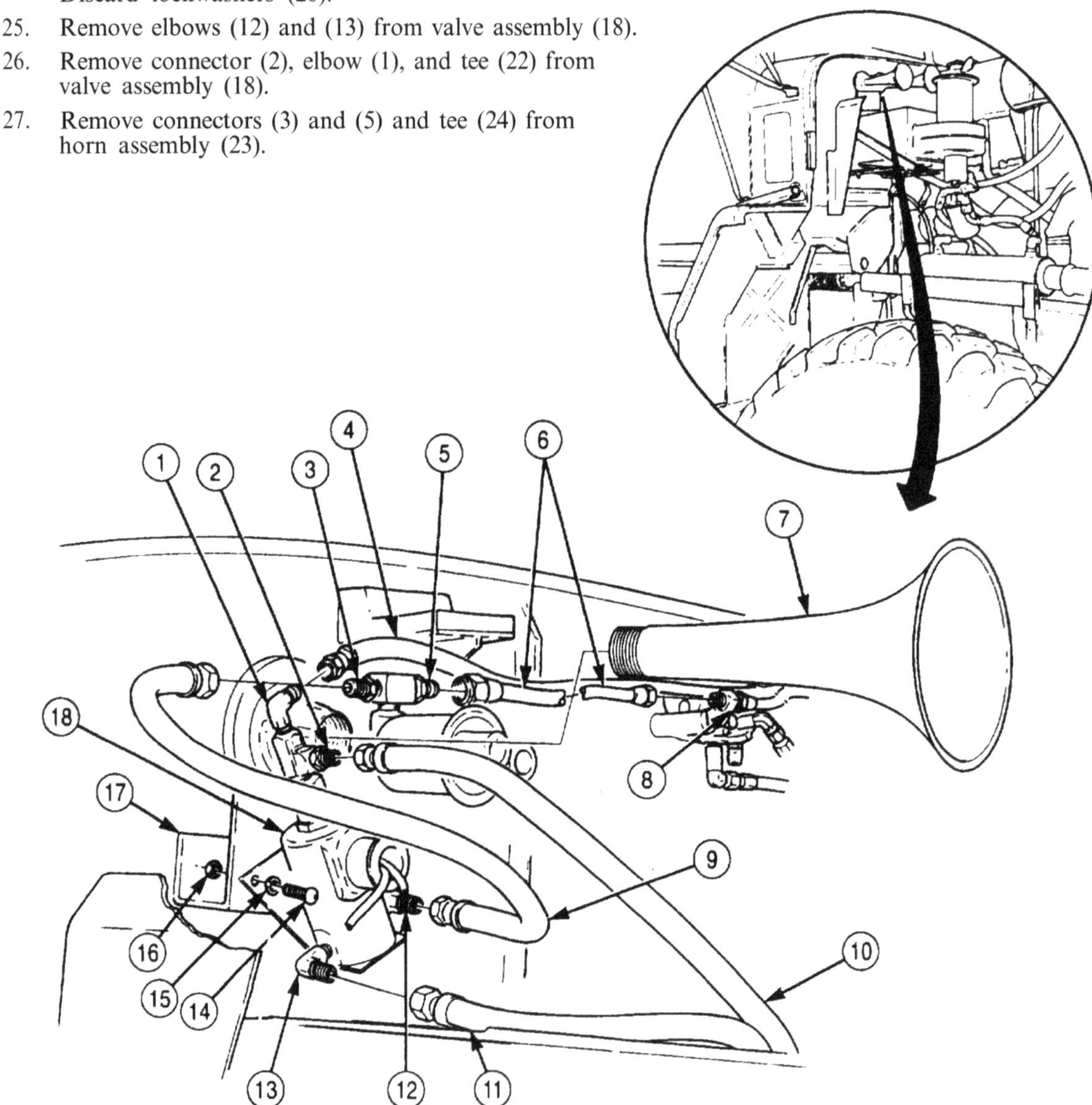

3-447. ENGINE EXHAUST BRAKE MODIFICATION KIT (M939/A1) REPLACEMENT (Contd)

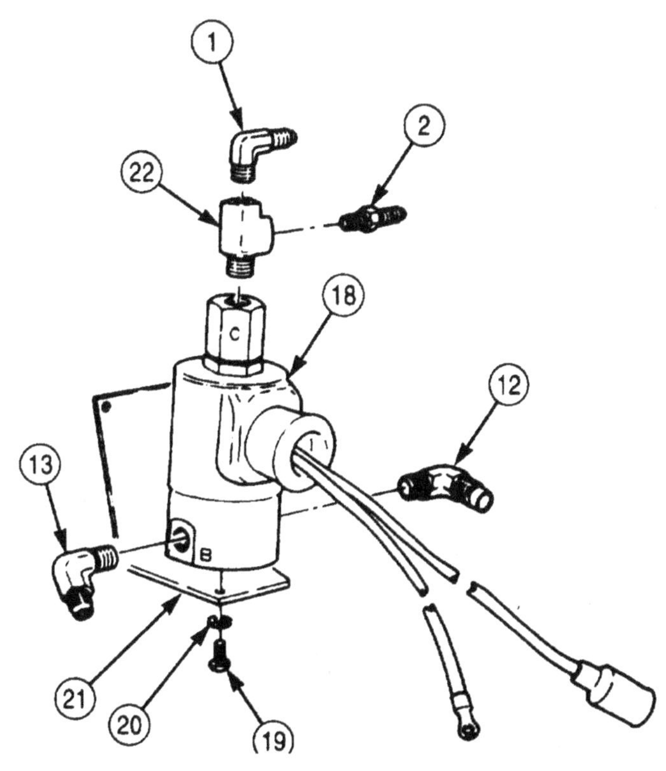

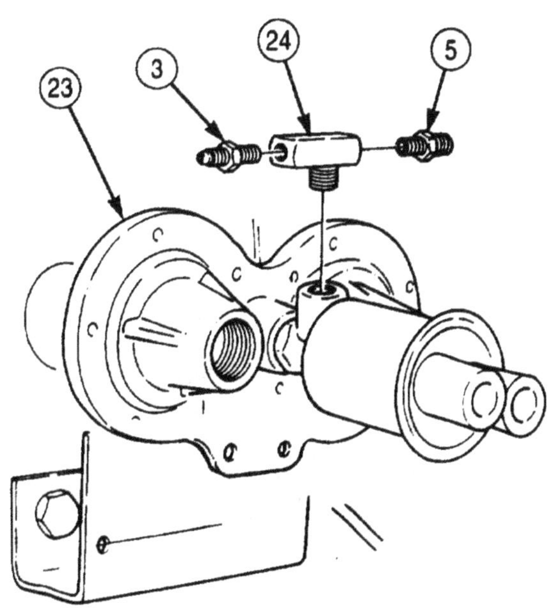

3-447. ENGINE EXHAUST BRAKE MODIFICATION KIT (M939/A1) REPLACEMENT (Contd)

28. Remove any shims (1) installed between exhaust brake assembly (6) and bracket assembly.
29. Remove clamp (10) and hose assembly (11) from oil dipstick tube (2).
30. Remove hose assembly (5) from quick-release valve (7) and exhaust brake assembly (6).
31. Remove two locknuts (9), screws (8), clamps (3), and hose assembly (11) from oil dipstick tube (2). Discard locknuts (9).
32. Remove nut (18), screw (19), coupling (20), and gasket (17) from rear exhaust pipe (16) and pipe assembly (12). Discard gasket (17).
33. Remove nut (21), clamp (14), pipe assembly (12), and gasket (13) from exhaust brake assembly (15). Discard gasket (13).

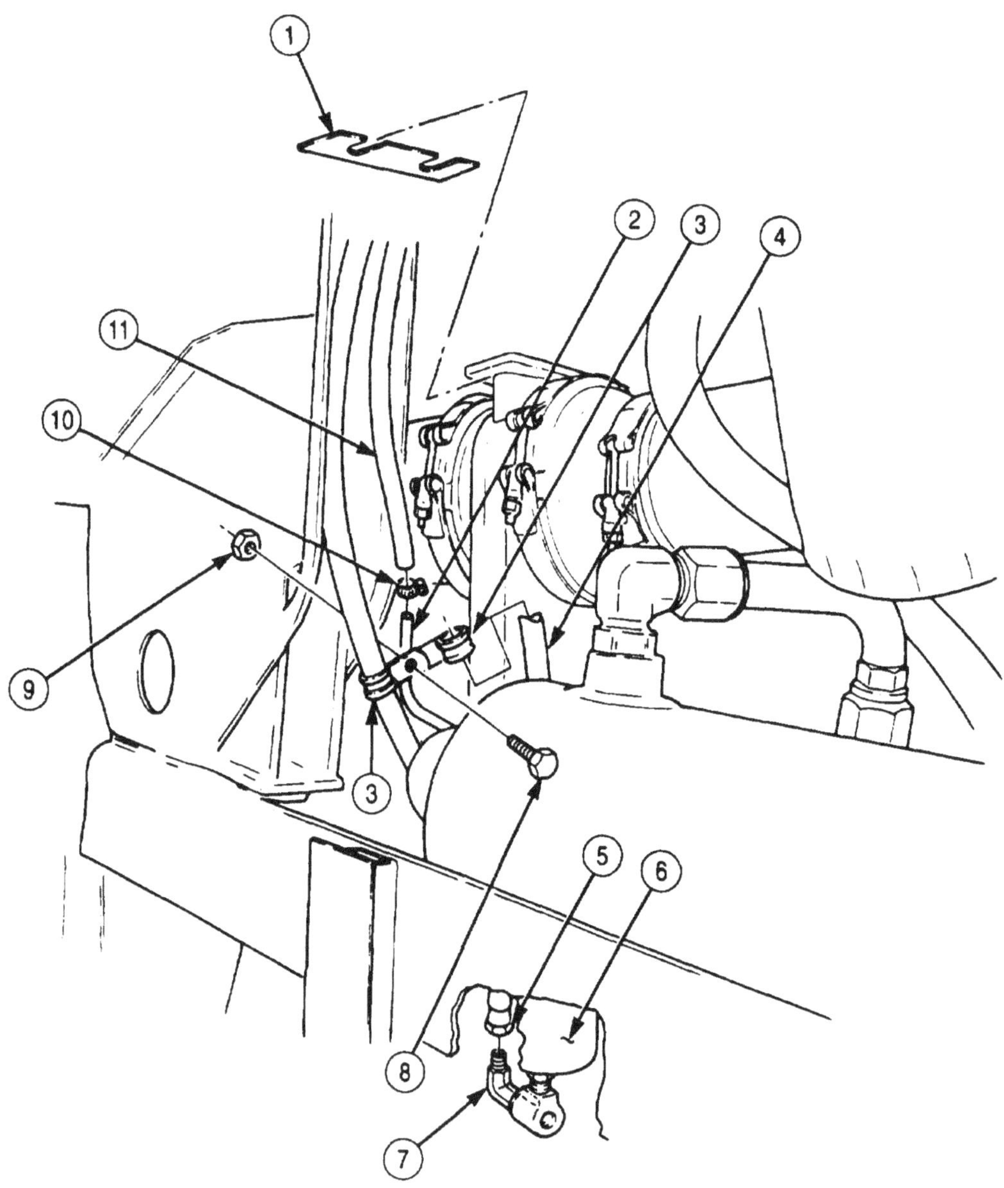

3-447. ENGINE EXHAUST BRAKE MODIFICATION KIT (M939/A1) REPLACEMENT (Contd)

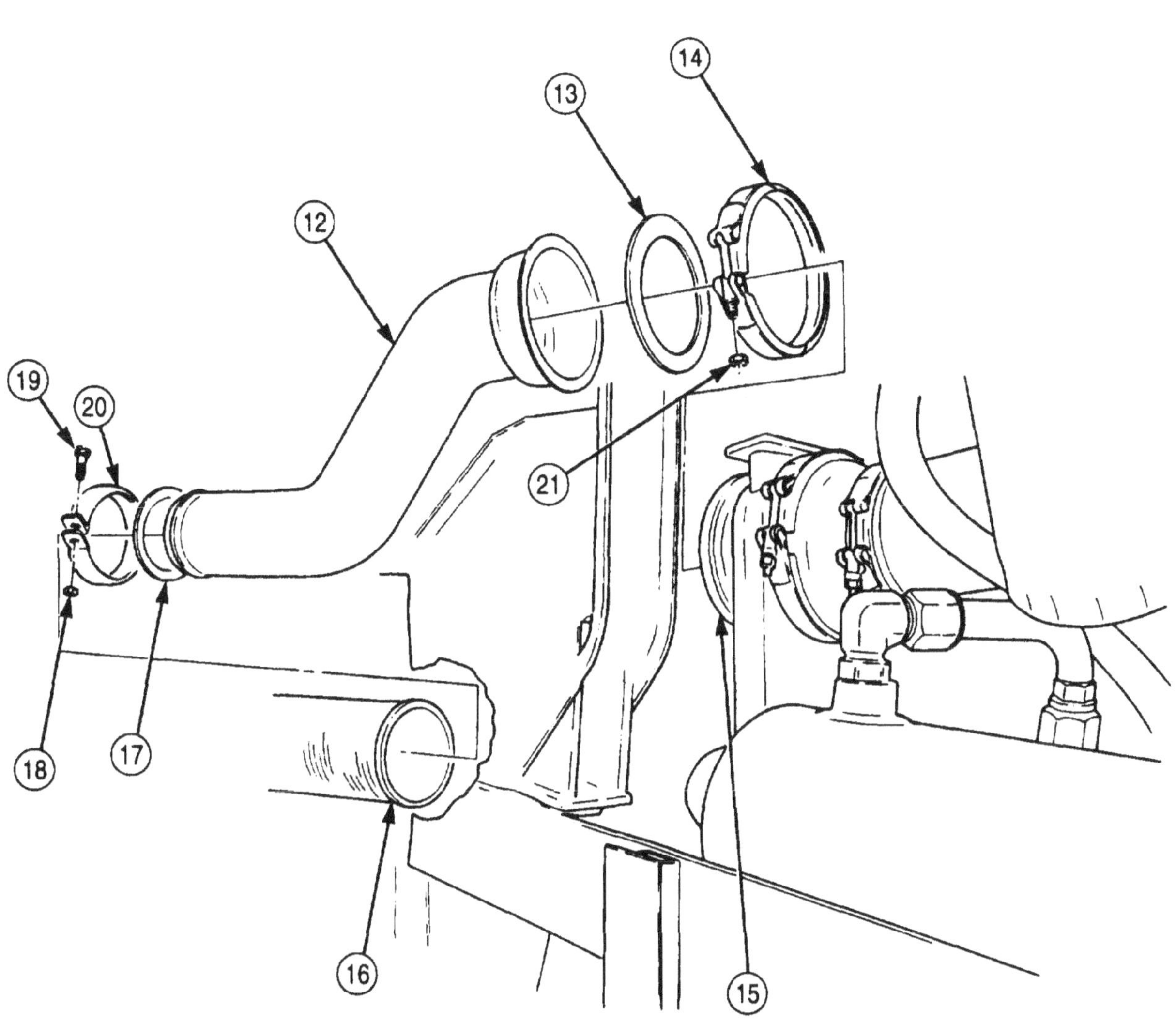

3-447. ENGINE EXHAUST BRAKE MODIFICATION KIT (M939/A1) REPLACEMENT (Contd)

34. Remove nut (6), coupling (5), and gasket (7) from adapter (4) and exhaust manifold (8). Discard gasket (7).
35. Remove two screws (12), washers (11). bracket assembly (10), and exhaust brake assembly (13) from engine block (9).
36. Remove nut (14), clamp (2), adapter (4). and gasket (1) from exhaust brake assembly (13). Discard gasket (1).
37. Remove two screws (16), washers (15), and exhaust brake assembly (13) from bracket assembly (10).
38. Remove elbow (19) from quick-release valve (20) and exhaust brake assembly (13).
39. Remove tube (17) and elbow (18) from exhaust brake assembly (13).

b. Installation

NOTE

- Apply sealant to male threads of fittings and leave first two starter threads clear of sealant.
- It may be necessary to make adjustments to fittings for alignment purposes.
- Do not tighten exhaust components until exhaust system is completely assembled except for steps 5 and 7.
- Ensure antiseize tape is applied to any screws attached to engine assembly.

1. Install elbow (18) and tube (17) on exhaust brake assembly (13).
2. Install elbow (19) on quick-release valve (20) and exhaust brake assembly (13).
3. Install exhaust brake assembly (13) on bracket assembly (10) with two washers (15) and screws (16).
4. Install new gasket (1) and adapter (4) on exhaust brake assembly (13) with adapter (4) reference line (3) at 3 O'clock position as viewed from driver's seat and install clamp (2) and nut (14).
5. Install exhaust brake assembly (13) and bracket assembly (10) on engine block (9) with two washers (11) and screws (12).
6. Install new gasket (7) and coupling (5) on adapter (4) and exhaust manifold (8) with nut (6).

3-447. ENGINE EXHAUST BRAKE MODIFICATION KIT (M939/A1) REPLACEMENT (Contd)

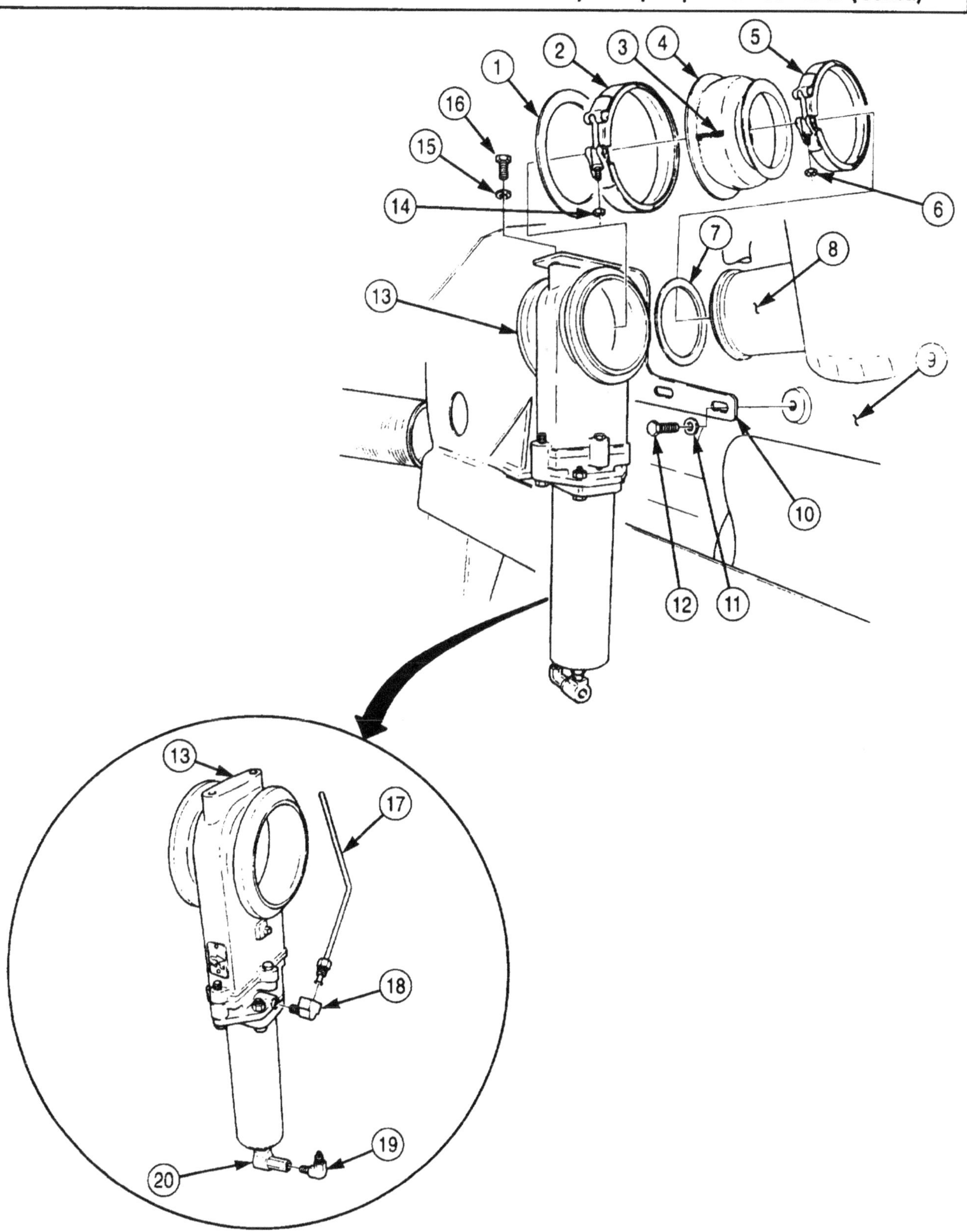

3-447. ENGINE EXHAUST BRAKE MODIFICATION KIT (M939/A1) REPLACEMENT (Contd)

7. Install new gasket (1) and pipe assembly (10) on exhaust brake assembly (4) with clamp (2) and nut (3).
8. Install new gasket (9) and pipe assembly (10) on rear exhaust pipe (5) with coupling (8), screw (7), and nut (6).

NOTE

If gap exists between exhaust brake assembly and bracket assembly, install shims before tightening hardware.

9. Tighten all exhaust brake assembly hardware, exhaust clamps, and related components.
10. Install two hose clamps (14) and hose assembly (11) on oil dipstick tube (15) with two screws (18) and new locknuts (19).
11. Install hose assembly (17) on quick-release valve (16) and exhaust brake assembly (4).
12. Position clamp (20) over hose assembly (11) and install hose assembly (11) on tube (13). Tighten clamp (20).
13. Install shims (12) between exhaust brake assembly (4) and bracket assembly.

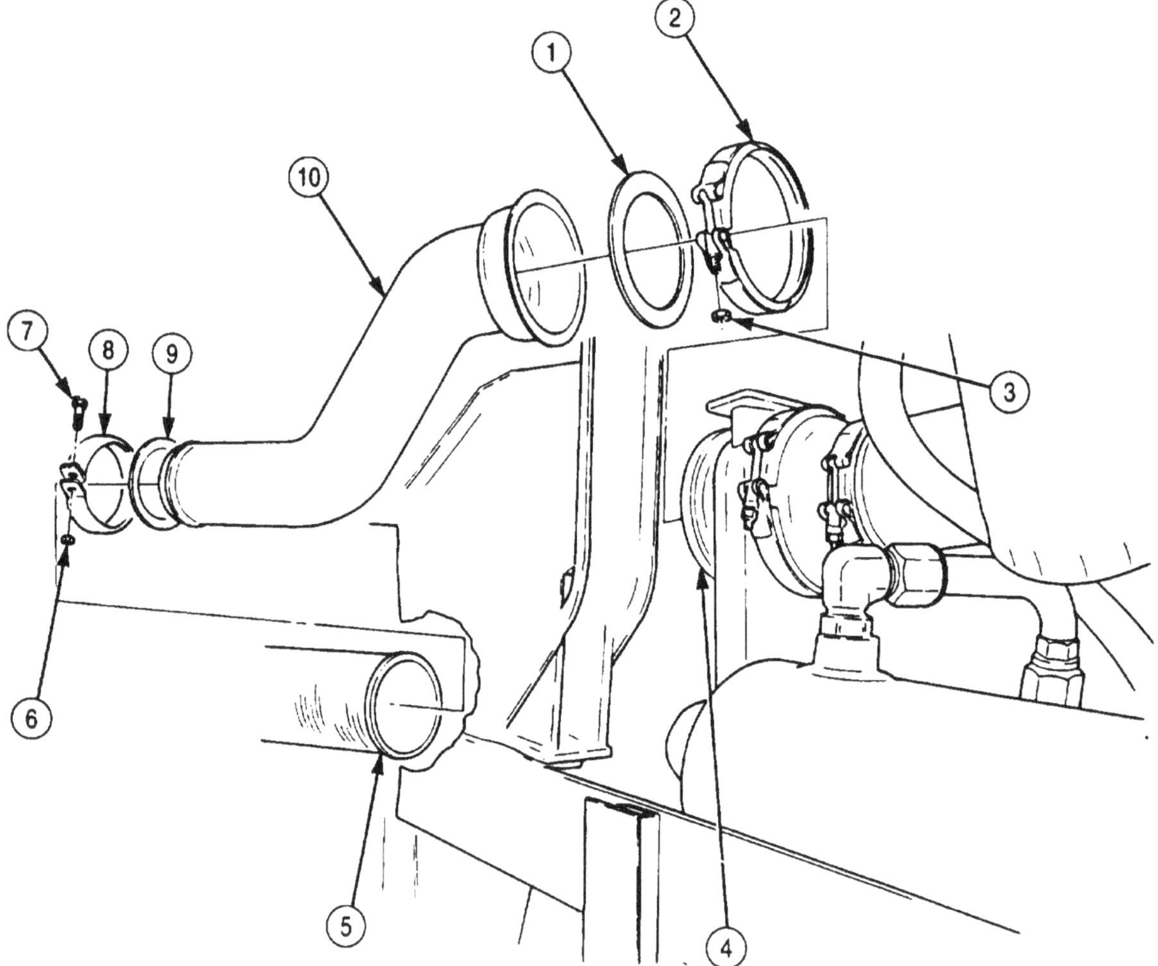

3-447. ENGINE EXHAUST BRAKE MODIFICATION KIT (M939/A1) REPLACEMENT (Contd)

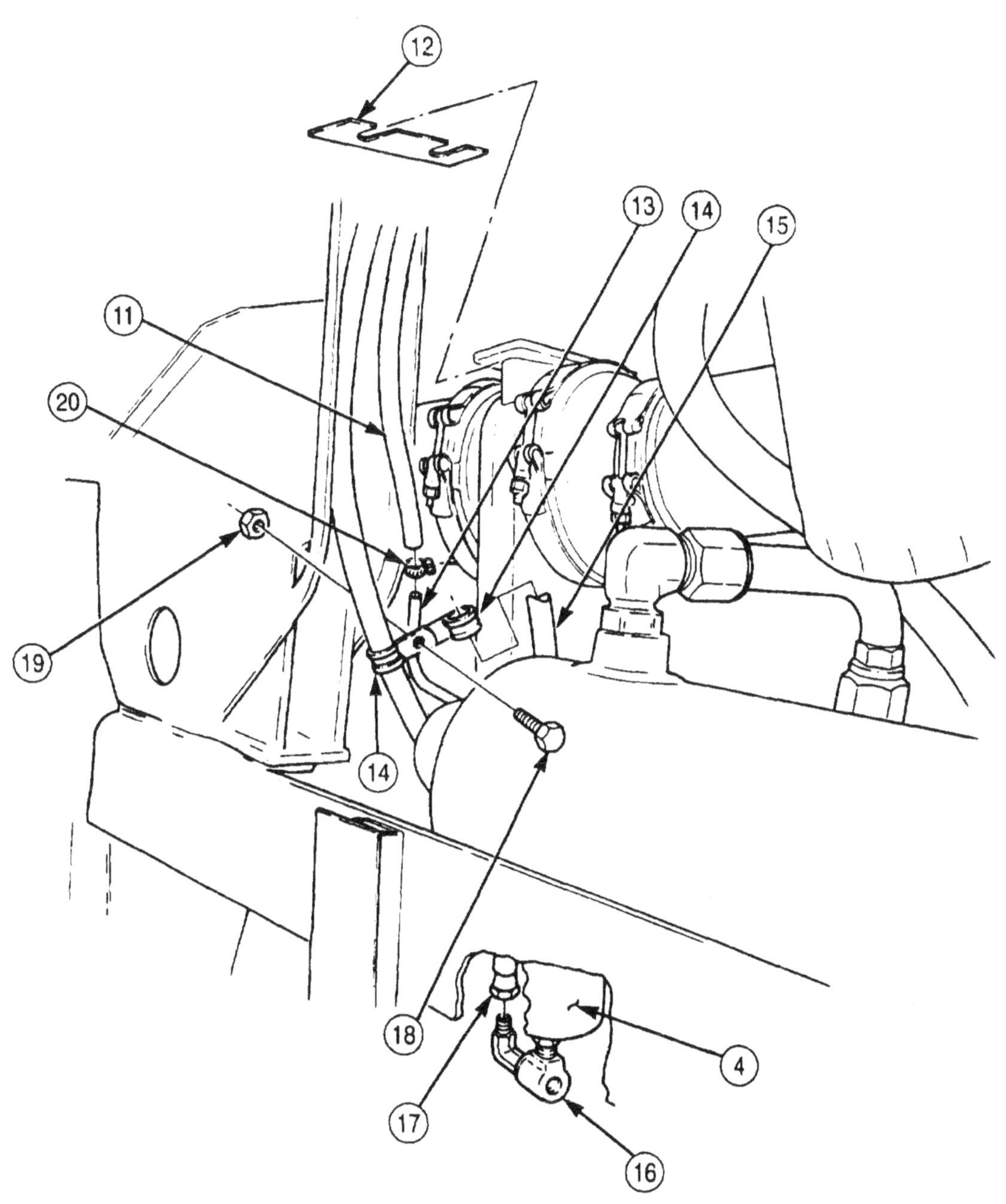

3-447. ENGINE EXHAUST BRAKE MODIFICATION KIT (M939/A1) REPLACEMENT (Contd)

14. Install tee (11) and connectors (10) and (12) on horn assembly (13).
15. Install tee (9), elbow (1), and connector (2) on valve assembly (3).
16. Install elbows (4) and (8) on valve assembly (3).
17. Install bracket (7) on valve assembly (3) with two new lockwashers (6) and screws (5).
18. Install horn assembly (13) (para. 3-103).
19. Install born bell (16) on horn assembly (13).
20. Install valve assembly (3) on horn bracket (24) with washer (22), screw (21), and nut (23).
21. Install hose assembly (18) on elbow (4) and connector (10).
22. Install hose assembly (20) on elbow (8).
23. Install hose assembly (19) on connector (2).
24. Install hose assembly (14) on elbow (1).
25. Install air supply line (15) on connectors (12) and (17).

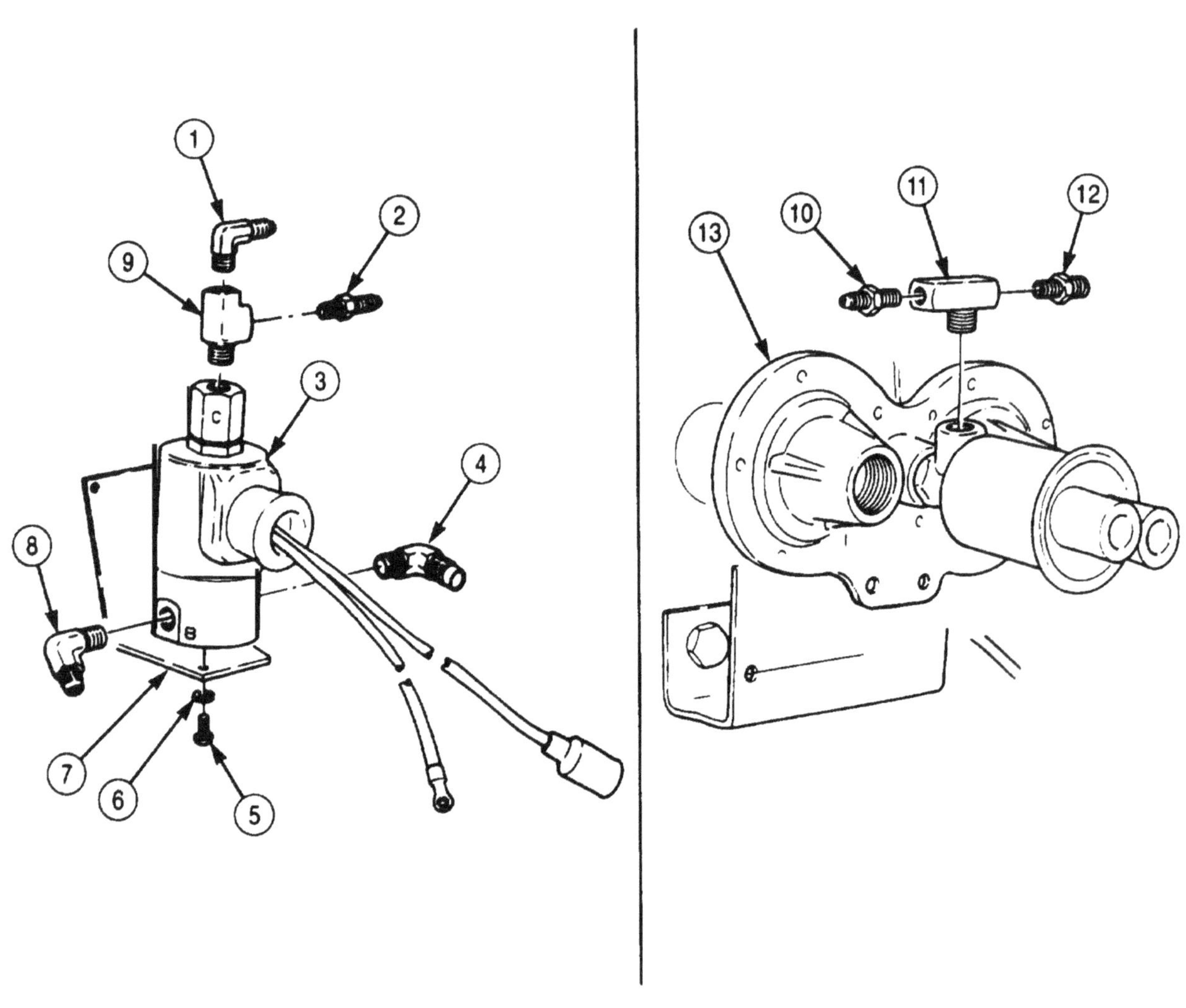

3-447. ENGINE EXHAUST BRAKE MODIFICATION KIT (M939/A1) REPLACEMENT (Contd)

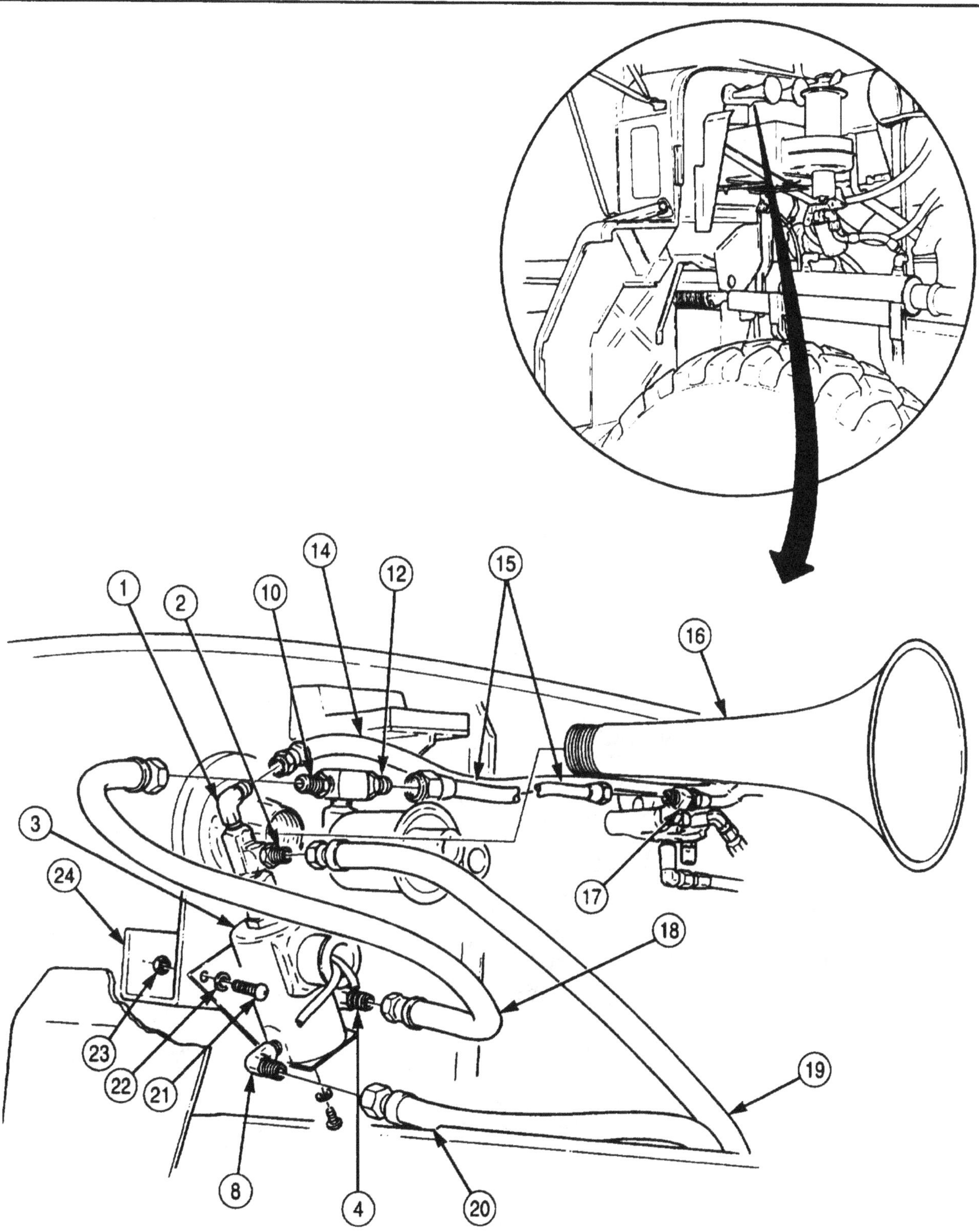

3-447. ENGINE EXHAUST BRAKE MODIFICATION KIT (M939/A1) REPLACEMENT (Contd)

26. Install valve assembly ground lead (2) on screw (1) with nut (3).
27. Connect cable (4) to valve assembly lead (6) and route cable (4) through firewall.
28. Connect elbow (22) and hose assembly (21) to intake manifold (20).
29. Loosen two fuel pump mounting screws (17) and install bracket assembly (15) on fuel pump (18).
30. Install lever switch bracket (16) on bracket assembly (15) with two washers (9) and screws (8). Do not tighten screws (8).
31. Install T-bolt (13) and lever (14) on fuel pump throttle control lever (10) with washer (12) and new locknut (11).
32. Position fuel pump throttle control lever (10) so it touches lever switch bracket (16).
33. Tighten fuel pump mounting screws (17) and screws (8).
34. Connect cable (7) to lead (19) and secure with tie-down strap.
35. Route cable (4) through firewall and connect to lead (19). Secure with new tiedown strap (5).

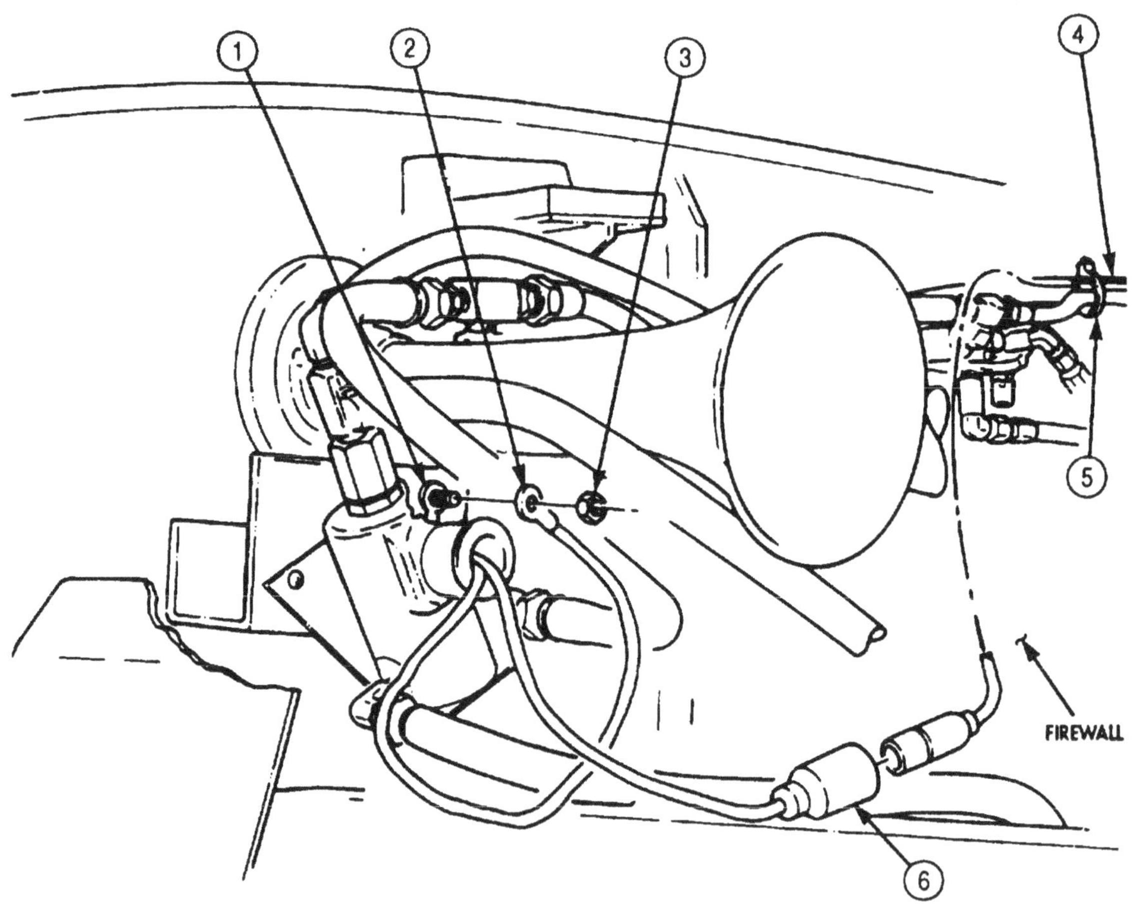

3-447. ENGINE EXHAUST BRAKE MODIFICATION KIT (M939/A1) REPLACEMENT (Contd)

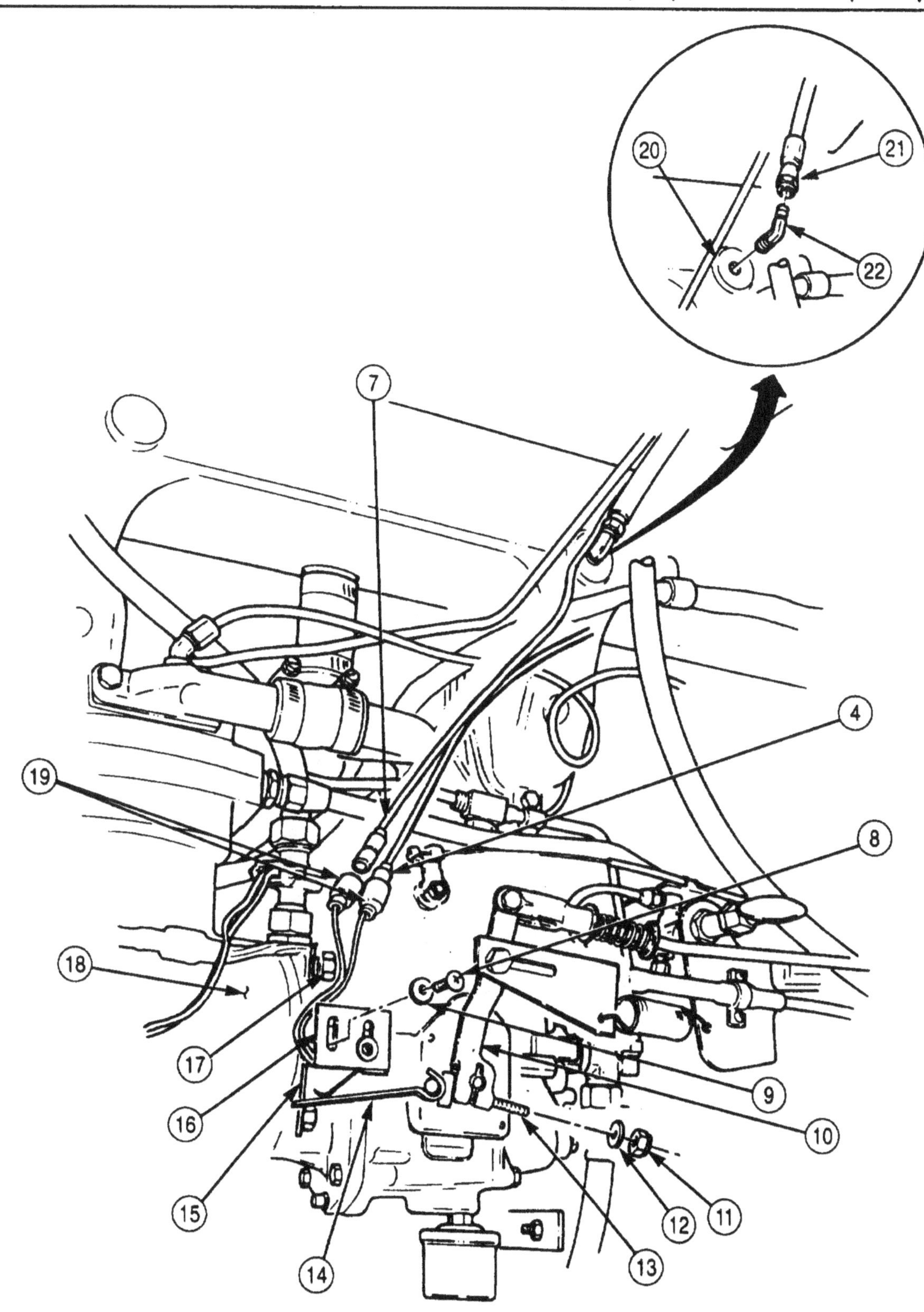

3-447. ENGINE EXHAUST BRAKE MODIFICATION KIT (M939/A1) REPLACEMENT (Contd)

NOTE

Instrument cluster may have to be removed for switch installation (para. 3-83).

36. Install toggle switch (4) and plate (5) on instrument cluster (3) with washer (1) and nut (2).

NOTE

Ensure that toggle switch is in OFF position.

37. Disconnect lead (9) from circuit breaker (11).
38. Connect lead assembly (10) to circuit breaker (11).
39. Connect lead (9) to lead assembly (10).
40. Connect two toggle switch leads (8) to cable (12) and lead assembly (10) behind instrument panel (7).
41. Secure cable (12) to wiring harness (6) with new tiedown strap (13).
42. Install instrument cluster (3) if required (para. 3-83).

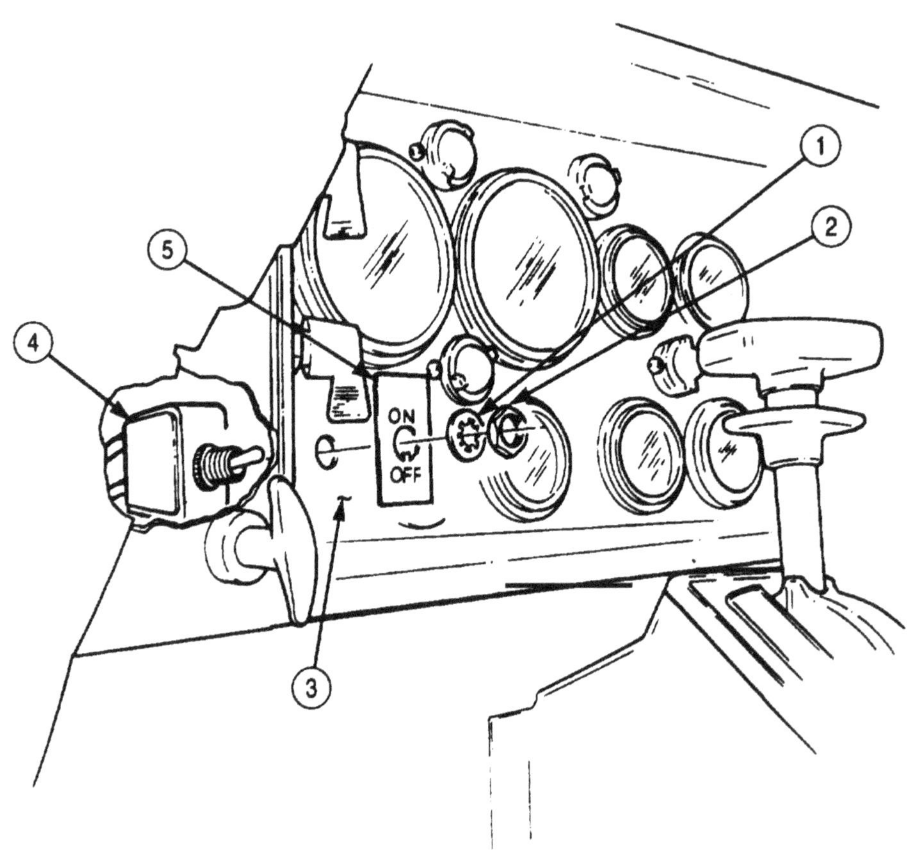

3-447. ENGINE EXHAUST BRAKE MODIFICATION KIT (M939/A1) REPLACEMENT (Contd)

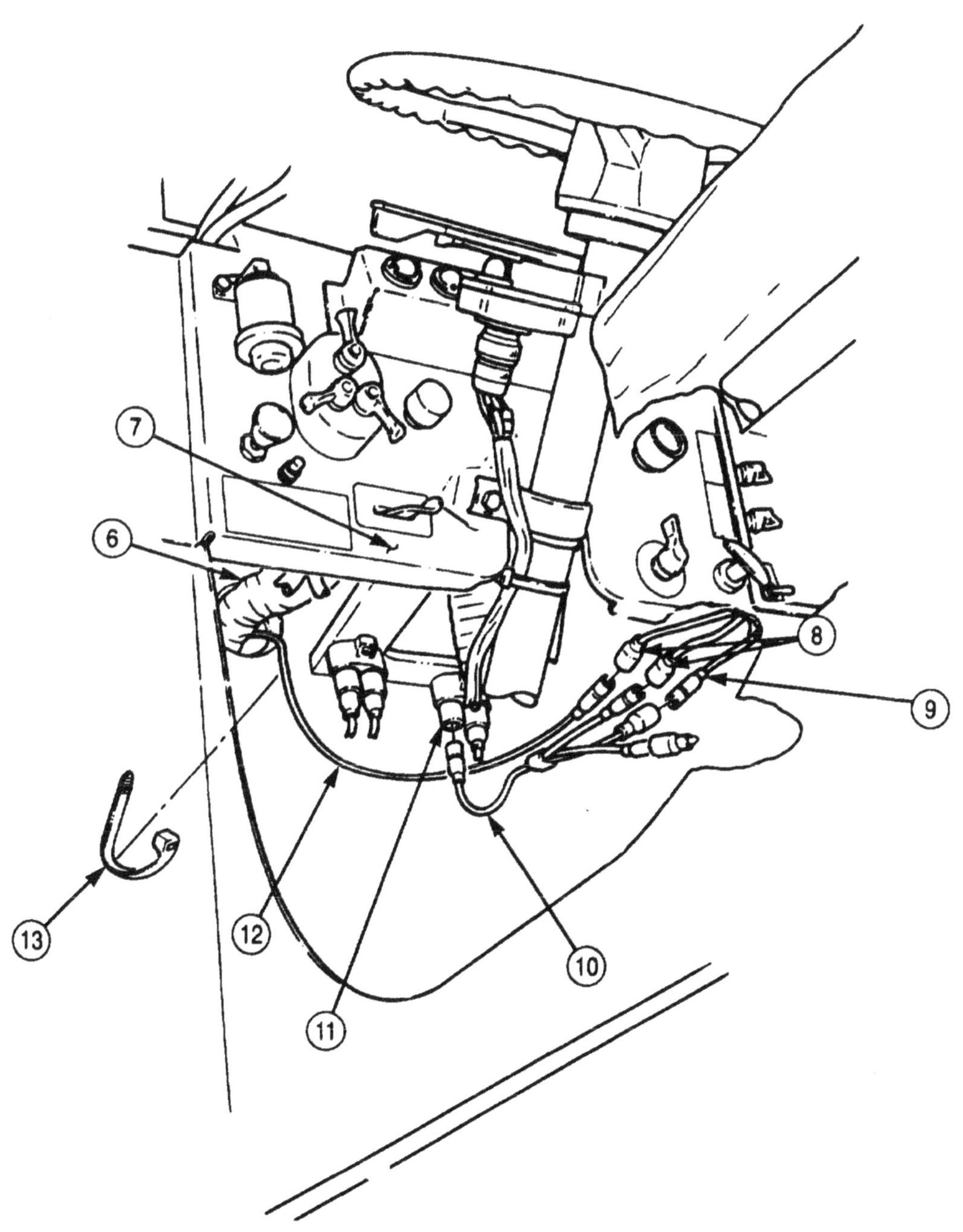

FOLLOW-ON TASKS:
- Connect battery ground cables (para. 3-126).
- Install left and right engine splash shields (TM 9-2320-272-10).
- Install right hood bumper (para. 3-274).
- Install exhaust pipe heat shield if removed (para. 3-49).

3-448. ENGINE EXHAUST BRAKE MODIFICATION KIT (M939A2) REPLACEMENT

THIS TASK COVERS:

a. Removal

b. Installation

INITIAL SETUP:

APPLICABLE MODELS

M939A2

TOOLS

General mechanic's tool kit (Appendix E, Item 1)

MATERIALS/PARTS

Lockwasher (Appendix D, Item 345)
Lockwasher (Appendix D, Item 415)
Tiedown straps (Appendix D, Item 697)
Two lockwashers (Appendix D, Item 346)
Six lockwashers (Appendix D, Item 371)
Two lockwashers (Appendix D, Item 347)
Two gaskets (Appendix D, Item 242)
Two gaskets (Appendix D, Item 170)
Sealant (Appendix C, Item 63)
Antiseize tape (Appendix C, Item 72)

REFERENCES (TM)

TM 9-2320-272-10
TM 9-2320-272-24P

EQUIPMENT CONDITION

- Parking brake set (TM 9-2320-272-10).
- Exhaust pipe heat shield removed if needed (para. 3-49).
- Right hood bumper removed (para. 3-274).
- Remove left and right engine splash shields removed (TM 9-2320-272-10).
- Air reservoir drained (TM-2520-272-10).
- Battery ground cables disconnected (para. 3-126).

NOTE

- Transmission oil cooler oil filter assembly may have to be removed to reposition heater (para. 3-139).
- Remove four screws securing personnel heater mounting bracket to firewall. Reposition and secure heater for installation of kit components (para. 3-292).

a. Removal

NOTE

- Remove all tiedown straps as required and note location and position for installation.
- Tag all leads for installation.

1. Remove two toggle switch leads (1) from cable (5) and lead assembly (3).
2. Remove lead (2) from lead assembly (3).
3. Remove lead assembly (3) from circuit breaker (7).
4. Remove tiedown strap (6) and cable (5) from wiring harness (8) and pull through firewall (4). Discard tiedown strap (6).
5. Connect lead (2) to circuit breaker (7).

NOTE

- Instrument cluster may have to be removed for switch removal (para. 3-83).
- Ensure that toggle switch is in OFF position.

6. Remove nut (9), washer (10), plate (11), and toggle switch (13) from instrument cluster (12).

3-448. ENGINE EXHAUST BRAKE MODIFICATION KIT (M939A2) REPLACEMENT (Contd)

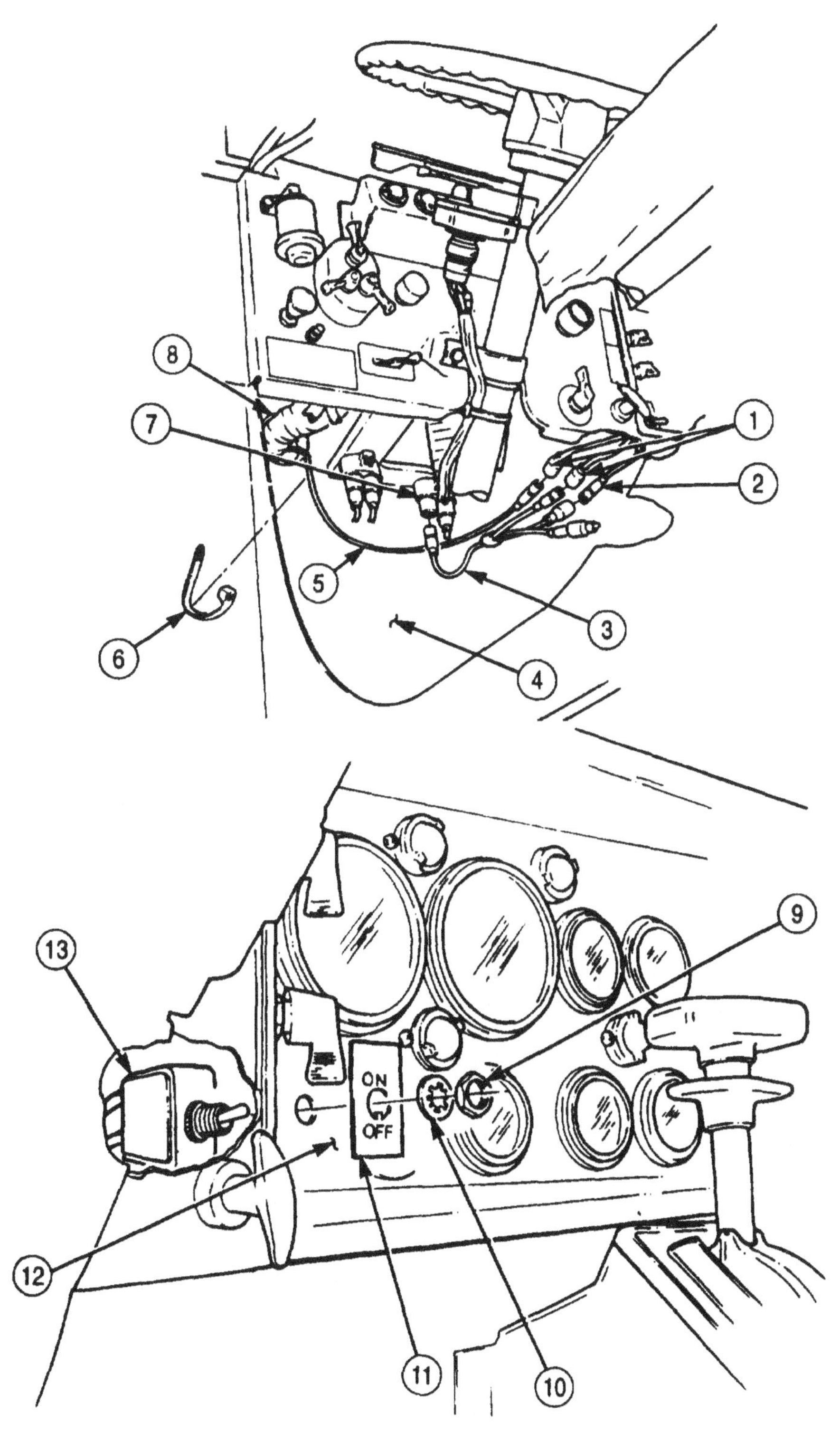

3-448. ENGINE EXHAUST BRAKE MODIFICATION KIT (M939A2) REPLACEMENT (Contd)

7. Disconnect cable (3) from lead (4) of lever switch bracket (5).
8. Disconnect cable (2) from lead (1) of lever switch bracket (5).
9. Disconnect accelerator linkage from throttle control lever (para. 3-43).
10. Remove two screws (7), washers (6), and lever switch bracket (5) from bracket assembly (8).
11. Remove screw (14), ball joint (12), lockwasher (13), and throttle control lever (15) from control lever (16). Discard lockwasher (13).
12. Remove two screws (11), lockwashers (10), clamp (9), and bracket assembly (8) from engine assembly (17) and bracket (18). Discard lockwashers (10).
13. Connect accelerator linkage to throttle control lever (para. 3-43).

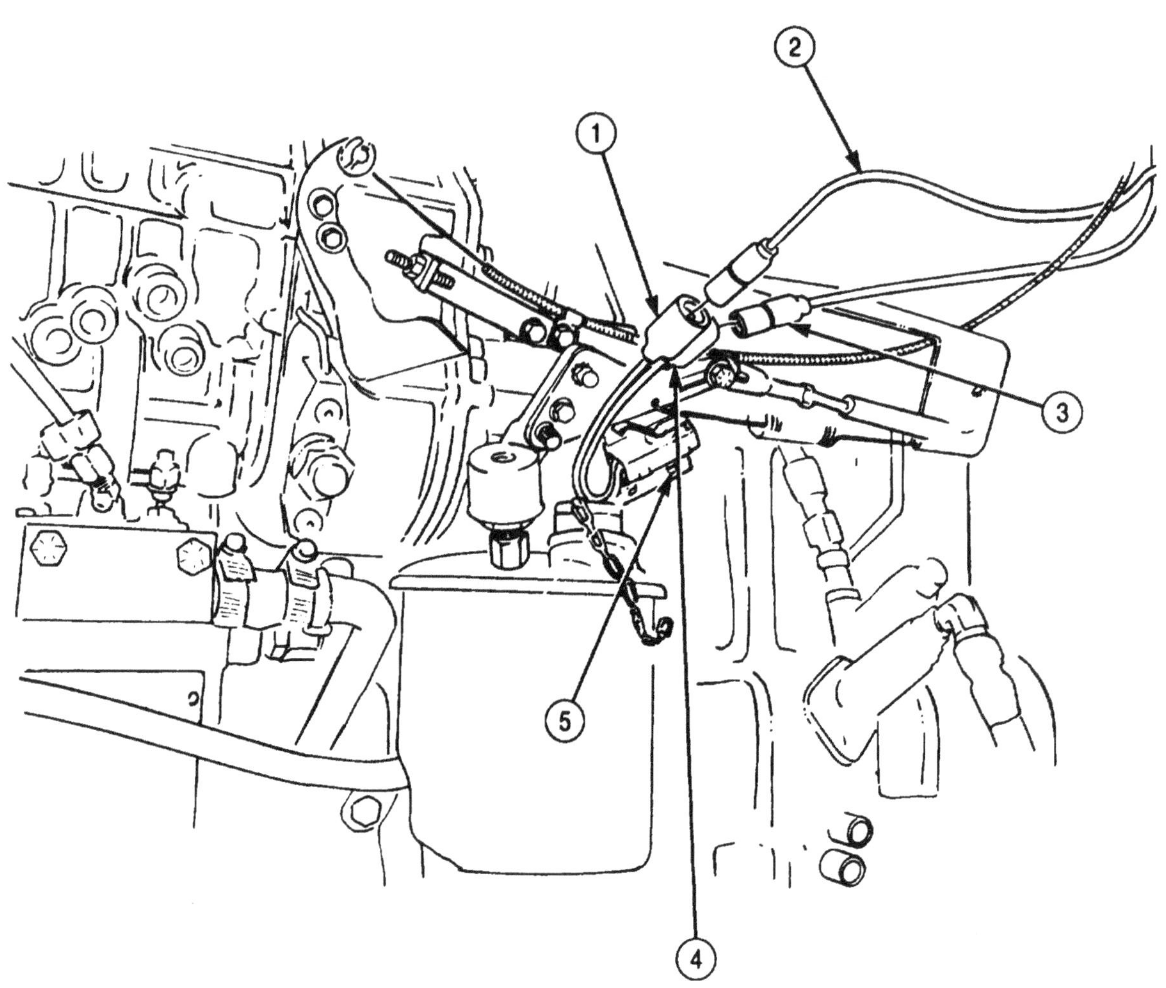

3-448. ENGINE EXHAUST BRAKE MODIFICATION KIT (M939A2) REPLACEMENT (Contd)

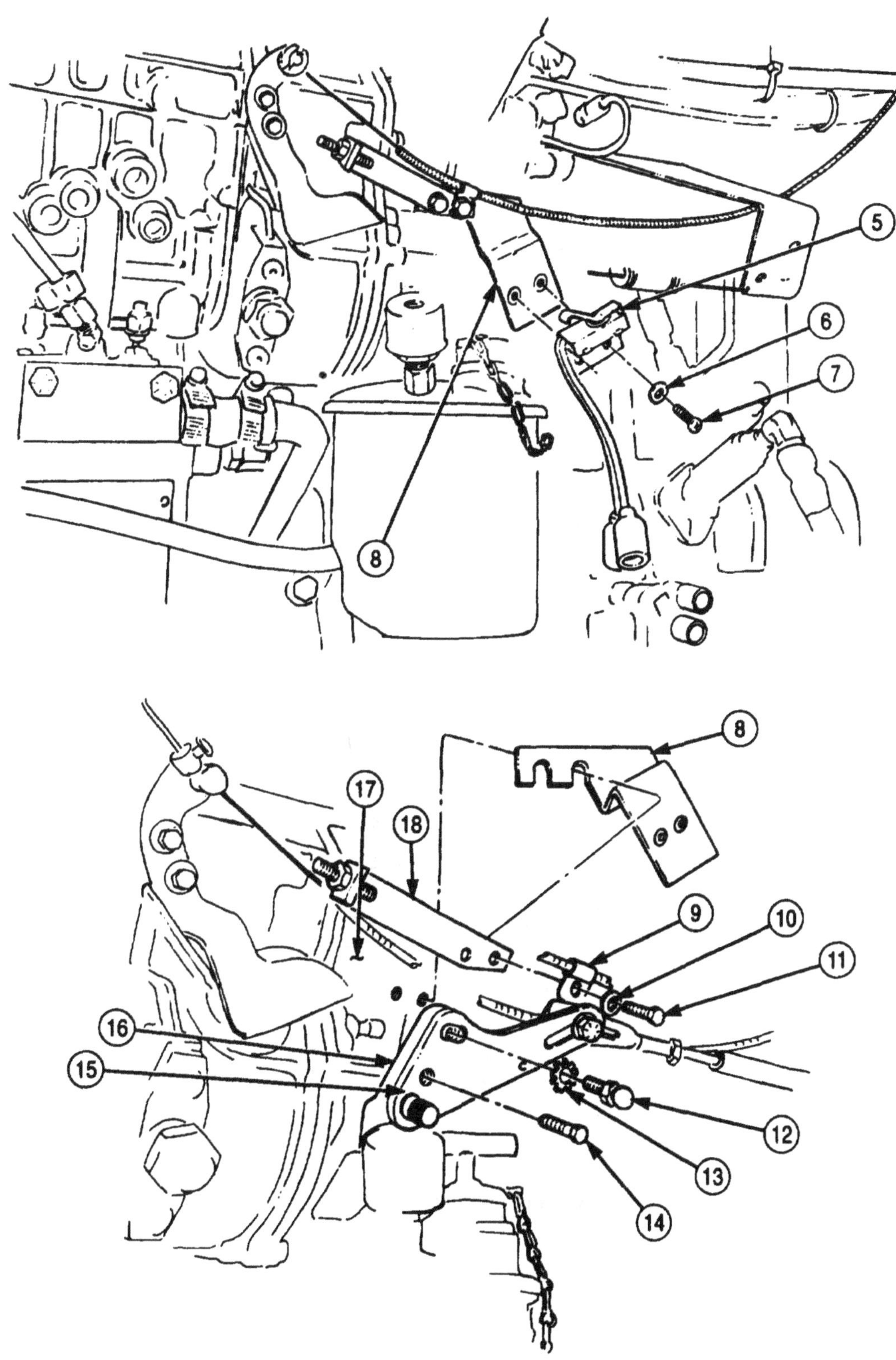

3-448. ENGINE EXHAUST BRAKE MODIFICATION KIT (M939A2) REPLACEMENT (Contd)

14. Disconnect cable (3) from valve assembly lead (4) and remove tiedown strap (2) and cable (3) from firewall (1). Discard tiedown strap (2).
15. Remove hose assembly (7) from elbow (6) and elbow (5).
16. Remove hose assembly (13) from elbow (14) and elbow (10).
17. Remove hose assembly (9) from elbow (8).
18. Remove hose assembly (12) from elbow (11).

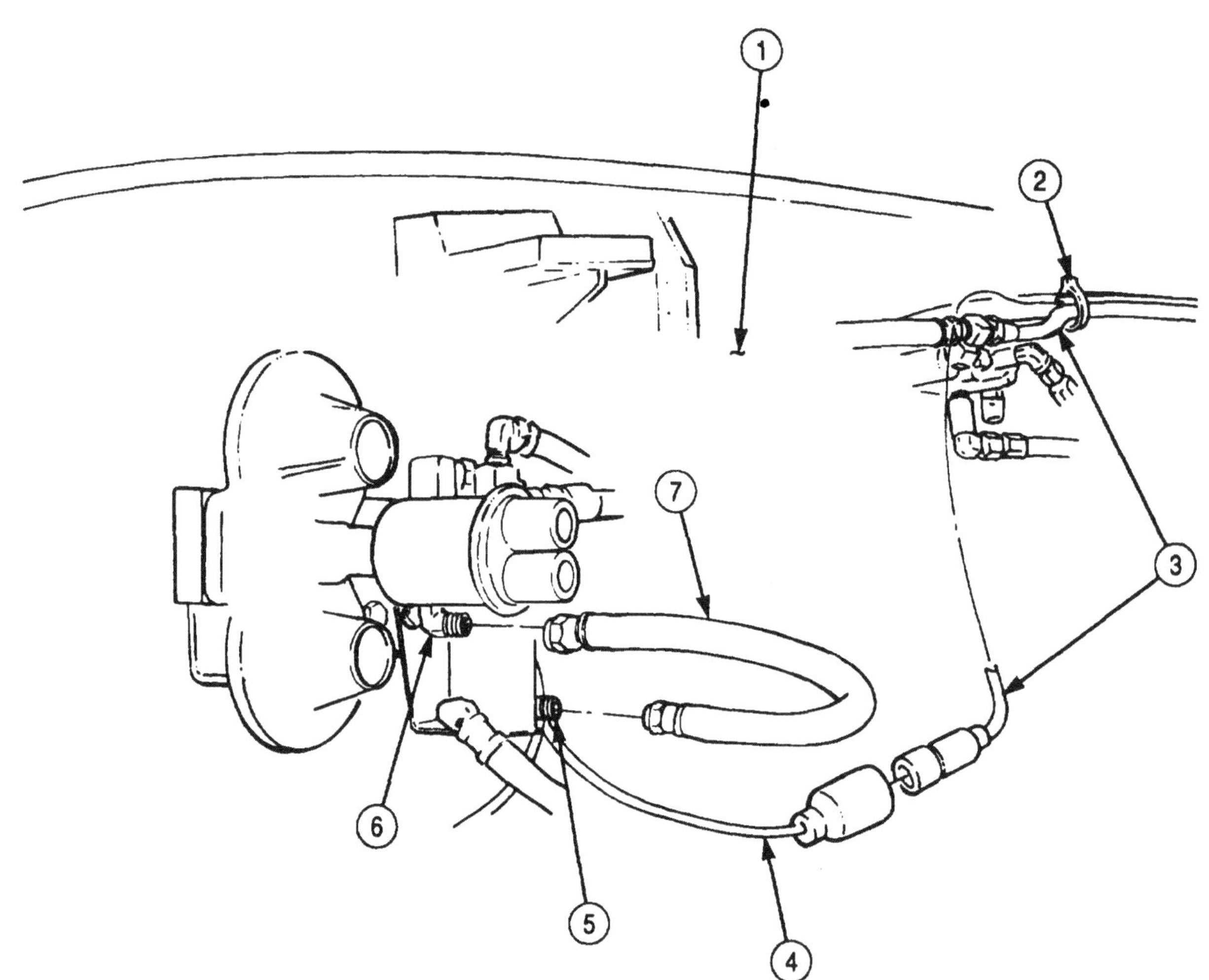

AIR HORN BELLS NOT SHOWN FOR CLARITY.

3-448. ENGINE EXHAUST BRAKE MODIFICATION KIT (M939A2) REPLACEMENT (Contd)

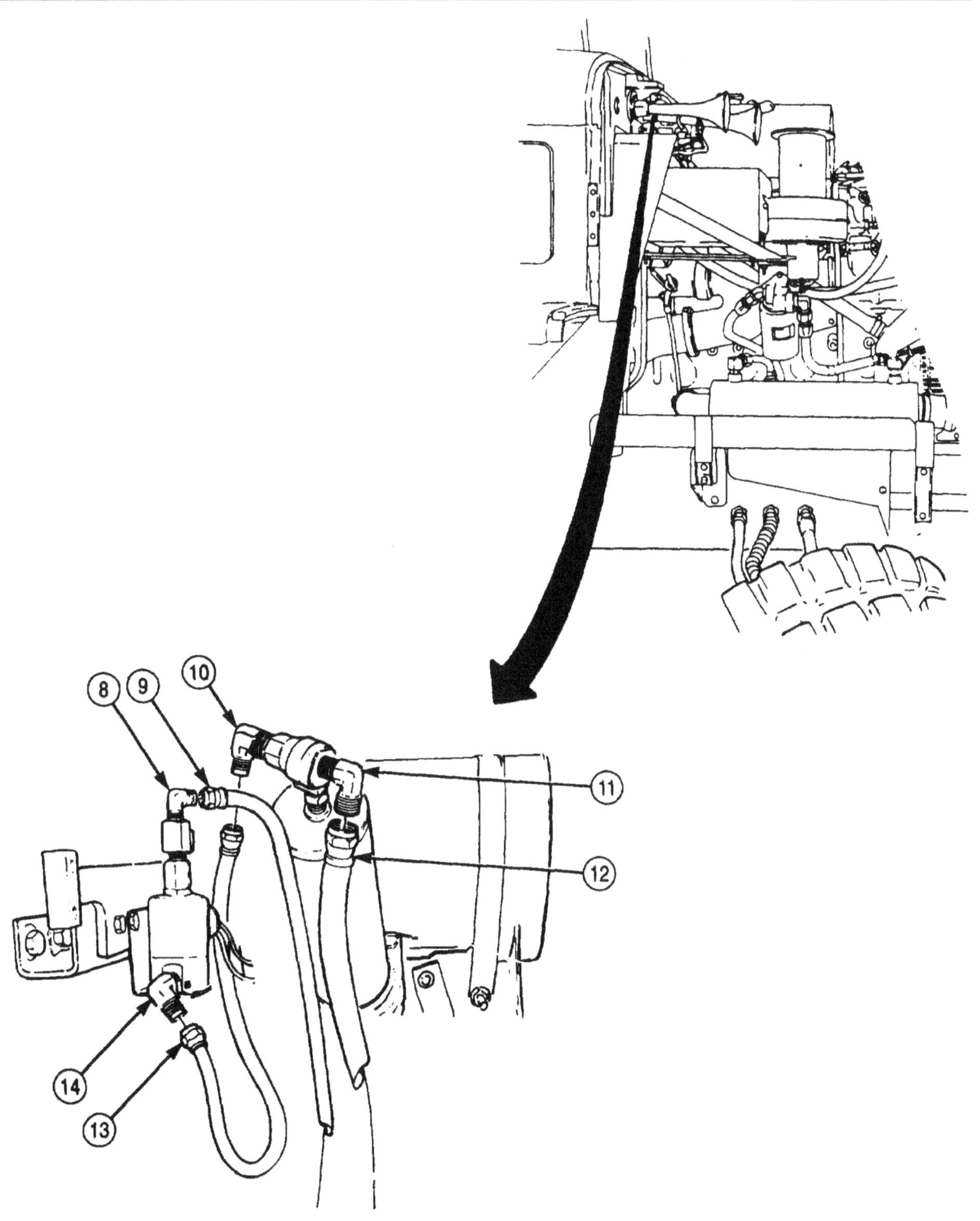

3-448. ENGINE EXHAUST BRAKE MODIFICATION KIT (M939A2) REPLACEMENT (Contd)

19. Remove hose (5) from connector (8).
20. Loosen two nuts (2) and remove air horn supply tube (3) from elbow (1) and tee (4).
21. Remove two nuts (24), lockwashers (25), screws (9), and horn assembly (10) from support brackets (12). Discard lockwashers (25).
22. Remove elbows (1) and (29) and tee (28) from horn assembly (10).
23. Remove nut (20), lockwasher (19), screw (15), lockwasher (16), valve assembly ground wire (14), and harness ground wire (13) from bracket (18) and support bracket (12). Discard lockwashers (19) and (16).
24. Remove nut (26), lockwasher (27), screw (17), and bracket (18) from support bracket (12) and horn support bracket (21). Discard lockwasher (27).
25. Remove two screws (35), lockwashers (34), and bracket (18) from valve assembly (32). Discard lockwashers (34).
26. Remove elbows (33) and (36) from valve assembly (32).
27. Remove connector (8), elbow (31), and tee (30) from valve assembly (32).
28. Remove nut (22), lockwasher (23), screw (11), and support bracket (12) from horn support bracket (21). Discard lockwasher (23).
29. Loosen clamp (6) and remove hose (5) from tube (7).

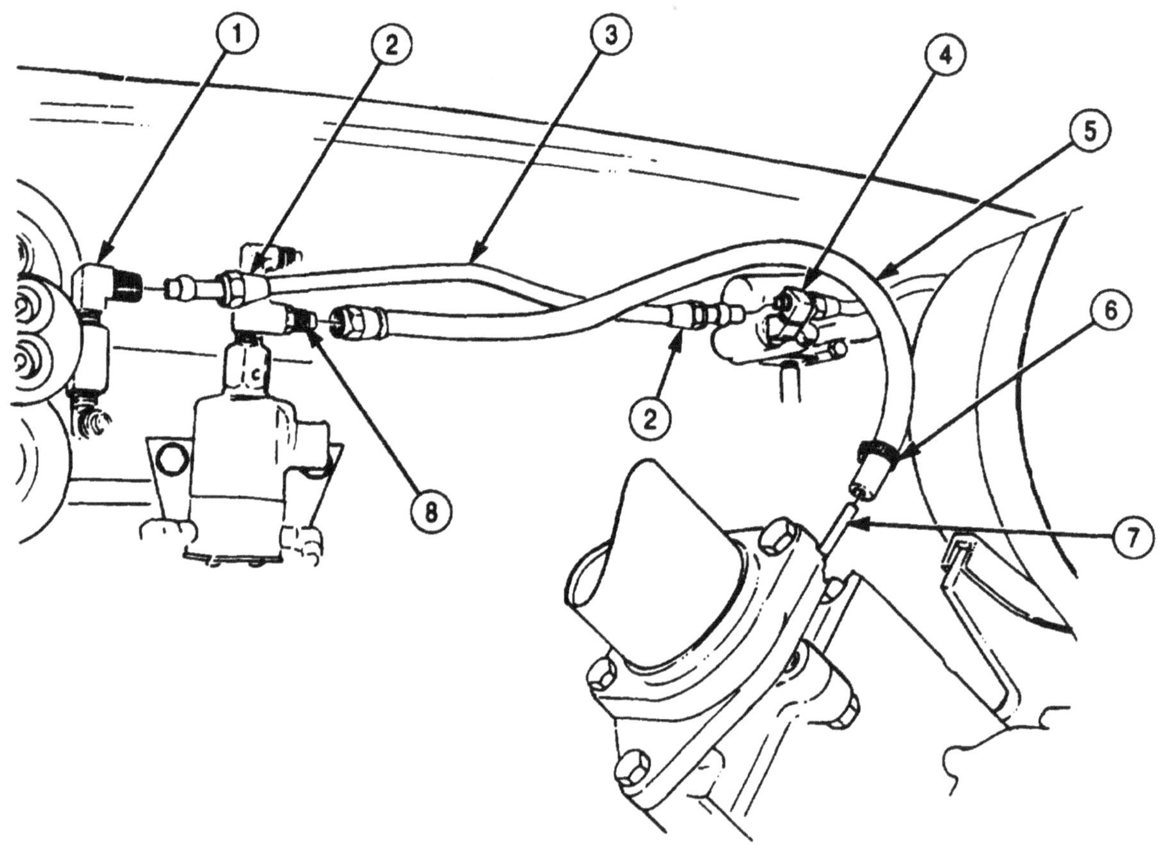

3-448. ENGINE EXHAUST BRAKE MODIFICATION KIT (M939A2) REPLACEMENT (Contd)

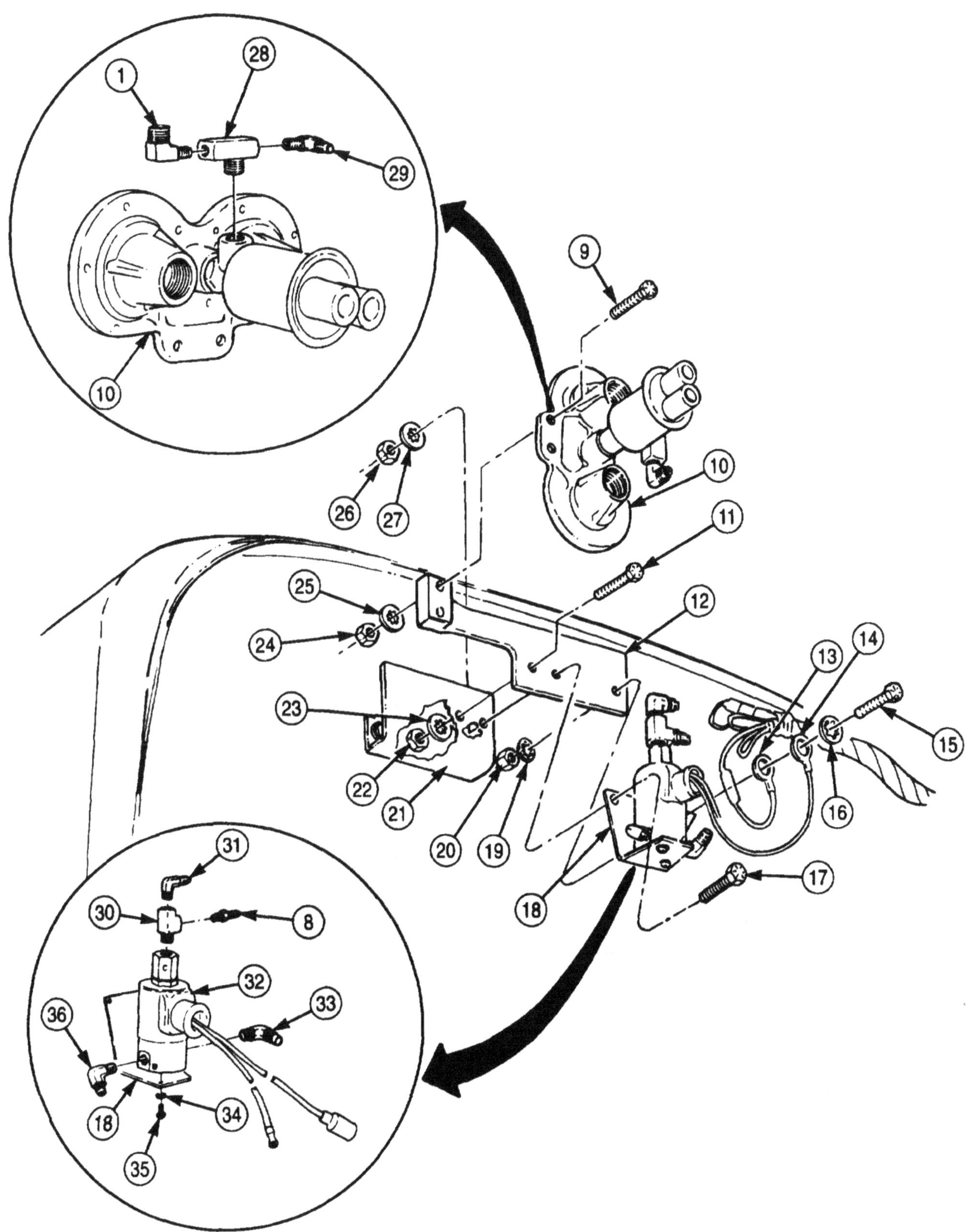

3-448. ENGINE EXHAUST BRAKE MODIFICATION KIT (M939A2) REPLACEMENT (Contd)

30. Remove nut (4), screw (3), and surge tank support (1) from surge tank mounting bracket (2).
31. Remove screw (5), lockwasher (6), and surge tank support (1) from exhaust manifold (7). Discard lockwasher (6).
32. Remove nut (17), screw (19), coupling (20), pipe assembly (16), and gasket (21) from rear exhaust pipe (18). Discard gasket (21).
33. Remove nut (22), clamp (12), gasket (11), and pipe assembly (16) from exhaust brake assembly (10). Discard gasket (11).
34. Remove clamp (14), gasket (23), and adapter assembly (13) from turbocharger (15). Discard gasket (23).
35. Remove two screws (35), four washers (36), cylinder support assembly (25), and exhaust brake assembly (10) from surge tank mounting bracket (2).
36. Remove nut (29), clamp (28), adapter assembly (13), and gasket (30) from exhaust brake assembly (10). Discard gasket (30).
37. Remove two screws (26), washers (27), and cylinder support assembly (25) from exhaust brake assembly (10).
38. Remove elbow (31) from outlet port (32) of quick-release valve (33).
39. Remove elbow (34) from inlet port (24) of quick-release valve (33).
40. Remove tube (8) and elbow (9) from exhaust brake assembly (10).
41. Disconnect rear exhaust pipe from muffler (para. 3-49).

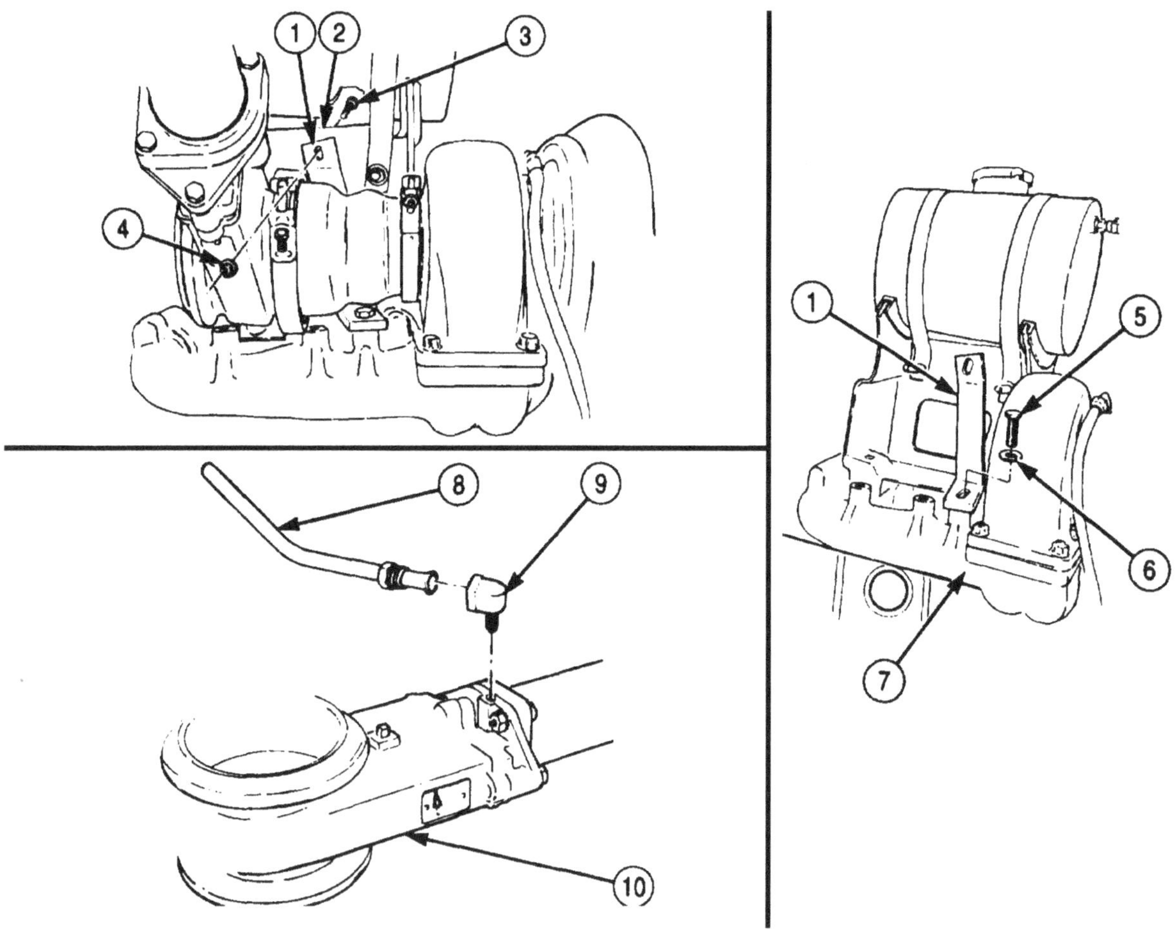

3-448. ENGINE EXHAUST BRAKE MODIFICATION KIT (M939A2) REPLACEMENT (Contd)

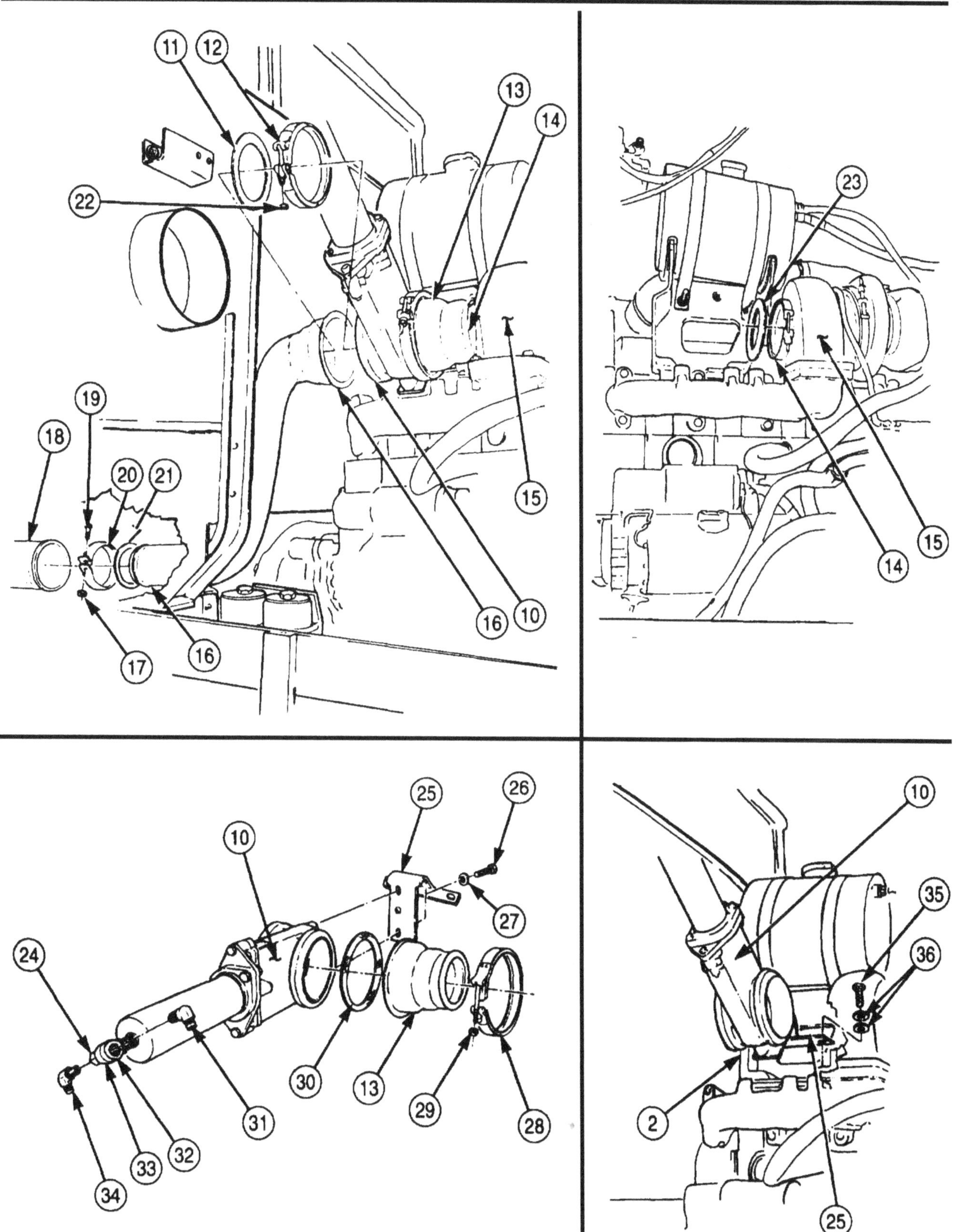

3-448. ENGINE EXHAUST BRAKE MODIFICATION KIT (M939A2) REPLACEMENT (Contd)

42. Disconnect right front brake vent hose (13) from connector (9).
43. Remove hoses (14) and (5) and elbows (12) and (6) from tee (11).
44. Remove tee (11), bushing (10), connector (9), and tee (8) from brake hose coupling (7)

b. Installation

NOTE

● Apply sealant to male threads of fittings and leave first two starter threads clear of sealant.

*It may be necessary to make adjustments to fittings for alignment purposes.

● Do not tighten exhaust components until exhaust system is completely assembled.

● Ensure antiseize tape is applied to any screws attached to engine assembly.

1. Install tee (8) in brake hose coupling (7).
2. Install connector (9) in tee (8).
3. Install bushing (10) and tee (11) in tee (8).
4. Install elbows (6) and (12) into tee (11).
5. Install hoses (5) and (14) in elbows (6) and (12).
6. Connect right front brake vent line (13) to connector (9).
7. Install surge tank support (1) on exhaust manifold (4) with new lockwasher (3) and screw (2). Do not tighten bolt (2).

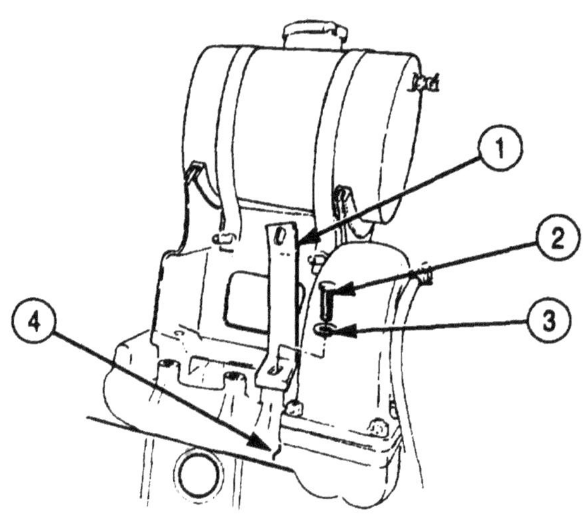

3-448. ENGINE EXHAUST BRAKE MODIFICATION KIT (M939A2) REPLACEMENT (Contd)

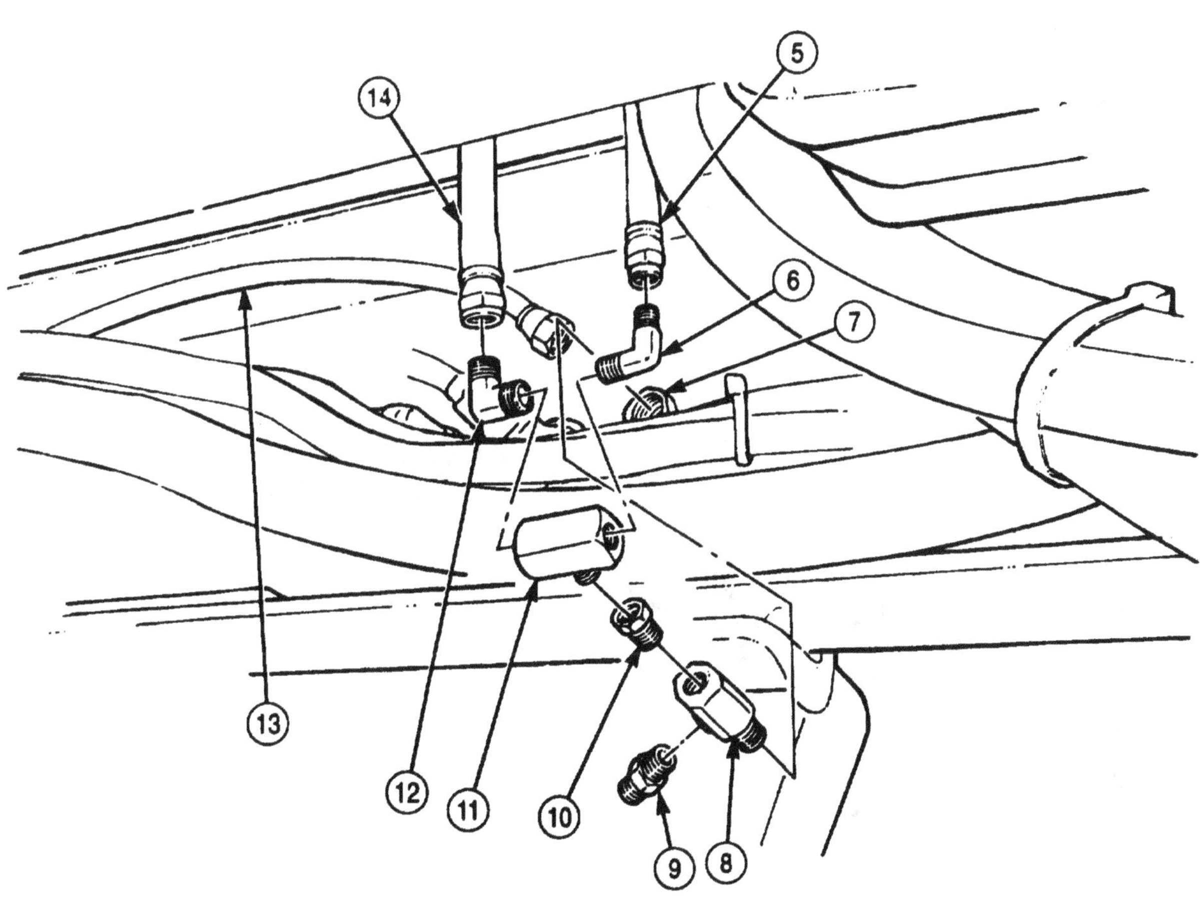

3-448. ENGINE EXHAUST BRAKE MODIFICATION KIT (M939A2) REPLACEMENT (Contd)

8. Install elbow (15) and tube (14) on exhaust brake assembly (13).
9. Install elbow (10) in inlet port (11) of quick-release valve (12).
10. Install elbow (8) in outlet port (9) of quick-release valve (12).
11. Install cylinder support assembly (1) on exhaust brake assembly (13) with two washers (2) and screw (3).
12. Install new gasket (7) and adapter assembly (6) on exhaust brake assembly (13) with clamp (4) and nut (5).
13. Install exhaust brake assembly (13) and cylinder support assembly (1) on surge tank mounting bracket (21) with four washers (20) and two screws (19).
14. Install new gasket (16) and adapter assembly (6) on turbocharger (18) with clamp (17).
15. Install new gasket (22) and pipe assembly (24) on exhaust brake assembly (13) with clamp (23) and nut (30).
16. Install new gasket (29) and rear exhaust pipe (26) on pipe assembly (24) with coupling (28), screw (27), and nut (25).
17. Connect rear exhaust pipe to muffler (para. 3-49).

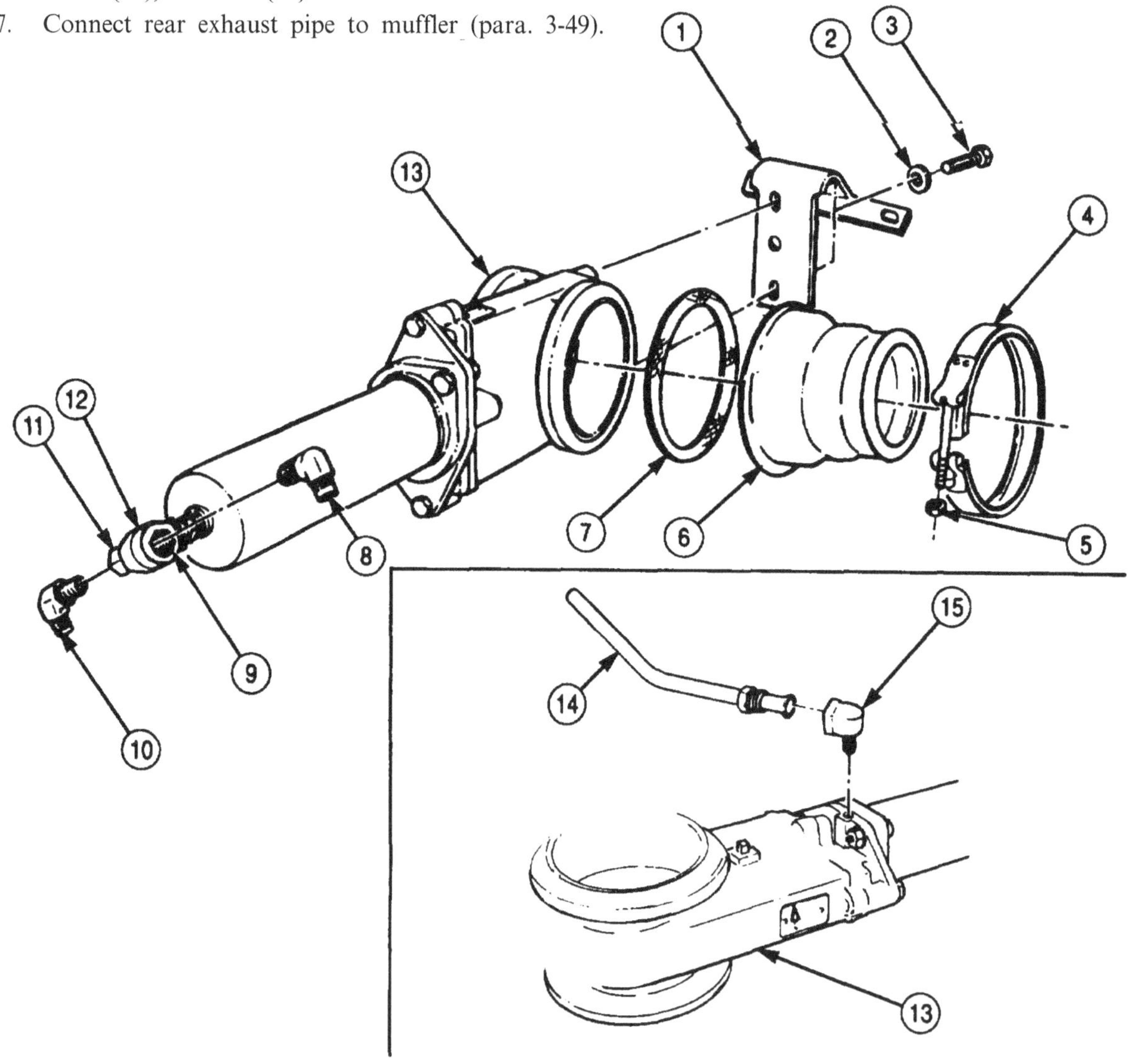

3-448. ENGINE EXHAUST BRAKE MODIFICATION KIT (M939A2) REPLACEMENT (Contd)

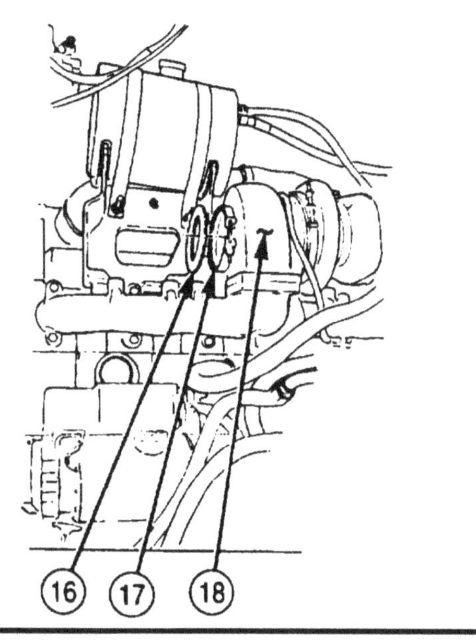

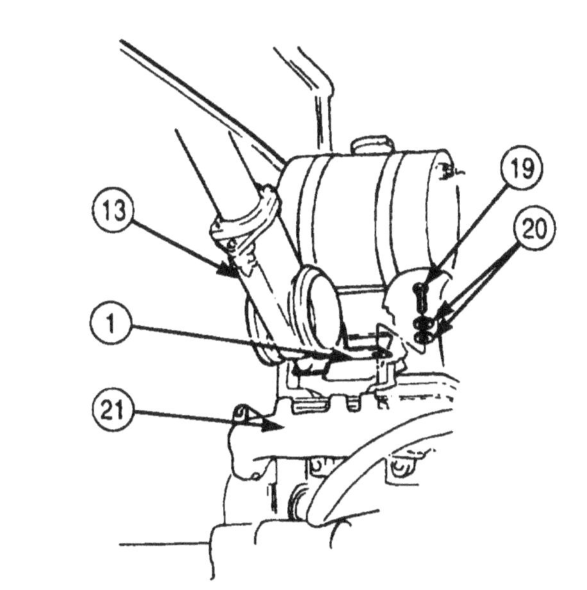

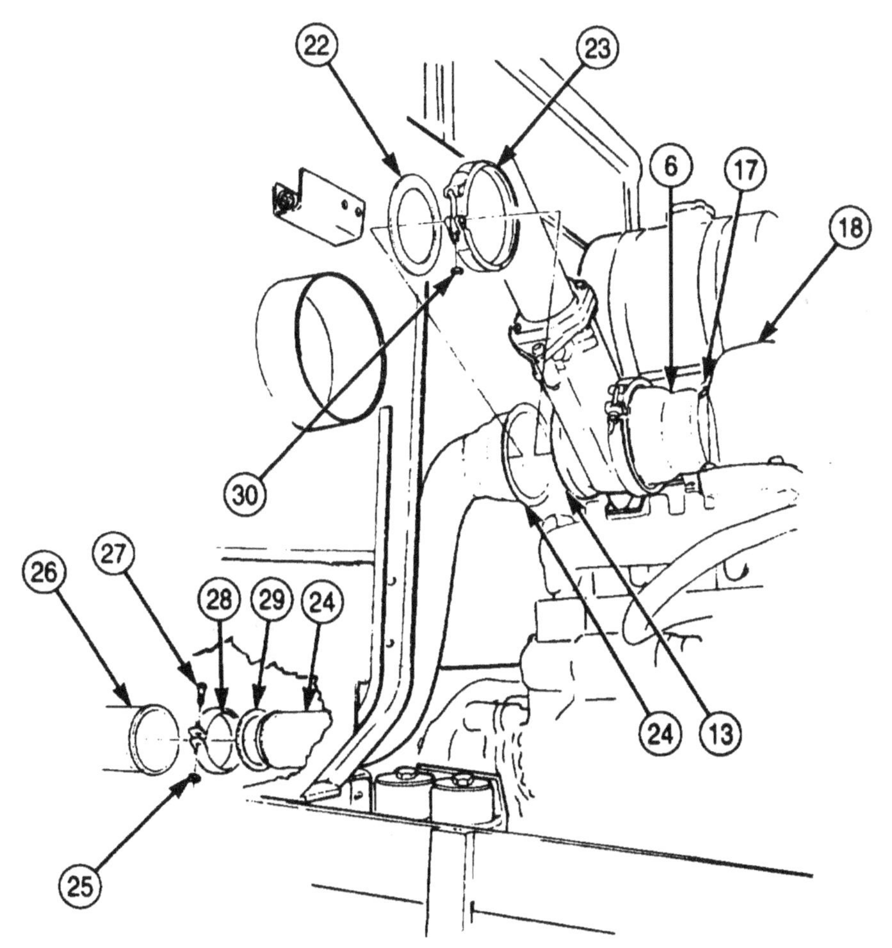

3-448. ENGINE EXHAUST BRAKE MOMDICATION KIT (M939A2) REPLACEMENT (Contd)

18. Install surge tank support (1) on surge tank mounting bracket (2) with screw (3) and nut (4).
19. Tighten all exhaust clamps and screws.
20. Position clamp (33) over hose assembly (31) and slide hose assembly (31) over tube (34) on exhaust brake assembly (35) and tighten clamp (33).
21. Install air horn supply tube (30) on governor assembly (32).
22. Install support bracket (14) on horn support bracket (26) with screw (13), new lockwasher (24), and nut (25). Do not tighten nut (25).
23. Install tee (42), elbow (37), and connector (36) in valve assembly (20).
24. Install elbows (39) and (41) in valve assembly (20).
25. Install bracket (21) on valve assembly (20) with two new lockwashers (39) and screws (40).
26. Install valve assembly (20) and bracket (21) on support bracket (14) and horn support bracket (26) with screw (19), new lockwasher (11), and nut (10). Do not tighten nut (10).
27. Align harness ground wire (15) and valve assembly ground wire (18) with remaining hole on valve assembly (20) and align bracket (21) with support assembly (14).
28. Install harness ground wire (15) and valve assembly ground wire (18) on support bracket (14) with screw (17), new lockwashers (16) and (22), and nut (23).
29. Tighten nuts (25), (10), and (23).
30. Install tee (6) in inlet port (8) of horn assembly (9).
31. Install elbows (5) and (7) in tee (6).
32. Install horn assembly (9) on support bracket (14) with two screws (12), new lockwashers (28), and nuts (27).
33. Install air horn supply tube in elbow (5) on horn assembly (9) and tighten nuts (29).
34. Install hose assembly (31) in connector (36) on valve assembly (20).

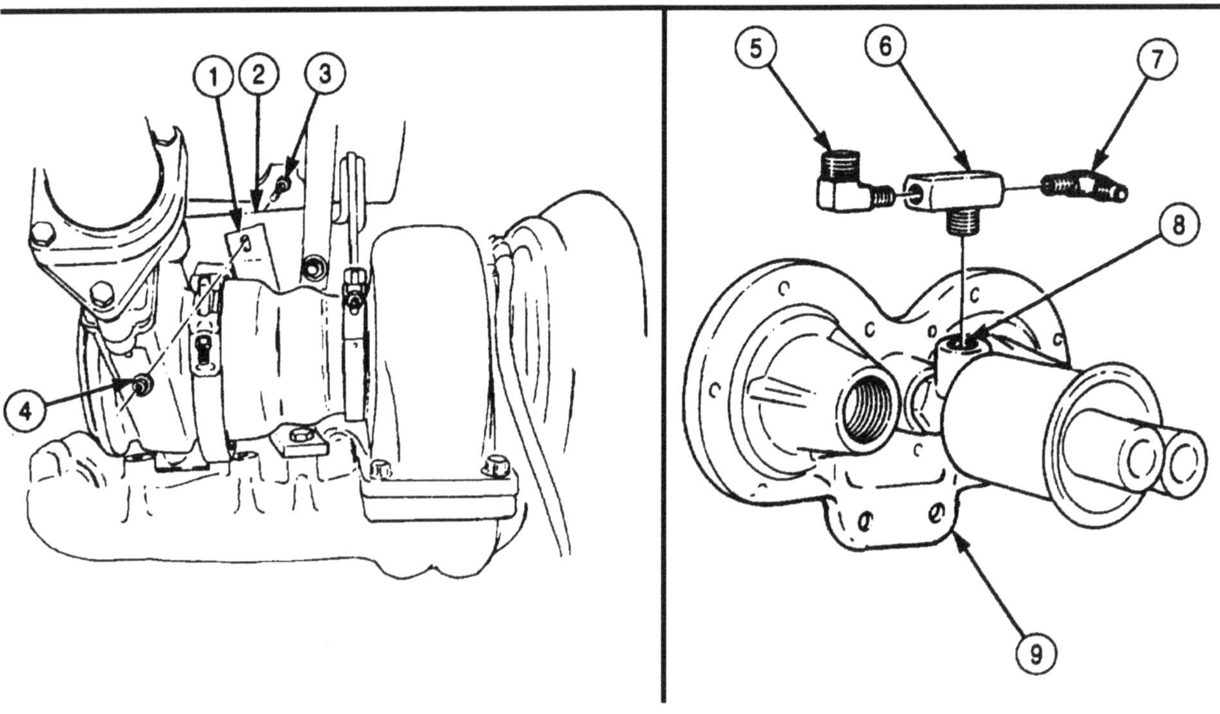

3-448. ENGINE EXHAUST BRAKE MODIFICATION KIT (M939A2) REPLACEMENT (Contd)

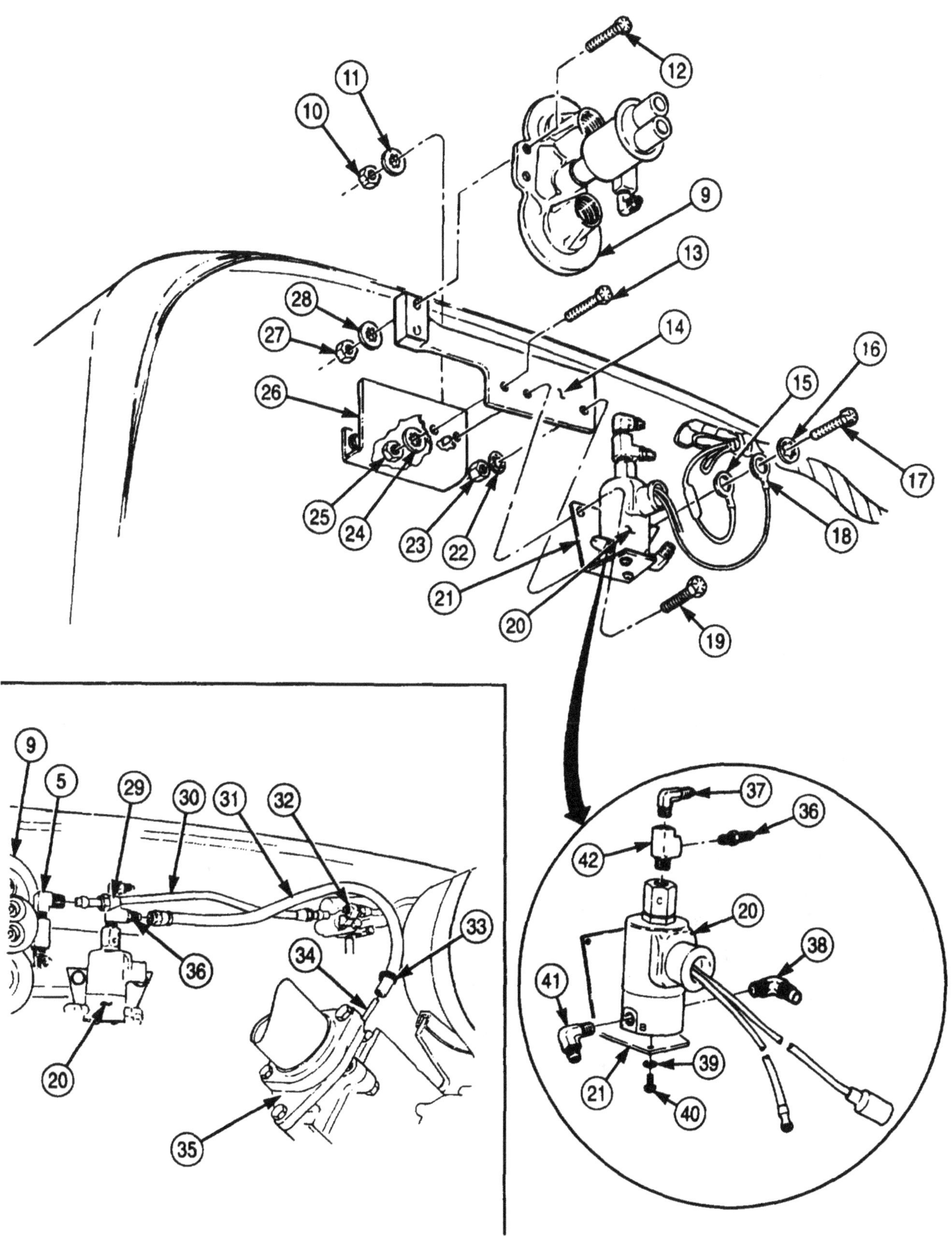

3-448. ENGINE EXHAUST BRAKE MODIFICATION KIT (M939A2) REPLACEMENT (Contd)

35. Install hose assembly (5) on elbow (3) on exhaust brake assembly (4).
36. Install hose assembly (1) on elbow (9) on valve assembly (8).
37. Install hose assembly (6) on elbow (2) on exhaust brake assembly (4) and elbow (7) on valve assembly (8).
38. Install hose assembly (10) on elbow (15) in valve assembly (8) and elbow (16) in horn assembly (17).
39. Connect cable (13) to valve assembly lead (14) and route cable (13) through firewall (11). Secure with new tiedown strap (12).
40. Disconnect accelerator linkage from throttle control lever (para. 3-43).
41. Install bracket assembly (19) and bracket (18) on engine assembly (27) with two new lock-washers (20) and screws (21).
42. Install throttle control lever (25) on control lever (26) with new lockwasher (23), ball joint (22), and screw (24).

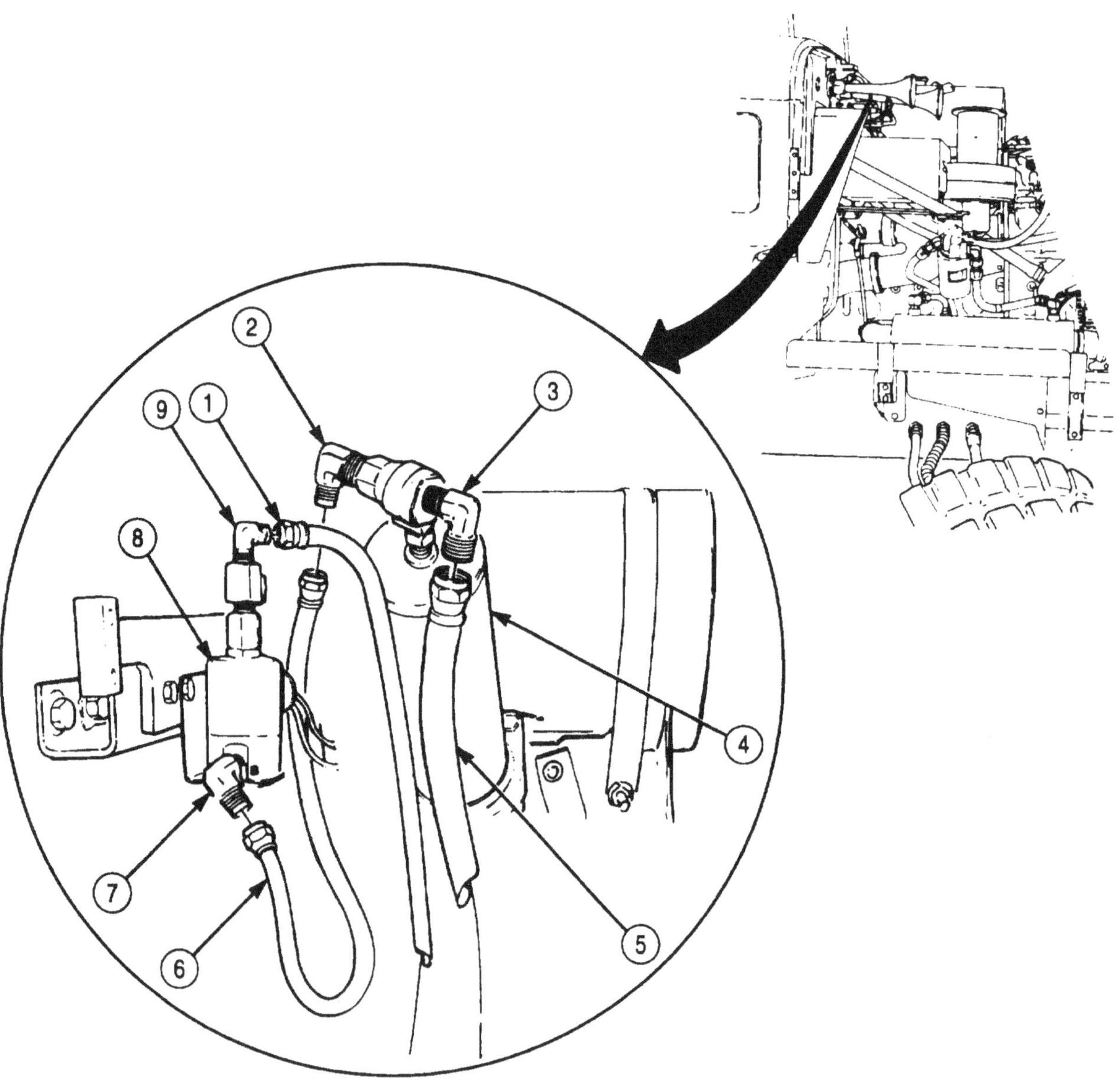

3-448. ENGINE EXHAUST BRAKE MODIFICATION KIT (M939A2) REPLACEMENT (Contd)

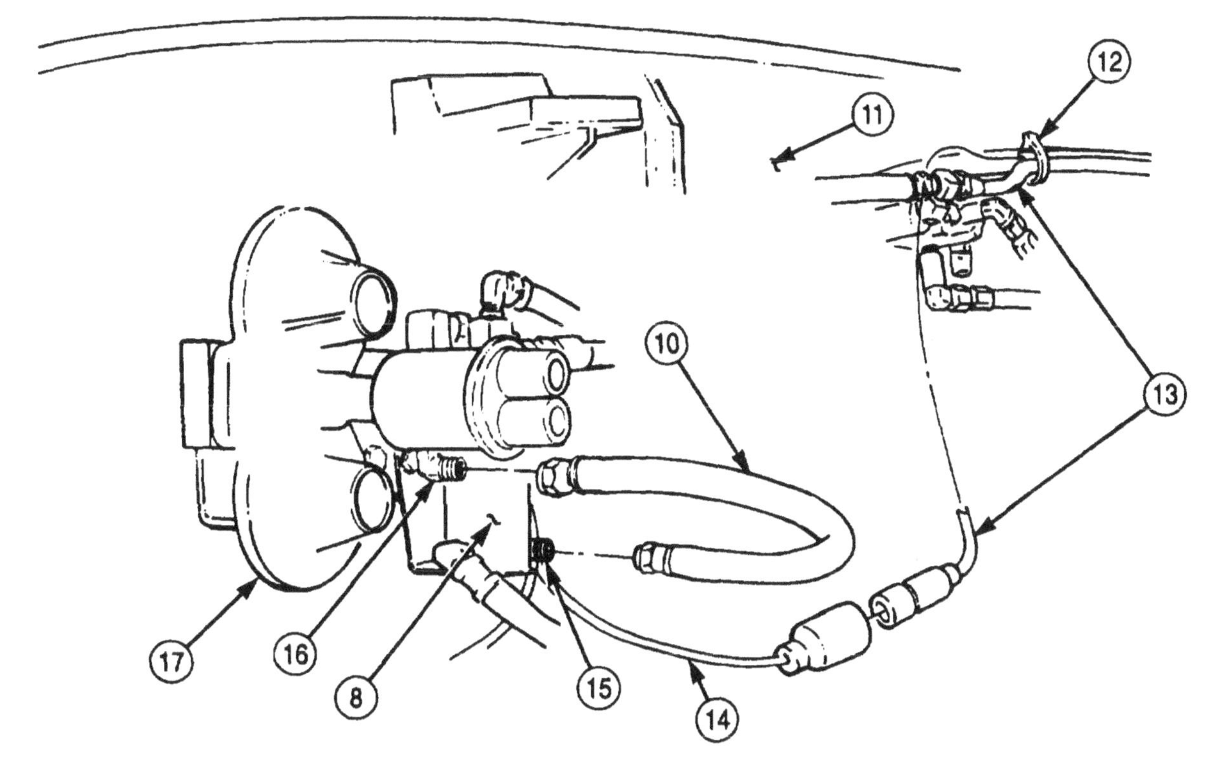

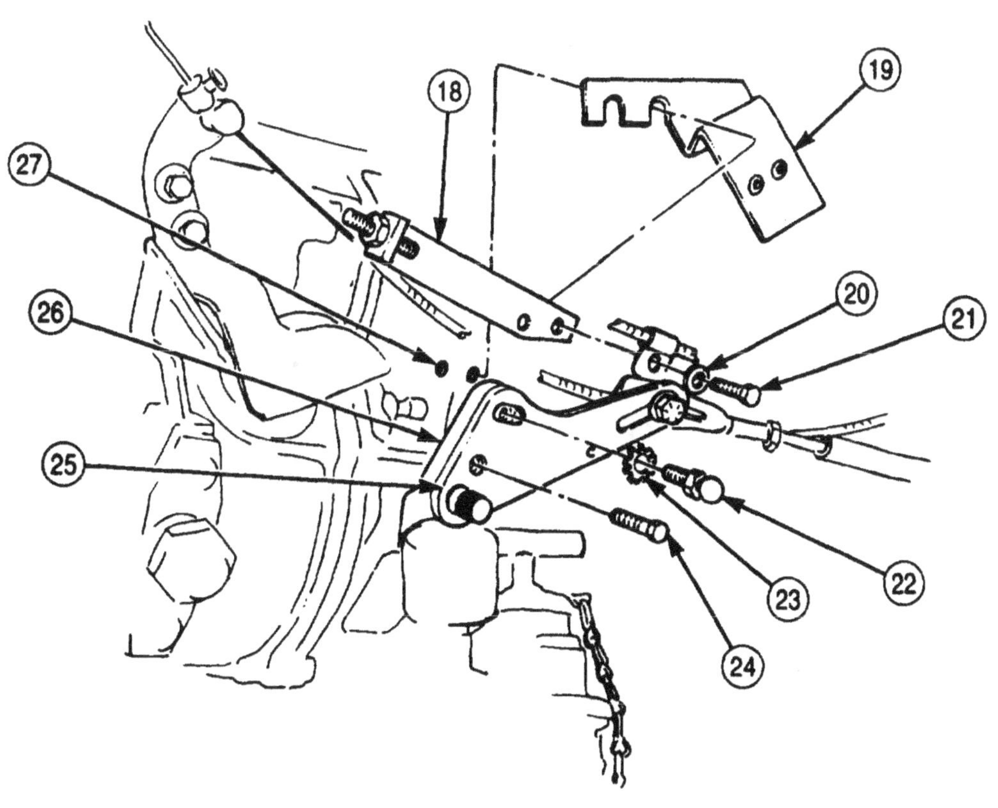

3-448. ENGINE EXHAUST BRAKE MODIFICATION KIT (M939A2) REPLACEMENT (Contd)

43. Install lever switch bracket (2) on bracket assembly (5) with two washers (3) and screws (4). Do not tighten screws (4).
44. Connect cable (7) to lead (6) of lever switch bracket (2).
45. Connect cable (8) to lead (10) of lever switch bracket (2) and route cable (8) through firewall (9).

NOTE

- Instrument cluster may have to be removed for switch installation (para. 3-83).
- Ensure that toggle switch is in OFF position.

46. Install toggle switch (15) and plate (14) on instrument cluster (11) with washer (13) and nut (12).
47. Disconnect lead (17) from circuit breaker (19).
48. Connect lead assembly (18) to circuit breaker (19).
49. Connect cable (8) to wiring harness (21) with new tiedown strap (20).
50. Connect lead (17) to lead assembly (18).
51. Connect toggle switch leads (16) to cable (8) and lead assembly (18).
52. Install instrument cluster if necessary (11) (para. 3-83).

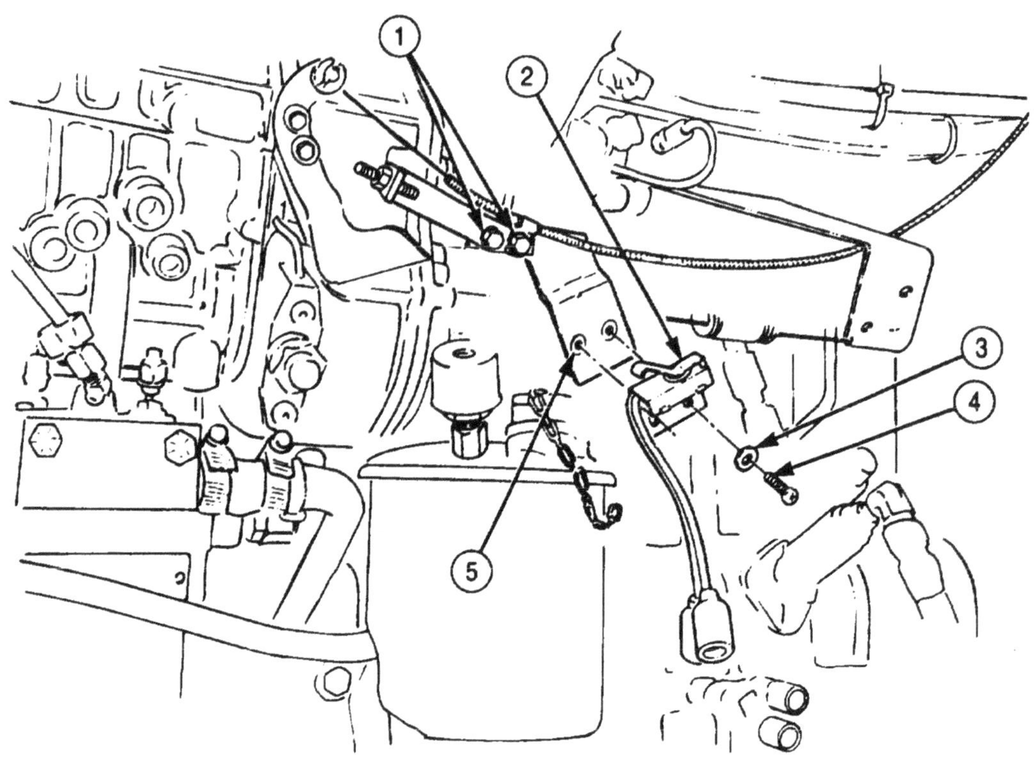

3-448. ENGINE EXHAUST BRAKE MODIFICATION KIT (M939A2) REPLACEMENT (Contd)

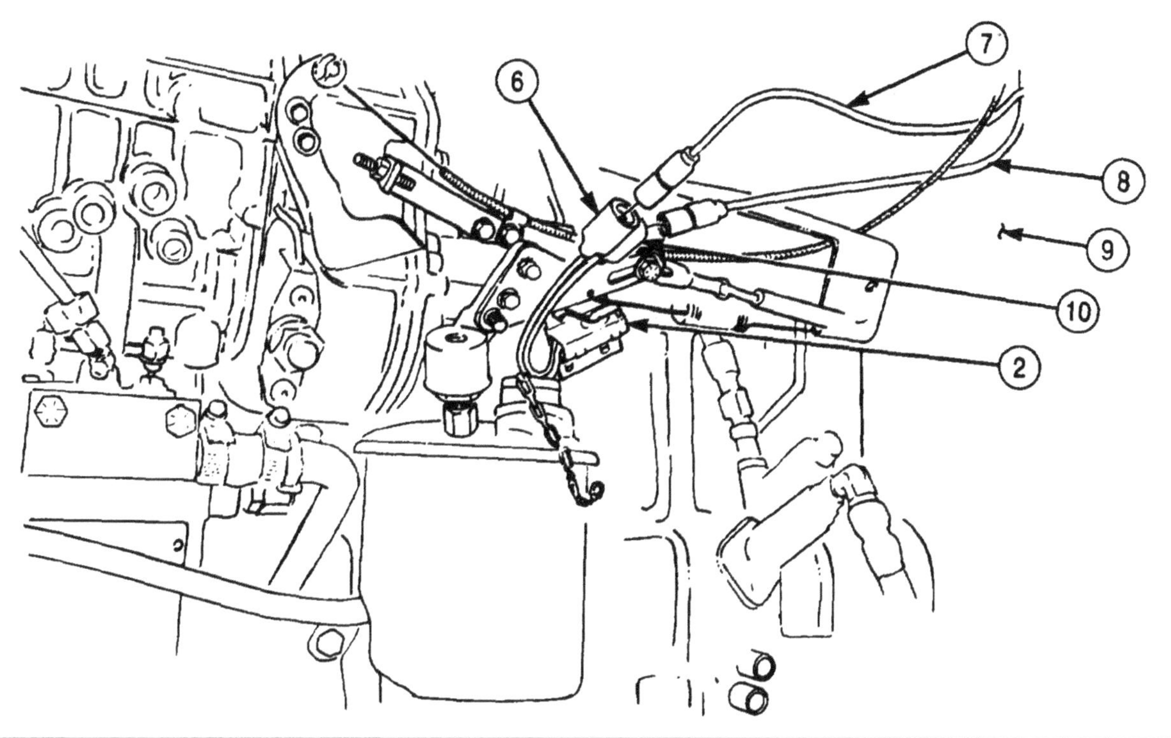

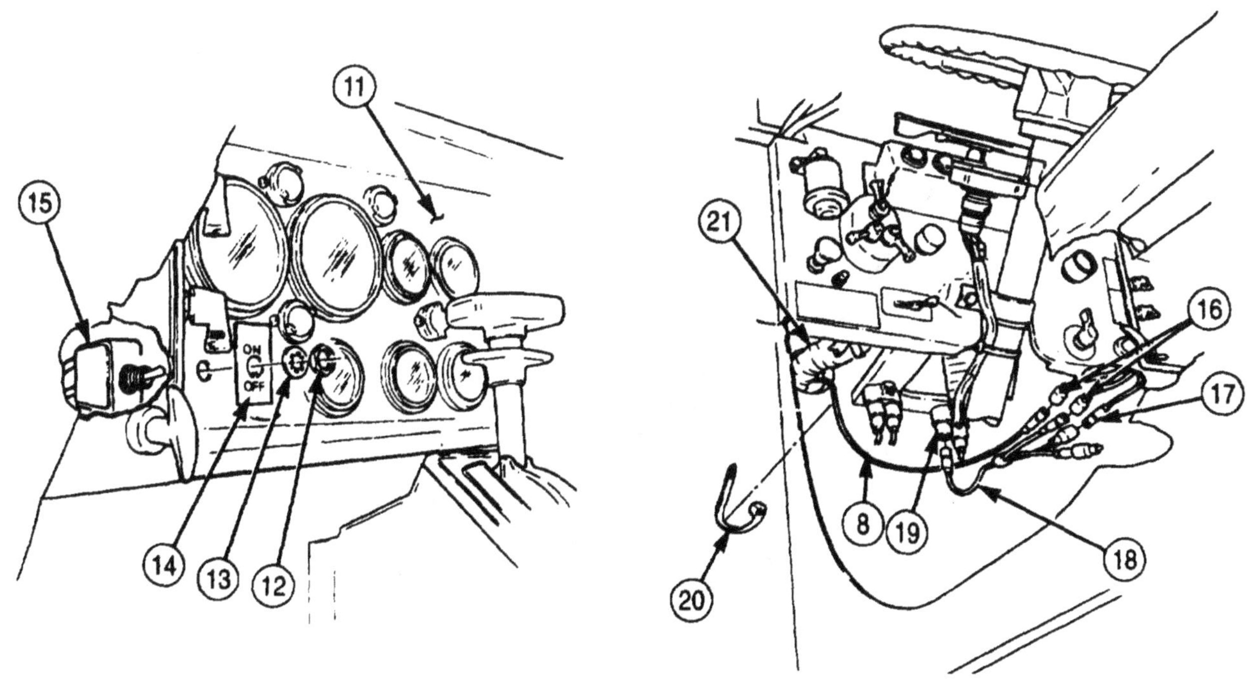

FOLLOW-ON TASKS:
- Install right hood bumper (para. 3-274).
- Install exhaust pipe heat shield if removed (para. 3-49).
- Install left and right engine splash shields (TM 9-2320-272-10).
- Connect battery ground cables (para. 3-126).

3-449. EXHAUST HEAT SHIELD ACCESSORY KIT REPLACEMENT

THIS TASK COVERS:

a. Removal

b. Installation

INITIAL SETUP:

APPLICABLE MODELS
M929/A1/A2, M930/A1/A2

TOOLS
General mechanic's tool kit (Appendix E, Item 1)

MATERIALS/PARTS
Two locknuts (Appendix D, Item 329)

REFERENCES (TM)
TM 9-2320-272-10
TM 9-2320-272-24P

EQUIPMENT CONDITION
Parking brake set (TM 9-2320-272-10).

a. Removal

1. Remove two locknuts (1) and shield (7) from U-bolt (4). Discard locknuts (1).
2. Remove two nuts (2), washers (3), clamp (6), and U-bolt (4) from exhaust pipe (5).

b. Installation

1. Install U-bolt (4) and clamp (6) on exhaust pipe (5) with two washers (3) and nuts (2).
2. Install shield (7) on U-bolt (4) with two new locknuts (1).

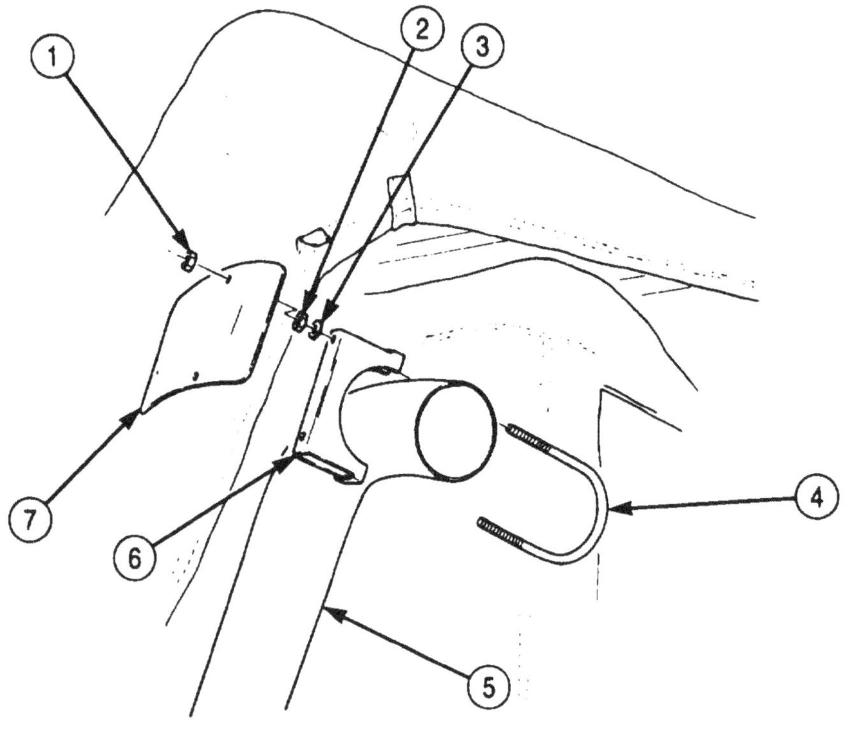

3-450. VAN HANDRAIL MODIFICATION KIT (M934/A1/A2) REPLACEMENT

THIS TASK COVERS:

a. Ladder Handrail Guide Removal
b. Ladder Handrail Guide Installation
c. Ladder Handrail Hangers Removal
d. Ladder Handrail Hangers Installation
e. Van Door Grab Handles Removal
f. Van Door Grab Handles Installation
g. Door Check Spacer Removal
h. Door Check Spacer Installation

INITIAL SETUP:

APPLICABLE MODELS
M934/A1/A2

TOOLS
General mechanic's tool kit (Appendix E, Item 1)

MATERIALS/PARTS
Four locknuts (Appendix D, Item 306)
Lockwasher (Appendix D, Item 400)

REFERENCES (TM)
TM 9-2320-272-10
TM 9-2320-272-24P

EQUIPMENT CONDITION
- Parking brake set (TM 9-2320-272-10).
- Ladders removed (TM 9-2320-272-10).

NOTE

All ladder handrail guides are replaced the same way.

a. Ladder Handrail Guide Removal

1. Remove four locknuts (6), washers (2), screws (3), washers (2), and handrail guide (5) from ladder (4). Discard locknuts (6).
2. Remove screw (1) and pin assembly (7) from ladder (4).

b. Ladder Handrail Guide Installation

1. Install handrail guide (5) on ladder (4) with four washers (2), screws (3), washers (2), and new locknuts (6).
2. Install pin assembly (7) on ladder (4) with screw (1).

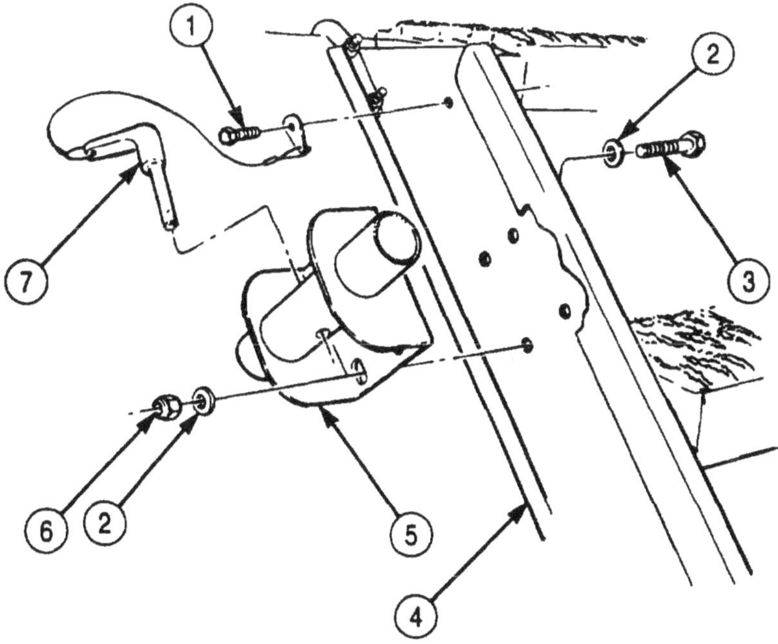

3-450. VAN HANDRAIL MODIFICATION KIT (M934/A1/A2) REPLACEMENT (Contd)

NOTE

Left and right van rear door handrail guides are replaced the same way. This procedure is for left van rear door handrail hangers.

c. Ladder Handrail Hangers Removal

1. Release two straps (2) and remove handrail (3) from upper (1) and lower (4) support brackets.
2. Remove four screws (6) and upper support bracket (1) from van rear door (5).
3. Remove four screws (7) and lower support bracket (4) from van rear door (5).
4. Remove eight screws (8), four footman loops (9), and two straps (2) from van rear door (5).

d. Ladder Handrail Hangers Installation

1. Install four footman loops (9) and two straps (2) on van rear door (5) with eight screws (8).
2. Install lower bracket (4) on van rear door (5) with four screws (7).
3. Install upper bracket (1) on van rear door (5) with four screws (6).
4. Position handrail (3) on upper (1) and lower (4) support brackets and secure with two straps (2).

3-450. VAN HANDRAIL MODIFICATION KIT (M934/A1/A2) REPLACEMENT (Contd)

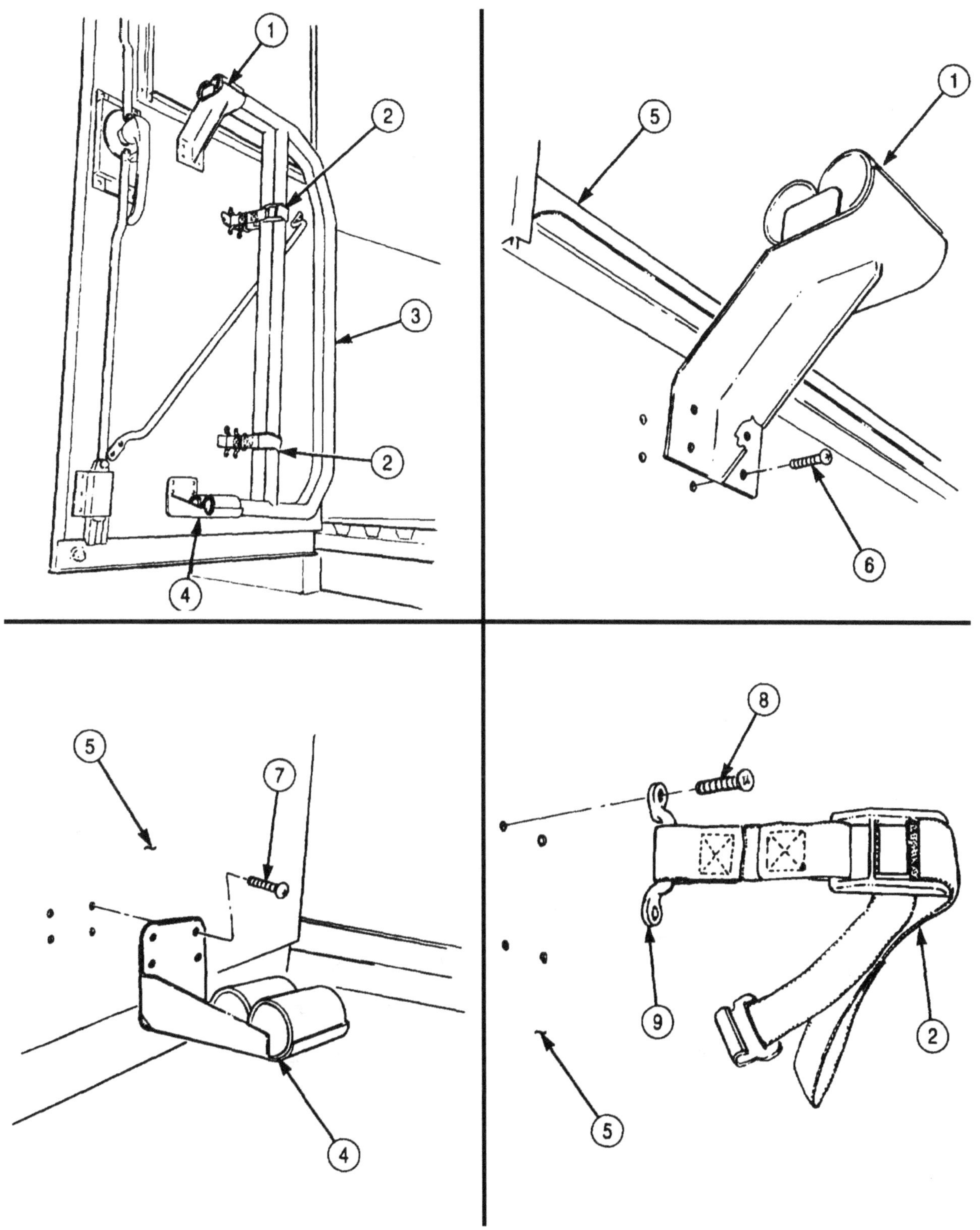

3-450. VAN HANDRAIL MODIFICATION KIT (M934/A1/A2) REPLACEMENT (Contd)

NOTE

Left and right van side door grab handles and swivel grab handles are replaced the same way. This procedure is for left van side door grab handle.

e. Van Door Grab Handles Removal

1. Remove four screws (2) and grab handle (3) from van rear door (1).
2. Remove ten screws (6), two swivel mounts (5), and swivel grab handle (4) from van side door (7).

f. Van Door Grab Handles Installation

1. Install grab handle (3) on van rear door (1) with four screws (2).
2. Install swivel mount (5) at bottom of van side door (7) with five screws (6).
3. Position swivel grab handle (4) in swivel mount (5).
4. Insert swivel mount (5) on swivel grab handle (4) and install on van side door (7) with five screws (6).

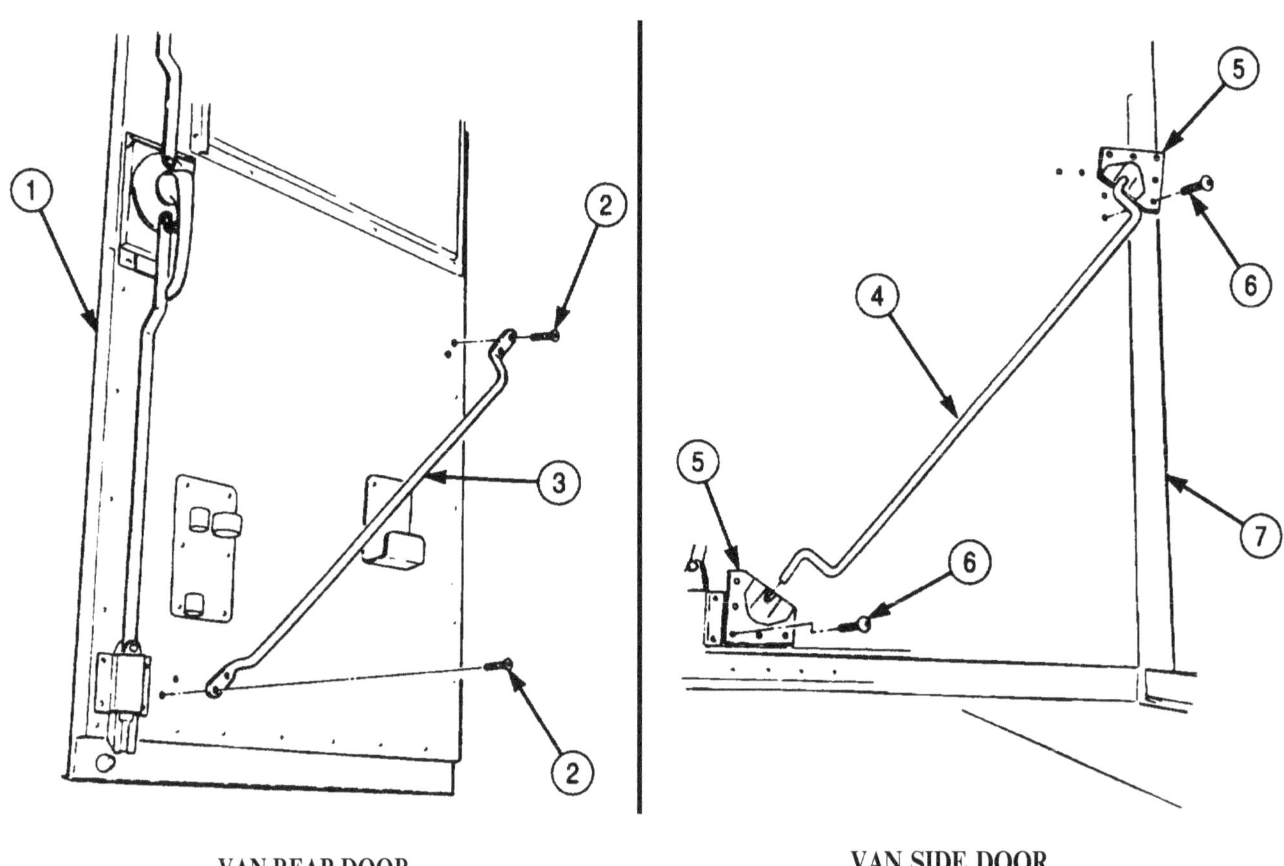

VAN REAR DOOR

VAN SIDE DOOR

3-450. VAN HANDRAIL MODIFICATION KIT (M934/A1/A2) REPLACEMENT (Contd)

g. Door Check Spacer Removal

1. Remove nut (6) from rod end (2) of door check (3).
2. Remove nut (7), lockwasher (8), and arm bracket (5) from screw (1). Discard lockwasher (8).
3. Remove arm bracket (5) and spacer (4) from rod end (2) of door check (3).

h. Door Check Spacer Installation

1. Install arm bracket (5) on rod end (2) with nut (6). Do not tighten nut (6).
2. Install spacer (4) and arm bracket (5) on rod end (2).
3. Install arm bracket (5) on screw (1) with new lockwasher (8) and nut (7).
4. Tighten nut (6).

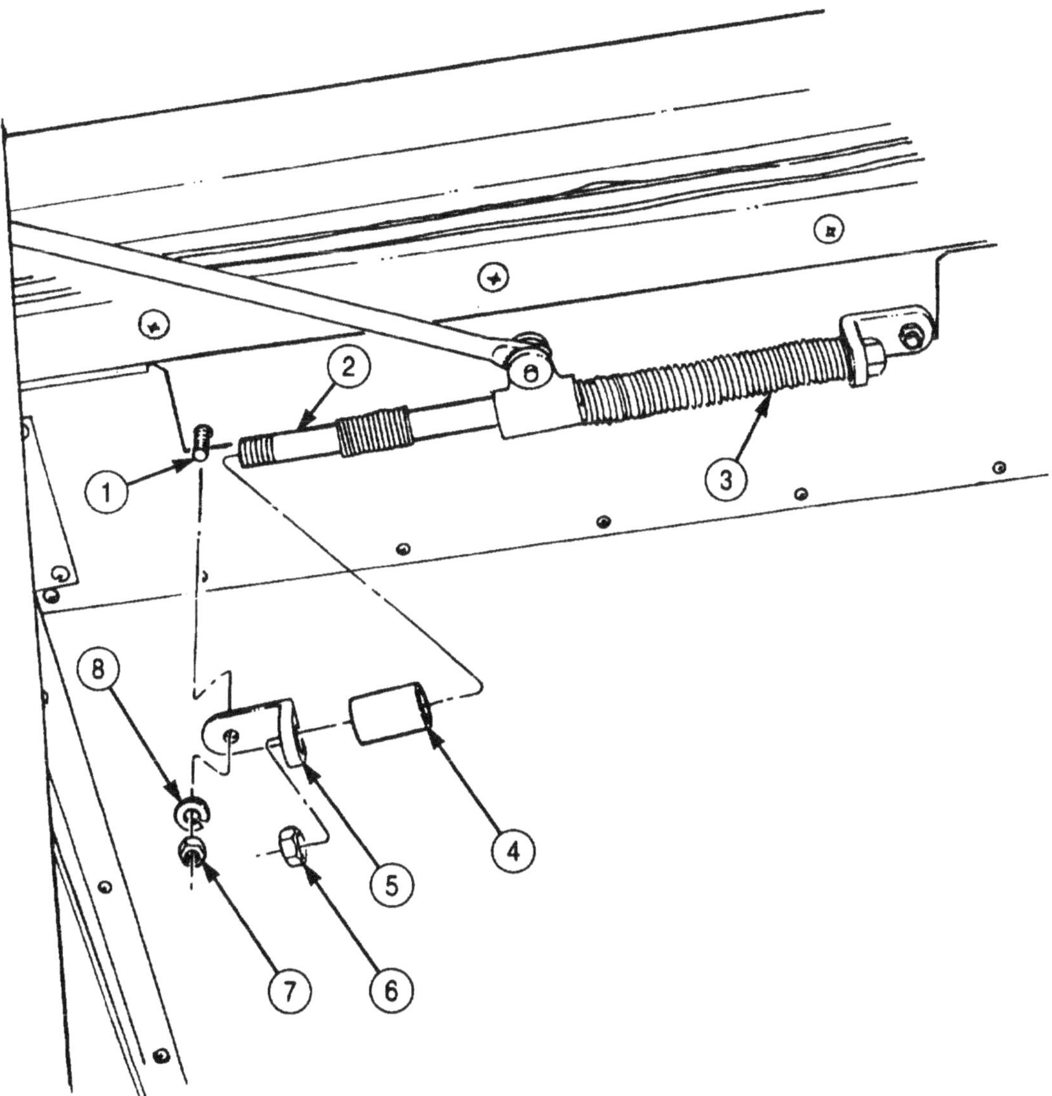

FOLLOW-ON TASK: Install ladders (TM 9-2320-272-10).

Section XVI. CENTRAL TIRE INFLATION SYSTEM (CTIS) MAINTENANCE

3-451. CENTRAL TIRE INFLATION SYSTEM (CTIS) MAINTENANCE INDEX

PARA. NO.	TITLE	PAGE NO.
3-452.	Fabrication of Air Lines	3-1232
3-453.	Pressure Transducer Replacement	3-1232
3-454.	Pneumatic Controller and Relief Valve Maintenance	3-1236
3-455.	Relief Safety Valve Maintenance	3-1240
3-456.	Front Wheel Valve Maintenance	3-1244
3-457.	Rear Wheel Valve Maintenance	3-1246
3-458.	Wheel Valve Filter Replacement	3-1248
3-459.	Wheel Valve Maintenance	3-1250
3-460.	Front Hubs Repair	3-1252
3-461.	Rear Hubs Repair	3-1258
3-462.	Hub Air Seal Leak Test	3-1264
3-463.	Rear Axle Air Manifold Maintenance	3-1268
3-464.	Air Dryer and Check Valve Maintenance	3-1270
3-465.	Air Dryer Filter Replacement	3-1276
3-466.	Water Separator Maintenance	3-1278
3-467.	Pressure Switch Replacement	3-1280
3-468.	Electronic Control Unit (ECU) Replacement	3-1282
3-469.	Amber Warning Light Replacement	3-1283
3-470.	CTIS Wiring Harness Replacement	3-1284
3-471.	Speed Signal Generator Replacement	3-1288

3-452. FABRICATION OF AIR LINES

For fabrication of air lines, refer to TM 9-243.

3-453. PRESSURE TRANSDUCER REPLACEMENT

THIS TASK COVERS:

a. Removal b. Installation

INITIAL SETUP:

APPLICABLE MODELS
M939A2

TOOLS
General mechanic's tool kit (Appendix E, Item 1)

MATERIALS/PARTS
Three lockwashers (Appendix D, Item 345)
Two cotter pins (Appendix D, Item 83)

REFERENCES (TM)
TM 9-2320-272-10
TM 9-2320-272-24P

EQUIPMENT CONDITION
- Parking brake set (TM 9-2320-272-10).
- Air reservoirs drained (TM 9-2320-272-10).

a. Removal

NOTE

Perform steps 1 through 7 for vehicles equipped with PTO.

1. Remove six screws (5) and access cover (6) from shift panel (2).
2. Remove two nuts (10), screws (14), clamp (13), PTO control cable (9), and spacer (12) from bracket (11).
3. Remove cotter pin (4), washer (3), clevis pin (8), and clevis (7) from control lever (1). Discard cotter pin (4).
4. Remove two nuts (19), screws (24), clamp (23), front winch control cable (22), and spacer (21) from bracket (20).
5. Remove cotter pin (15), washer (16), clevis pin (25), and clevis (17) from shift lever (18). Discard cotter pin (15).
6. Remove three screws (27), lockwashers (26), and PTO control panel assembly (2) from mounting plate (30) and cab floor (31). Discard lockwashers (26).
7. Install three screws (27) on mounting plate (30) and cab floor (31) to hold pneumatic controller (32) in place.

NOTE

Perform steps 8 and 9 for vehicles without PTO.

8. Remove three screws (34), lockwashers (33), and cover (35) from mounting plate (30) and cab floor (31).
9. Install three screws (34) on mounting plate (30) and cab floor (31) to hold pneumatic controller (32) in place.
10. Disconnect wire (28) from pressure transducer (29).
11. Remove pressure transducer (29) from pneumatic controller (32).

3-453. PRESSURE TRANSDUCER REPLACEMENT (Contd)

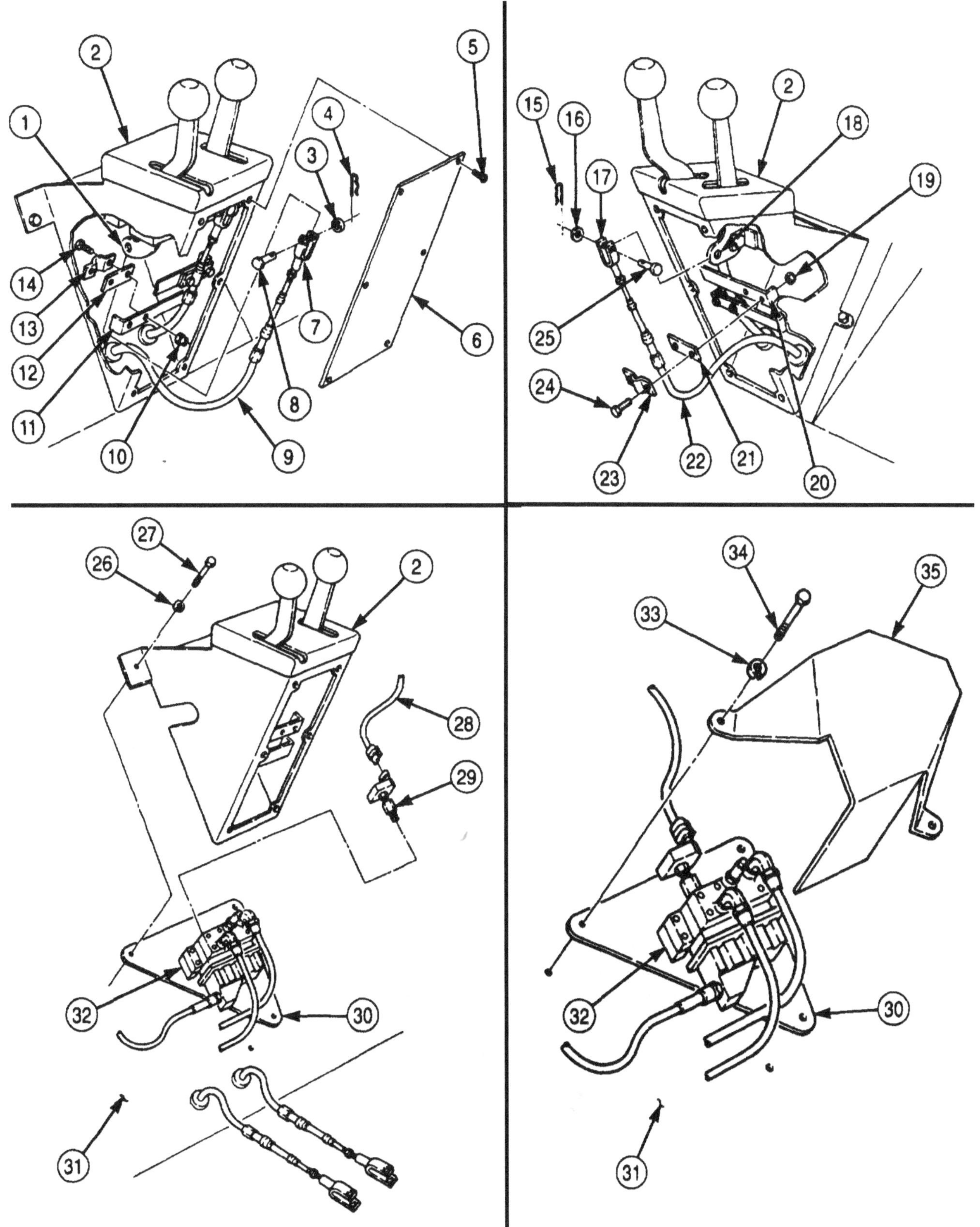

3-453. PRESSURE TRANSDUCER REPLACEMENT (Contd)

b. Installation

1. Install pressure transducer (11) on pneumatic controller (5) and connect wire (10) to pressure transducer (11).

NOTE

Perform steps 2 and 3 for vehicles without PTO.

2. Remove three screws (1) from mounting plate (3) and cab floor (4).
3. Install cover (2) on mounting plate (3) and cab floor (4) with three new lockwashers (6) and screws (1).

NOTE

Perform steps 4 through 10 for vehicles equipped with PTO.

4. Remove three screws (8) from mounting plate (3) and cab floor (4).
5. Install PTO control panel assembly (9) on mounting plate (3) and cab floor (4) with three new lockwashers (7) and screws (8).
6. Install clevis (21) on shift lever (14) with clevis pin (13), washer (22), and new cotter pin (12).
7. Install front winch control cable (18) on bracket (16) with spacer (17), clamp (20), two screws (19), and nuts (15).
8. Install clevis (28) on shift lever (23) with clevis pin (29), washer (24), and new cotter pin (25).
9. Install PTO control cable (30) on bracket (32) with spacer (33), clamp (34), two screws (35), and nuts (31).
10. Install access cover (27) on shift panel (9) with six screws (26).

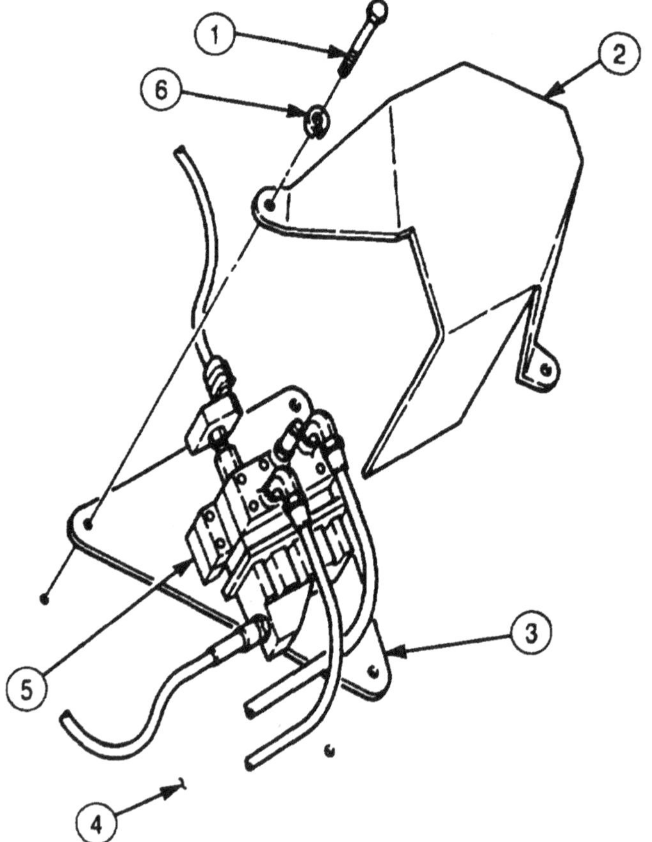

3-453. PRESSURE TRANSDUCER REPLACEMENT (Contd)

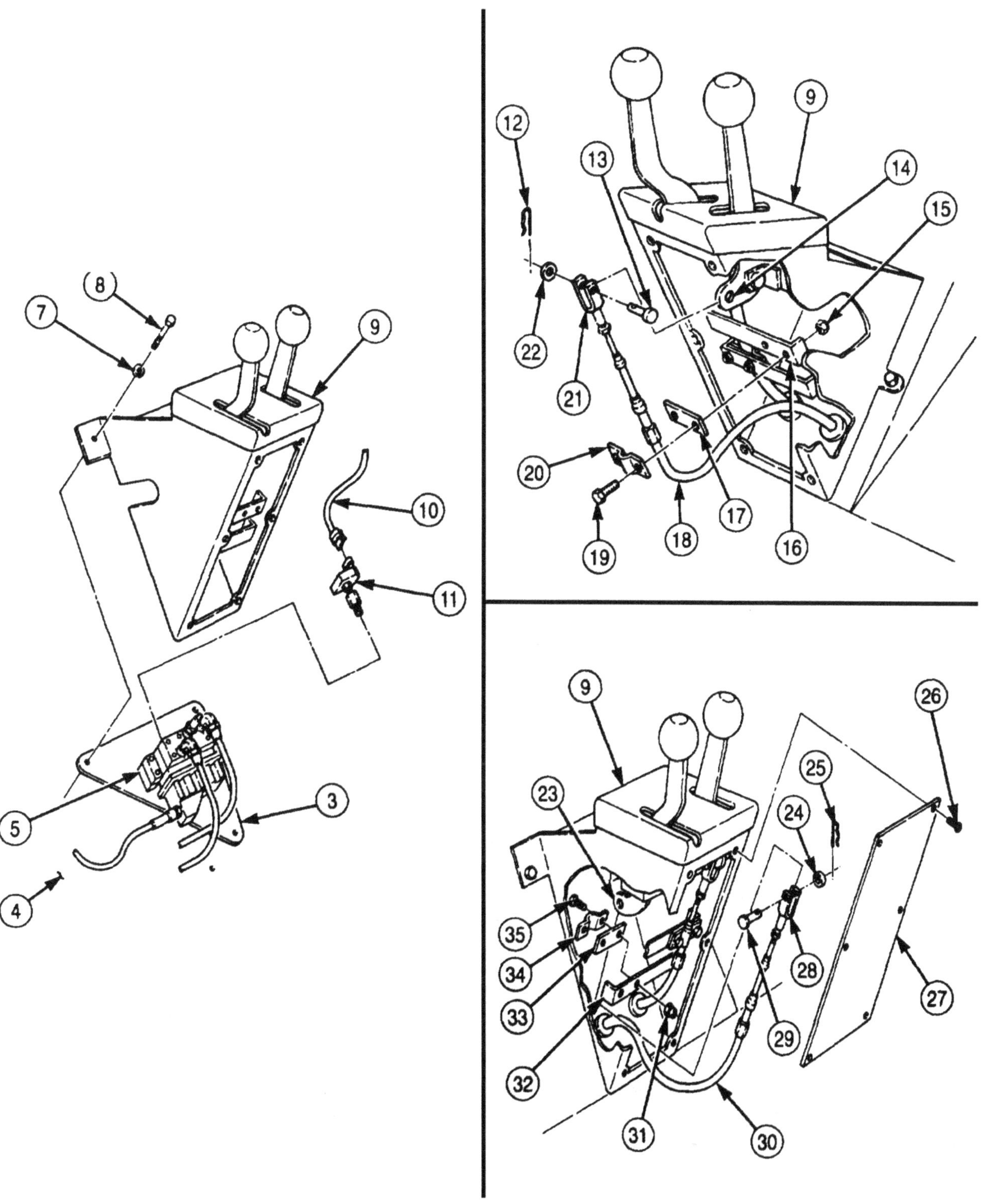

FOLLOW-ON TASK: Start engine (TM 9-2320-272-10) and check for leaks and proper CTIS operation.

3-454. PNEUMATIC CONTROLLER AND RELIEF VALVE MAINTENANCE

THIS TASK COVERS:

a. Removal
b. Disassembly
c. Cleaning and Inspection
d. Assembly
e. Installation

INITIAL SETUP:

APPLICABLE MODELS
M939A2

TOOLS
General mechanic's tool kit (Appendix E, Item 1)
Multimeter (Appendix E, Item 86)

MATERIALS/PARTS
Relief valve kit (Appendix D, Item 531)
Petrolatum (Appendix C, Item 53)
Antiseize tape (Appendix C, Item 72)

REFERENCES (TM)
TM 9-2320-272-10
TM 9-2320-272-24P

EQUIPMENT CONDITION
- Air reservoirs drained (TM 9-2320-272-10).
- Pressure transducer removed (para. 3-453).

GENERAL SAFETY INSTRUCTIONS
When cleaning with compressed air, wear eyeshields and ensure source pressure does not exceed 30 psi (207 kPa).

a. Removal

1. Disconnect connector (11) from solenoid receptacle (10).
2. Disconnect air lines (5) and (7) from elbows (4) and (8).
3. Remove relief valve (6) and elbows (4) and (8) from pneumatic controller (9).
4. Remove four screws (3), pneumatic controller (9), and two spacers (2) from mounting plate (13).
5. Remove three screws (1) and mounting plate (13) from cab floor (12).

b. Disassemble

1. Remove nut (21) from solenoid receptacle (10), and remove six screws (20), solenoid protector (19), and gasket (17) from cover plate (15). Discard solenoid protector (19) and gasket (17).

NOTE

Mark wiring harnesses to correspond to C, D, and S marks on base plate.

2. Remove three nuts (18), wiring harness (22), three gaskets (23), and seats (24) from studs (25) on cover plate (15).
3. Remove six screws (27) and base plate (26) from valve body (14).
4. Remove four screws (16) and cover plate (15) from valve body (14).

NOTE

Mark location and position of valve cartridges for installation.

5. Using access through orifice passages (32), carefully pry out valve cartridge (28) and two valve cartridges (29) from valve body ports (31). Discard valve cartridges (28) and (29).
6. Remove three O-rings (30) and five O-rings (33) from valve body (14). Discard O-rings (30) and (33).

3-454. PNEUMATIC CONTROLLER AND RELIEF VALVE MAINTENANCE (Contd)

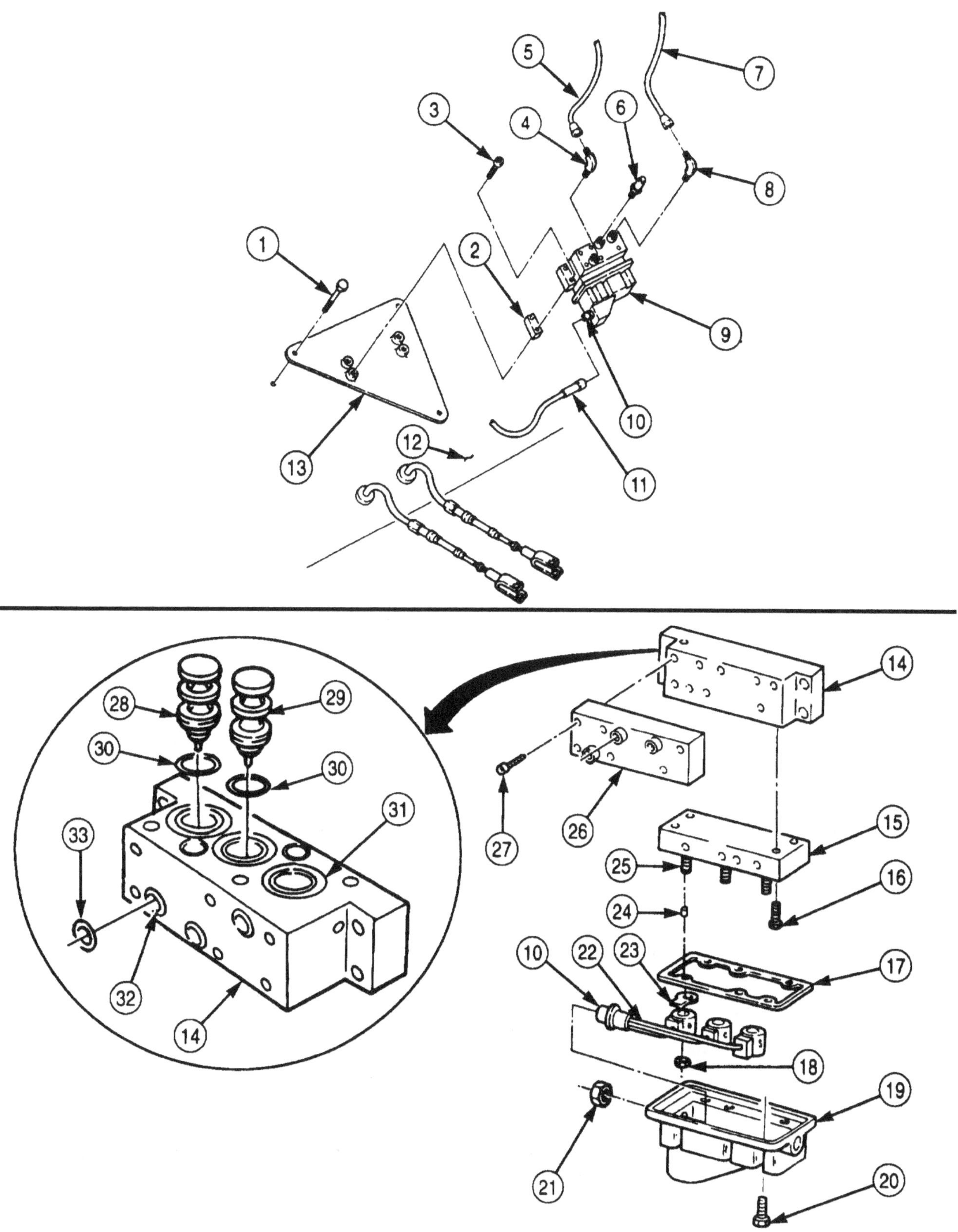

3-454. PNEUMATIC CONTROLLER AND RELIEF VALVE MAINTENANCE (Contd)

c. Cleaning and Inspection

1. For general cleaning instructions, refer to para. 2-14.
2. For general inspection instructions, refer to para. 2-15.

WARNING

Eyeshields must be worn when cleaning with compressed air. Compressed air source will not exceed 30 psi (207 kPa). Failure to do so may result in injury to personnel.

3. Dry all parts and clear all passages with compressed air.
4. Inspect base plate (15), cover plate (3), and valve body (2) for cracks and stripped threads. Replace if damaged.
5. Inspect relief valve (27) for bends, cracks, and stripped threads. Replace relief valve (27) if bent, cracked, or threads are stripped.
6. Inspect wiring harness (6) for broken wires or cracked insulation. Replace wiring harness (6) if wires are broken or insulation is cracked.

NOTE

Perform step 7 if condition of wiring harness as inspected in step 6 is satisfactory.

7. Using table 3-2, Electric Connections, check wiring harness (6) for continuity as follows:
   a. Check for continuity between pin A and pin E.
   b. Check for continuity between pin A and pin F.
   c. Check for continuity between pin B and pin D. Replace wiring harness (6) if continuity check a, b, or c fails.

Table 3-2. Electrical Connections.

ELECTRICAL CONNECTIONS		
PIN	WIRE	LABEL
A	Common wire from inflate and deflate valve	—
B	Common wire from exhaust valve	—
D	Hot wire from exhaust valve	C-CONTROL
E	Hot wire from exhaust valve	D-DEFLATE
F	Hot wire from exhaust valve	S-SUPPLY

d. Assembly

1. Install three new O-rings (18) on valve body ports (19) and five new O-rings (20) on valve body ports (21).
2. Apply a thin coat of petrolatum to two new valve cartridges (17) and new valve cartridge (16), and install valve cartridges (16) and (17) on valve body (2).
3. Install cover plate (3) on valve body (2) with four screws (4).
4. Install base plate (15) on valve body (2) with six screws (1).

NOTE

Ensure marked wiring harness ends correspond to C, D, and S marks on base plate.

5. Install three new seats (13), new gaskets (12), and wiring harness (6) on cover plate studs (14) with three nuts (7).

3-454. PNEUMATIC CONTROLLER AND RELIEF VALVE MAINTENANCE (Contd)

6. Install new gasket (5) on new solenoid protector (9).
7. Route solenoid receptacle (11) through hole in solenoid protector (9), and install receptacle (11) on solenoid protector (9) with nut (10).
8. Install solenoid protector (9) on cover plate (3) with six screws (8).

e. Installation

1. Install two spacers (23) and pneumatic controller (30) on mounting plate (33) with four screws (24).
2. Install mounting plate (33) with pneumatic controller (30) on cab floor (32) with three screws (22).

NOTE

Wrap all male threads with antiseize tape before installation.

3. Install elbows (25) and (29) and relief valve (27) on pneumatic controller (30).
4. Connect lines (26) and (28) to elbows (25) and (29).
5. Connect connector (31) to solenoid receptacle (11).

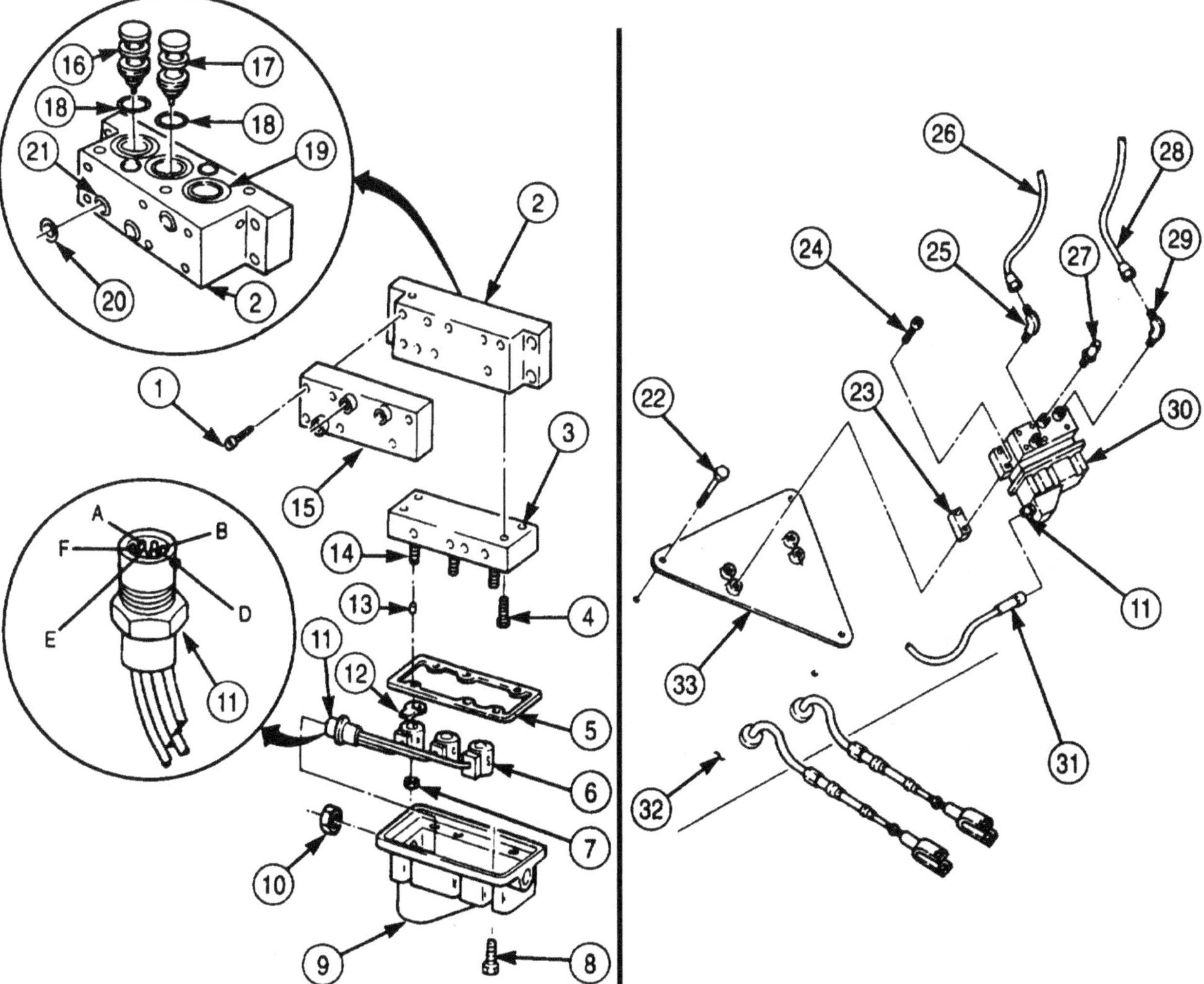

FOLLOW-ON TASKS:
- Install pressure transducer (para. 3-453).
- Start engine (TM 9-2320-272-10) and check for air leaks and proper CTIS operation.

3-455. RELIEF SAFETY VALVE MAINTENANCE

THIS TASK COVERS:

a. Removal
b. Disassembly
c. Cleaning and Inspection
d. Assembly
e. Installation

INITIAL SETUP:

APPLICABLE MODELS
M939A2

TOOLS
General mechanic's tool kit (Appendix E, Item 1)

MATERIALS/PARTS
O-ring (Appendix D, Item 488)
Two sleeves (Appendix D, Item 656)
Two sleeves (Appendix D, Item 655)
Cap and plug set (Appendix C, Item 14)
Antiseize tape (Appendix C, Item 72)

REFERENCES (TM)
TM 9-2320-272-10
TM 9-2320-272-24P

EQUIPMENT CONDITION
- Parking brake set (TM 9-2320-272-10).
- Air resevoirs drained (TM 9-2320-272-10).

GENERAL SAFETY INSTRUCTIONS
Eyeshields must be worn when releasing compressed air.

a. Removal

WARNING

Air pressure may create airborne debris. Use eye protection, or injury to personnel may result.

NOTE

- There are three relief safety valves, one for each axle. Relief safety valves are mounted the same and perform the same function. This procedure covers maintenance of one relief valve.
- When removing air lines, plug ends and tag for installation.
- Perform steps 1 and 2 for front axle relief safety valve.

1. Disconnect air lines (22), (23), and (29) from connectors (30) and (28) and elbow (26).
2. Remove sleeve (24), elbow (26), connectors (30) and (28), and reducers (25) and (27) from relief safety valve (9). Discard sleeve (24).

NOTE

Perform steps 3 and 4 for forward-rear axle relief safety valve.

3. Disconnect air lines (12), (16), (21), and (19) from tee (14) and elbows (11) and (18).
4. Remove sleeves (13) and (15), tee (14), elbows (11) and (18), and reducers (20) and (17) from relief safety valve (9). Discard sleeves (13) and (15).

NOTE

Perform steps 5 and 6 for rear-rear axle relief safety valve.

5. Disconnect air lines (10), (5), and (8) from elbows (1), (3), and (7).
6. Remove sleeve (4), elbows (1), (7), and (3), and reducers (2) and (6) from relief safety valve (9). Discard sleeve (4).
7. Remove two nuts (34), washers (33), screws (31), washers (32), and relief safety valve (9) from bracket (39).
8. Remove nut (41), washer (40), screw (37), washer (36), and bracket (39) from frame (38). Remove spacer (35) (M936A2 vehicles only).

3-455. RELIEF SAFETY VALVE MAINTENANCE (Contd)

b. Disassembly

1. Remove clamp (45), duckbill (44), and hose (43) from adapter (42).
2. Remove adapter (42) and O-ring (46) from relief safety valve (9). Discard O-ring (46).

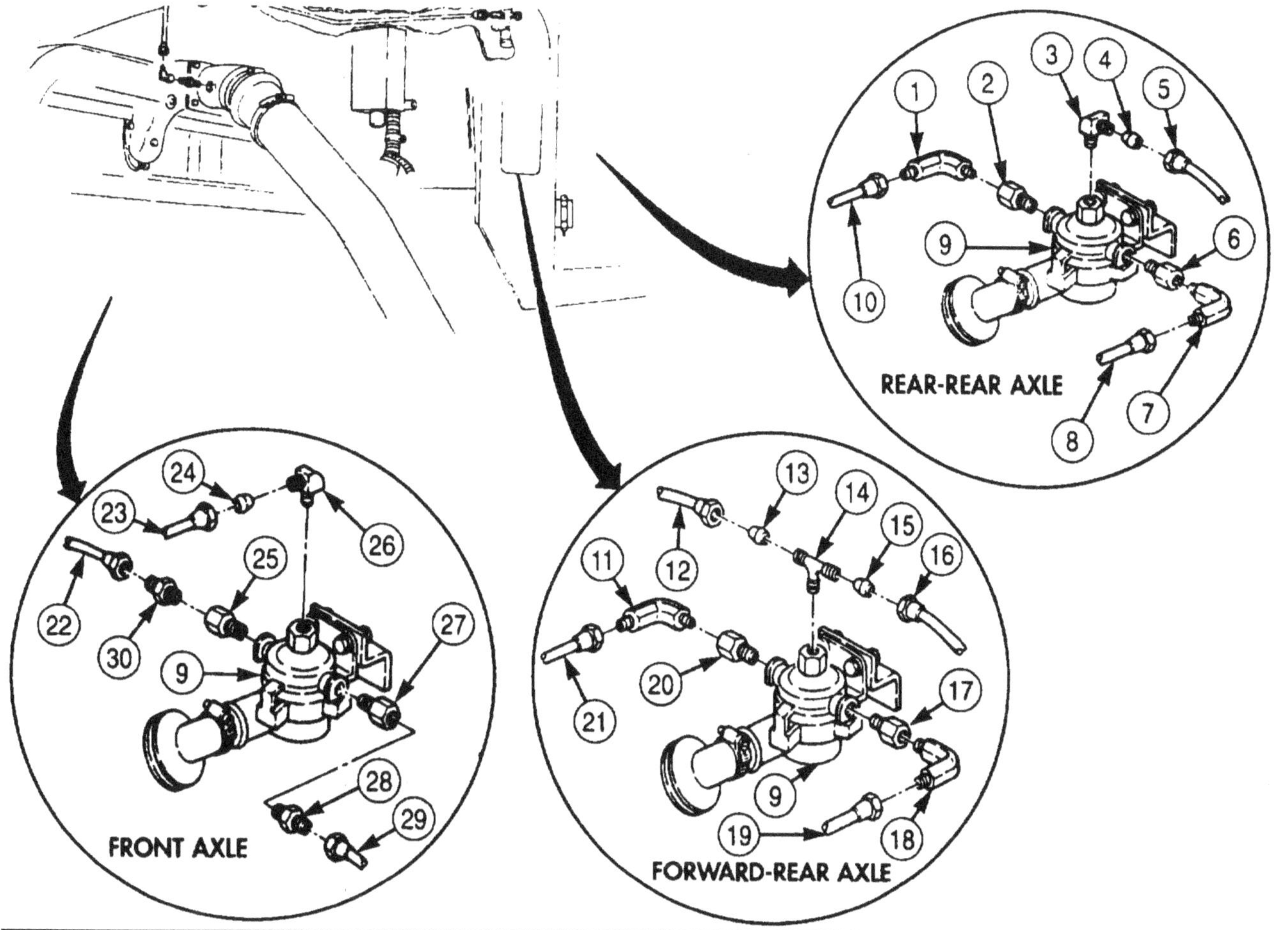

REAR-REAR AXLE

FRONT AXLE

FORWARD-REAR AXLE

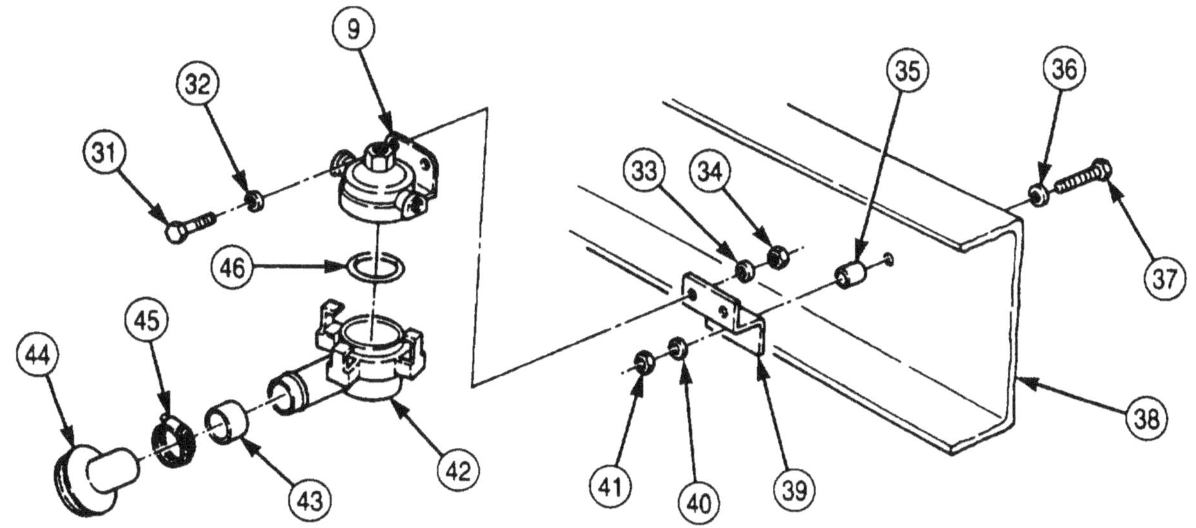

3-455. RELIEF SAFETY VALVE MAINTENANCE (Contd)

c. Cleaning and Inspection

1. For general cleaning instructions, refer to para. 2-14.
2. For general inspection instructions, refer to para 2-15.
3. Inspect relief safety valve (3) and adapter (13) for cracks, breaks, or stripped threads. Replace relief safety valve (3) and/or adapter (13) if damaged.
4. Inspect hose (14) and duckbill (15) for cracks, tears, or deterioration. Replace hose (14) and/or duckbill (15) if damaged.

NOTE

M936A2 model vehicles have spacer.

5. Inspect spacer (6), bracket (10), and clamp (16) for cracks, bends, and excessive corrosion. Replace spacer (6), bracket (10), or clamp (16) if damaged.
6. Inspect all fittings for stripped threads. Replace if threads are stripped.

d. Assembly

1. Install hose (14) and duckbill (15) on adapter (13) with clamp (16).
2. Install new O-ring (17) and adapter (13) on relief safety valve (3).

e. Installation

1. Install bracket (10) on frame (9) with washer (7), screw (8), washer (11), and nut (12). Install spacer (6) (M936A2 vehicles only).
2. Install safety relief valve (3) on bracket (10) with two washers (2), screws (1), washers (4), and nuts (5).

NOTE

- Wrap all male pipe threads with antiseize tape before installation.
- Perform steps 3 through 5 for rear-rear axle exhaust valve.

3. Install reducers (19) and (23) and elbows (18), (24), and (20) on relief safety valve (3).
4. Connect air lines (26) and (25) to elbows (18) and (24).
5. Insert new sleeve (21) on air line (22), and connect air line (22) to elbow (20).

NOTE

Perform steps 6 through 8 for forward-rear axle relief safety valve.

6. Install reducers (36) and (33), elbows (27) and (34), and tee (30) on relief safety valve (3).
7. Connect air lines (37) and (35) on elbows (27) and (34).
8. Insert new sleeves (29) and (31) on air lines (28) and (32), and connect air lines (28) and (32) to tee (30).

NOTE

Perform steps 9 through11 for front axle relief safety valve.

9. Install reducers (41) and (43), connectors (46) and (44), and elbow (42) on relief safety valve (3).
10. Connect air lines (38) and (45) to connectors (46) and (44).
11. Insert new sleeve (40) on air line (39), and connect air line (39) to elbow (42).

3-455. RELIEF SAFETY VALVE MAINTENANCE (Contd)

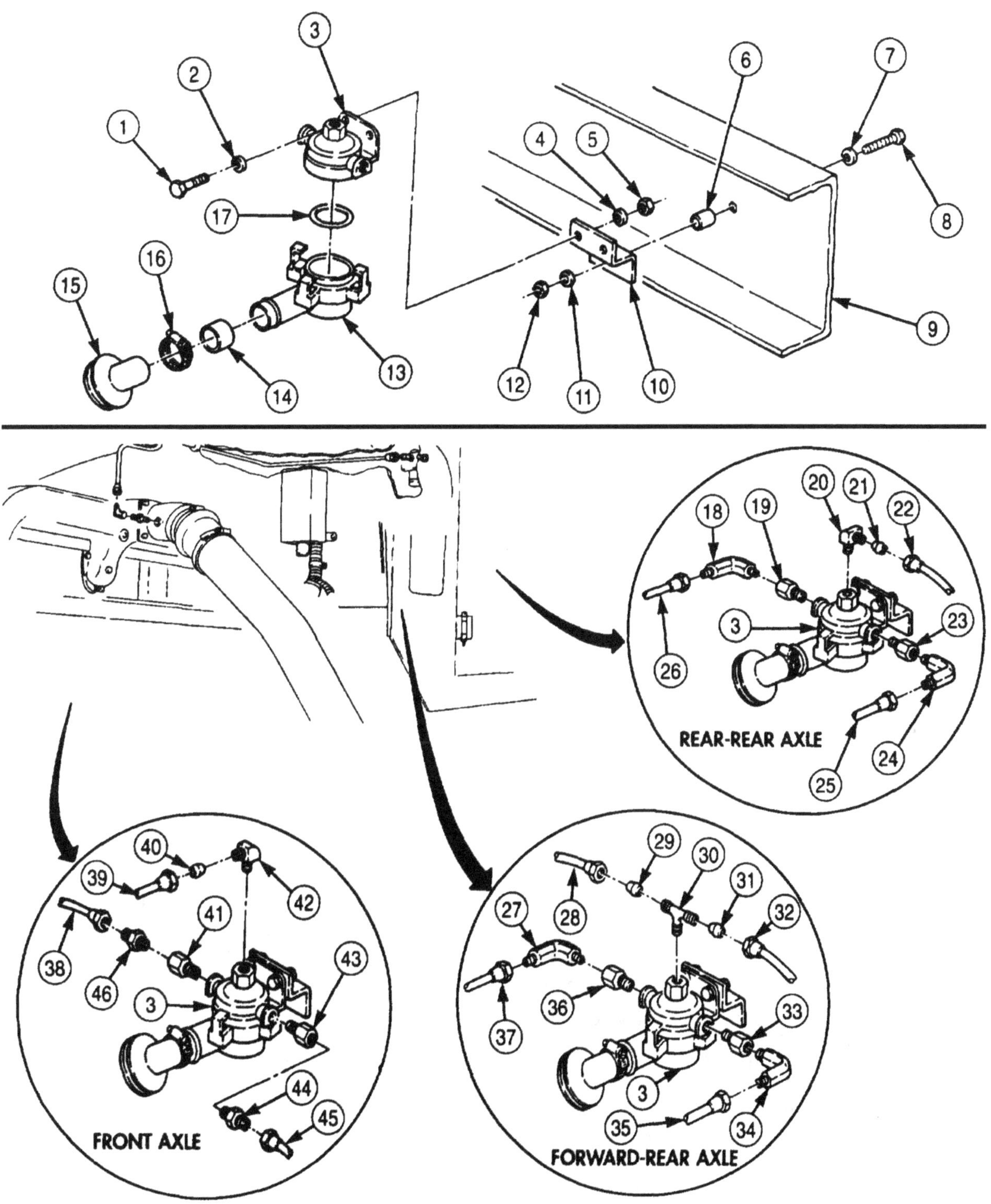

FOLLOW-ON TASK: Start engine (TM 9-2320-272-10) and check for air leaks and proper CTIS operation.

3-456. FRONT WHEEL VALVE MAINTENANCE

THIS TASK COVERS:

a. Removal
b. Cleaning and Inspection
c. Installation

INITIAL SETUP:

APPLICABLE MODELS
M939A2

TOOLS
General mechanic's tool kit (Appendix E, Item 1)

MATERIALS/PARTS
Three locknuts (Appendix D, Item 334)
Three lockwashers (Appendix D, Item 400)
O-ring (Appendix D, Item 437)
O-ring (Appendix D, Item 487)
Four locknuts (Appendix D, Item 273)
Antiseize tape (Appendix C, Item 72)

REFERENCES (TM)
TM 9-2320-272-10
TM 9-2320-272-24P

EQUIPMENT CONDITION
- Parking brake set (TM 9-2320-272-10).
  Air reservoirs drained (TM 9-2320-272-10).

GENERAL SAFETY INSTRUCTIONS
Eyeshields must be worn when releasing compressed air.

a. Removal

WARNING

Air system components are subject to high pressure. Always relieve pressure before loosening or removing air system components.

1. Remove cap (2) and valve stem (3) from tank valve (5), and allow tire (18) to deflate completely. Install valve stem (3) and cap (2) on tank valve (5).
2. Remove two locknuts (27), washers (26), screws (28), washers (29), shield (1), and spacer (25) from studs (16) and axle flange (20). Discard locknuts (27).
3. Remove hose (15) from elbow (7) and turret valve (17).
4. Loosen nut (14) and remove air manifold (22) from wheel valve (4).
5. Remove screw (24), washer (23), air manifold (22), and O-ring (21) from axle flange (20) and air tube (19). Discard O-ring (21).
6. Remove two locknuts (8), washers (9), washers (10), and wheel valve (4) from studs (16). Discard locknuts (8).
7. Remove three locknuts (13), lockwashers (12), and bracket (11) from wheel valve (4). Discard locknuts (13) and lockwashers (12).
8. Remove elbow (7) and O-ring (6) from wheel valve (4). Discard O-ring (6).

NOTE

For repair of wheel valves, refer to para. 3-461.

b. Cleaning and Inspection

1. For general cleaning instructions, refer to para. 2-14.
2. For general inspection instructions, refer to para. 2-15.
3. Inspect wheel valve (4) for cracks, leaks, and stripped threads. Replace or repair wheel valve (4) if cracked, leaking, or threads are stripped.
4. Inspect bracket (11), shield (1), air manifold (22), hose (15), spacer (25), elbow (7), and turret valve (17) for cracks, bends, or stripped threads. Replace parts if damaged.

3-456. FRONT WHEEL VALVE MAINTENANCE (Contd)

c. Installation

NOTE

Wrap all male pipe threads with antiseize tape before installation.

1. Install hose (15) on turret valve (17).
2. Install new O-ring (21) on air tube (19).
3. Install new O-ring (6) and elbow (7) on wheel valve (4). Finger-tighten elbow (7).
4. Install bracket (11) on wheel valve (4) with three new lockwashers (12) and new locknuts (13).

NOTE

Ensure air manifold is properly seated on wheel before tightening.

5. Connect air manifold (22) on wheel valve (4). Finger-tighten nut (14), but allow for movement.
6. Install wheel valve (4) with bracket (11) on two studs (16) with washers (10), washers (9), and new locknuts (8). Finger-tighten locknuts (8).
7. Install screw (24) and washer (23) on air manifold (22) and axle flange (20). Finger-tighten screw (24).
8. Tighten two locknuts (8), screw (24), and nut (14).
9. Tighten elbow (7) to align with hose (15) and connect hose (15) to elbow (7).

NOTE

Spacer has long and short end. Install short end of spacer on flange of air manifold.

10. Install spacer (25) and shield (1) on axle flange (20) with two washers (29) and screws (28).
11. Install two washers (26) and new locknuts (27) on studs (16).

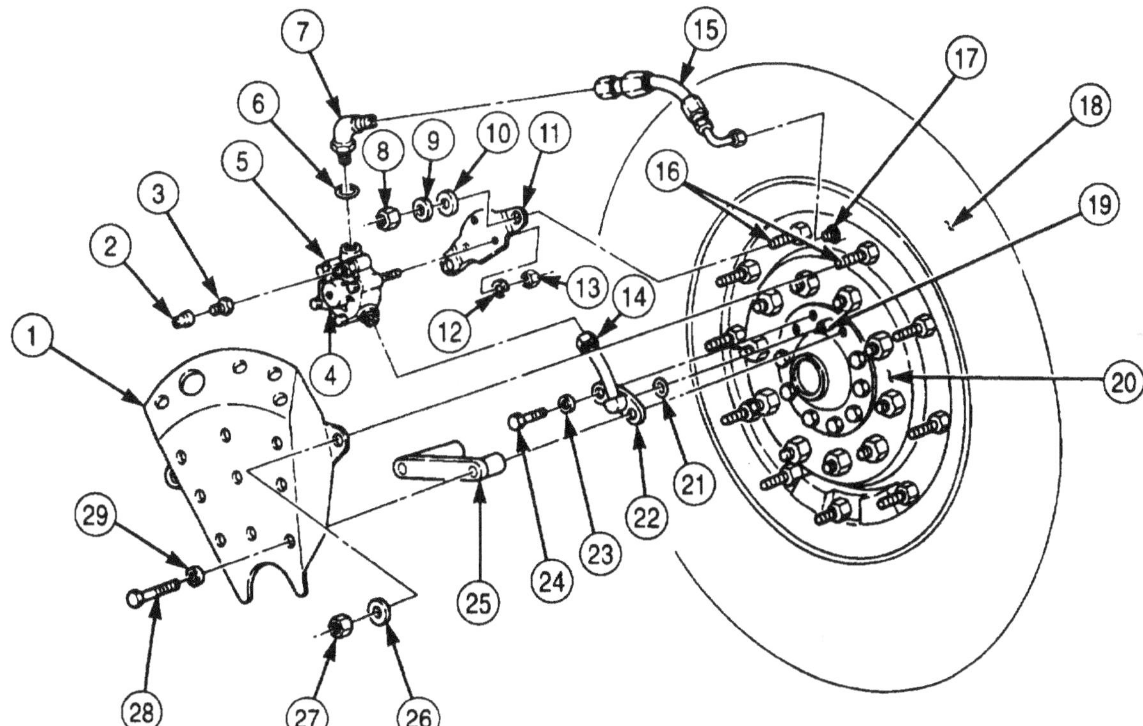

FOLLOW-ON TASK: Start engine (TM 9-2320-272-10), operate CTIS, allow tire to inflate, and check for leaks.

3-457. REAR WHEEL VALVE MAINTENANCE

THIS TASK COVERS:

a. Removal
b. Cleaning and Inspection
c. Installation

INITIAL SETUP:

APPLICABLE MODELS
M939A2

TOOLS
General mechanic's tool kit (Appendix E, Item 1)

MATERIALS/PARTS
Two locknuts (Appendix D, Item 334)
O-ring (Appendix D, Item 437)
Antiseize tape (Appendix C, Item 72)

REFERENCES (TM)
TM 9-2320-272-10
TM 9-2320-272-24P

EQUIPMENT CONDITION
- Parking brake set (TM 9-2320-272-10).
- Air reservoirs drained (TM 9-2320-272-10).

GENERAL SAFETY INSTRUCTIONS
Eyeshields must be worn when releasing compressed air.

a. Removal

WARNING

Air system components are subject to high pressure. Always relieve pressure before loosening or removing air system components.

1. Remove cap (11) and valve stem (12) from tank valve (13), and allow tire (17) to deflate completely.
2. Remove two nuts (6), washers (7), screws (10), washers (9), and support bracket (8) from bracket (3) and axle flange (14).
3. Remove hose (5) from connector (4) and tube (15).
4. Remove wheel valve (21) from connector (20).
5. Remove connector (20), O-ring (19), and washer (18) from hub (16). Discard O-ring (19).
6. Remove connector (4) from wheel valve (21).
7. Remove two locknuts (1), washers (2), and bracket (3) from wheel valve (21). Discard locknuts (1).

b. Cleaning and Inspection

1. For general cleaning instructions, refer to 2-14.
2. For general inspection instructions, refer to para. 2-15.
3. Inspect wheel valve (21) for cracks, leaks, and stripped threads. Replace wheel valve (21) if cracked or leaking. Repair stripped threads.
4. Inspect bracket (3), support bracket (8), connectors (4) and (20), hose (5), tube (15), and hub (16) for cracks, bends, and stripped threads. Replace part(s) if damaged.

3-457. REAR WHEEL VALVE MAINTENANCE (Contd)

c. Installation

1. Install bracket (3) on wheel valve (21) with two washers (2) and new locknuts (1).

NOTE

Wrap all male pipe threads with antiseize tape before installation.

2. Install connector (4) on wheel valve (21)
3. Install washer (18), new O-ring (19), and connector (20) on hub (16).
4. Install wheel valve (21) on connector (20).
5. Install hose (5) on connector (4) and tube (15).
6. Install support bracket (8) on axle flange (14) and bracket (3) with two washers (9), screws (10), washers (7), and nuts (6).
7. Install valve stem (12) and cap (11) on tank valve (13).

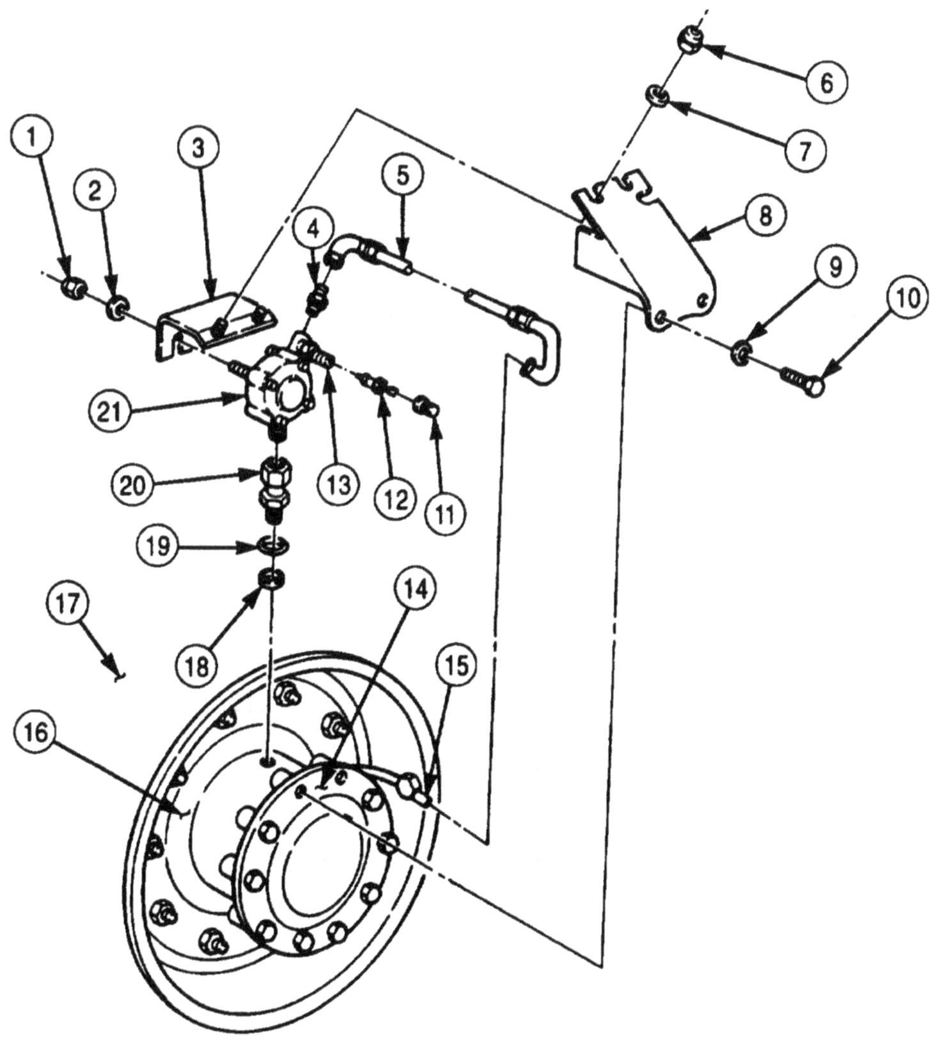

FOLLOW-ON TASK: Start engine (TM 9-2320-272-10), operate CTIS, allow tire to inflate, and check for leaks.

3-458. WHEEL VALVE FILTER REPLACEMENT

THIS TASK COVERS:

a. Removal **b. Installation**

INITIAL SETUP:

APPLICABLE MODELS
M939A2

TOOLS
General mechanic's tool kit (Appendix E, Item 1)

MATERIALS/PARTS
Filter (Appendix D, Item 122)
O-ring (Appendix D, Item 437)
Two locknuts (Appendix D, Item 334)

REFERENCES (TM)
TM 9-2320-272-10
TM 9-2320-272-24P

EQUIPMENT CONDITION
- Parking brake set (TM 9-2320-272-10).
- Air reservoirs drained (TM 9-2320-272-10).

GENERAL SAFETY INSTRUCTIONS
Eyeshields must be worn when releasing compressed air.

a. Removal

WARNING

Air system components are subject to high pressure. Always relieve pressure before loosening or removing air system components.

NOTE

Replacement of wheel valve filter is the same for front and rear wheel valves. This procedure is for front wheel valves.

1. Remove cap (2) and valve stem (3) from wheel valve (4) and allow tire (10) to deflate completely
2. Remove two locknuts (14), washers (13), screws (15), washers (16), shield (1), and spacer (12) from studs (9) and axle flange (11). Discard locknuts (14).
3. Disconnect hose (8) from elbow (7).
4. Remove elbow (7), O-ring (6), and filter (5) from wheel valve (4). Discard O-ring (6) and filter (5).

b. Installation

1. Install new filter (5), new O-ring (6), and elbow (7) on wheel valve (4).
2. Connect hose (8) to elbow (7).
3. Install spacer (12) and shield (1) on studs (9) and axle flange (11) with two washers (16), screws (15), washers (13), and new locknuts (14)
4. Install valve stem (3) and cap (2) on wheel valve (4).

3-458. WHEEL VALVE FILTER REPLACEMENT (Contd)

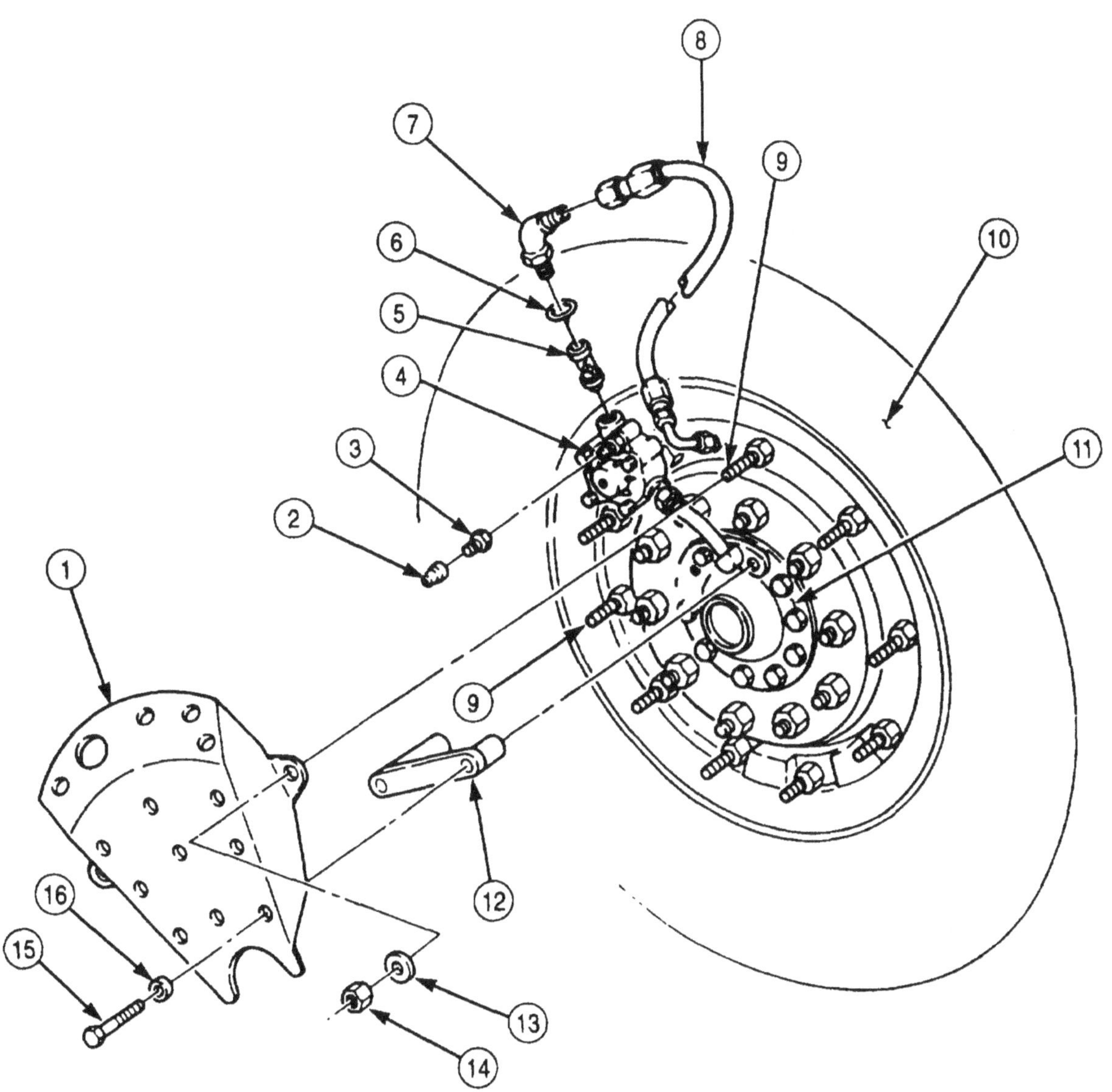

FOLLOW-ON TASK: Start engine (TM 9-2320-2762-10), operate CTIS, allow tire to inflate, and check for leaks.

3-459. WHEEL VALVE MAINTENANCE

THIS TASK COVERS:

a. Disassembly
b. Cleaning and Inspection
c. Assembly

INITIAL SETUP:

APPLICABLE MODELS
M939A2

TOOLS
General mechanic's tool kit (Appendix E, Item 1)

MATERIALS/PARTS
Relay valve kit (Appendix D, Item 530)
Filter (Appendix D, Item 122)
Antiseize tape (Appendix C, Item 72)

REFERENCES (TM)
TM 9-2320-272-10
TM 9-2320-272-24P

EQUIPMENT CONDITION
Wheel valve removed (para. 4-456 or 4-457).

GENERAL SAFETY INSTRUCTIONS
Eyeshields must be worn during disassembly.

a. Disassembly

1. Remove filter (9) from valve body (10). Discard filter (9).

WARNING

Wheel valve cover is under spring tension. Wear eye protection when disassembling wheel valve, or injury to personnel may result.

2. Holding cover (4) in place, remove four nuts (12), screw-assembled lockwasher (1), and three screw-assembled lockwashers (2) from base (11), valve body (10), and cover (4). Discard screw-assembled lockwashers (1) and (2).
3. Carefully remove cover (4), spring (5), plug (6), diaphragm (7), and base (11) from valve body (10). Discard spring (5) and diaphragm (7).
4. Remove ball (3) from cover (4). Discard ball (3).
5. Remove tank valve (8) from valve body (10).

b. Cleaning and Inspection

1. For general cleaning instructions, refer to para. 2-14.
2. For general inspection instructions, refer to para. 2-15.
3. Inspect base (11), valve body (10), cover (4), and tank valve (8) for cracks and stripped threads. Replace part(s) if damaged.

c. Assembly

NOTE

Wrap male threads with antiseize tape before installation.

1. Install tank valve (8) on valve body (10).
2. Install new ball (3) in cover (4).
3. Position base (11), new diaphragm (7), plug (6), new spring (5), and cover (4) on valve body (10). Carefully hold in place.
4. Install three new screw-assembled lockwashers (2), new screw-assembled lockwasher (1), and four nuts (12) on cover (4), valve body (10), and base (11).
5. Install new filter (9) in valve body (10).

3-459. WHEEL VALVE MAINTENANCE (Contd)

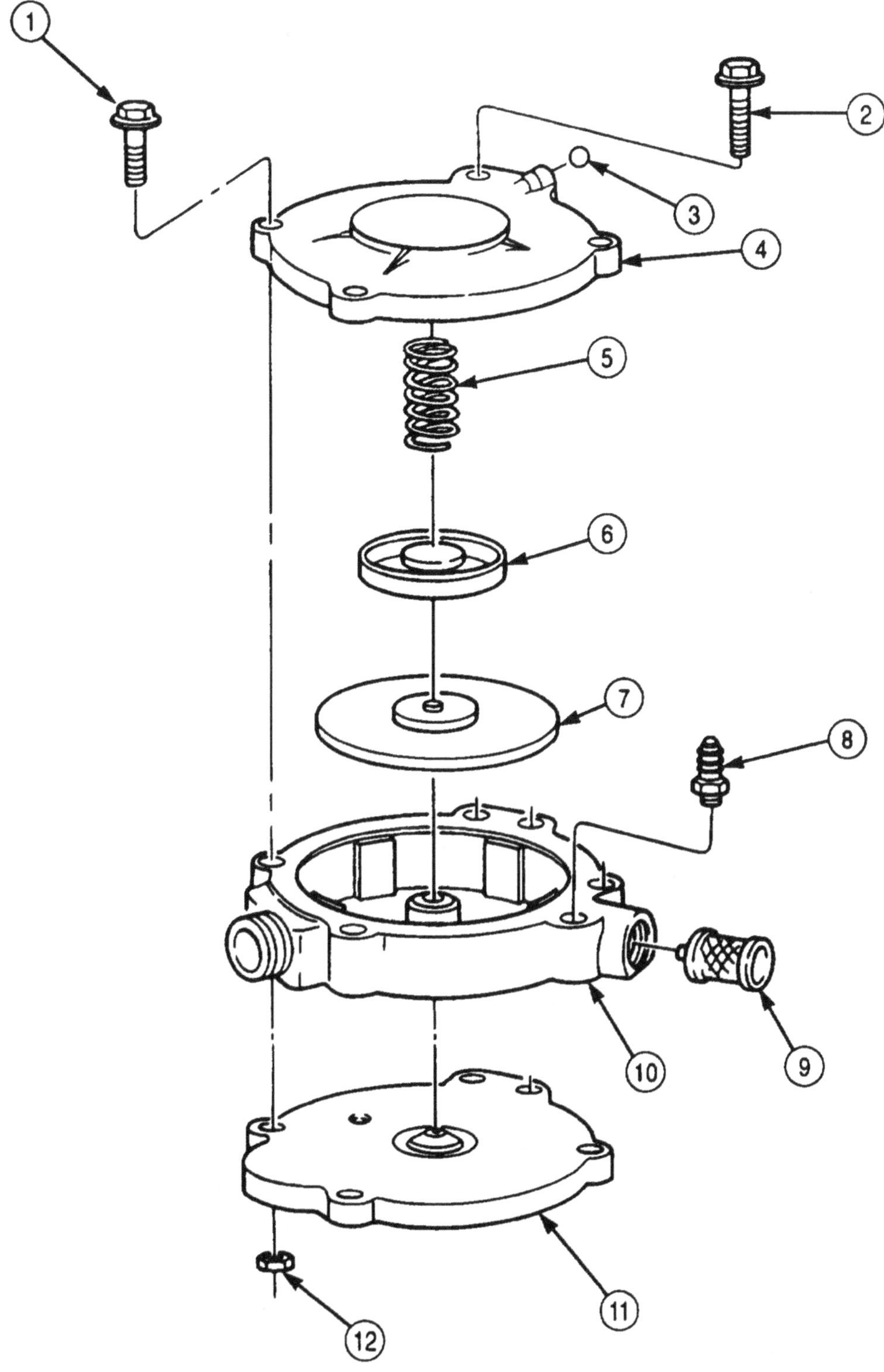

FOLLOW-ON TASK: Install wheel valve (para. 4-456 or 4-457).

3-460. FRONT HUBS REPAIR

THIS TASK COVERS:

a. Removal
b. Cleaning and Inspection
c. Repair
d. Installation

INITIAL SETUP:

APPLICABLE MODELS
M939A2

SPECIAL TOOLS
Bearing punch (Appendix E, Item 12)
Air seal installer (Appendix E, Item 5)

TOOLS
General mechanic's tool kit (Appendix E, Item 1)
Torque wrench (Appendix E, Item 144)

MATERIALS/PARTS
Oil seal (Appendix D, Item 629)
Seal (Appendix D, Item 630)
Oil seal (Appendix D, Item 497)
Ferrule (Appendix D, Item 118)
Gasket (Appendix D, Item 246)
GAA grease (Appendix C, Item 28)
Sealant (Appendix C, Item 30)
Abrasive cloth (Appendix C, Item 20)
Drycleaning solvent (Appendix C, Item 71)

REFERENCES (TM)
TM 9-214
TM 9-2320-272-10
TM 9-2320-272-24P

EQUIPMENT CONDITION
- Rear wheels chocked (TM 9-2320-272-10).
- Front wheel removed (TM 9-2320-272-10).
- Front wheel valve removed (para. 3-456).

GENERAL SAFETY INSTRUCTIONS
- When cleaning with compressed air, wear eyeshields and ensure source pressure does not exceed 30 psi (207 kPa).
- Drycleaning solvent is flammable and toxic. Do not use near an open flame.
- Keep fire extinguisher nearby when using drycleaning solvent.

a. Removal

1. Remove brake drum (9) from hub (2).
2. Remove ten screws (10) and washers (11) from drive flange (7).
3. Install two screws (10) on threaded holes (8) of drive flange (7). Tighten screws (2) until drive flange (7) separates from hub (2).
4. Remove drive flange (7) from hub (2) and remove two screws (10) from drive flange (7).

NOTE

Tag inner and outer bearing for installation.

5. Remove outer bearing nut (6) and washer (5) from spindle (1).
6. Remove nut (4) and outer bearing (3) from spindle (1).

NOTE

Assistant will help with step 7.

7. Remove hub (2) from spindle (1).
8. Remove snapring (14) and two air seals (13) from hub (2). Discard air seals (13).
9. Remove screw (15), ferrule (20), tee (12), and gasket (19) from hub (2). Discard ferrule (20) and gasket (19).
10. Remove hub seal (18) and inner bearing (17) from rear of hub (2). Discard hub seal (18).

3-460. FRONT HUBS REPAIR (Contd)

b. Cleaning and Inspection

1. For general cleaning instructions, refer to para. 2-14.
2. For general inspection instructions, refer to para. 2-15.
3. Inspect inner (17) and outer (3) bearings in accordance with the TM 9-214. Replace bearings (17) or (3) and bearing cups if either is damaged.
4. Inspect hub (2), drive flange (7), and plastic sleeving (16) for cracks, grooves, scores, and elongated holes. Replace part(s) if cracked or holes are elongated. Notify your supervisor if grooved or scored.
5. Inspect seal surface of spindle (1) for extensive rust, scratches, and grooves. Repair spindle (1) if rusted, scratched or grooved.

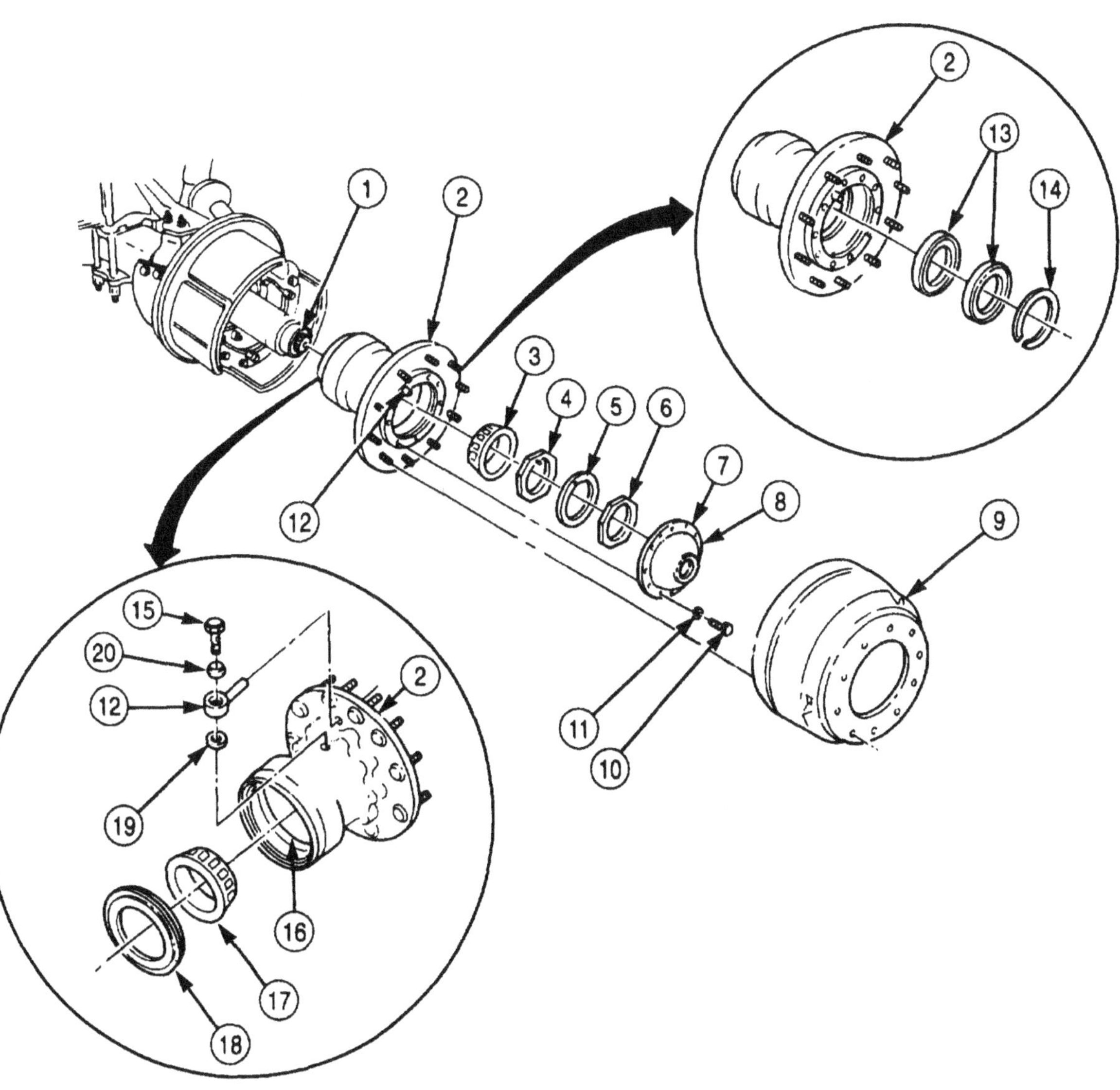

3-460. FRONT HUBS REPAIR (Contd)

c. Repair

1. Using bearing punch, remove outer bearing cup (4) from hub (5) by tapping alternately on outer edge of cup (4). Discard outer bearing cup (4).
2. Using bearing punch, remove inner bearing cup (1) from hub (5) by tapping alternately on outer edge of cup (1). Discard inner bearing cup (1).
3. Remove plastic sleeve (2) from hub (5) by splitting and prying out of hub (5). Discard plastic sleeve (2).

NOTE

Bearing and bearing cup must be replaced as a match set.

4. Install new outer bearing cup (4) in hub (5). Ensure outer bearing cup (4) is seated properly.
5. Install new plastic sleeve (2) on hub (5).
6. Install new inner bearing cup (1) in hub (5). Ensure inner bearing cup (1) is seated properly.

NOTE

- Perform steps 7 and 8 if stud failed inspection.
- Use left-hand threaded studs for hubs mounted on the left-hand side of the vehicle, and right-hand threaded studs for hubs mounted on the right side of the vehicle.

7. Remove damaged studs (3) from hub (5). Discard studs (3).
8. Install new studs (3) on hub (5), as required.
9. Using abrasive cloth, remove rust from seal surface of spindle (7).
10. Disconnect air line (6) from steering knuckle (8).

WARNING

Eyeshields must be worn when cleaning with compressed air. Compressed air source will not exceed 30 psi (207 kPa). Failure to do so may result in injury to personnel.

11. Using external air source, blow dirt and debris from surface of spindle (7) by forcing air thorough steering knuckle (8).

WARNING

Drycleaning solvent is flammable and toxic. Do not use near an open flame and always have a fire extinguisher nearby when solvents are used. Use only in well-ventilated places, wear protective clothing, and dispose of cleaning rags in approved container. Failure to do this will result in injury to personnel and/or damage to equipment.

12. Clean entire surface of spindle (7) with drycleaning solvent,
13. Connect air line (6) to steering knuckle (8).

CAUTION

Avoid filing too deeply into seal surface of spindle. Excessive filing may result in permanent damage to spindle.

14. Remove deep scratches by carefully filing seal surface of spindle (7) in a circular motion.
15. Using abrasive cloth, remove light surface scratches from seal surface of spindle (7).

NOTE

It is unnecessary to remove the circular depression of a groove; only the raised portion or lip.

3-460. FRONT HUBS REPAIR (Contd)

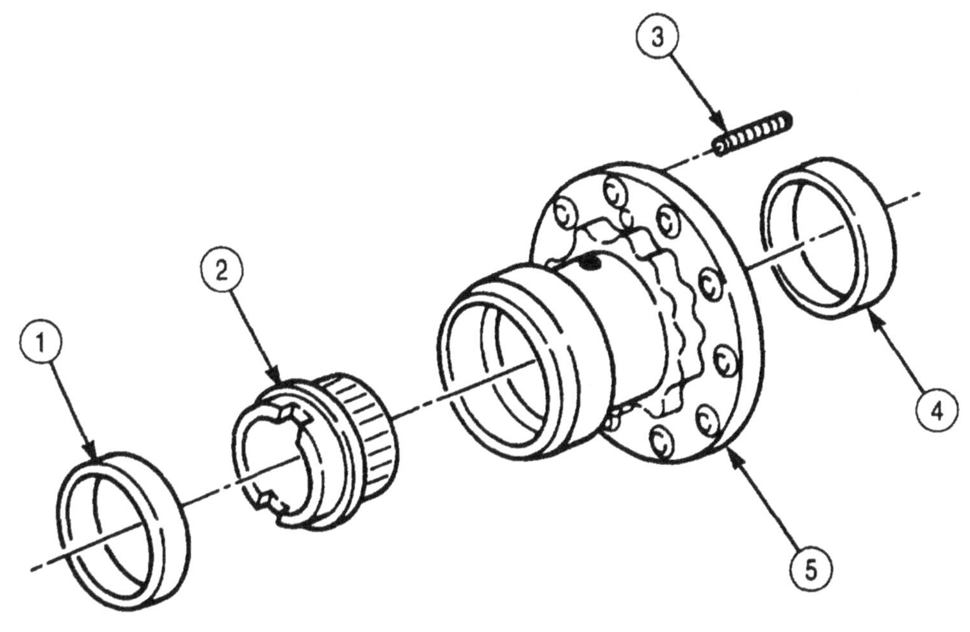

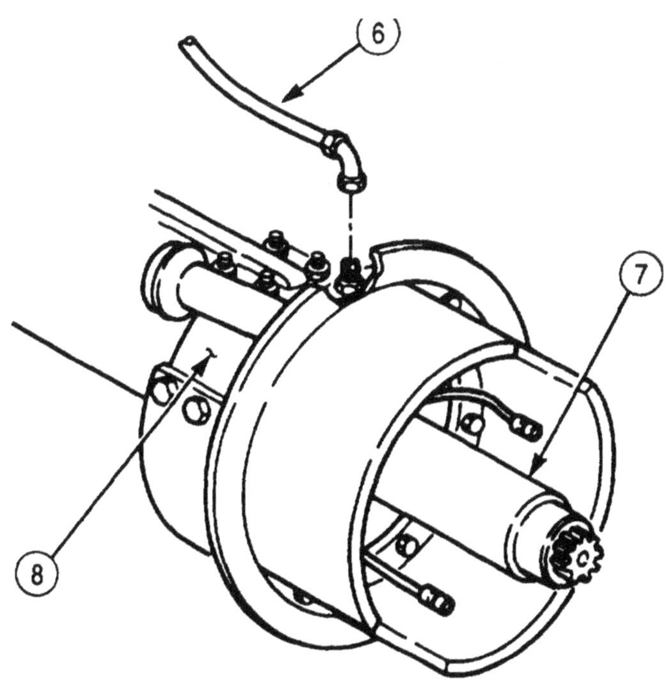

3-460. FRONT HUBS REPAIR (Contd)

d. Installation

1. Using long end of air seal installer, install new air seal (14) on hub (2), with spring facing out. Ensure air seal (14) is seated properly in hub (2).
2. Using short end of seal installer, install new air seal (15) on hub (2), with spring facing out. Ensure air seal (15) is seated properly in hub (2).
3. Install snapring (16) on hub (2).
4. Install new gasket (20), tee (21), new ferrule (22), and screw (17) on hub (2). Finger-tighten screw (17).
5. Align drive flange (9) with tee (21) and install drive flange (9) on hub (2) with two screws (11).
6. Tighten screw (17) 35 lb-ft (48 N•m).
7. Install two screws (11) in threaded holes (8) of drive flange (9).
8. Remove two screws (11) and drive flange (9) from hub (2).

CAUTION

Ensure bearings are matched with bearing cups before installation. Damage to bearings and/or bearing cups may result.

9. Install inner bearing (18) and new hub seal (19) on hub (2).

CAUTION

Do not allow hub assembly to slide directly over threaded end of spindle. Damage to air seals may result.

NOTE

Assistant will help with step 10.

10. Install hub (2), outer bearing (3), and bearing adjusting nut (5) on spindle (13).
11. While rotating hub (2), tighten bearing adjusting nut (5) 50 lb-ft (68 N•m).
12. Back out bearing adjusting nut (5) 1/8 to 1/4 turn so washer (6) can be aligned with slot in spindle (13) and adjusting nut locking pin (4).
13. Install washer (6) and nut (7) on spindle (13). Tighten nut (7) 250-400 lb-ft (339-542 N•m).
14. Apply sealant to sealing surfaces of hub (2) and drive flange (9).
15. Align drive flange (9) with tee (21), and install drive flange (9) on drive shaft (1) and hub (2) with ten washers (10) and screws (11). Tighten screws (11) 60-100 lb-ft (81-136 N•m).
16. Install brake drum (12) on hub (2).

3-460. FRONT HUBS REPAIR (Contd)

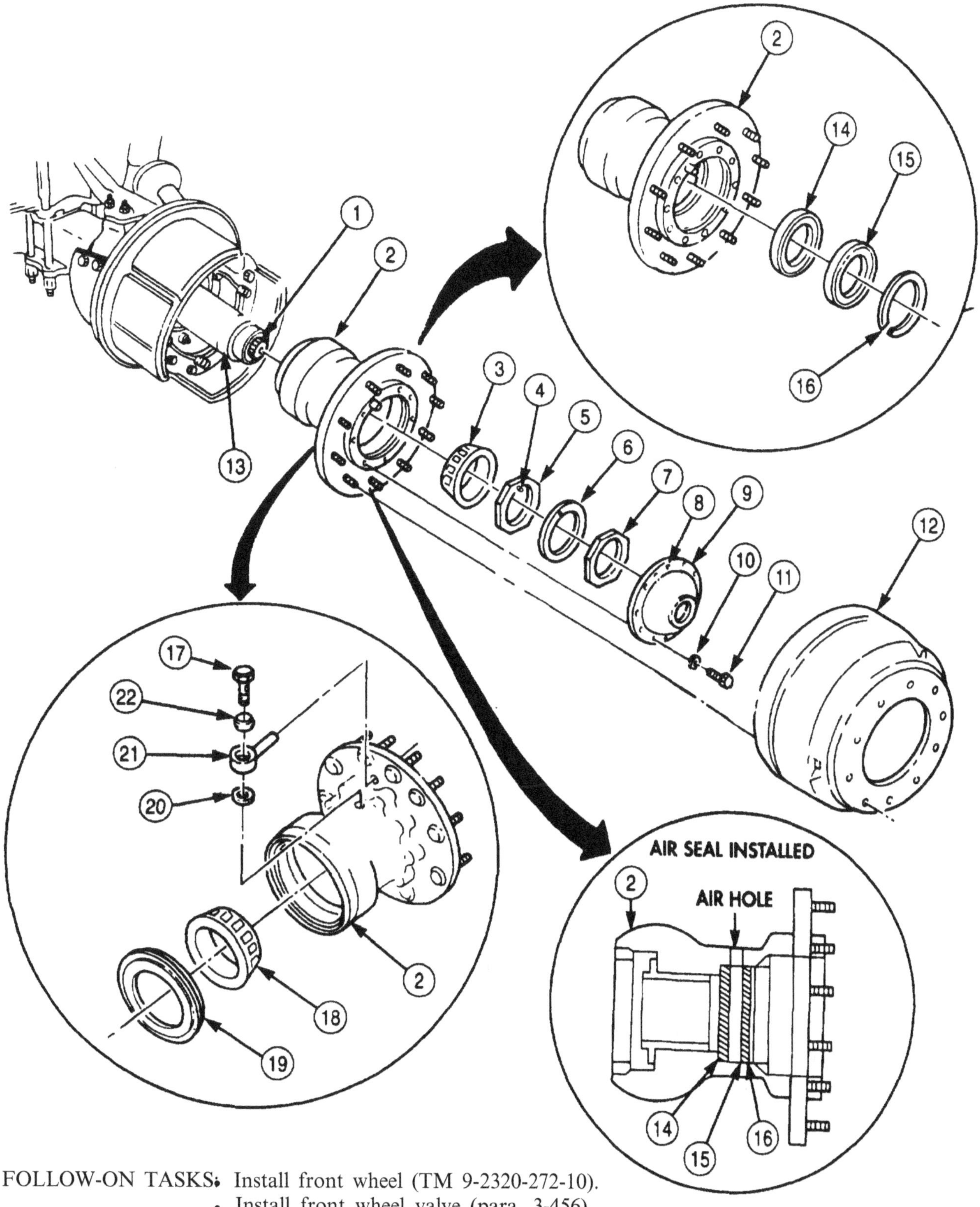

AIR SEAL INSTALLED

AIR HOLE

FOLLOW-ON TASKS:
- Install front wheel (TM 9-2320-272-10).
- Install front wheel valve (para. 3-456).
- Perform hub air seal leak test (para. 3-462).

3-461. REAR HUBS REPAIR

THIS TASK COVERS:

a. Removal
b. Cleaning and Inspection
c. Repair
d. Installation

INITIAL SETUP:

APPLICABLE MODELS
M939A2

SPECIAL TOOLS
Bearing punch (Appendix E, Item 12)
Air seal installer (Appendix E, Item 5)

TOOLS
General mechanic's tool kit (Appendix E, Item 1)
Torque wrench (Appendix E, Item 145)

MATERIALS/PARTS
Seal (Appendix D, Item 629)
Seal (Appendix D, Item 630)
Hub seal (Appendix D, Item 497)
Ten lockwashers (Appendix D, Item 350)
GAA grease (Appendix C, Item 28)
Sealant (Appendix C, Item 30)
Abrasive cloth (Appendix C, Item 20)
Drycleaning solvent (Appendix C, Item 71)

REFERENCES (TM)
TM 9-2320-272-10
TM 9-2320-272-24P

EQUIPMENT CONDITION
- Front wheels chocked (TM 9-2320-272-10).
- Spring brake caged (TM 9-2320-272-10).
- Rear wheel(s) removed (TM 9-2320-272-10).

GENERAL SAFETY INSTRUCTIONS
- When cleaning with compressed air, wear eyeshields and ensure source pressure does not exceed 30 psi (207 kPa).
- Drycleaning solvent is flammable and toxic. Do not use near an open flame.
- Keep fire extinguisher nearby when using drycleaning solvent.

a. Removal

1. Remove brake drum (10) from hub (6).
2. Remove ten screws (9), lockwashers (8), and axle shaft (7) from hub (6). Discard lockwashers (8).

NOTE

Tag inner and outer bearings for installation.

3. Remove outer bearing nut (1) and washer (2) from spindle (5).
4. Remove bearing adjusting nut (3) and outer bearing (4) from spindle (5).

CAUTION

Do not allow hub assembly to slide directly over threaded end of spindle. Damage to air seals may result.

NOTE

Assistant will help with step 5.

5. Remove hub (6) from spindle (5).
6. Remove hub seal (13) and inner bearing (14) from hub (6). Discard hub seal (13).
7. Remove snapring (11) and two air seals (12) from hub (6). Discard air seals (12).

3-461. REAR HUBS REPAIR (Contd)

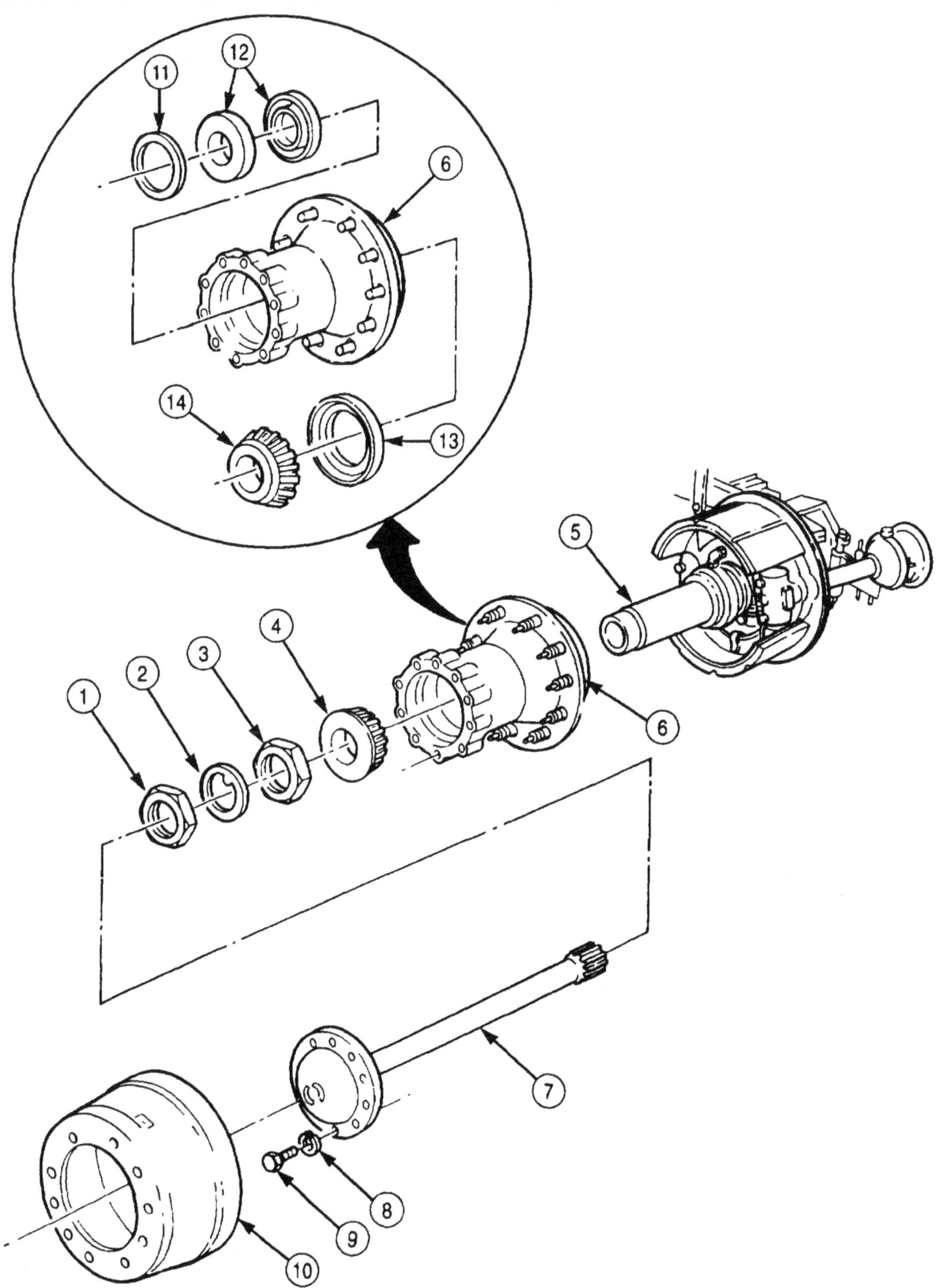

3-461. REAR HUBS REPAIR (Contd)

b. Cleaning and Inspection

1. For general cleaning instructions, refer to para. 2-14.
2. For general inspection instructions, refer to para. 2-15.

NOTE

Bearings and bearing cups must be replaced as a matched set.

3. Inspect inner bearing cup (2) and outer bearing cup (5) for cracks, chips, or excessive wear. Replace bearing(s) and bearing cup(s) if damaged (task cl.
4. Inspect plastic sleeve (1) for excessive wear. Replace plastic sleeve (1) if worn (task c).
5. Inspect studs (3), oil slinger (4), and hub (6) for stripped threads, chips, and cracks. Replace damaged parts (task c).
6. Inspect seal surface of spindle (9) for rust, scratches, and grooves.
   a. If rusted, disconnect air line (7) from air manifold fitting (8).

WARNING

Eyeshields must be worn when cleaning with compressed air. Compressed air source will not exceed 30 psi (207 kPa). Failure to do so may result in injury to personnel.

   b. Using external air source, blow dirt and debris from seal surface of spindle (9) by forcing air through air manifold fitting (8).

WARNING

Drycleaning solvent is flammable and toxic. Do not used near an open flame and always have a fire extinguisher nearby when solvents are used. Use only in well-ventilated places, wear protective clothing, and dispose of cleaning rags in approved container. Failure to do this will result in injury to personnel and/or damage to equipment.

   c. Thoroughly clean entire surface of spindle (9) with drycleaning solvent.
   d. Connect air line (7) to air manifold fitting (8).

CAUTION

Avoid filing too deeply in seal surface of spindle. Excessive filing may result in permanent damage to spindle.

   e. Remove deep scratches by carefully tiling seal surface of spindle (9) in a circular motion.

NOTE

It is unnecessary to remove the circular depression of a groove; only the raised portion or lip.

   f. Using abrasive cloth, remove light surface scratches from seal surface of spindle (9).
   g. If excessively rusted, scratched, or grooved, notify DS maintenance for rear axle replacement.

c. Repair

1. Using bearing punch, remove outer bearing cup (5) from hub (6) by tapping alternately on outer edges of outer bearing cup (5). Discard outer bearing cup (5).
2. Using bearing punch, remove inner bearing cup (2) from hub (6) by tapping alternately on outer edge of inner bearing cup (2).
3. Remove plastic sleeve (1) from hub (6) by splitting and prying plastic sleeve (1) from hub (6). Discard plastic sleeve (1).

3-461. REAR HUBS REPAIR (Contd)

NOTE

Bearings and bearing cups must be replaced as a matched set.

4. Install new outer bearing cup (5) on hub (6). Ensure outer bearing cup (5) is properly seated on hub (6).
5. Install new plastic sleeve (1) on hub (6).
6. Install new inner bearing cup (2) on hub (6). Ensure inner bearing cup (2) is seated properly on hub (6).

NOTE

- Perform steps 7 through 9 if slinger or studs failed inspection.
- Use left-hand threaded studs for hub mounted on the left side of vehicle, and right-hand threaded studs for hubs mounted on the right side of vehicle.

7. Remove studs (3) and slinger (4) from hub (6).
8. Remove damaged studs (3) from slinger (4) and hub (6). Discard studs (3) and slinger (4).
9. Position new slinger (4) on hub (6), and install with new studs (3).

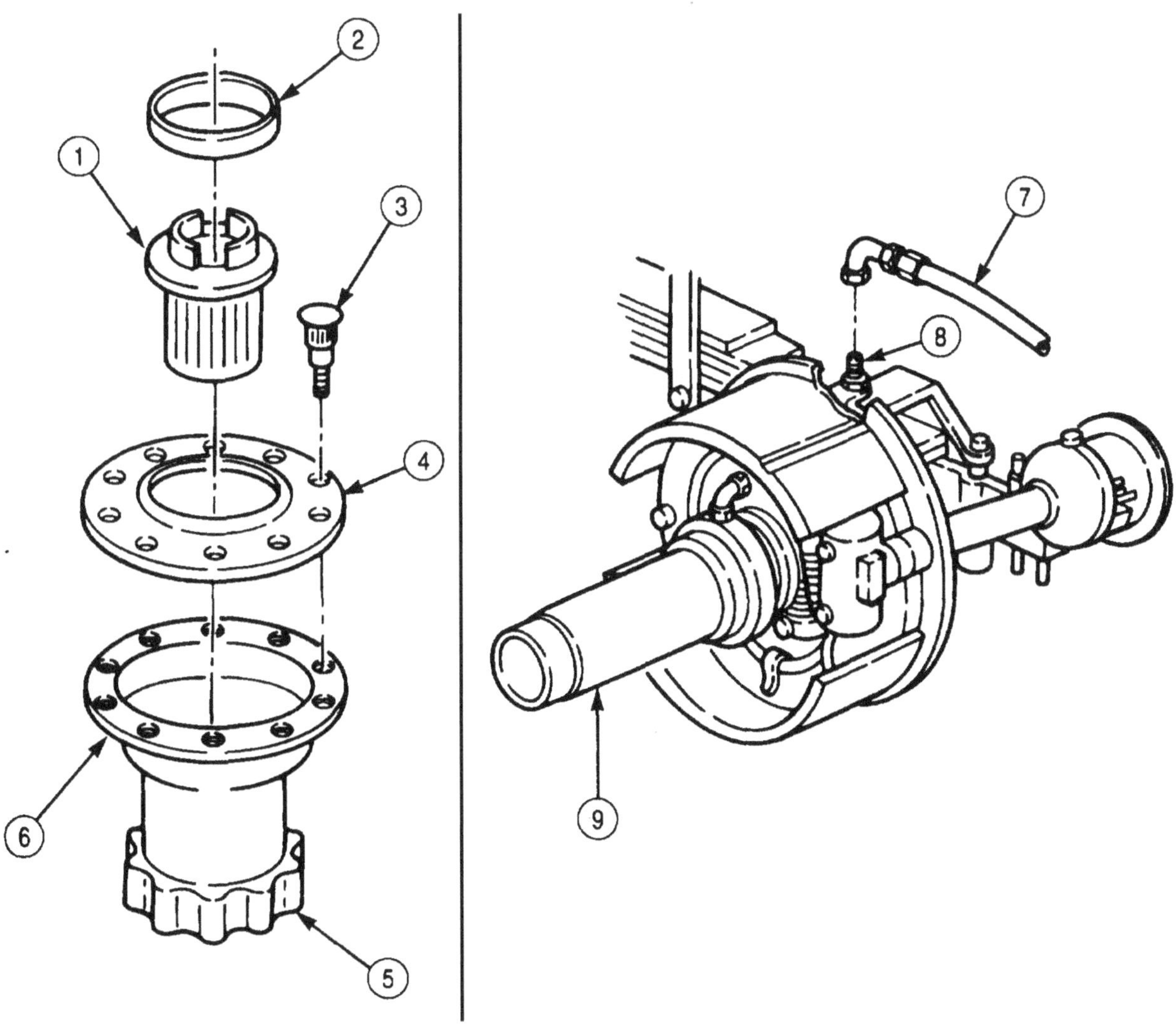

3-461. REAR HUBS REPAIR (Contd)

d. Installation

NOTE

Ensure double-lip seal is placed outward.

1. Using long end of air seal installer, install new air seal (12) on hub (6), with spring facing out, until seated in hub (6).
2. Using short end of air seal installer, install new air seal (12) on hub (6), with spring facing inward, until seated in hub (6).
3. Install snapring (11) on hub (6).

CAUTION

Ensure bearings are matched with bearing cups before installation, or damage to bearings may result.

NOTE

- Pack inner and outer bearings with GAA grease prior to installation.
- Pack inner rubber section of hub seal with GAA grease.

4. Install inner bearing (14) and new hub seal (13) on hub (6).

CAUTION

Do not allow hub assembly to slide directly over threaded end of spline. Damage to air seals may result.

NOTE

Assistant will help with step 5.

5. Install hub (6), outer bearing (4), and bearing adjusting nut (3) on spindle (5).
6. While rotating hub (6), tighten bearing adjusting nut (3) 50 lb-ft (68 N•m).
7. Back out bearing adjusting nut (3)1/8 to 1/4 turn so washer (2) can be aligned with slot on spindle (5) and locking pin on adjusting nut (3).
8. Install washer (2) and outer bearing nut (1) on spindle (5). Tighten outer bearing nut (1) 250-400 lb-ft (339-542 N•m).
9. Apply sealant to mating surface of hub (6) and inside of flange on axle shaft (7).
10. Install axle shaft (7) on hub (6) with ten new lockwashers (8) and screws (9). Tighten screws (9) 60-100 lb-ft (81-136 N•m).
11. Install brake drum (10) on hub (6).

3-461. REAR HUBS REPAIR (Contd)

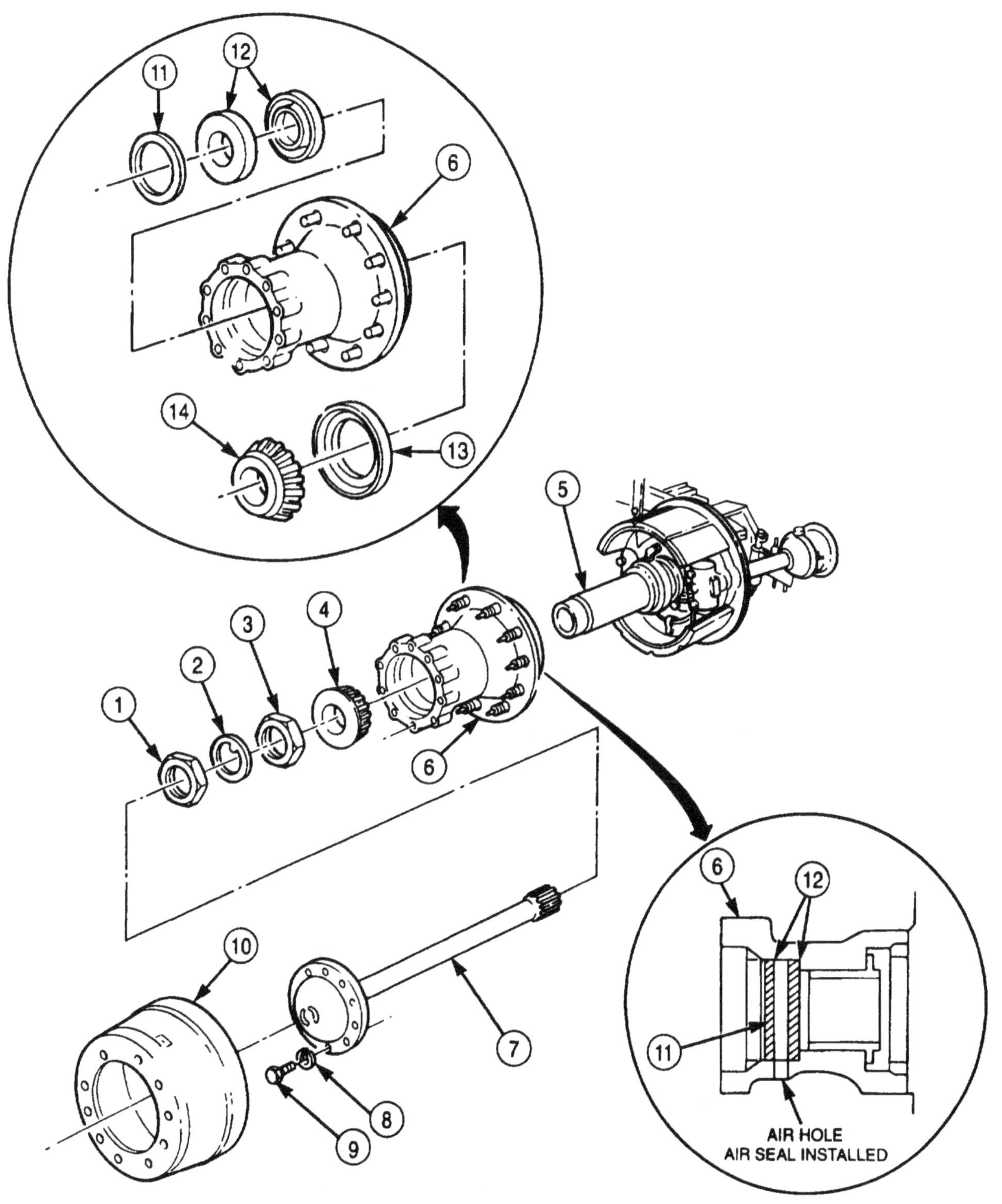

FOLLOW-ON TASKS: • Install rear wheels (TM 9-2320-272-10).
• Perform hub air seal leak test for rear axle (para. 3-462).
• Uncage spring brake (TM 9-2320-272-10).

3-462. HUB AIR SEAL LEAK TEST

THIS TASK COVERS:

a. Front Hub Leak Test **b. Rear Hub Leak Test**

INITIAL SETUP:

APPLICABLE MODELS
M939A2

SPECIAL TOOLS
Test kit (Appendix E, Item 137)
Air gauge (Appendix E, Item 4)

TOOLS
General mechanic's tool kit (Appendix E, Item 1)

MATERIALS/PARTS
Antiseize tape (Appendix C, Item 72)

EQUIPMENT CONDITION
- Front wheel valve removed (para. 3-456).
- Rear wheel valve removed (para. 3-457).

GENERAL SAFETY INSTRUCTIONS
When working with compressed air, wear eyeshields and ensure source pressure does not exceed 30 psi (207 kPa).

a. Front Hub Leak Test

NOTE

Air seals may not retain prescribed pressure immediately after installation. Pressure will be maintained after seals have been in operation and have seated on spindle.

1. Install O-ring (8) and air manifold (9) on drive flange (7) with washer (5) and screw (4).
2. Install short end of spacer (3) on air manifold (9) with washer (2) and screw (1).
3. Install plug (6) on air manifold (9).
4. Remove hose (11) from bulkhead fitting (10) and install air pressure gauge on hose (11).

WARNING

Eyeshields must be worn when working with compressed air. Compressed air source will not exceed 30 psi (207 kPa). Failure to do so may result in injury to personnel.

CAUTION

Ensure external air source is known to be as clean and dry as air supplied by vehicle. If in doubt, use the vehicle's air supply. Contamination may cause damage to CTIS.

5. Apply air pressure to front hub until air pressure gauge reads 80 psi (552 kPa).
6. Observe air pressure gauge for one minute. Pressure should not drop below 70 psi (483 kPa). If air pressure dropped below 70 psi (483 kPa), check connections using soapsud method.
7. If connections are tight, replace hub air seals (para. 3-460 or 3-461).

WARNING

Air pressure will create airborne debris. Use eye protection. Failure to do so may result in injury to personnel.

8. If air pressure remained at 70 psi (483 kPa) or above, release air pressure through valve on air pressure gauge.
9. Remove air pressure gauge from hose (11) and connect hose (11) to bulkhead fitting (10).

3-462. HUB AIR SEAL LEAK TEST (Contd)

NOTE

Wrap all male threads with antiseize tape before installation.

10. Remove screw (4), washer (5), air manifold (9), and O-ring (8) from drive flange (7).
11. Remove plug (6) from air manifold (9).
12. Remove screw (1), washer (2), and spacer (3) from air manifold (9).

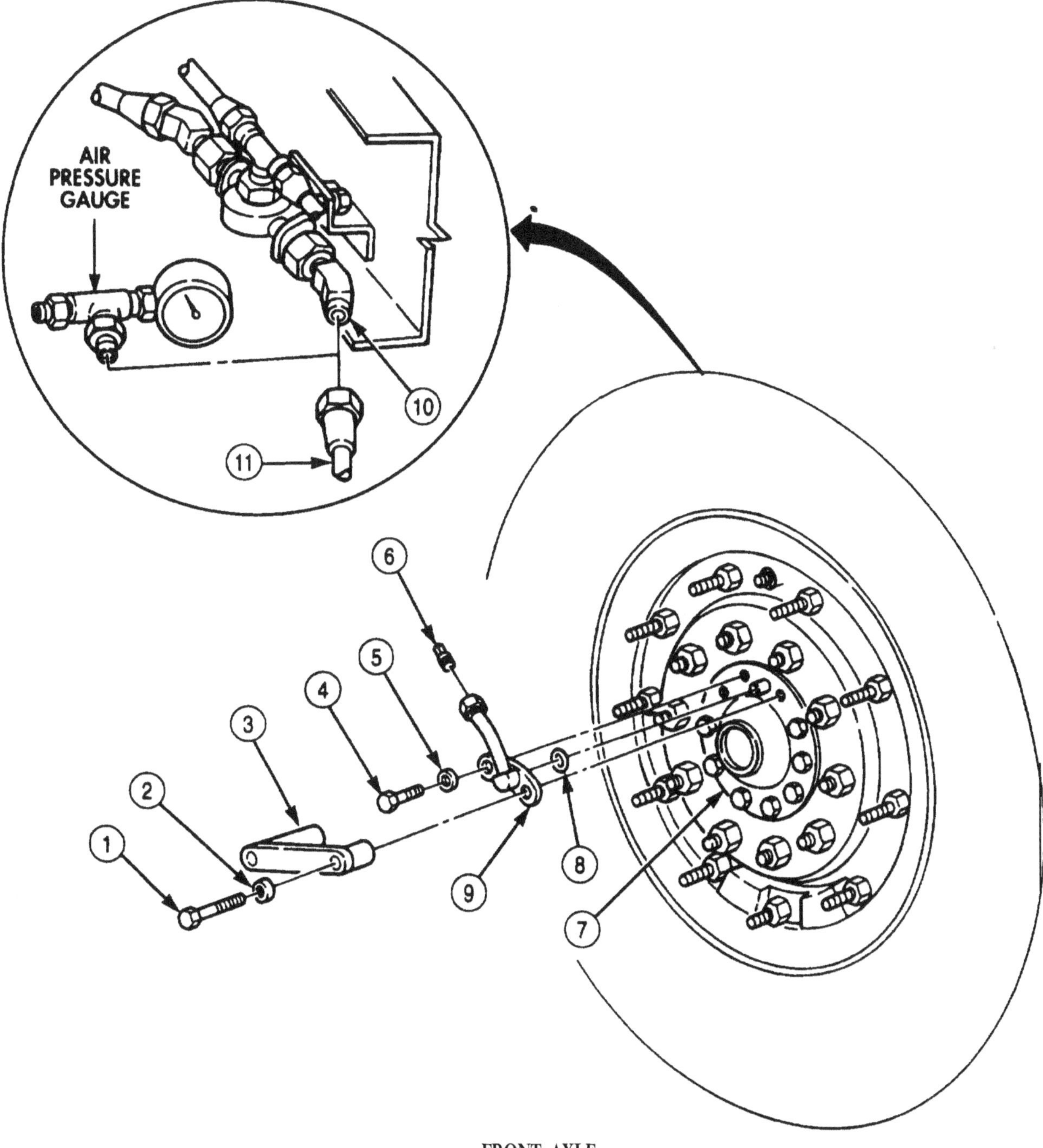

FRONT AXLE

3-462. HUB AIR SEAL LEAK TEST (Contd)

b. Rear Hub Leak Test

NOTE

Air seals may not retain prescribed pressure immediately after installation. Pressure will be maintained after seals have been in operation and have seated on spindle.

1. Install washer (4), O-ring (3), and rear hub connector (1) on hub (5).
2. Install plug (2) on connector (1).
3. Disconnect hose (7) from quick-exhaust valve elbow (6) and install pressure gauge on hose (7).

WARNING

Eyeshields must be worn when working with compressed air. Compressed air source will not exceed 30 psi (207 kPa). Failure to do so may result in injury to personnel.

CAUTION

Ensure external air source is known to be as clean and dry as air supplied by vehicle. If in doubt, use the vehicle's air supply. Contamination may cause damage to CTIS.

4. Apply air pressure to rear hub (5) until air pressure gauge reads 80 psi (552 kPa).
5. Observe air pressure gauge for one minute. Pressure gauge should not drop below 70 psi (483 kPa). If air dropped below 70 psi (483 kPa), check connections for leaks using soapsud method.
6. If connections are tight, replace hub air seals (para. 3-460 or 3-461).

WARNING

Air pressure will create airborne debris. Use eye protection. Failure to do so may result in injury to personnel.

7. If air pressure remained at 70 psi (483 kPa) or above, release air pressure through valve on air pressure gauge.

NOTE

Wrap all male threads with antiseize tape before installation.

8. Remove air pressure gauge from hose (7), and connect hose (7) to quick-exhaust valve elbow (6).
9. Remove plug (2) from rear hub connector (1).
10. Remove connector (1), O-ring (3), and washer (4) from hub (5).

3-462. HUB AIR SEAL LEAK TEST (Contd)

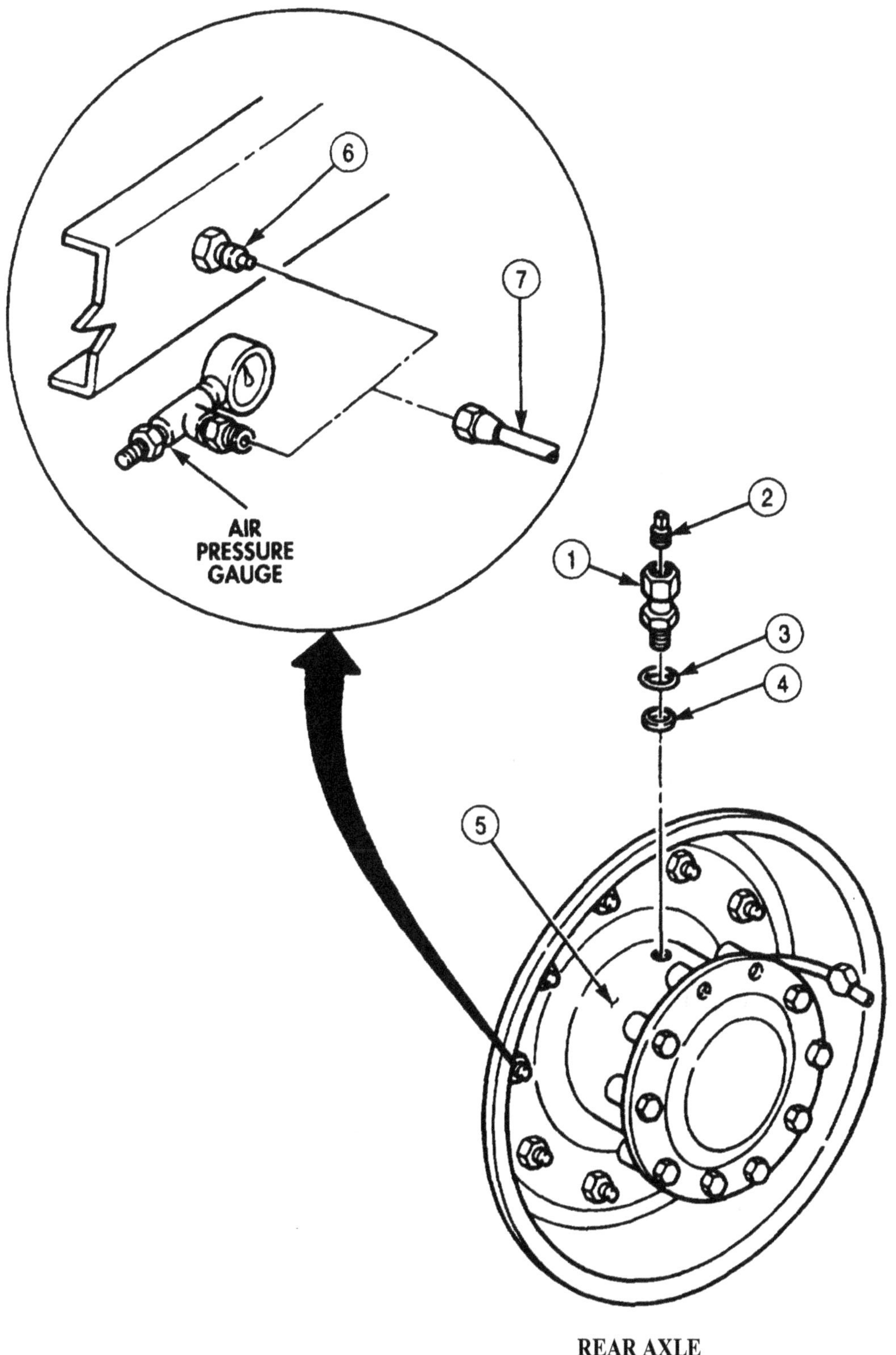

REAR AXLE

FOLLOW-ON TASKS: • Install front wheel valve (para. 3-456).
• Install rear wheel valve (para. 3-457).

3-463. REAR AXLE AIR MANIFOLD MAINTENANCE

THIS TASK COVERS:

a. Removal
b. Cleaning and Inspection
c. Installation

INITIAL SETUP:

APPLICABLE MODELS
M939A2

TOOLS
General mechanic's tool kit (Appendix E, Item 1)

MATERIALS/PARTS
Two O-rings (Appendix D, Item 488)
Antiseize tape (Appendix C, Item 72)
Sealing compound (Appendix C, Item 66)

EQUIPMENT CONDITION
Rear hub removed (para. 3-461).

a. Removal

1. Disconnect air line (11) from fitting (12), and remove fitting (12) from air manifold (13).
2. Remove two screws (8), washers (9), air manifold (13), and O-ring (10) from dust shield (4). Discard O-ring (10).
3. Disconnect nut (2) and tube (3) from elbow (1) and dust shield (4).
4. Remove elbow (1) and O-ring (7) from adapter ring (6). Discard O-ring (7).
5. Remove adapter ring (6) from spindle (5).

b. Cleaning and Inspection

1. For general cleaning instructions, refer to para. 2-14.
2. For general inspection instructions, refer to para. 2-15.
3. Inspect fitting (12), air manifold (13), tube (3), elbow (1), and adapter ring (6) for stripped threads, cracks, bends, and excessive wear. Replace parts if damaged.

c. Installation

1. Apply a thin coat of sealing compound to mating surfaces of spindle (5) and adapter ring (6), and install adapter ring (6) on spindle (5).

NOTE

Wrap all male threads with antiseize tape before installation.

2. Install new O-ring (7) and elbow (1) on adapter ring (6).
3. Insert tube (3) through hole in dust shield (4) and install tube (3) on elbow (1) with nut (2).
4. Install new O-ring (10) and air manifold (13) on dust shield (4) with two washers (9) and screws (8).
5. Install fitting (12) on air manifold (13).
6. Connect air line (11) to fitting (12).

3-463. REAR AXLE AIR MANIFOLD MAINTENANCE (Contd)

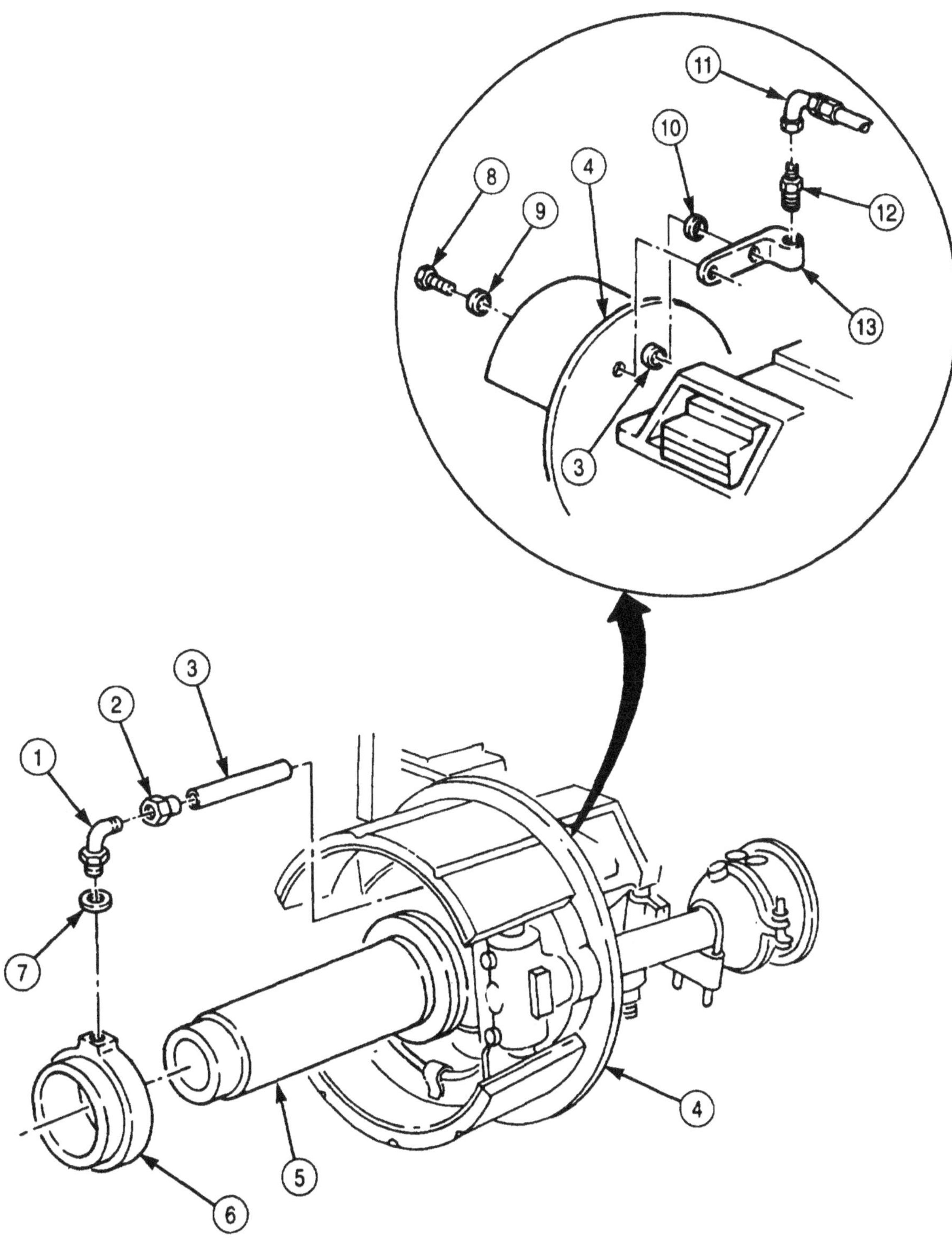

FOLLOW-ON TASK: Install rear hub (para. 3-461).

3-464. AIR DRYER AND CHECK VALVE MAINTENANCE

THIS TASK COVERS:

a. Removal
b. Disassembly
c. Cleaning and Inspection
d. Assembly
e. Installation

INITIAL SETUP:

APPLICABLE MODELS
M939A2

TOOLS
General mechanic's tool kit (Appendix E, Item 1)

MATERIALS/PARTS
Filter element kit (Appendix D, Item 124)
Four locknuts (Appendix D, Item 294)
Terminal (Appendix D, Item 681)
Connector (Appendix D, Item 41)
Two lockwashers (Appendix D, Item 400)
Tiedown straps (Appendix D, Item 696)
Cap and plug set (Appendix C, Item 14)
Antiseize tape (Appendix C, Item 72)
Lint-free cloth (Appendix C, Item 21)

REFERENCES (TM)
TM 9-2320-272-10
TM 9-2320-272-24P

EQUIPMENT CONDITION
- Parking brake set (TM 9-2320-272-10).
- Air reservoirs drained (TM 9-2320-272-10).

GENERAL SAFETY INSTRUCTIONS
- When cleaning with compressed air, wear eyeshields and ensure source pressure.
- Eyeshields must be worn when releasing compressed air.

---

a. Removal

WARNING

Air system components are subject to high pressure. Always relieve pressure before loosening or removing air system components. Failure to do so may result in injury to personnel.

NOTE

- Note location and position of tiedown straps for installation.
- When removing air lines, plug ends and tag for installation.

1. Remove all tiedown straps (13) from air lines (12), (16), and (18). Discard tiedown straps (13).
2. Disconnect lead (26) from connector (25).
3. Remove nut (22) and ground wire (21) from screw (23).
4. Remove air line (18) and elbow (17) from air dryer (14).
5. Remove air line (12) and connector (11) from adapter (10).
6. Remove air line (16) and elbow (8) from adapter (9).
7. Remove two locknuts (1), washers (2), screws (15), washers (2), and dryer (14) from bracket (5). Discard locknuts (1).
8. Remove two locknuts (7), washers (6), screws (3), and bracket (5) from frame (4). Discard locknuts (7).

NOTE

Perform steps 9 and 10 for M931A2 and M932A2 model vehicles.

3-464. AIR DRYER AND CHECK VALVE MAINTENANCE (Contd)

9. Remove two locknuts (34), washers (35), screws (32), washers (33), and air dryer (14) from bracket (29). Discard locknuts (34).
10. Remove two locknuts (27), washers (28), screws (31), and bracket (29) from frame (30). Discard locknuts (27).

NOTE

Perform steps 11 and 12 if replacing check valve.

11. Remove air line (16) and elbow (19) from check valve (20).
12. Remove check valve (20) from relief valve (24).

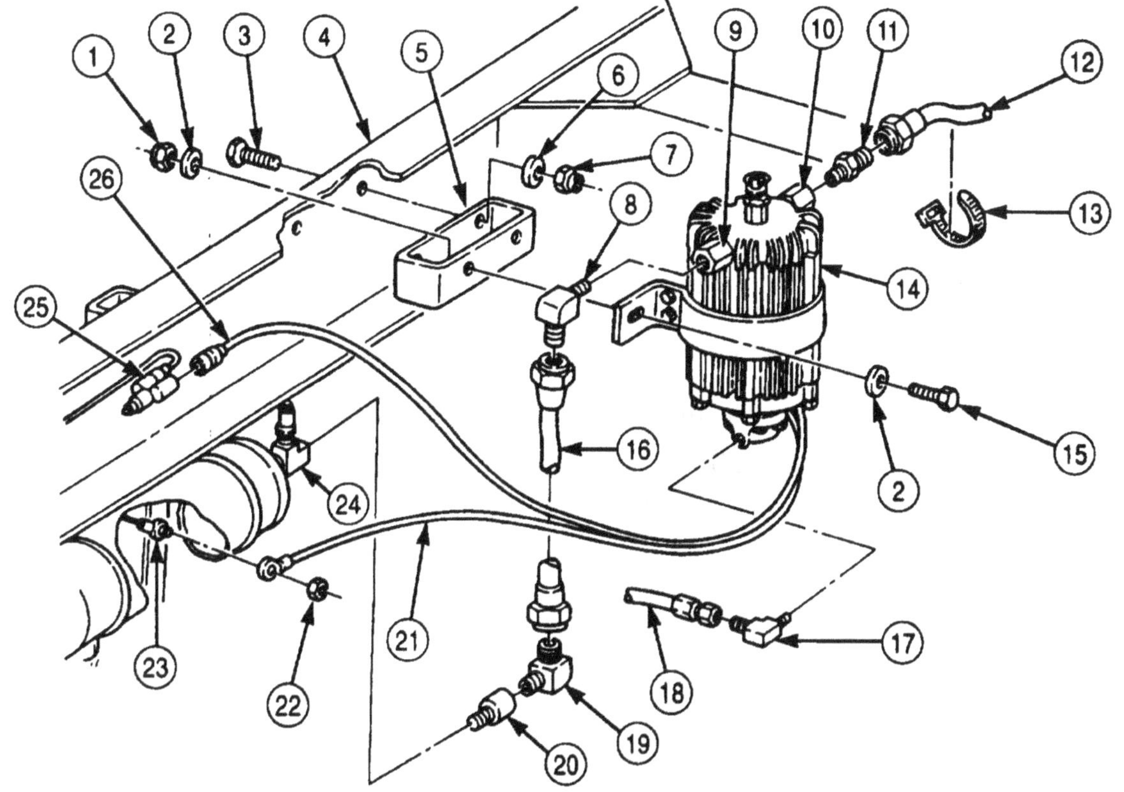

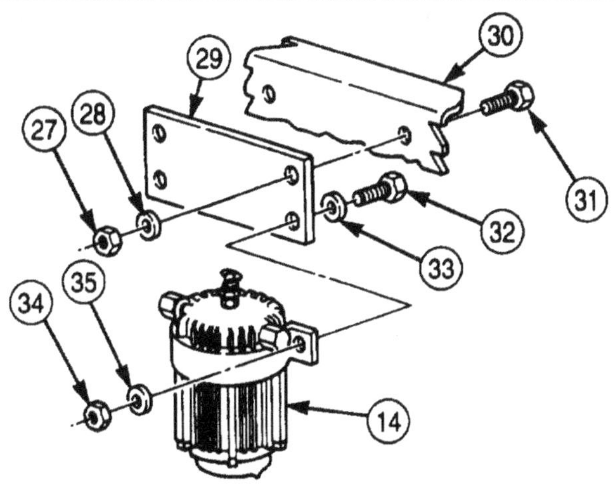

M931A2 AND M932A2

3-464. AIR DRYER AND CHECK VALVE MAINTENANCE (Contd)

b. Disassembly

NOTE

Note location of bracket on air dryer shell for installation.

1. Remove two nuts (6), lockwashers (5), screws (8), spacers (4), and bracket (7) from air dryer shell (13). Discard lockwashers (5).
2. Remove two adapters (3), O-rings (2), and safety valve (1) from shell (13). Discard O-rings (2).
3. Remove four screws (10), brake valve (9), O-ring (11), and filter (12) from shell (13). Discard O-ring (11) and filter (12).

c. Cleaning and Inspection

WARNING

Eyeshields must be worn when cleaning with compressed air. Compressed air source will not exceed 30 psi (207 kPa). Failure to do so may result in injury to personnel.

CAUTION

Do not clean air dryer assembly with liquid solvent. Damage to internal components may result.

1. Clean all parts with compressed air and wipe with lint-free cloth.
2. Inspect shell (13) and brake valve (9) for cracks and stripped threads. Replace shell (13) or brake valve (9) if cracked or threads are stripped.
3. Inspect safety valve (1) and adapters (3) for stripped threads. Replace safety valve (1) or adapters (3) if threads are stripped.
4. Inspect bracket (7) for cracks, bends, and excessive corrosion. Replace bracket (7) if cracked, bent, or excessively corroded.

d. Assembly

1. Install new filter (12) in shell (13).
2. Install new O-ring (11) and brake valve (9) on shell (13) with four screws (10).

NOTE

Wrap all male pipe threads with antiseize tape before installation.

3. Install safety valve (1), two new O-rings (2), and adapters (3) on shell (13).
4. Install bracket (7) on shell (13) with two screws (8), spacers (4), new lockwashers (5), and nuts (6).

3-464. AIR DRYER AND CHECK VALVE MAINTENANCE (Contd)

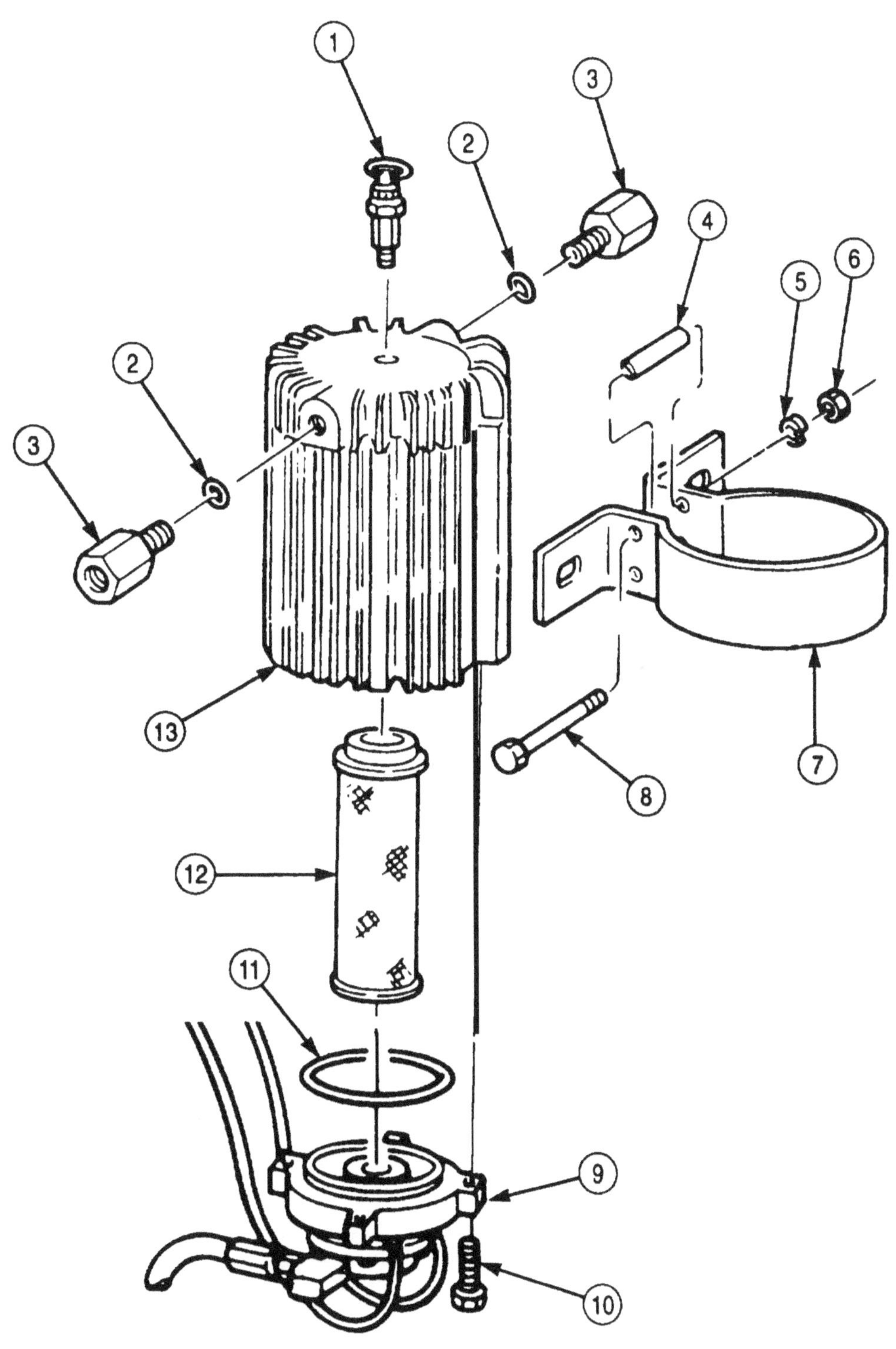

3-464. AIR DRYER AND CHECK VALVE MAINTENANCE (Contd)

e. Installation

NOTE

Wrap all male pipe threads with antiseize tape before installation.

Perform steps 1 and 2 if check valve was removed.

1. Install check valve (20) on relief valve (25).
2. Install elbow (19) on check valve (20), and connect air line (16) to elbow (19).
3. Install bracket (5) on frame (4) with two screws (3), washers (6), and new locknuts (7).
4. Install air dryer (14) on bracket (5) with two washers (2), screws (15), washers (2), and new locknuts (1).

NOTE

Perform steps 5 and 6 for M931A2 and M932A2 model vehicles.

5. Install plate (31) on frame (32) with two screws (33), washers (30), and new locknuts (29).
6. Install air dryer (14) on plate (31) with two washers (35), screws (34), washers (35), and new locknuts (36).
7. Install elbow (8) on adapter (9), and connect air line (16) to elbow (8).
8. Install connector (11) and air line (12) on adapter (10).
9. Install elbow (17) and air line (18) on air dryer (14).

NOTE

Perform steps 10 and 11 only if new air dryer is being installed.

10. Cut ground wire (21) to length and attach new terminal (23) to ground wire (21).
11. Cut lead (28) to length and attach new connector (27) to lead (28).
12. Install ground wire (21) on screw (24) with nut (22).
13. Connect lead (27) to connector (26).
14. Install new tiedown straps (13) on air lines (12), (16), and (18) as previously noted.

3-464. AIR DRYER AND CHECK VALVE MAINTENANCE (Contd)

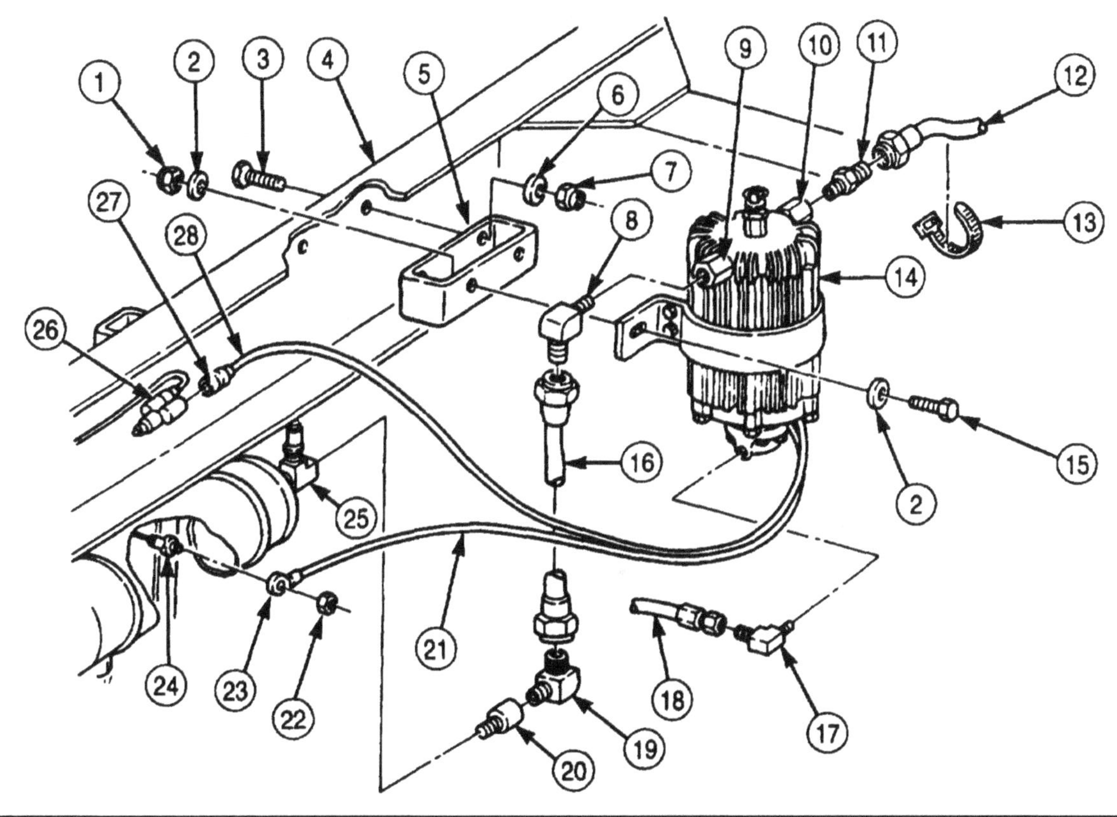

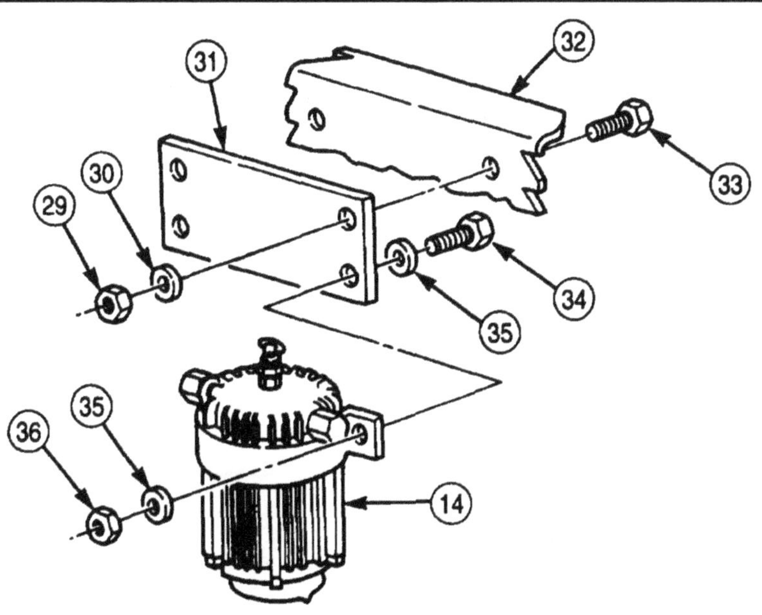

M931A2 AND M932A2

FOLLOW-ONTASK: Start engine (TM 9-2320-272-10) and check for air leaks and proper CTIS operation.

3-465. AIR DRYER FILTER REPLACEMENT

THIS TASK COVERS:

a. Removal b. Installation

INITIAL SETUP:

APPLICABLE MODELS
M939A2

TOOLS
General mechanic's tool kit (Appendix E, Item 1)

MATERIALS/PARTS
Filter element kit (Appendix D, Item 124)

REFERENCES (TM)
TM 9-2320-272-10
TM 9-2320-272-24P

EQUIPMENT CONDITION
- Parking brake set (TM 9-2320-272-10).
- Air reservoirs drained (TM 9-2320-272-10).

GENERAL SAFETY INSTRUCTIONS
Eyeshields must be worn when releasing compressed air.

a. Removal

WARNING

Air system components are subject to high pressure. Always relieve pressure before loosening or removing air system components. Failure to do so may result in injury to personnel.

1. Remove four screws (3), brake valve (2), and filter element (5) from air dryer shell (1). Discard filter element (5).
2. Remove O-ring (4) from brake valve (2). Discard O-ring (4).

b. Installation

1. Install new O-ring (4) on brake valve (2).
2. Install new filter element (5) on air dryer shell (1), and position brake valve (2) under filter element (5).
3. Install brake valve (2) on air dryer shell (1) with four screws (3).

3-465. AIR DRYER FILTER REPLACEMENT (Contd)

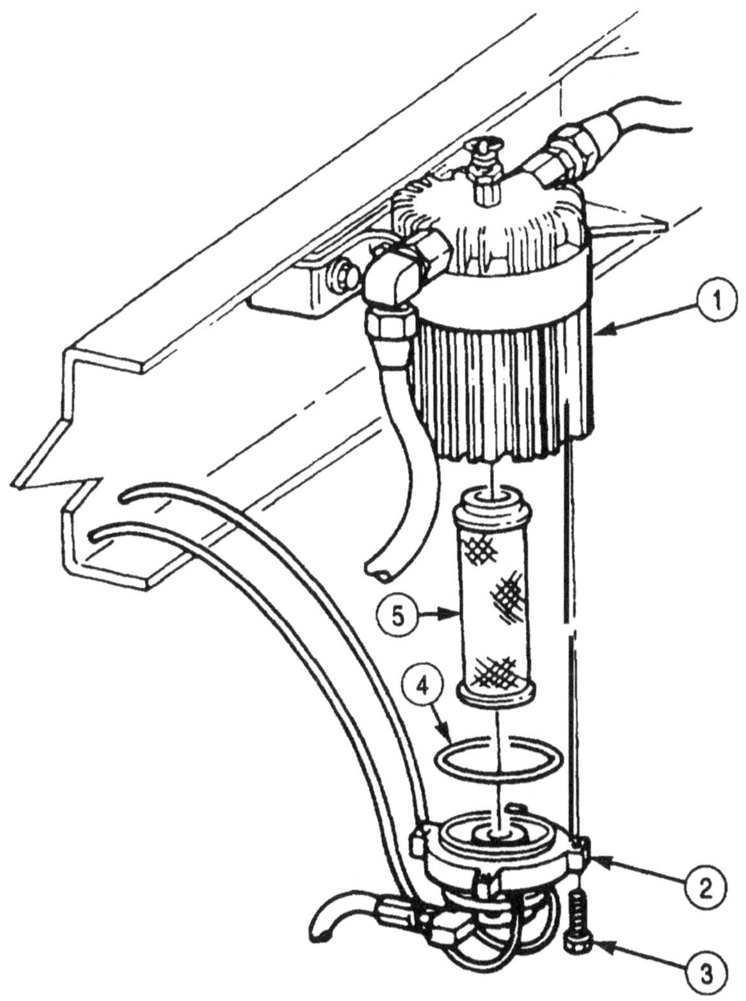

FOLLOW-ON TASK: Start engine (TM 9-2320-272-10) and check for leaks.

3-466. WATER SEPARATOR MAINTENANCE

THIS TASK COVERS:

a. Removal
b. Disassembly
c. Cleaning and Inspection
d. Assembly
e. Installation

INITIAL SETUP:

APPLICABLE MODELS
M939A2

TOOLS
General mechanic's tool kit (Appendix E, Item 1)

MATERIALS/PARTS
Fluid pressure kit (Appendix 128)
Crease (Appendix C, Item 32)

REFERENCES (TM)
TM 9-2320-272-10
TM 9-2320-272-24P

EQUIPMENT CONDITION
- Parking brake set (TM 9-2320-272-10).
- Air reservoirs drained (TM 9-2320-272-10).

GENERAL SAFETY INSTRUCTIONS
When cleaning with compressed air, wear eyeshields and ensure source pressure does not exceed 30 psi (207 kPa).

a. Removal

1. Disconnect two air lines (1) from adapters (2) on water separator (3).
2. Remove two adapters (2) from water separator (3).

b. Disassembly

1. Remove body (4) from bowl (12).
2. Remove stud (9), filter (8), louver (7), gasket (6), and O-ring (5) from body (4). Discard filter (8), gasket (6), and O-ring (5).
3. Remove automatic drain valve (10) and O-ring (11) from bowl (12). Discard O-ring (11).

c. Cleaning and Inspection

1. For general inspection instructions, refer to para. 2-15.
2. Clean bowl (12) with warm water and all other parts with soap and warm water.

WARNING

Eyeshields must be worn when cleaning with compressed air. Compressed air source will not exceed 30 psi (207 kPa). Failure to do so may result in injury to personnel.

3. From inside body (4), blow out internal passages with compressed air.

d. Assembly

1. Coat lip of automatic drain valve (10) with grease and install new O-ring (11) and automatic drain valve (10) in bowl (12).
2. Install new O-ring (5), new gasket (6), louver (7), and new filter (8) on body (4) with stud (9).
3. Install body (4) on bowl (12).

3-466. WATER SEPARATOR MAINTENANCE (Contd)

e. Installation

1. Install two adapters (2) on water separator (3).
2. Connect two air lines (1) to adapters (2).

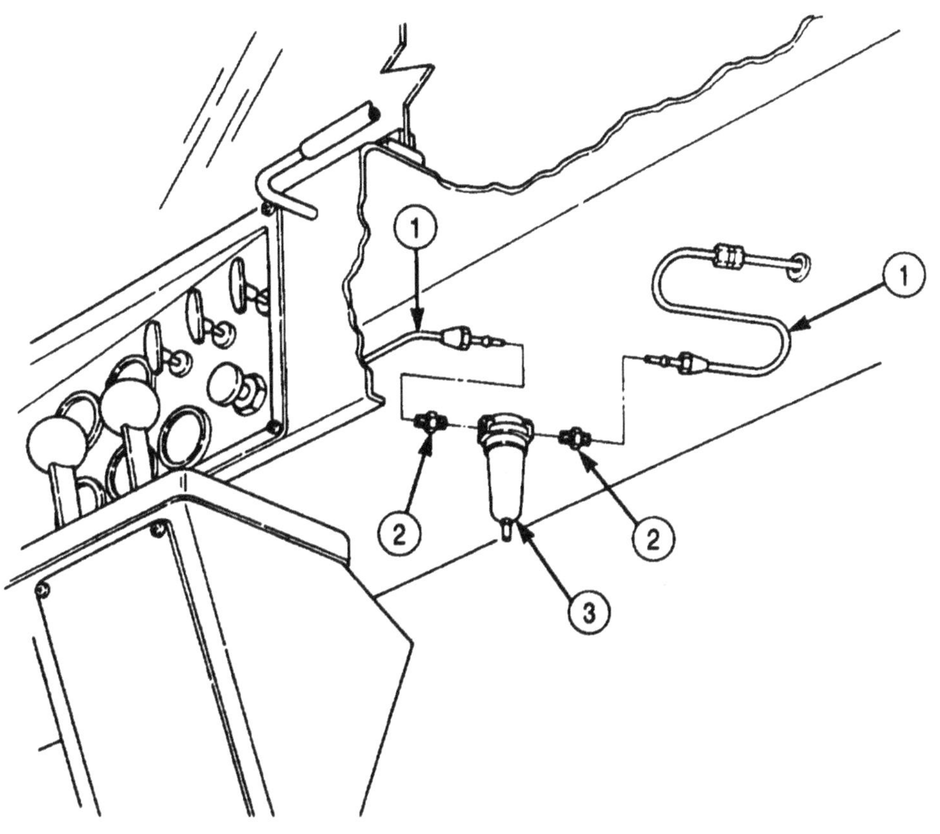

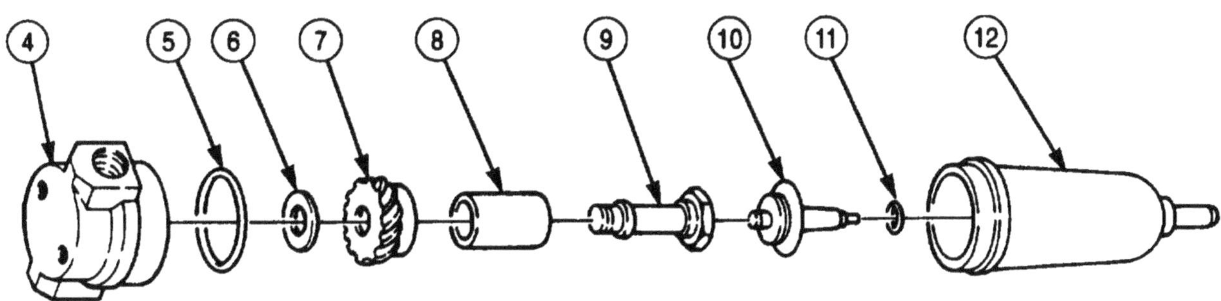

FOLLOW-ON TASK: Start engine (TM 9-2320-272-10) and check for leaks.

3-467. PRESSURE SWITCH REPLACEMENT

THIS TASK COVERS:

a. Removal **b. Installation**

INITIAL SETUP:

APPLICABLE MODELS
M939A2

TOOLS
General mechanic's tool kit (Appendix E, Item 1)

MATERIALS/PARTS
Tiedown straps (Appendix D, Item 690)
Antiseize tape (Appendix C, Item 72)

REFERENCES (TM)
TM 9-2320-272-10
TM 9-2320-272-24P

EQUIPMENT CONDITION
- Parking brake set (TM 9-2320-272-10)
- Air reservoirs drained (TM 9-2320-272-10).
- Battery ground cables disconnected (para. 3-126).

GENERAL SAFETY INSTRUCTIONS
- Eyeshields must be worn when releasing compressed air.
- Air reservoirs must be drained before removing air system components.

a. Removal

WARNING

- Release all air pressure before loosening or removing air system component(s).
- Eyeshields must be worn when releasing compressed air. Failure to do so may result in injury to personnel.

1. Remove tiedown straps (3) from pressure switch lead (4). Discard tiedown straps (3).
2. Disconnect pressure switch (2) from wiring harness (1).
3. Remove pressure switch lead (4) from elbow (5).

b. Installation

NOTE

Wrap male threads with antiseize tape before installation.

1. Install pressure switch lead (4) on elbow (5).
2. Connect pressure switch (2) to wiring harness (1).
3. Install new tiedown straps (3) on pressure switch lead (4).

3-467. PRESSURE SWITCH REPLACEMENT (Contd)

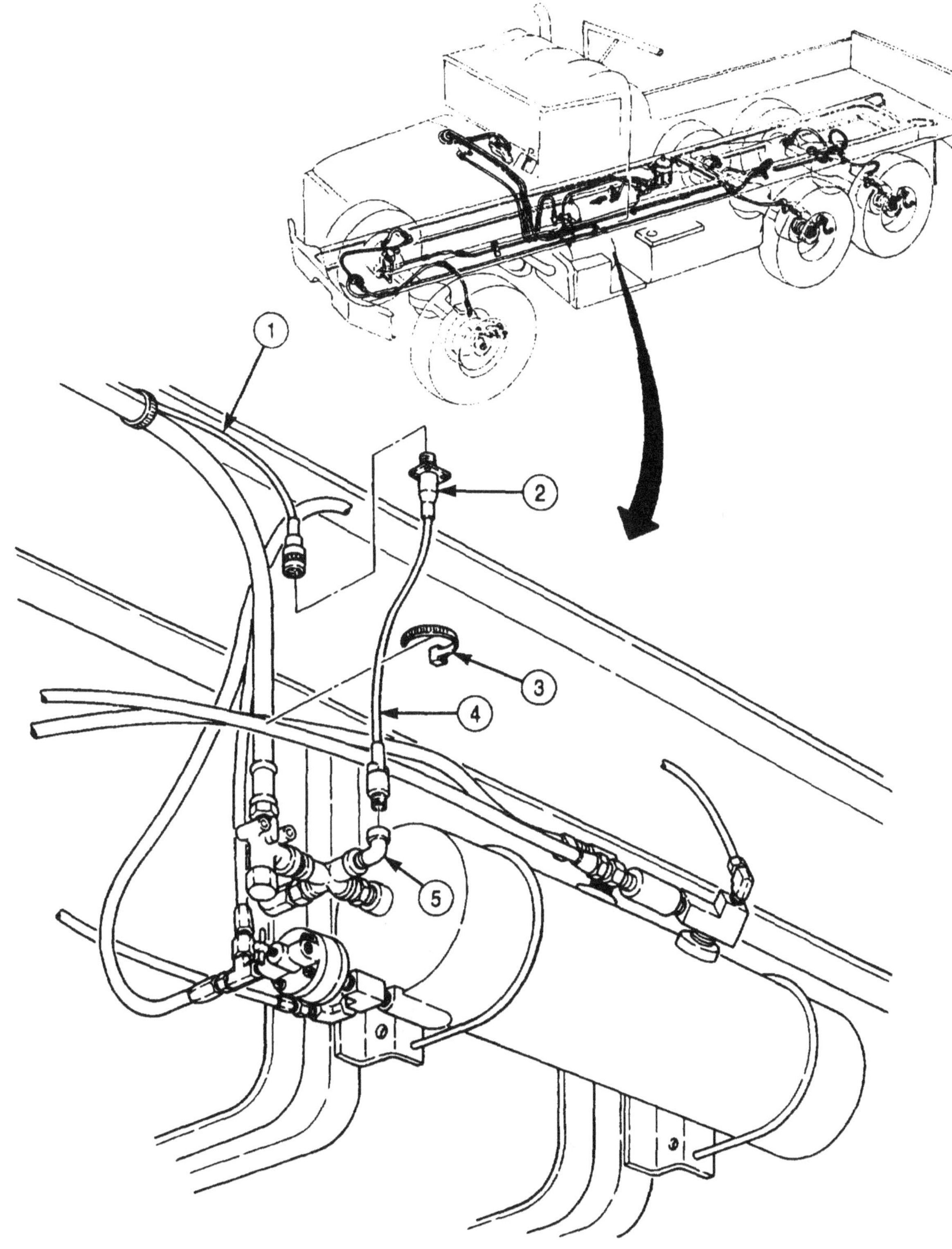

FOLLOW-ON TASKS:
- Connect battery ground cables (para. 3-126).
- Start engine (TM 9-2320-272-10) and check for air leaks.

3-468. ELECTRONIC CONTROL UNIT (ECU) REPLACEMENT

THIS TASK COVERS:

a. Removal **b. Installation**

INITIAL SETUP:

APPLICABLE MODELS
M939A2

TOOLS
General mechanic's tool kit (Appendix E, Item 1)

REFERENCES (TM)
TM 9-2320-272-10
TM 9-2320-272-24P

EQUIPMENT CONDITION
- Parking brake set (TM 9-2320-272-10).
- Battery ground cables disconnected (para. 3-126).

CAUTION

Use extreme care when handling ECU. Failure to do so may result in damage to internal components.

a. Removal

1. Disconnect CTIS wiring harness connector (1) from ECU receptacle (5).
2. Remove three screws (3) and ECU (4) from transmission shifter tower (2).

b. Installation

1. Install ECU (4) on transmission shifter tower (2) with three screws (3).
2. Connect CTIS wiring harness connector (1) to ECU receptacle (5).

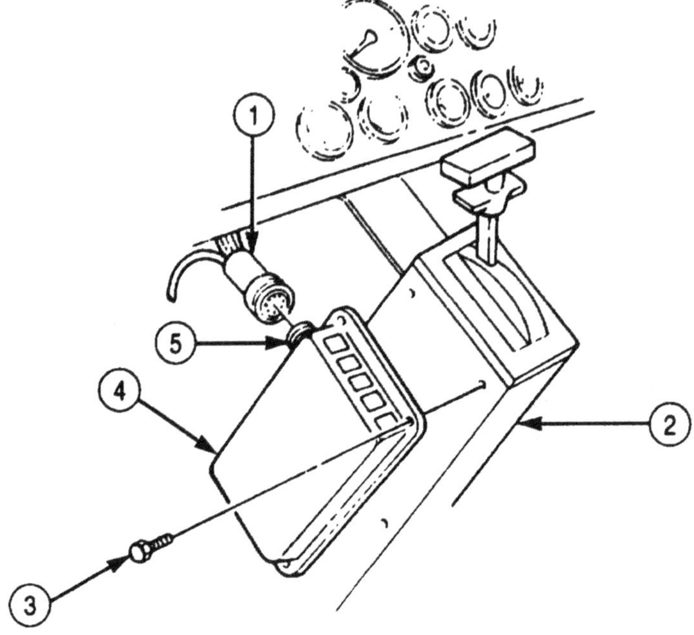

FOLLOW-ON TASKS:
- Connect battery ground cables (para. 3-126).
- Start engine (TM 9-2320-272-10) and check CTIS for proper operation.

3-469. AMBER WARNING LIGHT REPLACEMENT

THIS TASK COVERS:

a. Removal **b. Installation**

INITIAL SETUP:

APPLICABLE MODELS
M939A2

TOOLS
General mechanic's tool kit (Appendix E, Item 1)

MATERIALS/PARTS
Lockwasher (Appendix D, Item 418)

REFERENCES (TM)
TM 9-2320-272-10
TM 9-2320-272-24P

EQUIPMENT CONDITION
- Parking brake set (TM 9-2320-272-10).
- Battery ground cables disconnected (para. 3-126).

a. Removal

1. Disconnect electrical lead (8) from wiring harness (9).
2. Remove screw (6), ground strap (7), lockwasher (5), and J-nut (2) from bracket (3) and instrument cluster (1). Discard lockwasher (5).
3. Remove two nuts (10), screws (4), and amber warning light (11) from instrument cluster (1).

b. Installation

1. Install amber warning light (11) on instrument cluster (1) with two screws (4) and nuts (10).
2. Install new lockwasher (5) and ground strap (7) on bracket (3) and instrument cluster (1) with J-nut (2) and screw (6).
3. Connect electrical lead (8) to wiring harness (9).

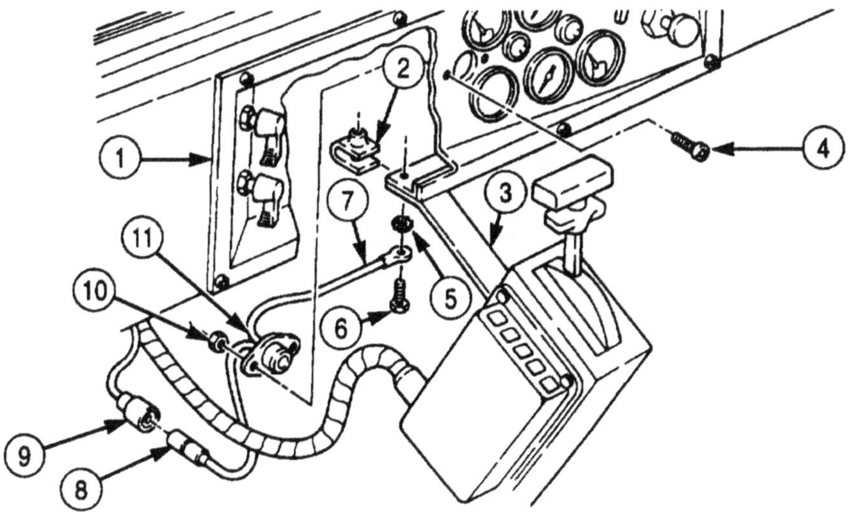

FOLLOW-ON TASK: Connect battery ground cables (para. 3-126).

3-470. CTIS WIRING HARNESS REPLACEMENT

THIS TASK COVERS:

a. Removal **b. Installation**

INITIAL SETUP:

APPLICABLE MODELS
M939A2

TOOLS
General mechanic's tool kit (Appendix E, Item 1)

MATERIALS/PARTS
Grommet (Appendix D, Item 252)
Tiedown straps (Appendix D, Item 690)

REFERENCES (TM)
TM 9-2320-272-10
TM 9-2320-272-24P

EQUIPMENT CONDITION
- Parking brake set (TM 9-2320-272-10).
- Battery ground cables disconnected (para. 3-126).

a. Removal

NOTE

- When removing CTIS wiring harness, note location of all tiedown straps and protective covering for installation.
- Tag all wiring harness leads and note routing for installation.

1. Disconnect wiring harness connector (12) from ECU receptacle (9).
2. Disconnect wiring harness connector (10) from amber warning light wire (11).
3. Disconnect wiring harness connector (18) from power and ground cable (17).
4. Disconnect wiring harness connector (15) from blackout wire (16).
5. Disconnect wiring harness connector (20) from pressure transducer (19).
6. Disconnect wiring harness connector (22) from pneumatic controller solenoid receptacle (21).
7. Disconnect wiring harness connector (1) from speed signal generator wire (23) on transfer case (5).
8. Disconnect wire (4) from pressure switch wire (24) located on right frame rail (2) above wet tank (3).
9. Remove screw (26), washer (27), and clamp (28) from wiring harness (7) and bracket (25).
10. Remove screw (30), washer (29), and clamp (31) from wiring harness (7) and firewall (13).
11. Remove wiring harness (7) from frame rails (2) and (8) and crossmember (6).
12. Carefully pull wiring harness (7) through grommet (14) on firewall (13).

NOTE

Perform step 13 if grommet is damaged.

13. Remove grommet (14) from firewall (13). Discard grommet (14).

3-470. CTIS WIRING HARNESS REPLACEMENT (Contd)

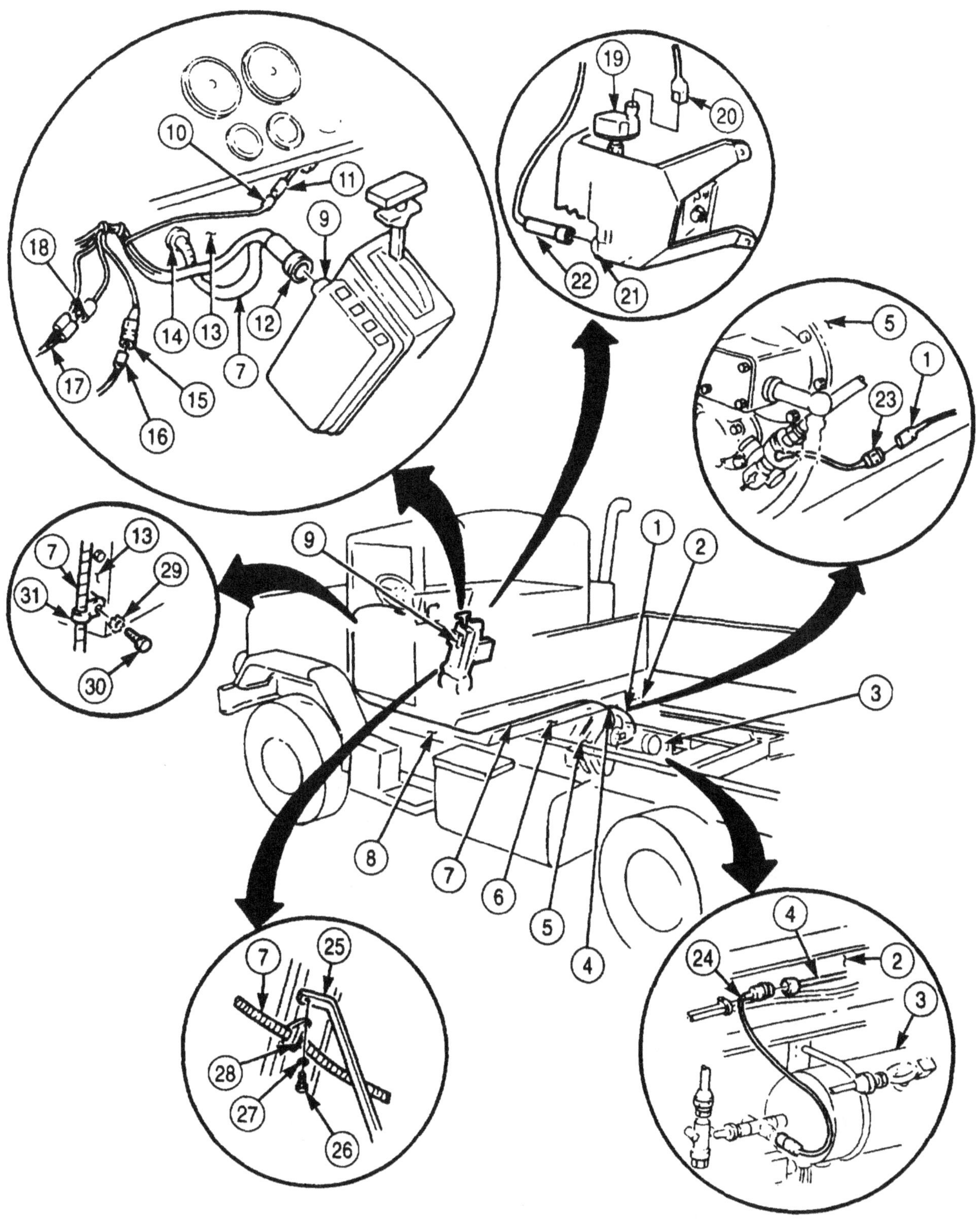

3-470. CTIS WIRING HARNESS REPLACEMENT (Contd)

b. Installation

CAUTION

Use care when routing CTIS wiring harness. Snagging may result, and forceful pulling will cause damage to harness.

NOTE

- Perform step 1 if grommet was removed.
- Route CTIS wiring harness and install tiedown straps as noted in removal.

1. Install new grommet (14) on firewall (13).
2. Insert wiring harness (7) through grommet (14) on firewall (13), and pull wiring harness (7) into engine compartment.
3. Route wiring harness (7) along frame rail (8), crossmember (6), and frame rail (2).
4. Connect wiring harness connector (1) to speed signal generator wire (23) on transfer case (5).
5. Connect wiring harness connector (4) to pressure switch wire (24) located above wet tank (3) at right frame rail (2).
6. Position clamp (28) on wiring harness (7) and install clamp (28) on bracket (25) with washer (27) and screw (26).
7. Position clamp (31) on wiring harness (7) and install clamp (31) on firewall (13) with washer (29) and screw (30).
8. Connect wiring harness connector (22) to pneumatic controller solenoid receptacle (21).
9. Connect wiring harness connector (19) to pressure transducer (20).
10. Connect wiring harness connector (15) to blackout wire (16).
11. Connect wiring harness connector (18) to power and ground cable (17).
12. Connect wiring harness connector (10) to amber warning light wire (11).
13. Connect wiring harness connector (12) to ECU receptacle (9).

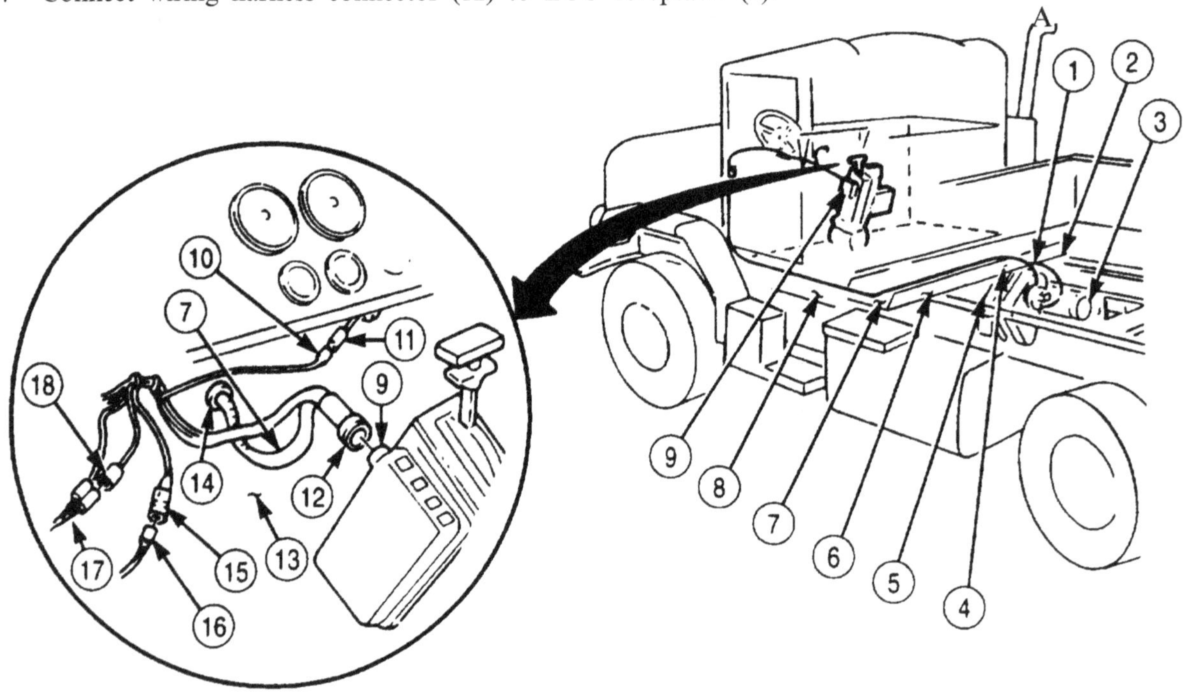

3-470. CTIS WIRING HARNESS REPLACEMENT (Contd)

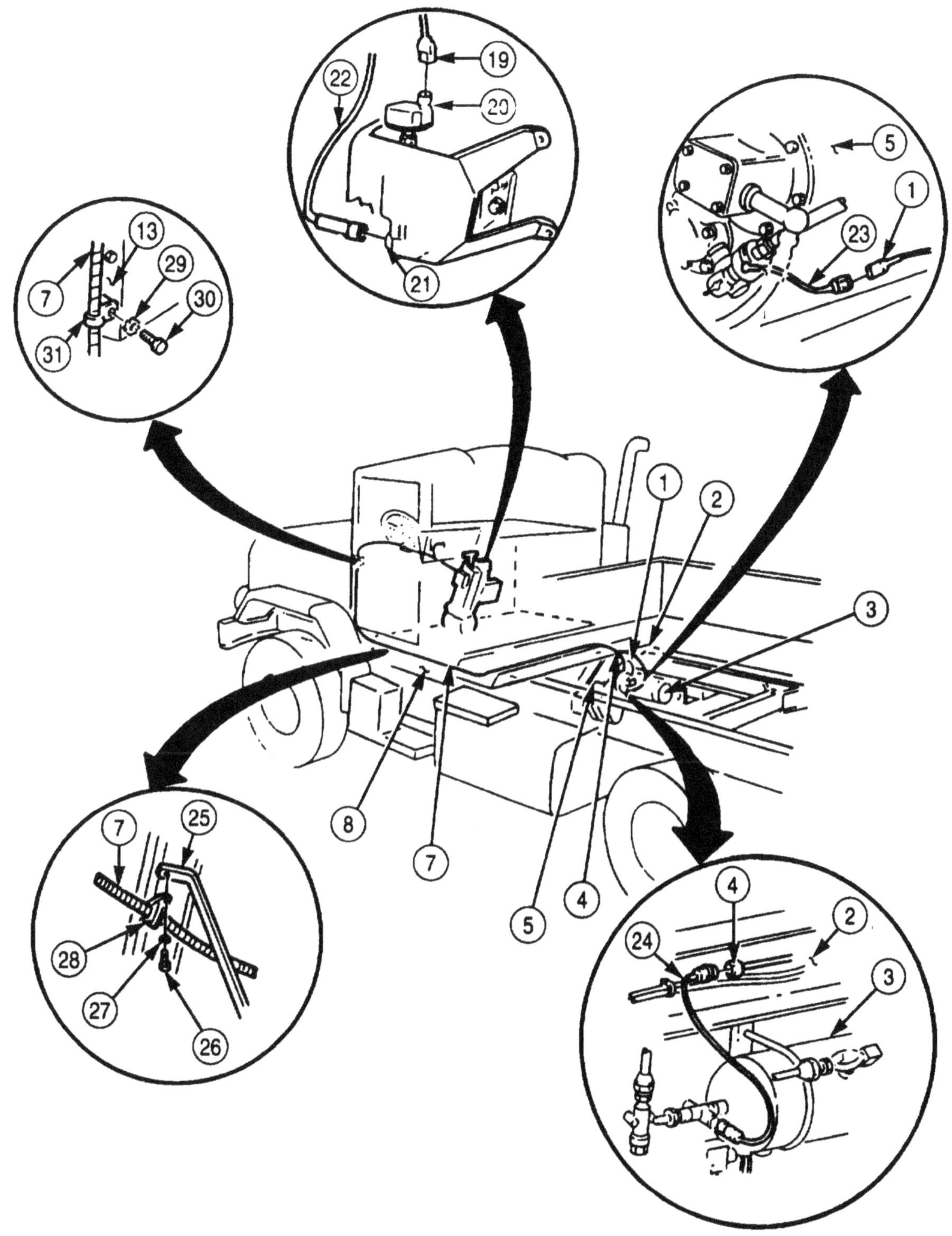

FOLLOW-ON TASKS:
- Connect battery ground cables (para. 3-126).
- Start engine and check CTIS for proper operation (TM 9-2320-272-10).

3-471. SPEED SIGNAL GENERATOR REPLACEMENT

THIS TASK COVERS:

a. Removal **b. Installation**

INITIAL SETUP:

APPLICABLE MODELS
M939A2

TOOLS
General mechanic's tool kit (Appendix E, Item 1)

REFERENCES (TM)
TM 9-2320-272-10
TM 9-2320-272-24P

EQUIPMENT CONDITION
- Parking brake set (TM 9-2320-272-10).
- Battery ground cables disconnected (para. 3-126).

a. Removal

NOTE

Gasket and drive pins are provisioned with speedometer drive adapter.

1. Disconnect speed signal generator electrical lead (5) from wiring harness (4).
2. Disconnect speedometer cable (1) from speed signal generator (3), and remove gasket (2) from speedometer cable (1).
3. Remove speed signal generator (3) and drive pin (6) from speedometer drive adapter (7).

b. Installation

NOTE

When installing drive pin, ensure end with double tab is facing speed signal generator.

1. Align tab of new drive pin (6) with slot of speedometer drive adapter (7), and install drive pin (6) on speedometer drive adapter (7).
2. Install speed signal generator (3) on drive pin (6) and speedometer drive adapter (7).
3. Install new gasket (2) on speedometer cable (l), and connect speedometer cable (1) to speed signal generator (3).
4. Connect speed signal generator electrical lead (5) to wiring harness (4).

3-471. SPEED SIGNAL GENERATOR REPLACEMENT (Contd)

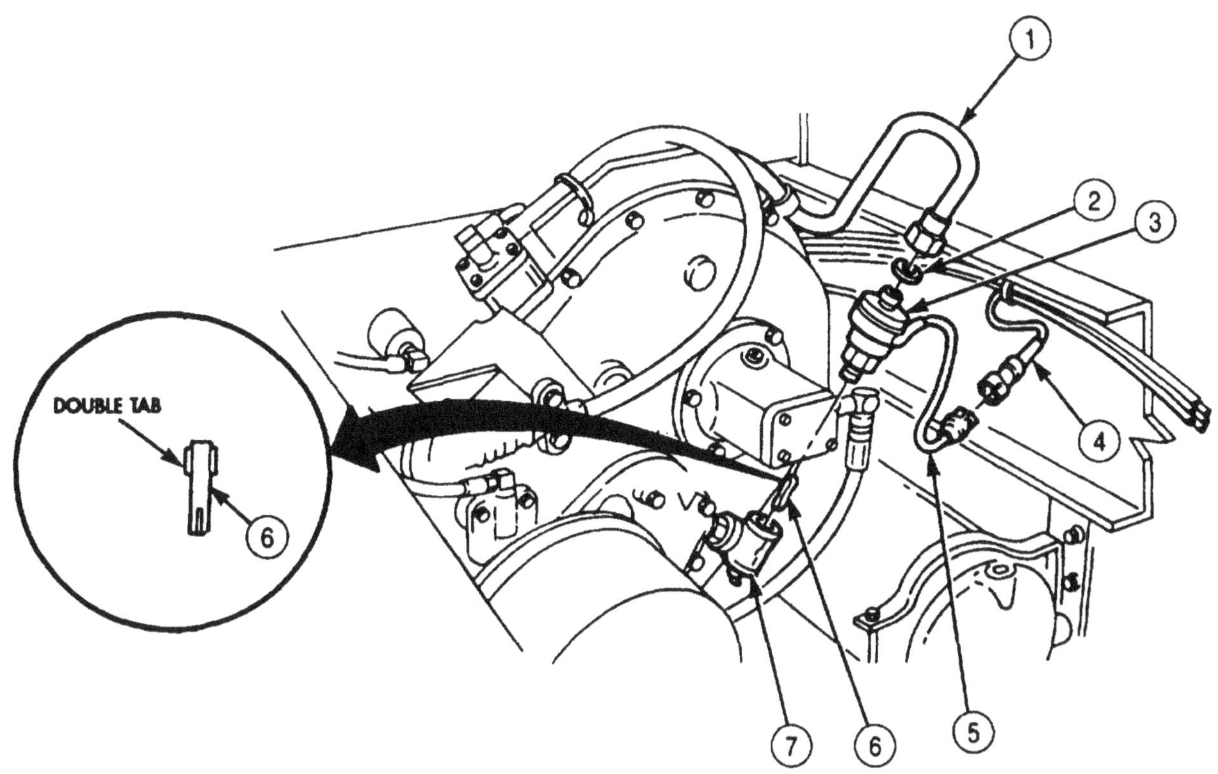

FOLLOW-ON TASK: Connect battery ground cables (para. 3-126).

CHAPTER 4
DIRECT SUPPORT (DS) MAINTENANCE

Section I. Direct. Support (DS) and General Support (GS) Mechanical Troubleshooting (page 4-1)
Section II. Power Plant Maintenance (page 4-35)
Section III. Engine Maintenance (M939/A1) (page 4-117)
Section IV. Engine Maintenance (M939A2) (page 4-271)
Section V. Cooling System Maintenance (page 4-352)
Section VI. Electrical System Maintenance (page 4-388)
Section VII. Transmission Maintenance (page 4-465)
Section VIII. Transfer Case Maintenance (page 4-504)
Section IX. Front and Rear Axle Maintenance (page 4-548)
Section X. Compressed Air and Brake System Maintenance (page 4-584)
Section XI. Power Steering System Maintenance (page 4-605)
Section XII. Frame Maintenance (page 4-680)
Section XIII. Body, Cab, and Hood Maintenance (page 4-691)
Section XIV. Special Purpose Bodies Maintenance (page 4-715)
Section XV. Winch, Hoist, and Power Takeoff Maintenance (page 4-808)
Section XVI. Special Purpose Kits Maintenance (page 4-1002)

4-1. GENERAL

This section provides troubleshooting and maintenance instructions for maintaining M939, M939A1, and M939A2 series vehicles at the Direct Support (DS) maintenance level. A description of the differences between models can be found in chapter 1 of TM 9-2320-272-24-1. Procedures found in this chapter pertain to all models within a vehicle series unless otherwise designated.

Section I. DIRECT SUPPORT (DS) AND GENERAL SUPPORT (GS) MECHANICAL TROUBLESHOOTING

4-2. SCOPE

a. This section provides diagnostic and corrective procedures for the mechanical system at the Direct Support (DS) maintenance level. These procedures cannot give all the answers or correct all vehicle malfunctions encountered. However, these procedures are an organized step-by-step approach to a problem that will direct tests and inspections toward the source of a problem and successful solution. Information in this section is for use in conjunction with, and as a supplement to, troubleshooting procedures found in TM 9-2320-272-24-1.

b. Do the easiest things first. Most problems are easily corrected.

c. Doublecheck before disassembly. The source of most problems can be traced to more than one part in a system.

d. Before attempting to correct a problem, diagnose the cause of the problem. Do not allow the same failure to occur again.

e. The following index will help you find specific problems and solutions in table 4-1, Mechanical Troubleshooting.

MECHANICAL TROUBLESHOOTING INDEX

MALFUNCTION NO.	MALFUNCTION	TROUBLESHOOTING PROCEDURE PAGE
	ENGINE	
1.	Engine will not crank	4-4
2.	Engine cranks, but fails to start.	4-4
3.	Engine starter motor operates, but does not engage flywheel ring gear.	4-4
4.	Engine runs, but misfires	4-5
5.	Engine stops	4-5
6.	Poor acceleration and/or lack of power	4-5
7.	Low oil pressure	4-6
8.	Contaminated engine oil	4-7
9.	Excessive oil consumption.	4-7
10.	Air cleaner element clogged with oil	4-7
11.	Engine knocks (mechanical noise)	4-7
12.	Excessive engine vibration	4-7
13.	Excessive gear noise	4-7
	FUEL SYSTEM	
14.	Engine idle rough, erratic.	4-7
15.	Excessive fuel consumption.	4-8
16.	Engine surges at idle speed.	4-8
17.	Engine surges at all speeds.	4-8
18.	Excessive exhaust smoke at idle, and under load.	4-8
19.	Engine fails to stop	4-8
20.	Engine misses	4-8
21.	Low power or loss of power.	4-9
22.	Engine overspeed	4-9
23.	Engine lubricating oil diluted by fuel	4-9
24.	Engine fuel knocks	4-9
	COOLING SYSTEM	
25.	Engine overheats.	4-9
26.	Loss of coolant.	4-9
27.	Contaminated coolant.	4-9
	TRANSMISSION	
28.	Transmission shifts occur at too high a speed.	4-9
29.	Transmission shifts occur at too low a speed	4-9
30.	Low main operating pressure at all shift ranges	4-10
31.	Clutch slippage in all forward and/or reverse gears	4-10
32.	Excessive vehicle creep (first and reverse)	4-10
33.	Vehicle moves in neutral.	4-10
34.	No response to shift lever movement	4-10
35.	Oil thrown out of oil tiller tube.	4-10
36.	Oil leaking into converter housing	4-10
37.	Rough shifting	4-11
38.	Transmission oil dirty, foamy, and/or milky	4-11
39.	Oil leak at output shaft.	4-11
	TRANSFER CASE	
40.	Transfer case will not shift into gear	4-11
41.	Transfer case will not stay in gear	4-12
42.	Excessive noise during operation	4-12
43.	Lubrication leaks	4-12

MECHANICAL TROUBLESHOOTING INDEX (Contd)

MALFUNCTION NO.	MALFUNCTION	TROUBLESHOOTING PROCEDURE PAGE
	AIR COMPRESSOR	
44.	Low air pressure (no air leaks, governor properly adjusted and operative)	4-12
45.	Air compressor passes excessive oil (excessive oil bled from air reservoirs)	4-12
46.	Air compressor does not unload (air governor adjusted and operative)	4-12
47.	Air compressor head leaking water	4-13
	STEERING	
48.	Steering wheel hard to turn	4-13
49.	Excessive power steering pump noise	4-13
50.	Oil leaking from steering pump	4-13
51.	Steering gear leaking oil	4-13
52.	Excessive play at steering wheel	4-13
	FRONT/REAR AXLES	
53.	Excessive play (backlash)	4-14
54.	Excessive noise	4-14
55.	Lubricant leaking	4-14
	DUMP BODY	
56.	Dump raises to full dump position, but does not power down	4-14
57.	Dump body does not hold in raised position	4-14
58.	Hoist assembly does not raise dump body	4-14
59.	Dump hoist is inoperative or operates slowly	4-15
	MEDIUM WRECKER CRANE	
60.	Boom does not respond properly when boom control levers are moved	4-21
61.	All boom controls operate abnormally	4-32
	Swing control operates abnormally; all other controls operate normally	4-32
63.	Crowd control operates abnormally; all other controls operate normally	4-32
64.	Hoist control operates abnormally; all other controls operate normally	4-32
65.	Boom control operates abnormally; all other controls operate normally	4-32
	POWER TAKEOFF	
66.	Noisy power takeoff	4-33
67.	Power takeoff slips out of gear	4-33
68.	Lubricant leaking	4-33
	FRONT WINCH	
69.	Front winch is inoperative or operates slowly	4-33

Table 4-1. Mechanical Troubleshooting.

MALFUNCTION TEST OR INSPECTION CORRECTIVE ACTION

ENGINE

1. ENGINE WILL NOT CRANK

Step 1. Check for mechanical or hydraulic seizure. Remove fuel injectors before attempting crankshaft rotation test (para. 4-12 or 4-55).

a. Try to rotate engine manually using engine barring tool.

b. If crankshaft will not rotate, go to step 2.

c. If crankshaft rotates, check if liquid is discharged.

d. If liquid is coolant, replace cylinder head(s) (para. 4-12 or 4-41).

e. If liquid is fuel, replace fuel injectors (para. 4-32 or 4-55).

Step 2. Engine must be removed and inspected for extent of internal damage (para. 4-6).

END OF TESTING!

2. ENGINE CRANKS, BUT FAILS TO START

Step 1. Check for defective fuel shutoff valves (M939/A1).

Remove, inspect, and replace fuel shutoff valve if necessary (para. 4-36 or 4-37).

Step 2. Check for broken fuel supply pump driveshaft.

a. Remove tachometer cable from fuel pump, crank engine, and observe if tachometer driveshaft end of pump is rotating.

b. If driveshaft does not rotate, replace fuel pump (para. 4-35 or 4-57).

Step 3. Check fuel injector and valve adjustment (para. 4-33 or 4-56).

Step 4. Check for dirty, plugged, and damaged fuel injectors.

If dirty, plugged, or damaged, clean or replace fuel injectors as necessary (para. 4-32 or 4-55).

WARNING

Diesel fuel is flammable. Do not perform fuel system procedures near open flame. Injury to personnel may result.

NOTE

Perform step 5 for M939A2 series vehicles.

Step 5. Check fuel injection pump for proper operation.

a. Loosen two injector tubes on fuel pump injectors.

b. Crank engine and visually inspect fuel delivery.

c. If no fuel is present, replace or repair fuel injection pump (para. 4-57).

END OF TESTING!

3. ENGINE STARTER MOTOR OPERATES, BUT DOES NOT ENGAGE FLYWHEEL RING GEAR

Check starter drive gear and flywheel ring gear for broken and missing teeth.

Table 4-1. Mechanical Troubleshooting (Contd).

MALFUNCTION TEST OR INSPECTION CORRECTIVE ACTION

a. Remove starter motor (para. 3-82).

b. If drive gear teeth are broken or missing, replace or repair starter (para. 3-82 or TM 9-2920-243-34).

c. Using engine barring tool, turn engine and inspect flywheel ring gear teeth through starter motor opening in engine block.

d. If teeth are broken or missing, replace flywheel ring gear (para. 4-15).

END OF TESTING!

4. ENGINE RUNS, BUT MISFIRES

Step 1. Perform malfunction 2.

Step 2. Check for damaged pushrods, camshaft, and tappets.

If damaged, replace (para. 4-20).

END OF TESTING!

5. ENGINE STOPS

Step 1. Check for evidence of engine overheating.

a. If no overheating is evident, perform malfunction 1.

b. If overheating is evident, proceed to step 2.

Step 2. Check cooling system.

a. Check for defective radiator fan clutch drive (TM 9-2320-272-10).

b. Install fan clutch override bolt (TM 9-2320-272-10) and start engine.

c. If engine starts and overheats, replace water pump (para. 3-68 or 3-69).

d. If engine does not start, perform malfunction 2.

END OF TESTING!

6. POOR ACCELERATION AND/OR LACK OF POWER

Step 1. Perform malfunction 2.

WARNING

When performing engine stall check, place wheel chocks in front of wheels to prevent vehicle from rolling. Failure to do so may cause injury to personnel and damage to equipment.

NOTE

Perform step 2 for M939A2 engines, model number up to 44629589.

Step 2. Perform engine stall check.

a. Operate vehicle to stabilize engine and transmission at normal operating temperature.

b. Check tachometer for proper operation.

Table 4-1. Mechanical Troubleshooting (Contd).

MALFUNCTION
TEST OR INSPECTION
CORRECTIVE ACTION

c. Ensure front and rear wheels are chocked. Apply parking and service brakes.

d. Place transmission gearshift lever in DRIVE position, press accelerator to floor, and record time it takes engine to reach 1,450-1,550 rpm.

CAUTION

Stop engine stall test if engine oil temperature exceeds 250°F (121°C) or transmission oil temperature exceeds 290°F (143°C) or damage to engine may result.

e. If engine does not reach limits within 20 seconds, perform FF152 fuel pump adjustment (para. 4-57).

f. Place transmission gearshift lever in NEUTRAL position, raise and maintain engine speed at 1,000 rpm to allow engine and transmission oil to cool.

END OF TESTING!

7. LOW OIL PRESSURE

Check engine oil pressure with pressure gauge.

a. If pump pressure is low, replace oil pump (para. 4-21 or 4-49).

b. If pump pressure is correct at oil pump, camshaft and crankshaft bearings are worn. Replace or repair engine (para. 4-6).

END OF TESTING!

8. CONTAMINATED ENGINE OIL

Step 1. If fuel is present in engine oil, adjust or replace fuel injectors (para. 4-33, 4-55, or 4-56).

Step 2. If coolant is present in engine oil:

a. Check fuel pump for damage.

If damaged, replace fuel pump (para. 4-35 or 4-57).

NOTE

During engine operation, oil pressure will be higher than coolant pressure. A leak in oil cooler will show oil in coolant. However, after engine is shut down, pressure in coolant system will cause coolant to leak into engine oil.

b. Check cylinder head for loose or missing expansion plugs.

If loose or missing, replace expansion plugs as necessary (para. 4-12).

c. Check cylinder head gasket for cracks and damage.

If cracked or damaged, replace cylinder head gasket (para. 4-12 or 4-41).

d. Check cylinder head for cracked coolant passages.

If coolant passages are cracked, replace cylinder head (para. 4-12 or 4-41).

e. Check cylinder liners for cracks.

If cracked, replace cylinder liners. Notify General Support maintenance.

f. Check engine block for cracked water passages.

If cracked, replace engine block (para. 5-7 or 5-28).

END OF TESTING!

Table 4-1. Mechanical Troubleshooting (Contd).

MALFUNCTION TEST OR INSPECTION CORRECTIVE ACTION

9. EXCESSIVE OIL CONSUMPTION

Step 1. Check for external oil leakage.

Replace or repair as necessary.

Step 2. Replace engine (para. 4-4 or 4-5).

END OF TESTING!

10. AIR CLEANER ELEMENT CLOGGED WITH OIL

Step 1. Clean air cleaner housing and replace element (para. 3-15).

Step 2. Check fluid levels in transmission, transfer case, and fuel tank(s) for signs of overfilling, and correct as necessary (LO 9-2320-272-12).

Step 3. Trace source of oil through vent line tubes on air inlet stack and check related components.

END OF TESTING!

11. ENGINE KNOCKS (MECHANICAL NOISE)

Replace engine (para. 4-4 or 4-5).

END OF TESTING!

12. EXCESSIVE ENGINE VIBRATION

Step 1. Check for loose or damaged vibration damper.

If loose, tighten. If damaged, replace (para. 4-14 or 4-42).

Step 2. Check for loose or damaged engine mounts (M939A2).

If loose, tighten. If damaged, replace (para. 4-39).

Step 3. Check for loose or damaged flywheel or flywheel ring gear.

If loose, tighten. If damaged, replace (para. 4-15, 4-16, or 4-43).

Step 4. Check fuel injectors for proper operation and adjustment (para. 4-33, 4-34, or 4-56).

END OF TESTING!

13. EXCESSIVE GEAR NOISE

Step 1. Check for loose or damaged flywheel or flywheel ring gear.

If loose, tighten. If damaged, replace (para. 4-15, 4-16, or 4-43).

Step 2. Check backlash of all accessory drive gears (para. 4-21, 4-49, 5-8, 5-9, 5-27, or 5-29).

Step 3. Replace engine (para. 4-5).

END OF TESTING!

FUEL SYSTEM

14. ENGINE IDLE ROUGH, ERRATIC

Step 1. Check for loose or damaged engine mounts (M939A2).

If loose, tighten. If damaged, replace (para. 4-39).

Step 2. Perform malfunction 2.

END OF TESTING!

Table 4-1. Mechanical Troubleshooting (Contd).

MALFUNCTION
TEST OR INSPECTION
CORRECTIVE ACTION

15. EXCESSIVE FUEL CONSUMPTION

Check fuel injectors for proper operation and adjustment (para. 4-33, 4-34, or 4-56).

END OF TESTING!

16. ENGINE SURGES AT IDLE SPEED

Perform malfunction 2.

END OF TESTING!

17. ENGINE SURGES AT ALL SPEEDS

Replace or repair fuel pump (para. 4-35 or 4-57).

END OF TESTING!

18. EXCESSIVE EXHAUST SMOKE AT IDLE, AND UNDER LOAD

Step 1. Perform malfunction 2.

Step 2. Check cylinder head for damaged and burned valves.

Replace or repair cylinder head (para. 4-12 or 4-41).

Step 3. Check cylinder liners for scoring and wear.

If scored or worn, replace cylinder liners (para. 5-7 or 5-28).

Step 4. Check pistons and piston rings for wear and damage.

If worn or damaged, replace as necessary (para. 5-10 or 5-26).

END OF TESTING!

19. ENGINE FAILS TO STOP

NOTE

This malfunction pertains to M939 and M939A1 series vehicles.

Check fuel shutoff solenoid for proper operation.

If defective, replace fuel shutoff valve (para. 4-36 or 4-37).

END OF TESTING!

20. ENGINE MISSES

Step 1. Perform malfunction 2.

Step 2. Check cylinder head for damaged and burned valves.

Replace or repair cylinder head (para. 4-12 or 4-41).

END OF TESTING!

21. LOW POWER OR LOSS OF POWER

Step 1. Perform malfunction 2.

Step 2. Check cylinder head for damaged and burned valves.

Replace or repair cylinder head (para. 4-12 or 4-41).

END OF TESTING!

Table 4-1. Mechanical Troubleshooting (Contd).

MALFUNCTION TEST OR INSPECTION CORRECTIVE ACTION

22. ENGINE OVERSPEED

Replace or repair fuel pump (para. 4-35 or 4-57).

END OF TESTING!

23. ENGINE LUBRICATING OIL DILUTED BY FUEL

Perform malfunction 2.

END OF TESTING!

24. ENGINE FUEL KNOCKS

Step 1. Check fuel injectors for proper operation and adjustment, Notify General Support maintenance.

Step 2. Check valve and injector timing adjustment (para. 4-32 or 4-56).

NOTE

Perform step 3 for M939A2 series vehicles.

Step 3. Check fuel injection pump timing. Notify General Support maintenance.

END OF TESTING!

COOLING SYSTEM

25. ENGINE OVERHEATS

Step 1. Check for proper operation of fan drive clutch. Replace fan drive clutch if malfunctioning (para. 4-61 or 4-62).

Step 2. Check water pump for wear or damage. Repair or replace water pump if worn or damaged (para. 4-63 or 4-29).

END OF TESTING!

26. LOSS OF COOLANT

Perform malfunction 8, step 2.

END OF TESTING!

27. CONTAMINATED COOLANT

Perform malfunction 8, step 2.

END OF TESTING!

TRANSMISSION

28. TRANSMISSION SHIFTS OCCUR AT TOO HIGH A SPEED

Step 1. Check governor pressure (para. 5-60).

If malfunction is not corrected, proceed to step 2.

Step 2. Adjust modulator linkage (para. 3-145).

If malfunction is not corrected, replace transmission (para. 4-71).

END OF TESTING!

29. TRANSMISSION SHIFTS OCCUR AT TOO LOW A SPEED

Perform malfunction 1.

END OF TESTING!

Table 4-1. Mechanical Troubleshooting (Contd).

MALFUNCTION
    TEST OR INSPECTION
        CORRECTIVE ACTION

30. LOW MAIN OPERATING PRESSURE AT ALL SHIFT RANGES

Perform transmission pressure test (para. 5-60).

If oil pressure is still low, replace transmission (para. 4-71).

END OF TESTING!

31. CLUTCH SLIPPAGE IN ALL FORWARD AND/OR REVERSE GEARS

Replace transmission (para. 4-71).

END OF TESTING!

32. EXCESSIVE VEHICLE CREEP (FIRST AND REVERSE)

Check and adjust engine idle (para. 3-42 or 3-43).

END OF TESTING!

33. VEHICLE MOVES IN NEUTRAL

Check shift range selector (para. 4-73).

a. Adjust shift selector linkage (para. 4-75).

b. If malfunction continues, replace transmission (para. 4-71).

END OF TESTING!

34. NO RESPONSE TO SHIFT LEVER MOVEMENT

Check if shift linkage is disconnected or broken.

a. Connect or replace linkage if necessary (para. 4-75).

b. If linkage is sound, replace transmission (para. 4-71).

END OF TESTING!

35. OIL THROWN OUT OF OIL FILLER TUBE

Step 1. Check transmission oil level (table 2-1).

Drain or fill to proper level.

Step 2. Check and clean breather vent.

If worn or damaged, replace breather vent (para. 3-136).

Step 3. Disconnect vent line at breather, start engine, and check for compressed air coming through vent line.

If air is present, perform compressed air and brake troubleshooting (para. 2-23).

Step 4. Replace transmission (para. 4-71).

END OF TESTING!

36. OIL LEAKING INTO CONVERTER HOUSING

Step 1. Check if leaking engine oil.

If leaking engine oil, replace engine rear crankshaft seal (para. 4-17 or 4-45).

Step 2. Check if leaking transmission fluid.

If leaking transmission fluid, replace transmission (para. 4-71).

END OF TESTING!

Table 4-1. Mechanical Troubleshooting (Contd).

MALFUNCTION
TEST OR INSPECTION
CORRECTIVE ACTION

37. ROUGH SHIFTING

Step 1. Check manual selector for proper operation.

If defective, replace manual selector (para. 4-73).

Step 2. Check shift cable for binding and proper adjustment.

Replace or adjust shift cable as necessary (para. 4-75).

Step 3. Check if modulator is sticking.

Replace modulator if necessary (para. 3-145).

Step 4. Check if modulator cable is kinked or out of adjustment.

Replace or adjust modulator cable as necessary (para. 3-145).

END OF TESTING!

38. TRANSMISSION OIL DIRTY, FOAMY, AND/OR MILKY

NOTE

Dirt/grit in transmission oil indicates oil needs to be changed (step 1). Foaminess indicates contamination by air (step 2) or water (step 3). Milkiness indicates contamination by water (step 3).

Step 1. Check for dirt/grit.

a. Perform transmission service procedure (para. 3-133).

b. Inspect all external transmission fittings for looseness. If loose, tighten.

c. Replace transmission (para. 4-71).

Step 2. Check for excessive foaming.

Ensure transmission fluid is at proper level (TM 9-2320-272-10). Drain or fill fluid as necessary (LO 9-2320-272-12).

Step 3. Check for coolant in transmission fluid.

If coolant is present in transmission oil, replace transmission oil cooler (para. 3-140 or 3-141).

END OF TESTING!

39. OIL LEAK AT OUTPUT SHAFT

Check oil seal at output shaft flange for wear or damage.

If worn or damaged, replace seal and output flange (para. 4-78).

TRANSFER CASE

40. TRANSFER CASE WILL NOT SHIFT INTO GEAR

Step 1. Check shift linkage for proper adjustment.

Adjust shift linkage if necessary (para. 4-88).

Step 2. Check shift linkage for bends, breaks, or disconnections.

Repair or replace shift linkage (para. 4-88).

Step 3. If malfunction continues, replace transfer case (para. 4-94 or 4-95).

END OF TESTING!

Table 4-1. Mechanical Troubleshooting (Contd).

MALFUNCTION TEST OR INSPECTION CORRECTIVE ACTION

41. TRANSFER CASE WILL NOT STAY IN GEAR

Step 1. Check shift linkage adjustment.

Adjust shift linkage if necessary (para. 4-88).

Step 2. Remove interlock air cylinder (para. 4-83) and check if pushrod is binding.

a. If binding in interlock air cylinder or transfer case, clean ports and pushrod, and reinstall.

b. If malfunction continues, go to step 3.

Step 3. Replace transfer case (para. 4-94 or 4-95).

END OF TESTING!

42. EXCESSIVE NOISE DURING OPERATION

Replace transfer case (para. 4-94 or 4-95).

END OF TESTING!

43. LUBRICATION LEAKS

Step 1. Check for defective seals or gaskets.

If defective, replace seals or gaskets (para. 4-93 or 5-63).

Step 2. Check for cracked transfer case.

Replace transfer case (para. 4-94 or 4-95).

END OF TESTING!

AIR COMPRESSOR

44. LOW AIR PRESSURE (NO AIR LEAKS, GOVERNOR PROPERLY ADJUSTED AND OPERATIVE)

Step 1. Remove unloader valve and unloader valve spring from compressor and inspect for wear and damage.

Replace worn unloader valve spring (para. 4-31 or 4-52).

Step 2. If pressure is still low, replace air compressor (para. 3-206 or 4-31).

END OF TESTING!

45. AIR COMPRESSOR PASSES EXCESSIVE OIL (EXCESSIVE OIL BLED FROM AIR RESERVOIRS)

Drain air reservoirs (TM 9-2320-272-10) and check for evidence of oil.

a. If oil is present, replace air compressor (para. 3-206 or 4-31).

b. On M939A2 series vehicles, service air dryer filter element (para. 3-465).

END OF TESTING!

46. AIR COMPRESSOR DOES NOT UNLOAD (AIR GOVERNOR ADJUSTED AND OPERATIVE)

Remove unloader valve and unloader valve spring from compressor and inspect for wear and damage.

Replace worn unloader valve spring (para. 4-31 or 4-52).

END OF TESTING!

Table 4-1. Mechanical Troubleshooting (Contd).

MALFUNCTION
TEST OR INSPECTION
CORRECTIVE ACTION

47. AIR COMPRESSOR HEAD LEAKING WATER

Step 1. Ensure screws securing head are tightened properly.
a. Tighten screws to 30-35 lb-ft (41-48 N•m).
b. If head is still leaking water, go to step 2.

Step 2. Check head and head cover for cracks.
a. Replace head or head cover if cracked (para. 4-31 or 4-52).
b. If head still leaks, go to step 3.

Step 3. Ensure head fittings are correctly installed.
If not, install fittings correctly (para. 4-31 or 4-52).

END OF TESTING!

STEERING

48. STEERING WHEEL HARD TO TURN

Step 1. Perform steering pump pressure test (para. 4-123 or 4-124).
If pressure is abnormal, repair or replace steering pump (para. 3-236, 4-126, or 4127).

Step 2. Check for broken piston rings in power steering assist cylinder.
a. Replace broken piston or piston rings (para. 4-121).
b. If steering wheel is still hard to turn, go to step 3.

Step 3. Check front axle for security of mounting (para. 4-99).
If not secure, tighten.

END OF TESTING!

49. EXCESSIVE POWER STEERING PUMP NOISE

Replace or repair defective power steering pump (para. 3-236, 4-126, or 4-127).

END OF TESTING!

50. OIL LEAKING FROM STEERING PUMP

Replace or repair defective power steering pump (para. 3-236, 4-126, or 4-127).

END OF TESTING!

51. STEERING GEAR LEAKING OIL

Check steering gear seals for wear (para. 4-119).
Replace worn steering gear seals (para. 4-120).

END OF TESTING!

52. EXCESSIVE PLAY AT STEERING WHEEL

Step 1. Check steering gear (para. 4-123, 4-124, or 4-125) for loose mounting.
a. Tighten loose mounting bolts.
b. If excessive play continues, go to step 2.

Step 2. Check steering gear for proper adjustment.
a. Adjust steering gear (para. 4-123, 4-124, or 4-125).
b. If excessive play continues, replace steering gear (para. 4-117 or 4-118).

END OF TESTING!

Table 4-1. Mechanical Troubleshooting (Contd).

MALFUNCTION
TEST OR INSPECTION
CORRECTIVE ACTION

FRONT/REAR AXLES

53. EXCESSIVE PLAY (BACKLASH)

Replace or repair differential carrier assembly (para. 4-100).

END OF TESTING!

54. EXCESSIVE NOISE

Replace differential carrier assembly (para. 4-100). Notify General Support maintenance for repair.

END OF TESTING!

55. LUBRICANT LEAKING

Step 1. Check for worn seals.

Replace seals if worn (para. 4-102 or 4-103).

Step 2. Check for cracked differential housing.

Replace or repair differential carrier assembly (para. 4-100).

END OF TESTING!

DUMP BODY

56. DUMP BODY RAISES TO FULL DUMP POSITION, BUT DOES NOT POWER DOWN

Step 1. Check control linkage adjustment.

a. Adjust improperly adjusted linkage (para. 4-148).

b. If linkage is not at fault, go to step 2.

Step 2. Check for hydraulic system leaks.

a. If leakage exists, replace gaskets.

b. If no leaks exist, replace control valve (para. 4-151).

END OF TESTING!

57. DUMP BODY DOES NOT HOLD IN RAISED POSITION

Step 1. Check for hydraulic system leaks.

a. Replace leaking components and gaskets (para. 4-147, 4-150, or 4-151).

b. If no leaks exist, go to step 2.

Step 2. Check control linkage adjustment.

a. Adjust improperly adjusted linkage (para. 4-148).

b. If linkage is satisfactory, replace control valve (para. 4-151).

END OF TESTING!

58. HOIST ASSEMBLY DOES NOT RAISE DUMP BODY

Inspect for leaks in hydraulic system.

Replace leaking components (para. 4-147, 4-150, or 4-151).

END OF TESTING!

Table 4-1. Mechanical Troubleshooting (Contd).

MALFUNCTION
TEST OR INSPECTION
CORRECTIVE ACTION

59. DUMP HOIST IS INOPERATIVE OR OPERATES SLOWLY

WARNING

- When dump body is in a raised position, oil in hydraulic system is under pressure. Any movement of control valve or leakage at the hydraulic cylinder lines or connections will cause dump body to drop to the subframe. Never work under dump body unless safety braces are properly positioned. Failure to do so may cause injury or death.
- Do not loosen or take off any lines or hoses while the engine is running with the PTO engaged. Failure to do so may cause injury.

Before performing the troubleshooting procedure, complete the following preliminary inspections:

1. Check hydraulic reservoir oil level. If low, fill to top (LO 9-2320-272-12).
2. Check hydraulic oil for contamination (TB 43-0211).
3. Replace hydraulic oil filter (para. 3-336).
4. Check pump, cylinders, control valve, lines, and hoses for deterioration and leaks. Tighten any loose connections. Replace any component where either class II or class III leaks are found (chapter 4, section XIV).
5. Check that PTO control lever moves freely and fully engages and disengages the PTO. If PTO will not engage, replace PTO (para. 4-211).
6. Check that dump control lever moves freely. Inspect control cable adjustment at valve. If valve spool will not fully shift, replace control valve (para. 4-151).
7. Ensure latch releases dump hoist when dump body is raised. If latch fails to release dump hoist, lubricate latch (LO 9-2320-272-12). If latch still fails to release, replace latch.
8. Ensure latch secures dump hoist to subframe when dump body is resting on subframe. Lubricate latch assembly (LO 9-2320-272-12). If latch fails to secure dump body, replace latch.

CAUTION

When disconnecting hydraulic oil lines and hoses, plug all openings to prevent dirt from entering and causing internal parts damage.

NOTE

Have drainage container ready to catch excess oil.

Step 1. Remove hydraulic pump-to-control valve line and check for blockage (para. 4-147).

a. If blockage is present, replace and retest dump operation.

b. If blockage or obstruction is not present, proceed to step 2.

Table 4-1. Mechanical Troubleshooting (Contd).

MALFUNCTION TEST OR INSPECTION CORRECTIVE ACTION

Step 2. Prepare hydraulic system for test.

NOTE

- Perform steps a. and b. for M930/A1/A2 model vehicles.
- Tag all hoses for installation.

a. Disconnect front winch input hose (2) and output hose (1) from elbows (5) and (6) at control valve (7).

b. Install appropriate caps (3) and plugs (4) on elbows (5) and (6) and hoses (1) and (2) to block hydraulic oil flow.

c. Disconnect supply hose (11) and return hose (10) from elbows (17) and (18) at control valve (7).

d. Install appropriate caps (8) and plugs (9) on elbows (17) and (18) and hoses (10) and (11).

e. Disconnect line (14) from adapters (13) and (15).

f. Remove adapter (15) from elbow (16).

g. Install appropriate nipple (29) and tee (19) on elbow (16).

h. Install appropriate connector (23), reducer (24), nut (25), flowmeter (Appendix E, Item 169) (20), and connector (21) on tee (19).

i. Install appropriate hose (22) on connector (21) and adapter (13) at winch pump (12).

j. Install appropriate connector (27) and hose (26) on pressure gauge (Appendix E, Item 168) (28).

k. Install pressure gauge (28) with hose (26) on reducer (24).

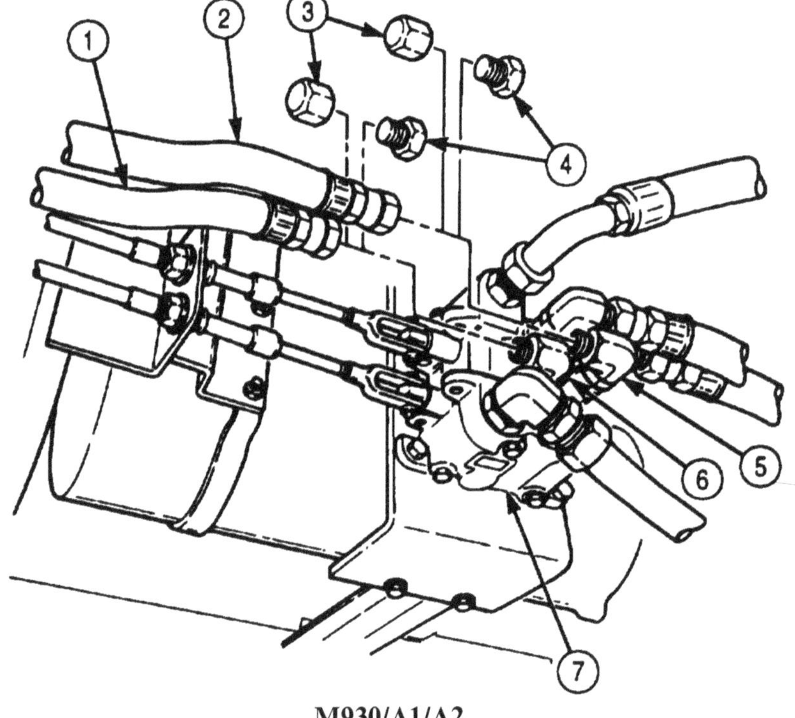

M930/A1/A2

Table 4-1. Mechanical Troubleshooting (Contd).

MALFUNCTION TEST OR INSPECTION CORRECTIVE ACTION

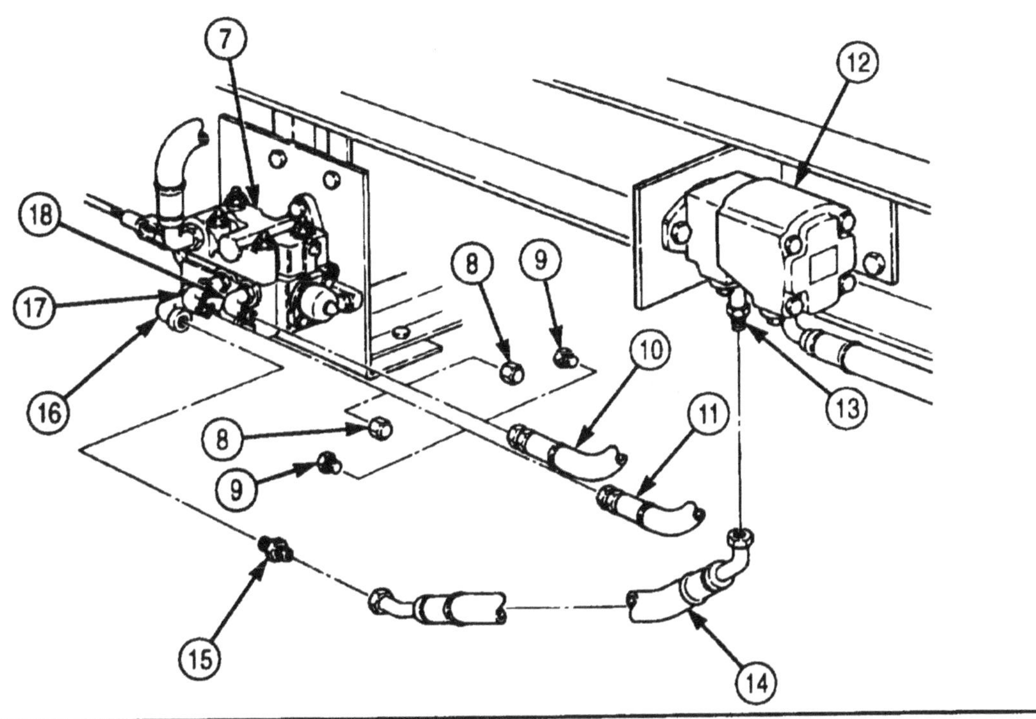

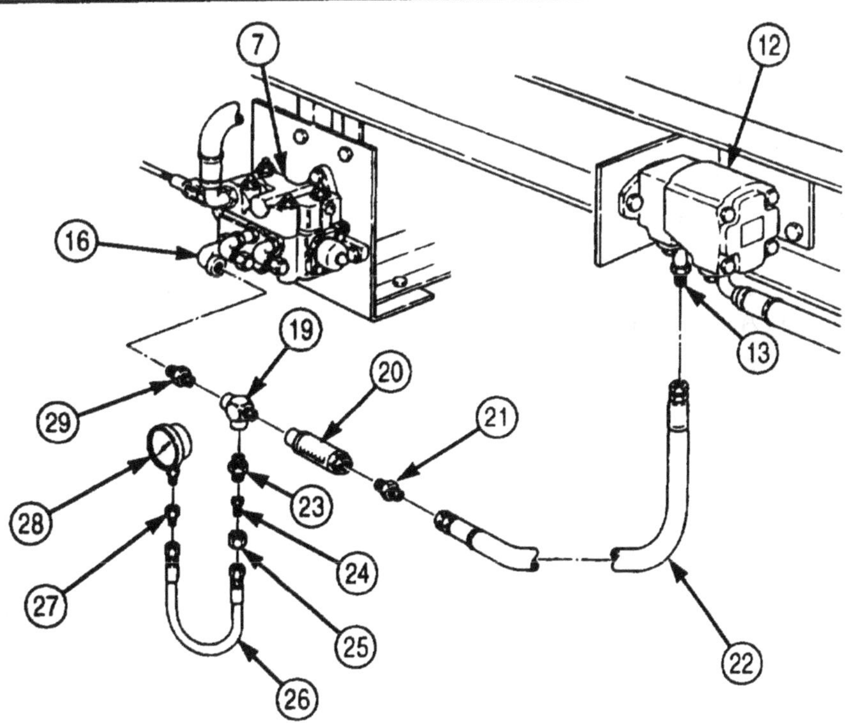

Table 4-1. Mechanical Troubleshooting (Contd).

MALFUNCTION TEST OR INSPECTION CORRECTIVE ACTION

Step 3. If hydraulic system oil temperature is greater than 32°F (0°C), proceed to step 5.

Step 4. If hydraulic system oil temperature is 32°F (0°C) or colder, start engine (TM 9-2320-272-10), engage PTO, allow hydraulic oil to circulate for 15 minutes with dump control in the neutral position, set idle at 800 rpm, and proceed to step 6.

Step 5. Start engine and set idle at 800 rpm (TM 9-2320-272-10).

Step 6. Engage PTO and place dump lever in raised position.

Step 7. Record pressure and flow readings.

a. M929/A1/A2 normal readings:

Flow rate - 5 gpm (18.9 Lpm) or greater

Pressure - 1,400-1,600 psi (9,653-11,032 kPa)

(1) If readings are normal, proceed to step 10.

(2) If pressure is less than 1,600 psi (11,032 kPa), and flow rate is less than 5 gpm (18.9 Lpm), proceed to step 9.

(3) If pressure is greater than 1,600 psi (11,032 kPa), proceed to step 10.

(4) If pressure is less than 1,400 psi (9,653 kPa), replace control valve (para. 4-151). Repeat steps 2 through 7.

b. M929/A1/A2 normal readings:

Flow rate - 5 gpm (18.9 Lpm) or greater

Pressure - 1,650-1,850 psi (11,377-12,756 kPa)

(1) If readings are normal, proceed to step 10.

(2) If pressure is less than 1,850 psi (12,756 kPa), and flow rate is less than 5 gpm (18.9 Lpm), proceed to step 9.

(3) If pressure is greater than 1,850 psi (12,756 kPa), proceed to step 10.

(4) If pressure is less than 1,650 psi (11,377 kPa), replace control valve (para. 4-151). Repeat steps 2 through 7.

Step 8. Remove test kit components:

a. Remove air pressure gauge (13) with hose (11) from reducer (9).

b. Remove hose (11) and connector (12) from air pressure gauge (13).

c. Remove hose (7) from connector (4) and adapter (5) at winch pump (6).

d. Remove connector (4), flowmeter (3), reducer (9), nut (10), and connector (8) from tee (2).

e. Remove tee (2) and nipple (14) from elbow (15) at control valve (1).

Table 4-1. Mechanical Troubleshooting (Contd).

MALFUNCTION TEST OR INSPECTION CORRECTIVE ACTION

f. Install adapter (22) on elbow (23).

g. Connect line (21) on adapter (22) at control valve (1) and adapter (20) at winch pump (6).

h. Remove caps (16) and plugs (17) from elbows (24) and (25) and hoses (18) and (19).

i. Connect supply line (19) and return line (18) to elbows (24) and (25) at control valve (1).

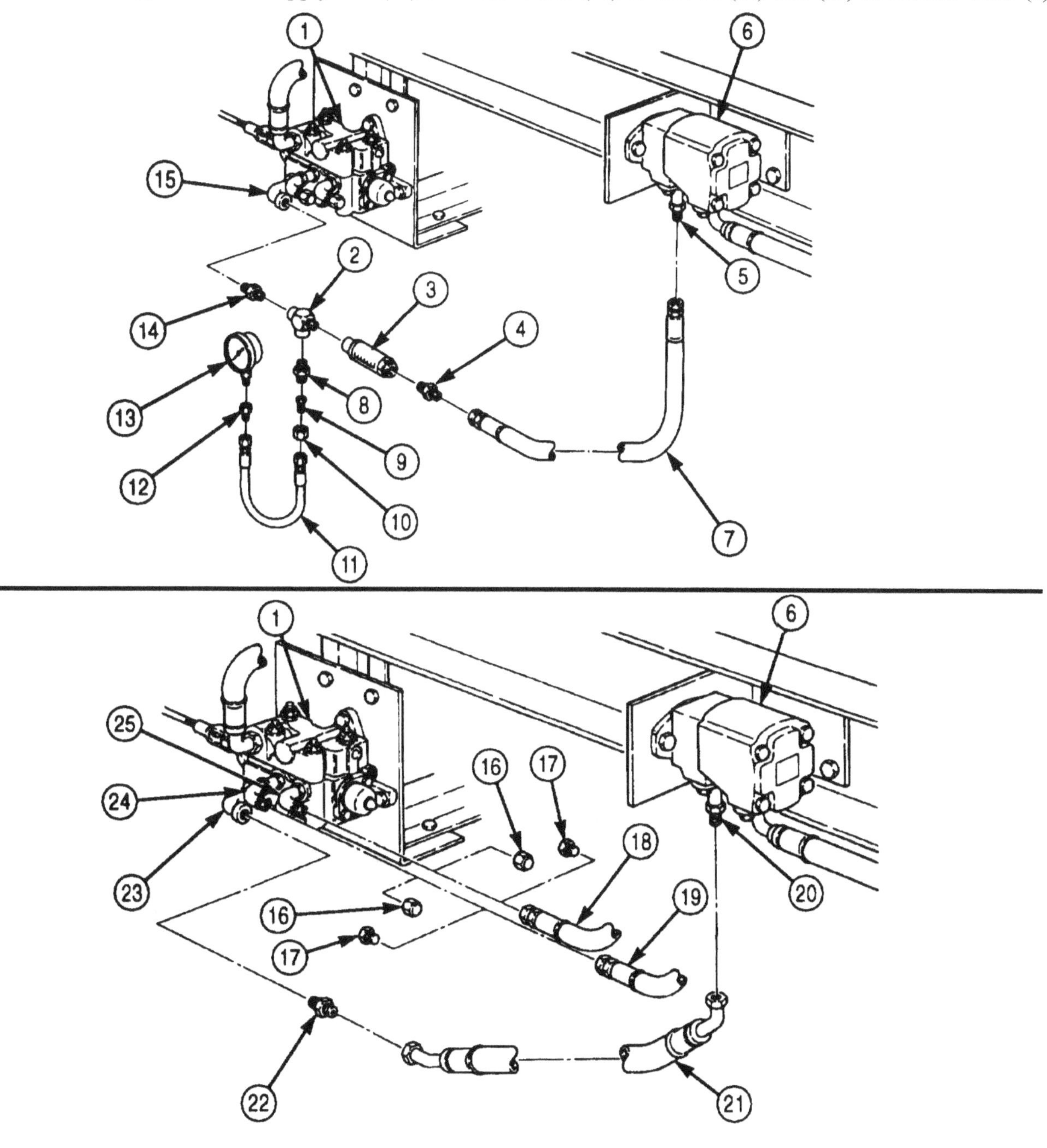

Table 4-1. Mechanical Troubleshooting (Contd).

MALFUNCTION TEST OR INSPECTION CORRECTIVE ACTION

NOTE

Perform steps j. and k. for M930/A1/A2 model vehicles.

j. Remove caps (3) and plugs (4) from elbows (5) and (6) and hoses (1) and (2).

k. Connect front winch input hose (2) and output hose (1) to elbows (6) and (5).

Step 9. Remove hydraulic pump suction line (para. 4-147) and inspect for bends, kinks, or restrictions.

a. If bends, kinks, or restrictions are present, replace hydraulic pump suction line (para. 4-147).

b. If restrictions are not present, replace hydraulic pump (para. 4-150).

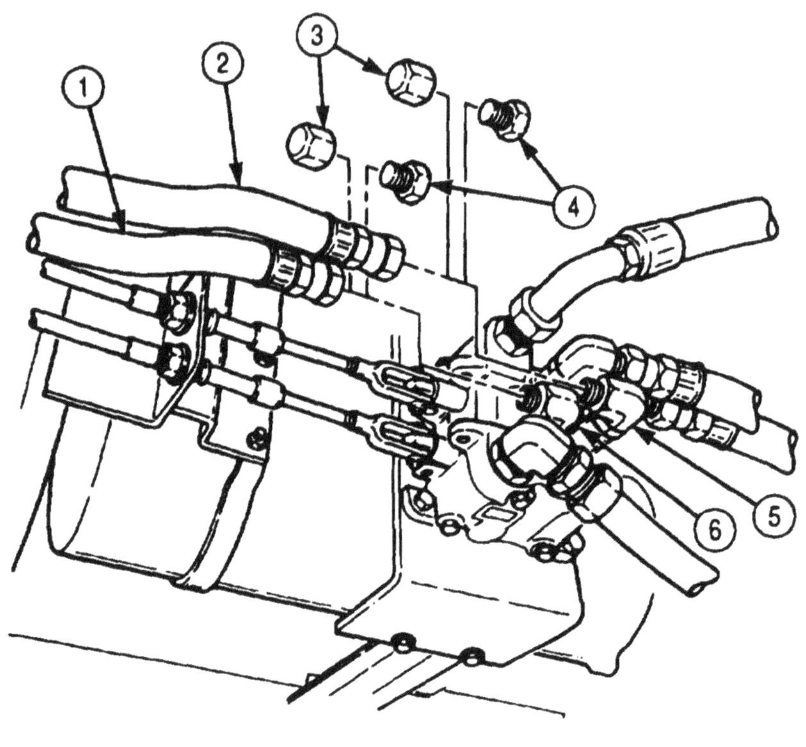

M930/A1/A2

Table 4-1. Mechanical Troubleshooting (Contd).

MALFUNCTION TEST OR INSPECTION CORRECTIVE ACTION

Step 10. Inspect hydraulic line between control valve and hydraulic reservoir for kinks, bends, or restrictions.

If bends, kinks, or restrictions are present, replace hydraulic line (para. 4-147).

Step 11. Inspect dump cylinder hydraulic lines for bends, kinks, or restrictions.

a. If bends, kinks, or restrictions are present, replace dump cylinder hydraulic lines (para. 4-147).

b. If bends, kinks, or restrictions are not present, replace dump cylinder (para. 5-105).

END OF TESTING!

MEDIUM WRECKER CRANE

WARNING

- When adjusting pressure relief valve settings, set at specified pressures only. Failure to do so could result in injury to personnel or damage to equipment.
- Do not loosen or remove any lines or hoses while the engine is running. Doing so may result in injury to personnel.

60. BOOM DOES NOT RESPOND PROPERLY WHEN BOOM CONTROL VALVE LEVERS ARE MOVED

Step 1. Visually inspect reservoir filter.

If damaged, replace reservoir filter (para. 5-336).

Step 2. Check hydraulic oil for contamination (LO 9-2320-272-12).

Clean hydraulic oil as required (TB 43-0211).

Step 3. Ensure hydraulic reservoir is filled to proper level (LO 9-2320-272-12).

Step 4. Ensure PTO lever fully engages and disengages PTO (TM 9-2320-272-10).

a. If PTO lever does not engage PTO, check control cable adjustment and adjust as required (para. 4-205).

b. If control cable is adjusted and lever will not engage PTO, replace transfer case PTO (para. 4-207).

Step 5. Ensure hand throttle cable is adjusted properly.

Adjust hand throttle cable as required (para. 3-45).

Step 6. Ensure engine governor maintains an engine speed of 1,200-1,300 rpm, as indicated by vehicle tachometer when PTO is engaged, transfer case is in neutral, transmission selector is in drive 5th position, and hand throttle is pulled out fully.

a. If governor does not operate properly, ensure fuel pump governor lever is pushed into the full fuel position by the air-actuated piston. Adjust lever adjustment screw and/or air cylinder piston as required (para. 4-35).

b. If air cylinder piston does not extend, inspect air system components. If air system components are damaged, replace (chapter 4, section X).

c. If air cylinder operates, but governor speed cannot be adjusted, replace fuel pump (para. 4-35 or 4-57).

Step 7. Inspect hydraulic pump, motors, valves, lines, and hoses for leaks or loose connections.

Tighten any loose connections and repair any leaks (TM 9-243).

Table 4-1. Mechanical Troubleshooting (Contd).

MALFUNCTION TEST OR INSPECTION CORRECTIVE ACTION

CAUTION

Cap or plug all openings immediately after disconnecting lines and hoses to prevent contamination. Remove caps or plugs prior to installation. Failure to do so may result in damage to equipment.

Step 8. Prepare hydraulic system for test.

a. Remove twelve screws (1), lockwashers (2), and rear winch control valve covers (3) and (4) from wrecker (5). Discard lockwashers (2).

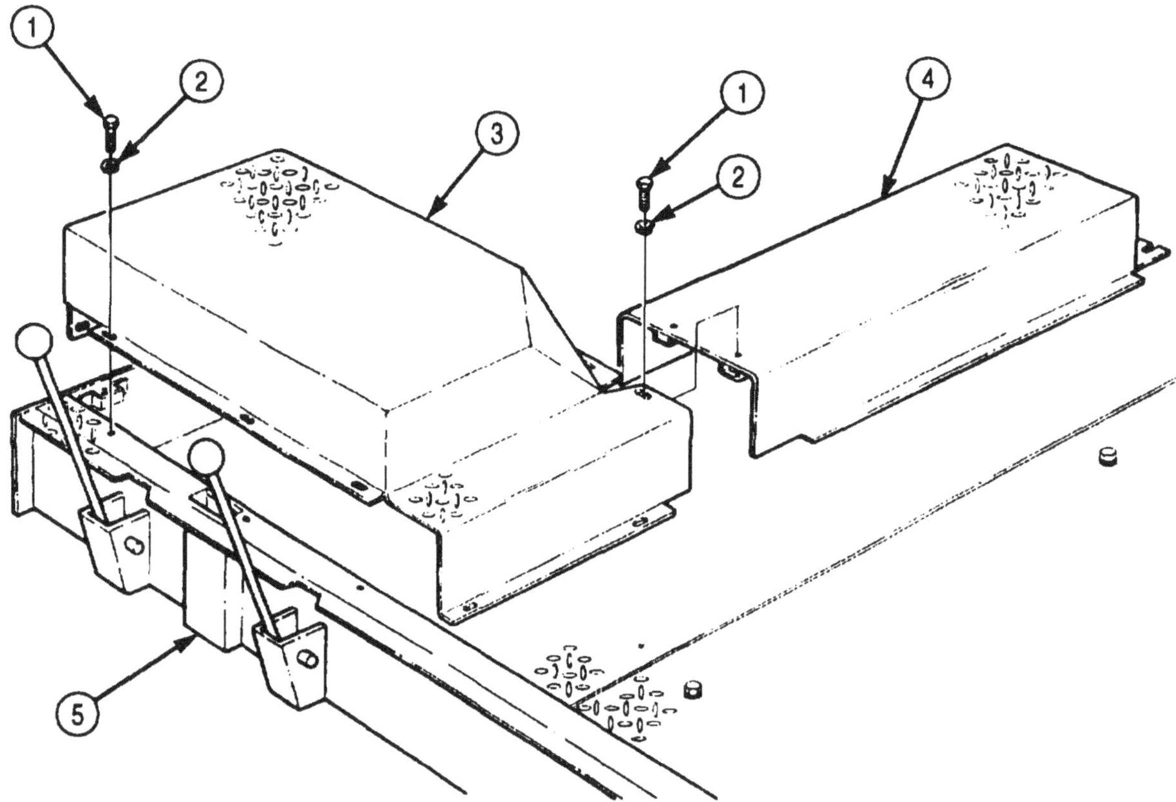

Table 4-1. Mechanical Troubleshooting (Contd).

MALFUNCTION TEST OR INSPECTION CORRECTIVE ACTION

NOTE

Have drainage container ready to catch oil.

b. With the help of an assistant, hold rear winch control lever in WIND position and crowd control lever in EXTEND position.

c. Remove two nuts (9) and U-bolt (7) from tube (6) and bracket (10).

d. Remove tube (6) from rear winch directional control valve (11) and tee (8).

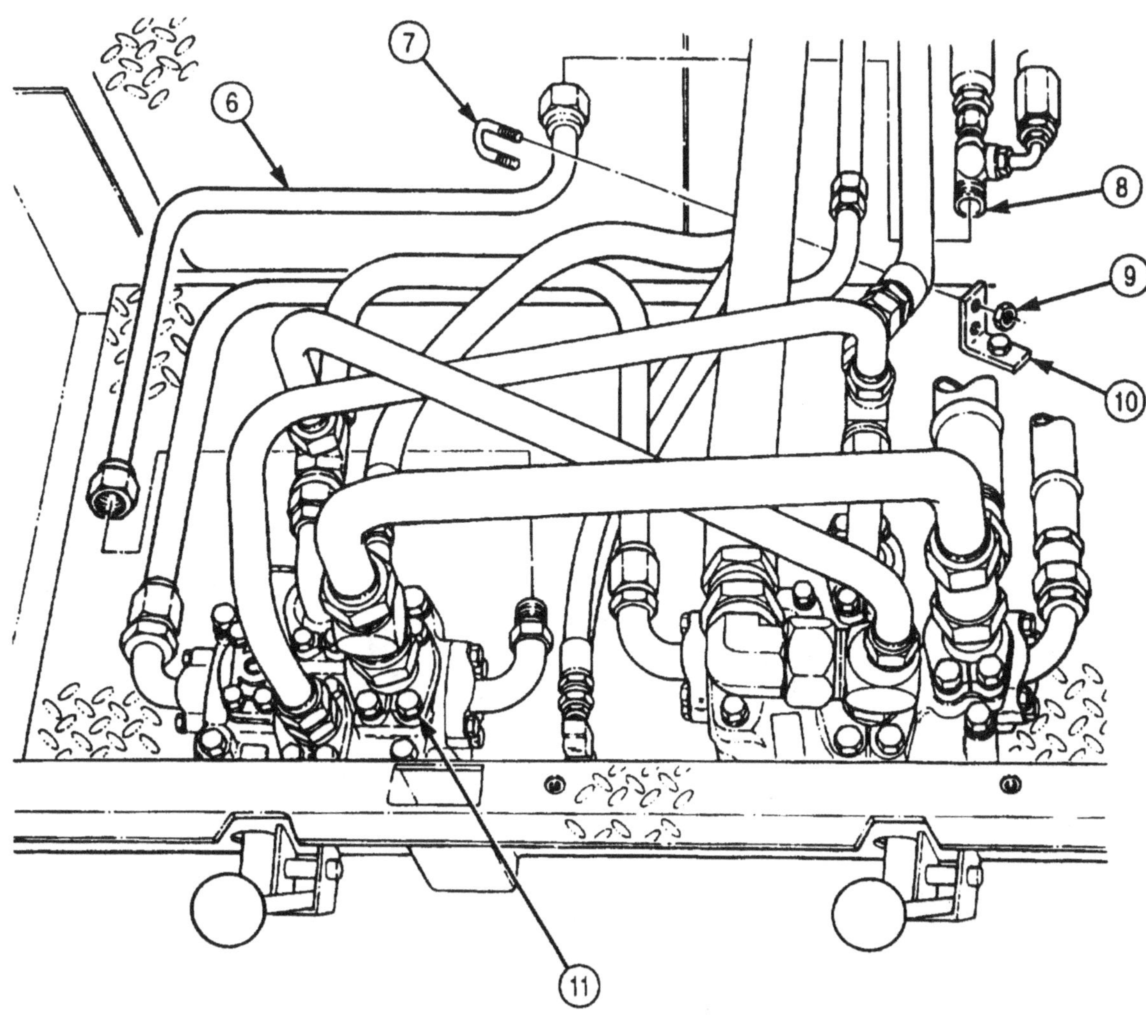

Table 4-1. Mechanical Troubleshooting (Contd).

MALFUNCTION TEST OR INSPECTION CORRECTIVE ACTION

e. Install appropriate hose assembly (1) on rear winch control valve elbow (8).

f. Install appropriate elbow (7), reducer (6), and connector (4) on flowmeter (Appendix E, Item 170) (3), then install flowmeter (3) on tee (5).

g. Install appropriate reducer (9) and adapter (2) on flowmeter (3).

h. Connect hose assembly (1) to adapter (2).

Step 9. Install pressure gauge (Appendix E, Item 168) on boom control valve.

a. Remove five screws (10) and mounting plate (11) from gondola (13).

b. Place crowd control lever (15) in EXTEND position.

NOTE

Have drainage container ready to catch oil.

c. Remove plug (12) from directional control valve (14).

d. Install appropriate connector (19), hose (18), connector (17), and pressure gauge (16) on directional control valve (14).

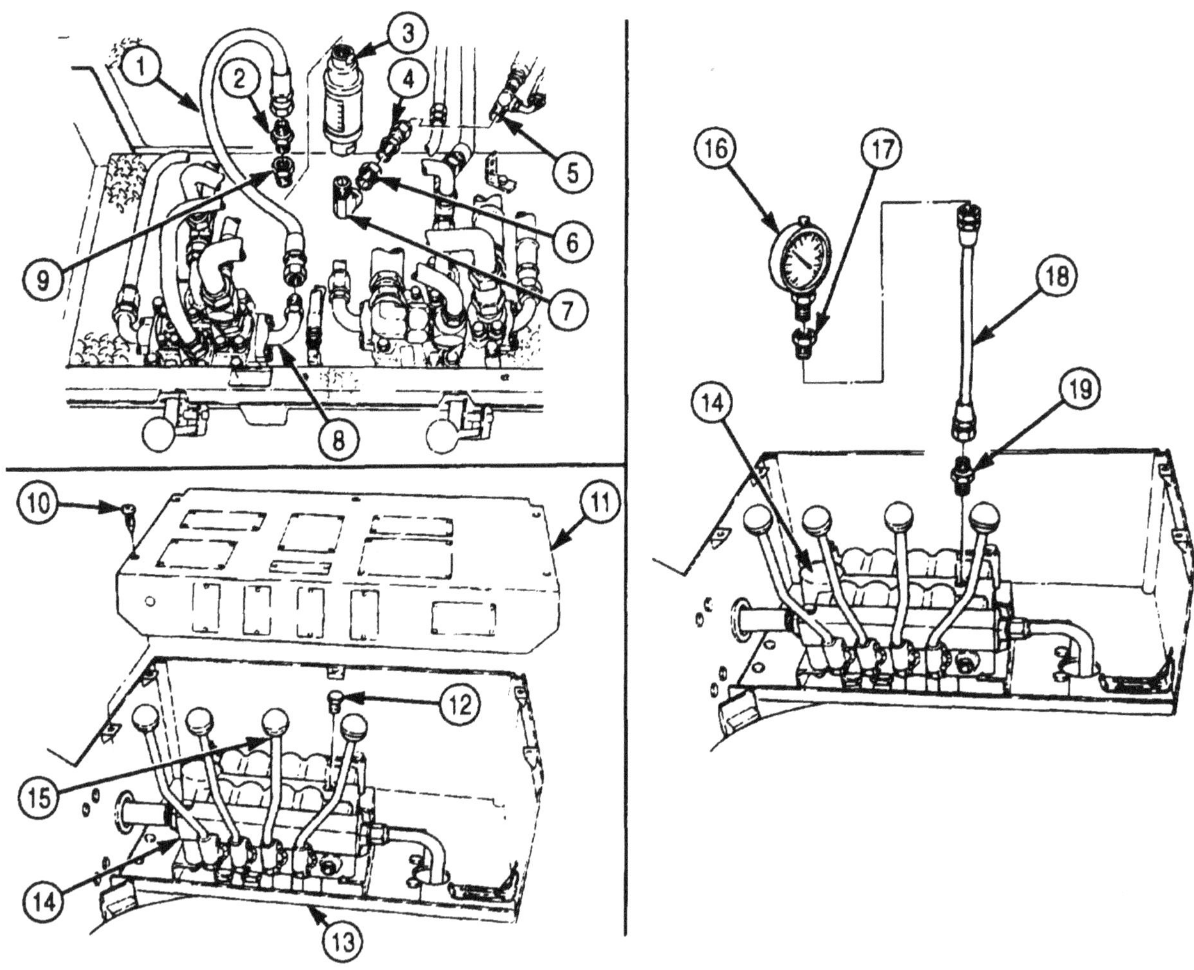

Table 4-1. Mechanical Troubleshooting (Contd).

MALFUNCTION TEST OR INSPECTION CORRECTIVE ACTION

Step 10. Check hydraulic oil temperature.

a. If oil temperature is greater than 32°F (0°C), proceed to step 11.

b. If oil temperature is 32°F (0°C) or cooler, start engine (TM 9-2320-272-10), engage crane hydraulic system (TM 9-2320-272-10), and allow oil to circulate for 15 minutes.

Step 11. Check boom relief pressure and flow. Start engine and engage crane hydraulic system (TM 9-2320-272-10). Ensure boom is secured in travel position and fully retracted. Move crown lever to the retract position. Record flow and pressure readings. Shut down engine. Proper flow rate is 65 gpm (246 Lpm) or greater. Proper boom relief pressure is 1,325-1,375 psi (9,136-9,480 kPa).

a. If readings are normal, proceed to steps 16 and 17.

b. If pressure is 1,500 psi (10,343 kPa) or less, and flow rate is less than 65 gpm (246 Lpm), proceed to step 12.

c. If pressure is greater than 1,500 psi (10,343 kPa), proceed to step 14.

d. If pressure is greater than 1,325 psi (9,136 kPa), and flow rate is 65 gpm (246 Lpm) or greater, proceed to step 15.

e. If pressure is normal, and flow rate is 65 gpm (246 Lpm) or greater, perform malfunctions 56 through 59.

Step 12. Check pump suction line for restrictions.

a. If pump suction line is not restricted, proceed to step 13.

b. If pump suction line is restricted, clear restrictions. Check boom relief pressure and flow rate (step 11).

Step 13. Check air pressure being delivered to transmission 5th gear lockup pressure switch.

a Remove hose assembly (21) from tee (22).

b. Install appropriate hose (20), reducer (23), and pressure gauge (Appendix E, Item 168) (16) on tee (22).

c. Start engine, place transfer case in neutral, and engage transfer case PTO (TM 9-2320-272-10). Record pressure.

d. If no air pressure is being delivered to pressure switch, inspect air lines for restrictions or leaks. If air lines are restricted or have leaks, clear restrictions or repair leaks (TM 9-243). Recheck boom relief pressure (step 11).

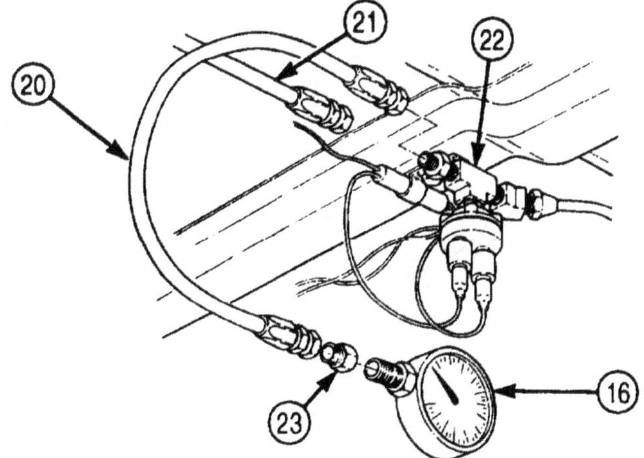

Table 4-1. Mechanical Troubleshooting (Contd).

MALFUNCTION TEST OR INSPECTION CORRECTIVE ACTION

e. If pressure being delivered to pressure switch is 5 psi (34 kPa) or greater, replace pressure switch (para. 3-138). Recheck boom relief pressure (step 11).

f. If flow rate is less than 65 gpm (246 Lpm), replace transmission solenoid valve (para. 4-80). Recheck flow rate (step 11). If flow rate is less than 65 gpm (246 Lpm), replace hydraulic pump (para. 4-176).

g. If flow rate is normal, proceed to steps 16 and 17.

h. Remove pressure gauge (4), reducer (5), and hose (1) from tee (3).

i. Connect hose (2) to tee (3).

Step 14. Check boom relief pressure (step 11) and adjust as required (para. 4-204).

a. If boom relief pressure is lowered, check flow rate (step 11).

b. If flow rate is 65 gpm (246 Lpm) or greater, perform malfunctions 56 through 59 as required.

c. If flow rate is less than 65 gpm (246 Lpm), perform step 12.

d. If boom pressure cannot be lowered, check for restrictions in reservoir return line.

e. If no restrictions are found, replace boom control valve (para. 4-204). Recheck boom relief pressure (step 11).

f. If restrictions are found, clear restrictions. Recheck boom relief pressure (step 11).

Step 15. Check boom relief pressure (step 11) and adjust as required (para. 4-204).

a. If boom relief pressure can be raised, perform malfunction 60, steps 16 and 17.

b. If boom pressure cannot be raised, remove flowmeter and pressure gauge from boom control valve (steps 16 and 17).

c. Install pressure gauge on rear winch control valve (step 18), and check pressure at rear winch control valve (step 19).

Step 16. Remove flowmeter.

NOTE

Ensure engine is off.

a. Remove adapter (9) from tee (10) and reducer (11).

b. Remove reducer (11), elbow (12), flowmeter (8), reducer (14), and connector (7) from hose assembly (6).

c. Remove hose assembly (6) from rear winch control valve elbow (13).

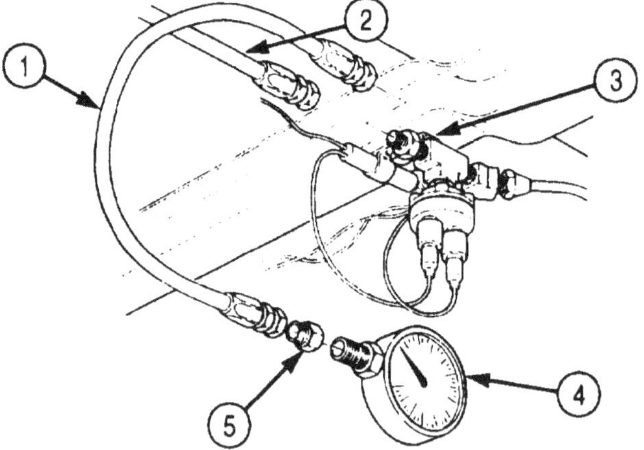

Table 4-1. Mechanical Troubleshooting (Contd).

MALFUNCTION TEST OR INSPECTION CORRECTIVE ACTION

d. Install tube (15) on rear winch control valve elbow (13) and tee (10).

e. Install tube (15) on bracket (18) with U-bolt (16) and two nuts (17).

f. Install rear winch control valve covers (21) and (22) on wrecker (23) with twelve new lockwashers (20) and screws (19).

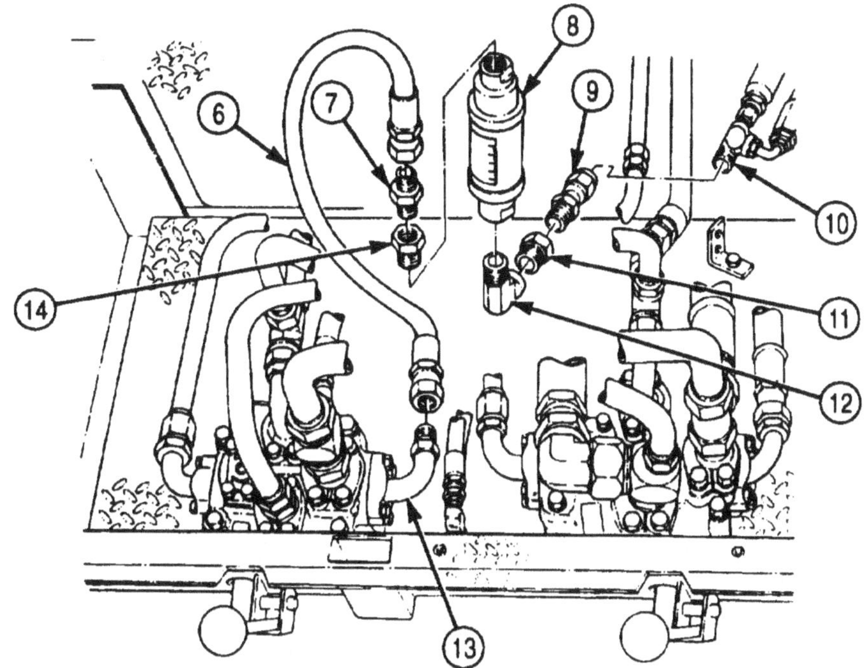

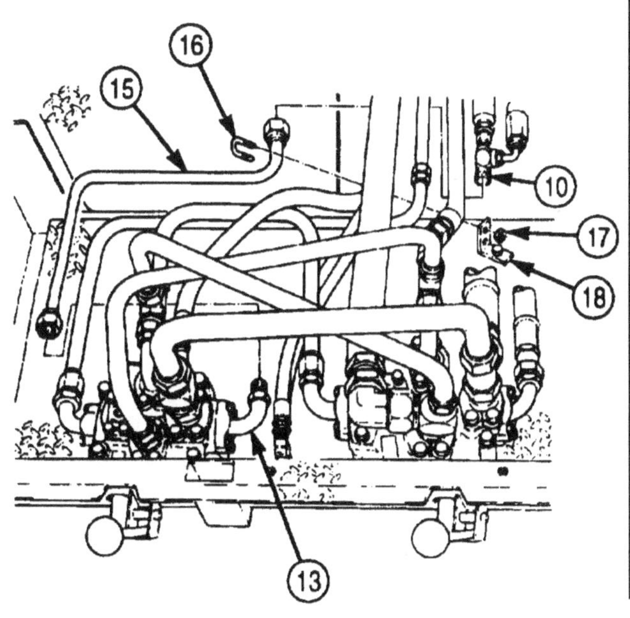

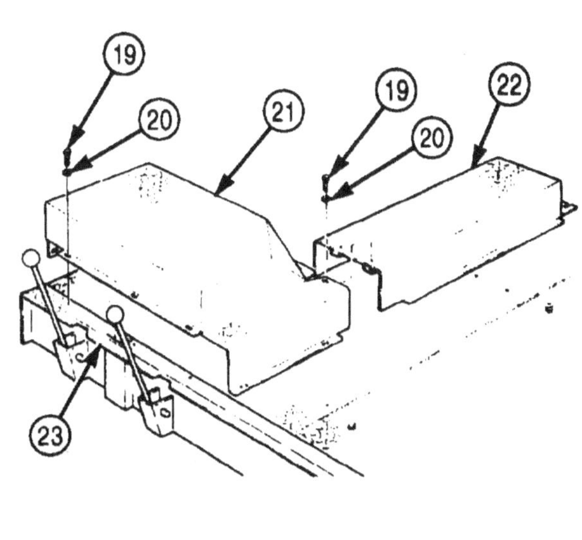

Table 4-1. Mechanical Troubleshooting (Contd).

MALFUNCTION TEST OR INSPECTION CORRECTIVE ACTION

Step 17. Remove pressure gauge from boom control valve.

NOTE

Ensure engine is off.

a. Remove pressure gauge (1), connector (2), hose assembly (3), and connector (4) from boom control valve (5).

b. Install plug (8) in boom control valve (5).

c. Install mounting plate (7) on gondola (9) with five screws (6).

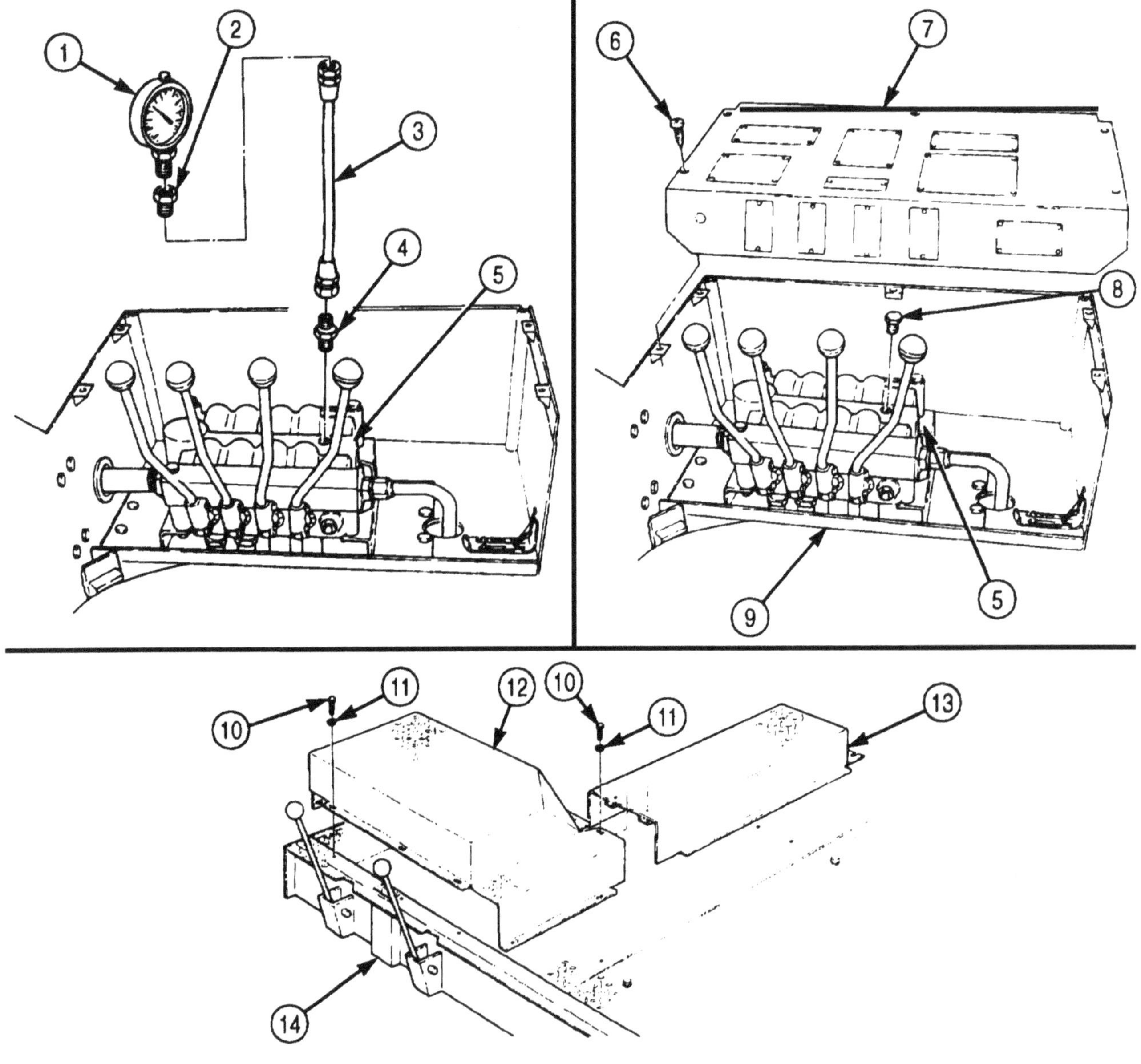

Table 4-1. Mechanical Troubleshooting (Contd).

MALFUNCTION TEST OR INSPECTION CORRECTIVE ACTION

Step 18. Install pressure gauge (Appendix E, Item 168) on rear winch directional control valve.

NOTE

Ensure engine is off.

a. Remove twelve screws (10), lockwashers (11), and control valve covers (12) and (13) from wrecker (14). Discard lockwashers (11).

NOTE

- If gauge port cover plate assembly or rear winch control valve is equipped with pipe plug, proceed to step d. If gauge port cover plate assembly is not equipped with pipe plug, proceed to step b.
- Have drainage container ready to catch oil.

b. Remove four screws (15), lockwashers (16), cover plate assembly (19), and O-ring (17) from rear winch control valve (18). Discard O-ring (17) and lockwashers (16).

c. Install new O-ring (17) and cover plate assembly (19) with plug (23) on rear winch control valve (18) with four new lockwashers (16) and screws (15).

d. Remove plug (23) from cover plate assembly (19).

e. Install appropriate connector (24), hose assembly (20), connector (22), and pressure gauge (21) on rear winch control valve (18).

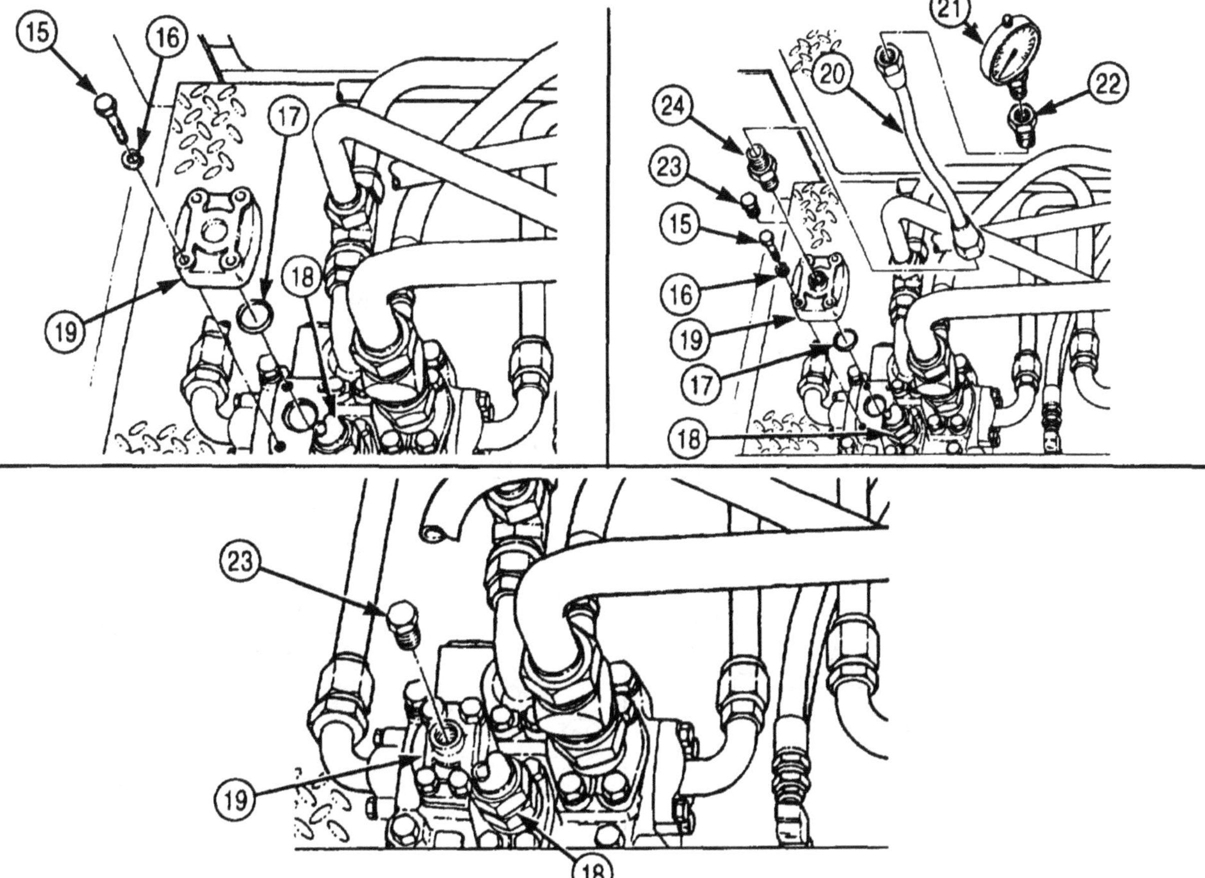

Table 4-1. Mechanical Troubleshooting (Contd).

MALFUNCTION TEST OR INSPECTION CORRECTIVE ACTION

f. Remove hose assembly (1) from tee (4).

g. Install appropriate plug (2) in hose assembly (1) and cap (3) on tee (4).

Step 19. Check rear winch control valve pressure. Start engine, place transfer case in neutral, engage crane hydraulic system, place torque selector in low torque, high-speed position, and move rear winch control valve to UNWIND (TM 9-2320-272-10). Record pressure. Shut down engine. Proper pressure is 1,600-1,700 psi (11,032-11,722 kPa).

a. If rear winch control valve pressure is 1,600-1,700 psi (11,032-11,722 kPa), replace boom control valve (para. 4-201), and check and adjust boom relief pressure (steps 20 through 23).

b. If rear winch control valve pressure is less than 1,600 psi (11,032 kPa), adjust rear winch control valve pressure. Remove relief valve adjusting screw cap, loosen jamnut, and turn adjusting screw to obtain correct pressure reading.

c. If relief pressure cannot be adjusted, replace rear winch control valve (para. 4-179), check and adjust rear winch control valve pressure, adjust boom relief pressure (steps 20 through 23), and replace boom control valve (para. 4-201) if boom relief pressure cannot be adjusted properly.

d. If rear winch control valve pressure can be adjusted, adjust boom relief pressure (steps 20 through 23).

e. If rear winch control valve pressure is greater than 1,700 psi (11,722 kPa), adjust rear winch control valve pressure. Remove relief valve adjusting screw cap, loosen jamnut, and turn adjusting screw to obtain correct pressure reading.

f. If rear winch control valve pressure can be adjusted, replace boom control valve (para. 4-201), and check and adjust boom relief pressure (steps 20 through 23).

g. If rear winch control valve pressure cannot be adjusted, replace rear winch control valve (para. 4-179), check and adjust rear winch control valve pressure, replace boom control valve (para. 4-201), and check and adjust boom relief pressure (steps 20 through 23).

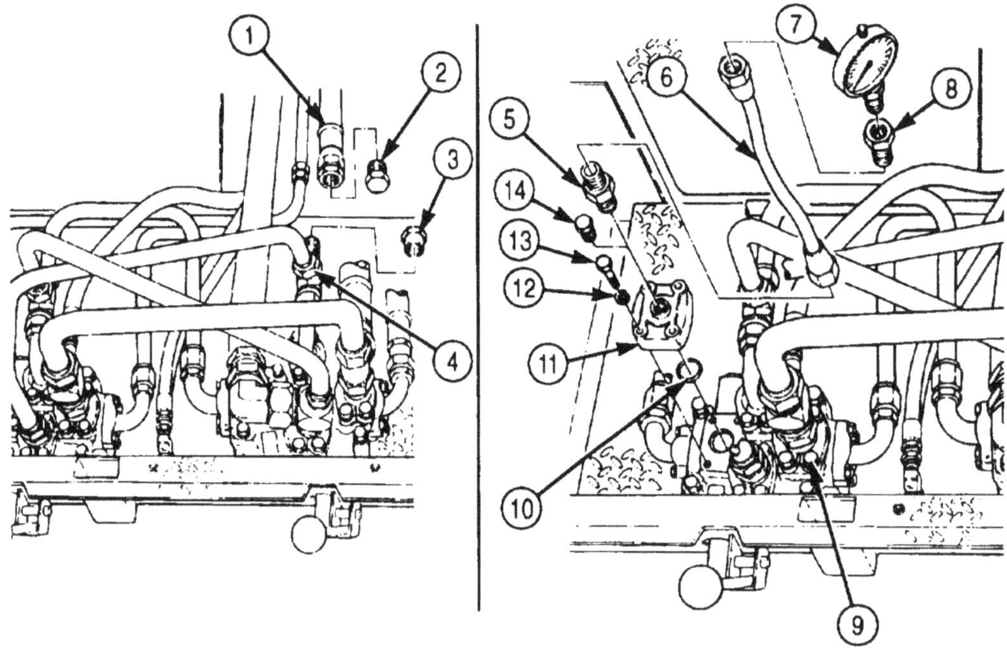

Table 4-1. Mechanical Troubleshooting (Contd).

MALFUNCTION TEST OR INSPECTION CORRECTIVE ACTION

Step 20. Remove pressure gauge from rear winch directional control valve.

NOTE

- Ensure engine is off.
- Have drainage container ready to catch oil.

a. Remove plug (2) from hose assembly (1) and cap (3) from tee (4).

b. Install hose assembly (1) on tee (4).

c. Remove pressure gauge (7), connector (8), hose (6), and connector (5) from rear winch control valve (9).

d. Install plug (14) in cover plate (11).

NOTE

If gauge port cover plate of rear winch control valve is equipped with a pipe plug, proceed to step g. If gauge port cover plate is not equipped with a pipe plug, proceed to step e.

e. Remove four screws (13), lockwashers (12), cover plate assembly (11), and O-ring (10) from rear winch control valve (9). Discard O-ring (10) and lockwashers (12).

f. Install new O-ring (10) and cover plate assembly (11) on rear winch control valve (9) with four new lockwashers (12) and screws (13).

g. Install rear winch control valve covers (17) and (18) on wrecker (19) with twelve new lockwashers (16) and screws (15).

Step 21. Install pressure gauge (Appendix E, Item 168) on boom control valve (step 9).

Step 22. Check boom relief pressure (step 11), and adjust as required (para. 4-204).

Step 23. Remove pressure gauge from boom control valve (step 17).

END OF TESTING!

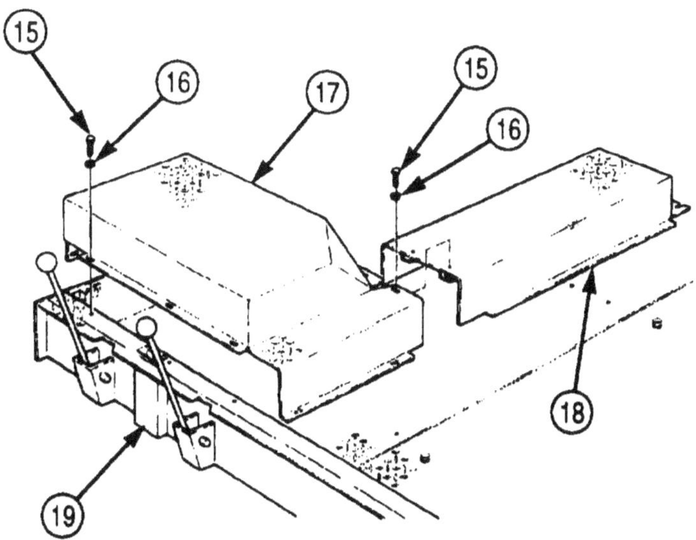

Table 4-1. Mechanical Troubleshooting (Contd).

MALFUNCTION TEST OR INSPECTION CORRECTIVE ACTION

61. ALL BOOM CONTROLS OPERATE ABNORMALLY

Step 1. Perform malfunction 60, steps 1 through 7.

Step 2. Check reservoir return lines for blockage.

a. If blockage is found, clear lines.

b. If no blockage is found, replace boom control valve (para. 4-201).

Step 3. Check boom relief pressure (malfunction 60, step 11).

END OF TESTING!

62. WING CONTROL OPERATES ABNORMALLY; ALL OTHER CONTROLS OPERATE NORMALLY

Step 1. Perform malfunction 60, steps 1 through 7.

Step 2. Check swing motor hydraulic lines for blockage.

a. If blockage is found, clear lines.

b. If no blockage is found, replace swing motor (para. 4-199).

Step 3. Check swing control operation.

If swing control operates abnormally, replace boom control valve (para. 4-201).

Step 4. Check boom relief pressure (malfunction 60, step 11).

END OF TESTING!

63. CROWD CONTROL OPERATES ABNORMALLY; ALL OTHER CONTROLS OPERATE NORMALLY

Step 1. Perform malfunction 60, steps 1 through 7.

Step 2. Check boom extension cylinder hydraulic lines for blockage.

a. If blockage is found, clear lines.

b. If no blockage is found, replace boom extension cylinder (para. 4-196).

Step 3. Check crowd control operation.

If crowd control operates abnormally, replace boom control valve (para. 4-201).

Step 4. Check boom relief pressure (malfunction 60, step 11).

END OF TESTING!

64. HOIST CONTROL OPERATES ABNORMALLY; ALL OTHER CONTROLS OPERATE NORMALLY

Step 1. Perform malfunction 60, steps 1 through 7.

Step 2. Check hoist winch motor hydraulic lines for blockage.

a. If blockage is found, clear lines.

b. If no blockage is found, replace hoist winch motor (para. 4-199).

Step 3. Check hoist control operation. If hoist control operates abnormally, replace boom control valve (para. 4-201).

Step 4. Check boom relief pressure (malfunction 60, step 11).

END OF TESTING!

65. BOOM CONTROL OPERATES ABNORMALLY; ALL OTHER CONTROLS OPERATE NORMALLY

Step 1. Perform malfunction 60, steps 1 through 7.

Step 2. Check boom elevation cylinder hydraulic lines for blockage.

a. If blockage is found, clear lines.

b. If no blockage is found, replace elevating cylinders (para. 4-193).

Step 3. Check boom control operation. If boom control operates abnormally, replace boom control valve (para. 4-201).

Table 4-1. Mechanical Troubleshooting (Contd).

MALFUNCTION TEST OR INSPECTION CORRECTIVE ACTION

Step 4. Check boom relief pressure (malfunction 60, step 11).

END OF TESTING!

POWER TAKEOFF

66. NOISY POWER TAKEOFF

Remove and repair PTO.

a. If transfer case PTO (para. 4-207 and para. 4-208).

b. If transmission PTO (para. 4-211 and para. 4-212).

END OF TESTING!

67. POWER TAKEOFF SUPS OUT OF GEAR

Step 1. Adjust shift linkage.

a. If transfer case PTO (para. 4-205 and 4-206).

b. If transmission PTO (para. 4-209 or 4-210).

Step 2. If malfunction continues, replace poppet springs.

a. If transfer case PTO (para. 4-208).

b. If transmission PTO, replace PTO (para. 4-211).

END OF TESTING!

68. LUBRICANT LEAKING

Check for defective gaskets and seals.

Replace defective gaskets or seals.

a. If transfer case PTO (para. 4-208).

b. If transmission PTO (para. 4-212).

END OF TESTING!

FRONT WINCH

69. FRONT WINCH IS INOPERATIVE OR OPERATES SLOWLY

CAUTION

When disconnecting hydraulic oil lines and hoses, plug all openings to prevent dirt from entering and causing internal parts damage.

NOTE

Have drainage container ready to catch excess oil.

Step 1. Remove hydraulic pump to control valve line and check for blockage (para. 4-177).

a. If blockage is present, remove obstruction and reevaluate winch operation.

b. If blockage is not present, go to step 2.

Table 4-1. Mechanical Troubleshooting (Contd).

MALFUNCTION TEST OR INSPECTION CORRECTIVE ACTION

Step 2. Perform malfunction 59, step 2.

Step 3. If hydraulic system oil temperature is greater than 32°F (0°C), proceed to step 5.

Step 4. If hydraulic system oil temperature is 32°F (0°C) or colder, start engine, engage PTO (TM 9-2320-272-10), and allow hydraulic oil to circulate 15 minutes with winch control in neutral position, and proceed to step 5.

Step 5. Start engine and set idle at 800 rpm (TM 9-2320-272-10).

Step 6. Engage PTO and winch lever in the wind position (TM 9-2320-272-10).

Step 7. Record pressure and flow readings.

Normal readings:

Flow rate - 5 gpm (18.9 Lpm) or greater.

Pressure - 1,650-1,850 psi (11,377-12,756 kPa).

a. If readings are normal, go to step 10.

b. If reading is less than 1,850 psi (12,756 kPa), and flow rate is less than 5 gpm (18.9 Lpm), proceed to step 8.

c. If pressure is greater than 1,850 psi (12,756 kPa), go to step 10.

d. If pressure is less than 1,650 psi (11,377 kPa), replace hydraulic control valve (para. 4-201).

Step 8. Perform malfunction 59, step 8.

Step 9. Check pump suction line for restrictions. If no restrictions are present, replace hydraulic pump (para. 4-176).

Step 10. Check lines between control valve and reservoir for restrictions. If no restrictions are present, replace control valve (para. 4-178).

Step 11. Check winch motor lines for restrictions. If no restrictions are present, replace control valve (para. 4-178).

END OF TESTING!

Section II. POWER PLANT MAINTENANCE

4-3. POWER PLANT INDEX

PARA. NO.	TITLE	PAGE NO.
4-4.	Power Plant (M939/A1) Replacement	4-35
4-5.	Power Plant (M939A2) Replacement	4-67
4-6.	Engine and Container Replacement	4-86
4-7.	Preparing Replacement Engine for Installation in Vehicle	4-91
4-8.	Starting Repaired or Replaced Engine	4-95
4-9.	Engine Mounting on Repair Stand	4-100

4-4. POWER PLANT (M939/A1) REPLACEMENT

THIS TASK COVERS:

a. Preliminary Disconnections
b. Removal
c. Installation

INITIAL SETUP:

APPLICABLE MODELS

M939/A1

SPECIAL TOOLS

Engine and transmission sling (Appendix E, Item 4)

TOOLS

General mechanic's tool kit (Appendix E, Item 1)
Torque wrench (Appendix E, Item 144)
Lifting device
Chains

MATERIALS/PARTS

O-ring (Appendix D, Item 454)
Cotter pin (Appendix D, Item 62)
Tiedown strap (Appendix D, Item 684)
Lockwasher (Appendix D, Item 401)
Lockwasher (Appendix D, Item 403)
Cotter pin (Appendix D, Item 66)
Two locknuts (Appendix D, Item 283)
Two locknuts (Appendix D, Item 294)
Locknut (Appendix D, Item 416)
Locknut (Appendix D, Item 276)
Two lockwashers (Appendix D, Item 382)
Five lockwashers (Appendix D, Item 350)
Ten lockwashers (Appendix D, Item 377)
Lockwasher (Appendix D, Item 379)
Lockwasher (Appendix D, Item 376)
Lockwasher (Appendix D, Item 364)
Gasket sealant (Appendix C, Item 30)
Twine (Appendix C, Item 77)
Cap and plug set (Appendix C, Item 14)

PERSONNEL REQUIRED

Two

REFERENCES (TM)

LO 9-2320-272-12
TM 9-2320-272-10
TM 9-2320-272-24P

4-4. POWER PLANT (M939/A1) REPLACEMENT (Contd)

INITIAL SETUP (COND):

EQUIPMENT CONDITION

- Parking brake set (TM 9-2320-272-10).
- Battery ground cables disconnected (para. 3-126).
- Air reservoirs drained (TM 9-2320-272-10).
- Engine oil drained (para. 3-5).
- Transmission oil drained (para. 3-133).
- Hood removed (para. 3-275).
- Radiator drained and removed (para. 3-59).
- Coolant hoses and tubes removed (para. 3-54).
- Front exhaust pipe removed (para. 3-50).
- Engine oil dipstick and tube removed (para. 3-3).
- Air intake pipe and hump hoses removed (para. 3-14).
- Transmission PTO-to-hydraulic pump drive shaft removed, if equipped (para. 3-334).
- Transmission-to-transfer case propeller shaft removed (para. 3-148).
- Surge tank removed (3-61).
- Radiator fan blade assembly removed (para. 3-72).

GENERAL SAFETY INSTRUCTIONS

- Lifting device must have a capacity greater than the combined weight of the engine and transmission.
- All personnel must stand clear during lifting operations.
- Do not put hands between frame and engine supports during lifting operations.
- Do not detach chain from engine until engine is supported.
- Engine must be securely mounted on repair stand.
- Never start a new or repaired engine without performing run-in starting procedures (para. 4-8).

a. Preliminary Disconnections

NOTE

If a special purpose kit is installed on vehicle, refer to chapter 4, section XV, and make necessary disconnections.

1. Disconnect wires (3) and (2) from water temperature sending unit (4) and engine high temperature sending unit (1).
2. Remove hose clamp (8) and hose (6) from water heater shutoff (7). Tie hose (6) clear of engine (5).
3. Remove hose clamp (10) and hose (9) from water heater shutoff (11). Tie hose (9) clear of engine (5).
4. Remove two screw-assembled lockwashers (27) and terminal cover (26) from alternator (28). Discard screw-assembled lockwashers (27).
5. Remove two screws (21), lockwashers (20), and retainer (19) from alternator (28). Discard lockwashers (20).
6. Remove screw (14), lockwasher (13), and wire (12) from alternator (28). Discard lockwasher (13).

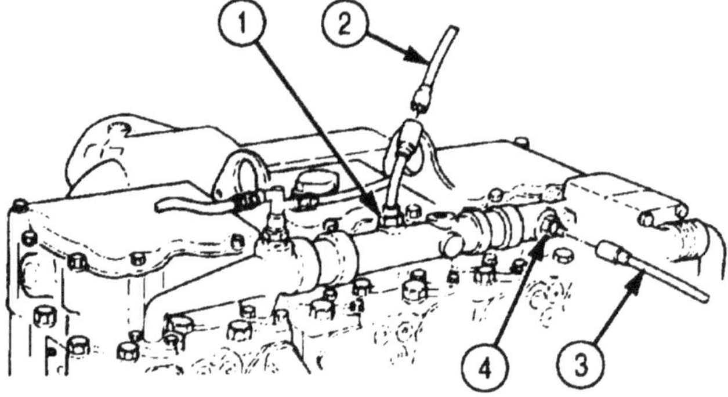

4-4. POWER PLANT (M939/A1) REPLACEMENT (Contd)

NOTE

Sealant must be removed before removing wires.

7. Remove nut (18), lockwasher (17), washer (16), and wire (15) from alternator (28). Discard lockwasher (17).
8. Remove nut (22), lockwasher (23), washer (24), and wire (25) from alternator (28). Discard lockwasher (23).
9. Disconnect plug (30) from connector (31).
10. Remove tiedown strap (29) from wires (12) and (15). Discard tiedown strap (29).

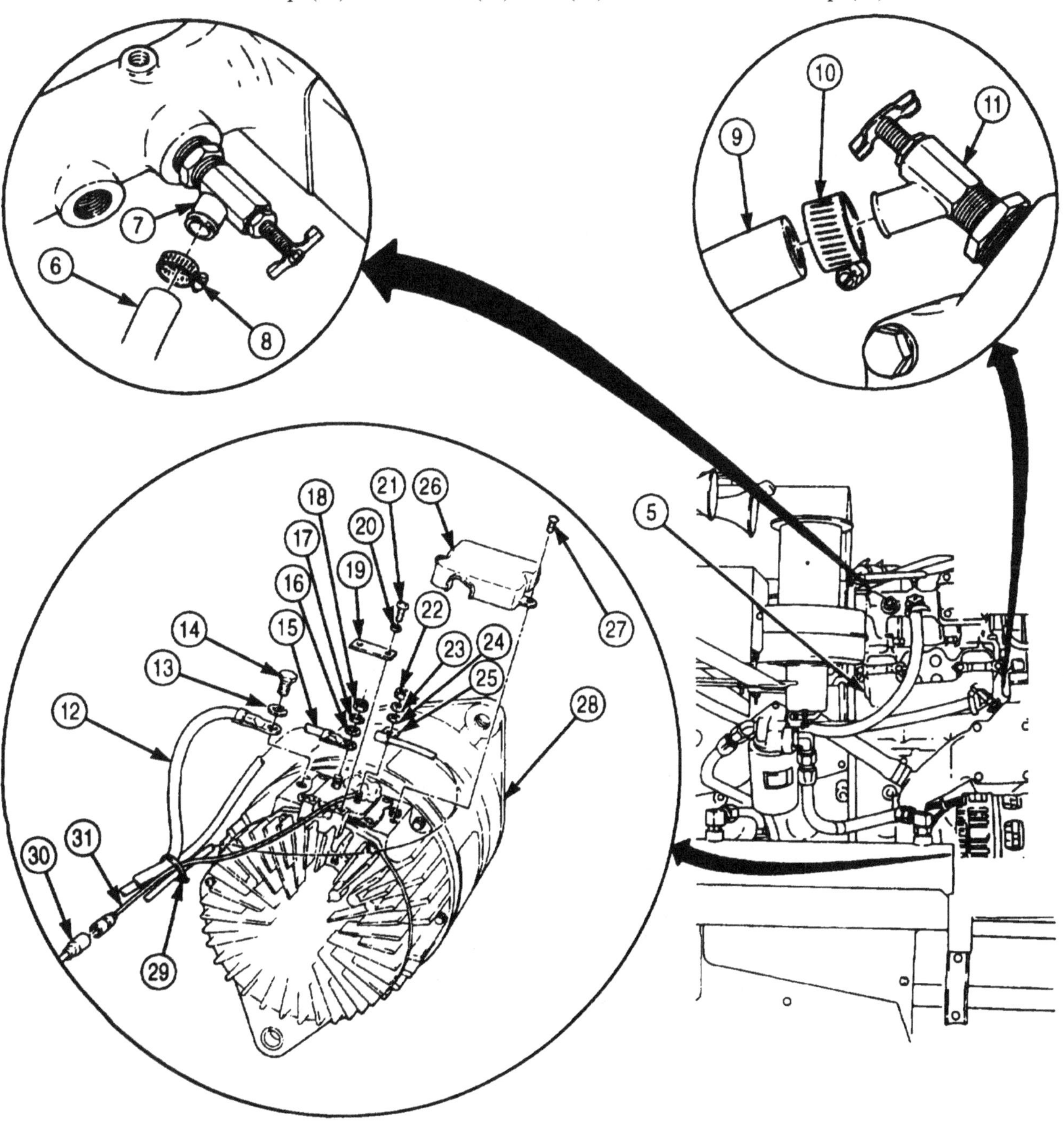

4-4. POWER PLANT (M939/A1) REPLACEMENT (Contd)

NOTE

Plug all openings and tag lines for installation.

11. Disconnect hose (1) from elbow (2) and tie hose (1) to engine (3).
12. Disconnect hose (4) from transmission oil cooler (5) and tie hose (4) to engine (3).
13. Disconnect hose (8) from fitting (7) and tie hose (8) to engine (3).
14. Loosen clamp (10), disconnect hose (9) from reservoir (6), and tie hose (9) to engine (3).

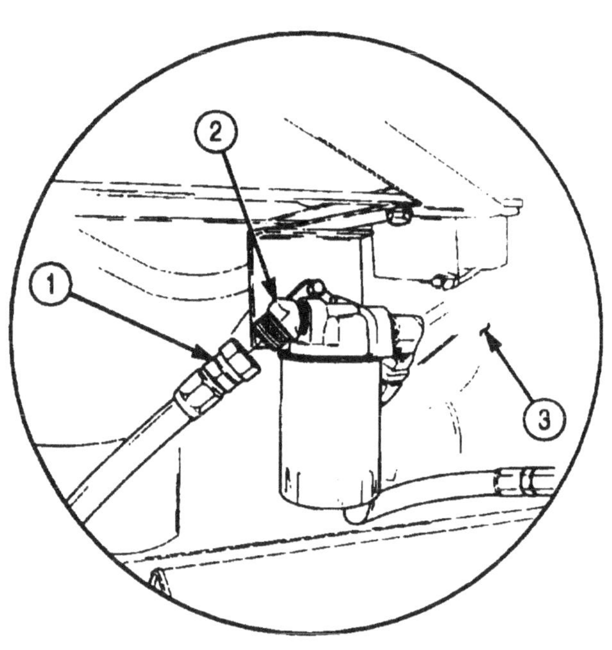

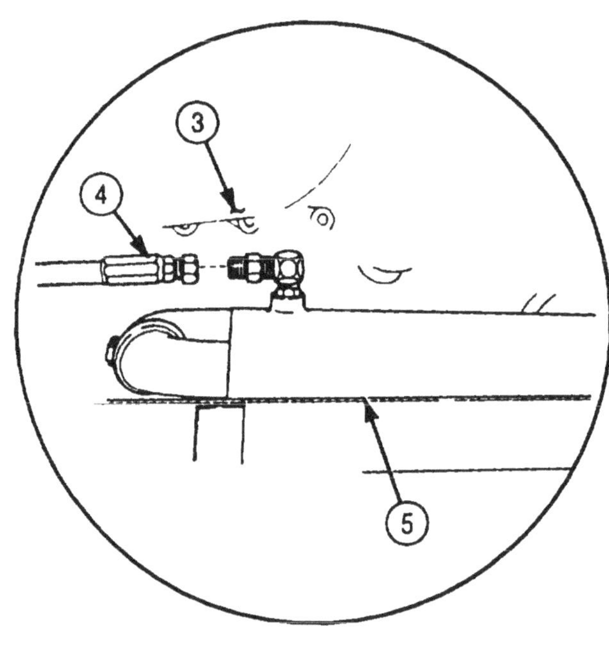

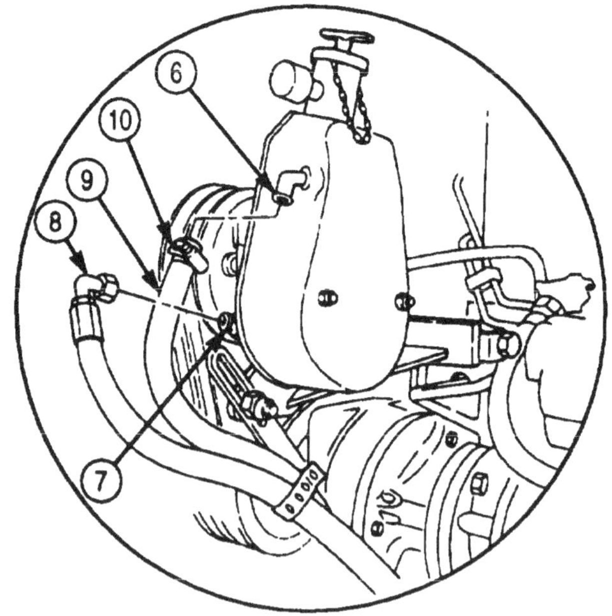

4-4. POWER PLANT (M939/A1) REPLACEMENT (Contd)

15. Remove screw (19) and connector (11) from emergency stop cable (13).
16. Remove nut (18), screw (15), washer (14), and clamp (16) from bracket (17), and pull cable (13) through swivel block (12).
17. Install connector (11) on cable (13) with screw (19).
18. Install clamp (16) on bracket (17) with screw (15), washer (14), and nut (18).
19. Remove nut (22) and two wires (21) from fuel shutoff solenoid (20).
20. Remove locknut (24), screw (26), and accelerator rod (25) from throttle lever (23), and tie accelerator rod (25) clear of engine. Discard locknut (24).
21. Remove locknut (27), screw (36), and link (28) from throttle lever (23). Discard locknut (27).
22. Remove return spring (35), two nuts (32), screws (34), cable clamp (33), and shim (31) from fuel primer pump bracket (30).
23. Remove modulator cable (29) from fuel primer pump bracket (30), and tie cable (29) clear of engine.

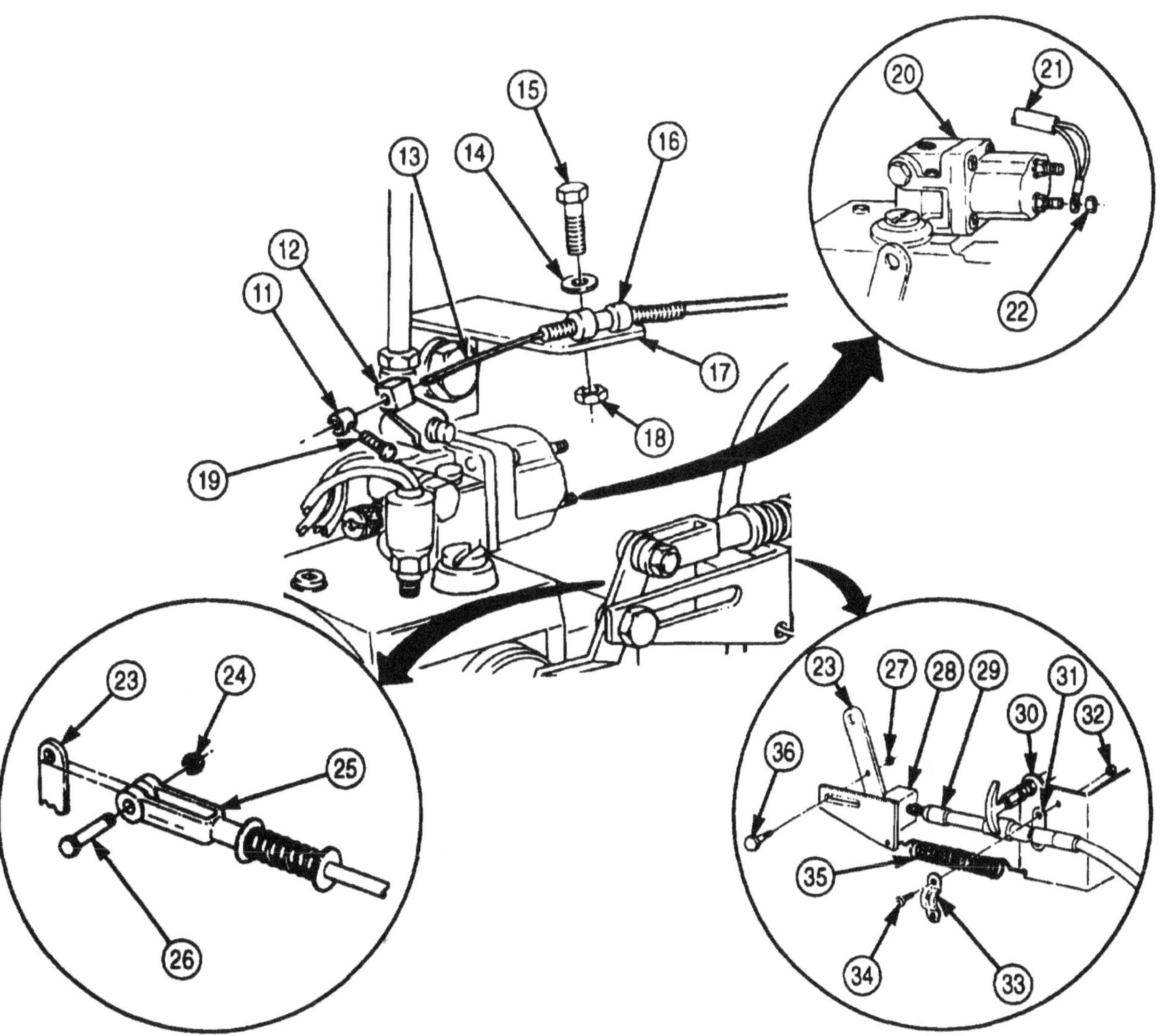

4-4. POWER PLANT (M939/A1) REPLACEMENT (Contd)

24. Disconnect air line (2) from air governor (1).

NOTE

Perform step 25 for M936/A1 model vehicles only.

25. Disconnect air line (3) from VS governor (4).

NOTE

Step 26 does not apply to M936/A1 model vehicles.

26. Disconnect connector (6) from fuel pressure transducer (7).
27. Disconnect fuel line (13) from fuel pump (12).
28. Disconnect air line (11) from air compressor (14).
29. Disconnect connector (15) from oil pressure sending unit (16).
30. Disconnect tachometer drive cable (5) from tachometer pulse sender (8).
31. Disconnect tachometer pulse sender connector (9) from pulse sender harness (10).

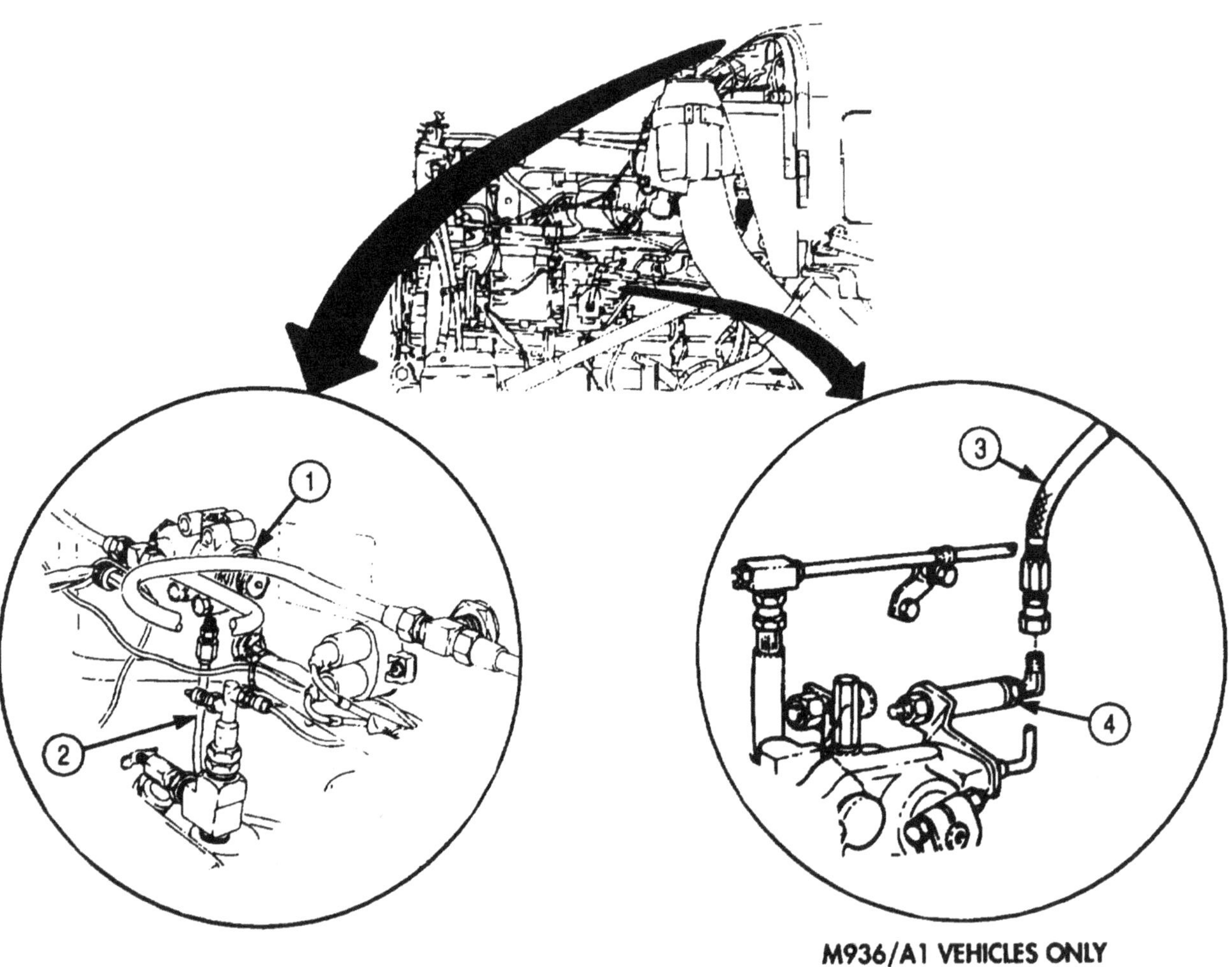

4-4. POWER PLANT (M939/A1) REPLACEMENT (Contd)

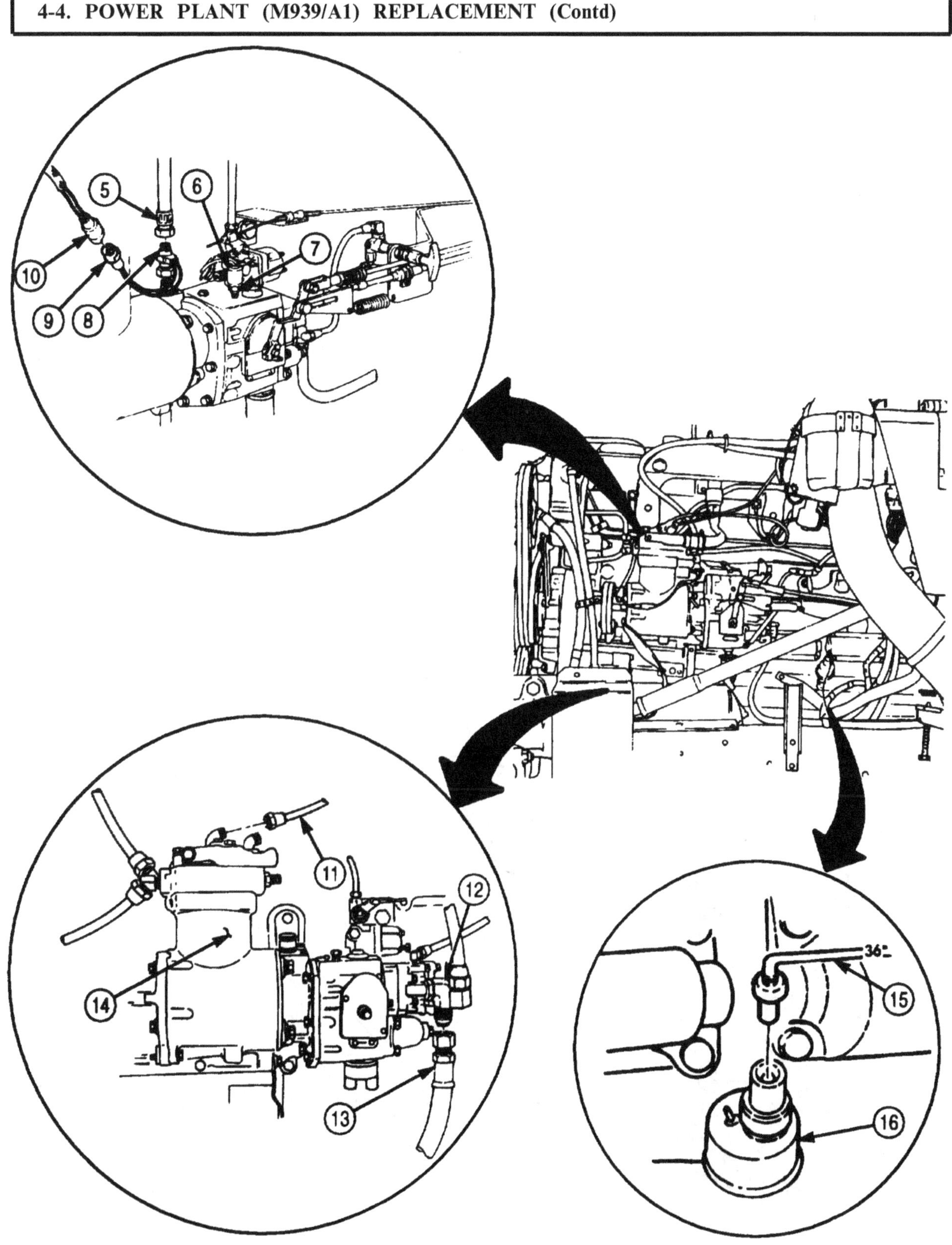

4-4. POWER PLANT (M939/A1) REPLACEMENT (Contd)

32. Remove two screws (8), washers (9), clamps (7), (5), and (10), and air line (4) from air intake manifold (3).
33. Remove tachometer cable (6), speedometer cable (11), and harness (12) from air intake manifold (3). Tie cables (6) and (11) and harness (12) clear of engine.
34. Disconnect two connectors (2) from ether start switch (1).

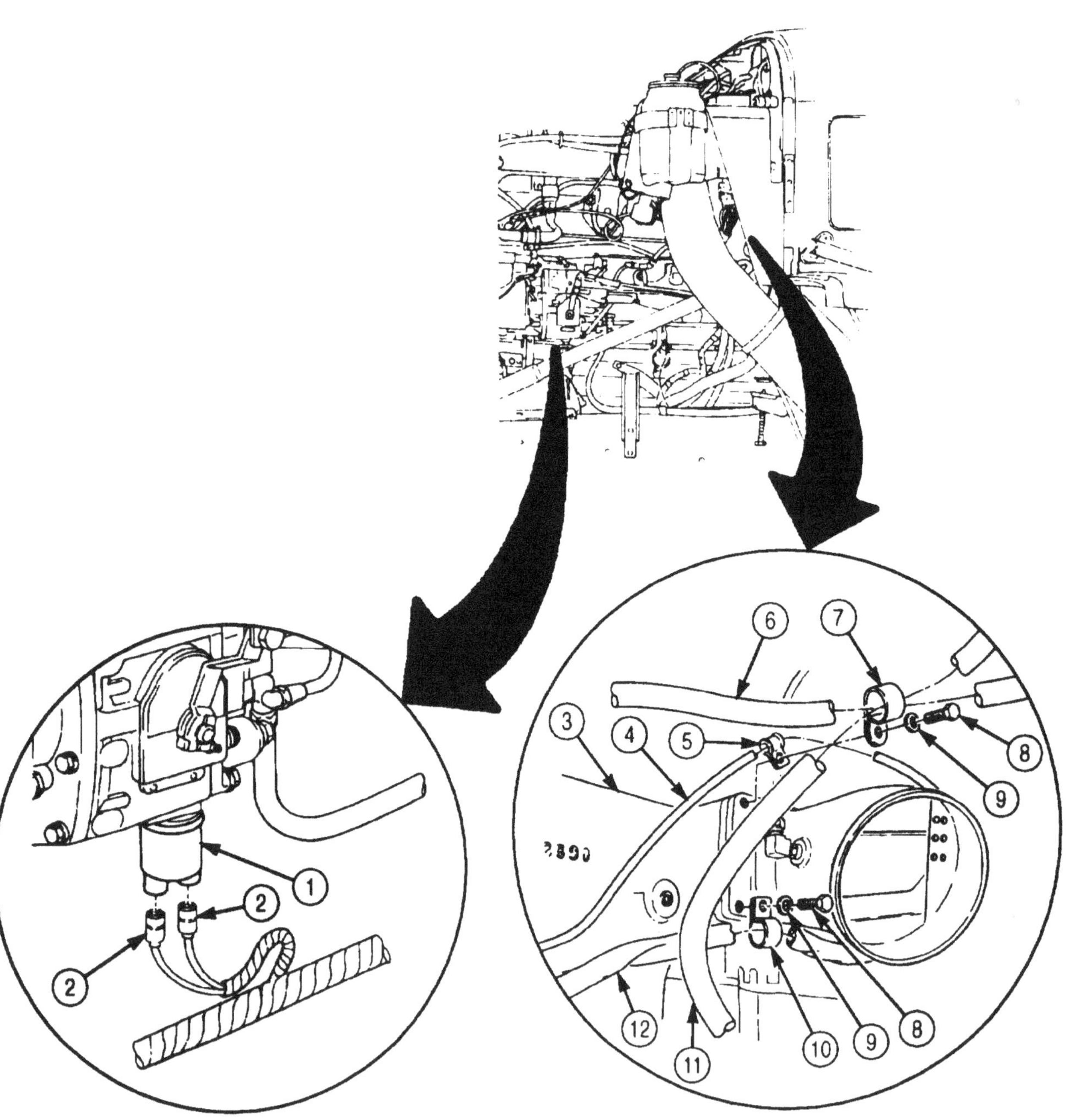

4-4. POWER PLANT (M939/A1) REPLACEMENT (Contd)

35. Disconnect fuel line (14) from fuel pump return hose (13). Tie fuel line (14) clear of engine (15).
36. Disconnect ether supply line (16) from ether start safety valve (17) and ether cylinder valve (22). Tie ether supply line (16) clear of engine (15).
37. Disconnect ether cylinder relief line (18) from safety valve (17) and atomizer (19). Tie either cylinder relief line (18) clear of engine (15).
38. Disconnect tube (20) from fitting (21). Tie tube (20) clear of engine (15).

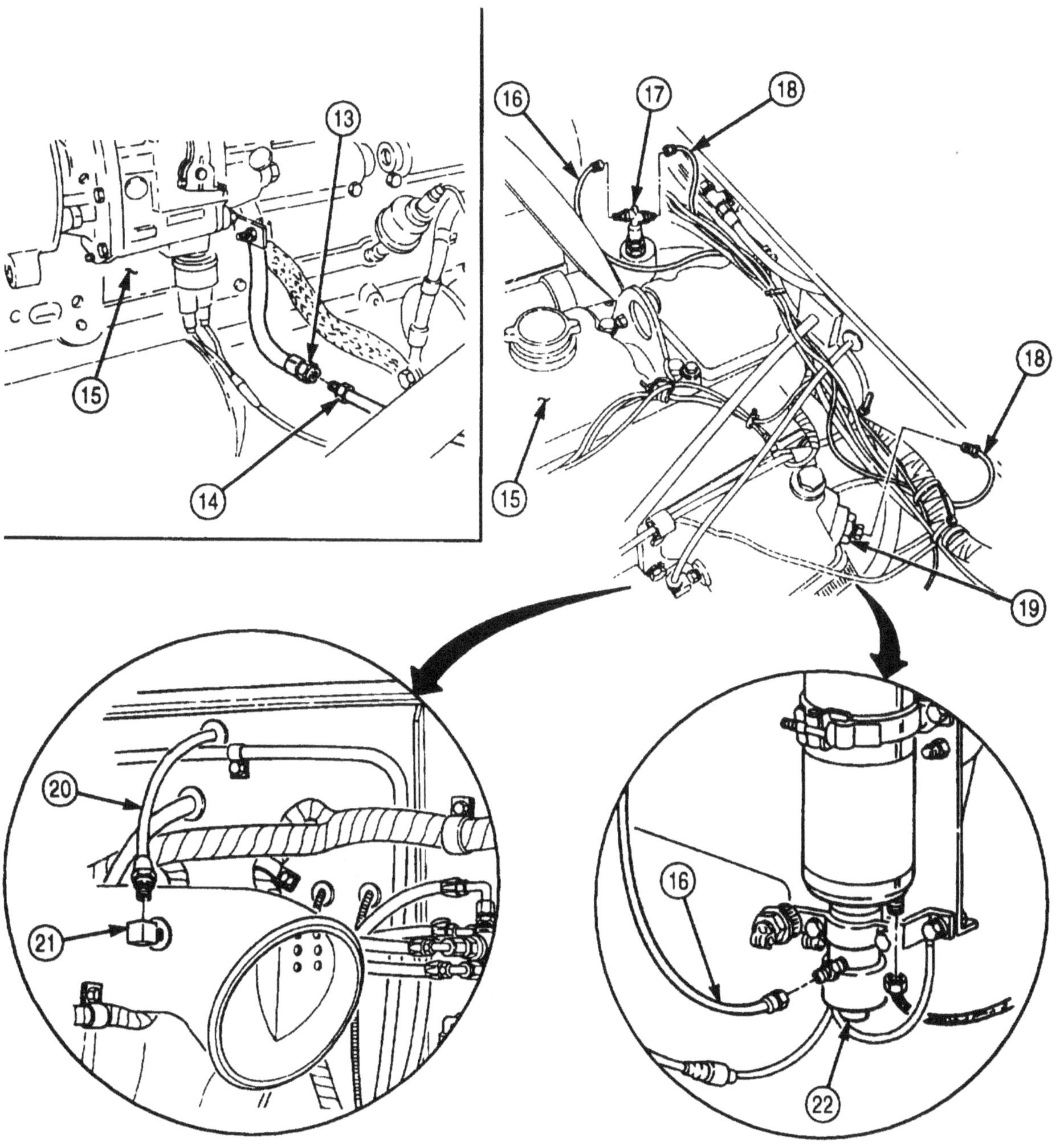

4-4. POWER PLANT (M939/A1) REPLACEMENT (Contd)

39. Remove nut (12), washer (13), and wires (14) and (11) from starter solenoid (4). Tie wires (14) and (11) clear of engine.
40. Remove screw (3), clip (2), and wire (1) from starter solenoid (4). Tie wire (1) clear of engine.
41. Remove nut (8), washer (9), wires (10) and (7), and ground strap (6) from starter motor (5). Tie wires (10) and (7) clear of engine.
42. Disconnect air line (15) from fitting (16).

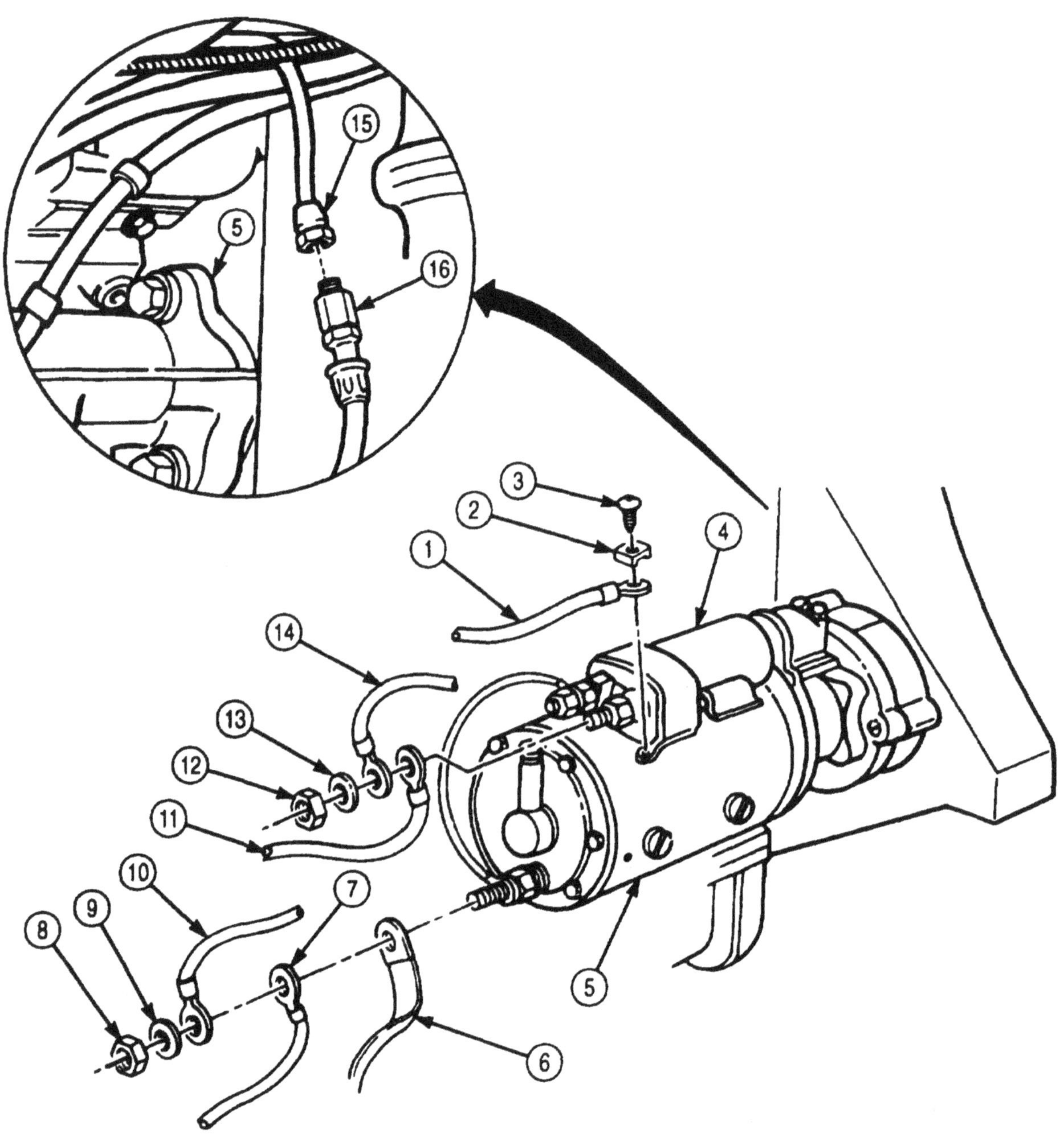

4-4. POWER PLANT (M939/A1) REPLACEMENT (Contd)

NOTE

Access for steps 43 through 45 is through door in cab floor.

43. Disconnect wire (17) from transmission oil temperature sending unit (23). Tie wire (17) clear of transmission.
44. Disconnect wires (17) and (20) from two clips (18) on transmission flange (19). Tie wires (17) and (20) clear of transmission.
45. Remove transmission dipstick (22) from tube (21). Plug opening in tube (21).

NOTE

Perform step 46 and 47 only on vehicles equipped with a transmission Power Takeoff (PTO).

46. Remove two nuts (24), screws (28), retaining strap (27), and spacer plate (26) from PTO cable (29) and bracket (25).
47. Remove cotter pin (33), washer (32), cable pin (30), and PTO cable (29) from PTO select lever (31). Discard cotter pin (33), and tie PTO cable (29) clear of transmission.

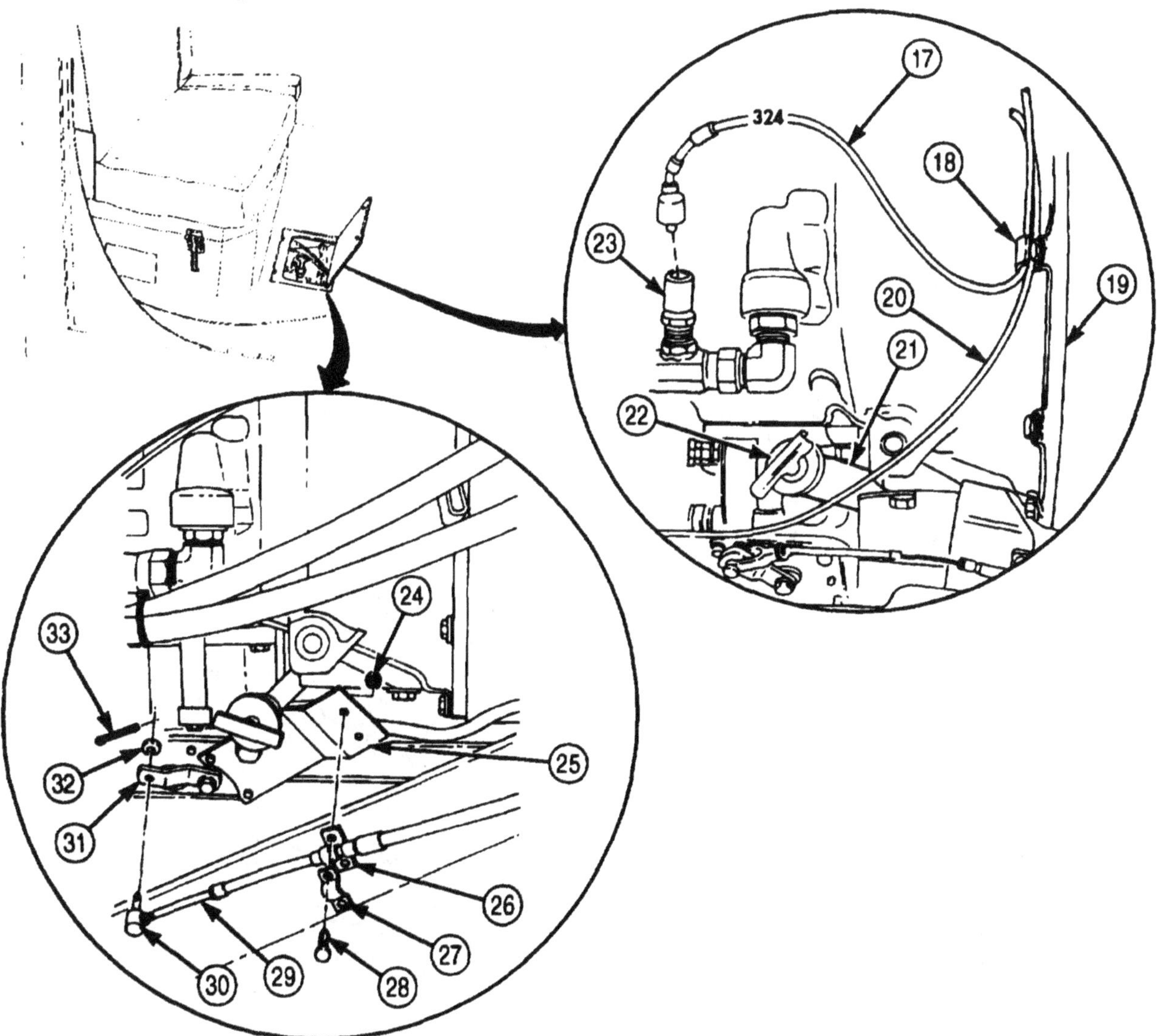

4-4. POWER PLANT (M939/A1) REPLACEMENT (Contd)

48. Disconnect breather vent line (3) from adapter elbow (2). Tie breather vent line (3) clear of transmission (1).
49. Remove two locknuts (5), U-bolt (7), and shim (6) from support bracket (4) and cable (8). Discard locknuts (5).
50. Remove cotter pin (10) and cable pin (9) from shift lever (11). Discard cotter pin (10) and tie cable (8) clear of transmission (1).
51. Disconnect wire (13) to 5th gear lock-in solenoid (14) from connector (15). Tie wire (13) to transmission (1).
52. Disconnect transorb diode wire (12) from connector (15).
53. Remove nut (18), screw (27), clamps (17) and (26), modulator cable (16), and speedometer cable (25) from transmission bracket (19). Tie modulator cable (16) and speedometer cable (25) clear of transmission (1).
54. Remove screw (22), bracket (21), modulator (24), and O-ring (23) from transmission (1). Discard O-ring (23).
55. Remove screw (20) and bracket (19) from transmission (1).
56. Disconnect two wires (28) from neutral start switch (29). Tie wires (28) clear of transmission (1).

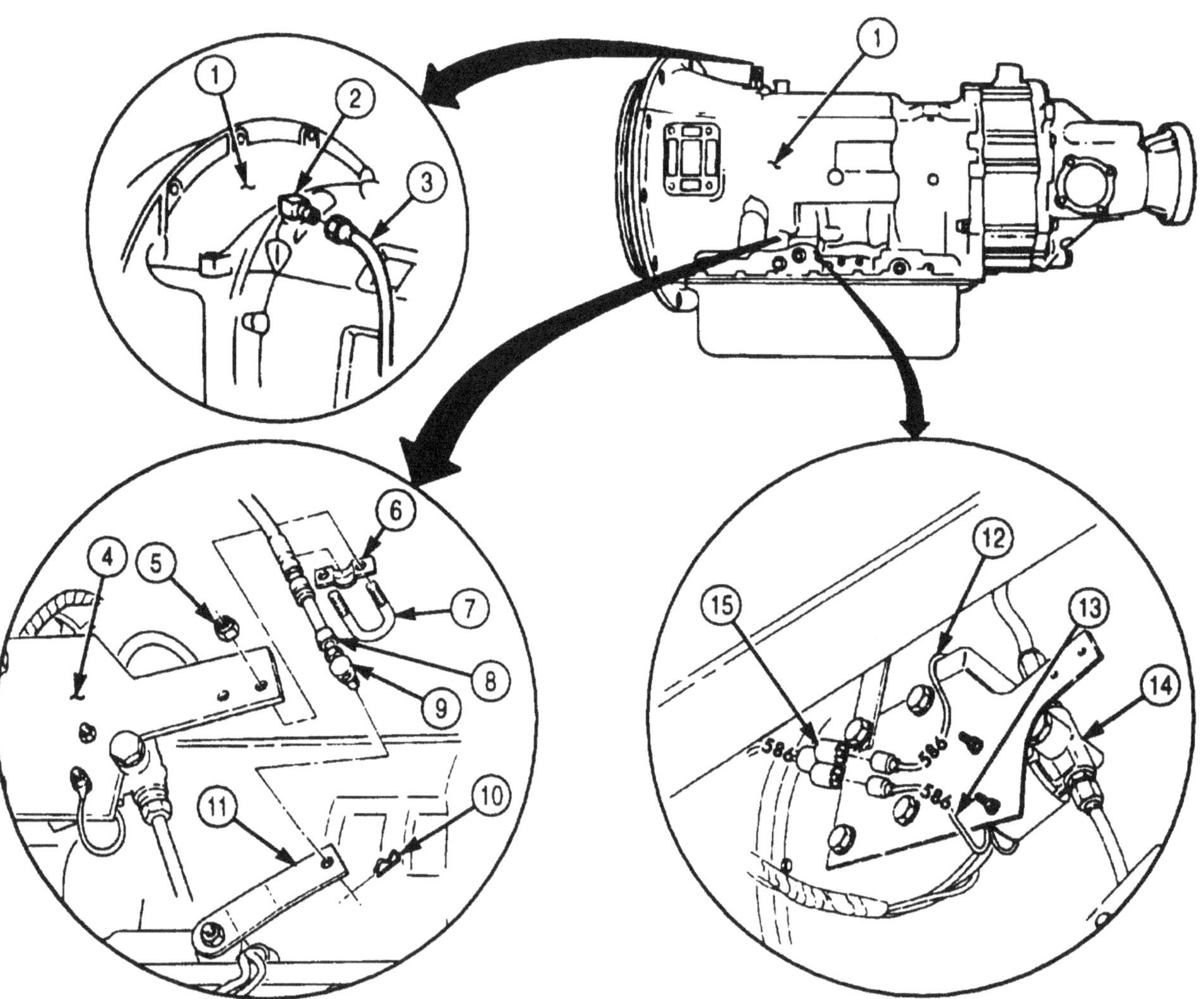

4-4. POWER PLANT (M939/A1) REPLACEMENT (Contd)

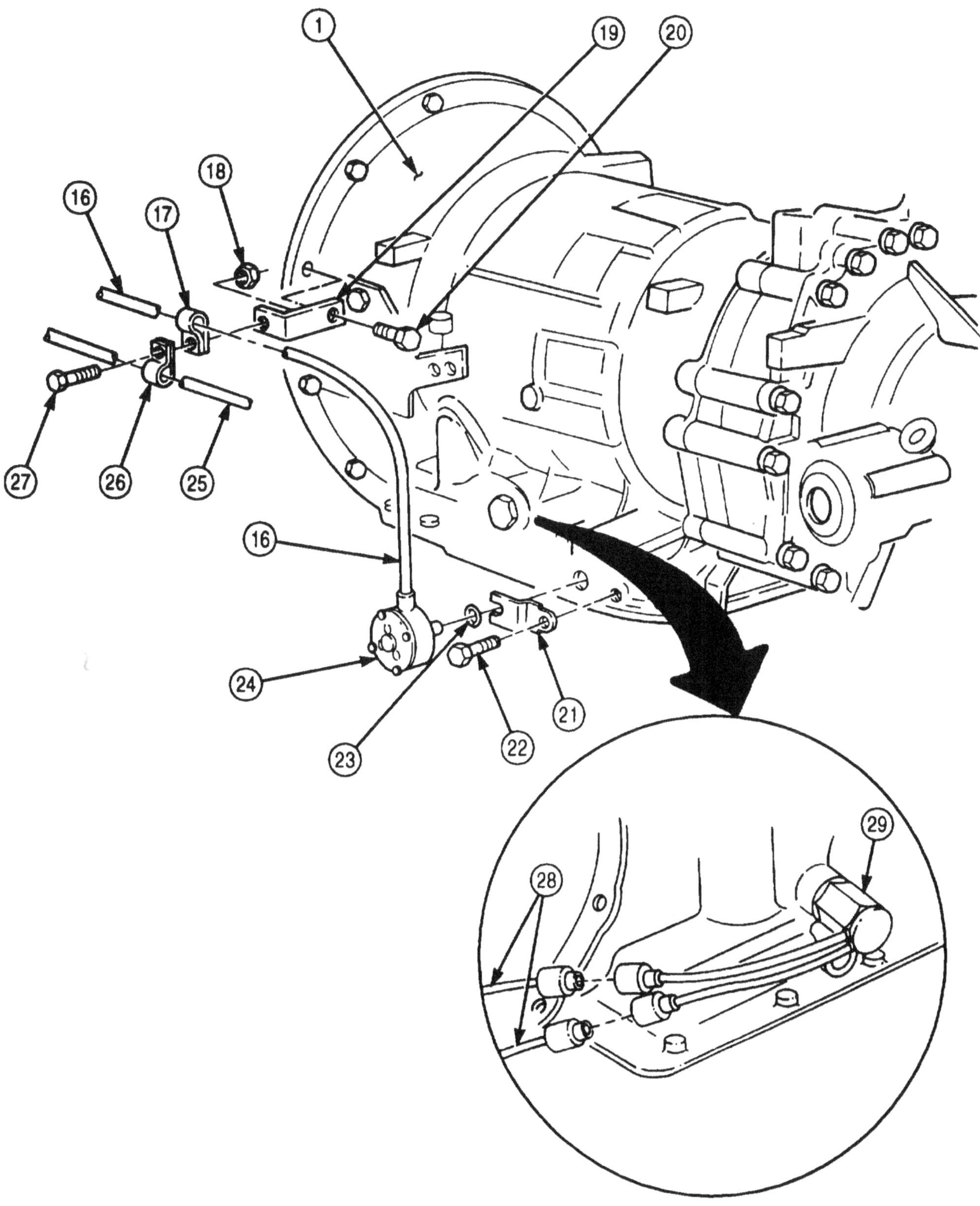

4-4. POWER PLANT (M939/A1) REPLACEMENT (Contd)

57. Remove screw (7), lockwasher (6), washer (5), ground strap (4), and lockwasher (3) from air compressor (2). Discard lockwashers (6) and (3). Tie ground strap (4) clear of engine (1).
58. Remove screw-assembled washer (9), ground wire (10), ground strap (11), and lockwasher (12) from air intake manifold (8). Discard lockwasher (12). Tie ground strap (11) and ground wire (10) clear of engine (1).

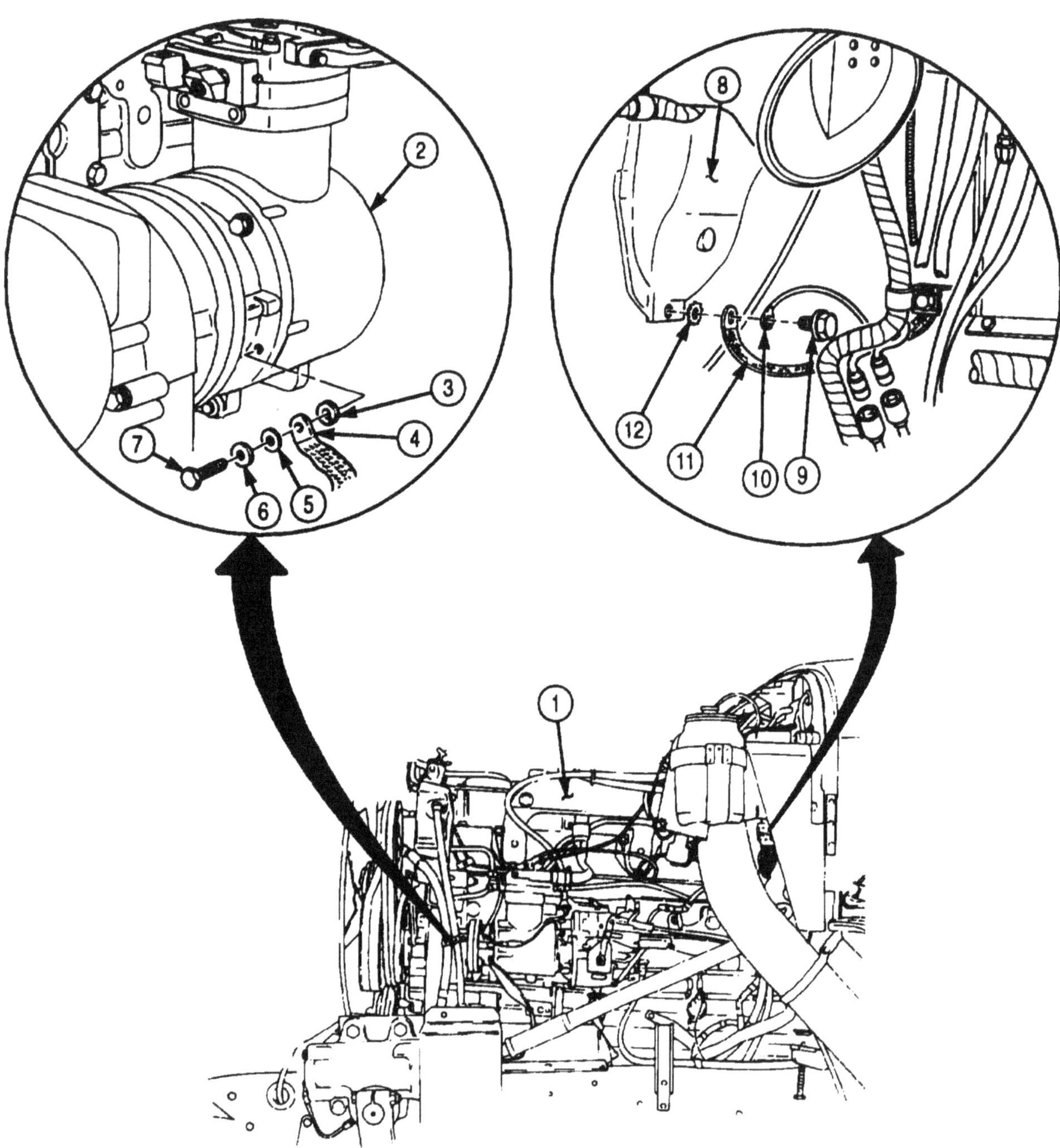

4-4. POWER PLANT (M939/A1) REPLACEMENT (Contd)

NOTE

Step 59 is performed only on vehicles equipped with front winch with level wind.

59. Remove four screws (14), lockwashers (15), and winch level wind (13) from winch (16). Discard lockwashers (15). Place winch level wind (12) clear of vehicle.

b. Removal

1. Remove two screws (17) and lockwashers (18) from transmission rear support bracket (19). Discard lockwashers (18).

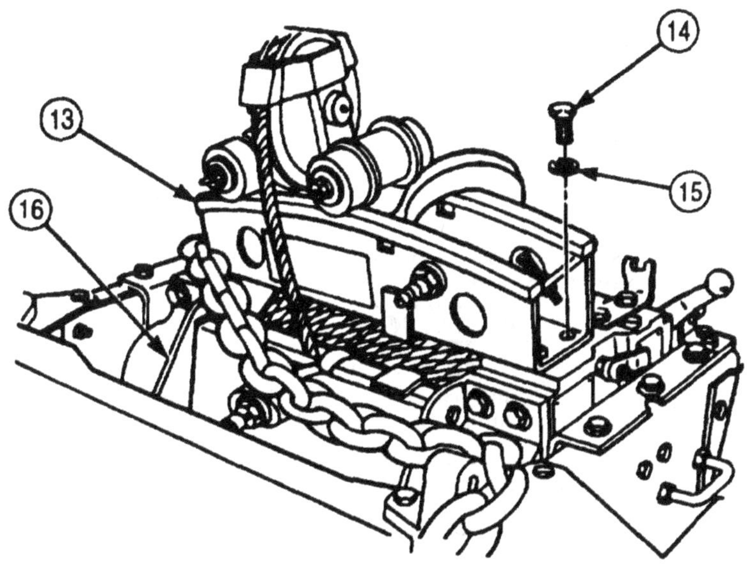

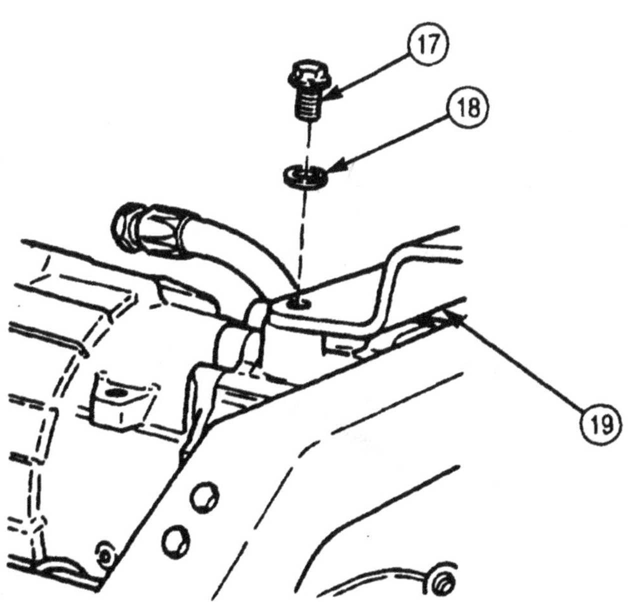

4-4. POWER PLANT (M939/A1) REPLACEMENT (Contd)

NOTE

Front of cab must be raised 4 in. (102 mm) to permit engine oil pan sump to clear the front axle differential housing. Two permanently mounted jack screws under the left and right cab A-posts permit raising front of cab.

2. Remove two locknuts (5), washers (4), and rubber cushions (3) from cab A-post support brackets (2). Discard locknuts (5).
3. Turn left and right jack screws (6) until A-posts (1) are approximately 4 in. (102 mm) above support brackets (2).

WARNING

Lifting device must have a weight capacity greater than the combined weight of the engine and transmission to prevent injury or death to personnel and damage to equipment.

4. Attach the adjustable end of an adjustable chain hoist to hoist hook and each adjustable chain hoist hook to engine lifting eyes (8). Raise hoist until all slack is removed from adjustable chain hoist, ensuring that hoist does not support weight of engine (7).
5. Remove two screws (11), lockwashers (10), five screws (13), lockwashers (14), trunnion cap (9), and trunnion mount (12) from frame crossmember (15). Discard lockwashers (10) and (14).
6. Remove two nuts (20), lockwashers (21), screws (16), and washers (17) from left and right engine supports (19) and flywheel housing (18). Discard lockwashers (21).

WARNING

- All personnel not participating in engine removal must stand clear during hoisting operations. A snapped cable, or swinging or shifting load, may cause injury to personnel.
- Do not use hands to free engine of hangups. Use tanker or prybars to avoid injury to personnel.

CAUTION

Always remove engine slowly. Lift out of chassis in short lifts and closely observe all engine and transmission attachments during removal to prevent damage to equipment.

NOTE

- Mechanic will direct all hoisting operations in steps 7 through 10 while assistant operates hoist or assists mechanic.
- Chain hoist must be adjusted to lower rear of transmission downward so engine is suspended at approximately a 15-20 degree angle to clear front axle.

7. Using socket wrench, adjust engine angle to 15-20 degrees with adjustable chain hoist.
8. Lift engine (7) slowly upward until clear of vehicle.

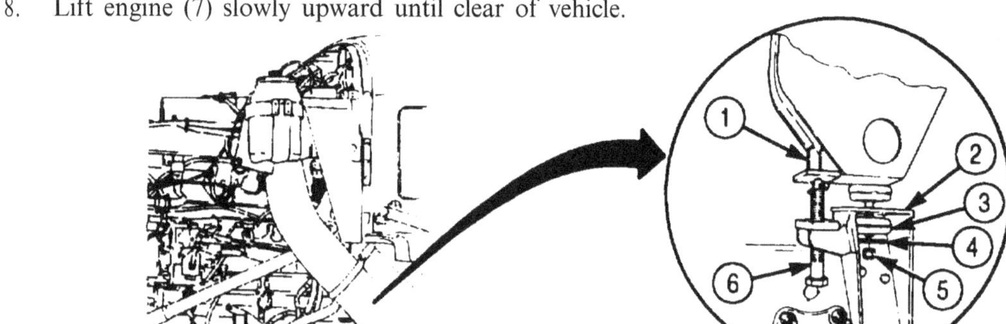

4-4. POWER PLANT (M939/A1) REPLACEMENT (Contd)

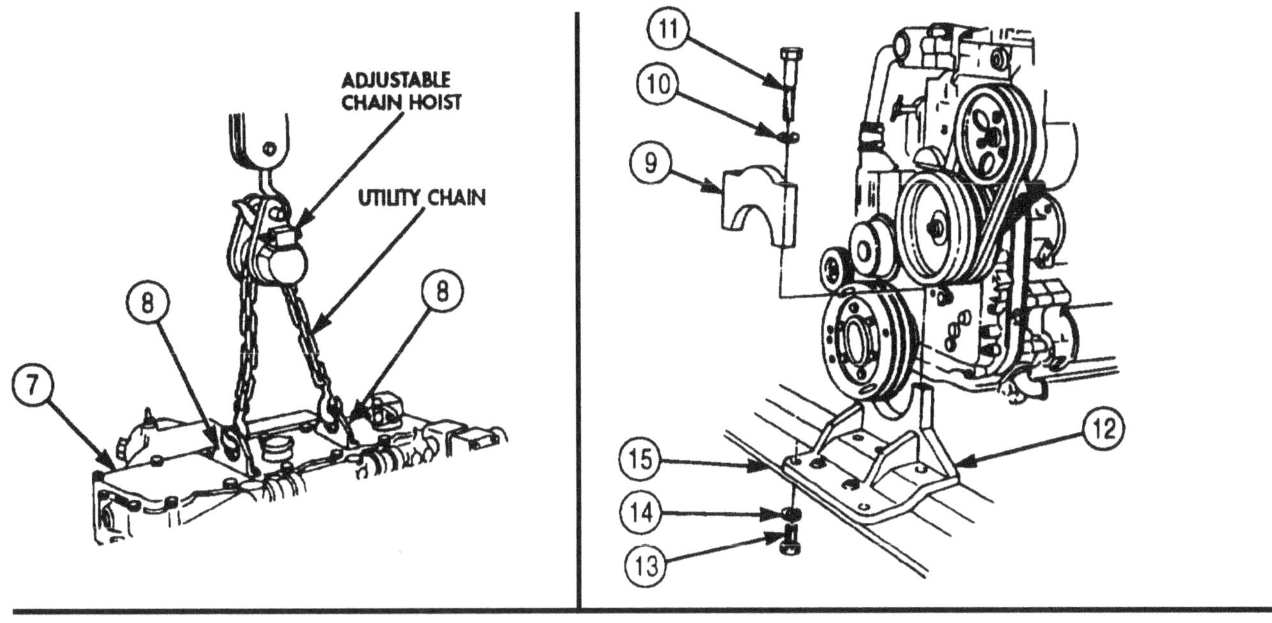

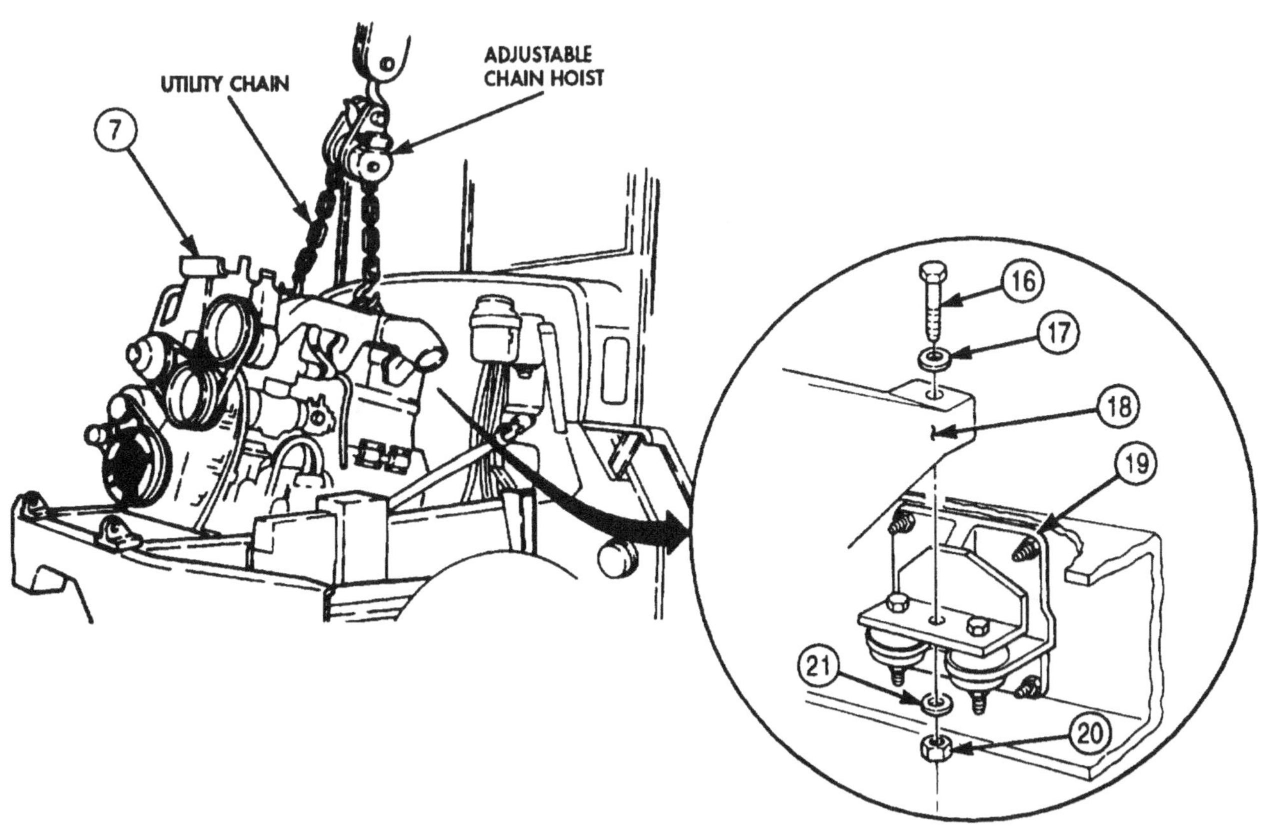

4-4. POWER PLANT (M939/A1) REPLACEMENT (Contd)

NOTE

Adjust chain hoist so engine and transmission are level for placement on trestles.

9. Lower engine (2) and transmission (3) onto engine transporter. Ensure two trestles of the transporter are positioned under each side of front gearcase cover (5) and two trestles are positioned under each end of flywheel (4).

WARNING

Do not detach chain hoist from engine until engine is equally distributed and is stable on transporter. An improperly supported engine may cause injury to personnel. Mounting engine on transport stand is solely temporary and is not intended as a supporting requirement for engine repair. When repairing engine, use repair stand.

10. Disconnect adjustable chain hoist and utility chains from two engine lifting eyes (1).

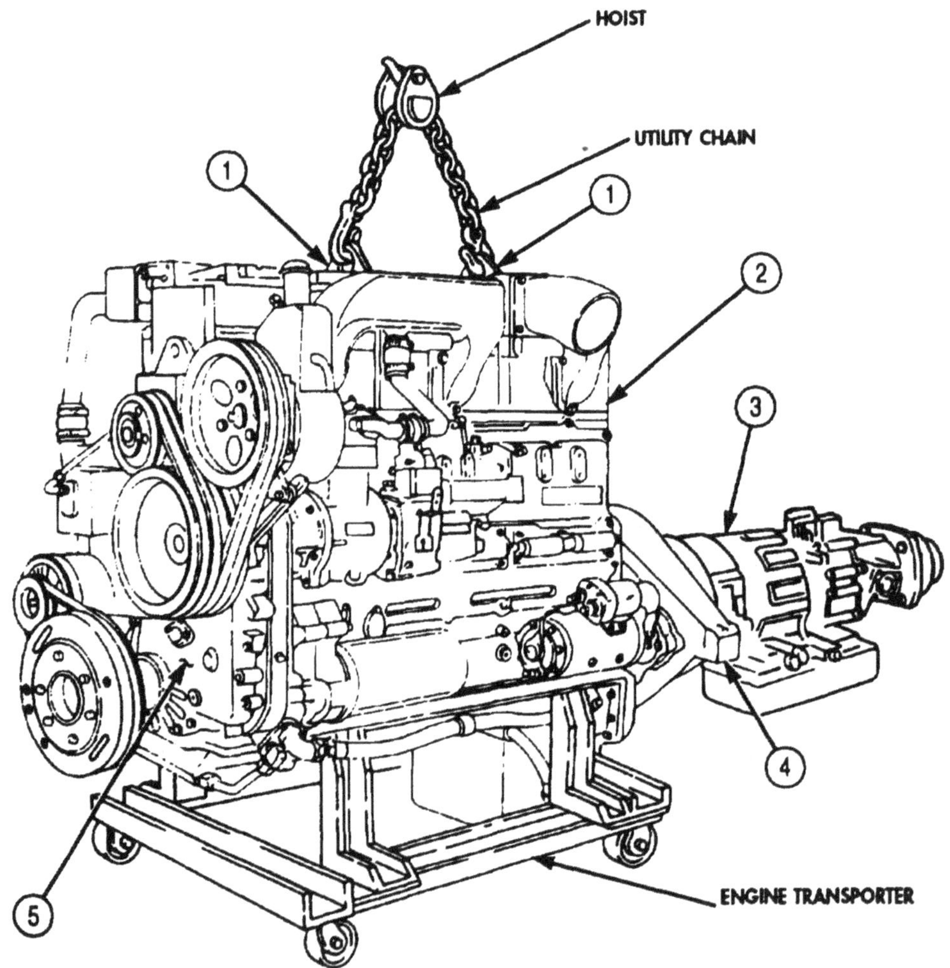

4-4. POWER PLANT (M939/A1) REPLACEMENT (Contd)

c. Installation

NOTE

When replacing engine, the transmission will be installed on new engine as outlined in para. 4-4.

1. Attach adjustable end of an adjustable chain hoist and chain to hoist hook and both chain hooks tc engine lifting eyes (1). Raise lifting device until all slack is removed from chain, ensuring lifting device does not support weight of engine (2).

WARNING

Personnel not participating in engine removal must stand clear during hoisting operations. A snapped cable, or swinging or shifting load, may cause injury to personnel.

CAUTION

Lower engine and transmission into chassis carefully and closely observe all engine and transmission components to prevent damage to equipment. Always remove engine slowly.

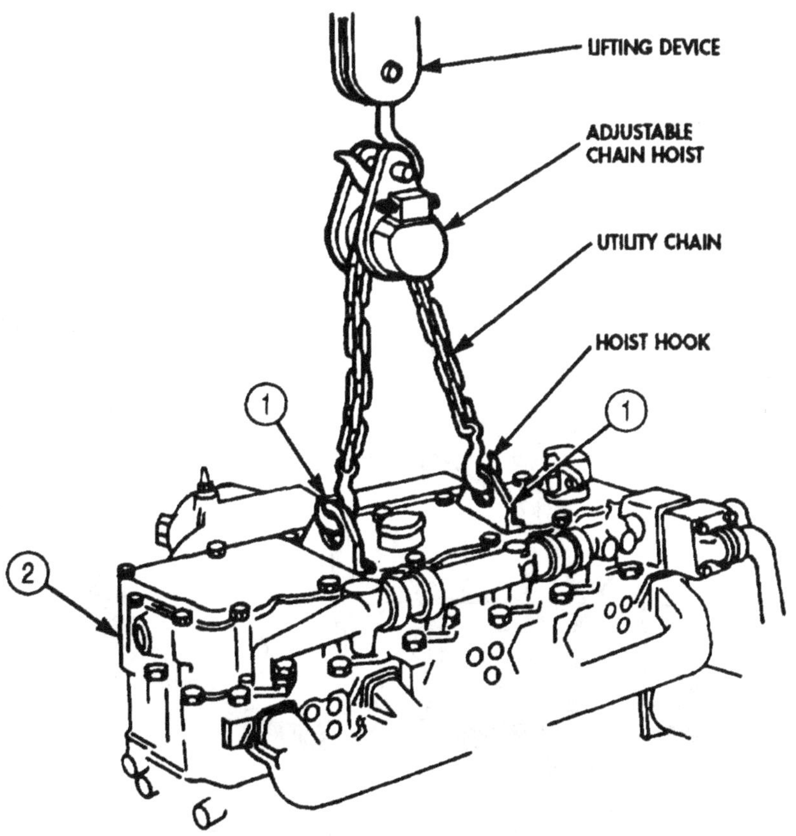

4-4. POWER PLANT (M939/A1) REPLACEMENT (Contd)

NOTE

- For steps 2 through 4, if engine is removed in the field, an additional assistant will be needed to operate the wrecker crane. Shop removal of engine requires a mechanic and one assistant if overhead hoist is available.
- Chain hoist must be adjusted so transmission points downward at approximately a 15-20 degree angle for engine to clear front axle.

2. Install trunnion mount (8) on frame crossmember (11) with five new lockwashers (10) and screws (9). Do not tighten screws (9).
3. Raise engine (1) and transmission (2) over front bumper (4) directly above engine compartment (3).
4. Using socket wrench, adjust engine angle to 15-20 degrees with adjustable ratchet on lifting device.
5. Slowly lower engine (1) and transmission (2) into engine compartment (3).
6. Lower engine (1) and transmission (2) until resting on trunnion mount (8) and left and right rear engine supports (15).
7. Using a drift pin, align holes in left and right rear engine supports (15) and engine flywheel housing (14), and install two washers (13), screws (12), new lockwashers (17), and nuts (16). Tighten nuts (16) 140-160 lb-ft (190-217 N·m).

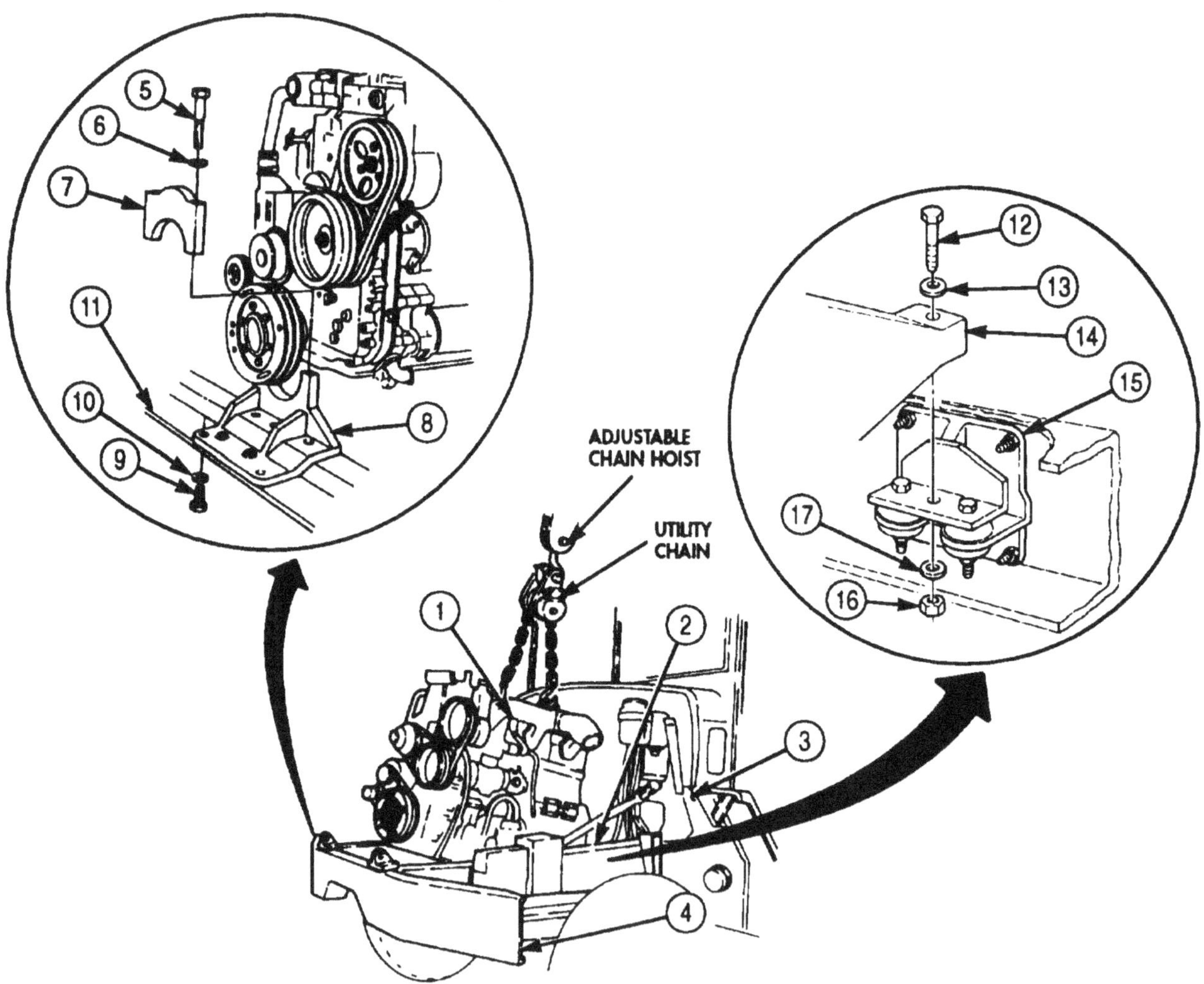

4-4. POWER PLANT (M939/A1) REPLACEMENT (Contd)

8. Install transmission (18) on rear support bracket (21) with two new lockwashers (20) and screws (19). Tighten screws (19) 75-85 lb-ft (102-115 N·m).
9. Turn left and right jack screws (27) until rubber cushions (28) of A-posts (22) rest on A-post support brackets (23). Ensure jack screws (27) are turned all the way down in A-post support brackets (23).
10. Install cab A-posts (22) on each A-post support bracket (23) with rubber cushions (24), washer (25), and locknut (26).
11. Install trunnion cap (7) on trunnion mount (8) with two new lockwashers (6) and screws (5). Tighten screws (5) 150 lb-ft (203 N·m).
12. Tighten screws (9) 65-75 lb-ft (88-102 N·m).

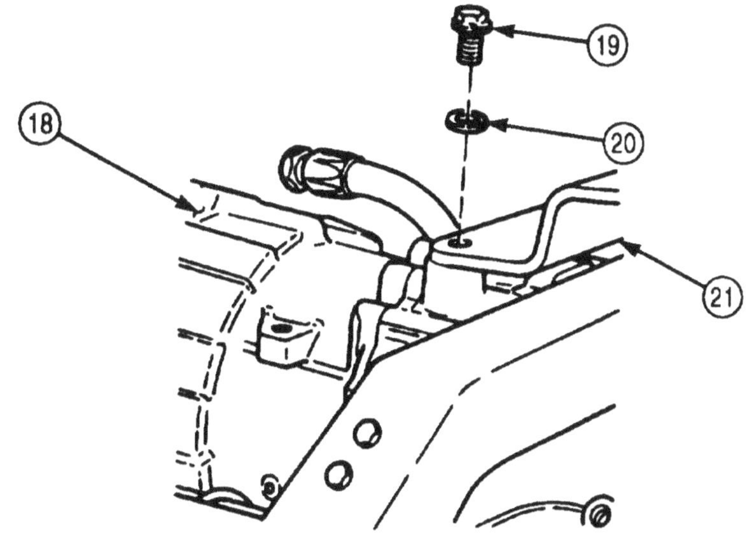

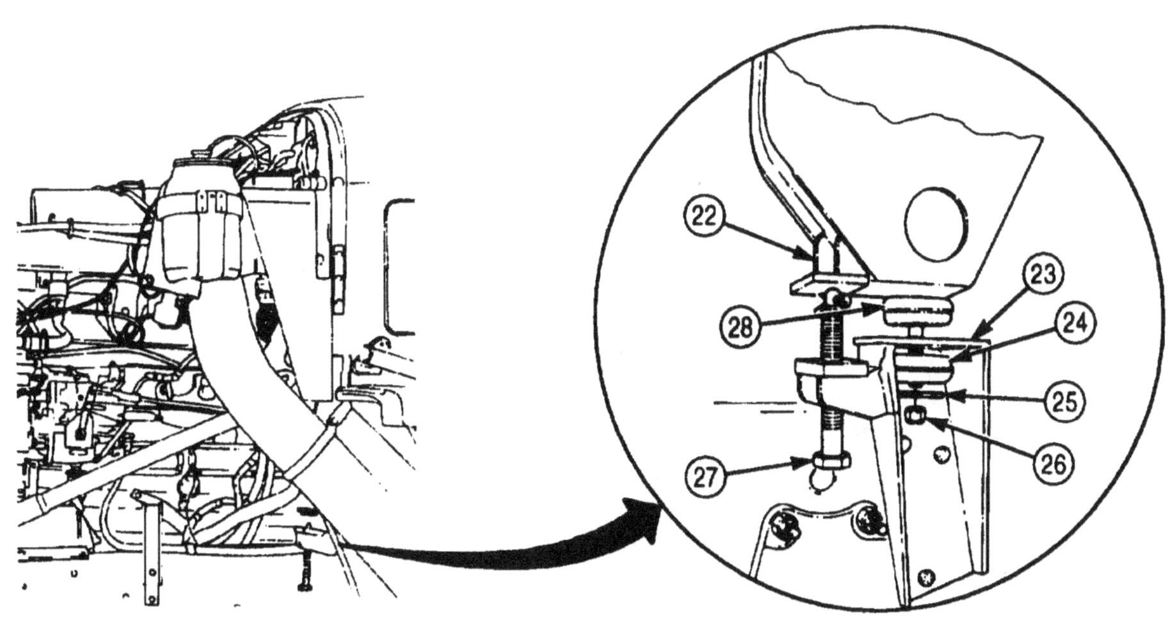

4-4. POWER PLANT (M939/A1) REPLACEMENT (Contd)

13. Disconnect lifting device, adjustable chain hoist, and chain from two engine lifting eyes (1) and hoist hook.

NOTE

Step 14 applies only to vehicles equipped with a front winch with level wind.

14. Install winch level wind (2) on winch (5) with four new lockwashers (4) and screws (3). Tighten screws (3) 70-90 lb-ft (95-122 N·m).

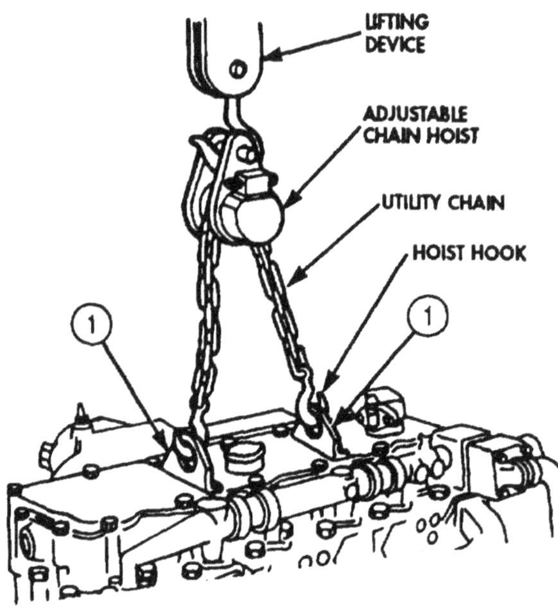

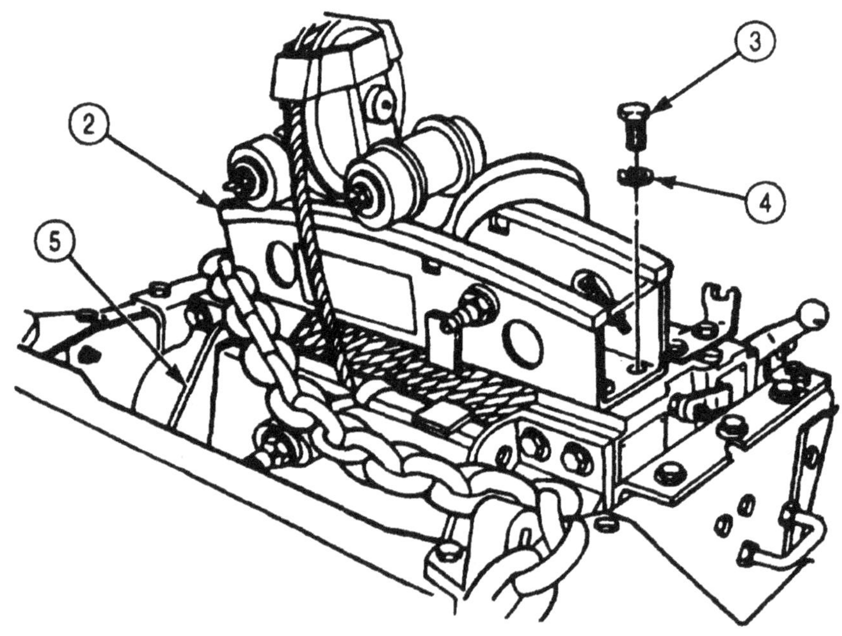

4-4. POWER PLANT (M939/A1) REPLACEMENT (Contd)

15. Connect two wires (7) to neutral start switch (8) on transmission (6).
16. Connect wires (10) and (11) to 5th gear lock-in connector (9).

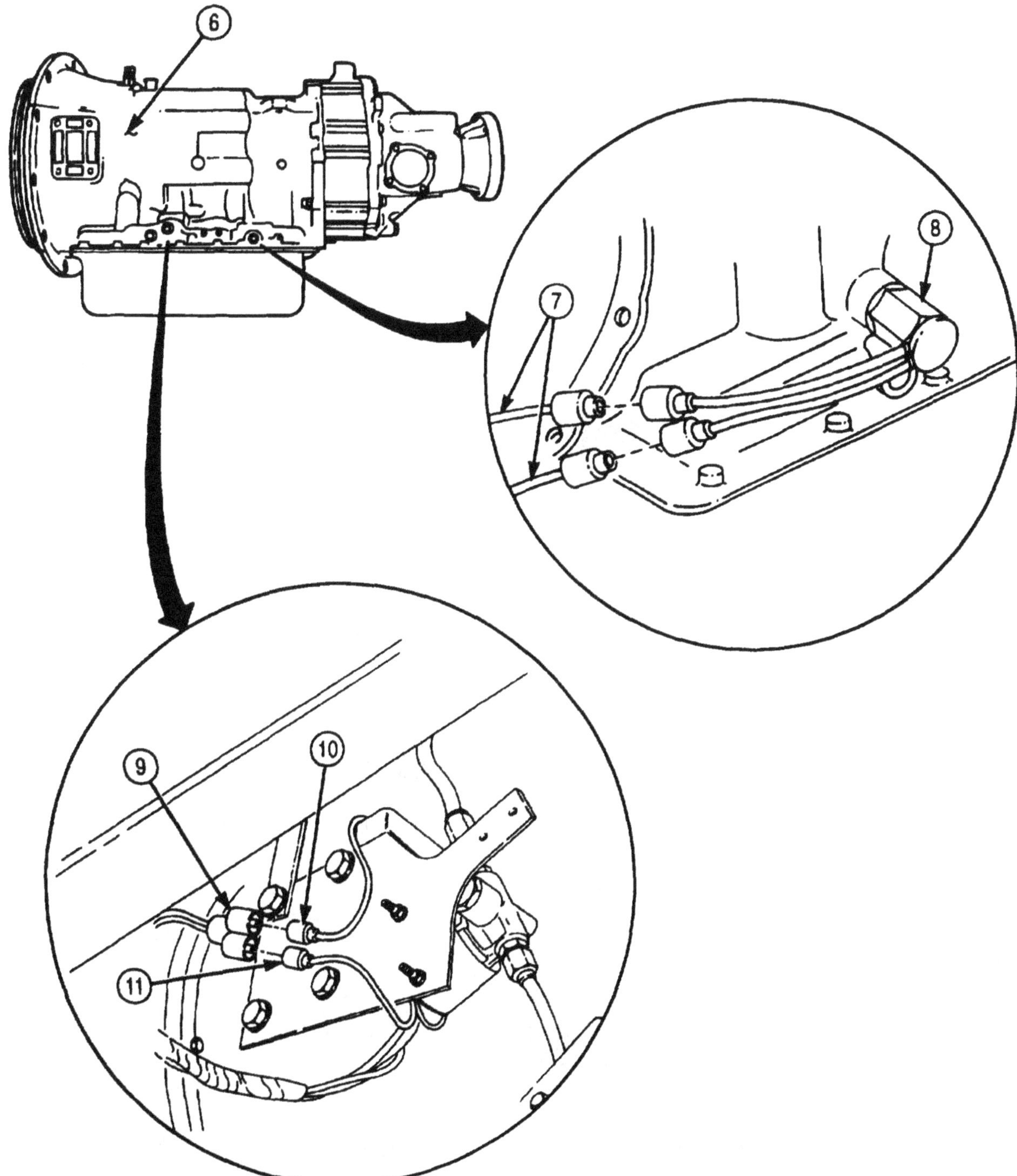

4-4. POWER PLANT (M939/A1) REPLACEMENT (Contd)

17. Connect breather vent line (3) to transmission adapter elbow (2).
18. Install transmission shift cable (19) and cable pin (20) on shift lever (22) with new cotter pin (21). Install transmission shift cable (19) on support bracket (15) with shim (17), U-bolt (18), and two new locknuts (16).
19. Install bracket (7), new O-ring (9), and modulator (10) on transmission (1) with screw (8).
20. Install bracket (5), and modulator cable (11) with clamp (14) and speedometer cable (12) with clamp (14) on transmission (1) with screws (6) and (13) and nut (4).

NOTE

- Step 21 applies only to vehicles equipped with a transmission PTO.
- Transmission connections listed in steps 21 through 24 can be made through access door in cab floor.

21. Install PTO cable (29) and cable pin (30) on select lever (31) with washer (32) and new cotter pin (23), and PTO cable (29) on bracket (25) with spacer plate (26), retaining strap (27), two screws (28), and nuts (24).
22. Install transmission oil dipstick (37) in dipstick tube (36).
23. Install wires (33) and (35) in two spring tension clips (34).
24. Connect wire (33) to transmission oil temperature sending unit (38).

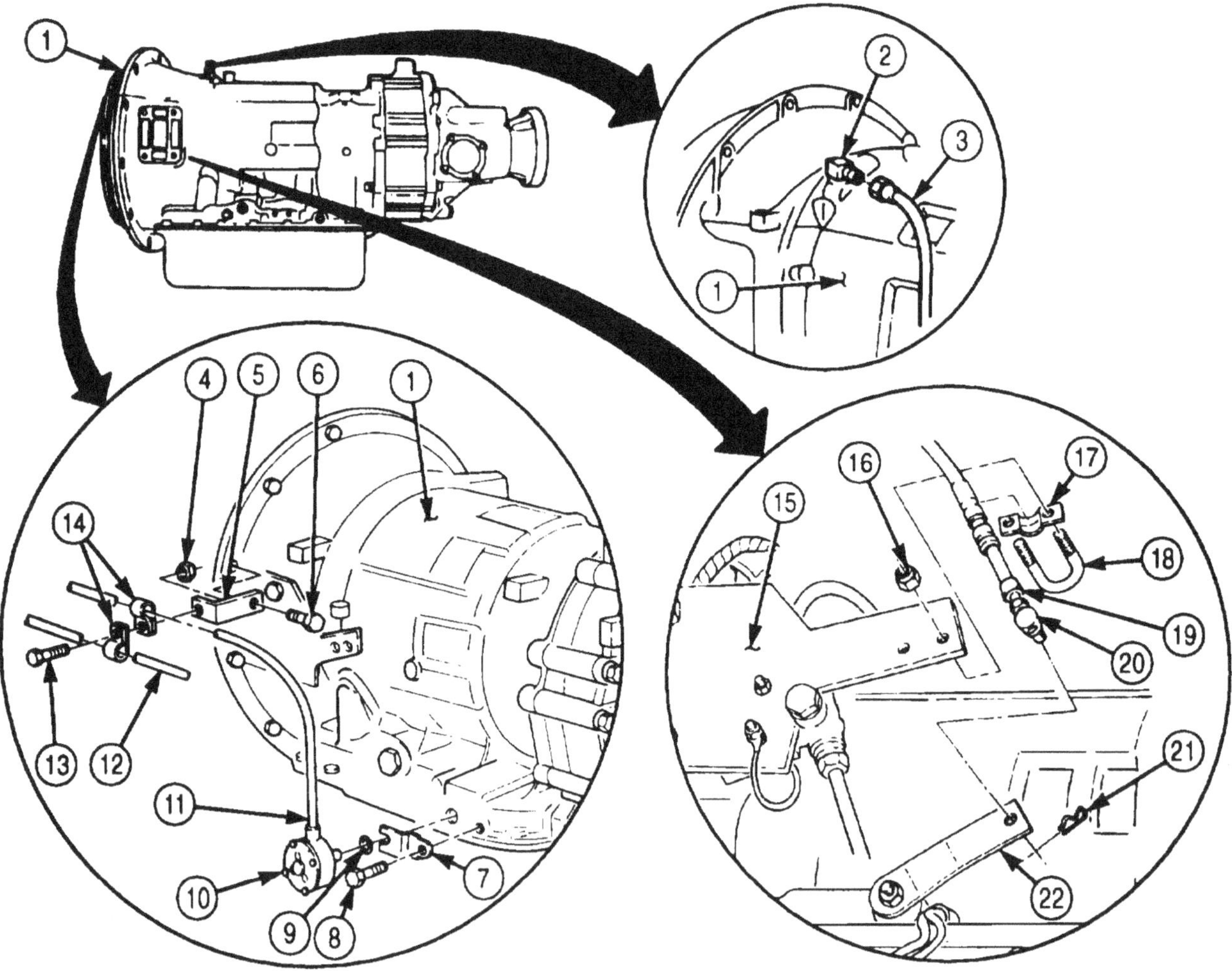

4-4. POWER PLANT (M939/A1) REPLACEMENT (Contd)

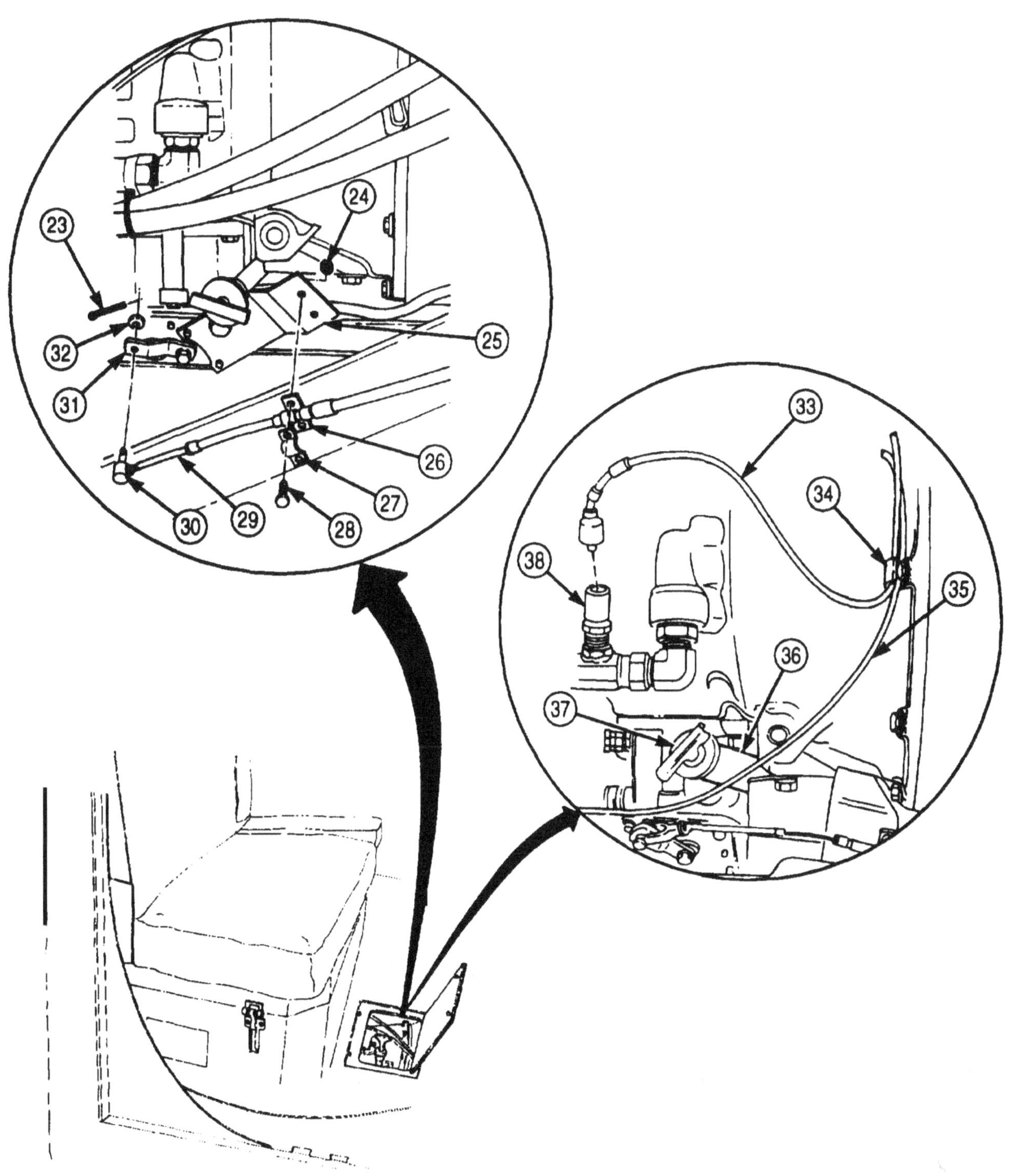

4-4. POWER PLANT (M939/A1) REPLACEMENT (Contd)

25. Install ground strap (3) on air compressor (1) with new lockwasher (2), washer (4), new lockwasher (5), and screw (6).
26. Install ground strap (10) and ground wire (9) on air intake manifold (7) with new lockwasher (11) and screw-assembled washer (8).
27. Connect air line (12) to fitting (13).
28. Install ground strap (20) and wires (21) and (24) on terminal post (19) with washer (23) and nut (22).
29. Install wires (14) and (25) on starter solenoid (18) with washer (27) and nut (26).
30. Install wire (15) on starter solenoid (18) with clip (17) and screw (16).

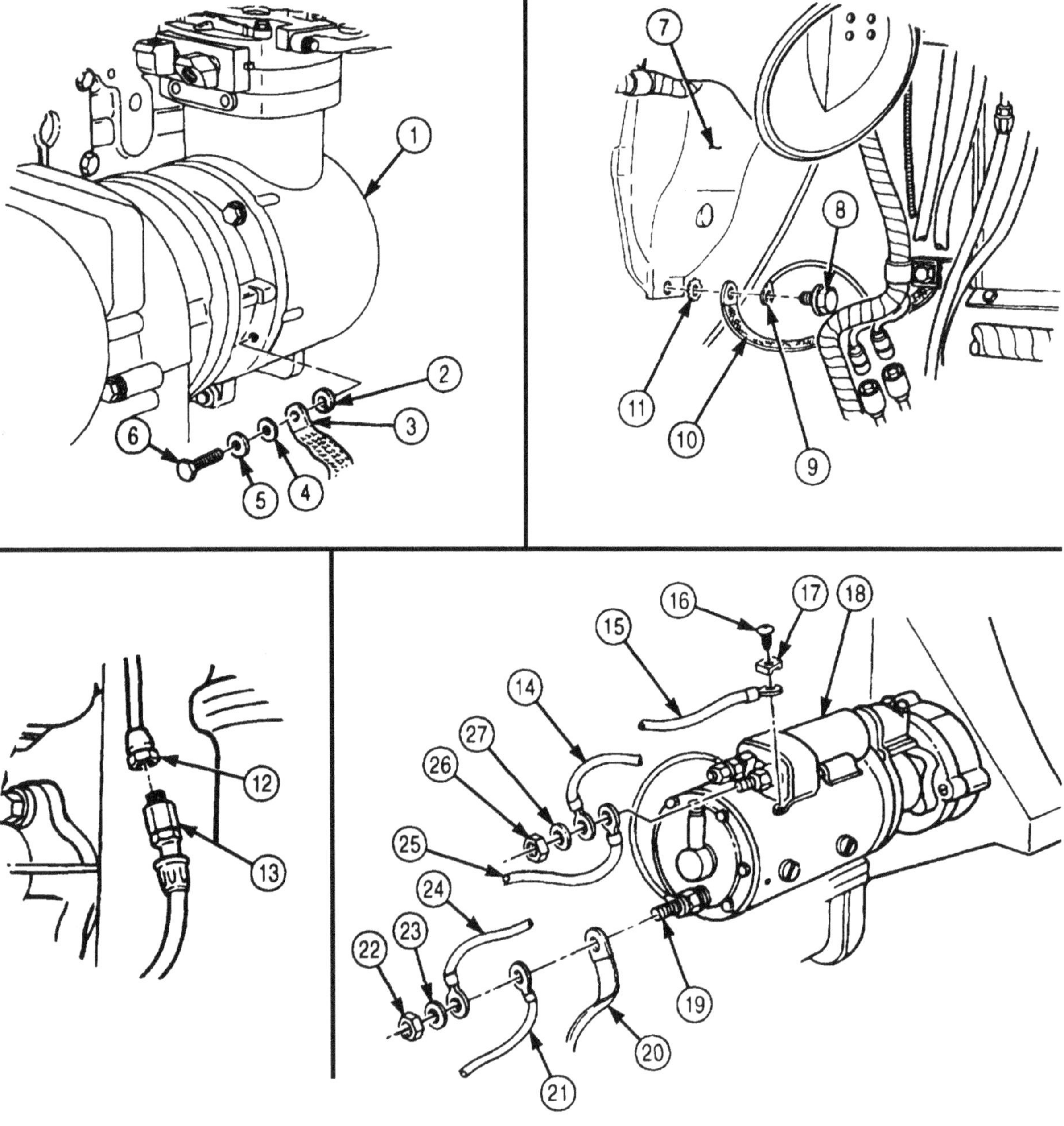

4-4. POWER PLANT (M939/A1) REPLACEMENT (Contd)

31. Connect air line (29) to air governor (28).
32. Connect air line (31) to compressor (30).

NOTE

Perform step 33 on M936/A1 vehicles only.

33. Connect air line (40) to VS governor (41).
34. Connect connector (36) to fuel pressure transducer (37).
35. Connect fuel line (32) to rear of fuel pump (33).
36. Connect tachometer drive cable (34) to tachometer pulse sender (35).
37. Connect harness (38) to tachometer pulse sender connector (39).
38. Connect wire (42) to oil pressure sending unit (43).

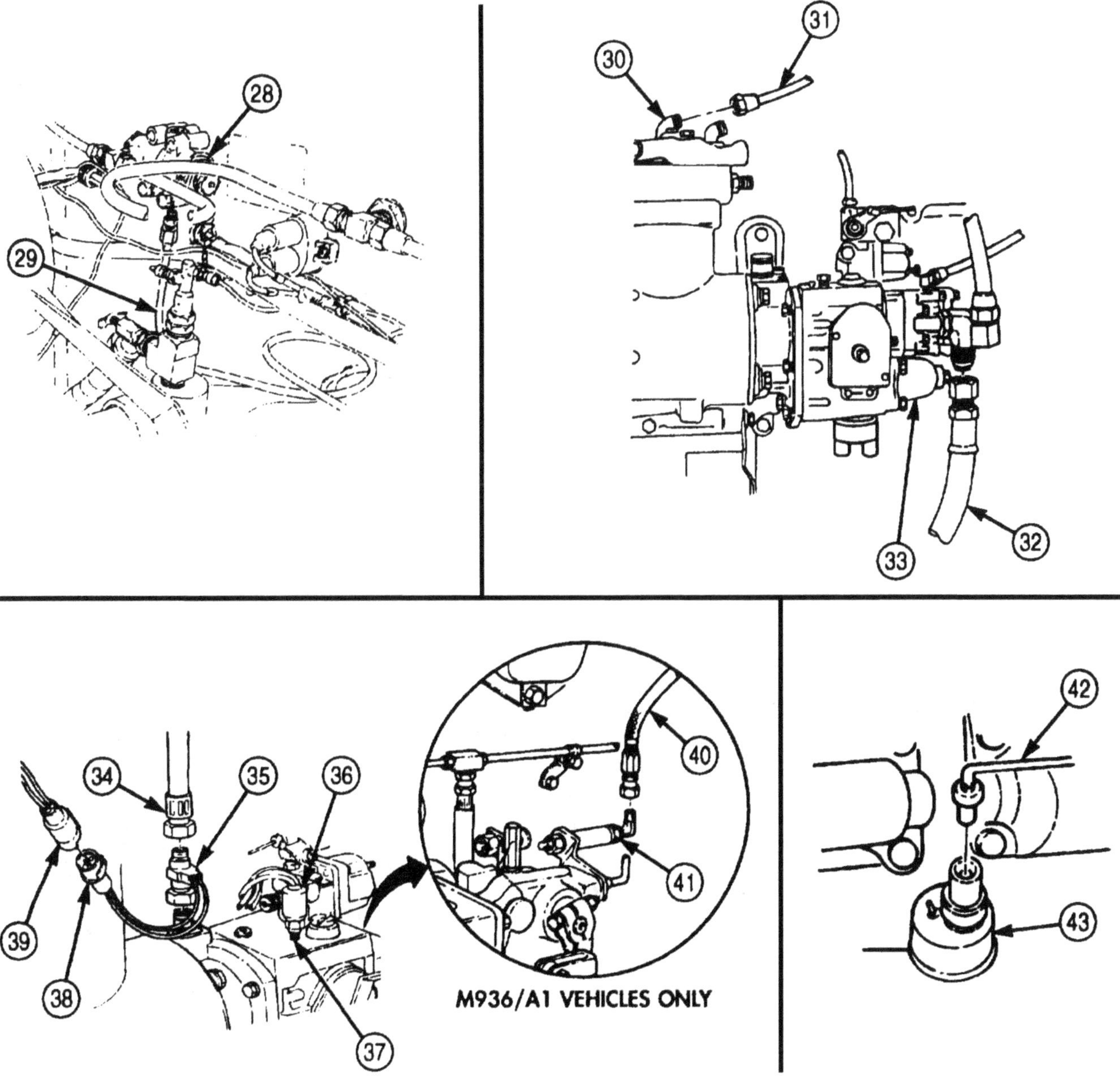

4-4. POWER PLANT (M939/A1) REPLACEMENT (Contd)

39. Connect fuel line (1) to fuel pump return hose (2).
40. Connect atomizer line (7) to ether start safety valve (6) and ether atomizer (8).
41. Connect line (5) from ether cylinder valve (9) to ether start safety valve (6).
42. Connect tube (10) to fitting (11).
43. Connect two wires (4) to ether start switch (3).
44. Connect air line (13), tachometer cable (14), speedometer cable (20), and harness (21) to air intake manifold (12) with clamps (15), (16), and (19), two washers (18), and screws (17).

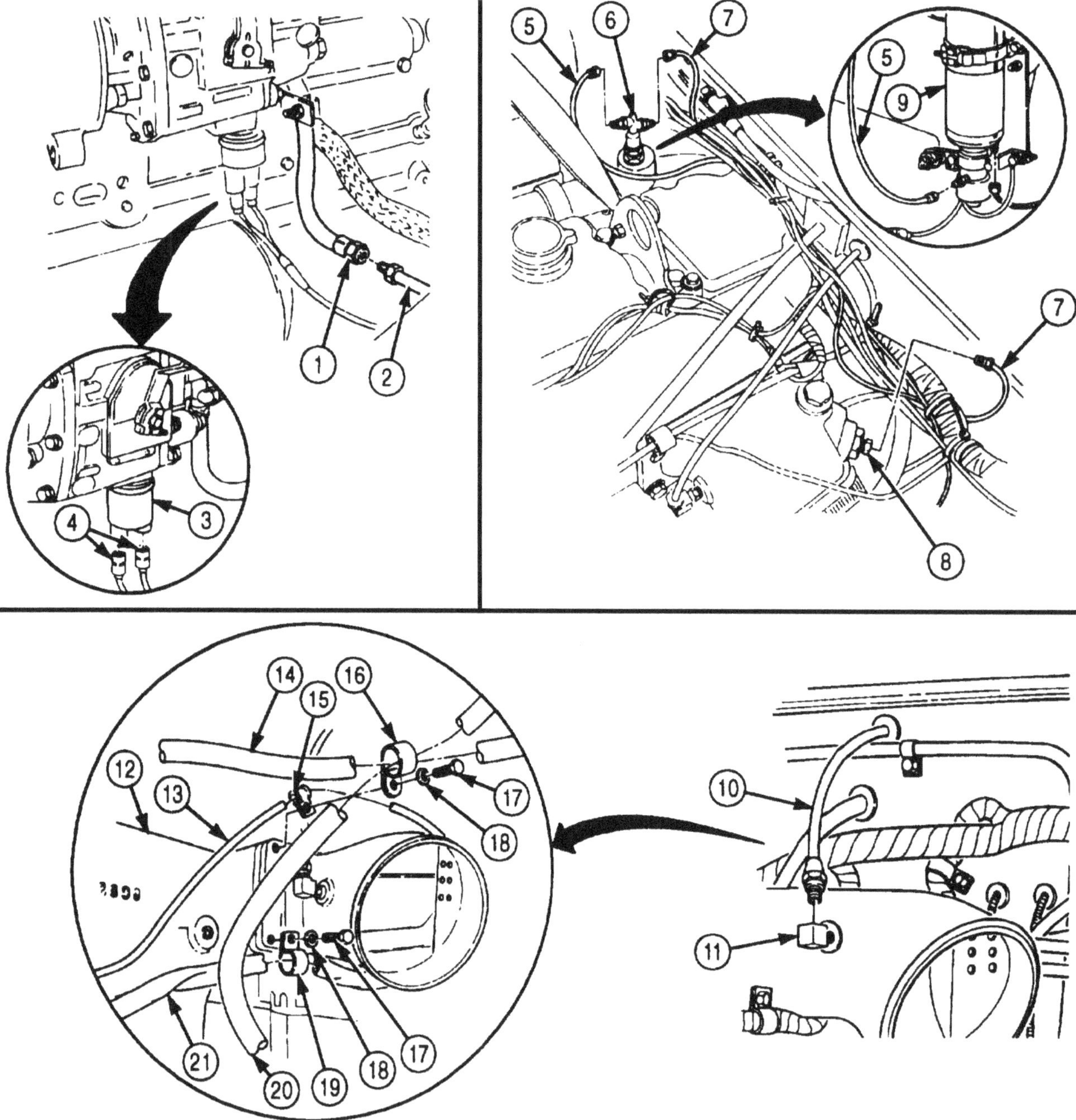

4-4. POWER PLANT (M939/A1) REPLACEMENT (Contd)

45. Install emergency stop cable (23) in swivel block (29).
46. Install connector (22) on cable (23) and tighten connector screw (30).
47. Install cable (23) on clamp bracket (28) with clamp (26), washer (24), screw (25), and nut (27).
48. Install wires (32) and (33) on fuel shutoff solenoid (31) with nut (34).
49. Install pump throttle lever (35) on accelerator rod (37) with screw (38) and new locknut (36).
50. Install pump throttle lever (35) on link (47) with screw (48) and new locknut (39).
51. Install modulator cable (42) on fuel primer pump bracket (43) with shim (40), clamp (45), two screws (46), and nuts (41).
52. Connect return spring (44) to link (47) and fuel primer pump bracket (43).

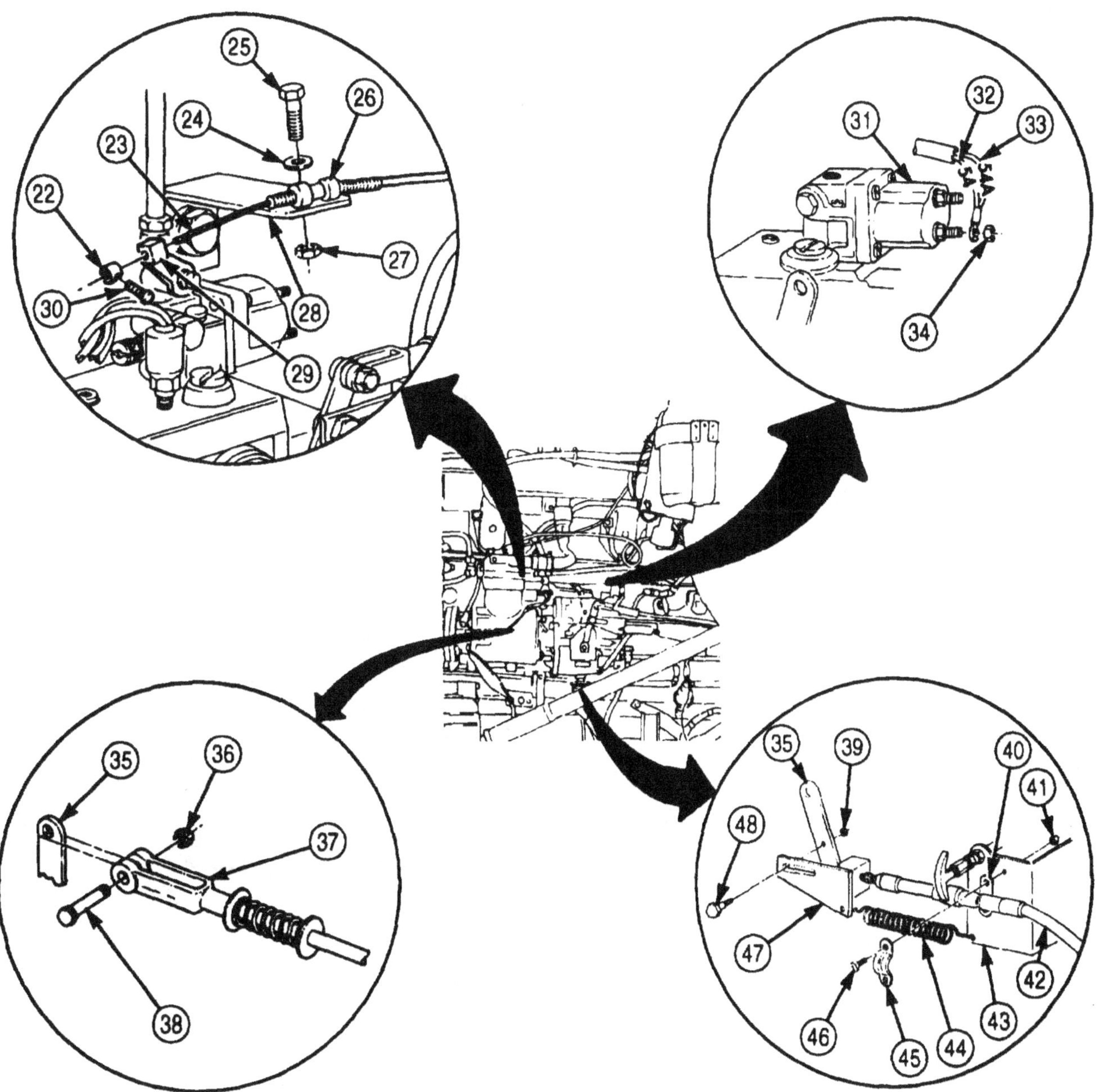

4-4. POWER PLANT (M939/A1) REPLACEMENT (Contd)

53. Connect hose (1) to elbow (2).
54. Connect hose (3) to transmission oil cooler (4).
55. Connect hose (8) to fitting (7).
56. Install return hose (9) on flange (6) of power steering pump reservoir (5) and tighten clamp (10).

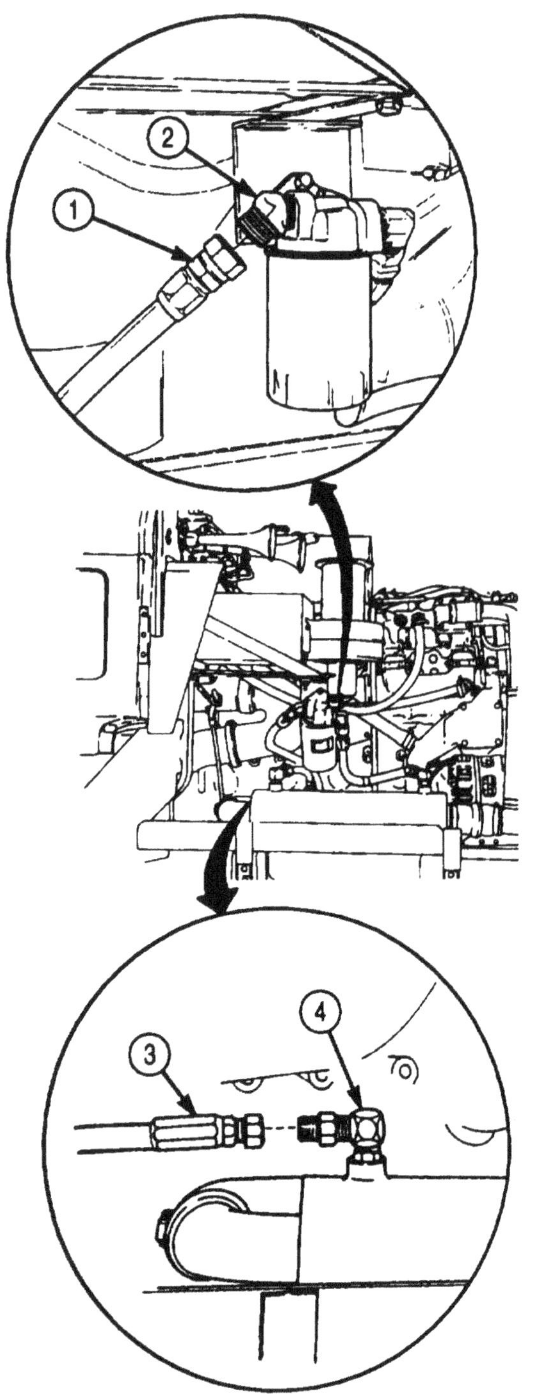

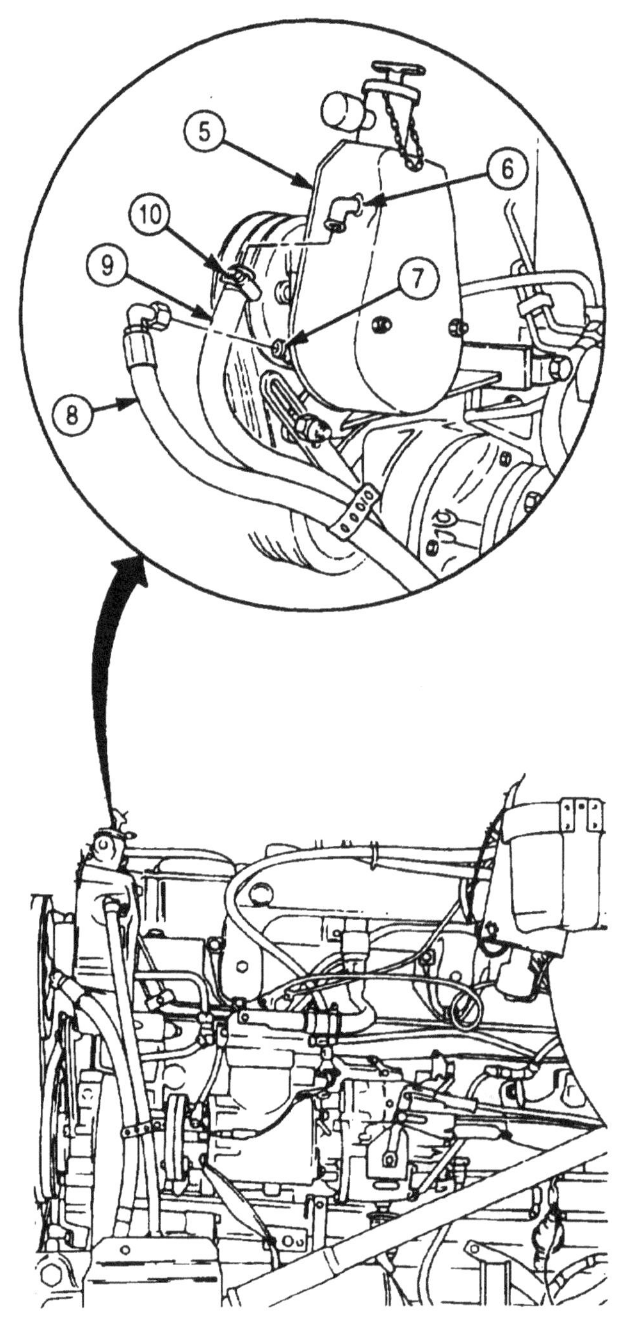

4-4. POWER PLANT (M939/A1) REPLACEMENT (Contd)

57. Install hose (13) on water heater shutoff valve (11) with clamp (12).
58. Install hose (14) on water heater shutoff valve (16) with clamp (15).
59. Install wire (29) on alternator (30) with washer (28), new lockwasher (27), and nut (26).
60. Install wire (17) on alternator (30) with washer (18), new lockwasher (19), and nut (20).
61. Install wire (34) on alternator (30) with new lockwasher (35) and screw (36).
62. Install wire retaining strap (21) on alternator (30) with two new lockwashers (22) and screws (23).
63. Seal wires (29) and (17) completely with adhesive sealant, and install terminal cover (24) on alternator (30) with two new screw-assembled lockwashers (25).
64. Connect wire (32) to connector (31).
65. Install new tiedown strap (33) around all wires.

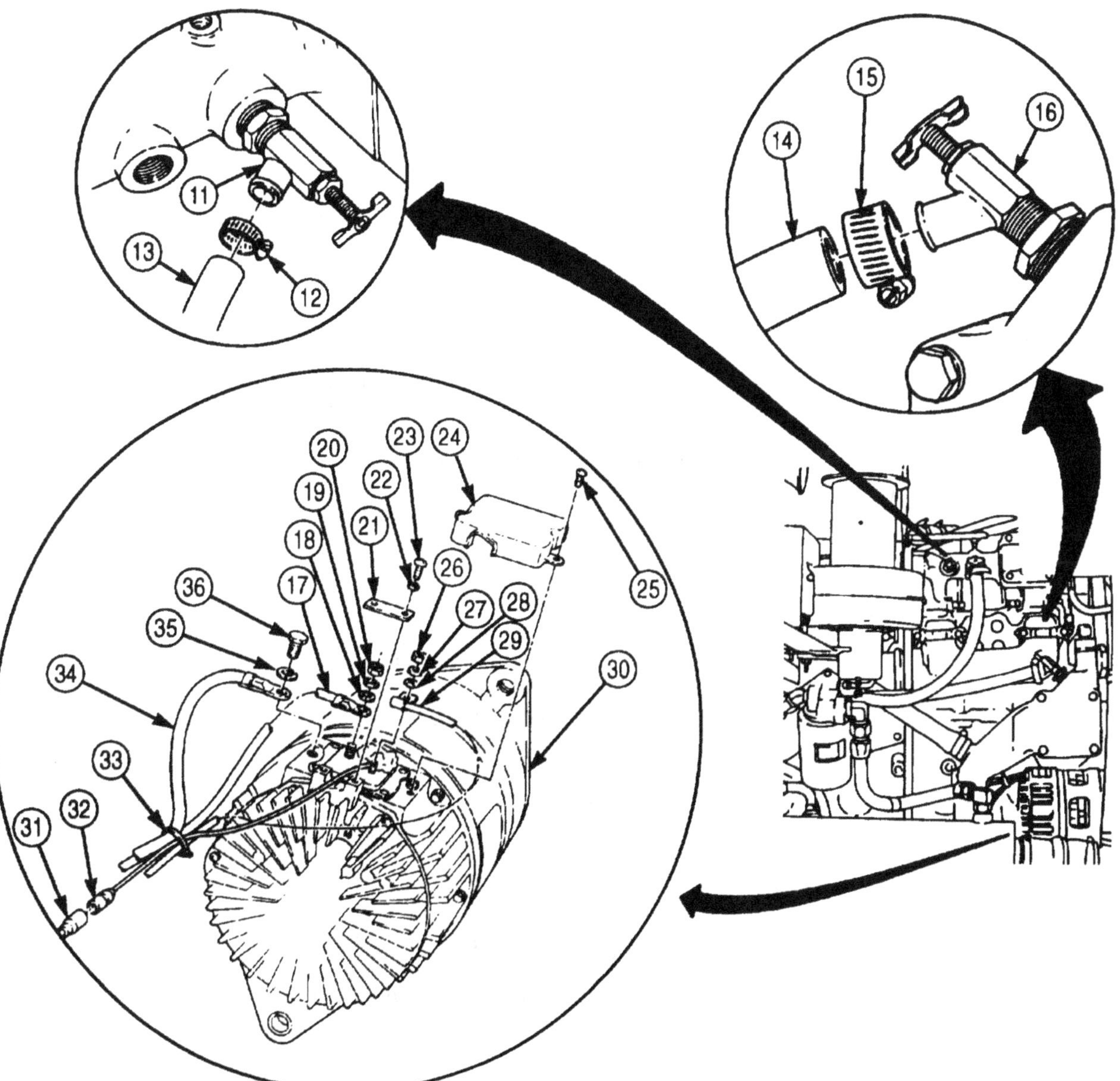

4-4. POWER PLANT (M939/A1) REPLACEMENT (Contd)

66. Connect wire (2) to engine high-temperature sending unit (1).
67. Connect wire (3) to water temperature sending unit (4).

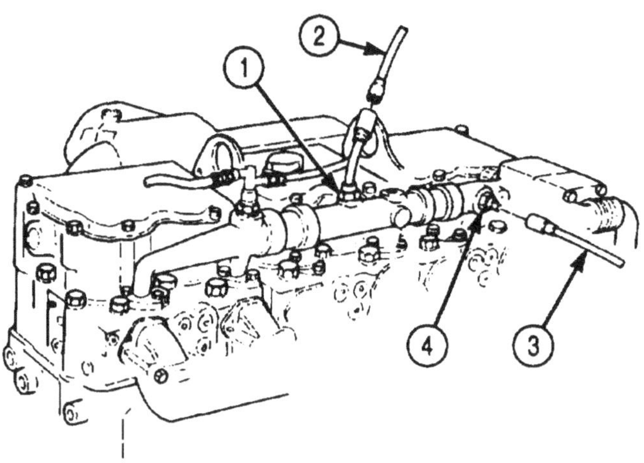

FOLLOW-ON TASKS:
- Install transmission-to-transfer case propeller shaft (para. 3-148).
- Install transmission PTO-to-hydraulic pump driveshaft, if equipped (para. 3-334).
- Install air intake pipe and hump hoses (para. 3-14).
- Install engine oil dipstick and tube (para. 3-3).
- Install front exhaust pipe (para. 3-50).
- Install radiator fan blade assembly (para. 3-72).
- Install radiator (para. 3-59).
- Install surge tank (3-61).
- Install coolant hoses and tubes (para. 3-54).
- Install hood assembly (para. 3-275).
- Fill steering system to proper oil level (LO 9-2320-272-12).
- Fill cooling system to proper coolant level (para. 3-53).
- Fill engine to proper oil level (para. 3-5).
- Fill transmission to proper oil level (para. 3-133).
- Close air reservoir drainvalves (TM 9-2320-272-10).
- Connect battery ground cables (para. 3-126).
- Adjust modulator cable (para. 3-145).
- Adjust throttle control cable (para. 3-45).
- Adjust emergency stop control cable (para. 3-44).
- Adjust accelerator linkage (para. 3-42).

CAUTION

Never start a new or repaired engine without performing run-in starting procedures (para. 4-8).

- Perform engine run-in starting procedures (para. 4-8).
- Start engine (TM 9-2320-272-10), allow air pressure to build up to normal operating range, and check for leaks. Road test vehicle.

4-5. POWER PLANT (M939A2) REPLACEMENT

THIS TASK COVERS:

a. Preliminary Disconnections
b. Removal
c. Installation

INITIAL SETUP:

APPLICABLE MODELS
M939A2

TOOLS
General mechanic's tool kit (Appendix E, Item 1)
Torque wrench (Appendix E, Item 144)
Lifting device
Chains

MATERIALS/PARTS
Two Iockpins (Appendix D, Item 271)
Tiedown strap (Appendix D, Item 692)
Cotter pin (Appendix D, Item 62)
Lockwasher (Appendix D, Item 416)
Lockwasher (Appendix D, Item 401)
Lockwasher (Appendix D, Item 403)
Two lockwashers (Appendix D, Item 364)
Four lockwashers (Appendix D, Item 416)
Two locknuts (Appendix D, Item 276)
Four locknuts (Appendix D, Item 306)
Tiedown strap (Appendix D, Item 696)
Lockwasher (Appendix D, Item 379)
Gasket sealant (Appendix C, Item 30)
Antiseize tape (Appendix C, Item 72)
Twine (Appendix C, Item 77)
Cap and plug set (Appendix C, Item 14)

PERSONNEL REQUIRED
Two

REFERENCES (TM)
LO 9-2320-272-12
TM 9-2320-272-10
TM 9-2320-272-24P

EQUIPMENT CONDITION
- Parking brake set (TM 9-2320-272-10).
- Air reservoirs drained (TM 9-2320-272-10).
- Battery ground cables disconnected (para. 3-126).
- Transmission oil drained (para. 3-133).
- Engine oil drained (para. 3-5).
- Power steering system drained (LO 9-2320-272-12).
- Hood removed (para. 3-275).
- Upper radiator hose and bracket removed (para. 3-58).
- Transmission PTO-to-hydraulic pump propeller shaft removed, if equipped (para. 3-334).
- Transmission-to-transfer case propeller shaft removed (para. 3-148).

GENERAL SAFETY INSTRUCTIONS
- All personnel must stand clear during lifting operations.
- Do not put hands between frame and engine supports during lifting operations.
- Do not detach chain from engine until engine is supported.
- Lifting device must have a capacity greater than the combined weight of the engine and transmission.
- Engine must be securely mounted on repair stand.

a. Preliminary Disconnections

CAUTION

- Cap or plug all openings after disconnecting lines and hoses to prevent contamination. Failure to do so may cause damage to equipment,
- Tie all loose connections away from engine. Failure to do so may cause damage to equipment.

NOTE

- If a special purpose kit is installed on vehicle, refer to chapter 4, section XV, and make necessary disconnections and connections.
- Tag all loose leads, lines, and tubes for installation.

4-5. POWER PLANT (M939A2) REPLACEMENT (Contd)

1. Disconnect tachometer cable (2) from tachometer drive (3) on front of engine (1).
2. Remove tiedown strap (12) and tachometer cable (2) from air intake tube (10). Discard tiedown strap (12).
3. Remove clamp (6) and air indicator tube (5) from air intake tube adapter (7).
4. Remove air intake tube adapter (7) from intake tube (10).
5. Remove two nuts (9), bracket (8), and U-bolt (18) from air intake tube (10) and bracket (16).
6. Remove clamps (4) and (11) and air intake tube (10) from pipe (14) and turbocharger (13).
7. Remove two screws (19), washers (17), bracket (16), and two spacers (15) from engine (1).
8. Remove tiedown strap (24) from cables (23), air line (28), and temperature sensor leads (27). Discard tiedown strap (24).
9. Disconnect connector (25) from temperature sensor connector (26).
10. Remove screw (34), washer (33), ground strap (32), lockwasher (31), and temperature sensor (30) from engine (1). Discard lockwasher (31).
11. Disconnect air line (28) from air compressor line (29).
12. Disconnect connector (42) from throttle control solenoid connector (41).
13. Disconnect connector (40) from transfer pump connector (39).
14. Disconnect fuel supply line (37) from transfer pump (38).
15. Disconnect connector (36) from oil pressure sending unit (35).
16. Disconnect ether supply tube (20) from ether atomizer nozzle (43).
17. Remove two screws (22), clamps (21), and cables (23) from engine (1).

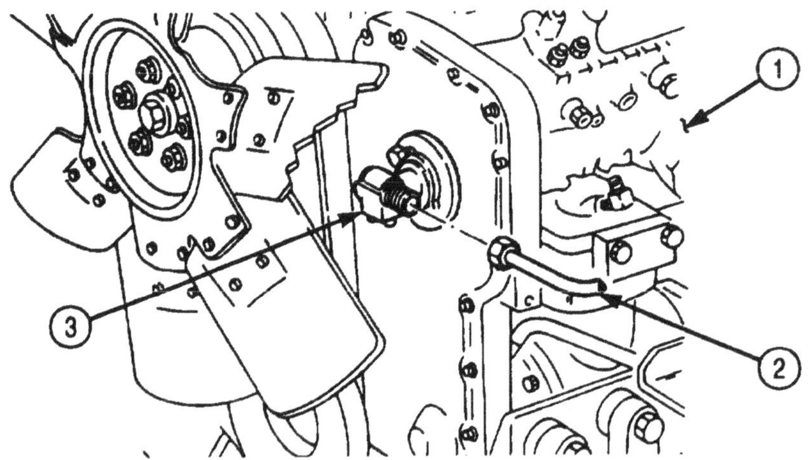

4-5. POWER PLANT (M939A2) REPLACEMENT (Contd)

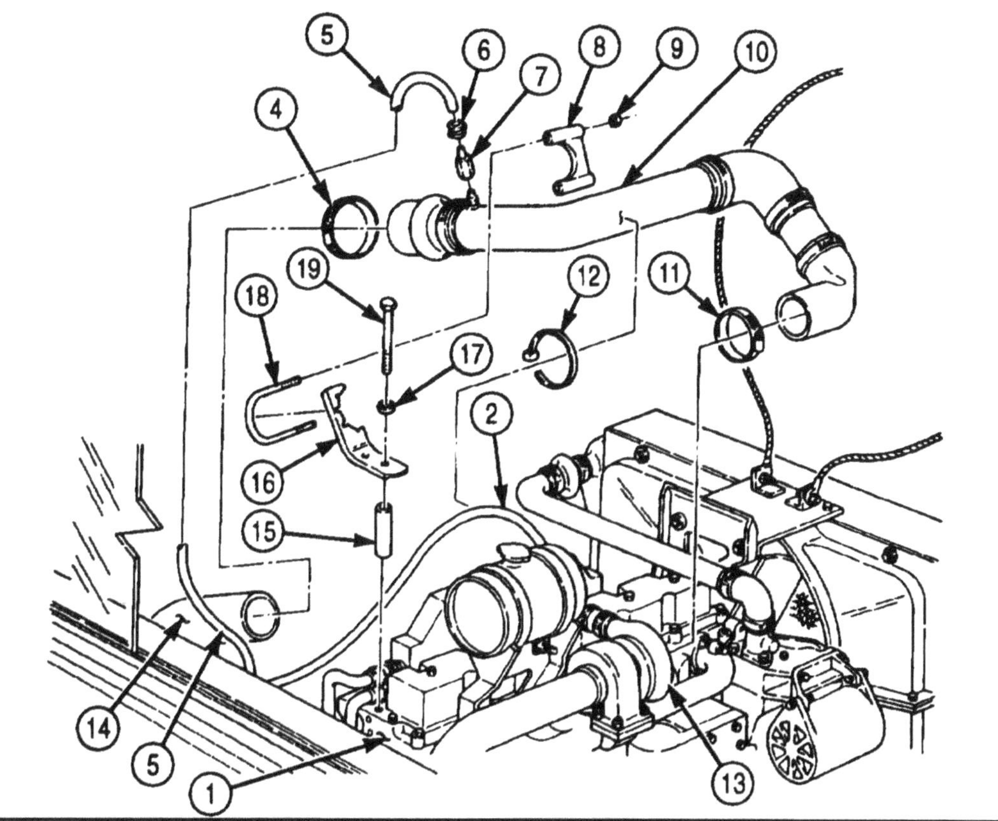

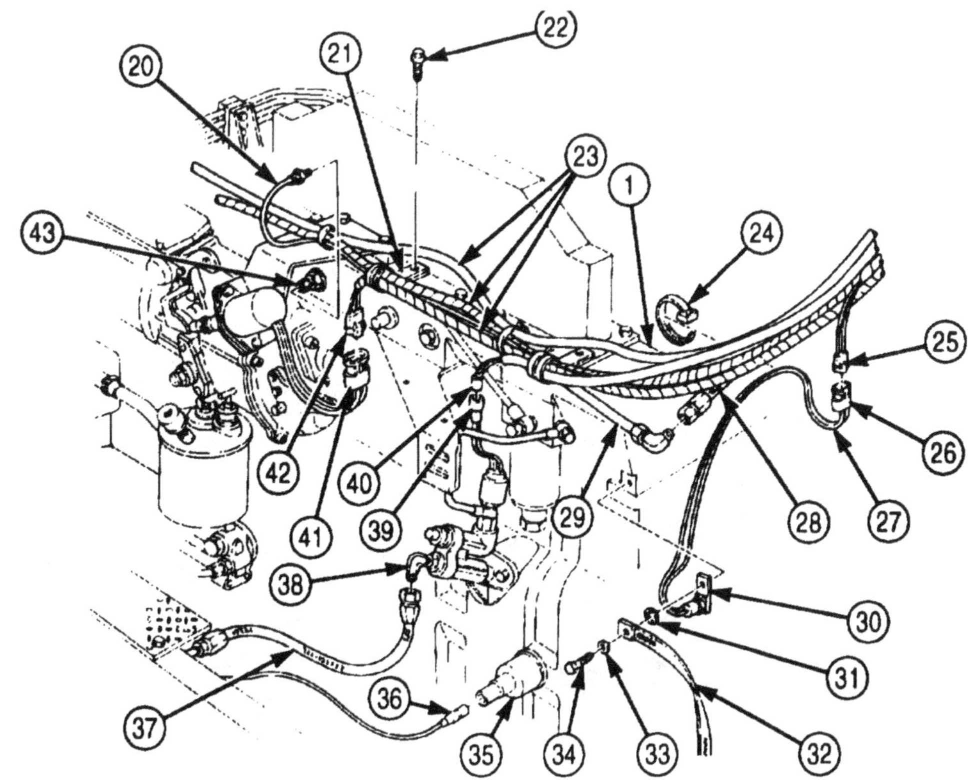

4-5. POWER PLANT (M939A2) REPLACEMENT (Contd)

18. Loosen clamp (20) and remove fuel return line (19) from fuel pump (1).
19. Remove screw (17), bracket (18), and fuel return line (19) from gearcase housing (21).
20. Full sleeve (15) away from socket (16), and remove accelerator linkage (14) from throttle control lever (2).
21. Remove spring (11) from modulator cable (8) and bracket (3).
22. Remove cotter pin (6), washer (7), and modulator cable (8) from throttle control lever (2). Discard cotter pin (6).
23. Remove two locknuts (4), U-bolt (10), two shims (5), and modulator cable (8) from bracket (3). Discard locknuts (4).
24. Remove nut (13), washer (12), and screw (9) from modulator cable (8).
25. Remove cotter pin (30), washer (31), and emergency stop cable (26) from fuel pump bracket (22). Discard cotter pin (30).
26. Remove screw (23), sleeve (24), and screw (25) from emergency stop cable (26).
27. Remove screw (29), clamp (27), and emergency stop cable (26) from bracket (28).

NOTE

Have drainage container ready to catch oil.

28. Disconnect power steering inlet line (35) from power steering pump (33).

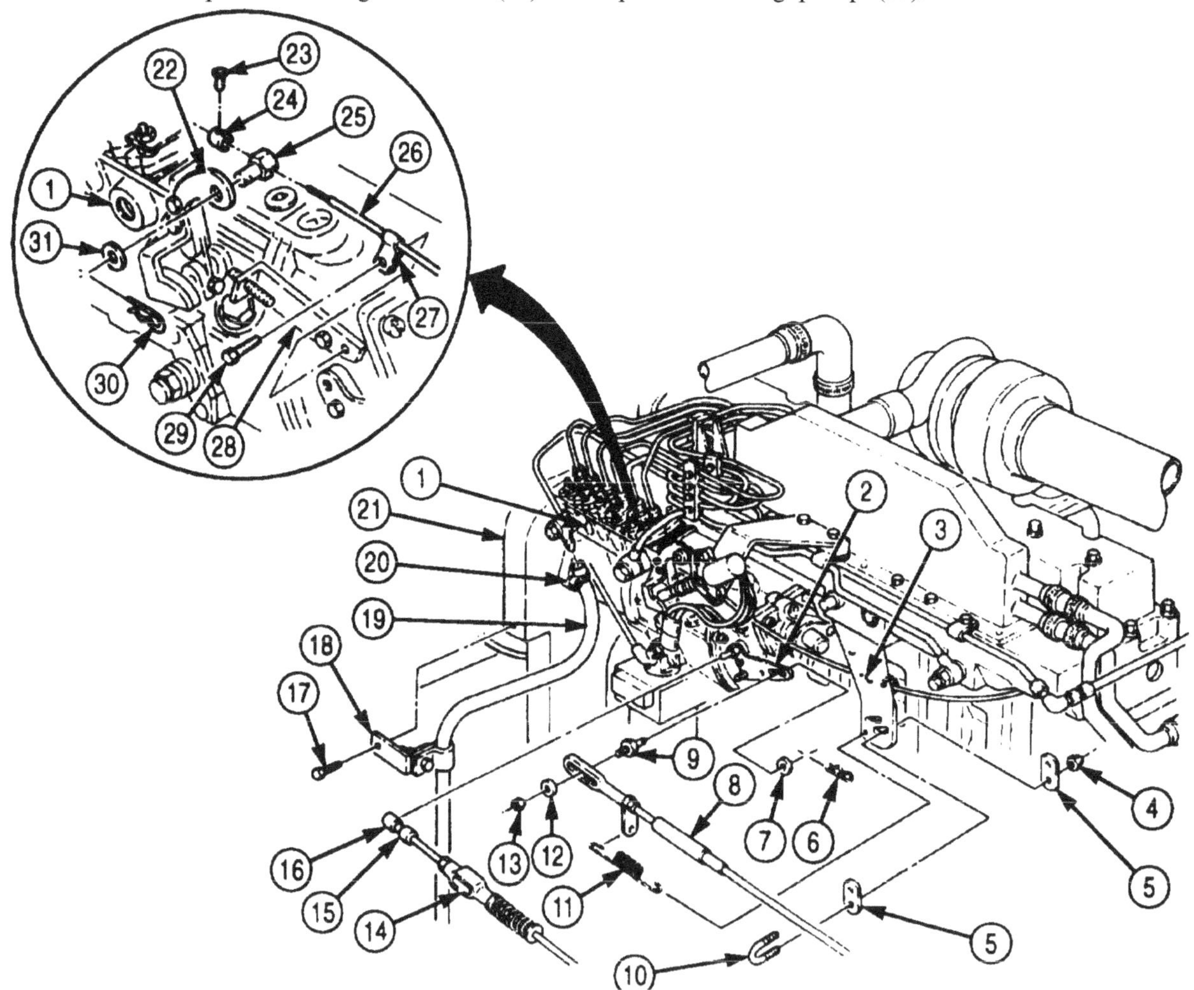

4-5. POWER PLANT (M939A2) REPLACEMENT (Contd)

29. Loosen clamp (39) and remove outlet line (40) from power steering pump (33).
30. Disconnect air line (41) from air compressor (32).
31. Remove screw (38), washer (37), clamp (36), and air line (41) from engine (34).
32. Loosen two screws (42) and screw (43), and rotate alternator (64) upward.

NOTE

All sealant must be removed prior to removing wires. Tag all wires for installation.

33. Remove two screw-assembled lockwashers (52) and terminal cover (51) from alternator (64). Discard screw-assembled lockwashers (52).
34. Remove screw (53), lockwasher (56), and negative wire (44) from alternator (64). Discard lockwasher (56).
35. Remove nut (47), lockwasher (46), and accessory wire (45) from alternator (64). Discard lockwasher (46).
36. Remove nut (50), lockwasher (49), and positive wire (48) from alternator (64). Discard lockwasher (49).
37. Disconnect lead (55) from AC wire (54).
38. Remove screw (60), clamp (62), and harness (59) from oil cooler housing (63).
39. Remove screw (57), clamp (58), and harnesses (59) and (61) from engine (34).
40. Disconnect lead (71) from temperature sending unit (65).
41. Remove screw (68), washer (67), ground strap (66), three ground wires (69), and lockwasher (70) from engine (34). Discard lockwasher (70).

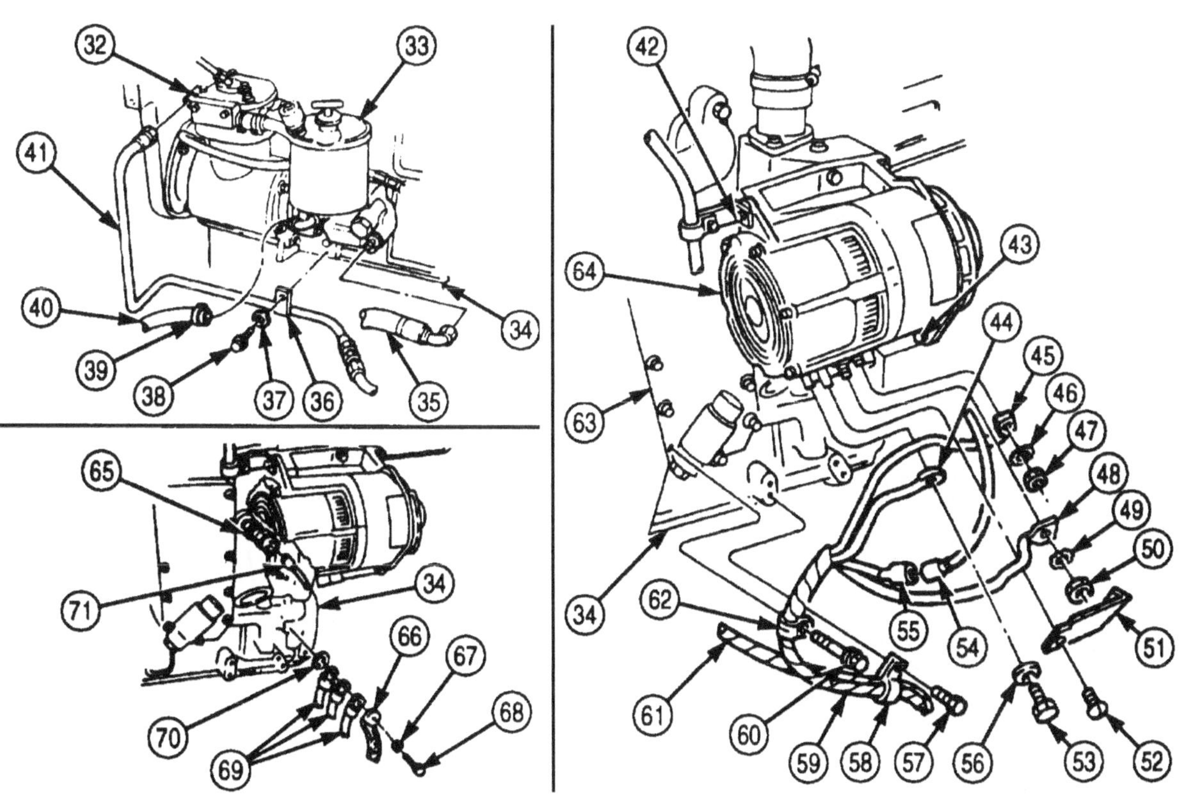

4-5. POWER PLANT (M939A2) REPLACEMENT (Contd)

42. Remove nut (22), lockwasher (21), and wire (4) from starter solenoid (20). Discard lockwasher (21).
43. Remove nut (8), lockwasher (9), and wires (10) and (11) from starter solenoid (20). Discard lockwasher (9).
44. Remove nut (18), lockwasher (19), and wire (17) from starter solenoid (20). Discard lockwasher (19).
45. Remove nut (12), lockwasher (13), and wires (14) and (15) from starter (16). Discard lockwasher (13).

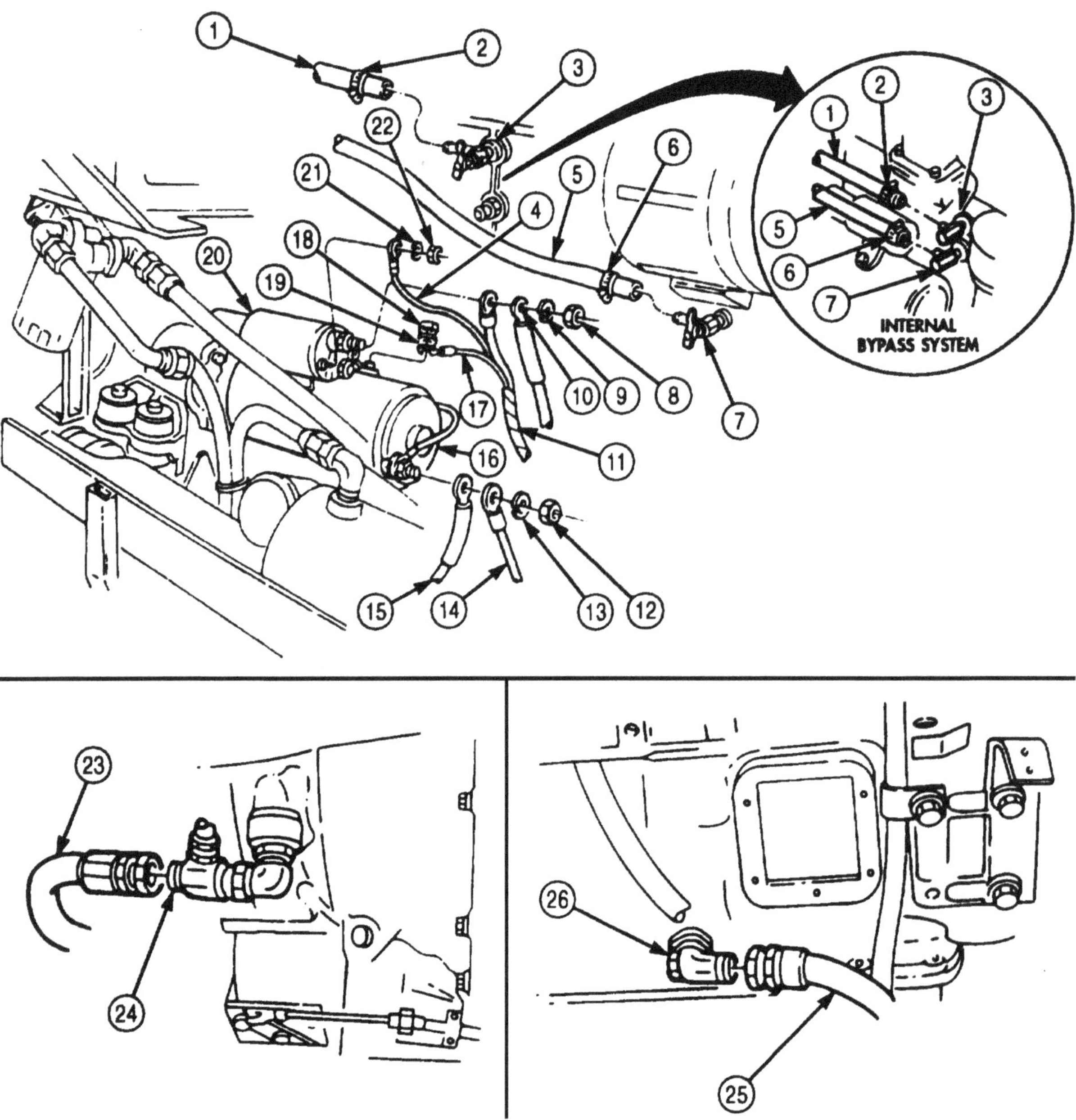

4-5. POWER PLANT (M939A2) REPLACEMENT (Contd)

46. Loosen clamps (2) and (6) and remove hoses (1) and (5) from fittings (3) and (7).
47. Disconnect oil cooler return hose (23) from temperature transmitter adapter (24).
48. Disconnect transmission supply hose (25) from lubrication valve adapter (26).
49. Remove nut (29), screw (34), clamp (33), and hose (32) from bracket (30).
50. Remove clamps (53) and (35) and hose (32) from surge tank fitting (27) and tube (48).

NOTE

Perform steps 51 and 52 for internal bypass systems.

51. Remove two clamps (42) and hose (43) from fitting (52) and thermostat canister (41).
52. Remove two clamps (44), thermostat canister (41), and hose (45) from tube (38).

NOTE

Perform step 53 for external bypass systems.

53. Remove two clamps (42) and hose (43) from fitting (54) and tube (38).
54. Remove two clamps (36), hump hose (37), two clamps (39), hump hose (40), and tube (38) from outlet tube (31) and transmission oil cooler (51).
55. Remove two clamps (50). elbow (49), two clamps (47), elbow (46), and tube (48) from outlet port (28) and transmission oil cooler (51).

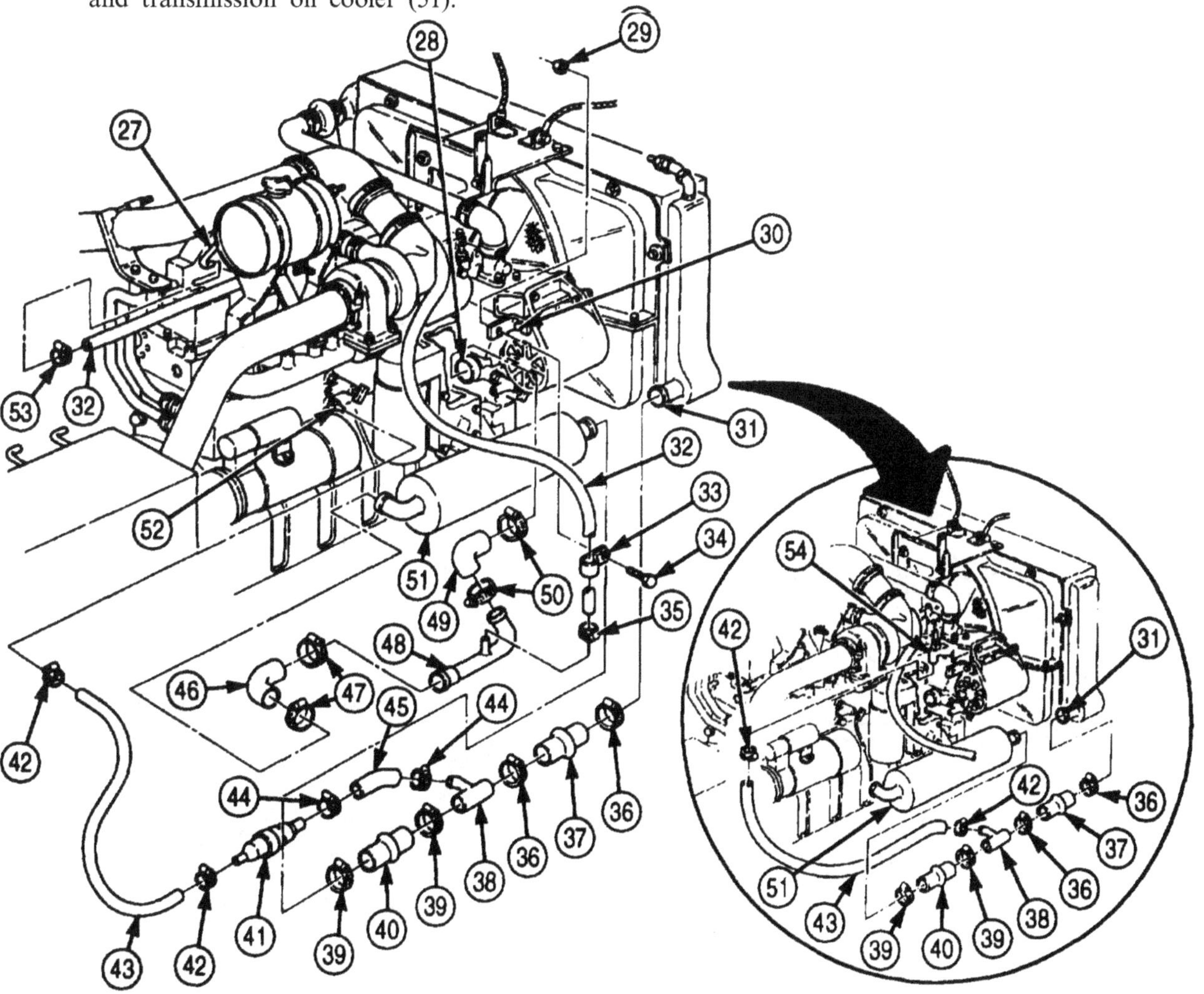

4-5. POWER PLANT (M939A2) REPLACEMENT (Contd)

56. Disconnect breather vent tube (3) from elbow (4).
57. Disconnect lead (2) from transmission temperature sending unit (1).
58. Disconnect two leads (16) from neutral safety switch wires (17).
59. Disconnect lead (9) from solenoid connector (10).
60. Remove cotter pin (15) and transmission shift cable (12) from shift lever (18). Discard cotter pin (15).
61. Remove two nuts (11), U-bolt (14), spacer (13), and transmission shift cable (12) from transmission adapter plate (19).
62. Remove nut (20), lockwasher (21), ground wire (22), screw (6), clamp (5), and modulator cable (7) from bracket (8). Discard lockwasher (21).
63. Remove and drain radiator (para. 3-60).
64. Remove surge tank and bracket (para. 3-62).

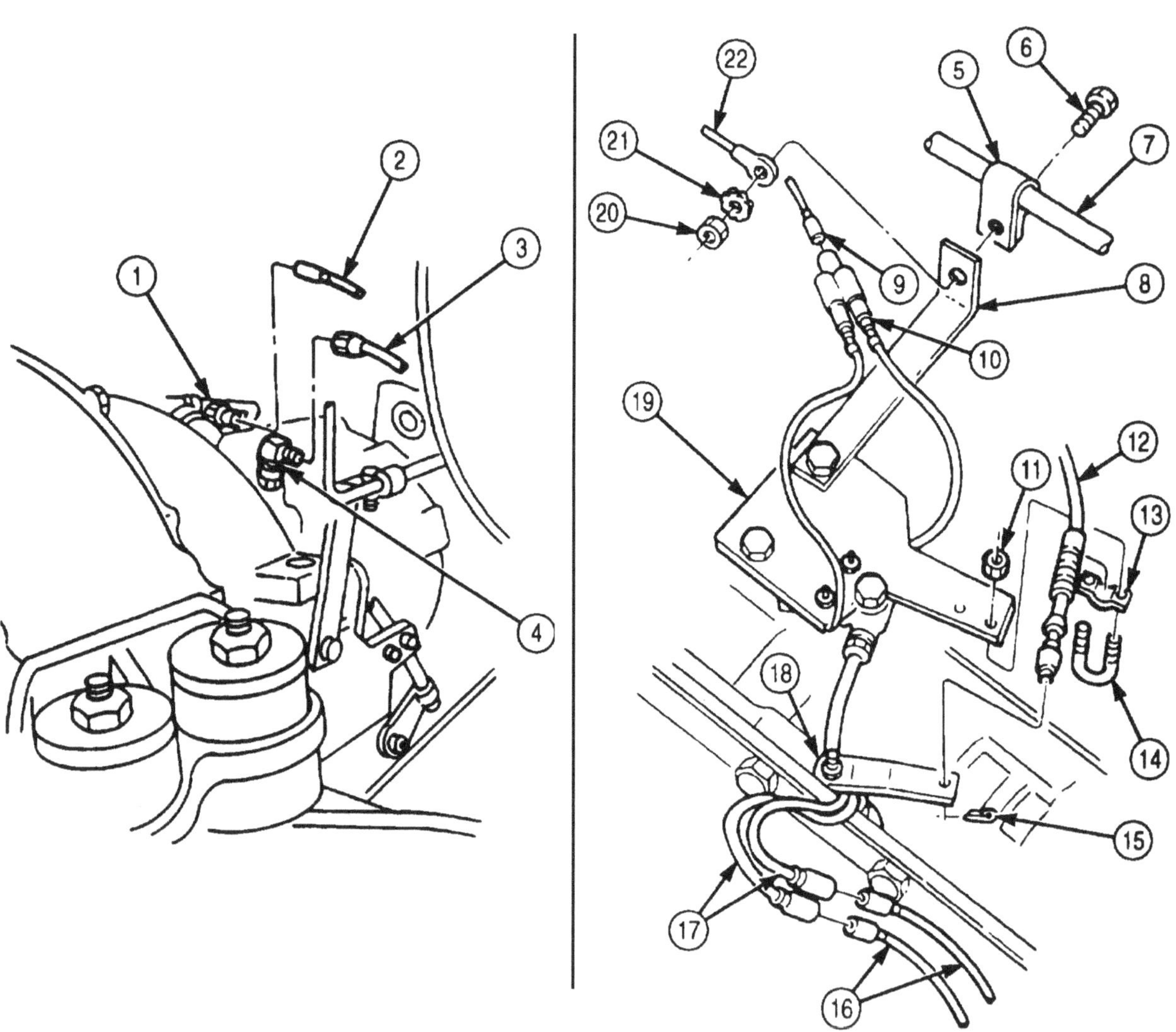

4-5. POWER PLANT (M939A2) REPLACEMENT (Contd)

b. Removal

1. Install utility chain on two lifting brackets (23).
2. Install lifting device on chain. Raise lifting device enough to remove slack from chain.
3. Remove three screws (28), lockwashers (29), and transmission mounting bracket (31) from transmission (32) and isolator (30). Discard lockwashers (29).
4. Remove two locknuts (36), washers (35), and back plate (34) from studs (37) and crossmember (33). Discard locknuts (36).
5. Remove two screws (25), washers (24), and isolators (26) from right engine mount bracket (27).
6. Remove two locknuts (38), washers (39), screws (40), and isolators (42) from left engine mount bracket (41). Discard locknuts (38).

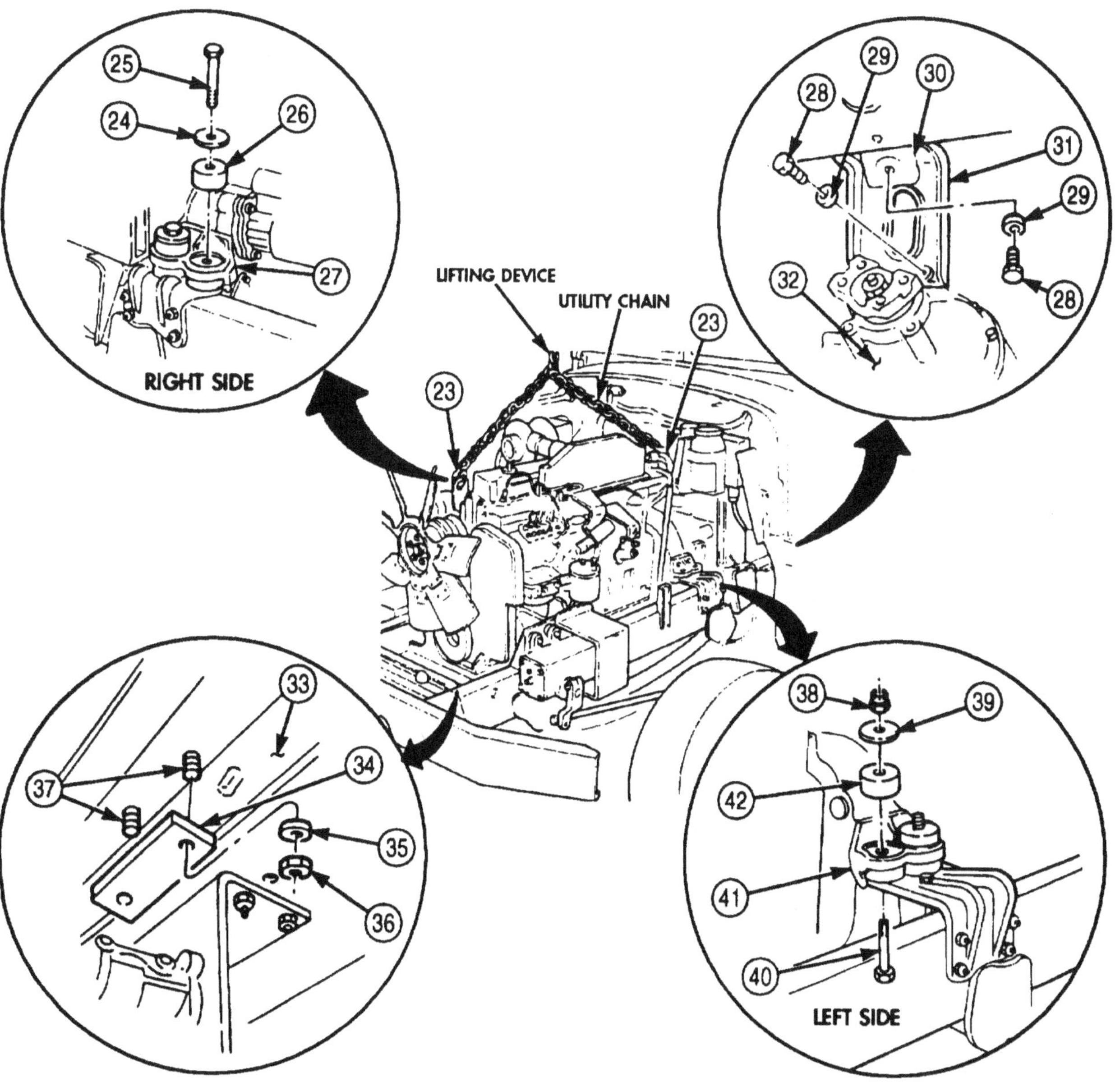

4-5. POWER PLANT (M939A2) REPLACEMENT (Contd)

WARNING

- Lifting device must have a weight capacity greater than the combined weight of the engine and transmission to prevent injury or death to personnel and damage to equipment.
- All personnel must stand clear during lifting operations. A snapped cable, or swinging or shifting load, may result in injury or death to personnel.
- Do not put hands between frame and engine supports during lifting operations. Use prybar to adjust position of engine during lifting operations. Failure to do so may result in injury or death to personnel.

CAUTION

Always remove engine slowly. Lift out of chassis in short lifts and closely observe all engine and transmission attachments during removal to prevent damage to equipment.

NOTE

Engine must be adjusted constantly during lifting operations.

7. Remove engine (3) and transmission (2) from vehicle.
8. Remove six isolators (4) and two washers (7) from crossmember (8) and left (6) and right (5) engine mounts.
9. Remove two screws (14), washers (9), and radiator brackets (13) from front engine mount (15).
10. Remove two screws (10), washers (11), and isolators (12) from front engine mount (15).

WARNING

- Do not detach chain from engine until engine is supported. Failure to do so may result in injury or death to personnel.
- Engine must be securely mounted on repair stand. Failure to do so may result in death or injury to personnel.

11. Remove lifting device and utility chain from two engine lifting brackets (1).

1. Install chain on two engine lifting brackets (1).
2. Install lifting device on chain. Raise lifting device enough to remove slack from chain.
3. Install two isolators (12), washers (11), and screws (10) on front engine mount (15).
4. Install two radiator brackets (13) on front engine mount (15) with two washers (9) and screws (14).

WARNING

- All personnel must stand clear during lifting operations. A snapped cable, or swinging or shifting load, may result in injury or death to personnel.
- Do not put hands between frame and engine supports during lowering operations. Use prybar to adjust position of engine lowering operations. Failure to do so may result in injury or death to personnel.

NOTE

Chain must be adjusted so transmission points downward to allow clearance of engine over front axle.

5. Raise engine (3) and transmission (2) and position in vehicle.

4-5. POWER PLANT (M939A2) REPLACEMENT (Contd)

CAUTION

Do not lower engine completely. Ensure all attaching components are clear and free from obstructions. Failure to do so may cause damage to engine components.

6. Position two washers (7) and six isolators (4) on crossmember (8) and left (6) and right (5) engine mounts.
7. Slowly lower engine (3) and transmission (2) into vehicle.
8. Level engine (3) and align screws (10) with holes in crossmember (8).
9. Lower engine (3) until it rests on crossmember (8) and left (6) and right (5) engine mounts.

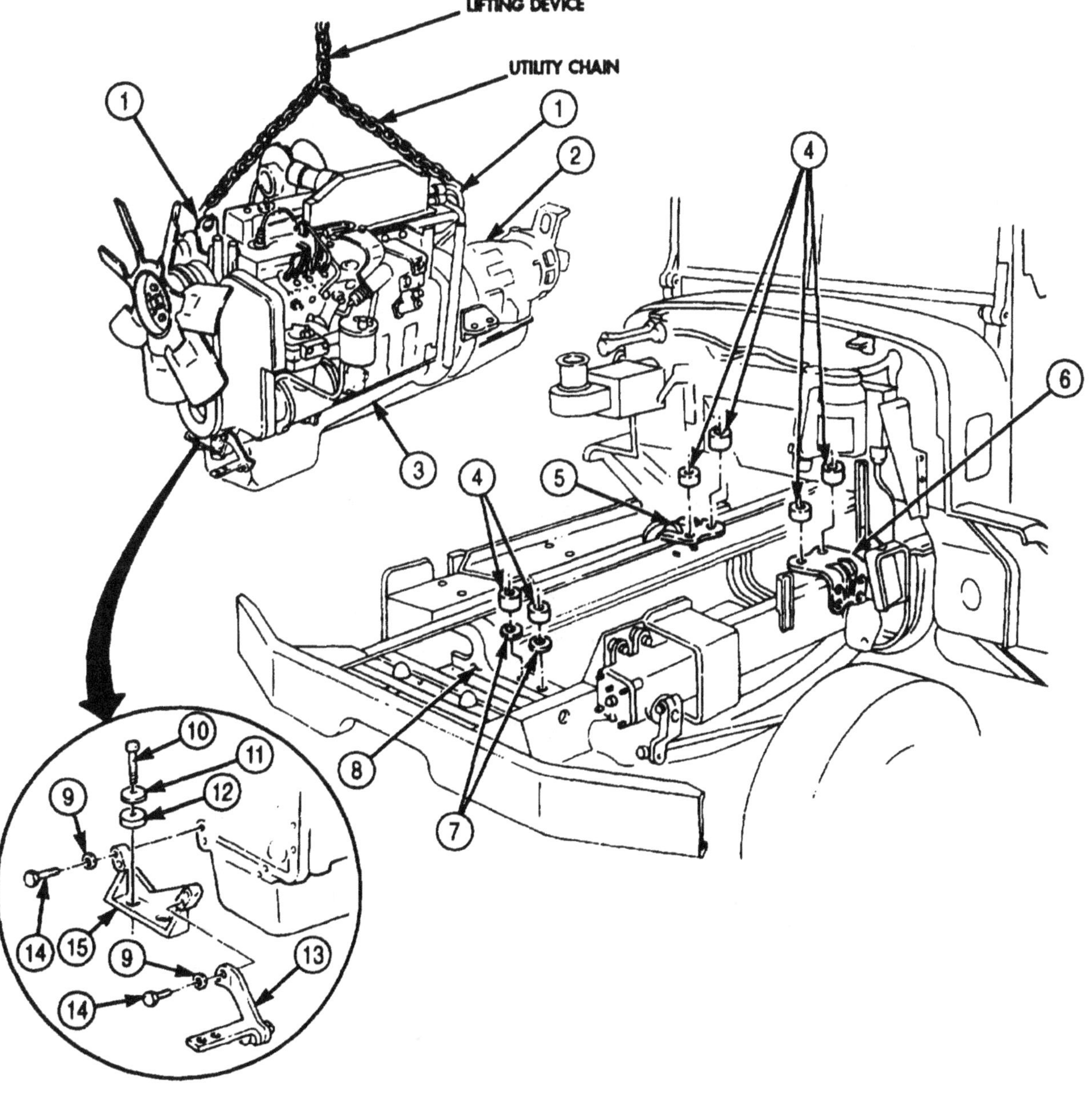

4-5. POWER PLANT (M939A2) REPLACEMENT (Contd)

10. Install two screws (20), isolators (22), washers (19), and locknuts (18) on left engine mounting bracket (21). Finger-tighten locknuts 18).
11. Install two isolators (4), washers (2), and screws (3) on right engine mounting bracket (5). Finger-tighten screws (3).
12. Install backing plate (14) on crossmember (13) and screws (17) with two washers (15) and new locknuts (16). Finger-tighten locknuts (16).
13. Install transmission mounting bracket (8) on transmission (11) with two new lockwashers (12) and screws (6).
14. Raise transmission (11) and install transmission mounting brackets (8) on isolator (7) with new lockwasher (9) and screw (10).
15. Tighten locknuts (18) 120-140 lb-ft (163-190 N•m).
16. Tighten screws (3) 120-140 lb-ft (163-190 N•m).
17. Tighten locknuts (16) 75-85 lb-ft (102-115 N•m).
18. Remove lifting device and utility chain from two engine lifting brackets (1).

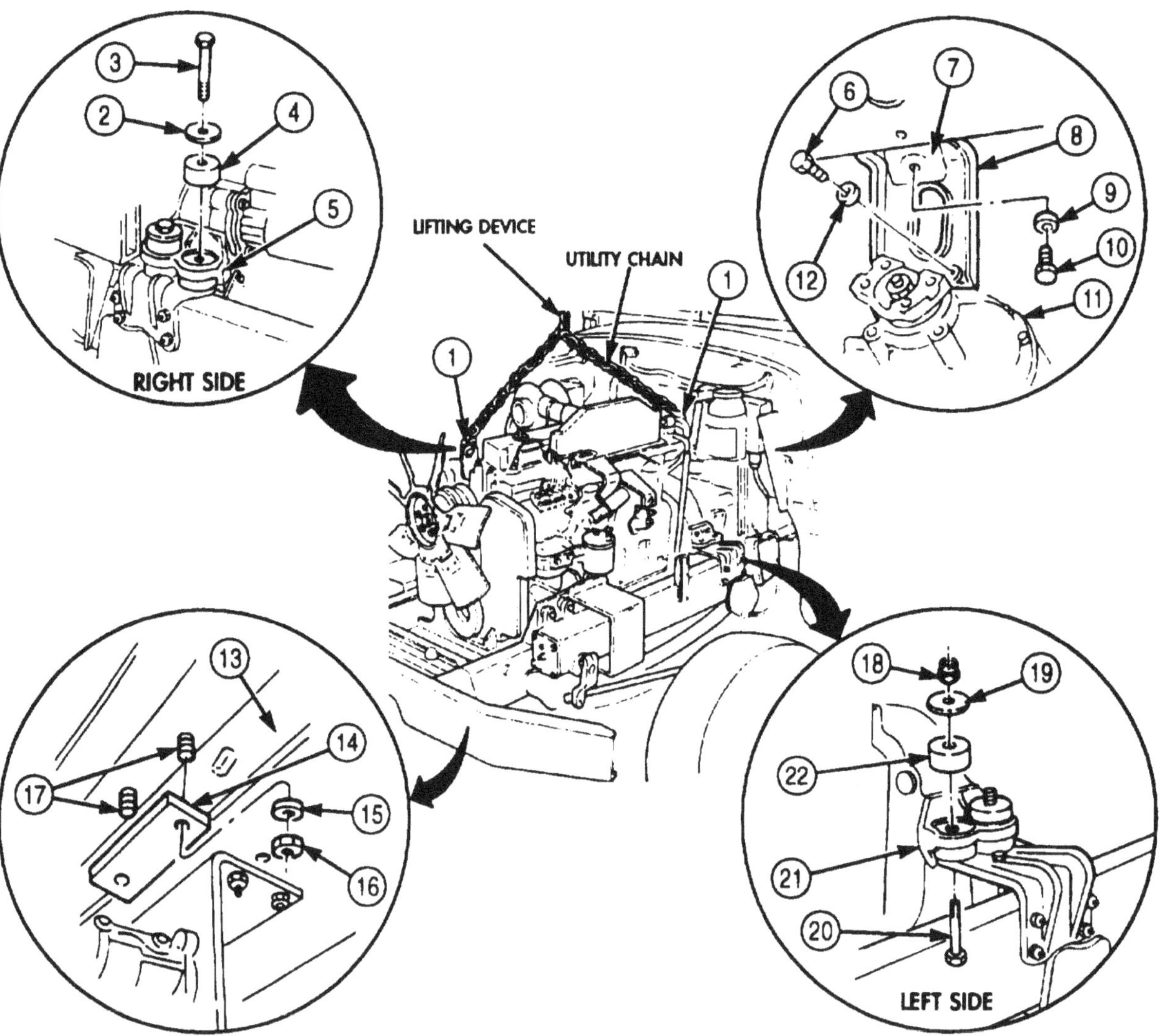

4-5. POWER PLANT (M939A2) REPLACEMENT (Contd)

19. Install transmission shift cable (30) on shift lever (35) with new cotter pin (34).
20. Install transmission shift cable (30) on transmission adapter plate (33) with spacer (31), U-bolt (32), and two nuts (29).
21. Connect two leads (36) to neutral safety switch wires (37).
22. Connect lead (40) to solenoid connector (39).
23. Install modulator cable (28), clamp (26), and ground wire (25) on bracket (38) with screw (27), new lockwasher (24), and nut (23).
24. Connect lead (42) to transmission temperature sending unit (41).

NOTE

Wrap all male pipe threads with antiseize tape before installation.

25. Install breather vent tube (43) on elbow (44).
26. Connect oil cooler return hose (47) on temperature transmitter adapter (48).
27. Connect transmission supply hose (46) on lubrication valve adapter (45).

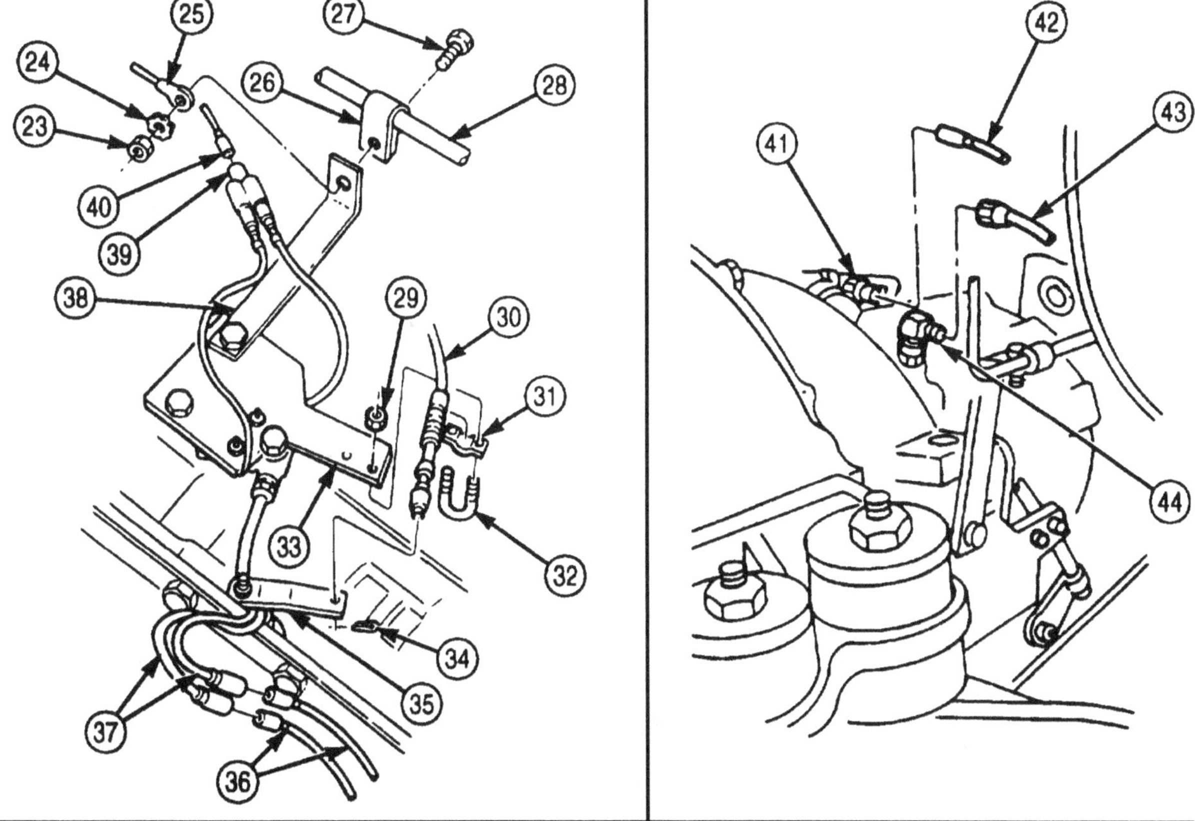

4-5. POWER PLANT (M939A2) REPLACEMENT (Contd)

28. Install wires (14) and (13) on starter (15) with new lockwasher (12) and nut (11).
29. Install wires (9) and (10) on starter solenoid (17) with new lockwasher (8) and nut (7).
30. Install wire (16) on starter (15) with new lockwasher (18) and nut (19).
31. Install wire (4) on starter solenoid (17) with new lockwasher (20) and nut (21).
32. Connect water heater hoses (1) and (5) on valves (3) and (6) and tighten two clamps (2).
33. Install new lockwasher (24), three ground wires (28), and ground strap (25) on engine (23) with washer (27) and screw (26).
84. Route lead (29) over engine (23) and connect to temperature sending unit (22).
35. Connect lead (43) to AC wire (42).

NOTE

Ensure terminals are clean prior to installing wires.

36. Install accessory wire (33) on alternator (52) with new lockwasher (34) and nut (35). Tighten nut (35) 20-25 lb-ft (2-3 N•m).
37. Install negative wire (32) on alternator (52) with new lockwasher (44) and screw (41). Tighten screw (41) 82-102 lb-ft (9-12 N•m).
38. Install positive wire (36) on alternator (52) with new lockwasher (37) and nut (38). Tighten nut (38) 45-55 lb-ft (5-6 N•m).

NOTE

Completely seal wires and inside of terminal cover with adhesive sealant.

39. Install terminal cover (39) on alternator (52) with two new screw-assembled lockwashers (40).
40. Install harnesses (47) and (49) and clamp (46) on engine (23) with screw (45).
41. Install harness (47) and clamp (50) on oil cooler housing (51) with screw (48).
42. Rotate alternator (52) upward and tighten screw (30).
43. Tighten two screws (31) 35 lb-ft (47 N-m).

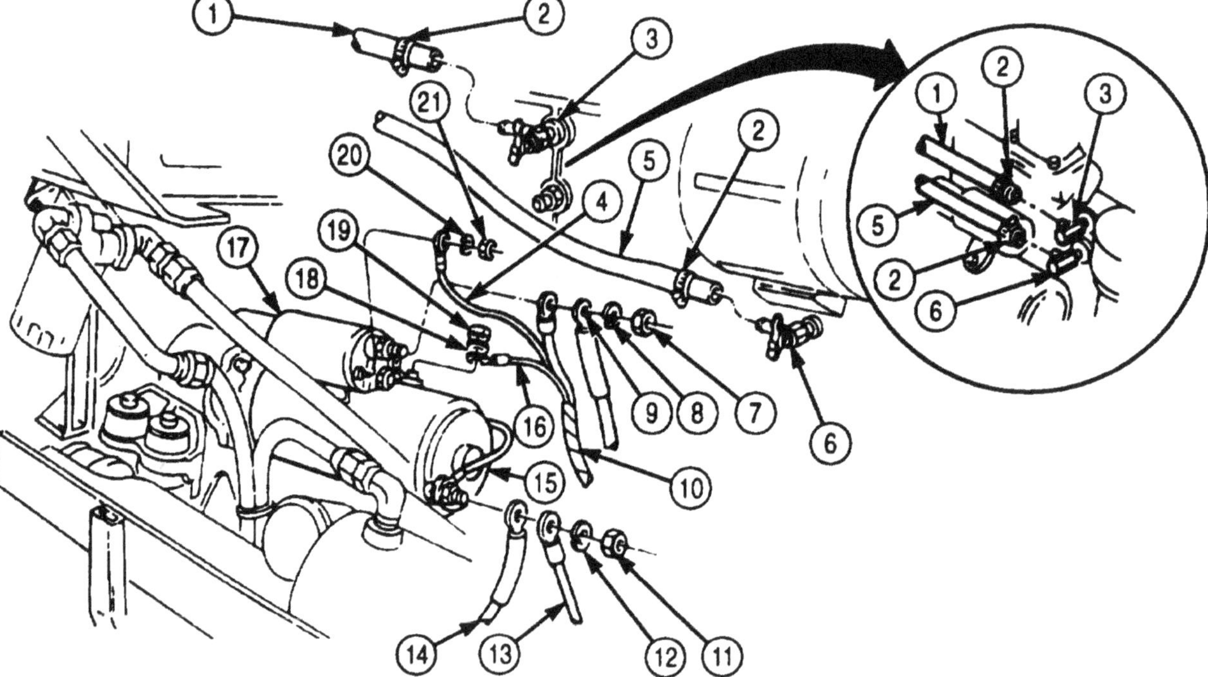

4-5. POWER PLANT (M939A2) REPLACEMENT (Contd)

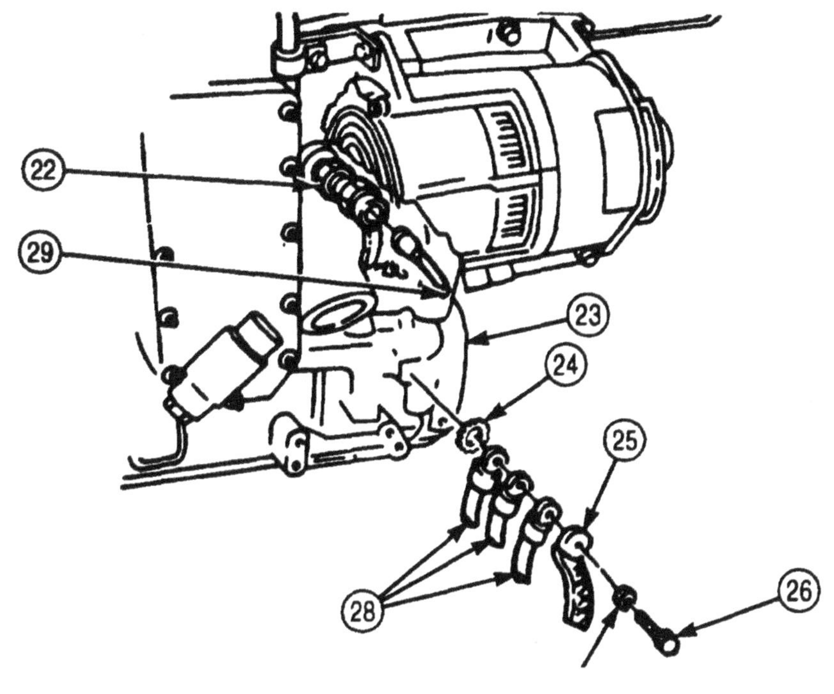

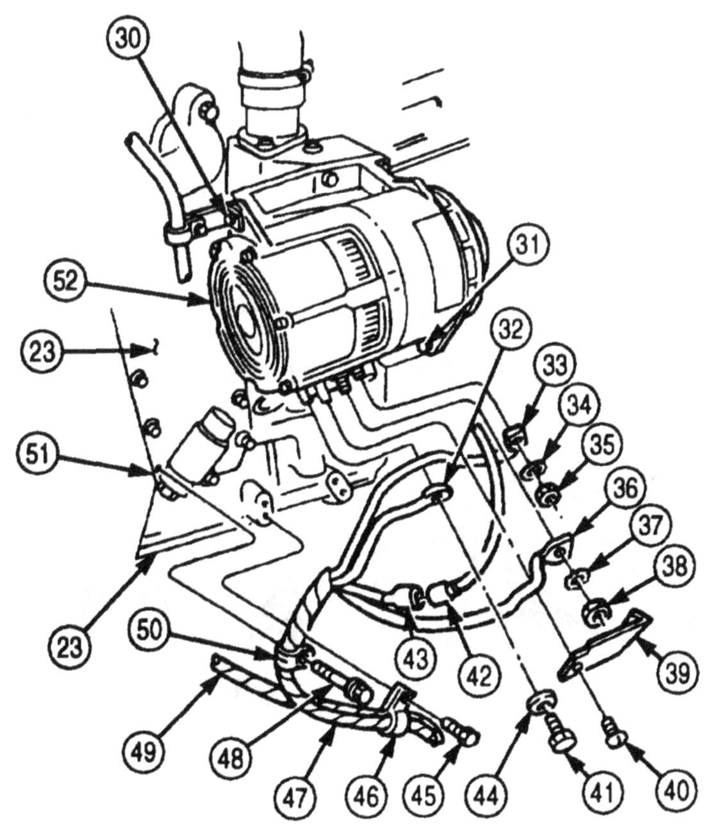

4-5. POWER PLANT (M939A2) REPLACEMENT (Contd)

NOTE

Wrap all male pipe threads with antiseize tape before installation.

44. Connect air line (10) on air compressor (1).
45. Install air line (10) on engine (3) with clamp (5), washer (6), and screw (7).
46. Install power steering inlet (4) and outlet (9) lines on power steering pump (2) and tighten clamp (8).
47. Install emergency stop cable (36) on screw (35) with sleeve (34) and screw (33).
48. Install emergency stop cable (36) on fuel pump bracket (32) with washer (41) and new cotter pin (40).
49. Install emergency stop cable (36) on fuel pump bracket (38) with clamp (37) and screw (39).
50. Install screw (19) on modulator cable (18) with washer (22) and nut (23).
51. Install modulator cable (18) and screw (19) on throttle control lever (12) with washer (17) and new cotter pin (16).
52. Install surge tank and bracket (para. 3-62).
53. Install radiator (para. 3-60).
54. Pull sleeve (25) away from socket (26) and install accelerator linkage (24) on throttle control lever (12).
55. Install modulator cable (18) on bracket (13) with two shims (15), U-bolt (20), and new locknuts (14).
56. Install spring (21) on modulator cable (18) and bracket (13).
57. Install fuel return line (29) on fuel pump (11) and tighten clamp (30).
58. Install fuel return line (29) and bracket (28) on gearcase housing (31) with screw (27).

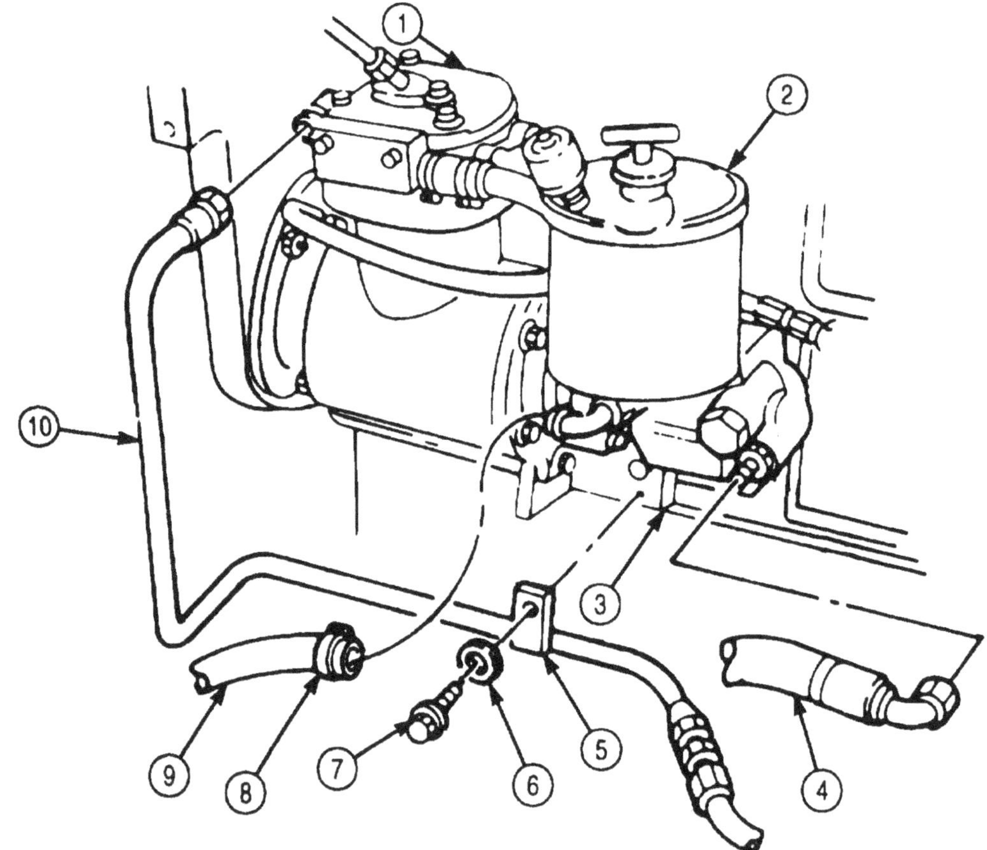

4-5. POWER PLANT (M939A2) REPLACEMENT (Contd)

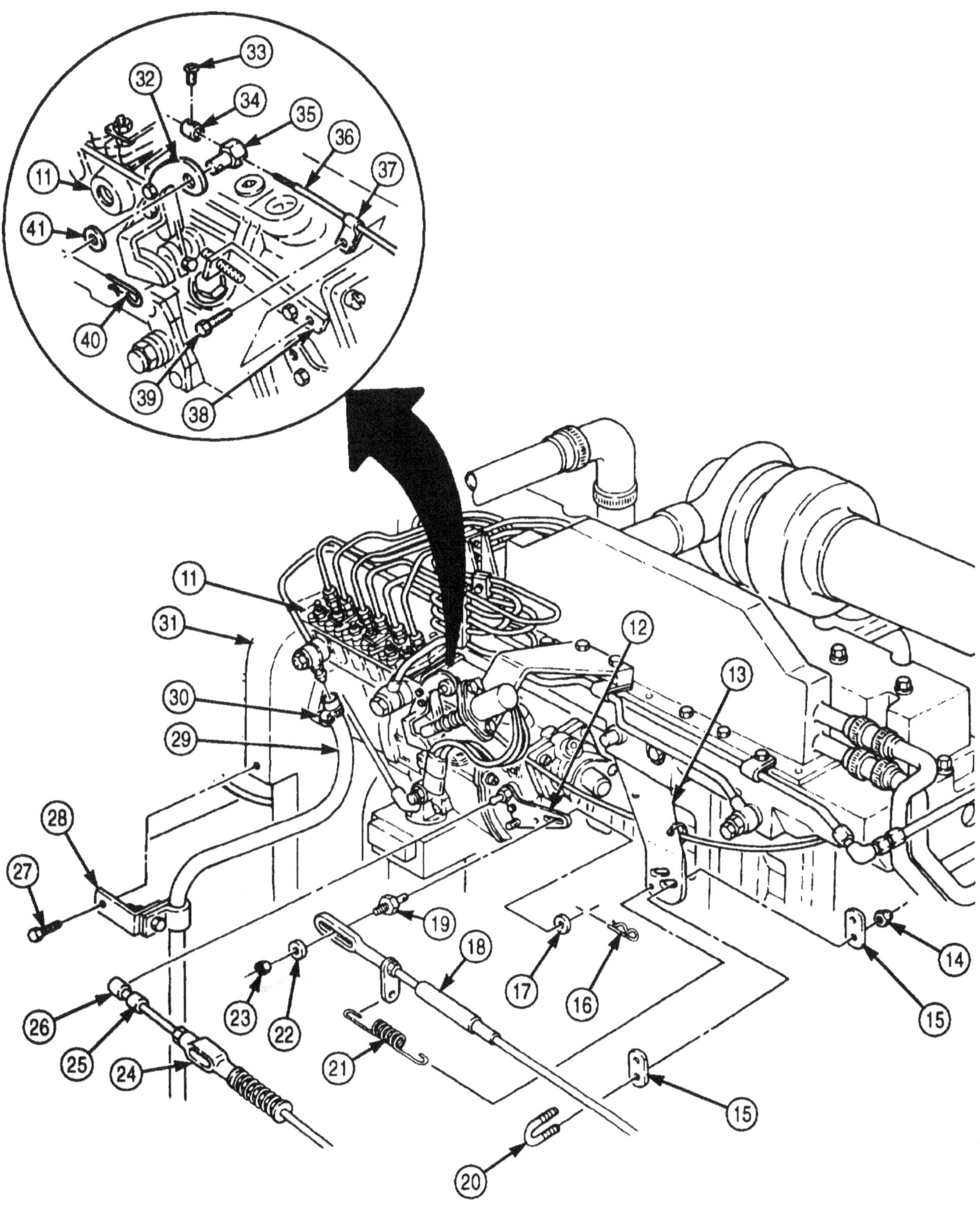

4-5. POWER PLANT (M939A2) REPLACEMENT (Contd)

59. Connect ether supply tube (20) to ether atomizer nozzle (42).
60. Connect connector (35) to oil pressure sending unit (34).
61. Connect fuel supply line (36) to fuel transfer pump (37).
62. Connect connector (39) to fuel transfer pump (38).
63. Connect connector (41) to throttle control solenoid connector (40).
64. Connect air line (27) to air compressor line (28).
65. Install temperature sensor (29), new lockwasher (30), and ground strap (31) on engine (11) with washer (32) and screw (33).
66. Connect wire (25) to temperature sensor wire (26).
67. Install two spacers (13) and bracket (14) on engine (11) with two washers (16) and screws (17).
68. Install air intake tube (7) on turbocharger (10) and pipe (12) with clamps (1) and (8).
69. Install bracket (5) and U-bolt (18) on air intake tube (7) and bracket (14) with two nuts (6).
70. Install air intake tube adapter (4) on air intake tube (7).
71. Install air indicator tube (2) on air intake tube adapter (4) with clamp (3).
72. Connect tachometer cable (15) on tachometer drive (19).
73. Secure tachometer cable (15) to air intake tube (7) with new tiedown strap (9).
74. Install cables (24) on engine (11) with two clamps (21) and screws (22).
75. Install new tiedown strap (23) on cables (24) and temperature sensor wire (26).

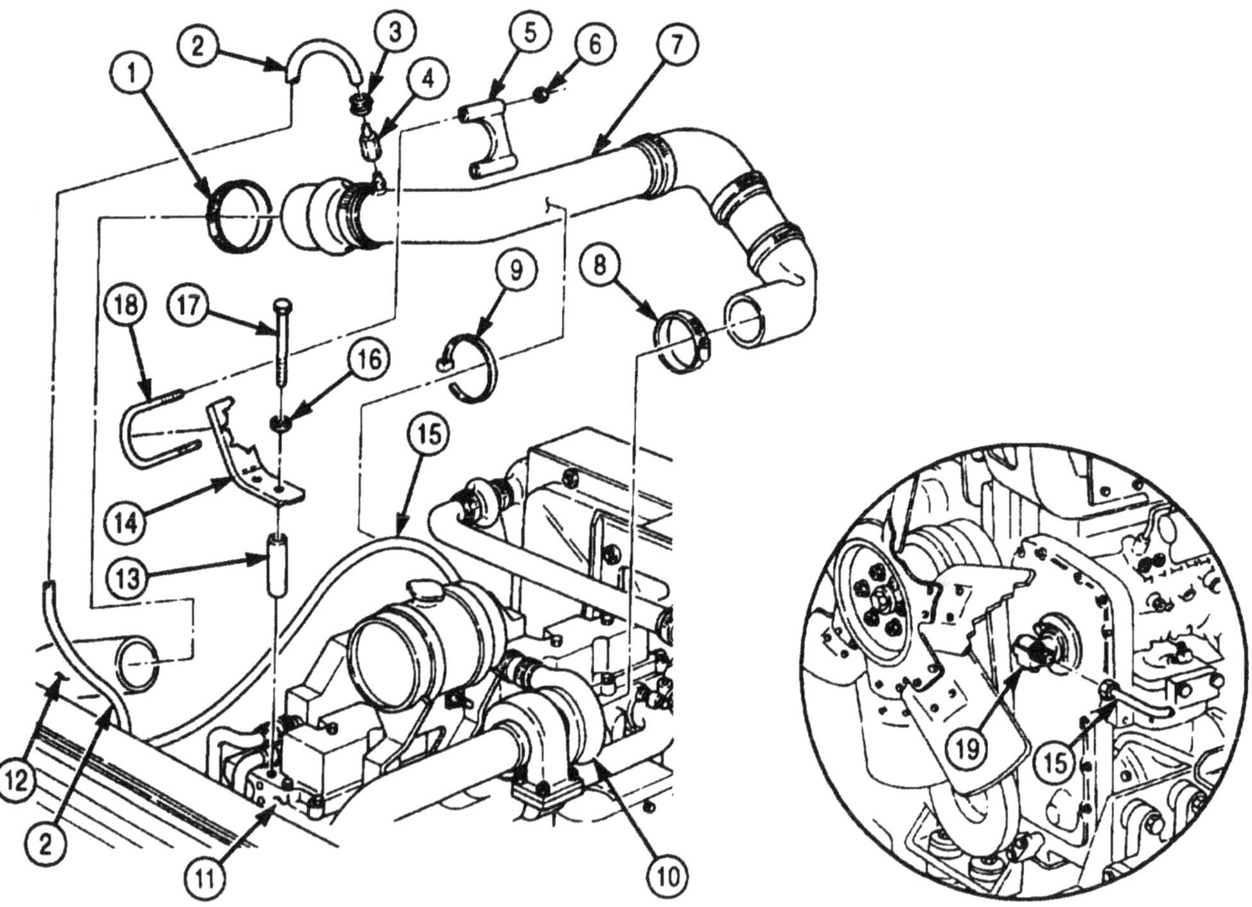

4-5. POWER PLANT (M939A2) REPLACEMENT (Contd)

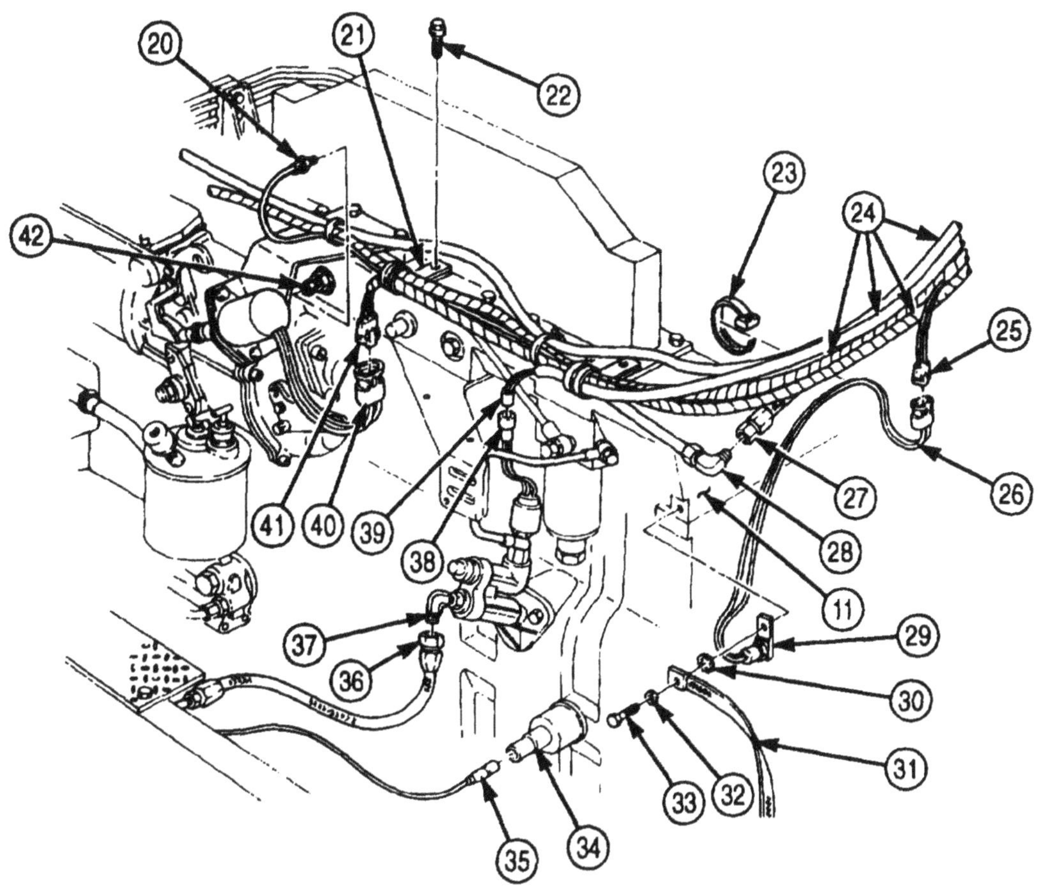

FOLLOW-ON TASKS:
- Install transmission-to-transfer case propeller shaft (para. 3-148).
- Install transmission PTC-to-hydraulic pump propeller shaft, if equipped (para. 3-334).
- Install upper radiator hose and bracket (para. 3-58).
- Install hood assembly (para. 3-275).
- Fill steering system to proper oil level (LO 9-2320-272-12).
- Fill engine to proper oil level (para. 3-5).
- Fill transmission to proper oil level (para. 3-133).
- Close air reservoirs drainvalve (TM 9-2320-272-10).
- Connect battery ground cables (para. 3-126).
- Adjust modulator cable (para. 3-145).
- Adjust accelerator linkage (para. 3-42).
- Fill coolant system to proper coolant level (para 3-53).

CAUTION

Never start a new or repaired engine without performing run-in starting procedures (para. 4-7).

- Perform engine run-in starting procedures (para. 4-8).
- Start engine (TM 9-2320-272-10), allow air pressure to build up to normal operating range, and check for leaks. Road test vehicle.

4-6. ENGINE AND CONTAINER REPLACEMENT

THIS TASK COVERS:

a Removal b. Installation

INITIAL SETUP:

APPLICABLE MODELS
All

TOOLS
General mechanic's tool kit (Appendix E, Item 1)
Lifting device
Utility chain

MATERIALS/PARTS
Forty-two lockwashers (Appendix D, Item 392)
Four lockwashers (Appendix D, Item 377)

REFERENCES (TM)
TM 9-2320-272-24P

GENERAL SAFETY INSTRUCTIONS
- All personnel must stand clear during lifting operations.
- Engine container pressure must be released before opening container.

WARNING

Engine container is pressurized. Ensure pressure is released before opening container. Failure to do so may cause injury or death to personnel.

a. Removal

1. Remove forty-two nuts (2), lockwashers (3), screws (4), and upper container section (1) from lower container section (5). Discard lockwashers (3).
2. Remove two screws (11), lockwashers (14), and trunnion cap (13) from front trunnion mount (12) and engine (6). Discard lockwashers (14).
3. Remove two screws (15), lockwashers (17), and washers (19) from rear trunnion mount (18) and flywheel housing (16). Discard lockwashers (17).

NOTE

Perform step 4 for late model engines.

4. Loosen two hose clamps (9) and remove breather tube (8) from breather (7) and elbow (10).

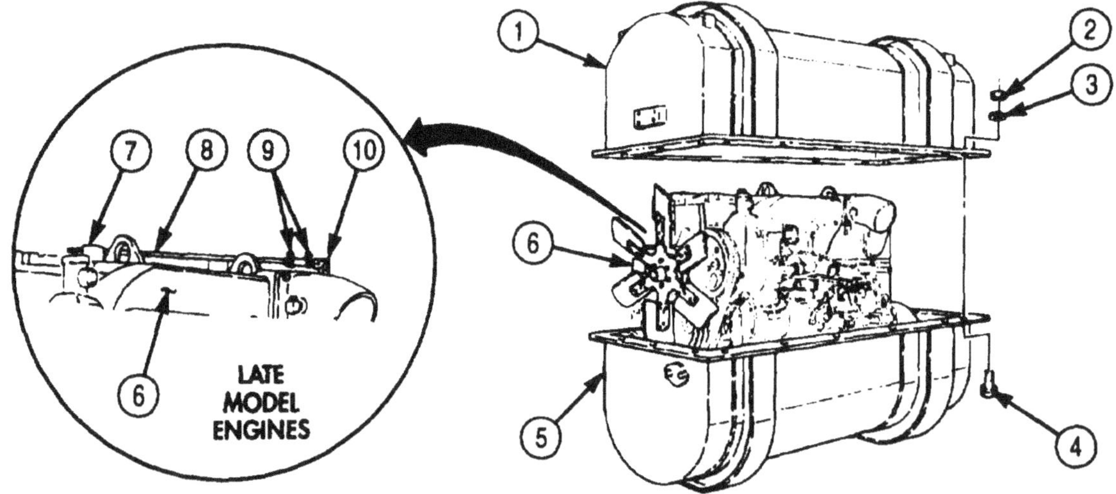

4-6. ENGINE AND CONTAINER REPLACEMENT (Contd)

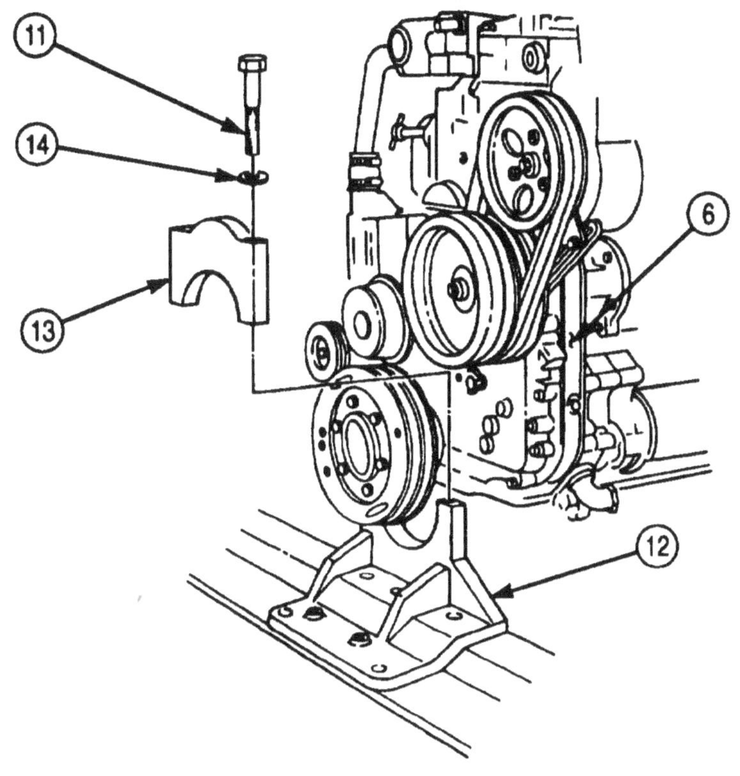

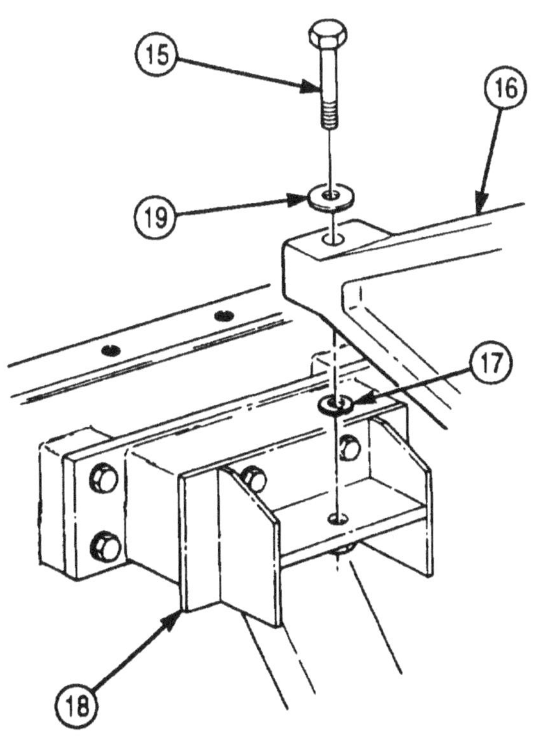

4-6. ENGINE AND CONTAINER REPLACEMENT (Contd)

5. Install utility chain and lifting device on two engine lifting eyes (1).

WARNING

All personnel must stand clear during lifting operations. A snapped cable, or swinging or shifting load, may result in injury or death to personnel.

NOTE

Assistant will help with steps 6 and 7.

6. Remove engine (2) from lower container section (3).
7. Lift engine (2) onto transporter stand.
8. Remove utility chain and lifting device from engine lifting eyes (1).

NOTE

Prepare engine for installation.

1. Install utility chain and lifting device on two engine lifting eyes (1).

WARNING

All personnel must stand clear during lifting operations. A snapped cable, or swinging or shifting load, may result in injury or death to personnel.

NOTE

Assistant will help with steps 2 and 3.

2. Remove engine (2) from transporter stand and install in lower container section (31, ensuring all mounting holes are aligned.
3. Remove chain and lifting device from two engine lifting eyes (1).

4-6. ENGINE AND CONTAINER REPLACEMENT (Contd)

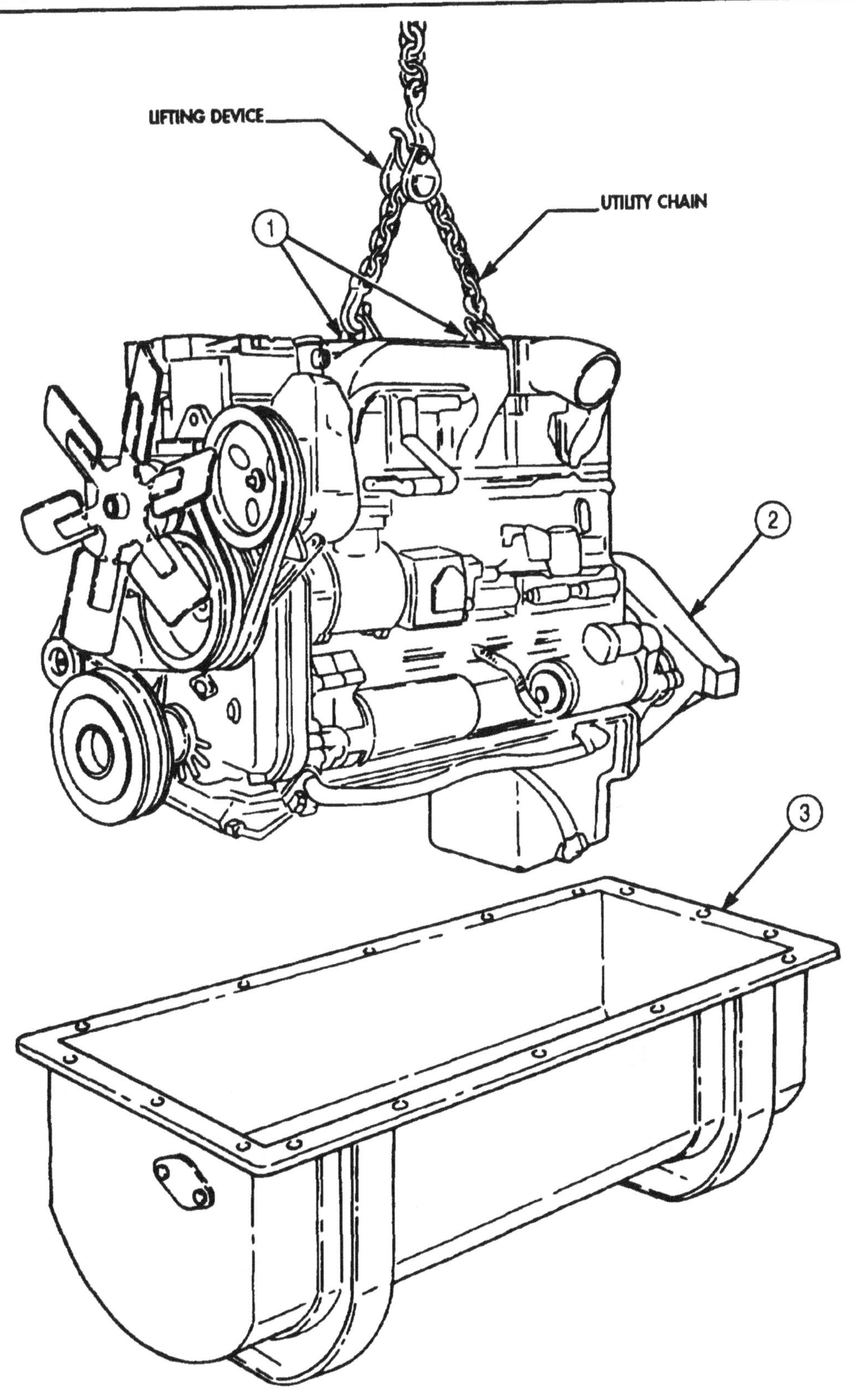

4-6. ENGINE AND CONTAINER REPLACEMENT (Contd)

NOTE

Perform step 4 for late model engines.

4. Install breather tube (8) on breather (7) and elbow (10) and tighten two hose clamps (9).
5. Install flywheel housing (12) on rear trunnion mount (13) with two washers (15), lockwashers (14), and screws (11).
6. Install trunnion cap (18) on front trunnion mount (17) and engine (6) with two lockwashers (19) and screws (16).
7. Install upper container section (1) on lower container section (5) with forty-two screws (4), lockwashers (3), and nuts (2). Tighten nuts (2) 85-105 lb-ft (115-142 N•m).

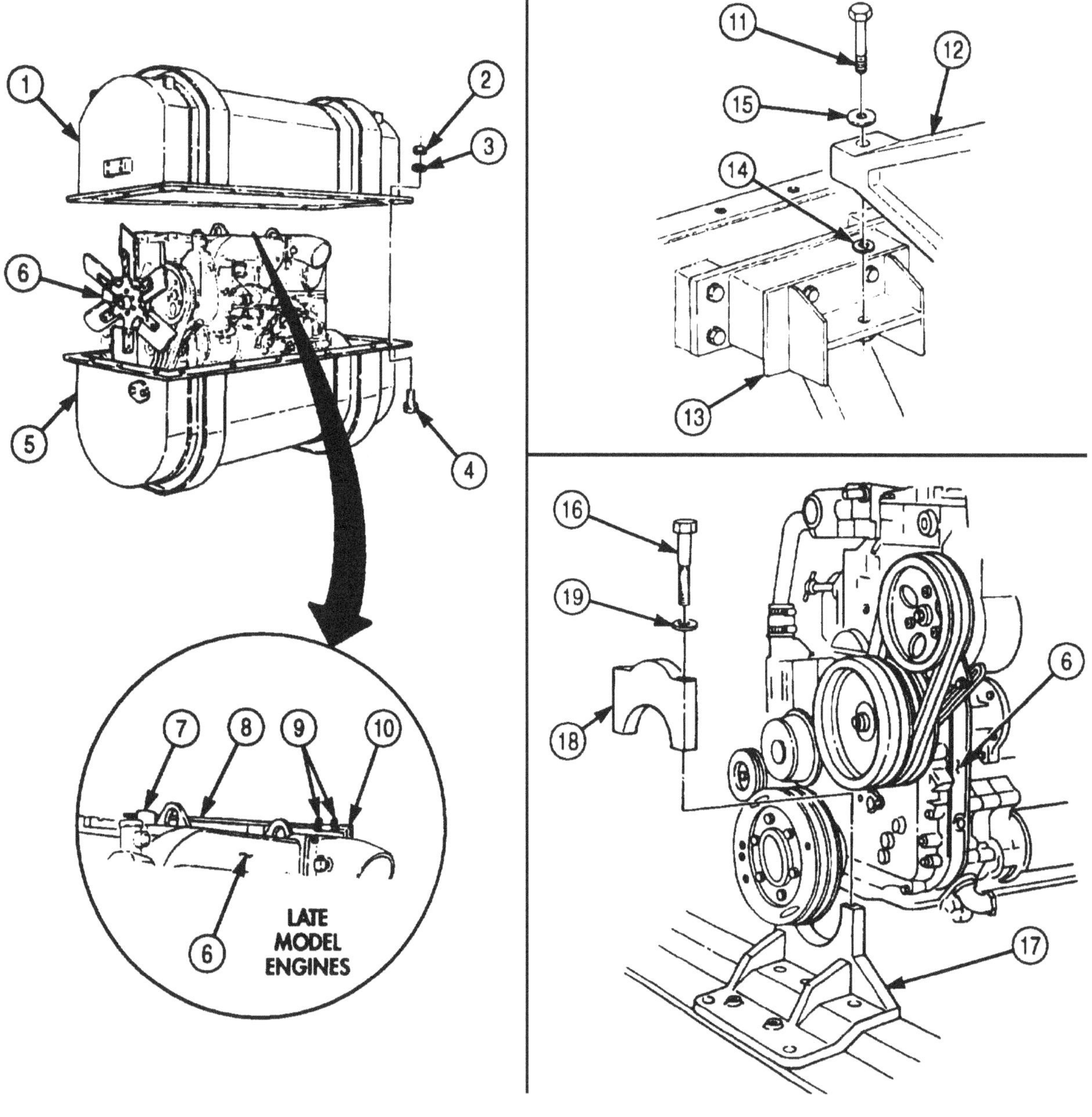

4-7. PREPARING REPLACEMENT ENGINE FOR INSTALLATION IN VEHICLE

THIS TASK COVERS:

a. Removal

b. Installation

INITIAL SETUP:

APPLICABLE MODELS
All

TOOLS
General mechanic's tool kit (Appendix E, Item 1)

MATERIALS/PARTS
Six lockwashers (Appendix D, Item 416)
Three locknuts (Appendix D, Item 313)
Antiseize tape (Appendix C, Item 72)
Gap and plug set (Appendix C, Item 14)

REFERENCES (TM)
TM 9-2320-272-24P

EQUIPMENT CONDITIONS
Engine and transmission removed from vehicle (para. 4-4 or para. 4-5).

a. Removal

CAUTION

Plug all open ports to prevent dirt or contamination from entering engine.

NOTE

Perform this task when preparing removed engine for installation in vehicle.

1. Remove shutoff valve (1) and adapter (2) from engine oil cooler (3).
2. Remove shutoff valve (7) and adapter (6) from water manifold (4).
3. Remove high-temperature switch (5) from water manifold (4).
4. Remove water temperature sending unit (8) from water manifold (4).

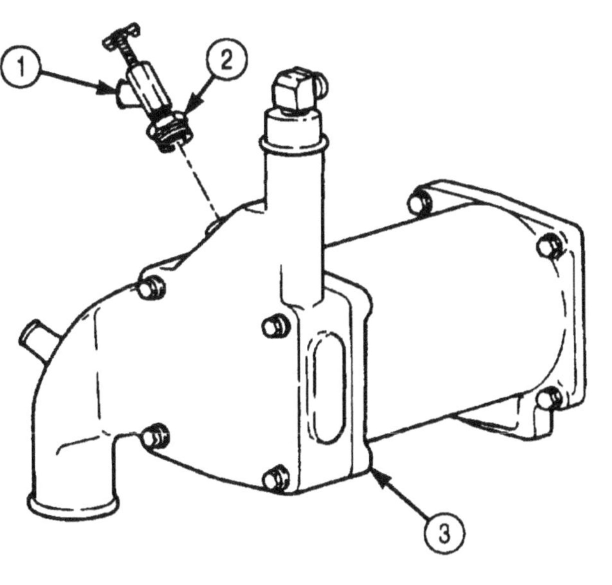

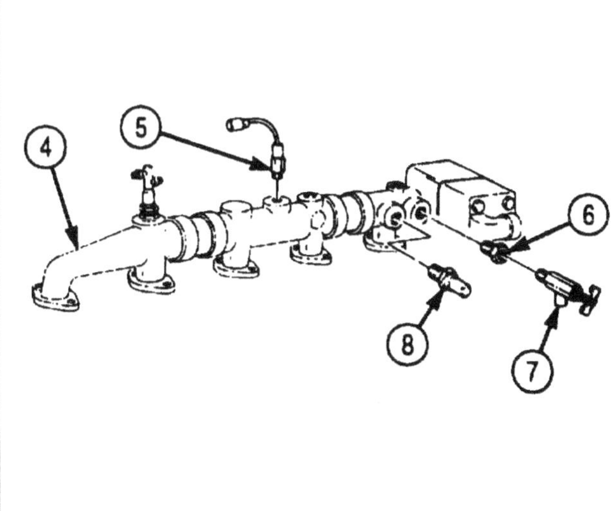

4-7. PREPARING REPLACEMENT ENGINE FOR INSTALLATION IN VEHICLE (Contd)

5. Remove two screws (4), lockwashers (5), washers (3), clamps (l), and transmission oil cooler lines (2) from access plate (6). Discard lockwashers (5).
6. Install two washers (3), new lockwashers (5), and screws (4) on access plate (6). Tag oil cooler lines (2) for installation.
7. Disconnect air line (7) from evaporator bottle (9).
8. Remove three locknuts (10), screws (8), evaporator bottle (9), and bracket (12) from primer pump bracket (11). Discard locknuts (10).
9. Disconnect two air lines (19) from elbows (18).
10. Remove two elbows (18) from air compressor (22).
11. Disconnect air line (21) from fitting (20).
12. Remove two screws (16), lockwashers (17), clamps (14), air line (15), and primer pump bracket (11) from engine block (13). Discard lockwashers (17).

b. Installation

NOTE

Wrap all male pipe threads with antiseize tape before installation.

1. Install primer pump bracket (11), clamps (14), and air line (15) on engine block (13) with two lockwashers (17) and screws (16).
2. Connect air line (21) to fitting (20).
3. Install two elbows (18) on air compressor (22).
4. Connect two air lines (19) to elbows (18).
5. Install evaporator bottle (9) and bracket (12) on primer pump bracket (11) with three screws (8) and locknuts (10).
6. Connect air line (7) to evaporator bottle (9).
7. Remove two screws (4), lockwashers (5), and washers (3), from access plate (6). Discard lockwashers (5).
8. Install two transmission oil cooler lines (2) on access plate (6) with two clamps (1), washers (3), lockwashers (5), and screws (4).
9. Install water temperature sending unit (5) on water manifold (1).

4-7. PREPARING REPLACEMENT ENGINE FOR INSTALLATION IN VEHICLE (Contd)

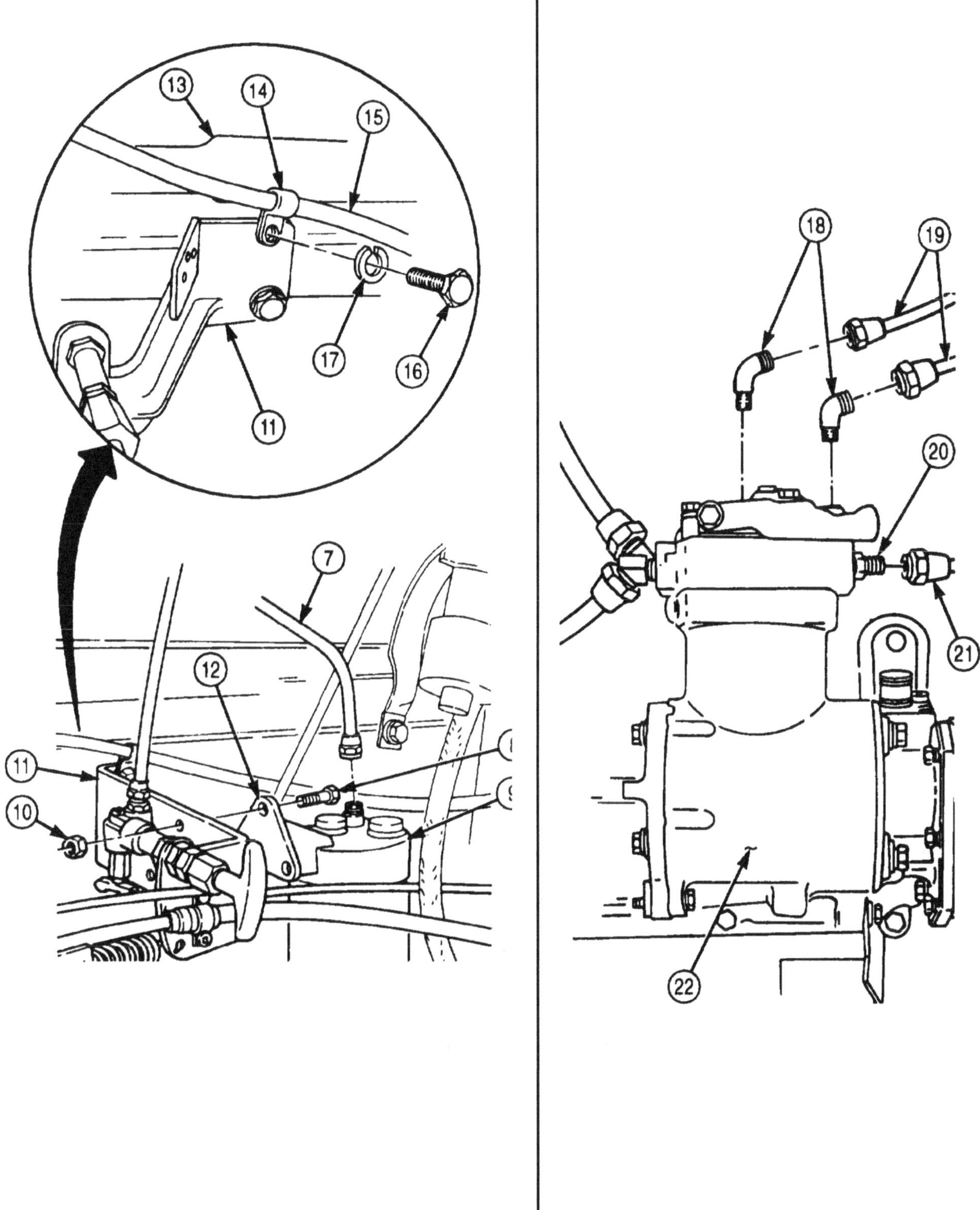

4-7. PREPARING REPLACEMENT ENGINE FOR INSTALLATION IN VEHICLE (Contd)

10. Install high-temperature switch (2) on water manifold (1).
11. Install adapter (3) and shutoff valve (4) on water manifold (1).
12. Install adapter (7) and shutoff valve (6) on oil cooler (8).

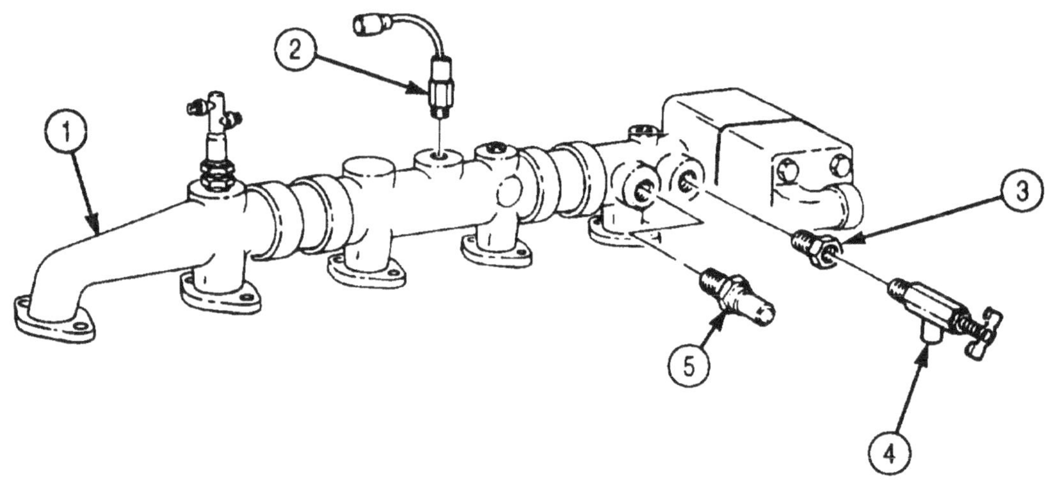

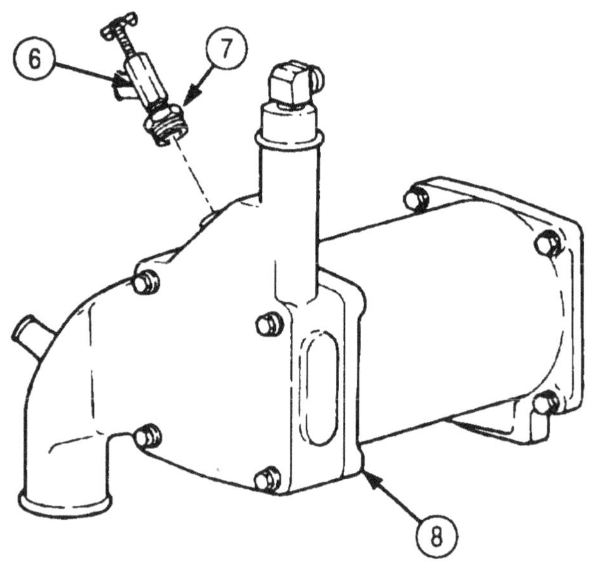

FOLLOW-ON TASK: Install engine and transmission (para. 4-4 or 4-5).

4-8. STARTING REPAIRED OR REPLACEMENT ENGINE

THIS TASK COVERS:

a. Priming Lubrication System (M939/A1)
b. Out-of-Chassis Run-In (M939/A1)
c. In-Chassis Run-In

INITIAL SETUP:

APPLICABLE MODELS
All

SPECIAL TOOLS
Priming pump (M939/A1 (Appendix E, Item 94)
Spring pack adjusting tool (M939/A1) (Appendix E, Item 130)

TOOLS
General mechanic's tool kit (Appendix E, Item 1)

MATERIALS/PARTS
Special seal (M939/A1) (Appendix D, Item 612)

PERSONNEL REQUIRED
TWO

REFERENCES (TM)
LO 9-2320-272-12
TM 9-2320-272-10
TM 9-2320-272-24P

GENERAL SAFETY INSTRUCTIONS
Engine compartment must be clear of tools and other materials before starting engine.

CAUTION

Engine lubrication system cannot be primed through bypass filter.

1. Remove pipe plug (3) from gearcase cover flange (1) on left side of engine (2).
2. Connect oil priming pump to orifice (4) in gearcase cover flange (1).
3. Prime until 30 psi (207 kPa) of pressure is obtained.

CAUTION

Do not crank engine continuously for more that 30 seconds. Wait 2-5 minutes before repeating to prevent starter motor damage.

4. Close fuel shutoff valve and crank engine 15 seconds while maintaining 15 psi (103 kPa) pump pressure.
5. Disconnect oil priming pump and replace pipe plug (3) in gearcase cover flange (1). Tighten pipe plug (3) 60-70 (81-95 N•m).
6. Check oil level (TM 9-2320-272-10).

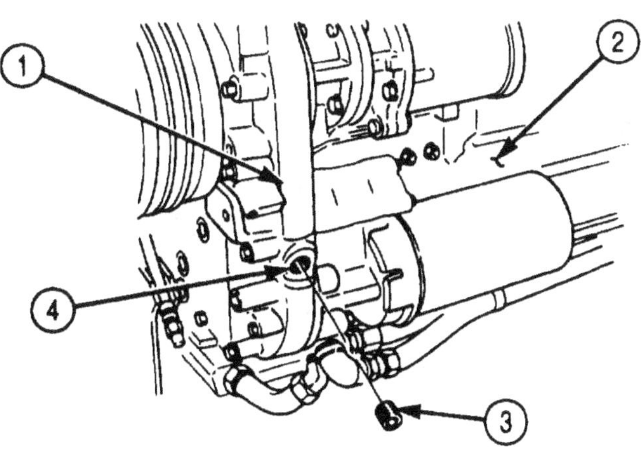

4-8. STARTING REPAIRED OR REPLACEMENT ENGINE (Contd)

b. Out-of-Chassis Run-In

NOTE

If engine dynamometer is available, follow manufacturer's instructions. Ensure dynamometer used is torque-rated at 685 lb-ft (929 N•m). If a dynamometer is not available, proceed to task c. of this paragraph.

1. Connect engine to dynamometer.
2. Start and run engine until oil temperature is 180°F (124°C).
3. Stop engine, inspect oil level, and check for leaks (TM 9-2320-272-10).

NOTE

Coolant should not exceed 195°F (90°C) or drop below 175°F (79°C) during engine load operation.

4. Start and run engine at 1,575 rpm, 125 hp, for 15 minutes.
5. Run engine at 2,100 r-pm, 188 hp, for 15 minutes.
6. Run engine at 2,100 rpm, 213 hp, for 15 minutes.
7. Run engine at 2,100 rpm, 225 hp, for 15 minutes.
8. Run engine at 2,100 rpm, 240 hp, for 15 minutes.
9. Remove engine load and idle until temperature drops.
10. Stop engine, inspect oil level, and check for leaks (refer to TM 9-2320-272-10).
11. Disconnect engine from dynamometer.

c. In-Chassis Run-In

WARNING

Ensure engine compartment is clear of tools and other materials before starting engine. Failure to do so may cause injury to personnel.

NOTE

Steps 1 through 6 apply to M939A2 only; steps 7 through 13 apply to M939fAl only.

1. Inspect engine coolant and oil levels (TM 9-2320-272-10).
2. Pull emergency stop control (4) out all the way to prevent engine from starting while priming.

CAUTION

- Do not operate M939A2 series vehicle starters continuously for more than 30 seconds at a time. Wait two minutes between periods of starter operation. Failure to do so may cause damage to equipment.
- If oil pressure drops below 10 psi (68 kPa) or rises sharply above 30 psi (207 kPa), stop engine and correct as required.

3. Crank engine until oil pressure registers on oil pressure gauge (3).
4. Push emergency stop control (4) in all the way.
5. Start and run engine at 1,000-2,000 rpm as indicated on tachometer (1) for 30 minutes, observing oil pressure (3) and water temperature (2) gauges for proper ranges (TM 9-2320-272-10).
6. Inspect engine for leaks (TM 9-2320-272-10).

4-8. STARTING REPAIRED OR REPLACEMENT ENGINE (Contd)

7. Start and idle engine at 800-1,000 rpm for 5-10 minutes, observing oil pressure (TM 9-2320-272-10). Reset throttle and proceed to step 14.

CAUTION

If oil pressure drops below 10 psi (68 kPa), or rises sharply above 30 psi (207 kPa), stop engine and correct as required.

8. Stop engine and inspect coolant and oil levels (TM 9-2320-272-10).
9. Inspect engine for leaks (TM 9-2320-272-10).
10. Start and run engine at 1/4-1/2 throttle until water temperature gauge (2) reaches 165°-195°F (73°-90°C).
11. Stop engine and inspect coolant and oil levels (TM 9-2320-272-10).
12. Inspect engine for leaks (TM 9-2320-272-10).
13. Start and idle engine at 800-1,000 rpm as indicated on tachometer (1) for 5-10 minutes, observing oil pressure gauge (3) for proper range (TM 9-2320-272-10).

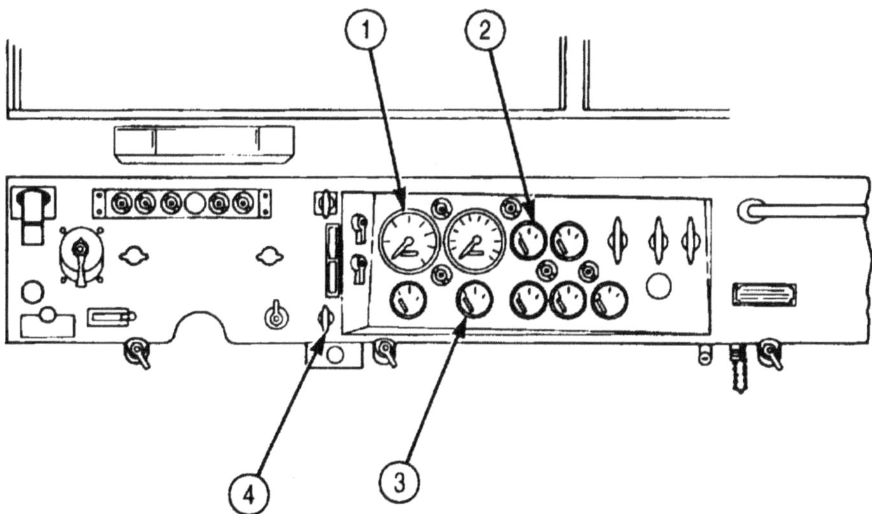

4-8. STARTING REPAIRED OR REPLACEMENT ENGINE (Contd)

NOTE

Coolant should not exceed 195°F (90°C) or drop below 175°F (79°C).

14. Check engine idle speed. If idle speed is not 625 ± 25 rpm on M939/A1 series vehicles, perform steps 15 through 20. If idle speed is not 550-650 rpm on M939A2 series vehicles, perform steps 21 through 25.
15. Break special seal (2) and remove pipe plug (3) from governor spring pack cover (4). Discard special seal (2).
16. Using spring pack adjusting tool, turn adjusting screw (5) in to increase idle speed; out to decrease idle speed. Correct idle speed is 625 2 25 rpm.
17. Install pipe plug (3) on spring pack cover (4).
18. Thread wire of new special seal (2) through pipe plug (3) and two hex-head cover screws (1), and twist wire of special seal (2) until secure.

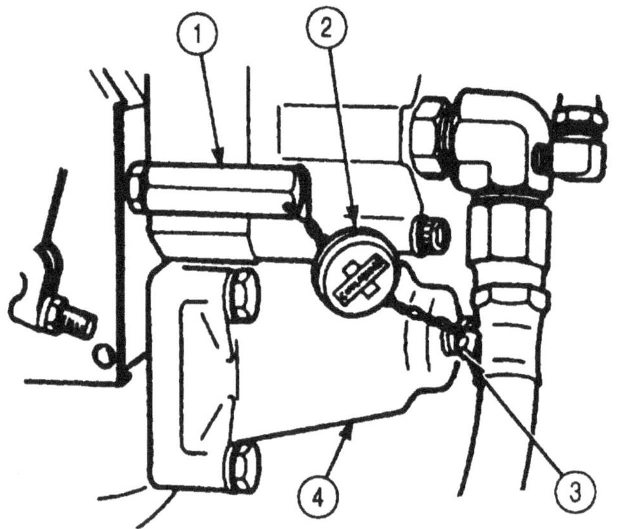

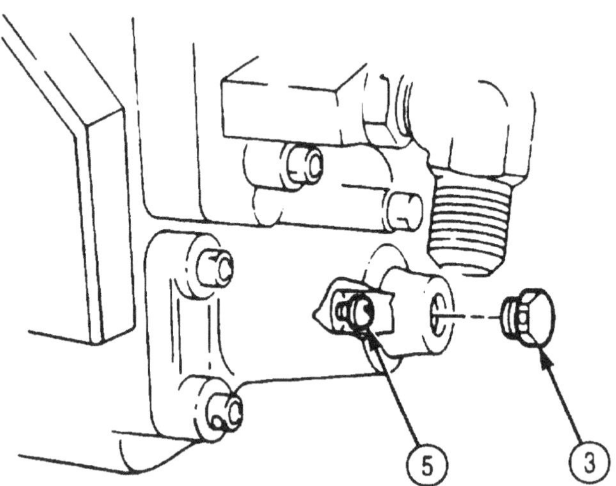

4-8. STARTING REPAIRED OR REPLACEMENT ENGINE (Contd)

19. Step engine and inspect coolant and oil levels (TM 9-2320-272-10).
20. Inspect engine for leaks (TM 9-2320-272-10).
21. Loosen locknut (8) on accelerator linkage (7).
22. Turn throttle rod (6) clockwise to increase idle speed; counterclockwise to decrease idle speed.
23. Tighten locknut (8) on accelerator linkage (7).
24. Step engine and inspect coolant and oil levels (TM 9-2320-272-10).
25. Inspect engine for leaks (TM 9-2320-272-10).

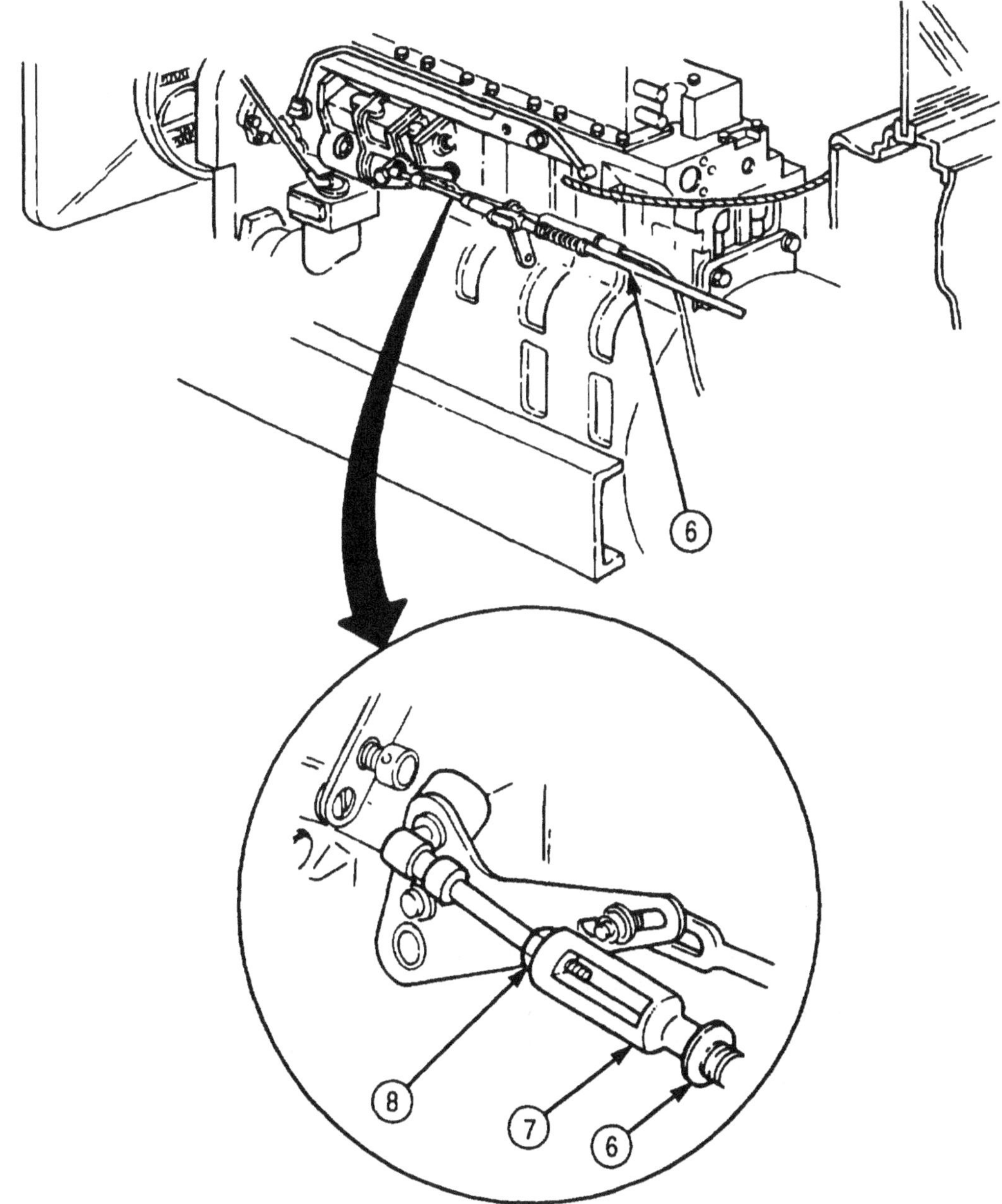

FOLLOW-ON TASK: Change engine oil after first 500 mi (805 km) (LO 9-2320-272-12).

4-9. ENGINE MOUNTING ON REPAIR STAND

THIS TASK COVERS:

a. Installation (M939/A1)
b. Installation (M939A2)
c. Removal (M939A2)
d. Removal (M939/A1)

INITIAL SETUP:

APPLICABLE MODELS
All

SPECIAL/TOOLS
Engine repair stand (Appendix E, Item 44)
Engine barring tool (M939A2) (Appendix E, Item 42)

TOOLS
General mechanic's tool kit (Appendix E, Item 1)

MATERIALS/PARTS
Gasket (M939A2) (Appendix D, Item 163)
Gasket (M939A2) (Appendix D, Item 139)
Gasket (Appendix D, Item 167)
Two seal washers (M939A2) (Appendix D, Item 644)
Banjo seal (M939A2) (Appendix D, Item 9)
Three screw-assembled lockwashers (M939/A1) (Appendix D, Item 585)
Three lockwashers (M939/A1) (Appendix D, Item 382)
Eleven lockwashers (M939/A1) (Appendix D, Item 354)
Two locknuts (M939/A1) (Appendix D, Item 281)
Locknut (M939/A1) (Appendix D, Item 327)
Gasket (M939/A1) (Appendix D, Item 156)
Gasket (M939/A1) (Appendix D, Item 155)
Twelve key washers (M939/A1) (Appendix D, Item 267)

MATERIALS/PARTS (Contd)
Six gaskets (M939/A1) (Appendix D, Item 168)
Packing sleeve (M939/A1) (Appendix D, Item 21)
Gasket (M939/A1) (Appendix D, Item 180)
Four lockwashers (M939/A1) (Appendix D, Item 416)
Antiseize tape (M939A2) (Appendix C, Item 72)
Gasket sealant (M939A2) (Appendix C, Item 30)

REFERENCES (TM)
TM 9-2320-272-24P

EQUIPMENT CONDITION
- Engine and transmission removed (M939/A1) (para. 4-4).
- Engine and transmission removed (M939A2) (para. 4-5).
- Transmission removed (para. 4-72).
- Engine steam-cleaned (para. 2-15).

GENERAL SAFETY INSTRUCTIONS
- Tilt rear of transmission slightly downward to prevent torque converter from sliding off.
- Engine must be securely mounted on repair stand.
- Ensure engine weight is evenly distributed.
- Lifting device must have a capacity greater than the combined weight of the engine and transmission.
- All personnel must stand clear during lifting operations.

NOTE

Spacer will be present on late model engines only.

1. Remove screw (1), lockwasher (2), washer (3), screw (9), lockwasher (8), washer (71, adjustable link (5), spacer (6), and washer (4) from alternator (18). Discard lockwashers (2) and (8).
2. Remove two screws (16), locknuts (11), washers (12), alternator belt (10), and alternator (18) from engine block (19). Discard locknuts (11).
3. Remove four screws (15), lockwashers (14), washers (13), and mounting bracket (17) from engine block (19). Discard lockwashers (14).

4-9. ENGINE MOUNTING ON REPAIR STAND (Contd)

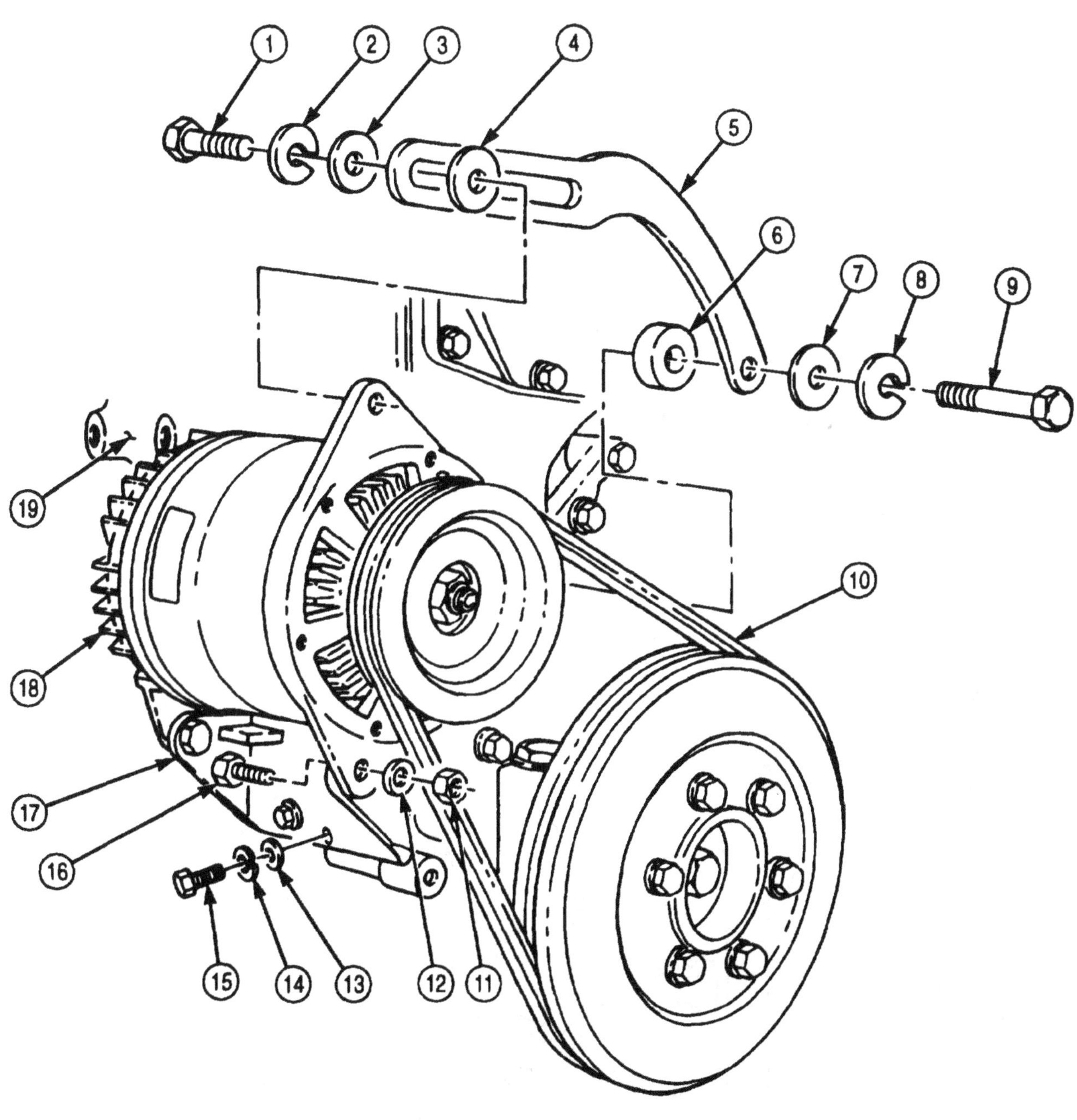

4-9. ENGINE MOUNTING ON REPAIR STAND (Contd)

4. Remove three screw-assembled lockwashers (1) and starter (2) from flywheel housing (3). Discard screw-assembled lockwashers (1).
5. Remove gasket (6), spacer (5), and gasket (4) from flywheel housing (3). Discard gaskets (6) and (4) and clean remains of gaskets (6) and (4) from mating surfaces.
6. Remove locknut (15), washer (16), screw (19), and washer (18) from power steering pump bracket (10) and adjustable link (17). Discard locknut (15).
7. Remove two screws (13), lockwashers (12), and washers (11) from power steering pump bracket (10). Discard lockwashers (12).
8. Remove two drivebelts (7) from pulley (8).
9. Remove power steering pump bracket (10) and power steering pump (9) from engine bracket (14).
10. Remove four screws (24), lockwashers (25), washers (23), engine access cover (22), and gasket (21) from engine block (20). Discard lockwashers (25) and gasket (21) and clean remains of gasket (21) from mating surfaces.
11. Remove screw (30), lockwasher (31), and breather tube mounting bracket (29) from engine block (20). Discard lockwasher (31).
12. Loosen hose clamp (27) and remove breather tube (26) from breather hose (28).

NOTE

Perform step 14 for late model engines only.

13. Loosen hose clamps (33) and (35) and remove breather tube (34) from breather (32) and elbow (36).

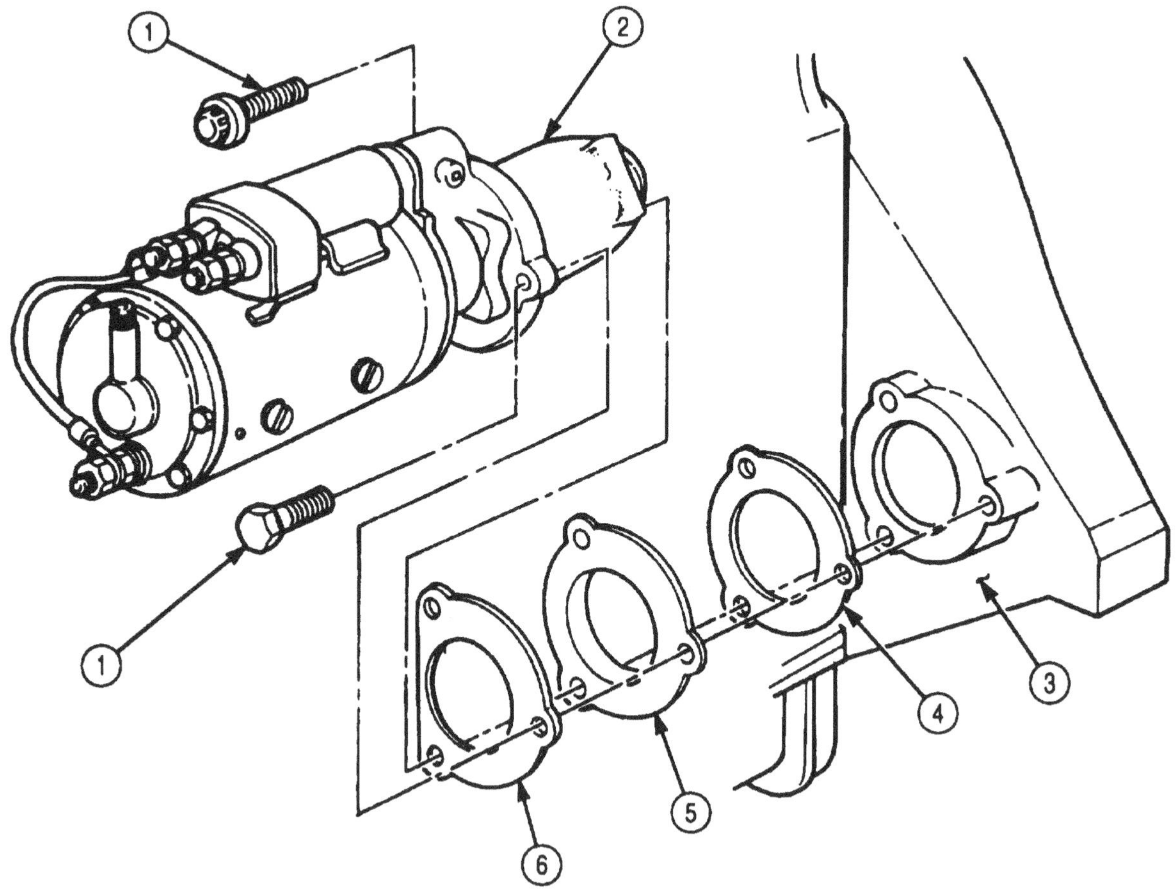

4-9. ENGINE MOUNTING ON REPAIR STAND (Contd)

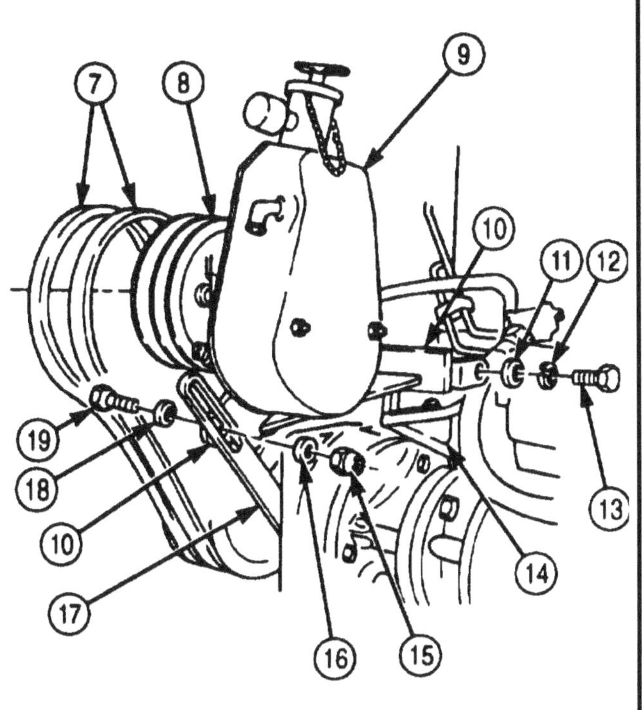

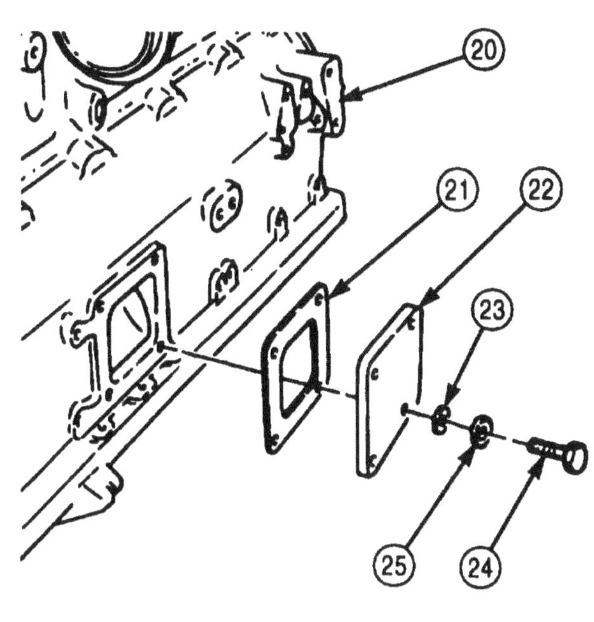

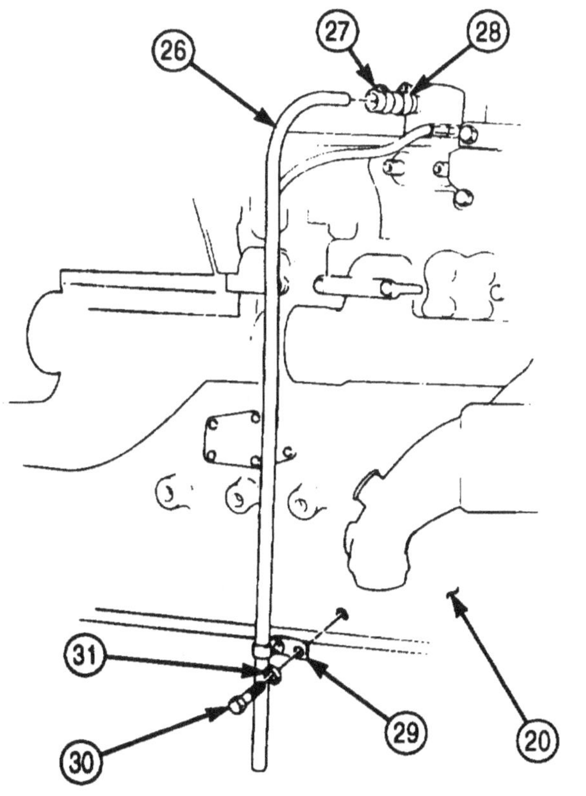

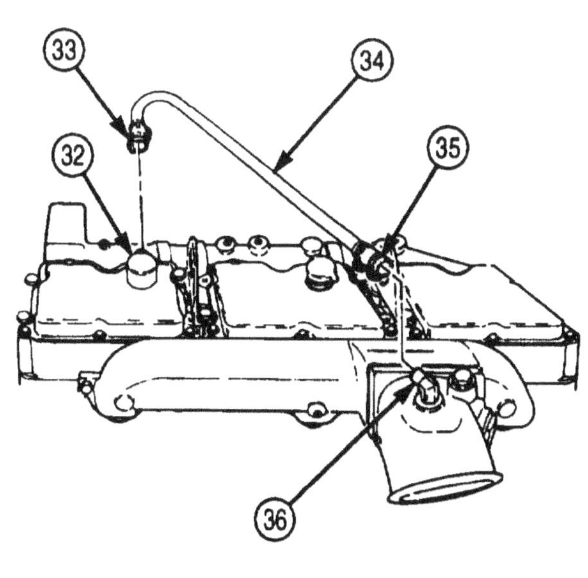

4-9. ENGINE MOUNTING ON REPAIR STAND (Contd)

14. Remove twelve screws (4), locktabs (5), six clamps (6), exhaust manifold (2), and six gaskets (3) from three cylinder heads (1). Discard locktabs (5) and gaskets (3).
15. Disconnect air compressor coolant line (9) and packing sleeve (8) from oil cooler elbow (7). Discard packing sleeve (8).
16. Remove five screws (13), lockwashers (12), oil cooler (14), and gasket (11) from cylinder block (10). Discard lockwashers (12) and gasket (11).

WARNING

- Lifting device must have a weight capacity greater than the combined weight of the engine and transmission to prevent injury or death to personnel and damage to equipment.
- All personnel must stand clear during lifting operations. A snapped cable, or swinging or shifiing load, may result in injury or death to personnel.
- Engine must be securely mounted on repair stand. Failure to do so may result in injury or death to personnel.

17. Attach utility chain to two engine lifting eyes (20).
18. Attach lifting device to utility chain.

NOTE

Assistant will help with step 20.

19. Position engine (16) and two spacers (15) against engine stand, and install with four washers (17) and bolts (18). Lower two bolts are installed in engine access holes (19).
20. Remove lifting device and utility chain.

NOTE

- For disassembly, cleaning, inspection and reassembly of oil cooler, refer to para. 4-23.
- For disassembly, cleaning, inspection and reassembly of power steering pump, refer to para. 4-126 or 4-127.

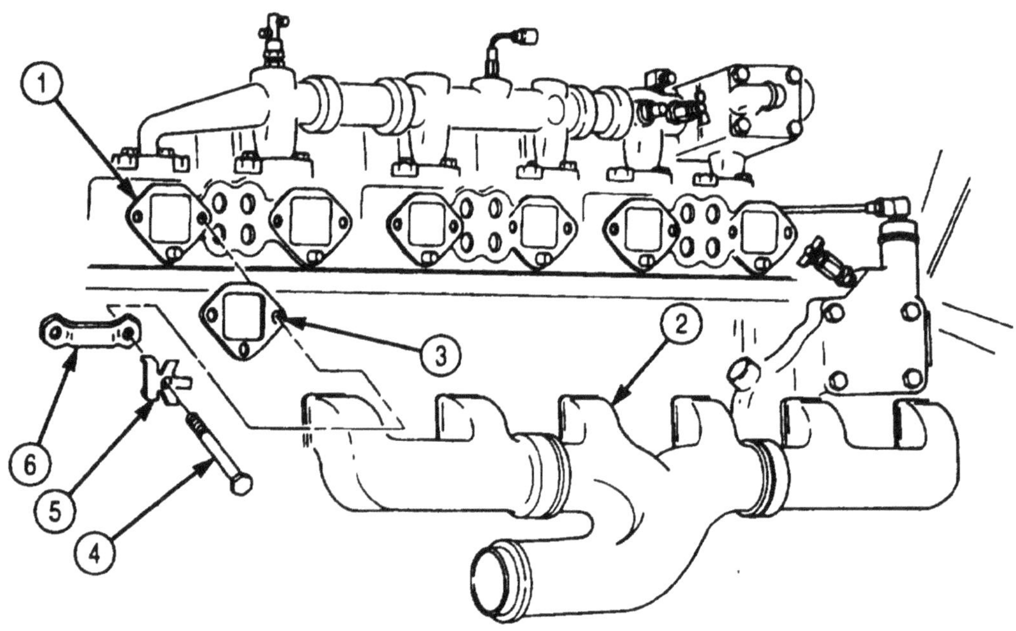

4-9. ENGINE MOUNTING ON REPAIR STAND (Contd)

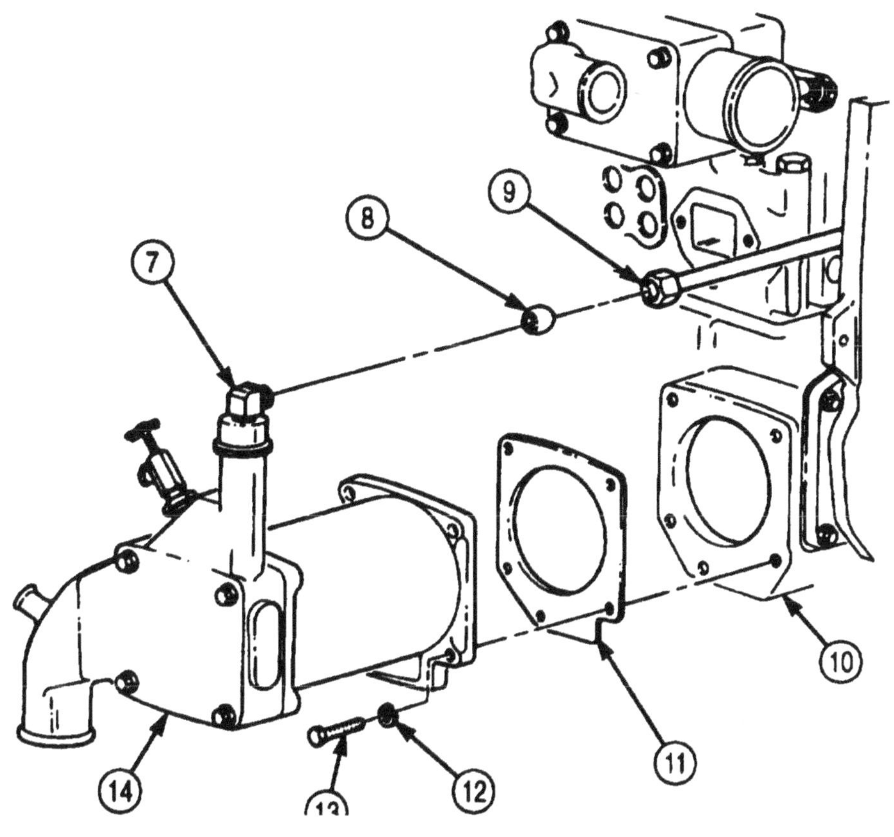

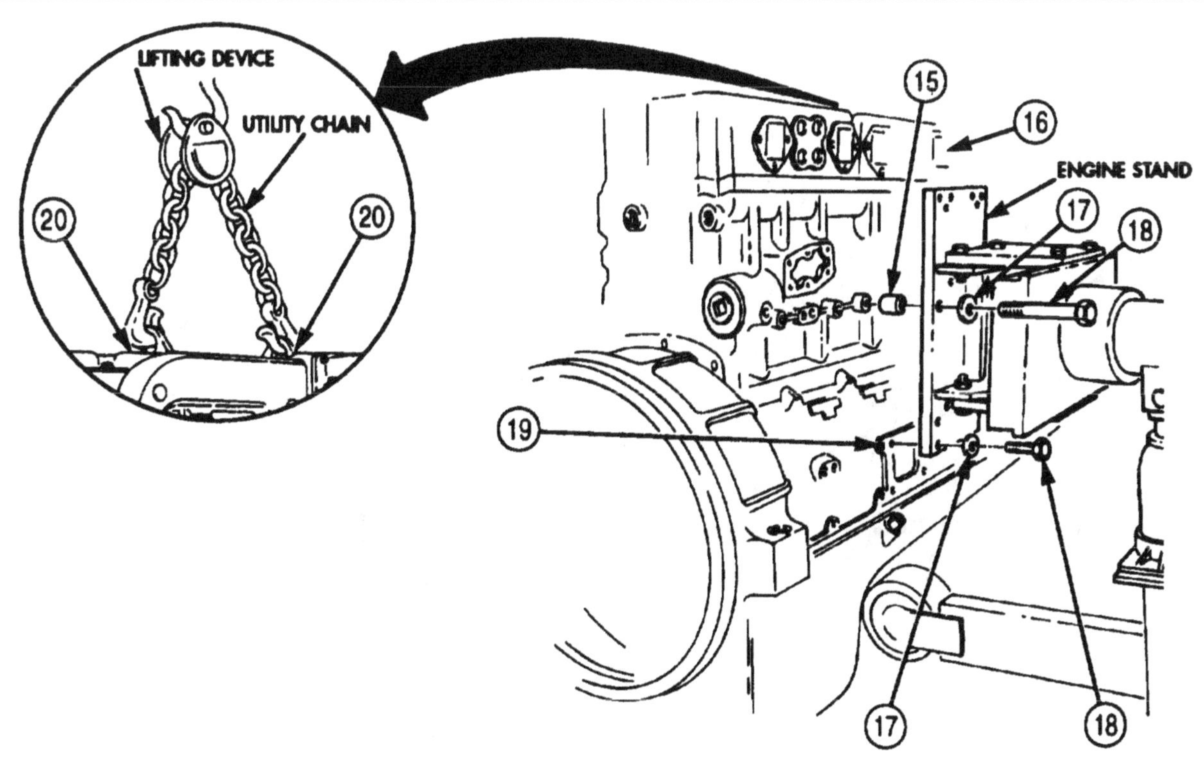

4-9. ENGINE MOUNTING ON REPAIR STAND (Contd)

b. Installation (M939A2)

1. Remove screw (5), fuel return manifold (4), and banjo seal (6) from screw (3). Discard banjo seal (6).
2. Remove screw (3) and two washers (2) from engine block (1). Discard washers (2).
3. Remove two screws (9), fuel transfer pump (8), fuel supply tube (7), and gasket (10) from engine block (1). Discard gasket (10).
4. Remove two screws (14), washers (15), power steering pump (13), and gasket (12) from air compressor (11). Discard gasket (12).
5. Remove oil sampling valve (19) from engine block (1).
6. Remove oil pressure sending unit (18) from engine block (1).
7. Disconnect nut (23) from elbow (22) and move oil line (28) aside.
8. Remove elbow (22) from engine block (1).
9. Remove screw (21) from air compressor (11).
10. Loosen two clamps (17) and remove air inlet line (20) from fitting (16) and air compressor (11).
11. Remove screw (24) and washer (25) from air compressor (11).
12. Remove screw (26) and bracket (27) from engine block (1).

WARNING

- Lifting device must have a weight capacity greater than the combined weight of the engine and transmission to prevent injury or death to personnel and damage to equipment.
- All personnel must stand clear during lifting operations. A snapped cable, or swinging or shifting load, may result in injury or death to personnel.
- Engine must be securely mounted on repair stand. Failure to do so may result in injury or death to personnel.

13. Attach utility chains and suitable lifting device to two engine lifting brackets (29).
14. Using six bolts (30), install engine block (1) on engine stand. Tighten six bolts (30) securely.
15. Remove lifting device from two engine lifting brackets (29).

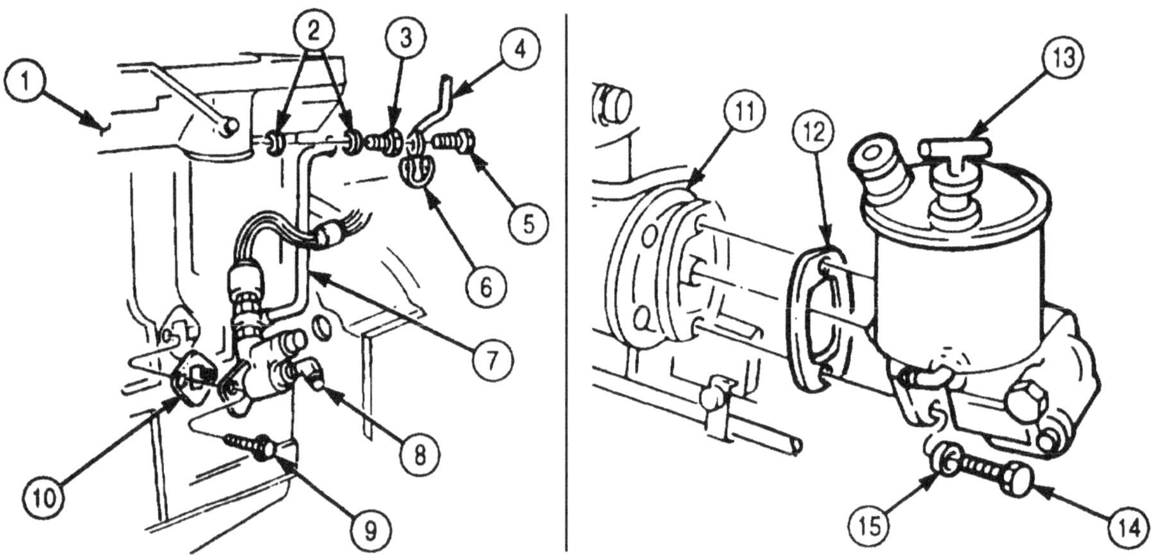

4-9. ENGINE MOUNTING ON REPAIR STAND (Contd)

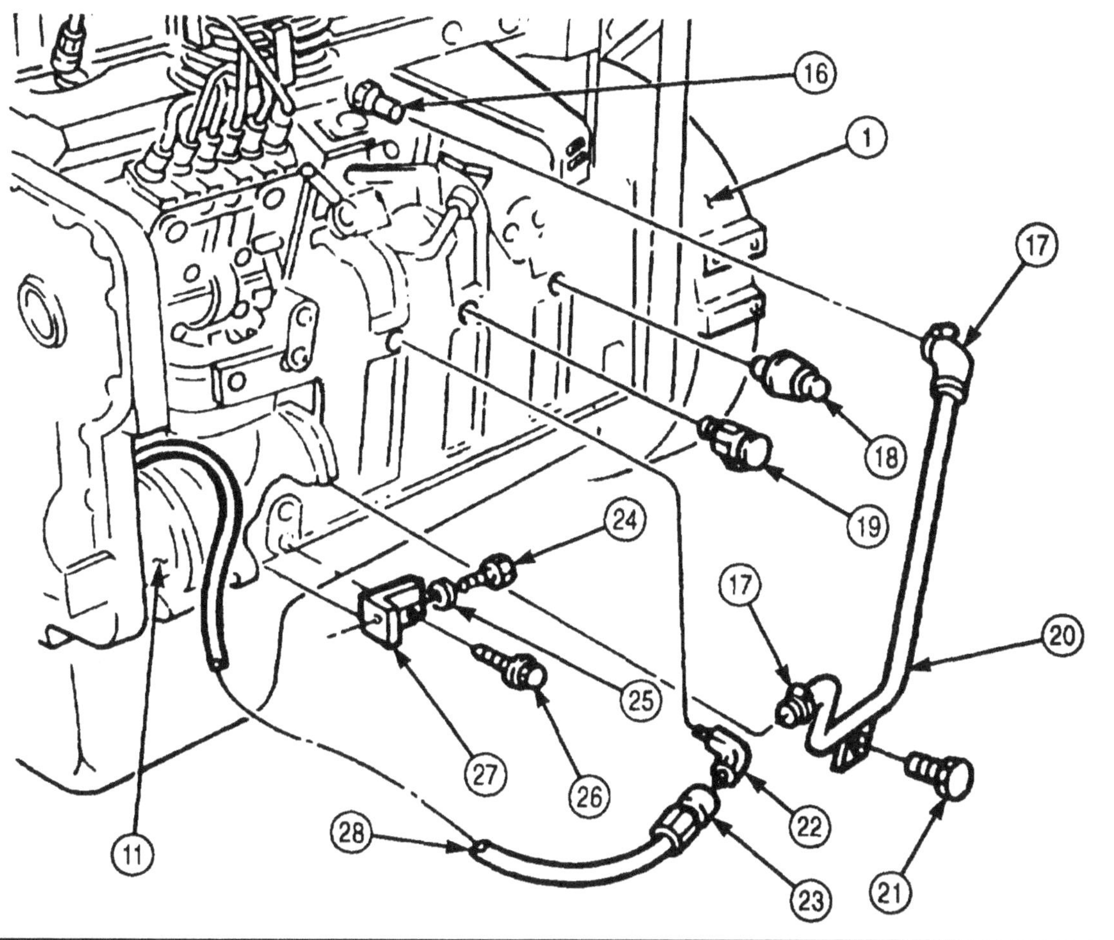

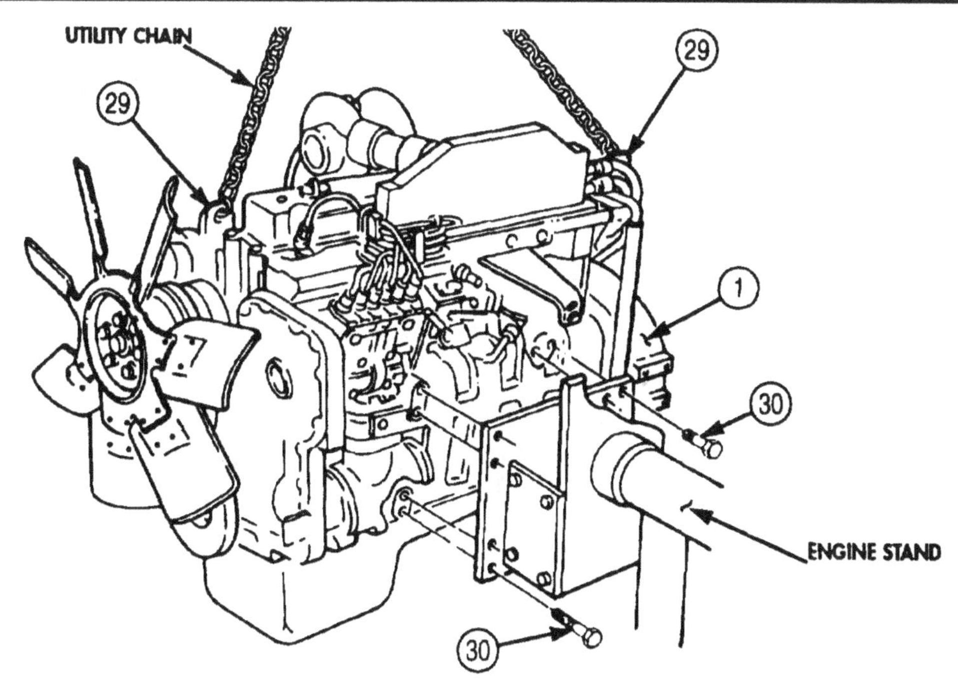

4-9. ENGINE MOUNTING ON REPAIR STAND (Contd)

c. Removal (M939A2)

WARNING

All personnel must stand clear during lifting operations. A snapped cable, or swinging or shifting load, may cause death or injury to personnel.

1. Install utility chains and lifting device on two lifting brackets (1).
2. Remove six bolts (3) and engine block (2) from engine stand.

WARNING

Ensure engine weight is evenly distributed. Failure to do so may result in injury or death to personnel.

3. Lower engine block (2) from engine stand.
4. Install bracket (15) on engine block (2) with screw (14).
5. Install bracket (15) on air compressor (17) with washer (13) and screw (12).
6. Install air inlet line (8) on air compressor (17) and fitting (4) with screw (9) and tighten clamps (5).

NOTE

Wrap male threads of elbow with antiseize tape.

7. Install elbow (10) on engine block (2).
8. Install air compressor oil line (16) on elbow (10) and tighten nut (11).
9. Apply antiseize tape to threads of oil pressure sending unit (6) and install on engine block (2).
10. Apply antiseize tape to threads of oil sampling valve (7) and install on engine block (2).

NOTE

Ensure all matting surfaces are clean before installing new gaskets.

11. Install new gasket (18) and power steering pump (19) on air compressor (17) with two washers (21) and screws (20).
12. Install new gasket (30) and fuel transfer pump (28) on engine block (2) with two screws (29).
13. Install fuel supply tube (27) on engine block (2) with two new washers (22) and screw (23).
14. Install fuel return manifold (24) and new banjo seal (26) on screw (23) with screw (25).

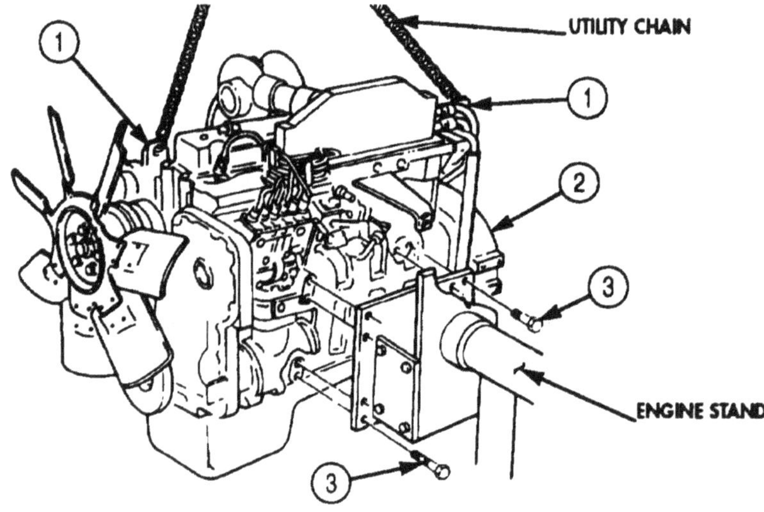

4-9. ENGINE MOUNTING ON REPAIR STAND (Contd)

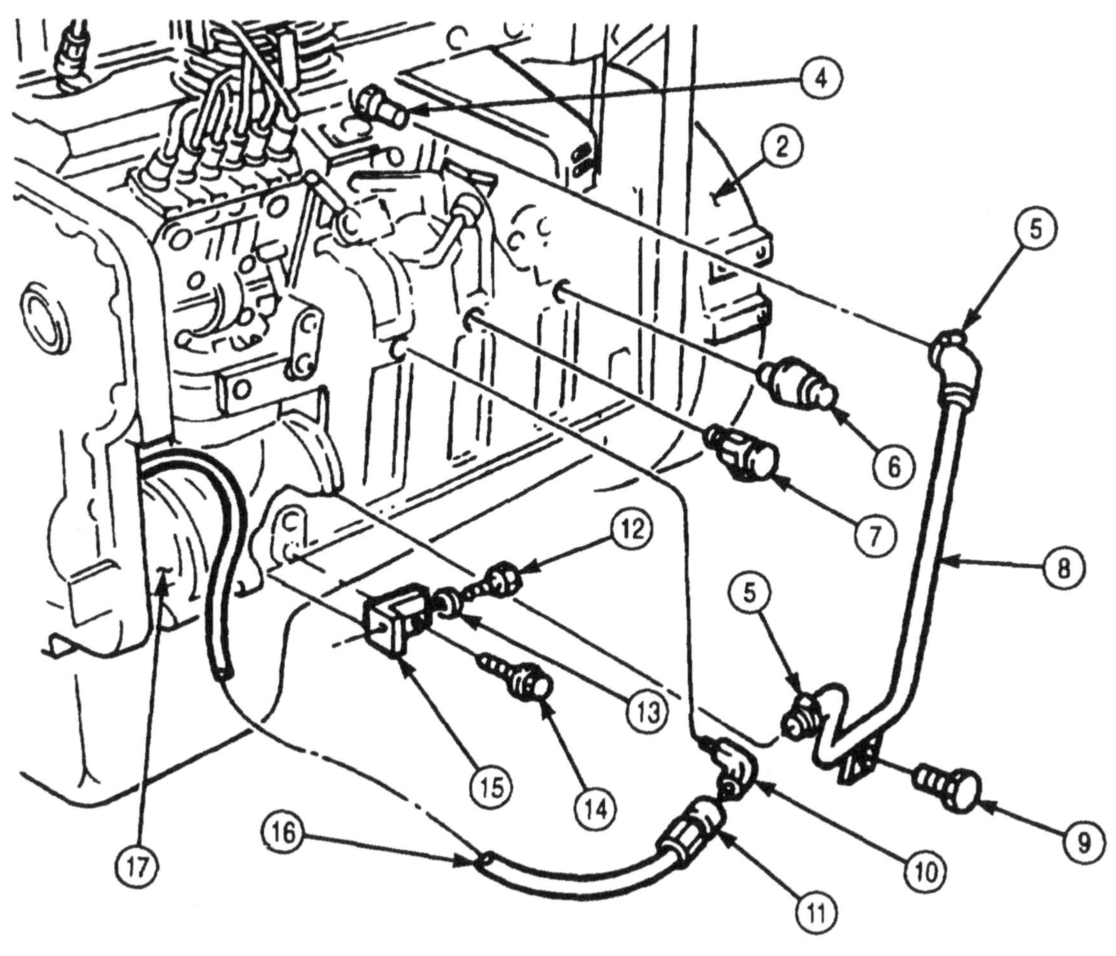

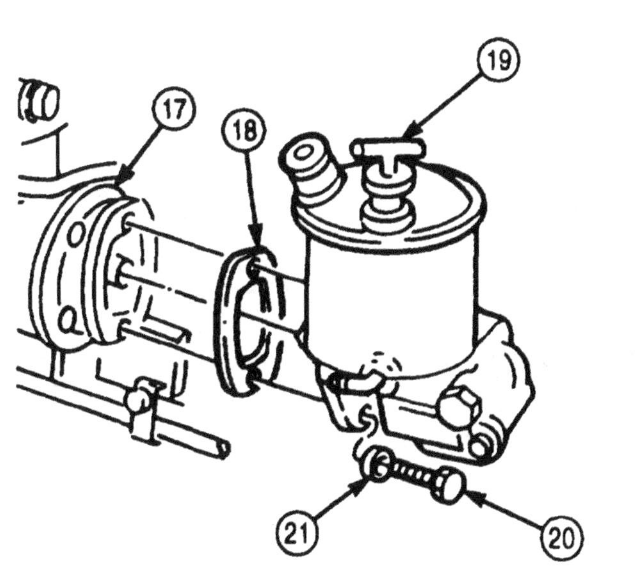

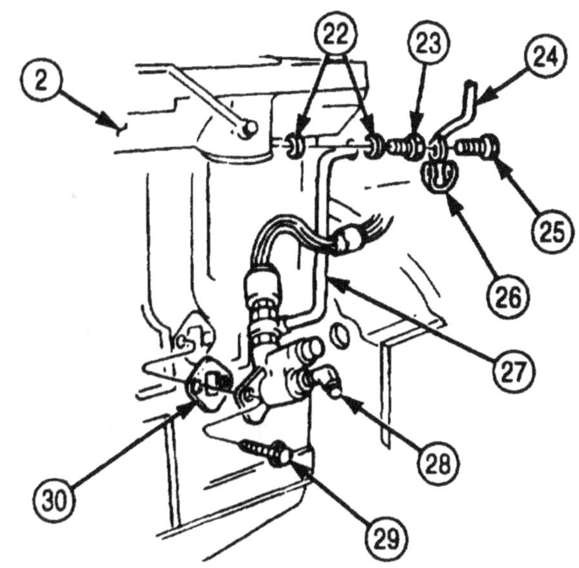

4-9. ENGINE MOUNTING ON REPAIR STAND (Contd)

d. Removal (M939/A1)

WARNING

- Lifting device must have a weight capacity greater than the combined weight of the engine and transmission to prevent injury or death to personnel and damage to equipment.
- All personnel must stand clear during hoisting operations. A snapped cable, or swinging or shifting load, may result in injury or death to personnel.

1. Attach utility chain and lifting device to two engine lifting eyes (5) and remove excess slack from utility chains.

NOTE

Assistant will help with steps 1 and 2.

2. Remove four bolts (2), washers (1), engine (4), and two spacers (3) from engine stand.
3. Position and mount engine (4) on engine transporter.
4. Install oil cooler (6) and new gasket (11) on cylinder block (10) with five new lockwashers (12) and screws (13).
5. Connect air compressor coolant line (9) and new packing sleeve (8) to oil cooler elbow (7).
6. Install exhaust manifold (15) and six new gaskets (16) on three cylinder heads (14) with six clamps (19), twelve new locktabs (18), and screws (17).

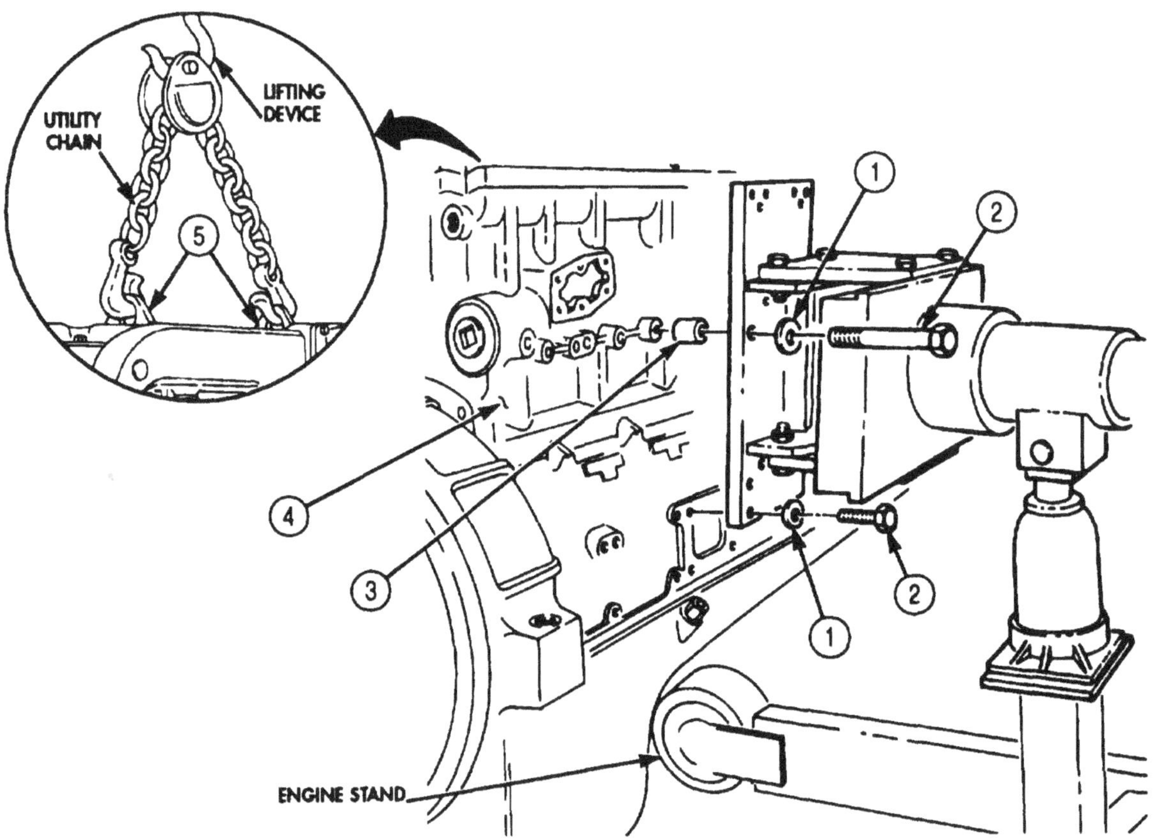

4-9. ENGINE MOUNTING ON REPAIR STAND (Contd)

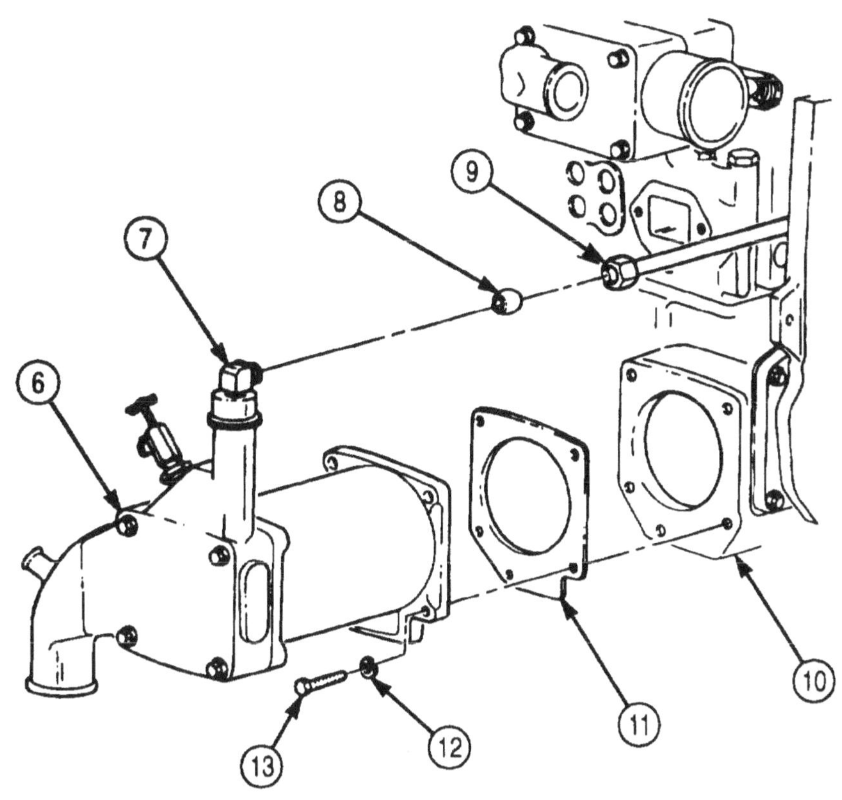

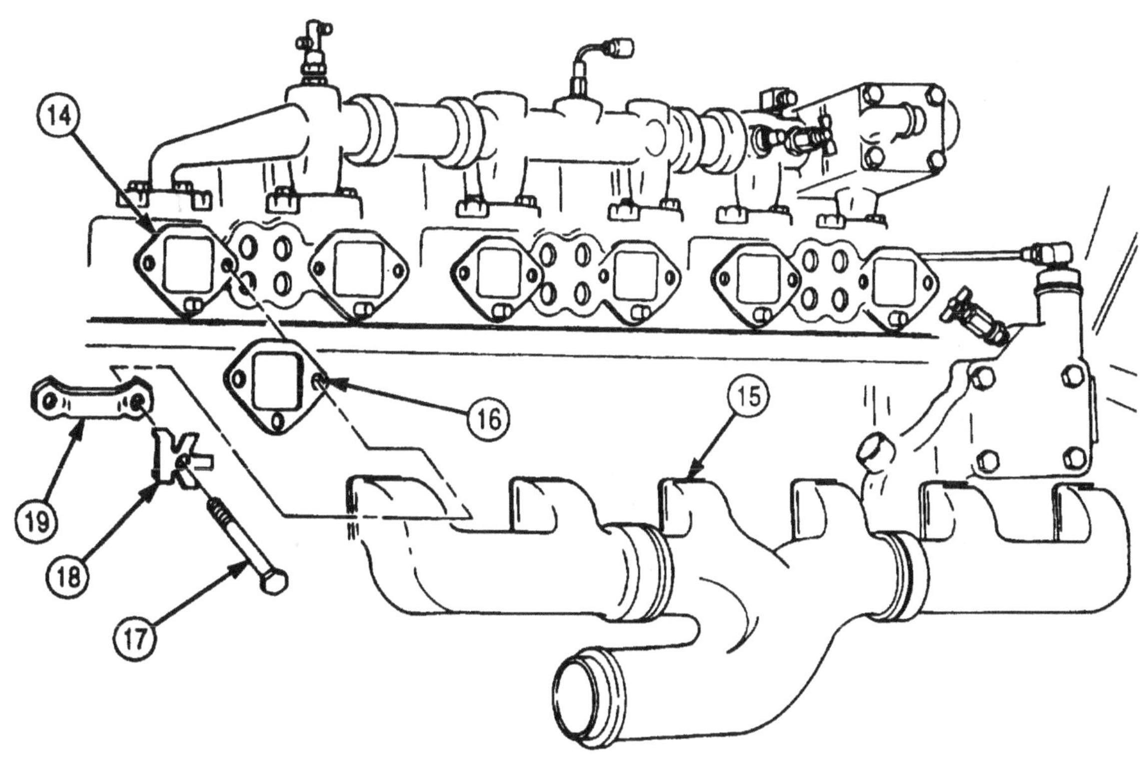

4-9. ENGINE MOUNTING ON REPAIR STAND (Contd)

NOTE

Perform steps 7 and 8 for early model engines and step 9 for late model engines.

7. Install breather tube (1) on breather hose (3) and tighten hose clamp (2).
8. Install breather tube mounting bracket (5) on engine block (4) with new lockwasher (7) and screw (6).
9. Install breather tube (10) on breather (8) and elbow (11) and tighten two hose clamps (9).
10. Install new gasket (12) and engine access cover (13) on engine block (4) with four washers (14), new lockwashers (15), and screws (16).
11. Install power steering pump bracket (23) and power steering pump (19) on engine bracket (24) with two washers (20), new lockwashers (21), and screws (22).
12. Install two drivebelts (17) on pulley (18).
13. Install adjustable link (27) on power steering pump bracket (23) with washer (28), screw (29), washer (26), and new locknut (25).

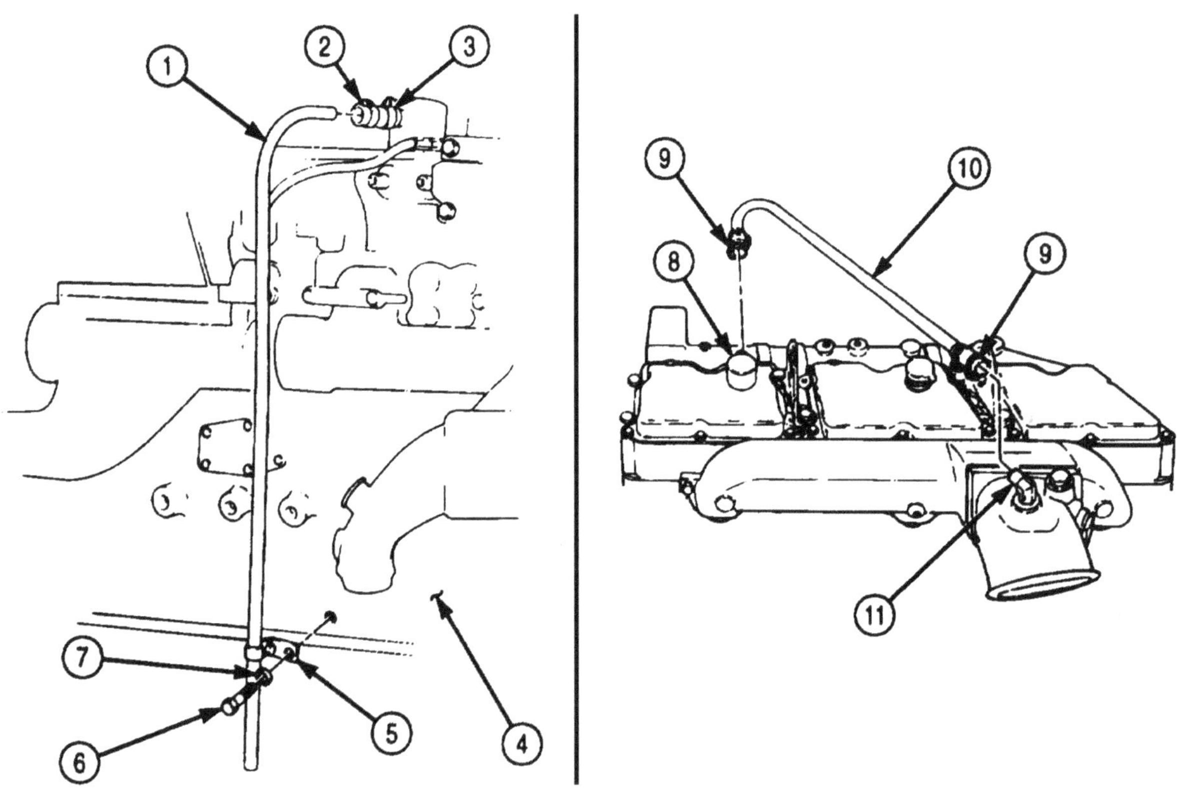

4-9. ENGINE MOUNTING ON REPAIR STAND (Contd)

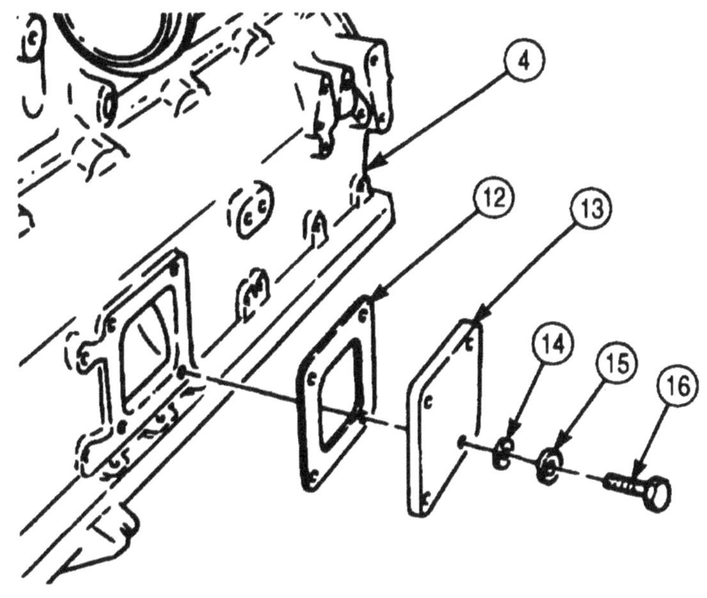

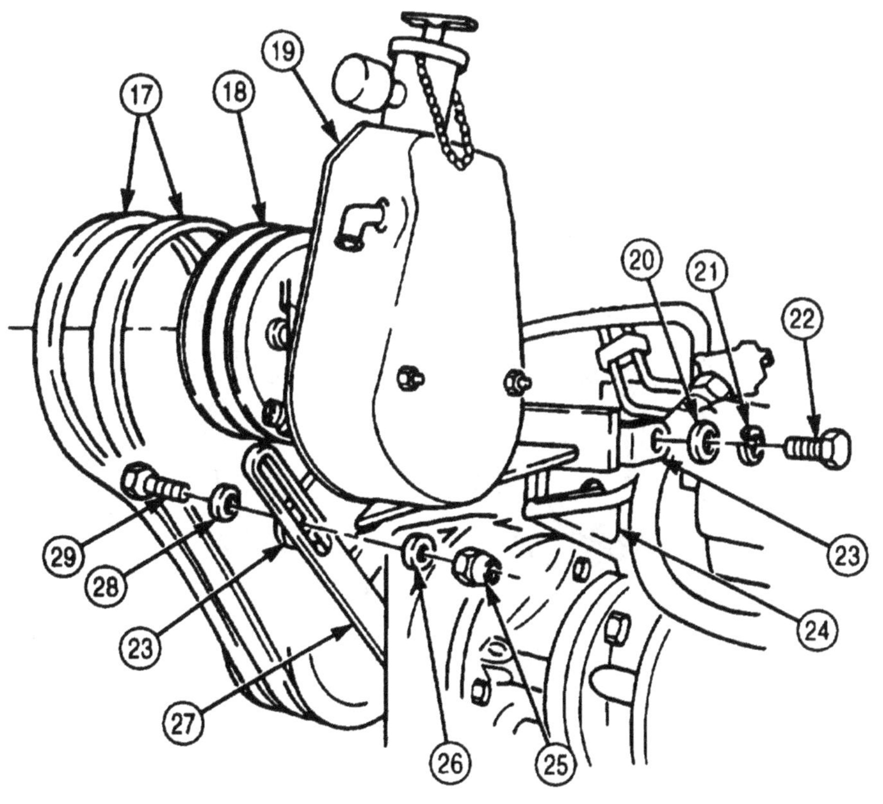

4-9. ENGINE MOUNTING ON REPAIR STAND (Contd)

14. Install new gasket (4), spacer (5), new gasket (6), and starter (2) on flywheel housing (3) with two new screw-assembled lockwashers (7), and new screw-assembled lockwasher (1).
15. Install mounting bracket (23) on engine block (26) with four washers (20), new lockwashers (21), and screws (22).
16. Install alternator belt (17) and alternator (25) on mounting bracket (23) with two screws (24), washers (19), and new locknuts (18).

NOTE

Spacer will be present on late model engines only.

17. Install washer (12), spacer (16), and adjustable link (11) on engine (26) with washer (10), new lockwasher (9), screw (8), washer (131, new lockwasher (14), and screw (15).

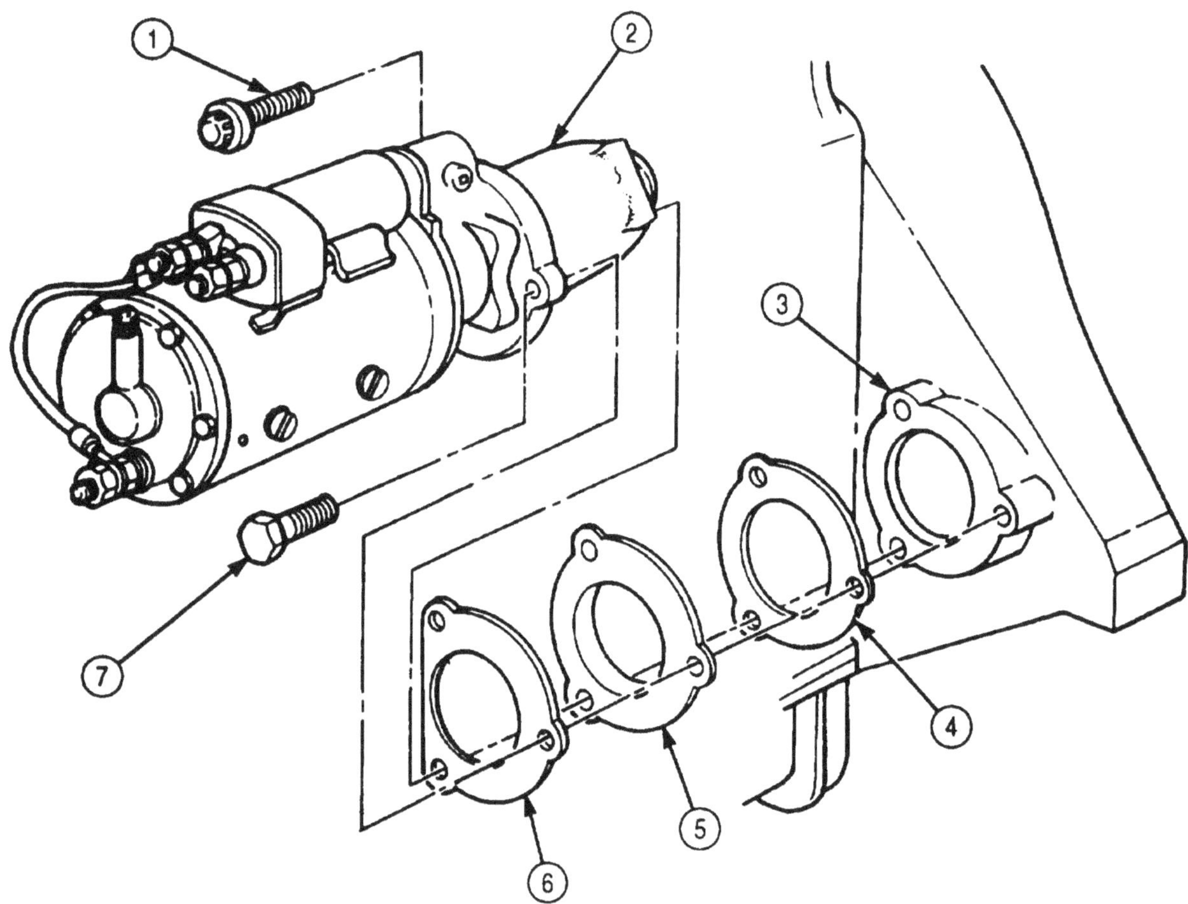

4-9. ENGINE MOUNTING ON REPAIR STAND (Contd)

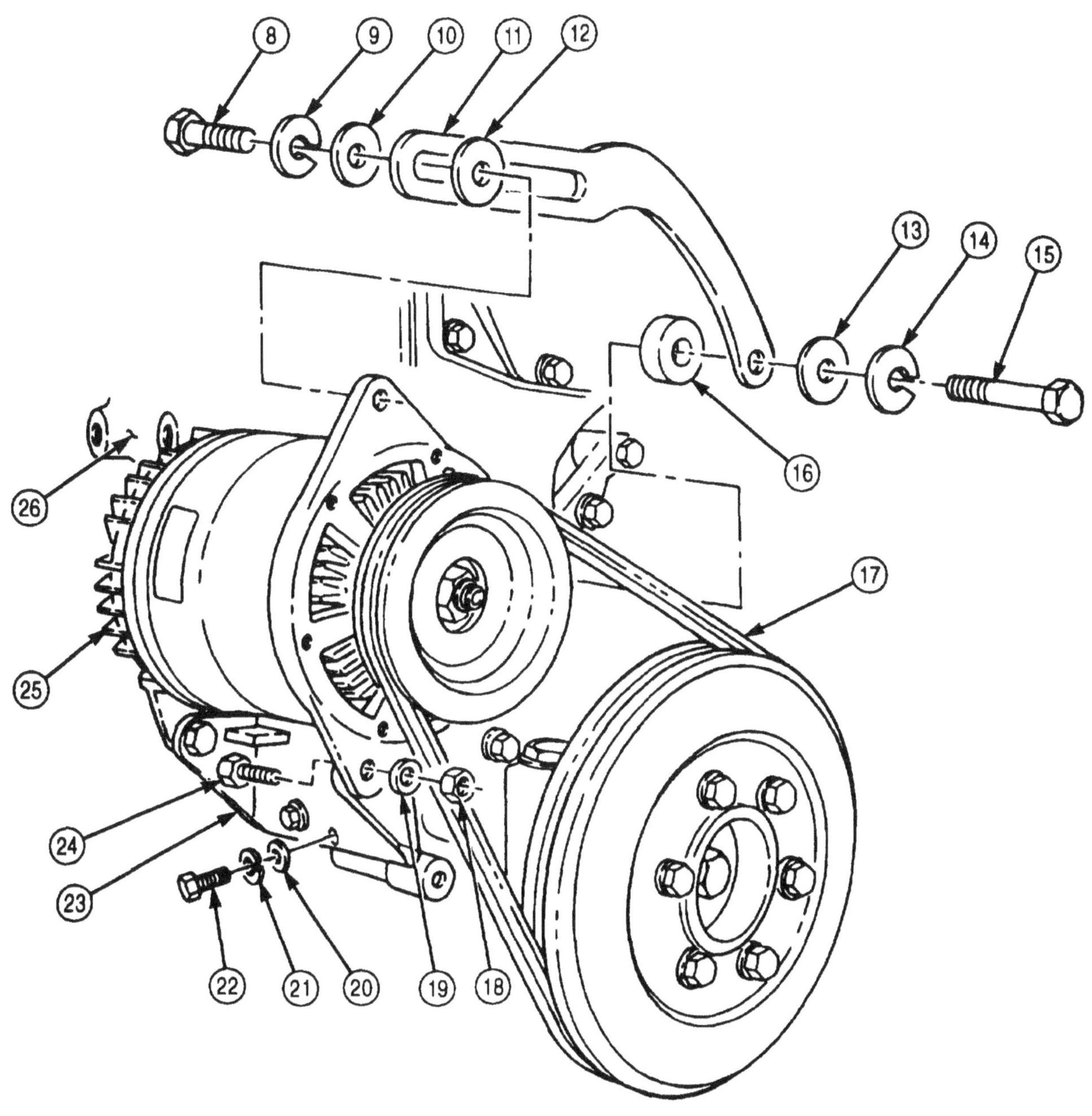

FOLLOW-ON TASKS:
- Install transmission (para. 4-72).
- Install engine and transmission (M939/A1) (para. 4-4).
- Install engine and transmission (M939A2) (para. 4-5).

Section III. ENGINE (M939/A1) MAINTENANCE

4-10. ENGINE (M939/A1) MAINTENANCE INDEX

PARA. NO.	TITLE	PAGE NO.
4-11.	Engine Lift Eyes Replacement	4-118
4-12.	Cylinder Head Replacement	4-120
4-13.	Engine Access Cover Replacement	4-124
4-14.	Crankshaft Vibration and Damper Replacement	4-126
4-15.	Flywheel Ring Gear Maintenance	4-129
4-16.	Flywheel Housing Maintenance	4-130
4-17.	Crankshaft Rear Cover and Seal Replacement	4-136
4-18.	Front Gearcase Cover Maintenance	4-138
4-19.	Cam Follower Housing Maintenance	4-144
4-20.	Rocker Lever Housing and Push Tubes Maintenance	4-154
4-21.	Engine Oil Pump Replacement	4-164
4-22.	Engine Oil Pan Maintenance	4-168
4-23.	Engine Oil Cooler Maintenance	4-174
4-24.	Engine Air Intake Manifold Maintenance	4-178
4-25.	Engine Exhaust Manifold Maintenance	4-184
4-26.	Engine Accessory Drive and Pulley Maintenance	4-186
4-27.	Engine Water Manifold Maintenance	4-194
4-28.	Water Header Plates Replacement	4-198
4-29.	Water Pump Maintenance	4-200
4-30.	Fuel Crossover Connectors and Crossheads Replacement	4-210
4-31.	Air Compressor Maintenance	4-214
4-32.	Engine Injector Timing Instructions	4-236
4-33.	Valve and Injector Adjustment (Dial Indicator Method)	4-248
4-34.	Injector Plunger and Valve Adjustments (Torque Method)	4-254
4-35.	Fuel Pump Replacement	4-260
4-36.	Fuel Pump Shutoff Valve Replacement	4-266
4-37.	Fuel Pump Shutoff Valve (M936) Replacement	4-268

4-11. ENGINE LIFT EYES REPLACEMENT

THIS TASK COVERS:

a. Removal b. Installation

INITIAL SETUP:

APPLICABLE MODELS
M939/A1

TOOLS
General mechanic's tool kit (Appendix E, Item 1)

MATERIALS/PARTS
Locknut (Appendix D, Item 288)

REFERENCES (TM)
TM 9-2320-272-10
TM 9-2320-272-24P

EQUIPMENT CONDITION
- Parking brake set (TM 9-2320-272-10).
- Left and right splash shields removed (TM 9-2320-272-10).

a. Removal

1. Remove locknut (3) and screw (2) from surge tank support (1) and rear lift eye (4). Discard locknut (3).
2. Remove four screws (6) and two lift eyes (4) and (7) from rocker lever housings (5).

b. Installation

1. Install two lift eyes (4) and (7) on rocker lever housings (5) with four screws (6).
2. Install surge tank support (1) on rear lift eye (4) with screw (2) and new locknut (3).

4-11. ENGINE LIFT EYES REPLACEMENT (Contd)

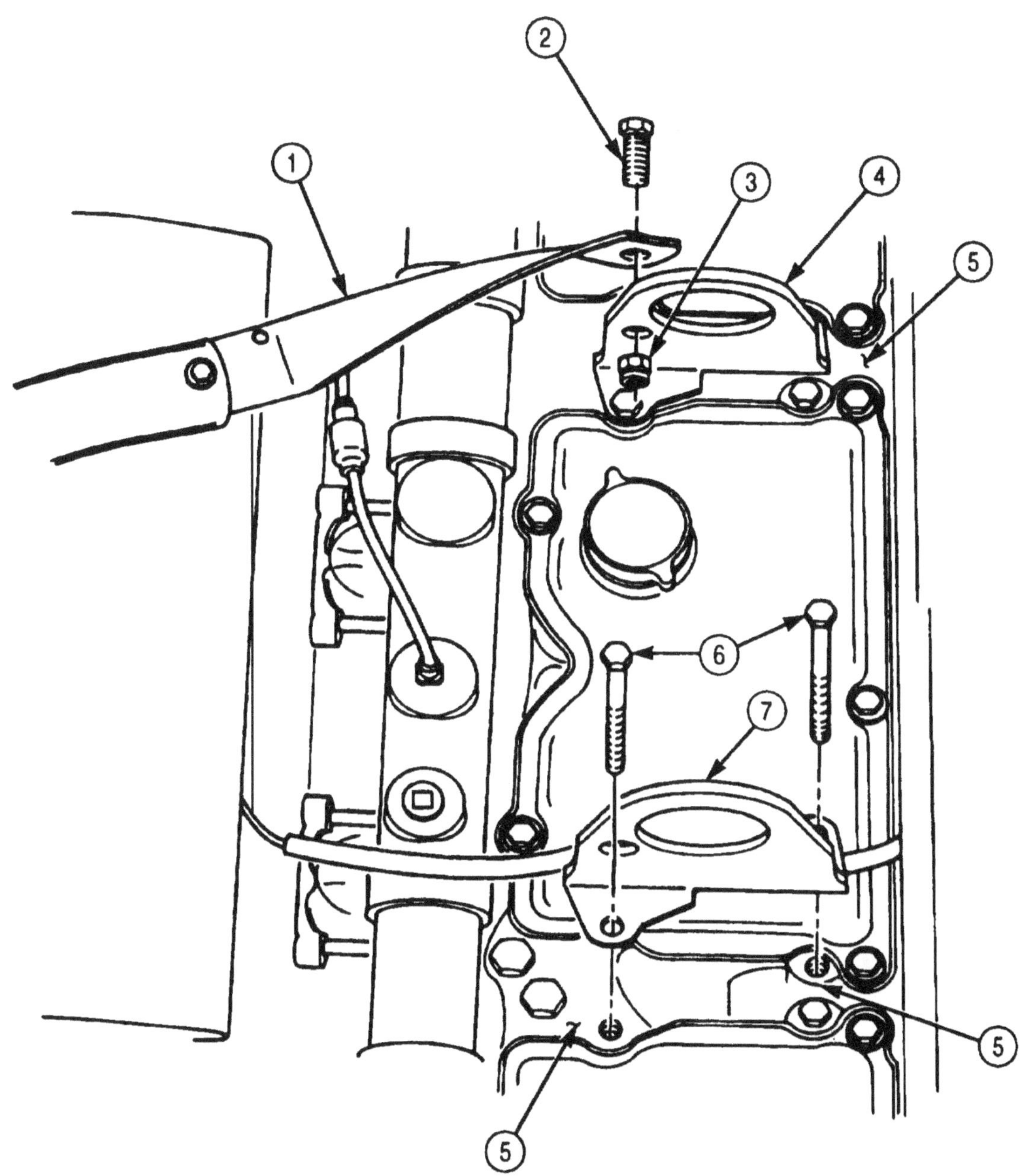

FOLLOW-ON TASK: Install left and right splash shield (TM 9-2320-272-10).

4-12. CYLINDER HEAD REPLACEMENT

THIS TASK COVERS:

a. Removal b. Installation

INITIAL SETUP:

APPLICABLE MODELS
M939/A1

SPECIAL TOOLS
Head holding fixture (Appendix E, Item 7)
Cleaning brush (Appendix E, Item 28)

TOOLS
General mechanic's tool kit (Appendix E, Item 1)
Torque wrench (Appendix E, Item 145)

MATERIALS/PARTS
Cylinder head gasket (Appendix D, Item 94)

REFERENCES (TM)
TM 9-2320-272-24P

EQUIPMENT CONDITION
- Water manifold removed (para. 4-27).
- Exhaust manifold removed (para. 4-25).
- Fuel crossover connectors and valve crossheads removed (para. 4-30).
- Injectors removed (para. 4-32).
- Engine fuel supply and return tubes removed (para. 3-17).

GENERAL SAFETY INSTRUCTIONS
- Compressed air source will not exceed 30 psi (207 kPa).
- Eyeshields must be worn when cleaning with compressed air.

NOTE

All three cylinder heads are replaced the same way. Only one is covered in this paragraph.

a. Removal

1. Following "outside-in" loosening sequence, remove twelve screws (2) and washers (3) from cylinder head (4).

NOTE

- Assistant will help with step 2.
- Tag cylinder head for installation.

2. Remove cylinder head (4) from cylinder block (1) by lifting straight up.
3. Remove cylinder head gasket (5) from cylinder block (1). Discard cylinder head gasket (5) and clean gasket remains from mating surfaces.

4-12. CYLINDER HEAD REPLACEMENT (Contd)

WARNING

- Compressed air source will not exceed 30 psi (207 kPa).When cleaning with compressed air, eyeshields must be worn. Failure to wear eyeshields may result in injury to personnel.
- Drycleaning solvent is flammable and will not be used near open flame. Use only in well-ventilated places and keep fire extinguisher nearby. Failure to do so may result in injury to personnel.

4. Place cylinder head (4) in cylinder head holding fixture, steam-clean cylinder head (4), and dry with compressed air.

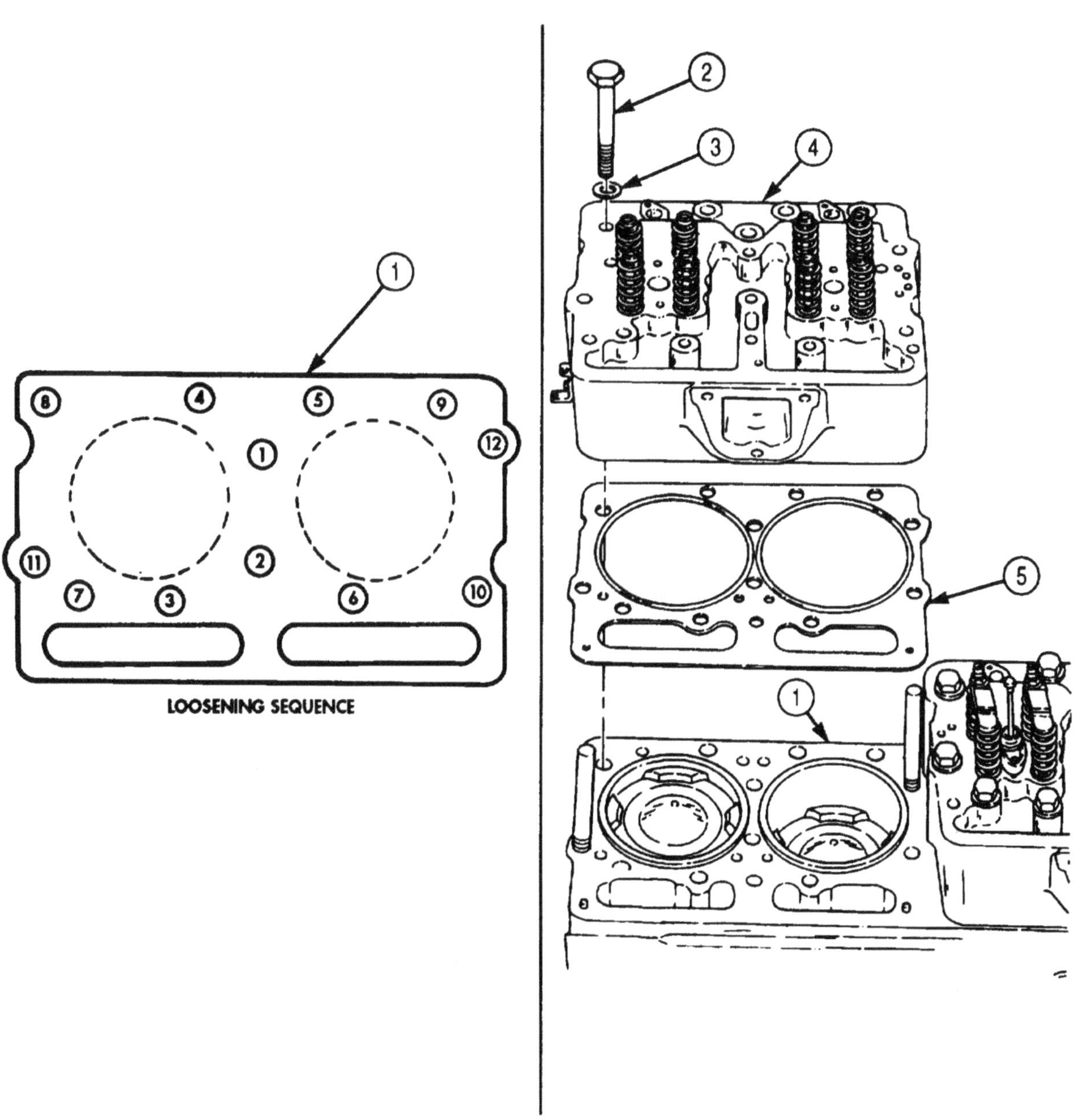

4-12. CYLINDER HEAD REPLACEMENT (Contd)

b. Installation

CAUTION

If old cylinder head is being installed, ensure it is installed in the same location.

NOTE

Assistant will help with steps 1 and 2.

1. Install new cylinder head gasket (5) on cylinder block (1). Ensure the word TOP is facing up.
2. Install cylinder head (4) on cylinder block (1) with twelve washers (3) and screws (2). Tighten screws (2) 25 lb-ft (34 N-m) in sequence shown.
3. Tighten screws (2) 80-100 lb-ft (109-136 N•m) in sequence shown.
4. Tighten screws (2) 180-200 lb-ft (244-271 N•m) in sequence shown.
5. Tighten screws (2) 280-300 lb-ft (380-407 N•m) in sequence shown.

4-12. CYLINDER HEAD REPLACEMENT (Contd)

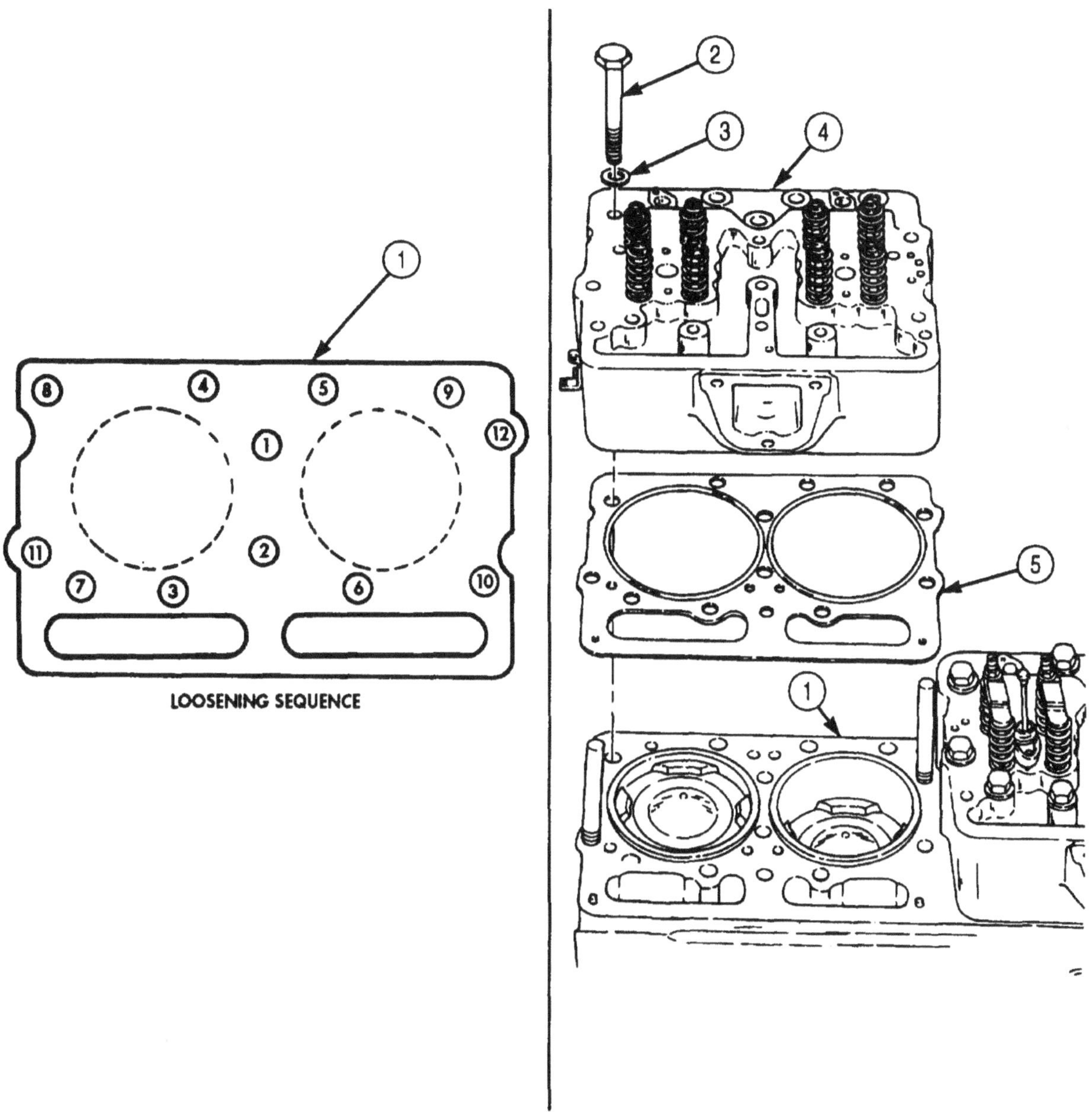

FOLLOW-ON TASKS:
- Install water manifold (para. 4-27).
- Install fuel injectors (para. 4-32).
- Install fuel crossover connectors and valve crossheads (para. 4-30).
- Install engine fuel supply and return tubes (para. 3-17).
- Install exhaust manifold (para. 4-25).

4-13. ENGINE ACCESS COVER REPLACEMENT

THIS TASK COVERS:

a. Removal
b. Installation

INITIAL SETUP:

APPLICABLE MODELS
M939/A1

TOOLS
General mechanic's tool kit (Appendix E, Item 1)

MATERIALS/PARTS
Engine access cover gasket (Appendix D, Item 167)
Four lockwashers (Appendix D, Item 416)

REFERENCES (TM)
LO 9-2320-272-12
TM 9-2320-272-10
TM 9-2320-272-24P

EQUIPMENT CONDITION
- Parking brake set (TM 9-2320-272-10).
- Right splash shield removed (TM 9-2320-272-10).
- Engine oil drained (LO 9-2320-272-12).

a. Removal

1. Remove two screws (9), lockwashers (10), washers (8), and clamps (11) from engine access cover (2) and move transmission cooler lines (7) to one side. Discard lockwashers (10).
2. Remove two screws (4), lockwashers (5), washers (3), engine access cover (2), and gasket (1) from cylinder block (6). Discard lockwashers (5) and gasket (1).

b. Installation

1. Install new gasket (1) and engine access cover (2) on cylinder block (6) with two washers (3), new lockwashers (5), and screws (4).
2. Install transmission cooler lines (7) on engine access cover (2) with two clamps (11), washers (8), new lockwashers (10), and screws (9).

4-13. ENGINE ACCESS COVER REPLACEMENT (Contd)

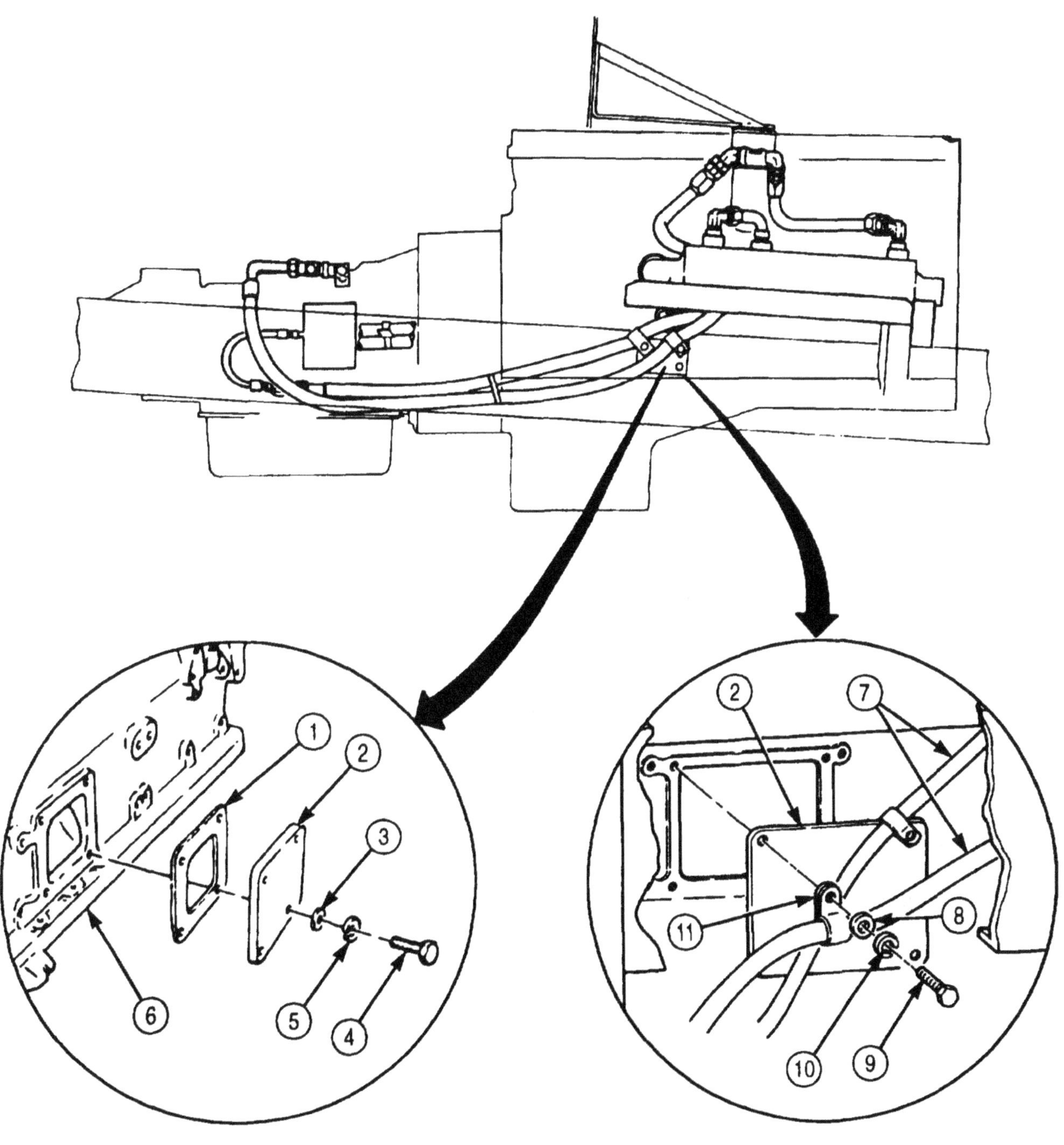

FOLLOW-ON TASKS: Install right splash shield (TM 9-2320-272-10).
- Fill engine to proper oil level (LO 9-2320-272-12).

4-14. CRANKSHAFT VIBRATION AND DAMPER REPLACEMENT

THIS TASK COVERS:

a. Runout and Wobble Check
b. Removal
c. Inspection
d. Installation

INITIAL SETUP:

APPLICABLE MODELS
M939/A1

SPECIAL TOOLS
Barring tool (Appendix E, Item 8)

TOOLS
General mechanic's tool kit (Appendix E, Item 1)
Dial indicator (Appendix E, Item 36)

MATERIALS/PARTS
Six lockwashers (Appendix D. Item 350)

REFERENCES (TM)
TM 9-2320-272-24P

EQUIPMENT CONDITION
- Radiator fan shroud removed (para. 3-63).
- Radiator fan blade removed (para. 3-72).
- Alternator belts removed (para. 3-78).

a. Runout and Wobble Check

NOTE

Perform steps 1 through 3 to check for runout.

1. Mount dial indicator and holding fixture on front gearcase cover (2).
2. Position dial arm on vibration damper (1) at surface (3) and zero-dial indicator.
3. Using barring tool, rotate vibration damper (1) and take reading. If reading exceeds 0.025 in. (0.63 mm), replace vibration damper (1).

NOTE

- Perform steps 4 through 6 to check for runout.
- Crankshaft must be kept at front or rear limit of thrust while checking vibration damper for wobble.

4. Using mounted dial indicator, position dial arm on vibration damper (1) at surface (4) and zero-dial indicator.
5. Using barring tool, rotate vibration damper (1) and take reading. If reading exceeds 0.030 in. (0.76 mm), replace vibration damper (1).
6. Remove dial indicator and fixture.

b. Removal

Remove six screws (5), lockwashers (6), and vibration damper (1) from crankshaft flange (7). Discard lockwashers (6).

4-14. CRANKSHAFT VIBRATION AND DAMPER REPLACEMENT (Contd)

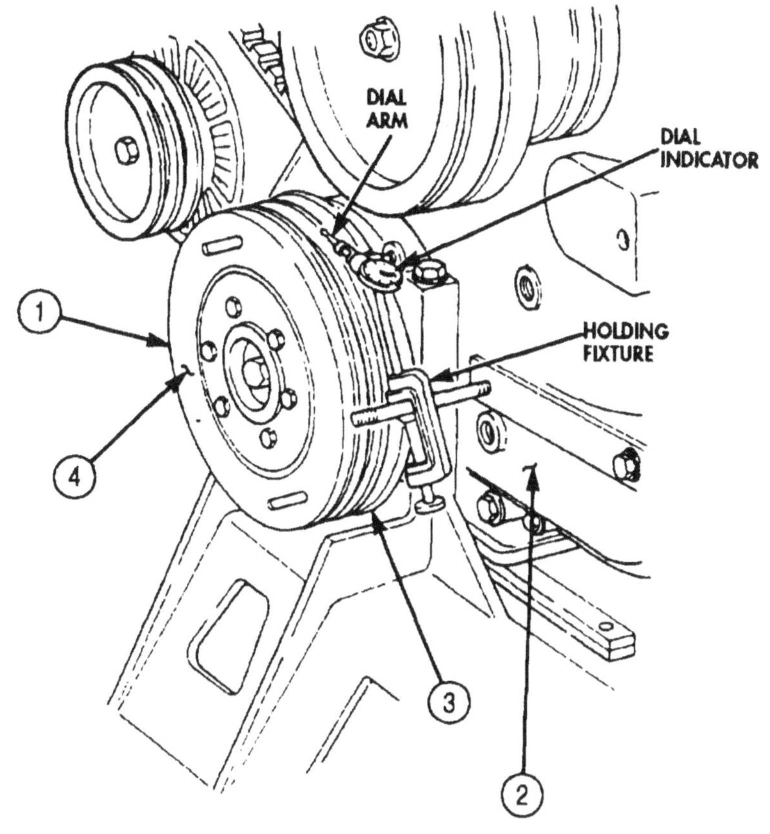

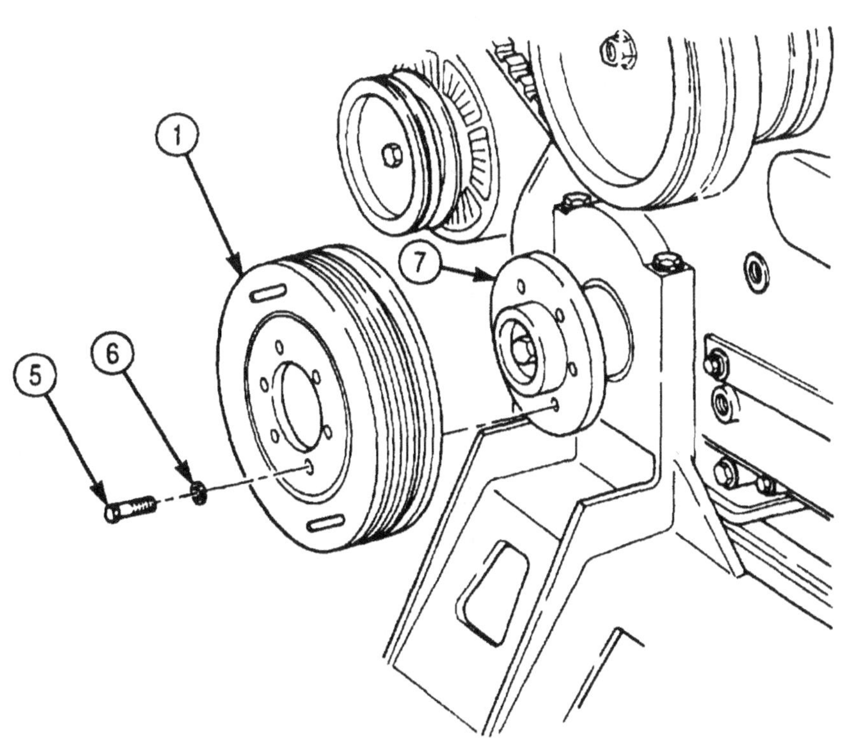

4-14. CRANKSHAFT VIBRATION AND DAMPER REPLACEMENT (Contd)

c. Inspection

Inspect vibration damper (1) to ensure mark (3) on hub (4) aligns with mark (2) on member (5). If alignment marks (2) and (3) are not within 0.062 in. (1.59 mm), replace vibration damper (1).

d. Installation

Install vibration damper (1) on crankshaft flange (8) with six new lockwashers (7) and screws (6). Tighten screws (6) 55-60 lb-ft (75-81 N•m).

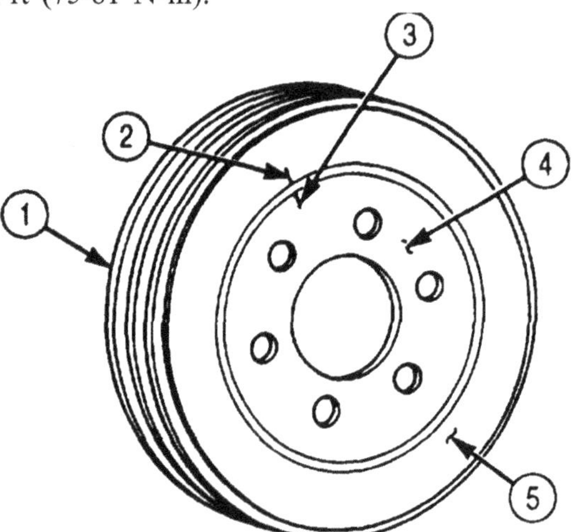

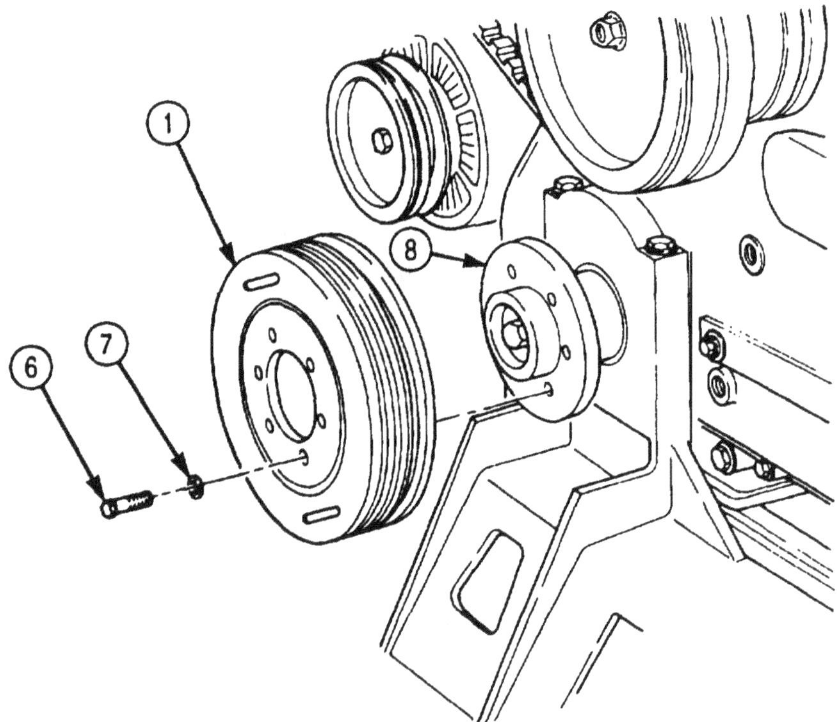

FOLLOW-ON TASKS:
- Install alternator belts (para. 3-78).
- Install radiator fan blade (para. 3-72).
- Install radiator fan shroud (para. 3-63).

4-15. FLYWHEEL RING GEAR MAINTENANCE

THIS TASK COVERS:

a. Removal　　b. Installation

INITIAL SETUP:

APPLICABLE MODELS
M939/A1

TOOLS
General mechanic's tool kit (Appendix E, Item 1)

REFERENCES (TM)
TM 9-2320-272-24P

EQUIPMENT CONDITION
Engine mounted on repair stand (para. 4-9).

a. Removal

WARNING

Support ring gear when removing flywheel screws. Ring gear may fail if not supported, causing injury to personnel.

CAUTION

Lock ring gear to prevent crankshaft from turning before removing screws.

NOTE

Flexplate and ring gear will be an assembly.

Remove six screws (1), washers (2), clutch spacer (3), flexplate (4), ring gear (5), and adapter plate (6) from crankshaft rear flange (7).

b. Installation

Install adapter plate (6), ring gear (5), flexplate (4), and clutch spacer (3) on crankshaft rear flange (7) with six washers (2) and screws (1). Tighten screws (1) alternately 200-220 lb-ft (271-298 N-m).

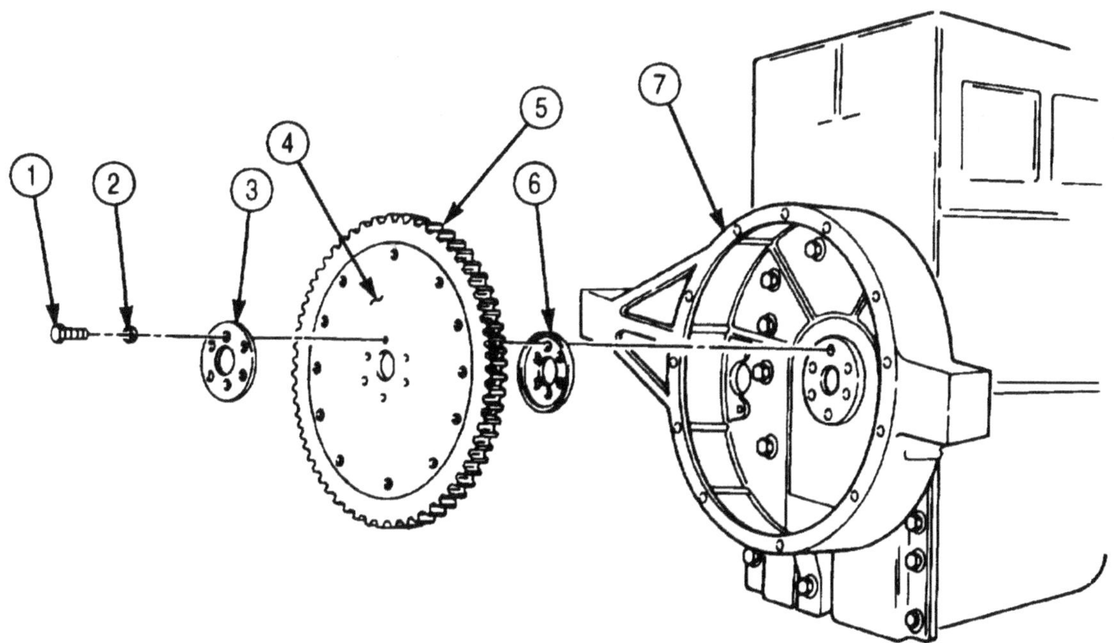

4-16. FLYWHEEL HOUSING MAINTENANCE

THIS TASK COVERS:

a. Removal
b. Cleaning and Inspection
c. Installation

INITIAL SETUP:

APPLICABLE MODELS
M939/A1

TOOLS
General mechanic's tool kit (Appendix E, Item 1)
Dial indicator (Appendix E, Item 36)
Soft-faced hammer

MATERIALS/PARTS
Nine lockwashers (Appendix D, Item 392)
Camshaft bore cork gasket (Appendix D, Item 32)

REFERENCES (TM)
TM 9-2320-272-24P

EQUIPMENT CONDITION
- Flywheel ring gear removed (para. 4-15).
- Engine oil pan removed (para. 4-22).

a. Removal

1. Remove nine screws (6), lockwashers (5), washers (4), and packing (3) from flywheel housing (1). Discard lockwashers (5).
2. Remove six screws (10) and lockwashers (9) from oil pan (8). Discard lockwashers (9).
3. Tap around side of flywheel housing (1) with wood block and soft-faced hammer, and remove flywheel housing (1) from dowel pins (7) in cylinder block (2).
4. Remove camshaft bore cork gasket (11) from flywheel housing (1). Discard gasket (11) and clean gasket remains from mating surfaces.

NOTE

Perform step 5 if installing new flywheel housing.

5. Remove dowel pins (7) from cylinder block (2).

b. Cleaning and Inspection

1. Clean flywheel housing (1) (para. 2-14). Ensure gasket mating surfaces are clean.
2. Inspect flywheel housing (1) (para. 2-15). Discard flywheel housing (1) if defective.
3. Inspect dowel pins (7) for damage and wear. Replace dowel pins (7) if damaged or outside diameter is worn more than 0.5005 in. (12.71 mm).

4-16. FLYWHEEL HOUSING MAINTENANCE (Contd)

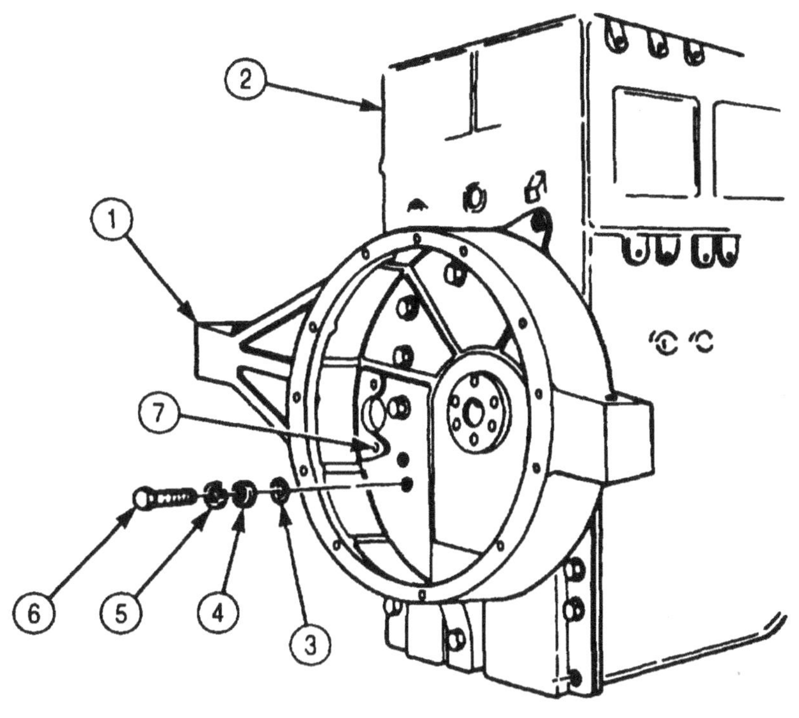

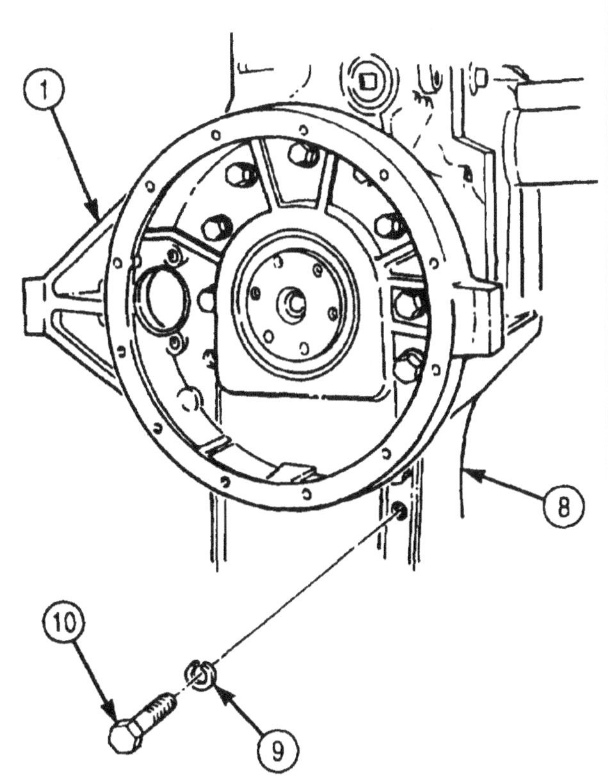

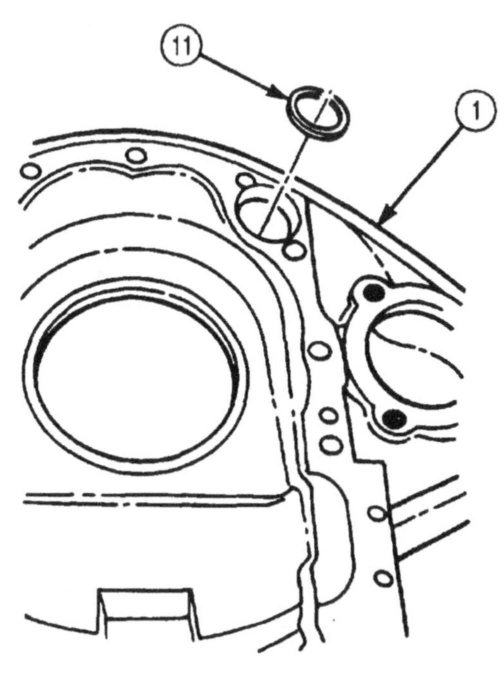

4-16. FLYWHEEL HOUSING MAINTENANCE (Contd)

c. Installation

1. Install new camshaft bore cork gasket (1) in flywheel housing (2).

NOTE

Perform step 2 when installing new flywheel housing.

2. Install two dowel pins (8) in engine block (3).
3. Install flywheel housing (2) over dowel pins (8) and seat against engine block (3).
4. Install nine packings (4), washers (5), new lockwashers (6), and screws (7) on flywheel housing (2) and finger-tighten screws (7).

NOTE

The flywheel bore must be centered to crankshaft rotation. Follow steps 5 through 18, using a dial indicator with dial gauge attachment.

5. Attach dial indicator attachment (10) to flange of crankshaft (9) and position dial indicator plunger against bore face (11).
6. Mark flywheel housing (2) with chalk marks at 12, 3, 6, and 9 o'clock positions and position dial indicator plunger at 3 o'clock position by rotating crankshaft (9). Set dial indicator to zero.

NOTE

Perform steps 7 and 8 to check flywheel housing vertical runout.

7. Rotate crankshaft (9) from 3 o'clock through 6 o'clock to 9 o'clock position. Record highest reading.
8. Rotate crankshaft (9) from 9 o'clock through 12 o'clock and back to 3 o'clock position. Record highest reading. The highest reading determines the up or down direction the flywheel housing (2) must be adjusted.

NOTE

Perform steps 9 through 12 to adjust flywheel housing vertical runout.

9. Rotate crankshaft (9) and dial indicator attachment (10) to point on bore face (11) where highest reading was recorded.
10. Set dial indicator to read one-half of total highest reading and slightly loosen flywheel housing mounting screws (12).
11. Using a soft-faced hammer, carefully tap flywheel housing (2) opposite the dial indicator until dial indicator reads zero.
12. Once dial indicator reads zero, tighten flywheel housing mounting screws (12) finger-tight. Flywheel housing (2) is now centered vertically. Do not remove dial indicator.

NOTE

Perform steps 13 through 15 to check flywheel housing horizontal runout.

13. Rotate crankshaft (9) so dial indicator is positioned at 12 o'clock, Set dial indicator to zero.
14. Rotate crankshaft (9) and check readings at 3 o'clock and 6 o'clock positions. Record highest reading.
15. Rotate crankshaft (9) and check reading at 9 o'clock and back to 12 o'clock position. Record highest reading. The highest reading recorded will indicate the direction the flywheel housing (2) must be moved to obtain correct alignment center.

4-16. FLYWHEEL HOUSING MAINTENANCE (Contd)

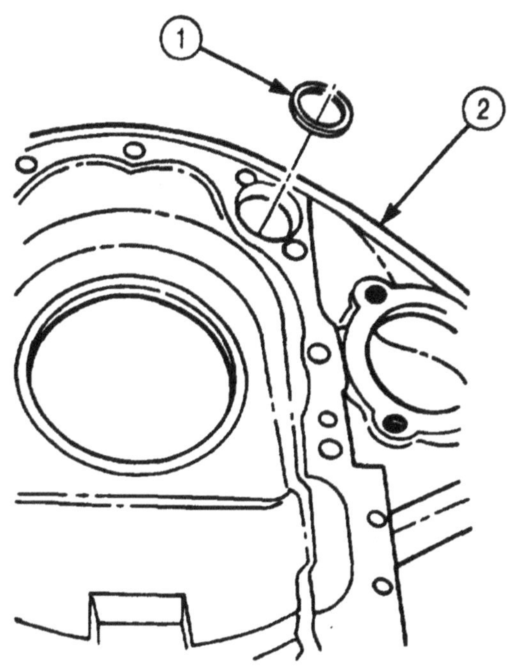

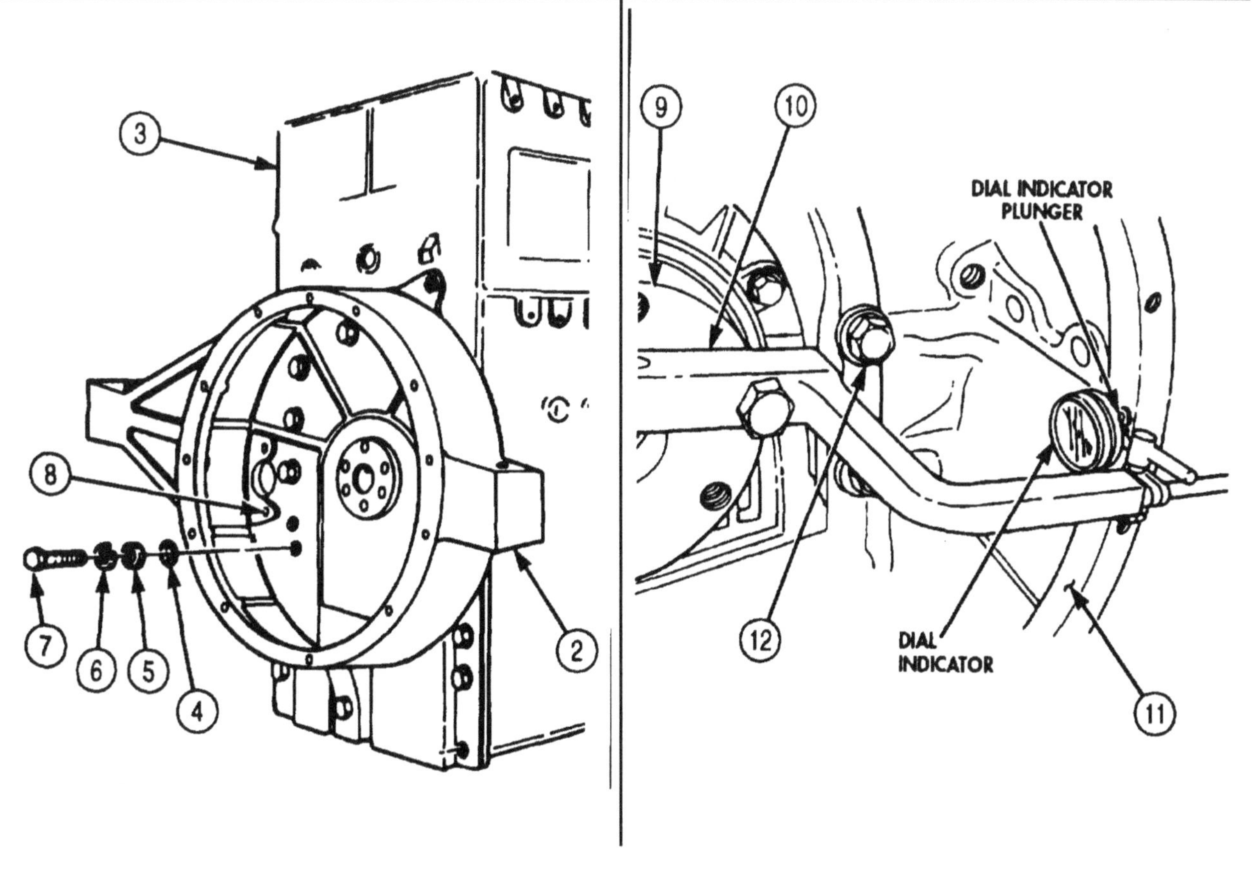

4-16. FLYWHEEL HOUSING MAINTENANCE (Contd)

NOTE

Perform steps 16 through 18 to adjust flywheel housing horizontal runout.

16. Rotate crankshaft (2) and set dial indicator at point where highest reading was recorded on bore face (3).
17. Set dial indicator to one-half of total highest reading and slightly loosen flywheel housing mounting screws (1).
18. Tap flywheel housing (4) opposite dial indicator with soft-faced hammer until dial indicator reads zero. Tighten nine flywheel housing mounting screws (1) 140-16 lb-ft (190-217 N•m) in sequence shown.
19. Reposition dial indicator against flywheel housing face (6).

NOTE

Perform steps 20 through 22 to check total flywheel housing runout.

20. Push crankshaft (2) forward to take up end play.
21. Set dial indicator to zero, rotate crankshaft (2), and read total runout on dial indicator. Total runout must not exceed 0.068 in. (0.20 mm). If runout is within specifications, flywheel housing (4) is properly positioned.
22. Remove dial indicator and dial indicator attachment (5).

NOTE

Perform steps 23 and 24 if dowel pins were removed.

23. Using drill and reaming fixture, ream dowel pin holes (9) to next oversize.
24. Install two new dowel pins (7) in flywheel housing (4) and engine block (8). Ensure dowel pins (7) are even with or 0.010 in. (0.25 mm) below the flywheel housing (4).

4-16. FLYWHEEL HOUSING MAINTENANCE (Contd)

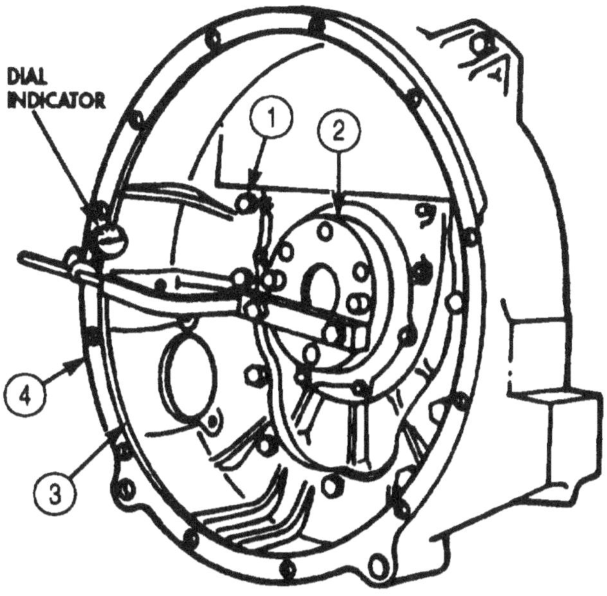

DIAL INDICATOR
1
2
4
3

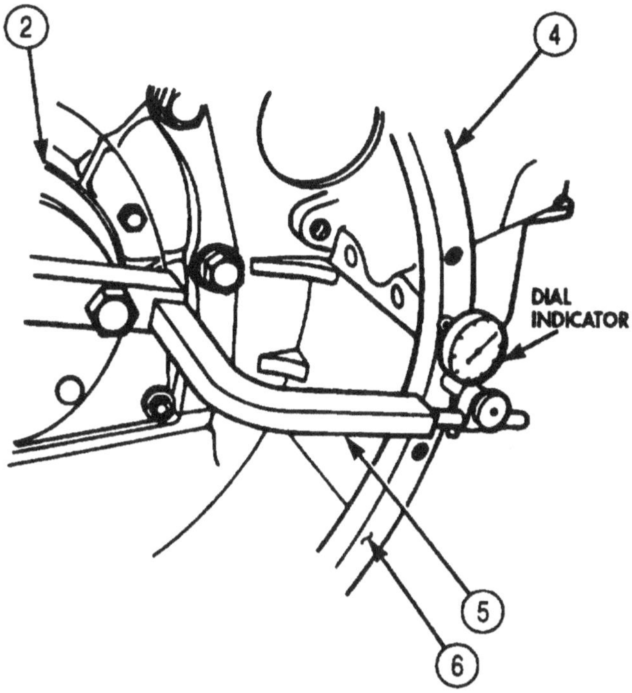

2
4
DIAL INDICATOR
5
6

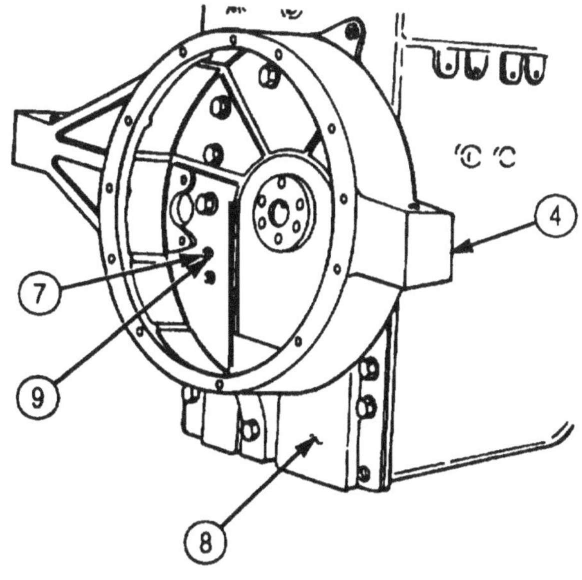

4
7
9
8

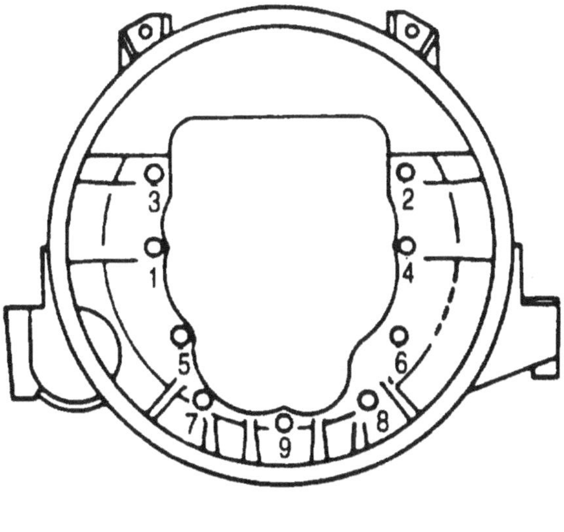

3
2
1
4
5
6
7
9
8

TIGHTENING SEQUENCE

FOLLOW-ON TASKS: • Install engine oil pan (para. 4-22).
• Install flywheel ring gear (para. 4-15).

4-17. CRANKSHAFT REAR COVER AND SEAL REPLACEMENT

THIS TASK COVERS:

a. Removal b. Installation

INITIAL SETUP:

APPLICABLE MODELS
M939/A1

TOOLS
General mechanic's tool kit (Appendix E, Item 1)

MATERIALS/PARTS
Rear cover plate gasket (Appendix D, Item 526)
Rear cover plate seal (Appendix D, Item 527)
O-ring (Appendix D, Item 439)
Fourteen screw-assembled lockwashers (Appendix D, Item 588)

REFERENCES (TM)
TM 9-2320-272-24P

EQUIPMENT CONDITION
Flywheel housing removed (para. 4-16).

a. Removal

1. Remove four screw-assembled lockwashers (2) from oil pan (1). Discard screw-assembled lockwashers (2).
2. Remove ten screw-assembled lockwashers (4), O-ring retainer (7), retainer gasket (6), cover seal (5), cover plate (3), and cover plate gasket (9). Discard seal (5), O-ring retainer gasket (6), and plate gasket (9).
3. Remove O-ring (8) from O-ring retainer (7). Discard O-ring (8).

b. Installation

1. Install new cover plate gasket (9) and cover plate (3) on cylinder block (10) with four new screw-assembled lockwashers (4). Do not tighten.
2. Install O-ring retainer (7) and gasket (6) over end of crankshaft (11) with six new screw-assembled lockwashers (4). Do not tighten.

NOTE

Perform steps 3 and 4 to check cover plate runout.

3. Position dial indicator on end of crankshaft (11) and indicator arm (12) to outside edge of O-ring retainer (7).
4. Rotate crankshaft (11) and read dial indicator to align cover plate (3). Cover plate (3) must be aligned within 0.005 in. (0.127 mm).
5. Tighten ten screw-assembled lockwashers (4) 24-29 lb-ft (33-39 N-m).
6. Install new rear cover seal (5) into rear cover plate (3) over rear end of crankshaft (11).
7. Install new O-ring (8) over O-ring retainer (7).
8. Install four screw-assembled lockwashers (2) in oil pan (1). Alternately tighten screw-assembled lockwashers (2) 15-20 lb-ft (20-27N•m).

4-17. CRANKSHAFT REAR COVER AND SEAL REPLACEMENT (Contd)

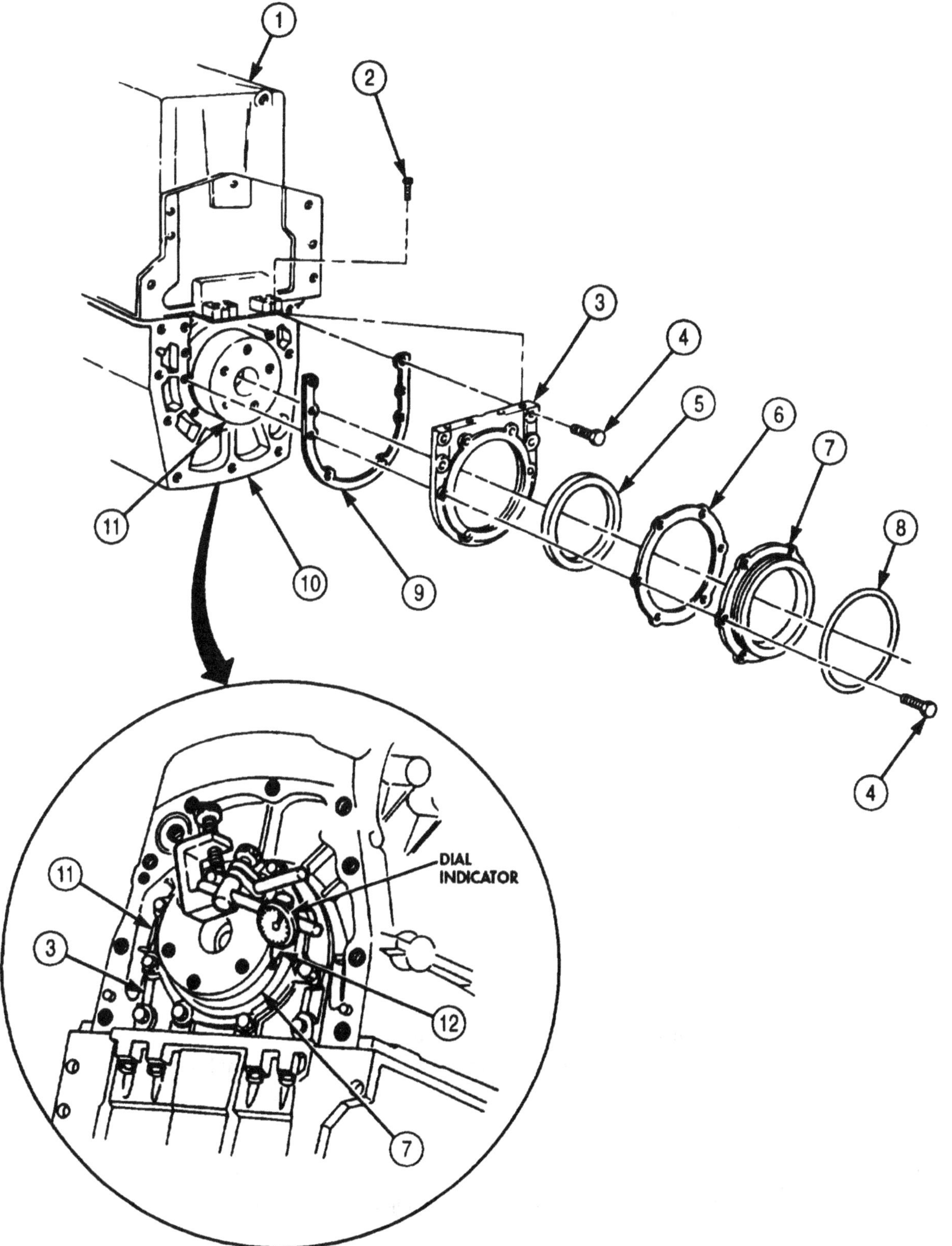

FOLLOW-ON TASK: Install flywheel housing (para. 4-16).

4-18. FRONT GEARCASE COVER MAINTENANCE

THIS TASK COVERS:

a. Removal
b. Disassembly
c. Cleaning and Inspection
d. Assembly
e. Installation

INITIAL SETUP:

APPLICATION MODELS
M939/A1

TOOLS
General mechanic's tool kit (Appendix E, Item 1)
Mechanical puller (Appendix E, Item 103)
Outside micrometer (Appendix E, Item 80)
Torque wrench (Appendix E, Item 145)
Arbor press
Mandrel

MATERIALS/PARTS
Sixteen lockwashers (Appendix D, Item 354)
Oil seal (Appendix D, Item 498)
Oil seal (Appendix D, Item 499)
Gasket (Appendix D, Item 169)
Shims (Appendix D, Item 651)
O-ring (Appendix D, Item 440)
Grease (Appendix C, Item 28)
Lubricating oil (Appendix C, Item 50)
Crocus cloth (Appendix C, Item 20)

REFERENCES (TM)
TM 9-2320-272-24P

EQUIPMENT CONDITION
- Engine mounted on repair stand (para. 4-9).
- Engine accessory drive pulley removed (para. 4-26).
- Water pump removed (para. 4-29).

a. Removal

1. Remove screw (16) and washer (17) from crankshaft flange (18).
2. Install screw (16) in crankshaft (19) three turns.
3. Using puller, remove crankshaft flange (18) from crankshaft (19).
4. Remove screw (16) from crankshaft (19).

NOTE

- Gearcase mounting screws are of different lengths. Tag screws for installation.
- Gearcase cover is mounted with screw-assembled washers for late model engines.

5. Remove screw (6), lockwasher (7), washer (5), and adjusting link (4) from gearcase cover (3). Discard lockwasher (7).
6. Remove four screws (12), washers (11), and brace (13) from engine block (1).
7. Remove fifteen screws (10), lockwashers (9), washers (8), gearcase cover (3), and gasket (2) from engine block (1). Discard lockwashers (9) and gasket (2).

b. Disassembly

1. Remove crankshaft seal (14) from gearcase cover (3). Discard crankshaft seal (14).
2. Remove accessory drive seal (15) from gearcase cover (3). Discard accessory drive seal (15).

4-18. FRONT GEARCASE COVER MAINTENANCE (Contd)

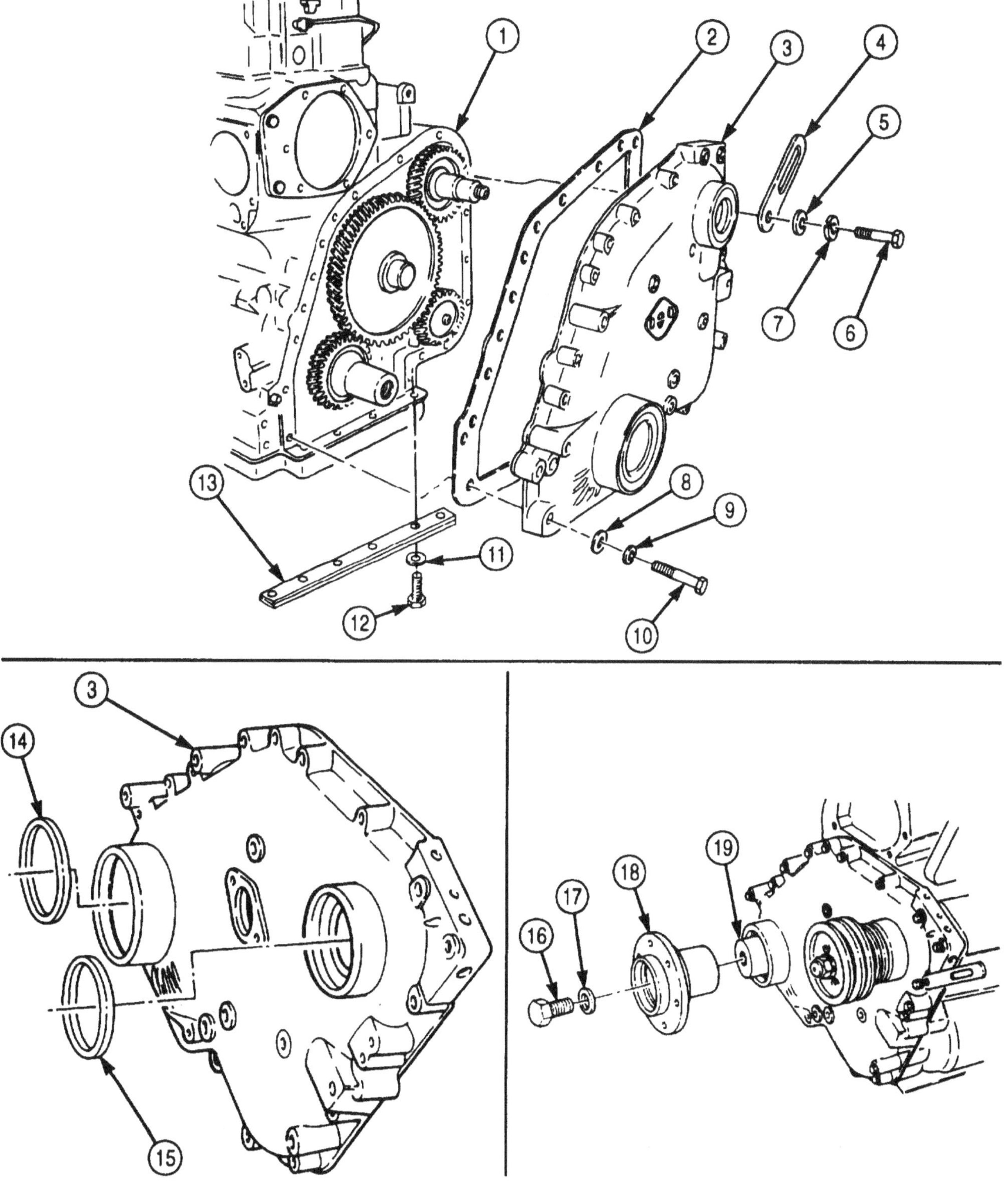

4-18. FRONT GEARCASE COVER MAINTENANCE (Contd)

c. Cleaning and Inspection

1. Clean gearcase cover (1) (para. 2-7).
2. Remove all gasket material from mating surfaces on gearcase cover (1).
3. Inspect gearcase cover (1) (para. 2-8). Replace gearcase cover (1) if any defect is noted.
4. Using surface plate, lap gasket mating surfaces to remove high and low area to ensure flatness and good sealing.
5. Check seal bores (5) and (6) for nicks, burrs, and gouges. Small nicks, burrs, and gouges can be smoothed out with crocus cloth.
6. Check bushing (7) in seal bore (6) for scoring and pitting. Discard bushing (7) if scored or pitted.
7. Check bushing (7) for wear using inside micrometer. Discard bushing (7) if inside diameter is more than 1.571 in. (39.90 mm).
8. Measure outside diameter of accessory driveshaft. Bushing-to-driveshaft clearance must be between 0.003 and 0.007 in. (0.08 and 0.18 mm). Use undersize bushings to obtain proper clearance.
9. Inspect bushing (8) for cracks or scoring. Replace bushing (8) if cracked or scored.
10. Measure bushing (8) outside diameter. If less than 4.745 in. (12.52 mm), replace bushing (8).
11. Inspect two dowel pins (3) in engine block (4) for cracks, breaks, or burrs. Replace dowel pins (3) if cracked or broken. Repair if burred (para. 2-9).
12. Inspect two dowel pin holes (2) in gearcase cover (1) for elongated holes. Replace gearcase cover (1) if holes are elongated.

d. Assembly

NOTE

Do not perform steps 1 through 3 unless installing new bushings.

1. Using arbor press and mandrel, remove bushing (7) from gearcase cover (1) and install new bushing (7) in gearcase cover (1).
2. Using puller, remove bushing (8) from gearcase cover (1).
3. Cut new bushing (8) with outside diameter 4.747-4.750 in. (120.57-120.65 mm). Using arbor press and mandrel, install bushing (8) on gearcase cover (1) with chamfer edge of bushing (8) toward gearcase cover (1).

NOTE

- Do not install mandatory replacement oil seals until gearcase cover is to be reassembled to engine. This prevents collection of dirt.
- Lightly coat all contacting parts of shafts, seals, and O-ring with lubricating oil before installation.

4. Using proper oil seal installation tool, install new oil seals (9) and (10) in gearcase cover (1).

4-18. FRONT GEARCASE COVER MAINTENANCE (Contd)

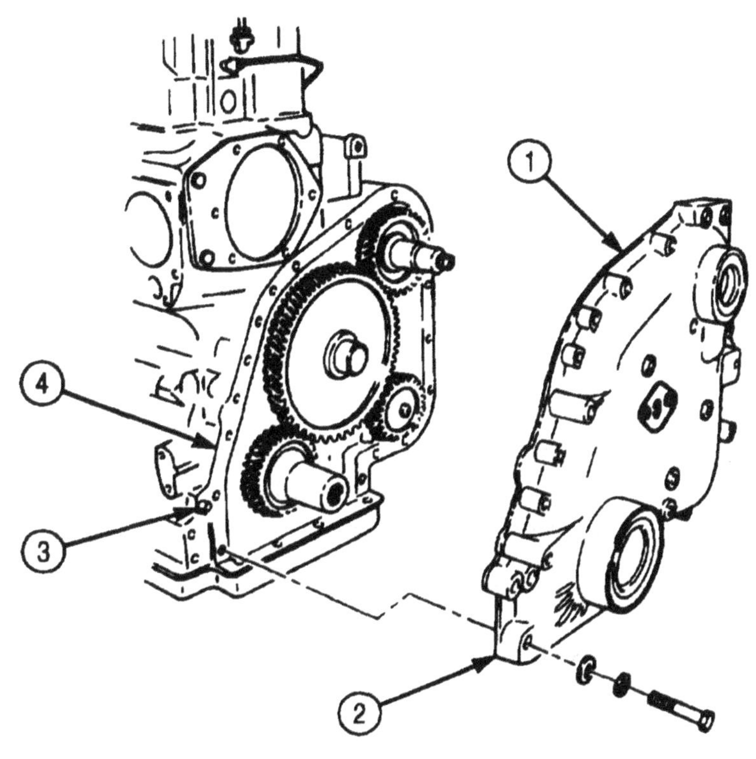

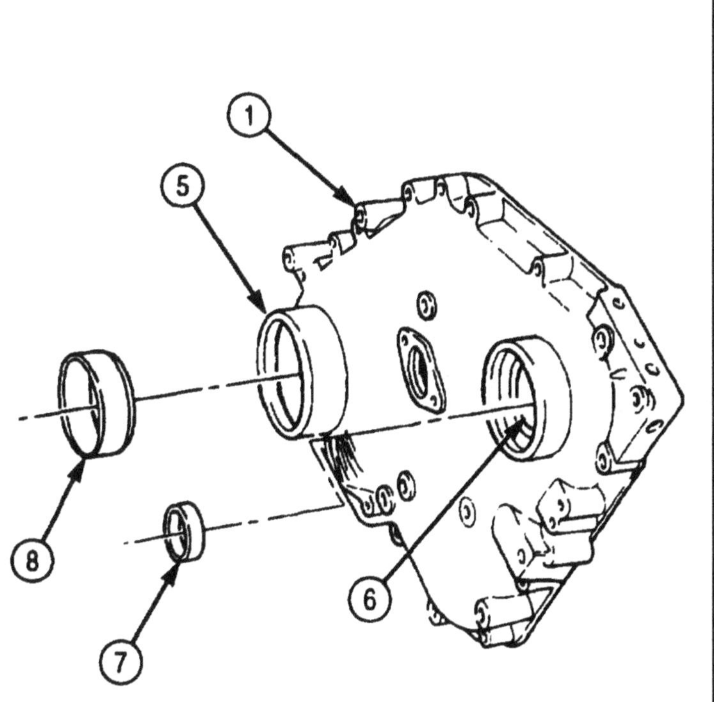

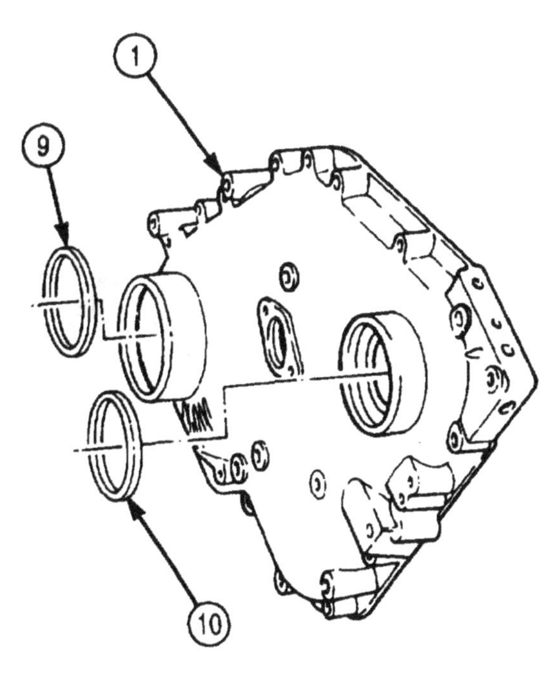

4-18. FRONT GEARCASE COVER MAINTENANCE (Contd)

e. Installation

NOTE

Perform step 1 only if camshaft and gear have been removed.

1. Remove two screws (1), lockwashers (2), washers (3), thrust plate (4), shims (5), and O-ring (6) from gearcase cover (7). Discard lockwashers (2).
2. Using GAA grease to hold gasket (11) in place, position new gasket (11) on engine block (9).

NOTE

Gearcase cover is mounted with screw-assembled washers for late model engines.

3. Install gearcase cover (7) on engine block (9) with fifteen washers (16), new lockwashers (17), and screws (18). Check gearcase cover (7) alignment at engine block (9) with straightedge (10). Tighten screws (18) 50 lb-ft (61-68 N•m).
4. Install adjusting link (12) on gearcase cover (7) with new washer (13), lockwasher (14), and screw (15).
5. Install brace (21) on oil pan (22) and gearcase cover (7) with four washers (19) and screws (20). Tighten screws (20) 35-40 lb-ft (48-54 N-m).

NOTE

Perform steps 6 through 8 to check camshaft thrust plate.

6. Remove O-ring (6) and shims (5) and push thrust plate (4) against camshaft (8).
7. Measure dimension between thrust plate (4) and gearcase cover (7) with feeler gauge, and install enough shims (5) to provide 0.001-0.005 in. (0.025-0.120 mm) end play.
8. Install new O-ring (6), new shims (5), and thrust plate (4) on gearcase cover (7) with two new lockwashers (2), washers (3), and screws (1).
9. Coat inside of crankshaft flange (25) with engine oil and install on crankshaft (26) with washer (24) and screw (23). Tighten screw (23) 180-200 lb-ft (244-271 N-m).

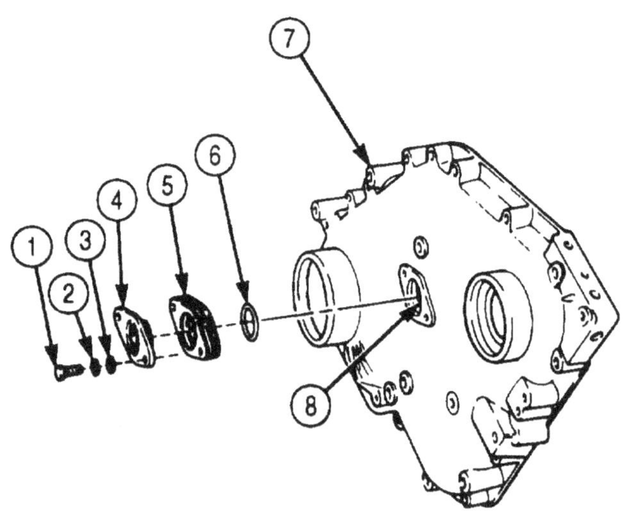

4-18. FRONT GEARCASE COVER MAINTENANCE (Contd)

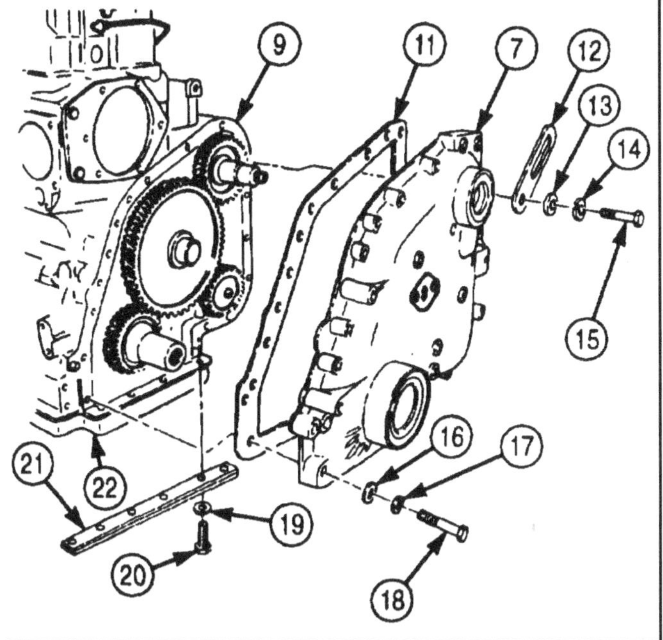

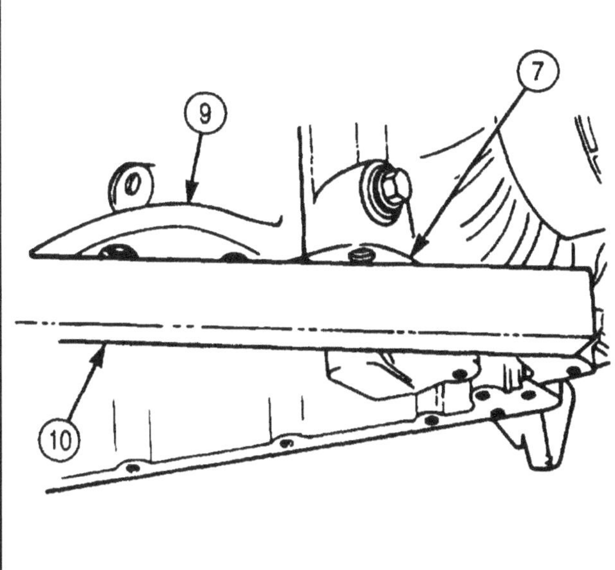

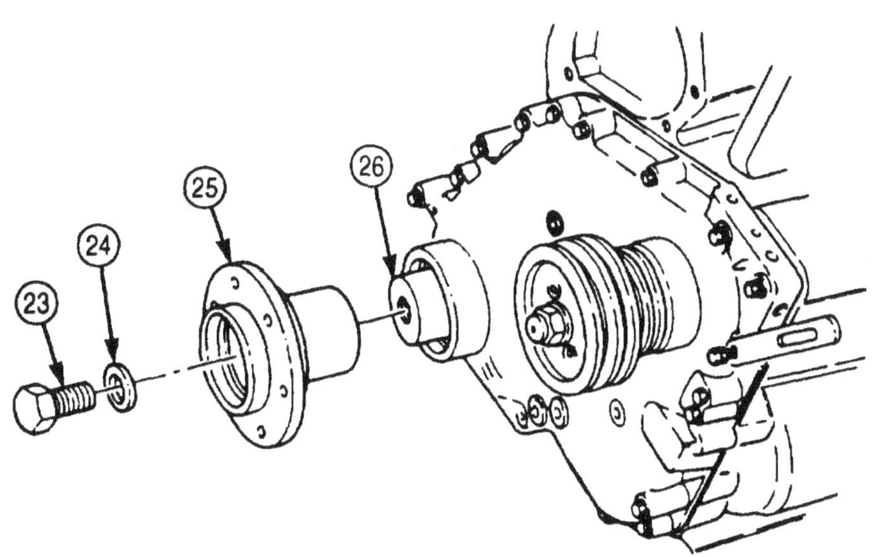

FOLLOW-ON TASKS:
- Dismount engine from repair stand (para. 4-9).
- Install engine accessory drive pulley (para. 4-26).
- Install water pump (para. 4-29).

4-19. CAM FOLLOWER HOUSING MAINTENANCE

THIS TASK COVERS:

a. Removal
b. Disassembly
c. Cleaning and Inspection
d. Assembly
e. Installation

INITIAL SETUP:

APPLICABLE MODELS
M939/A1

TOOLS
General mechanic's tool kit (Appendix E. Item 1)
Inside micrometer (Appendix E; Item 82)
Outside micrometer (Appendix E, Item 80)
Telescoping depth gauge (Appendix E, Item 136)
60-degree angle cutter (Appendix E, Item 6)
Feeler gauge, 0.006-in.
Soft-faced hammer
Arbor press
Mandrel

MATERIALS/PARTS
Two cup-plugs (Appendix D, Item 89)
Six screw-assembled lockwashers (Appendix D, Item 586)
Cam follower housing gasket kit (Appendix D, Item 534)
Lubricating oil (Appendix C, Item 50)
Prussian blue (Appendix C, Item 54)
Sealing compound (Appendix C, Item 68)
Drycleaning solvent (Appendix C, Item 71)

REFERENCES (TM)
TM 9-2320-272-24P

EQUIPMENT CONDITION
- Fuel pump removed (cylinders 3 and 4) (para. 4-35).
- Air compressor removed (cylinders 1 and 2) (para. 4-31).
- Rocker lever housings and push tubes removed (para. 4-20).

GENERAL SAFETY INSTRUCTIONS
- Keep fire extinguisher nearby when using drycleaning solvent.
- Drycleaning solvent is flammable and toxic. Do not use near an open flame.

NOTE

- Maintenance procedure for all three cam follower housings are the same.
- Cam follower bushings are mounted with screw-assembled lockwashers on late model engines.

a. Removal

NOTE

Removal of fuel line bracket is only required for number 2-3 cam follower bushing.

1. Remove six screw-assembled lockwashers (5) and fuel line bracket (6) from cam follower bushing (4). Discard screw-assembled lockwashers (5).

4-19. CAM FOLLOWER HOUSING MAINTENANCE (Contd)

CAUTION

Do not discard cam follower housing gaskets before measuring total thickness of each gasket. Total thickness of all gaskets is critical for sealing of cam followers on camshaft for correct injector and valve timing. Damage to engine will result if gasket spacing is not correct when installing cam followers.

NOTE

Tag cam follower housing for installation,

2. Carefully pry cam follower housing (4) from dowel pins (1) and remove from engine block (2).
3. Remove cam follower gaskets (3) from engine block (2) and measure thicknesses using micrometer. Record readings. Discard gaskets (3) after readings are recorded.
4. Clean remains of cam follower gasket (3) from engine block (2) mating surfaces.

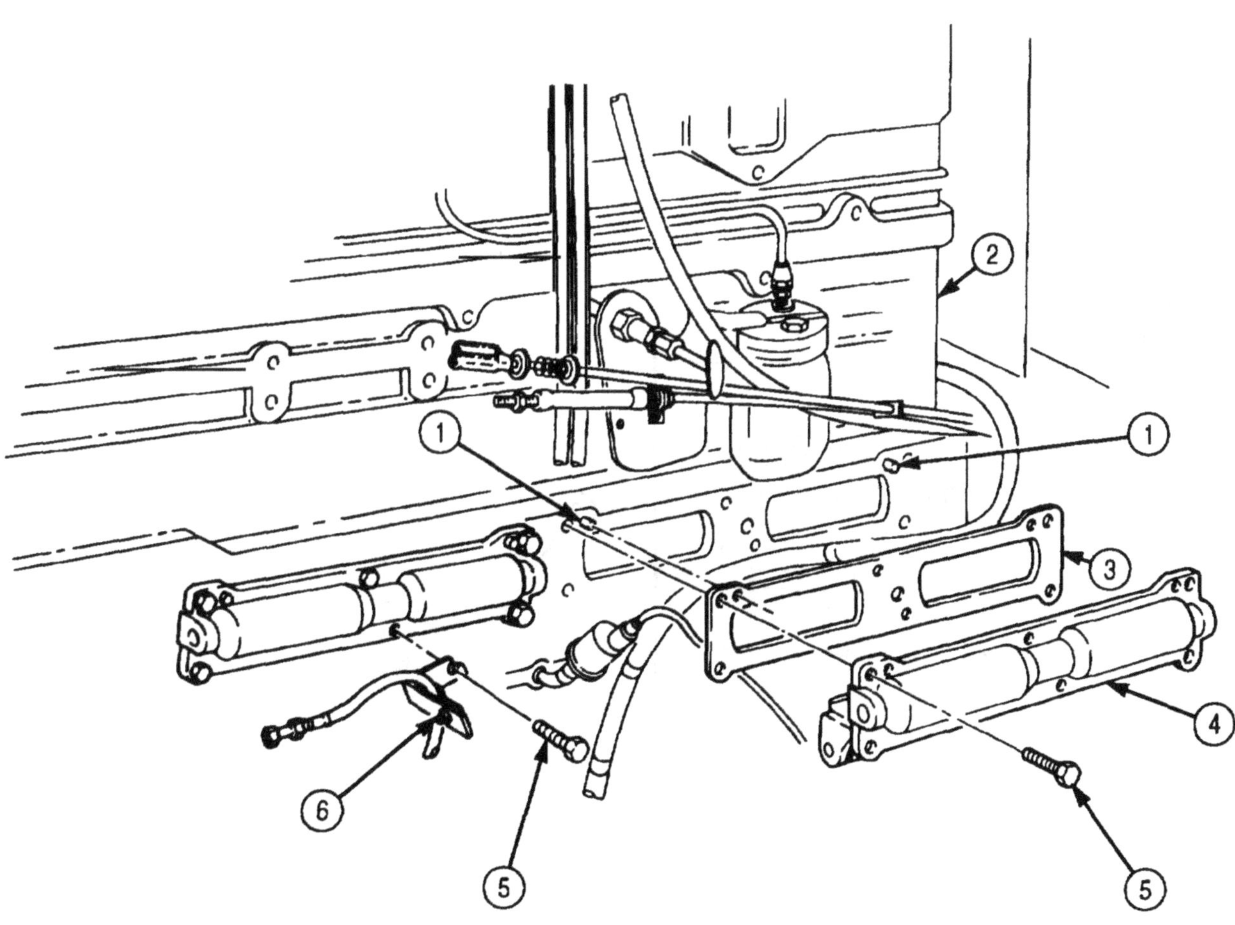

4-19. CAM FOLLOWER HOUSING MAINTENANCE (Contd)

b. Disassembly

1. Remove two shaft lockscrews (1) from cam follower housing (2).
2. Remove cup-plug (3) from cam follower housing (2) with center punch. Discard cup-plug (3).

NOTE

Before performing step 3, mark each cam follower lever with its location for installation.

3. Using arbor press and mandrel, remove cup-plug (3), two lever shafts (4), and six cam follower levers (5) from cam follower housing (2). Discard cup-plug (3).

NOTE

- Use drift to remove retainer pin.
- Use arbor press and mandrel to remove roller.

4. Remove retainer pin (6), roller pin (9), and roller (10) from cam follower lever (8).

NOTE

Perform steps 5 and 6 only if push tube insert or cam follower bushing are found to be defective in task c.

5. Using arbor press and mandrel, remove cam follower lever bushing (11) from cam follower lever (8) with center punch.
6. Using center punch, remove push tube insert (7) from cam follower lever (8).

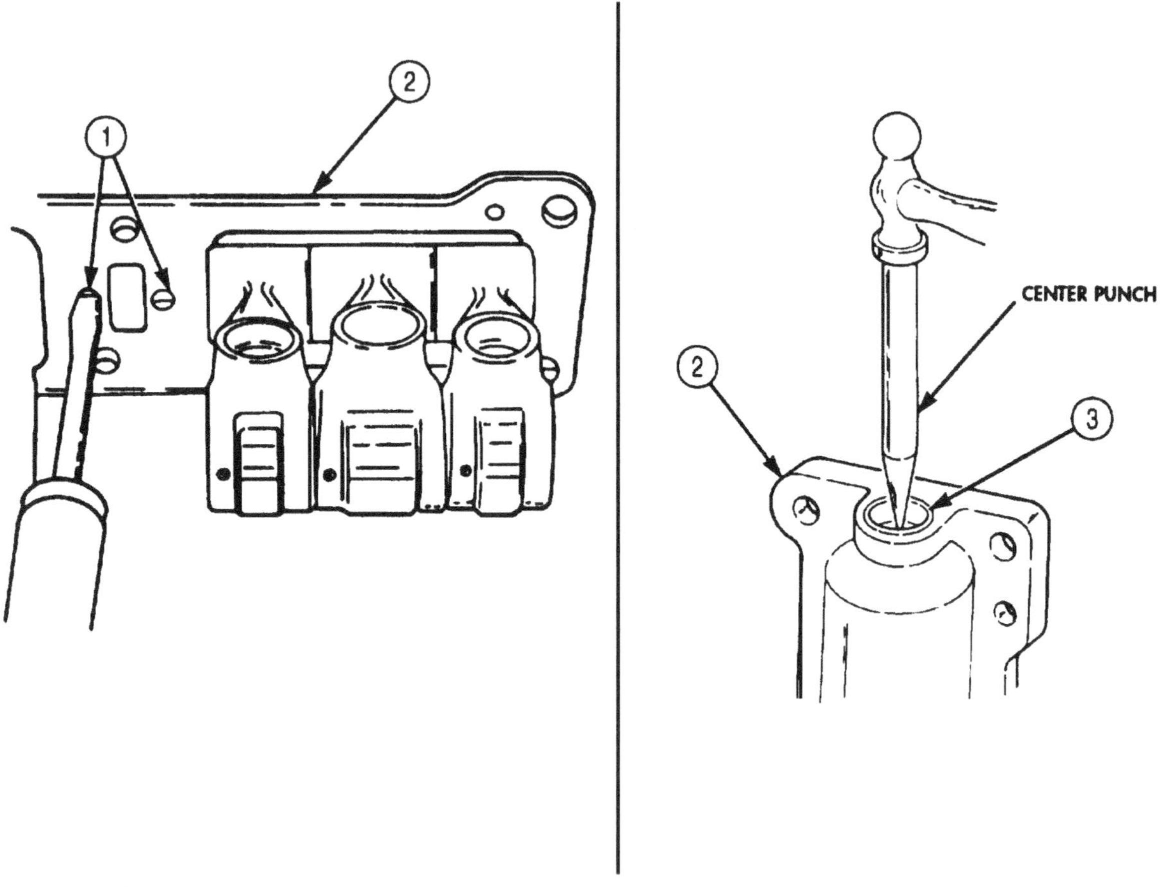

4-19. CAM FOLLOWER HOUSING MAINTENANCE (Contd)

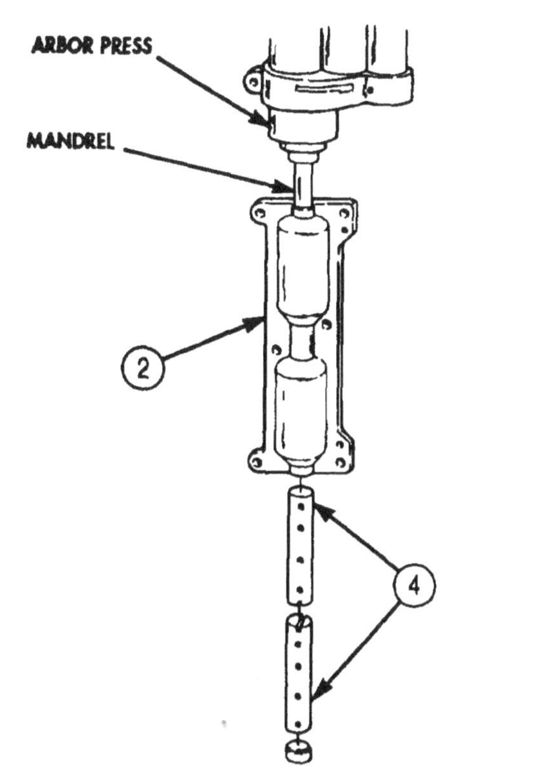

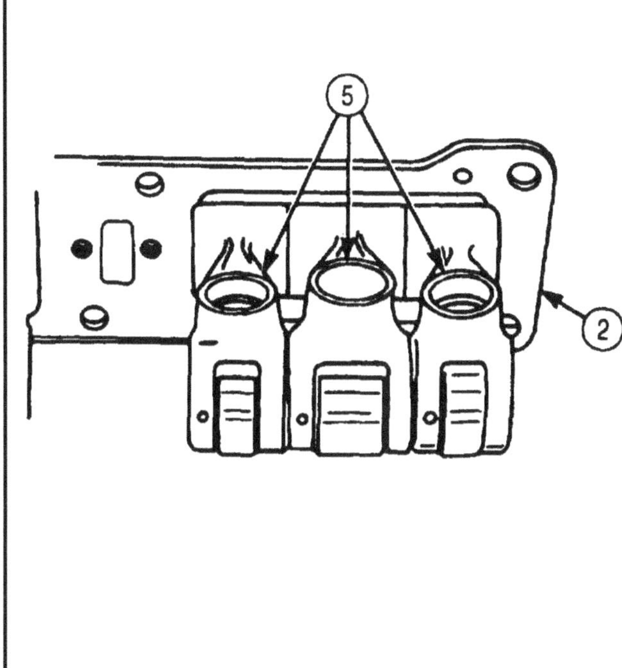

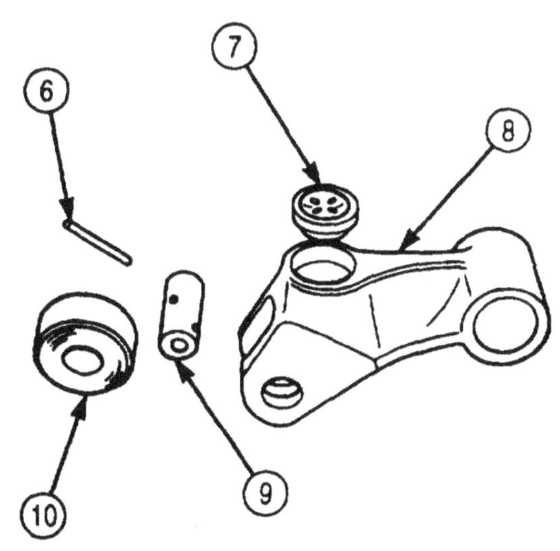

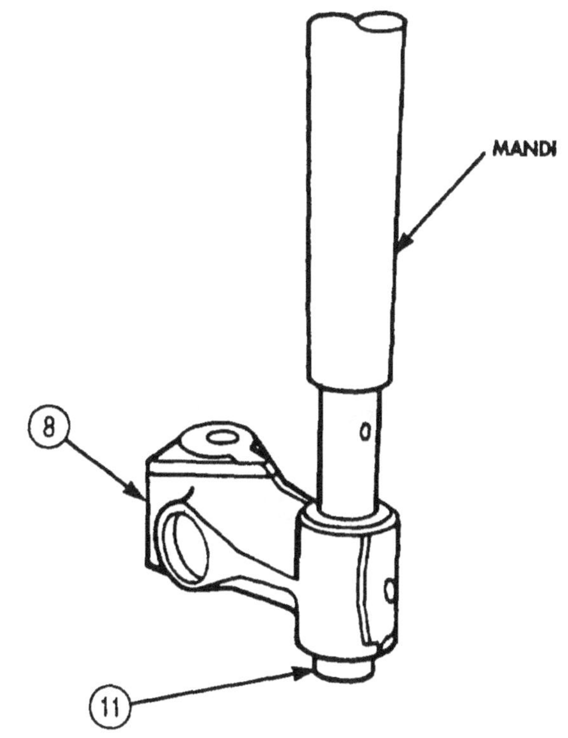

4-19. CAM FOLLOWER HOUSING MAINTENANCE (Contd)

c. Cleaning and Inspection

WARNING

Drycleaning solvent is flammable and toxic. Do not use near open flame and always have a tire extinguisher nearby when solvents are used. Use only in well-ventilated places, wear protective clothing, and dispose of cleaning rags in approved container. Failure to do this may result in injury or death to personnel and/or damage to equipment.

1. Clean cam follower housing (1) with drycleaning solvent and inspect for breaks and cracks. Replace if broken or cracked.
2. Clean cam follower shaft (2) with drycleaning solvent and inspect for breaks, cracks, or out-of-round condition. Replace if broken, cracked, or out-of-round. Replace if cam follower shaft (2) outer diameter is less than 0.748 in. (19.02 mm).
3. Measure all bearing surfaces with outside micrometer.
4. Clean cam follower lever (3) with drycleaning solvent and inspect for breaks and cracks. Replace if broken or cracked.
5. Inspect cam follower lever bushing (4) for breaks, cracks, or out-of-round condition. Replace if broken, cracked, or out-of-round.
6. Measure inner diameter of cam follower lever bushing (2) with telescoping gauge. Replace if inner diameter is more than 0.752 in. (19.10 mm).
7. Coat ball end of new push tube (5) with Prussian blue and place into push tube insert (6). Rotate ball end of push tube (5) to check push tube insert (6) for wear. Replace push tube insert (6) if wear area is not 80 percent blued.
8. Inspect cam follower roller pin (7) for breaks, cracks, or out-of-round condition. Replace if broken, cracked, or out-of-round.
9. Measure outside diameter of cam follower roller pin (7) with outside micrometer. Replace if outer diameter is less than 0.497 in. (12.62 mm).
10. Inspect exhaust and intake valve cam roller (8) for breaks, cracks, or out-of-round condition. Replace if broken, cracked, or out-of-round.
11. Set telescoping gauge to 0.503 in. (12.83 mm) and place into inner diameter of exhaust and intake valve cam roller (8). Replace if telescoping gauge slides into exhaust and intake valve cam roller (8).
12. Measure outer diameter of exhaust and intake valve cam rollers (8) with micrometer. Replace exhaust and intake valve cam rollers (8) if outer diameter is less than 1.248 in. (31.71 mm).

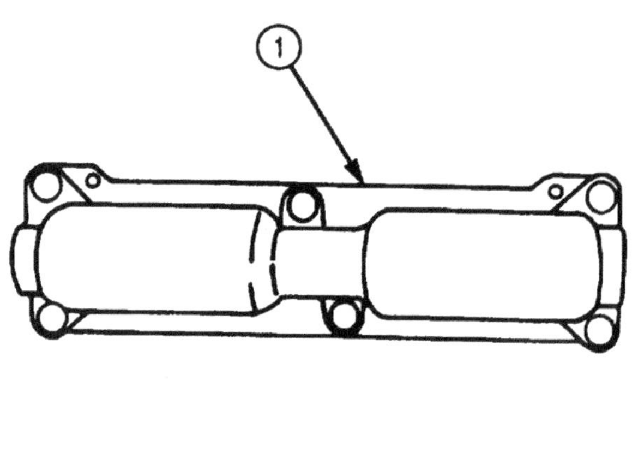

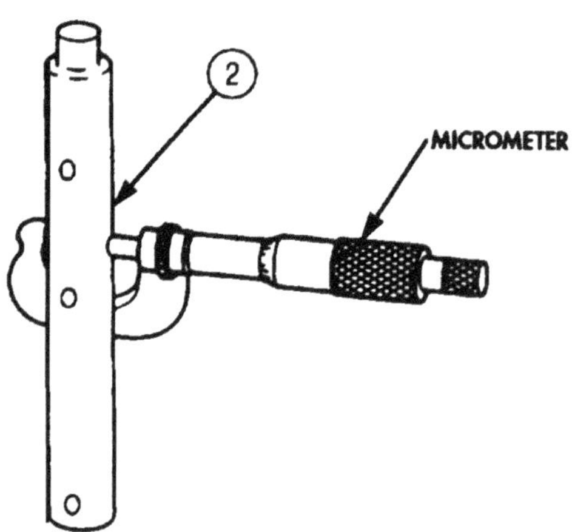

4-19. CAM FOLLOWER HOUSING MAINTENANCE (Contd)

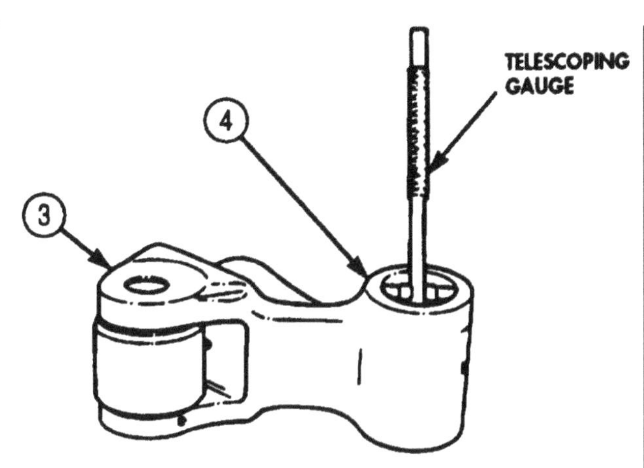

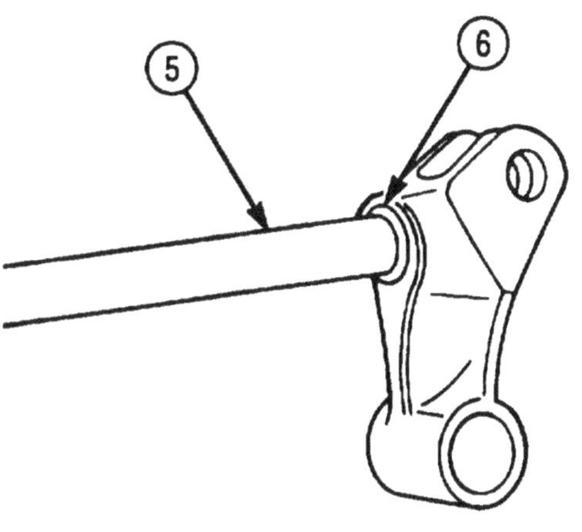

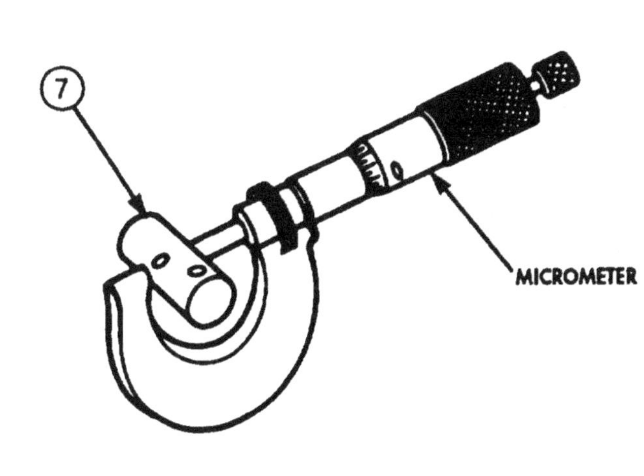

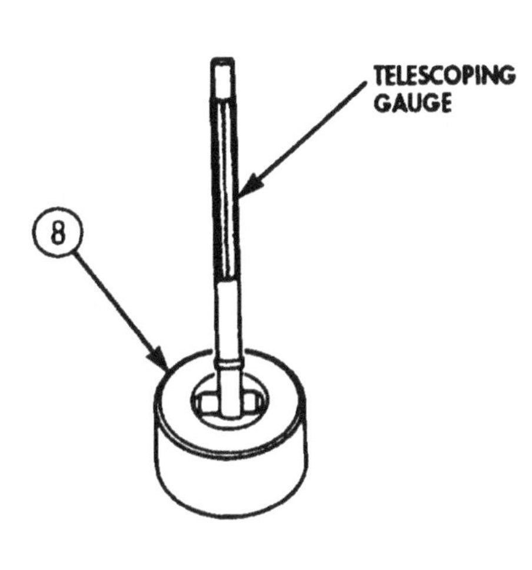

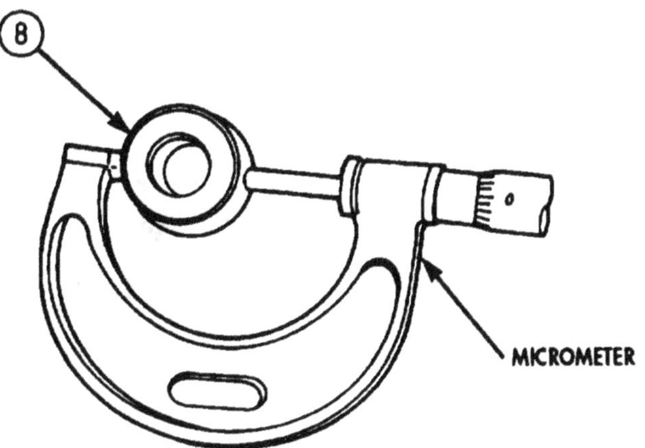

4-19. CAM FOLLOWER HOUSING MAINTENANCE (Contd)

d. Assembly

NOTE

- Apply light coat of oil to all parts before assembly.
- If new push tube insert is being installed, a new push tube must also be used.

1. Install push tube insert (1) into cam follower lever (2). Ensure push tube insert (1) is securely seated in cam follower lever (2).
2. Install cam follower lever bushing (3) into cam follower lever (2) with arbor press and mandrel. Align oil hole (4) in cam follower lever bushing (3) with oil hole (5) in cam follower lever (2).
3. Chamfer each end of cam follower lever bushing (3) with 60-degree angle cutter. Clean all metal chips from cam follower lever bushing (3) surfaces.
4. Place cam roller (8) into cam follower lever (2) with 0.006 in. (0.15 mm) feeler gauge between cam follower lever (2) and cam roller (8).
5. Install roller pin (6) through cam roller (8) with arbor press and mandrel and align roller pin (6) hole with hole in cam follower lever (2).
6. Install roller pin (6) on cam follower lever (2) with retainer pin (7).
7. Position six cam follower levers (2) into cam follower housing (9) in marked locations, and slide cam follower shaft (11) through cam follower housing (9) and six cam follower levers (2).
8. Align screw hole in cam follower housing (9) with screw hole in cam follower shaft (11) with temporary dummy screw (10) to prevent lockscrew (12) breakage when cup-plugs (13) are installed.

CAUTION

When pressing two cup-plugs into housing, press in only until plugs are flush with cam follower housing to avoid lockscrew breakage.

9. Coat two new cup-plugs (13) with sealing compound and press into cam follower housing (9) with arbor press and mandrel. Press only until flush with cam follower housing (9).
10. Remove dummy screw (10) from cam follower housing (9).
11. Install two shaft lockscrews (12) through cam follower housing (9) into cam follower shaft (11).

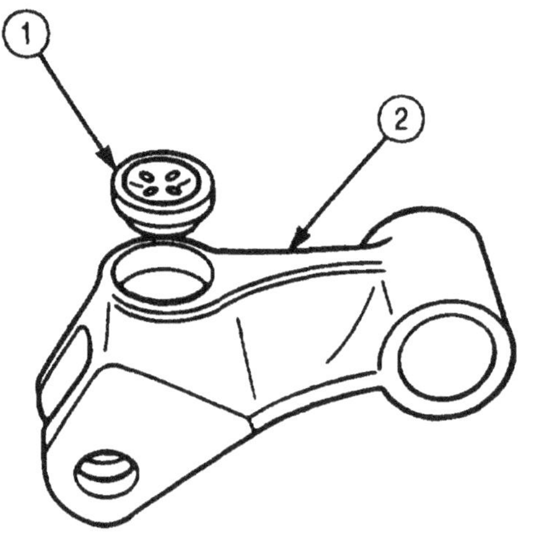

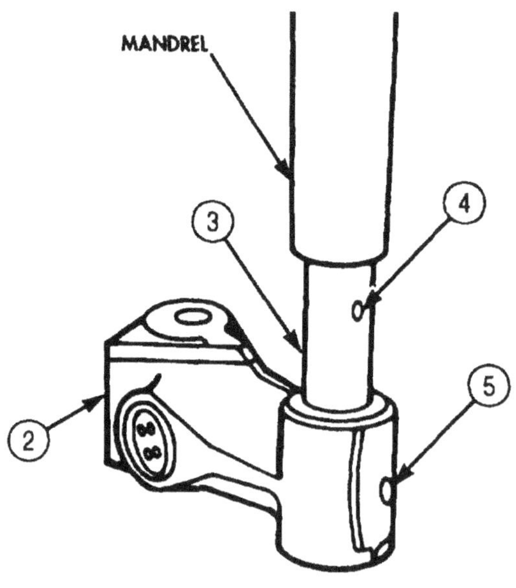

4-19. CAM FOLLOWER HOUSING MAINTENANCE (Contd)

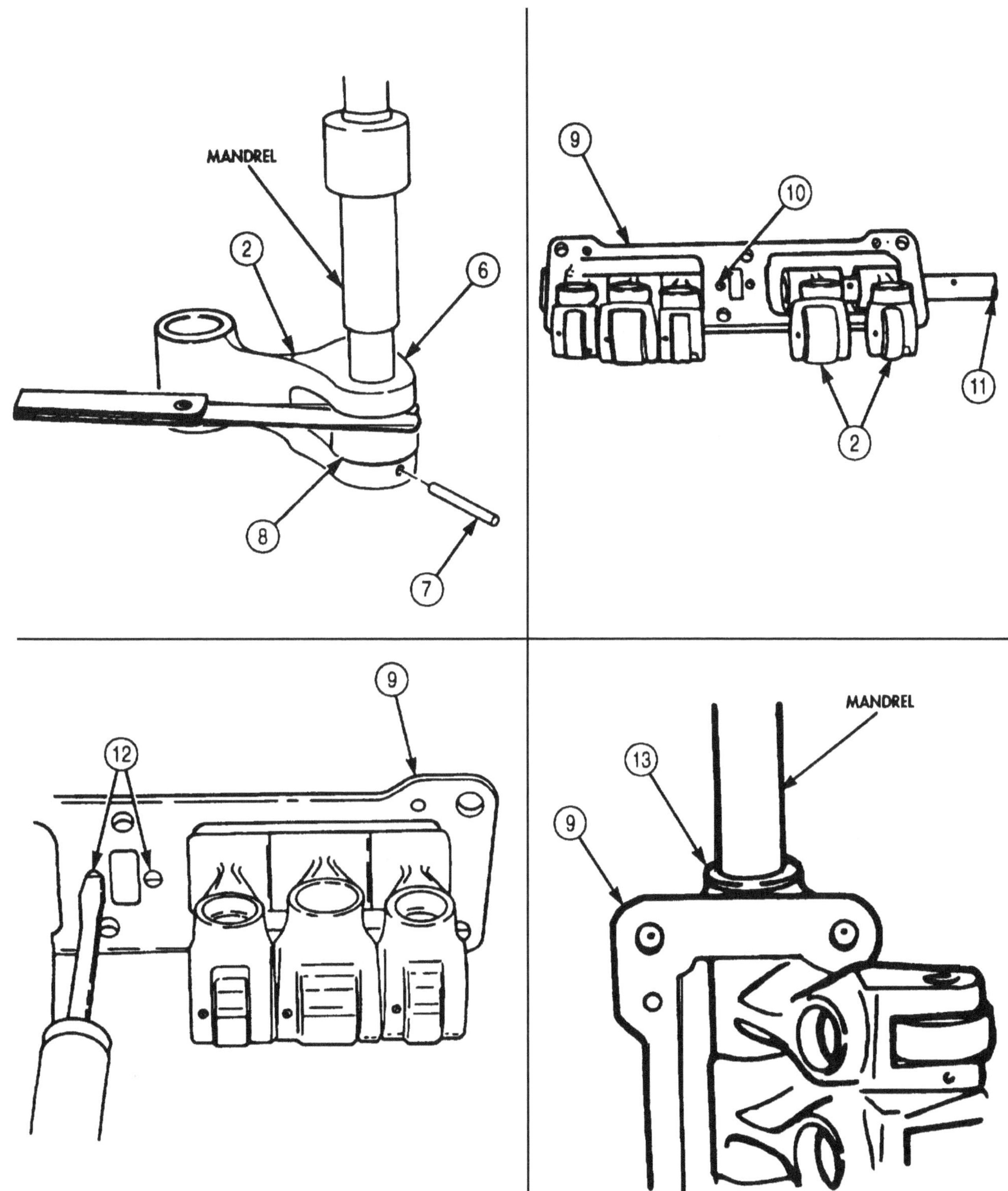

4-19. CAM FOLLOWER HOUSING MAINTENANCE (Contd)

e. Installation

CAUTION

- Before installation, ensure cam follower bushing levers have been lubricated (para. 2-18).
- If old cam follower housing assemblies are being installed, ensure they are installed in the same location from which they were removed.

NOTE

Cam follower housings are mounted with screw-assembled lockwashers on late model engines.

1. Check recorded measurements of cam follower housing gaskets (3) removed from engine block (2). Measurements should be 0.014-0.125 in. (0.36-3.2 mm).
2. Measure new cam follower housing gaskets (3) to ensure they measure exact thickness of originals.
3. Position cam follower housing gaskets (3) on engine block (2) with seals facing outward over dowels (1).
4. Position cam follower assembly (4) on cam follower housing gasket (3) over dowels (1) and seat against engine block (2). Lightly tap cam follower assembly (4) with soft-faced hammer.

NOTE

Installation of fuel line brackets is only required for No. 2-3 cam follower housing.

5. Install fuel line bracket (6) and six new screw-assembled lockwashers (5) on cam follower assembly (4).
6. Tighten screw-assembled lockwashers (5) 15 lb-ft (20 N•m) in sequence shown.
7. Tighten screw-assembled lockwashers (5) 30-35 lb-ft (41-48 N•m) in sequence shown.

4-19. CAM FOLLOWER HOUSING MAINTENANCE (Contd)

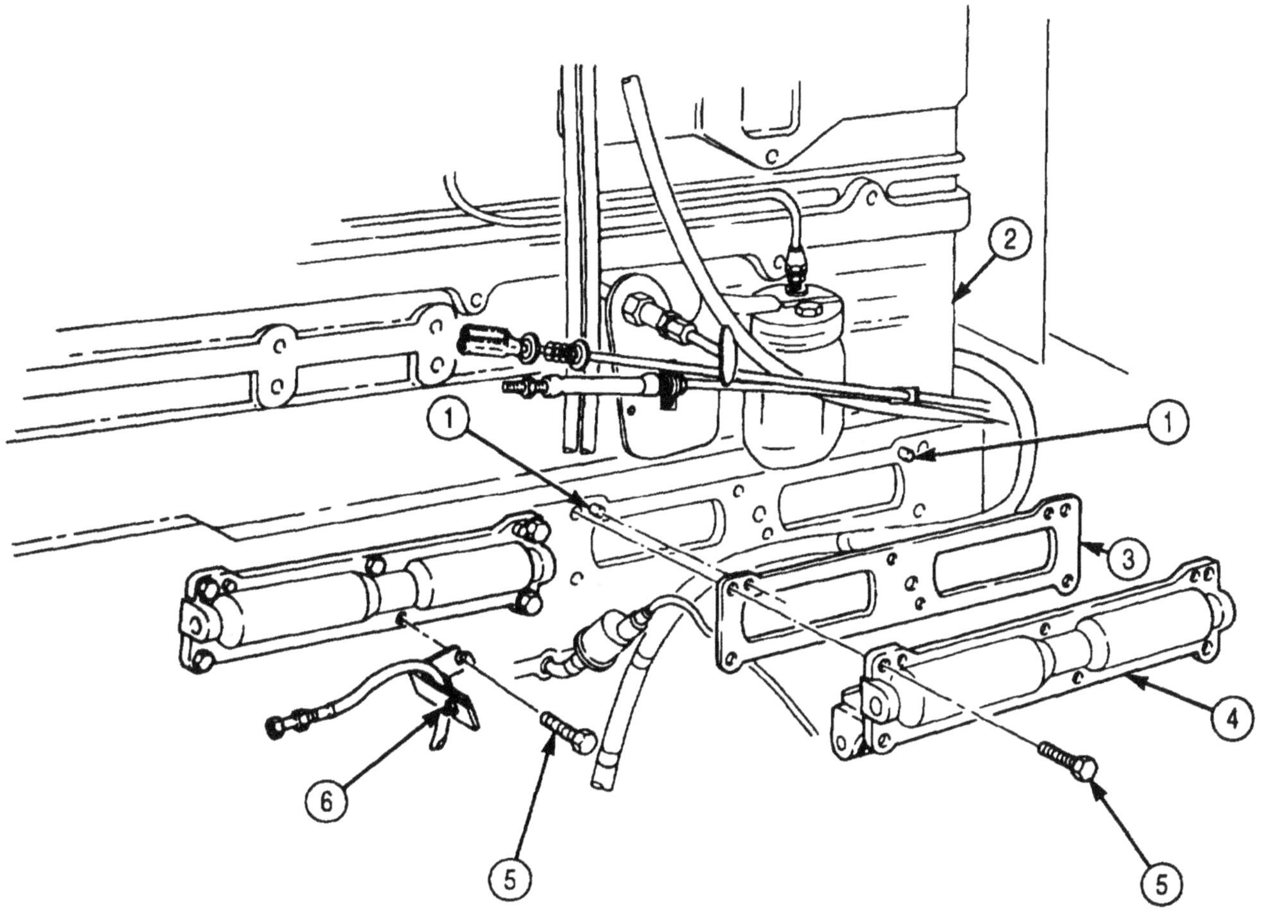

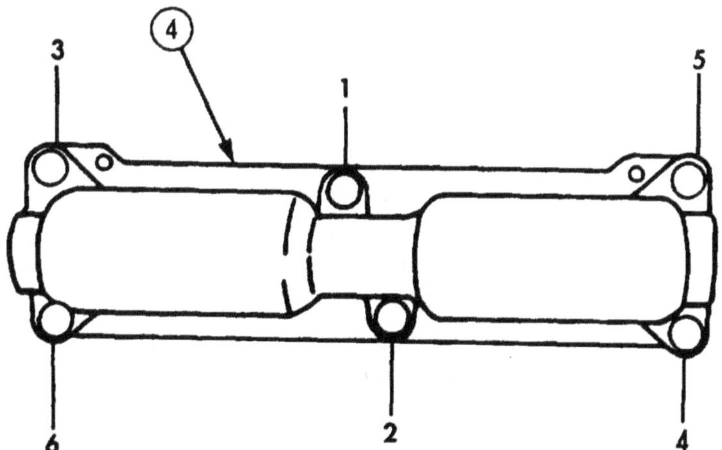

TIGHTENING SEQUENCE

FOLLOW-ON TASKS:
- Install rocker lever housings and push tubes (para. 4-20).
- Install air compressor (if removed) (para. 4-31).
- Install fuel pump (if removed) (para. 4-35).

4-20. ROCKER LEVER HOUSING AND PUSH TUBES MAINTENANCE

THIS TASK COVERS:

a. Removal
b. Disassembly
c. Cleaning and Inspection
d. Assembly
e. Installation

INITIAL SETUP:

APPLICABLE MODELS
M939/A1

SPECIAL TOOLS
Rocker lever bushing block and mandrel (Appendix E, Item 116)

TOOLS
General mechanic's tool kit (Appendix E, Item 1)
Inside micrometer (Appendix E, Item 82)
Outside micrometer (Appendix E, Item 80)
Radius gauge, 1/4-in. (Appendix E, Item 136)
Bore gauge (Appendix E, Item 136)
Torque wrench (Appendix E, Item 144)
Brass drift
Arbor press

MATERIALS/PARTS
Three gaskets (Appendix D, Item 187)
Two locknuts (Appendix D, Item 294)
Six O-rings (Appendix D, Item 473)
Three gaskets (Appendix D, Item 172)
Lubrication oil (Appendix C, Item 50)
Prussian blue (Appendix C, Item 54)
Drycleaning solvent (Appendix C, Item 71)

PERSONNEL REQUIRED
TWO

REFERENCES (TM)
TM 9-2320-272-10
TM 9-2320-272-24P

EQUIPMENT CONDITION
- Parking brake set (TM 9-2320-272-10).
- Engine lift eyes removed (para. 4-11).
- Intake manifold removed (para. 4-24).
- Engine access cover fin cab) removed (para. 4-13).

GENERAL SAFETY INSTRUCTIONS
- Keep fire extinguisher nearby when using drycleaning solvent.
- Drycleaning solvent is flammable and toxic. Do not use near an open flame.

a. Removal

NOTE

Perform step 1 for removing rocker lever housing cover closest to firewall.

1. Remove two locknuts (5), washers (4), and rubber cushions (3) from left and right cab A-posts (1). Discard locknuts (5).

NOTE

Assistant will help with step 2.

2. Turn left and right screw jacks (6) to raise cab above frame (2) 4-5 in. (102-127 mm).
3. Loosen hose clamps (10) and remove hose (9) from breather (8).

NOTE

Perform step 4 for late model engine only.

4. Loosen two clamps (14) and remove breather tube (15) from breather (8) and elbow (16).

NOTE

Rocker lever housings are mounted with screw-assembled lockwashers on late model engine.

5. Remove fifteen screws (11), three rocker covers (7), and gaskets (12) from rocker lever housings (13). Discard gaskets (12) and clean gasket remains from mating surfaces.

4-20. ROCKER LEVER HOUSING AND PUSH TUBES MAINTENANCE (Contd)

6. Loosen eighteen adjusting screw locknuts (22) on eighteen rocker levers (18).
7. Loosen eighteen adjusting screws (23) on eighteen rocker levers (18) two turns.

NOTE

Tag rocker lever housings and push tubes for installation.

8. Remove eighteen screws (17), upper radiator support bracket (24), three rocker lever housings (13), gaskets (20), and eighteen push tubes (21) from three cylinder heads (19). Discard gaskets (20) and clean gasket remains from mating surfaces of three cylinder heads (19).

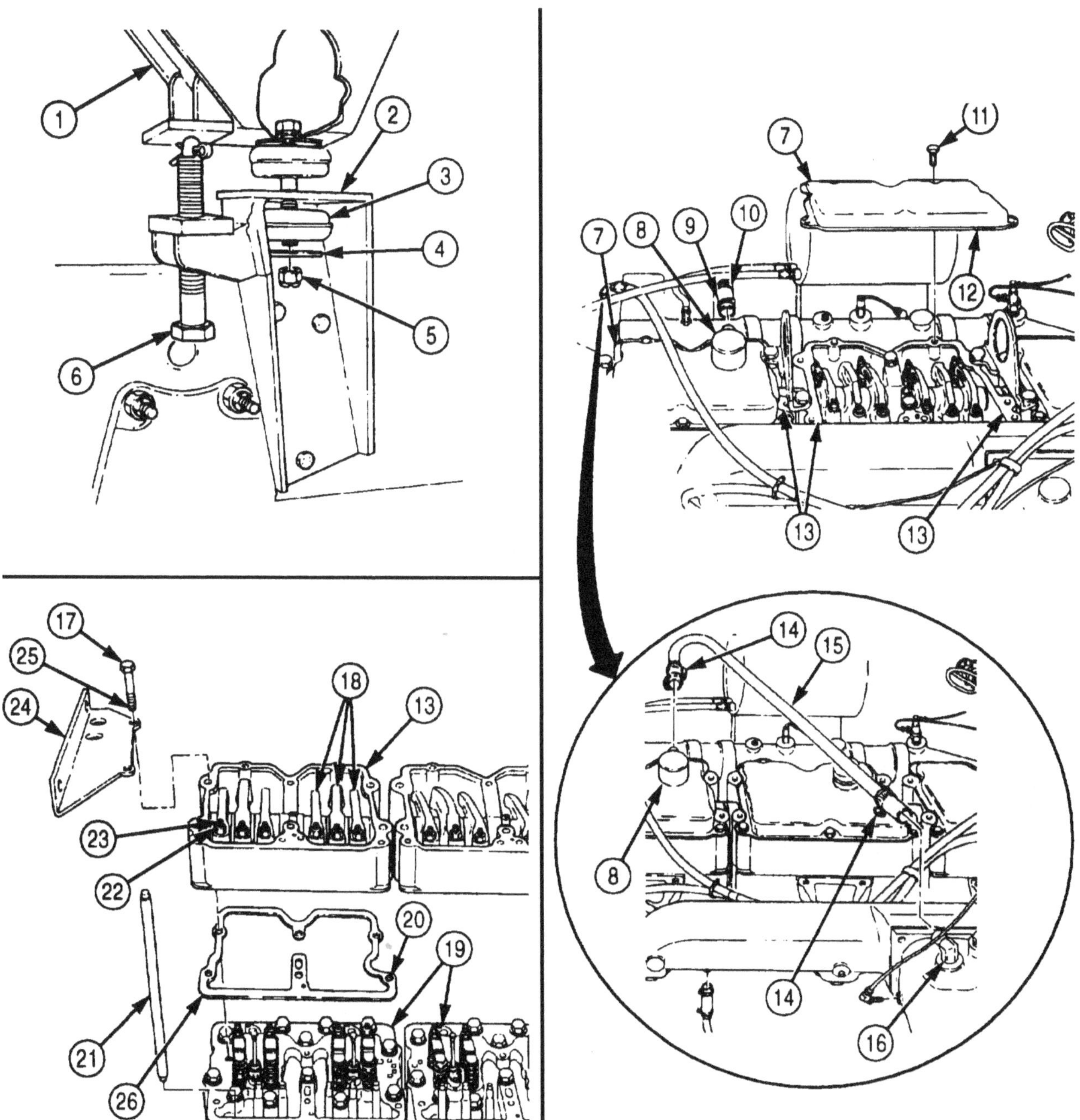

4-20. ROCKER LEVER HOUSING AND PUSH TUBES MAINTENANCE (Contd)

b. Disassembly

1. Remove shaft retaining setscrew (8) from rocker lever housing (9).

CAUTION

When removing rocker lever shaft, center the brass drift and lightly tap on shaft. Striking the housing or levers will cause damage.

NOTE

- Mark locations of rocker levers for proper installation during assembly.
- Mark direction of rocker lever shaft for proper installation during assembly.

2. Remove rocker lever shaft (7) from rocker lever housing (9) with brass drift.
3. Remove two O-rings (6) from rocker lever shaft (7). Discard O-rings (6).
4. Remove two exhaust rocker levers (3), injector rocker levers (4), and intake rocker levers (5) from rocker lever housing (9).
5. Remove six rocker lever adjusting screws (2) and nuts (1) from rocker levers (3), (4), and (5).

c. Cleaning and Inspection

1. Clean rocker lever bushings (13) and inspect for cracks and pitting. Discard bushing (13) if pitted or cracked.
2. Measure inner diameter of rocker lever bushings (13) at several points. Replace rocker lever bushings (13) if inner diameter exceeds 1.129 in. (28.66 mm).

NOTE

Steps 3 and 4 are performed only if bushings are to be replaced.

3. Using arbor press and mandrel, remove rocker lever bushings (13).

CAUTION

Ensure new bushing oil holes are properly aligned to oil passages in rocker levers. Failure to do so will cause lubrication failure and severe engine damage.

4. Using arbor press and mandrel, install new rocker lever bushings (13) in rocker levers (3), (4), and (5). Ensure oil holes (12) are aligned with rocker lever oil passages (14).
5. Clean intake rocker lever (5), exhaust rocker lever (3), and injector rocker lever (4), and inspect for breaks, cracks, and plugged oil passages. Replace if broken, cracked, or if oil passages (14) are plugged.
6. Coat new injector link (11) ball end with Prussian blue and place ball end into socket seat (10) of injector rocker lever (4).
7. Using hand pressure, rotate injector link (11) and check socket seat (10) wear area. If socket seat (10) wear area is not 80 percent blued, replace socket seat (10).

4-20. ROCKER LEVER HOUSING AND PUSH TUBES MAINTENANCE (Contd)

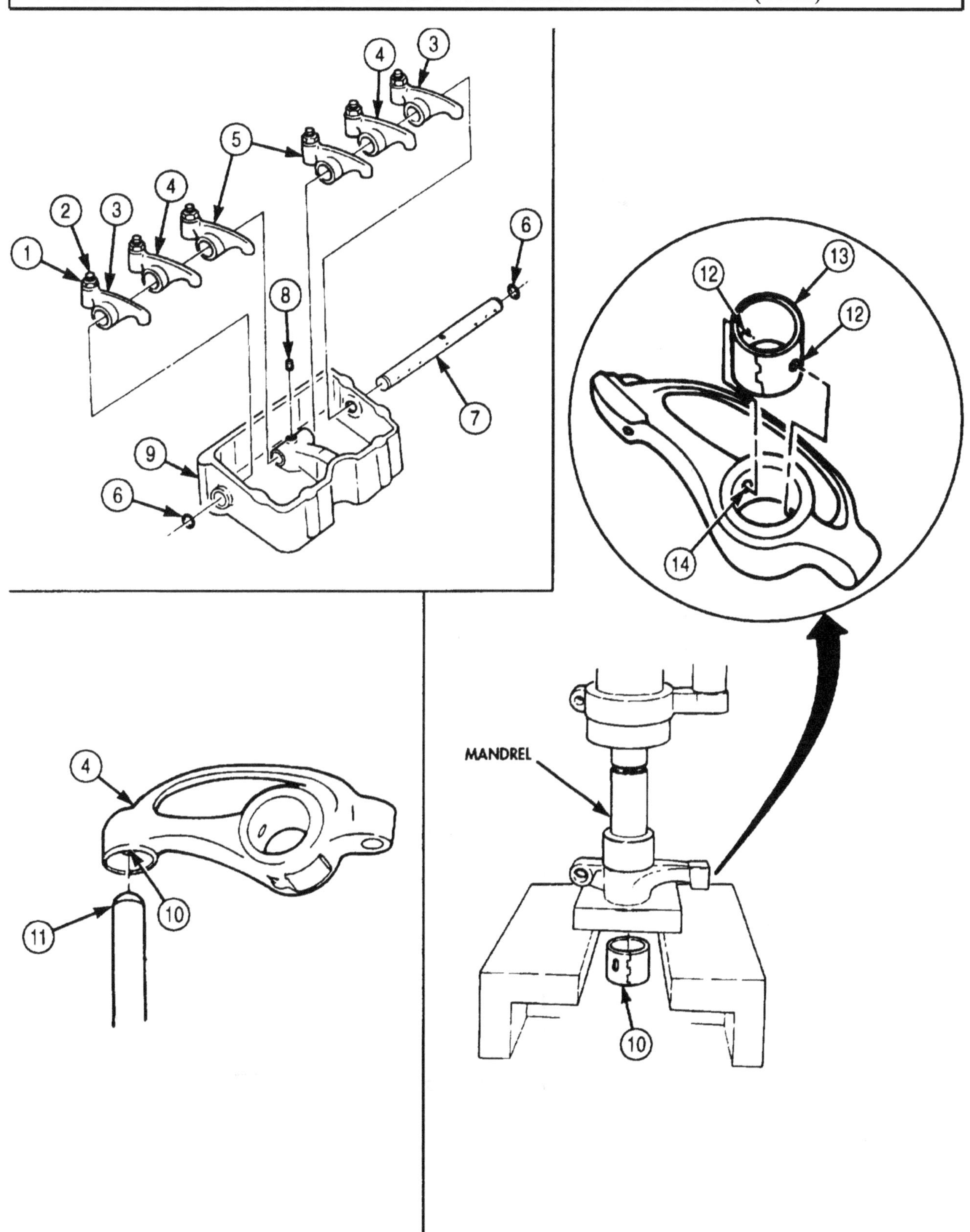

4-20. ROCKER LEVER HOUSING AND PUSH TUBES MAINTENANCE (Contd)

8. Clean intake (1), exhaust (2), and injector (3) push tubes and inspect for cracks and bends. Replace if cracked or bent.
9. Coat ball end (7) of new adjusting screw (4) with Prussian blue and place ball end (7) into tube socket (5).

NOTE

Perform steps 10 through 19 for intake, exhaust, and injector push tubes.

10. Using hand pressure, rotate new adjusting screw (4) and check tube socket (5) wear area. Replace if tube socket (5) wear area is not 80 percent blued.
11. Clean six valve adjusting screws (4) and move new nut (6) full length of adjusting screw (4) by hand. Replace adjusting screw (4) if new nut (6) binds on threads.
12. Inspect ball end (7) of six valve adjusting screws (4) for flat spots with 1/4-in. (6.35-mm) radius gauge. Discard valve adjusting screws (4) if flat spots are noted.
13. Clean rocker lever housing (8) and inspect for cracks and breaks. Replace if cracked or broken.
14. Inspect shaft bore (9) of rocker lever housing (8) for scratches. Replace rocker lever housing (8) if shaft bore (9) is scratched.
15. Using bore gauge, measure inner diameter of shaft bore (9) of rocker lever housing (8) at several points for wear. Replace rocker lever housing (8) if inner diameter is more than 1.125 in. (28.56 mm).

NOTE

Perform step 16 if plugs are to be replaced.

16. Remove rocker lever plugs (12) from rocker lever shaft (11). Discard rocker lever plugs (12) if cracked, pitted, or threads are damaged.

WARNING

Drycleaning solvent is flammable and toxic. Do not use near open flame and always have a fire extinguisher nearby when solvents are used. Use only in well-ventilated places, wear protective clothing, and dispose of cleaning rags in approved container. Failure to do this may result in injury or death to personnel and/or damage to equipment.

17. Clean oil passages (10) of rocker lever shaft (11) with drycleaning solvent and brush and inspect. Replace rocker lever shaft (11) if scratched.
18. Measure outer diameter of rocker lever shaft (11) at several points for wear. Replace if outer diameter is less than 1.122 in. (28.50 mm).
19. Install rocker lever plugs (12) on rocker lever shaft (11). Tighten plugs (12) 50-70 lb-ft (68-95 N-m). Mark end of rocker lever plug to indicate setscrew hole (13) position.

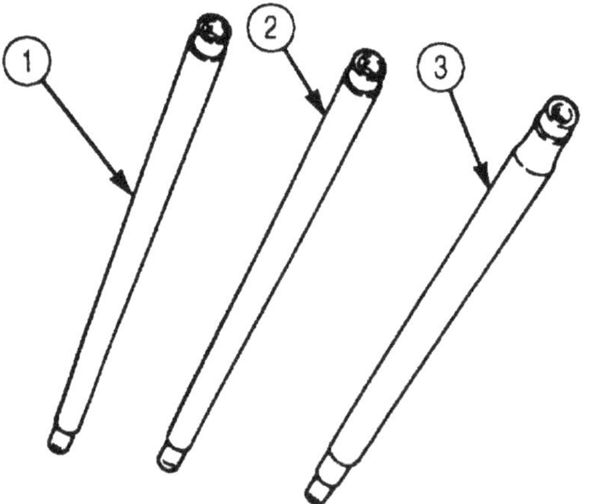

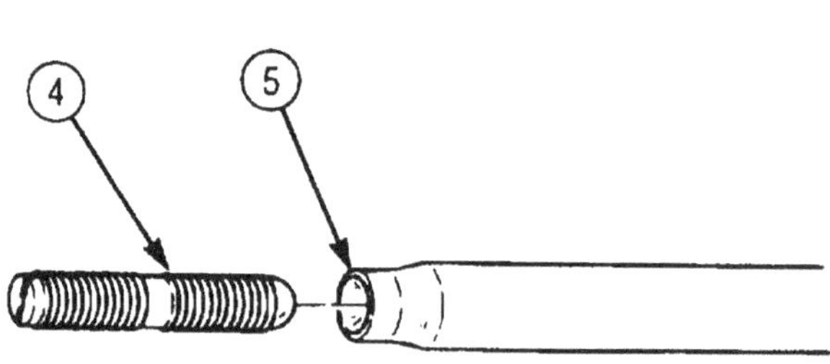

4-20. ROCKER LEVER HOUSING AND PUSH TUBES MAINTENANCE (Contd)

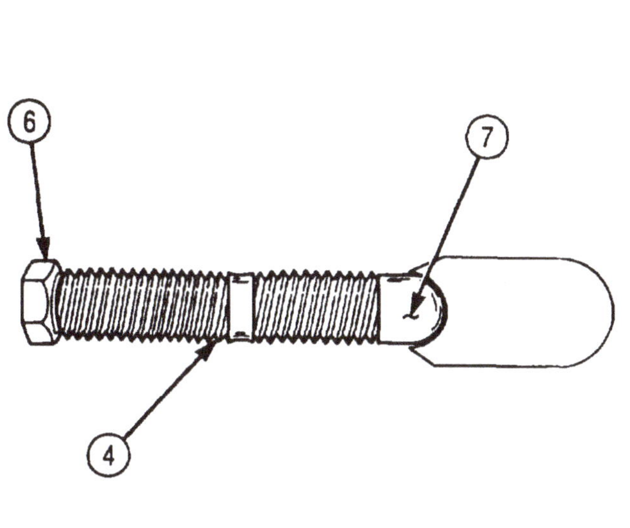

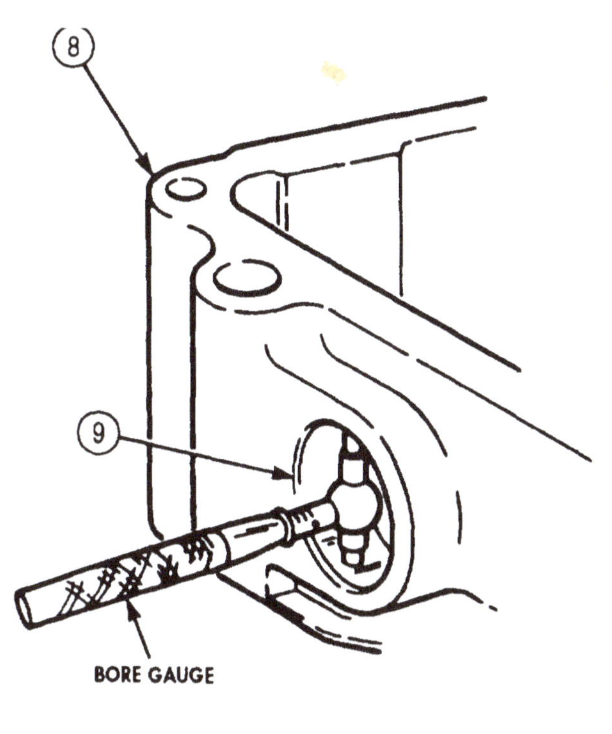

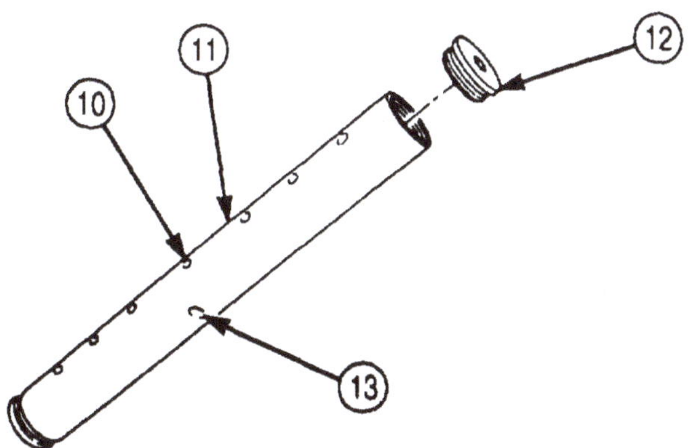

4-20 ROCKER LEVER HOUSING AND PUSH TUBES MAINTENANCE (Contd)

d. Assembly

NOTE

Apply light coat of lubricating oil to all parts before assembly.

1. Install one rocker lever adjusting screw (2) and nut (1) in rocker levers (3), (4), and (5).

CAUTION

Ensure rocker lever shaft oil holes are properly aligned to oil passages in rocker lever housing. Failure to do this will cause lubrication failure and engine damage.

2. Install two exhaust rocker levers (3), intake rocker levers (5), and injector rocker levers (4) in rocker lever housing (10).
3. Install rocker lever shaft (8) through one end of rocker lever housing (10) and through rocker levers (3), (4), and (5) and center of rocker lever housing (10).
5. Install first new O-ring (6) and next set of rocker levers (3), (4), and (5) in rocker lever housing (10).
6. Install second new O-ring (6) on rocker lever shaft (8).
7. Complete installation of rocker lever shaft (8) through rocker lever housing (10).
8. Align hole (7) in rocker lever shaft (8) with hole in rocker lever housing (10) and install setscrew (9).

e. Installation

NOTE

- Rocker lever housings are mounted with screw-assembled washers for late model engines.
- Do not mix push tubes during installation. The injector tube is the largest, and is positioned between the intake and exhaust push tubes. Intake and exhaust push tubes are identical.
- Ensure push tubes remain seated during rocker lever housing installation.

1. Install eighteen push tubes (12) through cylinder head (11) and into socket seat (14) on cam follower (13).

4-20. ROCKER LEVER HOUSING AND PUSH TUBES MAINTENANCE (Contd)

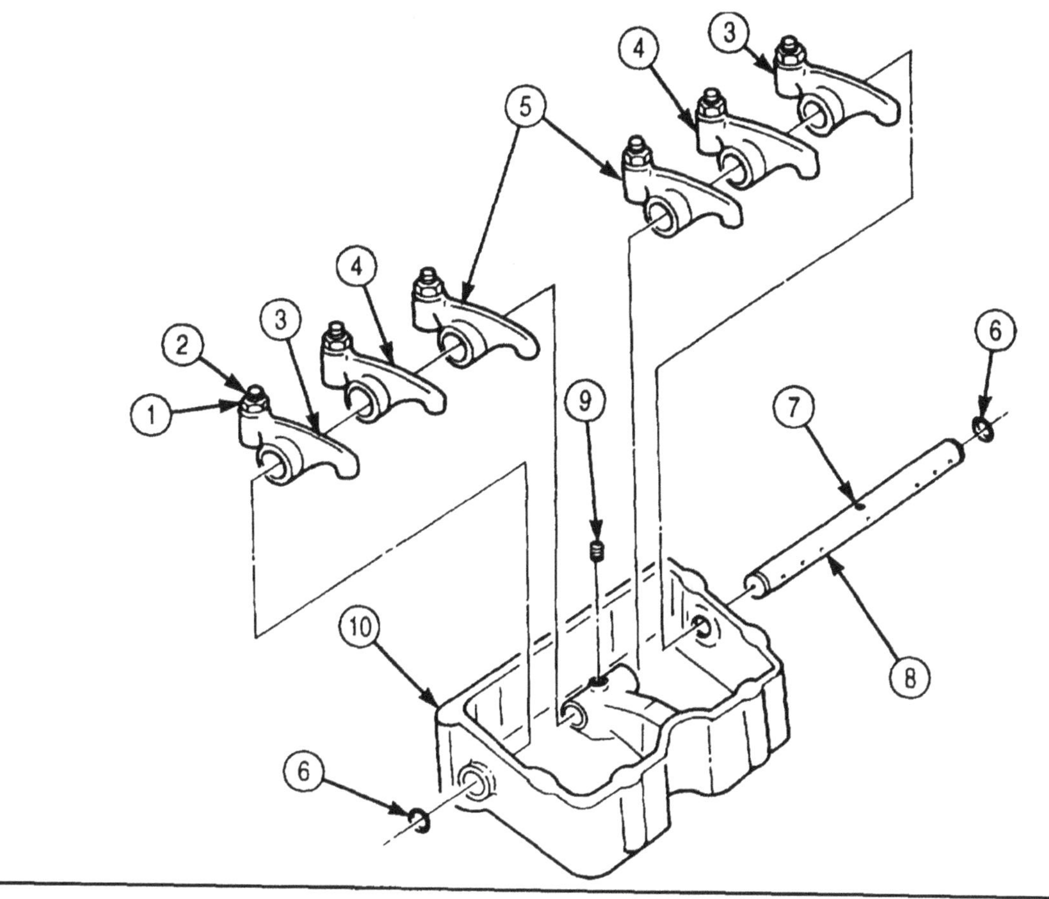

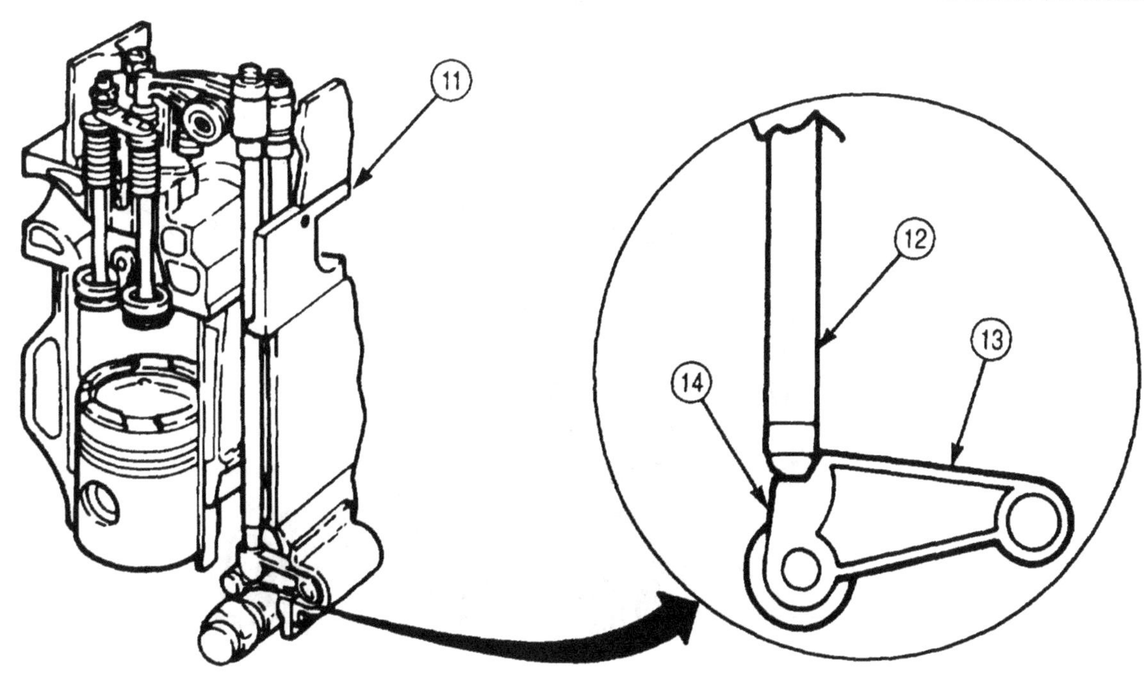

4-20. ROCKER LEVER HOUSING AND PUSH TUBES (M939/A1) MAINTENANCE (Contd)

2. Install three new gaskets (7) and rocker lever housings (6) on cylinder heads (8) and align push tubes (9) with rocker levers.
3. Install upper radiator support bracket (1) and two lifting eyes (3) on rocker lever housings (6) with six screws (2). Finger-tighten screws (2).

NOTE

Rocker lever housings are mounted using screw-assembled lockwashers on late model engine.

4. Install twelve washers (5) and screws (4) on rocker lever housings (6) and cylinder heads (8). Tighten screws (2) and (4) 50-65 lb-ft (68-88 N•m) in sequence shown.
5. Install three new gaskets (17) and rocker lever housing covers (18) on three rocker lever housings (6) with fifteen screws (16). Tighten screws (16) 10-15 lb-ft (14-20 N•m).
6. Connect breather hose (20) to breather (19) with hose clamp (21).

NOTE

Perform step 7 for late model engine only.

7. Connect breather hose (23) to breather (19) and elbow (24) with two clamps (22).

NOTE

- Perform steps 8 for no. three rocker lever housing cover only.
- Assistant will help with step 8.

8. Turn left and right screw jacks (15) at the same time and lower cab on frame (11). Ensure clearance between cab A-post (10) and jack screw (15) is 1-2 in. (25-50 mm).
9. Install two rubber cushions (12) on left and right cab A-posts (10) with two washers (13) and new locknuts (14).

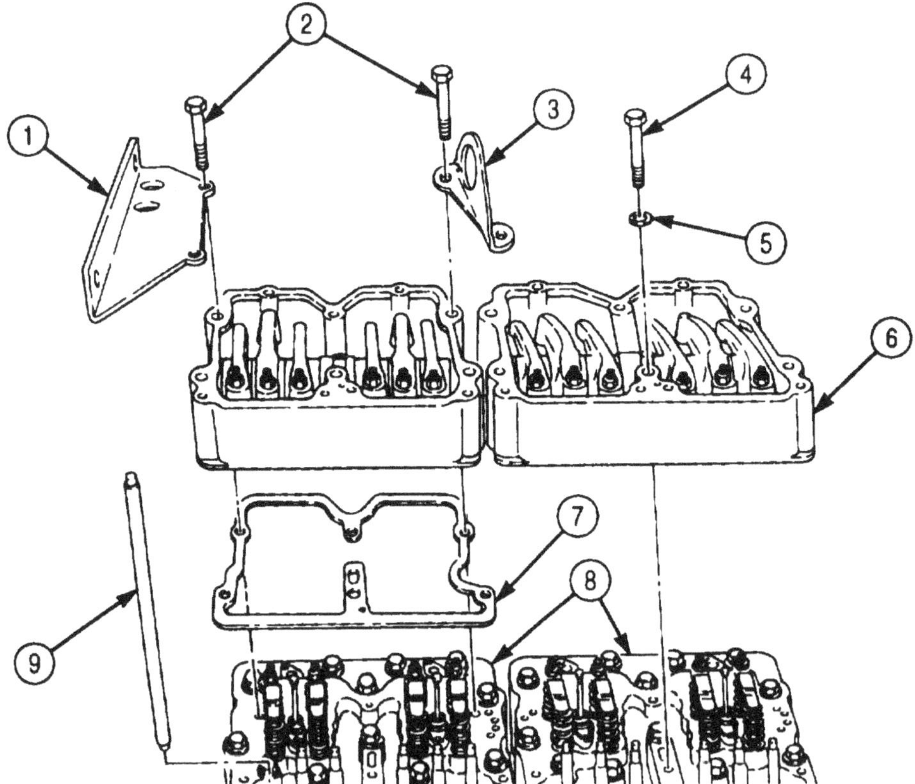

4-20. ROCKER LEVER HOUSING AND PUSH TUBES MAINTENANCE (Contd)

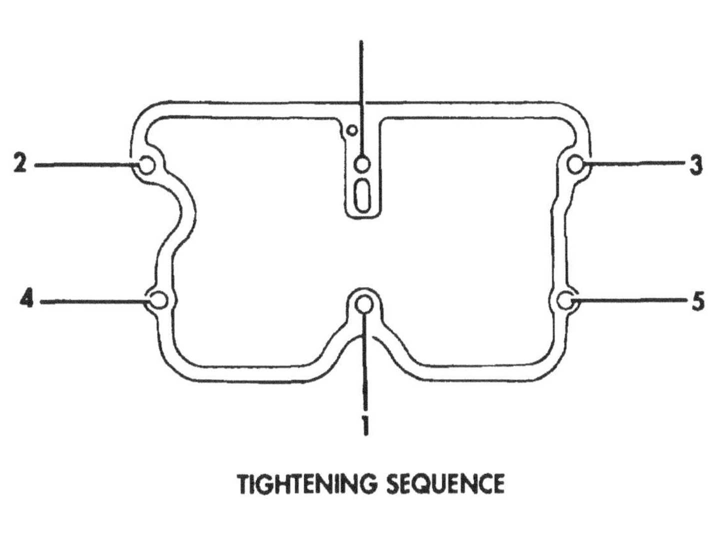

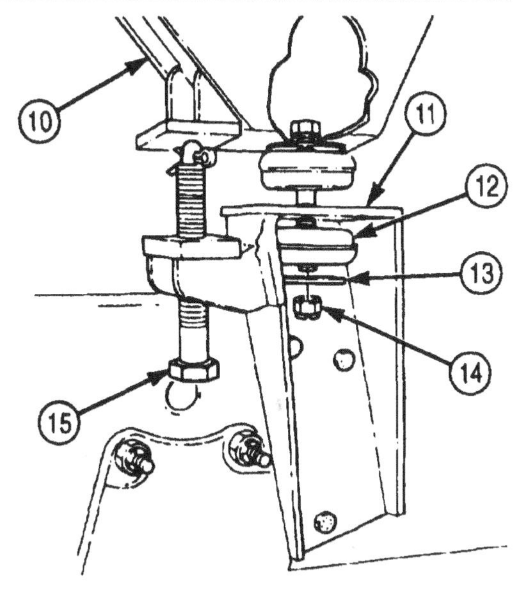

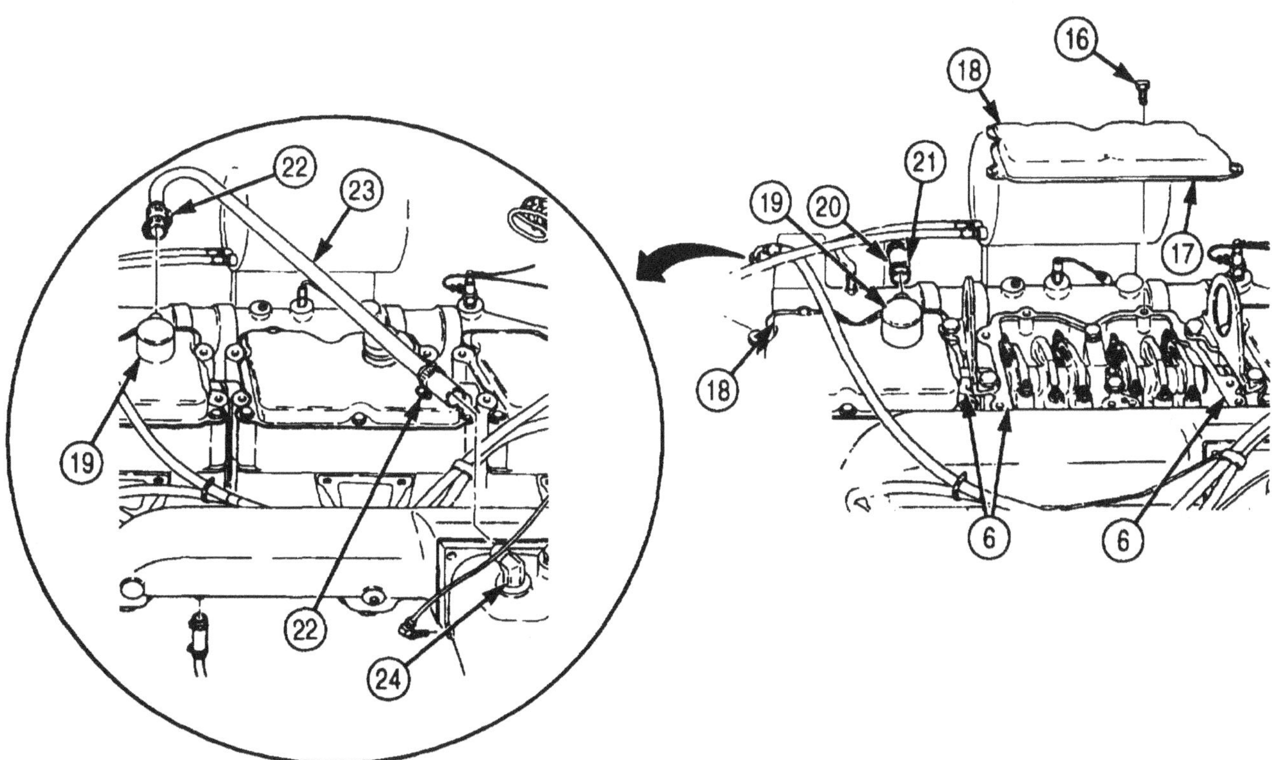

FOLLOW-ON TASKS:
- Adjust valve and injector (dial indicator method) (para. 4-33).
- Install engine access cover (in cab) (para. 4-13).
- Install intake manifold (para. 4-24).
- Install engine lift eyes (para. 4-11).

4-21. ENGINE OIL PUMP REPLACEMENT

THIS TASK COVERS:

a. Gear Backlash Test
b. Removal
c. Installation

INITIAL SETUP:

APPLICABLE MODELS
M939/A1

TOOLS
General mechanic's tool kit (Appendix E, Item 1)
Dial indicator (Appendix E, Item 36)
Torque wrench (Appendix E, Item 144)

MATERIALS/PARTS
Oil pump flange gasket (Appendix D, Item 176)
Six lockwashers (Appendix D, Item 354)
Two lockwashers (Appendix D, Item 411)
Two lockwashers (Appendix D, Item 382)

REFERENCES (TM)
TM 9-2320-272-24P

EQUIPMENT CONDITION
- Engine oil filter removed (para. 3-5).
- Engine front gearcase cover removed (task a. only) (para. 4-18).

a. Gear Backlash Test

1. Place dial indicator on engine block flange (2). Ensure anvil is positioned against drive gear tooth (1). Turn drive gear (3) clockwise until tight.
2. Zero-dial indicator index line and turn drive gear (3) counterclockwise until tight. Note amount of movement of dial indicator hand. Normal range of movement is 0.004-0.016 in. (0.10-0.40 mm). Replace drive gear (3) if less than 0.002 in. (0.05 mm).

b. Removal

CAUTION

Mounting and assembly screws are of different sizes and lengths. Screws must be tagged for installation. Misplaced screws can damage parts.

1. Remove two screws (8) and lockwashers (7) from oil pump body flange (16). Discard lockwashers (7).
2. Completely loosen screw (13) and lockwasher (15) from oil pump body flange (16). Screw (13) is removed during next step. Discard lockwasher (15).
3. Remove screws (6) and (10) and lockwashers (5) and (11) from filter head (9). Discard lockwashers (5) and (11).
4. Remove oil pump (12) and gasket (4) from front gearcase (14). Discard gasket (4) and clean gasket remains from mating surfaces of oil pump (12) and front gearcase (14).

4-21. ENGINE OIL PUMP REPLACEMENT (Contd)

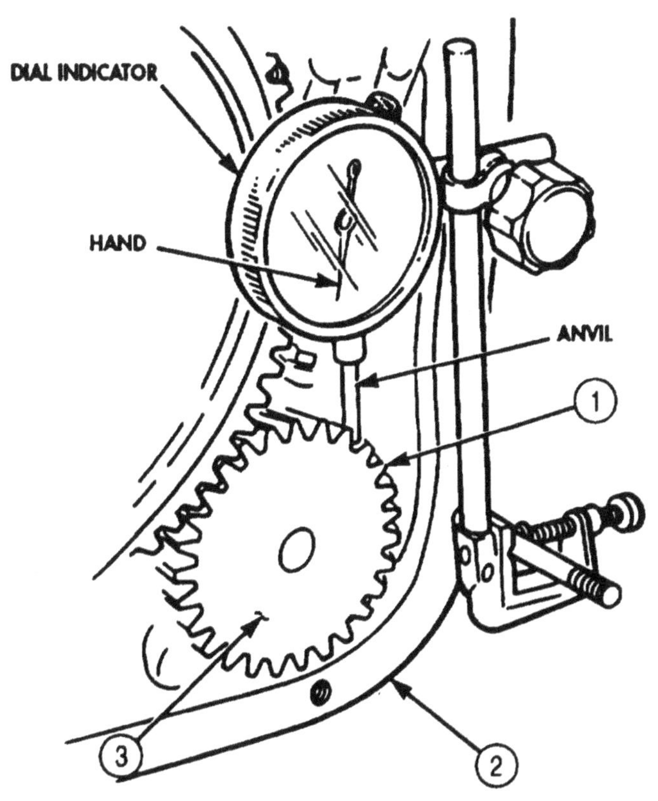

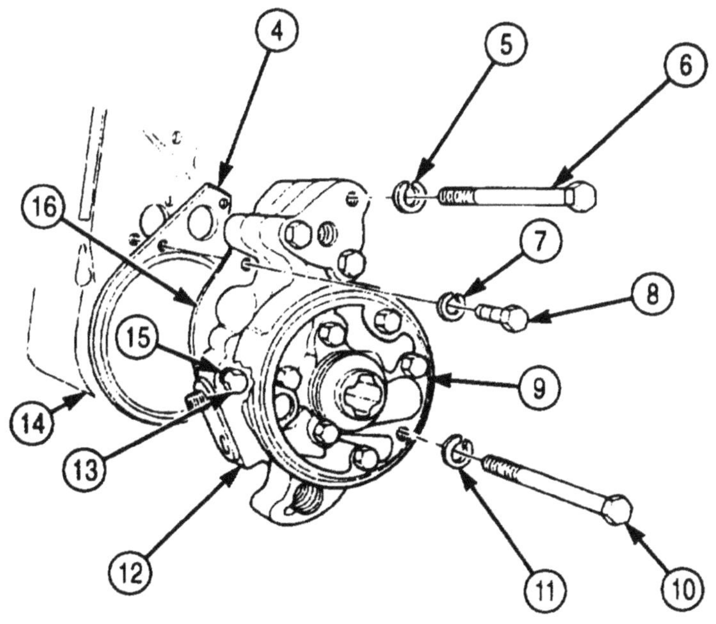

4-21. ENGINE OIL PUMP REPLACEMENT (Contd)

c. Installation

CAUTION

Mounting screws are of different sizes. Do not force screws to bottom. Misplaced screws can damage parts.

1. Position new gasket (1) and oil pump (9) on gearcase (11) and install new lockwashers (2) and (8) and screws (3) and (7) through oil filter head (6) into gearcase (11).
2. Install three new lockwashers (4) and screws (5) through oil pump flange (13) into gearcase (11).
3. Install screw (10) and new lockwasher (12) through oil pump flange (13) into gearcase (11).
4. Tighten screws (3), (7), and (10) 35-46 lb-ft (48-62 N•m).

NOTE

After installation, repeat gear backlash test, task a.

4-21. ENGINE OIL PUMP REPLACEMENT (Contd)

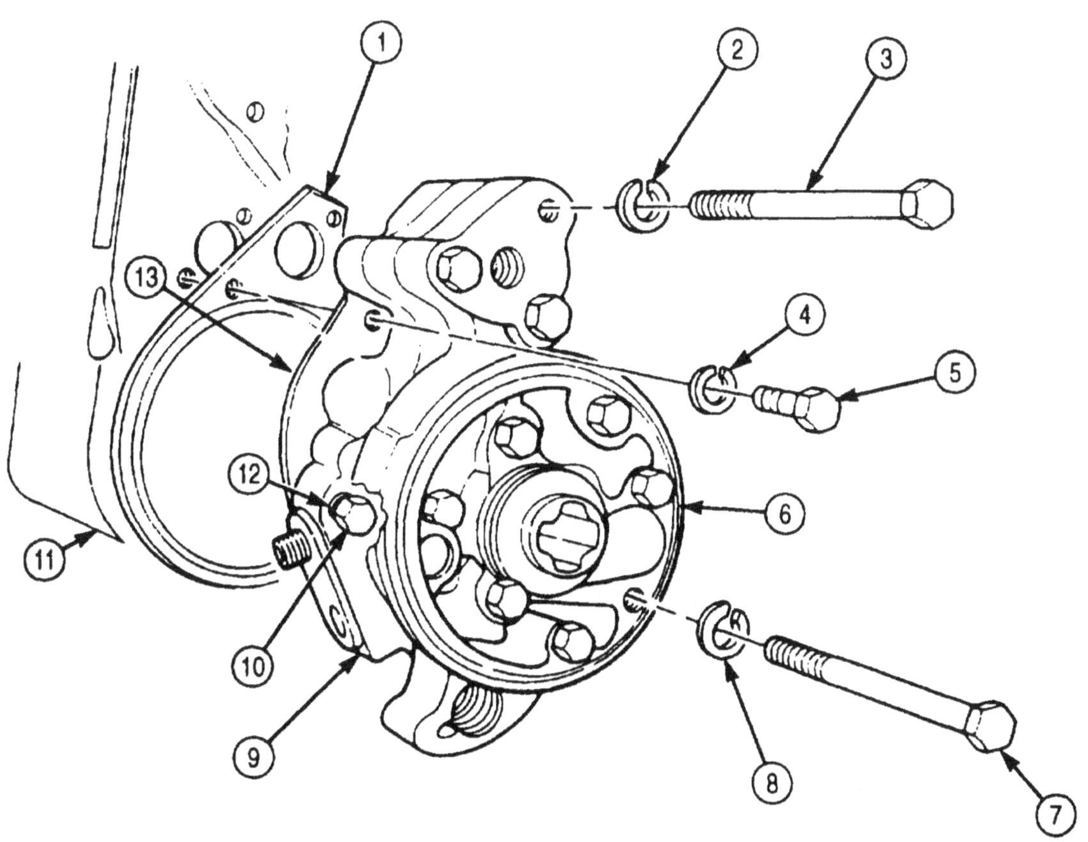

FOLLOW-ON TASKS:
- Install front gearcase cover (para. 4-18).
- Install engine oil filter (para. 3-5).

4-22. ENGINE OIL PAN MAINTENANCE

THIS TASK COVERS:

a. Removal
b. Disassembly
c. Cleaning and Inspection
d. Assembly
e. Installation

INITIAL SETUP:

APPLICABLE MODELS
M939/A1

TOOLS
General mechanic's tool kit (Appendix E, Item 1)
Torque wrench (Appendix E, Item 146)

MATERIALS/PARTS
Three lockwashers (Appendix D, Item 364)
Oil pan gasket (Appendix D, Item 493)
Aerator gasket (Appendix D, Item 177)
Gasket (Appendix D, Item 493)
Suction flange gasket (Appendix D, Item 178)
Sealing compound (Appendix C, Item 62)

REFERENCES (TM)
TM 9-2320-272-24P
LO 9-2320-272-12

EQUIPMENT CONDITION
- Front sump tube removed (para. 3-9).
- Oil dipstick tube removed (para. 3-3).
- Engine oil drained (para. 3-5).

a. Removal

1. Disconnect oil return hose (6) from aerator (3).
2. Disconnect oil pickup hose (2) from suction flange (5).
3. Remove two screws (7), washers (8), clamps (1), oil return hose (6), and pickup hose (2) from oil pan (4).
4. Remove six screws (15) and washers (14) from oil pan (4) and flywheel housing (10).
5. Remove four screws (13) and washers (12) from rear of oil pan (4).
6. Remove thirty screws (17), washers (16), four screws (19), washers (20), brace (18), oil pan (4), and oil pan gasket (11) from engine block (21), and front gearcase cover (9). Discard oil pan gasket (11) and clean gasket remains from mating surfaces of brace (18). oil pan (4), engine block (21), and front gearcase cover (9).

4-22. ENGINE OIL PAN MAINTENANCE (Contd)

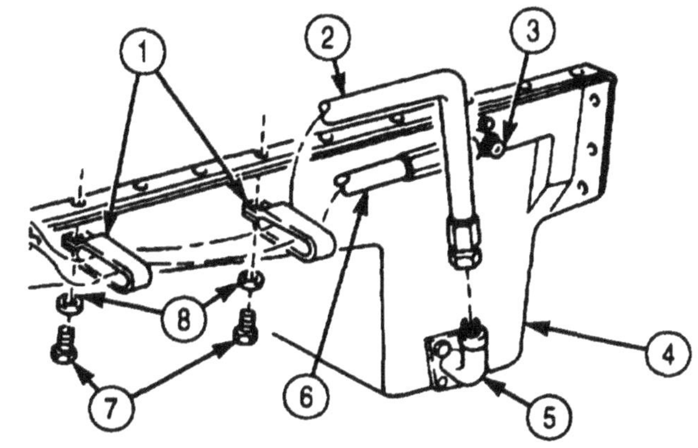

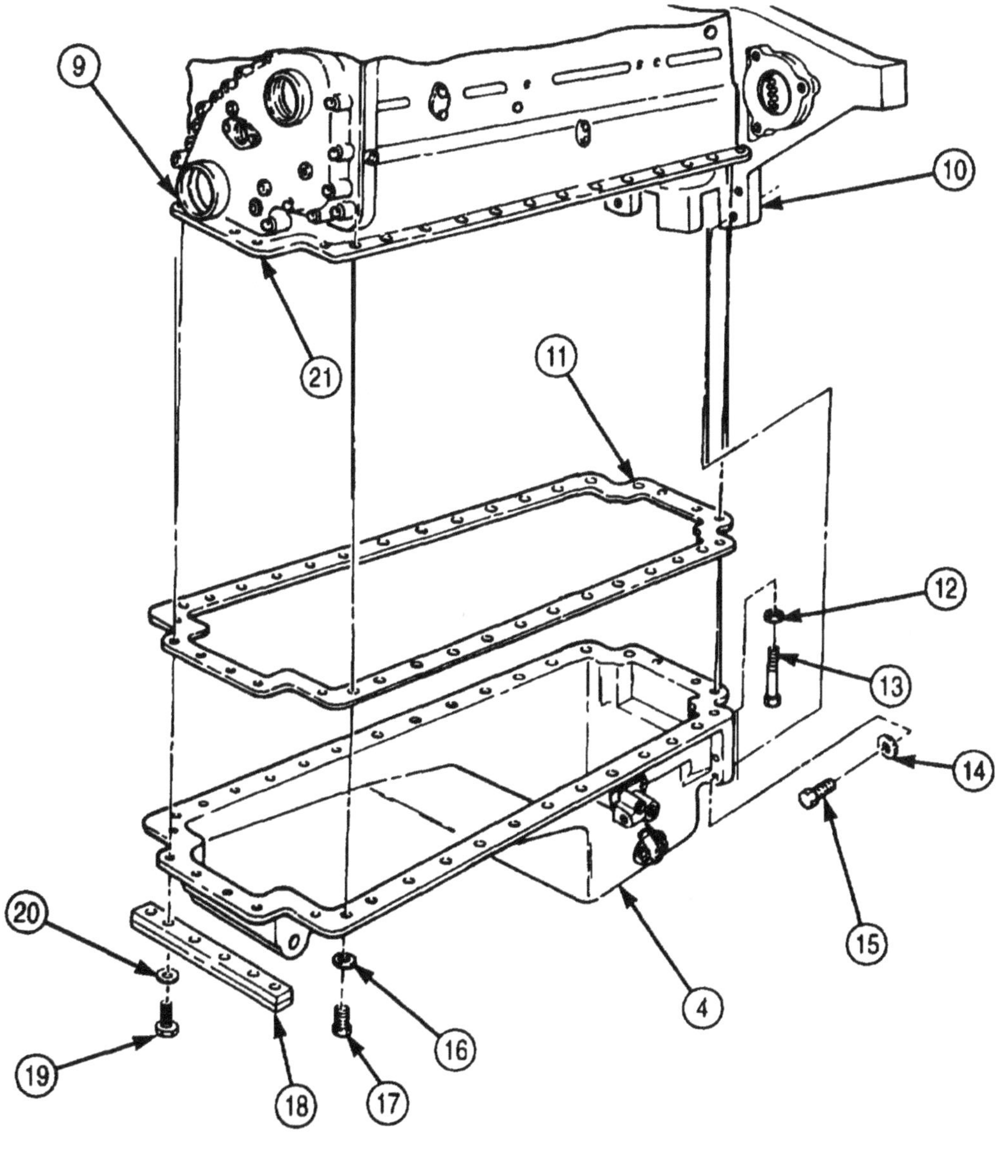

4-22. ENGINE OIL PAN MAINTENANCE (Contd)

b. Disassembly

1. Remove four screws (1), strainer screen (2), drainplug (4), and gasket (3) from oil pan (14). Discard gasket (3).
2. Remove three screws (9), lockwashers (8), washers (7), aerator (6), and gasket (5) from oil pan (14). Discard lockwashers (8) and aerator gasket (5) and clean gasket remains from oil pan mating surfaces.
3. Remove two screws (10), washers (11), suction flange (12), and gasket (13) from oil pan (14). Discard gasket (13) and clean gasket remains from oil pan mating surfaces.

c. Cleaning and Inspection

1. Clean oil pan (14), aerator (6), drainplug (4), suction flange (12), and screen (2) (para. 2-14).
2. Inspect oil pan (14), aerator (6), drainplug (4), suction flange (12), and screen (2) (para. 2-15).
3. Check oil pan (14) for cracks, screen (2) for damage or tears, and oil pan (14), aerator (6), drainplug (4), suction flange (12), and screen (2) for damaged threads and uneven gasket mating surfaces. Replace oil pan (14) if damaged or cracked. Replace screen (2) if damaged or torn. Repair or replace oil pan (14) if threads are damaged.

d. Assembly

1. Install strainer screen (2) on oil pan (14) with four screws (1).
2. Install new gasket (3) and drainplug (4) on oil pan (14).
3. Apply thin coating of sealing compound to new aerator gasket (5).
4. Install aerator gasket (5) and aerator (6) on oil pan (14) with three washers (7), new lockwashers (8), and screws (9). Tighten screws (9) 10-12 lb-ft (14-16 N•m).
5. Apply a thin coating of sealing compound to new suction flange gasket (13).
6. Install suction flange gasket (13) and suction flange (12) on oil pan (14) with two washers (11) and screws (10). Tighten screws (10) 19-22 lb-ft (26-30 4-168N•m).

4-22. ENGINE OIL PAN MAINTENANCE (Contd)

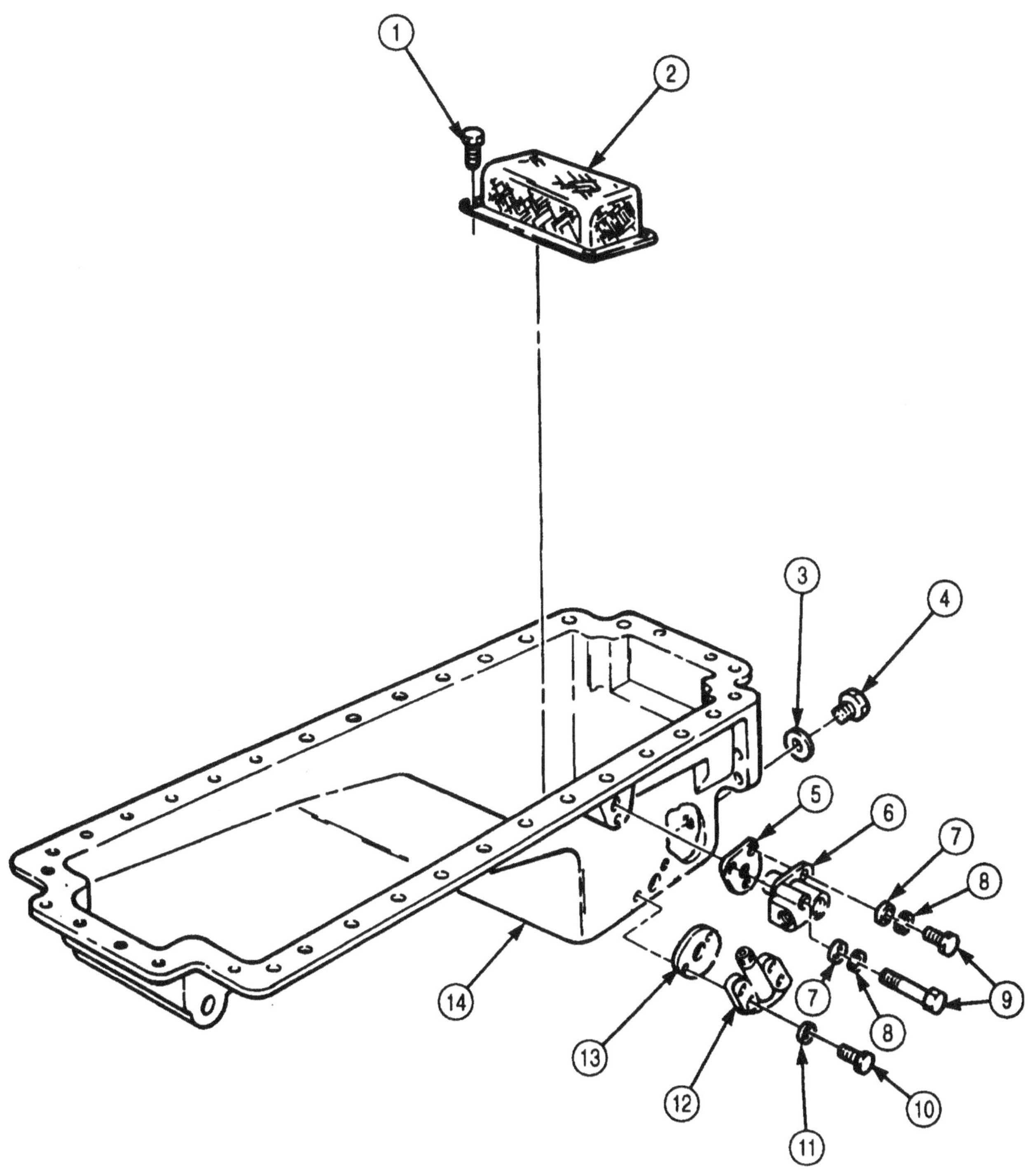

4-22. ENGINE OIL PAN MAINTENANCE (Contd)

e. Installation

NOTE

Engine oil pan is mounted using screw-assembled washers on late model engines.

1. Install new engine oil pan gasket (3) and oil pan (8) on engine block (14) and front gearcase cover (1) with brace (11), four washers (13), screws (12), thirty washers (9), and screws (10). Do not tighten screws (12) and (10).
2. Install rear of oil pan (8) on flywheel housing (2) with four washers (4) and screws (5). Do not tighten screws (5).
3. Install new engine oil pan gasket (3) and oil pan (8) on flywheel housing (2) with six washers (6) and screws (7). Tighten screws (7) alternately 70-80 lb-ft (95-109 N•m).
4. Alternately tighten screws (12) 35-40 lb-ft (20-54 N•m).
5. Alternately tighten screws (10) 15-40 lb-ft (20-54 N•m).
6. Connect oil pickup hose (16) to suction flange (18).
7. Connect oil return hose (19) to aerator (17).
8. Install oil pickup hose (16) and return hose (19) on oil pan (8) with two clamps (15), washers (21), and screws (20). Tighten screws (20) 35-40 lb-ft (48-54 N•m).

4-22. ENGINE OIL PAN MAINTENANCE (Contd)

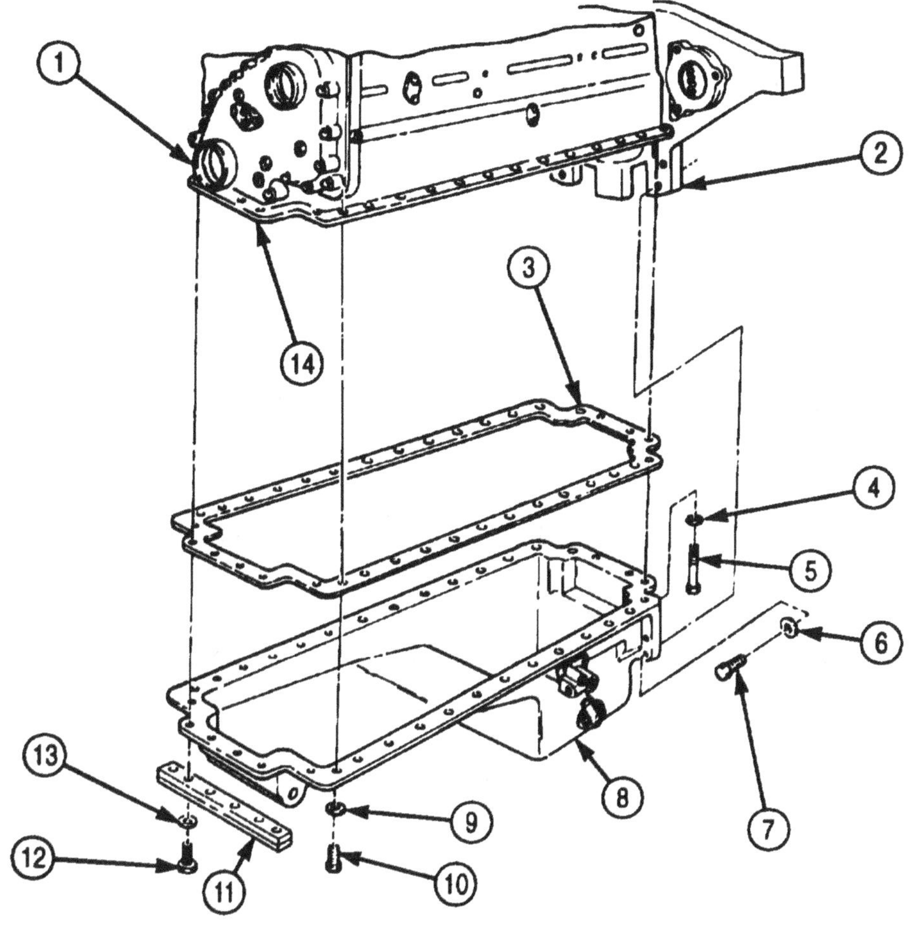

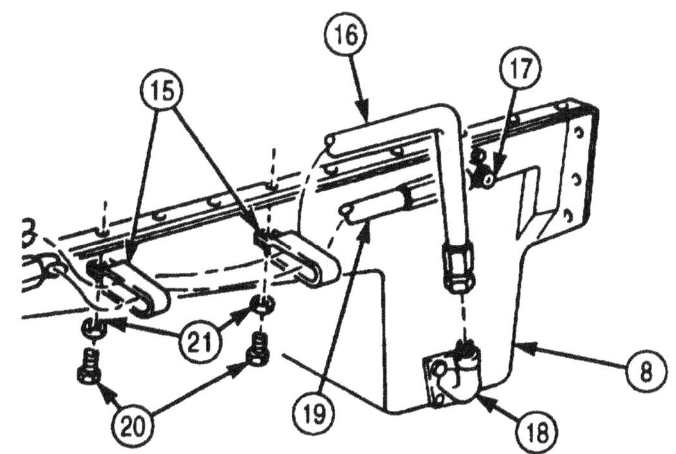

FOLLOW-ON TASKS:
- Install front sump tube (para. 3-9).
- Install oil dipstick tube (para. 3-3).
- Refill engine oil (LO 9-2320-272-12).

4-23. ENGINE OIL COOLER MAINTENANCE

THIS TASK COVERS:

a. Removal
b. Disassembly
c. Cleaning and Inspection
d. Assembly
e. Installation

INITIAL SETUP:

APPLICABLE MODELS
M939/A1

TOOLS
General mechanic's tool kit (Appendix E, Item 1)

MATERIALS/PARTS
Bushing (Appendix D, Item 21)
Nine lockwashers (Appendix D, Item 354)
Gasket (Appendix D, Item 179)
Gasket (Appendix D, Item 180)
Two O-rings (Appendix D, Item 441)
Two retaining rings (Appendix D, Item 537)
Lubricating oil (Appendix C, Item 50)
Antiseize tape (Appendix C, Item 72)
Drycleaning solvent (Appendix C, Item 71)

REFERENCES (TM)
TM 9-2320-272-10
TM 9-2320-272-24P

EQUIPMENT CONDITION
- Parking brake set (TM 9-2320-272-10).
- Right splash shield removed (TM 9-2320-272-10).
- Cooling system drained (para. 3-53).

GENERAL SAFETY INSTRUCTIONS
- Keep fire extinguisher nearby when using drycleaning solvent.
- Drycleaning solvent is flammable and toxic. Do not use near an open flame.
- When cleaning with compressed air, wear eyeshields and ensure source pressure does not exceed 30 psi (207 kPa).

a. Removal

1. Loosen hose clamp (23) and disconnect heater hose (21) from shutoff valve (24).
2. Remove shutoff valve (24) and adapter (22) from oil cooler (17).
3. Loosen two hose clamps (19) and disconnect surge tank-to-oil cooler hose (20) and transmission cooler-to-oil cooler hose (18) from oil cooler (17).
4. Disconnect air compressor coolant line (12) and bushing (11) from oil cooler elbow (1). Discard bushing (11).
5. Remove five screws (16), lockwashers (15), oil cooler (17), and gasket (14) from engine block (13). Discard lockwashers (15) and gasket (14). Clean gasket remains from engine block (13) and oil cooler (17) mating surfaces.

b. Disassembly

1. Remove four screws (2), lockwashers (3), cooler housing (6), and gasket (9) from oil cooler end cover (10). Discard lockwashers (3) and gasket (9), and clean gasket remains from mating surfaces of end cover (10) and cooler housing (6).
2. Remove pipe plug (5), elbow (1), two retaining rings (8), O-rings (7), and oil cooler element (4) from oil cooler end cover (10). Discard retaining rings (8) and O-rings (7).

4-23. ENGINE OIL COOLER MAINTENANCE

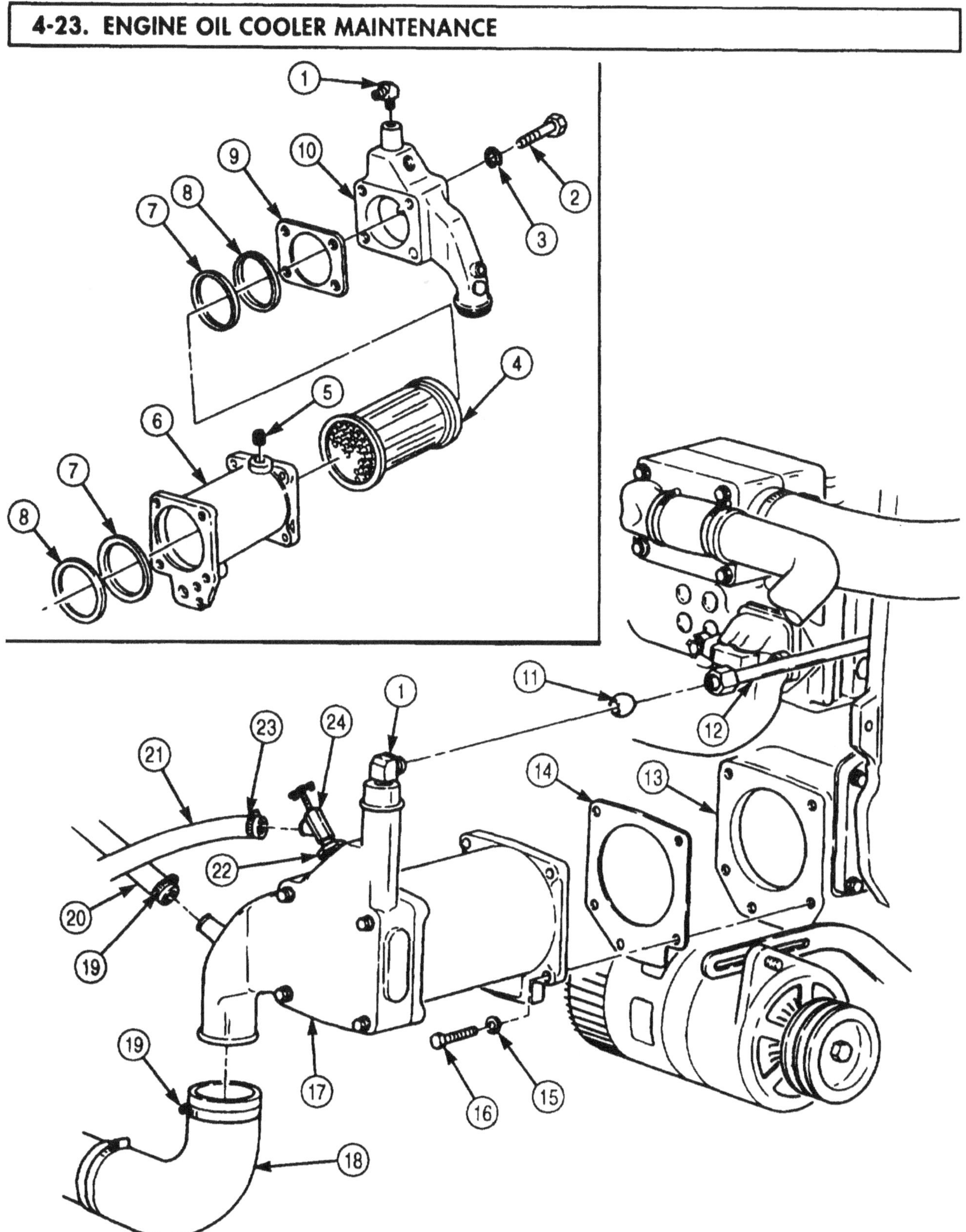

4-23. ENGINE OIL COOLER MAINTENANCE

c. Cleaning and Inspection

WARNING

Drycleaning solvent is flammable and toxic. Do not use near open flame and always have a fire extinguisher nearby when solvents are used. Use only in well-ventilated places, wear protective clothing, and dispose of cleaning rags in approved container. Failure to do this may result in injury or death to personnel and/or damage to equipment.

NOTE

To prevent hardening and drying of foreign substances, clean cooler element as soon as possible after removal.

1. Clean cooler housing (6) with drycleaning solvent and inspect for cracks and stripped threads. Replace cooler housing (6) if cracked or threads are stripped.
2. Soak and flush oil cooler element (4) with drycleaning solvent and inspect for broken and cracked welds. Replace cooler element (4) if broken or cracked.

WARNING

Eyeshields must be worn when cleaning with compressed air. Compressed air source will not exceed 30 psi (207 kPa). Failure to do so may result in injury to personnel.

3. Immerse oil cooler element (4) in water, apply 30 psi (207 kPa) air pressure, and plug opposite end. Replace oil cooler element (4) if air bubbles are observed.

d. Assembly

1. Align index marks on oil cooler element (4) and install in cooler housing (6).

NOTE

Apply engine oil to lubricate O-rings before installation.

2. Install two new O-rings (7) and new retaining rings (8) in each end of cooler housing (6).
3. Wrap end of elbow (1) with antiseize tape and install in end cover (10).
4. Wrap pipe plug (5) with sealing tape and install in cooler housing (6).
5. Install new gasket (9) and end cover (10) on cooler housing (6) with four new lockwashers (3) and screws (2). Tighten screws (2) 30-35 lb-ft (41-47 N•m).

e. Installation

1. Install new gasket (14) and oil cooler (17) on engine block (13) with five new lockwashers (15) and screws (16). Tighten screws (16) 30-35 lb-ft (40-47 N•m).
2. Connect new bushing (11) and air compressor coolant line (12) to oil cooler elbow (1).
3. Install adapter (22) and shutoff valve (24) on oil cooler (17).
4. Connect surge tank-to-oil cooler hose (20) and transmission cooler-to-oil cooler hose (18) to oil cooler (17) with two clamps (19).
5. Connect heater hose (21) to shutoff valve (24) with hose clamp (23).

4-23. ENGINE OIL COOLER MAINTENANCE

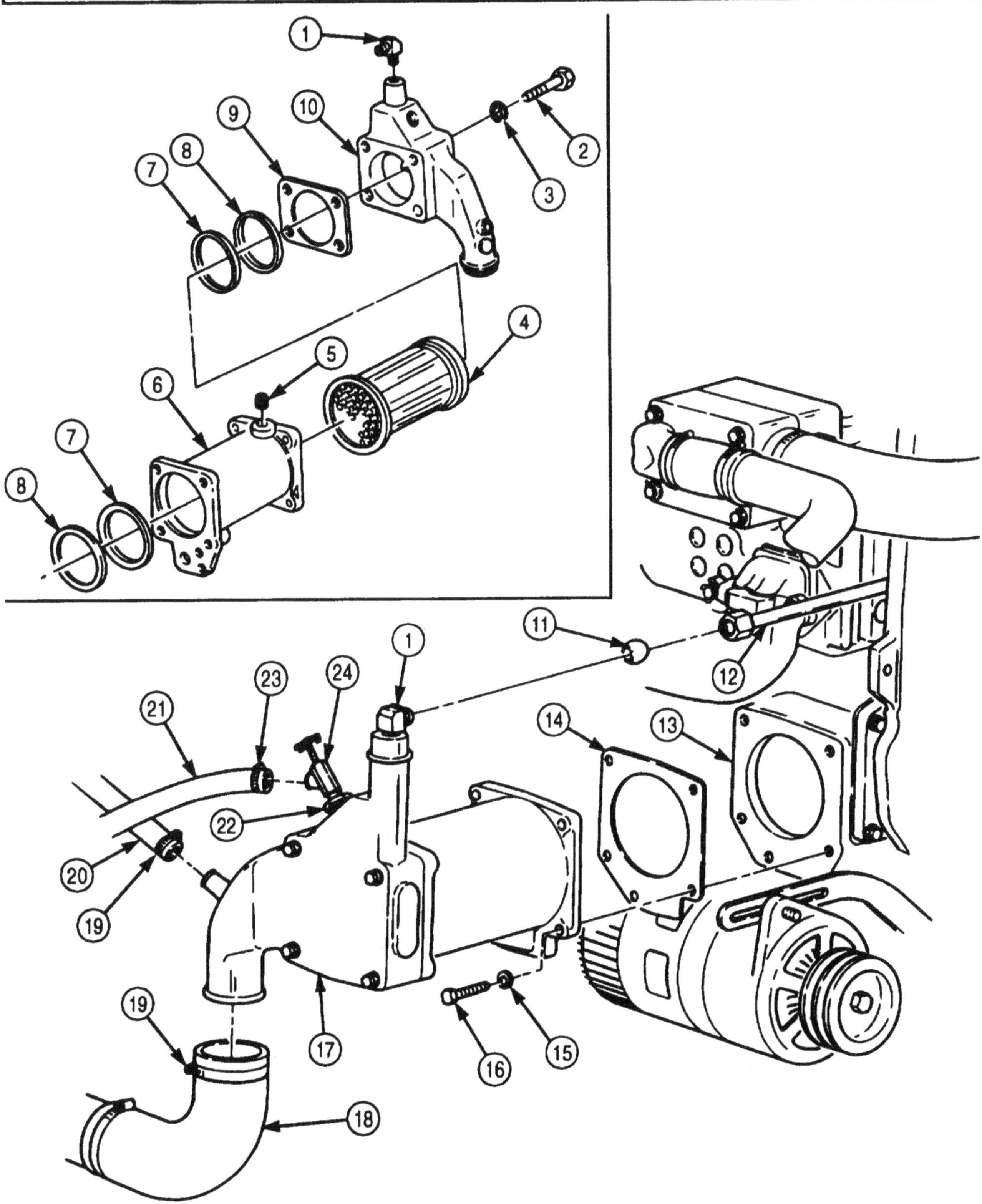

FOLLOW-ON TASKS: • Fill cooling system to proper level (para. 3-53).
• Install right splash shield (TM 9-2320-272-10).

4-24. ENGINE AIR INTAKE MANIFOLD MAINTENANCE

THIS TASK COVERS:

a Removal
b. Disassembly
c. Cleaning and Inspection
d. Assembly
e. Installation

INITIAL SETUP:

APPLICABLE MODELS
M939/A1

TOOLS
General mechanic's tool kit (Appendix E, Item 1)

MATERIALS/PARTS
Gasket (Appendix D, Item 182)
Lockwasher (Appendix D, Item 379)
Three gaskets (Appendix D, Item 181)
Cap and plug set (Appendix C, Item 14)
Antiseize tape (Appendix C, Item 72)

REFERENCES (TM)
TM 9-2320-272-10
TM 9-2320-272-24P

EQUIPMENT CONDITION
- Air intake pipe and hump hose removed (para. 3-14).
- Air compressor air intake tube removed (para. 4-31).
- Ether atomizer removed (para. 3-37).
- Crankcase breather tube removed (para. 3-2).

GENERAL SAFETY INSTRUCTIONS
When cleaning with compressed air, wear eyeshields and ensure source pressure does not exceed 30 psi (207 kPa).

a. Removal

CAUTION

Cover or plug all openings to prevent dirt from entering and damaging engine components.

1. Loosen nut (27) and disconnect air tube (8) from air compressor (28) and elbow (26).
2. Loosen nut (14) and disconnect air tube (8) from air governor (13) and adapter (15).
3. Remove screw (11), washer (10), two clamps (9), and air tube (8) from engine access cover (12).
4. Remove screw (23), wire clamp (24), and wiring harness (25) from lower left side of air intake connector (2) and intake manifold (3).

NOTE

Wire clamp will remain on wiring harness.

5. Disconnect hose (20) from elbow (21).
6. Remove elbow (21) and air cleaner indicator filter (22) from intake manifold (3).

NOTE

Tag cable ground strap for installation.

7. Remove screw (18), washer (17), lockwasher (16), and cable ground strap (19) from intake manifold (3). Discard lockwasher (16).

NOTE

- Air intake manifold is mounted with screw-assembled washers on late model engines. Screws, lockwashers, and washers From early model engines will be discarded and replaced with screw-assembled washers.
- Perform step 8 only if clamps are on vehicle.

8. Remove screw (30), two clamps (29), and fuel tubes (31) from cylinder heads (1).

4-24. ENGINE AIR INTAKE MANIFOLD MAINTENANCE (Contd)

9. Remove eight screws (4), lockwashers (5), washers (6), intake manifold (3), and three gaskets (7) from cylinder heads (1). Discard lockwashers (5) and gaskets (7). Discard screws (4), lockwashers (5), and washers (6) from early model engines. Clean gasket remains from intake manifold and cylinder head mating surfaces.

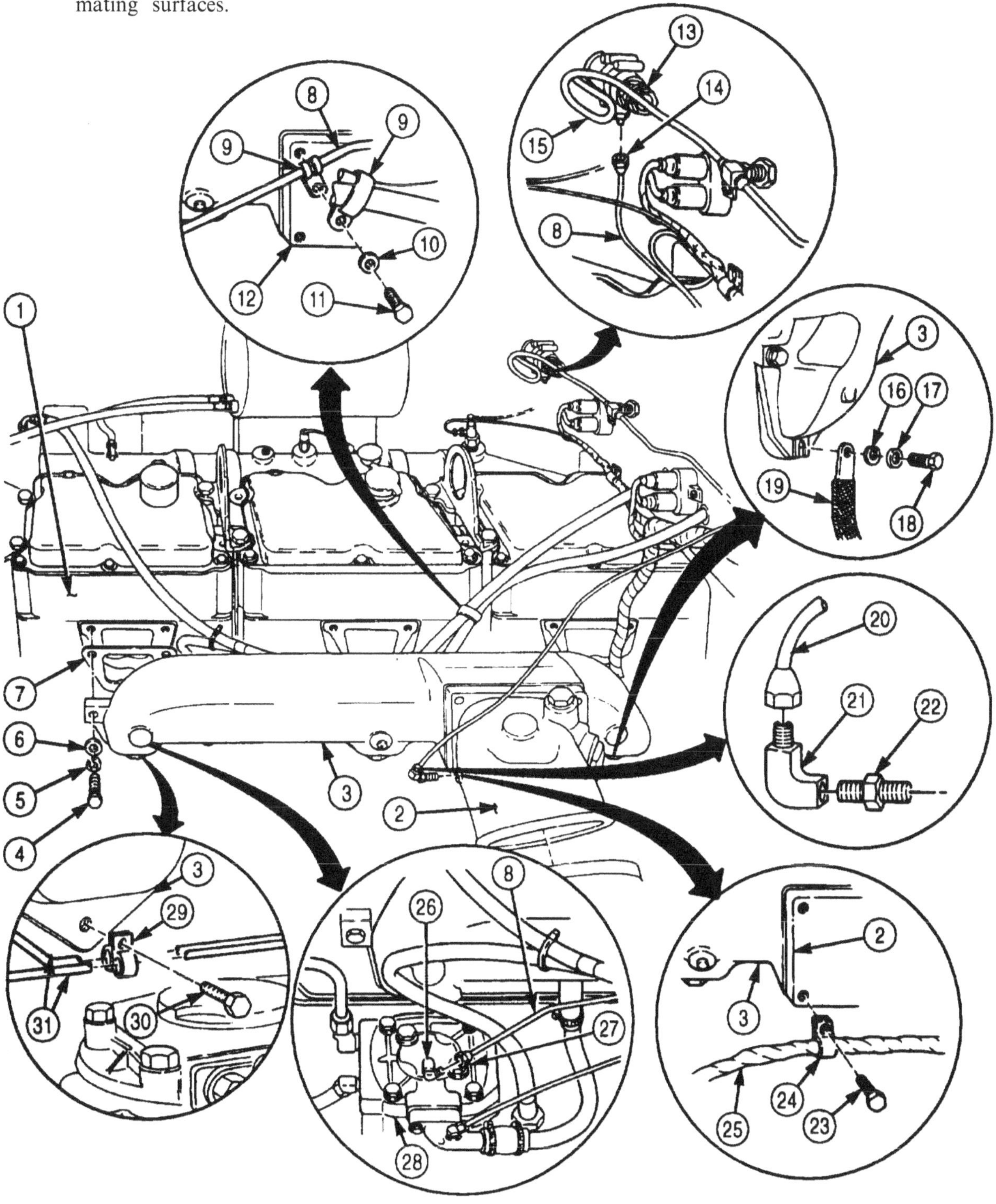

4-24. ENGINE AIR INTAKE MANIFOLD MAINTENANCE (Contd)

b. Disassembly

1. Remove pipe plug (10) from intake manifold (1).
2. Remove adapter (4) and plug (3) from air intake connector (8).

NOTE

- Air connector is mounted with screw-assembled washers on late model engines. Screws, lockwashers, and washers from early model engines will be discarded and replaced with screw-assembled washers.
- Perform step 3 for late model engines.

3. Remove elbow (2) from air intake connector (8).
4. Remove two screws (6), lockwashers (7), washers (5), air intake connector (8), and gasket (9) from intake manifold (1). Discard lockwashers (7) and gasket (9). Clean gasket remains from air intake connector and intake manifold mating surfaces.

c. Cleaning and Inspection

1. Brush and clean intake manifold (1) and air intake connector (8) and inspect for breaks, cracks, and elongated holes. Replace if broken, cracked, or if holes are elongated (para. 4-1).

WARNING

Eyeshields must be worn when cleaning with compressed air. Compressed air source will not exceed 30 psi (207 kPa). Failure to do so may result in injury to personnel.

2. Clean internal passages of intake manifold (1) and air intake connector (8) with compressed air and inspect threaded holes, screws, pipe plugs, and adapter for stripped or crossed threads. Repair or replace if threaded parts have stripped or crossed threads (para. 4-1).

d. Assembly

NOTE

Male pipe threads must be wrapped with antiseize tape before installation.

1. Install adapter (4) and plug (3) in air intake connector (8).
2. Install pipe plug (10) in intake manifold (1).

NOTE

Perform step 3 for late model engines.

3. Install elbow (2) on air intake connector (8).
4. Install air intake connector (8) and new gasket (9) on intake manifold (1) with two screws (6), lockwashers (7), and washers (5). Tighten screws (5) 25-30 lb-ft (34-41 N•m).

4-24. ENGINE AIR INTAKE MANIFOLD MAINTENANCE (Contd)

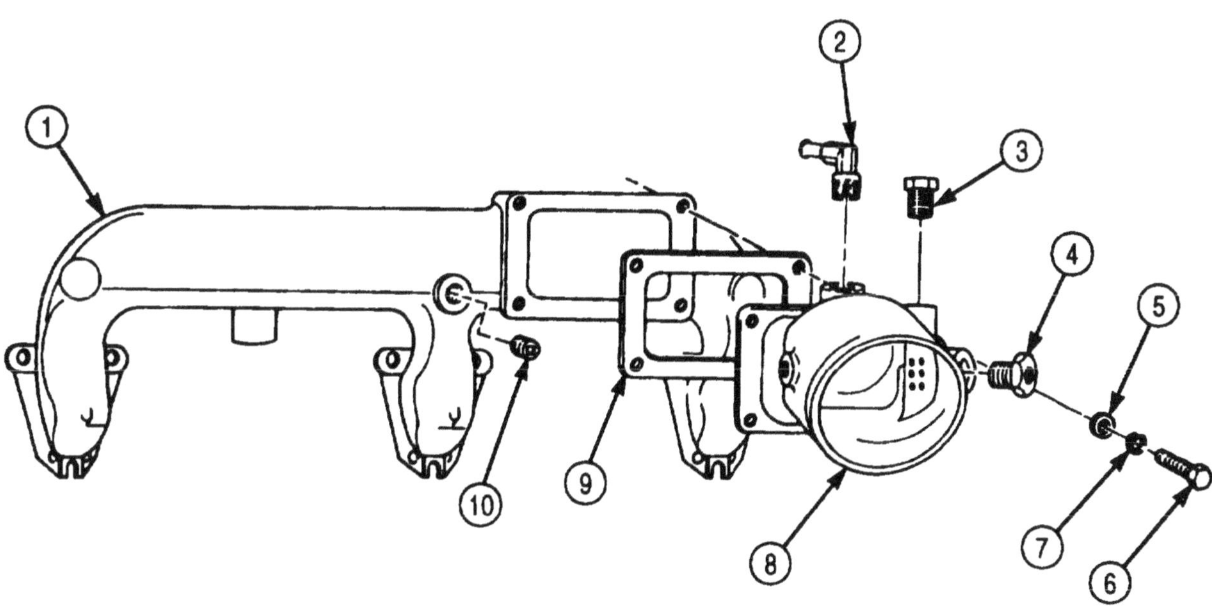

4-24. ENGINE AIR INTAKE MANIFOLD MAINTENANCE (Contd)

e. Installation

NOTE

Wrap male pipe threads with antiseize tape before installation.

1. Install intake manifold (3) and three new gaskets (5) on cylinder heads (1) with eight screw-assembled washers (4). Tighten screw-assembled washers (4) 25-30 lb-ft (34-41 N•m).

NOTE

Perform step 2 only if clamps were removed previously.

2. Install two clamps (27) and fuel tubes (29) on intake manifold (3) with screw (28). Tighten screw (28) 20-25 lb-ft (27-34 N•m).
3. Install cable ground strap (17) on intake manifold (3) with washer (14), new lockwasher (15), and screw (16). Tighten screw (16) 25-30 lb-ft (34-41 N•m).
4. Install wire harness (23) and clamp (22) on lower left side of air intake connector (2) with screw (21). Tighten screw (21) 25-30 lb-ft (34-41 N•m).
5. Install elbow (19) and air cleaner indicator filter (20) on left side of air intake connector (2).
6. Connect hose (18) to elbow (19).
7. Connect air tube (6) and nut (12) to adapter (13) on air governor (11).
8. Connect air tube (6) and nut (25) to elbow (24) on air compressor (26).
9. Position two clamps (7) to screw hole in top left of engine access cover (10) and install air tube (6) and two clamps (7) on air intake connector (2) with washer (8) and screw (9). Tighten screw (9) 25-30 lb-ft (34-41 N•m).

4-24. ENGINE AIR INTAKE MANIFOLD MAINTENANCE (Contd)

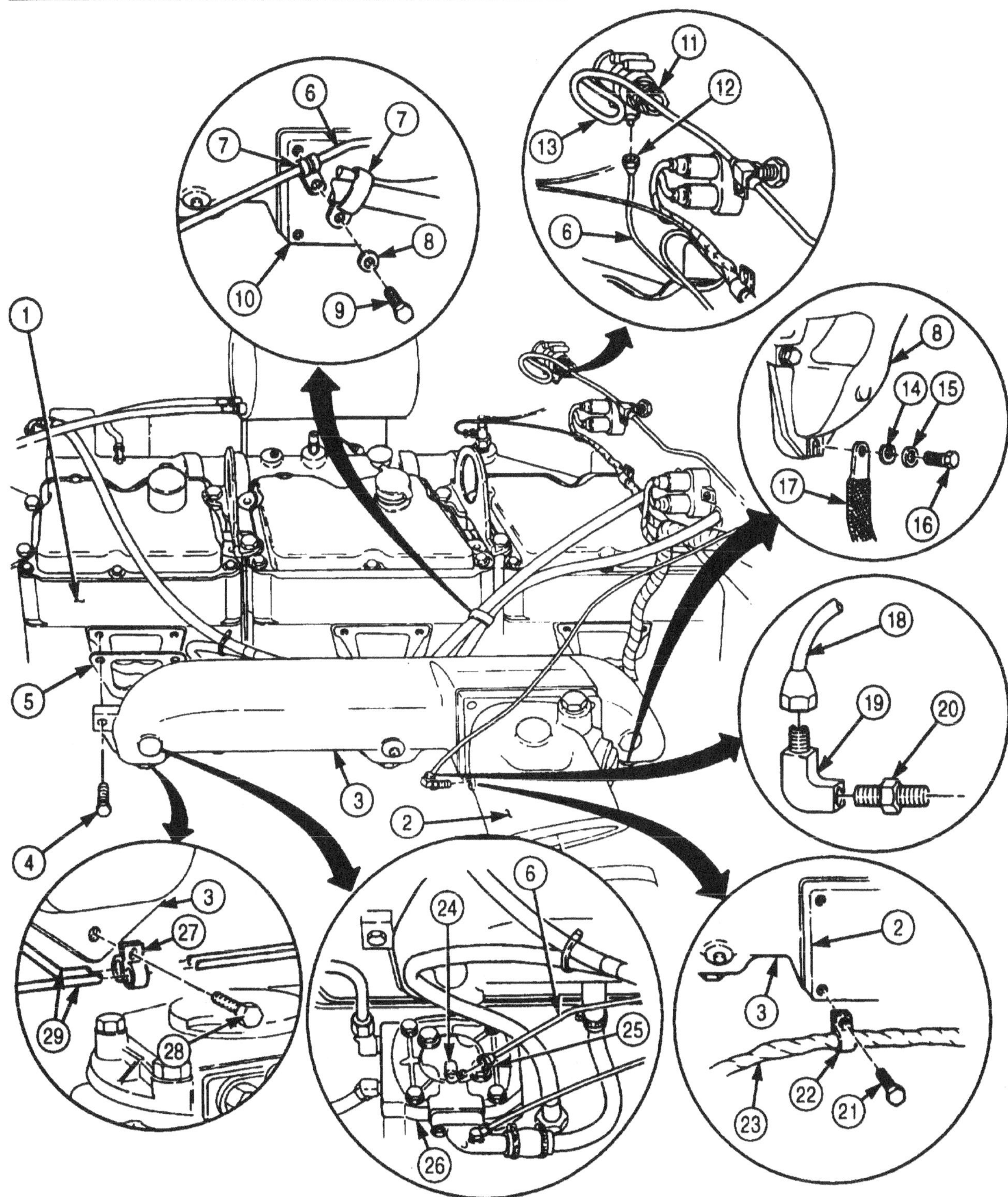

FOLLOW-ON TASKS:
- Install ether atomizer (para. 3-37).
- Install air compressor air intake tube (para. 4-31).
- Install air intake pipe and pump hose (para. 3-14).
- Install crankcase breather tube (para. 3-2).

4-25. ENGINE EXHAUST MANIFOLD MAINTENANCE

THIS TASK COVERS:

a. Removal b. Installation

INITIAL SETUP:

APPLICABLE MODELS

M939/A1

TOOLS

General mechanic's tool kit (Appendix E, Item 1)

MATERIALS/PARTS

Eight key washers (Appendix D, Item 267)
Six gaskets (Appendix D, Item 168)
Gasket (Appendix D, Item 175)

REFERENCES (TM)

TM 9-2320-272-10
TM 9-2320-272-24P

EQUIPMENT CONDITION

- Parking brake set (TM 9-2320-272-10).
- Crankcase breather tube and mounting bracket removed (para. 3-2).
- Surge tank removed (para. 3-61).

a. Removal

1. Remove manifold coupling clamp (4) from front exhaust pipe (5). Using soft-faced hammer, tap clamp (4).
2. Separate front exhaust pipe (5) from exhaust manifold (2) and remove gasket (3) from front exhaust pipe (5). Discard gasket (3).
3. Loosen screw (17) and nut (15) on dipstick tube (16).
4. Loosen clamp (12) and remove heater inlet hose (13) and clamp (12) from shutoff valve (11) on water manifold (1).
5. Unlock eight key washers (7) and remove eight screws (6), key washers (7), and four clamps (8) from exhaust manifold (2). Discard locktabs (7).
6. Position oil dipstick bracket (14) out of the way.
7. Remove exhaust manifold (2) and six gaskets (9) from three cylinder head dowels (10). Discard gaskets (9). Separate manifold (2) sections.

b. Installation

1. Position six new gaskets (9) and exhaust manifold (2) on three cylinder head dowels (10) for installation.
2. Install four clamps (8), eight new key washers (7), screws (6), and oil dipstick bracket (14) on exhaust manifold (2).
3. Reposition oil dipstick bracket (14) for installation.
4. Connect heater inlet hose (13) to shutoff valve (11) on water manifold (1) and tighten clamp (12).
5. Tighten screw (17) and nut (15) on dipstick tube (16).
6. Install new gasket (3) between exhaust manifold (2) and front exhaust pipe (5).
7. Position manifold coupling clamp (4) on front exhaust pipe (5)

4-25. ENGINE EXHAUST MANIFOLD MAINTENANCE (Contd)

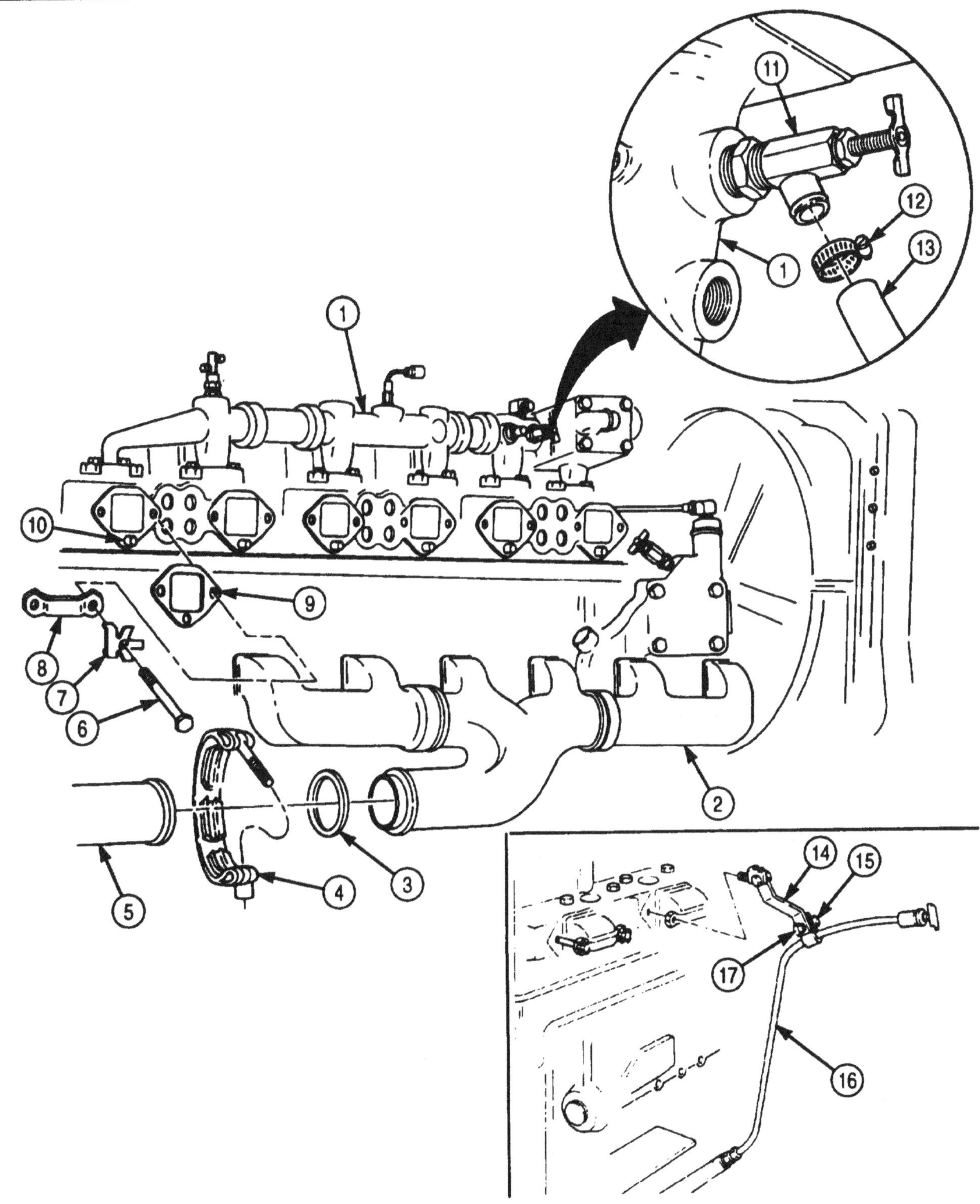

FOLLOW-ON TASKS: • Install surge tank (para. 3-61).
• Install crankcase breather tube and mounting bracket (para. 3-2).

4-26. ENGINE ACCESSORY DRIVE AND PULLEY MAINTENANCE

THIS TASK COVERS:

a. Accessory Drive Pulley Removal
b. Accessory Drive Removal
c. Accessory Drive Disassembly
d. Accessory Drive Cleaning and Inspection
e. Accessory Drive Assembly
f. Accessory Drive Installation
g. Accessory Drive Pulley Installation

INITIAL SETUP:

APPLICABLE MODELS

M939/A1

SPECIAL TOOLS

Mechanical puller kit (Appendix E, Item 102)

TOOLS

General mechanic's tool kit (Appendix E, Item 1)
Dial indicator (Appendix E, Item 36)
Inside micrometer (Appendix E, Item 82)
Outside micrometer (Appendix E, Item 80)
Torque wrench (Appendix E, Item 144)
Arbor press
Soft-faced hammer

MATERIALS/PARTS

Five screw-assembled lockwashers (Appendix D, Item 589)
Gasket (Appendix D, Item 243)
Gasket (Appendix D, Item 183)
Dowel pin (Appendix D, Item 97)
Antiseize tape (Appendix C, Item 72)
GAA grease (Appendix C, Item 28)

REFERENCES (TM)

TM 9-2320-272-24P

EQUIPMENT CONDITION

- Air compressor removed (para. 4-31).
- Power steering drivebelts removed (para. 3-230).
- Radiator removed (para. 3-59).
- Water pump drivebelt removed (para. 3-67).

GENERAL SAFETY INSTRUCTIONS

- Keep fire extinguisher nearby when using drycleaning solvent.
- Drycleaning solvent is flammable and toxic. Do not use near an open flame.
- When cleaning with compressed air, wear eyeshields and ensure source pressure does not exceed 30 psi (207 kPa).

a. Accessory Drive Pulley Removal

1. Remove nut (6) and washer (5) from accessory drive pulley (2).
2. Using mechanical puller, remove accessory drive pulley (2) from accessory driveshaft (4).
3. Remove gasket (1) from accessory drive pulley (2). Discard gasket (1).

NOTE

Perform step 4 only if dowel pin is damaged.

4. Remove dowel pin (3) from accessory driveshaft (4). Discard dowel pin (3).

b. Accessory Drive Removal

1. Remove five screw-assembled lockwashers (9) from accessory drive front flange (10). Discard screw-assembled lockwashers (9).
2. Using soft-faced hammer, loosen accessory drive housing (8) and gasket (7) from engine block gearcase (11) and remove housing (8) and gasket (7) from gearcase (11). Discard gasket (7) and clean gasket remains from mating surfaces of accessory drive housing and engine block gearcase.

4-26. ENGINE ACCESSORY DRIVE AND PULLEY MAINTENANCE (Contd)

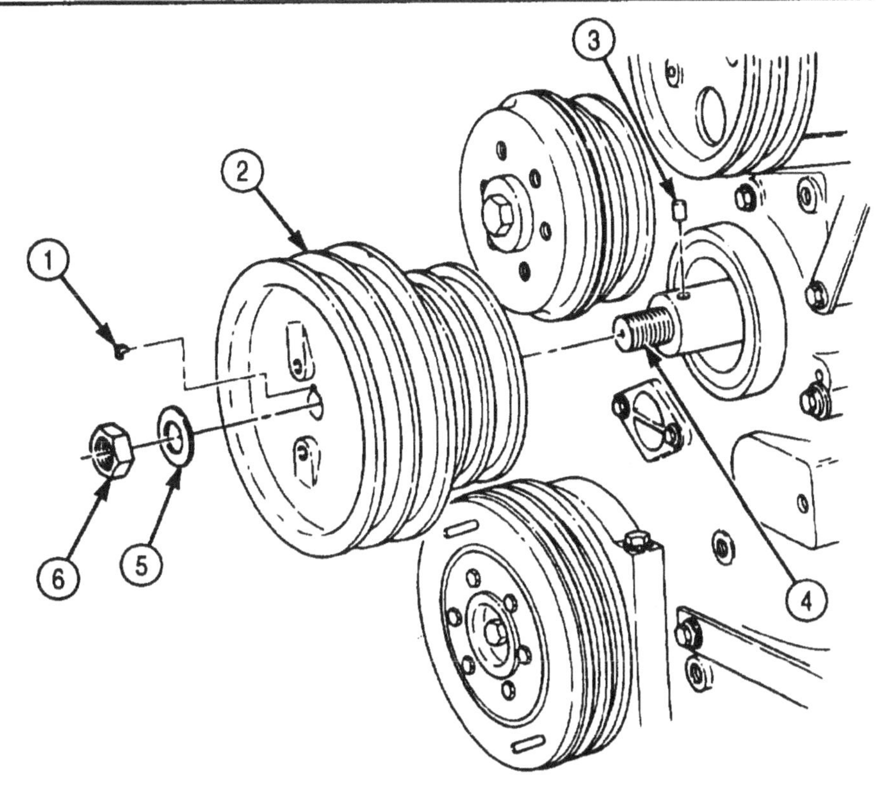

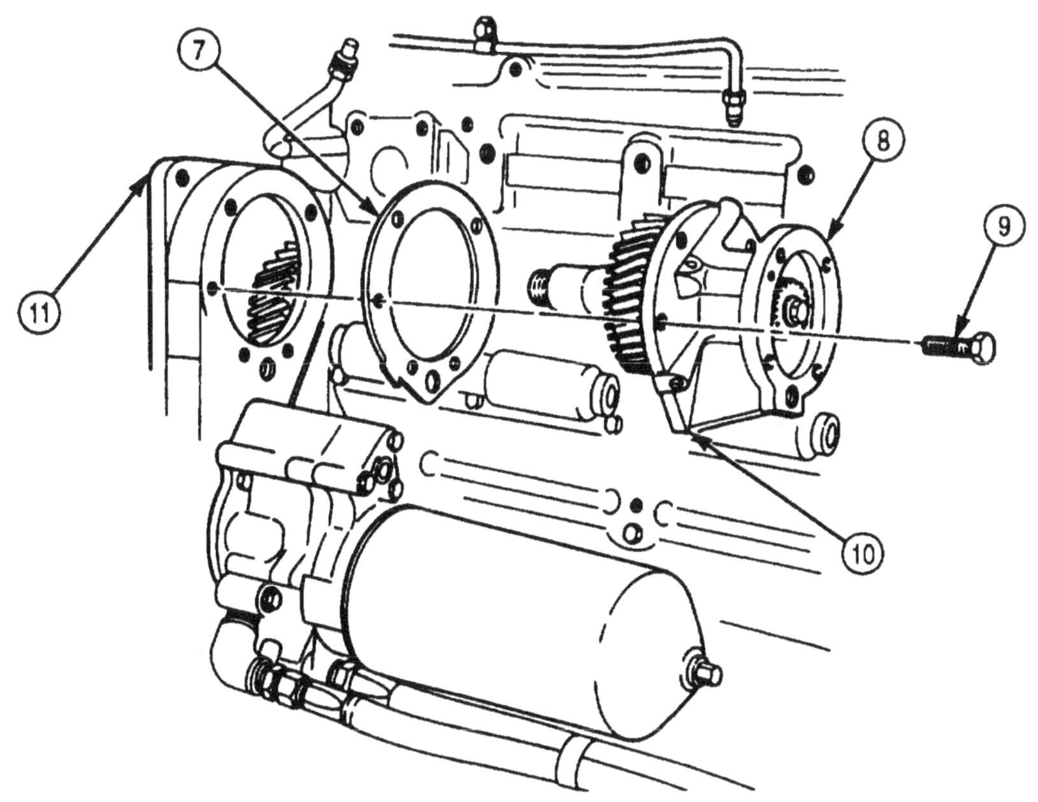

4-26. ENGINE ACCESSORY DRIVE AND PULLEY MAINTENANCE (Contd)

c. Accessory Drive Disassembly

1. Using dial indicator gauge, measure end play of driveshaft (3). Note end play measurement for reference.
2. Remove screw (11) and washer (12) from driveshaft (3).
3. Using arbor press and mandrel, press driveshaft (3) through coupling halfshaft (10).
4. Remove driveshaft (3), drive gear (l), thrust washers (5) and (8), and washer (9) from accessory drive housing (7) and slide through bushing (6) in accessory drive housing (7).
5. Using arbor press and mandrel, press driveshaft (3) through drive gear (1).

NOTE

Do not remove dowel pins unless damaged.

6. Remove dowel pins (2) and (4) from driveshaft (3).

d. Accessory Drive Cleaning and Inspection

WARNING

Drycleaning solvent is flammable and toxic. Do not use near open flame and always have a fire extinguisher nearby when solvents are used. Use only in well-ventilated places, wear protective clothing, and dispose of cleaning rags in approved container. Failure to do this may result in injury or death to personnel and/or damage to equipment.

Eyeshields must be worn when cleaning with compressed air. Compressed air source will not exceed 30 psi (207 kPa). Failure to do so may result in injury to personnel.

1. Clean accessory drive housing (7) with drycleaning solvent and blow out passages and bore with compressed air,
2. Inspect accessory drive housing (7) for breaks and cracks. Replace if broken or cracked.
3. Inspect bushing (6) on accessory drive housing (7) for pitting, galling, and cracks. Replace bushing (6) if pitted, galled, or cracked.
4. Check inside diameter of bushing (6) at both ends. Replace bushing (6) if either measurement is greater than 1.321 in. (33.6 mm).

NOTE

Perform steps 5 and 6 only if bushing must be replaced.

5. Using arbor press and mandrel, press bushing (6) from accessory drive housing (7).
6. Using arbor press and mandrel, install new bushing (6) in accessory drive housing (7) flush with face.
7. Clean driveshaft (3) with drycleaning solvent.
8. Inspect driveshaft (3) for breaks, cracks, and galling. Replace if broken, cracked, or galled.
9. Inspect driveshaft (3) for stripped or crossed threads. Repair or replace driveshaft (3) if threads are stripped or crossed (para. 4-1).
10. Check driveshaft (3) outside diameter at bushing (5) location. Replace driveshaft (3) if outside diameter is less than 1.310 in. (33.27 mm).
11. Inspect dowel pin holes (13) of driveshaft (3). Discard driveshaft (3) if holes (13) are enlarged.
12. Using drycleaning solvent, clean drive gear (1) and coupling halfshaft (10).

4-26. ENGINE ACCESSORY DRIVE AND PULLEY MAINTENANCE (Contd)

13. Inspect drive gear (1) and coupling halfshaft (10) for breaks, cracks, and galling in bore. Replace if cracked, broken, or bore shows galling.
14. Inspect drive gear (1) and coupling halfshaft (10) for chipped and broken teeth. Replace drive gear (1) and coupling halfshaft (10) if teeth are broken or chipped.
15. Inspect thrust washers (5) and (8) for cracks and scoring. Replace if cracked, broken, or scored.
16. Replace both thrust washers (5) and (8) if end play measured in task b., step 1, is greater than 0.012 in. (0.3 mm).
17. Inspect two dowel pins (2) and (4) for burrs or cracks. Replace if cracked or burred.

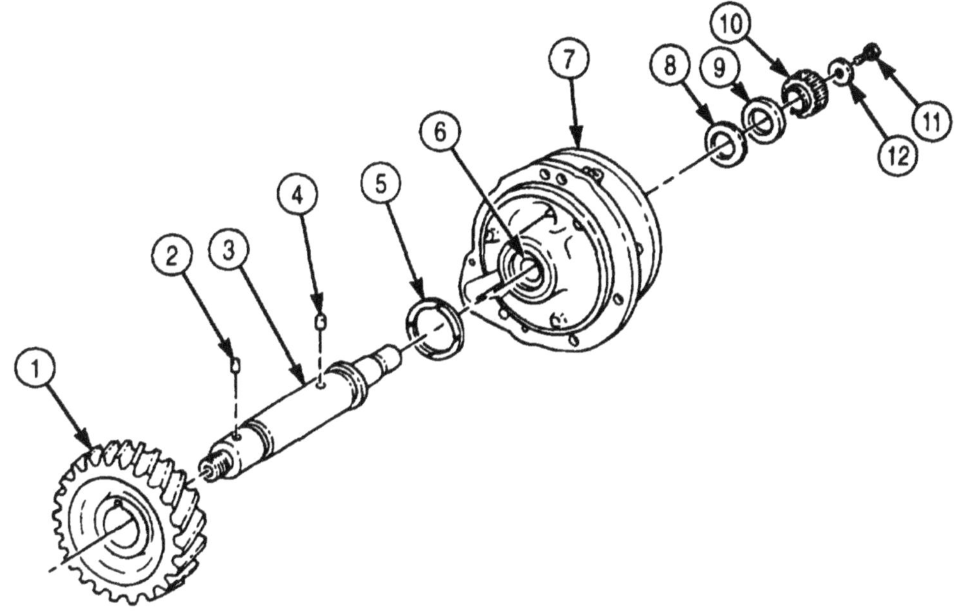

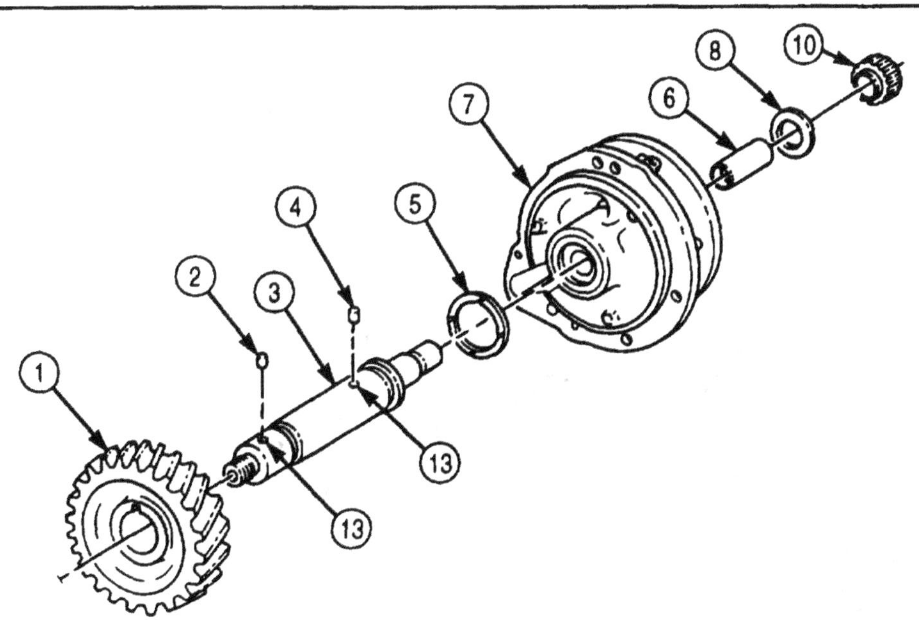

4-26. ENGINE ACCESSORY DRIVE AND PULLEY MAINTENANCE (Contd)

e. Accessory Drive Assembly

1. Install new dowel pin (4) in hole (14) of driveshaft (3).

NOTE

Remaining dowel pin is installed with accessory drive pulley.

2. Align slot (1) in bore of drive gear (2) with dowel pin (4) and press drive gear (2) on driveshaft (3) over dowel pin (4). Ensure drive gear (2) is seated on shoulder (13).
3. Install large thrust washer (5) on front of accessory drive housing (7).
4. Install driveshaft (3) and drive gear (2) through thrust washer (5) and bushing (6) in accessory drive housing (7).
5. Install thrust washer (8) over driveshaft (3) and position in accessory drive housing (7). Ensure grooved side of thrust washer (8) faces away from accessory drive housing (7).
6. Install washer (9) on driveshaft (3) against thrust washer (8).

NOTE

Ensure flat end of coupling halfshaft faces away from accessory drive housing.

7. Press coupling halfshaft (10) on driveshaft (3) until flush with end.
8. Measure driveshaft (3) end play. End play should be 0.002-0.012 in. (0.05-0.26 mm). If end play is not within limits, press driveshaft (3) through coupling halfshaft (10) to obtain proper end play.
9. Install washer (11) and screw (12) on driveshaft (3). Tighten screw (12) 30-35 lb-ft (41-47 N•m).

f. Accessory Drive Installation

NOTE

If accessory drive gear and camshaft gear are not properly aligned, valve, injector, and compressor timing will be incorrect.

1. Remove pipe plug (20) from front gearcase cover (18).
2. Rotate crankshaft (17) to No. 1 piston Top Dead Center (TDC) firing stroke.
3. Rotate crankshaft (17) 90 degrees past TDC.
4. Install accessory drive housing (23) and new gasket (22) on engine block gearcase (21) with five new screw-assembled lockwashers (24). Ensure timing marks (19) on accessory drive gear (15) and camshaft gear (16) align. Tighten screw-assembled lockwashers (24) 40-45 lb-ft (54-61 N•m).
5. Wrap pipe plug (20) threads with antiseize tape and install in front gearcase cover (18).

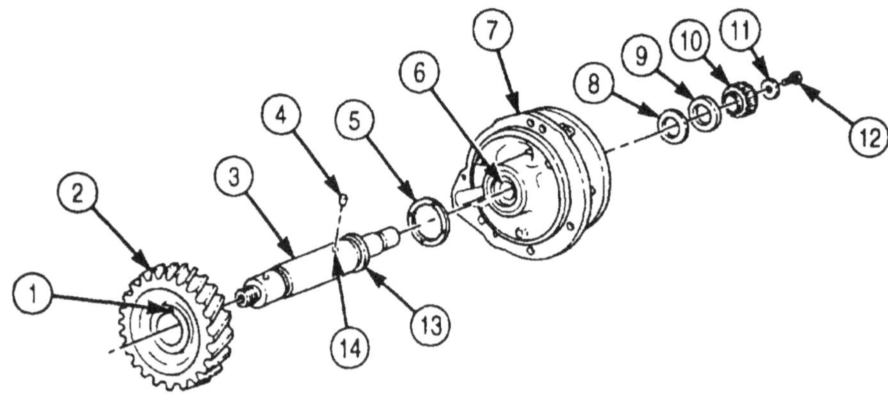

4-26. ENGINE ACCESSORY DRIVE AND PULLEY MAINTENANCE (Contd)

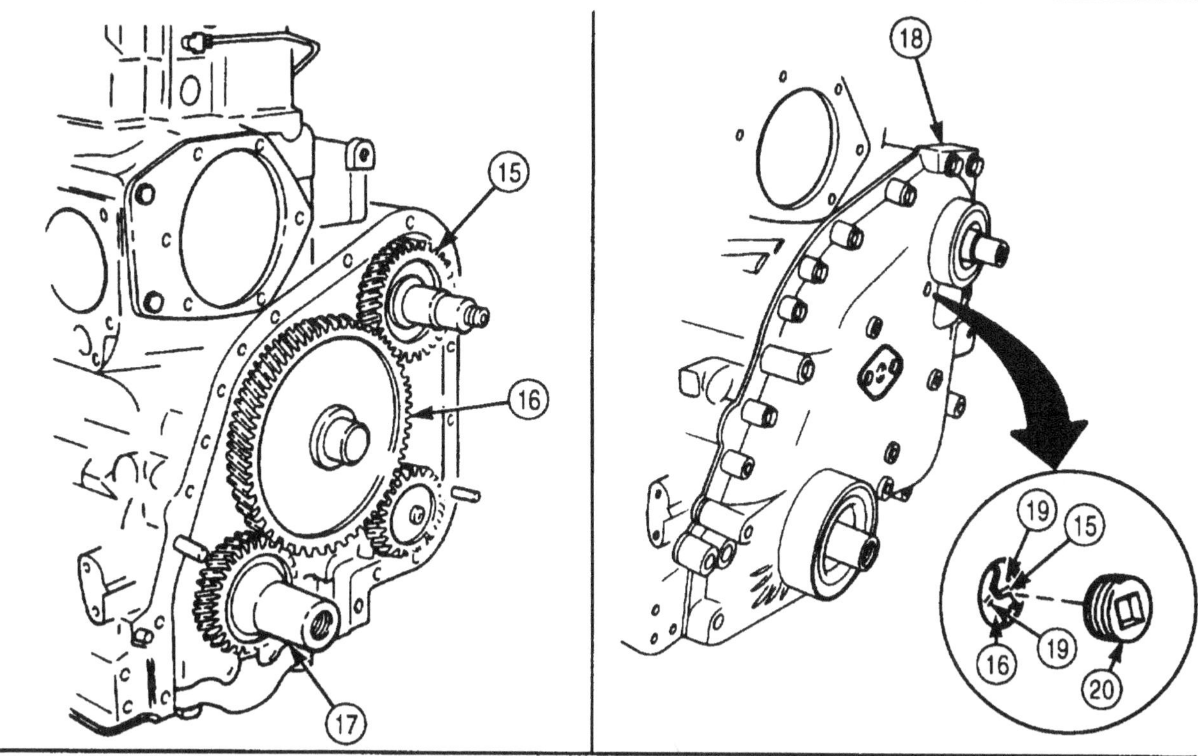

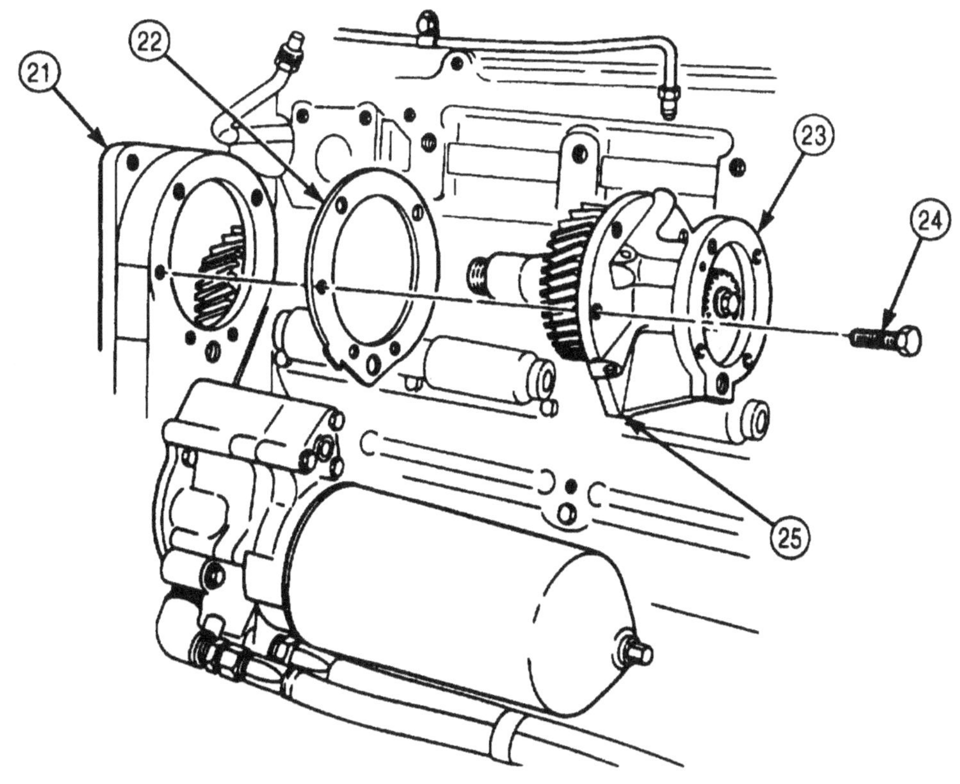

4-26. ENGINE ACCESSORY DRIVE AND PULLEY MAINTENANCE (Contd)

g. Accessory Drive Pulley Installation

1. Apply a light coat of GM grease to accessory driveshaft (4).

NOTE

Perform step 2 if dowel pin was removed.

2. Install new dowel pin (3) in accessory driveshaft (4).
3. Install new gasket (1) in keyway of accessory drive pulley (2).
4. Install nut (6) and washer (5) in accessory drive pulley (2).

4-26. ENGINE ACCESSORY DRIVE AND PULLEY MAINTENANCE (Contd)

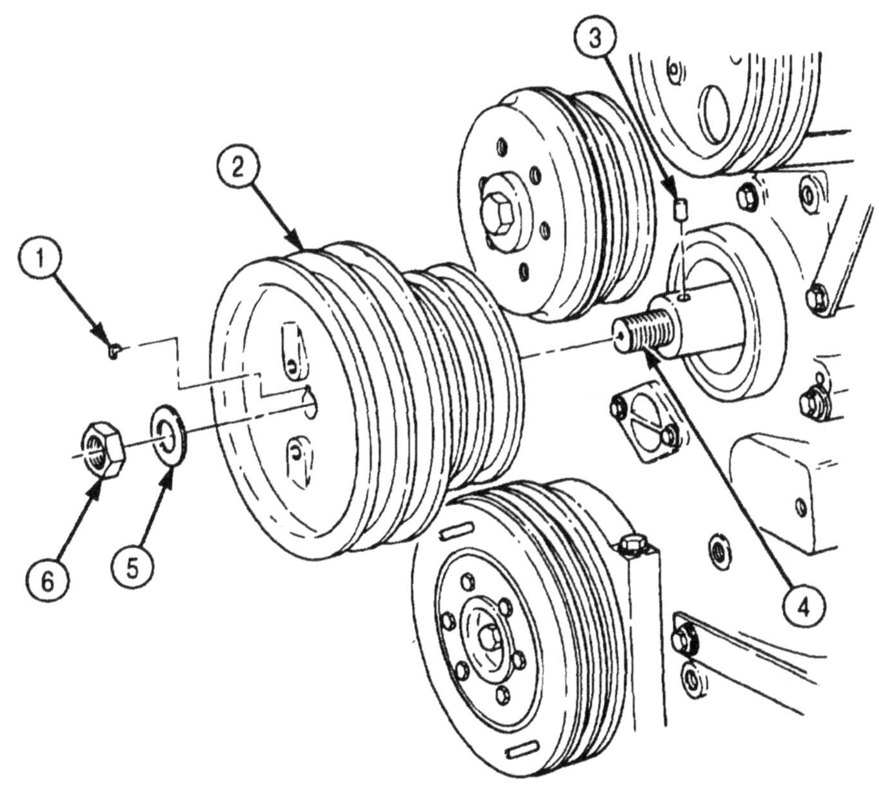

FOLLOW-ON TASKS:
- Install air compressor (para. 4-31).
- Install water pump drivebelt (para. 3-67).
- Install radiator (para. 3-59).
- Install power steering drivebelts (para. 3-230).

4-27. ENGINE WATER MANIFOLD MAINTENANCE

THIS TASK COVERS:

a. Removal
b. Disassembly
c. Cleaning and Inspection
d. Assembly
e. Installation

INITIAL SETUP:

APPLICABLE MODELS
M939/A1

TOOLS
General mechanic's tool kit (Appendix E, Item 1)
Wire brush

MATERIALS/PARTS
Six O-rings (Appendix D, Item 443)
Four O-rings (Appendix D, Item 442)
Twelve screw-assembled lockwashers (Appendix D, Item 590)
GAA grease (Appendix C, Item 28)
Antiseize tape (Appendix C, Item 72)
Cap and plug set (Appendix C, Item 14)

REFERENCES (TM)
TM 9-2320-272-10
TM 9-2320-272-24P

EQUIPMENT CONDITION
- Surge tank removed (para. 3-61).
- Engine crankcase breather draft tube removed (para. 3-2).
- Thermostat and housing removed (para. 3-65).
- Fan drive clutch actuator removed (para. 3-73).

GENERAL SAFETY INSTRUCTIONS
Wear eyeshields during cleaning procedure.

a. Removal

NOTE

Have container ready to catch coolant.

1. Loosen hose clamp (7) and disconnect personnel heater inlet hose (8) from water manifold shutoff drainvalve (6).
2. Disconnect connector (5) from water temperature sending unit (4) at water manifold (12).
3. Disconnect ether cylinder-to-safety valve line (3) and safety valve-to-atomizer line (1) from ether start safety valve (2).

NOTE

Clean area around water manifold to prevent dirt or debris from entering cylinder head water ports when water manifold is removed.

4. Remove twelve screw-assembled lockwashers (13) and manifold (12) from cylinder heads (11). Discard screw-assembled lockwashers (13).
5. Remove six O-rings (9) from cylinder head water ports (10). Discard O-rings (9).
6. Plug openings on six open cylinder head water ports (10) to prevent dirt from entering water ports (10).

b. Disassembly

1. Remove rear water manifold section (14) and O-ring (15) from rear coupling (16). Discard O-ring (15).
2. Remove rear coupling (16) and O-ring (17) from center water manifold section (18). Discard O-ring (17).
3. Remove center water manifold section (18) and O-ring (19) from front coupling (20). Discard O-ring (19).
4. Remove front coupling (20) and O-ring (21) from front water manifold section (22). Discard O-ring (21).
5. Remove ether start safety valve (2) from rear water manifold section (14).
6. Remove water temperature sending unit (4) and water manifold shutoff drainvalve (6) from front water manifold section (22).

4-27. ENGINE WATER MANIFOLD MAINTENANCE (Contd)

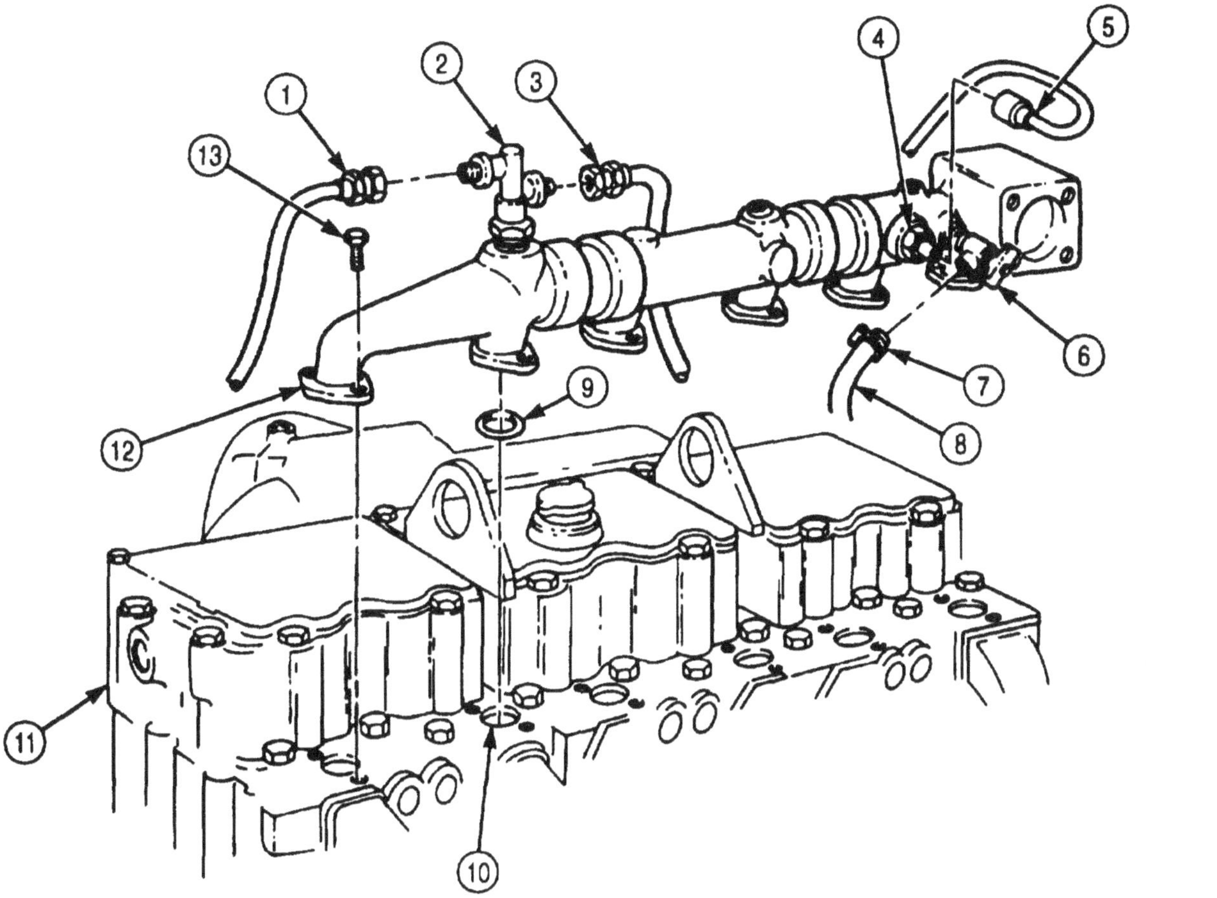

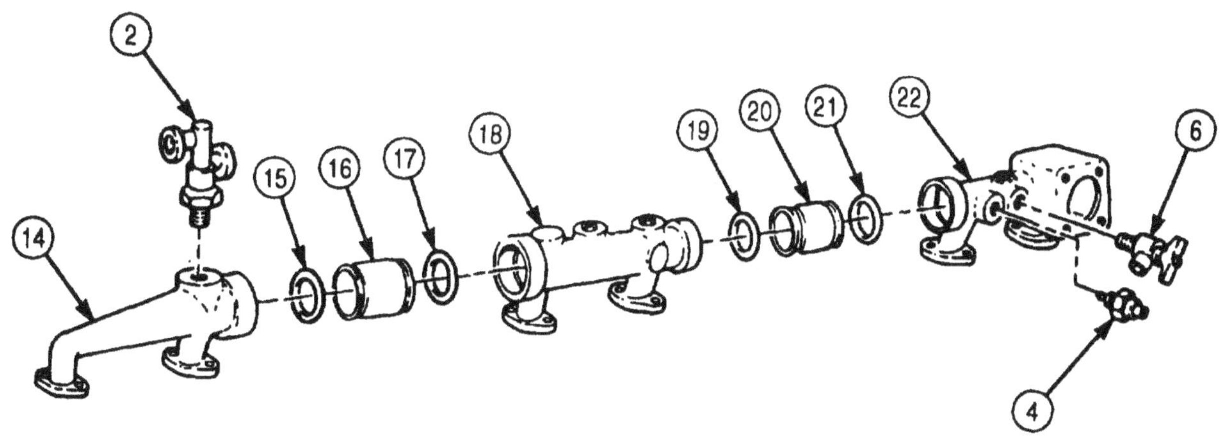

4-27. ENGINE WATER MANIFOLD MAINTENANCE (Contd)

c. Cleaning and Inspection

WARNING

Eyeshields must be worn during cleaning procedure. Failure to wear eyeshields may result in injury to personnel.

1. Clean water manifold sections (1), (6), and (10), and couplings (4) and (8) with wire brush and inspect. Replace if broken or cracked.
2. Check heater shutoff drainvalve (11) for proper opening and closing. Replace if defective.
3. Test water temperature sending unit (12) (para. 3-93). Replace if defective.

d. Assembly

1. Wrap male threaded ends of water temperature sending unit (12) and heater shutoff drainvalve (11) with antiseize tape and install in front water manifold section (10).
2. Apply light coat of GAA grease to new O-ring (9) and install on front coupling (8).
3. Install one end of front coupling (8) in bore of front water manifold section (10) until O-ring (9) is seated.
4. Apply light coat of GAA grease to new O-ring (7) and install on front coupling (8).
5. Install bore of center water manifold section (6) over end of front coupling (8) until seated against O-ring (7).
6. Apply light coat of GAA grease to new O-ring (5) and install on rear coupling (4).
7. Install one end of rear coupling (4) in bore of center water manifold section (6) until O-ring (5) is seated.
8. Wrap male threaded end of ether start safety valve (2) with antiseize tape and install in rear water manifold section (1).
9. Apply light coat of GM grease to new O-ring (3) and install on rear coupling (4).
10. Install rear water manifold section (1) bore over end of rear coupling (4) until seated against O-ring (3).

e. Installation

NOTE

Ensure all cylinder head water ports are unplugged.

1. Apply light coat of GAA grease to six new O-rings (18) and install in each cylinder head water port (19).
2. Install water manifold (21) on cylinder heads (20) with twelve new screw-assembled lockwashers (22). Alternately tighten screw-assembled lockwashers (22) 30-35 lb-ft (41-47 N•m).
3. Install ether cylinder on safety valve line (14) and safety valve-to-atomizer line (13) on ether start safety valve (2) at same points where disconnected.
4. Connect connector (15) to water temperature sending unit (12).
5. Connect personnel heater inlet hose (17) to heater shutoff drainvalve (11) with hose clamp (16).

4-27. ENGINE WATER MANIFOLD MAINTENANCE (Contd)

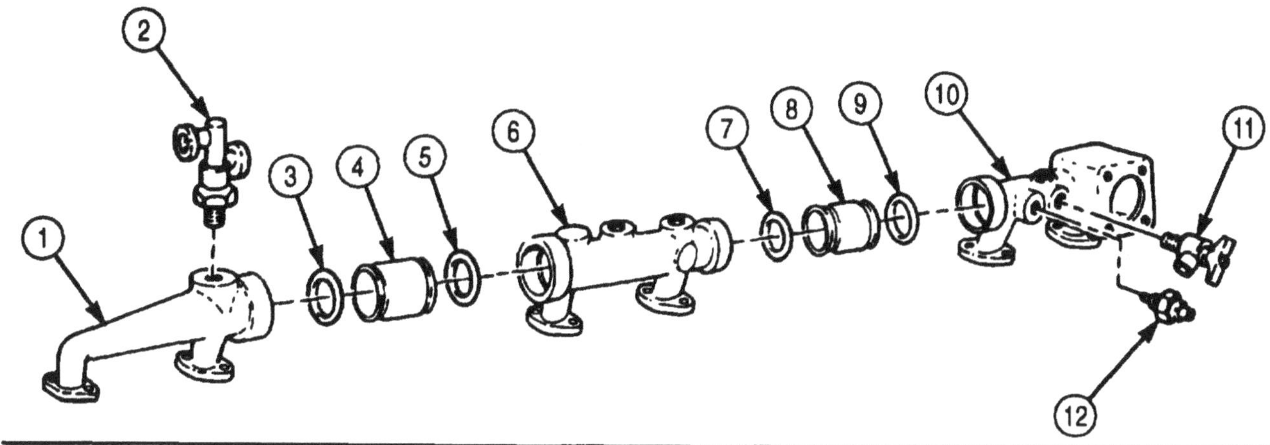

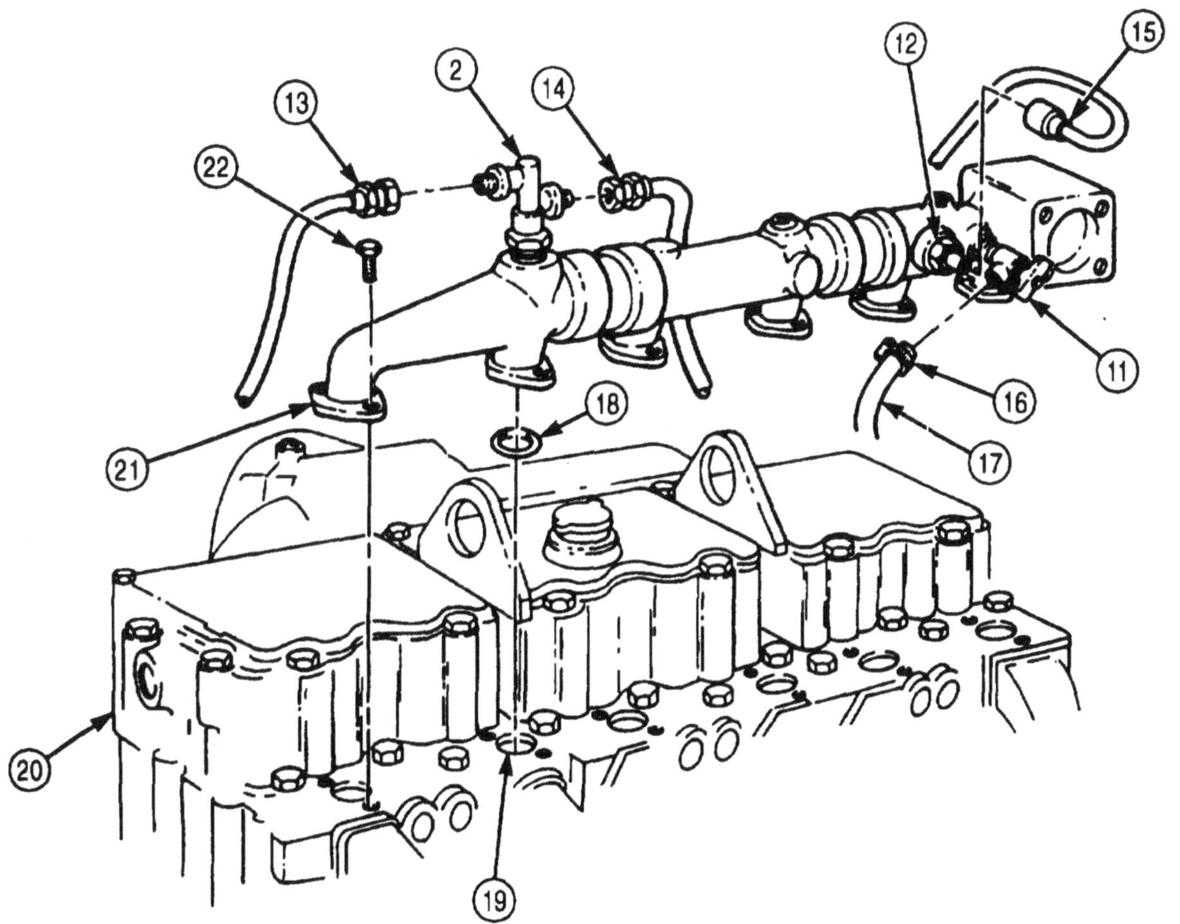

FOLLOW-ON TASKS:
- Install fan drive clutch actuator (para. 3-73).
- Install thermostat and housing (para. 3-65).
- Install engine crankcase breather draft tube (para. 3-2).
- Install surge tank (para. 3-61).
- Fill cooling system to proper level (para. 3-53).
- Start engine (TM 9-2320-272-10) and check for leaks.

4-28. WATER HEADER PLATES REPLACEMENT

THIS TASK COVERS:

a. Removal b. Installation

INITIAL SETUP:

APPLICABLE MODELS
M939/A1

TOOLS
General mechanic's tool kit (Appendix E, Item 1)

MATERIALS/PARTS
Two gaskets (Appendix D, Item 184)

REFERENCES (TM)
TM 9-2320-272-10
TM 9-2320-272-24P

EQUIPMENT CONDITION
- Parking brake set (TM 9-2320-272-10).
- Cooling system drained (para. 3-53).

NOTE

Water header plates are mounted with screw-assembled lockwashers on late model engines.

a. Removal

1. Remove twelve screws (1), two water header plates (2), and gaskets (3) from cylinder block (4). Discard gaskets (3).
2. Clean gasket remains from mating surface of cylinder block (4) and water header plate (2).

b. Installation

Install two new gaskets (3) and two water header plates (2) on cylinder block (4) with twelve screws (1).

4-28. WATER HEADER PLATES REPLACEMENT (Contd)

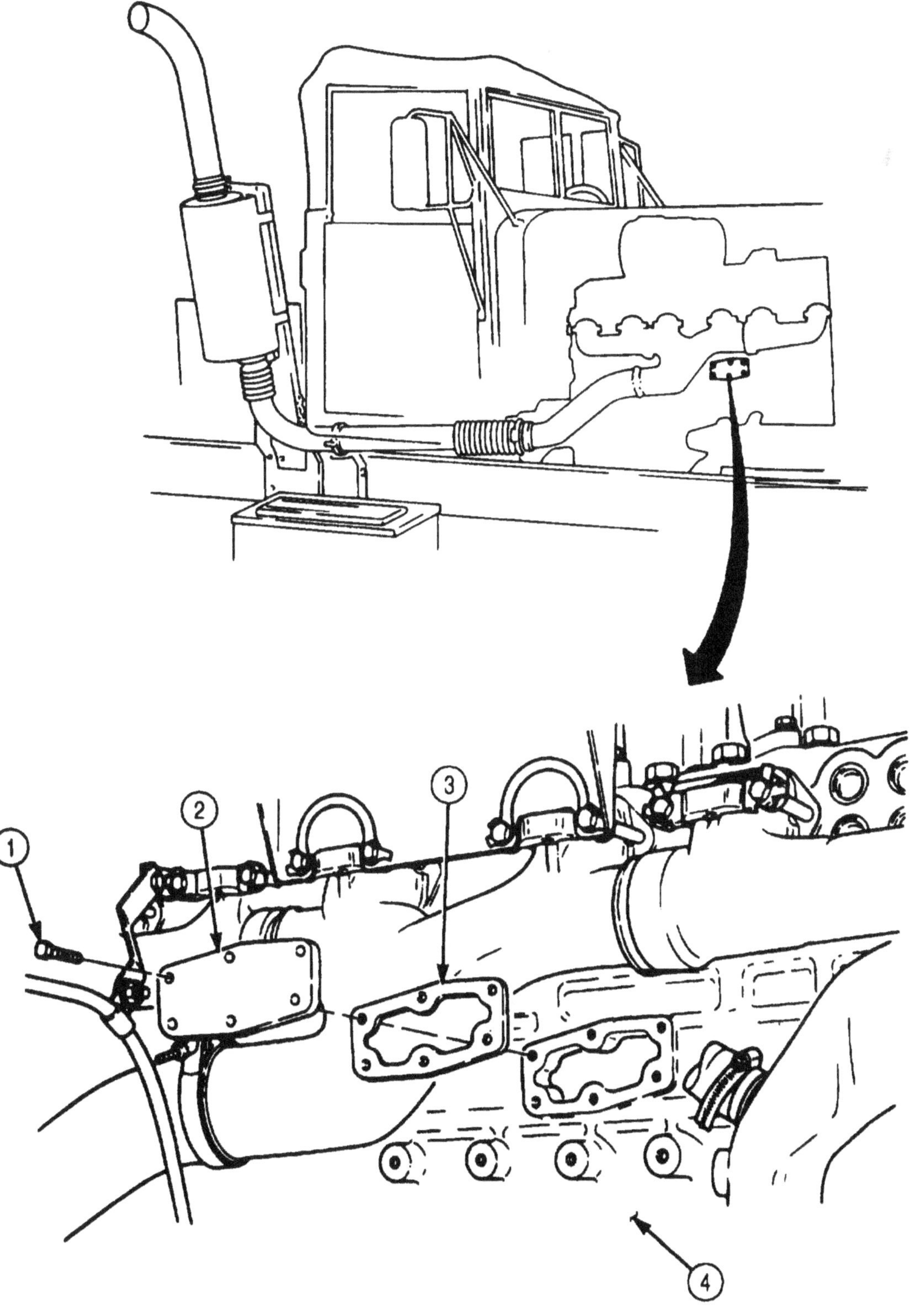

FOLLOW-ON TASK: Fill cooling system to proper level (para. 3-53).

4-29. WATER PUMP MAINTENANCE

THIS TASK COVERS:

a. Removal
b. Disassembly
c. Cleaning and Inspection
d. Assembly
e. Installation

INITIAL SETUP:

APPLICABLE MODELS

M939/A1

TOOLS

General mechanic's tool kit (Appendix E, Item 1)
Mechanical wedge puller (Appendix E, Item 101)
Outside micrometer (Appendix E, Item 80)
Snap gauge (Appendix E, Item 123)
Torque wrench (Appendix E, Item 146)
Arbor press
Wire brush
Mandrel driver

MATERIALS/PARTS

Two lockwashers (Appendix D, Item 350)
Gasket (Appendix D, Item 153)
Water pump seal assembly (Appendix D, Item 635)
Snapring (Appendix D, Item 660)
O-ring (Appendix D, Item 425)
Snapring (Appendix D, Item 662)
Eight lockwashers (Appendix D, Item 349)
Two relief fittings (Appendix D, Item 532)
GAA grease (Appendix C, Item 28)
Detergent (Appendix C, Item 27)

REFERENCES (TM)

TM 9-2320-272-10
TM 9-2320-272-24P
TM 9-214

EQUIPMENT CONDITION

- Cooling system drained (para. 3-53).
- Fan drive clutch removed (para. 3-75).
- Water pump drivebelt removed (para. 3-67).
- Alternator adjusting link removed (para. 3-79).

GENERAL SAFETY INSTRUCTIONS

When cleaning with compressed air, wear eyeshields and ensure source pressure does not exceed 30 psi (207 kPa).

a. Removal

1. Remove screw (2), lockwasher (3), hose clamp (4), and spacer (5) from engine bracket (6). Discard lockwasher (3).
2. Remove screw (1), lockwasher (12), and washer (11) from engine bracket (6). Discard lockwasher (12).
3. Remove six screws (9) and lockwashers (8) from support bracket (7). Discard lockwashers (8).
4. Remove support bracket (7) from engine block (10).
5. Remove water pump body (16) and O-ring (15) from water pump support (14). Discard O-ring (15).
6. Remove two screws (17) and lockwashers (18) from engine block (10). Discard lockwashers (18).
7. Remove water pump support (14) and gasket (13) from engine block (10). Discard gasket (13).

4-29. WATER PUMP MAINTENANCE (Contd)

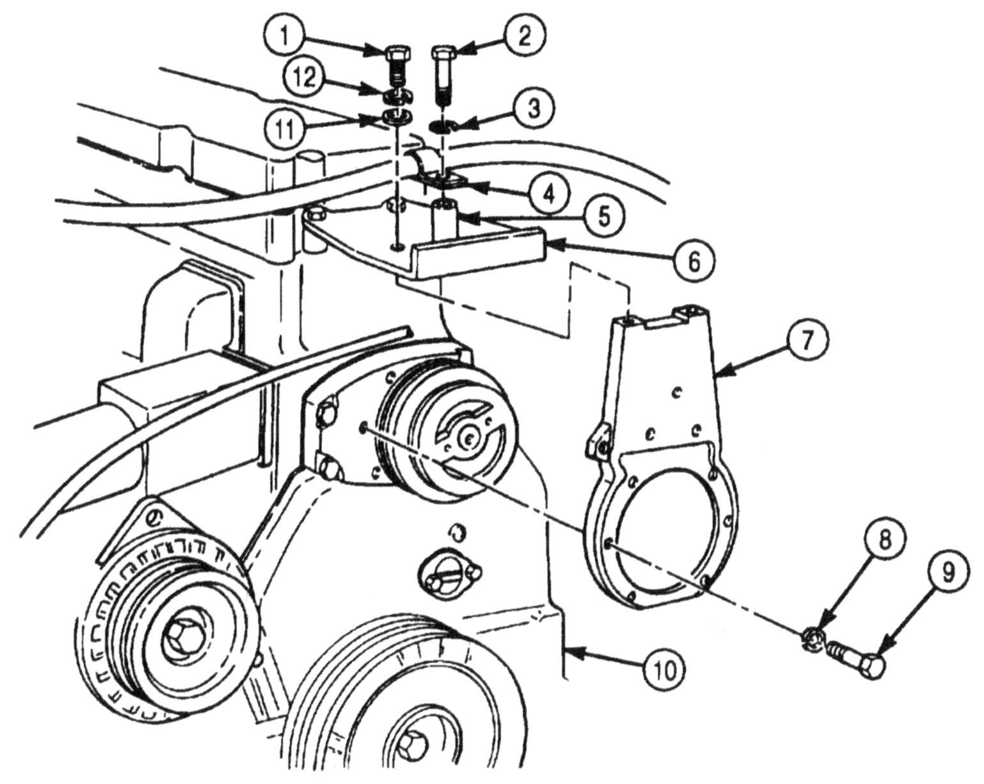

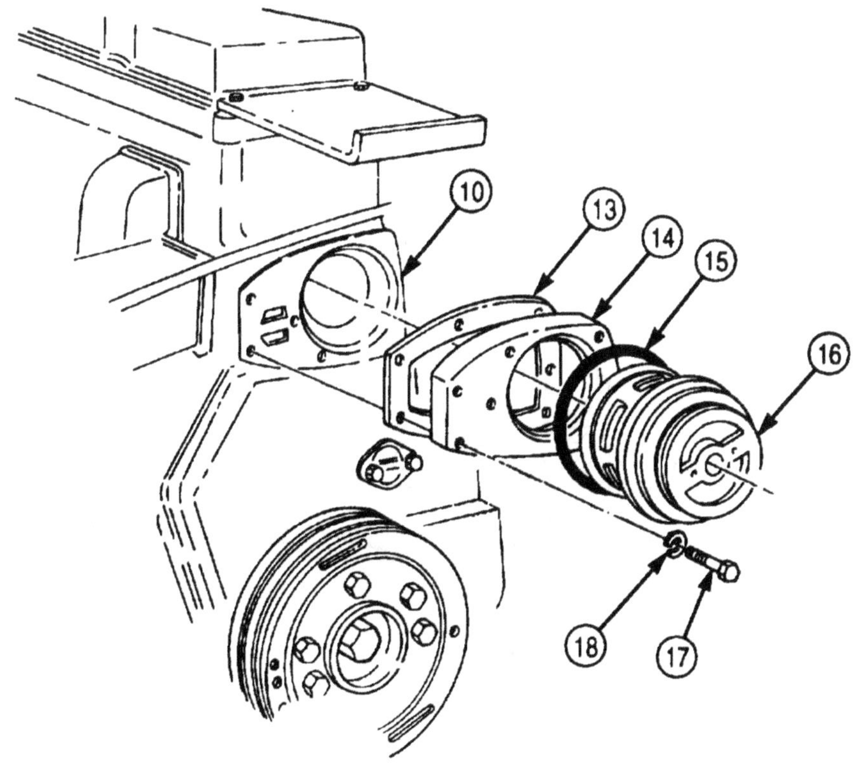

4-29. WATER PUMP MAINTENANCE (Contd)

b. Disassembly

1. Using mechanical wedge puller, remove drive pulley (2) from shaft (1) of water pump body (3).
2. Using puller, remove water pump impeller (4) from shaft (1).
3. Remove water pump seal (5) from shaft (1). Discard water pump seal (5).
4. Remove snapring (7) from ring groove (8) inside bore (6) of water pump body (3). Discard snapring (7).
5. Place water pump body (3) on arbor press with bore (6) facing downward, and support water pump body (3) with two blocks of wood.
6. Remove shaft assembly (9) from water pump body (3) by applying pressure on impeller end of shaft (9) and press shaft (9) down and out of bore (6).

NOTE

Bearing assemblies come out mounted on shalt.

7. Using mechanical wedge puller clamped between grease grooves (13) on spacer (12), remove bearing (11) and spacer (12) from shaft (9)
8. Remove snapring (14) from snapring groove (15) on shaft (9). Discard snapring (14).
9. Remove bearing (10) by pressing shaft (9) down through bearing (10).
10. Press water pump seal (16) out at impeller end of water pump body (3). Discard water pump seal (16).
11. Remove relief fitting (17) from water pump body (3). Do not discard plug (17) unless damaged.
12. Remove two vent relief fittings (18) from water pump body (3). Discard relief fittings (18).

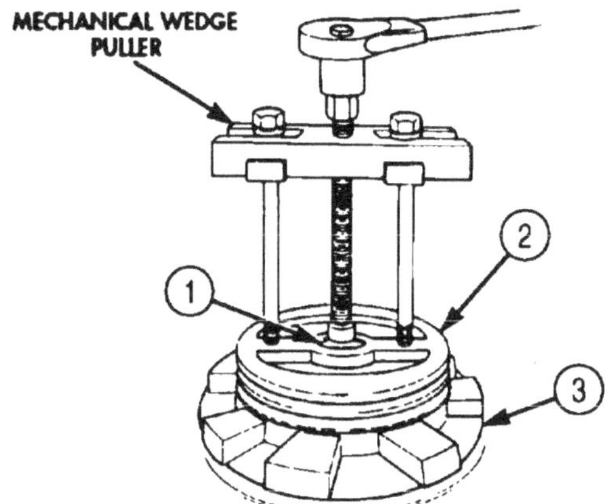

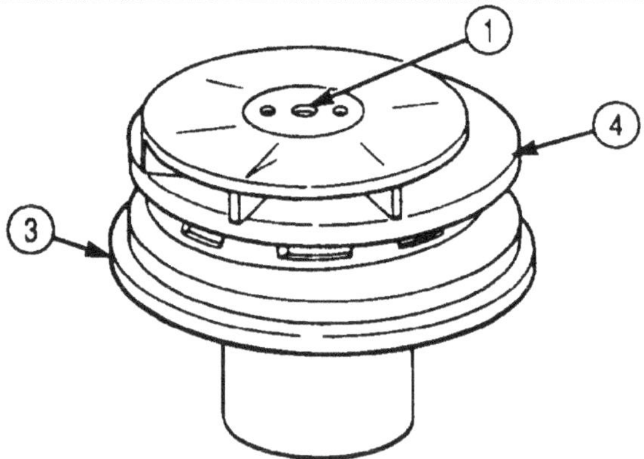

4-29. WATER PUMP MAINTENANCE (Contd)

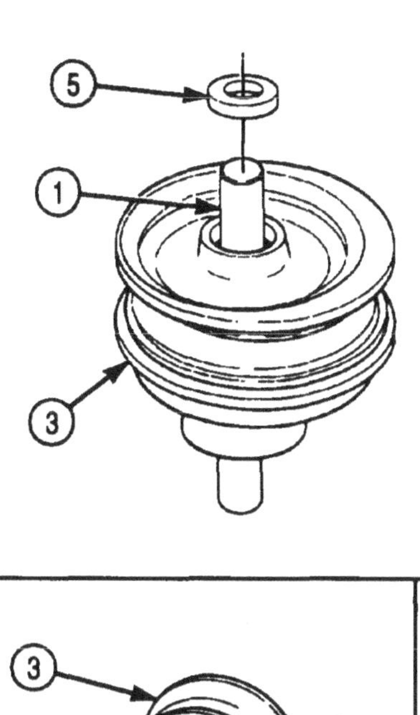

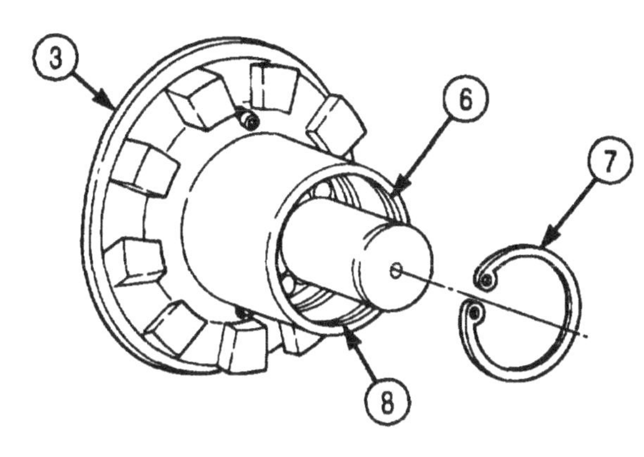

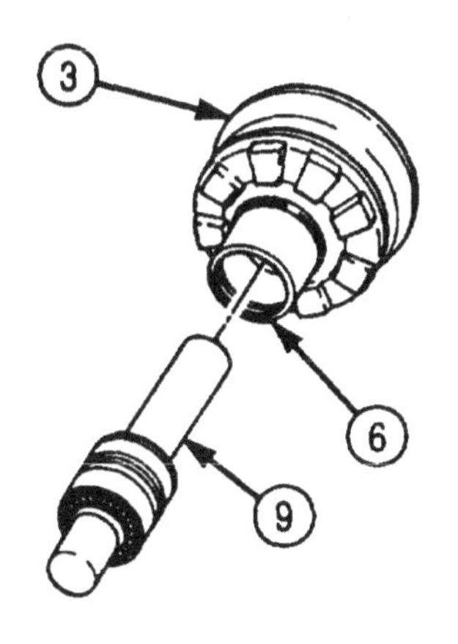

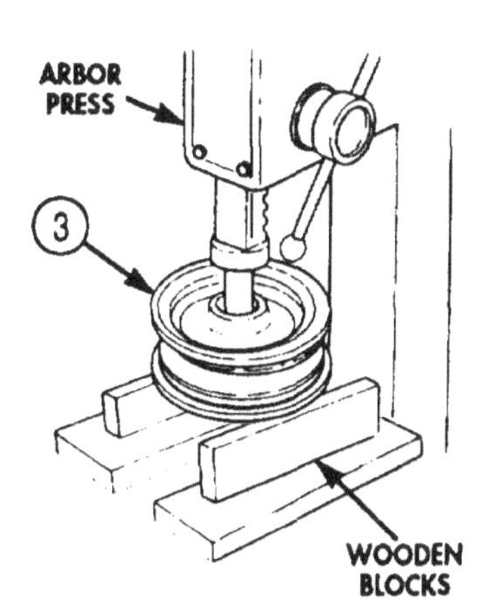

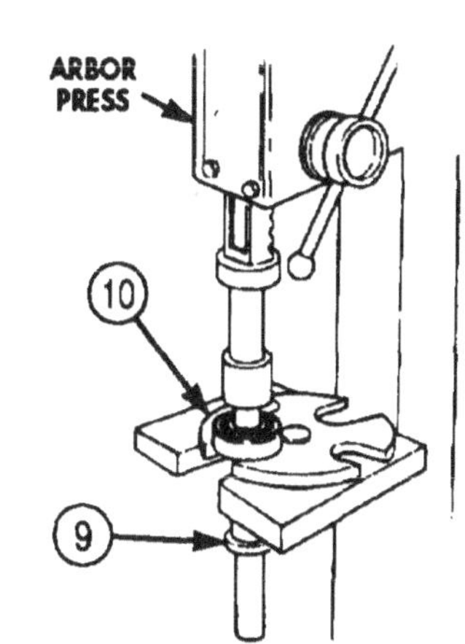

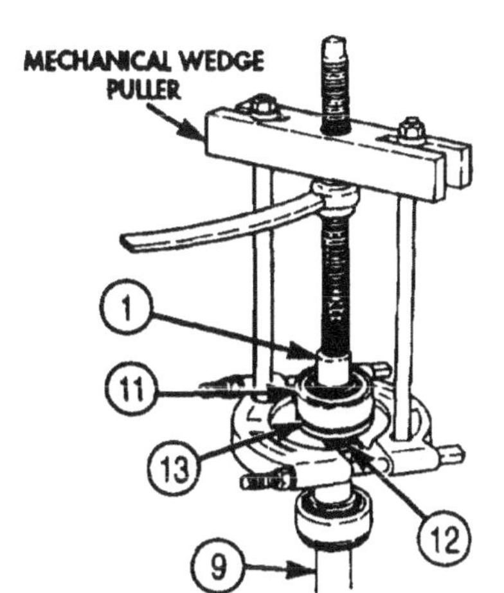

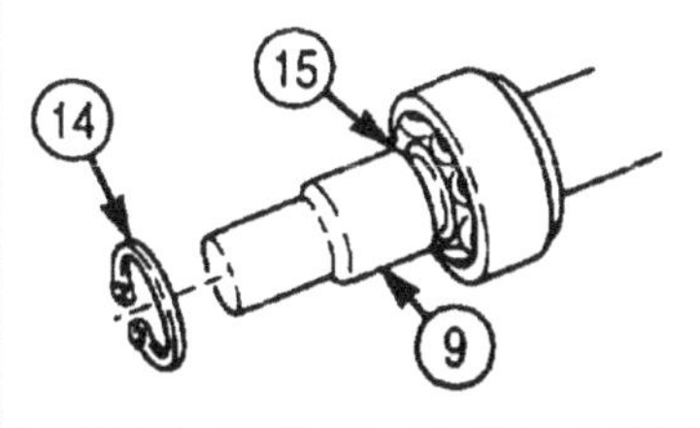

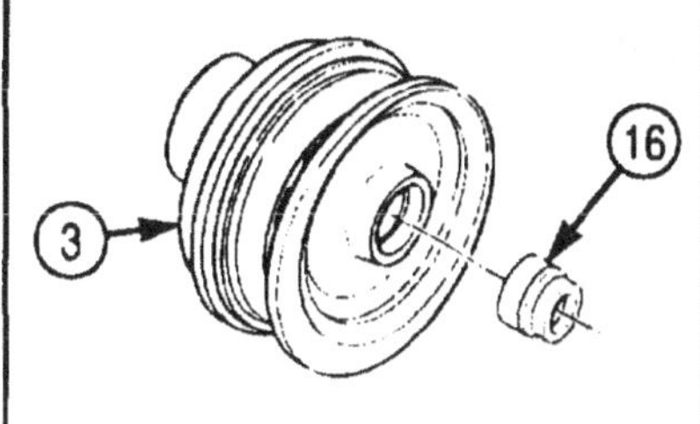

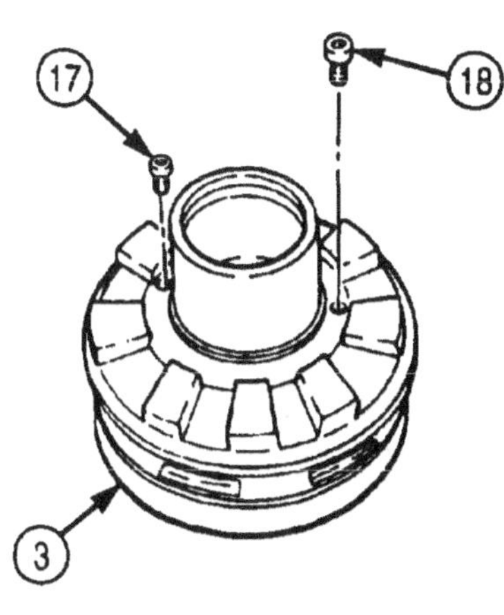

4-29. WATER PUMP MAINTENANCE (Contd)

c. Cleaning and Inspection

WARNING

Eyeshields must be worn when cleaning with compressed air. Compressed air source will not exceed 30 psi (207 kPa). Failure to do so may result in injury to personnel.

CAUTION

Rotate bearings very slowly while cleaning. Do not spin bearing races with compressor air when drying bearings. Serious damage will result.

NOTE

Do not service bearings in dirty surroundings.

1. Clean bearings (4) and (5), water pump body (6), shaft (3), and water pump impeller (8) (para. 2-14 and TM 9-214).
2. Clean gasket remains from water pump support (2) mating surfaces and engine block (1).
3. Inspect bearings (4) and (5) (para. 2-15). See TM 9-214 for additional inspection standards.
4. Check bearings (4) and (5) for heat discoloration, pits, scored ball or rollers, breaks, cracks, splits, dents, rust, or corrosion. Replace bearings (4) and/or (5) if any of these defects are noted.
5. Inspect spacer (7) for cracks and galls. Replace spacer (7) if cracked or galled.
6. Inspect water pump impeller (8) for cracks and heavy corrosion. Replace if cracked. Clean with wire' brush if corroded.
7. Inspect shaft (3) for cracks, scores, and galls. Replace shaft (3) if cracked, scored, or galled.
8. Inspect water pump body (6) for cracks, pits, and heavy corrosion. Replace water pump body (6) if cracked, pitted, or corroded.

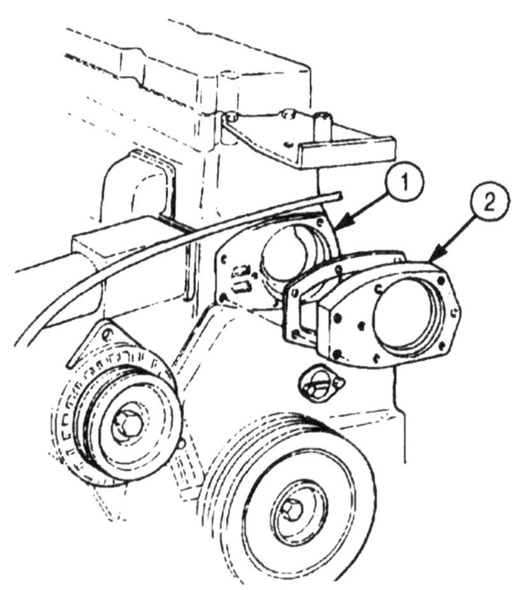

4-29. WATER PUMP MAINTENANCE (Contd)

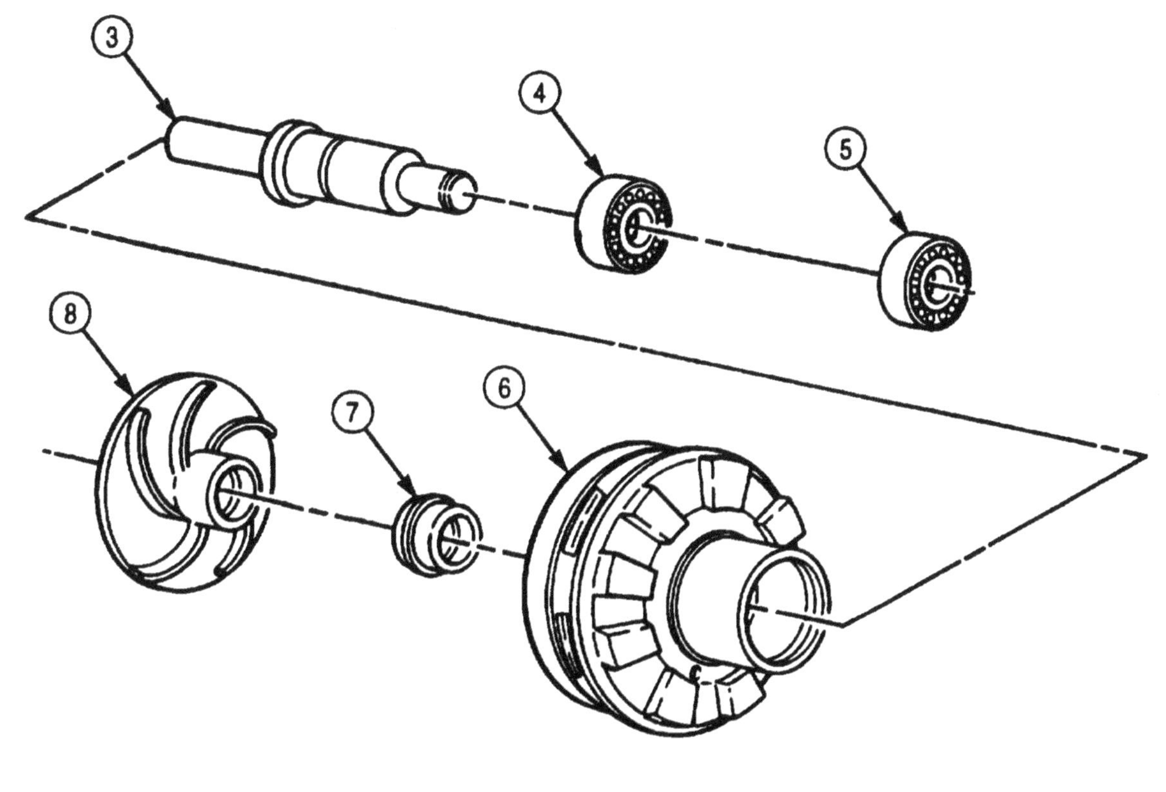

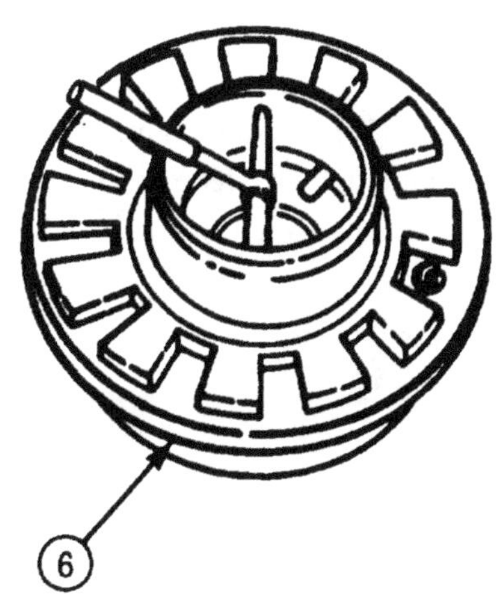

4-29. WATER PUMP MAINTENANCE (Contd)

d. Assembly

NOTE

New replacement bearings are installed as they are removed from packages, without cleaning or repacking. Original bearings, approved by inspection for reuse, must be packed after cleaning and inspection.

1. Apply light coat of clean GAA grease to shaft (1).
2. Pack bearing (3) with GAA grease (TM 9-214).
3. Using arbor press and mandrel, press shaft (1) down through bearing (3) until shoulder (2) bottoms on bearing (3).
4. Install snapring (4) in ring groove (5) on shaft (1).
5. Install spacer (7) on shaft (1) and seat against bearing (3).
6. Press bearing (6) over shaft (1) with arbor press and mandrel until seated against spacer (7).
7. Coat new water pump seal (9) with detergent to ease installation. Ensure carbon face on new water pump seal (9) is free of grease.
8. Using mandrel, install new water pump seal (9) at impeller end of water pump body (8).
9. Place water pump body (8) on arbor press with bore (10) facing upward.
10. Position shaft assembly (1) in water pump body (8) and align straight with bore (10).
11. Press shaft assembly (1) into water pump body (8) with mandrel.
12. Install new snapring (11) in bore (10) of snapring groove (12).
13. Install new relief fitting (17) in hole (18) on water pump body (8).

CAUTION

Water pumps are lubricated only after rebuild. The lubrication passages are normally kept plugged to prevent over-greasing which can cause seal blowout and damage to water pump.

NOTE

- The following procedure must be maintained as described to prevent water pump damage.
- Grease fitting is installed in water pump body passage temporarily to fill water pump cavity with lubricant.

14. Install a grease fitting (14) in passage (13) of water pump body (8).
15. Fill water pump body (8) with 0.60-0.70 cu in. (0.31-0.37 oz) of GM grease. Ensure grease does not come in contact with shaft (15).
16. Remove grease fitting (14) from passage (13) of water pump body (8). Discard grease fitting (14).
17. Install new plug (16) in passage (13) and tighten securely.

4-29. WATER PUMP MAINTENANCE (Contd)

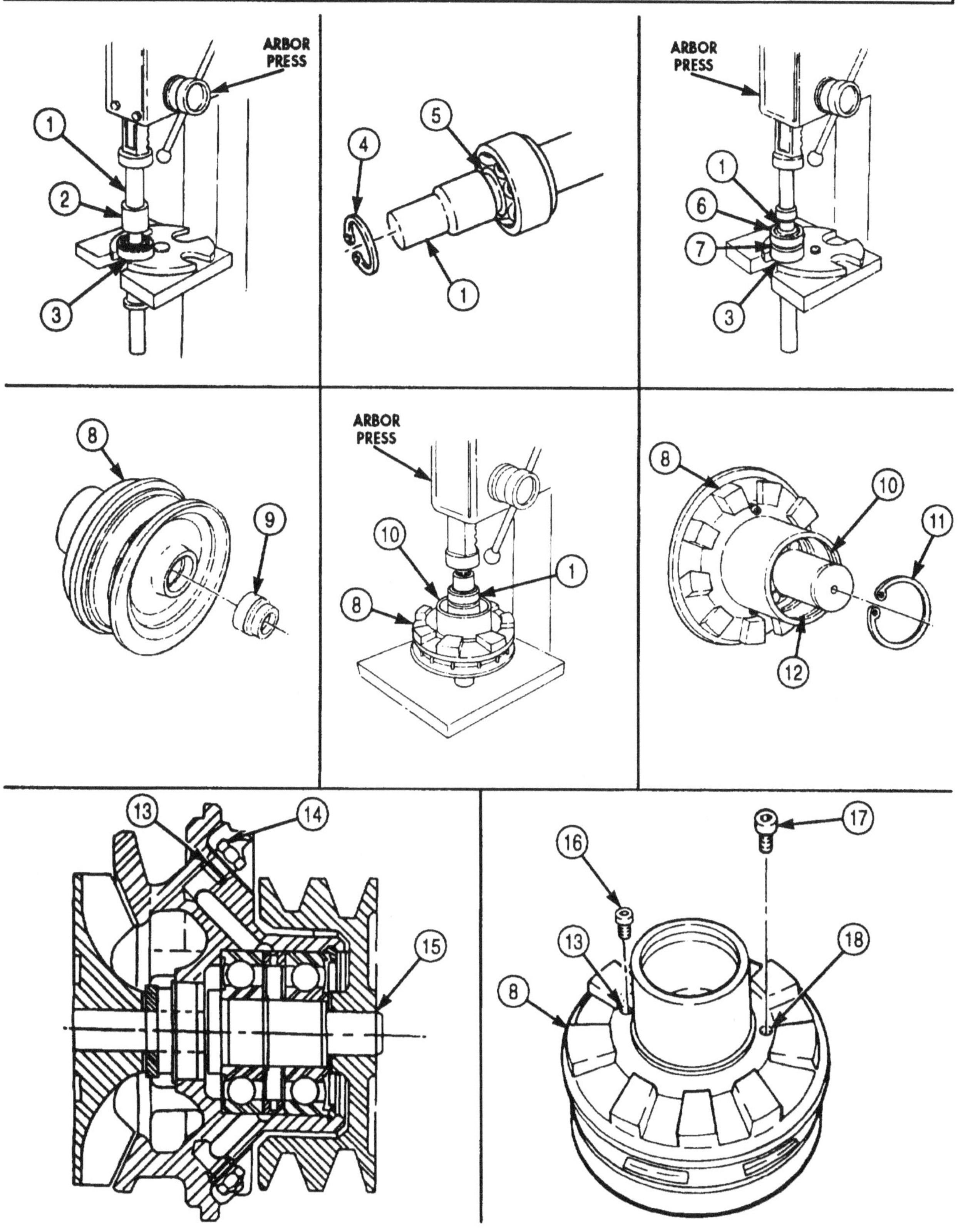

4-29. WATER PUMP MAINTENANCE (Contd)

18. Support water pump body (2) directly on impeller end of shaft (3) on arbor press.
19. Position drive pulley (1) and press on shaft (3) until seated.
20. Place water pump body (2) on arbor press with pulley drive (1) face down.
21. Apply coating of detergent to water pump seal (4) and install water pump seal (4) on shaft (3) with stainless steel surface facing upward.

NOTE

Minimum press fit between shaft and impeller is 0.001 in. (0.03 mm).

22. Position water pump impeller (5) and, using mandrel driver, press on shaft (3).

e. Installation

1. Apply light coat of clean GM grease to both sides of new gasket (7).
2. Position new gasket (7) on mating surface of water pump support (8) and align to screw holes.
3. Position water pump support (8) on engine block (6) and align to screw holes.
4. Install water pump support (8) on engine block (6) with two new lockwashers (11) and screws (10). Do not tighten screws (10) at this time.
5. Install new O-ring (9) and water pump body (2) on water pump support (8).
6. Tighten two screws (10) 30 lb-ft (41 N•m).
7. Install support bracket (17) on water pump support (8) with six new lockwashers (18) and screws (19). Tighten screws (19) 30 lb-ft (41 N•m).
8. Install top of support bracket (17) to upper radiator support bracket (16) with spacer (15), washer (20), two new lockwashers (13), screws (12), and hose clamp (14). Tighten all parts securely.

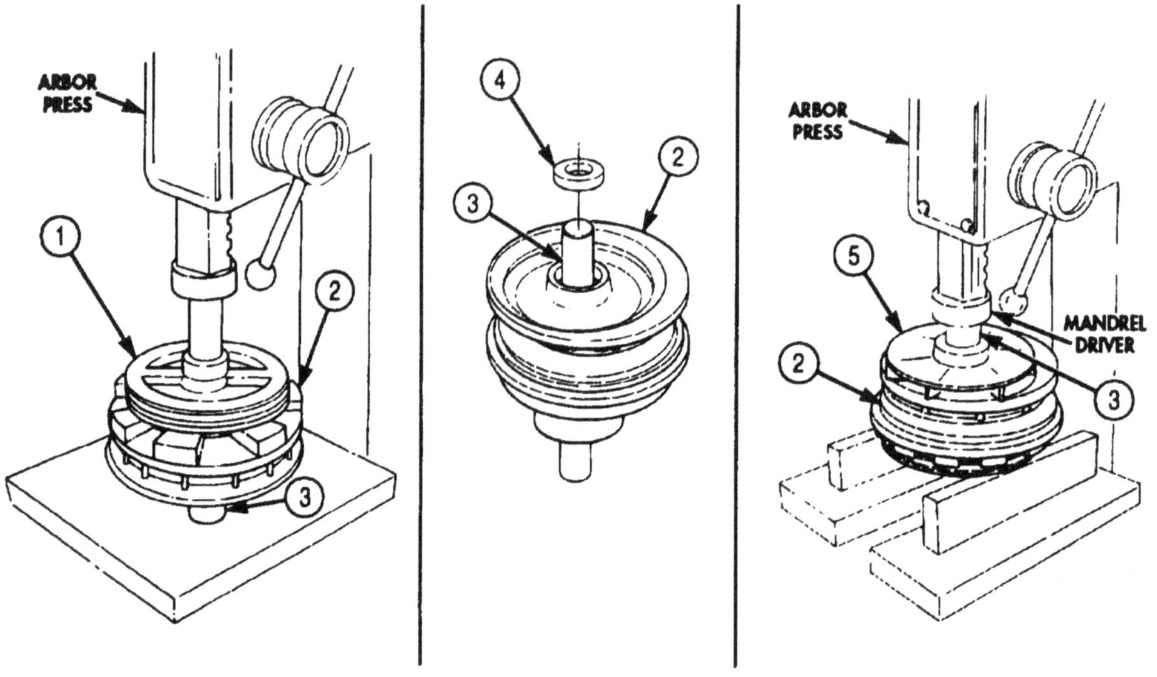

4-29. WATER PUMP MAINTENANCE (Contd)

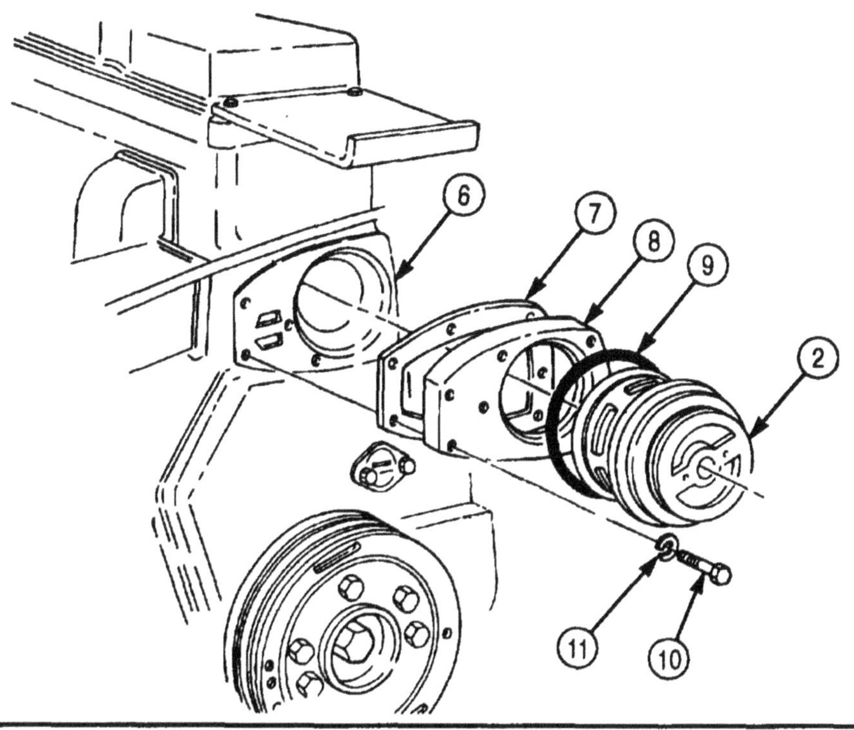

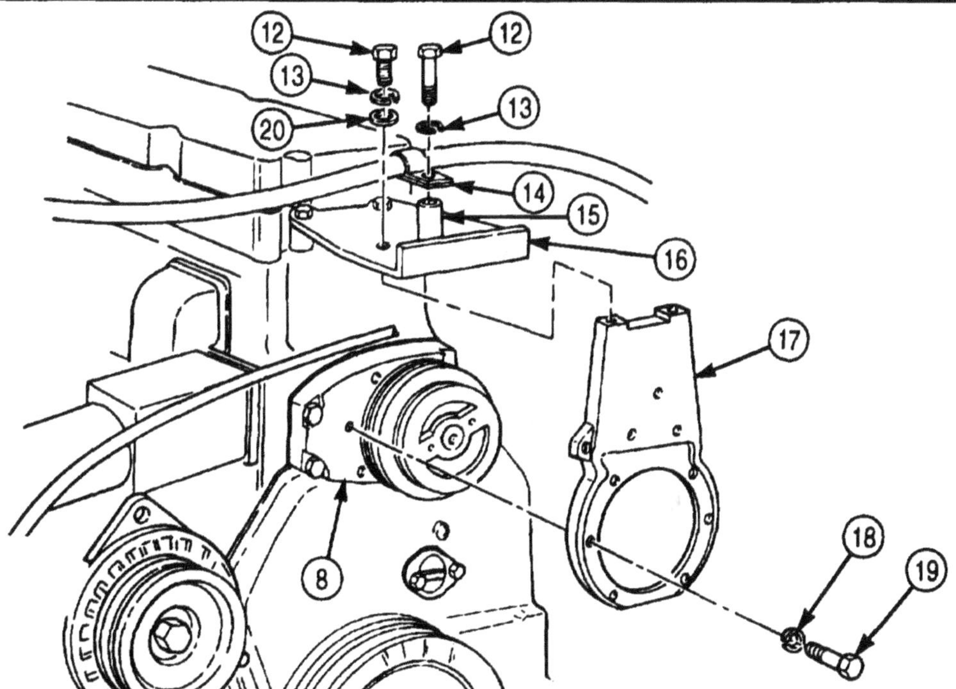

FOLLOW-ON TASKS:
- Install and adjust water pump drivebelt (para. 3-67).
- Install fan drive clutch (para. 3-75).
- Fill cooling system (para. 3-53).
- Install alternator adjusting link (para. 3-79).
- Start engine (TM 9-2320-272-10) and check for coolant leaks.

4-30. FUEL CROSSOVER CONNECTORS AND CROSSHEADS REPLACEMENT

THIS TASK COVERS:

a. Removal
b. Inspection
c. Installation and Adjustment

INITIAL SETUP:

APPLICABLE MODELS
M939/A1

SPECIAL TOOLS
Torque wrench adapter (Appendix E, Item 143)

TOOLS
General mechanic's tool kit (Appendix E, Item 1)
Dial indicator (Appendix E, Item 36)
Torque wrench (Appendix E, Item 146)

MATERIALS/PARTS
Eight screw-assembled lockwashers (Appendix D, Item 578)
Eight O-rings (Appendix D, Item 438)
Lubricating oil (Appendix C, Item 50)

REFERENCES (TM)
TM 9-2320-272-10
TM 9-2320-272-24P

EQUIPMENT CONDITION
- Parking brake set (TM 9-2320-272-10).
- Left and right splash shields removed (TM 9-2320-272-10).
- Rocker lever housings and push tubes removed (para. 4-20).

NOTE

Fuel crossover connectors are mounted using screw-assembled lockwashers on late model engines.

a. Removal

1. Remove eight screw-assembled lockwashers (1), two crossover connectors (4), and eight O-rings (3) from cylinder heads (2). Discard O-rings (3) and screw-assembled lockwashers (1).

NOTE

Tag all crossheads for installation.

2. Loosen twelve crosshead adjusting nuts (6) on twelve valve crossheads (7).
3. Remove twelve valve crossheads (7) from cylinder heads (2).

b. Inspection

1. Inspect twelve valve crossheads (7) for cracks, breaks, or scoring. Replace if cracked, broken, or scored.
2. Inspect adjusting screws (5) for damaged threads. Replace adjusting screws (5) if threads are damaged.

4-30. FUEL CROSSOVER CONNECTORS AND CROSSHEADS REPLACEMENT (Contd)

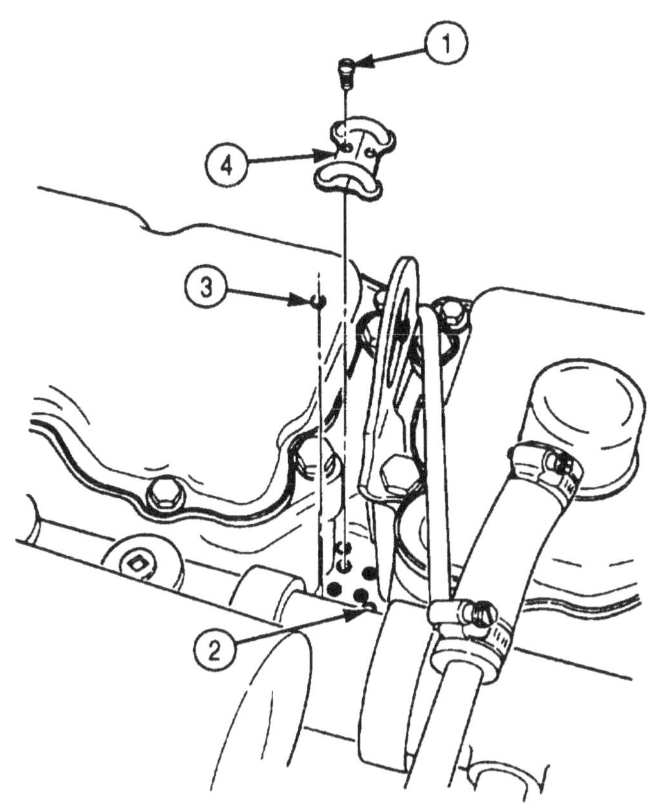

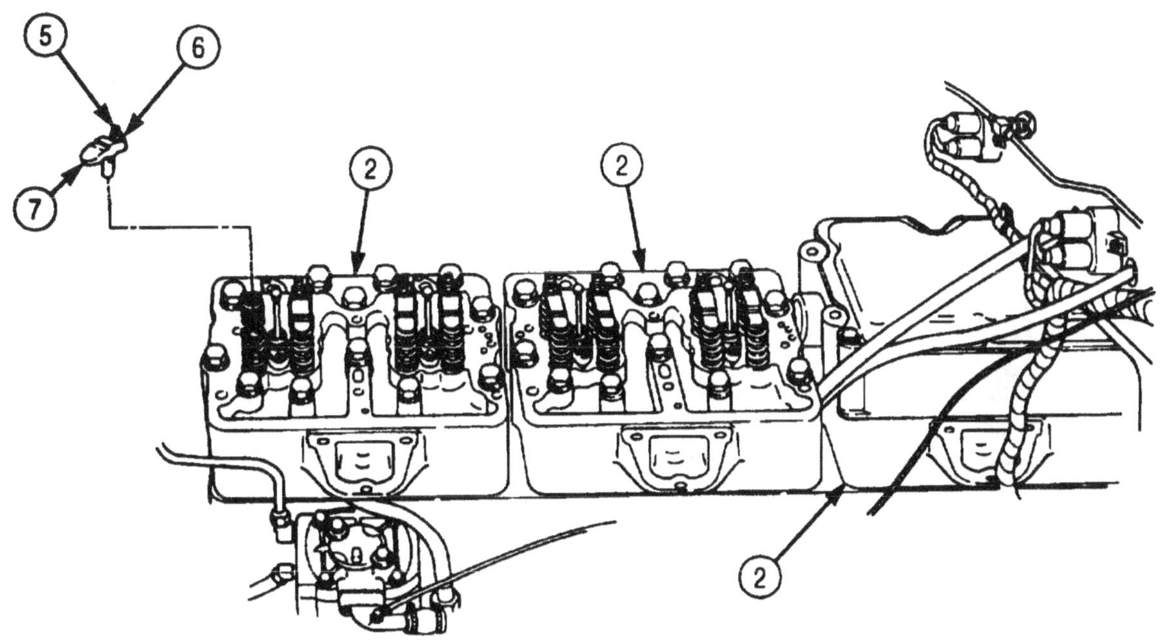

4-30. FUEL CROSSOVER CONNECTORS AND CROSSHEADS REPLACEMENT (Contd)

c. Installation and Adjustment

1. Loosen valve crosshead adjusting nuts (3) and back out adjusting screw (2) one full turn.
2. Apply light film of clean engine oil to twelve valve crossheads (1) and install on guide (6) of cylinder head (4). Ensure adjusting screw (2) faces toward manifold side of engine.
3. Hold valve crosshead (1) down with finger pressure so it contacts valve stem (5) on side opposite adjusting screw (2).

NOTE

- It may be necessary to loosen locknut to complete step 4.
- Ensure adjusting screw is lightly seated.

4. Turn adjusting screw (2) down until it just touches valve stem (5).
5. Set up dial indicator over center of valve crosshead (1) and zero-dial indicator while pressing down on valve crosshead (1).
6. Hold valve crosshead (1) down lightly and turn adjusting screw (2) in until dial indicator reads 0.025-0.040 in. (0.64-0.80 mm).
7. Tighten locknuts (3) 22-26 lb-ft (30-35 N•m).

NOTE

If minimum clearance reading on dial indicator is not 0.025 in. (0.64 mm), advance adjusting screw 1/3 of one hex on new valve crossheads and guides or 1/2 hex on old crossheads and guides, retighten locknut, and check clearance.

8. Apply light coat of clean engine oil to eight new O-rings (8) and insert in fuel crossover connector bores on cylinder heads (4).
9. Install two crossover connectors (9) on cylinder heads (4) with eight new screw-assembled lockwashers (7). Tighten screw-assembled lockwashers (7) 34-38 lb-in. (3.8-4.3 N•m).

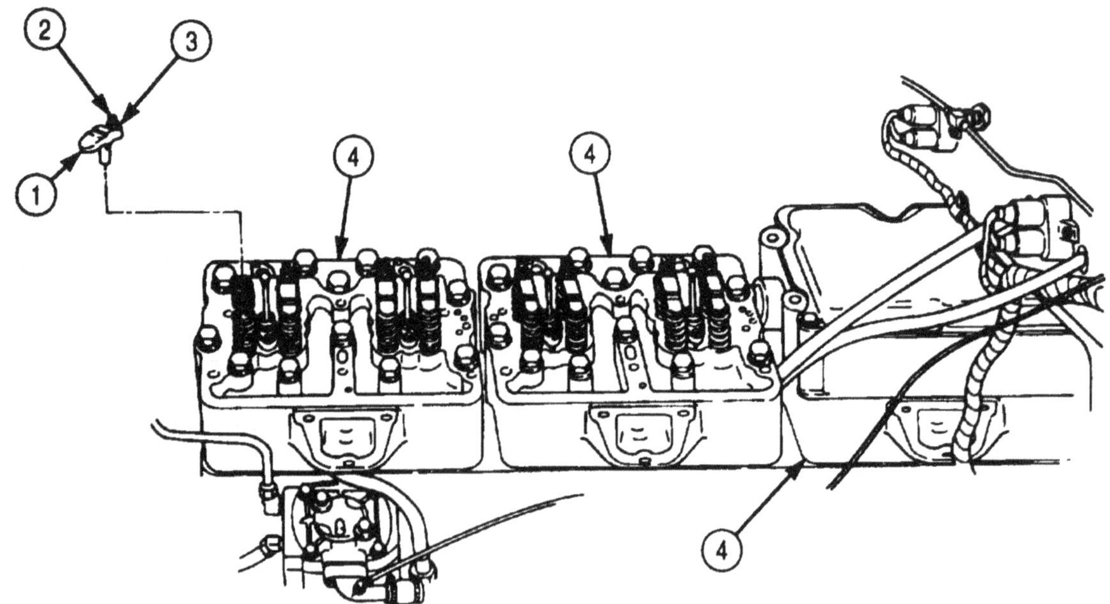

4-30. FUEL CROSSOVER CONNECTORS AND CROSSHEADS REPLACEMENT (Contd)

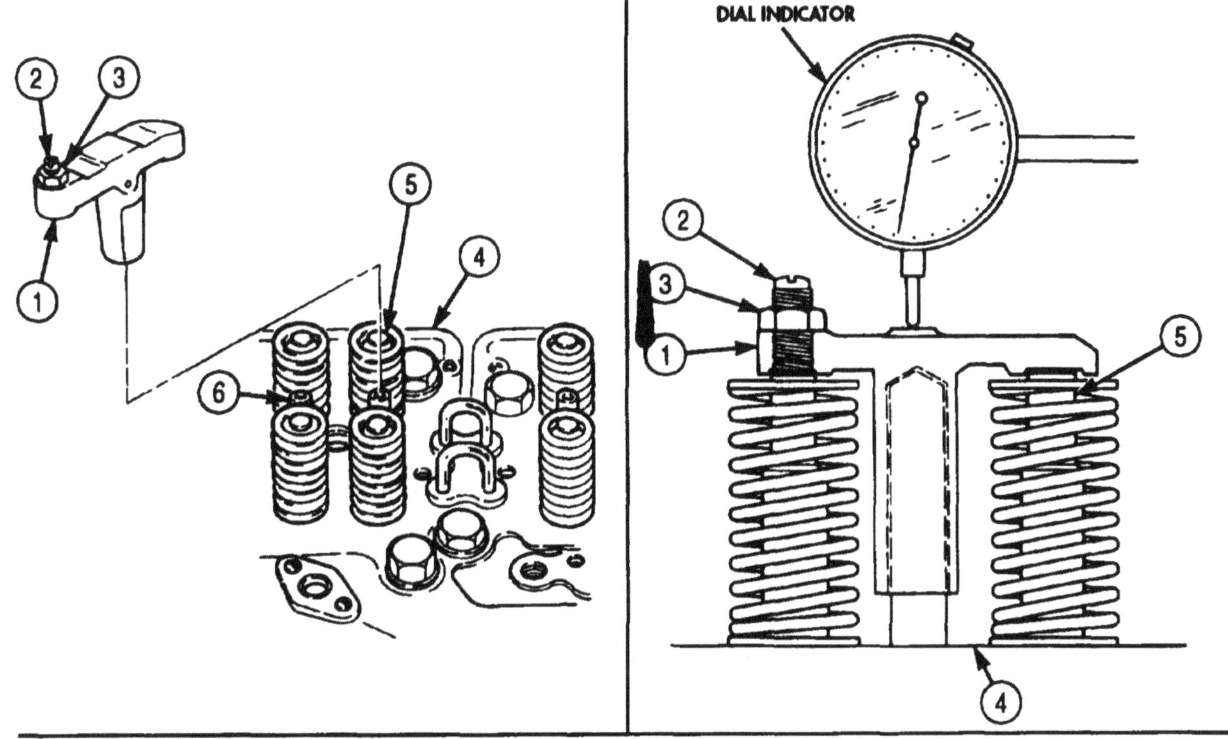

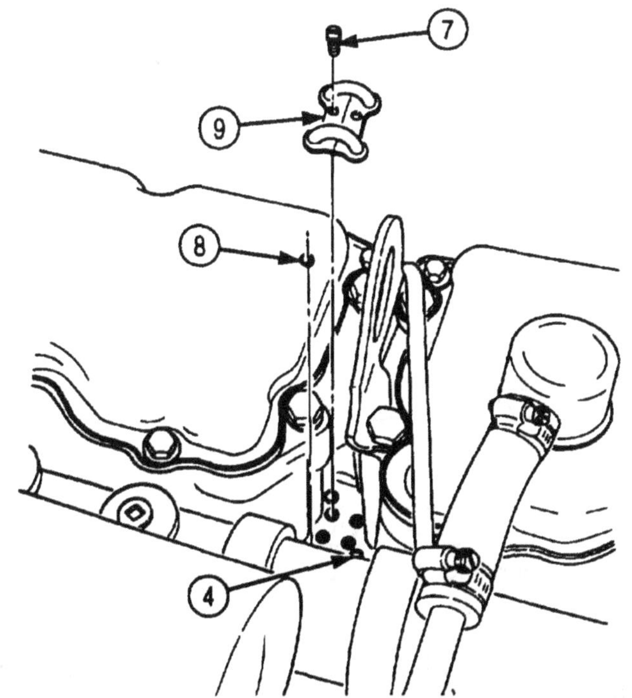

FOLLOW-ON TASKS:
- Install rocker lever housings and push tubes (para. 4-20).
- Install left and right splash shields (TM 9-2320-272-10).

4-31. AIR COMPRESSOR MAINTENANCE

THIS TASK COVERS:

a. Removal
b. Disassembly
c. Cleaning
d. Inspection and Repair
e. Assembly
f. Installation and Timing

INITIAL SETUP:

APPLICABLE MODELS
M939/A1

SPECIAL TOOLS
Piston ring expander (Appendix E, Item 96)
Mounting plate (Appendix E, Item 85)
Engine barring tool (Appendix E, Item 43)

TOOLS
General mechanic's tool kit (Appendix E, Item 1)
Torque wrench (Appendix E, Item 146)
Depth micrometer (Appendix E, Item 81)
Outside micrometer (Appendix E, Item 80)
Inside micrometer (Appendix E, Item 82)
Dial bore gauge (Appendix E, Item 136)
Spring tester (Appendix E, Item 131)
Dial indicator gauge (Appendix E, Item 36)
Ring compressor (Appendix E, Item 32)
Telescoping gauge (Appendix E, Item 136)
Arbor press
Vise
Feeler gauge

MATERIALS/PARTS
Twelve lockwashers (Appendix D, Item 364)
Housing support gasket (Appendix D, Item 185)
Four bushings (Appendix D, Item 21)
Repair kit (Appendix D, Item 534)
Repair kit (Appendix D, Item 535)
Gasket (Appendix D, Item 186)
Two screw-assembled lockwashers (Appendix D, Item 591)
Antiseize compound (Appendix C, Item 10)
Drycleaning solvent (Appendix C, Item 71)
Lubricating oil (Appendix C, Item 50)
Antiseize tape (Appendix C, Item 72)

REFERENCES (TM)
TM 9-2320-272-10
TM 9-2320-272-24P

EQUIPMENT CONDITION
- Air reservoirs drained (TM 9-2320-272-10).
- Left splash shield removed (TM 9-2320-272-10).
- Fuel pump removed (para. 4-35).
- Power steering pump removed (para. 3-236).

GENERAL SAFETY INSTRUCTIONS
- Do not disconnect air lines before draining air reservoirs.
- Drycleaning solvent is flammable and toxic. Do not use near open flame.
- Keep fire extinguisher nearby when using drycleaning solvent.
- Hold down unloader valve assembly during removal.
- When cleaning with compressed air, wear eyeshield and ensure source pressure does not exceed 30 psi (207 kPa).

a. Removal

WARNING

Do not disconnect air lines before draining air reservoirs. Small parts under pressure may shoot out with high velocity, causing injury to personnel.

1. Disconnect coolant outlet line (1) from elbow (5) on air compressor (2) and elbow (15) on engine oil cooler (16). Discard bushings (14).
2. Remove elbow (5) from air compressor (2).

4-31. AIR COMPRESSOR MAINTENANCE (Contd)

3. Remove nut (13), lockwasher (12), washer (11), screw (7), washer (8), and clamp (9) with coolant outlet line (1) from bracket (10). Discard lockwasher (12).
4. Remove coolant inlet line (4) from elbow (3) on air compressor (2) and water pump fitting (6). Discard inserts (14).
5. Remove elbow (3) from air compressor (2).
6. Remove four screw-assembled washers (18) and power steering pump pivot bracket (19) from engine (17).

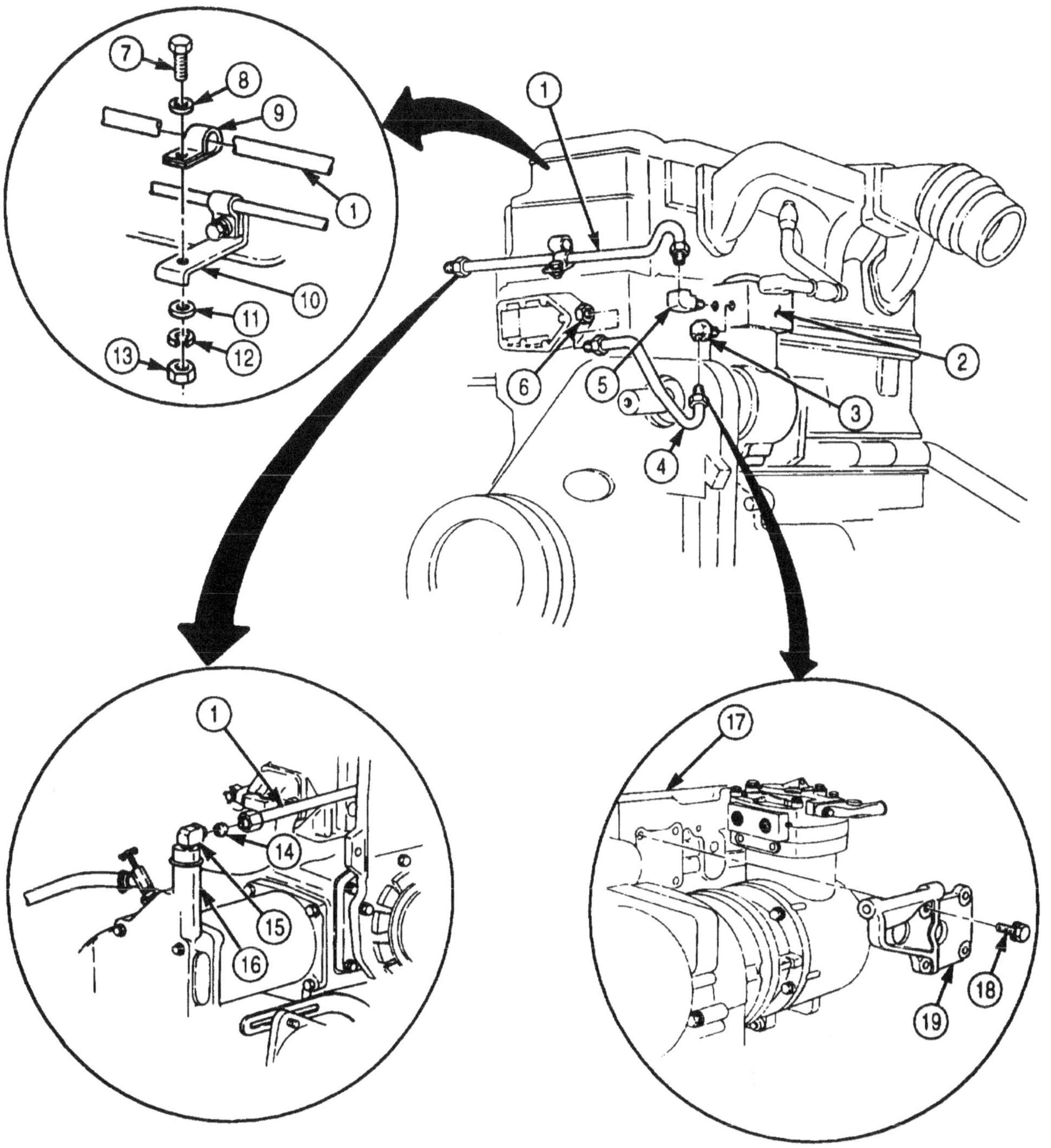

4-31. AIR COMPRESSOR MAINTENANCE (Contd)

7. Disconnect governor air line (3) from elbow (2) on unloader valve body (1).
8. Disconnect alcohol evaporator air line (4) from elbow (10) on compressor inlet (9).
9. Disconnect supply reservoir air line (6) from adapter (5) on compressor head (11).
10. Remove elbow (10) from compressor inlet (9), elbow (2) from valve body (1), and adapters (5) from compressor head (11).
11. Loosen two clamps (8) and remove air intake tube (7) from compressor inlet (9).
12. Remove screw-assembled washer (20) and ground cable (19) from air compressor (18).
13. Remove two screws (21) from accessory drive housing (13).
14. Remove two nuts (15), lockwashers (16), washers (17), screws (12), and washers (17) from accessory drive housing (13). Discard lockwashers (16).
15. Remove screw-assembled lockwashers (23) and (27) and washers (22) and (26) from bracket (25) and engine (24). Discard screw-assembled lockwashers (23) and (27).
16. Loosen screw (28), rotate bracket (25) downward, and remove air compressor (18) and gasket (14) from accessory drive housing (13). Discard gasket (14). Clean gasket remains from mating surfaces.
17. Remove drive coupling (29) from accessory drive housing (13).

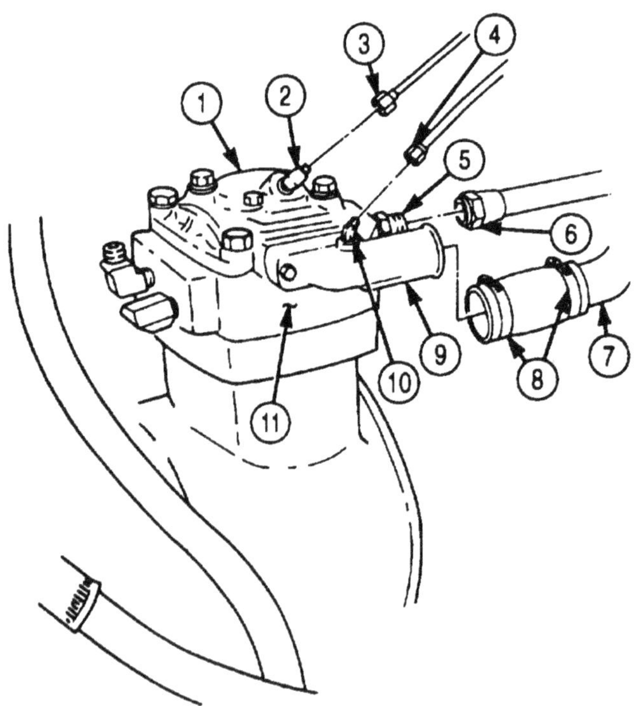

4-31. AIR COMPRESSOR MAINTENANCE (Contd)

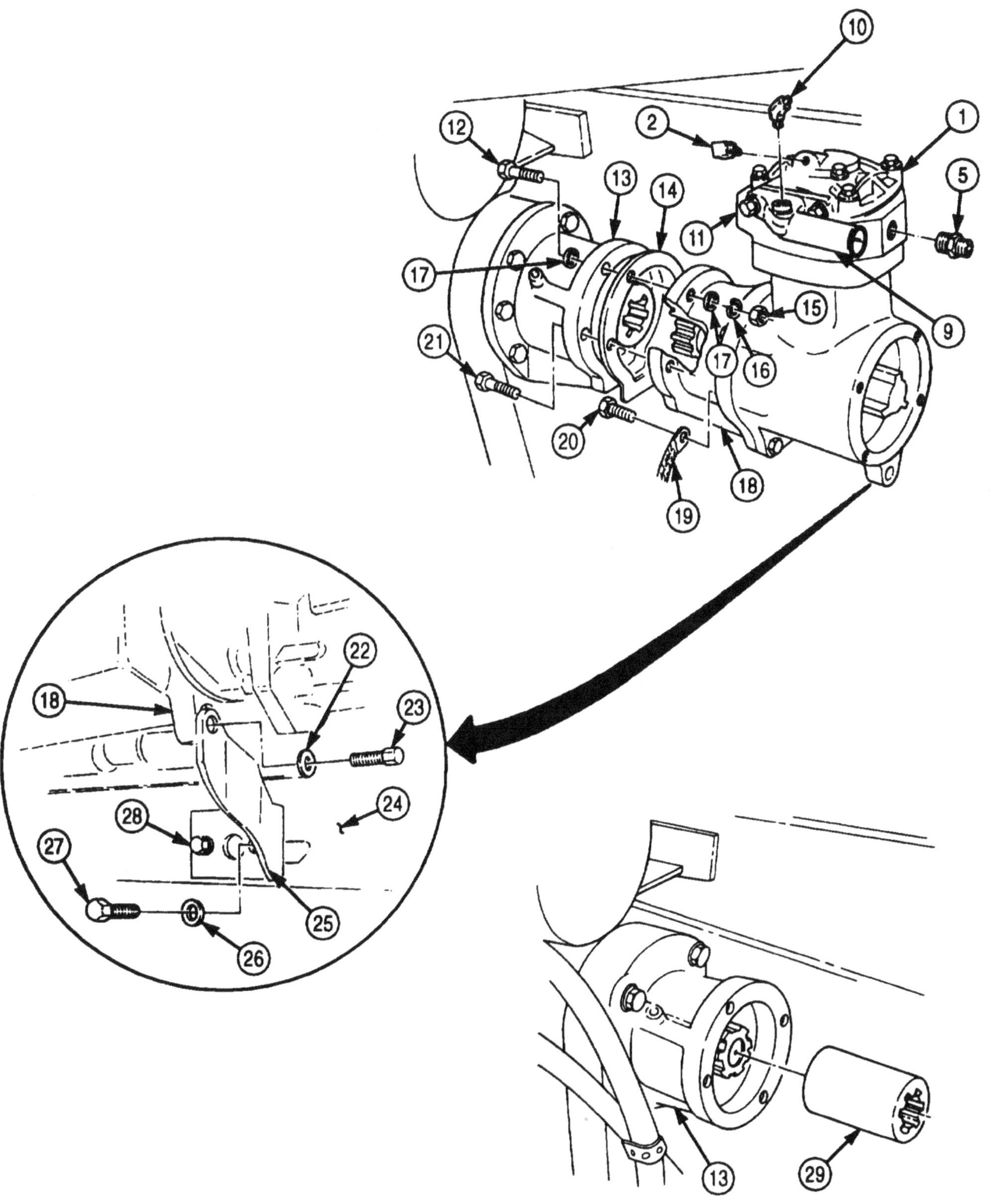

4-31. AIR COMPRESSOR MAINTENANCE (Contd)

b. Disassembly

1. Install mounting plate on air compressor (9) with two screws (8).
2. Install mounting plate and air compressor (9) in vise.
3. Remove two screws (11), washers (12), lockwashers (10), compressor inlet (13), and gasket (14) from cylinder head cover (7). Discard lockwashers (10) and gasket (14).

WARNING

Unloader valve assembly must be held down during removal. Small parts under pressure may shoot out, causing injury to personnel.

4. Remove two screws (4), washers (5), unloader valve assembly (1), unloader valve spring (2), intake valve assembly (3), and intake valve spring (6) from cylinder head cover (7).
5. Remove unloader valve cap (18), packing seal (17), and O-ring (16) from unloader valve body (15). Discard packing seal (17) and O-ring (16).
6. Remove intake valve seat (19) and disc valve (20) from intake assembly valve (3).
7. Remove four screws (21), washers (23), lockwashers (22), cylinder head cover (7), and gasket (24) from cylinder head (31). Discard lockwashers (22) and gasket (24).
8. Remove cylinder head (31) and gasket (32) from crankcase (33). Discard gasket (32).
9. Using arbor press, remove exhaust valve seat (26) from the bottom side of cylinder head (31).
10. Remove O-ring (25), packing seal (27), and exhaust valve (28) from exhaust valve seat. (26). Discard O-ring (25) and packing seal (27).
11. Remove wear plate (30) and exhaust valve spring (29) from cylinder head (31).
12. Remove six screws (34), lockwashers (35), and four washers (38) from crankcase (33) and front support (37). Discard lockwashers (35).

CAUTION

Use care when removing crankshaft from connecting rod. Crankshaft must be 90° before or after Top Dead Center (TDC) of piston for ease of removal. Failure to do so may damage connecting rod and crankshaft.

13. Remove front support (37), gasket (36), and crankshaft (41) from crankcase (33) as an assembly. Discard gasket (36).
14. Remove pipe plug (42) from crankshaft (41).
15. Using arbor press, remove drive coupling (39), thrust washer (40), and crankshaft (41) from front support (37).

4-31. AIR COMPRESSOR MAINTENANCE (Contd)

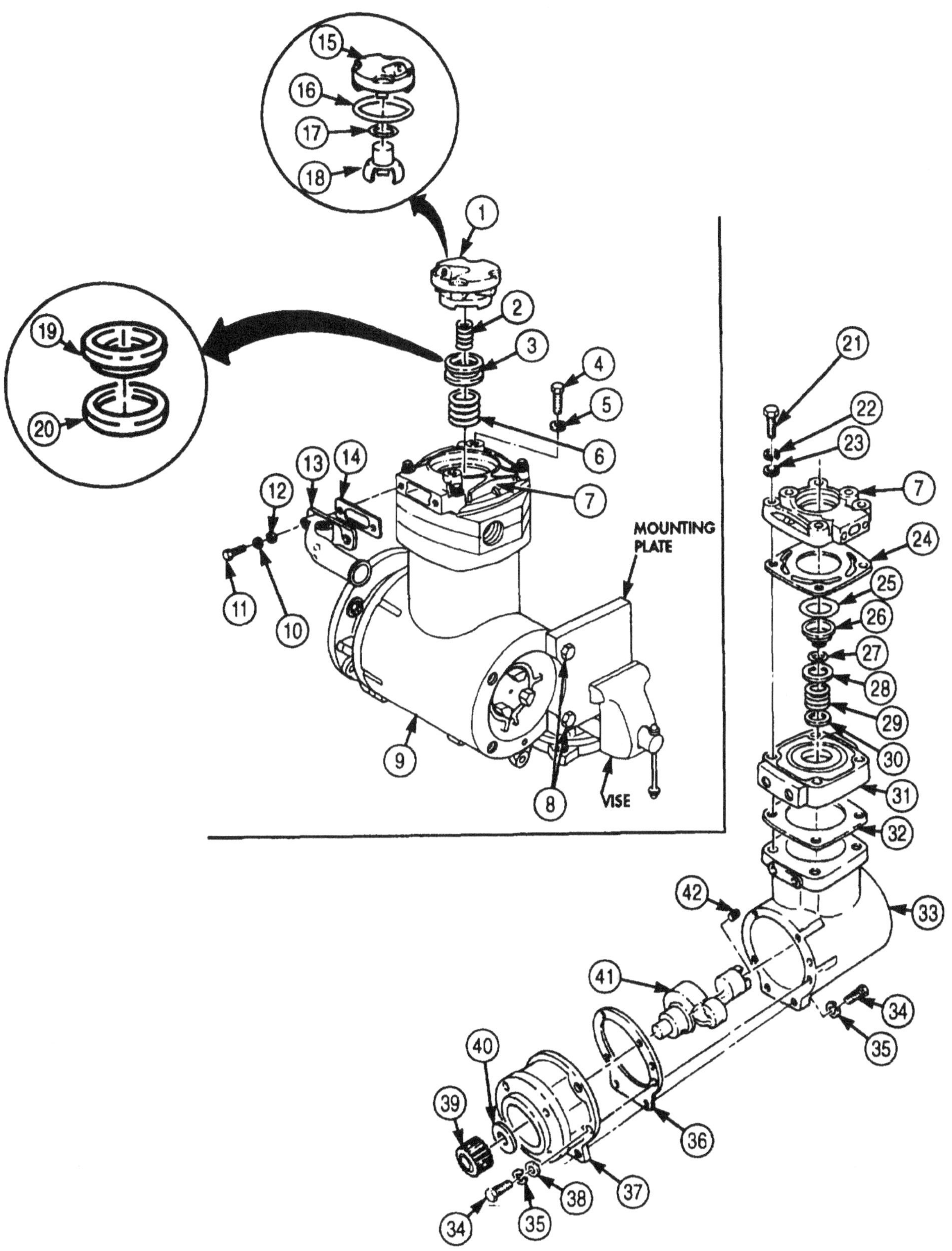

4-31. AIR COMPRESSOR MAINTENANCE (Contd)

16. Remove piston (7) and connecting rod (8) from crankcase (9).

NOTE

Tag compression and oil rings for assembly.

17. Remove compression rings (1) and (2), oil ring expander (4), and two oil rings (3) from piston (7). Discard compression rings (1) and (2), oil rings (3), and oil ring expander (4).

CAUTION

If piston pin cannot be removed from piston by hand pressure, place piston in hot water to expand piston pin bore to allow removal. Driving pin from piston may cause damage to piston.

18. Remove two retaining rings (5), piston pin (6), and connecting rod (8) from piston (7).
19. Remove crankcase (9) and mounting plate from vise.
20. Remove two screws (10) and mounting plate from crankcase (9).

c. Cleaning

WARNING

- Drycleaning solvent is flammable and toxic. Do not used near an open flame and always have a fire extinguisher nearby when solvents are used. Use only in well-ventilated places, wear protective clothing, and dispose of cleaning rags in approved container. Failure to do this will result in injury to personnel and/for damage to equipment.
- Eyeshields must be worn when cleaning with compressed air. Compressed air source will not exceed 30 psi (207 kPa). Failure to do so may result in injury to personnel.

1. Immerse all air compressor parts in drycleaning solvent. Remove all carbon deposits, rust, and scale. Use compressed air to dry parts.
2. Ensure that interior drilled oil passages (12) in crankshaft (11) are thoroughly cleaned.

CAUTION

Do not use screwdriver or scraper to remove carbon and scales. This may cause damage to sealing surfaces.

3. Clean all gasket remains from mating surfaces using drycleaning solvent.

d. Inspection and Repair

1. Using depth micrometer, measure exhaust valve (13) seat height. Replace if seat height is less than 0.485 in. (12.32 mm).
2. Measure intake valve (14) seat height. Replace if seat height is less than 0.270 in. (6.86 mm).
3. Measure intake valve surface (15). Surface must be flat within 0.001 in. (0.03 mm) total micrometer reading. If not, replace.

4-31. AIR COMPRESSOR MAINTENANCE (Contd)

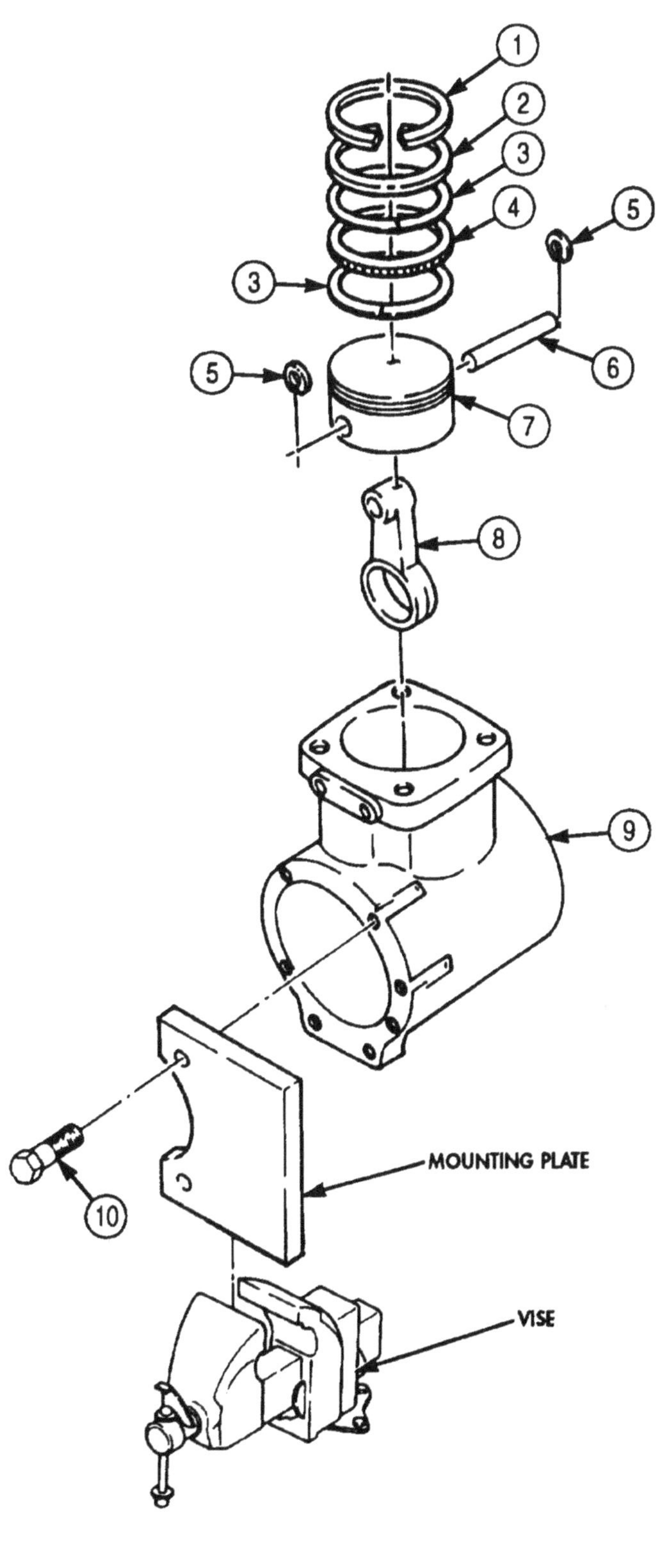

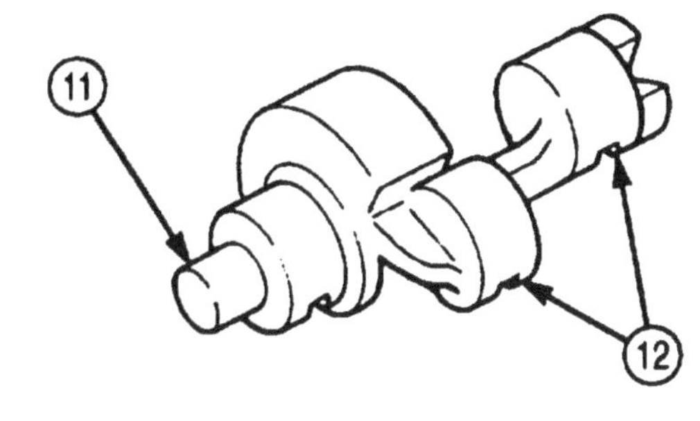

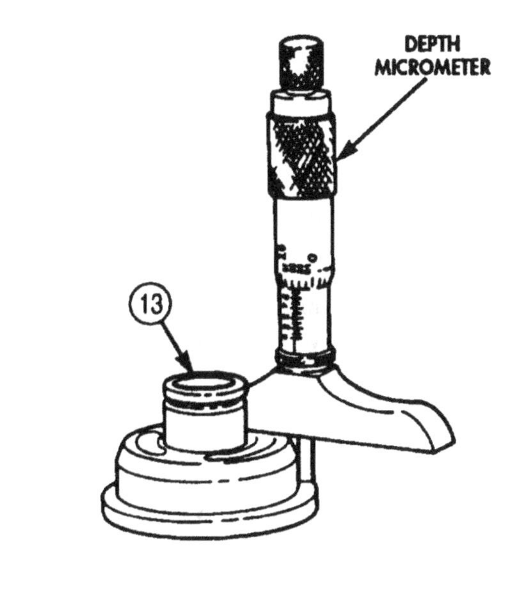

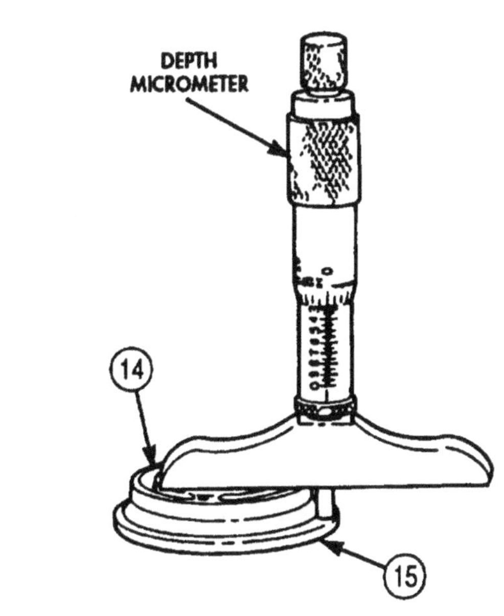

4-31. AIR COMPRESSOR MAINTENANCE (Contd)

4. Inspect narrow part of unloader cap (1) for scoring and excessive wear. Inspect unloader cap (1) seating surface for distortion, pitting, and excessive wear. If scored, distorted, pitted, or excessively worn, replace unloader cap (1).
5. Inspect cylinder head cover (5) and cylinder (6) for cracks, breaks, and bends. Replace cylinder head cover (5) and/or cylinder head (6) if cracked, broken, or bent.
6. Inspect valve springs for wear using spring tester and dial gauge.
   a. Replace unloader valve spring (2) if worn past 12.00 lb. (5.4 kg).
   b. Replace intake valve spring (3) if worn past 0.55 lb (0.25 kg).
   c. Replace exhaust valve spring (4) if worn past 8.00 lb. (3.63 kg).
7. Inspect crankshaft (9) for scratches, scoring, and excessive wear using outside micrometer:
   a. Replace crankshaft (12) if scratched or worn.
   b. Replace crankshaft if journal (10) if worn past 1.871 in. (47.52 mm).
   c. Replace crankshaft if journal (8) if worn past 1.933 in. (49.10 mm).
   d. Replace crankshaft if journal (7) if worn past 1.871 in. (47.52 mm).
8. Inspect front support (13) for scratches, scoring, and breaks and measure outside diameter of thrust support flange (11) for wear using outside micrometer. Replace front support (13) if scratched or broken or if thrust support flange (13) measures less than 1.287 in. (32.69 mm).
9. Measure thickness of thrust bearing (15). Replace if thickness is less than 0.240 in. (6.10 mm).
10. Inspect support bearing (12) for scoring, wear, or other damage and measure inside diameter using micrometer. Replace support bearing (12) if scored, damaged, or inside diameter is more than 1.887 in. (47.69 mm).
11. Inspect drive gear (14) for wear, scoring, and damage. Replace drive gear (14) if worn, scored, or damaged.

4-31. AIR COMPRESSOR MAINTENANCE (Contd)

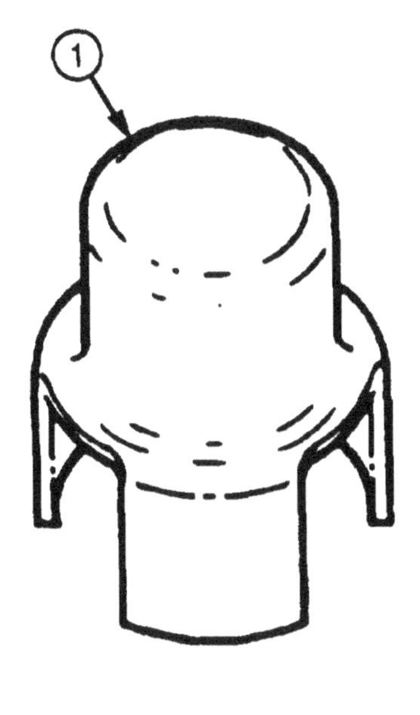

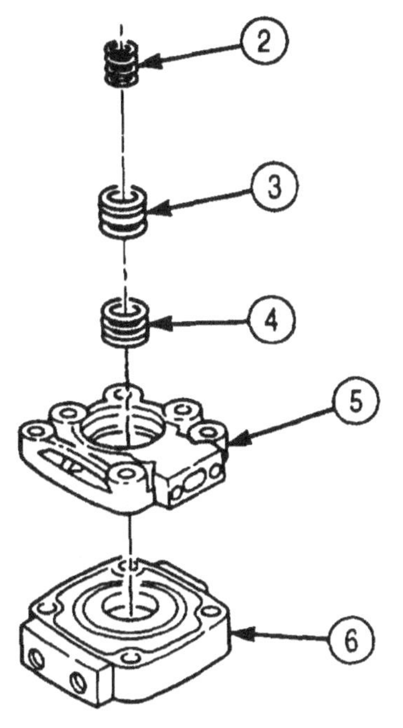

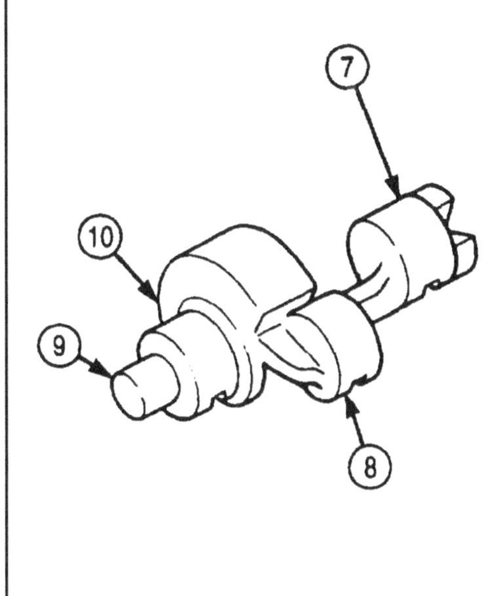

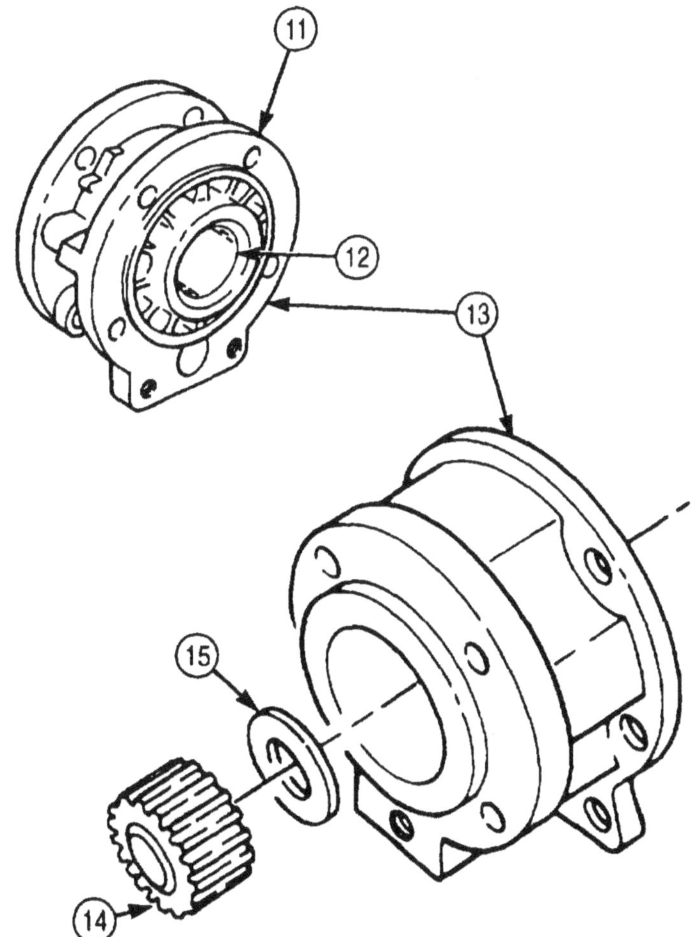

4-31. AIR COMPRESSOR MAINTENANCE (Contd)

12. Inspect connecting rod bushing (5) for scoring and damage. Replace connecting rod bushing (5) if scored or damaged.
13. Inspect connecting rod (4) for bends or twists and inspect connecting rod bores (7) and (6) for wear.
    a. Replace connecting rod (4) if bend exceeds 0.002 in. (0.05 mm) or if twist exceeds 0.004 in. (0.10 mm).
    b. Replace connecting rod (4) if inner diameter of connecting rod bore (7) exceeds 0.689 in. (17.50 mm).
    c. Replace connecting rod (4) if inner diameter of connecting rod bore (6) exceeds 1.935 in. (49.15 mm).
14. Measure outside diameter of piston pin (2) for wear. Replace if outside diameter is less than 0.687 in. (17.45 mm).
15. Inspect piston (1) for scoring, cracks, and damage and measure piston skirt (3) and pin bore (8) diameters.
    a. Replace piston (1) if scored, cracked, or damaged.
    b. Replace piston (1) if piston skirt (3) diameter is less than 3.617 in. (91.86 mm) at 70°F (21°C).
    c. Replace piston (1) if piston bore (8) diameter is more than 0.689 in. (17.50 mm) at 70°F (21°C).
16. Install new compression ring (9) in compression ring groove (10) of piston (1). Insert 0.009 in. (0.23 mm) feeler gauge between compression ring (9) and groove (10). Replace piston (1) if compression ring (9) can be pressed below piston surface with feeler gauge in place.
17. Seat new compression ring (12) in unworn portion of cylinder bore (11) and measure ring end gap with feeler gauge. Gap should be between 0.010-0.020 in. (0.25-0.51 mm). Use different sized ring (12) if necessary.

NOTE

The maximum crankcase bore is 3.6285 in. (92.16 mm). The maximum out-of-round specification is 0.0014 in. (0.036 mm).

18. Inspect cylinder bore (11) for scoring. Burnish cylinder bore (11), as required, to remove scoring.
19. Using inside micrometer, measure cylinder bore (11) for out-of-roundness. If out-of-round measurement is not within specifications, hone bore to receive 0.010, 0.020, or 0.030 in. (0.25, 0.51, or 0.76 mm) oversize piston (1) and ring (9).
20. Measure inside diameter of crankcase bushing (14). Replace crankcase bushing (14) if measurement exceeds 1.878 in. (47.70 mm).

4-31. AIR COMPRESSOR MAINTENANCE (Contd)

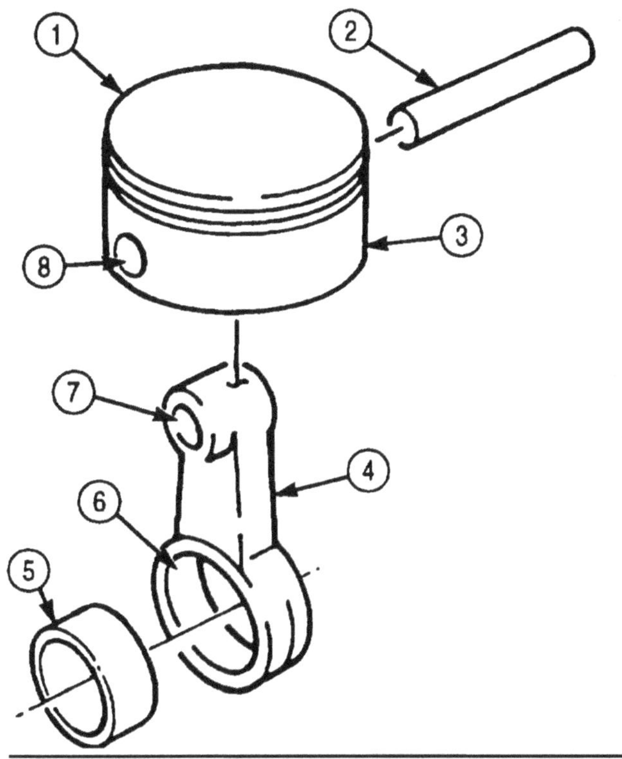

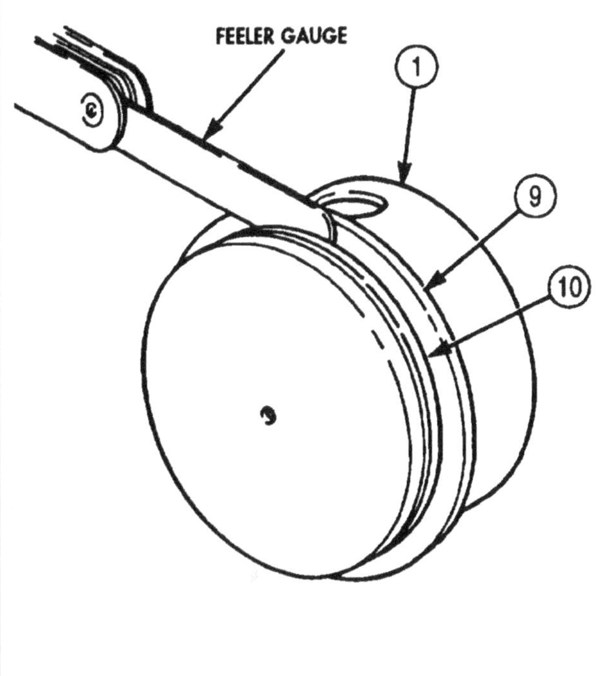

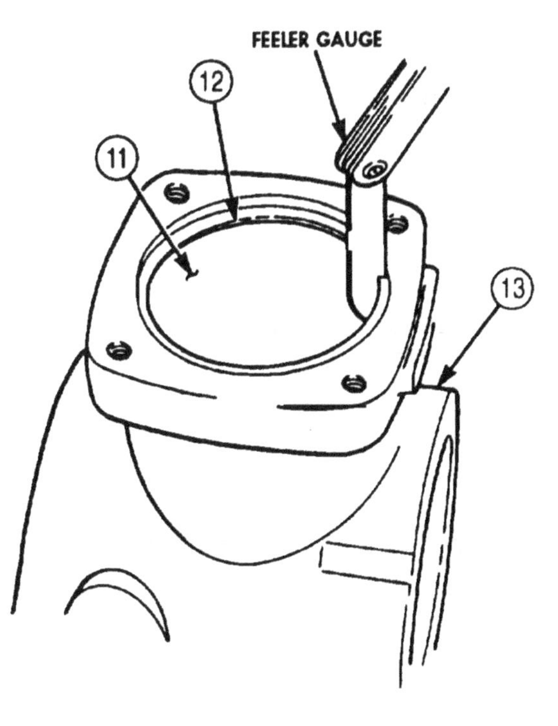

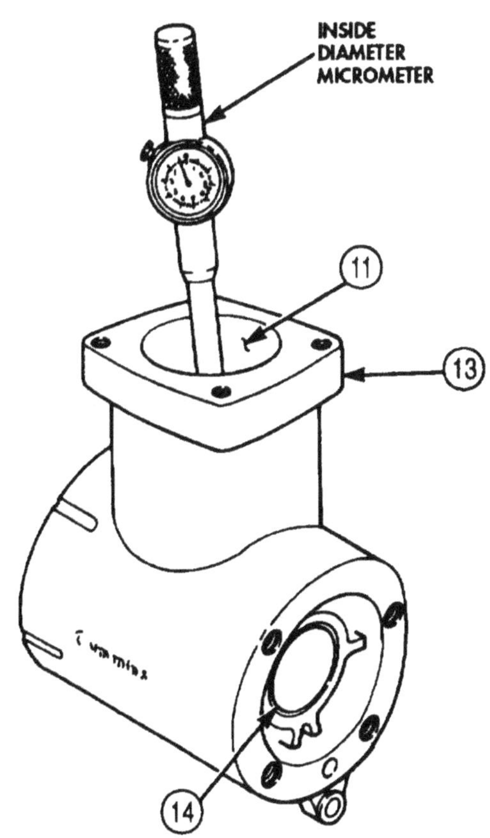

4-31. AIR COMPRESSOR MAINTENANCE (Contd)

e. Assembly

1. Install retaining ring (11) in piston pin bore (10).
2. Place piston end of connecting rod (9) in piston (8).

CAUTION

Do not drive piston pin into bore. Damage to piston will result.

3. Install piston pin (7) through piston pin bore (10) and connecting rod (9) until seated against retaining ring (11).

CAUTION

- If piston pin cannot be installed in piston by hand pressure, place piston in hot water to expand piston pin bore to allow removal. Driving pin in piston may cause damage to piston.
- Coat all metal parts with a thin film of lubricating oil prior to assembly.

4. Install retaining ring (6) in piston pin bore (10).

NOTE

- Compression rings and oil ring gaps must be installed 180° apart. If rings are not staggered 180°, leaks will develop during the compression stage resulting in decreased efficiency.
- Ensure ends of compression rings do not overlap.

5. Install oil ring (4) and two compression rings (3) and (5) in bottom piston ring groove.

NOTE

Ensure compression ring is installed with the word TOP toward crown of piston.

6. Install compression ring (1) and intermediate ring (2) on piston (8) and stagger gaps 180° apart. Lubricate with clean lubricating oil.
7. Install mounting plate on crankcase (13) with two screws (14).
8. Position mounting plate and crankcase (13) in vise.
9. Install ring compressor on piston and connecting rod assembly (12).

CAUTION

Do not force piston into cylinder. This can damage rings and/or cylinder wall.

10. Lubricate piston and connecting rod assembly (12) and crankcase (13) with clean lubricating oil.
11. Install piston and connecting rod assembly (12) in crankcase (13) cylinder bore.
12. Lubricate crankshaft (18) and install in support (17).
13. Install pipe plug (19) in crankshaft (18).
14. Install thrust bearing (16) in support (17) on crankshaft (18) end. Ensure notches are toward crankshaft (18).
15. Using arbor press, install drive gear (15) on crankshaft support (17) with protruding end of drive gear (15) toward crankshaft (18).

4-31. AIR COMPRESSOR MAINTENANCE (Contd)

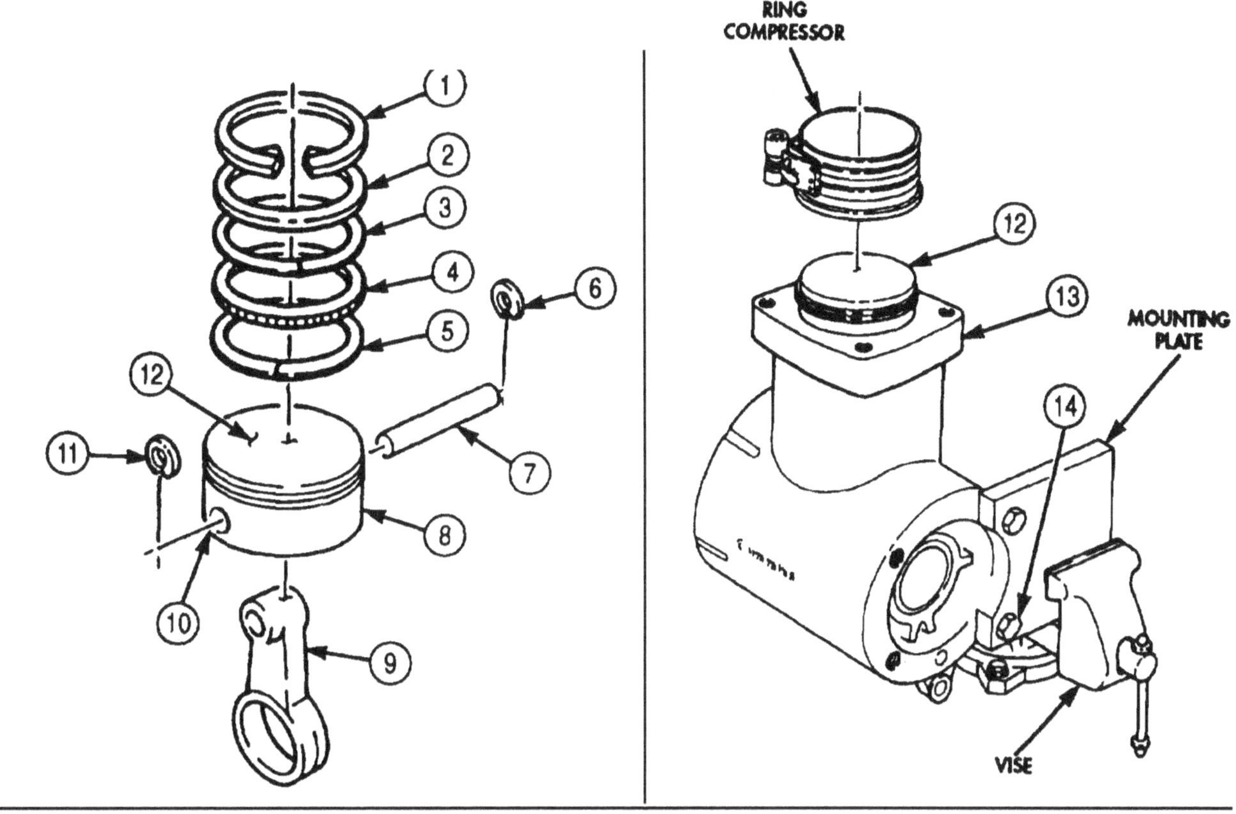

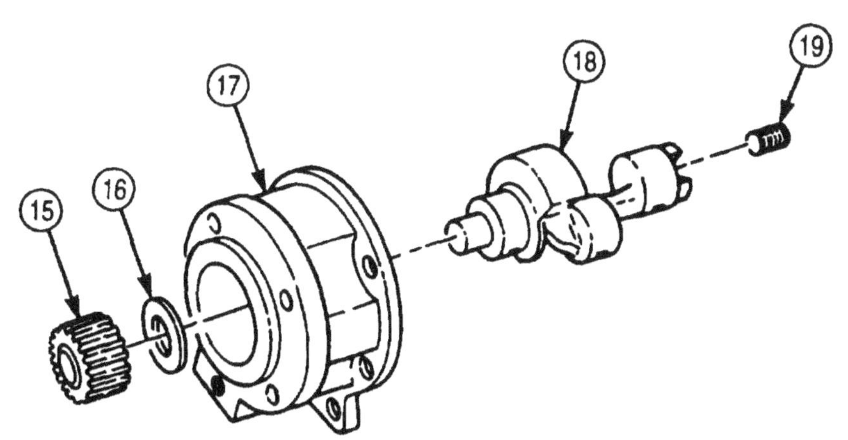

4-31. AIR COMPRESSOR MAINTENANCE (Contd)

NOTE

Perform steps 16 and 17 to install crankshaft, support assembly, and new gasket in crankcase.

16. Position piston (2) at 90 degrees before or after TDC and install crankshaft (7) in connecting rod (9) journal.
17. Install new gasket (8) and support assembly (6) on crankcase (1) with four washers (5), six screws (3), and lockwashers (4). Tighten screws (3) 30-35 lb-ft (41-48 N•m).
18. Install exhaust valve (18), new O-ring (15), and new packing seal (17) on valve seat (16).
19. Using thumb pressure, press valve seat (16) into position in cylinder head (21), install wear plate (20), exhaust valve spring (19), and valve seat (16) in cylinder head (21).
20. Install new gasket (22) and cylinder head (21) on crankcase (1).
21. Install new gasket (14) and cylinder head cover (13) on cylinder head (21) with four new lockwashers (11), washers (12), and screws (10).

4-31. AIR COMPRESSOR MAINTENANCE (Contd)

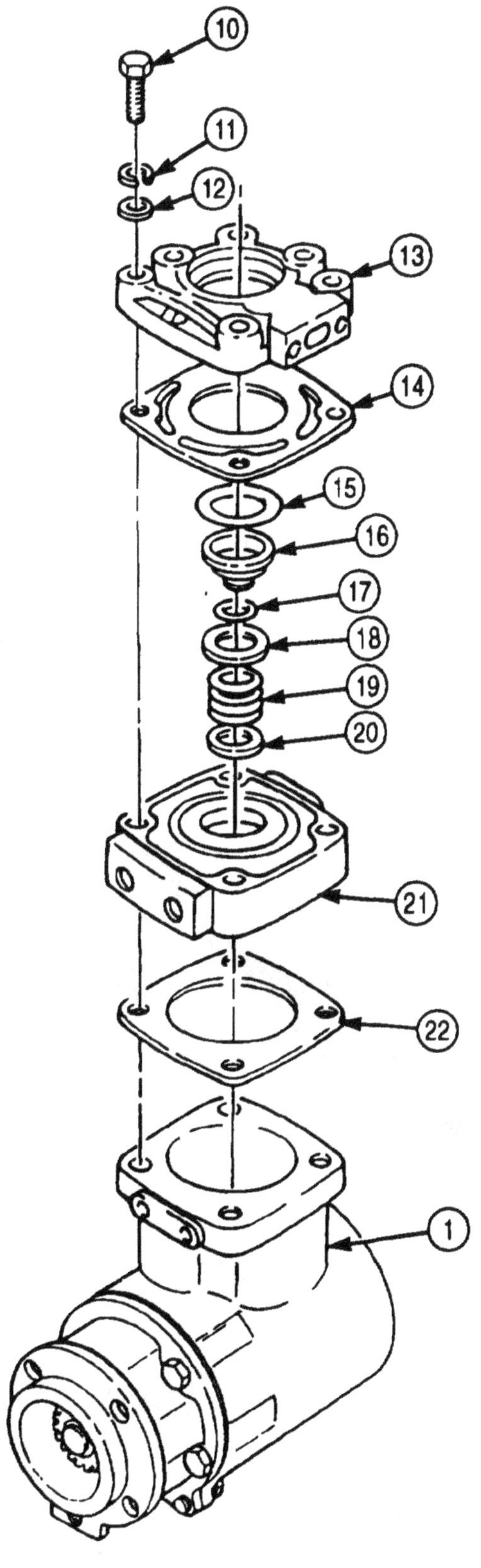

4-31. AIR COMPRESSOR MAINTENANCE (Contd)

22. Install new O-ring (9), new packing seal (10), and unloader valve (11) in unloader cap (8).
23. Install intake valve (13) on intake valve seat (12).
24. Install intake valve seat assembly (4) and intake valve spring (5) in cylinder head cover (6).
25. Install unloader spring (3) and unloader cap assembly (2) in cylinder head cover (6) with two washers (7) and screws (1).
26. Install new gasket (18) and air inlet connection (17) on cylinder head cover (6) with two new lockwashers (15), washers (16), and screws (14).
27. Remove two screws (20) and mounting plate from crankcase (19).
28. Remove mounting plate from vise.

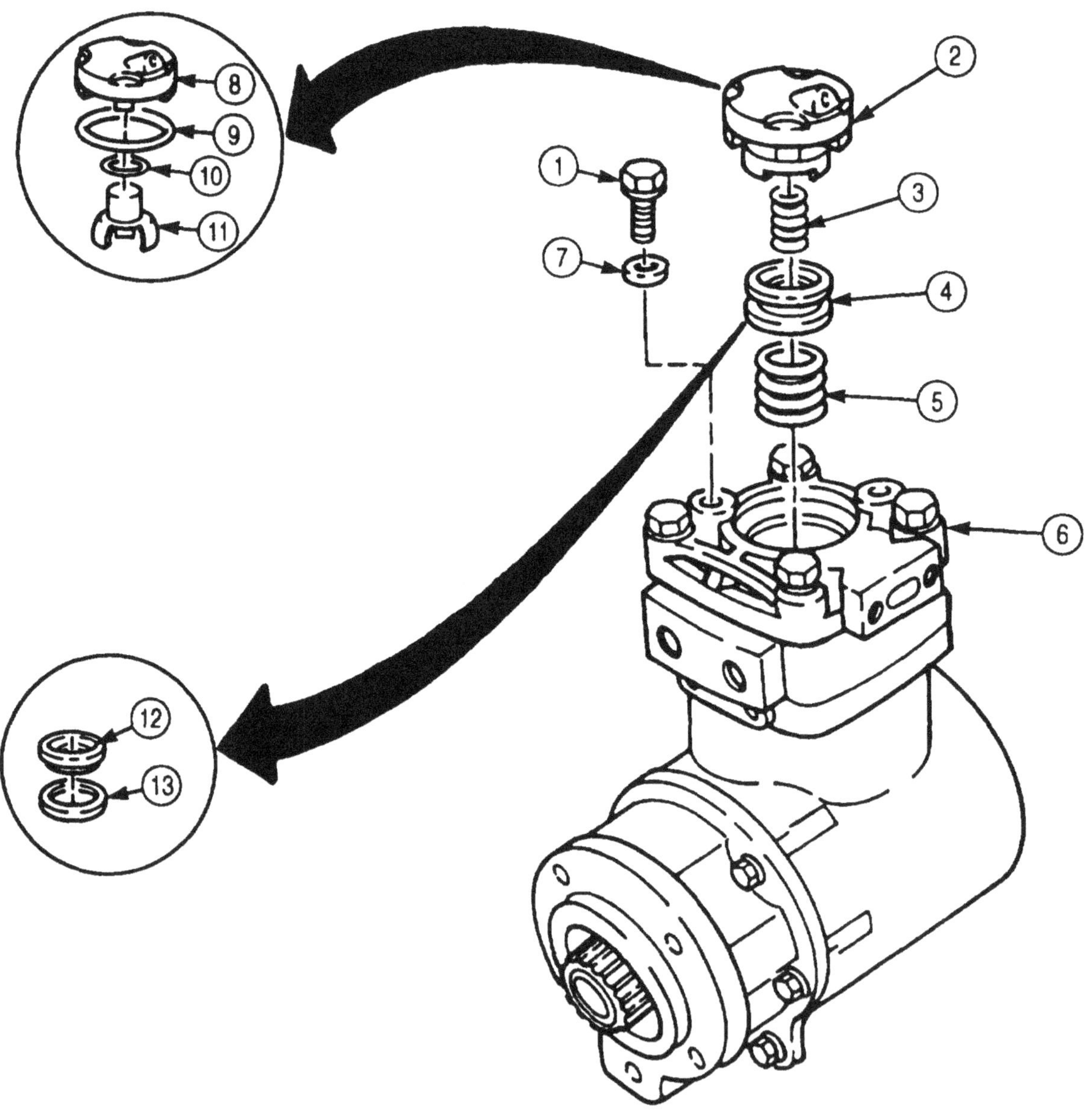

4-31. AIR COMPRESSOR MAINTENANCE (Contd)

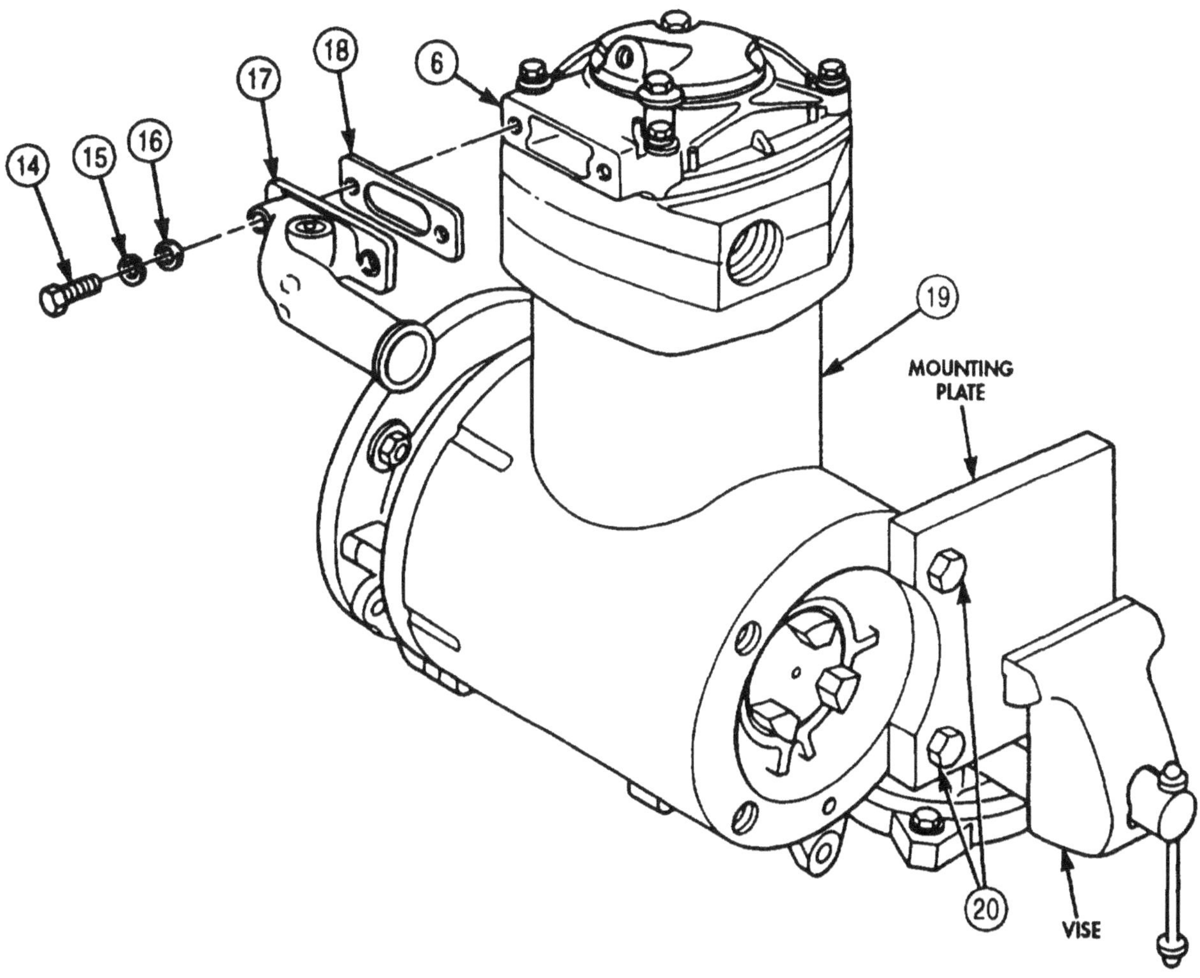

4-31. AIR COMPRESSOR MAINTENANCE (Contd)

f. Installation and Timing

1. Install drive coupling (2) in accessory drive housing (1) over accessory driveshaft (3).

NOTE

- If installing a new air compressor, use fitting from old compressor. Clean all male pipe threads and wrap with sealing tape before installation.
- Ensure oil passage holes in new gasket are open, and align with oil passages before installation.

2. Align new gasket (6) on accessory drive housing (1).
3. Using engine barring tool, rotate engine to 1-6 valve set position index mark (4).
4. Set air compressor timing mark pointing between 9 and 10 o'clock looking at coupling end and install air compressor (10) on accessory drive housing (1) with two screws (5), washers (7), screws (13), new lockwashers (9), washers (7), and nuts (8).
5. Install ground strap (11) on air compressor (10) with screw (12).
6. Rotate bracket (17) upward and install on air compressor (10) and engine (16) with new screw-assembled lockwashers (15) and (19) and washers (14) and (18). Tighten screw (20).

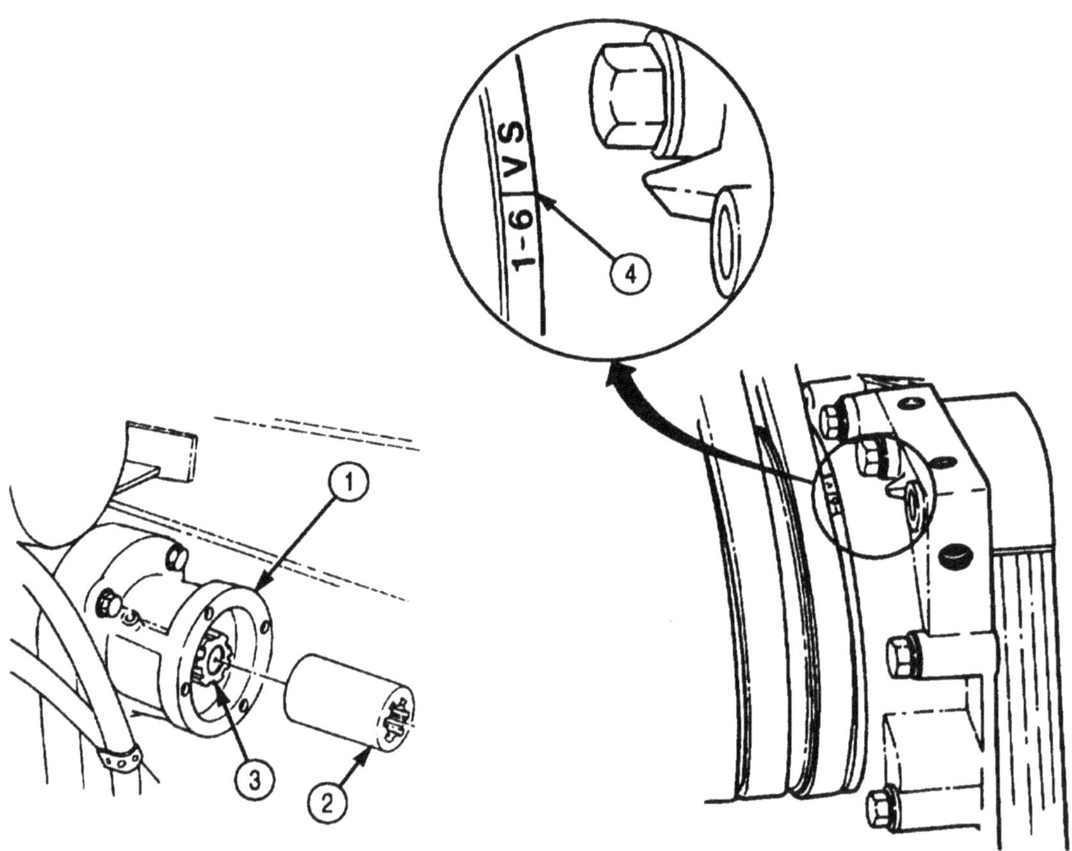

4-31. AIR COMPRESSOR MAINTENANCE (Contd)

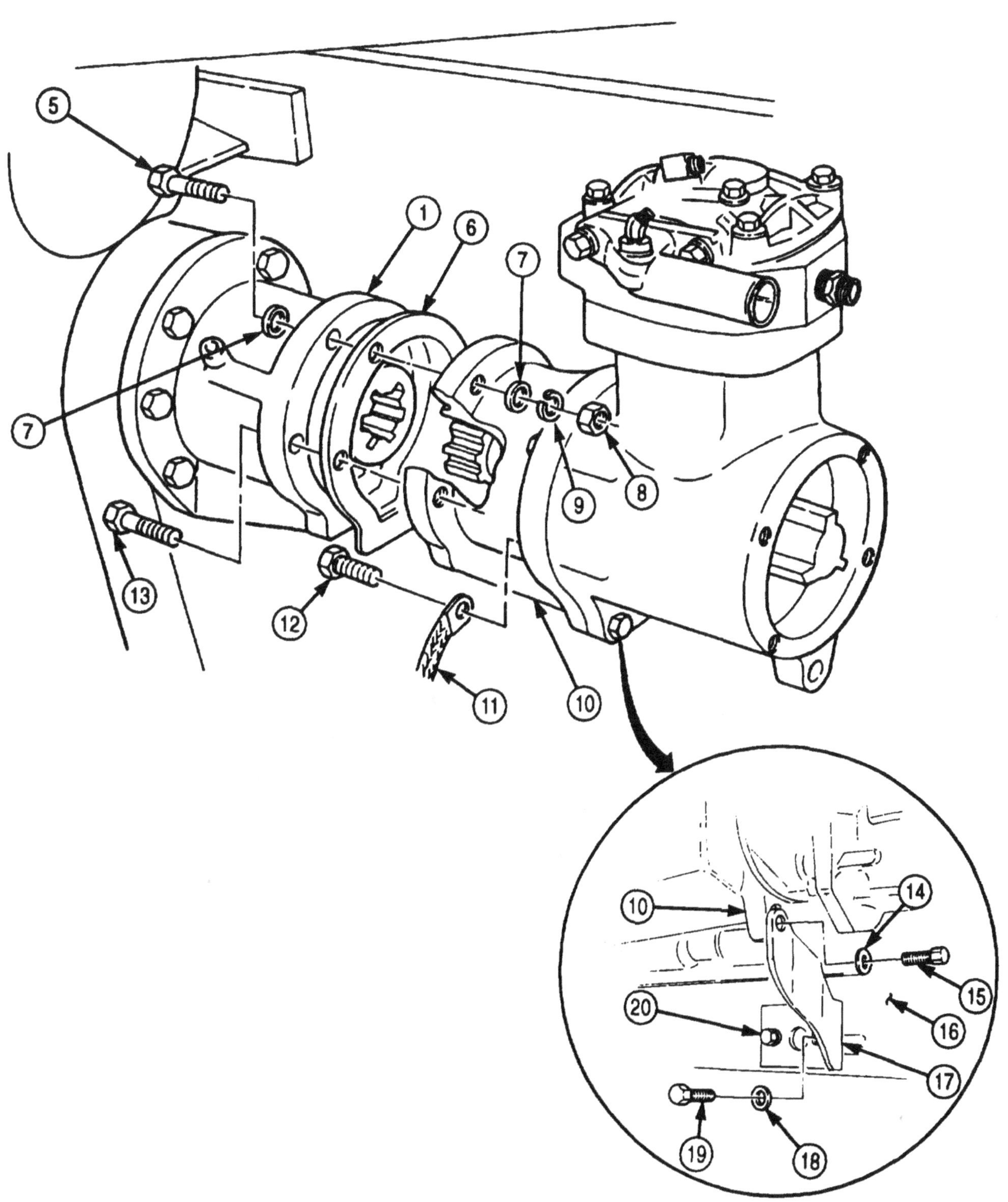

4-31. AIR COMPRESSOR MAINTENANCE (Contd)

7. Install elbows (6) and (8) on air compressor (5).
8. Install elbows (21) and (29) on unloader valve body (20) and compressor inlet (28).
9. Install adapter (24) on compressor head (30).
10. Using new inserts (10) on each end of coolant outlet line (4), install coolant outlet line (4) on elbow (8) on air compressor (5) and elbow (12) on engine oil cooler (11).
11. Install coolant outlet line (4) on bracket (16) with clamp (15), washer (14), screw (13), washer (17), lockwasher (18), and nut (19).
12. Install power steering pump pivot bracket (3) on engine (1) with four screw-assembled washers (2).
13. Using new inserts (10) on each end of coolant inlet line (7), install coolant inlet line (7) on elbow (6) and water pump fitting (9).
14. Connect air intake tube (26) to compressor inlet (28) and tighten two clamps (27).
15. Connect alcohol evaporator air line (23) to elbow (29).
16. Connect governor air line (22) to elbow (21).
17. Connect supply reservoir air line (25) to adapter (24).

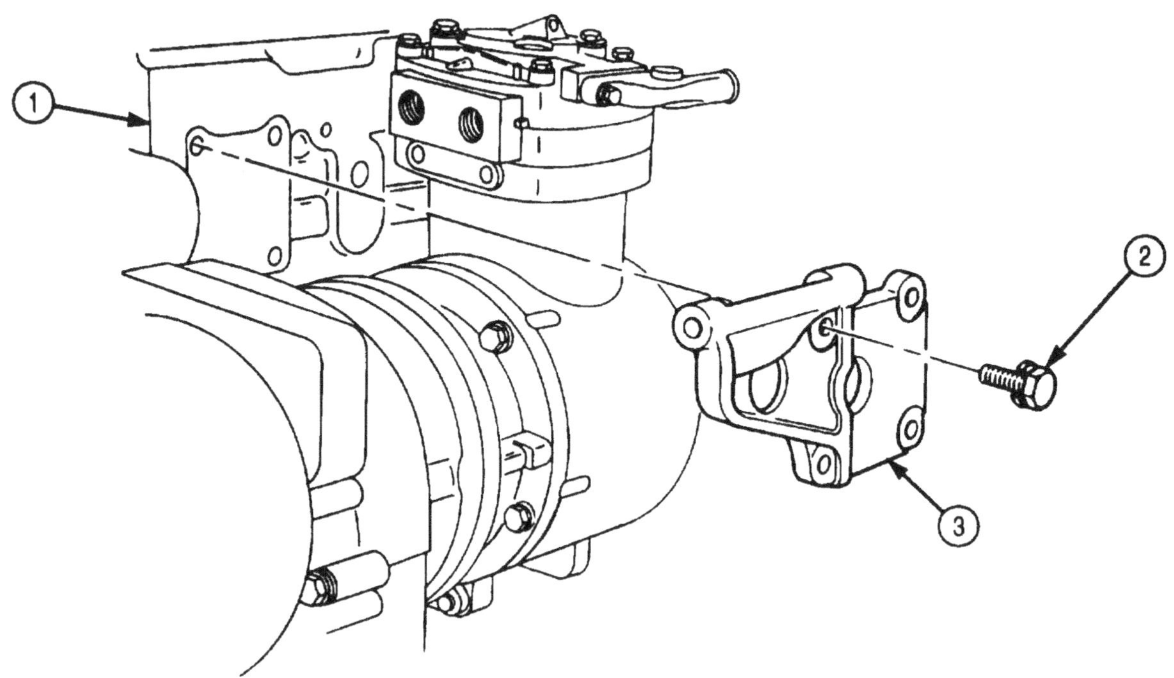

4-31. AIR COMPRESSOR MAINTENANCE (Contd)

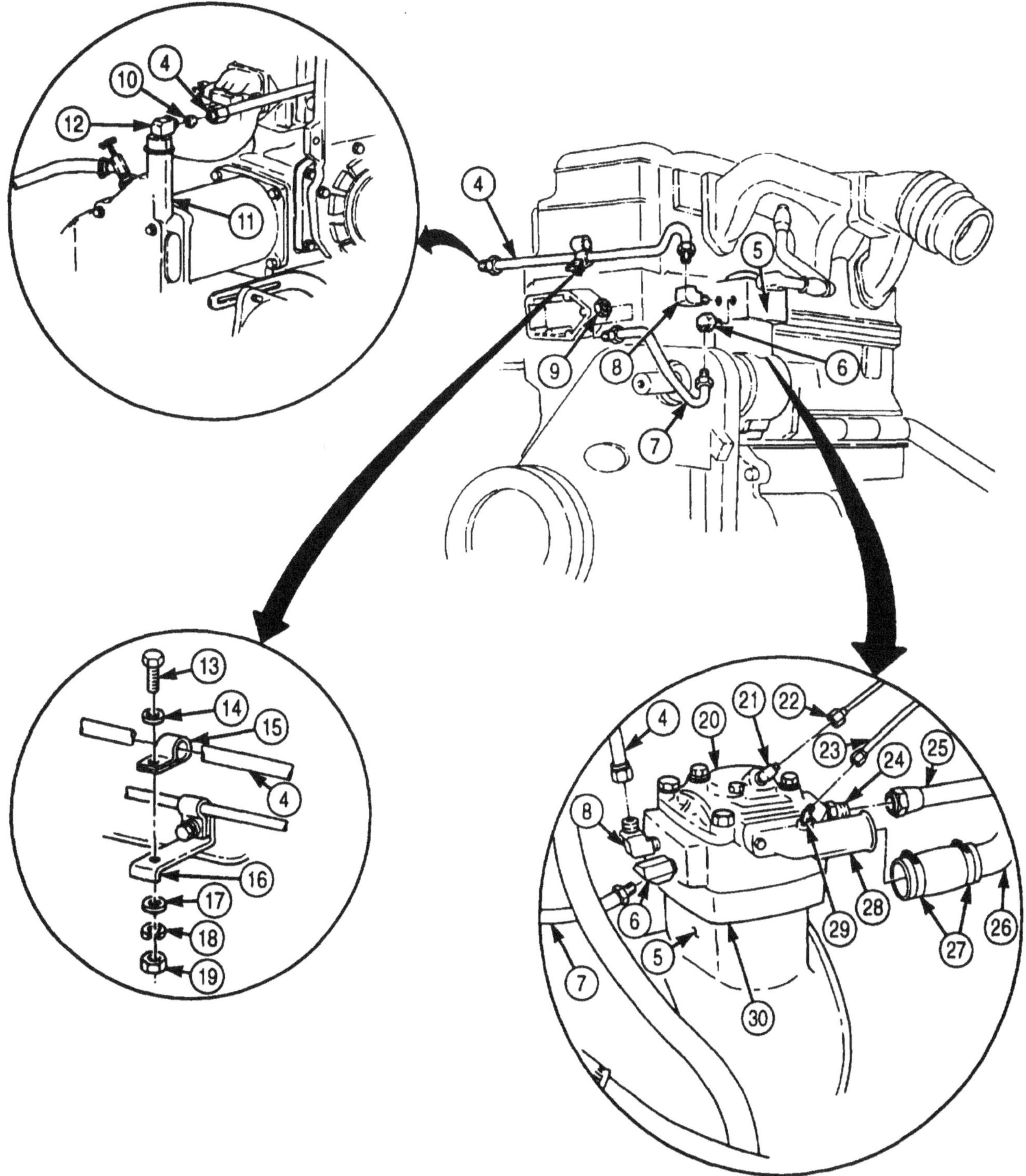

FOLLOW-ON TASKS: • Install fuel pump (para. 4-35).
- Install power steering pump (para. 3-236).
- Install left splash shield (TM 9-2320-272-10).
- Start engine (TM 9-2320-272-10) and check air compressor for leaks and proper operation.

4-32. ENGINE INJECTOR TIMING INSTRUCTIONS

THIS TASK COVERS:

a. Rocker Lever Housing Covers Removal
b. Rocker Lever Housing and Push Tubes Removal
c. Valve Crossheads Removal
d. Fuel Injectors Removal
e. General Instructions
f. Timing Tool Setup
g. Injector Timing
h. Fuel Injectors Installation
i. Valve Crossheads Installation and Adjustment
j. Rocker Lever Housing and Push Tubes Installation
k. Rocker Lever Housing Covers Installation

INITIAL SETUP:

APPLICABLE MODELS
M939/A1

SPECIAL TOOLS
Dial indicator (Appendix E, Item 36)
Spring compressor (Appendix E, Item 129)
Injector timing fixture (Appendix E, Item 76)

TOOLS
General mechanic's tool kit (Appendix E, Item 1)
Torque wrench (Appendix E, Item 144)
Torque wrench adapter (Appendix E, Item 143)

MATERIALS/PARTS
Four gaskets (Appendix D, Item 172)
Gasket (Appendix D, Item 187)
Lubricating oil (Appendix C, Item 50)

REFERENCES (TM)
TM 9-2320-272-24P

EQUIPMENT CONDITION
Engine mounted on repair stand (para. 4-9).

a. Rocker Lever Housing Covers Removal

NOTE

- Rocker lever housing covers are mounted with screw-assembled washers on late model engines.
- All rocker lever housing covers are removed the same way. This task covers removal of center housing cover.

1. Remove five screws (2), washers (3), rocker lever housing cover (1), and gasket (4) from rocker lever housing (5). Discard gasket (4).
2. Clean gasket remains from mating surfaces of rocker lever housing cover (1), rocker lever housing (5), and washers (3).

b. Rocker Lever Housing and Push Tubes Removal

1. Loosen eighteen locknuts (14) on rocker levers (16).
2. Turn eighteen adjusting screws (15) on rocker levers (16) two turns.
3. Remove six screws (7), two lifting eyes (8), and upper radiator support bracket (6) from rocker lever housings (5).

NOTE

- Rocker lever housings are mounted with screw-assembled washers for late model engines.
- Tag rocker lever housings for installation.

4. Remove twelve screws (9), washers (10), three rocker lever housings (5), and gaskets (13) from cylinder heads (11). Discard gaskets (13) and clean gasket remains from rocker lever housing (5) and cylinder head (11) mating surfaces.

4-32. ENGINE INJECTOR TIMING INSTRUCTIONS (Contd)

CAUTION

Each cylinder has an exhaust push tube, intake push tube, and injector push tube. It is important each push tube be tagged for installation in the same location.

5. Remove eighteen push tubes (12) from cylinder heads (11).

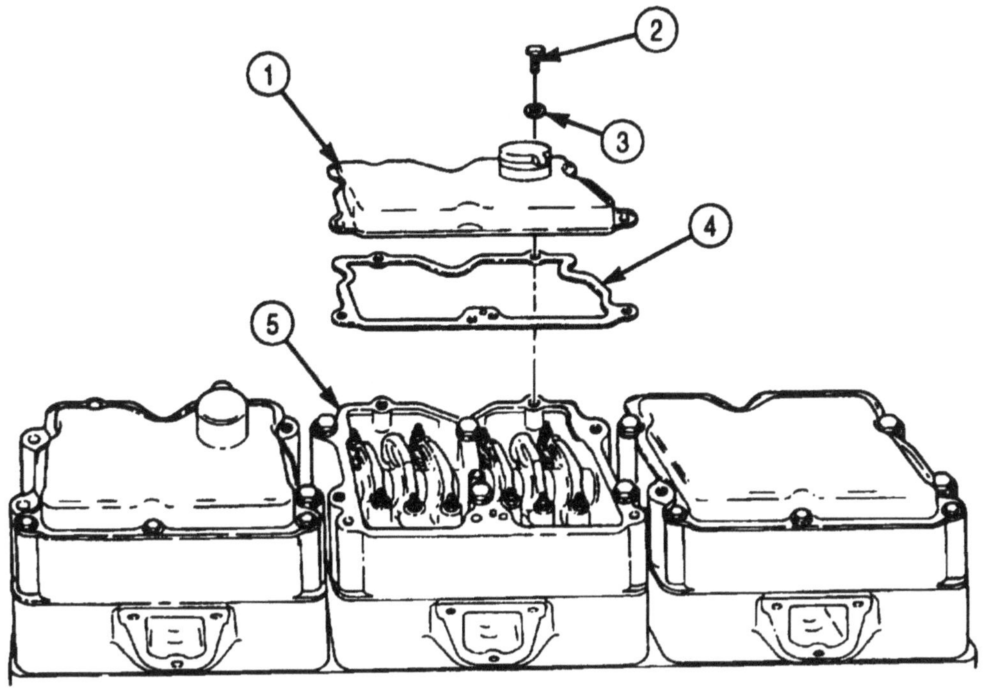

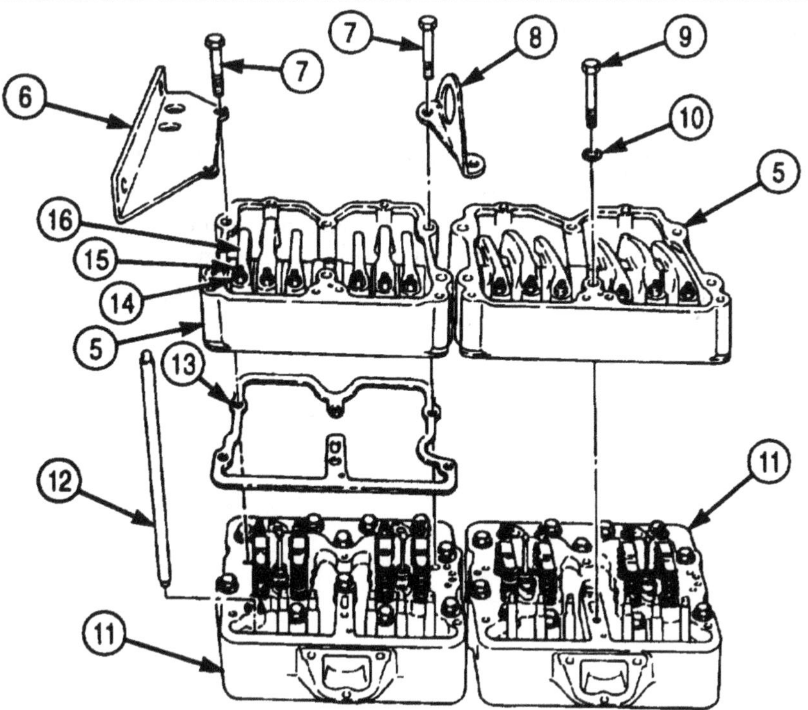

4-32. ENGINE INJECTOR TIMING INSTRUCTIONS (Contd)

c. Valve Crossheads Removal

NOTE

Tag all crossheads for installation.

1. Loosen twelve crosshead adjusting nuts (2) on crossheads (1).
2. Remove twelve crossheads (1) from cylinder heads (3).

d. Fuel Injectors Removal

CAUTION

- Do not turn injector upside down after removal. Plunger will fall out and be damaged.
- Do not damage injector tip during handling. Ensure injectors and plungers are not intermixed. Always number injectors according to the cylinder head from which they were removed.

NOTE

Top stop and non-top stop injectors are removed the same way. This task covers non-top stop injectors.

1. Remove injector link (4) from injector (6).
2. Remove two screws (8) from injector retaining clamp (5).
3. Remove retaining clamp (5), washer (7), and injector (6) from cylinder head (3). Keep injector (6) in safe place.

e. General Instructions

NOTE

- This timing procedure is for engines mounted on test stand only.
- The precise timing of the injector push tube travel with corresponding piston travel is accomplished by using the injector timing fixture to measure travel of these two parts.
- Adjustments to injection timing are made by altering the thickness of cam follower gaskets.
- The injection timing may be advanced or retarded by adding or removing cam follower gaskets.
- The timing operation is performed on only one cylinder on each head; cylinder No. 6 cannot be timed with engine in vehicle.
- Ensure camshaft is pushed back against rear of cylinder block for zero end play.

4-32. ENGINE INJECTOR TIMING INSTRUCTIONS (Contd)

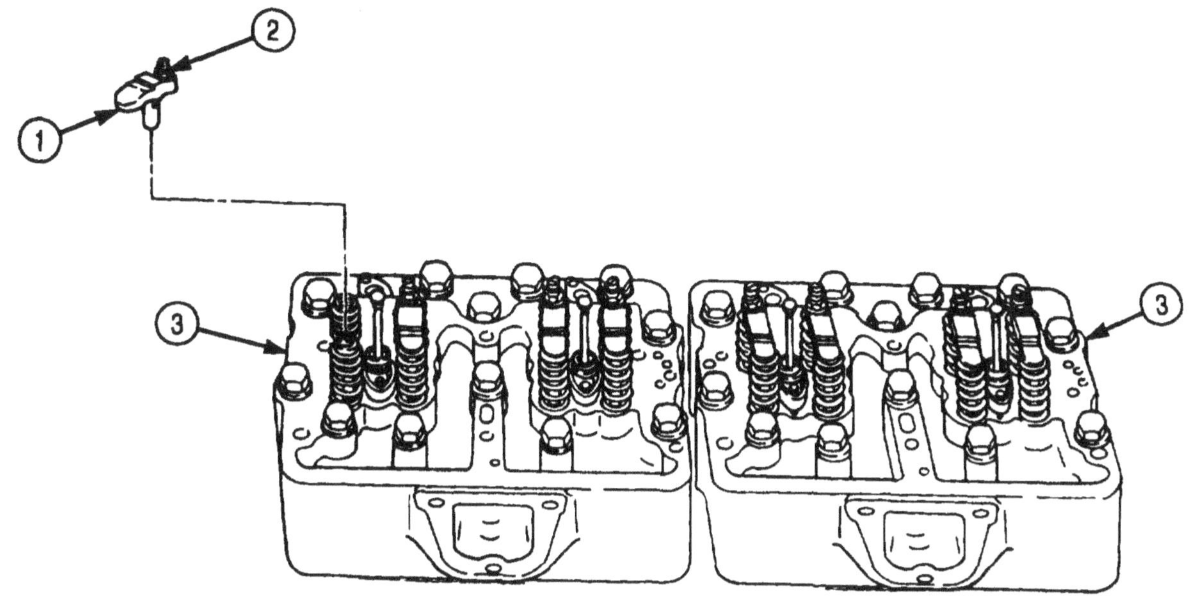

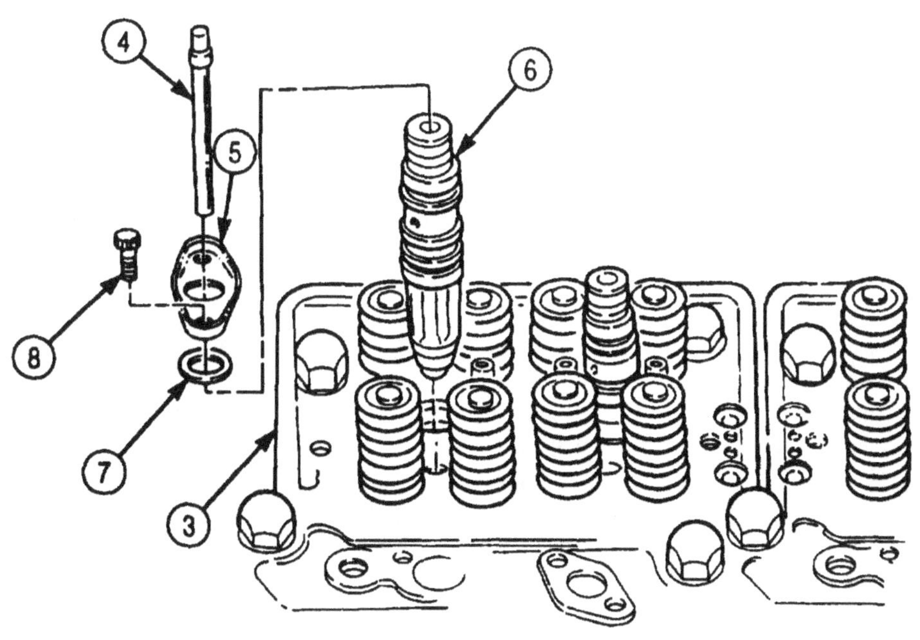

4-32. ENGINE INJECTOR TIMING INSTRUCTIONS (Contd)

f. Timing Tool Setup

NOTE

Each of the two dial indicators used in timing the engine must have a total travel of at least 0.250 in. (6.35 mm).

1. Position timing tool fixture in injector sleeve (4).
2. Engage rod (5) of push tube indicator in injector push tube socket.
3. Install timing tool fixture by tightening knurled holddowns (1) and (2) evenly by hand. Ensure timing tool fixture is straight on cylinder head (3).

g. Injector Timing

1. Loosen dial indicator supports on timing tool fixture.

NOTE

Perform steps 2 through 5 for timing procedure 1.

2. Rotate crankshaft (7) on engine (6) in direction of engine (6) rotation to TDC. Piston travel plunger will be near full upward position.
3. Adjust both dial indicators on timing tool fixture to their fully compressed position. To prevent damage, raise both dial indicators approximately 0.020 in. (0.5 mm), and lock in place with setscrew.
4. Rotate crankshaft (7) back and forth to ensure piston is precisely at TDC on compression stroke.

NOTE

Both dial indicators move in the same direction when piston is on compression stroke.

5. TDC is indicated by maximum clockwise position of the piston travel pointer. Turn the piston travel dial indicator face to align zero with the pointer. Lock piston travel dial indicator face with thumbscrews.

NOTE

Perform steps 6 and 7 for timing procedure 2.

6. Rotate crankshaft (7) in direction of engine rotation of 90 degrees After Top Dead Center (ATDC). Piston travel plunger will be near bottom of its travel.
7. Turn push tube travel dial indicator face to align zero with pointer. Lock face with thumbscrew.

NOTE

Perform step 8 for timing procedure 3.

8. Rotate crankshaft (7) in opposite direction of engine rotation (backwards) through TDC, then 45 degrees farther to Before Top Dead Center (BTDC) to remove gear train lash.

NOTE

Perform steps 9 through 11 for timing procedure 4.

9. Turn crankshaft (7) in direction of engine rotation until piston follower rod on timing tool fixture is in contact with indicator stem.
10. Move crankshaft (7) very slowly until dial indicator reads 0.0032 in. (0.0812 mm) before zero. Piston travel piston is now positioned at 19 degrees BTDC.

NOTE

This position is actually 0.2032 in. (5.1612 mm) before "0" because the dial indicator pointer has rotated twice (over 0.200 in.) (5.08 mm) as the crankshaft was moved to 45 degrees BTDC (timing procedure 3).

4-32. ENGINE INJECTOR TIMING INSTRUCTIONS (Contd)

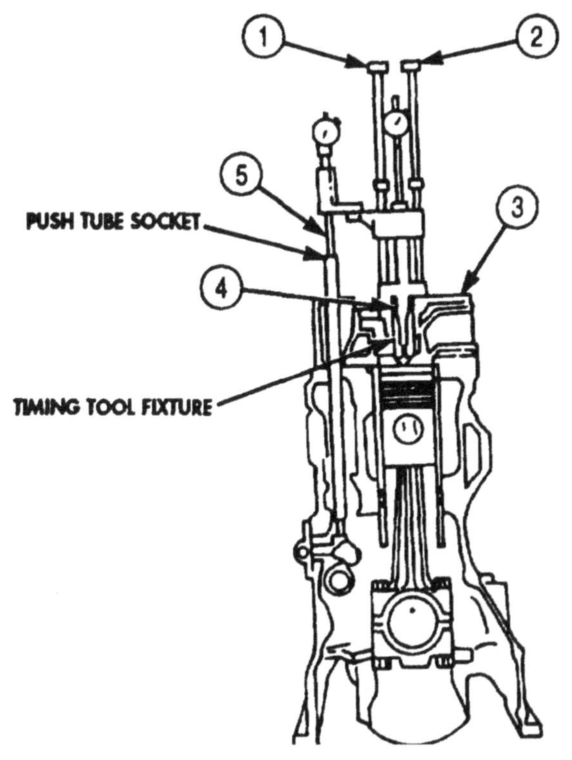

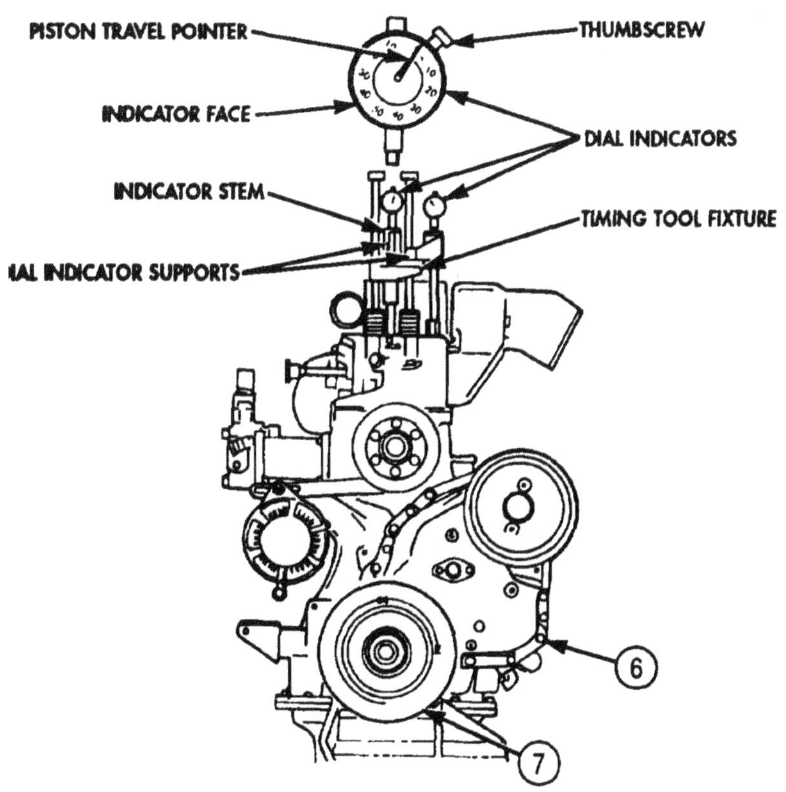

4-32. ENGINE INJECTOR TIMING INSTRUCTIONS (Contd)

11. Read right hand push tube travel indicator. If push tube travel is not 0.0290 in. (0.74 mm), continue to step 12.

NOTE

Perform the following steps before making changes in cam follower gaskets to correct injection timing.

12. Ensure cam screws in cam follower housing are tightened to specifications (para. 4-20).
13. Recheck positioning of piston and push tube indicators. Ensure indicators are not bottoming or binding.
14. Carefully recheck piston and push tube indicators TDC (steps 2 through 5).
15. If required, advance injector timing by adding cam follower gaskets, or retard injector timing by removing cam follower gaskets (para. 4-19).
16. Remove timing fixture from engine (1).

h. Fuel Injectors Installation

CAUTION

Ensure no foreign objects have fallen into cylinder head through injector bore.

NOTE

- If injector condition is unknown, or has been disassembled, it must be calibrated before installation (para. 4-33).
- Top stop and non-top stop injectors are installed the same way. This task covers non-top stop injectors,

1. Apply clean engine oil to lubricate three injector (3) O-rings (5).
2. Start injector (3) into injector bore (6). Align screen on fuel inlet hole with exhaust side of cylinder head (7).
3. Install spring compressor on cylinder head (7) and place over injector plunger (4).
4. Give spring compressor a quick push to seat injector (3). A click will be heard when injector (3) seats properly.
5. Position retaining ring (8) and clamp plate (10), with counterbore up, over injector (3) and start two screws (9). Do not tighten screws (9).
6. Carefully insert injector link (2) in injector (3) and tighten screws (9) incrementally in 4 lb-ft (5 N•m) steps to 11-12 lb-ft (15-16 N•m).
7. Raise injector link (2) 1/3 its length and allow to fall back into injector (3). If link binds or sticks, loosen screws (9) and retighten as described in step 6.

4-32. ENGINE INJECTOR TIMING INSTRUCTIONS (Contd)

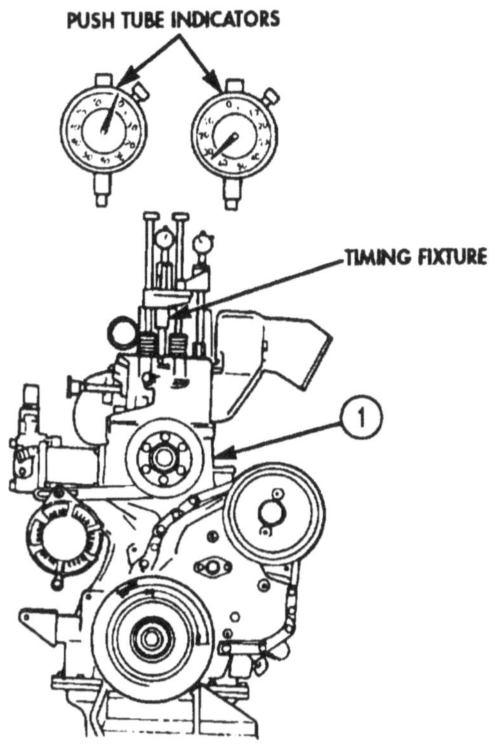

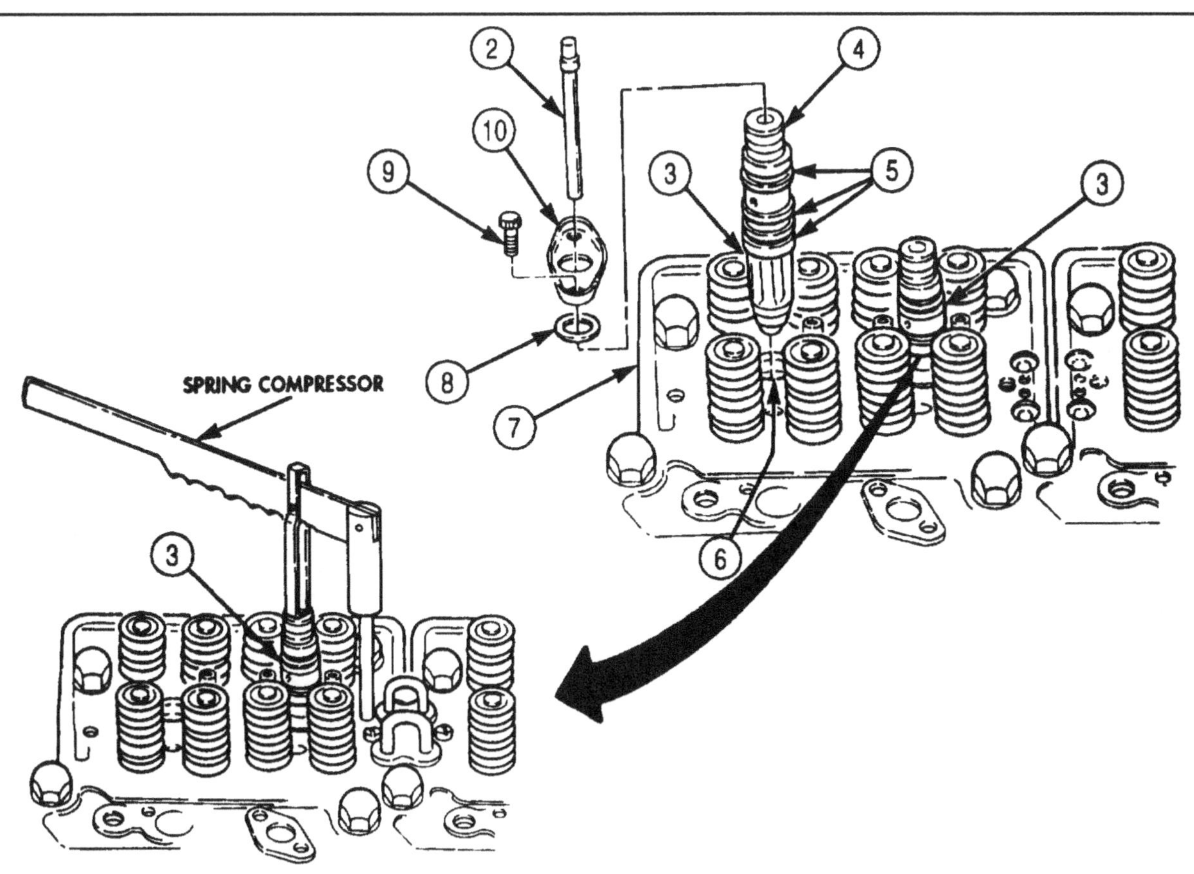

4-32. ENGINE INJECTOR TIMING INSTRUCTIONS (Contd)

i. Valve Crossheads Installation and Adjustment

1. Loosen valve crosshead locknut (5) and back out adjusting screw (4) one full turn.
2. Apply light film of clean engine oil to coat twelve valve crossheads (6).
3. Install twelve valve crossheads (6) on guide of cylinder head (7) with adjusting screw (5) facing toward exhaust manifold side of engine.

NOTE

It may be necessary to loosen valve crosshead locknut to complete step 4.

4. Hold crosshead (6) down with finger pressure so it contacts valve stem (9) on side opposite adjusting screw (4) and turn adjusting screw (4) down until it just touches valve stem (8).

NOTE

Ensure adjusting screw is just lightly seated.

5. Set up dial indicator over center of crosshead (6) and zero-dial indicator while pressing down on valve crosshead (6).
6. Hold crosshead (6) down lightly and turn adjusting screw (4) until dial indicator reads 0.025-0.040 in. (0.64-0.80 mm).
7. Using torque wrench adapter, tighten valve crosshead locknuts (5) 22-26 lb-ft (30-35 N•m).

NOTE

If clearance in step 6 is not at least 0.025 in. (0.64 mm), advance adjusting screw 1/3 of one hex on new crossheads and guides or 1/2 of one hex on old crossheads and guides, retighten locknut, and check clearance.

j. Rocker Lever Housing and Push Tubes Installation

CAUTION

- Do not mix push tubes during installation. The injector tube is the largest, and is positioned between the intake and exhaust push tubes. Intake and exhaust push tubes are identical.
- Seating push tube lower ball ends into cam follower socket seats is critical. Several visual checks must be made during installation to ensure tubes remain properly seated.

1. Install two exhaust valve push tubes (1), injector push tubes (2), and intake valve push tubes (3) by passing ball end (10) of each down through opening in cylinder head (7) and into socket seat (12) on cam follower (111.

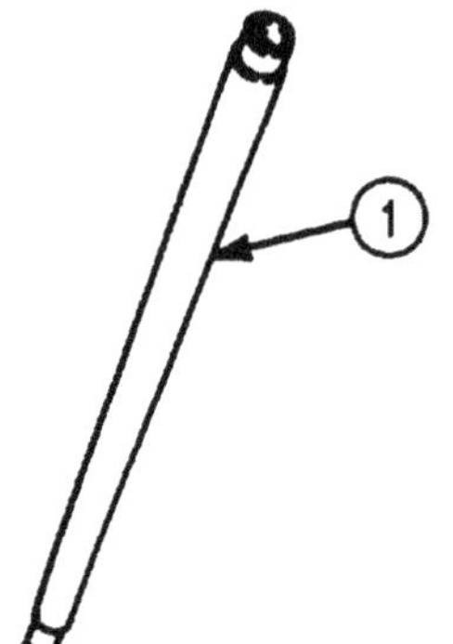

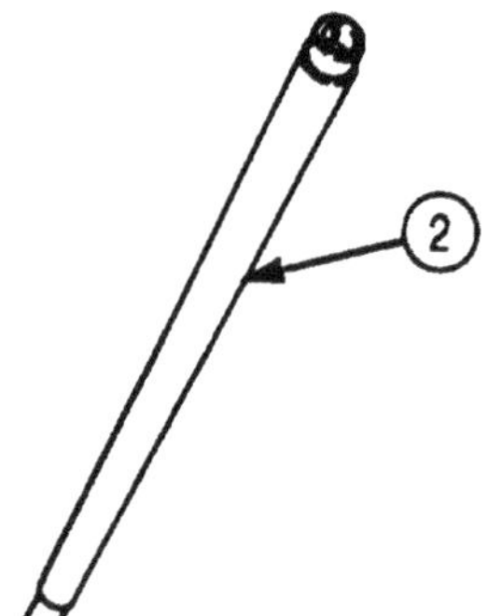

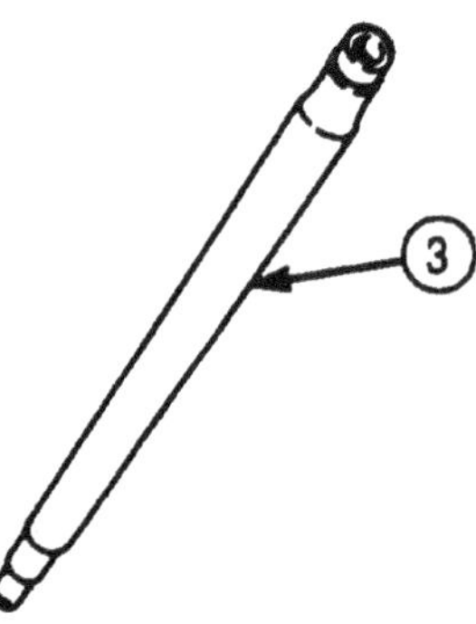

4-32. ENGINE INJECTOR TIMING INSTRUCTIONS (Contd)

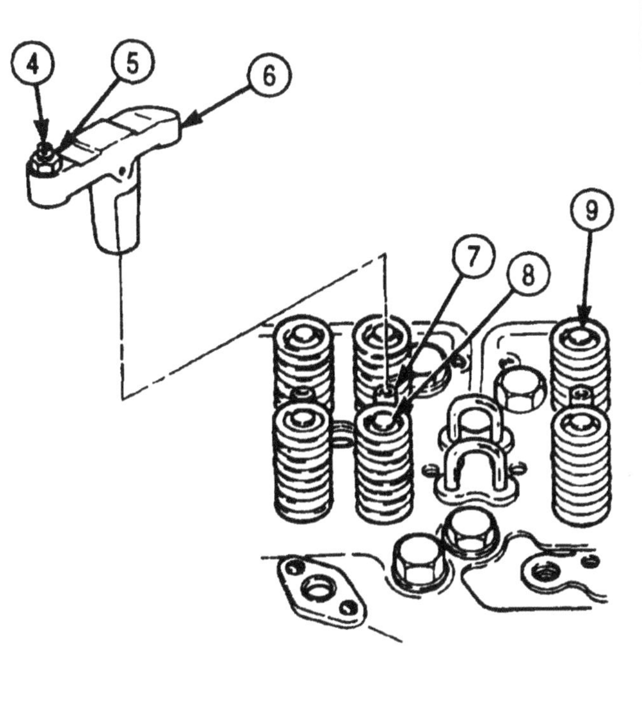

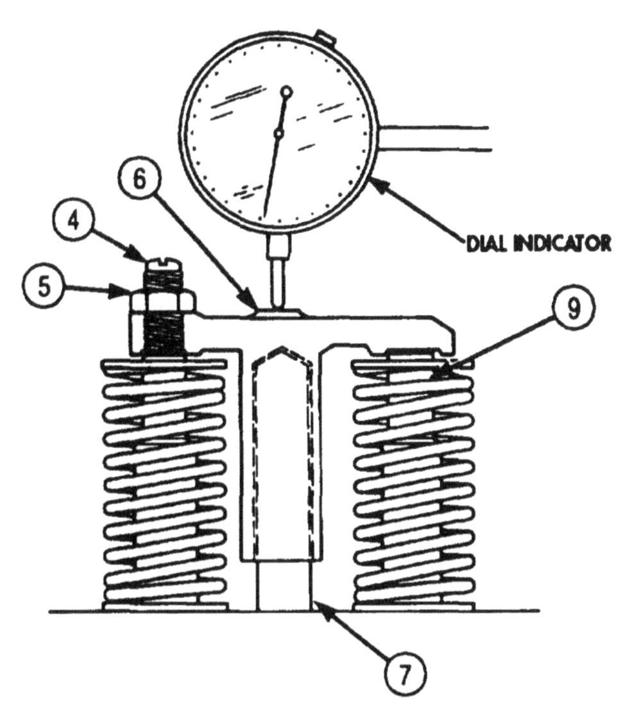

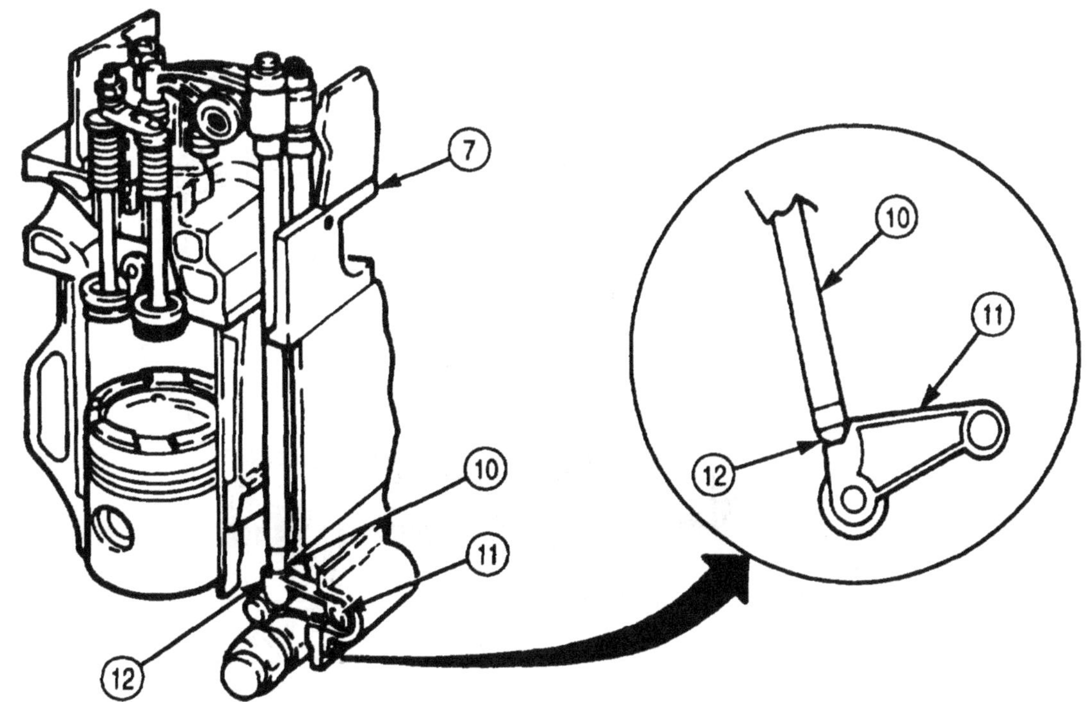

4-32. ENGINE INJECTOR TIMING INSTRUCTIONS (Contd)

NOTE

- Ensure push tubes remain seated during rocker lever housing installation.
- Rocker lever housings are mounted with screw-assembled washers on late model engines.

2. Loosen eighteen screws (9) on three rocker lever housings (6).
3. Position three new gaskets (8) and rocker lever housings (6) on cylinder heads (7) and install with twelve washers (5) and screws (4). Do not tighten screws (4).
4. Install two lifting eyes (3) and upper radiator support bracket (1) on rocker lever housings (6) with six screws (2). Tighten screws (2) and (4) 55-65 lb-ft (75-88 N•m) in sequence shown.

k. Rocker Lever Housing Covers Installation

NOTE

- Rocker lever housing covers are mounted with screw-assembled washers on late model engines.
- All rocker lever covers are installed the same way. This task covers the installation of the center housing cover.

Install new housing cover gasket (13) and rocker lever housing cover (12) on rocker lever housing (6) with five washers (11) and screws (10).

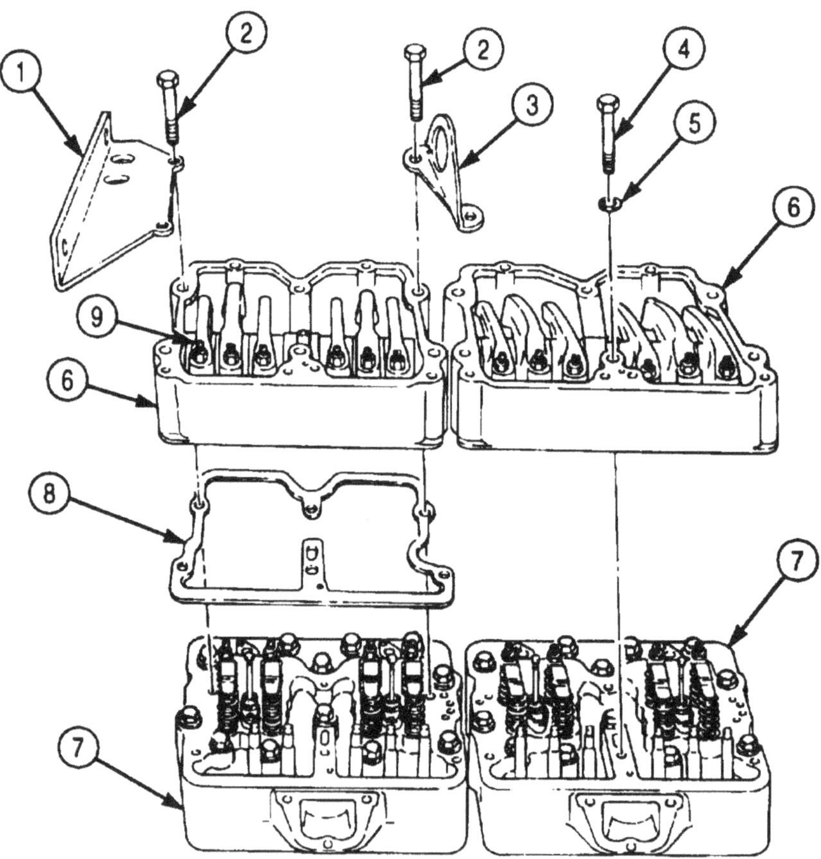

4-32. ENGINE INJECTOR TIMING INSTRUCTIONS (Contd)

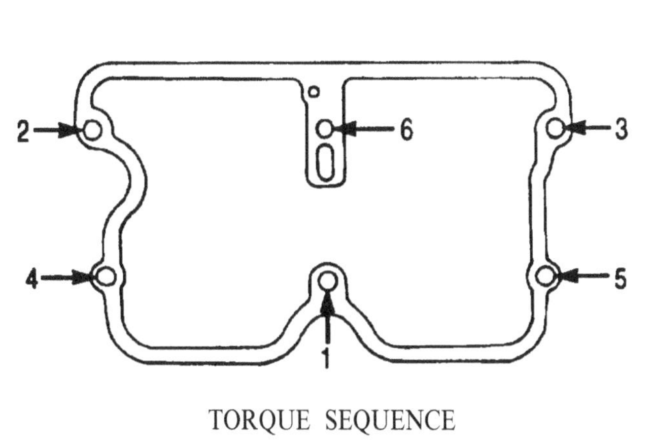

TORQUE SEQUENCE

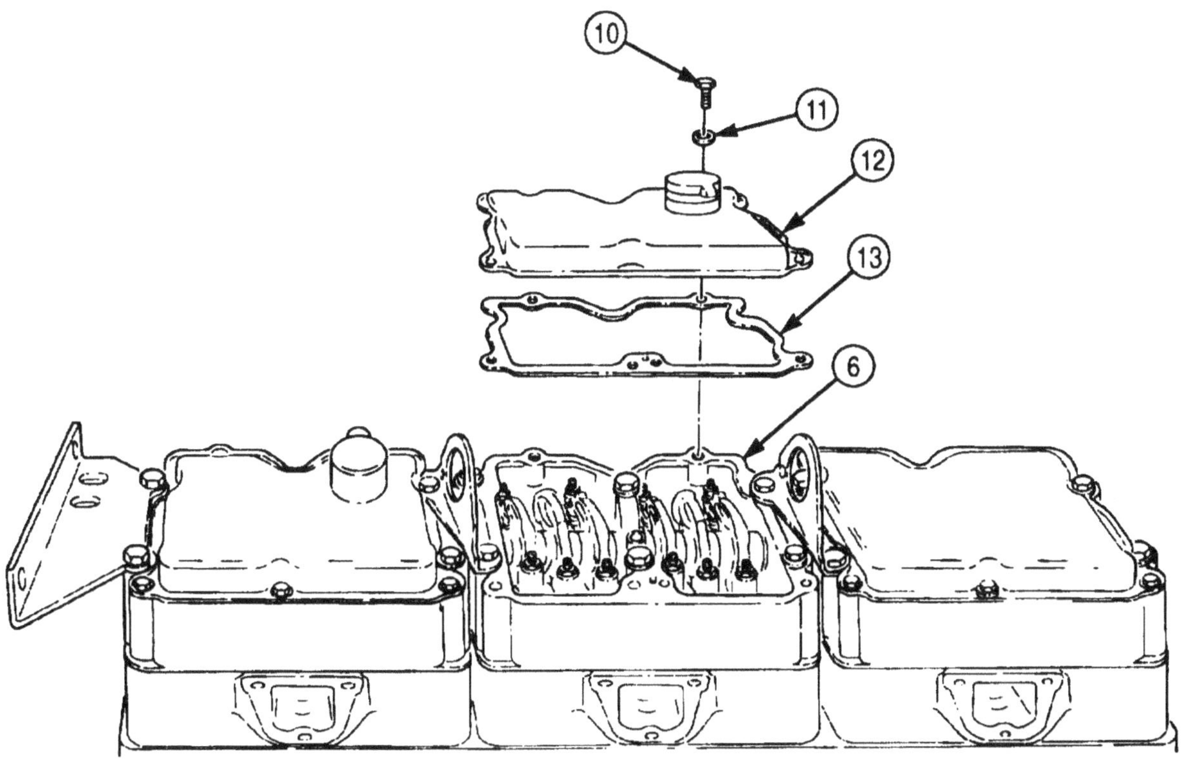

FOLLOW-ON TASK: Remove engine from repair stand (para. 4-9).

4-33. VALVE AND INJECTOR ADJUSTMENT (DIAL INDICATOR METHOD)

THIS TASK COVERS:

a. Checking Plunger Free 'Ravel **b. Injector and Valve Adjustments**

INITIAL SETUP:

APPLICABLE MODELS
M939/A1

SPECIAL TOOLS
Injector and valve adjustment kit (Appendix E, item 69)
Barring tool (Appendix E, Item 8)
Torque wrench adapter (Appendix E, Item 143)

TOOLS
General mechanic's tool kit (Appendix E, Item 1)
Torque wrench (Appendix E, Item 144)
Torque wrench adapter (Appendix E, Item 143)
Dial indicator (Appendix E, Item 36)
Feeler gauge

REFERENCES (TM)
TM 9-2320-272-10
TM 9-2320-272-24P

EQUIPMENT CONDITION
- Rocker lever housing covers removed (in-vehicle) (para. 4-20).
- Valve crossheads adjusted (para. 4-30).
- Rocker lever housing covers removed (para. 4-20).
- Fuel shutoff handle pulled (TM 9-2320-272-10).
- Battery ground cables disconnected (para. 3-126).

GENERAL SAFETY INSTRUCTIONS
If task is being performed while engine is in vehicle, ensure fuel shutoff handle is pulled and battery ground cables are disconnected to prevent engine starting.

a. Checking Plunger Free Travel

WARNING

If task is being performed while engine is in vehicle, ensure fuel shutoff handle is pulled and battery ground cables are disconnected to prevent engine starting. Failure to do this may result in injury to personnel.

CAUTION

This procedure is for non-top stop injectors. It is used to prevent excessive loading of the injector actuating systems and possible failure.

1. Loosen adjusting screw (2) and locknut (3) on injector rocker levers (4).
2. Install dial indicator, fixture, and extension arm on housing (6) at water manifold (1) side.
3. Position dial indicator extension arm on top of injector plunger (5) and set dial to zero.
4. Using engine barring tool to rotate crankshaft, rotate engine and record the total free travel amount of each plunger (5). If plunger (5) free travel exceeds 0.206 in. (5.23 mm) on any cylinder, the torque method of adjustment must be used (para. 4-34).
5. If free travel of all injector plungers (5) checks within limits, reset dial indicator on No. 3 cylinder.

b. Injector and Valve Adjustments

NOTE

1 Before adjusting injectors and valves, check whether rocker housings are cast iron or aluminum so correct clearance setting tolerances are used.

4-33. VALVE AND INJECTOR ADJUSTMENT (DIAL INDICATOR METHOD) (Contd)

NOTE

- During rebuild, injectors and valves are "cold set" with temperature of oil and components within 10oF (-12°C) of ambient air or room temperature. Final "hot set" adjustments must be made when engine is at operating temperature. When warming engine for hot adjustment, ensure rocker cover is installed.
- The injector and valve adjusting procedures below and in the referenced table of specifications require that injectors be adjusted before the valves.
- Perform this procedure for non-top stop injectors.

1. Using engine barring tool, rotate crankshaft pulley until the mark 1-6 VS (8) is aligned with pointer (9) on gearcase cover (7).

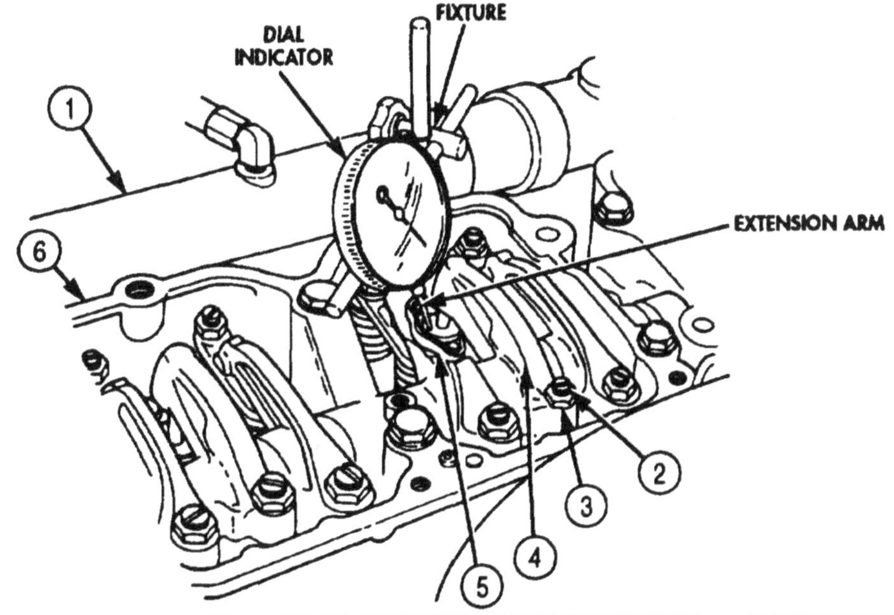

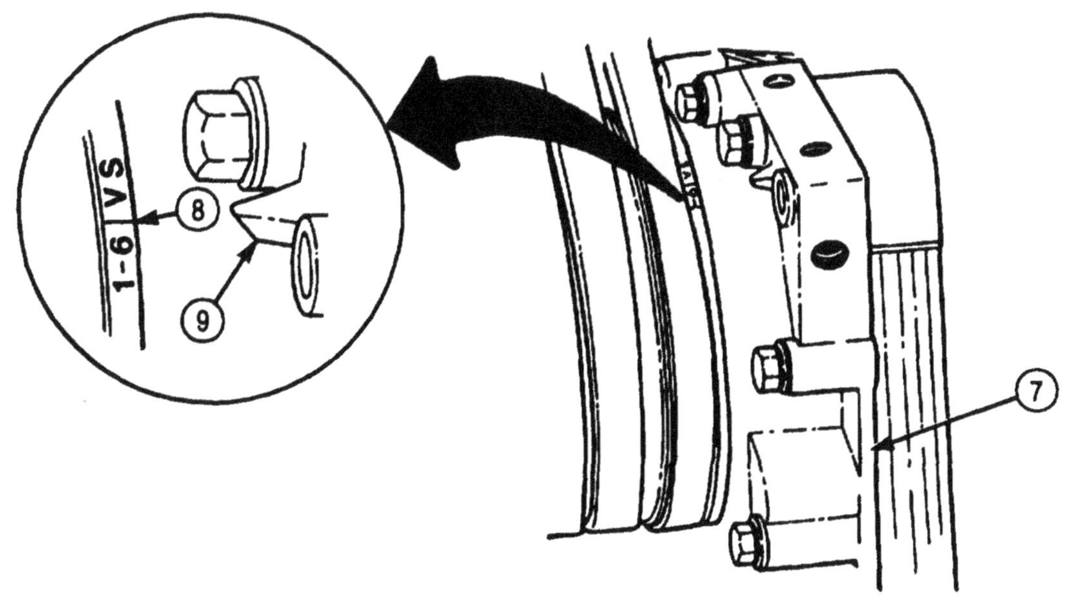

4-33. VALVE AND INJECTOR ADJUSTMENT (DIAL INDICATOR METHOD) (contd)

NOTE

Both valve rocker levers for cylinder No. 5 must be free (valves closed). Injector plunger for cylinder No. 3 must be at top of travel before beginning adjustments.

2. Shake intake (9) and exhaust (10) rocker levers on No. 5 cylinder (8) by hand to ensure they are free (in closed position). Ensure valve springs are not compressed.
3. If injector plunger (5) on No. 3 cylinder (1) is not at top of travel, rotate crankshaft 360° and realign marks 1-6 VS with the pointer.
4. Using wrench to hold locknuts (6), turn adjusting screws (7) on injector rocker levers (2) down until link pin (4) in plunger (5) contacts lever cup (3). Advance adjusting screw (7) 15° to squeeze oil from lever cup. Loosen adjusting screw (7) several turns.
6. Position dial indicator extension arm from fixture on top of cylinder (14) injector plunger (5).
6. Using rocker lever actuator (12), press injector lever (13) down toward fuel injector until injector plunger (5) is bottomed.
7. Release lever (11) and allow injector plunger (5) to rise, then press to bottom again. Ensure injector plunger (5) is held bottomed.
8. Set dial indicator to zero, release injector plunger (5), and allow dial needle to rise. Dial must show travel of 0.169-0.171 in. (4.29-4.34 mm).
9. Turn adjusting screw (7) until dial indicator reads 0.170 in. (4.32 mm) for aluminum housing.
10. Tighten locknut (6) on adjusting screw (7) 40-45 lb-ft (54-61 N•m)
11. Press injector plunger (5) with rocker lever actuator several times to check adjustment reading.

4-33. VALVE AND INJECTOR ADJUSTMENT (DIAL INDICATOR METHOD) (Contd)

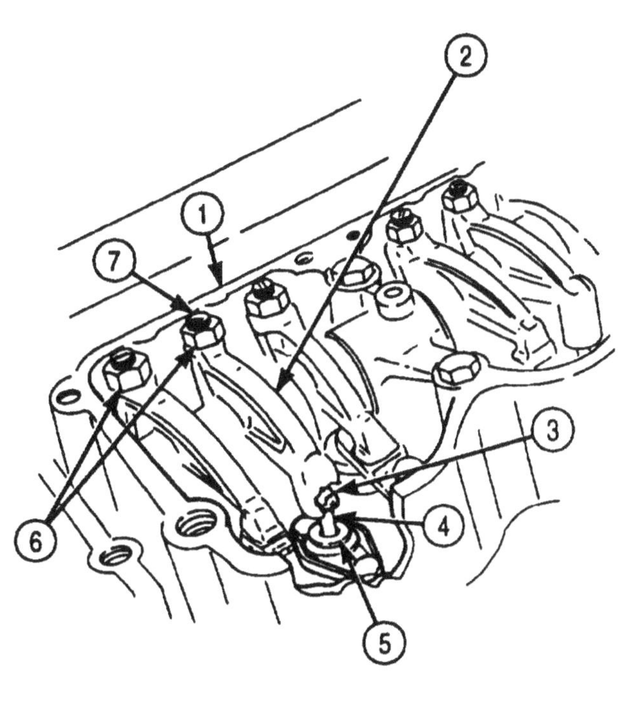

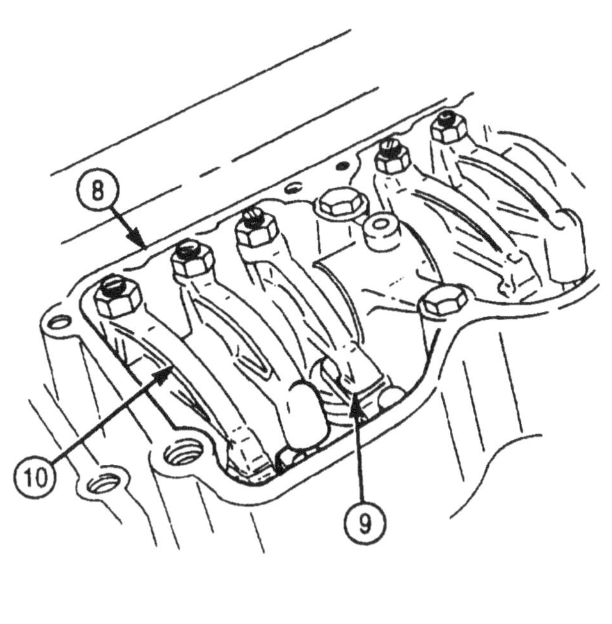

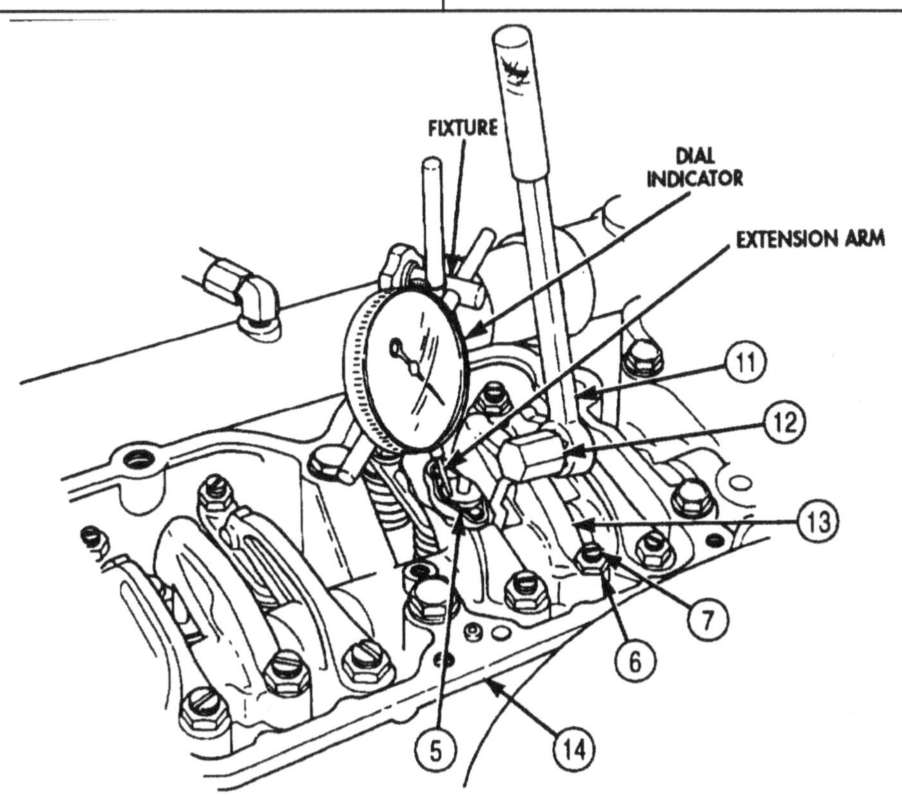

4-33. VALVE AND INJECTOR ADJUSTMENT (DIAL INDICATOR METHOD) (Contd)

Table 4-2. Engine Firing Order.	
Right hand rotation	1-5-3-6-2-4

NOTE

Preceding steps cover injector adjustments for No. 3 cylinder. Follow table 4-3 for the remaining five injectors.

Table 4-3. Injector and Valve Set Position.

ROTATION DIRECTION	PULLEY POSITION	ADJUST CYLINDER	
		INJECTOR	VALVE
Start	1-6 VS	3	5
Advance to	2-5 VS	6	3
Advance to	3-4 vs	2	6
Advance to	1-6 VS	4	2
Advance to	2-5 VS	1	4
Advance to	3-4 VS	5	1

NOTE: Two complete revolutions of pulley are required to adjust all injectors and valves.

Table 4-4. Uniform Plunger Travel Adjustment Limits.

OIL TEMP.	INJECTOR PLUNGER TRAVEL			VALVE CLEARANCE			
ADJUST VALVE		RECHECK LIMIT		INTAKE		EXHAUST	
IN.	(MM)	IN.	(MM)	IN.	(MM)	IN.	(MM)
Aluminum Rocker Housing							
Cold 0.170	(4.32)	0.169-0.171	(4.29-4.34)	0.011	(0.28)	0.023	(0.58)

NOTE

With engine position at 1-6 VS mark, No. 3 cylinder has been adjusted, and now valves in No. 5 cylinder must be adjusted.

12. Loosen locknut (1) on intake rocker lever (4) of No. 5 cylinder (3) and back out adjusting screw (2).
13. Insert feeler gauge between rocker lever (4) nose and crosshead. See table 4-4 for valve clearance settings.
14. Slowly turn down adjusting screw (2) with screwdriver until rocker lever (4) nose touches feeler gauge.
15. Check clearance by removing and inserting feeler gauge. There will be a slight drag on gauge when clearance is correct.
16. Ensure adjusting screw (2) does not move by holding it firmly in position while tightening locknut (1) 40-45 lb-ft (54-61 N•m).
17. Repeat steps 12 through 16 to adjust exhaust rocker lever (5). See table 4-4 for valve clearance settings.
18. Advance pulley (6) so pointer (8) is at next timing mark (7) and perform injector and valve adjustments indicated in tables 4-2 and 4-3.
19. Advance pulley (6) to next timing mark (7) after each set of injector and valve adjustments until all timing is completed.

4-33. VALVE AND INJECTOR ADJUSTMENT (DIAL INDICATOR METHOD) (Contd)

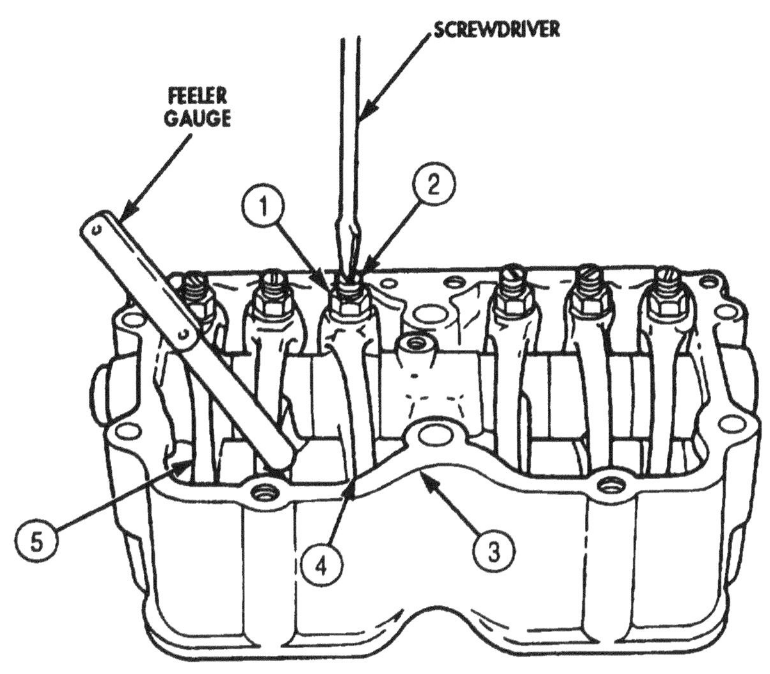

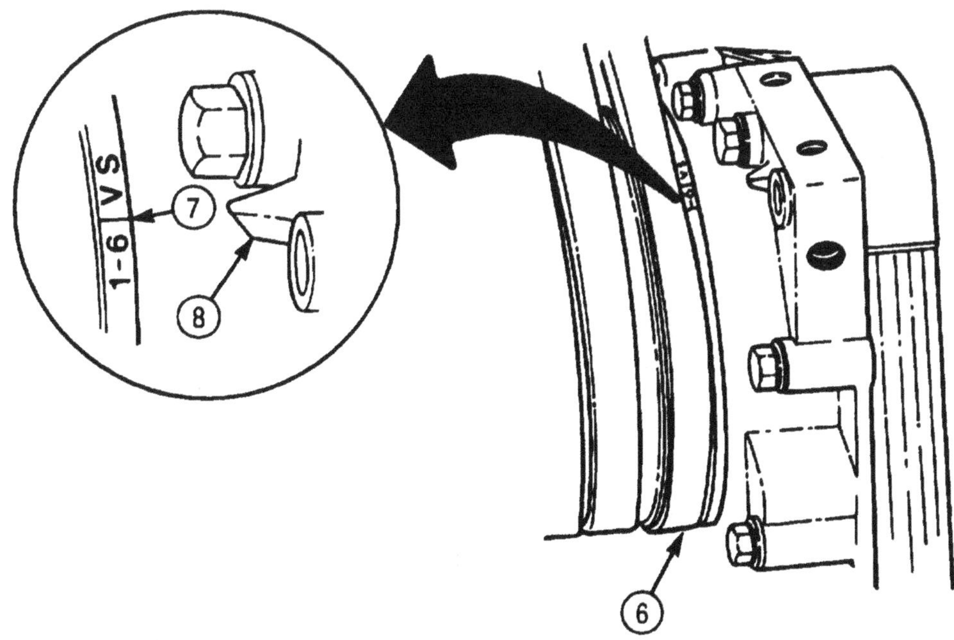

FOLLOW-ON TASKS:
- Install rocker lever housing covers (in-vehicle) (para. 4-20).
- Install rocker lever housing covers (para. 4-32).
- Connect battery ground cables (para. 3-126).
- Push in fuel shutoff handle (TM 9-2320-272-10).
- Start engine (TM 9-2320-272-10), run until normal operating temperature is reached, and check for leaks.

4-34. INJECTOR PLUNGER AND VALVE ADJUSTMENTS (TORQUE METHOD)

THIS TASK COVERS:

a. Pre-adjustment Setup

b. Injector and Valve Adjustment

INITIAL SETUP:

APPLICABLE MODELS

M939/A1

SPECIAL TOOLS

Barring tool (Appendix E, Item 8)

TOOLS

General mechanic's tool kit (Appendix E, Item 1)
Torque wrench (Appendix E, Item 144)
Torque wrench adapter (Appendix E, Item 143)
Torque wrench (Appendix E, Item 142)
Wire gauge
Feeler gauge

REFERENCES (TM)

TM 9-2320-272-10
TM 9-2320-272-24P

EQUIPMENT CONDITION

- Valve crossheads adjusted (para. 4-30).
- Rocker lever housing covers removed (para. 4-32).
- Rocker lever housing covers removed (in-vehicle) (para. 4-20).
- Fuel shutoff handle pulled (TM 9-2320-272-10)
- Battery ground cables disconnected (para. 3-126).

GENERAL SAFETY INSTRUCTIONS

If task is being performed while engine is in vehicle, ensure fuel shutoff handle is pulled and battery ground cables are disconnected to prevent engine starting.

a. Pre-adjustment Setup

WARNING

If task is being performed while engine is in vehicle, ensure fuel shutoff handle is pulled and battery ground cables are disconnected to prevent engine starting. Failure to do this may result in injury to personnel.

NOTE

- Injector plungers are adjusted before valves are adjusted.
- Loosening all injector rocker lever adjusting screws and locknuts will help indicate difference between cylinders that have been adjusted and those cylinders still needing adjustment.

1. Loosen six locknuts (3) on injector rocker levers (1) and adjusting screws (2) one full turn.
2. Rotate engine in gearcase (4) in direction of operating rotation until timing marks 1-6 VS on accessory drive pulley (5) align with pointer (6).

4-34. INJECTOR PLUNGER AND VALVE ADJUSTMENTS (TORQUE METHOD) (Contd)

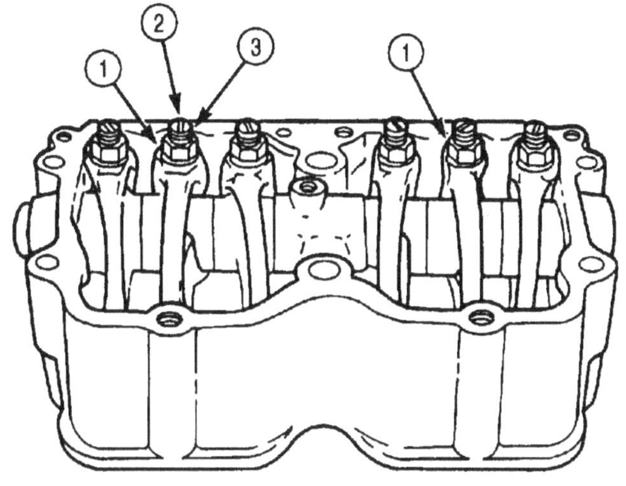

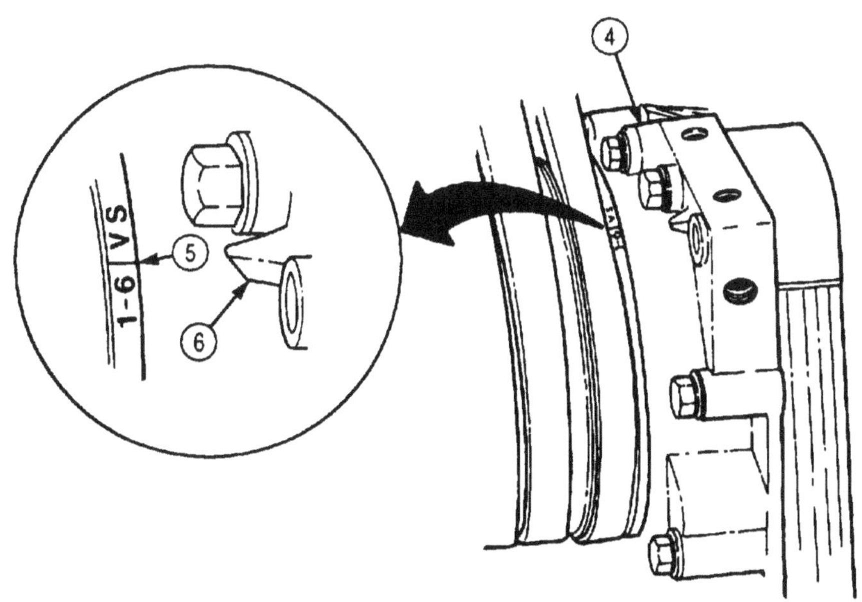

4-34. INJECTOR PLUNGER AND VALVE ADJUSTMENTS (TORQUE METHOD) (Contd)

b. Injector and Valve Adjustment

NOTE

- Before adjusting injectors and valves, check if rocker housings are cast iron or aluminum, so that the correct clearance setting tolerances listed in tables 4-5 and 4-6 are used.
- During rebuild, injectors and valves are "cold set" with oil and component parts temperature within 10° (-12°C) of ambient air or room temperature.
- The injector and valve adjusting procedures below and the referenced "table of specifications" require that the injectors be adjusted before the valves.

1. Loosen locknut (3) on injector rocker lever (1) and hold locknut (3) with wrench.
2. Turn adjusting screw (2) down until top of plunger (4) contacts cup (5).
3. Advance adjusting screw (2) 15 degrees to squeeze oil from cup (5).
4. Back adjusting screw (2) out one full turn. Ensure spring retainer (8) is against adjusting screw (6) of injector (7).

NOTE

Use torque wrench to adjust injectors. Set the torque wrench on value required and pull to zero. Break adjusting screw loose each time, and pull to torque value shown in each tightening pass.

5. Tighten adjusting screw (2), making two or three passes with torque wrench, to values listed in table 4-5.

NOTE

Perform step 6 for top-stop injectors.

6. Tighten adjusting screw (2) 5-6 lb-in. (0.6-0.7 N-m).
7. Tighten locknut (3) 40-45 lb-ft (54-61 N•m).

Table 4-5. Injector Adjustment.

COLD SET
Aluminum Rocker Housing
72 lb-in. (8.1 N·m)

4-34. INJECTOR PLUNGER AND VALVE ADJUSTMENTS (TORQUE METHOD) (Contd)

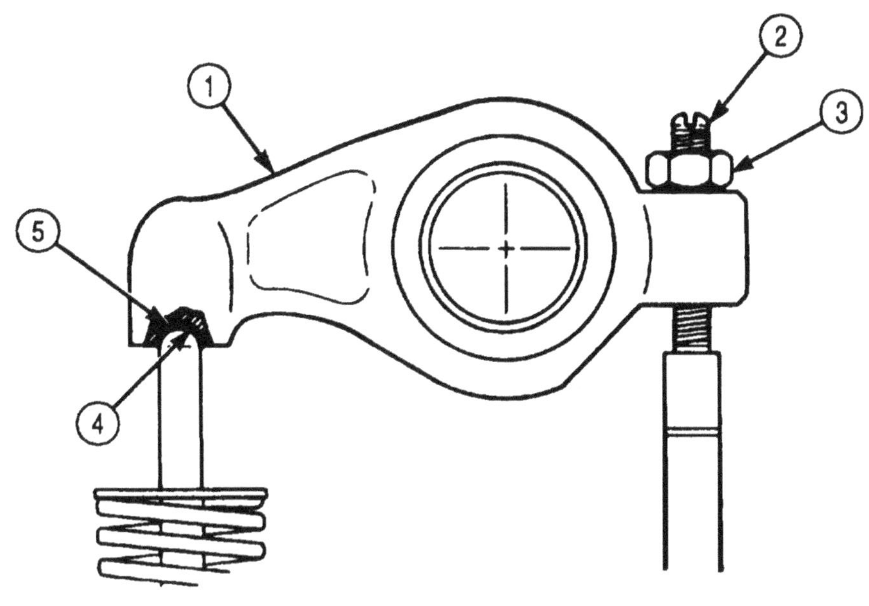

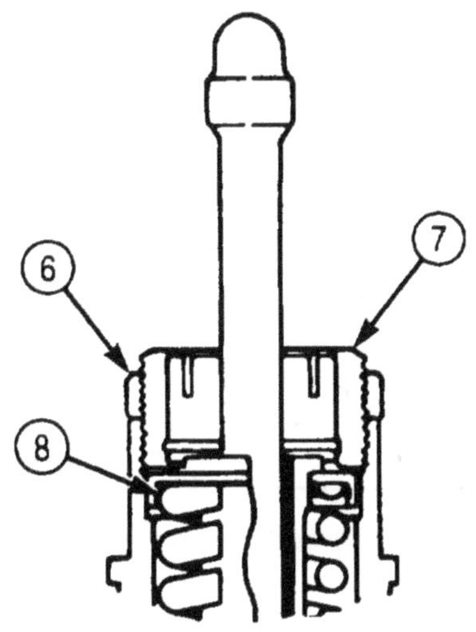

4-34. INJECTOR PLUNGER AND VALVE ADJUSTMENTS (TORQUE METHOD) (Contd)

CAUTION

Before checking or setting valve clearances, ensure crossheads are adjusted.

NOTE

- Crossheads operate two valves with one rocker lever. Crosshead adjustments are necessary to ensure equal operation of each valve.
- The same engine position used in adjusting injectors is used for setting intake and exhaust valves.

8. Loosen locknut (7) attached to adjusting screw (1) on crosshead (6).
9. Back out adjusting screw (1) on crosshead (6) one turn.
10. Apply light finger pressure to rocker lever contact surface (2) to maintain crosshead (6) contact with valve stem (4) opposite adjusting screw (1).
11. Turn adjusting screw (1) down until it touches valve stem (5).
12. Tighten locknut (7) 22-26 lb-ft (30-35 N•m).
13. Hold adjusting screw (1) with screwdriver and tighten locknuts (7) 25-30 lb-ft (34-41 N•m).
14. Check crosshead (6) valve spring retainer (3) clearance with wire gauge. At this point, there must be a minimum clearance of 0.020 in. (0.51 mm).
15. Loosen locknut (8) on intake rocker lever (11) and back out adjusting screw (9).
16. Insert feeler gauge between rocker lever (11) nose and crosshead (6) on cylinder (10). Refer to table 4-6 for valve clearance settings.
17. Using screwdriver, slowly turn down adjusting screw (9) until intake rocker lever (11) nose touches feeler gauge.
18. Check intake rocker lever (11) clearance by removing and inserting feeler gauge. There will be a slight drag on feeler gauge when clearance is correct.
19. Hold adjusting screw (9) firmly in position on cylinder (10) to prevent adjusting screw (9) motion, and tighten locknut (8) 40-45 lb-ft (54-61 N•m).
20. Repeat steps 15 through 19 to adjust exhaust rocker lever (12).
21. Rotate pulley (13) so pointer (15) is at next timing mark (14) and perform injector and valve adjustments for cylinder indicated by VS mark (rocker levers are loose).
22. Rotate pulley (13) after each set of injector and valve adjustments until all timing is completed.

Table 4-6. Valve Clearance Inch (MM) (Torque Method).

Intake valves	Exhaust Valves
Cold Set	Cold Set
Aluminum Rocker Housing	
0.014 in. (0.36 mm)	0.027 in. (0.79 mm)

4-34. INJECTOR PLUNGER AND VALVE ADJUSTMENTS (TORQUE METHOD) (Contd)

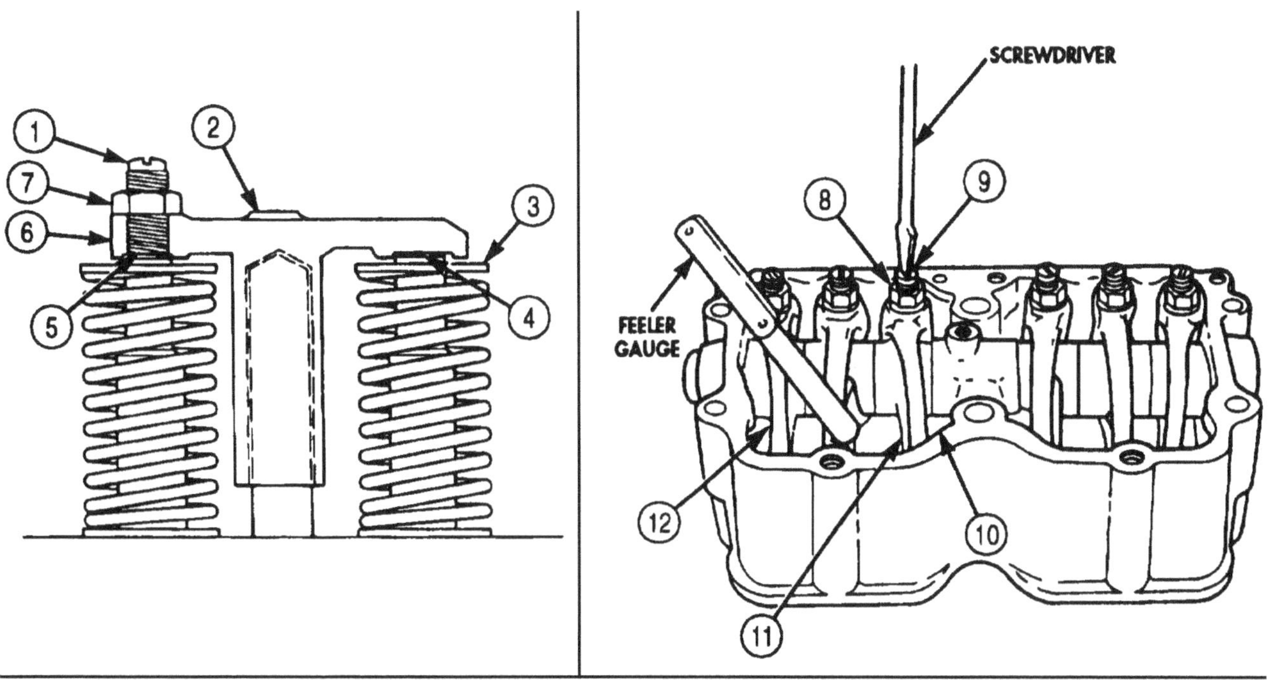

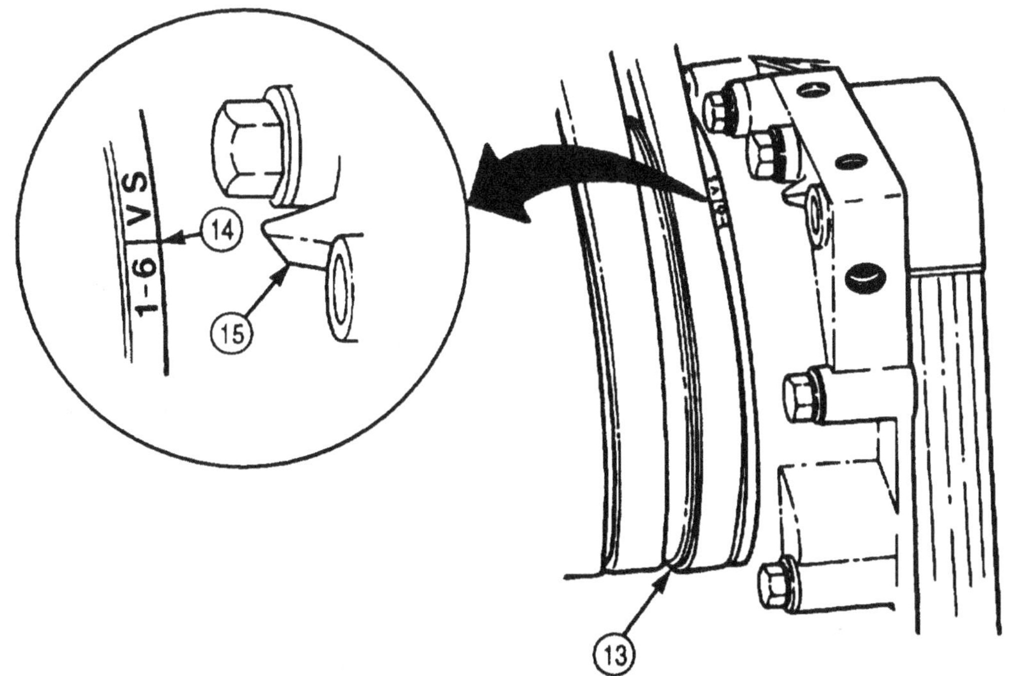

FOLLOW-ON TASKS: Install rocker lever housing covers (para. 4-32).

- Install rocker lever housing covers (in-vehicle) (para. 4-20).
- Connect battery ground cables (para. 3-126).
- Push in fuel shutoff handle (TM 9-2320-272-10).
- Start engine (TM 9-2320-272-10), run until normal operating temperature is reached, and check for leaks.

4-35. FUEL PUMP REPLACEMENT

THIS TASK COVERS:

a. Removal
b. Installation
c. On-Engine Adjustments

INITIAL SETUP:

APPLICABLE MODELS
M939/A1

TOOLS
General mechanic's tool kit (Appendix E, Item 1)

MATERIALS/PARTS
Three locknuts (Appendix D, Item 276)
Gasket (Appendix D, Item 138)
Lockwasher (Appendix D, Item 382)
Cap and plug set (Appendix C, Item 14)
Lubricating oil (Appendix C, Item 50)
Antiseize tape (Appendix C, Item 72)

REFERENCES (TM)
TM 9-2320-272-10
TM 9-2320-272-24P

EQUIPMENT CONDITION
- Parking brake set (TM 9-2320-272-10).
- Hood raised and secured (TM 9-2320-272-10).
- Left splash shield removed (TM 9-2320-272-10).
- Air reservoir drained (TM 9-2320-272-10).

GENERAL SAFETY INSTRUCTIONS
- Fuel pump body must be thoroughly cleaned before disconnecting any attaching components.
- Do not disconnect air line before draining air reservoirs.
- Hearing protection must be worn when engine is running.

a. Removal

1. Disconnect lead (1) from fuel pressure transducer (11) and remove fuel pressure transducer (11) from fuel pump (10).
2. Disconnect fuel line (4) from fuel shutoff valve (6).
3. Remove nut (8) and wires (7) from terminal (9).
4. Loosen screw (2) and remove cable (5) from fuel shutoff lever (3).
5. Remove locknut (14) and screw (20) and disconnect rod clevis (13) from pump throttle lever (12). Discard locknut (14).
6. Remove return spring (17) from modulator control link (16) and modulator cable clamp bracket (18).
7. Remove locknut (22) and (15), screws (24) and (19), and rod clevis (23) from pump throttle lever (21). Discard locknuts (22) and (15).

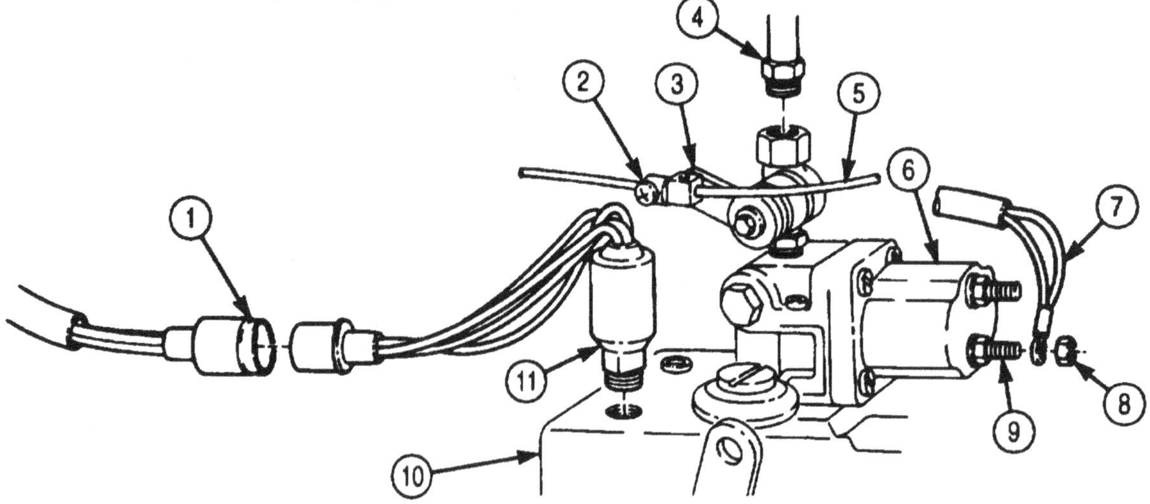

4-35. FUEL PUMP REPLACEMENT (Contd)

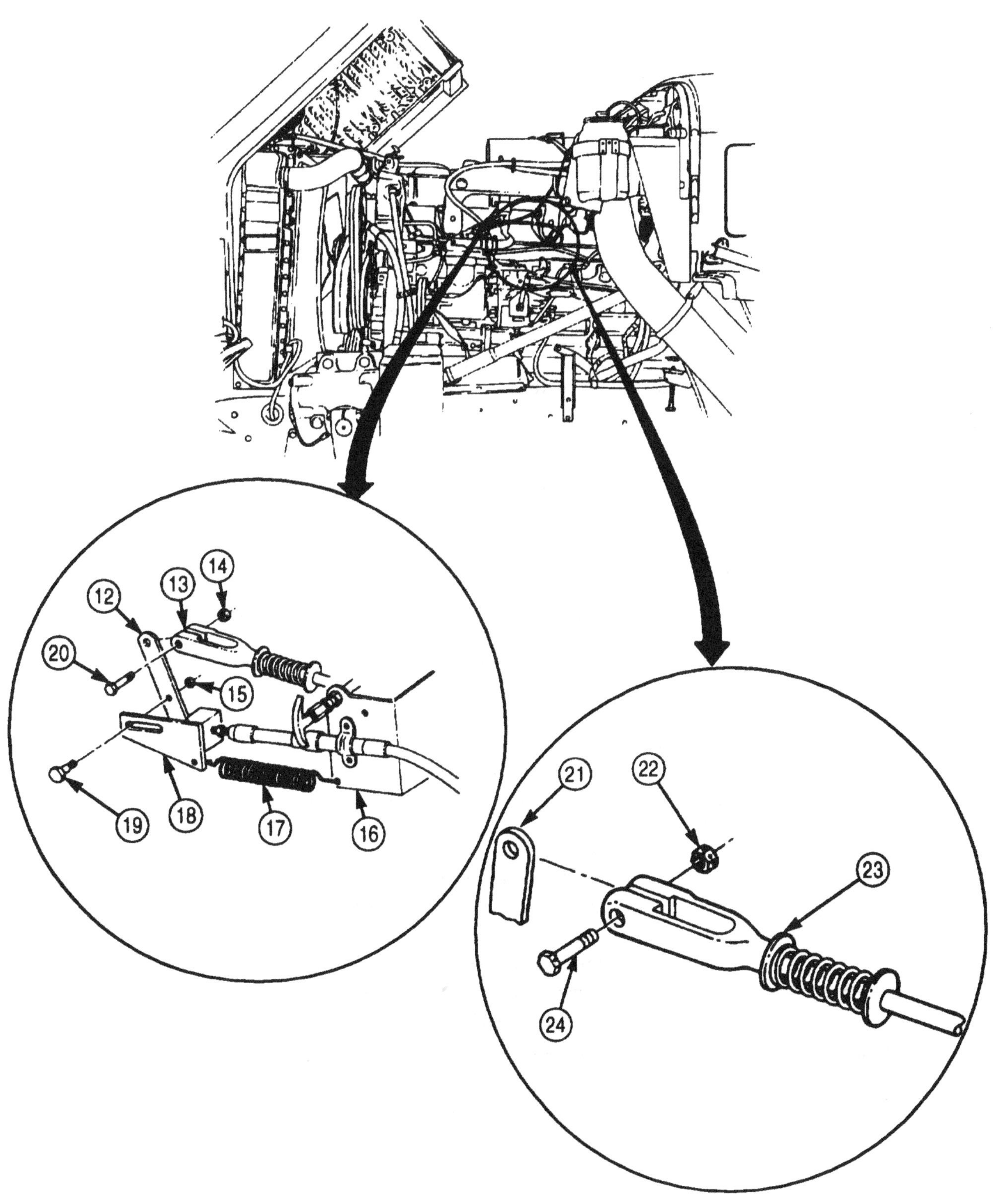

4-35. FUEL PUMP REPLACEMENT (Contd)

8. Disconnect tachometer pulse sender (10) from tachometer drive housing (9).
9. Disconnect primer pump fuel line (4), inlet fuel line (5), and fuel outlet line (3) from inlet adapter elbow (6) and fuel outlet line fitting (2).
10. Disconnect two connectors (7) from ether start fuel pressure switch (8) and remove ether start fuel pressure switch (8) from bottom of fuel pump (1).

NOTE

Steps 11 through 13 apply to fuel pump with VS governor only.

11. Remove two screws (16) and air cylinder bracket (17) from VS governor (11).

WARNING

Do not disconnect air line before draining air reservoirs. Small parts under pressure may shoot out with high velocity, causing injury to personnel.

12. Remove air line (15) and adapter (14) from air cylinder (13).
13. Remove nut (12) and air cylinder (17) from bracket (12).
14. Remove three screws (24), washers (25), screw (23), lockwasher (22). washer (21), and fuel pump (1) from air compressor (18). Discard lockwasher (32).
15. Remove rubber spider coupling (20) and fuel pump gasket (19) from air compressor (18). Discard gasket (19). Clean gasket remains from mating surfaces of air compressor (18) and rubber spider coupling (20).

b. Installation

1. Place new gasket (19) and rubber spider coupling (20) on air compressor (18).
2. Align fuel pump drive (26) with rubber spider coupling (20) on air compressor (18) and install fuel pump (1) on air compressor (18) with three washers (25). screws (24), washer (21), new lockwasher (22), and screw (23).

NOTE

Squirt clean oil into pump through adapter elbow hole. This aids fuel pickup and provides pump lubrication on initial start.

3. Connect tachometer pulse sender (10) to tachometer drive housing (9).
4. Connect primer pump fuel line (4), inlet fuel line (5), and fuel outlet line (3) to inlet adapter elbow (6) and fuel outlet line fitting (2).
5. Install ether start fuel pressure switch (8) on fuel pump (1).
6. Connect two connectors (7) to ether start fuel pressure switch (8).
7. Place fuel shutoff control cable (32) through hole in shutoff lever clamp (29). Ensure shutoff lever (30) is in the forward position and tighten screw (28).
8. Connect fuel line (31) to fuel shutoff valve (33).
9. Connect two wires (34) to shutoff valve terminal (36) with nut (35).
10. Install fuel pressure transducer (37) in fuel pump (1) and connect to lead (27).

NOTE

Steps 11 through 13 apply to fuel pump with VS governor only.

11. Install air cylinder bracket (17) on VS governor (11) with two screws (16).
12. Install air cylinder (13) on bracket (17) with nut (12).
13. Apply antiseize tape to male threads of adapter (14) and air line (15) and install on air cylinder (13).

4-35. FUEL PUMP REPLACEMENT (Contd

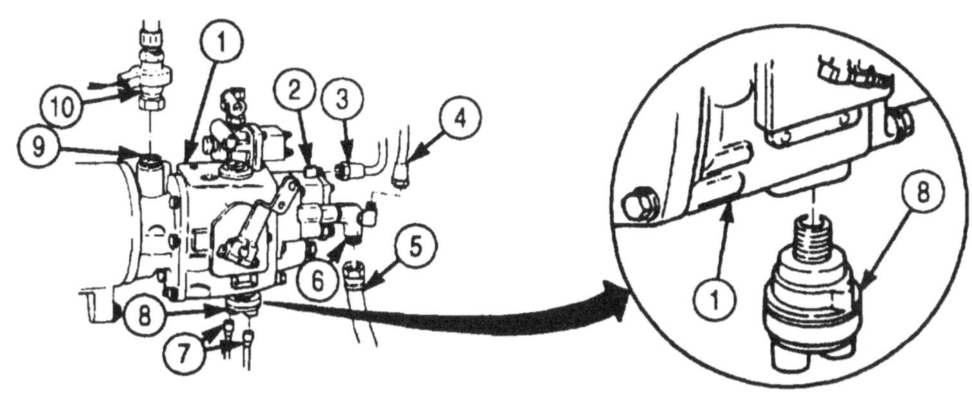

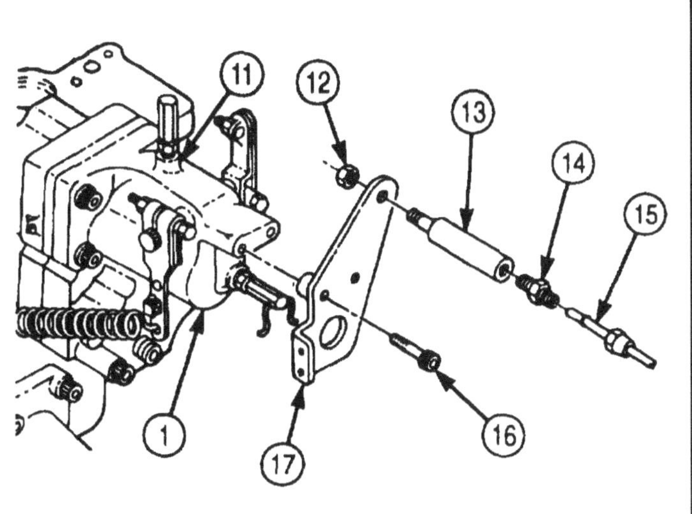

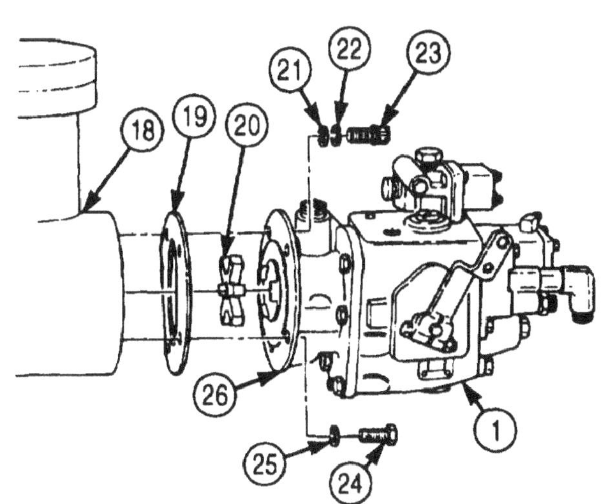

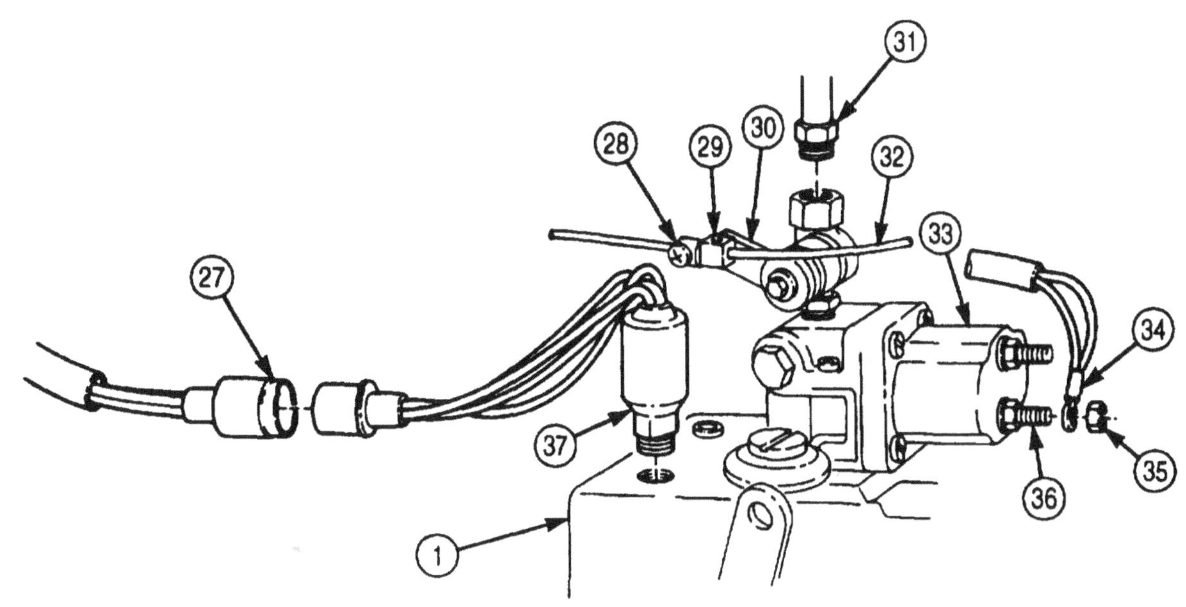

4-35. FUEL PUMP REPLACEMENT (Contd)

14. Connect rod clevis (2) to throttle lever (1) with screw (9) and new locknut (3).
15. Connect modulator link (7) to throttle lever (1) with screw (8) and new locknut (4).
16. Connect modulator return spring (6) to modulator link (7) and cable clamp bracket (5).
17. Connect rod clevis (12) to throttle lever (10) with screw (13) and new locknut (11).

c. On-Engine Adjustments

CAUTION

Do not change pump settings made during calibration.

1. Remove nut (16), lockwasher (17), washer (18), and screw (19) from throttle lever (1) and fuel pump (14). Discard lockwasher (17).
2. Slide throttle lever (1) off splined throttle shaft (20).
3. Turn splined throttle shaft (20) clockwise until resting against idle adjusting screw (15).
4. Slide throttle lever (1) on splined throttle shaft (20) and install with screw (19), washer (18), new lockwasher (17), and nut (16). Do not tighten nut (16).
5. Prime fuel system and allow to warm up to operating temperature (TM 9-320-272-10).
6. Check idle speed. If idle speed is not 600-650 rpm, stop engine and check linkage adjustment.
7. When idle speed is correct, stop engine and tighten nut (16).

WARNING

Hearing protection must be worn by mechanic when engine is running. Noise levels produced by this vehicle exceed 85dB, which may cause injury to personnel.

NOTE

Steps 8 through 17 apply to fuel pump with vs governor only.

8. Start engine (TM 9-2320-272-10) and allow air system to reach normal operating pressure.
9. Engage crane drive control lever (TM 9-2320-272-10).
10. Engage transfer power takeoff lever (TM 9-2320-272-10)
11. Remove spring (26) from throttle lever (1) and bracket (25).
12. Loosen nut (21) and adjusting screw (22).
13. Place throttle lever (23) forward in full throttle position and hold. Turn adjusting screw (22) until it touches air cylinder piston shaft (24).
14. Release throttle lever (23), hold adjusting screw (22), and tighten nut (21).
15. Install spring (26) on throttle lever (1) and bracket (25).
16. Disengage transfer PTO lever (TM 9-2320-272-10).
17. Disengage crane drive control lever (TM 9-2320-272-10).

4-35. FUEL PUMP REPLACEMENT (Contd)

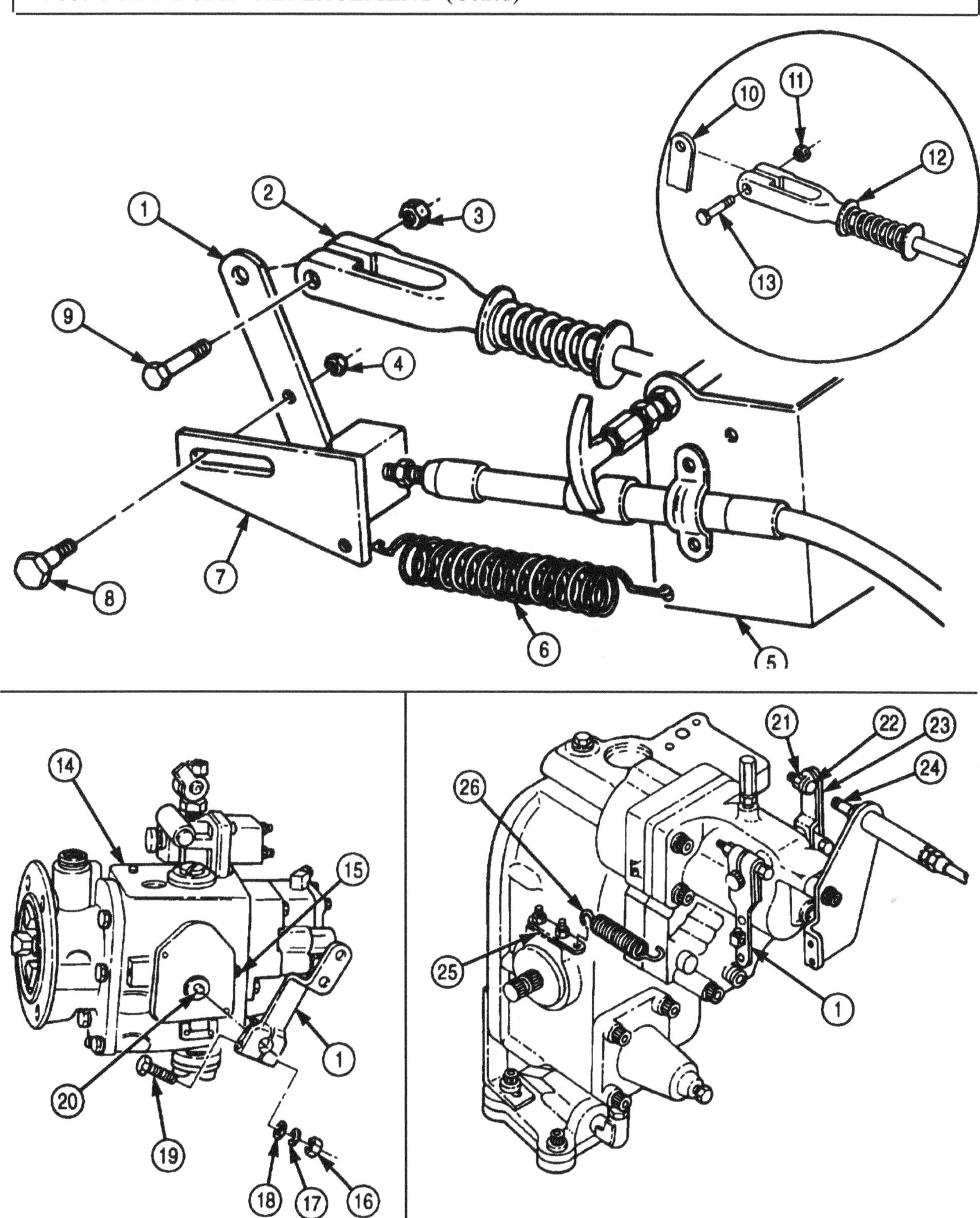

FOLLOW-ON TASK: Install left splash shield (TM 9-2320-272-10).

4-36. FUEL PUMP SHUTOFF VALVE REPLACEMENT

THIS TASK COVERS:

a. Manual Shutoff Valve Removal
b. Fuel Pump Shutoff Valve Removal
c. Fuel Pump Shutoff Valve Installation
d. Manual Shutoff Valve Installation

INITIAL SETUP:

APPLICABLE MODELS
M939/A1 (except M936)

TOOLS
General mechanic's tool kit (Appendix E, Item 1)

MATERIALS/PARTS
O-ring (Appendix D, Item 475)
Two lockwashers (Appendix D, Item 407)
Cap and plug set (Appendix C, Item 14)

REFERENCES (TM)
TM 9-2320-272-10
TM 9-2320-272-24P

EQUIPMENT CONDITION
- Hood raised and secured (TM 9-2320-272-10).
- Left splash shield removed (TM 9-2320-272-10).

a. Manual Shutoff Valve Removal

1. Disconnect fuel line (3) from manual fuel shutoff valve (4).
2. Loosen screw (15), remove clip (1), and pull fuel shutoff control cable (5) until free of shutoff lever (2).
3. Remove manual fuel shutoff valve (4) from fuel shutoff valve (12).

b. Fuel Pump Shutoff Valve Removal

1. Remove nut (10) and wires (9) from terminal (11).
2. Remove two screws (6), lockwashers (7), and washers (8) from shutoff valve (12). Discard lockwashers (7).
3. Remove fuel shutoff valve (12) and O-ring (13) from fuel pump (14). Plug openings in fuel pump (14). Discard O-ring (13).

c. Fuel Pump Shutoff Valve Installation

1. Install new O-ring (13) and fuel shutoff valve (12) on fuel pump (14) with two washers (8), new lockwashers (7), and screws (6).
2. Install wires (9) on terminal (11) with nut (10).

d. Manual Shutoff Valve Installation

1. Install manual fuel shutoff valve (4) in fuel shutoff valve (12).
2. Install control cable (5) on shutoff lever (2) with clip (1) and screw (15). Ensure shutoff lever (2) is in forward position.
3. Connect fuel line (3) to manual fuel shutoff valve (4).

4-36. FUEL PUMP SHUTOFF VALVE REPLACEMENT (Contd)

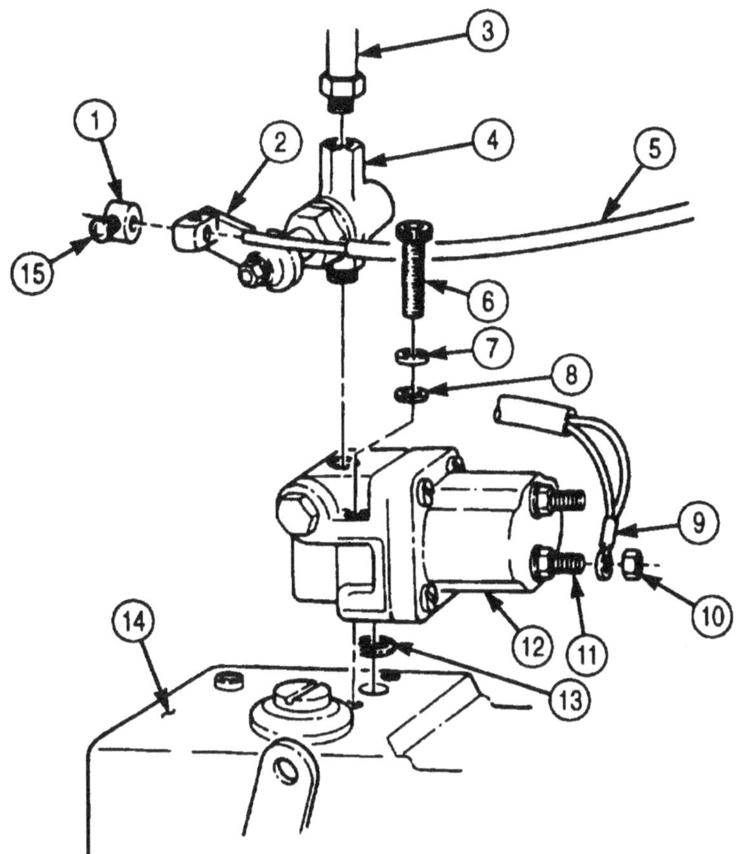

FOLLOW-ON TASKS:
- Start engine (TM 9-2320-272-10) and check fuel pump shutoff valves for proper operation.
- Install left splash shield (TM 9-2320-272-10).

4-37. FUEL PUMP SHUTOFF VALVE (M936) REPLACEMENT

THIS TASK COVERS:

a. Removal **b. Installation**

INITIAL SETUP:

APPLICABLE MODELS
M936

TOOLS
General mechanic's tool kit (Appendix E, Item 1)

MATERIALS/PARTS
O-ring (Appendix D, Item 445)
Two lockwashers (Appendix D, Item 407)
Cap and plug set (Appendix C, Item 14)

REFERENCES (TM)
TM 9-2320-272-10
TM 9-2320-272-24P

EQUIPMENT CONDITION
- Hood raised and secured (TM 9-2320-272-10).
- Left splash shield removed (TM 9-2320-272-10).

a. Removal

1. Disconnect fuel line (2) from manual fuel shutoff valve (1).
2. Remove nut (7) and wires (6) from terminal (8).
3. Loosen screw (5) on clip (3) and remove clip (3) and fuel shutoff control cable (4) from manual fuel shutoff valve (1).
4. Remove manual fuel shutoff valve (1) from fuel shutoff valve (11).
5. Remove two screws (14), lockwashers (13), washers (12), fuel shutoff valve (11), and O-ring (9) from fuel pump (10). Plug openings in fuel pump (10) and discard O-ring (9) and lockwashers (13).

b. Installation

1. Unplug openings in fuel pump (10) and install new O-ring (9) and fuel shutoff valve (11) on fuel pump (10) with two new lockwashers (13), washers (12), and screws (14).
2. Install manual fuel shutoff valve (1) on fuel shutoff valve (11).
3. Ensuring shutoff lever is in forward position, install fuel shutoff control cable (4) on manual fuel shutoff valve (1) with clip (3) and tighten screw (5).
4. Install wires (6) on terminal (8) with nut (7).
5. Connect fuel line (2) to manual fuel shutoff valve (1).

4-37. FUEL PUMP SHUTOFF VALVE (M936) REPLACEMENT (Contd)

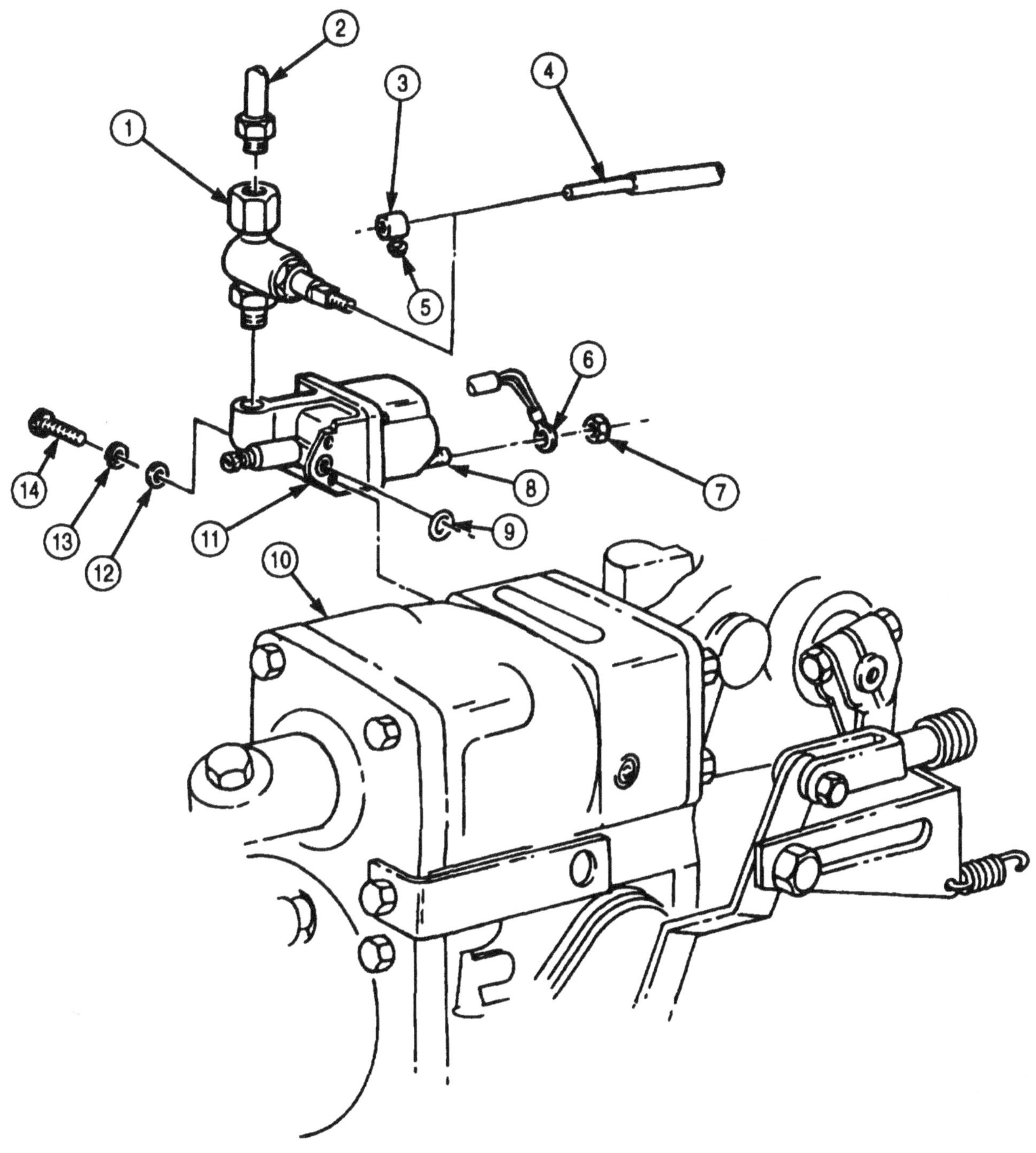

FOLLOW-ON TASKS:
- Start engine (TM 9-2320-272-10) and check fuel pump shutoff valve for proper operation.
- Install left splash shield (TM 9-2320-272-10).

INDEX

	Para	Page
A		
100-amp alternator harness replacement:		
Installation	3-427b	3-1098
Removal	3-427a	3-1096
100-amp alternator replacement:		
Installation	3-426b	3-1094
Removal	3-426a	3-1094
100-amp voltage regulator replacement:		
Installation	3-428b	3-1100
Removalea	3-428a	3-1100
A-frame kit maintenance:		
Inspection	3-414b	3-1048
Installation	3-414c	3-1048
Removal	3-414a	3-1046
Access step, spare tire carrier, cargo, replacement:		
Installation	3-260b	3-720
Removal	3-260a	3-720
Access step, spare tire carrier, dump and tractor, replacement:		
Installation	3-255b	3-711
Removal	3-255a	3-710
Accessory drive and pulley, engine, maintenance:		
Accessory drive assembly	4-26e	4-190
Accessory drive cleaning and inspection	4-26d	4-188
Accessory drive disassembly	4-26c	4-188
Accessory drive installation	427f	4-190
Accessory drive pulley installation	4-26g	4-192
Accessory drive pulley	4-26a	4-186
Accessory drive removal	4-26b	4-186
Adjustment, bearing, wheel:		
Front wheel bearing adjustment	3-225a	3-648
Rear wheel bearing adjustment	3-225b	3-650
Air compressor (M939/A1) maintenance:		
Assembly	4-31e	4-226
Cleaning	4-31c	4-220
Disassembly	4-31b	4-218
Inspection and repair	4-31d	4-220
Installation and timing	4-31f	4-232
Removal	4-31a	4-214
Air compressor (M939A2) maintenance:		
Assembly	4-52c	4-322
Cleaning and inspection	4-52b	4-318
Disassembly	4-52a	4-314
Air conditioner drain tube replacement:		
Installation	3-382b	3-938
Removal	3-382a	3-938
Air dryer and check valve maintenance:		
Assembly	3-464d	3-1272
Cleaning and inspection	3-464c	3-1272
Disassembly	3-464b	3-1272
Installation	3-464e	3-1274
Removal	3-464a	3-1270
Air dryer filter replacement:		
Installation	3-465b	3-1276
Removal	3-465a	3-1276
Air dryer kit (M923/A1/A2, M925/A1/A2, M927/A1/A2, M928/A1/A2, M934/A1/A2) replacement:		
Installation	3-412b	3-1026
Removal	3-412a	3-1022
Air dryer kit (M929/A1/A2, M930/A1/A2, M931/A1/A2, M932/A1/A2, M936/A1/A2) replacement:		
Installation	3-413b	3-1039
Removal	3-413a	3-1032
Air lines, fabrication	3-452	3-1232
Air manifold, rear axle, maintenance:		
Cleaning and inspection	3-463b	3-1268
Installation	3-463c	3-1268
Removal	3-463a	3-1268
Air seals, hub, leak test:		
Front hub leak test	3-462a	3-1264
Rear hub leak test	3-462b	3-1266
Air vent control assembly, fresh, replacement:		
Installation	3-293b	3-784
Removal	3-293a	3-784
Alternator, 100-amp, replacement:		
Installation	3-426b	3-1094
Removal	3-426a	3-1094
Amber warning light replacement:		
Installation	3-469b	3-1283
Removal	3-469a	3-1283

INDEX (Contd)

	Para	Page
Anchor post, toggle clamp, replacement:		
Installation	3-358b,	3-897
Removal	3-358a	3-897
Approach plates, fifth wheel, replacement:		
Installation	3-249b	3-698
Removal	3-249a	3-698
Arm, pitman, replacement moss):		
Installation	3-227b	3-653
Removal	3-227a	3-653
Arm, pitman, replacement (Sheppard):		
Installation	3-228b	3-654
Removal	3-228a	3-654
Assist cylinder, steering, replacement:		
Assembly	3-233c	3-664
Disassembly	3-233b	3-664
Installation	3-233d	3-666
Removal	3-233a	3-662
Travel adjustment	3-233e	3-666
Atmospheric fuel tank vent system kit replacement:		
Installation	3-441b	3-1141
Removal	3-441a	3-1140
Automatic brake (hoist winch) adjustment:		
Adjustment	3-383	3-940
Automatic throttle kit (M936/A1) replacement:		
Installation	3-440b	3-1136
Removal.a	3-440a	3-1134

B

	Para	Page
Backrest cushion, companion seat cushion, frame, and, replacement:		
Installation	3-286b	3-770
Removal	3-286a	3-770
Backrest cushion, driver's seat cushion, frame and, replacement:		
Installation	3-285b	3-768
Removal	3-285a	3-768
Baffles, seals, and plates, radiator, replacement:		
Installation	3-272b	3-738
Removal	3-272a	3-738
Battery box heater pad, engine coolant, replacement:		
Installation	3-404b	3-994
Removal	3-404a	3-994
Bearing adjustment, wheel:		
Front wheel bearing adjustment	3-225a	3-648
Rear wheel bearing adjustment	3-225b	3-650
Blackout circuit plungers, hinged roof-operated, replacement:		
Installation	3-377b	3-924
Removal	3-377a	3-924
Blackout light switch and 110-volt receptacle replacement:		
Installation	3-373b	3-916
Removal	3-373a	3-916
Blackout light switch, side and rear door maintenance:		
Installation	3-376b	3-922
Removal	3-376a	3-922
Blackout panel, window replacement:		
Installation	3-352b	3-890
Removal	3-352a	3-890
Bonnet handle and control rod replacement:		
Installation	3-361b	3-900
Removal	3-361a	3-900
Boom floodlight wire replacement:		
Installation	3-385b	3-946
Removal	3-385a	3-946
Bows, cargo body cover, replacement:		
Installation	3-341b	3-872
Removal	3-341a	3-872
Brake modification kit (M939A2), engine exhaust brake, replacement:		
Installation	3-447b	3-1194
Removal	3-447a	3-1186
Brake modification kit (M939A2), engine exhaust brake, replacement:		
Installation	3-448b	3-1214
Removal	3-448a	3-1204
Brake, (hoist winch), automatic, adjustment:		
Adjustment	3-383	3-940

INDEX (Contd)

	Para	Page
Brushguard, window, replacement:		
Installation	3-355b	3-893
Removal	3-355a	3-893
Bumper, hood, replacement:		
Installation	3-274b	3-741
Removal	3-274a	3-741
Bumperette replacement:		
Installation	3-246b	3-694
Removal	3-246a	3-694
Bumpers, ladder rack, replacement:		
Installation	3-366b	3-907
Removal	3-366a	3-907
Bumpers, rubber, side panel, replacement:		
Installation	3-367b	3-908
Removal	3-367a	3-908
Bumpers, tailgate, replacement:		
Installation	3-261b	3-721
Removal	3-261a	3-721
C		
Cab cowl vent screen and door replacement:		
Installation	3-299b	3-793
Removal	3-299a	3-793
Cab door catch replacement:		
Installation	3-321b	3-821
Removal	3-321a	3-821
Cab door check rod replacement:		
Installation	3-318b	3-817
Removal	3-318a	3-817
Cab door dovetail replacement:		
Installation	3-310b	3-809
Removal	3-310a	3-809
Cab door dovetail wedge replacement:		
Installation	3-309b	3-808
Removal	3-309a	3-808
Cab door glass maintenance:		
Adjustment	3-314c	3-813
Installation	3-314b	3-813
Removal	3-314a	3-813
Cab door hinge replacement:		
Installation	3-320b	3-820
Removal	3-320a	3-820
Cab door inspection hole cover replacement:		
Installation	3-312b	3-811
Removal	3-312a	3-811
Cab door lock replacement:		
Installation	3-313b	3-812
Removal	3-313a	3-812
Cab door regulator assembly replacement:		
Installation	3-317b	3-816
Removal	3-317a	3-816
Cab door replacement:		
Installation	3-319b	3-318
Removal	3-319a	3-818
Cab door weatherseal replacement:		
Installation	3-311b	3-810
Removal	3-311a	3-810
Cab grab handle replacement:		
Installation	3-288b	3-773
Removal	3-288a	3-773
Cab heat and defrost air ducting replacement:		
Installation	3-297b	3-791
Removal	3-297a	3-791
Cab hood stop bracket replacement:		
Installation	3-273b	3-740
Removal	3-273a	3-740
Cab insulation replacement:		
Installation	3-306b	3-804
Removal	3-306a	3-804
Cab mount, rear, replacement:		
Installation	3-305b	3-802
Removal	3-305a	3-802
Cab top seal and retainer replacement:		
Installation	3-316b	3-815
Removal	3-316a	3-815
Cab turnbuttons and lashing hooks replacement:		
Installation	3-303b	3-798
Removal	3-303a	3-798
Cab windshield hinge assembly replacement:		
Installation	3-280b	3-756
Removal	3-280a	3-756
Cab, mount, front, replacement:		
Installation	3-304b	3-800
Removal	3-304a	3-800
Cable, hinged roof and floor counterbalance, maintenance:		
Cable adjustment	3-356c	3-895
Installation	3-356b	3-894
Removal	3-356a	3-894

INDEX (Contd)

	Para	Page
Cable, hoist, winch, replacement:		
Installation	3-384b	3-944
Removal	3-384a	3-942
Cam follower housing maintenance:		
Assembly	4-19d	4-150
Cleaning and inspection	4-19c	4-148
Disassembly	4-19b	4-146
Installation	4-19e	4-152
Removal	4-19a	4-144
Cargo body cover bows replacement:		
Installation	3-341b	3-872
Removal	3-341a	3-872
Cargo spare tire carrier (M923, M925, M927, M928) replacement:		
Installation	3-258b	3-716
Removal	3-258a	3-716
Cargo spare tire carrier (M923A1/A2,M925A1/A2, M927A1/A2, M928A1/A2) replacement:		
Installation	3-259b	3-718
Removal	3-259a	3-718
Cargo spare tire carrier access step replacement:		
Installation	3-260b	3-720
Removal	3-260a	3-720
Cargo stowage box replacement:		
Installation	3-346b	3-880
Removal	3-346a	3-880
Cargo tailgate replacement:		
Installation	3-343b	3-876
Removal	3-343a	3-876
Cargo troop seat replacement:		
Installation	3-340b	3-870
Removal	3-340a	3-870
Cargo upper and lower wheel splash guard replacement:		
Installation	3-342b	3-874
Removal	3-342a	3-874
Check valve, air dryer and, maintenance:		
Assembly	3-464d	3-1272
Cleaning and inspection	3-464c	3-1272
Disassembly	3-464b	3-1272
Installation	3-464e	3-1274
Removal	3-464a	3-1270
Checks, door, replacement:		
Installation	3-365b	3-906
Removal	3-365a	3-906
Chemical agent alarm mounting bracket kit replacement:		
Chemical alarm wiring harness installation	3-417d	3-1064
Chemical alarm wiring harness removal	3-417a	3-1058
Detector and alarm bracket installation	3-417c	3-1062
Detector and alarm bracket removal	3-417b	3-1060
Chock anchors, front and rear field (M936/A1), replacement:		
Installation	3-244b	3-690
Removal	3-244a	3-690
Clamp, ladder locking, replacement:		
Installation	3-360b	3-899
Removal	3-360a	3-899
Clevis, winch cable, replacement:		
Installation	3-326b	3-832
Removal	3-326a	3-831
Clip, side locking pin retaining, replacement:		
Installation	3-344b	3-878
Removal	3-344a	3-878
Companion seat cushion, backrest cushion, and frame replacement:		
Installation	3-286b	3-770
Removal	3-286a	3-770
Compartment, map, replacement:		
Installation	3-287b	3-772
Removal	3-287a	3-772
Compressor, air, (M939/A1), maintenance:		
Assembly	4-31e	4-226
Cleaning	4-31c	4-220
Disassembly	4-31b	4-218
Inspection and repair	4-31d	4-220
Installation and timing	4-31f	4-232
Removal	4-31a	4-214
Compressor, air, (M939A2), maintenance:		
Assembly	4-52c	4-322
Cleaning and inspection	4-52b	4-318
Disassembly	4-52a	4-314
Control Unit, Electronic (ECU), replacement:		
Installation	3-468b	3-1282
Removal	3-468a	3-1282

INDEX (Contd)

	Para	Page
Controls, defrost and heat, replacement:		
Installation	3-294b	3-786
Removal	3-294a	3-786
Convoy warning light harness (M929/A1/A2, M930/A1/A2 replacement:		
Installation	3-434b	3-1118
Removal	3-434a	3-1116
Convoy warning light harness (M934/A1/A2) replacement:		
Installation	3-435b	3-1122
Removal	3-435a	3-1120
Convoy warning light harness replacement:		
Installation	3-433b	3-1114
Removal	3-433a	3-1114
Convoy warning light mount (M929/A1/A2, M930/A1/A2) replacement:		
Installation	3-432b	3-1112
Removal	3-432a	3-1112
Convoy warning light mount (M934/A1/A2) replacement:		
Installation	3-431b	3-1110
Removal	3-431a	3-1108
Convoy warning light mount replacement:		
Installation	3-430b	3-1106
Removal	3-430a	3-1104
Convoy warning light replacement:		
Installation	3-437b	3-1128
Removal	3-437a	3-1128
Convoy warning light resistor and leads replacement:		
Installation	3-436b	3-1126
Removal	3-436a	3-1124
Convoy warning light switch replacement:		
Installation	3-438b	3-1130
Removal	3-438a	3-1130
Coolant, heater engine, replacement:		
Installation	3-397b	3-974
Removal	3-397a	3-974
Cover, cab door inspection hole, replacement:		
Installation	3-312b	3-811
Removal	3-312a	3-811
Cover, crankshaft rear, seal and, replacement:		
Installation	4-17b	4-136
Removal	4-17a	4-136
Cover, engine access, replacement:		
Installation	4-13b	4-124
Removal	4-13a	4-124
Cover, housing assembly, replacement:		
Installation	3-333b	3-852
Removal	3-333a	3-852
Cover, kit, radiator, replacement:		
Installation	3-395b	3-970
Removal	3-395a	3-970
Crane hydraulic filter maintenance:		
Assembly	3-391d	3-962
Cleaning and inspection	3-391c	3-962
Disassembly	3-391b	3-960
Installation	3-391e	3-962
Removal	3-391a	3-960
Crane wiring harness replacement:		
Installation	3-386b	3-950
Removal	3-386a	3-948
Crankshaft rear cover and seal replacement:		
Installation	4-17b	4-136
Removal	4-17a	4-136
Crankshaft vibration and damper replacement:		
Inspection	4-14c	4-128
Installation	4-14d	4-128
Removal	4-14b	4-126
Runout and wobble check	4-14a	4-126
Crossheads, fuel crossover connectors and, replacement:		
Inspection	4-30b	4-210
Installation and adjustment	4-30c	4-212
Removal	4-30a	4-210
CTIS wiring harness replacement:		
Installation	3-470b	3-1286
Removal	3-470a	3-1284
Cylinder head replacement:		
Installation	4-12b	4-122
Removal	4-12a	4-120
Cylinder, steering assist, replacement:		
Assembly	3-233c	3-664
Disassembly	3-233b	3-664
Installation	3-233d	3-666
Removal	3-233a	3-662
Travel adjustment	3-233e	3-666

INDEX (Contd)

	Para	Page
D		
Damper, crankshaft vibration and, replacement:		
Inspection	4-14c	4-128
Installation	4-14d	4- 128
Removal	4-14b	4-126
Runout and wobble check	4-14a	4-126
Davit and pulley, swing, van, replacement:		
Installation	3-265b	3-726
Removal	3-265a	3-726
Davit chain and wire rope, van, replacement:		
Installation	3-264b	3-724
Removal	3-264a	3-724
Davit winch (M934A1/A2), van, replacement:		
Installation	3-266b	3-728
Removal	3-266a	3-728
Deck plate, fifth wheel, replacement:		
Installation	3-251b	3-700
Removal	3-251a	3-700
Deck plate, forward, replacement:		
Installation	3-392b	3-964
Removal	3-392a	3-964
Decontamination (M13) apparatus mounting bracket kit replacement:		
Installation (M923/A1/A2, M925/A1/A2, M927/A1/A2, M928/A1/A2)	3-419h	3-1076
Installation (M929/A1/A2, M930/A1/A2, M931/A1/A2, M932/A1/A2)	3-419B,	3-1074
Installation (M934/A1/A2)	3-419d	3-1074
Installation (M936/A1/A2)	3-419f	3-1076
Removal (M923/A1/A2, M925/A1/A2, M927/A1/A2, M928/A1/A2)	3-419g	3-1076
Removal (M929/A1/A2, M930/A1/A2, M931/A1/A2, M932/A1/A2)	3-419a	3-1074
Removal (M934/A1/A2)	3-419c	3-1074
Removal (M936/A1/A2)	3-419e	3-1076
Defrost and heat controls replacement:		
Installation	3-294b	3-786
Removal	3-294a	3-786
Direct and general support mechanical troubleshooting symptom index	4-2	4-2
Direct and general support mechanical troubleshooting	4-1	4-1
Diverter assembly replacement:		
Installation	3-295b	3-788
Removal	3-295a	3-788
Door catch, cab, replacement:		
Installation	3-321b	3-821
Removal	3-321a	3-821
Door checks replacement:		
Installation	3-365b	3-906
Removal	3-365a	3-906
Door handle and lock replacement:		
Installation	3-364b	3-905
Removal	3-364a	3-905
Door handle, outside, replacement:		
Installation	3-307b	3-806
Removal	3-307a	3-806
Door hinge and seal replacement:		
Installation	3-362b	3-902
Removal	3-362a	3-902
Door hinge, cab, replacement:		
Installation	3-320b	3-820
Removal	3-320a	3-820
Door regulator assembly, cab, replacement:		
Installation	3-317b	3-816
Removal	3-317a	3-816
Door, cab, replacement:		
Installation	3-319b	3-318
Removal	3-319a	3-818
Door, check rod, cab, replacement:		
Installation	3-318b	3-817
Removal	3-318a	3-817
Dovetail wedge, cab door, replacement:		
Installation	3-309b	3-808
Removal	3-309a	3-808
Dovetail, cab door, replacement:		
Installation	3-310b	3-809
Removal	3-310a	3-809
Drag link replacement:		
Installation	3-229b	3-655
Removal	3-229a	3-655

INDEX (Contd)

	Para	Page
Drain kit, hydraulic reservoir, replacement:		
Installation	3-444b	3-1152
Removal	3-444a	3-1152
Drain tube, air conditioner, replacement:		
Installation	3-382b	3-938
Removal	3-382a	3-938
Drivebelts, steering pump, maintenance (M939/A1):		
Adjustment	3-230a	3-656
Inspection	3-230c	3-656
Installation	3-230d	3-657
Removal	3-230a	3-656
Driver's seat cushion, backrest cushion, and frame replacement:		
Installation	3-285b	3-768
Removal	3-285a	3-768
Driver's seat frame and base maintenance:		
Inspection and repair	3-284b	3-766
Installation	3-284c	3-766
Removal	3-284a	3-764
Driver's seat replacement:		
Installation	3-283b	3-762
Removal	3-283a	3-762
Drum (M939/A1), front hub and, maintenance:		
Cleaning and inspection	3-223b	3-640
Installation	3-223d	3-640
Lubrication	3-223c	3-640
Removal	3-223a	3-638
Drum (M939/A1), rear hub and, maintenance:		
Cleaning and inspection	3-224b	3-644
Installation	3-224d	3-646
Lubrication	3-224c	3-644
Removal	3-224a	3-642
Ducting, cab heat and defrost, air, replacement:		
Installation	3-297b	3-791
Removal	3-297a	3-791
Ducting, fresh air inlet, replacement:		
Installation	3-296b	3-790
Removal	3-296a	3-790
Dump and tractor spare tire carrier access step replacement:		
Installation	3-255b	3-711
Removal	3-255a	3-710
Dump spare tire carrier (M929, M930) replacement:		
Installation	3-256b	3-712
Removal	3-256a	3-712
Dump spare tire carrier (M929A1/A2, M930A1/A2) replacement:		
Installation	3-257b	3-714
Removal	3-257a	3-713
Dump tailgate assembly replacement:		
Installation	3-348b	3-882
Removal	3-348a	3-882
Dump tailgate control linkage replacement:		
Installation	3-349b	3-886
Removal	3-349a	3-884

E

	Para	Page
Electronic Control Unit (ECU) replacement:		
Installation	3-468b	3-1282
Removal	3-468a	3-1282
Emergency/blackout light lamp and lens replacement:		
Installation	3-372b	3-915
Removal	3-372a	3-915
Engine access cover replacement:		
Installation	4-13b	4-124
Removal	4-13a	4-124
Engine accessory drive and pulley maintenance:		
Accessory drive assembly	4-26e	4-190
Accessory drive cleaning and inspection	4-26d	4-188
Accessory drive disassembly	4-26c	4-188
Accessory drive installation	4-27f	4-190
Accessory drive pulley installation	4-26g	4-192
Accessory drive pulley removal	4-26a	4-186
Accessory drive removal	4-26b	4-186
Engine air intake manifold maintenance:		
Assembly	4-24d	4-180
Cleaning and inspection	4-24c	4-180
Disassembly	4-24b	4-180
Installation	4-24e	4-182
Removal	4-24a	4-178

INDEX (Contd)

	Para	Page
Engine and container replacement:		
Installation	4-6b	4-88
Removal	4-6a	4-86
Engine coolant battery box heater pad replacement:		
Installation	3-404b	3-994
Removal	3-404a	3-994
Engine coolant heater control box replacement:		
Installation	3-399b	3-978
Removal	3-399a	3-978
Engine coolant heater harness (M939/A1) replacement:		
Installation	3-400b	3-980
Removal	3-400a	3-980
Engine coolant heater harness (M939A2) replacement:		
Installation	3-401b	3-982
Removal	3-401a	3-982
Engine coolant heater hose replacement:		
Installation	3-403b	3-990
Removal	3-403a	3-986
Engine coolant heater pump replacement:		
Installation	3-398b	3-976
Removal	3-398a	3-976
Engine coolant heater replacement:		
Installation	3-397b	3-974
Removal	3-397a	3-974
Engine coolant oil pan shroud and exhaust tube replacement:		
Installation	4-402b	3-984
Removal	4-402a	3-984
Engine exhaust brake modification kit (M939/A1) replacement:		
Installation	3-447b	3-1194
Removal	3-447a	3-1186
Engine exhaust brake modification kit (M939A2) replacement:		
Installation	3-448b	3-1214
Removal	3-448a	3-1204
Engine exhaust manifold maintenance:		
Installation	4-25b	4-184
Removal	4-25a	4-184
Engine hood maintenance:		
Adjustment	3-275c	3-746
Installation	3-275b	3-744
Removal	3-275a	3-742
Engine injector timing instructions:		
Fuel injectors installation	4-32h	4-242
Fuel injectors removal	4-32d	4-238
General instructions	4-32e	4-238
Injector timing	4-320	4-240
Rocker lever housing covers installation	4-32k	4-246
Rocker lever housing covers removal	4-32a	4-236
Rocker lever housing and push tubes removal	4-32b	4-236
Rocker lever housing and push tubes installation	4-32j	4-244
Timing toolsetup	4-32f	4-240
Valve crossheads installation and adjustment	4-32i	4-244
Valve crossheads removal	4-32c	4-238
Engine lift eyes replacement:		
Installation	4-11b	4-118
Removal	4-11a	4-118
Engine mounting on repair stand:		
Installation (M939/A1)	4-9a	4-100
Installation (M939A2)	4-9b	4-106
Removal (M939/A1)	4-9d	4-110
Removal (M939A2)	4-9c	4-108
Engine oil cooler maintenance:		
Assembly	4-23d	4-176
Cleaning and inspection	4-23c	4-176
Disassembly	4-23b	4-174
Installation	4-23e	4-176
Removal	4-23a	4-174
Engine oil pan maintenance:		
Assembly	4-22d	4-170
Cleaning and inspection	4-22c	4-170
Disassembly	422b	4-170
Installation	4-22e	4-172
Removal	4-22a	4-168
Engine oil pump replacement:		
Gear backlashtest	4-21a	4-164
Installation	4-21c	4-166
Removal	4-21b	4-164

INDEX (Contd)

	Para	Page
Engine water manifold maintenance:		
Assembly	4-27d	4-196
Cleaning and inspection	4-27c	4-196
Disassembly	4-27b	4-194
Installation	4-27e	4-196
Removal	4-27a	4-194
European mini-lighting kit replacement:		
Installation	3-439b	3-1132
Removal	3-439a	3-1132
Exhaust heat shield accessory kit replacement:		
Installation	3-449b	3-1224
Removal	3-449a	3-1224
Exhaust manifold, engine, maintenance:		
Installation	4-25b	4-184
Removal	4-25a	4-184
Expanding and retracting mechanism locks replacement:		
Installation	3-378b	3-926
Removal	3-378a	3-926

F

	Para	Page
Fabrication of air lines	3-452	3-1232
Fender extension, front, replacement:		
Installation	3-300b	3-794
Removal	3-300a	3-794
Fender splash shield replacement:		
Installation	3-301b	3-795
Removal	3-301a	3-795
Fifth wheel approach plates replacement:		
Installation	3-249b	3-698
Removal	3-249a	3-698
Fifth wheel deck plate replacement:		
Installation	3-251b	3-700
Removal	3-251a	3-700
Fifth wheel spacers replacement:		
Installation	3-252b	3-701
Removal	3-252a	3-701
Fifth wheel, tractor, replacement:		
Installation	3-248b	3-696
Removal	3-248a	3-696
Filer, air dryer, replacement:		
Installation	3-465b	3-1276
Removal	3-465a	3-1276
Filter, crane hydraulic, maintenance:		
Assembly	3-391d	3-962
Cleaning and inspection	3-391c	3-962
Disassembly	3-391b	3-960
Installation	3-391e	3-962
Removal	3-391a	3-960
Filter, power steering pump (M939A2), maintenance:		
Cleaning	3-237c	3-674
Filter installation	3-237e	3-674
Filter removal	3-237a	3-674
Reservoir installation	3-237d	3-674
Reservoir removal	3-237b	3-674
Filter, wheel valve, replacement:		
Installation	3-458b	3-1248
Removal	3-458a	3-1248
Filter, winch hydraulic oil reservoir, replacement:		
Installation	3-336b	3-858
Removal	3-336a	3-858
Fire extinguisher mounting bracket kit replacement:		
Installation	3-416b	3-1056
Removal	3-416a	3-1056
Fluorescent light tube replacement:		
Installation	3-371b	3-914
Removal	3-371a	3-914
Flywheel housing maintenance:		
Cleaning and inspection	4-16b	4-130
Installation	4-16c	4-132
Removal	4-16a	4-130
Flywheel ring gear maintenance:		
Installation	4-15b	4-129
Removal	4-15a	4-129
Forward deck plate replacement:		
Installation	3-392b	3-964
Removal	3-392a	3-964
Frame assembly, outer, windshield and, replacement:		
Installation	3-278b	3-752
Removal	3-278a	3-752
Frame assembly, windshield, replacement:		
Installation	3-279b	3-754
Removal	3-279a	3-754

INDEX (Contd)

	Para	Page
Frame extension, winch, replacement:		
Installation	3-245b	3-692
Removal	3-245a	3-692
Fresh air inlet ducting replacement:		
Installation	3-296b	3-790
Rsmoval	3-296a	3-790
Fresh air vent control assembly replacement:		
Installation	3-293b	3-784
Removal	3-293a	3-784
Front and rear field chock anchors (M936/A1) replacement:		
Installation	3-244b	3-690
Removal	3-244a	3-690
Front and rear lifting shackle and bracket replacement:		
Installation	3-241b	3-684
Removal	3-241a	3-684
Front bumper and plates replacement:		
Installation	3-243b	3-688
Removal	3-243a	3-688
Front cab mount replacement:		
Installation	3-304b	3-800
Removal	3-304a	3-800
Front fender extension replacement:		
Installation	3-300b	3-794
Removal	3-300a	3-794
Front gearcase cover maintenance:		
Assembly	4-18d	4-140
Cleaning and inspection	4-18c	4-140
Disassembly	4- 18b	4-138
Installation	4-18e	4-142
Removal	4-18a	4-138
Front hub and drum (M939/A1) maintenance:		
Cleaning and inspection	3-223b	3-640
Installation	3-223d	3-640
Lubrication	3-223c	3-640
Removal	3-223a	3-638
Front hubs repair:		
Cleaning and inspection	3-460b	3-1253
Installation	3-460d	3-1256
Removal	3-460a	3-1252
Repair	3-460c	3-1254
Front wheel valve maintenance:		
Cleaning and inspection	3-456b	3-1244
Installation	3-456c	3-1245
Removal	3-456a	3-1244
Front winch automatic brake adjustment:		
Adjustment	3-323b	3-826
Testing	3-323a	3-824
Front winch cable chain and hook replacement:		
Installation	3-325b	3-830
Removal	3-325a	3-830
Front winch cable replacement:		
Installation	3-327b	3-834
Removal	3-327a	3-834
Front winch drag brake adjustment:		
Adjustment	3-324b	3-828
Testing	3-324a	3-828
Front winch motor replacement:		
Installation	3-328b	3-838
Removal	3-328a	3-836
Front winch replacement:		
Installation	3-329b	3-842
Removal	3-329a	3-840
Fuel crossover connectors and crossheads replacement:		
Inspection	4-30b	4-210
Installation and adjustment	4-30c	4-212
Removal	4-30a	4-210
Fuel pump replacement:		
Installation	4-35b	4-262
On-engine adjustments	4-35c	4-264
Removal	4-35a	4-260
Fuel pump shutoff valve (M936) replacement:		
Installation	4-37b	4-268
Removal	4-37a	4-268
Fuel pump shutoff valve replacement:		
Fuel pump shutoff valve installation	4-36c	4-266
Fuel pump shutoff valve removal	4-36b	4-266
Manual shutoff valve installation	4-36d	4-266
Manual shutoff valve removal	4-36a	4-266

INDEX (Contd)

	Para	Page
Fuel tank vent system kit, atmospheric, replacement:		
Installation	3-441b	3-114b
Removal	3-441a	3-1140

G

	Para	Page
Gearcase cover, front, maintenance:		
Assembly	4-18d	4-140
Cleaning and inspection	4-18c	4-140
Disassembly	4-18b	4-138
Installation	4-18e	4- 142
Removal	4-18a	4-138
Generator, speed signal, replacement:		
Installation	3-471b	3-1288
Removal	3-471a	3-1288
Glass, cab door, maintenance:		
Adjustment	3-314c	3-813
Installation	3-314b	3-813
Removal	3-314a	3-813
Grab handle, cab, replacement:		
Installation	3-288b	3-773
Removal	3-288a	3-773
Grab handle, hood, replacement:		
Installation	3-271b	3-737
Removal	3-271a	3-737

H

	Para	Page
Hand airbrake air supply valve replacement:		
Installation	3-422b	3-1084
Removal	3-422a	3-1084
Hand airbrake controller valve replacement:		
Installation	3-423b	3-1086
Removal	3-423a	3-1086
Hand airbrake doublecheck valves replacement:		
Installation (forward-rear axle doublecheck valve)	3-424b	3-1088
Installation (rear-rear axle doublecheck valve)	3-424d	3-1090
Removal (forward-rear axle doublecheck valve)	3-424a	3-1088
Removal (rear-rear axle doublecheck valve)	3-424c	3-1090
Hand airbrake tractor protection valve replacement:		
Installation	3-425b	3-1092
Removal	3-425a	3-1092
Handle, and control rod, bonnet, replacement:		
Installation	3-361b	3-900
Removal	3-361a	3-900
Handle, and lock, door, replacement:		
Installation	3-364b	3-905
Removal	3-364a	3-905
Handle, inside door, window regulator and, replacement:		
Installation	3-308b	3-807
Removal	3-308a	3-807
Handrail modification kit, (M934/A1/A2), van, replacement:		
Door check spacer installation	3-450h	3-1229
Door check spacer removal	3-450g	3-1229
Ladder handrail guide installation	3-450b	3-1225
Ladder handrail guide removal	3-450a	3-1225
Ladder handrail hangers installation	3-450d	3-1226
Ladder handrail hangers removal	3-450c	3-1226
Van door grab handles installation	3-450f	3-1228
Van door grab handles removal	3-450e	3-1228
Hard top kit replacement:		
Installation	3-396b	3-972
Removal	3-396a	3-972
Harness (M939/A1), engine coolant heater, replacement:		
Installation	3-400b	3-980
Removal	3-400a	3-980
Harness (M939A2), engine coolant heater, replacement:		
Installation	3-401b	3-982
Removal	3-401a	3-982
Harness, 100-amp alternator, replacement:		
Installation	3-427b	3-1098
Removal	3-427a	3-1096
Harness, convoy warning light (M929/A1/A2, M930/A1/A21, replacement:		
Installation	3-434b	3-1118
Removal	3-434a	3-1116

INDEX (Contd)

	Para	Page
Harness, convoy warning light, replacement:		
Installation	3-433b	3-1114
Removal	3-433a	3-1114
Harness, swingfire heater, replacement:		
Installation	3-408b	3- 1012
Removal	3-408a	3-1010
Head, cylinder, replacement:		
Installation	4-12b	4-122
Removal	4-12a	4-120
Header plates, water, replacement:		
Installation	4-28b	4-198
Removal	4-28a	4-198
Heat shield accessroy, exhaust, replacement:		
Installation	3-449b	3-1224
Removal	3-449a	3-1224
Heater and exhaust, van, replacement:		
Installation	3-379b	3-930
Removal	3-379a	3-928
Heater control box, engine coolant, replacement:		
Installation	3-399b	3-978
Removal	3-399a	3-978
Heater fuel pump (M934A1/A2, van, replacement:		
Installation (M934)	3-381d	3-936
Installation (M934A1/A2)	3-381b	3-934
Removal (M934)	3-381c	3-936
Removal (M934A1/A2)	3-381a	3-934
Heater pad, swingfire heater battery box, replacement:		
Installation	3-410b	3-1018
Removal	3-410a	3-1018
Heater pump, engine coolant, replacement:		
Installation	3-398b	3-976
Removal	3-398a	3-976
Heater, electrical components, swingfire, replacement:		
Circuit breaker installation	3-407f	3-1006
Circuit breaker removal	3-407c	3-1006
Electrical connector installation	3-407h	3-1008
Electrical connector removal	3-407a	3-1004
Relay installation	3-407g	3-1008
Relay removal	3-406b	3-1004
Thermal switch installation	3-407e	3-1006
Thermal switch removal	3-407d	3-1006
Heater, mounting bracket and swingfire, replacement:		
Installation	3-406b	3-1002
Removal	3-406a	3-1000
Heater, personnel, hot water, replacement:		
Installation	3-292b	3-782
Removal	3-292a	3-782
Hinge, and seal, door, replacement:		
Installation	3-362b	3-902
Removal	3-362a	3-902
Hinged roof and floor counterbalance cable maintenance:		
Cable adjustment	3-356c	3-895
Installation	3-356b	3-894
Removal	3-356a	3-894
Hinged roof-operated blackout circuit plungers replacement:		
Installation	3-377b	3-924
Removal	3-377a	3-924
Hoist winch cable replacement:		
Installation	3-384b	3-944
Removal	3-384a	3-942
Hood bumper replacement:		
Installation	3-274b	3-741
Removal	3-274a	3-741
Hood grab handle replacement:		
Installation	3-271b	3-737
Removal	3-271a	3-737
Hood latch and bracket replacement:		
Installation	3-268b	3-732
Removal	3-268a	3-732
Hood retaining bracket replacement:		
Installation	3-247b	3-695
Removal	3-247a	3-695
Hood stop bracket, cab, replacement:		
Installation	3-273b	3-740
Removal	3-273a	3-740
Hood stop cables replacement:		
Installation	3-270b	3-736
Removal	3-270a	3-736

INDEX (Contd)

	Para	Page
Hood support bar and bracket replacement:		
Installation	3-269b	3-734
Removal	3-269a	3-734
Hood, engine, maintenance:		
Adjustment	3-275c	3-746
Installation	3-275b	3-744
Removal	3-275a	3-742
Hook, pintle, maintenance:		
Assembly	3-242d	3-687
Cleaning and inspection	3-242c	3-686
Disassembly	3-242b	3-686
Installation	3-242e	3-687
Removal	3-242a	3-686
Hook, side panel roof swivel, replacement:		
Installation	3-359b	3-898
Removal	3-359a	3-898
Hose chafe guard kit, hydraulic, replacement:		
Installation	3-443b	3-1150
Removal	3-443a	3-1148
Hose, heater inlet and outlet, personnel, replacement:		
Installation	3-291b	3-780
Removal	3-291a	3-780
Hose, heater, engine coolant, replacement:		
Installation	3-403b	3-990
Removal	3-403a	3-986
Hoses, power steering pump pressure and return, replacement (Ross):		
Installation	3-234b	3-668
Removal	3-234a	3-668
Hoses, power steering pump pressure and return, replacement (Sheppard):		
Installation	3-235b	3-670
Removal	3-235a	3-670
Hoses, steering assist cylinder, replacement:		
Installation	3-232b	3-660
Removal	3-232a	3-660
Housing assembly cover replacement:		
Installation	3-333b	3-852
Removal	3-333a	3-852
Housing, cam follower, maintenance:		
Assembly	4-19d	4-150
Cleaning and inspection	4-19c	4-148
Disassembly	4-19b	4-146
Installation	4-19e	4-152
Removal	4-19a	4-144
Housing, flywheel, maintenance:		
Cleaning and inspection	4-16b	4-130
Installation	4-16c	4-132
Removal	4-16a	4-130
Hub air seal leak test:		
Front hub leak test	3-462a	3-1264
Rear hub leak test	3-462b	3-1266
Hub and drum (M939/A1), front, maintenance:		
Cleaning and inspection	3-223b	3-640
Installation	3-223d	3-640
Lubrication	3-223c	3-640
Removal	3-223a	3-638
Hub and drum (M939/A1), rear, maintenance:		
Cleaning and inspection	3-224b	3-644
Installation	3-224d	3-646
Lubrication	3-224c	3-644
Removal	3-224a	3-642
Hubs, front, repair:		
Cleaning and inspection	3-460b	3- 1253
Installation	3-460d	3-1256
Removal	3-460a	3- 1252
Repair	3-460c	3-1254
Hubs, rear, repair:		
Cleaning and inspection	3-461b	3-1260
Installation	3-461d	3-1262
Removal	3-461a	3-1258
Repair	3-461c	3-1260
Hydraulic hose chafe guard kit replacement:		
Installation	3-443b	3-1150
Removal	3-443a	3-1148
Hydraulic reservoir drain kit replacement:		
Installation	3-444b	3-1152
Removal	3-444a	3-1152
Hydraulic reservoir shutoff modification kit replacement:		
Installation	3-445b	3-1156
Removal	3-445a	3-1154

INDEX (Contd)

	Para	Page
I		
Injector (dial indicator method), valve and, adjustment:		
Checking plunger free travel	4-33a	4-248
Injector and valve adjustments	4-33b	4-248
Injector plunger and valve adjustments (torque method):		
Injector and valve adjustment	4-34b	4-256
Pre-adjustment setup	4-34a	4-254
Inside telephone jack post replacement:		
Installation	3-374b	3-918
Removal	3-374a	3-918
Insulation, cab, replacement:		
Installation	3-306b	3-804
Removal	3-306a	3-804
K		
Kit, A-frame, maintenance:		
Inspection	3-414b	3-1048
Installation	3-414c	3-1048
Removal	3-414a	3-1046
Kit, air dryer (M923/A1/A2, M925/A1/A2, M927/A1/A2, M928/A1/A2, M934/A1/A2) replacement:		
Installation	3-412b	3-1026
Removal	3-412a	3-1022
Kit, air dryer (M929/A1/A2, M930/A1/A2, M931/A1/A2, M932/A1/A2, M936/A1/A2) replacement:		
Installation	3-413b	3-1039
Removal	3-413a	3-1032
Kit, hardtop, replacement:		
Installation	3-396b	3-972
Removal	3-396a	3-972
Kit, mud guard (M931/A1/A2, M932/A1/A2), replacement:		
Installation	3-420b	3-1078
Removal	3-420a	3-1078
Kit, troop seat and siderack (M929/A1/A2, M930/A1/A2) maintenance:		
Siderack assembly	3-429c	3- 1102
Siderack disassembly	3-429b	3- 1102
Troopseat assembly	3-429d	3- 1102
Troop seat disassembly	3-429a	3-1102
L		
Ladder locking clamp replacement:		
Installation	3-360b	3-899
Removal	3-360a	3-899
Ladder rack bumpers replacement:		
Installation	3-366b	3-907
Removal	3-366a	3-907
Lamp and lens, emergency/ blackout light, replacement:		
Installation	3-372b	3-915
Removal	3-372a	3-915
Latch and bracket, hood, replacement:		
Installation	3-268b	3-732
Removal	3-268a	3-732
Lift eyes, engine, replacement:		
Installation	4-11b	4-118
Removal	4-11a	4-118
Lifting shackle and bracket, front and rear, replacement:		
Installation	3-241b	3-684
Removal	3-241a	3-684
Light tube, fluorescent, replacement:		
Installation	3-371B	3-914
Removal	3-371a	3-914
Lightweight weapon station modification kit maintenance:		
Assembly	3-446c	3-1170
Disassembly	3-446b	3-1168
Installation	3-446d	3-1174
Removal	3-446a	3-1160
Link, drag, replacement:		
Installation	3-229b	3-655
Removal	3-229a	3-655
Lock, cab door, replacement:		
Installation	3-313b	3-812
Removal	3-313a	3-812
Lock, rear, side panel, replacement:		
Installation	3-368b	3-910
Removal	3-368a	3-909
Lock, side panel exterior, replacement:		
Installation	3-370b	3-912
Removal	3-370a	3-912
Lock, side panel front and hinged-type roof, replacement:		
Installation	3-369b	3-911
Removal	3-369a	3-911

INDEX (Contd)

	Para	Page
Locks, expanding and retracting mechanism, replacement:		
Installation	3-378b	3-926
Removal	3-378a	3-926

M

	Para	Page
Machine gun mounting kit maintenance:		
Assembly	3-418c	3-1070
Disassembly	3-418b	3- 1068
Installation	3-418d	3-1072
Removal	3-418a	3-1066
Manifold, engine air intake, maintenance:		
Assembly	4-24d	4-180
Cleaning and inspection	4-24c	4-180
Disassembly	4-24b	4 180
Installation	4-24e	4-182
Removal	4-24a	4-178
Manifold, engine water, maintenance:		
Assembly	4-27d	4-196
Cleaning and inspection	4-27c	4-196
Disassembly	4-27b	4-194
Installation	4-27e	4-196
Removal	4-27a	4-194
Map compartment replacement:		
Installation	3-287b	3-772
Removal	3-287a	3-772
Mechanical troubleshooting, direct and general support symptom index	4-2	4-2
Mini-lighting kit, European, replacement:		
Installation	3-439b	3-1132
Removal	3-439a	3-1132
Mirror, brace and, rearview, replacement:		
Installation	3-290b	3-779
Removal	3-290a	3-778
Mount, convoy warning light (M929/A1/A2, M930/A1/A2), replacement:		
Installation	3-432b	3-1112
Removal	3-432a	3-1112
Mount, convoy warning light (M934/A1/A2), replacement:		
Installation	3-431b	3-1110
Removal	3-431a	3-1108
Mount, convoy warning light, replacement:		
Installation	3-430b	3-1106
Removal	3-430a	3-1104
Mounting bracket kit, chemical agent alarm, replacement:		
Chemical alarm wiring harness installation	3-417d	3-1064
Chemical alarm wiring harness removal	3-417a	3-1058
Detector and alarm bracket installation	3-417c	3-1062
Detector and alarm bracket removal	3-417b	3-1060
Mounting bracket kit, decontamination (M13) apparatus, replacement:		
Installation (M923/A1/A2, M925/A1/A2, M927/A1/A2, M928/A1/A2	3-419h	3- 1076
Installation (M929/A1/A2, M930/A1/A2, M931/A1/A2,	3-419b	3-1074
M932/A1/A2)	3-419d	3-1074
Installation (M934/A1/A2)	3-419f	3-1076
Installation (M936/A1/A2)		
Removal (M923/A1/A2, M925/A1/A2, M927/A1/A2, M928/A1/A2)	3-419g	3-1076
Removal (M929/A1/A2, M930/A1/A2, M931/A1/A2, M932/A1/A2)	3-419a	3-1074
Removal (M934/A1/A2)	3-419c	3-1074
Removal (M936/A1/A2)	3-419e	3-1076
Mounting bracket kit, fire extinguisher, replacement:		
Installation	3-416b	3-1056
Removal	3-416a	3-1056
Mounting bracket, pioneer tool kit, replacement:		
Bracket installation (M923, M925, M927, M928)	3-415b	3-1052
Bracket installation (M929, M930)	3-415f	3-1054
Bracket installation (M929A1, M930A1)	3-415h	3- 1054
Bracket installation (M931, M932)	3-415d	3-1052
Bracket removal (M923, M925, M927, M928)	3-415a	3-1052
Bracket removal (M929, M930)	3-415e	3-1054

INDEX (Contd)

	Para	Page
Bracket removal (M929A1, M930A1)	3-415g	3-1054
Bracket removal (M931, M932)	3-415c	3-1052
Mounting kit, machine gun, maintenance:		
Assembly	3-418c	3-1070
Disassembly	3-418b	3-1068
Installation	3-418d	3-1072
Removal	3-418a	3-1066
Mounting kit, rifle, replacement:		
Installation on left door	3-421c	3-1082
Installation on dash and floor	3-421d	3-1083
Removal from dash and floor	3-421b	3-1081
Removal from left door	3-421a	3-1080
Mud guard kit (M931/A1/A2, M932/A1/A2) replacement:		
Installation	3-420b	3-1078
Removal	3-420a	3-1078

0

	Para	Page
Oil cooler, engine, maintenance:		
Assembly	4-23d	4-176
Cleaning and inspection	4-23c	4-176
Disassembly	4-23b	4-174
Installation	4-23e	4-176
Removal	4-23a	4-174
Oil pan shroud and exhaust tube, engine coolant replacement:		
Installation	4-402b	3-984
Removal	4-402a	3-984
Oil pan shroud and exhaust tube, swingfire heater, replacement:		
Installation	3-409b	3-1016
Removal	3-409a	3-1016
Oil pan, engine, maintenance:		
Assembly	4-22d	4-170
Cleaning and inspection	4-22c	4-170
Disassembly	4-22b	4-170
installation	4-22e	4-172
Removal	4-22a	4-168
Oil pump, engine, replacement:		
Gear backlashtest	4-21a	4- 164
Installation	4-21c	4-166
Removal	4-21b	4-164
Oil reservoir (M932/A1/A2), tractor winch hydraulic, replacement:		
Installation	3-338b	3-866
Removal	3-338a	3-864
Oil reservoir, winch hydraulic, replacement:		
Installation	3-337b	3-862
Removal	3-337a	3-860
Outside door handle replacement:		
Installation	3-307b	3-806
Removal	3-307a	3-806
Outside telephone jack post replacement:		
Installation	3-375b	3-920
Removal	3-375a	3-920

P

	Para	Page
Panel seals replacement:		
Installation	3-363b	3-904
Removal	3-363a	3-904
Personnel heater inlet and outlet hose replacement:		
Installation	3-291b	3-780
Removal	3-291a	3-780
Personnel hot water heater replacement:		
Installation	3-292b	3-782
Removal	3-292a	3-782
Pintle hook maintenance:		
Assembly	3-242d	3-687
Cleaning and inspection	3-242c	3-686
Disassembly	3-242b	3-686
Installation	3-242e	3-687
Removal	3-242a	3-686
Pioneer tool kit mounting bracket replacement:		
Bracket installation (M923, M925, M927, M928)	3-415b	3-1052
Bracket installation (M929, M930)	3-415f	3-1054
Bracket installation (M929A1, M930A1)	3-415h	3-1054
Bracket installation (M931, M932)	3-415d	3-1052
Bracket removal (M923, M925, M927, M928)	3-415a	3-1052
Bracket removal (M929, M930)	3-415e	3-1054

INDEX (Contd)

	Para	Page
Bracket removal (M929A1, M930A1)	3-415g	3-1054
Bracket removal (M931, M932)	3-415c	3-1052
Pitman arm replacement (Ross):		
Installation	3-227b	3-653
Removal	3-227a	3-653
Pitman arm replacement (Sheppard):		
Installation	3-228b	3-654
Removal	3-228a	3-654
Pitman, arm, replacement (Ross):		
Installation	3-227b	3-653
Removal	3-227a	3-653
Pitman, arm, replacement (Sheppard):		
Installation	3-228b	3-654
Removal	3-228a	3-654
Plates, front bumper and, replacement:		
Installation	3-243b	3-688
Removal	3-243a	3-688
Pneumatic controller and relief valve maintenance:		
Assembly	3-454d	3-1238
Cleaning and inspection	3-454c	3-1238
Disassembly	3-454b	3-1236
Installation	3-454e	3-1239
Removal	3-454a	3-1236
Power cable reel (M934A1/A2), van, replacement:		
Installation	3-380b	3-932
Removal	3-380a	3-932
Power plant (M939/A1) replacement:		
Installation	4-4c	4-53
Preliminary disconnections	4-4a	4-36
Removal	4-4b	4-49
Power plant (M939A2) replacement:		
Installation	4-5c	4-76
Preliminary disconnections	4-5a	4-67
Removal	4-5b	4-75
Power steering pump (M9391Al) maintenance:		
Assembly	3-236c	3-672
Disassembly	3-236b	3-672
Installation	3-236d	3-673
Removal	3-236a	3-672
Power steering pump filter (M939A2) maintenance:		
Cleaning	3-237c	3-674
Filter installation	3-237e	3-674
Filter removal	3-237a	3-674
Reservoir installation	3-237d	3-674
Reservoir removal	3-237b	3-674
Power steering pump filter and reservoir (M939A2) maintenance:		
Cleaning	3-237c	3-674
Filter installation	3-237e	3-674
Filter removal	3-237a	3-674
Reservoir installation	3-237d	3-674
Reservoir removal	3-237b	3-674
Power steering pump maintenance:		
Assembly	3-236c	3-673
Disassembly	3-236b	3-672
Installation	3-236d	3-673
Removal	3-236a	3-672
Power steering pump pressure and return hoses replacement (Ross):		
Installation	3-234b	3-668
Removal	3-234a	3-668
Power steering pump pressure and return hoses replacement (Ross):		
Installation	3-234b	3-668
Removal	3-234a	3-668
Power steering pump pressure and return hoses replacement (Sheppard):		
Installation	3-235b	3-670
Removal	3-235a	3-670
Power steering pump pressure and return hoses replacement (Sheppard):		
Installation	3-235b	3-670
Removal	3-235a	3-670
Preparing replacement engine for installation in vehicle:		
Installation	4-7b	4-92
Removal	4-7a	4-91
Pressure lines, steering gear-to-assist cylinder replacement:		
Installation	3-239b	3-680
Removal	3-239a	3-678
Pressure relief valve maintenance:		
Adjustment	3-389b	3-956
Testing	3-389a	3-956

INDEX (Contd)

	Para	Page
Pressure switch replacement:		
Installation	3-467b	3-1280
Removal	3-467a	3-1280
Pressure transducer replacement:		
Installation	3-453b	3-1234
Removal	3-453a	3-1232
Propeller shaft, PTO-to-hydraulic pump, transmission, replacement:		
Installation	3-334b	3-853
Removal	3-334a	3-853
FTO-to-hydraulic pump propeller shaft, transfer, replacement:		
Installation	3-393b	3-966
Removal	3-393a	3-966
Pump (M939/A1), power steering, maintenance:		
Assembly	3-236c	3-672
Disassembly	3-236b	3-672
Installation	3-236d	3-673
Removal	3-236a	3-672
Pump, fuel, replacement:		
Installation	4-35b	4-262
On-engine adjustments	4-35c	4-264
Removal	4-35a	4-260
Pump, steering, drivebelts, maintenance (M939/A1):		
Adjustment	3-230b	3-656
Inspection	3-230c	3-656
Installation	3-230d	3-657
Removal	3-230b	3-656
Pump, swingfire heater, replacement:		
Installation	3-405b	3-998
Removal	3-405a	3-996
Pump, water, maintenance:		
Assembly	4-29d	4-206
Cleaning and inspection	4-29c	4-204
Disassembly	4-29b	4-202
Installation	4-29e	4-208
Removal	4-29a	4-200
Pump, wrecker crane hydraulic, replacement:		
Installation	3-388b	3-954
Removal	3-388a	3-954
Push tubes, rocker lever housing and, maintenance:		
Assembly	4-20d	4-160
Cleaning and inspection	4-20c	4-156
Disassembly	4-20b	4-156
Installation	4-20e	4-160
Removal	4-20a	4-154

	Para	Page
R		
Radiator baffles, seals, and plates replacement:		
Installation	3-272b	3-738
Removal	3-272a	3-738
Radiator cover kit replacement:		
Installation	3-395b	3-970
Removal	3-395a	3-970
Rear axle air manifold maintenance:		
Cleaning and inspection	3-463b	3-1268
Installation	3-463c	3-1268
Removal	3-463a	3-1268
Rear cab mount replacement:		
Installation	3-305b	3-802
Removal	3-305a	3-802
Rear hub and drum (M939/A1) maintenance:		
Cleaning and inspection	3-224b	3-644
Installation	3-224d	3-646
Lubrication	3-224c	3-644
Removal	3-224a	3-642
Rear hubs repair:		
Cleaning and	3-461b	3-1260
Installation	3-461d	3-1262
Removal	3-461a	3-1258
Repair	3-461c	3-1260
Rear wheel valve maintenance:		
Cleaning and inspection	3-457b	3-1246
Installation	3-457c	3-1247
Removal	3-457a	3-1246
Rear winch adjustment:		
Cable tensioner adjustment	3-330b	3-844
Cable tensioner check	3-330a	3-844
Rear winch cable replacement:		
Installation	3-331b	3-846
Removal	3-331a	3-846
Rear winch replacement:		
Installation	3-332b	3-848
Removal	3-332a	3-847
Rearview mirror and brace replacement:		
Installation	3-290b	3-779
Removal	3-290a	3-778
Reflectors replacement:		
Installation	3-345b	3-879
Removal	3-345a	3-879

INDEX (Contd)

	Para	Page
Relief safety valve maintenance:		
Assembly	3-455d	3-1242
Cleaning and inspection	3-455c	3-1242
Disassembly	3-455b	3-1241
Installation	3-455e	3-1242
Removal	3-455a	3-1240
Reservoif shutoff modification kit, hydraulic, replacement:		
Installation	3-445b	3-1156
Removal	3-445a	3-1154
Reservoir, jet, and control, windshield wiper, replacement:		
Installation	3-281b	3-758
Removal	3-281a	3-758
Resistor, convoy warning light, leads and, replacement:		
Installation	3-436b	3-1126
Removal	3-436a	3-1124
Retaining bracket, hood, replacement:		
Installation	3-247b	3-695
Removal	3-247a	3-695
Retractable window regulator replacement:		
Installation	3-354b	3-892
Removal	3-354a	3-892
Retractable window replacement:		
Installation	3-351b	3-889
Removal	3-351a	3-889
Rifle mounting kit replacement:		
Installation on left door	3-421c	3-1082
Installation on dash and floor	3-421d	3-1083
Removal from dash and floor	3-421b	3-1081
Removal from left door	3-421a	3-1080
Ring gear, flywheel, maintenance:		
Installation	4-15b	4-129
Removal	4-15a	4-129
Rocker lever housing and push tubes maintenance:		
Assembly	4-20d	4-160
Cleaning and inspection	4-20c	4-156
Disassembly	4-20b	4-156
Installation	4-20e	4-160
Removal	4-20a	4-154
Rotation, wheel and, tire (M939):		
Front wheel installation	3-218f	3-621
Inner rear wheel installation	3-218d	3-619
Wheel and tire removal	3-218b	3-616
Inspection	3-218e	3-620
Outer rear wheel installation	3-218a	3-616
Wheel and tire rotation	3-218c	3-618
Rotation, wheel and, tire (M939A1/A2):		
Front wheel installation	3-219e	3-625
Inspection	3-219b	3-622
Rear wheel installation	3-219d	3-624
Wheel and tire removal	3-219a	3-622
Wheel and tire rotation	3-219c	3-624
S		
Screen, window, replacement:		
Installation	3-353b	3-891
Removal	3-353a	3-891
Seal and retainer, cab top, replacement:		
Installation	3-316b	3-815
Removal	3-316a	3-815
Seals, panel, replacement:		
Installation	3-363b	3-904
Removal	3-363a	3-904
Seat frame and base, driver's, maintenance:		
Inspection andrepair	3-284b	3-766
Installation	3-284c	3-766
Removal	3-284a	3-764
Seat, driver's, replacement:		
Installation	3-283b	3-762
Removal	3-283a	3-762
Seatbelt replacement:		
Companion seatbelts installation	3-289d	3-776
Companion seatbelts removal	3-289c	3-776
Driver's seatbelt installation	3-289b	3-774
Driver's seatbelt removal	3-289a	3-774
Separator, water, maintenance:		
Assembly	3-466d	3-1278
Cleaning and inspection	3-466c	3-1278
Disassembly	3-466b	3-1278
Installation	3-466e	3-1279
Removal	3-466a	3-1278

INDEX (Contd)

	Para	Page
Shutoff valve (M936), fuel pump, replacement:		
Installation	4-37b	4-268
Removal	4-37a	4-268
Shutoff valve, fuel pump, replacement:		
Fuel pump shutoff valve installation	4-36c	4-266
Fuel pump shutoff valve removal	4-36b	4-266
Manual shutoff valve installation	4-36d	4-266
Manual shutoff valve removal	4-36a	4-266
Side and rear door blackout light switch maintenance:		
Installation	3-376b	3-922
Removal	3-376a	3-922
Side locking pin retaining clip replacement:		
Installation	3-344b	3-878
Removal	3-344a	3-878
Side panel exterior lock replacement:		
Installation	3-370b	3-912
Removal	3-370a	3-912
Side panel front lock and hinged-type roof lock replacement:		
Installation	3-369b	3-911
Removal	3-369a	3-911
Side panel rear lock replacement:		
Installation	3-368b	3-910
Removal	3-368a	3-909
Side panel roof swivel hook replacement:		
Installation	3-359b	3-898
Removal	3-359a	3-898
Side panel rubber bumpers replacement:		
Installation	3-367b	3-908
Removal	3-367a	3-908
Side panel-to-roof toggle clamp replacement:		
Installation	3-357b	3-896
Removal	3-357a	3-896
Snubber valve assembly replacement:		
Installation	3-390b	3-958
Removal	3-390a	3-958
Spacers, fifth wheel, replacement:		
Installation	3-252b	3-701
Removal	3-252a	3-701
Spare tire carrier (M923, M925, M927, M928), cargo, replacement:		
Installation	3-258b	3-716
Removal	3-258a	3-716
Spare tire carrier (M923A1/A2,M925A1,/A2, M927A1/A2, M928A1,A2), cargo, replacement:		
Installation	3-259b	3-718
Removal	3-259a	3-718
Spare tire carrier (M929, M930), dump, replacement:		
Installation	3-256b	3-712
Removal	3-256a	3-712
Spare tire carrier (M929A1/A2, M930A1/A2), dump, replacement:		
Installation	3-257b	3-714
Removal	3-257a	3-713
Spare tire carrier (M931, M932), tractor, replacement:		
Installation	3-253b	3-702
Removal	3-253a	3-702
Spare tire carrier (M931A1/A2, M932A1/A2), tractor, replacement:		
Installation	3-254b	3-706
Removal	3-254a	3-704
Spare tire carrier (M9341, van, replacement	3-262	3-722
Spare tire carrier (M934A1/A2), van, replacement:		
Installation	3-263b	3-722
Removal	3-263a	3-722
Speed signal generator replacement:		
Installation	3-471b	3-1288
Removal	3-471a	3-1288
Splash guard, cargo upper and lower wheel, replacement:		
Installation	3-342b	3-874
Removal	3-342a	3-874
Splash shield, fender, replacement:		
Installation	3-301b	3-795
Removal	3-301a	3-795

INDEX (Contd)

	Para	Page
Starting repaired or replacement engine:		
In-chassis run-in	4-8c	4-96
Out-of-chassis run-in (M939/A1)	4-8b	4-96
Priming lubrication system (M939/A1)	4-8a	4-95
Steering assist cylinder hoses replacement:		
Installation	3-232b	3-660
Removal	3-232a	3-660
Steering assist cylinder hoses replacement:		
Installation	3-232b	3-660
Removal	3-232a	3-660
Steering assist cylinder replacement:		
Assembly	3-233c	3-664
Disassembly	3-233b	3-664
Installation	3-233d	3-666
Removal	3-233a	3-662
Travel adjustment	3-233e	3-666
Steering assist cylinder replacement:		
Assembly	3-233c	3-664
Disassembly	3-233b	3-664
Installation	3-233d	3-666
Removal	3-233a	3-662
Travel adjustment	3-233e	3-666
Steering assist cylinder stone shield replacement:		
Installation	3-231b	3-658
Removal	3-231a	3-658
Steering assist cylinder stone shield replacement:		
Installation	3-231b	3-658
Removal	3-231a	3-658
Steering gear stone shield replacement:		
Installation	3-238b	3-676
Removal	3-238a	3-676
Steering gear stone shield replacement:		
Installation	3-238b	3-676
Removal	3-238a	3-676
Steering gear-to-assist cylinder pressure lines replacement:		
Installation	3-239b	3-680
Removal	3-239a	3-678
Steering gear-to-assist cylinder pressure lines replacement:		
Installation	3-239b	3-680
Removal	3-239a	3-678
Steering pump drivebelts maintenance (M939/A1):		
Adjustment	3-230a	3-656
Inspection	3-230c	3-656
Installation	3-230d	3-657
Removal	3-230b	3-656
Steering pump drivebelts maintenance (M939/A1):		
Adjustment	3-230a	3-656
Inspection	3-230c	3-656
Installation	3-230d	3-657
Removal	3-230b	3-656
Steering wheel replacement:		
Installation	3-226b	3-652
Removal	3-226a	3-651
Steering wheel replacement:		
Installation	3-226b	3-652
Removal	3-226a	3-651
Steering, assist cylinder, replacement:		
Assembly	3-233c	3-664
Disassembly	3-233b	3-664
Installation	3-233d	3-666
Removal	3-233a	3-662
Travel adjustment	3-233e	3-666
Step, tailgate personnel, replacement:		
Installation	3-347b	3-881
Removal	3-347a	3-881
Steps, toolbox and, replacement:		
Installation	3-302b	3-796
Removal	3-302a	3-796
Stone shield, steering assist cylinder, replacement:		
Installation	3-231b	3-658
Removal	3-231a	3-658
Stone shield, steering gear, replacement:		
Installation	3-238b	3-676
Removal	3-238a	3-676
Stop bracket and latch, windshield, replacement:		
Installation	3-276b	3-747
Removal	3-276a	3-747
Stop cables, hood, replacement:		
Installation	3-270b	3-736
Removal	3-270a	3-736
Stowage, box cargo, replacement:		
Installation	3-346b	3-880
Removal	3-346a	3-880

INDEX (Contd)

	Para	Page
Stud, wheel rim, replacement:		
Installation	3-222b	3-636
Removal	3-222a	3-636
Support bar and bracket, hood, replacement:		
Installation	3-269b	3-734
Removal	3-269a	3-734
Swingfire heater and mounting bracket replacement:		
Installation	3-406b	3-1002
Removal	3-406a	3-1000
Swingtire heater battery box heater pad replacement:		
Installation	3-410b	3-1018
Removal	3-410a	3-1018
Swingfire heater electrical components replacement:		
Circuit breaker installation	3-407f	3-1006
Circuit breaker removal	3-407c	3-1006
Electrical connector installation	3-407h	3-1008
Electrical connector removal	3-407a	3-1004
Relay installation	3-407g	3-1008
Relay removal	3-407b	3-1004
Thermal switch installation	3-407e	3-1006
Thermal switch removal	3-407d	3-1006
Swingfire heater harness replacement:		
Installation	3-408b	3-1012
Removal	3-408a	3-1010
Swingfire heater oil pan shroud and exhaust tube replacement:		
Installation	3-409b	3- 10 16
Removal	3-409a	3-1016
Swingfire heater pump replacement:		
Installation	3-405b	3-998
Removal	3-405a	3-996
Swingfire heater water jacket replacement:		
Installation	3-411b	3-1020
Removal	3-411a	3-1020
Switch, blackout light, 100-volt, receptacle, and, replacement:		
Installation	3-373b	3-916
Removal	3-373a	3-916
Switch, convoy warning light, replacement:		
Installation	3-438b	3-1130
Removal	3-438a	3-1130
Switch, pressure, replacement:		
Installation	3-467b	3-1280
Removal	3-467a	3-1280
T		
Tailgate assembly, dump replacement:		
Installation	3-348b	3-882
Removal	3-348a	3-882
Tailgate bumpers replacement:		
Installation	3-261b	3-721
Removal	3-261a	3-721
Tailgate control linkage, dump, replacement:		
Installation	3-349b	3-886
Removal	3-349a	3-884
Tailgate personnel step replacement:		
Installation	3-347b	3-881
Removal	3-347a	3-881
Tailgate, cargo, replacement:		
Installation	3-343b	3-876
Removal	3-343a	3-876
Telephone jack post, inside, vreplacement:		
Installation	3-374b	3-918
Removal	3-374a	3-918
Telephone jack post, outside, replacement:		
Installation	3-375b	3-920
Removal	3-375a	3-920
Throttle kit (M936/A1), automatic, replacement:		
Installation	3-440b	3-1136
Removal	3-440a	3-1134
Tiedown kit, vehicle, replacement:		
Installation	3-442b	3-1146
Removal	3-442a	3-1146
Timing, engine injector, instructions:		
Fuel injectors installation	4-32h	4-242
Fuel injectors removal	4-32d	4-238
General instructions	4-32e	4-238
Injector timing	4-32g	4-240

INDEX (Contd)

	Para	Page
Rocker lever housing covers installation	4-32k	4-246
Rocker lever housing covers removal	4-32a	4-236
Rocker lever housing and push tubes removal	4-32b	4-236
Rocker lever housing and push tubes installation	4-32j	4-244
Timing tool setup	4-32f	4-240
Valve crossheads installation and adjustment	4-32i	4-244
Valve crossheads removal	4-32c	4-238
Tire and tube (M939) maintenance:		
Inspection	3-220b	3-628
Installation	3-220c	3-628
Removal	3-220a	3-626
Tire and wheel (M939A1/A2) maintenance:		
Assembly	3-221b	3-634
Disassembly	3-221a	3-630
Tire, wheel and, rotation (M939):		
Front wheel installation	3-218f	3-621
Inner rear wheel installation	3-218d	3-619
Wheel and tire removal	3-218a	3-616
Inspection	3-218b	3-616
Outer rear wheel installation	3-218e	3-620
Wheel and tire removal	3-218a	3-616
Wheel and tire rotation	3-218c	3-618
Tire, wheel and, rotation (M939A1/A2):		
Front wheel installation	3-219e	3-625
Inspection	3-219b	3-622
Rear wheel installation	3-219d	3-624
Wheel and tire removal	3-219a	3-622
Wheel and tire rotation	3-219c	3-624
Toggle clamp anchor post replacement:		
Installation	3-358b	3-897
Removal	3-358a	3-897
Toggle clamp, side panel-to-roof, replacement:		
Installation	3-357b	3-896
Removal	3-357a	3-896
Toolbox and steps replacement:		
Installation	3-302b	3-796
Removal	3-302a	3-796
Toolbox, spare tire carrier, tractor, replacement:		
Installation	3-250b	3-699
Removal	3-250a	3-699
Tractor fifth wheel replacement:		
Installation	3-248b	3-696
Removal	3-248a	3-696
Tractor spare tire carrier (M931, M932) replacement:		
Installation	3-253b	3-702
Removal	3-253a	3-702
Tractor spare tire carrier (M931A1/A2), M932A1/A2) replacement:		
Installation	3-254b	3-706
Removal	3-254a	3-704
Tractor spare tire carrier toolbox replacement:		
Installation	3-250b	3-699
Removal	3-250a	3-699
Tractor winch hydraulic oil reservoir (M932/A1/A2) replacement:		
Installation	3-338b	3-866
Removal	3-338a	3-864
Transducer, pressure, replacement:		
Installation	3-453b	3-1234
Removal	3-453a	3-1232
Transfer PTO-to-hydraulic pump propeller shaft replacement:		
Installation	3-393b	3-966
Removal	3-393a	3-966
Transmission PTO-to-hydraulic pump propeller shaft replacement:		
Installation	3-334b	3-853
Removal	3-334a	3-853
Transmission PTO-to-hydraulic pump propeller shaft universal joint maintenance:		
Assembly	3-335c	3-856
Disassembly	3-335a	3-854
Inspection	3-335b	3-854
Troop seat and siderack kit (M929/A1/A2, M930/A1,A2) maintenance:		
Siderack assembly	3-429c	3-1102
Siderack disassembly	3-429b	3-1102
Troop seat assembly	3-429d	3-1102
Troop seat disassembly	3-429a	3-1102

INDEX (Contd)

	Para	Page
Troop seat, cargo, replacement:		
Installation	3-340b	3-870
Removal	3-340a	3-870
Troubleshooting, direct and general support, mechanical symptom index	4-2	4-2
Tube (M939), tire and, maintenance:		
Inspection	3-220b	3-628
Installation	3-220c	3-628
Removal	3-220a	3-626
Turnbuttons and lashing hooks, cab, replacement:		
Installation	3-303b	3-798
Removal	3-303a	3-798

U

	Para	Page
Universal joint, PTO-to-hydraulic pump propeller shaft, transmission, maintenance:		
Assembly	3-335c	3-856
Disassembly	3-335a	3-854
Inspection	3-335b	3-854

V

	Para	Page
Valve and injector adjustment (dial indicator method):		
Checking plunger free travel	4-33a	4-248
Injector and valve adjustments	4-33b	4-248
Valve, air supply, hand airbrake, replacement:		
Installation	3-422b	3-1084
Removal	3-422a	3-1084
Valve, assembly, snubber, replacement:		
Installation	3-390b	3-958
Removal	3-390a	3-958
Valve, doublecheck, hand airbrake, replacement:		
Installation (forward-rear axle doublecheck valve)	3-424b	3-1088
Installation (rear-rear axle doublecheck valve)	3-424d	3-1090
Removal (forward-rear axle doublecheck valve)	3-424a	3-1088
Removal (rear-rear axle doublecheck valve)	3-424c	3- 1090
Valve, front wheel, maintenance:		
Cleaning and inspection	3-456b	3-1244
Installation	3-456c	3-1245
Removal	3-456a	3-1244
Valve, hand airbrake controller, replacement:		
Installation	3-423b	3-1086
Removal	3-423a	3-1086
Valve, hand airbrake tractor protection, valve replacement:		
Installation	3-425b	3-1092
Removal	3-425a	3-1092
Valve, injector plunger (torque method) and, adjustments:		
Injector and valve adjustment	4-34b	4-256
Pre-adjustment setup	4-34a	4-254
Valve, pneumatic controller and relief, maintenance:		
Assembly	3-454d	3-1238
Cleaning and inspection	3-454c	3-1238
Disassembly	3-454b	3-1236
Installation	3-454e	3-1239
Removal	3-454a	3-1236
Valve, pressure relief, maintenance:		
Adjustment	3-389b	3-956
Testing	3-389a	3-956
Valve, rear wheel, maintenance:		
Cleaning and inspection	3-457b	3-1246
Installation	3-457c	3-1247
Removal	3-457a	3-1246
Valve, relief safety, maintenance:		
Assembly	3-455d	3-1242
Cleaning and inspection	3-455c	3- 1242
Disassembly	3-455b	3-1241
Installation	3-455e	3- 1242
Removal	3-455a	3- 1240
Valve, wheel, maintenance:		
Assembly	3-459c	3- 1250
Cleaning and inspection	3-459b	3-1250
Disassembly	3-459a	3-1250
Van davit chain and wire rope replacement:		
Installation	3-264b	3-724
Removal	3-264a	3-724
Van davit winch (M934A1/A2) replacement:		
Installation	3-266b	3-728
Removal	3-266a	3-728

INDEX (Contd)

	Para	Page
Van handrail modification kit (M934/A1/A2) replacement:		
Door check spacer installation	3-450h	3-1229
Door check spacer removal	3-450g	3-1229
Ladder handrail guide installation	3-450b	3-1225
Ladder handrail guide	3-450a	3-1225
Ladder handrail hangers installation	3-450d	3-1226
Ladder handrail hangers removal	3-450c	3-1226
Van door grab handles installation	3-450f	3-1228
Van door grab handles removal	3-450e	3-1228
Van heater and exhaust replacement:		
Installation	3-379b	3-930
Removal	3-379a	3-928
Van heater fuel pump (M934A1/A2) replacement:		
Installation (M934)	3-381d	3-936
Installation (M934A1/A2)	3-381b	3-934
Removal (M934)	3-381c	3-936
Removal (M934A1/A2)	3-381a	3-934
Van power cable reel (M934A1/A2) replacement:		
Installation	3-380b	3-932
Removal	3-380a	3-932
Van rear door and side door window replacement:		
Installation	3-350b	3-888
Removal	3-350a	3-888
Van spare tire carrier (M934) replacement	3-262	3-722
Van spare tire carrier (M934A1/A2) replacement:		
Installation	3-263b	3-722
Removal	3-263a	3-722
Van swing davit and pulley replacement:		
Installation	3-265b	3-726
Removal	3-265a	3-726
Vehicle tiedown kit replacement:		
Installation	3-442b	3-1146
Removal	3-442a	3-1146
Vent door weatherseal replacement:		
Installation	3-298b	3-792
Removal	3-298a	3-792
Vent, screen and door, cab cowl, replacement:		
Installation	3-299b	3-793
Removal	3-299a	3-793
Voltage regulator, 100-amp, replacement:		
Installation	3-428b	3-1100
Removal	3-428a	3-1100
W		
Warning light, amber, replacement:		
Installation	3-469b	3-1283
Removal	3-469a	3-1283
Warning light, convoy, replacement:		
Installation	3-437b	3-1128
Removal	3-437a	3-1128
Washer hoses, windshield, replacement:		
Installation	3-282b	3-760
Removal	3-282a	3-760
Water header plates replacement:		
Installation	4-28b	4-198
Removal	4-28a	4-198
Water jacket, swingfire heater, replacement:		
Installation	3-411b	3-1020
Removal	3-411a	3-1020
Water pump maintenance:		
Assembly	4-29d	4-206
Cleaning and inspection	4-29c	4-204
Disassembly	4-29b	4-202
Installation	4-296	4-208
Removal	4-29a	4-200
Water separator maintenance:		
Assembly	3-466d	3-1278
Cleaning and inspection	3-466c	3-1278
Disassembly	3-466b	3-1278
Installation	3-466e	3-1279
Removal	3-466a	3-1278
Weapon station modification kit, lightweight, maintenance:		
Assembly	3-446c	3-1170
Disassembly	3-446b	3-1168
Installation	3-446d	3-1174
Removal	3-446a	3-1160
Weatherseal, cab door, replacement:		
Installation	3-311b	3-810
Removal	3-311a	3-810

INDEX (Contd)

	Para	Page
Weatherseal, vent door, replacement:		
Installation	3-298b	3-792
Removal	3-298a	3-792
Weatherstripping (cab door), window, replacement:		
Installation	3-315b	3-814
Removal	3-315a	3-814
Wheel and tire (M939) rotation:		
Front wheel installation	3-218f	3-621
Inner rear wheel installation	3-218d	3-619
Inspection	3-218b	3-616
Outer rear wheel installation	3-2 18e	3-620
Wheel and tire removal	3-218a	3-616
Wheel and tire rotation	3-218c	3-618
Wheel and tire (M939A1/A2) rotation:		
Front wheel installation	3-219e	3-625
Inspection	3-219b	3-622
Rear wheel installation	3-219d	3-624
Wheel and tire removal	3-219a	3-622
Wheel and tire rotation	3-219c	3-624
Wheel bearing adjustment:		
Front wheel bearing adjustment	3-225a	3-648
Rear wheel bearing adjustment	3-225b	3-650
Wheel rim stud replacement:		
Installation	3-222b	3-636
Removal	3-222a	3-636
Wheel valve filter replacement:		
Installation	3-458b	3-1248
Removal	3-458a	3-1248
Wheel valve maintenance:		
Assembly	3-459c	3-1250
Cleaning and inspection	3-459b	3-1250
Disassembly	3-459a	3-1250
Wheel, steering, replacement:		
Installation	3-226b	3-652
Removal	3-226a	3-651
Winch automatic brake, front, adjustment:		
Adjustment	3-323b	3-826
Testing	3-323a	3-824
Winch cable chain and hook, front, replacement:		
Installation	3-325b	3-830
Removal	3-325a	3-830
Winch cable clevis replacement:		
Installation	3-326b	3-832
Removal	3-326a	3-831
Winch cable, front, replacement:		
Installation	3-327b	3-834
Removal	3-327a	3-834
Winch cable, rear, replacement:		
Installation	3-331b	3-846
Removal	3-331a	3-846
Winch drag brake, front, adjustment:		
Adjustment	3-324b	3-828
Testing	3-324a	3-828
Winch frame extension replacement:		
Installation	3-245b	3-692
Removal	3-245a	3-692
Winch hydraulic oil reservoir filter replacement:		
Installation	3-336b	3-858
Removal	3-336a	3-858
Winch hydraulic oil reservoir replacement:		
Installation	3-337b	3-862
Removal	3-337a	3-860
Winch motor, front, replacement:		
Installation	3-328b	3-838
Removal	3-328a	3-836
Winch, front, replacement:		
Installation	3-329b	3-842
Removal	3-329a	3-840
Winch, rear, adjustment:		
Cable tensioner adjustment	3-330b	3-844
Cable tensioner check	3-330a	3-844
Winch, rear, replacement:		
Installation	3-332b	3-848
Removal	3-332a	3-847
Window blackout panel replacement:		
Installation	3-352b	3-890
Removal	3-352a	3-890
Window brushguard replacement:		
Installation	3-355b	3-893
Removal	3-355a	3-893
Window regulator and inside door handle replacement:		
Installation	3-308b	3-807
Removal	3-308a	3-807
Window regulator, retractable, replacement:		
Installation	3-354b	3-892
Removal	3-354a	3-892

INDEX (Contd)

	Para	Page
Window screen replacement:		
Installation	3-353b	3-891
Removal	3-353a	3-891
Window weatherstripping (cab door) replacement:		
Installation	3-315b	3-814
Removal	3-315a	3-814
Window, retractable, replacement:		
Installation	3-351b	3-889
Removal	3-351a	3-889
Window, van rear door and side door, replacement:		
Installation	3-350b	3-888
Removal	3-350a	3-888
Windshield and outer frame assembly replacement:		
Installation	3-278b	3-752
Removal	3-278a	3-752
Windshield frame assembly replacement:		
Installation	3-279b	3-754
Removal	3-279a	3-754
Windshield hinge, cab assembly, replacement:		
Installation	3-280b	3-756
Removal	3-280a	3-756
Windshield stop bracket and latch replacement:		
Installation	3-276b	3-747
Removal	3-276a	3-747
Windshield washer hoses replacement:		
Installation	3-282b	3-760
Removal	3-282a	3-760
Windshield wiper blade, wiper arm, and wiper motor replacement:		
Installation	3-277b	3-750
Removal	3-277a	3-748
Windshield wiper reservoir, jet, and control replacement:		
Installation	3-281b	3-758
Removal	3-281a	3-758
Wiper motor, windshield wiper arm, blade, wiper, and, replacement:		
Installation	3-277b	3-750
Removal	3-277a	3-748
Wire, boom, floodlight, replacement:		
Installation	3-385b	3-946
Removal	3-385a	3-946
Wiring harness, crane, replacement:		
Installation	3-386b	3-950
Removal	3-386a	3-948
Wiring harness, CTIS, replacement:		
Installation	3-470b	3-1286
Removal	3-470a	3-1284
Wrecker crane hydraulic hose and tube replacement:		
Installation	3-387b	3-952
Removal	3-387a	3-952
Wrecker crane hydraulic pump replacement:		
Installation	3-388b	3-954
Removal	3-388a	3-954

www.ingramcontent.com/pod-product-compliance
Lightning Source LLC
Chambersburg PA
CBHW080412030426
42335CB00020B/2427

* 9 7 8 1 9 5 4 2 8 5 6 4 4 *